AMMO & BALLISTICS 5

AMMO & BALLISTICS 5

Ballistic Data out to 1,000 Yards for over 190 Calibers and over 2,600 Factory Loads. Includes Data on All Factory Centerfire and Rimfire Cartridges for Rifles and Handguns.

by

Bob Forker

SAFARI PRESS INC

The trademark Safari Press ® is registered with the U.S. Patent and Trademark Office and with government trademark and patent offices in other countries.

Forker, Bob

Fifth edition

Safari Press

2013, Long Beach, California

ISBN 978-1-57157-402-2

Library of Congress Catalog Card Number: 2009926919

10 9 8 7 6 5 4 3 2 1

Printed in the United States of America

Readers wishing to receive the Safari Press catalog, featuring many fine books on big-game hunting, wingshooting, and sporting firearms, should write to Safari Press, P.O. Box 3095, Long Beach, CA 90803, USA. Tel: (714) 894-9080 or visit our Web site at www.safaripress.com.

Table of Contents

Data for Small and Medium Centerfire Rifle Cartridges

Data for Small and Medium Centerfire Rifle Cartridges (continued)

Data for Cartridges Suitable for Dangerous Game

Data for Pistol and Revolver Cartridges

Data for Rimfire Cartridges

Kevin Steele
Foreword

From 1987 to 2005 I served as the executive editor, editor, editorial director, and ultimately publisher of *Guns & Ammo* magazine. Now, that may seem a long time to many, but in truth, it's only a drop in the bucket compared to the author of this tome, Bob Forker. So I was pleased when I was asked to write the foreword for what is now the fifth edition of *Ammo and Ballistics,* and I decided to impart a bit about the back story that led up to the initial release of this title.

Bob's long tenure at *Guns & Ammo* began way back in 1964 at the invitation of then technical editor, Bob Hutton. Hutton was impressed with the job young engineer Forker had done in designing the Powley Computer. A Powley-what, you may ask? Well, the Powley computer allowed users to make ballistic calculations long before the advent of the personal computer. The initial concept and mathematical formulas had been worked out by a gentleman named Homer Powley. Forker assisted Powley by taking his data and incorporating it into a slide-rule-like device that simplified the process. All this was accomplished before Bill Gates left third grade.

So, at Bob Hutton's invitation, Forker began assisting him with various technical projects and eventually contributing to *Guns & Ammo* under his own byline. Then, in 1968, Bob accepted a position with Hughes Helicopters. That job would keep him out of the print industry for the next fifteen years. During that interim, Bob was kept busy at Hughes, where as an engineer he worked as a department manager in weapons systems and ordnance. Just a couple of his high-profile projects were the 30mm Chain Gun and the Apache attack helicopter weapon systems.

Fast forward to 1983. Forker is walking across the tarmac at Van Nuys airport when he trips over the feet of legendary *Guns & Ammo* publisher Tom Siatos who was fiddling with something beneath his pet Cessna. Knowing Tom as I did, my guess is he most likely jumped up red-faced and swearing at the clumsy oaf who had tripped over him. Of course, upon recognizing Forker, Siatos instantly transformed from red-faced former Marine sergeant to charming gentleman, and promptly invited Forker to lunch. This inevitably led to Forker coming back to write once again for *Guns & Ammo.*

Four years later I was asked by Siatos to join the *Guns & Ammo* team, and from then to the present day I have been honored to call Bob Forker my colleague and friend. During my years at *Guns & Ammo,* I can't even begin to recall how many hours Bob spent in my office discussing guns and ammunition. Those many discussions led to some great articles and projects, and ultimately this book.

Looking back, I know that just a couple of the ideas we came up with during those days included the Ballistic Pendulum, which Bob designed as a fairly simple but effective way of measuring and qualifying a given long-arms' recoil force. My all-time favorite was theorizing about how long a barrel had to be before it was too long for a bullet to exit. At the time, I was just kidding, but Bob took the notion to heart. Before long, he and his buddy, barrelmaker Bo Clerke, had an answer, at least for the .32 S&W cartridge: thirty feet!

Strangely enough, Bob was never very keen on wildcat cartridges, holding the belief that there was little a wildcatter could really do to create practical efficiencies over and above what the factories were capable of developing. This was ultimately proven true as time went on and new propellant developments allowed standard commercial cartridges from the .22 to the .458 to eclipse the velocity gains of many of the older, once popular wildcats.

But more than anything else, I believe that Bob's greatest contribution to the worlds of firearms and ammunition is this book you hold in your hand. *Ammo and Ballistics 5* is truly a comprehensive study of all the commercially loaded cartridges now in existence. Each factory load in every different bullet weight is catalogued and compared in an easy to use fashion. Velocity, energy, wind drift . . . all are presented with data out to an unprecedented 1,000 yards. Ballistic coefficients, pressure units, recoil factors—you name it and it's all here.

Believe me, I have used just about every reference source from reloading manuals to factory-load tables to ballistics programs to intensive and time-consuming online searches. You can find the data when you have to, but no other single source provides the depth of material and ease of use as does *Ammo and Ballistics.*

Here's just one example.

I recently had to decide upon a factory load to take with me on an ibex hunt in Spain. I wanted to use my pet Blaser R8 .30-06, and I wanted to know the flattest shooting factory load currently available. So, I picked up my copy of *Ammo and Ballistics 4*, went to the 30-06 section, and found sixteen factory loads using my choice of 165-grain bullets. Now, one might presuppose that the bullet with the highest BC would be the flattest shooting. But BC isn't everything because velocity is also a factor. With these variables, it could have taken a while to come up with the right answer, but *Ammo and Ballistics* provided it in less than a minute! (No, I'm not going to give you the answer, you'll have to buy this book and look it up for yourself!)

Bob Forker has been the technical guru for *Guns & Ammo* for almost fifty years. You can't get a better recommendation than that, and you won't find a better source for ballistics information than *Ammo and Ballistics.* For that reason alone, this book is worth every penny of the purchase price. Rest assured, you will never regret buying this book. It does an exceptional job of providing fast, accurate, and easy reference; it's so good that you'll wonder how you ever got along without it!

Kevin E. Steele
Publisher, *Petersen's HUNTING*
Publisher and Editor Emeritus, *Guns & Ammo*

Acknowledgments

We wish to thank all the people who have contributed to this database. The list is long, but without the cooperation of each and every one of these contributors the book would be less useful.

A-Square Company, Inc.
•
Aguila
•
ATK
•
Black Hills Ammunition, Inc.
•
CCI / Speer
•
Cor-Bon Bullet Co.
•
Dakota Arms, Inc.
•
Eley Ltd.
•
Federal Cartridge Company
•
Fiocchi
•
Hornady Manufacturing Company
•
Patria Lapua Oy
•
Lazzeroni Arms Company
•
Kynoch (Kynamco Limited)
•
MagTech
•
Norma Precision AB
•
Nosler, Inc.
•
PMC Ammunition
•
Remington Arms
•
Dynamit Nobel-RWS Inc.
•
Sellier & Bellot, USA
•
Ultramax Ammunition
•
Weatherby
•
Winchester Ammunition

A very special thank you goes to Bob Nosler of Nosler Bullets who has so generously allowed us to use their cartridge drawings in this book. Many of the drawings used were furnished by Nosler.

The ballistic calculations included in this book were done using Dr. Ken Oehler's (Oehler Research, Inc.) Ballistic Explorer program.

Mr. Ken Green, Director of Technical Affairs for SAAMI has been most generous with his assistance in establishing the performance specifications for new cartridges. Without his help the book would be less complete.

Introduction

There was a time when someone who needed ammunition went to his local hardware store and said he wanted a box of, let's say, .30-30 ammunition. That was it, he was lucky to find any ammo in the right caliber, let alone to have any choice of different bullets, brands, performance levels, etc. He took his box of cartridges and went hunting. Life was simple.

Today we "suffer" from just the opposite problem, for we have far more variations than anyone can remember. In the .30-30 Winchester alone, there are at least seventy-five combinations of bullet weight, brand, etc., available for retail sale in the U.S. That may be a problem, but it's a good class of problem. This book provides a way to see just what factory ammunition is available, in what caliber, in what bullet styles, and with enough detailed performance data so that you are able to make some meaningful comparisons among the selection choices.

As we started to collect these data, we immediately ran into the question of how much to include and where to cut off. That's not an easy decision, and the decisions we ultimately made are certain to not please everyone. We started with the fundamental decision to include only factory-made ammunition. That excludes custom reloaders and wildcat calibers. Since there is more than a little bit of factory ammunition sold in the U.S. that is not actually made here (Lapua, MagTech, and Norma come quickly to mind), we did decide to include foreign manufacturers who have a distribution system in place in this country. To keep the size of this book within reasonable bounds, we have limited the listings to calibers that are well known in the U.S.

As soon as you begin to get past the calibers that are standardized in the U.S. by the Small Arms and Ammunition Manufacturers' Institute (SAAMI), you get into a large gray area. Weatherby's cartridges provide an excellent example of this category. When the Weatherby line of cartridges was first introduced, you either bought your ammunition from the Weatherby factory or loaded your own. The concept

of proprietary cartridges furnished by custom gunmakers is not at all new. Just after 1900 in England, that practice was more the rule than the exception. The .375 Holland & Holland Magnum was once a proprietary number. Like the .375 H&H, over time, several of the Weatherby calibers have received SAAMI standardization. Ammunition for these calibers is now produced by independent ammunition makers. That certainly takes those calibers out of the "proprietary" designation. For other Weatherby calibers, the only factory ammunition that gets to the dealer's shelves still comes exclusively through Weatherby. That isn't bad; it is just the way things are. As of this writing, I believe that all Weatherby calibers that are not covered by SAAMI standards are controlled by CIP, the agency that oversees ammunition producers in Western Europe.

There are several U.S. companies who, relatively recently, have each developed their own lines of proprietary calibers. In general, these companies provide a source of factory loaded ammunition for their own calibers. While most of these cartridges have not received formal standardization in the U.S. or elsewhere, these folks are building their products under their own closely controlled conditions that quite often include outside testing by independent laboratories. Deciding which of these calibers to include in the listings wasn't easy. We did our best to make the book as inclusive as possible.

Having a comprehensive listing of what's offered is only the start of the ammunition selection process. The performance data allow you to compare velocity, energy, Taylor KO Index, bullet path, and wind drift for various ranges. With that information you can quickly see the trade-offs between light and fast bullets and the heavier and slower offerings.

In compiling this comparison data we have, whenever possible, used the manufacturer's own data. When the manufacturer was unable to supply specific data, we have used our best efforts to construct the missing items, using the best supporting information we could find.

Author's Notes

How to Select Ammunition

When you set out to buy ammunition, you usually know at least two things right at the start. You know what caliber you want, and you generally know what the ammunition will be used for—target practice or hunting. If the use is for target practice, the selection process is relatively easy because once the caliber has been selected, your primary concern is accuracy. All you need to do is test all the possible candidates in your caliber (and there won't be many) to see which performs best in your gun.

If the use is for hunting, the problem is considerably more complicated. The ultimate goal in the hunt is to obtain a reliable, clean kill. The first step in this is to get a hit. While a hit is certainly a necessary condition for a kill, after you get that hit the bullet has to finish the job. Up until about fifty years ago there was little done to design bullets specifically for the hunting application. Most of the attention that was paid to bullet performance was in the African calibers. Of course, the direct consequences (to the shooter) of poor bullet performance on an African hunt are much easier to visualize.

In the U.S., John Nosler developed the partition bullet and brought it into general usage. But Nosler's bullets were only available to reloaders, and as late as the mid-'60s they weren't being used by the major manufacturers to produce factory ammunition.

I can remember getting letters in that time period that asked if there were any bullets that got Nosler performance that didn't cost so much. After all, Nosler partition bullets cost about a dime each at that time. That seemed like a lot to folks who were used to paying no more than a nickel for the same weight bullet. What these folks didn't seem to realize was that it's the worst sort of economy to accept poor terminal performance to save a few cents per bullet. Thirty years ago it amazed me to see how many hunters couldn't understand that.

The results obtained with the Nosler and other premium performance hunting bullets that followed gradually convinced more and more hunters that these "expensive" bullets were a bargain when they produced good kills. But the benefits of these bullets were still limited to the hunters who were loading their own ammo. As time went on, the benefits of good bullets became clearer to an increasing proportion of hunters. Once they could see that there was a market established, it wasn't long before the custom reloaders began offering ammunition loaded with premium bullets.

When the ammo factories saw the custom reloaders cutting into their sales, they began to get the picture. Maybe there really was a market for ammunition loaded with premium bullets. From a tentative start with one premium loading, the main line factories are offering more and more choices of premium bullets. Today these choices include not only specialty bullets furnished by individual makers but also improved performance bullet designs produced by the ammo factories themselves. It is this increased offering of premium bullets that's mostly responsible for the quantum jump in the number of ammunition variations offered today.

As to the question of which premium bullet to use, that question goes beyond the scope of this book. There are now so many specialized forms of high-performance bullets incorporated in factory ammunition that giving meaningful advice as to which bullet might work best for your requirement just isn't possible without testing in your specific hunting conditions. It could very well be that several brands will give nearly identical performance in your application. That's a good class of problem. If you can't tell the difference between two or three brands, it really comes down to which is the easiest to locate.

The value of this book is that it can help to narrow the selection process down to a few promising candidates. With the list narrowed down to a manageable number, you can check with either the manufacturers or with gun shops in your area for more information to help you make your decision. My own recommendation is that when you find a bullet that works well for you, just stick with it. Unless you particularly enjoy trying new things, there's no point switching around if you are getting good results.

Within this book the listings are first divided into four sections: centerfire rifle cartridges, African cartridges, pistol cartridges, and rimfire cartridges. Within the separate parts, the listings are next divided by caliber and specific cartridge listings. Within a cartridge listing, the entries run in increasing order of bullet weight. Within a bullet weight, the listings are by bullet style, then velocity.

Find the cartridge you want, go to the desired bullet weight range, and then look through the listings to see just who makes what. Check the bullet description and the performance numbers to see if this is what you need. If you aren't quite satisfied, you might want to check listings for higher or lower bullet weights. If the performance you want isn't listed here, it probably isn't available as a factory product.

Standardization

It's only about 150 years since the breechloading guns firing ammunition contained in metal cartridge cases began to come into common use. The first of these cartridges in popular usage were rimfire rounds with about .50 caliber projectiles. Soon after, the centerfire primer system took over and continues today. All the early centerfire cartridges were loaded with black powder, but after about 1890 the nitro (smokeless) powders made their appearance and soon became the product of choice.

Smokeless powder brought the potential for greatly increased performance and a less critical cleaning requirement. This improved performance came at the price of significantly increased working pressures. As the government became more and more intrusive into our lives, the U.S. ammunition makers of the early 1900s recognized that sooner or later they were going to have to get together to adopt industry standards or be subject to government regulation of some type that was already beginning in Europe.

Ultimately, the Small Arms and Ammunition Manufacturers' Institute was formed in the U.S. SAAMI, as the group is called, is a voluntary organization. You don't have to be a member of SAAMI to manufacture ammunition in the U.S. Neither do members have to conform to SAAMI standards with their ammunition. For liability reasons, if for no other, a manufacturer would be foolish in the extreme to make a product that, at least in terms of chamber pressure, didn't follow the collective experience lead of SAAMI.

In addition to the velocity and chamber pressure standardization that SAAMI provides, this organization also distributes dimensional standards for both cartridges and chambers. It's easy to see the value of the dimensional standards. When you go out to buy ammunition for your gun, you want it to fit in that gun's chamber. This is an important function. Imagine the chaos that would result if every gun and every ammunition manufacturer worked to their own idea of the correct dimensions. Proprietary cartridges don't have this level of formal control (nor do they need it) until they cease to be exclusively proprietary and there is more than one maker of guns and ammunition for the caliber.

In Europe, most ammunition is controlled by an organization called Commission Internationale Permanente (CIP). CIP's approach to ammunition performance standardization seems to be that they will mandate maximum average chamber pressures but will allow individual manufacturers to load any bullet weight to any velocity desired so long as the maximum average pressure limits are observed. From the pure "blow up the gun" standpoint, controlling maximum pressures is all that matters. For European manufacturers, conforming to CIP standards is often a matter of law; it's not voluntary in many countries.

Unlike CIP, SAAMI specifications do control velocities and in doing so, specify the bullet weights that are associated with each of these standard velocities. The tolerance on the velocities called out in the specifications is a whopping 90 feet per second. That means that if a particular cartridge and bullet weight have a nominal velocity of 2,700 fps, the ammunition conforms to the standard if the average velocity for the lot is between 2,610 fps and 2,790 fps. No ammunition that I know of has shot-to-shot variations of 180 fps. That would be pretty awful stuff. But if a manufacturer makes up a run of ammunition that turns out to have an average velocity of, let's say, 2,615 fps when the nominal is 2,700 fps, and other characteristics (especially pressure) are nominal, the ammunition still conforms to the SAAMI standards.

Several ammunition companies have developed new high-energy propellants or highly specialized loading techniques that allow loading of some calibers to as much as 200 fps faster than the commonly advertised loading velocity levels without exceeding the allowable pressures. This high-performance ammunition offers a way to get near magnum performance from a "standard" rifle.

Energy vs. Momentum— The "Great" Debate

It started with hunters. There was and is a genuine difference of opinion between those folks who believe that high velocity is "everything" and the higher the striking velocity the better. They are opposed by another group who feels as strongly that as long as the striking velocity meets some minimum standard, the best measure of effectiveness is bullet weight and perhaps the size of the hole the bullet makes. It's pretty obvious that neither position, as simplified above, is a perfect measure of effectiveness. Within the last fifteen years or so, this debate has expanded from the hunting fields into the city streets and has become an important consideration in selection of ammunition for the self-defense application.

Because, for almost every specific factory cartridge, the user has a choice of bullet weight, this debate has become a problem for everyone who buys ammunition. You can characterize the debate as light and fast vs. heavy and slow. In physics terms, the debate comes down to whether effectiveness is more nearly comparable to energy than it is to momentum. Let's look at these terms a little more deeply.

Energy is defined as one-half of Mass times the square of Velocity. As long as we confine our discussion to the Earth (and I understand there's not much to hunt on the moon), we can think of Mass as being the same as weight. Momentum is defined as Mass times Velocity. When you look at these two definitions, you see that since Energy is proportional to Velocity squared and since Momentum is proportional to Velocity to the first power, Energy puts a premium in Velocity. That means that if you are on the side of light and fast, you are saying that Energy is the better measure of effectiveness.

As near as I can determine, the heavy and slow school was dominated in the early days by the professional British African hunters. This school of thought was responsible for pushing the African guns into some impressively large calibers. John "Pondoro" Taylor wanted a measure of effectiveness that he believed reflected his real world. He finally settled on a measure that is Bullet Weight (pounds) times Velocity (fps) times the Bullet Diameter (inches), and which was called the Taylor Knock Out Index. Taylor's KO Index needs some explanation. Taylor didn't think of the index as a measure of killing power but rather in the boxing sense of actually producing a short-term knockout of the animal. Because of this, he restricted the bullets to solids. His theory was that if the bullet didn't penetrate the skull on a brain shot, it still had the capability of stunning the animal for a few seconds so that the hunter could get off a second, more lethal shot.

Since the Taylor KO index represents a momentum value (times bullet diameter), it can be used as an alternate to the energy value. We have included it for all bullet styles for this purpose, although some bullets, especially varmint styles, will clearly not come close to meeting the original intent.

If you take either measure to extremes, you find that these measures both fall apart. The particles (electrons or some such little things) fired into patients to break up kidney stones are more energetic than bullets because they are traveling tens or even hundreds of thousands of feet per second. These highly energetic, but very light, particles break up the stones but don't kill the patient, clearly demonstrating that energy isn't a perfect measure of killing power. At the same time, a professional football quarterback can put more momentum (or

Taylor Index for that matter) on a thrown football than many guns can put onto their bullets. Since we don't lose a dozen or so wide receivers each weekend, momentum isn't a perfect measure of killing power either. Part of the explanation of this is because neither measure (energy or momentum) by itself, takes bullet construction into account. The terminal performance of the bullet on the target is at least (and possibly more) important than which measure you prefer to use. Generally the more effective cartridges have not only more energy but also more momentum, so it is not easy to find a clean demonstration of which measure is best.

When the military is looking at the effectiveness of weapon systems, one factor that gets into the effectiveness calculation is the probability of a kill for each shot. But the probability of a kill depends on two factors, the probability of a hit and the probability of a kill *given* a hit. The probability of a hit depends on many things, the accuracy of the shooter and his equipment and, of course, the accuracy of the bullet. Velocity probably plays a part in probability of a hit because higher velocity leads to flatter trajectory and lower wind drift.

The probability of a kill *given* a hit depends heavily on the bullet construction. It also depends on striking velocity. It is generally accepted that for a bullet to be effective it must penetrate deeply enough to reach a vital organ. It must also leave a large enough bullet path to provide a conduit for blood to "drain." This definition of how the bullet should perform leads us to our current crop of high-performance hollowpoint and controlled expansion bullets.

There are at least two applications that seem to contradict this conventional wisdom. The first comes along with light varmint hunting. In this application, the ability to get a hit isn't a foregone conclusion. A two-inch target at three hundred yards isn't all that easy to hit, and it takes an accurate bullet to even get a hit. At the same time, almost any hit with almost any bullet will produce a kill. Generally speaking, target bullets work well on light varmints. In this example, bullet performance after the hit isn't nearly as important as getting a hit in the first place. When we go through the probability of a kill equation, we find that in this application the hit probability is the controlling factor.

There is a class of heavy, and by implication, dangerous, game that is so tough skinned that conventional expanding bullets can't achieve the required penetration. Getting a hit is not the hard part but penetrating into a vital organ isn't easy (see the comments on Taylor's KO Index). For this application, hunters have concluded that solid bullets or bullets with very, very thick jackets, possibly with a core of compressed powdered tungsten, give the best performance. Within broad limits, accuracy doesn't matter.

Just as it isn't possible to generalize about the "best" bullet for every application, neither is it possible to come up with a single "best" caliber. In the final analysis, the shooter has to do his homework. He should look at the printed data and talk to other shooters to see what they are using. Then the final decision will be made when he tries his selection under his own field conditions. After all, performance in the field is the only thing that really matters.

Ballistics—Trajectory and All That Good Stuff

If you never take a shot beyond 100 yards or so, you can skip this section. But anyone, either hunter or target shooter, who is even thinking about firing at longer ranges should have at least a basic idea of why a bullet flies the way it does. Let's begin with the concept of a bullet that is fired in a more or less horizontal direction, the way 99.99 percent of all bullets are fired. After the bullet leaves the gun, it encounters two major external forces: One is gravity, and the other is the resistance of the air to the passage of the bullet. As you can no doubt guess, the air resistance acts to slow the bullet and gravity pulls it down. The big question is how much of each.

If we start with the gravity business, the classic comparison is that if a bullet were fired horizontally and another bullet were dropped from the height of the gun at the same time the gun was fired they would both reach the ground at the same time. That's true only in an over-simplified sense. In the real world, the fired bullet ends up getting some "lift" from the air in a fashion similar to a modern ski-jumper who gets some "lift" by laying out his body over the tips of the skis. This lift actually reduces the amount of bullet drop from what would be obtained in a place having no atmosphere (and therefore no air resistance), somewhere like the surface of the moon.

But air resistance plays a much more important role than simply reducing the amount of bullet drop. Air resistance starts slowing the bullet soon after it leaves the muzzle. The amount of slowing at any time during the flight depends on the speed of the bullet at that time. Near the muzzle the amount of slowing is relatively large, and, as the bullet moves downrange, the rate of slowing goes down. If you get out near the maximum range of the gun, or for bullets fired in a near-vertical direction, the bullet slows until reaching the top of the trajectory and then, as it starts down, starts speeding

up again. As the bullet continues downward, it encounters a more dense atmosphere and may actually start slowing again, even as it continues to fall. All that happens at ranges a long way beyond any range that is useful for either hunting or target shooting, and we don't need to concern ourselves with those very long trajectories.

Along about 1890, when smokeless powder was beginning to come into general military usage, the performance of military rifles and cannon took a quantum jump. The world's military powers all began a scientific study of the flight of projectiles. There were some data that existed, mostly empirical, but all of a sudden the ranges and times of flight to be considered were much larger than in the black-powder days.

It is interesting that at roughly the same time aviation began to develop heavier-than-air machines, first with gliders and then with powered aircraft. Since both ballistics and aerodynamics involve things flying through the air, you might wonder why the two sciences proceeded down very separate paths for the best part of seventy years. There's no explanation that satisfies me (and I've worked both sides of that street), but I suppose it boils down to the fact that the two groups didn't talk to each other. And to be fair, the velocity ranges of concern to each group were very different, indeed. Today, at the professional level, the speed of aeronautical things moving through the air has caught up with bullet speeds, and the science of ballistics is slowly moving closer to the aerodynamic approach.

For the average shooter, the older ballistics work is still valid and probably easier to use. Let's take a little look at how some of this science evolved and what it means to you today. The first big breakthrough came when the various countries began to conduct firing programs with "standard" projectiles. These programs were undoubtedly carried out separately and each country jealously protected its own data. The concept of a "standard" projectile was a big breakthrough. The problems resulting from trying to draw meaningful comparisons of data from numerous shots are hard enough if all the projectiles are alike. It isn't hard to understand that it becomes impossible if all the data are for different projectiles.

In the United States the work was led by a Capt. James M. Ingalls. Working with his own data and from earlier work by a Russian, Col. Mayevski, Ingalls produced a table that showed what happened to the speed of a bullet as it proceeded downrange. These data started with the highest velocity Ingalls could obtain from his test gun and documented the projectile's speed as a function of time. The tables also showed how far the bullet had traveled in the previous time increments.

Ingalls's second breakthrough came when he recognized that if the standard projectile was moving at, let's say, 2,000 feet per second, it didn't matter what the original launch speed was, the standard bullet was always going to slow down from 2,000 fps to 1,900 fps in the same elapsed time, and cover the same distance while it did so.

Those two insights allowed the use of Ingalls's tables to predict trajectory data for any gun firing the standard projectile, no matter what the muzzle velocity. Ingalls's standard projectile was one inch in diameter and weighed one pound. It was pretty clear that something more was needed if the concept was to be useful for more than just the standard projectile. Ingalls's third breakthrough came when he saw that he could "adjust" his tables by a factor that compensated for changes in the size, shape, and weight of the bullet. This factor became known as Ingalls's Ballistic Coefficient (C_1). If you knew the Ballistic Coefficient and muzzle velocity of a bullet, you could do a pretty fair job of predicting downrange performance. It wasn't long before some other bright lad found a way to predict the bullet drop when the downrange time, distance, and velocity factors were known.

Shortly after the turn of the century, a group called the Gavre Commission combined the ballistic work of Ingalls with that from other countries and produced some new and slightly improved tables. These new tables recognized that the velocity decay for a flat-based projectile was different from a boattail projectile. The different projectile shapes were assigned to tables known as G1, G2, G3, etc. (the "G" being for Gavre). With these new tables, the precision of the performance predictions was improved by a small amount. The multiple forms of ballistic coefficient generated a new class of problem. The ballistic coefficient for a projectile that used the G1 table could not be compared with the coefficient for a projectile that used the G2 table. That wasn't too much of a problem if you were working with only one or two different bullets, but it became a nightmare as the number of different bullet shapes increased.

Within about the last fifteen years or so, the various ammunition and bullet manufacturers have agreed that since any ballistic prediction is by its nature never absolutely perfect, they would standardize by using the equivalent of G1 coefficients for all bullets. The error that this approach introduces is quite small for all realistic ranges. As they say in aerospace, the results are "close enough for government work." By using one type of coefficient you can look at the coefficients for a variety of bullets and know that a coefficient of 0.400 is always "better" than a coefficient of 0.300. All the performance tables in this book are based on the G1 coefficients.

Maximum Effective and Point-Blank Ranges

There are two ballistic concepts that are much more important to hunters than to target shooters. These are Maximum Effective Range (MER) and Maximum Point-Blank Range (MPBR). These two ranges are not the same thing at all. MPBR is defined as the maximum range for which a gun will keep the bullets within a given distance of the aim point (without changing the sights or "holding off"). In its pure form it really doesn't depend on the ability of the shooter, only on his equipment.

MER, on the other hand, depends heavily on the shooter. By my definition, MER is the longest range at which a shooter can keep his bullets within a given size circle. To do this he can adjust sights, hold off, or use any other means at his disposal to get his hit. Any given shooter has a variety of MER's depending on the equipment he is using, the type of rest he has available, and what the weather, especially the wind, is doing.

Maximum Effective Range

Let's look at MER in more detail. I've done this test myself and with a variety of other shooters of different skill levels. Take a paper plate of any convenient size (6 or 8 inches diameter will do for a start) and staple the plate to a stick. We can start with a simple situation. Using any gun with iron sights from a freestanding position, how far out can you keep 9 out of 10 shots on the plate? If you are an "average" shot, you may have trouble keeping 9 out of 10 on the plate at 50 yards. Some people think that 4 out of 5 is a good enough hit percentage. What we are after here is to demonstrate a high probability that you can hit the vital zone with your first shot. This test is most valid if done using simulated hunting conditions, starting with a cold gun and without any shooting prior to your first shot. If 50 yards is too easy for you, then start increasing the range until you start to see misses. If you use a gun with a scope instead of the iron sights, you might move the plate out a bit farther still.

Now, let's take another step. If you fire from a good rest with a flat-shooting gun using a scope, you might be able to do the 9 out of 10 thing out to 200 or even 300 yards. If you aren't a practiced long-range shooter, it's unlikely you are going to be able to get out much beyond 300 yards. These tests are only for a stationary target. Could you do better on a moving target? Few of us can! If the target is presented at ranges other than the even 50- or 100-yard increments, the problem gets a lot harder, especially when the range starts to get to be much beyond 200 yards, and nearly impossible beyond 300 yards.

What does this test represent? The size of the plate represents the lethal area of deer-size game. If you can't reliably put your shots into the lethal area beyond any given range, then you shouldn't be taking a shot at game under those conditions. It is bad enough to miss completely but it is far worse to gut-shoot a buck and have it run off too fast and too far to track, only to die somewhere in the brush and be wasted. Maximum Effective Range depends far more on the shooter, his equipment, and the conditions under which the shot was taken than on the detail ballistics of the cartridge.

Maximum Point-Blank Range

The concept of MPBR is that for any given lethal zone diameter, and for any gun, there exists one sighting distance at which the gun can be "zeroed" that will cause the trajectory between the first crossing of the sight line (for practical purposes, starting at the muzzle) and the zero range to be above the sight line by no more than the radius of the lethal area. Furthermore, from the zero range on out to MPBR, the bullet's trajectory will be below the sight line. At MPBR the impact point will be one lethal zone radius BELOW the sight line. The idea is that out to the MPBR, the bullet will always be within the lethal radius from the sight line and the shooter shouldn't have to hold high (or low), but instead to always hold dead on. Using a .22-250 for an example, in a varmint application with a 2-inch diameter lethal circle, the bullet will be one inch high at 125 yards if you have the gun zeroed for 200 yards. The MPBR for this example is 230 yards. If you had a lethal circle that was 8 inches in diameter, the 4-inch high point would be reached at 170 yards with a 275-yard zero and a MPBR of 370 yards. You can see that the MPBR depends heavily on the size of the lethal circle.

The concept of MPBR is not without controversy. The controversy isn't about the concept itself or even the actual MPBR numbers; it is about what happens in

the field. I know two professional guides who are 180 degrees apart on this subject. One loves the idea. He says that he has the client demonstrate where the client's gun is zeroed before the hunt. The guide judges the range to the target and instructs the client to hold dead on if he (the guide) decides the range is within MPBR. The idea of Maximum Effective Range also plays into his decision to tell the client to shoot. For most shooters with modern guns, their Maximum Effective Range is shorter than the MPBR, sometimes very much shorter.

The second professional guide says that in his experience the idea of MPBR may be technically correct but is terrible in practice. No matter where he tells the client to hold, the client makes his own adjustment for range. Furthermore, the client almost always overestimates the range. The combination of holding high for a real range that is shorter (often much shorter) than the estimated range and shooting at a range shorter than the zero range (where the trajectory is above the sight line) results far too often in shooting over the back of the game.

What these two positions clearly say to me is that if the shooter can determine the range accurately, either by lots of

Maximum Point-Blank Range Scope Sight Use This Table for 2-Inch-Diameter Lethal Zone (Varmint Shooting)					
Ballistic Coefficient	Muzzle Velocity fps	First LOS Crossing yards	High Point yards	"Zero" Range yards	Max. Point Blank Range yards
0.200	2,000	25	70	110	125
	2,500	35	85	135	155
	3,000	40	105	160	185
	3,500	45	120	185	210
	4,000	50	135	210	235
0.300	2,000	25	75	115	130
	2,500	35	85	140	160
	3,000	40	105	165	190
	3,500	45	120	195	220
	4,000	50	140	220	250
0.400	2,000	25	75	115	135
	2,500	35	85	145	170
	3,000	40	110	170	200
	3,500	45	125	200	230
	4,000	55	140	225	260
0.500	2,000	25	75	120	135
	2,500	35	90	145	170
	3,000	40	110	175	230
	3,500	45	125	200	230
	4,000	55	145	230	265

practice estimating, by using a rangefinder, or by listening to his guide, AND if he knows where his gun will shoot at that range, AND if he can hold well enough, he will get good hits. If you don't start holding off, the MPBR concept puts an upper limit on the range at which you should attempt a shot. If you want to start holding off, you need to know both the range to the target and how high you must hold to get hits at that range. That knowledge doesn't come without lots and lots of practice. Finally, given a good hit, the rest depends on the quality of the bullet. If the bullet doesn't perform, everything else is wasted.

The tables nearby list the MPBR for bullets with assorted ballistic coefficients at different muzzle velocities, and for both 2-inch and 6-inch lethal areas. You may wonder why bullet weight isn't included in the tabulated data. The fact is that any bullet with a ballistic coefficient of, let's say 0.350, that launched at a given muzzle velocity (for instance 3,000 fps) will fly along the same trajectory path. Bullet weight is accounted for in the ballistic coefficient number because ballistic coefficient equals sectional density divided by the bullet's form (shape) factor. Sectional density is bullet weight divided by the bullet diameter squared. The caliber of the gun doesn't matter in the table but is also indirectly involved in two ways. Small calibers have lighter bullets that need better form factors to have the same ballistic coefficient as a heavier bullet. The smaller caliber guns generally produce higher velocities than the larger calibers. These tables are calculated using a sight height above the bore line of 1.5 inches, which is pretty standard for most scope-sighted rifles. A third table shows what happens if the gun has iron sights (0.9 inches above the bore line).

To use these tables, start with the table for the correct lethal zone diameter (2 inches or 6 inches). Once you find the approximate ballistic coefficient of your bullet and pick the line nearest to your muzzle velocity, you can read across the table to see the following data: the range at which the bullet first crosses the line of sight (going up), the distance to the high point of the trajectory, the range that the gun should be zeroed (where the bullet comes down to cross the line of sight for the second time, and finally the Maximum Point-Blank Range.

For example, let's look at MPBR for a .300 Winchester firing 180-grain Hornady ammunition. The Ballistic coefficient of the ammunition is 0.438 and the nominal muzzle velocity is 2,960 fps. We will use the chart for scope sights and a 6-inch lethal zone. In the 6-inch table, use a Ballistic Coefficient of

0.400. If you look at the listing for a BC of 0.500, you will find that large changes in the BC number don't make a huge difference in the MPBR data. Likewise, use the line for 3,000 fps. Again that's close enough to 2,960. You can then read that the first crossing is at 25 yards (all these data are rounded to the nearest 5 yards), the high point is at 140 yards, the zero range is 250 yards, and the MPBR is 290 yards.

If the gun had iron sights the numbers would be as follows:

first crossing .... 15 yards
high point .... 130 yards
zero range .... 240 yards
MPBR .... 285 yards

Maximum Point-Blank Range
Scope Sight
Use This Table for 6-Inch-Diameter Lethal Zone
(Large Game Shooting)

Ballistic Coefficient	Muzzle Velocity fps	First LOS Crossing yards	High Point yards	"Zero" Range yards	Max. Point Blank Range yards
0.200	2,000	20	90	155	180
	2,500	20	110	190	220
	3,000	25	130	225	260
	3,500	30	150	260	300
	4,000	35	165	290	335
0.300	2,000	20	95	165	195
	2,500	20	115	205	235
	3,000	25	140	240	280
	3,500	30	160	275	320
	4,000	35	180	310	360
0.400	2,000	20	95	170	200
	2,500	20	115	210	245
	3,000	25	140	250	290
	3,500	30	160	285	335
	4,000	35	185	325	375
0.500	2,000	20	100	175	205
	2,500	25	120	215	250
	3,000	25	140	255	300
	3,500	30	165	295	345
	4,000	35	190	330	390

You can see here that the sight height above the bore line does change the numbers a little. I wonder how many people can keep 9 out of 10 (or even 4 out of 5) in 6 inches at 285 yards using a gun with open iron sights. I'll wager there aren't many. I sure can't.

Maximum Point-Blank Range Iron Sights Use This Table for 6-Inch-Diameter Lethal Zone (Large Game Shooting)					
Ballistic Coefficient	Muzzle Velocity fps	First LOS Crossing yards	High Point yards	"Zero" Range yards	Max. Point Blank Range yards
0.200	2,000	10	85	150	175
	2,500	15	105	185	220
	3,000	15	125	220	255
	3,500	20	140	250	290
	4,000	20	160	280	325
0.300	2,000	10	85	160	185
	2,500	15	110	195	230
	3,000	15	130	235	275
	3,500	20	150	270	315
	4,000	20	170	300	355
0.400	2,000	10	90	165	195
	2,500	15	110	200	240
	3,000	15	130	240	285
	3,500	20	150	275	325
	4,000	20	175	310	370
0.500	2,000	10	90	165	200
	2,500	15	115	205	245
	3,000	15	130	245	290
	3,500	20	155	285	335
	4,000	25	175	320	380

There Is Always a Trade-Off

When we select ammunition for a specific application, there's always a trade-off. If we can have the luxury of selecting the very "best" gun for the application, the trade-offs are lessened. But few of us are fortunate to have guns for every possible situation. For certain applications, like varmint or target shooting, we can trade hunting bullet effectiveness for accuracy. For general hunting, the last little bit of accuracy, or for that matter the last little bit of ballistic coefficient, doesn't justify trading away any bullet performance.

You can begin to get the idea here. In the final analysis you, and only you, can decide which way to lean on each of these trades. You can get help from various sources including friends, gun shop employees, and especially from professional guides. You have to then evaluate just what the advice from each source is worth. The guide probably has the most to lose if he gives you bad advice. If your hunt isn't successful, you won't be back.

This book provides the technical information on cartridge performance you need to help you decide on what trades are best from the purely technical standpoint. We don't even attempt to take a stand on light and fast vs. heavy and slow. That's a fundamental choice you must make for yourself. Good luck and good shooting!

CAUTION and WARNING!

Common sense needs to be used when handling and discharging a firearm. **Always keep the following in mind:** Always point a firearm in a safe direction, and never point a firearm at another person. Treat all firearms as though they are loaded. Wear eye and hearing protection at all times when handling firearms. Only adults competent in handling firearms and ammunition should ever attempt to load or discharge a firearm.

Do not attempt to use this book to handload your own ammunition up to the bullet velocities listed on these pages. Your firearm may not be able to withstand the pressures generated by the loads listed in this book. If you aren't sure about your gun, consult a competent gunsmith. The handloading of ammunition and the discharging of a firearm should never be attempted without the supervision of an adult experienced in both handloading and firearms. Do not attempt to handload ammunition without knowing how to read signs of (excessive) pressure in both guns and ammunition. Keep these principles of safety in mind so as to provide a safe environment for everyone.

How to Use This Book

The listings in this book are divided into four sections: Small and Medium Cartridges, Dangerous Game Cartridges, Pistol and Revolver Cartridges, and Rimfire Cartridges. Within these four sections the listings begin with the smallest and end with the largest caliber. If you refer to the sample listing on page xxi, you will see that there is a ❶ description (with short historical notes) and a ❷ drawing of each caliber.

For each caliber we list a ❸ Relative Recoil Factor to give you some idea of the recoil the caliber will generate. Most shooters can handle the recoil of a .30-06 reasonably well with some practice; it has a relative recoil factor of 2.19. Should you carry a .700 Nitro Express, you will note that its relative recoil factor is 9.00, more than four times what a .30-06 generates! Relative recoil factor is based on the muzzle momentum of the bullet and the expelled powder gas for a typical loading. Below the relative recoil factor, you will find the **controlling agency for standardization of this ammunition.** In the Author's Notes there is a discussion of standardization and what it means to the shooter.

Item ❹ gives the standard performance numbers that have been established for this caliber. We list the Maximum Average Pressures obtained by both the Copper Crusher and the Transducer methods. Number ❺ gives two figures. The first is the standard barrel length the factories use to create their velocity figures. If you shoot a .300 Weatherby Magnum with a 22-inch barrel, do not be surprised if your chronograph shows velocities significantly below the factory figures provided here; the reason for this is that this caliber is tested in the factory with a 26-inch barrel. The second number is the twist rate of the rifling the factories use in the factory test barrels.

Next come the listings of all the factory loads currently available in this caliber. The listings start with the lightest bullet and progress **down** to the heaviest bullet available in factory loadings. Within each bullet weight there are listings for the bullet styles available. Within each bullet style, the listings run from the highest muzzle velocity to the lowest.

Under each specific loading you will find ❻ manufacturer, ❼ bullet weight, ❽ the manufacturer's name for his loading, and ❾ the factory stock number (in parentheses) that can be used to order that particular cartridge and load. The individual cartridge listings also provide ❿ velocity, ⓫ energy, and ⓬ Taylor's Knock-Out Index. (See "The 'Great' Debate" in "How to Select Ammunition" for a discussion of the significance of these factors.) ⓭ The figures for the category of "path • inches" show the bullet's position relative to the line of sight at ranges up to 1,000 yards (depending on the listings section). For small and medium centerfire listings, the figures assume a scope-sighted rifle that is set at a 200-yard "zero." For the dangerous-game calibers, the figures are based on iron sights and a 150-yard "zero." The rimfire listings are based on scope sights with a 100-yard "zero." Note that for handgun cartridges the "path • inches" listing is replaced by figures for mid-range trajectory height in inches. ⓮ The category of "wind • drift inches" shows how much the bullet is pushed off-course by a direct crosswind of 10 mph at ranges out to 1,000 yards. The G1 Ballistic Coefficient value ⓯ is useful for those shooters who want to calculate their own ballistic data.

Find the cartridge you want, locate the desired bullet weight, and then look through the entries to see who makes what. Check the performance numbers to see whether a particular load is what you need. If you aren't quite satisfied, you can check the listings for the next higher or lower bullet weight in this caliber. If you are just playing "What if," you can also try other calibers to see if you can find a load that would do your job. If you cannot find a load with the performance you want, it probably isn't available as a factory product. Please note that some listings in this book are followed by a legend, such as "(Discontinued in 2004)." These listings have dropped out of the manufacturer's catalog but may still be available at your local supplier.

.223 Remington

When the .223 Remington cartridge was adopted by the U.S. Army as M193 5.56mm Ball ammunition in 1964, that action ensured that the .223 would become the most popular .22 centerfire in the list. Every cartridge that becomes a U.S. Army standard, with the possible exception of the .30 Carbine, has gone on to a long and useful commercial life. Just look the .45-70, the .45 Colt, the .30-06, and the .45 ACP. The .223 case has been "necked" to every possible size, the TCU series of cartridges and the .30 Whisper being examples.

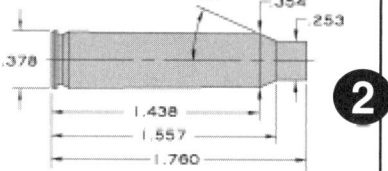

Even without the military application, the .223 Remington had plenty of potential to become popular. Based on the .222 and the .222 Remington Magnum, the .223 provides an excellent balance of accuracy and performance with good case and barrel life. It has become the standard by which all other .22s are judged.

Some guns chambered for the .223 have barrel twists as slow as 1 turn in 14 inches. That twist provides enough stability for bullets up to 55-grains but begins to be marginal above that level. Today there are bullets weighing as much as 77-grains available in loaded ammunition (80-grain bullets for the handloader). It takes a faster twist to stabilize these heavier bullets, and some gunmakers offer barrel twists as fast as 1 turn in 7 inches in this caliber. If you have one of the fast twist barrels, you may have problems with the very light varmint loads. The high velocities attained and the quick twist combine to produce a bullet spin rate that can literally rip thin jacketed bullets apart. Guns equipped with quick twist barrels will usually do better with 55-grain and heavier bullets.

Relative Recoil Factor = 0.80

Specifications:

Controlling Agency for Standardization of this Ammunition: SAAMI

Bullet Weight Grains	Velocity fps	Maximum Average Pressure Copper Crusher	Transducer
53	3305	52,000 cup	55,000 psi
55	3215	52,000 cup	55,000 psi
60	3200	52,000 cup	55,000 psi
64	3000	52,000 cup	55,000 psi

Standard barrel for velocity testing: 24 inches long—1 turn in 12-inch twist.

Availability:

Winchester 35-grain Ballistic Silvertip (S22336RLF) LF
G1 Ballistic Coefficient = 0.201

Distance • Yards	Muzzle	100	200	300	400	500	600	800	1000
Velocity • fps	3800	3251	2766	2330	1935	1586	1300	981	836
Energy • ft-lbs	1110	813	588	417	287	193	131	75	54
Taylor KO Index	4.3	3.6	3.1	2.6	2.2	1.8	1.5	1.1	0.9
Path • Inches	-1.5	+0.9	0.0	-5.5	-17.5	-38.9	-73.4	-215.1	-499.9
Wind Drift • Inches	0.0	1.1	4.9	11.8	22.7	39.0	62.0	129.7	219.2

Nosler 35-grain Ballistic Tip (06088) LF
G1 Ballistic Coefficient = 0.238

Distance • Yards	Muzzle	100	200	300	400	500	600	800	1000
Velocity • fps	3750	3214	2742	2315	1927	1587	1301	984	839
Energy • ft-lbs	1093	803	584	416	289	196	131	75	55
Taylor KO Index	4.2	3.6	3.1	2.6	2.2	1.8	1.5	1.1	0.9
Path • Inches	-1.5	+1.0	0.0	-5.6	-17.8	-39.5	-75.3	-216.7	-501.2
Wind Drift • Inches	0.0	1.1	4.8	11.8	22.7	38.9	61.7	128.8	217.6

Black Hills 36-grain Barnes Varmint Grenade (D223N15)
G1 Ballistic Coefficient = 0.149

Distance • Yards	Muzzle	100	200	300	400	500	600	800	1000
Velocity • fps	3750	3034	2421	1886	1439	1127	964	787	664
Energy • ft-lbs	1124	736	469	284	166	101	74	50	35
Taylor KO Index	4.3	3.5	2.8	2.2	1.7	1.3	1.1	0.9	0.8
Path • Inches	-1.5	+1.2	0.0	-7.2	-24.5	-58.9	-120.6	-365.2	-836.2
Wind Drift • Inches	0.0	1.6	7.0	17.6	35.7	63.4	100.4	194.0	313.0

Conversion Charts

After the first edition of Ammo & Ballistics was issued, we were surprised to discover how many readers wanted to use data in the metric system. In this edition we are providing conversion charts to aid these users. Most of the conversions are straightforward, but there is one point that should be borne in mind. It is easy to convert the velocities shown to meters per second. But when you do that you must remember that those metric velocities are correct at the ranges shown in YARDS. You can also convert the yardages into metric distances, and the results will be correct, but the increments will be 91.44 meters (not 100 meters). If you want metric velocity data for even 100-meter increments, the easiest way is to use the ballistic coefficient and the metric muzzle velocity with an external ballistics program (like Ken Oehler's Ballistic Explorer) to compute a new, all-metric table.

To accomplish these conversions, you have two choices. The charts give conversion values for even increments. If you want to find the metric equivalent of 2,750 fps, you add the numbers for 2,000 fps (609.57), 700 fps (213.35), and 50 fps (15.24). The result is 838.16 meters per second.

As an alternative, you can simply multiply the number to be converted by the conversion factor given. In the above example, this would be: 2,750 fps times .3048 equals 838.2 meters per second. In all conversions, you can expect small differences in the last digit due to round-off effects.

Medium Length Conversions
Foot/Meter Length Conversions
Feet times .3048 equals Meters
Meters times 3.281 equals Feet

Feet	Meters	Meters	Feet
1,000	304.8	1,000	3,281.0
900	274.3	900	2,952.9
800	243.8	800	2,624.8
700	213.4	700	2,296.7
600	182.9	600	1,968.6
500	152.4	500	1,640.5
400	121.9	400	1,312.4
300	91.4	300	984.3
200	61.0	200	656.2
100	30.5	100	328.1
90	27.4	90	295.3
80	24.4	80	262.5
70	21.3	70	229.7
60	18.3	60	196.9
50	15.2	50	164.1
40	12.2	40	131.2
30	9.1	30	98.4
20	6.1	20	65.6
10	3.0	10	32.8
9	2.7	9	29.5
8	2.4	8	26.2
7	2.1	7	23.0
6	1.8	6	19.7
5	1.5	5	16.4
4	1.2	4	13.1
3	0.9	3	9.8
2	0.6	2	6.6
1	0.3	1	3.3

Large Length Conversions
Yard/Meter Length Conversions
Yards times .9144 equals Meters
Meters times 1.094 equals Yards

Yards	Meters	Meters	Yards
1,000	914.4	1,000	1,094.0
900	823.0	900	984.6
800	731.5	800	875.2
700	640.1	700	765.8
600	548.6	600	656.4
500	457.2	500	547.0
400	365.8	400	437.6
300	274.3	300	328.2
200	182.9	200	218.8
100	91.4	100	109.4
90	82.3	90	98.5
80	73.2	80	87.5
70	64.0	70	76.6
60	54.9	60	65.6
50	45.7	50	54.7
40	36.6	40	43.8
30	27.4	30	32.8
20	18.3	20	21.9
10	9.1	10	10.9
9	8.2	9	9.8
8	7.3	8	8.8
7	6.4	7	7.7
6	5.5	6	6.6
5	4.6	5	5.5
4	3.7	4	4.4
3	2.7	3	3.3
2	1.8	2	2.2
1	0.9	1	1.1

Velocity Conversions
FPS/Meters Per Sec. Conversions
FPS times .3048 equals MPS
MPS times 3.281 equals FPS

FPS	MPS	MPS	FPS
4,000	1,219.20	2,000	6,562.0
3,000	914.40	1,000	3,281.0
2,000	609.60	900	2,952.9
1,000	304.80	800	2,624.8
900	274.32	700	2,296.7
800	243.84	600	1,968.6
700	213.36	500	1,640.5
600	182.88	400	1,312.4
500	152.40	300	984.3
400	121.92	200	656.2
300	91.44	100	328.1
200	60.96	90	295.3
100	30.48	80	262.5
90	27.43	70	229.7
80	24.38	60	196.9
70	21.34	50	164.1
60	18.29	40	131.2
50	15.24	30	98.4
40	12.19	20	65.6
30	9.14	10	32.8
20	6.10	9	29.5
10	3.05	8	26.2
9	2.74	7	23.0
8	2.44	6	19.7
7	2.13	5	16.4
6	1.83	4	13.1
5	1.52	3	9.8
4	1.22	2	6.6
3	0.91	1	3.3
2	0.61		
1	0.30		

Energy Conversions
Ft-Lbs times 1.356 equals Joules
Joules times .7376 equals Ft-Lbs

Ft-Lbs .... Joules	Joules ....Ft-Lbs
10,000 13,560	10,000 7,376
9,000 12,204	9,000 6,638
8,000 10,848	8,000 5,901
7,000 9,492	7,000 5,163
6,000 8,136	6,000 4,426
5,000 6,780	5,000 3,688
4,000 5,424	4,000 2,950
3,000 4,068	3,000 2,213
2,000 2,712	2,000 1,475
1,000 1,356	1,000 738
900 1,220	900 664
800 1,085	800 590
700 949	700 516
600 814	600 443
500 678	500 369
400 542	400 295
300 407	300 221
200 271	200 148
100 136	100 74
90 122	90 66
80 108	80 59
70 95	70 52
60 81	60 44
50 68	50 37
40 54	40 30
30 41	30 22
20 27	20 15
10 14	10 7

Weight Conversions
Grains times 0.06481 equals Grams
Grams times 15.43 equals Grains

Grains ... Grams	Grams ... Grains
1,000 ... 64.81	100 ... 1,543.0
900 ... 58.33	90 ... 1,388.7
800 ... 51.85	80 ... 1,234.4
700 ... 45.37	70 ... 1,080.1
600 ... 38.89	60 ... 925.8
500 ... 32.41	50 ... 771.5
400 ... 25.92	40 ... 617.2
300 ... 19.44	30 ... 462.9
200 ... 12.96	20 ... 308.6
100 ... 6.48	10 ... 154.3
90 ... 5.83	9 ... 138.9
80 ... 5.18	8 ... 123.4
70 ... 4.54	7 ... 108.0
60 ... 3.89	6 ... 92.6
50 ... 3.24	5 ... 77.2
40 ... 2.59	4 ... 61.7
30 ... 1.94	3 ... 46.3
20 ... 1.30	2 ... 30.9
10 ... 0.65	1 ... 15.4
9 ... 0.58	0.9 ... 13.9
8 ... 0.52	0.8 ... 12.3
7 ... 0.45	0.7 ... 10.8
6 ... 0.39	0.6 ... 9.3
5 ... 0.32	0.5 ... 7.7
4 ... 0.26	0.4 ... 6.2
3 ... 0.19	0.3 ... 4.6
2 ... 0.13	0.2 ... 3.1
1 ... 0.06	0.1 ... 1.5
0.9 ... 0.06	
0.8 ... 0.05	
0.7 ... 0.05	
0.6 ... 0.04	
0.5 ... 0.03	
0.4 ... 0.03	
0.3 ... 0.02	
0.2 ... 0.01	
0.1 ... 0.01	

Inch-millimeter Conversions
Inches times 25.40 equals millimeters
Millimeters times .03937 equals inches

in .... mm	mm ....in
1,000 25,400	10,000 ... 393.70
900 22,860	9,000 ... 354.33
800 20,320	8,000 ... 314.96
700 17,780	7,000 ... 275.59
600 15,240	6,000 ... 236.22
500 12,700	5,000 ... 196.85
400 10,160	4,000 ... 157.48
300 7,620	3,000 ... 118.11
200 5,080	2,000 ... 78.74
100 2,540	1,000 ... 39.37
90 2,286	900 ... 35.43
80 2,032	800 ... 31.50
70 1,778	700 ... 27.56
60 1,524	600 ... 23.62
50 1,270	500 ... 19.69
40 1,016	400 ... 15.75
30 762	300 ... 11.81
20 508	200 ... 7.87
10 254	100 ... 3.94
9 229	90 ... 3.54
8 203	80 ... 3.15
7 178	70 ... 2.76
6 152	60 ... 2.36
5 127	50 ... 1.97
4 102	40 ... 1.57
3 76	30 ... 1.18
2 51	20 ... 0.79
1 25	10 ... 0.39
0.9 23	9 ... 0.35
0.8 20	8 ... 0.31
0.7 18	7 ... 0.28
0.6 15	6 ... 0.24
0.5 13	5 ... 0.20
0.4 10	4 ... 0.16
0.3 8	3 ... 0.12
0.2 5	2 ... 0.08
0.1 3	1 ... 0.04

DISCLAIMER

The reader may notice certain discrepancies in spellings in this book. Unlike most books where consistency is paramount, the author and the publisher decided to reproduce the actual word usage for the cartridges exactly as they appear in the manufacturers' catalogs. This is where the discrepancies arise. One manufacturer may use ***Soft Nose*** as part of its description for the ammunition in its catalog, while another might use ***Softnose, soft nose, or softnose.*** (Safari Press's style is to combine the two words and lowercase it—softnose.) Furthermore, when you go to your local shop to buy any ammunition, you may find the word usage on the box differs slightly from what you will find in this book. This further discrepancy is the result of a manufacturer failing to replicate exactly the wording in its catalog to that found on its cartridge boxes. For the purpose of this book, we have used only those names supplied from the manufacturer's catalog, which may differ slightly from what is printed on the cartridge box or seen elsewhere.

A Note about Packaging, Older Loads, and Updates Made to Each Edition

Some years ago, a hunter in Southern California was preparing for a hunt in Africa when his companion for the trip called from Germany to ask him to pick up some ".460 Weatherby ammo in a local shop," as the friend could find none in Germany and he needed some for his safari. It will be a surprise to almost none of our readers that .460 Weatherby cartridges are not found in every gun shop in town, but our friend in Germany didn't know that. After a lot of calls, the hunter found one shop claiming to have a package of twenty. When the hunter arrived, the package actually contained only nineteen cartridges, and it was surmised that a shop employee had removed one ten years earlier for a cartridge collector! The original sticker price on that box was $18.95 (!), and it was barely readable because it was so old. Obviously the package had been sitting in that store for over two decades. What is the moral of this story? Not all ammo you buy today in a gun shop was delivered straight from the plant last month!

So, when you look over the performance and stock numbers listed in these pages, be aware that, depending on the date of manufacture, a lot may have changed between the time the package in your hand was manufactured and the data in this book were produced. First of all, always note the factory stock number (in parentheses) because the exact name of the load may change over time. For instance, Remington now loads and has loaded a 45-grain softpoint in .22 Hornet. Some years ago it was listed on the box as "45-grain PTD Softpoint," but the most current packaging we could find lists this as "45-grain PSP," yet the load and the factory stock number (R22HN1) are exactly the same.

We happen to shoot Remington's excellent ammo a lot, and so, maybe, we get to look at it more than some other brands, which is why we have taken yet another example from Remington. In the spring of 2009, Remington's Web site gave the following listing: the 300 Remington Ultra Magnum, loaded with a 150-grain Core-Lokt PSP bullet, was designated as RL300UM1 under "order number." However, when we got hold of a package of the same caliber, it showed an order number of R300UM1. Obviously, we cannot physically get hold of every package that the manufacturers produce for 2,000-some loads, and so the reader must keep in mind that small variations can occur. The order or stock numbers on the actual package, on the Web sites, and in the catalogs may differ slightly.

Second of all, things change, even if the name, the stock number, and the package are identical to the one made last year. One reason for the great popularity of this book is that Mr. Robert Forker does his homework for each edition—again and again! About 50 percent of all loads that were listed in the third edition and are now listed in the fourth edition have seen material changes in ballistic coefficients and/or velocity. This means that the table for that load altered in a significant manner. And that is only for the same loads listed from one edition to another. Additionally, we have eliminated dozens of loads no longer made and added scores of new loads brought onto the market since the fourth edition. In short the team of Bob Forker and Safari Press updates *Ammo & Ballistics* significantly each edition. Even if nothing appears to have changed on the package, chances are there have been changes to the performance of the rounds found in that package.

Classic Big-Game Cartridges

Ten Game-Getting Cartridges That Have Stood the Test of Time

by John Barsness

The .375 H&H has often been called the most versatile big-game round ever designed, particularly useful in Africa where you never know what you might run into next—the reason that about 40 percent of the author's kudus have been taken with the old .375.

"Classic" is one of the most flexible words in the history of the English language. A recent edition of *Webster's Unabridged Dictionary* lists twenty meanings for classic, several involving ancient Greece and Rome, but yet another is "automobiles made between about 1925 and 1948." This would be disputed by those who went to high school in the 1960s, and who firmly believe the 1956 Chevrolet to be the true classic.

In this essay the definition of classic will focus on centerfire cartridges introduced before World War II. If hunters still like and use a cartridge that's been around for most of a century, it qualifies for the seventh definition in Webster for the term classic: "of enduring interest, quality,

or style." Such cartridges have stuck around primarily because they really work (quality), but often they also evoke the history of hunting (enduring interest), and perhaps a certain style. The style might be due to well-known hunters who favored a particular cartridge, or even the simple poetry of the cartridge's name. Any hunter who misses the lilt in "seven by fifty-seven" probably shoots a rifle chambered for one of the more obscure alphabet magnums.

A cartridge may initially become popular because of the human yearning for novelty, especially if jump-started by advertising. However, initial popularity is often fueled by what one friend calls "churners," the rifle enthusiasts who constantly

The .257 Roberts is not just a highly capable rifle for game like pronghorn and mule deer, but it fits neatly into lightweight rifles like this Kimber 84M Classic Select Grade that weighs less than 6½ pounds with a scope.

search for the perfect cartridge. These guys are much like dogs—quintessential optimists who, despite previous experience, feel that somehow the next cartridge (or bowl of food) will change their lives.

The history of hunting cartridges is littered with "in-cartridges" abandoned by the churners after they discovered that the cartridges only shot bullets instead of a light-ray from beyond the known universe. The list of somnolent "in-cartridges" is long, ranging from the .303 Savage, essentially a .30-30 advertised in the 1890s as good for everything from ego to elephants, to the 7mm STW, which has now disappeared completely from the Remington lineup and apparently supplanted by the 7mm RUM.

Sometimes a classic cartridge receives an initial boost by being a military cartridge, providing lots of cheap rifles and ammo, but that isn't a guarantee of success, even in America. Observe the .30-40 Krag, a cartridge much admired by the philosophical opposite of the churner, those who prefer instant obsolescence to instant celebrity.

A true classic cartridge has survived because it has done its job, over and over and over again. It may not be among the top sellers right now, but, like the 1898 Mauser or Nosler Partition, it set the standard, casting a long, steady shadow over anything that follows. Classic cartridges are often favored by experienced hunters who accept the fact that rifles are relatively simple mechanisms for pushing bullets, and the rest of the deal is up to the shooter.

The "top ten list" is one of the classic forms of the English language, partly because it's less difficult to compose than an Elizabethan sonnet. Here's one hunter's top-ten list, from smallest- to largest-bore:

.257 Roberts

This choice was tough for a hunter who owns .25 caliber rifles chambered for cartridges from the .25-35 WCF to the .257 Weatherby Magnum, but the Roberts stands out. It fits neatly into cute little bolt actions such as the Kimber Model 84, yet still performs admirably on deer-size animals, a category that describes 90 percent of the world's big game. Many hunters think of it as a woman's and a kid's cartridge, but a lot of he-men carry a .257 because it works.

The case is the 7x57 Mauser necked down, and it was first developed as a wildcat for long-range woodchuck shooting in the 1920s by Ned Roberts, a well-known "gun crank" and writer. It was called the .25 Roberts until Remington introduced the .257 Remington-Roberts in 1935. Soon Winchester brought out an identical cartridge called the .257 Winchester-Roberts. Thankfully this hyphenation soon ended; otherwise, the hunting world would resemble a New England cocktail party full of interbred bluebloods such as the .257 Griffin & Howe-Roberts and .257 Thompson-Center-Roberts.

6.5x55 Swedish Mauser

Some hunters might instead pick the 6.5x54mm Mannlicher, perhaps because it was used by Margot Macomber to shoot her husband in Ernest Hemingway's famous short story. But the Mannlicher round is rarely seen any more, while the 6.5x55 remains fairly popular.

When handloaded with bullets in the 120-grain range, the 6.5x55 is very similar to the .257 Roberts, and I've used it to take pronghorns out to 400 yards. However, bullets up to 160 grains can also be used, making the 6.5x55 more suitable for game larger than deer. It has accounted for untold numbers of alg (the Swedish name for the animal Americans call moose), as well as many Scandinavian polar bears. North American elk hunters are more likely to use spitzer bullets in the

The .30-06 has been a worldwide big-game cartridge for over a century, and has been chambered in about every bolt action made, from the pre-WWII Model 70 Winchester (top) to the modern New Ultra Light Arms Model 24 (below).

The .416 Rigby was made famous by Robert Ruark; this CZ 550 Magnum was modified to balance much like Harry Selby's rifle, and it even weighs the same 9¼ pounds.

140-grain range rather than traditional bluntnose 155- to 160-grain bullets.

The popular name is deceiving, since the cartridge was jointly developed in the early 1890s by Sweden and Norway for military use. Sweden chambered the round in Mauser rifles, while Norway used Krag-Jorgensens, so it could also be inflicted with too many hyphens—6.5x55 Norwegian-Swedish-Mauser-Krag-Jorgensen, anyone?

.270 Winchester

In the early 1990s the Boone and Crockett Club sent me the list of which cartridges accounted for record animals taken during the previous few years, and the 1993 record book included my article on the data, "The Caliber of Record." The .270 appeared regularly in several categories, but led all others in bighorn sheep, desert sheep, and Coues deer. As I noted then, this could be seen almost as a fashion statement, honoring a well-known writer who really liked hunting both bighorn sheep and Coues deer.

Only three other commercial .270 cartridges have ever been introduced, but none have come close to the popularity of the original. A couple of years ago I was in a local sporting goods store when two guys asked the clerk if he had any .270 rifles. The clerk pulled a .270 WSM off the rack. The two guys said no, they wanted a regular .270 for one of their nephews, not some newfangled cartridge that cost twice as much and didn't kill elk any better. The store didn't have any "regular .270s" in stock, so the two guys stomped out.

7x57 Mauser

The 7x57 gained its original fame as a military cartridge, serving with distinction in the Spanish-American War, the Boer War, and the Mexican revolution. After that it gained fame among hunters as the favored elephant cartridge of "Karamojo" Bell.

Like the other rounds listed so far, the 7x57 is distinguished by excellent performance with relatively light recoil. I haven't slain any elephants with the 7x57, but I have used it at ranges out to 400 yards on pronghorns and springbok, and out to 300 yards on animals from kudu to moose. According to modern theory, the 7x57 doesn't shoot flat enough for pronghorns and is too small for moose, but none of the animals argued the point.

But hey, don't take my word for it. The 7x57 is also the favorite plains-game round of the most experienced African PH I've hunted with, 60-year-old Kevin Thomas, who grew up in what was then Rhodesia and has taken thousands of animals in a career as game ranger and guide. Shoot any animal from impala to eland in the right place with a 7x57, Kevin says, then load 'em in the Land Cruiser.

.30-06 Springfield

The .30-06 was the American military's answer to the 7x57. Hemingway used the .30-06 successfully during his Green Hills of Africa safari on Cape buffalo and black rhino, while my good friend Phil Shoemaker, an Alaskan master guide, has put down many wounded brown bears with the .30-06, while backing up clients who used bigger rifles. Phil also came up with my favorite .30-06 quote: "Anybody who claims the .30-06 isn't adequate for big game is unwittingly commenting on his own marksmanship."

The .30-06 is also one of the truly worldwide hunting cartridges. In the early 1990s I hunted red stag in the Czech Republic. One of my guides, Vaclav, was a Czech rifle loony and asked in accented English what my rifle was chambered for.

"Thirty-aught-six," I said. Vaclav shook his head, so I tried again. "Seven point six-two by sixty-three?"

He shook his head again, then said. "Let me haff a cartridge." He held the headstamp up to the sunlight and then smiled. "Ah, da Spring-field! Excellent!"

.300 Holland & Holland

This is the least popular round on this list, though the recently introduced Ruger No. 1 in .300 H&H is selling briskly. Nevertheless, the

.300 H&H deserves to be here, mostly because it was the original "belted" .300 magnum, the direct father of the .300 Weatherby and, thus, the grandfatherly inspiration of the .300 Winchester Magnum.

When H&H's ".30 Super" appeared in 1925, it merely duplicated the ballistics of the .30-06, in a larger case designed to keep pressures down in the hot tropics of the far-reaching British Empire. It wasn't until newer powders boosted ballistics in the 1930s that the round became known in America as the .300 H&H Magnum.

In 2000 Winchester introduced a belt-free .30-caliber magnum with a short, fat case advertised as the ultimate answer to any .300 magnum "problem." Supposedly the .300 WSM fed easier because of the lack of a useless belt and was much more efficient, whatever that means. The .300 WSM is a fine cartridge, and I've taken game from pronghorns to elk with it myself, but all it does is reproduce the ballistics of the .300 H&H, a 180-grain bullet at about 3,000 fps, a combination still hard to beat for 99 percent of the world's big game. Oh, and in a bolt action the .300 WSM doesn't feed nearly as slickly as the .300 H&H.

9.3x62 Mauser

Here we'll skip over a bunch of excellent "medium bores," including the .35 Whelen, for a German cartridge that until recently hardly any Americans knew existed. Essentially a .30-06 cartridge necked up to take .366-inch diameter bullets, the 9.3x62 Mauser was developed in 1905 by Otto Bock to provide a round that would function perfectly in unaltered 1898 Mauser actions, yet be capable of taking any game on earth.

It has been compared to both the .35 Whelen and .375 H&H, a fair assessment. When handloaded, it's a marvelously flexible cartridge. With modern powders, 250-grain bullets can safely be pushed

The .300 Holland & Holland not only set the standard for .300 magnum performance, but its case was the inspiration for the .300 Weatherby and .300 Winchester Magnums.

to over 2,600 fps, basically matching the 250-grain factory load of the .338 Winchester Magnum, while 286-grain bullets can reach 2,500 fps, nearly duplicating the .375 H&H. Five rounds fit in the magazine of almost any bolt action, always a virtue when hunting dangerous game. It will kill deer without much meat damage, yet it also works fine on grizzlies and Cape buffalo. Its only real rival is the 9.3x74R, a rimmed case duplicating the 9.3x62's ballistics in single-shot and double rifles.

The .375 Holland & Holland is often called the world's best all-round big-game cartridge. It was introduced in 1912, and even approaching its centennial is still extremely popular, especially in Alaska and Africa. Like the .300 H&H, it feeds very slickly, and is considered about ideal for lions and brown bears.

Some hunters claim the .375 isn't really enough for Cape buffalo, but I have noticed that most of these are Americans, and American hunting literature has a long tradition of transforming African game into super-beasts. Most of the PHs I've known, from Finn Aagaard to Kevin Thomas, regard the .375 as the very best Cape buffalo cartridge for a safari client, and most use it themselves when hunting buffalo. Kevin Thomas, in fact, prefers his .375 H&H to his .458 Lott when guiding buffalo hunters, so sometimes he finds himself in the curious position of finishing off bulls wounded by more powerful rifles.

.416 Rigby

There are relatively few cartridges instantly recognizable by their nonnumerical name. The Hornet, Swift, Roberts, and Springfield come to mind, but the only really big round is the .416 Rigby. It is also one of the few rounds associated firmly with one hunter, the African PH Harry Selby, who was made famous by Robert Ruark's book *Horn of the Hunter*.

Curiously, though Selby used his Rigby for close to half a century, it was a second choice both early and late. His first dangerous-game rifle was a big-bore double, but it was run over by a truck so Selby bought the only "big" rifle he could afford at a local store. Eventually, when he reached an age when most Americans would be retired, Selby shot the barrel out, so he returned it to Rigby for rebarreling. Since British gunsmiths are slower than a Greek tragedy, he bought a Winchester Model 70 in .458 Winchester Magnum to tide him over. By the time his .416 came back, three years later, Selby had grown used to the .458, so he sold the .416 to an American client.

I shot Harry Selby's .416 Rigby a number of years ago in Arkansas, where the owner lives. This is no mark of distinction, for half of American gunwriters have done the same thing. The rifle proved both relatively light and yet shooter-friendly. Since it had obviously been adequate for many huge animals, I eventually remodeled a CZ 550 Magnum to much the same dimensions.

While most modern hunters assume that the original ballistics of the .416 Rigby were a 400-grain bullet at 2,400 fps (probably because they've read that the .416 Remington Magnum duplicates the Rigby), the original velocity was actually around 2,300. This is how I load my rifle, mostly because that's where the ammo matches the fixed express sights, and it has worked just fine on both Cape and water buffalo. Some people feel compelled to hot-rod the Rigby, just because the extra case capacity is there, but the big virtue of the .416 Rigby is power combined with controllable recoil, as Harry Selby demonstrated for decades.

.45-70 Springfield

It's been claimed by at least one reputable historian of the Old West that more bison were killed during the hide-hunting days by the .45-70 "trapdoor" Springfield than any other cartridge and rifle, despite the notoriety of the Sharps. This is partly because trapdoors were plentiful and cheap after their introduction as the U.S. Army rifle in 1873, but it's also because a heavy .45-caliber bullet kills big game well, even at black-powder velocities.

The .45-70 adapted very well to the smokeless era. In modern rifles, smokeless powders are capable of driving 400-grain bullets at 2,000 fps. This moves the .45-70 into an entirely different class, capable of not just killing but stopping large game, the reason some Alaskan brown bear guides carry .45-70 lever actions. Yet the old round still works great on common game such as deer, black bears, and pigs. Unlike the .416 Rigby, it's still a cartridge of the common hunter, the main reason it has stuck around longer than any other classic big-game cartridge.

The .270 Winchester is still known as, perhaps, the perfect mule-deer cartridge. Eileen Clarke took this nice buck in the mountains of Montana using a New Ultra Light Arms Model 24 and a 130-grain Nosler Partition.

Data for Small and Medium Centerfire Rifle Cartridges

.17 Hornady Hornet

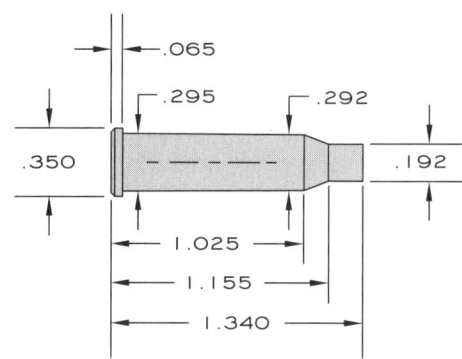

In 2011 Hornady decided to take advantage of the huge popularity of the .17 Hornady Magnum Rimfire (HMR) with a centerfire round with considerably better performance. (See the rimfire section.) The result of this is the .17 Hornady Hornet. Late in the year details began to emerge. Hornady's .17 is very similar to the .17 Ackley Hornet, which is at least 60 years old. The Hornady version is very slightly shorter than the Ackley and is SAAMI standardized, which the Ackley version never was. Owners of guns chambered for the .17 Ackley Hornet now have a source of factory ammunition. Time will tell if the current rage for small calibers will continue, but, if it does, this little Hornady should be a winner.

Relative Recoil Factor = 0.38

Specifications:

Controlling Agency for Standardization of this Ammunition: SAAMI

Bullet Weight Grains	Velocity fps	Maximum Average Pressure	
		Copper Crusher	Transducer
20	3,650	N/S	50,000 psi (estimated)

Standard barrel for velocity testing: 24 inches long

Availability:

Hornady 20-grain V-MAX (83005)
G1 Ballistic Coefficient = 0.185

Distance • Yards	Muzzle	100	200	300	400	500	600	800	1000
Velocity • fps	3650	3078	2574	2122	1721	1383	1135	903	775
Energy • ft-lbs	592	421	294	200	131	85	57	36	27
Taylor KO Index	1.8	1.5	1.3	1.1	0.9	0.7	0.6	0.5	0.4
Path • Inches	-1.5	+1.1	0.0	-6.4	-20.6	-46.8	-91.7	-270.1	-619.2
Wind Drift • Inches	0.0	1.3	5.6	13.8	27.0	46.9	74.8	151.8	249.7

The .17 Hornet is the latest entry in a recent spate of .17 caliber factory cartridges.

.17 Remington

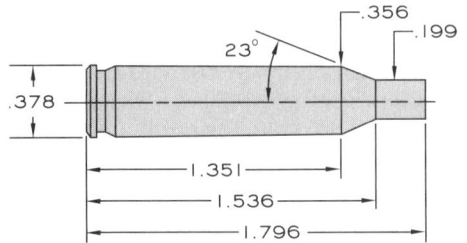

The .17 Remington was introduced in 1971. While it is based on the .222 Remington case, this design isn't just a .222 necked to .17 caliber. The case body was lengthened and the neck shortened to provide just a little more case volume than a simple necking down of the .222 would have created. The .17 Remington has never become as popular as some of its larger cousins, the .222 and .223 for instance, and as a result there are only two factory loadings available for this caliber.

Relative Recoil Factor = 0.45

Specifications:

Controlling Agency for Standardization of this Ammunition: SAAMI

Bullet Weight Grains	Velocity fps	Maximum Average Pressure	
		Copper Crusher	Transducer
25	4,000	52,000 cup	N/S

Standard barrel for velocity testing: 24 inches long—1 turn in 9-inch twist

Availability:

Remington 20-grain AccuTip-V (PRA17RA)

G1 Ballistic Coefficient = 0.185

Distance • Yards	Muzzle	100	200	300	400	500	600	800	1000
Velocity • fps	4250	3594	3028	2529	2081	1684	1352	983	826
Energy • ft-lbs	802	·574	407	284	192	126	81	43	30
Taylor KO Index	2.1	1.8	1.5	1.2	1.0	0.8	0.7	0.5	0.4
Path • Inches	-1.5	+0.6	0.0	-4.4	-14.4	-32.8	-64.3	-197.3	-463.4
Wind Drift • Inches	0.0	1.1	4.7	11.4	22.0	37.8	60.5	129.5	222.5

Remington 25-grain Hornady Hollow Point (R17R2)

G1 Ballistic Coefficient = 0.190

Distance • Yards	Muzzle	100	200	300	400	500	600	800	1000
Velocity • fps	4040	3428	2895	2420	1993	1617	1312	974	825
Energy • ft-lbs	906	652	465	325	221	145	96	53	38
Taylor KO Index	2.5	2.1	1.8	1.5	1.2	1.0	0.8	0.6	0.5
Path • Inches	-1.5	+0.8	0.0	-5.0	-15.9	-35.6	-68.8	-204.6	-486.2
Wind Drift • Inches	0.0	1.1	4.8	11.7	22.6	38.9	62.1	131.5	223.9

Prairie dogs remain a favorite target for varmint shooters all over the United States. (Safari Press Archives)

.204 Ruger

Though the .17 centerfires have been around for quite a while, there were not any factory centerfire 5mm's until 2004. There have been .20 caliber and 5mm (pretty much the same thing) wildcats for 50 years or so, but no factory wanted to go to the trouble. Ruger induced Hornady to design and develop this round that has already been chambered in two Ruger rifle styles. The cartridge designer started with the .222 Remington Magnum length rather than the .223 length, and that resulted in a case with enough volume to give this cartridge the second highest advertised muzzle velocity on the market today.

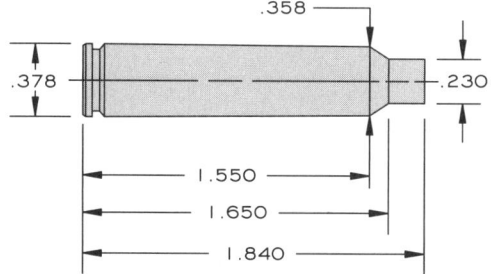

Relative Recoil Factor = 0.60

Specifications:

Controlling Agency for Standardization of this Ammunition: SAAMI

Bullet Weight Grains	Velocity fps	Maximum Average Pressure Copper Crusher	Transducer
Data pending.			

Availability:

Cor-Bon 26-grain Varmint Grenade (SD20426VG/20) LF (Lead Free)
G1 Ballistic Coefficient = 0.100

Distance • Yards	Muzzle	100	200	300	400	500	600	800	1000
Velocity • fps	4200	3070	2179	1469	1046	872	760	592	463
Energy • ft-lbs	1019	544	274	125	63	44	33	20	12
Taylor KO Index	3.2	2.3	1.7	1.1	0.8	0.7	0.6	0.4	0.3
Path • Inches	-1.5	+1.2	0.0	-8.9	-34.6	-91.8	-194.8	-599.9	-1413
Wind Drift • Inches	0.0	2.2	10.0	27.1	58.0	101.1	153.6	286.8	465.2

Hornady 30-grain NTX (83209) LF
G1 Ballistic Coefficient = 0.205

Distance • Yards	Muzzle	100	200	300	400	500	600	800	1000
Velocity • fps	4225	3632	3114	2652	2234	1856	1525	1077	892
Energy • ft-lbs	1189	879	646	468	332	230	155	77	53
Taylor KO Index	3.7	3.2	2.7	2.3	2.0	1.6	1.3	0.9	0.8
Path • Inches	-1.5	+0.6	0.0	-4.2	-13.4	-29.6	-56.1	-161.3	-385.5
Wind Drift • Inches	0.0	1.1	4.7	11.4	22.0	37.8	60.5	129.5	222.5

Hornady 32-grain V-MAX (83204)
G1 Ballistic C efficient = 0.210

Distance • Yards	Muzzle	100	200	300	400	500	600	800	1000
Velocity • fps	4225	3645	3137	2683	2272	1899	1569	1104	908
Energy • ft-lbs	1268	944	699	512	357	256	175	87	59
Taylor KO Index	3.9	3.4	2.9	2.5	2.1	1.8	1.5	1.0	0.8
Path • Inches	-1.5	+0.6	0.0	-4.1	-13.1	-29.0	-54.5	-155.0	-368.8
Wind Drift • Inches	0.0	1.0	4.1	9.8	18.7	31.6	49.7	106.2	187.8

Remington 32-grain AccuTip-V (PRA204A)
G1 Ballistic Coefficient = 0.207

Distance • Yards	Muzzle	100	200	300	400	500	600	800	1000
Velocity • fps	4225	3632	3114	2652	2234	1856	1529	1077	890
Energy • ft-lbs	1268	937	689	500	355	245	166	82	56
Taylor KO Index	3.9	3.4	2.9	2.1	1.5	1.7	1.4	1.0	0.8
Path • Inches	-1.5	+0.6	0.0	-4.1	-13.1	-28.9	-56.6	-166.3	-393.7
Wind Drift • Inches	0.0	1.0	4.2	10.1	19.2	32.6	51.5	110.2	194.2

Fiocchi 32-grain V-Max Polymer Tip BT (204HVA)
G1 Ballistic Coefficient = 0.210

Distance • Yards	Muzzle	100	200	300	400	500	600	800	1000
Velocity • fps	4125	3559	3061	2616	2212	1846	1521	1091	897
Energy • ft-lbs	1209	900	666	486	348	242	164	83	57
Taylor KO Index	3.8	3.3	2.9	2.4	2.1	1.7	1.4	1.0	0.8
Path • Inches	-1.5	+0.6	0.0	-4.4	-13.9	-30.6	-57.7	-164.4	-389.0
Wind Drift • Inches	0.0	1.0	4.2	10.1	19.3	32.6	51.4	109.6	192.3

Winchester 32-grain Ballistic Silvertip (SBST204R)
G1 Ballistic Coefficient = 0.206

Distance • Yards	Muzzle	100	200	300	400	500	600	800	1000
Velocity • fps	4050	3482	2984	2537	2132	1767	1451	1046	876
Energy • ft-lbs	1165	862	632	457	323	222	150	78	55
Taylor KO Index	3.8	3.2	2.8	2.4	2.0	1.6	1.4	1.0	0.8
Path • Inches	-1.5	+0.7	0.0	-4.6	-14.7	-32.6	-61.7	-177.4	-419.1
Wind Drift • Inches	0.0	1.0	4.4	10.6	20.2	34.4	54.4	115.7	200.9

Federal 32-grain Nosler Ballistic Tip (P204B)
G1 Ballistic Coefficient = 0.207

Distance • Yards	Muzzle	100	200	300	400	500	600	800	1000
Velocity • fps	4030	3470	2970	2520	2120	1760	1450	1046	877
Energy • ft-lbs	1155	855	625	450	320	220	149	78	55
Taylor KO Index	3.8	3.2	2.8	2.4	2.0	1.6	1.4	1.0	0.8
Path • Inches	-1.5	+0.7	0.0	-4.7	-14.9	-33.1	-62.1	-178.1	-420.0
Wind Drift • Inches	0.0	1.0	4.4	10.6	20.2	34.4	54.4	115.5	200.5

Federal 32-grain Speer TNT Green (P204D) LF
G1 Ballistic Coefficient = 0.160

Distance • Yards	Muzzle	100	200	300	400	500	600	800	1000
Velocity • fps	4030	3320	2710	2170	1710	1330	1077	855	724
Energy • ft-lbs	1155	980	520	335	205	125	82	52	37
Taylor KO Index	3.8	3.1	2.5	2.0	1.6	1.2	1.0	0.8	0.7
Path • Inches	-1.5	+0.9	0.0	-5.7	-19.1	-45.3	-89.8	-279.4	-659.8
Wind Drift • Inches	0.0	1.4	5.9	14.6	23.9	51.1	82.6	167.9	276.5

Nosler 32-grain Ballistic Tip (60089) LF
G1 Ballistic Coefficient = 0.228

Distance • Yards	Muzzle	100	200	300	400	500	600	800	1000
Velocity • fps	3800	3319	2890	2499	2140	1810	1517	1106	922
Energy • ft-lbs	1026	783	593	444	325	233	164	87	60
Taylor KO Index	3.5	3.1	2.7	2.3	2.0	1.7	1.4	1.0	0.9
Path • Inches	-1.5	+0.8	0.0	-5.0	-15.7	-34.1	-63.2	-173.4	-396.6
Wind Drift • Inches	0.0	1.0	4.2	9.9	18.9	31.9	49.9	104.8	182.7

Winchester 34-grain Hollow Point (X204R)
G1 Ballistic Coefficient = 0.168

Distance • Yards	Muzzle	100	200	300	400	500	600	800	1000
Velocity • fps	4025	3339	2751	2232	1775	1393	1119	878	743
Energy • ft-lbs	1223	842	571	376	238	146	95	58	42
Taylor KO Index	4.0	3.3	2.7	2.2	1.8	1.4	1.1	0.9	0.7
Path • Inches	-1.5	+0.8	0.0	-5.5	-18.1	-42.0	-86.8	-265.6	-603.9
Wind Drift • Inches	0.0	1.3	5.6	13.8	27.3	47.9	77.4	159.5	264.5

Federal 40-grain Sierra BlitzKing (P204A)
G1 Ballistic Coefficient = 0.287

Distance • Yards	Muzzle	100	200	300	400	500	600	800	1000
Velocity • fps	3750	3360	3010	2680	2380	2090	1830	1381	1086
Energy • ft-lbs	1220	980	785	625	490	380	290	165	102
Taylor KO Index	4.3	3.8	3.4	3.0	2.7	2.4	2.1	1.6	1.2
Path • Inches	-1.5	+0.8	0.0	-4.7	-14.1	-29.9	-53.6	-135.7	-292.0
Wind Drift • Inches	0.0	0.8	3.3	7.8	14.7	24.3	37.2	75.6	134.6

Hornady 40-grain V-MAX (83206)
G1 Ballistic Coefficient = 0.275

Distance • Yards	Muzzle	100	200	300	400	500	600	800	1000
Velocity • fps	3900	3482	3103	2755	2433	2133	1855	1384	1079
Energy • ft-lbs	1351	1077	855	674	526	404	306	170	103
Taylor KO Index	4.5	4.1	3.6	3.2	2.8	2.5	2.2	1.6	1.3
Path • Inches	-1.5	+0.7	0.0	-4.3	-13.2	-28.1	-50.6	-129.6	-282.7
Wind Drift • Inches	0.0	0.8	3.3	7.8	14.7	24.4	37.4	76.4	136.7

Remington 40-grain AccuTip Boat Tail (PRA204B)
G1 Ballistic Coefficient = 0.257

Distance • Yards	Muzzle	100	200	300	400	500	600	800	1000
Velocity • fps	3900	3451	3046	2677	2336	2021	1731	1264	1007
Energy • ft-lbs	1351	1058	824	636	485	363	266	142	90
Taylor KO Index	4.5	4.0	3.6	3.1	2.7	2.4	2.0	1.5	1.2
Path • Inches	-1.5	+0.7	0.0	-4.3	-13.2	-28.1	-54.2	-144.7	-327.9
Wind Drift • Inches	0.0	0.9	3.6	8.6	16.2	26.9	41.6	86.2	154.0

Nosler 40-grain Ballistic Tip (00505)

G1 Ballistic Coefficient = 0.239

Distance • Yards	Muzzle	100	200	300	400	500	600	800	1000
Velocity • fps	3625	3176	2771	2401	2059	1749	1476	1093	919
Energy • ft-lbs	1167	896	682	512	377	272	193	106	75
Taylor KO Index	4.2	3.7	3.2	2.8	2.4	2.0	1.7	1.3	1.1
Path • Inches	-1.5	+1.0	0.0	-5.5	-17.2	-37.1	-68.4	-185.7	-418.6
Wind Drift • Inches	0.0	1.0	4.2	10.2	19.3	32.6	50.9	106.1	183.3

Fiocchi 40-grain V-Max Polymer Tip BT (204HVB)

G1 Ballistic Coefficient = 0.275

Distance • Yards	Muzzle	100	200	300	400	500	600	800	1000
Velocity • fps	3700	3302	2939	2604	2293	2003	1734	1295	1035
Energy • ft-lbs	1216	968	767	602	457	356	267	149	95
Taylor KO Index	4.3	3.8	3.4	3.0	2.7	2.3	2.0	1.5	1.2
Path • Inches	-1.5	+0.8	0.0	-4.9	15.0	-31.7	-57.2	-147.1	-320.9
Wind Drift • Inches	0.0	0.8	3.5	8.4	15.7	26.2	40.2	82.5	148.2

Hornady 45-grain SP (83208)

G1 Ballistic Coefficient = 0.245

Distance • Yards	Muzzle	100	200	300	400	500	600	800	1000
Velocity • fps	3625	3188	2792	2428	2093	1787	1515	1121	937
Energy • ft-lbs	1313	1015	778	589	438	319	230	126	88
Taylor KO Index	4.8	4.2	3.7	3.2	2.7	2.3	2.0	1.5	1.2
Path • Inches	-1.5	+1.0	0.0	-5.5	-16.9	-36.3	-66.7	-179.0	-401.6
Wind Drift • Inches	0.0	1.0	4.1	9.8	18.7	31.4	49.0	101.8	176.8

Coyotes are easily killed with .22 centerfire rifles as was this Wyoming dog shot near Rawling. (Safari Press Archives)

.22 Hornet

The oldest of the centerfire .22s in use today is the .22 Hornet. The Hornet was developed from an old black-powder number called the .22 WCF (Winchester Center Fire). Winchester introduced the Hornet as a factory cartridge in the early 1930s. The initial success of the .22 Hornet probably came because there wasn't any competition. Hornets have a very mixed reputation for accuracy. Some guns do very well while others are terrible. Part of this seems to be the result of some strange chambering practices in factory guns produced in the U.S. Unlike the .218 Bee, the Hornet appears to be showing some new life. It's still an excellent little varmint cartridge when fired from a "good" gun as long as the range is limited to about 200 yards.

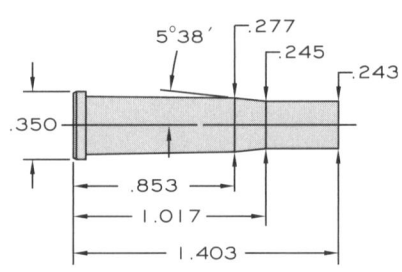

Relative Recoil Factor = 0.54

Specifications:

Controlling Agency for Standardization of this Ammunition: SAAMI

Bullet Weight Grains	Velocity fps	Maximum Average Pressure	
		Copper Crusher	Transducer
45	2655	43,000 cup	N/S
46	2655	43,000 cup	N/S

Standard barrel for velocity testing: 24 inches long —1 turn in 16-inch twist

Availability:

Federal 30-grain Speer TNT Green (P22D) LF
G1 Ballistic Coefficient = 0.091

Distance • Yards	Muzzle	100	200	300	400	500	600	800	1000
Velocity • fps	3150	2150	1390	990	830	720	628	479	365
Energy • ft-lbs	660	310	130	65	45	35	26	15	9
Taylor KO Index	3.0	2.1	1.3	1.0	0.8	0.7	0.6	0.5	0.4
Path • Inches	-1.5	+3.3	0.0	-22.8	-78.7	-179.7	-358.6	-1002	-2278
Wind Drift • Inches	0.0	3.5	17.2	46.2	87.9	139.4	201.3	361.7	585.8

Hornady 32-grain V-MAX (8302)
G1 Ballistic Coefficient = 0.109

Distance • Yards	Muzzle	100	200	300	400	500	600	800	1000
Velocity • fps	3070	2246	1571	1116	919	803	712	568	454
Energy • ft-lbs	732	392	192	97	66	50	39	25	16
Taylor KO Index	3.4	2.5	1.8	1.2	1.0	0.9	0.8	0.6	0.5
Path • Inches	-1.5	+2.9	0.0	-17.6	-62.9	-149.3	-290.0	-795.7	-1744
Wind Drift • Inches	0.0	2.9	13.9	37.1	72.6	117.1	169.9	302.4	478.3

Hornady 35-grain V-MAX (8302)
G1 Ballistic Coefficient = 0.109

Distance • Yards	Muzzle	100	200	300	400	500	600	800	1000
Velocity • fps	3070	2246	1571	1116	919	803	712	568	454
Energy • ft-lbs	732	392	192	97	68	50	39	25	16
Taylor KO Index	3.4	2.5	1.8	1.2	1.0	0.9	0.8	0.6	0.5
Path • Inches	-1.5	+2.9	0.0	-17.6	-62.8	-149.3	-289.9	-795.7	-1745
Wind Drift • Inches	0.0	2.9	13.9	37.1	72.6	117.1	169.9	302.4	478.3

Remington 35-grain AccuTip-V (PRA22HNA)
G1 Ballistic Coefficient = 0.109

Distance • Yards	Muzzle	100	200	300	400	500	600	800	1000
Velocity • fps	3100	2271	1591	1127	924	806	712	566	447
Energy • ft-lbs	747	401	197	99	66	51	39	25	16
Taylor KO Index	3.5	2.5	1.8	1.3	1.0	0.9	0.8	0.6	0.5
Path • Inches	-1.5	+3.0	0.00	-17.6	-62.5	-144.4	-273.8	-730.9	-1592
Wind Drift • Inches	0.0	2.9	13.7	36.6	72.0	116.6	169.4	302.0	477.7

Nosler 40-grain Ballistic Tip (02015)

G1 Ballistic Coefficient = 0.227

Distance • Yards	Muzzle	100	200	300	400	500	600	800	1000
Velocity • fps	2800	2444	2073	1738	1447	1217	1065	899	793
Energy • ft-lbs	721	530	382	268	186	132	101	72	56
Taylor KO Index	3.8	3.1	2.7	2.2	1.9	1.6	1.4	1.2	1.0
Path • Inches	-1.5	+2.3	0.0	-10.6	-33.1	-72.4	-135.1	-358.7	-754.7
Wind Drift • Inches	0.0	1.5	6.4	15.6	30.3	51.4	79.4	150.2	238.0

Federal 45-grain Jacketed Soft Point (P22B)

G1 Ballistic Coefficient = 0.145

Distance • Yards	Muzzle	100	200	300	400	500	600	800	1000
Velocity • fps	2690	2100	1590	1210	1000	890	808	677	573
Energy • ft-lbs	725	440	255	145	100	80	65	46	33
Taylor KO Index	3.9	3.0	2.3	1.7	1.4	1.3	1.2	1.0	0.8
Path • Inches	-1.5	+3.3	0.0	-17.6	-59.3	-134.2	-250.0	-657.1	-1372
Wind Drift • Inches	0.0	2.6	11.7	29.9	58.3	94.5	137.2	241.4	372.9

Remington 45-grain Pointed Soft Point (R22HN1

) G1 Ballistic Coefficient = 0.130

Distance • Yards	Muzzle	100	200	300	400	500	600	800	1000
Velocity • fps	2690	2042	1502	1128	948	840	756	622	513
Energy • ft-lbs	723	417	225	127	90	70	57	39	26
Taylor KO Index	3.9	2.9	2.2	1.6	1.4	1.2	1.1	0.9	0.7
Path • Inches	-1.5	+3.6	0.0	20.0	-66.9	-148.6	-275.7	-702.5	-1458
Wind Drift • Inches	0.0	2.9	13.5	34.8	66.7	106.4	153.3	268.4	416.3

Winchester 45-grain Soft Point (X22H1)

G1 Ballistic Coefficient = 0.130

Distance • Yards	Muzzle	100	200	300	400	500	600	800	1000
Velocity • fps	2690	2042	1502	1128	948	840	756	622	513
Energy • ft-lbs	723	417	225	127	90	70	57	39	26
Taylor KO Index	3.9	2.9	2.2	1.6	1.4	1.2	1.1	0.9	0.7
Path • Inches	-1.5	+3.6	0.0	-20.0	-66.9	-148.6	-275.7	-702.5	-1458
Wind Drift • Inches	0.0	2.9	13.5	34.8	66.7	106.4	153.3	268.4	416.3

Sellier & Bellot 45-grain SP (V330012U)

G1 Ballistic Coefficient = 0.124

Distance • Yards	Muzzle	100	200	300	400	500	600	800	1000
Velocity • fps	2346	1726	1251	998	869	777	701	575	473
Energy • ft-lbs	550	298	157	99	76	60	49	33	22
Taylor KO Index	3.4	2.5	1.8	1.4	1.3	1.1	1.0	0.8	0.7
Path • Inches	-1.5	+5.3	0.0	-27.7	-90.2	-198.7	-365.0	-927.1	-1920
Wind Drift • Inches	0.0	3.8	17.4	42.8	77.3	119.2	168.3	290.6	450.4

Remington 45-grain Hollow Point (R22HN2)

G1 Ballistic Coefficient = 0.130

Distance • Yards	Muzzle	100	200	300	400	500	600	800	1000
Velocity • fps	2690	2042	1502	1128	948	840	756	622	513
Energy • ft-lbs	723	417	225	127	90	70	57	39	26
Taylor KO Index	3.9	2.9	2.2	1.6	1.4	1.2	1.1	0.9	0.7
Path • Inches	-1.5	+3.6	0.0	20.0	-66.9	-148.6	-275.7	-702.5	-1458
Wind Drift • Inches	0.0	2.9	13.5	34.8	66.7	106.4	153.3	268.4	416.3

Sellier & Bellot 45-grain FMJ (V330002U)

G1 Ballistic Coefficient = 0.124

Distance • Yards	Muzzle	100	200	300	400	500	600	800	1000
Velocity • fps	2346	1726	1251	998	869	777	701	575	473
Energy • ft-lbs	550	298	157	99	76	60	49	33	22
Taylor KO Index	3.4	2.5	1.8	1.4	1.3	1.1	1.0	0.8	0.7
Path • Inches	-1.5	+5.3	0.0	-27.7	-90.2	-198.7	-365.0	-927.1	-1920
Wind Drift • Inches	0.0	3.8	17.4	42.8	77.3	119.2	168.3	290.6	450.4

Winchester 46-grain Hollow Point (X22H2)

G1 Ballistic Coefficient = 0.130

Distance • Yards	Muzzle	100	200	300	400	500	600	800	1000
Velocity • fps	2690	2042	1502	1128	948	840	756	622	513
Energy • ft-lbs	723	417	225	127	90	70	57	39	26
Taylor KO Index	3.9	2.9	2.2	1.6	1.4	1.2	1.1	0.9	0.7
Path • Inches	-1.5	+3.6	0.0	-20.0	-66.9	-148.6	-275.7	-702.5	-1458
Wind Drift • Inches	0.0	2.9	13.5	34.8	66.7	106.4	153.3	268.4	416.3

.218 Bee

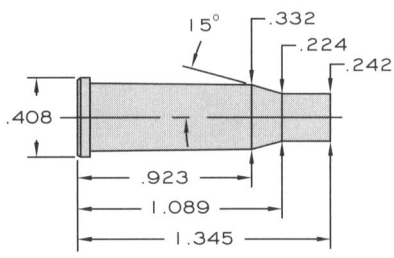

The .218 Bee can be thought of as a slight improvement over the .22 Hornet. Don't be fooled by the .218 designation. This cartridge uses standard 0.224-inch bullets, the same as all other centerfire .22s. Introduced in 1938, the Bee was created from the .25-20 Winchester case (32-20s can also be used) especially for the Model 65 Winchester (a lever-action rifle). It does have about 30 percent more case capacity than the Hornet, which allows the Bee to get about 50 fps more velocity from its 46-grain bullet. While the Bee could have been an excellent little cartridge in an accurate gun, and has found a following in single-shot guns, the factory loading has just about run out its lifetime, as evidenced by the fact that only Winchester makes a factory loading and only loads a single bullet choice. Today, there are just too many good centerfire .22s to spend very much time, money, or effort to build up a new rifle chambered for this cartridge, although TC barrels are available.

Relative Recoil Factor = 0.56

Specifications:

Controlling Agency for Standardization of this Ammunition: SAAMI

Bullet Weight Grains	Velocity fps	Maximum Average Pressure	
		Copper Crusher	Transducer
46	2,725	40,000 cup	N/S

Standard barrel for velocity testing: 24 inches long —1 turn in 16-inch twist

Availability:

Winchester 46-grain Hollow Point (X218B)

G1 Ballistic Coefficient = 0.130

Distance • Yards	Muzzle	100	200	300	400	500	600	800	1000
Velocity • fps	2760	2102	1550	1156	961	850	763	628	518
Energy • ft-lbs	778	451	245	136	94	74	60	40	27
Taylor KO Index	4.1	3.1	2.3	1.7	1.4	1.3	1.1	0.9	0.8
Path • Inches	-1.5	+3.4	0.0	-18.8	-63.6	-142.5	-265.4	-680.2	-1417
Wind Drift • Inches	0.0	2.8	13.0	33.6	65.4	104.4	151.3	265.9	413.0

European combination guns are among the most versatile weapons one can buy. This drilling has barrels for 12 gauge, .22 Hornet, and a .30-06. If you are buying such a gun, consider the rimmed centerfire cartridges as most suitable since they provide a very positive extraction. (Safari Press Archives)

.221 Remington Fireball

When Remington introduced the XP-100 bolt-action pistol in 1962, the company wanted a new cartridge to complete the package. The .221 Remington Fireball is that cartridge. It is noticeably shorter than the .222 Remington (0.3 inches to be exact) and, when compared with the .222, has the reduced performance you might expect from its smaller volume. The combination of the XP-100 pistol and the .221 Remington Fireball cartridge developed a reputation for fine accuracy. This cartridge dropped out of Remington's catalog a number of years ago, but in 2004 Remington decided to reopen the production line. If you have a .221 and are nearly out of ammo, you might seriously consider stocking up before production is stopped again.

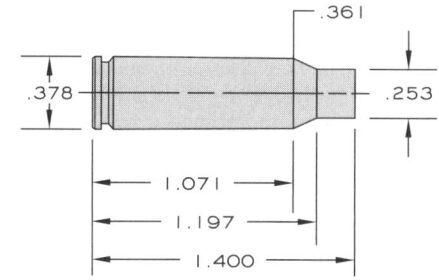

Relative Recoil Factor = 0.67

Specifications:

Controlling Agency for Standardization of this Ammunition: SAAMI

Bullet Weight Grains	Velocity fps	Maximum Average Pressure	
		Copper Crusher	Transducer
50	2,520	52,000 cup	N/S

Standard barrel for velocity testing: 10.5 inches long—1 turn in 12-inch twist*
*The ballistic data shown here were taken in a 24-inch barrel.

Availability:

Nosler 40-grain Ballistic Tip (04016)

G1 Ballistic Coefficient = 0.221

Distance • Yards	Muzzle	100	200	300	400	500	600	800	1000
Velocity • fps	3200	2761	2363	1998	1671	1390	1174	946	823
Energy • ft-lbs	910	677	496	355	248	172	123	79	60
Taylor KO Index	4.1	3.5	3.0	2.6	2.1	1.8	1.5	1.2	1.1
Path • Inches	-1.5	+1.6	0.0	-7.8	-24.5	-53.7	-101.1	-278.6	-611.3
Wind Drift • Inches	0.0	1.3	5.4	13.2	25.7	43.9	68.8	137.5	224.7

Remington 50-grain AccuTip-V Boat Tail (PRA221FB)

G1 Ballistic Coefficient = 0.238

Distance • Yards	Muzzle	100	200	300	400	500	600	800	1000
Velocity • fps	2995	2605	2247	1918	1622	1368	1170	954	837
Energy • ft-lbs	996	753	560	408	292	208	152	101	78
Taylor KO Index	4.8	4.2	3.6	3.1	2.6	2.2	1.9	1.5	1.3
Path • Inches	-1.5	+1.8	0.0	-8.8	-27.1	-59.0	-110.2	-295.4	-619.5
Wind Drift • Inches	0.0	1.3	5.4	13.2	25.4	43.2	67.2	132.5	215.2

Handgun hunting is a uniquely American sport, and many different guns and calibers have been developed for it over the last several decades.

.222 Remington

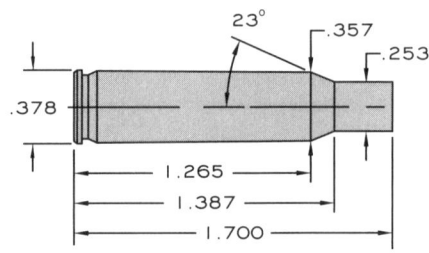

The .222 Remington, when introduced in 1950, was a whole new cartridge design. This new cartridge, which was an instant hit, would drive a 50-grain bullet in excess of 3,100 fps. The design combined a rimless 0.378-inch head diameter with a bottlenecked case, a conservative 23-degree shoulder, and a 0.3127-inch neck length to produce a cartridge that had all the elements needed to be superbly accurate. The .222 soon caught on as varmint rifle and for many years was the caliber of choice in benchrest competitions. Today, its popularity has been somewhat diluted by the .223 Remington (5.56 NATO) which is the U.S. standard infantry cartridge, but the .222 remains an excellent little cartridge. Most ammunition manufacturers load for this caliber. There are many ammunition choices available.

Relative Recoil Factor = 0.74

Specifications:

Controlling Agency for Standardization of this Ammunition: SAAMI

Bullet Weight Grains	Velocity fps	Maximum Average Pressure	
		Copper Crusher	Transducer
50–52	3,110	46,000 cup	50,000 psi
55	3,000	46,000 cup	50,000 psi

Standard barrel for velocity testing: 24 inches long—1 turn in 14-inch twist

Availability:

Hornady 40-grain V-MAX (8310)
G1 Ballistic Coefficient = 0.201

Distance • Yards	Muzzle	100	200	300	400	500	600	800	1000
Velocity • fps	3600	3074	2606	2184	1803	1474	1219	950	816
Energy • ft-lbs	1151	839	603	424	289	193	132	80	59
Taylor KO Index	4.6	3.9	3.3	2.8	2.3	1.9	1.6	1.2	1.0
Path • Inches	-1.5	+1.1	0.0	-6.2	-19.9	-44.4	-84.8	-244.1	-557.3
Wind Drift • Inches	0.0	1.2	5.2	12.6	24.5	42.2	67.0	137.6	228.9

Federal 40-grain Nosler Ballistic Tip (P222C)
G1 Ballistic Coefficient = 0.218

Distance • Yards	Muzzle	100	200	300	400	500	600	800	1000
Velocity • fps	3450	3000	2570	2190	1840	1530	1279	969	851
Energy • ft-lbs	1055	790	585	425	300	210	145	87	64
Taylor KO Index	5.5	4.8	4.1	3.5	2.9	2.4	2.0	1.6	1.4
Path • Inches	-1.5	+1.2	0.0	-6.5	-20.4	-44.7	-85.2	-239.6	-526.5
Wind Drift • Inches	0.0	1.2	4.9	11.9	23.0	39.1	61.6	126.6	211.6

Norma 40-grain V-MAX (15705)
G1 Ballistic Coefficient = 0.204

Distance • Yards	Muzzle	100	200	300	400	500	600	800	1000
Velocity • fps	3445	2936	2482	2070	1702	1390	1157	924	798
Energy • ft-lbs	1054	766	577	381	257	172	119	76	57
Taylor KO Index	4.4	3.8	3.2	2.6	2.2	1.8	1.5	1.2	1.0
Path • Inches	-1.5	+1.2	0.0	-7.0	-22.2	-49.6	-95.2	-272.0	-611.6
Wind Drift • Inches	0.0	1.3	5.5	13.5	26.3	45.4	71.9	145.0	237.9

Nosler 40-grain Ballistic Tip (05080)
G1 Ballistic Coefficient = 0.227

Distance • Yards	Muzzle	100	200	300	400	500	600	800	1000
Velocity • fps	3330	2879	2470	2097	1760	1465	1229	969	839
Energy • ft-lbs	0985	736	542	390	275	191	134	83	63
Taylor KO Index	4.3	3.7	3.2	2.7	2.3	1.9	1.6	1.2	1.1
Path • Inches	-1.5	+1.4	0.0	-7.1	-22.2	-48.7	-91.5	-253.7	-563.5
Wind Drift • Inches	0.0	1.2	5.2	12.5	24.1	41.2	64.8	131.5	217.5

Federal 43-grain Speer TNT Green (P222D) LF
G1 Ballistic Coefficient = 0.151

Distance • Yards	Muzzle	100	200	300	400	500	600	800	1000
Velocity • fps	3400	2750	2180	1680	1290	1050	921	762	646
Energy • ft-lbs	1361	887	557	333	196	129	100	68	49
Taylor KO Index	5.8	4.7	3.7	2.8	2.2	1.8	1.6	1.3	1.1
Path • Inches	-1.5	+1.6	0.0	-9.2	-31.4	-75.9	-147.9	-428.5	-950.6
Wind Drift • Inches	0.0	1.8	7.8	19.9	40.4	70.7	109.2	104.9	325.0

Sellier & Bellot 50-grain SP (V330122U)
G1 Ballistic Coefficient = 0.248

Distance • Yards	Muzzle	100	200	300	400	500	600	800	1000
Velocity • fps	3215	2821	2460	2127	1821	1548	1317	948	888
Energy • ft-lbs	1148	884	672	502	368	266	193	117	88
Taylor KO Index	5.1	4.5	3.9	3.4	2.9	2.5	2.1	1.5	1.4
Path • Inches	-1.5	+1.4	0.0	-7.2	-22.2	-47.8	-88.0	-235.4	-513.9
Wind Drift • Inches	0.0	1.1	4.7	11.4	21.8	36.9	57.5	116.8	195.3

Norma 50-grain Soft Point (15711)
G1 Ballistic Coefficient = 0.185

Distance • Yards	Muzzle	100	200	300	400	500	600	800	1000
Velocity • fps	3199	2679	2215	1800	1449	1181	1019	848	734
Energy • ft-lbs	1136	797	545	360	233	155	115	80	60
Taylor KO Index	5.1	4.3	3.5	2.9	2.4	1.9	1.6	1.4	1.2
Path • Inches	-1.5	+1.6	0.0	-8.9	-28.6	-65.2	-126.6	-357.3	-782.3
Wind Drift • Inches	0.0	1.5	6.7	16.6	32.8	56.9	88.8	170.4	271.7

Federal 50-grain Soft Point (222A)
G1 Ballistic Coefficient = 0.223

Distance • Yards	Muzzle	100	200	300	400	500	600	800	1000
Velocity • fps	3140	2710	2320	1960	1640	1370	1163	943	823
Energy • ft-lbs	1095	815	595	425	300	205	150	99	75
Taylor KO Index	5.0	4.3	3.7	3.1	2.6	2.2	1.9	1.5	1.3
Path • Inches	-1.5	+1.6	0.0	-8.2	-25.7	-56.5	-104.6	-286.6	-624.7
Wind Drift • Inches	0.0	1.3	5.5	13.4	26.0	44.4	69.6	138.2	225.0

Remington 50-grain Pointed Soft Point (R222R1)
G1 Ballistic Coefficient = 0.176

Distance • Yards	Muzzle	100	200	300	400	500	600	800	1000
Velocity • fps	3140	2600	2120	1700	1350	1110	971	813	700
Energy • ft-lbs	1095	750	500	320	200	135	105	73	54
Taylor KO Index	5.0	4.2	3.4	2.7	2.2	1.8	1.6	1.3	1.1
Path • Inches	-1.5	+1.9	0.0	-9.7	-31.6	-71.3	-145.1	-394.4	-829.3
Wind Drift • Inches	0.0	1.7	7.3	18.3	36.3	62.9	97.4	183.4	290.0

Winchester 50-grain Pointed Soft Point (X222R)
G1 Ballistic Coefficient = 0.175

Distance • Yards	Muzzle	100	200	300	400	500	600	800	1000
Velocity • fps	3140	2600	2120	1700	1350	1110	971	813	700
Energy • ft-lbs	1095	750	500	320	200	135	105	73	54
Taylor KO Index	5.0	4.2	3.4	2.7	2.2	1.8	1.6	1.3	1.1
Path • Inches	-1.5	+1.9	0.0	-9.7	-31.6	-71.3	-145.1	-394.4	-829.3
Wind Drift • Inches	0.0	1.7	7.3	18.3	36.3	62.9	97.4	183.4	290.0

Hornady 50-grain V-MAX (8316)
G1 Ballistic Coefficient = 0.242

Distance • Yards	Muzzle	100	200	300	400	500	600	800	1000
Velocity • fps	3345	2930	2551	2203	1883	1595	1350	1037	891
Energy • ft-lbs	1242	953	722	539	393	283	203	119	88
Taylor KO Index	5.4	4.7	4.1	3.5	3.0	2.6	2.2	1.7	1.4
Path • Inches	-1.5	+1.3	0.0	-6.7	-20.6	-44.4	-82.0	-221.2	-488.9
Wind Drift • Inches	0.0	1.1	4.6	11.1	21.3	36.0	56.2	115.2	194.2

Fiocchi 50-grain V-Max Polymer Tip BT (222HVA)
G1 Ballistic Coefficient = 0.223

Distance • Yards	Muzzle	100	200	300	400	500	600	800	1000
Velocity • fps	3200	2791	2414	2074	1769	1496	1270	1000	868
Energy • ft-lbs	1137	868	653	482	349	249	179	111	84
Taylor KO Index	5.1	4.5	3.9	3.3	2.8	2.4	2.0	1.6	1.4
Path • Inches	-1.5	+1.5	0.0	-7.5	-23.0	-49.7	-92.1	-248.5	-543.3
Wind Drift • Inches	0.0	1.2	5.0	11.9	22.9	38.9	60.8	122.9	203.9

Remington 50-grain AccuTip-V Boat Tail (PRV222RB)

G1 Ballistic Coefficient = 0.242

Distance • Yards	Muzzle	100	200	300	400	500	600	800	1000
Velocity • fps	3140	2744	2380	2045	1740	1471	1253	994	864
Energy • ft-lbs	1094	836	629	464	336	240	174	110	83
Taylor KO Index	5.0	4.4	3.8	3.3	2.8	2.4	2.0	1.6	1.4
Path • Inches	-1.5	+1.5	0.0	-7.8	-23.9	-51.7	-96.6	-261.6	-558.4
Wind Drift • Inches	0.0	1.2	5.0	12.2	23.4	39.6	61.6	124.0	204.8

Remington 50-grain Power-Lokt Hollow Point (R222R3)

G1 Ballistic Coefficient = 0.188

Distance • Yards	Muzzle	100	200	300	400	500	600	800	1000
Velocity • fps	3140	2635	2182	1777	1432	1172	1015	847	733
Energy • ft-lbs	1094	771	529	351	228	152	114	80	60
Taylor KO Index	5.0	4.2	3.5	2.8	2.3	1.9	1.6	1.4	1.2
Path • Inches	-1.5	+1.8	0.0	-9.2	-29.8	-67.1	-132.8	-363.1	-764.9
Wind Drift • Inches	0.0	1.6	6.8	16.8	33.1	57.2	89.2	170.3	271.0

Nosler 50-grain Ballistic Tip (05091)

G1 Ballistic Coefficient = 0.238

Distance • Yards	Muzzle	100	200	300	400	500	600	800	1000
Velocity • fps	3025	2631	2270	1940	1642	1385	1184	962	842
Energy • ft-lbs	1016	769	572	418	299	213	156	103	79
Taylor KO Index	4.8	4.2	3.6	3.1	2.6	2.2	1.9	1.5	1.3
Path • Inches	-1.5	+1.8	0.0	-8.6	-26.5	-57.4	-106.6	-286.0	-615.6
Wind Drift • Inches	0.0	1.3	5.4	13.1	25.2	42.8	66.8	132.2	215.1

A .222 Remington is a good choice for a roebuck hunt. This deer was taken in Scotland, where suppressors are legal and commonly used on hunting rifles. (Safari Press Archives)

Norma 50-grain Full Jacket (15715)

G1 Ballistic Coefficient = 0.192

Distance • Yards	Muzzle	100	200	300	400	500	600	800	1000
Velocity • fps	3199	2697	2247	1843	1496	1222	1047	868	753
Energy • ft-lbs	1136	809	562	378	249	166	122	84	63
Taylor KO Index	5.1	4.3	3.6	2.9	2.4	2.0	1.7	1.4	1.2
Path • Inches	-1.5	+1.6	0.0	-8.6	-27.6	-62.4	-120.4	-339.0	-742.8
Wind Drift • Inches	0.0	1.5	6.4	15.9	31.1	53.8	84.3	163.1	261.3

Norma 55-grain Oryx (15704)

G1 Ballistic Coefficient = 0.185

Distance • Yards	Muzzle	100	200	300	400	500	600	800	1000
Velocity • fps	3051	2549	2099	1700	1366	1127	988	830	721
Energy • ft-lbs	1137	794	538	353	228	155	119	84	64
Taylor KO Index	5.4	4.5	3.7	3.0	2.4	2.0	1.7	1.5	1.3
Path • Inches	-1.5	+2.0	0.0	-10.0	-32.2	-73.2	-141.6	-392.4	-845.0
Wind Drift • Inches	0.0	1.6	7.2	17.8	35.2	60.8	93.8	176.5	279.0

Lapua 55-grain FMJ (4315020)

G1 Ballistic Coefficient = 0.192

Distance • Yards	Muzzle	100	200	300	400	500	600	800	1000
Velocity • fps	2885	2522	2188	1880	1603	1365	1178	967	851
Energy • ft-lbs	1016	777	584	431	314	228	169	114	88
Taylor KO Index	5.1	4.4	3.9	3.3	2.8	2.4	2.1	1.7	1.5
Path • Inches	-1.5	+2.0	0.0	-9.3	-28.6	-61.4	-113.1	-297.9	-631.2
Wind Drift • Inches	0.0	1.3	5.4	13.2	25.3	42.7	66.2	129.8	210.7

Norma 55-grain FULL JACKET (15722)

G1 Ballistic Coefficient = 0.209

Distance • Yards	Muzzle	100	200	300	400	500	600	800	1000
Velocity • fps	2789	2366	1983	1640	1352	1139	1008	857	753
Energy • ft-lbs	950	684	480	329	223	158	124	90	69
Taylor KO Index	4.9	4.2	3.5	2.9	2.4	2.0	1.8	1.5	1.3
Path • Inches	-1.5	+2.4	0.0	-11.4	-35.8	-79.4	-149.8	-399.6	-839.4
Wind Drift • Inches	0.0	1.6	7.1	17.4	34.0	57.9	88.4	164.9	259.0

Norma 55-grain FULL JACKET (15721)

G1 Ballistic Coefficient = 0.209

Distance • Yards	Muzzle	100	200	300	400	500	600	800	1000
Velocity • fps	3084	2634	2225	1855	1530	1266	1084	899	784
Energy • ft-lbs	1162	847	605	420	289	196	144	98	75
Taylor KO Index	5.4	4.6	3.9	3.3	2.7	2.2	1.9	1.6	1.4
Path • Inches	-1.5	+1.7	0.0	-8.8	-28.0	-62.2	-118.1	-325.5	-705.6
Wind Drift • Inches	0.0	1.4	6.1	15.0	29.2	50.2	78.3	152.2	244.4

Norma 62-grain Soft Point (15716)

G1 Ballistic Coefficient = 0.214

Distance • Yards	Muzzle	100	200	300	400	500	600	800	1000
Velocity • fps	2887	2457	2067	1716	1413	1181	1035	872	765
Energy • ft-lbs	1148	831	588	405	275	192	148	1.5	80
Taylor KO Index	5.7	4.9	4.1	3.4	2.8	2.3	2.1	1.7	1.5
Path • Inches	-1.5	+2.1	0.0	-10.4	-32.8	-72.8	-139.5	-370.0	-763.3
Wind Drift • Inches	0.0	1.5	6.5	16.5	32.2	54.9	84.3	159.7	252.7

.222 Remington Magnum

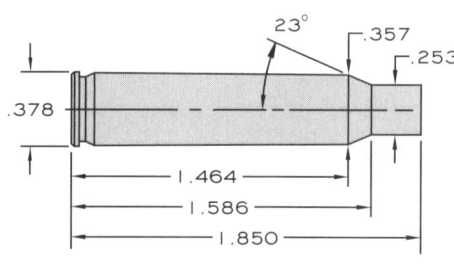

Remington's .222 Magnum is a cartridge that never really had a good chance to become popular. The .222 RM started life as an experimental military cartridge based on the .222 Remington case. Remington began to load for this "Improved" .222 in 1958, but the .223 Remington, which was adopted as a military cartridge, signed its death warrant in 1964. There's nothing wrong with the .222 Remington Magnum as a cartridge; in fact, the slightly greater volume and longer neck makes it a bit "better" than the .223 on a technical basis. But historically, a cartridge that's almost identical to a standard military cartridge is going nowhere. The only factory ammo available today comes from Nosler. The last catalog listing by Remington was in 1997.

Relative Recoil Factor = 0.79

Specifications:

Controlling Agency for Standardization for this Ammunition: SAAMI

Bullet Weight Grains	Velocity fps	Maximum Average Pressure	
		Copper Crusher	Transducer
55	3,215	50,000 cup	N/S

Standard barrel for velocity testing: 24 inches long—1 turn in 14-inch twist

Availability:

Nosler 50-grain Ballistic Tip (06088)

G1 Ballistic Coefficient = 0.238

Distance • Yards	Muzzle	100	200	300	400	500	600	800	1000
Velocity • fps	3340	2919	2534	2181	1859	1570	1326	1023	881
Energy • ft-lbs	1238	946	713	528	384	274	195	116	86
Taylor KO Index	5.3	4.7	4.1	3.5	3.0	2.5	2.1	1.6	1.4
Path • Inches	-1.5	+1.3	0.0	-6.8	-20.9	-45.2	-83.7	-227.1	-502.6
Wind Drift • Inches	0.0	1.1	4.7	11.4	21.8	36.9	57.7	118.2	198.5

Safety first! When introducing youngsters to shooting, you need to explain the basics. Here Wyoming outfitter Robb Wiley is making sure these young shooters understand the safety issues. (Safari Press Archives)

.223 Remington

When the .223 Remington cartridge was adopted by the U.S. Army as M193 5.56mm Ball ammunition in 1964, that action ensured that the .223 would become the most popular .22 centerfire in the list. Every cartridge that becomes a U.S. Army standard, with the possible exception of the .30 Carbine, has gone on to a long and useful commercial life. Just look at the .45-70, the .45 Colt, the .30-06, and the .45 ACP. The .223 case has been "necked" to every possible size, the TCU series of cartridges and the .30 Whisper being examples.

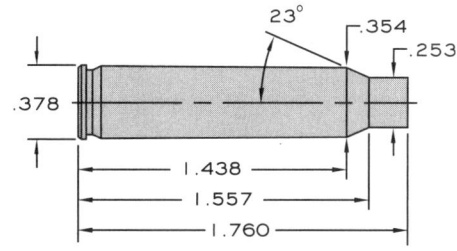

Even without the military application, the .223 Remington had plenty of potential to become popular. Based on the .222 and the .222 Remington Magnum, the .223 provides an excellent balance of accuracy and performance with good case and barrel life. It has become the standard by which all other .22s are judged.

Some guns chambered for the .223 have barrel twists as slow as 1 turn in 14 inches. That twist provides enough stability for bullets up to 55 grains but begins to be marginal above that level. Today there are bullets weighing as much as 77 grains available in loaded ammunition (80-grain bullets for the handloader). It takes a faster twist to stabilize these heavier bullets, and some gunmakers offer barrel twists as fast as 1 turn in 7 inches in this caliber. If you have one of the fast-twist barrels, you may have problems with the very light varmint loads. The high velocities attained and the quick twist combine to produce a bullet spin rate that can literally rip thin jacketed bullets apart. Guns equipped with quick-twist barrels will usually do better with 55-grain and heavier bullets.

Relative Recoil Factor = 0.80

Specifications:

Controlling Agency for Standardization of this Ammunition: SAAMI

Bullet Weight Grains	Velocity fps	Maximum Average Pressure	
		Copper Crusher	Transducer
53	3,305	52,000 cup	55,000 psi
55	3,215	52,000 cup	55,000 psi
60	3,200	52,000 cup	55,000 psi
64	3,000	52,000 cup	55,000 psi

Standard barrel for velocity testing: 24 inches long—1 turn in 12-inch twist

Availability:

Winchester 35-grain Ballistic Silvertip (S22336RLF) LF

G1 Ballistic Coefficient = 0.201

Distance • Yards	Muzzle	100	200	300	400	500	600	800	1000
Velocity • fps	3800	3251	2766	2330	1935	1586	1300	981	836
Energy • ft-lbs	1110	813	588	417	287	193	131	75	54
Taylor KO Index	4.3	3.6	3.1	2.6	2.2	1.8	1.5	1.1	0.9
Path • Inches	-1.5	+0.9	0.0	-5.5	-17.5	-38.9	-73.4	-215.1	-499.9
Wind Drift • Inches	0.0	1.1	4.9	11.8	22.7	39.0	62.0	129.7	219.2

Nosler 35-grain Ballistic Tip (06088) LF

G1 Ballistic Coefficient = 0.238

Distance • Yards	Muzzle	100	200	300	400	500	600	800	1000
Velocity • fps	3750	3214	2742	2315	1927	1587	1301	984	839
Energy • ft-lbs	1093	803	584	416	289	196	131	75	55
Taylor KO Index	4.2	3.6	3.1	2.6	2.2	1.8	1.5	1.1	0.9
Path • Inches	-1.5	+1.0	0.0	-5.6	-17.8	-39.5	-75.3	-216.7	-501.2
Wind Drift • Inches	0.0	1.1	4.8	11.8	22.7	38.9	61.7	128.8	217.6

Black Hills 36-grain Barnes Varmint Grenade (D223N15)

G1 Ballistic Coefficient = 0.149

Distance • Yards	Muzzle	100	200	300	400	500	600	800	1000
Velocity • fps	3750	3034	2421	1886	1439	1127	964	787	664
Energy • ft-lbs	1124	736	469	284	166	101	74	50	35
Taylor KO Index	4.3	3.5	2.8	2.2	1.7	1.3	1.1	0.9	0.8
Path • Inches	-1.5	+1.2	0.0	-7.2	-24.5	-58.9	-120.6	-365.2	-836.2
Wind Drift • Inches	0.0	1.6	7.0	17.6	35.7	63.4	100.4	194.0	313.0

Cor-Bon 36-grain Varmint Grenade (SD22336VG/20) LF
G1 Ballistic Coefficient = 0.149

Distance • Yards	Muzzle	100	200	300	400	500	600	800	1000
Velocity • fps	3600	2907	2311	1791	1366	1086	942	774	654
Energy • ft-lbs	1036	676	427	256	149	94	71	48	34
Taylor KO Index	4.1	3.3	2.7	2.1	1.6	1.3	1.1	0.9	0.8
Path • Inches	-1.5	+1.3	0.0	-8.0	-27.2	-65.2	-132.6	-394.2	-890.4
Wind Drift • Inches	0.0	1.7	7.4	18.7	37.9	67.0	104.9	200.1	320.0

Hornady 40-grain V-Max (8325)
G1 Ballistic Coefficient= 0.200

Distance • Yards	Muzzle	100	200	300	400	500	600	800	1000
Velocity • fps	3800	3249	2762	2324	1928	1578	1290	975	831
Energy • ft-lbs	1282	937	677	479	330	221	148	84	61
Taylor KO Index	6.1	5.2	4.4	3.7	3.1	2.5	2.1	1.6	1.3
Path • Inches	-1.5	+0.9	0.0	-5.5	-17.6	-39.2	-74.9	-217.5	-505.6
Wind Drift • Inches	0.0	1.2	5.0	11.9	22.9	39.4	62.6	131.0	221.2

Norma 40-grain V-MAX (15738)
G1 Ballistic Coefficient = 0.200

Distance • Yards	Muzzle	100	200	300	400	500	600	800	1000
Velocity • fps	3740	3196	2715	2281	1890	1546	1268	987	827
Energy • ft-lbs	1243	908	655	462	317	212	143	83	61
Taylor KO Index	4.8	4.1	3.5	2.9	2.4	2.0	1.6	1.3	1.1
Path • Inches	-1.5	+0.9	0.0	-5.7	-18.2	-40.6	-77.7	-225.2	-520.6
Wind Drift • Inches	0.0	1.2	5.0	12.1	23.4	40.2	63.9	133.0	223.5

Federal 40-grain Nosler Ballistic Tip (P223P)
G1 Ballistic Coefficient = 0.221

Distance • Yards	Muzzle	100	200	300	400	500	600	800	1000
Velocity • fps	3700	3210	2770	2370	2010	1680	1396	1038	880
Energy • ft-lbs	1215	915	680	500	360	250	173	96	69
Taylor KO Index	4.7	4.1	3.5	3.0	2.6	2.2	1.8	1.3	1.1
Path • Inches	-1.5	+1.0	0.0	-8.0	-17.3	-37.8	-71.6	-203.4	-458.9
Wind Drift • Inches	0.0	1.1	4.5	10.8	20.8	35.3	55.6	116.3	199.0

Nosler 40-grain Ballistic Tip (60001)
G1 Ballistic Coefficient = 0.221

Distance • Yards	Muzzle	100	200	300	400	500	600	800	1000
Velocity • fps	3700	3215	2784	2290	2030	1698	1418	1051	890
Energy • ft-lbs	1216	918	688	507	366	256	179	98	70
Taylor KO Index	4.7	4.1	3.6	3.1	2.6	2.2	1.8	1.3	1.1
Path • Inches	-1.5	+0.9	0.0	-5.5	-17.1	-37.3	-69.7	-193.6	-442.4
Wind Drift • Inches	0.0	1.0	4.4	10.7	20.4	34.6	54.0	113.8	195.4

Fiocchi 40-grain Hornady V-MAX Polymer Tip (223HVB)
G1 Ballistic Coefficient = 0.200

Distance • Yards	Muzzle	100	200	300	400	500	600	800	1000
Velocity • fps	3650	3117	2644	2217	1832	1497	1232	954	818
Energy • ft-lbs	1183	862	620	436	298	199	135	81	60
Taylor KO Index	4.7	4.0	3.4	2.8	2.3	1.9	1.6	1.2	1.0
Path • Inches	-1.5	+1.1	0.0	-6.1	-19.3	-43.1	-82.5	-238.3	-454.0
Wind Drift • Inches	0.0	1.2	5.1	12.5	24.2	41.7	66.2	136.6	227.8

Nosler 40-grain Ballistic Tip (60007) LF
G1 Ballistic Coefficient = 0.224

Distance • Yards	Muzzle	100	200	300	400	500	600	800	1000
Velocity • fps	3625	3146	2719	2330	1973	1654	1381	1035	881
Energy • ft-lbs	1167	879	657	482	346	243	170	95	69
Taylor KO Index	4.6	4.0	3.5	3.0	2.5	2.1	1.8	1.3	1.1
Path • Inches	-1.5	+1.0	0.0	-5.8	-18.0	-39.2	-73.2	-203.4	-462.5
Wind Drift • Inches	0.0	1.1	4.6	11.0	21.0	35.6	56.1	116.8	199.1

Black Hills 40-grain Hornady V-MAX (D223N11)
G1 Ballistic Coefficient = 0.200

Distance • Yards	Muzzle	100	200	300	400	500	600	800	1000
Velocity • fps	3600	3072	2803	2180	1798	1469	1210	945	812
Energy • ft-lbs	1150	839	602	422	287	192	130	79	59
Taylor KO Index	5.8	4.9	4.5	3.5	2.9	2.4	1.9	1.5	1.3
Path • Inches	-1.5	+1.1	0.0	-6.3	-20.0	-44.6	-85.6	-246.7	-563.3
Wind Drift • Inches	0.0	1.2	5.2	12.7	24.7	42.6	67.7	138.9	230.8

Cor-Bon 40-grain BlitzKing (SD22340BK/20)   G1 Ballistic Coefficient = 0.210

Distance • Yards	Muzzle	100	200	300	400	500	600	800	1000
Velocity • fps	3400	2919	2487	2093	1739	1434	1196	948	819
Energy • ft-lbs	1027	757	549	389	269	183	127	80	60
Taylor KO Index	4.4	3.7	3.2	2.7	2.2	1.8	1.5	1.2	1.0
Path • Inches	-1.5	+1.3	0.0	-7.0	-22.0	-48.7	-92.4	-260.4	-583.0
Wind Drift • Inches	0.0	1.2	5.3	12.9	25.1	43.0	68.0	137.7	227.1

Ultramax 40-grain Nosler Ballistic Tip (223R5)   G1 Ballistic Coefficient = 0.221

Distance • Yards	Muzzle	100	200	300	400	500	600	800	1000
Velocity • fps	3200	2762	2363	1999	1672	1391	1175	946	823
Energy • ft-lbs	910	678	496	355	248	172	123	80	60
Taylor KO Index	4.1	3.5	3.0	2.6	2.1	1.8	1.5	1.2	1.1
Path • Inches	-1.5	+1.6	0.0	-7.8	-24.5	-54.2	-102.7	-283.2	-605.4
Wind Drift • Inches	0.0	1.3	5.4	13.2	25.6	43.8	68.8	133.3	224.4

Ultramax 42-grain Frangible (223RF1)   G1 Ballistic Coefficient = 0.190

Distance • Yards	Muzzle	100	200	300	400	500	600	800	1000
Velocity • fps	3200	2694	2241	1835	1484	1210	1039	862	748
Energy • ft-lbs	955	677	469	314	205	137	101	69	52
Taylor KO Index	4.3	3.6	3.0	2.5	2.0	1.6	1.4	1.2	1.0
Path • Inches	-1.5	+1.7	0.0	-8.7	-27.9	-63.1	-122.0	-343.8	-753.3
Wind Drift • Inches	0.0	1.5	6.6	16.0	31.6	54.6	85.5	165.1	264.2

Federal 43-grain Speer TNT Green (P223R) LF   G1 Ballistic Coefficient = 0.152

Distance • Yards	Muzzle	100	200	300	400	500	600	800	1000
Velocity • fps	3600	2920	2330	1810	1390	1100	953	783	663
Energy • ft-lbs	1225	810	515	315	185	115	87	59	42
Taylor KO Index	5.0	4.0	3.2	2.5	1.9	1.5	1.3	1.1	0.9
Path • Inches	-1.5	+1.3	0.0	-7.9	-26.5	-63.4	-128.7	-383.0	-866.1
Wind Drift • Inches	0.0	1.4	6.3	15.5	30.8	53.9	85.7	169.2	274.0

Cor-Bon 45-grain Glaser Safety Slug (05400)   G1 Ballistic Coefficient = 0.135

Distance • Yards	Muzzle	100	200	300	400	500	600	800	1000
Velocity • fps	3375	2653	2034	1515	1146	961	854	702	585
Energy • ft-lbs	1138	703	413	229	131	92	73	49	34
Taylor KO Index	4.9	3.8	2.9	2.2	1.7	1.4	1.2	1.0	0.8
Path • Inches	-1.5	+1.8	0.0	-10.4	-36.5	-88.9	-178.8	-512.9	-1133
Wind Drift • Inches	0.0	2.0	9.1	23.6	48.4	83.5	126.4	232.2	366.6

Fiocchi 45-grain Frangible (223SFNT)   G1 Ballistic Coefficient = 0.202

Distance • Yards	Muzzle	100	200	300	400	500	600	800	1000
Velocity • fps	3300	2812	2373	1976	1623	1329	1117	908	787
Energy • ft-lbs	1087	790	562	390	263	176	125	82	62
Taylor KO Index	4.8	4.0	3.4	2.8	2.3	1.9	1.6	1.3	1.1
Path • Inches	-1.5	+1.5	0.0	-7.7	-24.5	-54.7	-104.7	-295.6	-655.4
Wind Drift • Inches	0.0	1.3	5.8	14.2	27.7	47.8	75.4	149.6	243.0

UMC (Remington) 45-grain Jacketed Hollow Point (L223R7)   G1 Ballistic Coefficient = 0.175

Distance • Yards	Muzzle	100	200	300	400	500	600	800	1000
Velocity • fps	3550	2953	2430	1968	1564	1245	1044	850	728
Energy • ft-lbs	1259	871	590	387	245	155	109	72	53
Taylor KO Index	5.2	4.4	3.6	3.0	2.4	2.0	1.6	1.3	1.1
Path • Inches	-1.5	+1.3	0.0	-7.2	-23.8	-55.4	-110.8	-319.8	-698.3
Wind Drift • Inches	0.0	1.4	6.3	15.5	30.8	53.9	85.7	169.2	274.0

USA (Winchester) 45-grain Jacketed Hollow Point (USA2232)   G1 Ballistic Coefficient = 0.173

Distance • Yards	Muzzle	100	200	300	400	500	600	800	1000
Velocity • fps	3600	3033	2533	2085	1687	1356	1119	896	769
Energy • ft-lbs	1295	919	641	434	284	184	125	80	59
Taylor KO Index	5.2	4.4	3.6	3.0	2.4	2.0	1.6	1.3	1.1
Path • Inches	-1.5	+1.2	0.0	-6.7	-21.4	-48.7	-97.4	-283.3	-623.5
Wind Drift • Inches	0.0	1.3	5.7	14.0	27.6	47.9	76.4	154.0	252.4

Remington 45-grain Jacketed Iron-Tin Core HP (DV223RA) LF G1 Ballistic Coefficient = 0.161

Distance • Yards	Muzzle	100	200	300	400	500	600	800	1000
Velocity • fps	3550	2911	2355	1865	1451	1150	987	811	691
Energy • ft-lbs	1251	847	554	347	210	132	97	66	48
Taylor KO Index	5.1	4.2	3.4	2.7	2.1	1.7	1.4	1.2	1.0
Path • Inches	-1.5	+1.3	0.0	-7.7	-25.6	-60.3	-121.0	-357.9	-807.5
Wind Drift • Inches	0.0	1.6	6.8	17.2	34.4	60.6	95.6	184.8	296.6

Norma 50-grain V-MAX (15739) G1 Ballistic Coefficient = 0.242

Distance • Yards	Muzzle	100	200	300	400	500	600	800	1000
Velocity • fps	3495	3065	2675	2317	1987	1688	1428	1073	911
Energy • ft-lbs	1357	1044	795	596	439	317	228	128	92
Taylor KO Index	5.6	4.9	4.3	3.7	3.2	2.7	2.3	1.7	1.5
Path • Inches	-1.5	+1.1	0.0	-6.0	-18.5	-40.0	-73.7	-199.0	-444.2
Wind Drift • Inches	0.0	1.0	4.4	10.5	20.0	33.7	52.6	109.0	186.4

UMC (Remington) 50-grain Jacketed Hollow Point (L223R8) G1 Ballistic Coefficient = 0.194

Distance • Yards	Muzzle	100	200	300	400	500	600	800	1000
Velocity • fps	3425	2899	2430	2010	1637	1327	1108	895	772
Energy • ft-lbs	1302	933	655	449	298	196	136	89	66
Taylor KO Index	5.5	4.6	3.9	3.2	2.6	2.1	1.8	1.4	1.2
Path • Inches	-1.5	+1.3	0.0	-7.3	-23.4	-53.2	-104.0	-296.6	-644.1
Wind Drift • Inches	0.0	1.4	5.8	14.3	28.0	48.5	76.9	153.5	250.2

Remington 50-grain AccuTip-V Boat Tail (PRA223RB) G1 Ballistic Coefficient = 0.242

Distance • Yards	Muzzle	100	200	300	400	500	600	800	1000
Velocity • fps	3410	2989	2605	2252	1928	1635	1381	1050	898
Energy • ft-lbs	1291	992	753	563	413	297	212	123	90
Taylor KO Index	5.5	4.8	4.2	3.6	3.1	2.6	2.2	1.7	1.4
Path • Inches	-1.5	+1.2	0.0	-6.4	-19.7	-42.4	-78.3	-211.6	-470.2
Wind Drift • Inches	0.0	1.1	4.5	10.8	20.7	35.0	54.7	112.7	191.3

Winchester 50-grain Ballistic Silvertip (SBST223) G1 Ballistic Coefficient = 0.239

Distance • Yards	Muzzle	100	200	300	400	500	600	800	1000
Velocity • fps	3410	2982	2593	2235	1907	1613	1363	1039	890
Energy • ft-lbs	1291	987	746	555	404	289	206	120	88
Taylor KO Index	5.5	4.8	4.1	3.6	3.1	2.6	2.2	1.7	1.4
Path • Inches	-1.5	+1.2	0.0	-6.4	-19.8	-42.8	-80.2	-220.8	-485.1
Wind Drift • Inches	0.0	1.1	4.6	11.1	21.2	35.8	55.8	115.0	194.5

American Eagle (Federal) 50-grain Jacketed Hollow Point (AE223G) G1 Ballistic Coefficient = 0.205

Distance • Yards	Muzzle	100	200	300	400	500	600	800	1000
Velocity • fps	3325	2840	2400	2010	1650	1360	1140	920	799
Energy • ft-lbs	1225	895	640	445	305	205	144	94	71
Taylor KO Index	5.3	4.5	3.8	3.2	2.6	2.2	1.8	1.5	1.3
Path • Inches	-1.5	+1.4	0.0	-7.5	-23.7	-52.8	-100.8	-284.5	-632.4
Wind Drift • Inches	0.0	1.3	5.6	13.8	26.8	46.2	73.0	145.7	237.8

Black Hills 50-grain Hornady V-Max (M223N7) G1 Ballistic Coefficient = 0.242

Distance • Yards	Muzzle	100	200	300	400	500	600	800	1000
Velocity • fps	3300	2909	2626	2173	1850	1562	1324	1024	882
Energy • ft-lbs	1231	940	708	524	380	271	195	116	86
Taylor KO Index	5.3	4.7	4.2	3.5	3.0	2.5	2.1	1.6	1.4
Path • Inches	-1.5	+1.3	0.0	-6.8	-21.1	-45.6	-85.8	-234.6	-5.9.6
Wind Drift • Inches	0.0	1.1	4.7	11.4	21.9	37.1	57.6	117.6	197.3

Fiocchi 50-grain V-MAX (223HVA) G1 Ballistic Coefficient = 0.242

Distance • Yards	Muzzle	100	200	300	400	500	600	800	1000
Velocity • fps	3300	2889	2513	2168	1851	1567	1326	1025	884
Energy • ft-lbs	1208	926	701	521	380	272	195	117	87
Taylor KO Index	5.3	4.6	4.0	3.5	3.0	2.5	2.1	1.6	1.4
Path • Inches	-1.5	+1.4	0.0	-6.9	-21.3	-45.9	-84.8	-229.0	-504.6
Wind Drift • Inches	0.0	1.1	4.7	11.3	21.7	36.8	57.5	117.4	197.1

Ultramax 50-grain TNT (223R9)
G1 Ballistic Coefficient = 0.223

Distance • Yards	Muzzle	100	200	300	400	500	600	800	1000
Velocity • fps	3100	2672	2282	1926	1608	1338	1138	928	810
Energy • ft-lbs	1067	793	578	412	287	199	144	96	73
Taylor KO Index	5.0	4.3	3.7	3.1	2.6	2.1	1.8	1.5	1.3
Path • Inches	-1.5	+1.7	0.0	-8.4	-26.5	-58.4	-110.8	-302.7	-641.0
Wind Drift • Inches	0.0	1.3	5.7	13.8	26.8	45.8	71.8	141.7	229.8

Ultramax 50-grain Nosler Ballistic Tip (223R4)
G1 Ballistic Coefficient = 0.238

Distance • Yards	Muzzle	100	200	300	400	500	600	800	1000
Velocity • fps	3100	2700	2334	1997	1692	1426	1214	974	850
Energy • ft-lbs	1067	810	605	443	318	226	164	105	80
Taylor KO Index	5.0	4.3	3.7	3.2	2.7	2.3	1.9	1.6	1.4
Path • Inches	-1.5	+1.6	0.0	-8.1	-25.0	-54.5	101.8	-275.9	-585.6
Wind Drift • Inches	0.0	1.2	5.2	12.6	24.4	41.4	64.6	129.0	211.4

Cor-Bon 50-grain Varmint Grenade (SD22350VG/20) LF
G1 Ballistic Coefficient = 0.183

Distance • Yards	Muzzle	100	200	300	400	500	600	800	1000
Velocity • fps	3000	2498	2049	1652	1326	1100	972	774	654
Energy • ft-lbs	999	693	466	303	195	134	105	74	56
Taylor KO Index	4.8	4.0	3.3	2.6	2.1	1.8	1.6	1.3	1.1
Path • Inches	-1.5	+2.1	0.0	-10.5	-33.9	-77.3	-149.3	-411.2	-881.6
Wind Drift • Inches	0.0	1.7	7.4	18.6	36.7	63.1	96.8	180.8	284.6

Fiocchi 50-grain FMJ Tundra "Green" (223TUNZ) LF
G1 Ballistic Coefficient = 0.210

Distance • Yards	Muzzle	100	200	300	400	500	600	800	1000
Velocity • fps	3300	2831	2407	2022	1677	1383	1159	932	809
Energy • ft-lbs	1209	890	643	454	312	212	149	96	73
Taylor KO Index	5.3	4.5	3.9	3.2	2.7	2.2	1.9	1.5	1.3
Path • Inches	-1.5	+1.4	0.0	-7.5	-23.7	-52.4	-99.5	-278.7	-617.8
Wind Drift • Inches	0.0	1.3	5.5	13.5	26.2	45.0	70.9	142.0	232.0

Black Hills 52-grain Match HP (M223N3)
G1 Ballistic Coefficient = 0.223

Distance • Yards	Muzzle	100	200	300	400	500	600	800	1000
Velocity • fps	3250	2810	2411	2046	1716	1430	1205	960	833
Energy • ft-lbs	1220	912	671	483	340	236	163	106	80
Taylor KO Index	5.4	4.7	4.0	3.4	2.9	2.4	2.0	1.6	1.4
Path • Inches	-1.5	+1.5	0.0	-7.5	-23.5	-51.6	-97.7	-270.6	-581.9
Wind Drift • Inches	0.0	1.2	5.3	12.8	24.8	42.2	66.4	133.5	219.6

Ultramax 52-grain HP (223R1)
G1 Ballistic Coefficient = 0.224

Distance • Yards	Muzzle	100	200	300	400	500	600	800	1000
Velocity • fps	3000	2586	2207	1862	1556	1301	1116	922	807
Energy • ft-lbs	1039	772	563	400	280	195	144	98	75
Taylor KO Index	5.0	4.3	3.7	3.1	2.6	2.2	1.9	1.5	1.3
Path • Inches	-1.5	+1.9	0.0	-9.1	-28.4	-62.6	-118.4	-319.0	-667.5
Wind Drift • Inches	0.0	1.4	5.9	14.3	27.8	47.4	73.8	143.8	231.4

Norma 53-grain SOFT POINT (15717)
G1 Ballistic Coefficient = 0.237

Distance • Yards	Muzzle	100	200	300	400	500	600	800	1000
Velocity • fps	3215	2804	2428	2082	1767	1490	1262	995	863
Energy • ft-lbs	1217	925	694	510	368	261	187	116	88
Taylor KO Index	5.5	4.8	4.1	3.5	3.0	2.5	2.1	1.7	1.5
Path • Inches	-1.5	+1.4	0.0	-7.4	-23.0	-49.7	-92.3	-250.0	-547.6
Wind Drift • Inches	0.0	1.2	5.0	12.0	23.1	39.3	61.5	124.4	206.2

Hornady 53-grain V-MAX (8025)
G1 Ballistic Coefficient = 0.290

Distance • Yards	Muzzle	100	200	300	400	500	600	800	1000
Velocity • fps	3465	3106	2775	2468	2180	1913	1668	1266	1030
Energy • ft-lbs	1413	1135	906	716	559	431	327	189	125
Taylor KO Index	5.9	5.3	4.7	4.2	3.7	3.2	2.8	2.1	1.7
Path • Inches	-1.5	+1.1	0.0	-5.6	-16.9	-35.6	-63.9	-162.2	-347.9
Wind Drift • Inches	0.0	0.9	3.6	8.6	16.1	26.7	41.0	83.6	146.6

Winchester 53-grain Hollow Point (X223RH)

G1 Ballistic Coefficient = 0.223

Distance • Yards	Muzzle	100	200	300	400	500	600	800	1000
Velocity • fps	3330	2882	2477	2106	1710	1475	1236	972	841
Energy • ft-lbs	1306	978	722	522	369	256	180	111	83
Taylor KO Index	5.6	4.9	4.2	3.6	3.0	2.5	2.1	1.6	1.4
Path • Inches	-1.5	+1.7	0.0	-7.4	-22.7	-49.1	-92.2	-257.0	-557.6
Wind Drift • Inches	0.0	1.2	5.1	12.4	23.9	40.8	64.2	130.4	215.9

Hornady 53-grain HP Steel (80274)

G1 Ballistic Coefficient = 0.218

Distance • Yards	Muzzle	100	200	300	400	500	600	800	1000
Velocity • fps	3315	2860	2448	2071	1732	1438	1206	958	830
Energy • ft-lbs	1293	962	705	505	353	243	171	108	81
Taylor KO Index	5.6	4.9	4.2	3.5	2.9	2.4	2.0	1.6	1.4
Path • Inches	-1.5	+1.4	0.0	-7.3	-22.7	-49.9	-94.0	-261.4	-580.0
Wind Drift • Inches	0.0	1.2	5.3	12.8	24.8	42.3	66.6	134.6	221.7

CorBon 53-grain DPX (DPX22353/20)

G1 Ballistic Coefficient = 0.256

Distance • Yards	Muzzle	100	200	300	400	500	600	800	1000
Velocity • fps	3000	2636	2299	1988	1704	1455	1249	1002	876
Energy • ft-lbs	1059	818	622	465	342	249	184	118	90
Taylor KO Index	5.1	4.5	3.9	3.4	2.9	2.5	2.1	1.7	1.5
Path • Inches	-1.5	+1.8	0.0	-8.4	-25.7	-55.0	-100.9	-266.0	-569.2
Wind Drift • Inches	0.0	1.2	5.0	12.1	23.2	39.2	60.8	121.4	199.5

PMC 53-grain HPWC (223HMA)

G1 Ballistic Coefficient = 0.218

Distance • Yards	Muzzle	100	200	300	400	500	600	800	1000
Velocity • fps	3330	2873	2460	2083	1742	1446	1212	960	832
Energy • ft-lbs	1305	971	712	510	357	246	173	109	82
Taylor KO Index	5.6	4.9	4.2	3.5	3.0	2.5	2.1	1.6	1.4
Path • Inches	-1.5	+1.4	0.0	-7.2	-22.5	-49.4	-93.0	-258.7	-574.9
Wind Drift • Inches	0.0	1.2	5.2	12.7	24.6	42.0	66.2	133.9	221.0

Sellier & Bellot 55-grain SP (V330212U)

G1 Ballistic Coefficient = 0.260

Distance • Yards	Muzzle	100	200	300	400	500	600	800	1000
Velocity • fps	3301	2917	2565	2238	1937	1662	1422	1087	927
Energy • ft-lbs	1331	1040	804	612	458	338	247	144	105
Taylor KO Index	5.8	5.1	4.5	3.9	3.4	2.9	2.5	1.9	1.6
Path • Inches	-1.5	+1.3	0.0	-6.6	-20.2	-43.2	-78.8	-207.9	-454.0
Wind Drift • Inches	0.0	1.0	4.3	10.4	24.2	33.2	51.6	105.6	179.7

Federal 55-grain Soft Point (223A)

G1 Ballistic Coefficient = 0.222

Distance • Yards	Muzzle	100	200	300	400	500	600	800	1000
Velocity • fps	3240	2800	2400	2040	1710	1420	1195	955	830
Energy • ft-lbs	1280	955	705	505	355	245	174	111	84
Taylor KO Index	5.7	4.9	4.2	3.6	3.0	2.5	2.1	1.7	1.5
Path • Inches	-1.5	+1.5	0.0	-7.6	-23.7	-51.9	-97.6	-269.2	-592.8
Wind Drift • Inches	0.0	1.2	5.3	12.9	25.0	42.8	67.2	134.9	221.5

Hornady 55-grain TAP-FPD (83278)

G1 Ballistic Coefficient = 0.257

Distance • Yards	Muzzle	100	200	300	400	500	600	800	1000
Velocity • fps	3240	2854	2500	2172	1871	1598	1364	1051	903
Energy • ft-lbs	1282	995	763	576	427	312	227	135	100
Taylor KO Index	5.7	5.0	4.4	3.8	3.3	2.8	2.4	1.8	1.6
Path • Inches	-1.5	+1.4	0.0	-7.0	-21.4	-45.9	-84.8	-227.8	-492.9
Wind Drift • Inches	0.0	1.1	4.5	10.9	20.8	35.1	54.6	111.4	187.9

Remington 55-grain Pointed Soft Point (R223R1)

G1 Ballistic Coefficient = 0.198

Distance • Yards	Muzzle	100	200	300	400	500	600	800	1000
Velocity • fps	3240	2750	2300	1910	1550	1270	1081	888	771
Energy • ft-lbs	1280	920	650	445	295	195	143	96	73
Taylor KO Index	5.7	4.8	4.0	3.4	2.7	2.2	1.9	1.6	1.4
Path • Inches	-1.5	+1.6	0.0	-8.2	-26.1	-58.3	-112.4	-316.6	-697.4
Wind Drift • Inches	0.0	1.4	6.1	15.0	29.3	50.6	79.6	155.9	251.4

Winchester 55-grain Pointed Soft Point (X223R)

G1 Ballistic Coefficient = 0.198

Distance • Yards	Muzzle	100	200	300	400	500	600	800	1000
Velocity • fps	3240	2750	2300	1910	1550	1270	1081	888	771
Energy • ft-lbs	1280	920	650	445	295	195	143	96	73
Taylor KO Index	5.7	4.8	4.0	3.4	2.7	2.2	1.9	1.6	1.4
Path • Inches	-1.5	+1.6	0.0	-8.2	-26.1	-58.3	-112.4	-316.6	-697.4
Wind Drift • Inches	0.0	1.4	6.1	15.0	29.3	50.6	79.6	155.9	251.4

Winchester 55-grain Pointed Soft Point TIN (X223RT) LF

G1 Ballistic Coefficient = 0.197

Distance • Yards	Muzzle	100	200	300	400	500	600	800	1000
Velocity • fps	3240	2747	2304	1905	1554	1270	1079	887	770
Energy • ft-lbs	1282	921	648	443	295	197	142	96	72
Taylor KO Index	5.7	4.8	4.1	3.4	2.7	2.2	1.9	1.6	1.4
Path • Inches	-1.5	+1.9	0.0	-8.5	-26.7	-57.6	-112.9	-318.1	-700.5
Wind Drift • Inches	0.0	1.4	6.1	15.0	29.5	50.9	80.0	156.5	252.2

Fiocchi 55-grain PSP (223B)

G1 Ballistic Coefficient = 0.235

Distance • Yards	Muzzle	100	200	300	400	500	600	800	1000
Velocity • fps	3230	2813	2434	2085	1767	1487	1256	990	859
Energy • ft-lbs	1273	966	723	530	381	270	193	120	90
Taylor KO Index	5.7	5.0	4.3	3.7	3.1	2.6	2.2	1.7	1.5
Path • Inches	-1.5	+1.5	0.0	-7.4	-22.9	-49.6	-92.2	-250.6	-550.3
Wind Drift • Inches	0.0	1.2	5.0	12.1	23.3	39.6	62.0	125.4	207.9

Black Hills 55-grain Soft Point (M223N2)

G1 Ballistic Coefficient = 0.245

Distance • Yards	Muzzle	100	200	300	400	500	600	800	1000
Velocity • fps	3200	2803	2438	2102	1795	1522	1293	1013	878
Energy • ft-lbs	1251	959	726	540	394	283	204	125	94
Taylor KO Index	5.6	4.9	4.3	3.7	3.2	2.7	2.3	1.8	1.5
Path • Inches	-1.5	+1.5	0.0	-7.4	-22.7	-49.1	-91.2	-247.3	-531.6
Wind Drift • Inches	0.0	1.1	4.8	11.7	22.4	37.8	59.0	119.7	199.2

Norma 55-grain Oryx (15719)

G1 Ballistic Coefficient = 0.185

Distance • Yards	Muzzle	100	200	300	400	500	600	800	1000
Velocity • fps	3117	2608	2152	1746	1401	1148	1000	837	725
Energy • ft-lbs	1187	831	566	372	240	161	122	86	64
Taylor KO Index	5.5	4.6	3.8	3.1	2.5	2.0	1.8	1.5	1.3
Path • Inches	-1.5	+1.8	0.0	-9.5	-30.6	-69.8	-135.0	-337.3	-819.2
Wind Drift • Inches	0.0	1.6	7.0	17.3	34.2	59.1	91.8	174.2	276.4

PMC 55-grain PSP (223B)

G1 Ballistic Coefficient = 0.264

Distance • Yards	Muzzle	100	200	300	400	500	600	800	1000
Velocity • fps	3100	2689	2314	1974	1817	1553	1333	1046	906
Energy • ft-lbs	1174	883	654	476	341	244	217	134	100
Taylor KO Index	5.5	4.7	4.1	3.5	3.2	2.7	2.3	1.8	1.6
Path • Inches	-1.5	+1.6	0.0	-7.6	-23.3	-49.6	-90.4	-237.1	-510.2
Wind Drift • Inches	0.0	1.1	4.6	11.1	21.2	35.7	55.4	112.0	187.1

Ultramax 55-grain Soft Point (223R3)

G1 Ballistic Coefficient = 0.230

Distance • Yards	Muzzle	100	200	300	400	500	600	800	1000
Velocity • fps	3000	2596	2226	1888	1586	1331	1140	936	820
Energy • ft-lbs	1099	823	605	435	307	216	159	107	82
Taylor KO Index	5.3	4.6	3.9	3.3	2.8	2.3	2.0	1.6	1.4
Path • Inches	-1.5	+1.8	0.0	-8.9	-27.8	-61.0	-114.7	-308.9	-647.2
Wind Drift • Inches	0.0	1.3	5.7	13.8	26.8	45.6	71.1	139.3	224.9

Cor-Bon 55-grain MPG (MPG22355/20) LF

G1 Ballistic Coefficient = 0.215

Distance • Yards	Muzzle	100	200	300	400	500	600	800	1000
Velocity • fps	3300	2860	2460	2094	1782	1472	1237	976	845
Energy • ft-lbs	1330	999	739	536	379	265	187	116	87
Taylor KO Index	5.8	5.0	4.3	3.7	3.1	2.6	2.2	1.7	1.5
Path • Inches	-1.5	+1.4	0.0	-7.2	-22.4	-48.9	-91.7	-252.6	-559.1
Wind Drift • Inches	0.0	1.2	5.1	12.4	23.9	40.7	64.0	129.6	214.5

Federal 55-grain Nosler Ballistic Tip (P223F)
G1 Ballistic Coefficient = 0.268

Distance • Yards	Muzzle	100	200	300	400	500	600	800	1000
Velocity • fps	3240	2870	2530	2220	1920	1660	1426	1095	934
Energy • ft-lbs	1280	1005	780	600	450	336	248	146	107
Taylor KO Index	5.7	5.1	4.5	3.9	3.4	2.9	2.5	1.9	1.6
Path • Inches	-1.5	+1.4	0.0	-6.8	-20.8	-44.2	-80.3	-209.8	-454.7
Wind Drift • Inches	0.0	1.0	4.3	10.3	19.6	32.8	50.9	103.8	176.4

Federal 55-grain Sierra GameKing Boat-Tail Hollow Point (P223E)
G1 Ballistic Coefficient = 0.250

Distance • Yards	Muzzle	100	200	300	400	500	600	800	1000
Velocity • fps	3240	2850	2490	2150	1850	1570	1337	1037	894
Energy • ft-lbs	1280	990	755	565	415	300	218	131	98
Taylor KO Index	5.7	5.0	4.4	3.8	3.3	2.8	2.4	1.8	1.6
Path • Inches	-1.5	+1.4	0.0	-7.1	-21.7	-46.7	-85.8	-229.1	-501.1
Wind Drift • Inches	0.0	1.1	4.6	11.2	21.4	36.1	56.3	114.7	192.5

Hornady 55-grain V-MAX (8327) – V-MAX w/Moly (83273)
G1 Ballistic Coefficient = 0.255

Distance • Yards	Muzzle	100	200	300	400	500	600	800	1000
Velocity • fps	3240	2850	2500	2172	1871	1598	1362	1052	905
Energy • ft-lbs	1282	995	763	576	427	312	227	135	100
Taylor KO Index	5.7	5.3	4.4	3.8	3.3	2.8	2.4	1.9	1.6
Path • Inches	-1.5	+1.4	0.0	-7.0	-21.4	-45.9	-84.2	-223.3	-487.3
Wind Drift • Inches	0.0	1.1	4.6	10.9	20.8	35.1	54.7	111.5	187.9

Remington 55-grain AccuTip-V Boat Tail (PRA223RC)
G1 Ballistic Coefficient = 0.255

Distance • Yards	Muzzle	100	200	300	400	500	600	800	1000
Velocity • fps	3240	2854	2500	2172	1871	1598	1363	1053	905
Energy • ft-lbs	1282	995	763	576	427	312	227	135	100
Taylor KO Index	5.7	5.0	4.4	3.8	3.3	2.8	2.4	1.9	1.6
Path • Inches	-1.5	+1.4	0.0	-7.1	-21.7	-46.3	-84.8	-227.9	-492.5
Wind Drift • Inches	0.0	1.1	4.6	10.9	20.8	35.1	54.6	111.4	187.7

Remington 55-grain Hollow Point Power-Lokt (R223R2)
G1 Ballistic Coefficient = 0.209

Distance • Yards	Muzzle	100	200	300	400	500	600	800	1000
Velocity • fps	3240	2773	2352	1969	1627	1341	1131	919	799
Energy • ft-lbs	1282	939	675	473	323	220	156	103	78
Taylor KO Index	5.7	4.9	4.1	3.5	2.9	2.4	2.0	1.6	1.4
Path • Inches	-1.5	+1.5	0.0	-7.9	-24.8	-55.1	-104.9	-293.0	-645.9
Wind Drift • Inches	0.0	1.3	5.7	14.0	27.2	46.7	73.5	145.8	237.0

Winchester 55-grain Ballistic Silvertip (SBST223B)
G1 Ballistic Coefficient = 0.267

Distance • Yards	Muzzle	100	200	300	400	500	600	800	1000
Velocity • fps	3240	2871	2531	2215	1923	1657	1422	1092	932
Energy • ft-lbs	1282	1006	782	599	451	335	247	146	106
Taylor KO Index	5.7	5.1	4.5	3.9	3.4	2.9	2.5	1.9	1.6
Path • Inches	-1.5	+1.4	0.0	-6.8	-20.8	-44.3	-81.1	-214.8	-463.8
Wind Drift • Inches	0.0	1.0	4.3	10.3	19.6	33.0	51.1	104.2	177.1

Black Hills 55-grain Barnes MPG (D223N16) LF
G1 Ballistic Coefficient = 0.215

Distance • Yards	Muzzle	100	200	300	400	500	600	800	1000
Velocity • fps	3200	2750	2342	1971	1639	1358	1148	931	811
Energy • ft-lbs	1250	924	670	475	328	225	161	106	80
Taylor KO Index	5.6	4.8	4.1	3.5	2.9	2.4	2.0	1.6	1.4
Path • Inches	-1.5	+1.6	0.0	-8.0	-25.0	-55.1	-104.3	-288.8	-633.7
Wind Drift • Inches	0.0	1.3	5.6	13.7	26.6	45.6	71.5	142.1	231.2

Federal 55-grain Barnes Triple-Shock X-Bullet (223A) LF
G1 Ballistic Coefficient = 0.215

Distance • Yards	Muzzle	100	200	300	400	500	600	800	1000
Velocity • fps	3200	2750	2350	1980	1650	1360	1148	931	811
Energy • ft-lbs	1250	925	670	475	330	225	161	106	80
Taylor KO Index	5.6	4.8	4.1	3.5	2.9	2.4	2.0	1.6	1.4
Path • Inches	-1.5	+1.6	0.0	-8.0	-25.1	-55.5	-104.3	-288.8	-633.7
Wind Drift • Inches	0.0	1.3	5.6	13.7	26.6	45.6	71.5	142.1	231.2

Black Hills 55-grain Barnes TSX (D223N17) LF

G1 Ballistic Coefficient = 0.210

Distance • Yards	Muzzle	100	200	300	400	500	600	800	1000
Velocity • fps	3200	2740	2324	1946	1610	1329	1125	917	798
Energy • ft-lbs	1250	917	660	463	316	216	155	103	78
Taylor KO Index	5.6	4.8	4.1	3.4	2.8	2.3	2.0	1.6	1.4
Path • Inches	-1.5	+1.6	0.0	-8.1	-25.5	-56.5	-107.3	-298.5	-655.0
Wind Drift • Inches	0.0	1.3	5.8	14.1	27.5	47.1	74.0	146.3	237.2

Cor-Bon 55-grain BlitzKing (SD22355BK/20)

G1 Ballistic Coefficient = 0.264

Distance • Yards	Muzzle	100	200	300	400	500	600	800	1000
Velocity • fps	3000	2646	2618	2014	1736	1488	1281	1021	890
Energy • ft-lbs	1099	855	656	496	368	271	200	127	97
Taylor KO Index	5.3	4.7	4.1	3.5	3.1	2.6	2.3	1.8	1.6
Path • Inches	-1.5	+1.8	0.0	-8.3	-25.2	-53.7	-98.0	-256.7	-548.3
Wind Drift • Inches	0.0	1.1	4.9	11.7	22.4	37.6	58.4	116.8	193.0

Ultramax 55-grain Nosler Ballistic Tip (223R7)

G1 Ballistic Coefficient = 0.267

Distance • Yards	Muzzle	100	200	300	400	500	600	800	1000
Velocity • fps	3000	2650	2325	2024	1748	1501	1293	1029	896
Energy • ft-lbs	1099	858	661	501	373	275	204	129	98
Taylor KO Index	5.3	4.7	4.1	3.6	3.1	2.6	2.3	1.8	1.6
Path • Inches	-1.5	+1.4	0.0	-8.2	-25.0	-53.4	-97.8	-257.7	-543.2
Wind Drift • Inches	0.0	1.1	4.8	11.5	22.0	37.0	57.4	150.7	190.2

Magtech 55-grain FMJ (223A)

G1 Ballistic Coefficient = 0.255

Distance • Yards	Muzzle	100	200	300	400	500	600	800	1000
Velocity • fps	3301	2909	2550	2219	1912	1635	1395	1068	914
Energy • ft-lbs	1331	1034	795	601	447	327	238	139	102
Taylor KO Index	5.8	5.1	4.5	3.9	3.4	2.9	2.5	1.9	1.6
Path • Inches	-1.5	+1.3	0.0	-3.7	-20.5	-43.9	-80.4	-213.5	-467.8
Wind Drift • Inches	0.0	1.0	4.4	10.6	20.3	34.2	53.2	108.9	184.6

Sellier & Bellot 55-grain FMJ (V330282U)

G1 Ballistic Coefficient = 0.260

Distance • Yards	Muzzle	100	200	300	400	500	600	800	1000
Velocity • fps	3283	2901	2550	2224	1924	1651	1412	1082	924
Energy • ft-lbs	1317	1028	794	604	452	333	244	143	104
Taylor KO Index	5.8	5.1	4.5	3.9	3.4	2.9	2.5	1.9	1.6
Path • Inches	-1.5	+1.3	0.0	-6.7	-20.5	-43.8	-79.9	-210.6	-459.6
Wind Drift • Inches	0.0	1.0	4.4	10.5	19.9	33.5	52.0	106.3	180.6

Norma 55-grain FULL JACKET (15726)

G1 Ballistic Coefficient = 0.209

Distance • Yards	Muzzle	100	200	300	400	500	600	800	1000
Velocity • fps	3250	2782	2360	1976	1636	1348	1137	922	801
Energy • ft-lbs	1290	946	631	478	327	222	158	104	78
Taylor KO Index	5.7	4.9	4.2	3.5	2.9	2.4	2.0	1.6	1.4
Path • Inches	-1.5	+1.5	0.0	-7.8	-24.6	-54.6	-103.9	-290.4	-640.4
Wind Drift • Inches	0.0	1.3	5.7	13.9	27.0	46.4	73.0	145.0	236.0

American Eagle (Federal) 55-grain FMJ Boattail (AE223)

G1 Ballistic Coefficient = 0.270

Distance • Yards	Muzzle	100	200	300	400	500	600	800	1000
Velocity • fps	3240	2870	2540	2220	1930	1670	1436	1101	939
Energy • ft-lbs	1280	1010	785	605	455	340	252	148	108
Taylor KO Index	5.7	5.1	4.5	3.9	3.4	2.9	2.5	1.9	1.7
Path • Inches	-1.5	+1.4	0.0	-6.8	-20.7	-43.9	-79.8	-207.9	-450.1
Wind Drift • Inches	0.0	1.0	4.3	10.2	19.4	32.5	50.3	102.6	174.8

Federal 55-grain Full Metal Jacket Boat-Tail (223B)

G1 Ballistic Coefficient = 0.270

Distance • Yards	Muzzle	100	200	300	400	500	600	800	1000
Velocity • fps	3240	2870	2540	2220	1930	1670	1436	1101	939
Energy • ft-lbs	1280	1010	785	605	455	340	252	148	108
Taylor KO Index	5.7	5.1	4.5	3.9	3.4	2.9	2.5	1.9	1.7
Path • Inches	-1.5	+1.4	0.0	-6.8	-20.7	-43.9	-79.8	-207.9	-450.1
Wind Drift • Inches	0.0	1.0	4.3	10.2	19.4	32.5	50.3	102.6	174.8

Fiocchi 55-grain FMJBT (223A)

G1 Ballistic Coefficient = 0.272

Distance • Yards	Muzzle	100	200	300	400	500	600	800	1000
Velocity • fps	3240	2877	2542	2231	1943	1679	1447	1110	945
Energy • ft-lbs	1281	1017	789	608	461	344	256	151	109
Taylor KO Index	5.6	4.9	4.3	3.7	3.2	2.7	2.3	1.8	1.6
Path • Inches	-1.5	+1.4	0.0	-6.8	-20.6	-43.6	-79.1	-205.6	-444.5
Wind Drift • Inches	0.0	1.0	4.2	10.1	19.2	32.1	49.7	101.3	172.7

Remington 55-grain Metal Case (R223R3)

G1 Ballistic Coefficient = 0.202

Distance • Yards	Muzzle	100	200	300	400	500	600	800	1000
Velocity • fps	3240	2759	2326	1933	1587	1301	1099	899	781
Energy • ft-lbs	1282	929	660	456	307	207	148	99	75
Taylor KO Index	5.7	4.9	4.1	3.4	2.8	2.3	1.9	1.6	1.4
Path • Inches	-1.5	+1.6	0.0	-8.1	-25.5	-57.0	-109.5	-307.6	-677.9
Wind Drift • Inches	0.0	1.4	5.9	14.6	28.5	49.1	77.3	152.1	246.0

UMC (Remington) 55-grain Metal Case (L223R3)

G1 Ballistic Coefficient = 0.202

Distance • Yards	Muzzle	100	200	300	400	500	600	800	1000
Velocity • fps	3240	2759	2326	1933	1587	1301	1099	899	781
Energy • ft-lbs	1282	929	660	456	307	207	148	99	75
Taylor KO Index	5.7	4.9	4.1	3.4	2.8	2.3	1.9	1.6	1.4
Path • Inches	-1.5	+1.6	0.0	-8.1	-25.5	-57.0	-109.5	-307.6	-677.9
Wind Drift • Inches	0.0	1.4	5.9	14.6	28.5	49.1	77.3	152.1	246.0

USA (Winchester) 55-grain Full Metal Jacket (USA223R1)

G1 Ballistic Coefficient = 0.255

Distance • Yards	Muzzle	100	200	300	400	500	600	800	1000
Velocity • fps	3240	2854	2499	2172	1869	1597	1363	1053	905
Energy • ft-lbs	1282	995	763	576	427	311	227	135	100
Taylor KO Index	5.7	5.0	4.4	3.8	3.3	2.8	2.4	1.9	1.6
Path • Inches	-1.5	+1.4	0.0	-7.0	-21.4	-45.9	-84.8	-227.9	-492.5
Wind Drift • Inches	0.0	1.1	4.6	10.9	20.8	35.1	54.6	111.4	187.7

Black Hills 55-grain Full Metal Jacket (D223N1)

G1 Ballistic Coefficient = 0.250

Distance • Yards	Muzzle	100	200	300	400	500	600	800	1000
Velocity • fps	3200	2810	2453	2132	1819	1548	1319	1029	890
Energy • ft-lbs	1250	965	735	735	550	404	293	213	129
Taylor KO Index	5.6	4.9	4.3	3.7	3.2	2.7	2.3	1.8	1.6
Path • Inches	-1.5	+1.5	0.0	-7.3	-22.4	-48.0	-88.3	-235.5	-513.3
Wind Drift • Inches	0.0	1.1	4.7	11.4	21.7	36.7	57.2	116.2	194.2

PMC 55-grain FMJ-BT (223A)

G1 Ballistic Coefficient = 0.268

Distance • Yards	Muzzle	100	200	300	400	500	600	800	1000
Velocity • fps	3200	2833	2493	2180	1893	1635	1406	1085	929
Energy • ft-lbs	1250	980	759	580	438	326	241	144	106
Taylor KO Index	5.6	5.0	4.4	3.8	3.3	2.9	2.5	1.9	1.6
Path • Inches	-1.5	+1.9	0.0	-7.1	-21.5	-45.5	-82.6	-215.8	-466.4
Wind Drift • Inches	0.0	1.1	4.7	11.4	21.7	36.7	57.2	116.2	194.2

Lapua 55-grain FMJ (4315040)

G1 Ballistic Coefficient = 0.250

Distance • Yards	Muzzle	100	200	300	400	500	600	800	1000
Velocity • fps	3130	2747	2395	2070	1770	1505	1284	1014	881
Energy • ft-lbs	1196	922	701	523	383	277	202	126	95
Taylor KO Index	5.5	4.8	4.2	3.6	3.1	2.6	2.3	1.8	1.6
Path • Inches	-1.5	+1.6	0.0	-7.7	-23.6	-50.6	-93.2	-248.2	-537.9
Wind Drift • Inches	0.0	1.1	4.9	11.7	22.5	38.0	59.2	119.3	198.0

Cor-Bon 55-grain FMJ (PM22355/50)

G1 Ballistic Coefficient = 0.225

Distance • Yards	Muzzle	100	200	300	400	500	600	800	1000
Velocity • fps	3000	2588	2211	1867	1562	1307	1121	925	811
Energy • ft-lbs	1099	818	597	426	298	209	154	105	80
Taylor KO Index	5.3	4.6	3.9	3.3	2.7	2.3	2.0	1.6	1.4
Path • Inches	-1.5	+1.9	0.0	-9.0	-28.2	-61.7	-115.7	-313.3	-673.2
Wind Drift • Inches	0.0	1.4	5.8	14.2	27.6	47.0	73.2	142.8	230.0

Ultramax 55-grain FMJ (223R2)

G1 Ballistic Coefficient = 0.240

Distance • Yards	Muzzle	100	200	300	400	500	600	800	1000
Velocity • fps	3000	2612	2256	1928	1633	1379	1180	961	842
Energy • ft-lbs	1099	834	622	454	326	232	170	113	87
Taylor KO Index	5.3	4.6	4.0	3.4	2.9	2.4	2.1	1.7	1.5
Path • Inches	-1.5	+1.8	0.0	-8.7	-26.9	-58.6	-109.3	-293.2	-615.8
Wind Drift • Inches	0.0	1.3	5.4	13.1	25.3	43.0	66.9	132.1	214.7

Federal 60-grain Nosler Partition (P223Q)

G1 Ballistic Coefficient = 0.227

Distance • Yards	Muzzle	100	200	300	400	500	600	800	1000
Velocity • fps	3160	2740	2350	2000	1680	1400	1187	955	833
Energy • ft-lbs	1330	1000	735	530	375	260	188	122	92
Taylor KO Index	6.1	5.3	4.5	3.8	3.2	2.7	2.3	1.8	1.6
Path • Inches	-1.5	+1.6	0.0	-8.0	-24.7	-54.0	-101.1	276.4	-603.6
Wind Drift • Inches	0.0	1.2	5.4	13.0	25.2	43.0	67.4	134.5	220.1

Hornady 60-grain TAP-FPD (83288)

G1 Ballistic Coefficient = 0.265

Distance • Yards	Muzzle	100	200	300	400	500	600	800	1000
Velocity • fps	3115	2754	2420	2110	1824	1567	1344	1051	909
Energy • ft-lbs	1293	1010	780	593	443	327	241	147	111
Taylor KO Index	6.0	5.3	4.6	4.1	3.5	3.0	2.6	2.0	1.7
Path • Inches	-1.5	+1.6	0.0	-7.5	-23.0	-48.9	-89.2	-233.8	-503.5
Wind Drift • Inches	0.0	1.1	4.6	11.0	21.0	35.4	54.8	111.0	185.8

Black Hills 60-grain Soft Point (M223N4)

G1 Ballistic Coefficient = 0.262

Distance • Yards	Muzzle	100	200	300	400	500	600	800	1000
Velocity • fps	3100	2735	2399	2087	1800	1542	1323	1039	900
Energy • ft-lbs	1281	997	767	580	432	317	233	144	108
Taylor KO Index	6.0	5.3	4.6	4.0	3.5	3.0	2.5	2.0	1.7
Path • Inches	-1.5	+1.6	0.0	-7.7	-23.4	-50.1	-92.0	-244.4	-520.6
Wind Drift • Inches	0.0	1.1	4.7	11.3	21.5	36.2	56.2	113.5	189.0

Ultramax 60-grain Nosler Partition (223R11)

G1 Ballistic Coefficient = 0.228

Distance • Yards	Muzzle	100	200	300	400	500	600	800	1000
Velocity • fps	3100	2684	2303	1955	1642	1375	1169	948	828
Energy • ft-lbs	1281	960	707	509	359	252	182	120	91
Taylor KO Index	6.0	5.2	4.4	3.8	3.2	2.6	2.2	1.8	1.6
Path • Inches	-1.5	+1.7	0.0	-8.3	-25.6	-56.8	-107.0	-291.3	-617.0
Wind Drift • Inches	0.0	1.3	5.5	13.3	25.8	44.0	68.7	136.3	221.9

Black Hills 60-grain Hornady V-MAX (M223N10)

G1 Ballistic Coefficient = 0.260

Distance • Yards	Muzzle	100	200	300	400	500	600	800	1000
Velocity • fps	3100	2733	2394	2080	1791	1533	1314	1033	896
Energy • ft-lbs	1281	995	764	576	428	313	230	142	107
Taylor KO Index	6.0	5.2	4.6	4.0	3.4	2.9	2.5	2.0	1.7
Path • Inches	-1.5	+1.6	0.0	-7.7	-23.5	-50.4	-92.7	-246.8	-525.6
Wind Drift • Inches	0.0	1.1	4.6	10.9	20.8	34.9	54.2	110.0	184.8

Winchester 60-grain PDX-1 Defender (S223RPDB)

G1 Ballistic Coefficient = 0.208

Distance • Yards	Muzzle	100	200	300	400	500	600	800	1000
Velocity • fps	2750	2329	1947	1608	1325	1121	997	850	747
Energy • ft-lbs	1007	723	505	344	234	167	132	96	74
Taylor KO Index	5.3	4.5	3.7	3.1	2.5	2.2	1.9	1.6	1.4
Path • Inches	-1.5	+2.5	0.0	-11.8	-37.2	-82.5	-155.5	-412.9	-863.8
Wind Drift • Inches	0.0	1.7	7.3	17.9	35.0	59.4	90.3	167.4	262.1

Remington 62-grain Core-Lokt Ultra Bonded (PRC223R4)

G1 Ballistic Coefficient = 0.235

Distance • Yards	Muzzle	100	200	300	400	500	600	800	1000
Velocity • fps	3100	2695	2324	1983	1676	1409	1200	966	843
Energy • ft-lbs	1323	1000	743	541	368	273	198	128	99
Taylor KO Index	6.2	5.3	4.6	3.9	3.3	2.8	2.4	1.9	1.7
Path • Inches	-1.5	+1.7	0.0	-8.2	-25.2	-54.8	-102.0	-275.7	-598.0
Wind Drift • Inches	0.0	1.2	5.3	12.9	24.8	42.1	65.8	131.2	214.7

Federal 62-grain Fusion (F223FS1)
G1 Ballistic Coefficient = 0.310

Distance • Yards	Muzzle	100	200	300	400	500	600	800	1000
Velocity • fps	3000	2700	2410	2150	1900	1670	1465	1151	982
Energy • ft-lbs	1240	1000	800	635	495	384	296	183	133
Taylor KO Index	6.0	5.4	4.8	4.3	3.8	3.3	2.9	2.3	1.9
Path • Inches	-1.5	+1.6	0.0	-7.7	-22.8	-48.0	-85.2	-213.7	-537.9
Wind Drift • Inches	0.0	1.0	4.1	9.6	18.2	30.3	46.4	93.2	158.1

Black Hills 62-grain Barnes TSX (D223N18) LF
G1 Ballistic Coefficient = 0.287

Distance • Yards	Muzzle	100	200	300	400	500	600	800	1000
Velocity • fps	3100	2766	2456	2167	1898	1652	1434	1115	954
Energy • ft-lbs	1323	1054	831	647	496	376	283	171	125
Taylor KO Index	6.2	5.5	4.9	4.3	3.8	3.3	2.8	2.2	1.9
Path • Inches	-1.5	+1.5	0.0	-7.3	-22.1	-46.6	-83.9	-214.8	-457.1
Wind Drift • Inches	0.0	1.0	4.2	10.1	19.1	31.9	49.2	99.5	168.6

Black Hills 62-grain Barnes TSX (D223N17) LF
G1 Ballistic Coefficient = 0.287

Distance • Yards	Muzzle	100	200	300	400	500	600	800	1000
Velocity • fps	3100	2729	2387	2070	1780	1521	1302	1026	892
Energy • ft-lbs	1323	1026	785	590	436	318	233	145	109
Taylor KO Index	5.5	4.8	4.2	3.6	3.1	2.7	2.3	1.8	1.6
Path • Inches	-1.5	+1.6	0.0	-7.8	-23.7	-50.7	-92.8	-245.2	-528.4
Wind Drift • Inches	0.0	1.1	4.8	11.5	22.0	37.1	57.6	116.2	193.2

Remington 62-grain Hollow Point Match (R223R6)
G1 Ballistic Coefficient = 0.205

Distance • Yards	Muzzle	100	200	300	400	500	600	800	1000
Velocity • fps	3025	2572	2162	1792	1471	1217	1051	878	767
Energy • ft-lbs	1260	911	643	442	298	204	152	106	81
Taylor KO Index	6.0	5.2	4.3	3.6	2.9	2.4	2.1	1.7	1.5
Path • Inches	-1.5	+1.9	0.0	-9.4	-29.9	-66.4	-126.9	-348.7	-751.3
sWind Drift • Inches	0.0	1.5	6.4	15.8	31.0	53.2	82.7	158.7	253.0

Cor-Bon 62-grain DPX (SD22362/20) LF
G1 Ballistic Coefficient = 0.250

Distance • Yards	Muzzle	100	200	300	400	500	600	800	1000
Velocity • fps	2750	2397	2071	1773	1508	1286	1122	940	832
Energy • ft-lbs	1041	791	591	433	313	228	173	122	95
Taylor KO Index	5.5	4.8	4.1	3.5	3.0	2.6	2.2	1.9	1.7
Path • Inches	-1.5	+2.3	0.0	-10.5	-32.1	-69.2	-127.3	-332.5	-695.0
Wind Drift • Inches	0.0	1.4	5.9	14.2	27.4	46.1	71.1	136.5	218.0

UMC (Remington) 62-grain Closed Tip Flat Base (L223R9)
G1 Ballistic Coefficient = 0.261

Distance • Yards	Muzzle	100	200	300	400	500	600	800	1000
Velocity • fps	3100	2734	2396	2083	1785	1537	1318	1036	898
Energy • ft-lbs	1323	1029	790	597	444	325	239	148	111
Taylor KO Index	6.2	5.4	4.8	4.1	3.5	3.0	2.6	2.1	1.8
Path • Inches	-1.5	+1.6	0.0	-7.7	-23.4	-50.1	-91.6	-241.0	-519.2
Wind Drift • Inches	0.0	1.1	4.7	11.3	21.6	36.4	56.5	114.0	190.1

USA (Winchester) 62-grain Full Metal Jacket (USA223R3)
G1 Ballistic Coefficient = 0.284

Distance • Yards	Muzzle	100	200	300	400	500	600	800	1000
Velocity • fps	3100	2762	2448	2155	1884	1636	1420	1104	946
Energy • ft-lbs	1323	1050	825	640	488	368	277	168	123
Taylor KO Index	6.2	5.5	4.9	4.3	3.7	3.2	2.8	2.2	1.9
Path • Inches	-1.5	+1.5	0.0	-7.4	-22.3	-47.1	-85.3	-221.4	-471.2
Wind Drift • Inches	0.0	1.0	4.3	10.2	19.4	32.4	50.0	101.1	171.1

American Eagle (Federal) 62-grain Full Metal Jacket Boat-Tail (AE223N)
G1 Ballistic Coefficient = 0.307

Distance • Yards	Muzzle	100	200	300	400	500	600	800	1000
Velocity • fps	3020	2710	2430	2160	1900	1670	1464	1148	978
Energy • ft-lbs	1225	1015	810	640	500	385	295	181	132
Taylor KO Index	6.0	5.4	4.8	4.3	3.8	3.3	2.9	2.3	1.9
Path • Inches	-1.5	+1.6	0.0	-7.6	-22.6	-47.4	-84.6	-212.8	-446.3
Wind Drift • Inches	0.0	1.0	4.1	9.7	18.3	30.4	46.7	93.8	159.3

Fiocchi 62-grain FMJ BT (223C)

G1 Ballistic Coefficient = 0.307

Distance • Yards	Muzzle	100	200	300	400	500	600	800	1000
Velocity • fps	3000	2694	2408	2139	1889	1659	1452	1140	975
Energy • ft-lbs	1238	999	798	630	491	379	290	179	131
Taylor KO Index	6.0	5.3	4.8	4.2	3.7	3.3	2.9	2.3	1.9
Path • Inches	-1.5	+1.7	0.0	-7.7	-23.0	-48.1	-86.0	-216.2	-453.1
Wind Drift • Inches	0.0	1.0	4.1	9.8	18.5	30.7	47.2	94.7	160.4

Ultramax 62-grain FMJ (223R10)

G1 Ballistic Coefficient = 0.260

Distance • Yards	Muzzle	100	200	300	400	500	600	800	1000
Velocity • fps	2925	2572	2244	1942	1667	1425	1229	994	872
Energy • ft-lbs	1178	911	694	519	383	280	208	136	105
Taylor KO Index	5.8	5.1	4.5	3.9	3.3	2.8	2.4	2.0	1.7
Path • Inches	-1.5	+1.9	0.0	-8.8	-27.0	-58.0	-106.8	-281.6	-588.3
Wind Drift • Inches	0.0	1.2	5.1	12.4	23.7	39.9	61.9	122.5	200.3

Federal 64-grain Soft Point (223L)

G1 Ballistic Coefficient = 0.257

Distance • Yards	Muzzle	100	200	300	400	500	600	800	1000
Velocity • fps	3050	2680	2340	2030	1740	1490	1275	1014	883
Energy • ft-lbs	1320	1020	780	585	430	315	231	146	111
Taylor KO Index	6.2	5.5	4.8	4.2	3.6	3.1	2.6	2.1	1.8
Path • Inches	-1.5	+1.7	0.0	-8.1	-24.6	-52.8	-96.7	-255.2	-548.5
Wind Drift • Inches	0.0	1.2	4.9	11.8	22.6	38.1	59.2	118.7	196.3

Winchester 64-grain Power Point (X223R2)

G1 Ballistic Coefficient = 0.258

Distance • Yards	Muzzle	100	200	300	400	500	600	800	1000
Velocity • fps	3020	2656	2320	2009	1724	1473	1265	1010	881
Energy • ft-lbs	1296	1003	765	574	423	308	227	145	110
Taylor KO Index	6.2	5.4	4.8	4.1	3.5	3.0	2.6	2.1	1.8
Path • Inches	-1.5	+1.7	0.0	-8.2	-25.1	-53.6	-98.6	-260.0	-557.1
Wind Drift • Inches	0.0	1.2	5.0	11.9	22.8	38.5	59.7	119.5	197.1

Winchester 64-grain Power Core 95/5 (X223LF) LF

G1 Ballistic Coefficient = 0.247

Distance • Yards	Muzzle	100	200	300	400	500	600	800	1000
Velocity • fps	3020	2641	2293	1971	1674	1425	1220	983	860
Energy • ft-lbs	1296	991	747	552	401	288	212	137	105
Taylor KO Index	6.2	5.4	4.7	4.0	3.4	2.9	2.5	2.0	1.8
Path • Inches	-1.5	+1.8	0.0	-8.4	-25.9	-55.8	-102.9	-273.9	-588.3
Wind Drift • Inches	0.0	1.2	5.2	12.6	24.1	40.8	63.4	126.2	206.6

Winchester 64-grain Power Max Bonded (X223R2BP)

G1 Ballistic Coefficient = 0.231

Distance • Yards	Muzzle	100	200	300	400	500	600	800	1000
Velocity • fps	3020	2616	2246	1908	1605	1347	1153	943	827
Energy • ft-lbs	1296	973	717	517	366	258	189	127	97
Taylor KO Index	6.2	5.4	4.6	3.9	3.3	2.8	2.4	1.9	1.7
Path • Inches	-1.5	+1.8	0.0	-8.8	-27.2	-59.2	-110.5	-298.4	-642.3
Wind Drift • Inches	0.0	1.3	5.6	13.6	26.3	44.8	69.9	137.4	222.6

Ultramax 68-grain HP (223R6)

G1 Ballistic Coefficient = 0.338

Distance • Yards	Muzzle	100	200	300	400	500	600	800	1000
Velocity • fps	2900	2627	2370	2128	1900	1689	1498	1192	1014
Energy • ft-lbs	1270	1042	848	684	545	431	339	215	155
Taylor KO Index	6.3	5.7	5.2	4.6	4.1	3.7	3.3	2.6	2.2
Path • Inches	-1.5	+1.8	0.0	-8.0	-23.7	-49.1	-86.9	-215.0	-443.6
Wind Drift • Inches	0.0	0.9	3.9	9.2	17.2	28.5	43.5	86.6	146.9

Black Hills 68-grain "Heavy" Match Hollow Point (M223N5)

G1 Ballistic Coefficient = 0.339

Distance • Yards	Muzzle	100	200	300	400	500	600	800	1000
Velocity • fps	2850	2581	2327	2088	1863	1656	1469	1174	1004
Energy • ft-lbs	1227	1006	818	658	524	414	326	208	152
Taylor KO Index	6.2	5.6	5.1	4.5	4.1	3.6	3.2	2.6	2.2
Path • Inches	-1.5	+1.9	0.0	-8.3	-24.6	-51.0	-90.2	-221.2	-453.2
Wind Drift • Inches	0.0	0.9	4.0	9.4	17.6	29.2	44.5	88.4	149.4

CorBon 69-grain BTHP (PM22369/20)

G1 Ballistic Coefficient = 0.338

Distance • Yards	Muzzle	100	200	300	400	500	600	800	1000
Velocity • fps	3000	2721	2459	2211	1978	1761	1562	1297	1037
Energy • ft-lbs	1379	1135	927	749	600	475	374	234	165
Taylor KO Index	6.6	6.0	5.4	4.9	4.4	3.9	3.4	2.9	2.3
Path • Inches	-1.5	+1.6	0.0	-7.4	-21.9	-45.4	-80.2	-198.2	-410.6
Wind Drift • Inches	0.0	0.9	3.7	8.7	16.4	27.1	41.3	82.4	141.2

Remington 69-grain MatchKing BTHP (RM223R1)

G1 Ballistic Coefficient = 0.336

Distance • Yards	Muzzle	100	200	300	400	500	600	800	1000
Velocity • fps	3000	2920	2457	2209	1975	1758	1556	1231	1033
Energy • ft-lbs	1379	1133	925	747	598	473	371	232	164
Taylor KO Index	6.6	6.0	5.4	4.9	4.4	3.9	3.4	2.7	2.3
Path • Inches	-1.5	+1.6	0.0	-7.4	-21.9	-45.4	-80.6	-199.3	-413.5
Wind Drift • Inches	0.0	0.9	3.7	8.8	16.5	27.3	41.6	83.1	142.3

Federal 69-grain Sierra MatchKing BTHP (GM223M)

G1 Ballistic Coefficient = 0.300

Distance • Yards	Muzzle	100	200	300	400	500	600	800	1000
Velocity • fps	2950	2640	2350	2080	1830	1600	1396	1103	952
Energy • ft-lbs	1335	1070	850	665	515	395	299	186	139
Taylor KO Index	6.5	5.8	5.2	4.6	4.0	3.5	3.1	2.4	2.1
Path • Inches	-1.5	+1.8	0.0	-8.1	-24.2	-50.9	-91.2	-230.7	-484.0
Wind Drift • Inches	0.0	1.0	4.3	10.3	19.5	32.5	50.0	100.0	168.1

Black Hills 69-grain Sierra BTHP MatchKing (M223N12)

G1 Ballistic Coefficient = 0.338

Distance • Yards	Muzzle	100	200	300	400	500	600	800	1000
Velocity • fps	2850	2580	2326	2086	1861	1653	1466	1171	1002
Energy • ft-lbs	1245	1020	829	667	531	419	329	210	154
Taylor KO Index	6.3	5.7	5.1	4.6	4.1	3.6	3.2	2.6	2.2
Path • Inches	-1.5	+1.9	0.0	-8.3	-24.7	-51.2	-90.6	-224.1	-461.1
Wind Drift • Inches	0.0	1.0	4.0	9.4	17.7	29.3	44.7	88.7	149.8

Lapua 69-grain Scenar (4315011)

G1 Ballistic Coefficient = 0.339

Distance • Yards	Muzzle	100	200	300	400	500	600	800	1000
Velocity • fps	2789	2525	2277	2044	1817	1614	1430	1151	992
Energy • ft-lbs	1199	983	800	644	506	399	315	203	151
Taylor KO Index	6.2	5.6	5.0	4.5	4.0	3.6	3.2	2.5	2.2
Path • Inches	-1.5	+2.0	0.0	-8.7	-25.9	-53.6	-94.8	-232.5	-475.1
Wind Drift • Inches	0.0	1.0	4.1	9.7	18.2	30.1	45.9	90.9	152.6

Fiocchi 69-grain Sierra MatchKing HPBT (223MKC)

G1 Ballistic Coefficient = 0.313

Distance • Yards	Muzzle	100	200	300	400	500	600	800	1000
Velocity • fps	2735	2444	2173	1928	1701	1497	1318	1069	937
Energy • ft-lbs	1145	915	723	569	443	344	266	175	135
Taylor KO Index	6.0	5.4	4.9	4.3	3.8	3.3	2.9	2.4	2.1
Path • Inches	-1.5	+2.2	0.0	-9.5	-28.3	-59.1	-105.5	-263.6	-543.1
Wind Drift • Inches	0.0	1.1	4.6	11.0	20.7	34.5	52.8	104.0	171.6

Hornady 75-grain TAP-FPD (80268)

G1 Ballistic Coefficient = 0.395

Distance • Yards	Muzzle	100	200	300	400	500	600	800	1000
Velocity • fps	2790	2561	2344	2137	1941	1757	1585	1294	1093
Energy • ft-lbs	1296	1092	915	760	627	514	418	279	199
Taylor KO Index	6.7	6.1	5.6	5.1	4.7	4.2	3.8	3.1	2.6
Path • Inches	-1.5	+1.9	0.0	-8.3	-24.2	-49.3	-85.8	-203.6	-405.0
Wind Drift • Inches	0.0	0.8	3.5	8.1	15.1	24.8	37.6	73.7	125.0

PMC 75-grain BTHP Match (223HMB)

G1 Ballistic Coefficient = 0.395

Distance • Yards	Muzzle	100	200	300	400	500	600	800	1000
Velocity • fps	2790	2561	2344	2137	1941	1757	1586	1295	1094
Energy • ft-lbs	1296	1092	915	760	627	514	419	279	199
Taylor KO Index	6.7	6.1	5.6	5.1	4.7	4.2	3.8	3.1	2.6
Path • Inches	-1.5	+1.9	0.0	-8.3	-24.1	-49.3	-85.7	-203.4	-404.4
Wind Drift • Inches	0.0	0.8	3.4	8.1	15.1	24.8	37.5	73.5	125.0

Hornady 75-grain BTHP (80264)
G1 Ballistic Coefficient = 0.395

Distance • Yards	Muzzle	100	200	300	400	500	600	800	1000
Velocity • fps	2930	2694	2470	2257	2055	1863	1685	1373	1143
Energy • ft-lbs	1429	1209	1016	848	703	578	473	314	218
Taylor KO Index	7.0	6.5	5.9	5.4	4.9	4.5	4.0	3.3	2.7
Path • Inches	-1.5	+1.2	0.0	-6.9	-20.7	-42.7	-76.5	-181.2	-360.4
Wind Drift • Inches	0.0	0.8	3.2	7.6	14.1	23.0	34.8	68.4	117.1

Ultramax 75-grain Boat-Tail Hollow Point Match (223R8)
G1 Ballistic Coefficient = 0.390

Distance • Yards	Muzzle	100	200	300	400	500	600	800	1000
Velocity • fps	2800	2568	2347	2137	1939	1752	1579	1286	1086
Energy • ft-lbs	1306	1098	918	761	626	511	416	276	197
Taylor KO Index	6.7	6.2	5.6	5.1	4.7	4.2	3.8	3.1	2.6
Path • Inches	-1.5	+1.9	0.0	-8.2	-24.1	-49.2	-85.8	-205.2	-411.8
Wind Drift • Inches	0.0	0.8	3.5	8.2	15.3	25.1	38.0	74.6	126.7

Hornady 75-grain Boat-Tail Hollow Point Match (8026)
G1 Ballistic Coefficient = 0.395

Distance • Yards	Muzzle	100	200	300	400	500	600	800	1000
Velocity • fps	2790	2561	2344	2137	1941	1757	1585	1294	1093
Energy • ft-lbs	1296	1092	915	760	627	514	418	279	199
Taylor KO Index	6.7	6.1	5.6	5.1	4.7	4.2	3.8	3.1	2.6
Path • Inches	-1.5	+1.9	0.0	-8.3	-24.2	-49.3	-85.8	-203.6	-405.0
Wind Drift • Inches	0.0	0.8	3.5	8.1	15.1	24.8	37.6	73.7	125.0

Black Hills 75-grain "Heavy" Match Hollow Point (M223N6)
G1 Ballistic Coefficient = 0.390

Distance • Yards	Muzzle	100	200	300	400	500	600	800	1000
Velocity • fps	2750	2520	2302	2094	1898	1714	1545	1260	1071
Energy • ft-lbs	1259	1058	883	731	600	489	397	265	191
Taylor KO Index	6.6	6.0	5.5	5.0	4.6	4.1	3.7	3.0	2.6
Path • Inches	-1.5	+2.0	0.0	-8.6	-25.1	-51.3	-89.4	-212.8	-424.2
Wind Drift • Inches	0.0	0.9	3.6	8.4	15.7	25.8	39.0	76.6	129.7

CorBon 77-grain HPBT (PM22377/20)
G1 Ballistic Coefficient = 0.350

Distance • Yards	Muzzle	100	200	300	400	500	600	800	1000
Velocity • fps	2800	2542	2298	2068	1852	1651	1470	1182	1013
Energy • ft-lbs	1341	1105	903	731	586	466	369	239	175
Taylor KO Index	6.9	6.3	5.7	5.1	4.6	4.1	3.6	2.9	2.5
Path • Inches	-1.5	+2.0	0.0	-8.6	-25.3	-52.2	-91.9	-223.9	-455.5
Wind Drift • Inches	0.0	0.9	3.9	9.3	17.4	28.8	43.8	86.7	146.2

Remington 77-grain MatchKing BTHP (RM223R3)
G1 Ballistic Coefficient = 0.362

Distance • Yards	Muzzle	100	200	300	400	500	600	800	1000
Velocity • fps	2788	2539	2303	2081	1871	1675	1496	1208	1031
Energy • ft-lbs	1329	1102	907	740	598	480	383	249	182
Taylor KO Index	6.9	6.3	5.7	5.1	4.6	4.1	3.7	3.0	2.5
Path • Inches	-1.5	+2.0	0.0	-8.5	-25.1	-51.7	-90.8	-219.4	-443.6
Wind Drift • Inches	0.0	0.9	3.8	9.0	16.8	27.7	42.2	83.2	140.7

Federal 77-grain Match (GM223M3)
G1 Ballistic Coefficient = 0.372

Distance • Yards	Muzzle	100	200	300	400	500	600	800	1000
Velocity • fps	2720	2480	2260	2040	1840	1650	1480	1203	1033
Energy • ft-lbs	1265	1055	870	710	580	465	375	248	182
Taylor KO Index	6.8	6.2	5.6	5.1	4.5	4.1	3.6	2.9	2.5
Path • Inches	-1.5	+2.1	0.0	-9.0	-26.1	-54.1	-94.4	-226.8	-455.5
Wind Drift • Inches	0.0	0.9	3.8	9.0	16.9	27.8	42.2	82.8	139.4

Black Hills 77-grain Sierra Matchking HP (M223N9)
G1 Ballistic Coefficient = 0.390

Distance • Yards	Muzzle	100	200	300	400	500	600	800	1000
Velocity • fps	2750	2520	2302	2094	1898	1714	1545	1260	1071
Energy • ft-lbs	1259	1086	906	750	616	503	408	272	196
Taylor KO Index	6.6	6.0	5.5	5.2	4.7	4.2	3.8	3.1	2.6
Path • Inches	-1.5	+2.0	0.0	-8.6	-25.1	-51.3	-89.4	-212.8	-424.2
Wind Drift • Inches	0.0	0.9	3.6	8.4	15.7	25.8	39.0	76.6	129.7

Fiocchi 77-grain Sierra MatchKing HPBT (223MKD)

G1 Ballistic Coefficient = 0.361

Distance • Yards	Muzzle	100	200	300	400	500	600	800	1000
Velocity • fps	2660	2418	2189	1972	1769	1576	1411	1150	999
Energy • ft-lbs	1209	999	819	665	535	425	341	226	171
Taylor KO Index	6.6	6.0	5.4	4.9	4.4	3.9	3.5	2.8	2.5
Path • Inches	-1.5	+2.2	0.0	-9.5	-28.1	-57.8	-101.5	-245.6	-494.7
Wind Drift • Inches	0.0	1.0	4.1	9.7	18.1	29.9	45.4	89.1	148.6

Nosler 77-grain Custom Competition (07010)

G1 Ballistic Coefficient = 0.340

Distance • Yards	Muzzle	100	200	300	400	500	600	800	1000
Velocity • fps	2600	2346	2106	1881	1673	1484	1319	1084	954
Energy • ft-lbs	1156	941	759	605	479	377	298	201	156
Taylor KO Index	6.4	5.8	5.2	4.6	4.1	3.7	3.3	2.7	2.4
Path • Inches	-1.5	+2.4	0.0	-10.3	-30.4	-63.1	-111.6	-273.4	-553.2
Wind Drift • Inches	0.0	1.1	4.5	10.7	20.2	33.4	50.9	99.3	163.1

While slower cartridges like the rimfires may be more ideal for the smallest animals, in Africa you never know what shot you might get. Craig Boddington used a .222 Remington to take this bat-eared fox at 200 yards. (Photo from Safari Rifles II by Craig Boddington, 2009, Safari Press Archives)

.22-250 Remington

The .22-250, which became a standardized factory cartridge in 1965, has a history that starts about 1915 when Charles Newton designed the .250 Savage. In the late 1920s and early 1930s, a number of people began experimenting with the .250 Savage necked to .22 caliber. One early number was the .220 Wotkyns Original Swift. J. E. Gebby and J. Bushnell Smith called their version the .22 Varminter and copyrighted the name. Most gunsmiths when chambering guns for the .22-caliber version of the .250 Savage simply called the cartridge the .22-250 to avoid any copyright troubles.

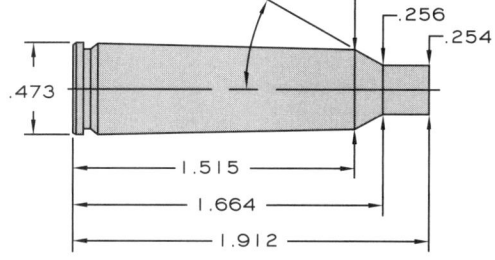

Remington's introduction of the .22-250 finally brought this fine cartridge its deserved recognition as a standardized factory caliber. For a given bullet weight, the .22-250 is only 50 fps slower than the .220 Swift. Some of the wildcat loadings from the 1950s produced velocities about 200 fps higher than today's factory standards. They also produced pressures to match, and that fact explains why the factories are sometimes slow to adopt wildcat designs and why they seldom achieve the velocity claims of the wildcat "inventors."

Relative Recoil Factor = 0.90

Specifications:

Controlling Agency for Standardization of this Ammunition: SAAMI

Bullet Weight Grains	Velocity fps	Maximum Average Pressure	
		Copper Crusher	Transducer
40	3,975	53,000 cup	65,000 psi
52	3,740	53,000 cup	65,000 psi
53–55	3,650	53,000 cup	65,000 psi
60	3,600	53,000 cup	65,000 psi

Standard barrel for velocity testing: 24 inches long—1 turn in 14-inch twist

Availability:

Winchester 35-grain Ballistic Silvertip (S22250RLF) LF
G1 Ballistic Coefficient = 0.201

Distance • Yards	Muzzle	100	200	300	400	500	600	800	1000
Velocity • fps	4350	3729	3189	2711	2280	1890	1548	1082	891
Energy • ft-lbs	1470	1081	790	571	404	278	186	91	62
Taylor KO Index	4.9	4.2	3.6	3.0	2.6	2.1	1.7	1.2	1.0
Path • Inches	-1.5	+0.5	0.0	-4.0	-12.7	-28.2	-53.6	-155.4	-375.0
Wind Drift • Inches	0.0	1.0	4.2	10.0	19.1	32.4	51.2	109.9	194.4

Nosler 35-grain Ballistic Tip (60006) LF
G1 Ballistic Coefficient = 0.204

Distance • Yards	Muzzle	100	200	300	400	500	600	800	1000
Velocity • fps	4200	3606	3089	2630	2213	1835	1505	1066	888
Energy • ft-lbs	1371	1010	741	538	381	262	176	88	61
Taylor KO Index	4.7	4.0	3.5	2.9	2.5	2.1	1.7	1.2	1.0
Path • Inches	-1.5	+0.6	0.0	-4.3	-13.6	-30.2	-57.2	-165.0	-394.0
Wind Drift • Inches	0.0	1.0	4.3	10.2	19.5	33.2	52.5	112.2	196.8

Cor-Bon 36-grain Varmint Grenade (SD2225036VG/20) LF
G1 Ballistic Coefficient = 0.149

Distance • Yards	Muzzle	100	200	300	400	500	600	800	1000
Velocity • fps	4300	3492	2815	2230	1722	1315	1059	836	701
Energy • ft-lbs	1478	975	634	398	237	138	90	56	39
Taylor KO Index	5.0	4.0	3.2	2.6	2.0	1.5	1.2	1.0	0.8
Path • Inches	-1.5	+0.7	0.0	-5.2	-17.5	-41.9	-86.6	-278.2	-670.0
Wind Drift • Inches	0.0	1.4	6.0	14.8	29.4	52.4	85.3	174.5	288.5

Black Hills 36-grain Barnes Varmint Grenade (1C22250BHGN3)
G1 Ballistic Coefficient = 0.149

Distance • Yards	Muzzle	100	200	300	400	500	600	800	1000
Velocity • fps	4250	3450	2780	2199	1696	1296	1048	832	698
Energy • ft-lbs	1444	952	618	378	230	134	88	55	39
Taylor KO Index	4.9	4.0	3.2	2.5	2.0	1.5	1.2	1.0	0.8
Path • Inches	-1.5	+0.7	0.0	-5.3	-18.0	-43.1	-89.1	-285.0	-683.3
Wind Drift • Inches	0.0	1.4	6.0	14.9	29.9	53.2	86.6	176.2	290.6

CorBon 40-grain BlitzKing (SD2225040/20)

G1 Ballistic Coefficient = 0.210

Distance • Yards	Muzzle	100	200	300	400	500	600	800	1000
Velocity • fps	4200	3623	3116	2664	2254	1882	1554	1096	904
Energy • ft-lbs	1567	1166	863	630	451	315	215	107	73
Taylor KO Index	5.4	4.6	4.0	3.4	2.9	2.4	2.0	1.4	1.2
Path • Inches	-1.5	+0.6	0.0	-4.2	-13.3	-29.4	-55.4	-157.7	-374.9
Wind Drift • Inches	0.0	1.0	4.1	9.9	18.9	31.9	50.3	107.3	189.4

Hornady 40-grain V-MAX (8335) – V-MAX w/Moly (83353)

G1 Ballistic Coefficient = 0.200

Distance • Yards	Muzzle	100	200	300	400	500	600	800	1000
Velocity • fps	4150	3553	3032	2568	2148	1771	1444	1035	866
Energy • ft-lbs	1529	1121	816	585	410	278	185	95	67
Taylor KO Index	5.3	4.5	3.9	3.3	2.7	2.3	1.8	1.3	1.1
Path • Inches	-1.5	+0.6	0.0	-4.5	-14.3	-31.7	-60.5	-176.1	-420.6
Wind Drift • Inches	0.0	1.0	4.4	10.6	20.4	34.8	55.2	118.0	205.0

Nosler 40-grain Ballistic Tip (60008) LF

G1 Ballistic Coefficient = 0.222

Distance • Yards	Muzzle	100	200	300	400	500	600	800	1000
Velocity • fps	3750	3258	2820	2422	2057	1722	1436	1058	893
Energy • ft-lbs	1249	943	706	521	376	263	183	99	71
Taylor KO Index	4.8	4.2	3.6	3.1	2.6	2.2	1.8	1.4	1.1
Path • Inches	-1.5	+0.9	0.0	-5.3	-16.6	-36.3	-37.8	-188.6	-432.8
Wind Drift • Inches	0.0	1.0	4.4	10.5	20.2	34.2	53.7	112.7	194.1

Federal 43-grain Speer TNT Green (P22250D)

G1 Ballistic Coefficient = 0.151

Distance • Yards	Muzzle	100	200	300	400	500	600	800	1000
Velocity • fps	4000	3250	2620	2070	1590	1220	1014	816	698
Energy • ft-lbs	1530	1010	655	405	240	145	98	64	45
Taylor KO Index	5.5	4.5	3.6	2.8	2.2	1.7	1.4	1.1	1.0
Path • Inches	-1.5	+0.9	0.0	-6.1	-20.6	-49.2	-101.2	-315.7	-740.5
Wind Drift • Inches	0.0	1.5	6.4	15.9	32.0	56.9	91.6	182.5	297.8

Remington 45-grain Jacketed Iron-Tin Core HP (DV2250RA) LF

G1 Ballistic Coefficient = 0.161

Distance • Yards	Muzzle	100	200	300	400	500	600	800	1000
Velocity • fps	4000	3293	2690	2159	1696	1320	1076	856	725
Energy • ft-lbs	1598	1084	723	466	287	174	116	73	53
Taylor KO Index	5.8	4.7	3.9	3.1	2.4	1.9	1.5	1.2	1.0
Path • Inches	-1.5	+0.9	0.0	-5.8	-19.1	-44.8	-90.5	-280.7	-661.3
Wind Drift • Inches	0.0	1.4	5.9	14.6	29.0	51.1	82.6	167.6	275.8

UMC (Remington) 45-grain Jacketed Hollow Point (L22503)

G1 Ballistic Coefficient = 0.173

Distance • Yards	Muzzle	100	200	300	400	500	600	800	1000
Velocity • fps	4000	3340	2770	2266	1819	1440	1158	901	765
Energy • ft-lbs	1598	1114	767	513	331	207	135	81	59
Taylor KO Index	5.8	4.8	4.0	3.3	2.6	2.1	1.7	1.3	1.1
Path • Inches	-1.5	+0.8	0.0	-5.4	-17.8	-40.8	-80.9	-247.7	-586.1
Wind Drift • Inches	0.0	1.3	5.4	13.3	26.2	45.7	73.6	152.5	254.0

USA (Winchester) 45-grain Jacketed Hollow Point (USA222502)

G1 Ballistic Coefficient = 0.175

Distance • Yards	Muzzle	100	200	300	400	500	600	800	1000
Velocity • fps	4000	3346	2781	2281	1837	1458	1175	909	772
Energy • ft-lbs	1598	1118	773	520	337	212	138	83	60
Taylor KO Index	5.8	4.8	4.0	3.3	2.6	2.1	1.7	1.3	1.1
Path • Inches	-1.5	+0.8	0.0	-5.4	-17.5	-40.2	-81.5	-248.1	-564.8
Wind Drift • Inches	0.0	1.2	5.4	13.1	25.7	44.8	72.2	149.9	250.1

Hornady 50-grain V-MAX (83366)

G1 Ballistic Coefficient = 0.242

Distance • Yards	Muzzle	100	200	300	400	500	600	800	1000
Velocity • fps	4000	3517	3086	2694	2334	2003	1702	1227	993
Energy • ft-lbs	1776	1373	1057	806	605	445	322	167	107
Taylor KO Index	6.4	5.6	4.9	4.3	3.7	3.2	2.7	2.0	1.6
Path • Inches	-1.5	+0.7	0.0	-4.3	-13.5	-29.1	-53.4	-143.2	-326.0
Wind Drift • Inches	0.0	0.9	3.7	8.8	16.7	27.9	43.3	90.5	161.6

Nosler 50-grain Ballistic Tip (05510)

G1 Ballistic Coefficient = 0.238

Distance • Yards	Muzzle	100	200	300	400	500	600	800	1000
Velocity • fps	3850	3376	2951	2564	2209	1884	1592	1155	949
Energy • ft-lbs	1645	1265	967	729	541	394	282	148	100
Taylor KO Index	6.2	5.4	4.7	4.1	3.5	3.0	2.5	1.8	1.5
Path • Inches	-1.5	+0.8	0.0	-4.8	-14.8	-32.2	-59.4	-160.9	-366.4
Wind Drift • Inches	0.0	0.9	4.0	9.4	17.9	30.1	46.9	98.1	172.8

UMC (Remington) 50-grain Jacketed Hollow Point (L22504)

G1 Ballistic Coefficient = 0.192

Distance • Yards	Muzzle	100	200	300	400	500	600	800	1000
Velocity • fps	3820	3245	2739	2286	1878	1523	1239	949	810
Energy • ft-lbs	1620	1169	833	580	392	258	170	100	73
Taylor KO Index	6.1	5.2	4.4	3.7	3.0	2.4	2.0	1.5	1.3
Path • Inches	-1.5	+0.9	0.0	-5.6	-18.0	-40.4	-78.3	-229.8	-535.7
Wind Drift • Inches	0.0	1.2	5.1	12.4	24.1	41.6	66.3	138.1	231.6

Winchester 50-grain Ballistic Silvertip (SBST22250)

G1 Ballistic Coefficient = 0.238

Distance • Yards	Muzzle	100	200	300	400	500	600	800	1000
Velocity • fps	3810	3341	2919	2536	2182	1859	1568	1140	941
Energy • ft-lbs	1611	1239	946	714	529	384	273	144	98
Taylor KO Index	6.1	5.4	4.7	4.1	3.5	3.0	2.5	1.8	1.5
Path • Inches	-1.5	+0.8	0.0	-4.9	-15.2	-32.9	-61.0	-165.4	-376.4
Wind Drift • Inches	0.0	1.0	4.0	9.6	18.2	30.6	47.7	99.8	175.2

Hornady 50-grain V-MAX (8336) – Moly (83363)

G1 Ballistic Coefficient = 0.242

Distance • Yards	Muzzle	100	200	300	400	500	600	800	1000
Velocity • fps	3800	3339	2925	2546	2198	1878	1589	1157	952
Energy • ft-lbs	1603	1238	949	720	536	392	281	149	101
Taylor KO Index	6.1	5.3	4.7	4.1	3.5	3.0	2.5	1.9	1.5
Path • Inches	-1.5	+0.8	0.0	-4.9	-15.2	-32.8	-60.4	-162.6	-368.5
Wind Drift • Inches	0.0	0.9	4.0	9.4	17.9	30.0	46.7	97.5	171.6

Remington 50-grain AccuTip (PRA2250RB)

G1 Ballistic Coefficient = 0.242

Distance • Yards	Muzzle	100	200	300	400	500	600	800	1000
Velocity • fps	3800	3339	2925	2546	2198	1878	1592	1158	950
Energy • ft-lbs	1603	1238	949	720	536	392	282	149	100
Taylor KO Index	6.1	5.3	4.7	4.1	3.5	3.0	2.5	1.9	1.5
Path • Inches	-1.5	+0.8	0.0	-4.9	-15.2	-32.8	-60.7	-166.0	-377.7
Wind Drift • Inches	0.0	0.9	3.9	9.4	17.8	29.9	46.6	97.3	171.3

Federal 50-grain Barnes Triple-Shock X-Bullet (P22250Q) LF

G1 Ballistic Coefficient = 0.199

Distance • Yards	Muzzle	100	200	300	400	500	600	800	1000
Velocity • fps	3750	3220	2730	2290	1890	1540	1265	965	825
Energy • ft-lbs	1585	1155	830	580	395	265	178	103	76
Taylor KO Index	6.0	5.2	4.4	3.7	3.0	2.5	2.0	1.5	1.3
Path • Inches	-1.5	+0.9	0.0	-5.6	-18.0	-40.7	-77.7	-225.6	-522.1
Wind Drift • Inches	0.0	1.2	5.0	12.1	23.5	40.4	64.2	133.5	153.2

Hornady 50-grain V-MAX (83366)

G1 Ballistic Coefficient = 0.242

Distance • Yards	Muzzle	100	200	300	400	500	600	800	1000
Velocity • fps	4000	3517	3086	2694	2334	2003	1702	1227	993
Energy • ft-lbs	1776	1373	1057	806	605	445	322	167	107
Taylor KO Index	6.4	5.6	4.9	4.3	3.7	3.2	2.7	2.0	1.6
Path • Inches	-1.5	+0.7	0.0	-4.3	-13.5	-29.1	-53.4	-143.2	-326.0
Wind Drift • Inches	0.0	0.9	3.7	8.8	16.7	27.9	43.3	90.5	161.6

Black Hills 50-grain Hornady V-MAX (1C22250BHGN1)

G1 Ballistic Coefficient = 0.242

Distance • Yards	Muzzle	100	200	300	400	500	600	800	1000
Velocity • fps	3700	3249	2842	2470	2127	1814	1535	1127	937
Energy • ft-lbs	1520	1172	897	678	503	365	262	141	98
Taylor KO Index	5.9	5.2	4.5	4.0	3.4	2.9	2.5	1.8	1.5
Path • Inches	-1.5	+0.9	0.0	-5.2	-16.2	-35.0	-64.4	-173.6	-391.8
Wind Drift • Inches	0.0	1.1	4.1	9.7	18.5	31.1	48.5	101.2	176.5

Norma 53-grain Soft Point (15733)

G1 Ballistic Coefficient = 0.237

Distance • Yards	Muzzle	100	200	300	400	500	600	800	1000
Velocity • fps	3707	3246	2830	2451	2106	1788	1507	1107	925
Energy • ft-lbs	1618	1240	943	707	522	376	267	191	144
Taylor KO Index	6.3	5.5	4.8	4.2	3.6	3.0	2.6	1.9	1.6
Path • Inches	-1.5	+0.9	0.0	-5.3	-16.4	-35.5	-65.9	-182.2	-411.7
Wind Drift • Inches	0.0	1.0	4.2	10.0	18.9	31.9	49.9	104.2	181.0

CorBon 53-grain DPX (DPX2225053/20)

G1 Ballistic Coefficient = 0.240

Distance • Yards	Muzzle	100	200	300	400	500	600	800	1000
Velocity • fps	3550	3111	2713	2349	2013	1709	1442	1077	912
Energy • ft-lbs	1484	1139	867	649	477	344	245	137	98
Taylor KO Index	6.0	5.3	4.6	4.0	3.4	2.9	2.4	1.8	1.5
Path • Inches	-1.5	+1.1	0.0	-5.8	-18.0	-38.8	-71.7	-194.2	-435.7
Wind Drift • Inches	0.0	1.0	4.3	10.4	19.8	33.4	52.2	108.4	186.1

Federal 55-grain Fusion (F22250FS1)

G1 Ballistic Coefficient = 0.157

Distance • Yards	Muzzle	100	200	300	400	500	600	800	1000
Velocity • fps	3680	2980	2410	1900	1470	1160	986	807	685
Energy • ft-lbs	1625	1085	705	440	265	165	119	80	57
Taylor KO Index	6.4	5.2	4.2	3.3	2.6	2.0	1.7	1.4	1.2
Path • Inches	-1.5	+1.2	0.0	-7.4	-24.9	-60.0	-117.9	-353.0	-803.1
Wind Drift • Inches	0.0	1.6	6.8	17.1	34.3	60.6	96.0	186.4	300.1

Sellier & Bellot 55-grain SBT (V330462U)

G1 Ballistic Coefficient = 0.250

Distance • Yards	Muzzle	100	200	300	400	500	600	800	1000
Velocity • fps	3680	3245	2852	2491	2158	1852	1577	1161	958
Energy • ft-lbs	1654	1287	994	758	569	419	304	165	112
Taylor KO Index	6.5	5.7	5.0	4.4	3.8	3.3	2.8	2.0	1.7
Path • Inches	-1.5	+0.9	0.0	-5.2	-16.1	-34.4	-63.1	-167.9	-375.9
Wind Drift • Inches	0.0	0.9	4.0	9.4	17.8	29.9	46.5	96.6	169.2

Remington 55-grain Pointed Soft Point (R22501)

G1 Ballistic Coefficient = 0.198

Distance • Yards	Muzzle	100	200	300	400	500	600	800	1000
Velocity • fps	3680	3137	2656	2222	1832	1493	1228	950	814
Energy • ft-lbs	1654	1201	861	603	410	272	184	110	81
Taylor KO Index	6.5	5.5	4.7	3.9	3.2	2.6	2.2	1.7	1.4
Path • Inches	-1.5	+1.0	0.0	-6.0	-19.1	-42.8	-82.1	-238.2	-548.2
Wind Drift • Inches	0.0	1.2	5.2	12.5	24.3	41.8	66.6	137.6	229.7

Winchester 55-grain Pointed Soft Point (X222501)

G1 Ballistic Coefficient = 0.198

Distance • Yards	Muzzle	100	200	300	400	500	600	800	1000
Velocity • fps	3680	3137	2656	2222	1832	1493	1228	950	814
Energy • ft-lbs	1654	1201	861	603	410	272	184	110	81
Taylor KO Index	6.5	5.5	4.7	3.9	3.2	2.6	2.2	1.7	1.4
Path • Inches	-1.5	+1.0	0.0	-6.0	-19.1	-42.8	-82.1	-238.2	-548.2
Wind Drift • Inches	0.0	1.2	5.2	12.5	24.3	41.8	66.6	137.6	229.7

Federal 55-grain Soft Point (22250A)

G1 Ballistic Coefficient = 0.223

Distance • Yards	Muzzle	100	200	300	400	500	600	800	1000
Velocity • fps	3650	3170	2730	2340	1980	1660	1384	1035	880
Energy • ft-lbs	1625	1225	915	670	480	335	234	131	95
Taylor KO Index	6.4	5.6	4.8	4.1	3.5	2.9	2.4	1.8	1.5
Path • Inches	-1.5	+1.0	0.0	-5.7	-17.8	-38.8	-72.5	-201.9	-460.3
Wind Drift • Inches	0.0	1.1	4.5	10.9	21.0	35.8	56.0	116.9	199.5

Norma 55-grain Oryx (15734)

G1 Ballistic Coefficient = 0.185

Distance • Yards	Muzzle	100	200	300	400	500	600	800	1000
Velocity • fps	3609	3041	2540	2091	1691	1358	1121	896	770
Energy • ft-lbs	1591	1130	788	534	349	225	154	98	72
Taylor KO Index	6.4	5.4	4.5	3.7	3.0	2.4	2.0	1.6	1.4
Path • Inches	-1.5	+1.2	0.0	-6.6	-21.3	-48.4	-94.5	-277.7	-634.2
Wind Drift • Inches	0.0	1.3	5.7	14.0	27.5	47.8	112.1	200.6	252.3

CorBon 55-grain BlitzKing (SD2225055/20)

G1 Ballistic Coefficient = 0.264

Distance • Yards	Muzzle	100	200	300	400	500	600	800	1000
Velocity • fps	3700	3205	2907	2560	2239	1941	1670	1236	1000
Energy • ft-lbs	1672	1318	1033	801	612	460	341	187	122
Taylor KO Index	6.5	5.6	5.1	4.5	3.9	3.4	2.9	2.2	1.8
Path • Inches	-1.5	+0.9	0.0	-5.0	-15.4	-32.7	-59.3	-154.8	-342.0
Wind Drift • Inches	0.0	0.9	3.7	8.8	16.6	27.6	42.7	88.2	155.8

Nosler 55-grain Ballistic Tip (60003)

G1 Ballistic Coefficient = 0.272

Distance • Yards	Muzzle	100	200	300	400	500	600	800	1000
Velocity • fps	3700	3295	2928	2589	2274	1983	1718	1279	1026
Energy • ft-lbs	1672	1326	1047	819	631	480	360	200	129
Taylor KO Index	6.5	5.8	5.2	4.6	4.0	3.5	3.0	2.3	1.8
Path • Inches	-1.5	+0.9	0.0	-4.9	-15.1	-32.0	-57.7	-149.0	-326.1
Wind Drift • Inches	0.0	0.8	3.6	8.5	15.9	26.5	40.8	84.0	148.7

Federal 55-grain Sierra GameKing BTHP (P22250B)

G1 Ballistic Coefficient = 0.250

Distance • Yards	Muzzle	100	200	300	400	500	600	800	1000
Velocity • fps	3650	3220	2830	2470	2140	1830	1558	1149	952
Energy • ft-lbs	1625	1265	975	745	560	410	296	161	111
Taylor KO Index	6.4	5.7	5.0	4.3	3.8	3.2	2.7	2.0	1.7
Path • Inches	-1.5	+0.9	0.0	-5.3	-16.4	-35.2	-64.4	-171.7	-384.0
Wind Drift • Inches	0.0	1.0	4.0	9.5	18.1	30.3	47.2	98.0	171.1

Fiocchi 55-grain V-Max Polymer Tip (22250HVD)

G1 Ballistic Coefficient = 0.267

Distance • Yards	Muzzle	100	200	300	400	500	600	800	1000
Velocity • fps	3680	3272	2900	2559	2240	1946	1676	1244	1006
Energy • ft-lbs	1654	1307	1027	799	613	462	343	189	124
Taylor KO Index	6.5	5.8	5.1	4.5	3.9	3.4	2.9	2.2	1.8
Path • Inches	-1.5	+0.9	0.0	-5.0	-15.4	-32.8	-59.6	-156.9	-349.0
Wind Drift • Inches	0.0	0.9	3.7	8.7	16.4	27.4	42.3	87.2	154.1

Federal 55-grain Sierra BlitzKing (P22250C)

G1 Ballistic Coefficient = 0.271

Distance • Yards	Muzzle	100	200	300	400	500	600	800	1000
Velocity • fps	3625	3230	2860	2530	2210	1930	1667	1244	1008
Energy • ft-lbs	1605	1270	1000	780	600	450	340	189	124
Taylor KO Index	6.4	5.7	5.0	4.5	3.9	3.4	2.9	2.2	1.8
Path • Inches	-1.5	+0.9	0.0	-5.2	-15.9	-33.7	-60.8	-157.3	-344.3
Wind Drift • Inches	0.0	0.9	3.7	8.7	16.4	27.4	42.3	86.9	153.2

Norma 55-grain FULL JACKET (15732)

G1 Ballistic Coefficient = 0.209

Distance • Yards	Muzzle	100	200	300	400	500	600	800	1000
Velocity • fps	3609	3094	2642	2233	1855	1527	1262	971	833
Energy • ft-lbs	1591	1174	855	610	424	288	196	116	85
Taylor KO Index	6.4	5.4	4.6	3.9	3.3	2.7	2.2	1.7	1.5
Path • Inches	-1.5	+1.1	0.0	-6.1	-19.2	-42.5	-80.8	-230.1	-525.2
Wind Drift • Inches	0.0	1.2	5.0	12.0	23.3	40.0	63.3	130.8	219.2

Hornady 60-grain SP (8039)

G1 Ballistic Coefficient = 0.264

Distance • Yards	Muzzle	100	200	300	400	500	600	800	1000
Velocity • fps	3530	3131	2767	2431	2119	1832	1573	1175	972
Energy • ft-lbs	1660	1306	1020	787	598	447	330	184	126
Taylor KO Index	6.8	6.0	5.3	4.7	4.1	3.5	3.0	2.3	1.9
Path • Inches	-1.5	+1.0	0.0	-5.6	-17.1	-36.4	-66.2	-173.2	-381.0
Wind Drift • Inches	0.0	0.9	3.9	9.3	17.6	29.5	45.6	94.1	164.1

Federal 60-grain Nosler Partition (P22250G)

G1 Ballistic Coefficient = 0.229

Distance • Yards	Muzzle	100	200	300	400	500	600	800	1000
Velocity • fps	3500	3050	2630	2260	1910	1610	1351	1026	877
Energy • ft-lbs	1630	1235	925	680	490	345	243	140	103
Taylor KO Index	6.7	5.9	5.0	4.3	3.7	3.1	2.6	2.0	1.7
Path • Inches	-1.5	+1.1	0.0	-6.2	-19.3	-41.9	-78.8	-220.2	-487.3
Wind Drift • Inches	0.0	1.1	4.6	11.2	21.4	36.4	57.2	118.2	200.0

CorBon 62-grain DPX (DPX2225062/20)

G1 Ballistic Coefficient = 0.250

Distance • Yards	Muzzle	100	200	300	400	500	600	800	1000
Velocity • fps	3500	3082	2702	2353	2030	1735	1475	1104	930
Energy • ft-lbs	1687	1308	1006	762	567	415	299	168	119
Taylor KO Index	6.9	6.1	5.4	4.7	4.0	3.4	2.9	2.2	1.8
Path • Inches	-1.5	+1.1	0.0	-5.9	-18.1	-38.8	-71.2	-190.2	-422.6
Wind Drift • Inches	0.0	1.0	4.2	10.1	19.2	32.2	50.2	103.8	178.8

Winchester 64-grain Power Point (X222502)

G1 Ballistic Coefficient = 0.253

Distance • Yards	Muzzle	100	200	300	400	500	600	800	1000
Velocity • fps	3500	3086	2708	2360	2033	1744	1482	1107	931
Energy • ft-lbs	1741	1353	1042	791	590	432	312	174	123
Taylor KO Index	7.2	6.3	5.5	4.8	4.2	3.6	3.0	2.3	1.9
Path • Inches	-1.5	+1.1	0.0	-5.9	-18.0	-38.6	-71.3	-193.1	-428.2
Wind Drift • Inches	0.0	1.0	4.2	10.0	19.0	32.0	49.8	103.0	177.7

The .243 bullet from a Ruger Model 77 performed perfectly for Finn Aagaard's son Harald on this whitetail, but Finn felt it lacks the bone-smashing penetration needed for game much over 200 pounds. (Photo from Guns and Hunting *by Finn Aagaard, 2012, Safari Press Archives)*

.225 Winchester

The .225 Winchester was introduced in 1964 as a replacement for the .220 Swift. It is classified as semirimless, rather than rimmed, and while it can be used in Model 70s and other box-magazine guns, it seems to have found the most favor in various single-shot actions. Remington's standardization of the .22-250 in 1965 may have dealt the .225 a severe, and nearly fatal, blow. Winchester had dropped this cartridge from its catalog but relisted it in 2004.

Relative Recoil Factor = 0.90

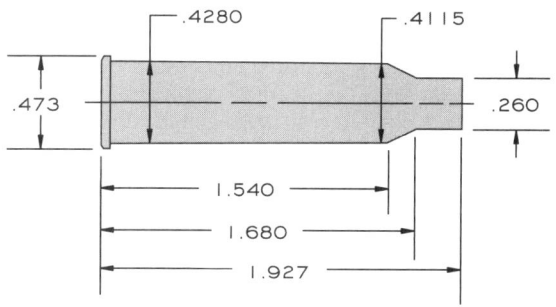

Specifications:

Controlling Agency for Standardization of this Ammunition: SAAMI

Bullet Weight Grains	Velocity fps	Maximum Average Pressure	
		Copper Crusher	Transducer
55	3,650	N/A	60,000 psi

Standard barrel for velocity testing: 24 inches long—1 turn in 14-inch twist

Availability:

Winchester 55-grain Pointed Soft Point (X2251)

G1 Ballistic Coefficient = 0.208

Distance • Yards	Muzzle	100	200	300	400	500	600	800	1000
Velocity • fps	3570	3066	2616	2208	1838	1514	1255	969	832
Energy • ft-lbs	1556	1148	836	595	412	280	192	115	85
Taylor KO Index	6.2	5.4	4.6	3.9	3.2	2.7	2.2	1.7	1.5
Path • Inches	-1.5	+1.1	0.0	-6.3	-19.8	-43.7	-84.1	-241.2	-534.8
Wind Drift • Inches	0.0	1.2	5.0	12.2	23.7	40.7	64.0	132.0	220.6

Finn Aagaard tests a Kimber Model 84 at the bench. (Photo from *Guns and Hunting* by Finn Aagaard, 2012, Safari Press Archives)

.224 Weatherby Magnum

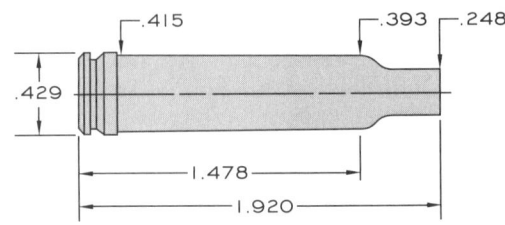

Weatherby's .224 Magnum, introduced in 1963, is a miniaturized version of the .300 Weatherby—belt, venturi shoulder, and all. This nifty little cartridge provides just a shade less performance than the current king of the centerfire .22s, the .22-250. At one time Norma loaded a 53-grain bullet in this case, but that loading was dropped a couple years ago. While Weatherby no longer lists new rifles in this caliber, they continue to provide this one loading.

Relative Recoil Factor = 0.90

Specifications:

Controlling Agency for Standardization of this Ammunition: CIP

Bullet Weight Grains	Velocity fps	Maximum Average Pressure	
		Copper Crusher	Transducer
N/S	N/S	3,800 bar	4,370 bar

Standard barrel for velocity testing: 26 inches long—1 turn in 14-inch twist

Availability:

Weatherby 55-grain Pointed-Expanding (H 224 55 SP)

G1 Ballistic Coefficient = 0.235

Distance • Yards	Muzzle	100	200	300	400	500	600	800	1000
Velocity • fps	3650	3192	2780	2403	2056	1741	1462	1081	910
Energy • ft-lbs.	1627	1244	944	705	516	370	261	143	101
Taylor KO Index	6.4	5.6	4.9	4.2	3.6	3.1	2.6	1.9	1.6
Path • Inches	-1.5	+1.0	0.0	-5.5	-17.1	-37.0	-69.2	-192.1	-432.0
Wind Drift • Inches	0.0	1.0	4.3	10.2	19.6	33.0	51.8	108.0	186.4

PH Larry McGillewie and Craig Boddington with a superb Cape grysbok, taken with a .243. Boddington has never actually taken a .243 to Africa, but it's an extremely useful cartridge for small antelope, so he has borrowed a few while there. (Photo from Safari Rifles II by Craig Boddington, 2009, Safari Press Archives)

.220 Swift

When it was introduced in 1935, the .220 Swift was the "swiftest" cartridge in the factory inventory. Today, the Swift still shares the "record" for .22s (4,250 fps with a 40-grain bullet). Much of what has been written about the Swift is a direct steal from the stories about King Arthur's sword—that is, mostly myth. The .220 Swift doesn't need any fiction. It's a very high-velocity .22, one suitable for all varmint applications. Part of the Swift's bad press has resulted from shooters using thin-jacketed varmint bullets on deer-size game. The results are seldom satisfactory, but it is unfair to blame the cartridge for the foolishness of a few uninformed people.

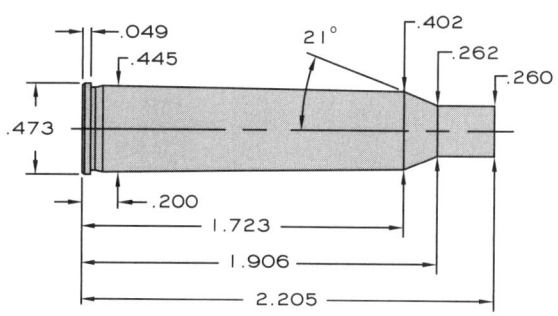

Relative Recoil Factor = 0.90

Specifications:

Controlling Agency for Ammunition Standardization: CIP

Bullet Weight Grains	Velocity fps	Maximum Average Pressure	
		Copper Crusher	Transducer
55	3,650	54,000 cup	N/A
60	3,600	54,000 cup	N/A

Standard barrel for velocity testing: 24 inches long—1 turn in 14-inch twist

Availability:

Federal 40-grain Nosler Ballistic Tip (P220B)
G1 Ballistic Coefficient = 0.221

Distance • Yards	Muzzle	100	200	300	400	500	600	800	1000
Velocity • fps	4250	3690	3200	2770	2370	2000	1674	1177	947
Energy • ft-lbs	1605	1210	910	680	500	355	249	123	80
Taylor KO Index	5.4	4.7	4.1	3.5	3.0	2.6	2.1	1.5	1.2
Path • Inches	-1.5	+0.5	0.0	-3.9	-12.5	-27.3	-51.2	-145.0	-342.6
Wind Drift • Inches	0.0	0.9	3.8	9.2	17.4	29.2	45.7	96.8	173.4

Hornady 40-grain V-Max Moly VX (83203)
G1 Ballistic Coefficient = 0.200

Distance • Yards	Muzzle	100	200	300	400	500	600	800	1000
Velocity • fps	4200	3597	3070	2602	2179	1796	1466	1044	871
Energy • ft-lbs	1566	1149	837	601	422	287	191	97	67
Taylor KO Index	5.4	4.6	3.9	3.3	2.8	2.3	1.9	1.3	1.1
Path • Inches	-1.5	+0.6	0.0	-4.3	-14.0	-30.8	-58.8	-171.1	-409.9
Wind Drift • Inches	0.0	1.0	4.4	10.5	20.1	34.3	54.3	116.2	202.8

Norma 50-grain Soft Point (15701)
G1 Ballistic Coefficient = 0.185

Distance • Yards	Muzzle	100	200	300	400	500	600	800	1000
Velocity • fps	4019	3395	2853	2371	1939	1562	1260	949	806
Energy • ft-lbs	1794	1280	904	624	418	271	176	100	72
Taylor KO Index	6.4	5.4	4.6	3.8	3.1	2.5	2.0	1.5	1.3
Path • Inches	-1.5	+0.8	0.0	-5.1	-16.5	-37.2	-72.5	-217.6	-516.6
Wind Drift • Inches	0.0	1.0	4.5	11.0	21.3	36.8	59.0	124.2	209.8

Winchester 50-grain Pointed Soft Point (X220S)
G1 Ballistic Coefficient = 0.200

Distance • Yards	Muzzle	100	200	300	400	500	600	800	1000
Velocity • fps	3870	3310	2816	2373	1972	1616	1318	985	837
Energy • ft-lbs	1663	1226	881	625	432	290	193	108	78
Taylor KO Index	6.2	5.3	4.5	3.8	3.2	2.6	2.1	1.6	1.3
Path • Inches	-1.5	+0.8	0.0	-5.2	-16.7	-37.1	-72.9	-215.0	-490.0
Wind Drift • Inches	0.0	1.1	4.8	11.6	22.4	38.4	61.1	128.4	218.1

Remington 50-grain Pointed Soft Point (R220S1)

G1 Ballistic Coefficient = 0.175

Distance • Yards	Muzzle	100	200	300	400	500	600	800	1000
Velocity • fps	3780	3158	2617	2135	1710	1357	1108	882	752
Energy • ft-lbs	1586	1107	760	506	325	204	136	86	63
Taylor KO Index	6.0	5.1	4.2	3.4	2.7	2.2	1.8	1.4	1.2
Path • Inches	-1.5	+1.0	0.0	-6.2	-20.1	-46.2	-94.2	-279.7	-623.6
Wind Drift • Inches	0.0	1.3	5.7	14.1	27.8	48.6	78.2	158.6	260.8

Hornady 50-grain V-MAX w/Moly (83213)

G1 Ballistic Coefficient = 0.242

Distance • Yards	Muzzle	100	200	300	400	500	600	800	1000
Velocity • fps	3850	3384	2965	2583	2232	1910	1617	1174	959
Energy • ft-lbs	1645	1271	976	741	553	405	290	153	102
Taylor KO Index	6.2	5.4	4.7	4.1	3.6	3.1	2.6	1.9	1.5
Path • Inches	-1.5	+0.8	0.0	-4.8	-14.8	-31.8	-58.6	-157.5	-357.5
Wind Drift • Inches	0.0	0.9	3.9	9.3	17.6	29.4	45.8	95.8	169.2

Nosler 50-grain Ballistic Tip (03077)

G1 Ballistic Coefficient = 0.238

Distance • Yards	Muzzle	100	200	300	400	500	600	800	1000
Velocity • fps	3900	3420	2990	2600	2242	1913	1617	1168	954
Energy • ft-lbs	1689	1299	993	751	558	406	290	152	101
Taylor KO Index	6.2	5.5	4.8	4.2	3.6	3.1	2.6	1.9	1.5
Path • Inches	-1.5	+0.7	0.0	-4.7	-14.5	-31.3	-57.8	-156.3	-356.7
Wind Drift • Inches	0.0	0.9	3.9	9.3	17.6	29.6	46.1	96.7	170.9

Federal 50-grain Sierra BlitzKing (P220C)

G1 Ballistic Coefficient = 0.249

Distance • Yards	Muzzle	100	200	300	400	500	600	800	1000
Velocity • fps	3530	3131	2767	2431	2119	1832	1573	1175	972
Energy • ft-lbs	1660	1306	1020	787	598	447	330	184	126
Taylor KO Index	6.8	6.0	5.3	4.7	4.1	3.5	3.0	2.3	1.9
Path • Inches	-1.5	+1.0	0.0	-5.6	-17.1	-36.4	-66.2	-173.2	-381.0
Wind Drift • Inches	0.0	0.9	3.9	9.3	17.6	29.5	45.6	94.1	164.1

Federal 52-grain Sierra MatchKing BTHP (P220V)

G1 Ballistic Coefficient = 0.225

Distance • Yards	Muzzle	100	200	300	400	500	600	800	1000
Velocity • fps	3830	3330	2890	2490	2120	1780	1488	1083	907
Energy • ft-lbs	1695	1280	965	715	520	365	256	136	95
Taylor KO Index	6.4	5.5	4.8	4.1	3.5	3.0	2.5	1.8	1.5
Path • Inches	-1.5	+0.8	0.0	-5.0	-15.7	-34.3	-63.9	-177.2	-408.2
Wind Drift • Inches	0.0	1.0	4.2	10.2	19.4	32.8	51.5	108.4	188.4

Norma 55-grain Oryx (15703)

G1 Ballistic Coefficient = 0.185

Distance • Yards	Muzzle	100	200	300	400	500	600	800	1000
Velocity • fps	3733	3184	2667	2205	1790	1438	1174	917	784
Energy • ft-lbs	1739	1238	869	594	391	253	168	103	75
Taylor KO Index	6.6	5.6	4.7	3.9	3.2	2.5	2.1	1.6	1.4
Path • Inches	-1.5	+1.0	0.0	-5.9	-19.2	-43.4	-84.7	-251.6	-583.7
Wind Drift • Inches	0.0	1.3	5.4	13.2	25.8	44.8	71.6	147.3	244.5

Hornady 55-grain V-MAX (8324) w/Moly (83243)

G1 Ballistic Coefficient = 0.255

Distance • Yards	Muzzle	100	200	300	400	500	600	800	1000
Velocity • fps	3680	3253	2867	2511	2183	1880	1605	1183	971
Energy • ft-lbs	1654	1292	1003	770	582	432	315	229	115
Taylor KO Index	6.5	5.7	5.0	4.4	3.8	3.3	2.8	2.1	1.7
Path • Inches	-1.5	+0.9	0.0	-5.2	-15.9	-34.0	-62.0	-163.9	-365.4
Wind Drift • Inches	0.0	0.9	3.9	9.2	17.4	29.2	45.3	93.9	154.9

Nosler 60-grain Partition (03049)

G1 Ballistic Coefficient = 0.228

Distance • Yards	Muzzle	100	200	300	400	500	600	800	1000
Velocity • fps	3600	3134	2714	2331	1980	1664	1393	1043	887
Energy • ft-lbs	1727	1309	982	724	523	369	258	145	105
Taylor KO Index	6.9	6.0	5.2	4.5	3.8	3.2	2.7	2.0	1.7
Path • Inches	-1.5	+1.0	0.0	-5.8	-18.0	-39.3	-73.2	-202.0	-457.8
Wind Drift • Inches	0.0	1.1	4.5	10.8	20.8	35.2	55.2	115.0	196.3

Hornady 60-grain Hollow Point (8122)

G1 Ballistic Coefficient = 0.264

Distance • Yards	Muzzle	100	200	300	400	500	600	800	1000
Velocity • fps	3600	3195	2826	2485	2169	1877	1611	1198	983
Energy • ft-lbs	1726	1360	1063	823	627	470	346	191	129
Taylor KO Index	6.9	6.1	5.4	4.8	4.2	3.6	3.1	2.3	1.9
Path • Inches	-1.5	+1.0	0.0	-5.3	-16.4	-34.8	-63.3	-165.5	-364.8
Wind Drift • Inches	0.0	0.9	3.8	9.1	17.2	28.7	44.4	91.7	160.9

Buck Buckner nailed this jack on the run near Tucson with Jack O'Connor's R-2 in 1997, fifty years after Jack last used it. (Photo from *Jack O'Connor* by Robert Anderson, 2008, Safari Press Archives)

.223 Winchester Super Short Magnum (WSSM)

Winchester's success with their Short Magnum cartridges led them to introduce a new series of Super Short Magnum cartridges beginning in 2003. The .223 WSSM is the .22-caliber version of that new series. The .223 WSSM has factory performance numbers that are slightly "better" than the .220 Swift but not up to what can be done with the wildcat .22-.284. It is a potent cartridge that lends itself to use in very short actions.

Relative Recoil Factor = 0.95

Specifications:

Controlling Agency for Standardization of this Ammunition: SAAMI

Bullet Weight Grains	Velocity fps	Maximum Average Pressure	
		Copper Crusher	Transducer
55	6,850	N/S	65,000 psi

Standard barrel for velocity testing: 24 inches long—1 turn in 12-inch twist

Availability:

Winchester 55-grain Ballisitic Silvertip (SBST223SS)

G1 Ballistic Coefficient = 0.276

Distance • Yards	Muzzle	100	200	300	400	500	600	800	1000
Velocity • fps	3850	3438	3064	2721	2402	2105	1831	1367	1071
Energy • ft-lbs.	1810	1444	1147	904	704	541	409	228	140
Taylor KO Index	6.8	6.1	5.4	4.8	4.2	3.7	3.2	2.4	1.9
Path • Inches	-1.5	+0.7	0.0	-4.4	-13.6	-28.8	-52.0	-138.1	-290.0
Wind Drift • Inches	0.0	0.8	3.4	7.9	14.9	24.6	37.8	77.4	138.3

Winchester 55-grain Pointed Soft Point (X223WSS)

G1 Ballistic Coefficient = 0.233

Distance • Yards	Muzzle	100	200	300	400	500	600	800	1000
Velocity • fps	3850	3367	2934	2541	2181	1851	1558	1129	934
Energy • ft-lbs.	1810	1384	1051	789	581	418	296	156	107
Taylor KOc Index	6.8	5.9	5.2	4.5	3.8	3.3	2.7	2.0	1.6
Path • Inches	-1.5	+0.8	0.0	-4.9	-15.1	-32.8	-60.7	-165.8	-379.6
Wind Drift • Inches	0.0	1.0	4.0	9.7	18.4	31.0	48.4	137.4	178.1

Winchester 64-grain Power Point (X223WSS1)

G1 Ballistic Coefficient = 0.235

Distance • Yards	Muzzle	100	200	300	400	500	600	800	1000
Velocity • fps	3600	3144	2732	2356	2011	1698	1428	1062	898
Energy • ft-lbs.	1841	1404	1061	786	574	410	290	160	115
Taylor KO Index	7.4	6.4	5.6	4.8	4.1	3.5	2.9	2.2	1.8
Path • Inches	-1.5	+1.0	0.0	-5.7	-17.7	-38.5	-72.0	-200.3	-448.9
Wind Drift • Inches	0.0	1.0	4.4	10.5	20.1	34.0	53.3	111.0	190.6

.243 Winchester

The .243 Winchester was the first (1955) of the spin-offs of the .308 cartridge after the .308's military adoption in 1954. There were few other 6mm cartridges to compete against. Except for the 6mm Navy, which was very much unloved and effectively obsolete, the 6mm size didn't go anywhere in the United States until after the announcement of the .243 Winchester. Today the 6mm calibers have taken over the middle ground between the .22s and the 7mms. Except for the .25-06, the .25-caliber guns are not doing well at all. The .243 is a truly versatile cartridge, one that's an excellent varmint caliber with the lighter bullets while retaining a very good capability against deer-size game with bullets in the 100-grain weight class.

Relative Recoil Factor = 1.25

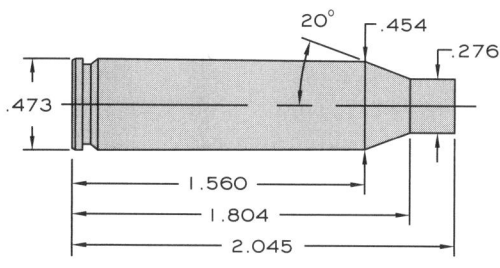

Specifications:

Controlling Agency for Standardization of this Ammunition: SAAMI

Bullet Weight Grains	Velocity fps	Maximum Average Pressure Copper Crusher	Transducer
75	3,325	52,000 cup	60,000 psi
80	3,325	52,000 cup	60,000 psi
85–87	3,300	52,000 cup	60,000 psi
100	2,950	52,000 cup	60,000 psi

Standard barrel for velocity testing: 24 inches long—1 turn in 10-inch twist

Availability:

Winchester 55-grain Ballistic Silvertip (SBST243)
G1 Ballistic Coefficient = 0.276

Distance • Yards	Muzzle	100	200	300	400	500	600	800	1000
Velocity • fps	3910	3493	3114	2766	2444	2144	1865	1393	1084
Energy • ft-lbs	1867	1489	1184	934	729	562	425	237	144
Taylor KO Index	7.4	6.5	5.8	5.1	4.5	3.9	3.4	2.5	2.0
Path • Inches	-1.5	+0.6	0.0	-4.3	-13.1	-27.8	-50.2	-128.4	-279.8
Wind Drift • Inches	0.0	0.8	3.5	8.2	15.4	25.6	39.5	81.2	144.9

Federal 55-grain Nosler Ballistic Tip (P243H)
G1 Ballistic Coefficient = 0.277

Distance • Yards	Muzzle	100	200	300	400	500	600	800	1000
Velocity • fps	3850	3440	3060	2720	2400	2110	1834	1370	1073
Energy • ft-lbs	1810	1445	1145	905	705	540	411	229	141
Taylor KO Index	7.4	6.6	5.8	5.2	4.6	4.0	3.5	2.6	2.0
Path • Inches	-1.5	+0.7	0.0	-4.4	-13.6	-28.9	-51.9	-132.8	-289.2
Wind Drift • Inches	0.0	0.8	3.4	7.9	14.8	24.6	37.7	77.1	137.8

Norma 58-grain V-MAX (16035)
G1 Ballistic Coefficient = 0.250

Distance • Yards	Muzzle	100	200	300	400	500	600	800	1000
Velocity • fps	3937	3476	3061	2683	2336	2014	1722	1252	999
Energy • ft-lbs	1997	1556	1207	928	703	523	382	202	129
Taylor KO Index	7.9	7.0	6.2	5.4	4.7	4.1	3.5	2.5	2.0
Path • Inches	-1.5	+0.7	0.0	-4.4	-13.7	-29.4	-53.8	-142.5	-320.8
Wind Drift • Inches	0.0	0.9	3.6	8.7	16.4	27.3	42.2	87.8	156.8

Hornady 58-grain V-MAX (8343)
G1 Ballistic Coefficient = 0.250

Distance • Yards	Muzzle	100	200	300	400	500	600	800	1000
Velocity • fps	3925	3465	3051	2674	2327	2007	1715	1247	997
Energy • ft-lbs	1984	1546	1199	921	697	519	379	200	128
Taylor KO Index	7.9	7.0	6.1	5.4	4.7	4.0	3.5	2.5	2.0
Path • Inches	-1.5	+0.7	0.0	-4.5	-13.8	-29.6	-54.2	-143.6	-323.2
Wind Drift • Inches	0.0	0.9	3.7	8.7	16.4	27.4	42.4	88.2	157.3

Black Hills 58-grain Hornady V-MAX (1C243BHGN6)

G1 Ballistic Coefficient = 0.250

Distance • Yards	Muzzle	100	200	300	400	500	600	800	1000
Velocity • fps	3800	3353	2950	2581	2241	1928	1644	1202	977
Energy • ft-lbs	1859	1448	1121	858	647	479	348	186	123
Taylor KO Index	7.7	6.8	5.9	5.2	4.5	3.9	3.3	2.4	2.0
Path • Inches	-1.5	+0.8	0.0	-4.8	-14.9	-32.0	-58.4	-155.3	-348.8
Wind Drift • Inches	0.0	0.9	3.8	9.0	17.1	28.6	44.4	92.3	163.9

Hornady 58-grain V-MAX Moly VX (83423)

G1 Ballistic Coefficient = 0.252

Distance • Yards	Muzzle	100	200	300	400	500	600	800	1000
Velocity • fps	3750	3308	2905	2544	2206	1895	1616	1193	966
Energy • ft-lbs	1811	1409	1090	833	627	463	336	180	120
Taylor KO Index	7.6	6.7	5.8	5.1	4.4	3.8	3.3	2.4	1.9
Path • Inches	-1.5	+0.8	0.0	-5.0	-15.4	-33.0	-60.6	-163.8	-370.0
Wind Drift • Inches	0.0	0.9	3.9	9.2	17.4	29.2	45.2	94.1	166.0

Black Hills 62-grain Barnes Varmint Grenade (1C243BHGN4) LF

G1 Ballistic Coefficient = 0.280

Distance • Yards	Muzzle	100	200	300	400	500	600	800	1000
Velocity • fps	3700	3337	2949	2618	2310	2023	1759	1319	1049
Energy • ft-lbs	1895	1506	1197	944	735	564	426	240	152
Taylor KO Index	8.0	7.2	6.3	5.6	5.0	4.4	3.8	2.8	2.3
Path • Inches	-1.5	+0.8	0.0	-4.9	-14.8	-31.4	-56.4	-144.3	-313.2
Wind Drift • Inches	0.0	0.8	3.5	8.2	15.4	25.6	39.3	80.4	142.6

Cor-Bon 62-grain Varmint Grenade (SD24362VG/20) LF

G1 Ballistic Coefficient = 0.280

Distance • Yards	Muzzle	100	200	300	400	500	600	800	1000
Velocity • fps	3600	3217	2866	2542	2239	1958	1701	1278	1030
Energy • ft-lbs	1785	1425	1131	890	691	528	399	225	146
Taylor KO Index	7.7	6.9	6.2	5.5	4.8	4.2	3.7	2.8	2.2
Path • Inches	-1.5	+0.9	0.0	-5.2	-15.8	-33.4	-60.1	-153.9	-333.7
Wind Drift • Inches	0.0	0.8	3.6	8.5	15.9	26.5	40.8	83.4	147.3

Fiocchi 70-grain PSP (243SPB)

G1 Ballistic Coefficient = 0.269

Distance • Yards	Muzzle	100	200	300	400	500	600	800	1000
Velocity • fps	3400	3020	2671	2347	2047	1771	1524	1151	964
Energy • ft-lbs	1796	1417	1108	856	651	487	361	206	144
Taylor KO Index	8.3	7.3	6.5	5.7	5.0	4.3	3.7	2.8	2.3
Path • Inches	-1.5	+1.2	0.0	-6.1	-18.5	-39.2	-71.2	-185.4	-404.4
Wind Drift • Inches	0.0	1.0	4.0	9.6	18.1	30.3	47.0	96.4	166.6

Federal 70-grain Nosler Ballistic Tip (P243F)

G1 Ballistic Coefficient = 0.309

Distance • Yards	Muzzle	100	200	300	400	500	600	800	1000
Velocity • fps	3450	3100	2800	2510	2240	1980	1744	1342	1079
Energy • ft-lbs	1850	1505	1220	980	780	610	473	280	181
Taylor KO Index	8.4	7.5	6.8	6.1	5.4	4.8	4.2	3.3	2.6
Path • Inches	-1.5	+1.0	0.0	-5.5	-16.5	-34.6	-61.5	-153.4	-323.4
Wind Drift • Inches	0.0	0.8	3.4	8.0	15.0	24.8	37.9	76.5	134.5

Cor-Bon 70-grain BlitzKing (SD24370/20)

G1 Ballistic Coefficient = 0.300

Distance • Yards	Muzzle	100	200	300	400	500	600	800	1000
Velocity • fps	3400	3058	2741	2446	2169	1911	1674	1282	1043
Energy • ft-lbs	1797	1454	1168	930	731	568	436	255	169
Taylor KO Index	8.3	7.4	6.7	5.9	5.3	4.6	4.1	3.1	2.5
Path • Inches	-1.5	+1.1	0.0	-5.8	-17.4	-36.4	-65.0	-163.6	-347.6
Wind Drift • Inches	0.0	0.8	3.6	8.4	15.8	26.2	40.2	81.6	142.8

Remington 75-grain AccuTip-V Boat Tail (PRA243WB)

G1 Ballistic Coefficient = 0.330

Distance • Yards	Muzzle	100	200	300	400	500	600	800	1000
Velocity • fps	3375	3065	2775	2504	2248	2008	1782	1395	1122
Energy • ft-lbs	1897	1564	1282	1044	842	671	529	324	210
Taylor KO Index	8.8	8.0	7.2	6.5	5.9	5.2	4.6	3.6	2.9
Path • Inches	-1.5	+1.1	0.0	-5.6	-16.8	-35.0	-61.8	-156.2	-319.6
Wind Drift • Inches	0.0	0.8	3.2	7.6	14.2	23.5	35.8	71.6	125.4

PMC 75-grain HP (243HVA)
G1 Ballistic Coefficient = 0.294

Distance • Yards	Muzzle	100	200	300	400	500	600	800	1000
Velocity • fps	3400	3051	2729	2428	2147	1886	1646	1256	1027
Energy • ft-lbs	1925	1556	1240	982	768	592	451	263	176
Taylor KO Index	8.9	7.9	7.1	6.3	5.6	4.9	4.3	3.3	2.7
Path • Inches	-1.5	+1.1	0.0	-5.8	-17.6	-36.9	-66.1	-167.2	-357.2
Wind Drift • Inches	0.0	0.9	3.6	8.6	16.2	26.9	41.4	84.2	147.1

Hornady 75-grain HP (8040)
G1 Ballistic Coefficient = 0.294

Distance • Yards	Muzzle	100	200	300	400	500	600	800	1000
Velocity • fps	3340	2996	2677	2380	2102	1844	1609	1230	1014
Energy • ft-lbs	1857	1494	1193	943	736	566	431	252	171
Taylor KO Index	8.7	7.8	7.0	6.2	5.5	4.8	4.2	3.2	2.6
Path • Inches	-1.5	+1.2	0.0	-6.1	-18.3	-38.5	-68.9	-174.6	-372.6
Wind Drift • Inches	0.0	0.9	3.7	8.8	16.6	27.6	42.5	86.4	150.3

Norma 75-grain V-MAX (16004)
G1 Ballistic Coefficient = 0.330

Distance • Yards	Muzzle	100	200	300	400	500	600	800	1000
Velocity • fps	3330	3023	2736	2467	2214	1975	1754	1373	1109
Energy • ft-lbs	1847	1522	1247	1014	816	650	512	314	205
Taylor KO Index	8.7	7.9	7.1	6.4	5.8	5.1	4.6	3.6	2.9
Path • Inches	-1.5	+1.1	0.0	-5.8	-17.4	-36.0	-63.7	-156.2	-323.5
Wind Drift • Inches	0.0	0.8	3.3	7.8	14.5	23.9	36.4	73.0	127.5

Federal 80-grain Soft Point (243AS)
G1 Ballistic Coefficient = 0.366

Distance • Yards	Muzzle	100	200	300	400	500	600	800	1000
Velocity • fps	3330	3050	2790	2540	2310	2090	1881	1509	1219
Energy • ft-lbs	1970	1655	1380	1150	945	775	628	405	264
Taylor KO Index	9.2	8.5	7.7	7.1	6.4	5.8	5.2	4.2	3.4
Path • Inches	-1.5	+1.1	0.0	-5.6	-16.6	-34.2	-59.6	-142.7	-287.8
Wind Drift • Inches	0.0	0.7	3.0	6.9	12.8	21.0	31.8	62.9	109.4

Remington 80-grain Pointed Soft Point (R243W1)
G1 Ballistic Coefficient = 0.256

Distance • Yards	Muzzle	100	200	300	400	500	600	800	1000
Velocity • fps	3350	2955	2593	2259	1951	1670	1427	1085	924
Energy • ft-lbs	1993	1551	1194	906	676	495	362	209	152
Taylor KO Index	9.3	8.2	7.2	6.3	5.4	4.6	4.0	3.0	2.6
Path • Inches	-1.5	+1.2	0.0	-6.5	-19.8	-42.4	-77.3	-204.9	-449.9
Wind Drift • Inches	0.0	1.0	4.4	10.4	19.9	33.4	51.7	106.1	180.9

Winchester 80-grain Pointed Soft Point (X2431)
G1 Ballistic Coefficient = 0.256

Distance • Yards	Muzzle	100	200	300	400	500	600	800	1000
Velocity • fps	3350	2955	2593	2259	1951	1670	1427	1085	924
Energy • ft-lbs	1993	1551	1194	906	676	495	362	209	152
Taylor KO Index	9.3	8.2	7.2	6.3	5.4	4.6	4.0	3.0	2.6
Path • Inches	-1.5	+1.2	0.0	-6.5	-19.8	-42.4	-77.3	-204.9	-449.9
Wind Drift • Inches	0.0	1.0	4.4	10.4	19.9	33.4	51.7	106.1	180.9

Hornady 80-grain GMX (80456) LF
G1 Ballistic Coefficient = 0.300

Distance • Yards	Muzzle	100	200	300	400	500	600	800	1000
Velocity • fps	3425	3080	2760	2463	2184	1924	1690	1293	1049
Energy • ft-lbs	2083	1684	1358	1077	847	657	507	297	195
Taylor KO Index	9.5	8.6	7.7	6.8	6.1	5.3	4.7	3.6	2.9
Path • Inches	-1.5	+1.1	0.0	-5.7	-17.1	-35.9	-63.9	-160.7	-341.6
Wind Drift • Inches	0.0	0.8	3.5	8.4	15.7	26.0	39.8	80.7	141.5

Remington 80-grain Power-Lokt Hollow Point (R243W2)
G1 Ballistic Coefficient = 0.256

Distance • Yards	Muzzle	100	200	300	400	500	600	800	1000
Velocity • fps	3350	2955	2593	2259	1951	1670	1427	1085	924
Energy • ft-lbs	1993	1551	1194	906	676	495	362	209	152
Taylor KO Index	9.3	8.2	7.2	6.3	5.4	4.6	4.0	3.0	2.6
Path • Inches	-1.5	+1.2	0.0	-6.5	-19.8	-42.4	-77.3	-204.9	-449.9
Wind Drift • Inches	0.0	1.0	4.4	10.4	19.9	33.4	51.7	106.1	180.9

CorBon 80-grain T-DPX (DPX24385/220)

G1 Ballistic Coefficient = 0.350

Distance • Yards	Muzzle	100	200	300	400	500	600	800	1000
Velocity • fps	3200	2918	2653	2403	2167	1944	1737	1378	1123
Energy • ft-lbs	1933	1607	1329	1090	886	714	569	358	238
Taylor KO Index	9.4	8.6	7.8	7.1	6.4	5.7	5.1	4.1	3.3
Path • Inches	-1.5	+1.3	0.0	-6.2	-18.6	-38.3	-67.2	-162.9	-332.8
Wind Drift • Inches	0.0	0.8	3.3	7.7	14.3	23.5	35.8	71.2	123.7

Remington 80-grain Copper Solid Tipped (PCS243WB) LF

G1 Ballistic Coefficient = 0.299

Distance • Yards	Muzzle	100	200	300	400	500	600	800	1000
Velocity • fps	3350	3011	2696	2403	2128	1872	1638	1255	1029
Energy • ft-lbs	1993	1610	1291	1025	804	622	477	280	188
Taylor KO Index	9.3	8.1	7.2	6.4	5.7	5.0	4.4	3.3	2.6
Path • Inches	-1.5	+0.8	0.0	-4.8	-14.5	-30.6	-54.8	-138.5	-297.5
Wind Drift • Inches	0.0	0.8	3.3	7.9	14.7	24.4	37.3	75.9	134.8

Nosler 85-grain Partition (60002)

G1 Ballistic Coefficient = 0.315

Distance • Yards	Muzzle	100	200	300	400	500	600	800	1000
Velocity • fps	3225	2911	2618	2343	2086	1847	1626	1263	1041
Energy • ft-lbs	1963	1599	1294	1036	821	643	499	301	205
Taylor KO Index	9.5	8.6	7.7	6.9	6.2	5.4	4.8	3.7	3.1
Path • Inches	-1.5	+1.3	0.0	-6.4	-19.2	-39.9	-71.0	-176.6	-370.1
Wind Drift • Inches	0.0	0.9	3.6	8.6	16.1	26.6	40.7	82.0	142.3

Federal 85-grain Sierra GameKing BTHP

G1 Ballistic Coefficient = 0.315

Distance • Yards	Muzzle	100	200	300	400	500	600	800	1000
Velocity • fps	3300	2980	2680	2410	2140	1900	1672	1295	1058
Energy • ft-lbs	2055	1675	1360	1090	870	680	528	317	211
Taylor KO Index	9.7	8.8	7.9	7.1	6.3	5.6	4.9	3.8	3.1
Path • Inches	-1.5	+1.2	0.0	-6.0	-18.2	-37.7	-67.3	-167.3	-351.0
Wind Drift • Inches	0.0	0.8	3.5	8.3	15.6	25.8	39.4	79.4	138.5

Black Hills 85-grain Barnes TSX (1C243BHGN4) LF

G1 Ballistic Coefficient = 0.333

Distance • Yards	Muzzle	100	200	300	400	500	600	800	1000
Velocity • fps	3200	2904	2628	2367	2122	1891	1678	1318	1080
Energy • ft-lbs	1932	1592	1303	1058	850	675	532	328	220
Taylor KO Index	9.4	8.6	7.8	7.0	6.3	5.6	5.0	3.9	3.2
Path • Inches	-1.5	+1.3	0.0	-6.4	-19.0	-39.3	-69.5	-170.4	-352.1
Wind Drift • Inches	0.0	0.8	3.4	8.1	15.2	25.0	38.2	76.4	132.7

Federal 85-grain Barnes Triple-Shock X-Bullet (P243K) LF

G1 Ballistic Coefficient = 0.333

Distance • Yards	Muzzle	100	200	300	400	500	600	800	1000
Velocity • fps	3200	2900	2630	2370	2120	1890	1676	1316	1078
Energy • ft-lbs	1935	1590	1305	1055	850	675	531	327	219
Taylor KO Index	9.4	8.6	7.8	7.0	6.3	5.6	4.9	3.9	3.2
Path • Inches	-1.5	+1.3	0.0	-6.3	-19.0	-39.2	-69.5	-170.6	-352.8
Wind Drift • Inches	0.0	0.8	3.4	8.1	15.2	25.1	38.3	76.6	133.0

Nosler 90-grain CT (10123)

G1 Ballistic Coefficient = 0.410

Distance • Yards	Muzzle	100	200	300	400	500	600	800	1000
Velocity • fps	3200	2958	2729	2511	2304	2107	1920	1580	1300
Energy • ft-lbs	2046	1748	1488	1260	1061	887	737	499	338
Taylor KO Index	10.0	9.2	8.5	7.8	7.2	6.6	6.0	4.9	4.1
Path • Inches	-1.5	+1.2	0.0	-5.9	-17.4	-35.4	-61.2	-143.3	-281.9
Wind Drift • Inches	0.0	0.7	2.8	6.4	11.9	19.3	29.1	56.8	97.7

Nosler 90-grain E-Tip (10123 and 60087) LF

G1 Ballistic Coefficient = 0.410

Distance • Yards	Muzzle	100	200	300	400	500	600	800	1000
Velocity • fps	3200	2958	2729	2511	2304	2107	1920	1580	1300
Energy • ft-lbs	2046	1748	1488	1260	1061	887	737	499	338
Taylor KO Index	10.0	9.2	8.5	7.8	7.2	6.6	6.0	4.9	4.1
Path • Inches	-1.5	+1.2	0.0	-5.9	-17.4	-35.4	-61.2	-143.3	-281.9
Wind Drift • Inches	0.0	0.7	2.8	6.4	11.9	19.3	29.1	56.8	97.7

Remington 90-grain Swift Scirocco Bonded (PRSC243WA)

G1 Ballistic Coefficient = 0.390

Distance • Yards	Muzzle	100	200	300	400	500	600	800	1000
Velocity • fps	3120	2871	2636	2411	2199	1997	1806	1467	1204
Energy • ft-lbs	1946	1647	1388	1162	966	797	652	430	290
Taylor KO Index	9.7	9.0	8.2	7.5	6.9	6.2	5.6	4.6	3.8
Path • Inches	-1.5	+1.4	0.0	-6.4	-18.8	-38.3	-66.8	-158.7	-318.5
Wind Drift • Inches	0.0	0.7	3.0	7.0	13.0	21.3	32.2	63.3	109.3

Lapua 90-grain Naturalis (N316201) LF

G1 Ballistic Coefficient = 0.230

Distance • Yards	Muzzle	100	200	300	400	500	600	800	1000
Velocity • fps	2986	2583	2214	1877	1576	1323	1134	934	819
Energy • ft-lbs	1782	1334	980	704	496	350	257	174	134
Taylor KO Index	9.3	8.1	6.9	5.9	4.9	4.1	3.5	2.9	2.6
Path • Inches	-1.5	+1.9	0.0	-9.0	-28.1	-61.2	-114.4	-308.3	-661.7
Wind Drift • Inches	0.0	1.3	5.6	14.0	27.0	46.0	71.6	140.0	225.9

Lapua 90-grain FMJ (4316052)

G1 Ballistic Coefficient = 0.378

Distance • Yards	Muzzle	100	200	300	400	500	600	800	1000
Velocity • fps	2904	2659	2427	2206	1998	1801	1620	1309	1096
Energy • ft-lbs	1686	1413	1117	973	798	649	524	343	240
Taylor KO Index	9.1	8.3	7.6	6.9	6.2	5.6	5.1	4.1	3.4
Path • Inches	-1.5	+1.7	0.0	-7.6	-22.4	-46.0	-80.3	-191.9	-385.3
Wind Drift • Inches	0.0	0.8	3.4	8.1	15.0	24.7	37.4	73.8	126.2

Hornady 95-grain SST (80463)

G1 Ballistic Coefficient = 0.355

Distance • Yards	Muzzle	100	200	300	400	500	600	800	1000
Velocity • fps	3185	2908	2648	2402	2169	1950	1690	1293	1048
Energy • ft-lbs	2139	1784	1478	1217	992	802	643	407	271
Taylor KO Index	10.5	9.6	8.4	7.9	7.2	6.4	5.8	4.6	3.7
Path • Inches	-1.5	+1.3	0.0	-6.3	-18.6	-38.4	-67.3	-162.5	-330.6
Wind Drift • Inches	0.0	0.8	3.2	7.6	14.1	23.2	35.3	70.1	121.7

Remington 95-grain AccuTip-V Boat Tail (PRA243WA)

G1 Ballistic Coefficient = 0.355

Distance • Yards	Muzzle	100	200	300	400	500	600	800	1000
Velocity • fps	3120	2847	2590	2347	2118	1902	1699	1353	1111
Energy • ft-lbs	2053	1710	1415	1162	946	763	609	386	260
Taylor KO Index	10.3	9.4	8.5	7.7	7.0	6.3	5.6	4.5	3.7
Path • Inches	-1.5	+1.3	0.0	-6.6	-19.5	-40.2	-70.7	-172.1	-353.5
Wind Drift • Inches	0.0	0.8	3.3	7.8	14.6	24.0	36.4	72.4	125.3

Winchester 95-grain XP3 (SXP243W)

G1 Ballistic Coefficient = 0.411

Distance • Yards	Muzzle	100	200	300	400	500	600	800	1000
Velocity • fps	3100	2864	2641	2428	2225	2032	1848	1518	1253
Energy • ft-lbs	2027	1730	1471	1243	1044	871	720	486	331
Taylor KO Index	10.2	9.4	8.7	8.0	7.3	6.7	6.1	5.0	4.1
Path • Inches	-1.5	+1.4	0.0	-6.4	-18.7	-38.0	-65.8	-154.2	-303.7
Wind Drift • Inches	0.0	0.7	2.9	6.7	12.4	20.2	30.4	59.5	102.2

Hornady 95-grain SST (80464)

G1 Ballistic Coefficient = 0.355

Distance • Yards	Muzzle	100	200	300	400	500	600	800	1000
Velocity • fps	3010	2744	2493	2255	2031	1820	1625	1297	1079
Energy • ft-lbs	1911	1588	1310	1073	870	699	558	355	245
Taylor KO Index	9.9	9.0	8.2	7.4	6.7	6.0	5.4	4.3	3.6
Path • Inches	-1.5	+1.0	0.0	-7.2	-21.2	-43.7	-76.8	-185.9	-378.3
Wind Drift • Inches	0.0	0.8	3.5	8.2	15.4	25.3	38.4	76.3	131.2

Federal 95-grain Fusion (F243FS1)

G1 Ballistic Coefficient = 0.375

Distance • Yards	Muzzle	100	200	300	400	500	600	800	1000
Velocity • fps	2980	2730	2490	2270	2060	1850	1665	1343	1115
Energy • ft-lbs	1875	1570	1310	1085	890	725	585	380	262
Taylor KO Index	9.8	9.0	8.2	7.5	6.8	6.1	5.5	4.4	3.7
Path • Inches	-1.5	+1.6	0.0	-7.2	-21.2	-43.2	-75.9	-181.6	-365.2
Wind Drift • Inches	0.0	0.8	3.3	7.8	14.6	23.9	36.3	71.7	123.1

Black Hills 95-grain Hornady SST (1C243HGN3)
G1 Ballistic Coefficient = 0.355

Distance • Yards	Muzzle	100	200	300	400	500	600	800	1000
Velocity • fps	2950	2687	2439	2205	1983	1776	1584	1265	1060
Energy • ft-lbs	1835	1523	1255	1025	829	665	529	338	237
Taylor KO Index	9.7	8.9	8.0	7.3	6.5	5.9	5.2	4.2	3.5
Path • Inches	-1.5	+1.7	0.0	-7.5	-22.2	-45.8	-80.6	-196.8	-403.4
Wind Drift • Inches	0.0	0.6	3.3	8.2	15.6	25.8	39.4	78.5	134.6

Fiocchi 95-grain SST Polymer Tip BT (243HSB)
G1 Ballistic Coefficient = 0.355

Distance • Yards	Muzzle	100	200	300	400	500	600	800	1000
Velocity • fps	2950	2658	2439	2204	1983	1775	1585	1267	1062
Energy • ft-lbs	1835	1523	1254	1025	829	665	530	339	238
Taylor KO Index	9.7	8.9	8.0	7.3	6.5	5.9	5.2	4.2	3.5
Path • Inches	-1.5	+1.7	0.0	-7.5	-22.2	-45.8	-80.5	-195.0	-396.7
Wind Drift • Inches	0.0	0.9	3.6	8.4	15.8	26.0	39.6	78.6	134.6

Hornady 95-grain SST InterLock (80464)
G1 Ballistic Coefficient = 0.355

Distance • Yards	Muzzle	100	200	300	400	500	600	800	1000
Velocity • fps	2950	2687	2439	2205	1983	1776	1584	1265	1060
Energy • ft-lbs	1835	1523	1255	1025	829	665	529	338	237
Taylor KO Index	9.7	8.9	8.0	7.3	6.5	5.9	5.2	4.2	3.5
Path • Inches	-1.5	+1.7	0.0	-7.5	-22.2	-45.8	-80.6	-196.8	-403.4
Wind Drift • Inches	0.0	0.6	3.3	8.2	15.6	25.8	39.4	78.5	134.6

Winchester 95-grain Ballistic Silvertip (SBST243A)
G1 Ballistic Coefficient = 0.400

Distance • Yards	Muzzle	100	200	300	400	500	600	800	1000
Velocity • fps	3100	2854	2626	2410	2203	2007	1819	1486	1223
Energy • ft-lbs	2021	1719	1455	1225	1024	850	698	466	316
Taylor KO Index	10.2	9.4	8.7	7.9	7.3	6.6	6.0	4.9	4.0
Path • Inches	-1.5	+1.4	0.0	-6.4	-18.9	-38.4	-66.8	-157.3	-311.6
Wind Drift • Inches	0.0	0.7	3.0	6.9	12.9	20.9	31.5	61.7	106.3

Federal 95-grain Nosler Ballistic Tip (P243J)
G1 Ballistic Coefficient = 0.377

Distance • Yards	Muzzle	100	200	300	400	500	600	800	1000
Velocity • fps	3025	2770	2540	2310	2100	1890	1701	1372	1133
Energy • ft-lbs	1930	1625	1355	1125	925	755	610	397	271
Taylor KO Index	10.0	9.1	8.4	7.6	6.9	6.2	5.6	4.5	3.7
Path • Inches	-1.5	+1.5	0.0	-6.9	-20.4	-41.6	-73.1	-174.7	-351.0
Wind Drift • Inches	0.0	0.8	3.2	7.6	14.2	23.3	35.3	69.7	120.0

Norma 95-grain FULL JACKET (16037)
G1 Ballistic Coefficient = 0.351

Distance • Yards	Muzzle	100	200	300	400	500	600	800	1000
Velocity • fps	3199	2918	2654	2405	2170	1948	1741	1383	1127
Energy • ft-lbs	2159	1797	1487	1221	994	801	640	403	268
Taylor KO Index	10.5	9.6	8.8	7.9	7.2	6.4	5.7	4.6	3.7
Path • Inches	-1.5	+1.3	0.0	-6.3	-18.5	-38.2	-67.1	-162.4	-331.4
Wind Drift • Inches	0.0	0.8	3.3	7.7	14.2	23.4	35.6	70.8	123.0

Fiocchi 100-grain PSP (243SPD)
G1 Ballistic Coefficient = 0.350

Distance • Yards	Muzzle	100	200	300	400	500	600	800	1000
Velocity • fps	3200	2918	2651	2399	2165	1947	1738	1380	1125
Energy • ft-lbs	2274	1892	1561	1279	1041	842	671	423	281
Taylor KO Index	11.1	10.1	9.2	8.3	7.5	6.8	6.0	4.8	3.9
Path • Inches	-1.5	+1.3	0.0	-6.2	-18.5	-38.2	-67.2	-162.7	-332.2
Wind Drift • Inches	0.0	0.8	3.3	7.7	14.3	23.5	35.7	71.1	123.4

Norma 100-grain Soft Point (16003)
G1 Ballistic Coefficient = 0.258

Distance • Yards	Muzzle	100	200	300	400	500	600	800	1000
Velocity • fps	2986	2624	2290	1981	1701	1453	1249	1002	877
Energy • ft-lbs	1980	1529	1165	871	643	469	346	223	171
Taylor KO Index	10.4	9.1	7.9	6.9	5.9	5.0	4.3	3.5	3.0
Path • Inches	-1.5	+1.8	0.0	-8.5	-25.8	-55.3	-101.4	-266.9	-570.4
Wind Drift • Inches	0.0	1.2	5.0	12.1	23.2	39.1	60.7	121.0	198.9

Lapua 100-grain Soft Point (4316056)

G1 Ballistic Coefficient = 0.479

Distance • Yards	Muzzle	100	200	300	400	500	600	800	1000
Velocity • fps	2986	2788	2598	2417	2242	2075	1916	1621	1369
Energy • ft-lbs	1980	1726	1499	1297	1117	956	815	584	416
Taylor KO Index	10.4	9.7	9.0	8.4	7.8	7.2	6.7	5.6	4.8
Path • Inches	-1.5	+1.5	0.0	-6.6	-19.3	-38.8	-66.4	-152.0	-292.2
Wind Drift • Inches	0.0	0.6	2.6	5.9	11.0	17.7	26.3	51.1	86.7

Federal 100-grain Sierra GameKing BTSP (P243C)

G1 Ballistic Coefficient = 0.430

Distance • Yards	Muzzle	100	200	300	400	500	600	800	1000
Velocity • fps	2960	2740	2530	2330	2150	1960	1791	1484	1237
Energy • ft-lbs	1945	1670	1425	1210	1020	855	713	489	340
Taylor KO Index	10.3	9.5	8.8	8.1	7.5	6.8	6.2	5.2	4.3
Path • Inches	-1.5	+1.6	0.0	-7.0	-20.4	-41.3	-71.4	-166.3	-325.1
Wind Drift • Inches	0.0	0.7	2.9	6.8	12.6	20.5	30.8	60.0	102.5

PMC 100-grain BTSP (243HIA)

G1 Ballistic Coefficient = 0.405

Distance • Yards	Muzzle	100	200	300	400	500	600	800	1000
Velocity • fps	2960	2728	2508	2299	2099	1910	1732	1418	1178
Energy • ft-lbs	1945	1653	1397	1173	978	810	666	447	308
Taylor KO Index	10.3	9.5	8.7	8.0	7.3	6.6	6.0	4.9	4.1
Path • Inches	-1.5	+1.6	0.0	-7.1	-20.9	-42.5	-73.7	-173.6	-343.3
Wind Drift • Inches	0.0	0.7	3.1	7.2	13.4	22.0	33.2	85.0	111.4

Federal 100-grain Nosler Partition (P243E)

G1 Ballistic Coefficient = 0.386

Distance • Yards	Muzzle	100	200	300	400	500	600	800	1000
Velocity • fps	2850	2610	2390	2170	1970	1780	1603	1302	1094
Energy • ft-lbs	1805	1515	1265	1045	860	700	571	377	266
Taylor KO Index	9.9	9.1	8.3	7.5	6.8	6.2	5.6	4.5	3.8
Path • Inches	-1.5	+1.8	0.0	-8.0	-23.3	-47.6	-82.8	-197.3	-394.4
Wind Drift • Inches	0.0	0.8	3.4	8.1	15.1	24.8	37.5	73.7	125.7

Federal 100-grain Soft Point (243B)

G1 Ballistic Coefficient = 0.358

Distance • Yards	Muzzle	100	200	300	400	500	600	800	1000
Velocity • fps	2960	2700	2450	2220	1990	1790	1600	1280	1070
Energy • ft-lbs	1945	1615	1330	1090	880	710	569	364	254
Taylor KO Index	10.3	9.4	8.5	7.7	6.9	6.2	5.6	4.4	3.7
Path • Inches	-1.5	+1.6	0.0	-7.9	-22.0	-45.4	-79.4	-192.2	-390.4
Wind Drift • Inches	0.0	0.8	3.5	8.4	15.6	25.7	39.0	77.4	132.7

Hornady 100-grain BTSP Interlock (8046)

G1 Ballistic Coefficient = 0.406

Distance • Yards	Muzzle	100	200	300	400	500	600	800	1000
Velocity • fps	2960	2728	2508	2299	2099	1910	1734	1420	1179
Energy • ft-lbs	1945	1653	1397	1174	979	810	667	448	309
Taylor KO Index	10.3	9.5	8.7	8.0	7.3	6.6	6.0	4.9	4.1
Path • Inches	-1.5	+1.6	0.0	-7.2	-21.0	-42.8	-73.6	-173.4	-342.8
Wind Drift • Inches	0.0	0.7	3.1	7.2	13.4	22.0	33.1	64.9	111.2

Remington 100-grain CoreLokt Ultra Bonded (PRC243WC)

G1 Ballistic Coefficient = 0.373

Distance • Yards	Muzzle	100	200	300	400	500	600	800	1000
Velocity • fps	2960	2709	2471	2246	2033	1832	1644	1324	1102
Energy • ft-lbs	1945	1629	1356	1120	917	745	600	390	270
Taylor KO Index	10.3	9.4	8.6	7.8	7.1	6.4	5.7	4.6	3.8
Path • Inches	-1.5	+1.5	0.0	-7.3	-21.6	-44.3	-77.5	-186.7	-379.0
Wind Drift • Inches	0.0	0.8	3.4	8.0	14.8	24.5	36.8	73.3	125.2

Remington 100-grain Core-Lokt Pointed Soft Point (R243W3)

G1 Ballistic Coefficient = 0.356

Distance • Yards	Muzzle	100	200	300	400	500	600	800	1000
Velocity • fps	2960	2697	2449	2215	1993	1786	1594	1274	1065
Energy • ft-lbs	1945	1615	1332	1089	882	708	564	360	252
Taylor KO Index	10.3	9.4	8.5	7.7	6.9	6.2	5.5	4.4	3.7
Path • Inches	-1.5	+1.6	0.0	-7.5	-22.0	-45.4	-79.8	-193.1	-392.8
Wind Drift • Inches	0.0	0.8	3.6	8.4	15.7	25.9	39.3	78.0	133.7

Winchester 100-grain Power-Point (X2432)

G1 Ballistic Coefficient = 0.356

Distance • Yards	Muzzle	100	200	300	400	500	600	800	1000
Velocity • fps	2960	2697	2449	2215	1993	1786	1594	1274	1065
Energy • ft-lbs	1945	1615	1332	1089	882	708	564	360	252
Taylor KO Index	10.3	9.4	8.5	7.7	6.9	6.2	5.5	4.4	3.7
Path • Inches	-1.5	+1.6	0.0	-7.5	-22.0	-45.4	-79.8	-193.1	-392.8
Wind Drift • Inches	0.0	0.8	3.6	8.4	15.7	25.9	39.3	78.0	133.7

Sellier and Bellot 100-grain Soft Point (V330812U)

G1 Ballistic Coefficient = 0.249

Distance • Yards	Muzzle	100	200	300	400	500	600	800	1000
Velocity • fps	2904	2505	2161	1865	1581	1339	1154	951	836
Energy • ft-lbs	1877	1398	1040	774	555	398	296	201	155
Taylor KO Index	10.1	8.7	7.5	6.5	5.5	4.6	4.0	3.3	2.9
Path • Inches	-1.5	+2.0	0.0	-9.4	-28.9	-62.7	-116.7	-309.8	-643.8
Wind Drift • Inches	0.0	1.3	5.6	13.6	26.1	44.3	68.8	134.4	216.9

Winchester 100-grain Power Max Bonded (X243BP)

G1 Ballistic Coefficient = 0.372

Distance • Yards	Muzzle	100	200	300	400	500	600	800	1000
Velocity • fps	2960	2708	2470	2240	2031	1829	1642	1323	1101
Energy • ft-lbs	1945	1628	1355	1118	915	743	599	389	269
Taylor KO Index	10.3	9.4	8.6	7.8	7.1	6.3	5.7	4.6	3.8
Path • Inches	-1.5	+1.6	0.0	-7.3	-21.6	-44.3	-77.5	-185.8	-374.2
Wind Drift • Inches	0.0	0.8	3.4	8.0	14.9	24.4	37.1	73.2	125.6

Cor-Bon 115-grain DTAC (PM243115/20)

G1 Ballistic Coefficient = 0.360

Distance • Yards	Muzzle	100	200	300	400	500	600	800	1000
Velocity • fps	2900	2644	2401	2172	1955	1753	1567	1257	1058
Energy • ft-lbs	2148	1785	1473	1205	977	785	627	404	286
Taylor KO Index	11.6	10.6	9.6	8.7	7.8	7.0	6.3	5.0	4.2
Path • Inches	-1.5	+1.7	0.0	-7.8	-23.0	-47.3	-83.0	-200.6	-406.7
Wind Drift • Inches	0.0	0.9	3.6	8.5	16.0	23.0	39.9	79.1	134.9

6mm Remington

Remington's 6mm was introduced in 1963 as a replacement for a basically identical 1955 cartridge called the .244 Remington. Remington expected that cartridge (the .244) to be a formidable competitor for Winchester's .243, but its guns were built with a 1 turn in 12-inch twist (vs. the .243's 1 turn in 10 inches). This left the .244 unable to stabilize 100- and 105-grain bullets reliably. In 1963 the caliber was renamed the 6mm Remington, and the rifles were manufactured with a twist rate of 1 turn in 9 inches. The sales of this cartridge must be sagging because there is little new activity from the ammo manufacturers.

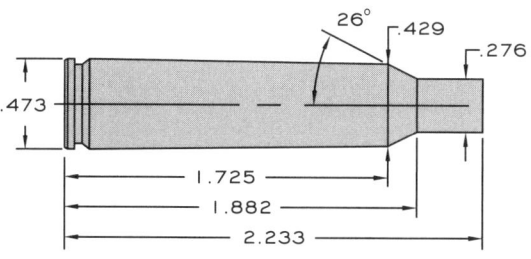

Relative Recoil Factor = 1.40

Specifications:

Controlling Agency for Standardization of this Ammunition: SAAMI

Bullet Weight Grains	Velocity fps	Maximum Average Pressure Copper Crusher	Transducer
80	3,400	52,000 cup	65,000 psi
90	3,175	52,000 cup	65,000 psi
100	3,090	52,000 cup	65,000 psi

Standard barrel for velocity testing: 24 inches long—1 turn in 9-inch twist

Availability:

Federal 80-grain Speer Hot-Cor SP (6AS)
G1 Ballistic Coefficient = 0.366

Distance • Yards	Muzzle	100	200	300	400	500	600	800	1000
Velocity • fps	3400	3120	2850	2600	2360	2140	1930	1550	1249
Energy • ft-lbs	2055	1725	1445	1200	995	815	662	427	277
Taylor KO Index	9.4	8.7	7.9	7.2	6.6	5.9	5.4	4.3	3.5
Path • Inches	-1.5	+1.0	0.0	-5.4	-15.8	-32.6	-56.8	-135.9	-273.8
Wind Drift • Inches	0.0	0.7	2.9	6.7	12.5	20.4	30.9	61.0	106.1

Federal 85-grain Barnes Triple Shock X-Bullet (P6D) LF
G1 Ballistic Coefficient = 0.333

Distance • Yards	Muzzle	100	200	300	400	500	600	800	1000
Velocity • fps	3350	3040	2760	2490	2240	2000	1778	1395	1124
Energy • ft-lbs	2120	1750	1435	1170	945	755	597	367	239
Taylor KO Index	9.9	9.0	8.1	7.3	6.6	5.9	5.2	4.1	3.3
Path • Inches	-1.5	+1.1	0.0	-5.7	-17.0	-35.3	-62.4	-152.7	-315.4
Wind Drift • Inches	0.0	0.8	3.2	7.6	14.2	23.4	35.7	71.4	124.8

Hornady 95-grain SST (81663)
G1 Ballistic Coefficient = 0.355

Distance • Yards	Muzzle	100	200	300	400	500	600	800	1000
Velocity • fps	3235	2955	2692	2443	2209	1987	1779	1417	1151
Energy • ft-lbs	2207	1841	1528	1259	1029	833	668	484	279
Taylor KO Index	10.7	9.7	8.9	8.1	7.3	6.6	5.9	4.7	3.8
Path • Inches	-1.5	+1.6	0.0	-6.1	-18.0	-37.0	-64.9	-156.6	-318.5
Wind Drift • Inches	0.0	0.8	3.2	7.4	13.8	22.7	34.5	68.5	119.0

Federal 100-grain Nosler Partition (P6C)
G1 Ballistic Coefficient = 0.386

Distance • Yards	Muzzle	100	200	300	400	500	600	800	1000
Velocity • fps	3100	2850	2610	2380	2170	1970	1780	1442	1185
Energy • ft-lbs	2135	1800	1515	1260	1045	860	703	462	312
Taylor KO Index	10.8	9.9	9.1	8.3	7.5	6.8	6.2	5.0	4.1
Path • Inches	-1.5	+1.4	0.0	-6.5	-19.2	-39.2	-68.2	-161.8	-323.1
Wind Drift • Inches	0.0	0.7	3.1	7.2	13.3	21.8	33.0	65.0	112.0

Federal 100-grain Hi-Shok Soft Point (6B)

G1 Ballistic Coefficient = 0.357

Distance • Yards	Muzzle	100	200	300	400	500	600	800	1000
Velocity • fps	3100	2830	2570	2330	2100	1890	1692	1349	1110
Energy • ft-lbs	2135	1775	1470	1205	985	790	636	404	273
Taylor KO Index	10.8	9.8	8.9	8.1	7.3	6.6	5.9	4.7	3.9
Path • Inches	-1.5	+1.4	0.0	-6.7	-19.8	-40.8	-71.5	-172.6	-350.9
Wind Drift • Inches	0.0	0.8	3.3	7.8	14.6	24.1	36.5	72.5	125.4

Hornady 100-grain BTSP InterLock (8166)

G1 Ballistic Coefficient = 0.405

Distance • Yards	Muzzle	100	200	300	400	500	600	800	1000
Velocity • fps	3100	2861	2634	2419	2231	2018	1832	1501	1237
Energy • ft-lbs	2134	1818	1541	1300	1068	904	746	500	340
Taylor KO Index	10.8	9.9	9.1	8.4	7.7	7.0	6.4	5.2	4.3
Path • Inches	-1.5	+1.4	0.0	-6.4	-18.8	-38.3	-66.3	-155.8	-307.9
Wind Drift • Inches	0.0	0.7	2.9	6.8	12.6	20.6	31.0	60.7	104.4

Winchester 100-grain Power Point (X6MMR2)

G1 Ballistic Coefficient = 0.356

Distance • Yards	Muzzle	100	200	300	400	500	600	800	1000
Velocity • fps	3100	2829	2573	2332	2104	1889	1689	1346	1107
Energy • ft-lbs	2133	1777	1470	1207	983	792	633	402	272
Taylor KO Index	10.8	9.8	8.9	8.1	7.3	6.6	5.9	4.7	3.8
Path • Inches	-1.5	+1.4	0.0	-6.7	-19.8	-40.8	-71.6	-173.0	-352.0
Wind Drift • Inches	0.0	0.8	3.3	7.9	14.7	24.1	36.7	72.8	125.9

(Left to right) .223 WSSM, .243 WSSM, .25 WSSM.

.243 Winchester Super Short Magnum (WSSM)

In the flurry of new cartridges in 2003, Winchester added a 6mm version of its WSSM, the .243. The .243 WSSM is interesting at least partly because the internal case volume is nearly identical to the "standard" .243 Winchester. That would suggest that the performance of the two cartridges should be essentially identical. But bullet-for-bullet the .243 WSSM outperforms its older cousin. Some experts have suggested that this is entirely due to the cartridge's shape. This overlooks the fact that the SSM uses a significantly higher working pressure level. Since there are no "old" guns chambered for .243 WSSM that have to be accommodated, Winchester can set the working pressure level at any number it thinks is appropriate.

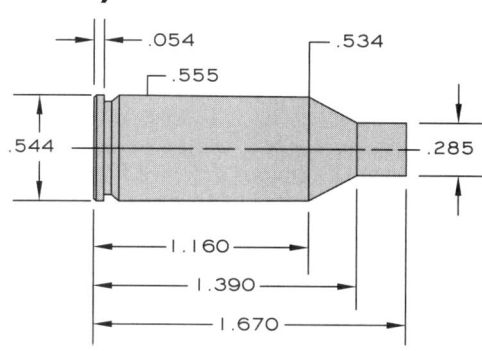

Relative Recoil Factor = 1.37

Specifications:

Controlling Agency for Standardization of this Ammunition: SAAMI

Bullet Weight Grains	Velocity fps	Maximum Average Pressure	
		Copper Crusher	Transducer
55	4,000	N/S	65,000 psi
95	3,250	N/S	65,000 psi

Standard barrel for velocity testing: 24 inches long—1 turn in 9-inch twist

Availability:

Winchester 55-grain Ballisitic Silvertip (SBST243SS)
G1 Ballistic Coefficient = 0.276

Distance • Yards	Muzzle	100	200	300	400	500	600	800	1000
Velocity • fps	4060	3628	3237	2880	2550	2243	1956	1463	1122
Energy • ft-lbs.	2013	1607	1280	1013	794	614	468	261	154
Taylor KO Index	7.8	6.9	6.2	5.5	4.9	4.3	3.7	2.8	1.9
Path • Inches	-1.5	+0.6	0.0	-3.9	-12.0	-25.5	-46.1	-118.7	-262.5
Wind Drift • Inches	0.0	0.8	3.2	7.5	14.0	23.0	35.2	71.8	129.0

Winchester 95-grain XP3 (SXP243WSS)
G1 Ballistic Coefficient = 0.411

Distance • Yards	Muzzle	100	200	300	400	500	600	800	1000
Velocity • fps	3150	2912	2686	2471	2266	2071	1884	1549	1276
Energy • ft-lbs.	2093	1788	1521	1287	1083	904	749	507	344
Taylor KOc Index	10.4	9.6	8.9	8.1	7.5	6.8	6.2	5.1	4.2
Path • Inches	-1.5	+1.3	0.0	-6.1	-18.0	-36.6	-63.4	-148.6	-292.4
Wind Drift • Inches	0.0	0.7	2.8	6.5	12.1	19.7	29.7	58.1	99.9

Winchester 95-grain Ballisitic Silvertip (SBST243SSA)
G1 Ballistic Coefficient = 0.410

Distance • Yards	Muzzle	100	200	300	400	500	600	800	1000
Velocity • fps	3250	2905	2674	2453	2244	2045	1857	1519	1248
Energy • ft-lbs.	2093	1780	1508	1270	1062	882	728	487	329
Taylor KOc Index	10.4	9.6	8.8	8.1	7.4	6.7	6.1	5.0	4.1
Path • Inches	-1.5	+1.3	0.0	-6.2	-18.2	-37.1	-64.3	-151.3	-299.4
Wind Drift • Inches	0.0	0.7	2.9	6.7	12.5	20.3	30.7	60.1	103.5

Winchester 100-grain Power-Point (X243WSS)
G1 Ballistic Coefficient = 0.356

Distance • Yards	Muzzle	100	200	300	400	500	600	800	1000
Velocity • fps	3110	2838	2583	2341	2112	1897	1696	1351	1110
Energy • ft-lbs.	2147	1789	1481	1217	991	799	639	405	274
Taylor KO Index	10.8	9.9	9.0	8.1	7.3	6.6	5.9	4.7	3.9
Path • Inches	-1.5	+1.4	0.0	-6.6	-19.7	-40.5	-71.2	-173.0	-355.1
Wind Drift • Inches	0.0	0.8	3.3	7.8	14.6	24.0	36.5	72.5	125.4

6.17mm (.243) Lazzeroni Spitfire

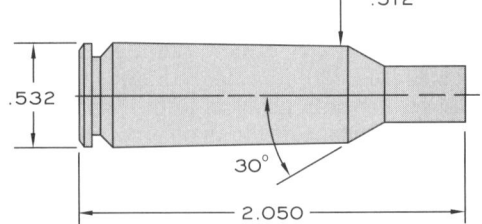

Lazzeroni has two lines of high-performance cartridges. The smallest of these is the 6.17 mm Spitfire. This is factory production ammunition, not a wildcat caliber (although like many cartridges it started as one). The Spitfire is a "short" magnum, suitable for adaptation to all actions designed for the .308 case. Performance levels of this and other Lazzeroni calibers are impressive.

Relative Recoil Factor = 1.37

Specifications:

Controlling Agency for Standardization of this Ammunition: Factory

Availability:

Lazzeroni 85-grain Nosler Partition (617SF085P)

G1 Ballistic Coefficient = 0.404

Distance • Yards	Muzzle	100	200	300	400	500	600	800	1000
Velocity • fps	3618	3316	3036	2772	2523	2287	2064	1660	1328
Energy • ft-lbs	2471	2077	1704	1450	1202	987	805	520	333
Taylor KO Index	10.7	9.8	9.0	8.2	7.4	6.7	6.1	4.9	3.9
Path • Inches	-1.5	+0.8	0.0	-4.6	-13.8	-28.4	-49.7	-118.9	-239.3
Wind Drift • Inches	0.0	0.6	2.7	6.3	11.7	19.1	28.8	56.7	98.8

A South African springbok with its characteristic "ruff," which displays for some time after death. (Photo from *Out of Bullets* by Johnny Chilton, 2011, Safari Press Archives)

.240 Weatherby Magnum

Weatherby's .240 Magnum fills the 6mm place in its extended family of cartridges. The .240 is a belted case (many people think that's what makes it a "magnum") with Weatherby's trademark venturi shoulder. If you look closely at the dimensions, you will see that this cartridge comes very, very close to duplicating the .30-06's dimensions, making it for talking purposes a 6mm-06. That's plenty of cartridge to give the .240 Weatherby the best performance on the 6mm list (with the exception of the Lazzeroni Spitfire's latest numbers). The .240 drives a 100-grain bullet about 300 fps faster than the 6mm Remington and 450 fps faster than the .243 Winchester with the same bullet. Rifles chambered for the .240 Weatherby have dropped off the current Weatherby inventory.

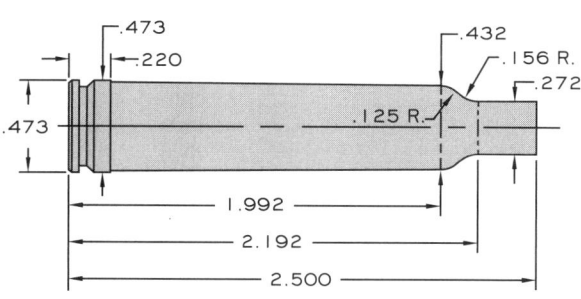

Relative Recoil Factor = 1.46

Specifications:

Controlling Agency for Standardization of this Ammunition: CIP

Bullet Weight Grains	Velocity fps	Maximum Average Pressure	
		Copper Crusher	Transducer
N/S	N/S	3,800 bar	4,370 bar

Standard barrel for velocity testing: 26 inches long—1 turn in 10-inch twist

Availability:

Weatherby 85-grain Barnes TSX (B 240 85 TSX)

G1 Ballistic Coefficient = 0.363

Distance • Yards	Muzzle	100	200	300	400	500	600	800	1000
Velocity • fps	3500	3222	2962	2717	2484	2264	2056	1674	1355
Energy • ft-lbs	2313	1961	1657	1394	1166	968	798	529	347
Taylor KO Index	10.3	9.5	8.7	8.0	7.3	6.7	6.1	4.9	4.0
Path • Inches	-1.5	+0.9	0.0	-4.9	-14.5	-29.8	-51.3	-122.6	-243.7
Wind Drift • Inches	0.0	0.6	2.6	6.2	11.4	18.6	28.0	54.8	94.9

Weatherby 87-grain SP (H 240 87 SP)

G1 Ballistic Coefficient = 0.328

Distance • Yards	Muzzle	100	200	300	400	500	600	800	1000
Velocity • fps	3523	3199	2898	2617	2352	2103	1873	1465	1164
Energy • ft-lbs	2399	1977	1622	1323	1069	855	678	415	262
Taylor KO Index	10.6	9.7	8.8	7.9	7.1	6.4	5.7	4.4	3.5
Path • Inches	-1.5	+0.7	0.0	-4.7	-15.3	-31.8	-56.2	-137.5	-284.8
Wind Drift • Inches	0.0	0.7	3.1	7.3	13.6	22.3	33.9	67.8	119.2

Weatherby 95-grain BST (N 240 95 BST)

G1 Ballistic Coefficient = 0.379

Distance • Yards	Muzzle	100	200	300	400	500	600	800	1000
Velocity • fps	3420	3146	2888	2645	2414	2195	1987	1611	1303
Energy • ft-lbs	2467	2087	1759	1475	1229	1017	833	548	358
Taylor KO Index	11.3	10.4	9.5	8.7	8.0	7.2	6.6	5.3	4.5
Path • Inches	-1.5	+1.2	0.0	-5.6	-15.4	-31.5	-54.9	-130.3	-260.1
Wind Drift • Inches	0.0	0.7	2.8	6.4	11.9	19.4	29.2	57.4	99.7

Weatherby 100-grain SP (H 240 100 SP)

G1 Ballistic Coefficient = 0.381

Distance • Yards	Muzzle	100	200	300	400	500	600	800	1000
Velocity • fps	3406	3134	2878	2637	2408	2190	1983	1610	1304
Energy • ft-lbs	2576	2180	1839	1544	1287	1065	874	576	379
Taylor KO Index	11.8	10.9	10.0	9.2	8.4	7.6	6.9	5.6	4.5
Path • Inches	-1.5	+0.8	0.0	-5.1	-15.5	-31.8	-55.3	-131.0	-261.2
Wind Drift • Inches	0.0	0.7	2.8	6.4	11.9	19.4	29.2	57.4	99.5

Weatherby 100-grain Nosler Partition (N 240 100 PT)

G1 Ballistic Coefficient = 0.385

Distance • Yards	Muzzle	100	200	300	400	500	600	800	1000
Velocity • fps	3406	3136	2882	2642	2415	2199	1996	1624	1318
Energy • ft-lbs	2576	2183	1844	1550	1294	1073	885	586	386
Taylor KO Index	11.8	10.9	10.0	9.2	8.4	7.6	6.9	5.6	4.6
Path • Inches	-1.5	+0.8	0.0	-5.0	-15.4	-31.6	-55.0	-130.0	-258.5
Wind Drift • Inches	0.0	0.7	2.7	6.4	11.7	19.1	28.8	56.5	97.9

Nosler 100-grain Partition (10051)

G1 Ballistic Coefficient = 0.384

Distance • Yards	Muzzle	100	200	300	400	500	600	800	1000
Velocity • fps	3125	2871	2632	2405	2189	1984	1792	1451	1189
Energy • ft-lbs	2169	1831	1538	1284	1064	874	713	467	314
Taylor KO Index	10.8	10.0	9.1	8.3	7.6	6.9	6.2	5.0	4.1
Path • Inches	-1.5	+1.4	0.0	-6.4	-18.8	-38.5	-67.1	-159.4	-318.6
Wind Drift • Inches	0.0	0.7	3.0	7.1	13.3	21.7	32.8	64.6	111.6

Three young hunters, one proud mother, and a happy outfitter. The American West offers some of the greatest hunting experiences available today anywhere on earth. (Safari Press Archives)

.25-20 Winchester

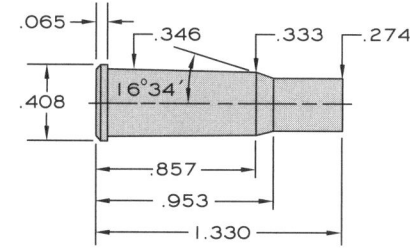

The .25-20 Winchester is an outgrowth of the .25-20 Single Shot. At the time of its introduction in 1893 for Winchester's Model 92 rifle, it was considered radical because of its "sharp" 16-degree shoulder. This caliber has almost reached the end of its commercial life.

Relative Recoil Factor = 0.57

Specifications:

Controlling Agency for Standardization of this Ammunition: SAAMI

Bullet Weight Grains	Velocity fps	Maximum Average Pressure	
		Copper Crusher	Transducer
86	1,445	28,000 cup	N/S

Standard barrel for velocity testing: 24 inches long—1 turn in 14-inch twist

Availability:

Remington 86-grain Soft Point (R25202)

G1 Ballistic Coefficient = 0.191

Distance • Yards	Muzzle	100	200	300	400	500	600	800	1000
Velocity • fps	1460	1194	1030	931	858	797	744	696	572
Energy • ft-lbs	407	272	203	165	141	122	106	81	63
Taylor KO Index	4.6	3.8	3.3	3.0	2.7	2.5	2.4	2.2	1.8
Path • Inches	-1.5	+1.4	0.0	-44.1	-128.3	-259.9	-430.6	-972.6	-1824
Wind Drift • Inches	0.0	4.0	15.7	33.7	56.7	84.5	116.8	196.3	297.0

Winchester 86-grain Soft Point (X25202)

G1 Ballistic Coefficient = 0.191

Distance • Yards	Muzzle	100	200	300	400	500	600	800	1000
Velocity • fps	1460	1194	1030	931	858	798	744	696	572
Energy • ft-lbs	407	272	203	165	141	122	106	81	63
Taylor KO Index	4.6	3.8	3.3	3.0	2.7	2.5	2.4	2.2	1.8
Path • Inches	-1.5	+1.4	0.0	-44.1	-128.3	-259.9	-430.6	-972.6	-1824
Wind Drift • Inches	0.0	4.0	15.7	33.7	56.7	84.5	116.8	196.3	297.0

Even with a copy of Ammo & Ballistics *at hand, you need to go to the range and make sure you understand how your rifle will shoot a particular load.* (Safari Press Archives)

.25-35 Winchester

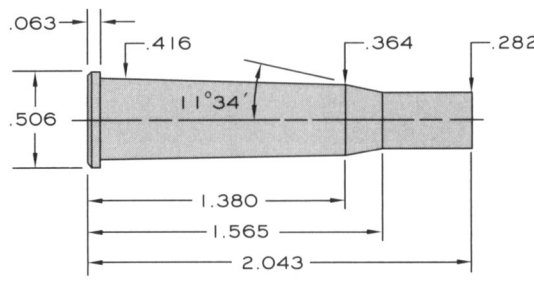

Introduced in 1895, the .25-35 Winchester has enough terminal performance at short ranges to be effective on deer-size game. The 117-grain bullet at 2,230 fps certainly isn't the same class as a .30-30, but this caliber continues to hang on in Winchester's catalog. This cartridge is also on its last legs, with only one loading offered.

Relative Recoil Factor = 1.18

Specifications:

Controlling Agency for Standardization of this Ammunition: SAAMI

Bullet Weight Grains	Velocity fps	Maximum Average Pressure	
		Copper Crusher	Transducer
117	2,210	37,000 cup	N/S

Standard barrel for velocity testing: 24 inches long—1 turn in 8-inch twist

Availability:

Winchester 117-grain Soft Point (X2535)

G1 Ballistic Coefficient = 0.214

Distance • Yards	Muzzle	100	200	300	400	500	600	800	1000
Velocity • fps	2230	1866	1545	1282	1097	984	906	791	700
Energy • ft-lbs	1292	904	620	427	313	252	213	163	127
Taylor KO Index	9.6	8.0	6.7	5.5	4.7	4.2	3.9	3.4	3.0
Path • Inches	-1.5	+4.3	0.0	-19.0	-59.2	-128.1	-231.9	-559.3	-1087
Wind Drift • Inches	0.0	2.2	9.6	23.5	44.5	71.8	104.3	182.1	276.8

Before each hunt, check the zero of your rifle even if you must do so with a makeshift target frame like in this case. Stick a few targets and some tape in your duffel, for there is almost always something lying about that can act like a target frame! (Safari Press Archives)

.250 Savage

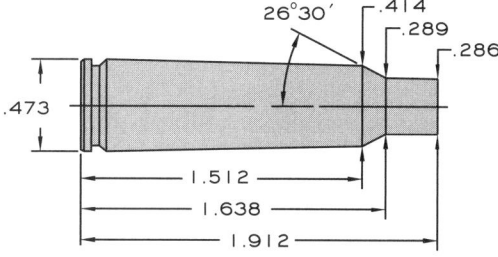

The .250 Savage is another cartridge that's nearing the end of a long and useful life. Announced in 1915, it was intended for the Savage 99 lever-action rifle. Because the 87-grain bullet would reach 3,000 fps (at least in the advertising literature), the cartridge became known as the .250-3000. While the introduction of the .243 Winchester has cut deeply into the popularity of the .250 Savage, it will live for many more years in the form of the .22-250, which was based on the .250 case.

Relative Recoil Factor = 1.27

Specifications:

Controlling Agency for Standardization of this Ammunition: SAAMI

Bullet Weight Grains	Velocity fps	Maximum Average Pressure	
		Copper Crusher	Transducer
87	3,010	45,000 cup	N/S
100	2,800	45,000 cup	N/S

Standard barrel for velocity testing: 24 inches long—1 turn in 14-inch twist

Availability:

Remington 100-grain Pointed Soft Point (R250SV)

G1 Ballistic Coefficient = 0.286

Distance • Yards	Muzzle	100	200	300	400	500	600	800	1000
Velocity • fps	2820	2504	2210	1936	1684	1461	1272	1029	903
Energy • ft-lbs	1765	1392	1084	832	630	473	359	235	181
Taylor KO Index	10.4	9.2	8.1	6.7	6.2	5.4	4.7	3.8	3.3
Path • Inches	-1.5	+2.0	0.0	-9.2	-27.7	-58.6	-106.6	-274.8	-568.7
Wind Drift • Inches	0.0	1.2	4.7	11.7	22.2	37.1	57.2	113.1	185.7

Winchester 100-grain Silvertip (X2503)

G1 Ballistic Coefficient = 0.255

Distance • Yards	Muzzle	100	200	300	400	500	600	800	1000
Velocity • fps	2820	2467	2140	1839	1569	1339	1162	961	847
Energy • ft-lbs	1765	1351	1017	751	547	398	300	205	159
Taylor KO Index	10.4	9.1	7.9	6.8	5.8	4.9	4.3	3.5	3.1
Path • Inches	-1.5	+2.1	0.0	-9.8	-29.4	-64.6	-119.2	-312.4	-644.6
Wind Drift • Inches	0.0	1.3	5.5	13.4	25.7	43.4	67.1	130.7	210.7

.257 Roberts

Remington adopted the .257 Roberts as a factory number in 1934 and Winchester followed a year later. The design of the .257 cartridge stemmed from work done as early as 1909 by Griffin & Howe, A. O. Niedner, and Major Ned Roberts. Their work culminated in the early '30s with a cartridge called the .25 Roberts. The .25 Roberts was a little different from the .257 Roberts, hence the two different descriptions. Both cartridges were certainly closely related to the 7x57 Mauser (as was the .30-06). The .257 Roberts is an "Oldie but Goodie" and still provides a useful capability, both as a varmint rifle and a gun for deer-size game.

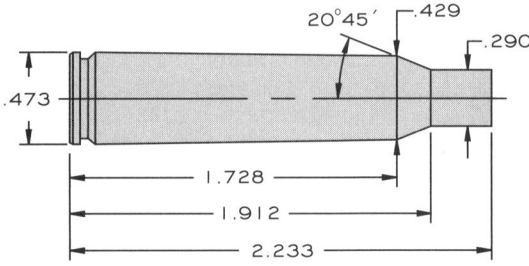

Relative Recoil Factor = 1.47

Specifications:

Controlling Agency for Standardization of this Ammunition: SAAMI

Bullet Weight Grains	Velocity fps	Maximum Average Pressure	
		Copper Crusher	Transducer
87	3,150	45,000 cup	54,000 psi
100	2,880	45,000 cup	54,000 psi
117	2,630	45,000 cup	54,000 psi
+ P Loads			
100	2,980	50,000 cup	58,000 psi
117	2,760	50,000 cup	58,000 psi

Standard barrel for velocity testing: 24 inches long—1 turn in 10-inch twist

Availability:

Nosler 100-grain Partition + P (11025)

G1 Ballistic Coefficient = 0.377

Distance • Yards	Muzzle	100	200	300	400	500	600	800	1000
Velocity • fps	3000	2749	2512	2286	2073	1873	1685	1360	1126
Energy • ft-lbs	1998	1678	1401	1160	954	779	630	411	282
Taylor KO Index	11.0	10.1	9.2	8.4	7.6	6.9	6.2	5.0	4.1
Path • Inches	-1.5	+1.6	0.0	-7.1	-20.8	-42.7	-74.5	-178.0	-357.4
Wind Drift • Inches	0.0	0.8	3.3	7.7	14.4	23.6	35.7	70.4	121.1

Nosler 100-grain Ballistic Tip + P (11056)

G1 Ballistic Coefficient = 0.393

Distance • Yards	Muzzle	100	200	300	400	500	600	800	1000
Velocity • fps	3000	2759	2530	2313	2107	1913	1729	1407	1165
Energy • ft-lbs	1998	1690	1422	1188	986	812	664	440	301
Taylor KO Index	11.0	10.1	9.3	8.5	7.7	7.0	6.3	5.2	4.3
Path • Inches	-1.5	+1.5	0.0	-7.0	-20.5	-41.8	-72.7	-172.1	-342.5
Wind Drift • Inches	0.0	0.8	3.1	7.4	13.7	22.4	33.8	66.4	114.1

Nosler 110-grain AccuBond + P (60010)

G1 Ballistic Coefficient = 0.425

Distance • Yards	Muzzle	100	200	300	400	500	600	800	1000
Velocity • fps	3000	2776	2564	2362	2169	1984	1810	1497	1245
Energy • ft-lbs	2198	1882	1606	1363	1149	961	801	543	379
Taylor KO Index	12.1	11.2	10.4	9.5	8.8	8.0	7.3	6.0	5.0
Path • Inches	-1.5	+1.5	0.0	-6.8	-19.9	-40.3	-69.6	-162.4	-317.9
Wind Drift • Inches	0.0	0.7	2.9	6.7	12.5	20.3	30.6	59.6	102.0

Nosler 110-grain AccuBond + P (11050)

G1 Ballistic Coefficient = 0.425

Distance • Yards	Muzzle	100	200	300	400	500	600	800	1000
Velocity • fps	2900	2682	2474	2275	2086	1905	1737	1435	1200
Energy • ft-lbs	2055	1757	1495	1265	1063	888	737	503	352
Taylor KO Index	11.7	10.8	10.0	9.2	8.4	7.7	7.0	5.8	4.8
Path • Inches	-1.5	+1.7	0.0	-7.4	-21.5	-43.6	-75.3	-175.8	-344.4
Wind Drift • Inches	0.0	0.7	3.0	7.1	13.1	21.4	32.2	62.8	107.1

Nosler 115-grain Ballistic Tip + P (11064)

G1 Ballistic Coefficient = 0.454

Distance • Yards	Muzzle	100	200	300	400	500	600	800	1000
Velocity • fps	2800	2599	2407	2224	2048	1882	1726	1443	1218
Energy • ft-lbs	2002	1725	1480	1262	1071	904	761	532	379
Taylor KO Index	11.8	11.0	10.2	9.4	8.6	7.9	7.3	6.1	5.1
Path • Inches	-1.5	+1.8	0.0	-7.8	-22.7	-45.8	-78.8	-182.0	-352.9
Wind Drift • Inches	0.0	0.7	3.0	6.9	12.8	20.8	31.2	60.5	102.7

Hornady 117-grain SST + P (81353)

G1 Ballistic Coefficient = 0.390

Distance • Yards	Muzzle	100	200	300	400	500	600	800	1000
Velocity • fps	2946	2705	2478	2265	2057	1863	1683	1368	1137
Energy • ft-lbs	2253	1901	1595	1329	1099	902	736	486	336
Taylor KO Index	12.0	11.0	10.1	9.2	8.4	7.6	6.8	5.6	4.6
Path • Inches	-1.5	+1.6	0.0	-7.3	-21.4	-43.8	-76.1	-80.7	-360.4
Wind Drift • Inches	0.0	0.8	3.2	7.6	14.2	23.2	35.1	69.0	118.3

Hornady 117-grain BTSP Interlock + P (8135)

G1 Ballistic Coefficient = 0.392

Distance • Yards	Muzzle	100	200	300	400	500	600	800	1000
Velocity • fps	2780	2550	2331	2122	1925	1740	1571	1281	1084
Energy • ft-lbs	2007	1689	1411	1170	963	787	641	427	306
Taylor KO Index	11.9	11.0	10.0	9.1	8.3	7.5	6.7	5.5	4.7
Path • Inches	-1.5	+1.9	0.0	-8.3	-24.4	-49.9	-86.9	-206.5	-411.3
Wind Drift • Inches	0.0	0.8	3.5	8.2	15.4	25.3	38.1	74.8	127.0

Remington 117-grain Core-Lokt Soft Point (R257)

G1 Ballistic Coefficient = 0.240

Distance • Yards	Muzzle	100	200	300	400	500	600	800	1000
Velocity • fps	2650	2291	1961	1663	1404	1199	1059	902	800
Energy • ft-lbs	1824	1363	999	718	512	373	291	211	166
Taylor KO Index	11.4	9.8	8.4	7.1	6.0	5.2	4.5	3.9	3.4
Path • Inches	-1.5	+2.6	0.0	-11.7	-36.1	-78.2	-144.9	-376.8	-779.8
Wind Drift • Inches	0.0	1.5	6.5	15.9	30.6	51.5	78.6	147.7	232.6

Winchester 117-grain Power Point +P (X257P3)

G1 Ballistic Coefficient = 0.241

Distance • Yards	Muzzle	100	200	300	400	500	600	800	1000
Velocity • fps	2780	2411	2071	1761	1488	1263	1102	925	816
Energy • ft-lbs	2009	1511	1115	806	576	415	316	222	173
Taylor KO Index	11.9	10.4	8.9	7.6	6.4	5.4	4.7	4.0	3.5
Path • Inches	-1.5	+2.6	0.0	-10.8	-33.0	-70.0	-130.8	-342.1	-701.0
Wind Drift • Inches	0.0	1.4	6.0	14.7	28.3	47.8	73.8	141.3	225.2

Federal 120-grain Nosler Partition + P (P257B)

G1 Ballistic Coefficient = 0.393

Distance • Yards	Muzzle	100	200	300	400	500	600	800	1000
Velocity • fps	2800	2570	2350	2140	1940	1760	1587	1294	1092
Energy • ft-lbs	2090	1760	1470	1220	1005	820	671	446	318
Taylor KO Index	12.3	11.3	10.4	9.4	8.5	7.8	7.0	5.7	4.8
Path • Inches	-1.5	+1.9	0.0	-8.2	-24.0	-49.2	-85.4	-202.7	-403.7
Wind Drift • Inches	0.0	0.8	3.5	8.1	15.2	24.9	37.6	73.8	125.5

.25 Winchester Super Short Magnum (WSSM)

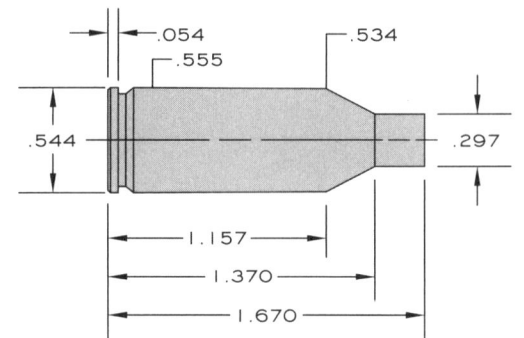

To complete their super-short line, at least for the moment, Winchester has introduced the .25 WSSM. The .25 WSSM's ballistic data are identical to the .25-06. The pressure numbers haven't been announced, but it is a good bet that the pressures will be in the 65,000-psi class.

Relative Recoil Factor = 1.57

Specifications:

Controlling Agency for Standardization of this Ammunition: SAAMI

Bullet Weight Grains	Velocity fps	Maximum Average Pressure Copper Crusher	Transducer
Data pending.			

Availability:

Winchester 85-grain Ballistic Silvertip (SBST25WSS)

G1 Ballistic Coefficient = 0.334

Distance • Yards	Muzzle	100	200	300	400	500	600	800	1000
Velocity • fps	3470	3156	2863	2589	2331	2088	1862	1462	1166
Energy • ft-lbs.	2273	1880	1548	1266	1026	823	654	404	257
Taylor KO Index	10.8	9.8	8.9	8.1	7.3	6.5	5.8	4.6	3.5
Path • Inches	-1.5	+1.0	0.0	-5.2	-15.7	-32.5	-57.5	-141.4	-294.2
Wind Drift • Inches	0.0	0.7	3.1	7.3	13.5	22.2	33.8	67.5	118.4

Winchester 110-grain AccuBond CT (S25WSCT)

G1 Ballistic Coefficient = 0.424

Distance • Yards	Muzzle	100	200	300	400	500	600	800	1000
Velocity • fps	3100	2870	2651	2442	2243	2053	1874	1548	1280
Energy • ft-lbs.	2347	2011	1716	1456	1228	1029	858	585	400
Taylor KOc Index	11.8	11.0	10.1	9.3	8.6	7.8	7.2	5.9	4.9
Path • Inches	-1.5	+1.3	0.0	-6.3	-18.5	-37.6	-64.9	-151.9	-299.4
Wind Drift • Inches	0.0	0.7	2.8	6.5	12.0	19.6	29.5	57.5	98.6

Winchester 115-grain Ballistic Silvertip (SBST25WSSA)

G1 Ballistic Coefficient = 0.447

Distance • Yards	Muzzle	100	200	300	400	500	600	800	1000
Velocity • fps	3060	2844	2639	2442	2254	2074	1905	1593	1330
Energy • ft-lbs.	2392	2066	1778	1523	1298	1099	927	648	452
Taylor KOc Index	12.9	12.0	11.1	10.3	9.5	8.8	8.0	6.7	5.6
Path • Inches	-1.5	+1.4	0.0	-6.4	-18.6	-37.7	-64.9	-150.2	-292.4
Wind Drift • Inches	0.0	0.6	2.7	6.2	11.4	18.6	27.9	54.0	92.2

Winchester 120-grain Positive Expanding Point (X25WSS)

G1 Ballistic Coefficient = 0.345

Distance • Yards	Muzzle	100	200	300	400	500	600	800	1000
Velocity • fps	2990	2717	2459	2216	1987	1773	1579	1254	1050
Energy • ft-lbs.	2383	1967	1612	1309	1053	838	664	419	294
Taylor KO Index	13.2	12.0	10.8	9.8	8.8	7.8	7.0	5.5	4.6
Path • Inches	-1.5	+1.6	0.0	-7.4	-21.8	-45.1	-79.7	-195.8	-404.1
Wind Drift • Inches	0.0	0.9	3.6	8.6	16.1	26.5	40.4	80.4	137.9

.25-06 Remington

The history of the conversion of wildcat cartridges to factory numbers is rather spotty. Part of this is because, until recently, wildcatters almost never had a numerical way to measure pressures and kept loading the charges (pressures) higher and higher until something looked like it was going to let go. If an ammo factory decided to make the wildcat cartridge into a standard caliber, the factory velocities (at rational pressures) were almost always substantially lower than the wildcatter's claims. In 1969 Remington made the .25-06 into a factory cartridge. As a wildcat, it had been around since the early 1920s. A few of today's .25-06 factory loadings list velocities for the 117-grain bullet in excess of 3,000 fps. By comparison, some old wildcat data list velocities well over 3,200 fps with the same bullets.

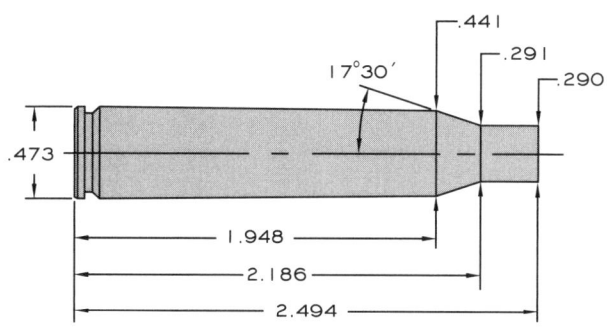

Relative Recoil Factor = 1.57

Specifications:

Controlling Agency for Standardization of this Ammunition: SAAMI

Bullet Weight Grains	Velocity fps	Maximum Average Pressure Copper Crusher	Transducer
87	3,420	53,000 cup	63,000 psi
90	3,420	53,000 cup	63,000 psi
100	3,210	53,000 cup	63,000 psi
117	2,975	53,000 cup	63,000 psi
120	2,975	53,000 cup	63,000 psi

Standard barrel for velocity testing: 24 inches long—1 turn in 10-inch twist

Availability:

Federal 85-grain Nosler Ballistic Tip (P2506G)

G1 Ballistic Coefficient = 0.329

Distance • Yards	Muzzle	100	200	300	400	500	600	800	1000
Velocity • fps	3550	3230	2930	2640	2380	2130	1896	1485	1178
Energy • ft-lbs	2380	1965	1615	1320	1070	855	679	416	262
Taylor KO Index	11.1	10.1	9.1	8.2	7.4	6.6	5.9	4.6	3.7
Path • Inches	-1.5	+0.9	0.0	-5.0	-15.0	-31.2	-55.0	-134.6	-278.5
Wind Drift • Inches	0.0	0.7	3.0	7.2	13.4	21.9	33.3	66.6	117.2

Winchester 85-grain Ballistic Silvertip (SBST2506A)

G1 Ballistic Coefficient = 0.334

Distance • Yards	Muzzle	100	200	300	400	500	600	800	1000
Velocity • fps	3470	3156	2864	2590	2332	2089	1862	1462	1166
Energy • ft-lbs	2273	1880	1548	1266	1026	824	654	404	257
Taylor KO Index	10.8	9.8	8.9	8.1	7.3	6.5	5.8	4.6	3.6
Path • Inches	-1.5	+1.0	0.0	-5.2	-15.7	-32.5	-57.5	-140.2	-289.0
Wind Drift • Inches	0.0	0.7	3.1	7.3	13.5	22.2	33.8	67.5	118.4

Winchester 90-grain Positive Expanding Point (X25061)

G1 Ballistic Coefficient = 0.260

Distance • Yards	Muzzle	100	200	300	400	500	600	800	1000
Velocity • fps	3440	3043	2680	2344	2034	1749	1498	1126	947
Energy • ft-lbs	2364	1850	1435	1098	827	611	449	254	179
Taylor KO Index	11.4	10.1	8.9	7.7	6.7	5.8	4.9	3.7	3.1
Path • Inches	-1.5	+1.4	0.0	-6.0	-18.4	-39.3	-71.5	-188.3	-414.3
Wind Drift • Inches	0.0	1.0	4.1	9.8	18.6	31.3	48.6	100.0	172.5

Nosler 100-grain Partition (60005)

G1 Ballistic Coefficient = 0.384

Distance • Yards	Muzzle	100	200	300	400	500	600	800	1000
Velocity • fps	3300	3036	2787	2552	2328	2117	1918	1557	1266
Energy • ft-lbs	2418	2046	1725	1446	1203	995	817	539	356
Taylor KO Index	12.1	11.1	10.2	9.4	8.5	7.8	7.0	5.7	4.6
Path • Inches	-1.5	+1.1	0.0	-5.6	-16.6	-34.0	-59.2	-140.1	-279.2
Wind Drift • Inches	0.0	0.7	2.8	6.6	12.3	20.0	30.2	59.4	102.6

Remington 100-grain Core-Lokt Pointed Soft Point (R25062)

G1 Ballistic Coefficient = 0.293

Distance • Yards	Muzzle	100	200	300	400	500	600	800	1000
Velocity • fps	3230	2893	2580	2287	2014	1762	1536	1181	989
Energy • ft-lbs	2316	1858	1478	1161	901	689	524	310	217
Taylor KO Index	12.0	10.6	9.5	8.4	7.4	6.5	5.6	4.3	3.6
Path • Inches	-1.5	+1.3	0.0	-6.6	-19.8	-41.7	-74.9	-190.2	-405.5
Wind Drift • Inches	0.0	0.9	3.9	9.3	17.6	29.2	45.0	91.2	157.2

Federal 100-grain Barnes Triple-Shock X-Bullet (P2506H)

G1 Ballistic Coefficient = 0.421

Distance • Yards	Muzzle	100	200	300	400	500	600	800	1000
Velocity • fps	3210	2970	2750	2540	2330	2140	1956	1619	1337
Energy • ft-lbs	2290	1965	1680	1430	1205	1015	850	582	397
Taylor KO Index	11.8	10.9	10.1	9.3	8.6	7.9	7.2	5.9	4.9
Path • Inches	-1.5	+1.2	0.0	-5.8	-17.0	-34.8	-59.9	-139.6	-273.1
Wind Drift • Inches	0.0	0.6	2.7	6.2	11.4	18.6	28.0	54.5	93.5

CorBon 100-grain T-DPX (DPX2506100/20)

G1 Ballistic Coefficient = 0.421

Distance • Yards	Muzzle	100	200	300	400	500	600	800	1000
Velocity • fps	3200	2964	2740	2528	2325	2131	1947	1611	1330
Energy • ft-lbs	2274	1951	1668	1419	1200	1009	842	576	393
Taylor KO Index	11.7	10.8	10.0	9.2	8.5	7.8	7.2	5.9	4.9
Path • Inches	-1.5	+1.2	0.0	-5.9	-17.2	-35.0	-60.4	-140.8	-275.5
Wind Drift • Inches	0.0	0.7	2.7	6.2	11.5	18.7	28.2	54.8	94.2

Federal 100-grain Nosler Ballistic Tip (P2506D)

G1 Ballistic Coefficient = 0.393

Distance • Yards	Muzzle	100	200	300	400	500	600	800	1000
Velocity • fps	3220	2970	2730	2500	2290	2080	1887	1538	1258
Energy • ft-lbs	2300	1955	1655	1390	1160	965	791	526	352
Taylor KO Index	11.8	10.9	10.0	9.2	8.4	7.6	6.9	5.6	4.6
Path • Inches	-1.5	+1.2	0.0	-5.9	-17.3	-35.5	-61.8	-145.8	-289.3
Wind Drift • Inches	0.0	0.7	2.8	6.7	12.4	20.2	30.4	59.7	103.0

Black Hills 100-grain Barnes Triple Shock (2C2506BHGN2)

G1 Ballistic Coefficient = 0.420

Distance • Yards	Muzzle	100	200	300	400	500	600	800	1000
Velocity • fps	3200	2964	2740	2527	2323	2129	1945	1609	1328
Energy • ft-lbs	2273	1951	1667	1418	1199	1007	840	575	391
Taylor KO Index	11.8	10.9	10.1	9.3	8.6	7.9	7.2	5.9	4.9
Path • Inches	-1.5	+1.2	0.0	-5.9	-17.2	-35.0	-60.4	-141.3	-278.2
Wind Drift • Inches	0.0	0.6	2.7	6.2	11.5	18.8	28.2	55.0	94.4

Winchester 90-grain Positive Expanding Point (X25061)

G1 Ballistic Coefficient = 0.260

Distance • Yards	Muzzle	100	200	300	400	500	600	800	1000
Velocity • fps	3440	3043	2680	2344	2034	1749	1498	1126	947
Energy • ft-lbs	2364	1850	1435	1098	827	611	449	254	179
Taylor KO Index	11.4	10.1	8.9	7.7	6.7	5.8	4.9	3.7	3.1
Path • Inches	-1.5	+1.4	0.0	-6.0	-18.4	-39.3	-71.5	-188.3	-414.3
Wind Drift • Inches	0.0	1.0	4.1	9.8	18.6	31.3	48.6	100.0	172.5

Nosler 100-grain E-Tip (60009) LF

G1 Ballistic Coefficient = 0.416

Distance • Yards	Muzzle	100	200	300	400	500	600	800	1000
Velocity • fps	3200	2961	2735	2521	2316	2121	1938	1598	1317
Energy • ft-lbs	2274	1947	1661	1411	1191	999	832	567	385
Taylor KO Index	11.7	10.9	10.0	9.3	8.5	7.8	7.1	5.9	4.8
Path • Inches	-1.5	+1.2	0.0	-5.9	-17.3	-35.1	-60.8	-141.8	-278.1
Wind Drift • Inches	0.0	0.6	2.7	6.3	11.7	19.0	28.5	55.6	95.6

Winchester 110-grain AccuBond CT (S2506CT)

G1 Ballistic Coefficient = 0.424

Distance • Yards	Muzzle	100	200	300	400	500	600	800	1000
Velocity • fps	3100	2870	2651	2442	2243	2053	1874	1548	1260
Energy • ft-lbs	2347	2011	1716	1456	1228	1029	858	585	400
Taylor KO Index	11.8	11.0	10.1	9.3	8.6	7.8	7.2	5.9	4.8
Path • Inches	-1.5	+1.3	0.0	-6.3	-18.5	-37.6	-64.9	-151.9	-299.4
Wind Drift • Inches	0.0	0.7	2.8	6.5	12.0	19.6	29.5	57.5	98.6

Federal 110-grain Nosler Ballistic Tip (P2506D)

G1 Ballistic Coefficient = 0.393

Distance • Yards	Muzzle	100	200	300	400	500	600	800	1000
Velocity • fps	3220	2970	2730	2500	2290	2080	1886	1537	1256
Energy • ft-lbs	2300	1955	1655	1390	1160	965	790	427	350
Taylor KO Index	11.8	10.9	10.0	9.1	8.4	7.6	6.9	5.6	4.6
Path • Inches	-1.5	+1.2	0.0	-5.9	-17.3	-35.5	-61.8	-145.9	-289.7
Wind Drift • Inches	0.0	0.7	2.8	6.7	12.4	20.2	30.5	59.7	103.6

Remington 115-grain Core-Lokt Ultra Bonded (PRC2506RA)

G1 Ballistic Coefficient = 0.380

Distance • Yards	Muzzle	100	200	300	400	500	600	800	1000
Velocity • fps	3000	2751	2516	2293	2081	1881	1693	1368	1131
Energy • ft-lbs	2298	1933	1616	1342	1106	903	732	478	327
Taylor KO Index	12.0	11.0	10.0	9.2	8.3	7.5	6.8	5.5	4.5
Path • Inches	-1.5	+1.4	0.0	-7.1	-20.7	-42.5	-74.2	-177.8	-359.4
Wind Drift • Inches	0.0	0.8	3.2	7.6	14.2	23.3	35.3	69.6	118.8

Winchester 115-grain Ballistic Silvertip (SBST2506)

G1 Ballistic Coefficient = 0.449

Distance • Yards	Muzzle	100	200	300	400	500	600	800	1000
Velocity • fps	3060	2844	2639	2442	2254	2074	1903	1590	1326
Energy • ft-lbs	2392	2066	1778	1523	1298	1099	925	646	449
Taylor KO Index	12.2	11.4	10.5	9.7	9.0	8.3	7.6	6.3	5.3
Path • Inches	-1.5	+1.3	0.0	-6.4	-18.6	-37.7	-65.0	-150.4	-293.0
Wind Drift • Inches	0.0	0.6	2.7	6.2	11.5	18.6	28.0	54.2	92.6

Federal 115-grain Nosler Partition (P2506E)

G1 Ballistic Coefficient = 0.391

Distance • Yards	Muzzle	100	200	300	400	500	600	800	1000
Velocity • fps	3030	2790	2550	2330	2120	1930	1744	1417	1170
Energy • ft-lbs	2345	1980	1665	1390	1150	945	777	513	350
Taylor KO Index	12.7	11.7	10.7	9.8	8.9	8.1	7.3	6.0	4.9
Path • Inches	-1.5	+1.5	0.0	-6.8	-20.1	-40.9	-71.3	-169.0	-336.7
Wind Drift • Inches	0.0	0.8	3.1	7.3	13.6	22.2	33.6	66.0	113.6

Black Hills 117-grain Hornady SST (1C2506BHGN3)

G1 Ballistic Coefficient = 0.390

Distance • Yards	Muzzle	100	200	300	400	500	600	800	1000
Velocity • fps	3150	2899	2662	2437	2223	2020	1828	1486	1218
Energy • ft-lbs	2577	2184	1841	1543	1284	1060	869	574	386
Taylor KO Index	13.5	12.5	11.4	10.5	9.5	8.7	7.9	6.4	5.2
Path • Inches	-1.5	+1.3	0.0	-6.2	-18.4	-37.5	-65.2	-154.4	-307.2
Wind Drift • Inches	0.0	0.7	3.0	6.9	12.8	21.0	31.7	62.4	107.6

Hornady 117-grain SST (81453)

G1 Ballistic Coefficient = 0.390

Distance • Yards	Muzzle	100	200	300	400	500	600	800	1000
Velocity • fps	3110	2861	2626	2403	2191	1989	1800	1463	1207
Energy • ft-lbs	2512	2127	1792	1500	1246	1028	842	556	375
Taylor KO Index	13.4	12.1	11.3	10.3	9.4	8.5	7.7	6.3	5.2
Path • Inches	-1.5	+1.6	0.0	-6.4	-18.9	-38.6	-67.2	-159.1	-316.8
Wind Drift • Inches	0.0	0.7	3.0	7.0	13.1	21.4	32.3	63.9	109.6

Federal 117-grain Sierra GameKing BTSP (P2506C)

G1 Ballistic Coefficient = 0.413

Distance • Yards	Muzzle	100	200	300	400	500	600	800	1000
Velocity • fps	3030	2800	2580	2370	2170	1980	1802	1481	1226
Energy • ft-lbs	2385	2035	1725	1455	1220	1015	843	570	391
Taylor KO Index	13.0	12.0	11.0	10.1	9.3	8.5	7.7	6.3	5.2
Path • Inches	-1.5	+1.5	0.0	-6.7	-19.7	-40.0	-69.2	-162.1	-319.2
Wind Drift • Inches	0.0	0.7	2.9	6.9	12.7	20.8	31.3	61.2	105.0

Federal 117-grain Soft Point (2506BS)

G1 Ballistic Coefficient = 0.360

Distance • Yards	Muzzle	100	200	300	400	500	600	800	1000
Velocity • fps	3030	2770	2520	2280	2060	1850	1656	1324	1097
Energy • ft-lbs	2385	1990	1650	1355	1105	890	713	456	313
Taylor KO Index	13.0	11.9	10.9	9.8	8.8	7.9	7.1	5.7	4.7
Path • Inches	-1.5	+1.5	0.0	-7.0	-20.8	-42.5	-74.9	-180.5	-366.2
Wind Drift • Inches	0.0	0.8	3.4	8.0	14.9	24.5	37.3	73.9	127.4

Hornady 117-grain BTSP InterLock (8145)

G1 Ballistic Coefficient = 0.391

Distance • Yards	Muzzle	100	200	300	400	500	600	800	1000
Velocity • fps	2990	2749	2520	2302	2096	1900	1716	1394	1155
Energy • ft-lbs	2322	1962	1649	1377	1141	938	765	505	347
Taylor KO Index	12.8	11.8	10.8	9.9	9.0	8.2	7.4	6.0	5.0
Path • Inches	-1.5	+1.6	0.0	-7.0	-20.7	-42.2	-73.5	-174.3	-347.4
Wind Drift • Inches	0.0	0.8	3.2	7.4	13.8	22.6	34.2	67.3	115.7

Hornady 117-grain SST InterLock (81454)

G1 Ballistic Coefficient = 0.390

Distance • Yards	Muzzle	100	200	300	400	500	600	800	1000
Velocity • fps	2990	2748	2519	2312	2093	1897	1713	1391	1152
Energy • ft-lbs	2322	1962	1648	1375	1138	935	762	503	345
Taylor KO Index	12.8	11.8	10.8	9.9	9.0	8.1	7.4	6.0	4.9
Path • Inches	-1.5	+1.7	0.0	-7.4	-21.5	-47.8	-73.7	-175.6	-352.7
Wind Drift • Inches	0.0	0.8	3.2	7.5	13.9	22.7	34.4	67.6	116.1

Remington 120-grain Pointed Soft Point Core-Lokt (R25063)

G1 Ballistic Coefficient = 0.363

Distance • Yards	Muzzle	100	200	300	400	500	600	800	1000
Velocity • fps	2990	2730	2484	2252	2032	1825	1635	1310	1090
Energy • ft-lbs	2383	1995	1644	1351	1100	887	713	458	316
Taylor KO Index	13.2	12.0	10.9	9.9	9.0	8.0	7.2	5.8	4.8
Path • Inches	-1.5	+1.6	0.0	-7.2	-21.4	-44.1	-76.9	-185.4	-375.5
Wind Drift • Inches	0.0	0.8	3.4	8.1	15.1	24.9	37.7	74.8	128.5

Winchester 120-grain Positive Expanding Point (X25062)

G1 Ballistic Coefficient = 0.347

Distance • Yards	Muzzle	100	200	300	400	500	600	800	1000
Velocity • fps	2990	2717	2459	2216	1987	1773	1578	1252	1047
Energy • ft-lbs	2382	1967	1612	1309	1053	838	663	418	292
Taylor KO Index	13.2	12.0	10.9	9.9	9.0	8.0	7.2	5.8	4.8
Path • Inches	-1.5	+1.5	0.0	-7.4	-21.8	-45.1	-76.8	-196.0	-404.8
Wind Drift • Inches	0.0	0.9	3.6	8.6	16.1	26.5	40.4	80.6	138.2

Federal 120-grain Fusion (F2506FS1)

G1 Ballistic Coefficient = 0.467

Distance • Yards	Muzzle	100	200	300	400	500	600	800	1000
Velocity • fps	2980	2780	2590	2400	2220	2050	1888	1590	1338
Energy • ft-lbs	2365	2055	1780	1535	1315	1120	950	674	477
Taylor KO Index	13.1	12.2	11.4	10.6	9.8	9.0	8.3	7.0	5.9
Path • Inches	-1.5	+1.5	0.0	-6.7	-19.5	-39.3	-67.5	-154.9	-297.8
Wind Drift • Inches	0.0	0.6	2.6	6.1	11.3	18.3	27.4	53.0	90.1

6.53mm (.257) Lazzeroni Scramjet

The 6.53 mm Scramjet is the .25-caliber entry in the Lazzeroni line of high-performance cartridges. The cases in the Lazzeroni line are not based on any existing cartridge. Three case-head sizes are used. (See the case drawings for details.) With a muzzle velocity of 3,750 fps, the 6.53 slightly exceeds the performance of the .257 Weatherby in this one specific loading.

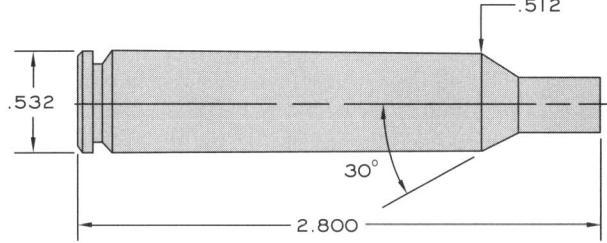

Relative Recoil Factor = 1.69

Specifications:

Controlling Agency for Standardization of this Ammunition: Factory

Availability:

Lazzeroni 100-grain Nosler Partition (653SJ100P)

G1 Ballistic Coefficient = 0.454

Distance • Yards	Muzzle	100	200	300	400	500	600	800	1000
Velocity • fps	3750	3501	3266	3044	2833	2631	2437	2076	1749
Energy • ft-lbs	3123	2722	2370	2058	1782	1537	1319	957	679
Taylor KO Index	13.8	12.9	12.0	11.2	10.4	9.7	8.9	7.6	6.4
Path • Inches	-1.5	+0.7	0.0	-3.9	-11.6	-23.7	-40.7	-93.5	-178.8
Wind Drift • Inches	0.0	0.5	2.0	4.7	8.6	13.9	20.7	39.4	66.7

This Chobe bushbuck was shot in Mozambique with a Ruger No. 1 in .25-06. It performed perfectly, as seen with this truly exceptional trophy. (Photo from *Safari Rifles II* by Craig Boddington, 2009, Safari Press Archives)

.257 Weatherby Magnum

If you want the highest velocity from any .25-caliber gun firing factory ammunition, the .257 Weatherby is the gun for you. Late in the WWII time frame, Roy Weatherby shortened a .300 magnum case and necked it to .25 caliber, thereby creating a case with significantly more volume than the .25-06. All other things being equal, a larger case volume translates directly into higher velocity. With the same bullet, the .257 Weatherby Magnum is about 200 fps faster than the .25-06 and at least 400 fps faster than the .257 Roberts. The .257 Weatherby is certainly the king of the .25-caliber hill.

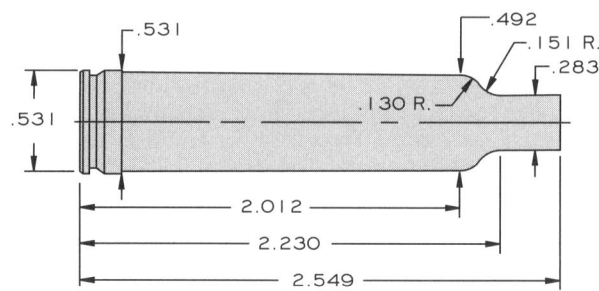

Relative Recoil Factor = 1.76

Specifications:

Controlling Agency for Standardization of this Ammunition: CIP

Bullet Weight Grains	Velocity fps	Maximum Average Pressure	
		Copper Crusher	Transducer
N/S	N/S	3,800 bar	4,370 bar

Standard barrel for velocity testing: 26 inches long—1 turn in 10-inch twist

Availability:

Weatherby 80-grain Barnes TTSX (B25780TTSX) LF
G1 Ballistic Coefficient = 0.378

Distance • Yards	Muzzle	100	200	300	400	500	600	800	1000
Velocity • fps	3870	3561	3274	3005	2753	2514	2290	1873	1511
Energy • ft-lbs	2661	2254	1906	1606	1348	1125	931	623	406
Taylor KO Index	11.4	10.5	9.6	8.8	8.1	7.4	6.7	5.5	4.4
Path • Inches	-1.5	+0.6	0.0	-3.9	-11.6	23.9	41.7	98.8.	196.2
Wind Drift • Inches	0.0	0.6	2.4	5.6	10.5	16.7	25.1	48.8	84.4

Weatherby 87-grain Pointed-Expanding (H 257 87 SP)
G1 Ballistic Coefficient = 0.323

Distance • Yards	Muzzle	100	200	300	400	500	600	800	1000
Velocity • fps	3825	3472	3147	2845	2563	2297	2051	1606	1255
Energy • ft-lbs	2826	2328	1913	1563	1269	1019	813	498	304
Taylor KO Index	12.2	11.1	10.1	9.1	8.2	7.3	6.6	5.1	4.0
Path • Inches	-1.5	+0.7	0.0	-4.2	-12.7	-26.6	-47.0	-115.0	-238.1
Wind Drift • Inches	0.0	0.7	2.9	6.7	12.4	20.4	30.9	61.5	108.6

Weatherby 100-grain Pointed-Expanding (H 257 100 SP)
G1 Ballistic Coefficient = 0.358

Distance • Yards	Muzzle	100	200	300	400	500	600	800	1000
Velocity • fps	3602	3298	3016	2750	2500	2266	2043	1638	1309
Energy • ft-lbs	2881	2416	2019	1515	1260	1040	927	596	380
Taylor KO Index	13.2	12.1	11.1	10.1	9.2	8.3	7.5	6.0	4.8
Path • Inches	-1.5	+0.8	0.0	-4.7	-14.0	-28.8	-50.5	-120.9	-244.1
Wind Drift • Inches	0.0	0.7	2.7	6.4	11.9	19.4	29.3	62.2	100.8

Weatherby 100-grain Spitzer (G257100SR)
G1 Ballistic Coefficient = 0.256

Distance • Yards	Muzzle	100	200	300	400	500	600	800	1000
Velocity • fps	3500	3091	2718	2375	2057	1766	1505	1126	944
Energy • ft-lbs	2721	2122	1641	1253	940	693	5.3	282	198
Taylor KO Index	12.9	11.3	10.0	8.7	7.6	6.5	5.5	4.1	3.5
Path • Inches	-1.5	+1.1	0.0	-5.8	-17.8	-38.2	-69.8	-184.9	-409.2
Wind Drift • Inches	0.0	1.0	4.1	9.8	18.6	31.3	48.6	100.4	173.7

Weatherby 100-grain Barnes TSX (B257100TSX) LF
G1 Ballistic Coefficient = 0.420

Distance • Yards	Muzzle	100	200	300	400	500	600	800	1000
Velocity • fps	3570	3312	3079	2840	2623	2416	2216	1850	1527
Energy • ft-lbs	2731	2437	2094	1793	1528	1296	1092	760	518
Taylor KO Index	14.4	13.4	12.4	11.5	10.6	9.8	9.0	7.5	6.2
Path • Inches	-1.5	+0.8	0.0	-4.5	-13.4	-27.3	-47.2	-109.6.	-213.1
Wind Drift • Inches	0.0	0.6	2.3	5.4	10.0	16.2	24.2	48.8	80.1

Nosler 100-grain Ballistic Tip (120)
G1 Ballistic Coefficient = 0.393

Distance • Yards	Muzzle	100	200	300	400	500	600	800	1000
Velocity • fps	3500	3230	2976	2737	2510	2294	2089	1712	1394
Energy • ft lbs	2721	2317	1968	1664	1399	1169	969	651	431
Taylor KO Index	12.9	11.9	10.9	10.0	9.2	8.4	7.7	6.3	5.1
Path • Inches	-1.5	+0.9	0.0	-4.8	-14.4	-29.4	-51.1	-120.1	-237.3
Wind Drift • Inches	0.0	0.6	2.6	6.0	11.0	18.0	27.0	52.7	91.0

Weatherby 110-grain Nosler AccuBond (N 257 110 ACB)
G1 Ballistic Coefficient = 0.418

Distance • Yards	Muzzle	100	200	300	400	500	600	800	1000
Velocity • fps	3460	3207	2969	2744	2529	2325	2130	1770	1458
Energy • ft-lbs	2925	2513	2154	1839	1563	1320	1109	765	519
Taylor KO Index	14.0	13.0	12.0	11.1	10.2	9.4	8.6	7.1	5.9
Path • Inches	-1.5	+0.9	0.0	-4.9	-14.4	-29.4	-50.8	-118.2	-230.6
Wind Drift • Inches	0.0	0.6	2.4	5.7	10.5	17.0	25.4	49.3	84.6

Nosler 110-grain AccuBond (60012)
G1 Ballistic Coefficient = 0.425

Distance • Yards	Muzzle	100	200	300	400	500	600	800	1000
Velocity • fps	3400	3155	2924	2705	2496	2297	2107	1755	1450
Energy • ft-lbs	2823	2431	2088	1787	1522	1289	1084	752	514
Taylor KO Index	13.7	12.7	11.8	10.9	10.1	9.3	8.5	7.1	5.9
Path • Inches	-1.5	+1.0	0.0	-5.1	-14.9	-30.3	-52.4	-121.5.	-236.4
Wind Drift • Inches	0.0	0.6	2.4	5.7	10.5	17.0	25.5	49.4	84.6

Weatherby 115-grain BST (N 257 115 BS)
G1 Ballistic Coefficient = 0.453

Distance • Yards	Muzzle	100	200	300	400	500	600	800	1000
Velocity • fps	3400	3170	2952	2745	2547	2357	2175	1837	1537
Energy • ft-lbs	2952	2566	2226	1924	1656	1419	1208	862	604
Taylor KO Index	14.4	13.4	12.5	11.6	10.8	10.0	9.2	7.8	6.5
Path • Inches	-1.5	+1.0	0.0	-5.0	-14.6	-29.6	-50.8	-116.7	-224.3
Wind Drift • Inches	0.0	0.6	2.3	5.3	9.8	15.8	23.7	45.4	77.2

Weatherby 117-grain Round Nose-Expanding (H 257 117 RN)
G1 Ballistic Coefficient = 0.256

Distance • Yards	Muzzle	100	200	300	400	500	600	800	1000
Velocity • fps	3402	2984	2595	2240	1921	1639	1389	1056	902
Energy • ft-lbs	3007	2320	1742	1302	956	690	502	290	212
Taylor KO Index	14.6	12.8	11.1	9.6	8.3	7.0	6.0	4.5	3.9
Path • Inches	-1.5	+1.2	0.0	-6.4	-19.7	-42.4	-78.1	-210.3	-466.4
Wind Drift • Inches	0.0	1.1	4.5	10.8	20.6	34.7	54.2	111.6	189.6

Weatherby 120-grain Nosler Partition (N 257 120 PT)
G1 Ballistic Coefficient = 0.392

Distance • Yards	Muzzle	100	200	300	400	500	600	800	1000
Velocity • fps	3305	3046	2801	2570	2350	2141	1944	1586	1293
Energy • ft-lbs	2910	2472	2091	1760	1471	1221	1088	671	446
Taylor KO Index	14.6	13.4	12.3	11.3	10.4	9.4	8.6	7.0	5.7
Path • Inches	-1.5	+1.1	0.0	-5.6	-16.4	-33.6	-58.3	-137.5	-272.8
Wind Drift • Inches	0.0	0.7	2.8	6.5	12.0	19.5	29.4	57.6	99.6

Nosler 120-grain Partition (12057)
G1 Ballistic Coefficient = 0.391

Distance • Yards	Muzzle	100	200	300	400	500	600	800	1000
Velocity • fps	3135	2886	2650	2426	2213	2011	1820	1481	1215
Energy • ft-lbs	2619	2219	1871	1569	1305	1078	883	584	393
Taylor KO Index	13.8	12.7	11.7	10.7	9.7	8.9	8.0	6.5	5.4
Path • Inches	-1.5	+1.3	0.0	-6.3	-18.5	-37.9	-65.9	-155.8	-310.0
Wind Drift • Inches	0.0	0.7	3.0	7.0	12.9	21.1	31.8	62.6	108.0

6.5x57 Mauser

Even Paul Mauser couldn't resist the temptation to neck down and neck up cartridge designs. The 6.5x57 is nothing more or less than a 7x57 necked to 6.5mm. The case can be made from .257 Roberts cases and vice versa. This is another cartridge that remains reasonably popular in Europe but is seldom seen in the U.S. The cartridge appeared in the mid-1890s, but there doesn't seem to be any record of it ever being adopted as a military number. Performance-wise, it is a hair more powerful than the 6.5x55 MS, but the difference isn't anything to write home about. Because of its early origins, the working pressures are held very much on the mild side.

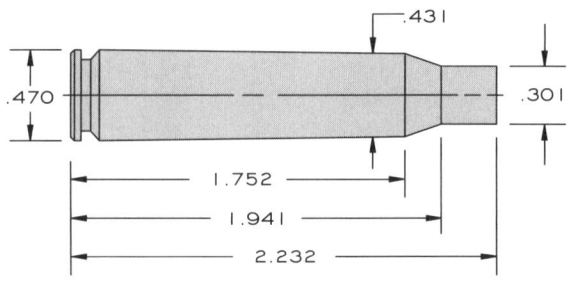

Relative Recoil Factor = 1.62

Specifications:

Controlling Agency for Standardization of this Ammunition: CIP

Bullet Weight Grains	Velocity fps	Maximum Average Pressure	
		Copper Crusher	Transducer
	N/S	(data not available)	

Standard barrel for velocity testing: N/S—1 turn in 7.87-inch twist

Availability:

Sellier & Bellot 131-grain SP (V330622U)   G1 Ballistic Coefficient = 0.343

Distance • Yards	Muzzle	100	200	300	400	500	600	800	1000
Velocity • fps	2543	2295	2060	1841	1638	1455	1296	1073	949
Energy • ft-lbs	1882	1532	1235	986	780	616	489	335	262
Taylor KO Index	12.6	11.3	10.2	9.1	8.1	7.2	6.4	5.3	4.7
Path • Inches	-1.5	+2.6	0.0	-10.8	-31.9	-66.0	-116.6	-284.8	-573.4
Wind Drift • Inches	0.0	1.1	4.6	11.0	20.6	34.1	51.8	100.6	164.3

6.5x55mm Swedish Mauser

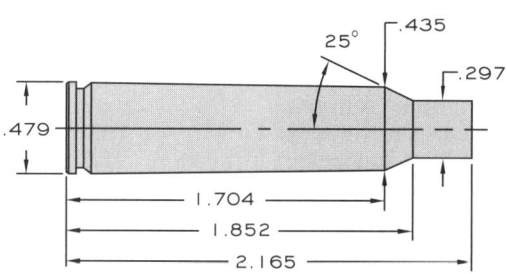

One of the oldest cartridges in the factory inventory, the 6.5x55mm Swedish Mauser was adopted as a military cartridge over 100 years ago. The cartridge has been very popular in Europe and still ranks somewhere like 15th in the list of American reloading die sales (only a couple places behind the .30-30). Performance-wise, this cartridge is virtually identical to the .257 Roberts. Because it is used in many military surplus guns with a wide range of strength characteristics, the factory specifications have been set very much on the mild side.

Relative Recoil Factor = 1.72

Specifications:

Controlling Agency for Standardization of this Ammunition: SAAMI

Bullet Weight Grains	Velocity fps	Maximum Average Pressure Copper Crusher	Transducer
160	2,380	46,000 cup	N/S

Standard barrel for velocity testing: 24 inches long—1 turn in 7.87-inch twist

Availability:

Federal 93-grain Sierra MatchKing BTHP (GM655M2)

G1 Ballistic Coefficient = 0.314

Distance • Yards	Muzzle	100	200	300	400	500	600	800	1000
Velocity • fps	2625	2350	2090	1850	1630	1430	1263	1041	921
Energy • ft-lbs	1425	1140	905	705	550	425	329	224	175
Taylor KO Index	9.2	8.2	7.3	6.5	5.7	5.0	4.4	3.7	3.2
Path • Inches	-1.5	+2.4	0.0	-10.3	-31.1	-64.8	-115.8	-288.3	-588.7
Wind Drift • Inches	0.0	1.2	4.9	11.6	21.9	36.4	55.7	108.5	176.6

Norma 100-grain Hollow Point (16527)

G1 Ballistic Coefficient = 0.321

Distance • Yards	Muzzle	100	200	300	400	500	600	800	1000
Velocity • fps	2625	2354	2100	1862	1645	1449	1281	1053	930
Energy • ft-lbs	1530	1232	980	771	601	466	365	246	192
Taylor KO Index	9.9	8.9	7.9	7.0	6.2	5.5	4.8	4.0	3.5
Path • Inches	-1.5	+2.4	0.0	-10.3	-30.7	-64.0	-113.9	-282.5	-576.0
Wind Drift • Inches	0.0	1.1	4.8	11.3	21.4	35.5	54.2	105.7	172.7

Lapua 100-grain Scenar HPBT (4316035)

G1 Ballistic Coefficient = 0.449

Distance • Yards	Muzzle	100	200	300	400	500	600	800	1000
VelocitLy fps	2625	2426	2235	2054	1881	1718	1566	1305	1115
Energy • ft-lbs	1530	1307	1110	937	786	655	545	378	276
Taylor KO Index	9.9	9.1	8.4	7.7	7.1	6.5	5.9	4.9	4.2
Path • Inches	-1.5	+2.2	0.0	-9.2	-26.6	-53.9	-93.0	-216.8	-422.7
Wind Drift • Inches	0.0	0.8	3.4	7.9	14.6	23.9	51.2	69.8	117.6

Lapua 100-grain FMJ (4316033)

G1 Ballistic Coefficient = 0.250

Distance • Yards	Muzzle	100	200	300	400	500	600	800	1000
Velocity • fps	2725	2374	2049	1752	1491	1273	1113	935	829
Energy • ft-lbs	1649	1251	932	682	494	360	275	194	152
Taylor KO Index	10.3	9.0	7.7	6.6	5.6	4.8	4.8	3.5	3.1
Path • Inches	-1.5	+2.4	0.0	-10.7	-32.8	-70.7	-130.1	-339.0	-706.8
Wind Drift • Inches	0.0	1.4	6.0	14.4	27.7	46.8	71.9	137.6	219.3

Lapua 108-grain Scenar HPBT (C316031)

G1 Ballistic Coefficient = 0.458

Distance • Yards	Muzzle	100	200	300	400	500	600	800	1000
Velocity • fps	3015	2811	2612	2418	2236	2062	1896	1592	1280
Energy • ft-lbs	2180	1895	1636	1402	1199	1020	862	608	393
Taylor KO Index	12.3	11.4	10.6	9.8	9.1	8.4	7.7	6.5	5.2
Path • Inches	-1.5	+1.5	0.0	-6.6	-19.1	-38.6	-66.3	-152.6	-294.4
Wind Drift • Inches	0.0	0.6	2.6	6.2	11.4	18.4	27.6	53.4	91.0

Lapua 108-grain Scenar HPBT (4316031)

G1 Ballistic Coefficient = 0.453

Distance • Yards	Muzzle	100	200	300	400	500	600	800	1000
Velocity • fps	2953	2748	2550	2358	2177	2003	1838	1539	1238
Energy • ft-lbs	2087	1811	1590	1333	1136	962	811	568	400
Taylor KO Index	12.0	11.2	10.4	9.6	8.9	8.2	7.5	6.3	5.0
Path • Inches	-1.5	+1.6	0.0	-6.9	-20.1	-40.7	-69.9	-161.3	-312.0
Wind Drift • Inches	0.0	0.7	2.8	6.4	11.8	19.2	28.9	56.0	95.3

Norma 120-grain Nosler BT (16522)

G1 Ballistic Coefficient = 0.430

Distance • Yards	Muzzle	100	200	300	400	500	600	800	1000
Velocity • fps	2822	2609	2407	2213	2030	1855	1690	1399	1176
Energy • ft-lbs	2123	1815	1544	1305	1098	917	761	522	369
Taylor KO Index	12.8	11.8	10.9	10.0	9.2	8.4	7.6	6.3	5.3
Path • Inches	-1.5	+1.8	0.0	-7.8	-22.7	-46.1	-79.7	-186.6	-367.2
Wind Drift • Inches	0.0	0.8	3.1	7.3	13.5	22.0	33.1	64.4	109.7

Norma 120-grain Full jacket (16542)

G1 Ballistic Coefficient = 0.428

Distance • Yards	Muzzle	100	200	300	400	500	600	800	1000
Velocity • fps	2690	2238	1833	1482	1212	1040	937	801	699
Energy • ft-lbs	1929	1336	897	587	391	288	234	171	130
Taylor KO Index	12.2	10.1	8.3	6.7	5.5	4.7	4.2	3.6	3.2
Path • Inches	-1.5	+2.8	0.0	-13.3	-42.9	-96.9	-181.9	-465.6	-941.7
Wind Drift • Inches	0.0	1.9	8.3	20.8	40.6	68.4	102.4	185.4	287.5

Lapua 123-grain Scenar HPBT (4316036)

G1 Ballistic Coefficient = 0.525

Distance • Yards	Muzzle	100	200	300	400	500	600	800	1000
Velocity • fps	3015	2836	2661	2490	2328	2172	2022	1742	1457
Energy • ft-lbs	2492	2204	1941	1699	1481	1298	1117	829	582
Taylor KO Index	14.0	13.2	12.3	11.6	10.8	10.1	9.4	8.1	6.8
Path • Inches	-1.5	+1.4	0.0	-6.3	-18.3	-36.7	-62.4	-140.9	-265.5
Wind Drift • Inches	0.0	0.6	2.3	5.3	9.7	15.7	23.4	44.6	75.1

Lapua 123-grain Scenar HPBT (C316032)

G1 Ballistic Coefficient = 0.520

Distance • Yards	Muzzle	100	200	300	400	500	600	800	1000
Velocity • fps	2930	2743	2571	2401	2241	2087	1939	1664	1391
Energy • ft-lbs	2337	2063	1811	1580	1372	1190	1027	757	550
Taylor KO Index	13.5	12.7	11.9	11.1	10.4	9.7	9.0	7.7	6.5
Path • Inches	-1.5	+1.6	0.0	-6.8	-19.7	-39.6	-67.4	-152.3	-287.9
Wind Drift • Inches	0.0	0.6	2.4	5.6	10.3	16.6	24.8	47.4	79.9

Federal 123-grain Sierra MatchKing BTHP (GM655M3)

G1 Ballistic Coefficient = 0.509

Distance • Yards	Muzzle	100	200	300	400	500	600	800	1000
Velocity • fps	2750	2570	2400	2240	2080	1930	1788	1525	1304
Energy • ft-lbs	2065	1810	1580	1370	1185	1020	873	636	465
Taylor KO Index	12.8	11.9	11.1	10.4	9.6	9.0	8.3	7.1	6.0
Path • Inches	-1.5	+1.9	0.0	-7.9	-22.9	-45.5	-77.9	-176.9	-338.1
Wind Drift • Inches	0.0	0.6	2.7	6.2	11.5	18.6	27.8	53.4	90.1

Lapua 123-grain Scenar HPBT (4316032)

G1 Ballistic Coefficient = 0.514

Distance • Yards	Muzzle	100	200	300	400	500	600	800	1000
Velocity • fps	2720	2548	2379	2217	2062	1914	1773	1515	1253
Energy • ft-lbs	2028	1779	1551	1347	1162	1001	859	627	430
Taylor KO Index	12.6	11.8	11.0	10.3	9.6	8.9	8.2	7.0	5.8
Path • Inches	-1.5	+1.9	0.0	-8.1	-23.3	-46.6	-79.5	-180.3	-342.3
Wind Drift • Inches	0.0	0.6	2.7	6.3	11.6	18.7	28.0	53.6	90.3

Lapua 123-grain Scenar HPBT (C316036)

G1 Ballistic Coefficient = 0.548

Distance • Yards	Muzzle	100	200	300	400	500	600	800	1000
Velocity • fps	2723	2559	2401	2249	1961	1826	1697	1576	1360
Energy • ft-lbs	2026	1789	1575	1382	1207	1050	911	679	505
Taylor KO Index	12.6	11.9	11.1	10.4	9.1	8.5	7.9	7.3	6.3
Path • Inches	-1.5	+1.9	0.0	-7.9	-22.8	-45.5	-77.2	-173.7	-326.6
Wind Drift • Inches	0.0	0.6	2.5	5.8	10.7	17.4	25.9	49.4	82.8

Norma 130-grain Hollow Point (16500) + (16513)

G1 Ballistic Coefficient = 0.548

Distance • Yards	Muzzle	100	200	300	400	500	600	800	1000
Velocity • fps	2723	2559	2401	2249	2102	1961	1826	1576	1380
Energy • ft-lbs	2141	1891	1665	1460	1276	1110	962	717	534
Taylor KO Index	13.4	12.5	11.8	11.0	10.3	9.6	9.0	7.7	6.8
Path • Inches	-1.5	+1.9	0.0	-7.9	-22.8	-45.5	-77.2	-173.7	-326.6
Wind Drift • Inches	0.0	0.6	2.5	5.8	10.7	17.4	25.9	49.4	82.8

Sellier and Bellot 131-grain SP (V330502A)

G1 Ballistic Coefficient = 0.341

Distance • Yards	Muzzle	100	200	300	400	500	600	800	1000
Velocity • fps	2602	2336	2098	1884	1677	1488	1323	1086	956
Energy • ft-lbs	1971	1590	1282	1034	818	645	510	343	266
Taylor KO Index	12.9	11.5	10.4	9.3	8.2	7.4	6.5	5.4	4.7
Path • Inches	-1.5	+2.4	0.0	-10.3	-30.4	-63.0	-111.6	-274.9	-554.5
Wind Drift • Inches	0.0	1.1	4.5	10.7	20.1	33.2	50.6	98.8	162.3

Lapua 139-grain Scenar HPBT (C316030)

G1 Ballistic Coefficient = 0.575

Distance • Yards	Muzzle	100	200	300	400	500	600	800	1000
Velocity • fps	2740	2568	2435	2289	2145	2009	1878	1634	1402
Energy • ft-lbs	2315	2062	1828	1615	1420	1246	1089	825	606
Taylor KO Index	14.4	13.6	12.8	12.0	11.2	10.5	9.8	8.6	7.3
Path • Inches	-1.5	+1.9	0.0	-7.7	-22.1	-44.1	-74.7	-167.0	-311.6
Wind Drift • Inches	0.0	0.6	2.4	5.5	10.1	16.2	24.2	45.9	76.9

Lapua 139-grain Scenar HPBT (4316030)

G1 Ballistic Coefficient = 0.575

Distance • Yards	Muzzle	100	200	300	400	500	600	800	1000
Velocity • fps	2620	2468	2320	2178	2040	1908	1782	1548	1317
Energy • ft-lbs	2117	1878	1660	1463	1285	1124	980	740	535
Taylor KO Index	13.7	12.9	12.2	11.4	10.7	10.0	9.3	8.1	6.9
Path • Inches	-1.5	+2.1	0.0	-8.6	-24.5	-48.8	-82.6	-184.8	-345.4
Wind Drift • Inches	0.0	0.6	2.5	5.9	10.8	18.0	25.8	49.2	82.1

Hornady 140-grain SST (85507)

G1 Ballistic Coefficient = 0.540

Distance • Yards	Muzzle	100	200	300	400	500	600	800	1000
Velocity • fps	2735	2569	2408	2254	2105	1961	1825	1572	1354
Energy • ft-lbs	2325	2051	1803	1579	1377	1196	1035	768	570
Taylor KO Index	14.4	13.6	12.7	11.9	11.1	10.4	9.6	8.3	7.1
Path • Inches	-1.5	+1.9	0.0	-7.9	-22.6	-45.2	-76.9	-173.2	-326.3
Wind Drift • Inches	0.0	0.6	2.5	5.9	10.8	17.5	26.1	49.9	83.8

Norma 140-grain Nosler Partition (16559)

G1 Ballistic Coefficient = 0.467

Distance • Yards	Muzzle	100	200	300	400	500	600	800	1000
Velocity • fps	2690	2500	2317	2142	1976	1817	1668	1402	1192
Energy • ft-lbs	2250	1943	1669	1427	1214	1027	865	611	442
Taylor KO Index	14.2	13.2	12.2	11.3	10.4	9.6	8.8	7.4	6.3
Path • Inches	-1.5	+2.0	0.0	-8.5	-24.6	-49.6	-85.2	-196.9	-381.6
Wind Drift • Inches	0.0	0.7	3.1	7.1	13.1	21.4	32.1	62.0	104.7

Federal 140-grain Soft Point (6555B)

G1 Ballistic Coefficient = 0.441

Distance • Yards	Muzzle	100	200	300	400	500	600	800	1000
Velocity • fps	2650	2450	2260	2080	1900	1740	1588	1323	1128
Energy • ft-lbs	2185	1865	1585	1340	1120	935	784	544	396
Taylor KO Index	14.0	12.9	11.9	11.0	10.0	9.2	8.4	7.0	6.0
Path • Inches	-1.5	+2.2	0.0	-9.0	-25.9	-52.8	-90.8	-211.4	-412.0
Wind Drift • Inches	0.0	0.8	3.3	7.8	14.4	23.4	35.3	68.5	115.5

Fiocchi 140-grain SST Polymer Tip BT (65HSA)

G1 Ballistic Coefficient = 0.358

Distance • Yards	Muzzle	100	200	300	400	500	600	800	1000
Velocity • fps	2625	2381	2150	1932	1729	1543	1378	1127	985
Energy • ft-lbs	2142	1762	1437	1160	929	740	590	395	302
Taylor KO Index	13.9	12.6	11.4	10.2	9.1	8.1	7.3	6.0	5.2
Path • Inches	-1.5	+2.3	0.0	-9.9	-29.1	-60.0	-105.5	-256.1	-515.8
Wind Drift • Inches	0.0	1.0	4.2	10.0	18.8	31.0	47.1	92.2	152.8

Nosler 140-grain AccuBond (60022)

G1 Ballistic Coefficient = 0.516

Distance • Yards	Muzzle	100	200	300	400	500	600	800	1000
Velocity • fps	2650	2479	2316	2158	2006	1860	1722	1417	1264
Energy • ft-lbs	2183	1910	1667	1448	1251	1075	922	673	496
Taylor KO Index	14.0	13.1	12.2	11.4	10.6	9.8	9.1	7.8	6.7
Path • Inches	-1.5	+2.1	0.0	-8.6	-24.6	-49.4	-84.2	-191.0	-362.5
Wind Drift • Inches	0.0	0.7	2.8	6.5	12.0	19.4	29.0	55.5	93.3

Remington 140-grain Pointed Soft Point Core-Lokt (R65SWE1)

G1 Ballistic Coefficient = 0.436

Distance • Yards	Muzzle	100	200	300	400	500	600	800	1000
Velocity • fps	2550	2353	2164	1984	1814	1655	1508	1258	1085
Energy • ft-lbs	2022	1720	1456	1224	1023	850	707	492	366
Taylor KO Index	13.5	12.4	11.4	10.5	9.6	8.7	8.0	6.7	5.8
Path • Inches	-1.5	+2.4	0.0	-9.8	-27.0	-57.8	-99.8	-232.8	-454.1
Wind Drift • Inches	0.0	0.8	3.5	8.3	15.4	25.2	37.2	73.4	122.9

Winchester 140-grain Soft Point (X6555)

G1 Ballistic Coefficient = 0.450

Distance • Yards	Muzzle	100	200	300	400	500	600	800	1000
Velocity • fps	2550	2359	2176	2002	1836	1680	1584	1285	1107
Energy • ft-lbs	2022	1731	1473	1246	1048	878	732	514	381
Taylor KO Index	13.5	12.5	11.5	10.6	9.7	8.9	8.4	6.8	5.9
Path • Inches	-1.5	+2.4	0.0	-9.7	-28.1	-56.8	-98.1	-227.7	-441.9
Wind Drift • Inches	0.0	0.8	3.4	8.0	14.8	24.2	36.4	70.4	118.0

This sika stag was taken in England's Dorset region with a 6.5x55 Swedish Mauser. The rifle is equipped with a silencer, which is legal for hunting in the United Kingdom. (Safari Press Archive)

Federal 140-grain Fusion (F6555FS1)

G1 Ballistic Coefficient = 0.441

Distance • Yards	Muzzle	100	200	300	400	500	600	800	1000
Velocity • fps	2530	2340	2150	1970	1800	1650	1504	1258	1087
Energy • ft-lbs	1990	1695	1435	1210	1010	845	703	492	367
Taylor KO Index	13.4	12.4	11.4	10.4	9.5	8.7	7.9	6.6	5.7
Path • Inches	-1.5	+2.4	0.0	-10.0	-28.9	-58.5	-100.9	-235.0	-457.3
Wind Drift • Inches	0.0	0.8	3.5	8.3	15.4	25.1	37.8	73.0	122.1

Lapua 140-grain Naturalis (N316101)

G1 Ballistic Coefficient = 0.323

Distance • Yards	Muzzle	100	200	300	400	500	600	800	1000
Velocity • fps	2625	2356	2104	1868	1650	1455	1287	1057	933
Energy • ft-lbs	2143	1727	1376	1085	847	658	515	348	271
Taylor KO Index	13.9	12.4	11.1	9.9	8.7	7.7	6.8	5.6	4.9
Path • Inches	-1.5	+2.4	0.0	-10.3	-30.6	-63.7	-113.4	-280.8	-572.2
Wind Drift • Inches	0.0	1.1	4.7	11.2	21.2	35.2	53.7	104.9	171.5

Fiocchi 142-grain Sierra MatchKing HPBT (65MKA)

G1 Ballistic Coefficient = 0.581

Distance • Yards	Muzzle	100	200	300	400	500	600	800	1000
Velocity • fps	2625	2474	2328	2187	2051	1919	1794	1561	1359
Energy • ft-lbs	2173	1930	1709	1508	1326	1162	1015	769	582
Taylor KO Index	14.1	13.2	12.5	11.7	11.1	10.3	9.6	8.4	7.3
Path • Inches	-1.5	+2.1	0.0	-8.5	-24.3	-48.4	-81.9	-183.0	-341.6
Wind Drift • Inches	0.0	0.6	2.5	5.8	10.6	17.1	25.5	48.4	80.8

Lapua 155-grain Mega (4316021)

G1 Ballistic Coefficient = 0.377

Distance • Yards	Muzzle	100	200	300	400	500	600	800	1000
Velocity • fps	2560	2332	2116	1912	1712	1547	1399	1145	1001
Energy • ft-lbs	2254	1871	1540	1257	1021	823	664	451	345
Taylor KO Index	15.0	13.6	12.4	11.2	10.0	9.0	8.2	6.7	5.9
Path • Inches	-1.5	+2.4	0.0	-10.2	-30.0	-61.6	-107.9	-258.2	-514.6
Wind Drift • Inches	0.0	1.0	4.1	9.7	18.2	30.0	45.4	88.3	146.3

Norma 156-grain Oryx SP (16562)

G1 Ballistic Coefficient = 0.348

Distance • Yards	Muzzle	100	200	300	400	500	600	800	1000
Velocity • fps	2559	2313	2081	1862	1660	1477	1317	1086	958
Energy • ft-lbs	2269	1854	1500	1202	955	756	601	409	318
Taylor KO Index	15.1	13.7	12.3	11.0	9.8	8.7	7.8	6.4	5.7
Path • Inches	-1.5	+2.5	0.0	-10.6	-31.2	-64.5	-113.8	-277.4	-558.3
Wind Drift • Inches	0.0	1.1	4.5	10.7	20.1	33.2	50.5	98.2	161.0

Norma 156-grain Alaska SP (16552)

G1 Ballistic Coefficient = 0.276

Distance • Yards	Muzzle	100	200	300	400	500	600	800	1000
Velocity • fps	2559	2250	1964	1701	1469	1273	1125	953	850
Energy • ft-lbs	2269	1755	1336	1002	748	562	439	315	250
Taylor KO Index	15.1	13.2	11.6	10.0	8.6	7.5	6.6	5.6	5.0
Path • Inches	-1.5	+2.7	0.0	-11.8	-35.6	-75.7	-137.3	-348.4	-712.3
Wind Drift • Inches	0.0	1.4	5.8	14.1	26.9	44.9	68.6	130.1	206.6

Norma 156-grain Vulcan HP (16556)

G1 Ballistic Coefficient = 0.354

Distance • Yards	Muzzle	100	200	300	400	500	600	800	1000
Velocity • fps	2560	2318	2090	1875	1674	1492	1332	1098	967
Energy • ft-lbs	2271	1862	1513	1218	971	771	615	418	324
Taylor KO Index	15.1	13.6	12.3	11.0	9.8	8.8	7.8	6.5	5.7
Path • Inches	-1.5	+2.5	0.0	-10.5	-30.9	-63.8	-112.4	-273.0	-548.6
Wind Drift • Inches	0.0	1.0	4.4	10.5	19.7	32.5	49.4	96.1	157.9

.260 Remington

Perhaps the story of the .260 Remington (introduced in 2001) should start out, "Where have you been?" That's because the .260 Remington is simply a 6.5 mm-.308. Since the .243 Winchester (1955) is a 6 mm-08 and Remington began offering the 7 mm-08 in 1980, the only logical reason the 6.5 mm-08 didn't develop sooner is that the 6.5mm caliber has only recently had a good selection of bullets for the reloader. Time will tell if the .260 Remington has what it takes to become a popular "standard."

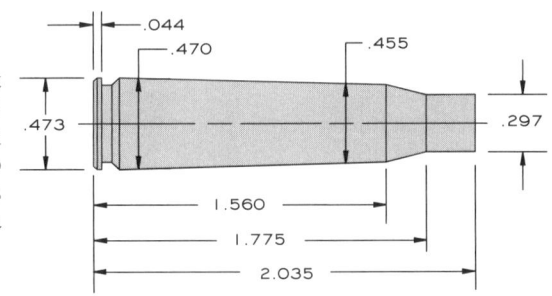

Relative Recoil Factor = 1.73

Specifications:

Controlling Agency for Standardization of this Ammunition: SAAMI

Bullet Weight Grains	Velocity fps	Maximum Average Pressure	
		Copper Crusher	Transducer
140	2,725	N/A	60,000 psi

Standard barrel for velocity testing: 24 inches long—1 turn in 9-inch twist

Availability:

Nosler 100-grain Partition (13017)
G1 Ballistic Coefficient = 0.326

Distance • Yards	Muzzle	100	200	300	400	500	600	800	1000
Velocity • fps	3200	2898	2615	2350	2100	1866	1650	1290	1060
Energy • ft-lbs	2274	1865	1509	1226	979	773	605	370	250
Taylor KO Index	12.1	10.9	9.9	8.9	7.9	7.0	6.2	4.9	4.0
Path • Inches	-1.5	+1.3	0.0	-6.4	-19.2	-39.9	-70.6	-174.1	-362.0
Wind Drift • Inches	0.0	0.8	3.5	8.3	15.6	25.8	39.4	79.0	137.1

Nosler 100-grain Ballistic Tip (13006)
G1 Ballistic Coefficient = 0.350

Distance • Yards	Muzzle	100	200	300	400	500	600	800	1000
Velocity • fps	3200	2918	2653	2403	2167	1944	1737	1378	1123
Energy • ft-lbs	2274	1891	1563	1283	1043	840	670	422	280
Taylor KO Index	12.1	11.0	10.0	9.1	8.2	7.3	6.6	5.2	4.2
Path • Inches	-1.5	+1.3	0.0	-6.2	-18.6	-38.3	-67.2	-162.9	-332.8
Wind Drift • Inches	0.0	0.8	3.3	7.7	14.3	23.5	35.8	71.2	123.7

Federal 120-grain Fusion (243FS1)
G1 Ballistic Coefficient = 0.391

Distance • Yards	Muzzle	100	200	300	400	500	600	800	1000
Velocity • fps	2950	2710	2480	2270	2060	1870	1688	1373	1141
Energy • ft-lbs	2320	1955	1640	1370	1130	930	760	502	347
Taylor KO Index	13.4	12.3	11.2	10.3	9.3	8.5	7.6	6.2	5.2
Path • Inches	-1.5	+1.6	0.0	-7.3	-21.4	-43.4	-75.8	-179.8	-358.3
Wind Drift • Inches	0.0	0.8	3.2	7.6	14.1	23.1	34.9	68.6	117.7

Federal 120-grain Nosler Ballistic Tip (P260B)
G1 Ballistic Coefficient = 0.418

Distance • Yards	Muzzle	100	200	300	400	500	600	800	1000
Velocity • fps	2950	2730	2510	2310	2110	1930	1755	1446	1203
Energy • ft-lbs	2320	1980	1680	1420	1190	990	821	557	386
Taylor KO Index	13.4	12.4	11.4	10.5	9.5	8.7	7.9	6.5	5.4
Path • Inches	-1.5	+1.6	0.0	-7.1	-20.8	-42.1	-73.1	-171.1	-336.3
Wind Drift • Inches	0.0	0.7	3.0	7.0	13.0	21.3	32.1	62.7	107.3

Federal 120-grain Barnes Triple-Shock X-Bullet (P260C) LF

G1 Ballistic Coefficient = 0.381

Distance • Yards	Muzzle	100	200	300	400	500	600	800	1000
Velocity • fps	2930	2690	2450	2240	2030	1830	1647	1333	1112
Energy • ft-lbs	2285	1920	1605	1330	1095	890	723	474	329
Taylor KO Index	13.3	12.2	11.1	10.1	9.2	8.3	7.5	6.0	5.0
Path • Inches	-1.5	+1.7	0.0	-7.6	-21.9	-44.7	-78.2	-186.6	-373.8
Wind Drift • Inches	0.0	0.8	3.4	7.9	14.7	24.1	36.5	71.9	123.2

Cor-Bon 120-grain T-DPX (DPX260120/20)

G1 Ballistic Coefficient = 0.450

Distance • Yards	Muzzle	100	200	300	400	500	600	800	1000
Velocity • fps	2900	2693	2496	2307	2127	1954	1791	1496	1255
Energy • ft-lbs	2241	1934	1661	1419	1205	1018	855	596	420
Taylor KO Index	13.1	12.2	11.3	10.4	9.6	8.8	8.1	6.8	5.7
Path • Inches	-1.5	+1.6	0.0	-7.2	-21.0	-42.5	-73.1	-169.1	-327.9
Wind Drift • Inches	0.0	0.7	2.8	6.6	12.3	20.0	30.0	58.2	99.0

Remington 120-grain AccuTip Boat Tail (PRA260RA)

G1 Ballistic Coefficient = 0.480

Distance • Yards	Muzzle	100	200	300	400	500	600	800	1000
Velocity • fps	2890	2697	2512	2334	2163	2000	1843	1558	1318
Energy • ft-lbs	2226	1938	1681	1451	1246	1065	906	647	463
Taylor KO Index	13.1	12.2	11.4	10.6	9.8	9.1	8.3	7.1	3.0
Path • Inches	-1.5	+1.9	0.0	-7.2	-20.7	-41.7	-71.5	-163.6	-313.4
Wind Drift • Inches	0.0	0.6	2.7	6.2	11.4	18.6	27.8	53.6	90.9

Cor-Bon 123-grain Scenar (PM260123/20)

G1 Ballistic Coefficient = 0.520

Distance • Yards	Muzzle	100	200	300	400	500	600	800	1000
Velocity • fps	2950	2769	2595	2428	2267	2111	1963	1686	1442
Energy • ft-lbs	2377	2095	1840	1610	1403	1218	1052	776	568
Taylor KO Index	13.7	12.8	12.0	11.3	10.5	9.8	9.1	7.8	6.7
Path • Inches	-1.5	+1.5	0.0	-6.7	-19.3	-38.7	-65.8	-148.8	-281.1
Wind Drift • Inches	0.0	0.6	2.4	5.5	10.1	16.4	24.4	46.7	78.7

Nosler 125-grain Partition (13050) + (60018)

G1 Ballistic Coefficient = 0.449

Distance • Yards	Muzzle	100	200	300	400	500	600	800	1000
Velocity • fps	2950	2741	2541	2350	2167	1992	1826	1525	1278
Energy • ft-lbs	2416	2085	1792	1533	1303	1102	926	646	453
Taylor KO Index	13.9	12.9	12.0	11.1	10.2	9.4	8.6	7.2	6.0
Path • Inches	-1.5	+1.6	0.0	-7.0	-20.2	-40.9	-70.4	-162.8	-315.5
Wind Drift • Inches	0.0	0.7	2.8	6.5	12.0	19.5	29.3	56.9	96.9

Nosler 130-grain AccuBond (13055)

G1 Ballistic Coefficient = 0.488

Distance • Yards	Muzzle	100	200	300	400	500	600	800	1000
Velocity • fps	2800	2613	2434	2262	2097	1939	1789	1516	1287
Energy • ft-lbs	2264	1972	1711	1478	1270	1086	924	663	479
Taylor KO Index	13.7	12.8	11.9	11.1	10.3	9.5	8.8	7.4	6.3
Path • Inches	-1.5	+1.8	0.0	-7.7	-22.2	-44.6	-76.2	-174.2	-333.2
Wind Drift • Inches	0.0	0.7	2.7	6.4	11.8	19.1	28.6	55.1	93.1

Cor-Bon 139-grain Scenar (PM260139/20)

G1 Ballistic Coefficient = 0.575

Distance • Yards	Muzzle	100	200	300	400	500	600	800	1000
Velocity • fps	2750	2593	2442	2295	2154	2017	1886	1641	1425
Energy • ft-lbs	2335	2076	1840	1626	1432	1256	1098	832	627
Taylor KO Index	14.4	13.6	12.8	12.0	11.3	10.6	9.9	8.6	7.5
Path • Inches	-1.5	+1.8	0.0	-7.7	-22.0	-43.8	-74.1	-165.6	-309.0
Wind Drift • Inches	0.0	0.6	2.4	5.5	10.0	16.2	24.0	45.7	76.4

Nosler 140-grain Partition (13063)

G1 Ballistic Coefficient = 0.486

Distance • Yards	Muzzle	100	200	300	400	500	600	800	1000
Velocity • fps	2725	2542	2367	2198	2037	1884	1737	1472	1254
Energy • ft-lbs	2308	2009	1741	1502	1290	1103	938	673	489
Taylor KO Index	14.4	13.4	12.5	11.6	10.8	9.9	9.2	7.8	6.6
Path • Inches	-1.5	+1.9	0.0	-8.1	-23.5	-47.2	-80.7	-184.5	-353.0
Wind Drift • Inches	0.0	0.7	2.8	6.6	12.2	19.8	29.6	57.0	96.3

Federal 140-grain Sierra GameKing (P260A)

G1 Ballistic Coefficient = 0.415

Distance • Yards	Muzzle	100	200	300	400	500	600	800	1000
Velocity • fps	2700	2490	2280	2090	1910	1730	1569	1293	1100
Energy • ft-lbs	2365	1920	1620	1360	1130	935	765	520	376
Taylor KO Index	14.3	13.1	12.0	11.0	10.1	9.1	8.3	6.8	5.8
Path • Inches	-1.5	+2.1	0.0	-8.8	-25.3	-51.9	-90.0	-211.6	-416.8
Wind Drift • Inches	0.0	0.8	3.4	8.1	15.0	24.6	37.1	72.3	122.2

Remington 140-grain Core-Lokt Ultra Bonded (PRC260RB)

G1 Ballistic Coefficient = 0.457

Distance • Yards	Muzzle	100	200	300	400	500	600	800	1000
Velocity • fps	2750	2554	2365	2185	2013	1849	1693	1417	1200
Energy • ft-lbs	2351	2027	1739	1484	1260	1063	891	625	448
Taylor KO Index	14.5	13.5	12.5	11.5	10.6	9.8	8.9	7.5	6.3
Path • Inches	-1.5	+1.9	0.0	-8.1	-23.6	-47.6	-81.9	-189.3	-366.5
Wind Drift • Inches	0.0	0.7	3.0	7.0	13.0	21.2	31.9	61.8	104.6

Remington 140-grain Core-Lokt Pointed Soft Point (R260R1)

G1 Ballistic Coefficient = 0.436

Distance • Yards	Muzzle	100	200	300	400	500	600	800	1000
Velocity • fps	2750	2544	2347	2158	1979	1812	1651	1371	1159
Energy • ft-lbs	2352	2011	1712	1488	1217	1021	847	584	417
Taylor KO Index	14.5	13.4	12.4	11.4	10.4	9.6	8.8	7.3	6.1
Path • Inches	-1.5	+1.9	0.0	-8.3	-24.0	-47.2	-83.9	-195.5	-381.6
Wind Drift • Inches	0.0	0.8	3.2	7.4	13.8	22.5	33.8	65.8	111.5

Remington 140-grain Core-Lokt Pointed Soft Point Managed Recoil (RL2601)

G1 Ballistic Coefficient = 0.435

Distance • Yards	Muzzle	100	200	300	400	500	600	800	1000
Velocity • fps	2360	2171	1991	1820	1660	1511	1377	1162	1026
Energy • ft-lbs	1731	1465	1232	1029	856	710	589	420	328
Taylor KO Index	12.5	11.5	10.5	9.6	8.8	8.0	7.3	6.1	5.4
Path • Inches	-1.5	+2.9	0.0	-11.7	-34.0	-68.9	-119.2	-278.3	-540.2
Wind Drift • Inches	0.0	1.0	4.0	9.4	17.4	28.4	42.6	81.7	134.2

Cor-Bon 142-grain HPBT (PM260142/20)

G1 Ballistic Coefficient = 0.480

Distance • Yards	Muzzle	100	200	300	400	500	600	800	1000
Velocity • fps	2700	2515	2337	2166	2003	1847	1700	1436	1223
Energy • ft-lbs	2299	1994	1722	1480	1265	1076	911	650	472
Taylor KO Index	14.5	13.5	12.5	11.6	10.7	9.9	9.1	7.7	6.5
Path • Inches	-1.5	+2.0	0.0	-8.4	-24.2	-48.7	-83.4	-191.2	-367.0
Wind Drift • Inches	0.0	0.7	2.9	6.8	12.6	20.5	30.8	59.4	100.2

6.5 Creedmore

This cartridge is another variant of a 6.5-.308. It has attained some popularity among long-range shooters shooting in practical style events. In 2008, Hornady announced that it would be building ammo for this cartridge. Performance-wise, it is difficult to tell this cartridge from the .260 Remington, but competition shooting is as much of a mind game as it is simply shooting skill, so you shoot what you think works best. With proper bullets, this cartridge is entirely suitable for light game.

Relative Recoil Factor = 1.73

Specifications:

Controlling Agency for Standardization of this Ammunition: SAAMI

Bullet Weight Grains	Velocity fps	Maximum Average Pressure Copper Crusher	Transducer
		N/A	60,000 psi

Standard barrel for velocity testing: 24 inches long—1 turn in 8-inch twist

Availability:

Hornady 120-grain GMX (81490)
G1 Ballistic Coefficient = 0.450

Distance • Yards	Muzzle	100	200	300	400	500	600	800	1000
Velocity • fps	3050	2850	2658	2475	2298	2129	1968	1668	1409
Energy • ft-lbs	2478	2163	1882	1631	1407	1208	1032	742	529
Taylor KO Index	13.8	12.9	12.0	11.2	10.4	9.6	8.9	7.5	6.4
Path • Inches	-1.5	+1.4	0.0	-6.3	-18.3	-36.9	-63.2	-144.4	-276.0
Wind Drift • Inches	0.0	0.6	2.5	5.8	10.6	17.1	25.6	49.3	83.6

Hornady 120-grain A-MAX (81492)
G1 Ballistic Coefficient = 0.465

Distance • Yards	Muzzle	100	200	300	400	500	600	800	1000
Velocity • fps	2910	2712	2522	2340	2166	1999	1839	1549	1306
Energy • ft-lbs	2256	1959	1695	1459	1250	1065	901	639	454
Taylor KO Index	13.2	12.3	11.4	10.6	9.8	9.0	8.3	7.0	5.9
Path • Inches	-1.5	+1.6	0.0	-7.1	-20.5	-41.4	-71.1	-163.2	-313.8
Wind Drift • Inches	0.0	0.6	2.7	6.3	11.6	18.9	28.3	54.6	92.8

Hornady 129-grain SST (81496)
G1 Ballistic Coefficient = 0.485

Distance • Yards	Muzzle	100	200	300	400	500	600	800	1000
Velocity • fps	2950	2756	2570	2392	2221	2057	1900	1612	1364
Energy • ft-lbs	2492	2175	1892	1639	1417	1212	1035	744	533
Taylor KO Index	14.4	13.4	12.5	11.6	10.8	10.0	9.2	7.8	6.6
Path • Inches	-1.5	+1.5	0.0	-6.8	-19.7	-39.5	-67.8	-154.9	-296.0
Wind Drift • Inches	0.0	0.6	2.6	6.0	11.0	17.8	26.6	51.2	86.7

Hornady 140-grain A-MAX (81494)
G1 Ballistic Coefficient = 0.585

Distance • Yards	Muzzle	100	200	300	400	500	600	800	1000
Velocity • fps	2710	2557	2409	2266	2128	1995	1867	1628	1417
Energy • ft-lbs	2283	2032	1804	1596	1408	1235	1084	824	625
Taylor KO Index	14.3	13.5	12.7	12.0	11.2	10.5	9.9	8.6	7.5
Path • Inches	-1.5	+1.9	0.0	-7.9	-22.6	-45.0	-76.1	-169.7	-316.2
Wind Drift • Inches	0.0	0.6	2.4	5.5	10.0	16.2	24.1	45.7	76.3

6.71mm (.264) Lazzeroni Phantom

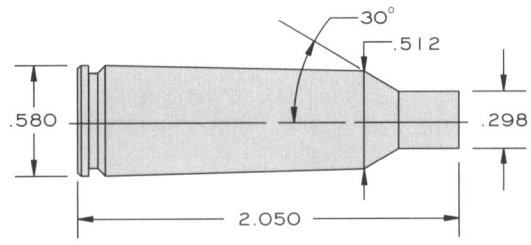

The second caliber in John Lazzeroni's short magnum series is his 6.71 Phantom. While it is difficult to make a perfect comparison, the factory loading for this round sets its performance at a level almost identical to the .264 Winchester Magnum.

Relative Recoil Factor = 1.79

Specifications:

Controlling Agency for Standardization of this Ammunition: Factory

Availability:

Lazzeroni 120-grain SP (671)

G1 Ballistic Coefficient = 0.525

Distance • Yards	Muzzle	100	200	300	400	500	600	800	1000
Velocity • fps	3312	3117	2930	2751	2579	2414	2255	1955	1681
Energy • ft-lbs	2923	2589	2289	2018	1773	1553	1356	1019	753
Taylor KO Index	15.0	14.1	13.3	12.5	11.7	10.9	10.2	8.8	7.6
Path • Inches	-1.5	+1.0	0.0	-5.1	-14.8	-29.7	-50.6	-113.9	-214.4
Wind Drift • Inches	0.0	0.5	2.0	4.7	8.6	13.8	20.4	38.9	65.2

Bill Quimby shot this springbok with a running shot near Kimberley, South Africa—a shot that surprised not only himself but also his hosts. (Photo from *Sixty Years a Hunter* by Bill Quimby, 2010, Safari Press Archives)

6.5mm Remington Magnum

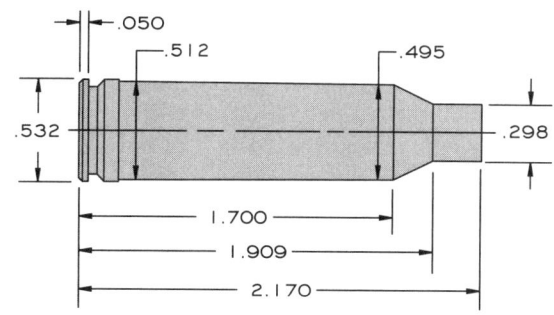

An early example of a short magnum, Remington's 6.5 RM was introduced in 1966. It was probably way before its time and handicapped by being introduced for an 18.5-inch barrel gun. It didn't last very long and dropped out of the ammo catalogs. There was nothing wrong with the cartridge. While it didn't have quite the performance of the .264 Winchester Magnum (due to a smaller case volume), it equaled the performance of two popular wildcats, the 6.5-06 and the 6.5-284, and avoided the .264's reputation for barrel burning. Recently Remington has restarted the manufacturing of this ammo, and Nosler is also offering a loading using their 125-grain Partition bullet.

Relative Recoil Factor = 1.72

Specifications:

Controlling Agency for Standardization of this Ammunition: SAAMI

Bullet Weight Grains	Velocity fps	Maximum Average Pressure	
		Copper Crusher	Transducer
120	3,195	53,000 cup	N/A

Standard barrel for velocity testing: 24 inches long—1 turn in 9-inch twist

Availability:

Remington 120-grain Core-Lokt Pointed Soft Point (R65MM2)

G1 Ballistic Coefficient = 0.323

Distance • Yards	Muzzle	100	200	300	400	500	600	800	1000
Velocity • fps	3210	2905	2621	2353	2102	1867	1647	1286	1057
Energy • ft-lbs	2745	2248	1830	1475	1177	929	723	441	298
Taylor KO Index	14.5	13.1	11.9	10.6	9.5	8.4	7.5	5.8	4.8
Path • Inches	-1.5	+1.3	0.0	-6.4	-19.1	-39.8	-70.4	-174.1	-362.7
Wind Drift • Inches	0.0	0.8	3.5	8.4	15.7	25.9	39.6	79.6	138.1

Nosler 125-grain Partition (14325)

G1 Ballistic Coefficient = 0.449

Distance • Yards	Muzzle	100	200	300	400	500	600	800	1000
Velocity • fps	3025	2812	2610	2416	2230	2053	1884	1576	1318
Energy • ft-lbs	2540	2196	1891	1620	1381	1170	985	690	482
Taylor KO Index	14.3	13.3	12.3	11.4	10.5	9.7	8.9	7.4	6.2
Path • Inches	-1.5	+1.4	0.0	-6.6	-19.1	-38.6	-66.4	-153.4	-297.0
Wind Drift • Inches	0.0	0.6	2.7	6.3	11.6	18.8	28.2	54.6	93.2

.264 Winchester Magnum

Winchester's .264 Magnum was touted to be a world beater when it was introduced in 1958. It never quite reached the popularity that was anticipated. This generally isn't because it lacked performance. The .264 WM could reach factory velocities well in excess of 3,000 fps even with the heavier bullet offerings. Nothing else in the 6.5mm class came close. Barrel life and a general shortage of good 6.5mm bullets for reloading didn't help its popularity. Winchester, who had previously dropped this caliber, reintroduced it into its catalog in 2001. Remington also makes factory ammo for the .264 Winchester Magnum, offering (like Winchester) only one loading, a 140-grain bullet. Nosler is now making six loadings.

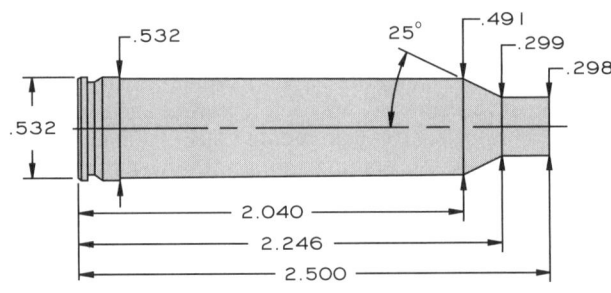

Relative Recoil Factor = 1.91

Specifications:

Controlling Agency for Standardization of this Ammunition: SAAMI

Bullet Weight Grains	Velocity fps	Maximum Average Pressure	
		Copper Crusher	Transducer
100	3,300	54,000 cup	64,000 psi
140	3,015	54,000 cup	64,000 psi

Standard barrel for velocity testing: 24 inches long—1 turn in 9-inch twist

Availability:

Nosler 100-grain Partition (14022)
G1 Ballistic Coefficient = 0.326

Distance • Yards	Muzzle	100	200	300	400	500	600	800	1000
Velocity • fps	3400	3084	2790	2514	2255	2011	1784	1392	1119
Energy • ft-lbs	2568	2112	1729	1404	1130	899	707	431	278
Taylor KO Index	12.8	11.6	10.5	9.5	8.5	7.6	6.7	5.2	4.2
Path • Inches	-1.5	+1.1	0.0	-5.6	-16.6	-34.6	-61.2	-150.4	-312.3
Wind Drift • Inches	0.0	0.8	3.2	7.7	14.3	23.6	35.9	72.0	126.2

Nosler 100-grain Ballistic Tip (14013)
G1 Ballistic Coefficient = 0.350

Distance • Yards	Muzzle	100	200	300	400	500	600	800	1000
Velocity • fps	3400	3105	2829	2570	2325	2093	1876	1490	1196
Energy • ft-lbs	2568	2141	1778	1467	1200	973	781	493	318
Taylor KO Index	12.8	11.7	10.7	9.7	8.8	7.9	7.1	5.6	4.5
Path • Inches	-1.5	+1.1	0.0	-5.4	-16.1	-33.3	-58.4	-140.9	-287.1
Wind Drift • Inches	0.0	0.7	3.0	7.1	13.1	21.6	32.7	64.9	113.3

Nosler 120-grain Ballistic Tip (14016)
G1 Ballistic Coefficient = 0.458

Distance • Yards	Muzzle	100	200	300	400	500	600	800	1000
Velocity • fps	3250	3030	2822	2622	2432	2249	2074	1750	1466
Energy • ft-lbs	2815	2448	2122	1833	1576	1348	1174	816	573
Taylor KO Index	14.7	13.7	12.8	11.9	11.0	10.2	9.4	7.9	6.6
Path • Inches	-1.5	+1.1	0.0	-5.5	-16.1	-32.3	-55.9	-128.4	-246.8
Wind Drift • Inches	0.0	0.6	2.4	5.6	10.2	16.6	24.8	47.7	81.2

Nosler 125-grain Partition (14072)
G1 Ballistic Coefficient = 0.449

Distance • Yards	Muzzle	100	200	300	400	500	600	800	1000
Velocity • fps	3200	2979	2768	2568	2376	2192	2016	1692	1412
Energy • ft-lbs	2843	2463	2128	1830	1587	1334	1128	795	554
Taylor KO Index	15.1	14.0	13.0	12.1	11.2	10.3	9.5	8.0	6.7
Path • Inches	-1.5	+1.2	0.0	-5.8	-16.8	-34.0	-58.5	-134.7	-260.2
Wind Drift • Inches	0.0	0.6	2.5	5.8	10.7	15.3	25.9	50.1	85.6

Nosler 130-grain AccuBond (14076) + (60019)

G1 Ballistic Coefficient = 0.488

Distance • Yards	Muzzle	100	200	300	400	500	600	800	1000
Velocity • fps	3100	2901	2710	2527	2352	2183	2022	1722	1458
Energy • ft-lbs	2775	2429	2121	1844	1597	1377	1180	856	614
Taylor KO Index	15.2	14.2	13.3	12.4	11.5	10.7	9.9	8.4	7.1
Path • Inches	-1.5	+1.3	0.0	-6.1	-17.6	-35.4	-60.5	-137.7	-262.4
Wind Drift • Inches	0.0	0.6	2.4	5.5	10.2	16.4	24.5	47.1	79.7

Remington 140-grain Core-Lokt Pointed Soft Point (R264W2)

G1 Ballistic Coefficient = 0.385

Distance • Yards	Muzzle	100	200	300	400	500	600	800	1000
Velocity • fps	3030	2782	2548	2326	2114	1914	1726	1397	1153
Energy • ft-lbs	2854	2406	2018	1682	1389	1139	926	607	414
Taylor KO Index	16.0	14.7	13.5	12.3	11.2	10.1	9.1	7.4	6.1
Path • Inches	-1.5	+1.5	0.0	-7.2	-20.8	-42.2	-72.0	-172.1	-347.0
Wind Drift • Inches	0.0	0.8	3.2	7.4	13.9	22.7	34.3	67.6	116.3

Winchester 140-grain Power Point (X2642)

G1 Ballistic Coefficient = 0.385

Distance • Yards	Muzzle	100	200	300	400	500	600	800	1000
Velocity • fps	3030	2782	2548	2326	2114	1914	1726	1397	1153
Energy • ft-lbs	2854	2406	2018	1682	1389	1139	926	607	414
Taylor KO Index	16.0	14.7	13.5	12.3	11.2	10.1	9.1	7.4	6.1
Path • Inches	-1.5	+1.5	0.0	-7.2	-20.8	-42.2	-72.0	-172.1	-347.0
Wind Drift • Inches	0.0	0.8	3.2	7.4	13.9	22.7	34.3	67.6	116.3

Nosler 140-grain Partition (14083)

G1 Ballistic Coefficient = 0.490

Distance • Yards	Muzzle	100	200	300	400	500	600	800	1000
Velocity • fps	2950	2758	2574	2398	2228	2065	1910	1623	1376
Energy • ft-lbs	2706	2365	2060	1787	1543	1326	1134	819	588
Taylor KO Index	15.6	14.6	13.6	12.7	11.8	10.9	10.1	8.6	7.3
Path • Inches	-1.5	+1.5	0.0	-6.8	-19.6	-39.5	-37.5	-153.9	-293.6
Wind Drift • Inches	0.0	0.6	2.5	5.9	10.8	17.6	26.2	50.5	85.5

Bill Quimby with his best Arizona pronghorn that was taken west of Flagstaff, circa 1992.
(Photo from *Sixty Years a Hunter* by Bill Quimby, 2010, Safari Press Archives)

6.8mm Remington SPC

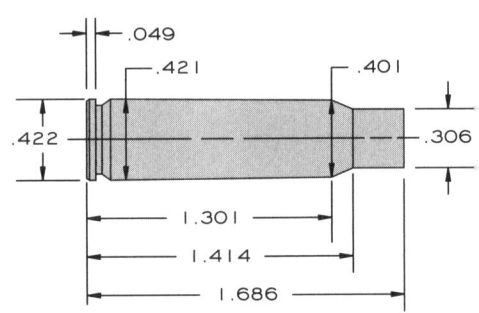

This is a relatively new cartridge, introduced in 2004, that was developed to fill the desire of various military units for a round that would allow heavier bullets to be fired in an M-16–type rifle. It is obvious that, as a minimum, the gun's barrel must be changed, but the conversion looks pretty easy. Based on the .30 Remington case, the 6.8mm Remington is considerably more potent than the .30 M1 Carbine and seems to be well suited to the assault rifle role. It is actually slightly more powerful than the 7.62x39 round. Its ultimate popularity will probably depend on what rifle selections become available.

Relative Recoil Factor = 1.41

Specifications:

Controlling Agency for Standardization of this Ammunition: SAAMI

Bullet Weight Grains	Velocity fps	Maximum Average Pressure Copper Crusher	Transducer
115	2,780	N/S	55,000 psi

Standard barrel for velocity testing: 24 inches long—1 turn in 10-inch twist

Availability:

Hornady 100-grain V-MAX (8346)

G1 Ballistic Coefficient = 0.370

Distance • Yards	Muzzle	100	200	300	400	500	600	800	1000
Velocity • fps	2570	2338	2118	1910	1716	1538	1379	1135	993
Energy • ft-lbs	1613	1335	1095	891	719	577	465	315	241
Taylor KO Index	11.1	10.1	9.1	8.2	7.4	6.6	5.9	4.9	4.3
Path • Inches	-1.5	+2.4	0.0	-10.2	-30.0	-61.6	-108.0	-260.0	-519.7
Wind Drift • Inches	0.0	1.0	4.2	9.9	18.5	30.5	46.2	90.1	149.0

Hornady 110-grain BTHP (8146)

G1 Ballistic Coefficient = 0.360

Distance • Yards	Muzzle	100	200	300	400	500	600	800	1000
Velocity • fps	2570	2332	2106	1893	1695	1514	1355	1115	979
Energy • ft-lbs	1613	1328	1083	875	702	560	448	304	234
Taylor KO Index	11.1	10.0	9.1	8.2	7.3	6.5	5.8	4.8	4.2
Path • Inches	-1.5	+2.5	0.0	-10.3	-30.4	-62.6	-110.0	-266.2	-534.1
Wind Drift • Inches	0.0	1.0	4.3	10.2	19.1	31.6	47.9	93.3	154.0

Hornady 110-grain V-MAX (8346)

G1 Ballistic Coefficient = 0.370

Distance • Yards	Muzzle	100	200	300	400	500	600	800	1000
Velocity • fps	2550	2319	2100	1893	1700	1524	1365	1126	988
Energy • ft-lbs	1588	1313	1077	875	706	567	455	310	238
Taylor KO Index	11.1	10.1	9.1	8.2	7.4	6.6	5.9	4.9	4.3
Path • Inches	-1.5	+2.5	0.0	-10.4	-30.6	-62.6	-110.1	-265.0	-529.5
Wind Drift • Inches	0.0	1.0	4.2	10.0	18.8	30.9	46.8	91.2	150.5

Hornady 110-grain BTHP (8146)

G1 Ballistic Coefficient = 0.360

Distance • Yards	Muzzle	100	200	300	400	500	600	800	1000
Velocity • fps	2550	2313	2088	1877	1680	1500	1341	1106	973
Energy • ft-lbs	1588	1306	1065	860	689	550	439	299	231
Taylor KO Index	11.1	10.1	9.1	8.2	7.3	6.5	5.8	4.8	4.2
Path • Inches	-1.5	+2.5	0.0	-10.7	-32.2	-67.8	-122.1	-307.6	-631.2
Wind Drift • Inches	0.0	1.0	4.4	10.4	19.4	32.0	48.6	94.5	155.4

Cor-Bon 110-grain T- DPX (DPX68110/20) LF

G1 Ballistic Coefficient = 0.350

Distance • Yards	Muzzle	100	200	300	400	500	600	800	1000
Velocity • fps	2400	2164	1942	1735	1548	1377	1235	1042	932
Energy • ft-lbs	1407	1144	922	735	584	463	372	265	212
Taylor KO Index	10.3	9.3	8.4	7.5	6.7	5.9	5.3	4.5	4.0
Path • Inches	-1.5	+3.0	0.0	-12.2	-36.1	-74.5	-131.4	-318.2	-632.9
Wind Drift • Inches	0.0	1.2	4.9	11.7	22.0	36.2	54.7	104.6	168.3

Remington 115-grain Core-Lokt Ultra Bonded (PRC68R4)

G1 Ballistic Coefficient = 0.295

Distance • Yards	Muzzle	100	200	300	400	500	600	800	1000
Velocity • fps	2625	2332	2058	1805	1574	1372	1206	1086	890
Energy • ft-lbs	1759	1389	1082	832	633	481	372	257	202
Taylor KO Index	11.9	10.6	9.4	8.2	7.2	6.2	5.5	4.9	4.1
Path • Inches	-1.5	+2.5	0.0	-10.7	-32.2	-67.8	-122.1	-307.6	-631.2
Wind Drift • Inches	0.0	1.2	5.2	12.5	23.8	39.6	60.7	117.4	189.5

Remington 115-grain Open Tip Match (R68R1)

G1 Ballistic Coefficient = 0.344

Distance • Yards	Muzzle	100	200	300	400	500	600	800	1000
Velocity • fps	2625	2373	2135	1911	1702	1513	1345	1100	965
Energy • ft-lbs	1759	1437	1163	932	740	584	462	309	238
Taylor KO Index	11.9	10.8	9.7	8.7	7.7	6.9	6.1	5.0	4.4
Path • Inches	-1.5	+2.3	0.0	-10.0	-29.6	-61.2	-108.2	-264.6	-535.4
Wind Drift • Inches	0.0	1.0	4.4	10.4	19.6	32.5	49.4	96.7	159.5

Remington 115-grain MatchKing BTHP (RM68R1)

G1 Ballistic Coefficient = 0.334

Distance • Yards	Muzzle	100	200	300	400	500	600	800	1000
Velocity • fps	2625	2365	2119	1889	1679	1484	1318	1080	950
Energy • ft-lbs	1759	1428	1147	911	717	562	443	298	231
Taylor KO Index	11.9	10.8	9.6	8.6	7.6	6.8	6.0	4.9	4.3
Path • Inches	-1.5	+2.4	0.0	-10.1	-30.1	-62.5	-110.6	-272.0	-552.2
Wind Drift • Inches	0.0	1.1	4.5	10.8	20.4	33.7	51.4	100.4	165.1

Cor-Bon 115-grain HPBT (PM68115/20)

G1 Ballistic Coefficient = 0.350

Distance • Yards	Muzzle	100	200	300	400	500	600	800	1000
Velocity • fps	2400	2164	1942	1735	1545	1377	1235	1042	932
Energy • ft-lbs	1471	1196	963	769	610	484	389	277	222
Taylor KO Index	11.0	9.9	8.9	7.9	7.1	6.3	5.6	4.8	4.3
Path • Inches	-1.5	+3.0	0.0	-12.2	-36.1	-74.5	-131.4	-318.2	-632.9
Wind Drift • Inches	0.0	1.2	4.9	11.7	22.0	36.2	54.7	104.6	168.3

Cor-Bon 115-grain HPBT Subsonic (PM68115S/20)

G1 Ballistic Coefficient = 0.350

Distance • Yards	Muzzle	100	200	300	400	500	600	800	1000
Velocity • fps	1000	947	903	864	830	798	769	716	670
Energy • ft-lbs	255	229	208	191	176	163	151	131	115
Taylor KO Index	4.6	4.3	4.1	3.9	3.8	3.6	3.5	3.3	3.1
Path • Inches	-1.5	+18.6	0.0	-61.2	-169.0	-327.3	-540.3	-1148	-2030
Wind Drift • Inches	0.0	1.5	5.8	12.8	22.4	34.6	49.3	86.5	134.6

UMC (Remington) 115-grain Metal Case (L68R2)

G1 Ballistic Coefficient = 0.292

Distance • Yards	Muzzle	100	200	300	400	500	600	800	1000
Velocity • fps	2625	2329	2053	1797	1565	1363	1199	998	887
Energy • ft-lbs	1759	1385	1076	825	625	474	367	255	201
Taylor KO Index	12.0	10.6	9.4	8.2	7.1	6.2	5.5	4.6	4.1
Path • Inches	-1.5	+2.1	0.0	-10.6	-32.4	-68.3	-122.9	-310.4	-637.0
Wind Drift • Inches	0.0	1.2	5.3	12.6	24.0	40.1	61.4	118.6	191.2

.270 Winchester

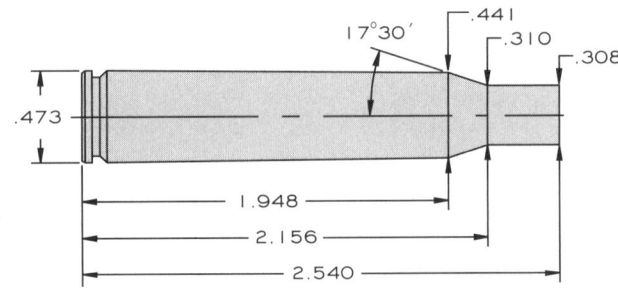

When Winchester took the .30-06 in 1925 and necked it to .270, I doubt if they even guessed that it would become one of the most popular nonmilitary calibers sold in the USA. The popularity of this cartridge is demonstrated by the fact that there are about 50 factory loadings available for the .270. That's a larger number than any other cartridge that is not a military standard. This popularity isn't an accident. The .270 drives 130-grain bullets to just over 3,000 fps, providing an excellent flat shooting capability that's bettered only by the magnums.

Relative Recoil Factor = 1.82

Specifications:

Controlling Agency for Standardization of this Ammunition: SAAMI

Bullet Weight Grains	Velocity fps	Maximum Average Pressure	
		Copper Crusher	Transducer
100	3,300	52,000 cup	65,000 psi
130	3,050	52,000 cup	65,000 psi
140	2,950	52,000 cup	65,000 psi
150	2,830	52,000 cup	65,000 psi
160	2,650	52,000 cup	65,000 psi

Standard barrel for velocity testing: 24 inches long—1 turn in 10-inch twist

Availability:

Cor-Bon 110-grain DPX (DPX270110/20)
G1 Ballistic Coefficient = 0.350

Distance • Yards	Muzzle	100	200	300	400	500	600	800	1000
Velocity • fps	3450	3152	2873	2611	2364	2130	1910	1519	1216
Energy • ft-lbs	2908	2427	2016	1665	1365	1108	891	563	361
Taylor KO Index	15.1	13.8	12.6	11.4	10.3	9.3	8.3	6.6	5.3
Path • Inches	-1.5	+1.0	0.0	-5.2	-15.6	-32.2	-56.5	-136.2	-277.2
Wind Drift • Inches	0.0	0.7	3.0	6.9	12.9	21.1	32.0	63.5	111.0

Federal 110-grain Barnes Tipped Triple-Shock (P270Q) LF
G1 Ballistic Coefficient = 0.380

Distance • Yards	Muzzle	100	200	300	400	500	600	800	1000
Velocity • fps	3400	3130	2870	2620	2390	2180	1975	1602	1297
Energy • ft-lbs	2825	2385	2010	1680	1400	1155	953	627	411
Taylor KO Index	14.8	13.6	12.5	11.4	10.4	9.5	8.6	7.0	5.6
Path • Inches	-1.5	+1.0	0.0	-5.3	-15.6	-32.1	-55.6	-131.9	-263.2
Wind Drift • Inches	0.0	0.7	2.8	6.4	12.0	19.5	29.4	57.8	1000.2

Norma 110-grain V-MAX (16940)
G1 Ballistic Coefficient = 0.370

Distance • Yards	Muzzle	100	200	300	400	500	600	800	1000
Velocity • fps	3215	2947	2695	2456	2230	2016	1814	1458	1187
Energy • ft-lbs	2525	2122	1774	1474	1215	993	804	520	344
Taylor KO Index	14.0	12.9	11.8	10.7	9.7	8.8	7.9	6.4	5.2
Path • Inches	-1.5	+1.2	0.0	-6.0	-17.9	-36.8	-64.2	-153.4	-309.0
Wind Drift • Inches	0.0	0.7	3.0	7.2	13.3	21.8	33.0	65.1	112.9

Remington 115-grain Core-Lokt Pointed Soft Point – Managed Recoil (RL270W2)
G1 Ballistic Coefficient = 0.295

Distance • Yards	Muzzle	100	200	300	400	500	600	800	1000
Velocity • fps	2710	2412	2133	1873	1636	1425	1248	1023	903
Energy • ft-lbs	1875	1485	1161	896	673	519	398	267	208
Taylor KO Index	12.3	11.0	9.7	8.5	7.4	6.5	5.7	4.7	4.1
Path • Inches	-1.5	+2.3	0.0	-10.0	-29.9	-62.9	-113.2	-286.3	-591.9
Wind Drift • Inches	0.0	1.2	5.0	11.9	22.6	37.8	58.0	113.5	184.9

Hornady 130-grain Interbond (80548)

G1 Ballistic Coefficient = 0.460

Distance • Yards	Muzzle	100	200	300	400	500	600	800	1000
Velocity • fps	3200	2984	2788	2582	2393	2213	2041	1721	1443
Energy • ft-lbs	2955	2570	2228	1924	1653	1414	1203	855	601
Taylor KO Index	16.5	15.4	14.4	13.4	12.4	11.4	10.5	8.9	7.5
Path • Inches	-1.5	+1.2	0.0	-5.7	-16.7	-33.7	-57.8	-132.7	-255.0
Wind Drift • Inches	0.0	0.6	2.4	5.6	10.4	16.8	25.2	48.5	82.6

Hornady 130-grain SST (80543)

G1 Ballistic Coefficient = 0.460

Distance • Yards	Muzzle	100	200	300	400	500	600	800	1000
Velocity • fps	3200	2984	2788	2582	2393	2213	2041	1721	1443
Energy • ft-lbs	2955	2570	2228	1924	1653	1414	1203	855	601
Taylor KO Index	16.3	15.2	14.2	13.1	12.2	11.3	10.4	8.8	7.3
Path • Inches	-1.5	+1.2	0.0	-5.7	-16.7	-33.7	-57.8	-132.7	-255.0
Wind Drift • Inches	0.0	0.6	2.4	5.6	10.4	16.8	25.2	48.5	82.6

Norma 130-grain Soft Point (16902)

G1 Ballistic Coefficient = 0.359

Distance • Yards	Muzzle	100	200	300	400	500	600	800	1000
Velocity • fps	3140	2868	2612	2370	2144	1928	1727	1379	1129
Energy • ft-lbs	2847	2375	1970	1622	1327	1074	861	549	368
Taylor KO Index	16.0	14.6	13.3	12.1	10.9	9.8	8.8	7.0	5.7
Path • Inches	-1.5	+1.4	0.0	-6.5	-19.2	-39.4	-69.1	-166.5	-337.9
Wind Drift • Inches	0.0	0.8	3.2	7.6	14.2	23.4	35.5	70.5	122.0

Nosler 130-grain AccuBond (60025)

G1 Ballistic Coefficient = 0.442

Distance • Yards	Muzzle	100	200	300	400	500	600	800	1000
Velocity • fps	3075	2856	2649	2450	2261	2079	1906	1591	1326
Energy • ft-lbs	2729	2354	2025	1733	1476	1248	1049	730	508
Taylor KO Index	15.9	14.7	13.7	12.7	11.7	10.7	9.8	8.2	6.8
Path • Inches	-1.5	+1.4	0.0	-6.4	-18.5	-37.5	-64.5	-149.2	-289.6
Wind Drift • Inches	0.0	0.6	2.7	6.2	11.5	18.7	28.0	54.4	92.9

Federal 130-grain Sierra GameKing BTSP (P270D)

G1 Ballistic Coefficient = 0.439

Distance • Yards	Muzzle	100	200	300	400	500	600	800	1000
Velocity • fps	3060	2840	2630	2430	2240	2060	1886	1571	1308
Energy • ft-lbs	2705	2325	1995	1705	1445	1220	1027	713	494
Taylor KO Index	15.7	14.6	13.5	12.5	11.5	10.6	9.7	8.1	6.7
Path • Inches	-1.5	+1.4	0.0	-6.4	-18.8	-38.1	-65.5	-151.8	-295.0
Wind Drift • Inches	0.0	0.7	2.7	6.3	11.7	19.0	28.5	55.4	94.7

Federal 130-grain Nosler Partition (P270P)

G1 Ballistic Coefficient = 0.416

Distance • Yards	Muzzle	100	200	300	400	500	600	800	1000
Velocity • fps	3060	2830	2610	2400	2200	2010	1831	1508	1248
Energy • ft-lbs	2705	2310	1965	1665	1400	1170	968	656	449
Taylor KO Index	15.7	14.6	13.4	12.3	11.3	10.3	9.4	7.8	6.4
Path • Inches	-1.5	+1.4	0.0	-6.5	-19.1	-41.8	-67.4	-157.6	-309.7
Wind Drift • Inches	0.0	0.7	2.9	6.7	12.4	20.3	30.6	59.7	102.4

Federal 130-grain Soft Point (270A)

G1 Ballistic Coefficient = 0.371

Distance • Yards	Muzzle	100	200	300	400	500	600	800	1000
Velocity • fps	3060	2880	2560	2330	2110	1900	1708	1373	1131
Energy • ft-lbs	2700	2265	1890	1585	1285	1043	842	544	370
Taylor KO Index	15.7	14.8	13.2	12.0	10.9	9.8	8.8	7.1	5.8
Path • Inches	-1.5	+1.5	0.0	-6.8	-20.0	-41.1	-71.9	-172.2	-347.1
Wind Drift • Inches	0.0	0.8	3.2	7.6	14.2	23.4	35.4	70.0	120.8

Hornady 130-grain SST (8054)

G1 Ballistic Coefficient = 0.460

Distance • Yards	Muzzle	100	200	300	400	500	600	800	1000
Velocity • fps	3060	2851	2651	2460	2277	2101	1934	1627	1365
Energy • ft-lbs	2702	2345	2028	1746	1496	1275	1080	764	538
Taylor KO Index	15.6	14.5	13.5	12.5	11.6	10.7	9.8	8.3	6.9
Path • Inches	-1.5	+1.4	0.0	-6.3	-18.5	-37.3	-64.0	-147.1	-283.3
Wind Drift • Inches	0.0	0.6	2.6	6.0	11.0	17.9	26.9	51.9	88.4

Hornady 130-grain SP InterLock (8055)

G1 Ballistic Coefficient = 0.410

Distance • Yards	Muzzle	100	200	300	400	500	600	800	1000
Velocity • fps	3060	2825	2603	2391	2188	1996	1816	1491	1232
Energy • ft-lbs	2702	2304	1955	1649	1382	1150	952	642	438
Taylor KO Index	15.7	14.5	13.4	12.3	11.3	10.3	9.3	7.7	6.3
Path • Inches	-1.5	+1.4	0.0	-6.6	-19.3	-39.2	-67.9	-159.3	-314.0
Wind Drift • Inches	0.0	0.7	2.9	6.8	12.6	20.6	31.1	60.9	104.7

PMC 130-grain SP (270HIA)

G1 Ballistic Coefficient = 0.409

Distance • Yards	Muzzle	100	200	300	400	500	600	800	1000
Velocity • fps	3060	2825	2603	2391	2188	1996	1815	1490	1231
Energy • ft-lbs	2702	2304	1955	1649	1382	1150	951	641	438
Taylor KO Index	15.8	14.6	13.4	12.3	11.3	10.3	9.4	7.4	6.4
Path • Inches	-1.5	+1.4	0.0	-6.6	-19.3	-39.2	-37.9	-159.4	-314.3
Wind Drift • Inches	0.0	0.7	2.9	6.8	12.7	20.7	31.2	61.0	104.7

Remington 130-grain AccuTip Boat Tail (PRA270WA)

G1 Ballistic Coefficient = 0.448

Distance • Yards	Muzzle	100	200	300	400	500	600	800	1000
Velocity • fps	3060	2845	2639	2442	2254	2076	1907	1595	1332
Energy • ft-lbs	2702	2335	2009	1721	1467	1243	1050	734	512
Taylor KO Index	15.7	14.6	13.6	12.6	11.6	10.7	9.8	8.2	6.9
Path • Inches	-1.5	+1.4	0.0	-6.4	-18.6	-37.7	-64.8	-149.8	-290.0
Wind Drift • Inches	0.0	0.6	2.6	6.2	11.4	18.6	27.8	53.9	92.0

Remington 130-grain Core-Lokt Pointed Soft Point (R270W2)

G1 Ballistic Coefficient = 0.336

Distance • Yards	Muzzle	100	200	300	400	500	600	800	1000
Velocity • fps	3060	2776	2510	2259	2022	1801	1599	1262	1051
Energy • ft-lbs	2702	2225	1818	1472	1180	936	738	460	319
Taylor KO Index	15.7	14.3	12.9	11.6	10.4	9.3	8.3	6.5	5.4
Path • Inches	-1.5	+1.5	0.0	-7.0	-20.9	-43.3	-76.6	-187.6	-386.6
Wind Drift • Inches	0.0	0.9	3.6	8.5	16.0	26.4	40.2	80.4	138.4

Federal 130-grain Fusion (F270FS1)

G1 Ballistic Coefficient = 0.400

Distance • Yards	Muzzle	100	200	300	400	500	600	800	1000
Velocity • fps	3050	2810	2580	2370	2160	1970	1787	1460	1206
Energy • ft-lbs	2685	2280	1925	1620	1350	1115	922	616	420
Taylor KO Index	15.7	14.5	13.3	12.2	11.2	10.2	9.2	7.5	6.2
Path • Inches	-1.5	+1.5	0.0	-6.6	-19.6	-39.8	-69.2	-163.0	-322.8
Wind Drift • Inches	0.0	0.7	3.0	7.0	13.0	21.3	32.1	63.0	108.3

Fiocchi 130-grain PSP (270SPB)

G1 Ballistic Coefficient = 0.409

Distance • Yards	Muzzle	100	200	300	400	500	600	800	1000
Velocity • fps	3010	2777	2557	2347	2147	1957	1778	1459	1209
Energy • ft-lbs	2614	2227	1887	1590	1330	1105	913	615	422
Taylor KO Index	15.5	14.3	13.2	12.1	11.1	10.1	9.2	7.5	6.2
Path • Inches	-1.5	+1.5	0.0	-6.8	-20.0	-40.7	-70.5	-165.6	-326.8
Wind Drift • Inches	0.0	0.7	3.0	7.0	13.0	21.2	31.9	62.5	107.3

Nosler 130-grain AccuBond (15405)

G1 Ballistic Coefficient = 0.435

Distance • Yards	Muzzle	100	200	300	400	500	600	800	1000
Velocity • fps	3000	2782	2574	2376	2186	2005	1834	1524	1270
Energy • ft-lbs	2599	2235	1913	1630	1380	1161	971	670	466
Taylor KO Index	15.3	14.2	13.1	12.1	11.1	10.2	9.3	7.8	6.5
Path • Inches	-1.5	+1.5	0.0	-6.8	-19.7	-39.9	-68.8	-159.8	-311.3
Wind Drift • Inches	0.0	0.7	2.8	6.6	12.1	19.8	29.7	57.7	98.6

Black Hills 130-grain Hornady SST (1C270BHGN3)

G1 Ballistic Coefficient = 0.460

Distance • Yards	Muzzle	100	200	300	400	500	600	800	1000
Velocity • fps	2950	2746	2551	2364	2185	2014	1851	1554	1306
Energy • ft-lbs	2513	2177	1879	1613	1378	1171	989	697	492
Taylor KO Index	15.2	14.2	13.2	12.2	11.3	10.4	9.6	8.0	6.7
Path • Inches	-1.5	+1.6	0.0	-6.9	-20.1	-40.5	-69.5	-160.1	-308.8
Wind Drift • Inches	0.0	0.6	2.7	6.3	11.6	18.9	28.4	54.9	93.4

Federal 130-grain Trophy Bonded Tip (P270TT4)

G1 Ballistic Coefficient = 0.439

Distance • Yards	Muzzle	100	200	300	400	500	600	800	1000
Velocity • fps	3200	2970	2760	2560	2360	2170	1933	1664	1384
Energy • ft-lbs	2955	2555	2200	1885	1610	1365	1147	800	553
Taylor KO Index	16.3	15.1	14.0	13.0	12.0	11.0	10.1	8.5	7.0
Path • Inches	-1.5	+1.2	0.0	-5.8	-16.9	-34.3	-59.1	-136.7	-265.1
Wind Drift • Inches	0.0	0.6	2.6	5.9	11.0	17.8	26.7	51.7	88.4

Hornady 130-grain GMX (8052) LF

G1 Ballistic Coefficient = 0.459

Distance • Yards	Muzzle	100	200	300	400	500	600	800	1000
Velocity • fps	3190	2976	2769	2573	2385	2202	2031	1712	1435
Energy • ft-lbs	2937	2553	2213	1911	1642	1404	1191	846	594
Taylor KO Index	16.2	15.1	14.1	13.1	12.1	11.2	10.3	8.7	7.3
Path • Inches	-1.5	+1.2	0.0	-5.5	-16.8	-33.9	-58.3	-133.8	-257.4
Wind Drift • Inches	0.0	0.6	2.4	5.7	10.5	16.9	25.4	48.9	83.3

Cor-Bon 130-grain DPX (DPX270130/20)

G1 Ballistic Coefficient = 0.430

Distance • Yards	Muzzle	100	200	300	400	500	600	800	1000
Velocity • fps	3100	2874	2659	2455	2259	2072	1895	1572	1305
Energy • ft-lbs	2775	2385	2042	1740	1472	1240	1037	714	492
Taylor KO Index	15.9	14.8	13.7	12.6	11.6	10.7	9.7	8.1	6.7
Path • Inches	-1.5	+1.4	0.0	-6.3	-18.4	-37.2	-64.2	-149.4	-291.4
Wind Drift • Inches	0.0	0.7	2.7	6.4	11.8	19.1	28.7	55.9	95.7

Federal 130-grain Trophy Bonded Tip (P270TT1)

G1 Ballistic Coefficient = 0.439

Distance • Yards	Muzzle	100	200	300	400	500	600	800	1000
Velocity • fps	3060	2840	2630	2430	2240	2060	1886	1571	1308
Energy • ft-lbs	2705	2325	1995	1705	1445	1220	1027	713	494
Taylor KO Index	15.7	14.6	13.5	12.5	11.5	10.6	9.7	8.1	6.7
Path • Inches	-1.5	+1.4	0.0	-6.4	-18.8	-38.1	-65.5	-151.8	-295.0
Wind Drift • Inches	0.0	0.7	2.7	6.3	11.7	19.0	28.5	55.4	94.7

Federal 130-grain Barnes Triple-Shock (P270L) LF

G1 Ballistic Coefficient = 0.439

Distance • Yards	Muzzle	100	200	300	400	500	600	800	1000
Velocity • fps	3060	2840	2630	2430	2240	2060	1886	1571	1308
Energy • ft-lbs	2705	2325	1995	1705	1445	1220	1027	713	494
Taylor KO Index	15.7	14.6	13.5	12.5	11.5	10.6	9.7	8.1	6.7
Path • Inches	-1.5	+1.4	0.0	-6.4	-18.8	-38.1	-65.5	-151.8	-295.0
Wind Drift • Inches	0.0	0.7	2.7	6.3	11.7	19.0	28.5	55.4	94.7

Federal 130-grain Nosler Ballistic Tip (P270F)

G1 Ballistic Coefficient = 0.435

Distance • Yards	Muzzle	100	200	300	400	500	600	800	1000
Velocity • fps	3060	2840	2630	2430	2230	2050	1878	1561	1299
Energy • ft-lbs	2700	2325	1990	1700	1440	1210	1018	704	487
Taylor KO Index	15.7	14.6	13.5	12.5	11.5	10.5	9.7	8.1	6.7
Path • Inches	-1.5	+1.4	0.0	-6.5	-18.8	-38.2	-65.8	-152.7	-297.3
Wind Drift • Inches	0.0	0.7	2.7	6.4	11.8	19.2	28.8	56.1	95.9

Remington 130-grain Swift Scirocco Bonded (PRSC270WA)

G1 Ballistic Coefficient = 0.433

Distance • Yards	Muzzle	100	200	300	400	500	600	800	1000
Velocity • fps	3060	2838	2627	2425	2232	2048	1872	1555	1293
Energy • ft-lbs	2702	2325	1991	1697	1438	1211	1012	698	482
Taylor KO Index	15.8	14.7	13.6	12.5	11.5	10.6	9.7	8.0	6.7
Path • Inches	-1.5	+1.4	0.0	-6.5	-18.8	-38.2	-66.0	-153.2	-298.6
Wind Drift • Inches	0.0	0.6	2.5	5.8	10.7	17.4	26.1	50.8	87.0

Remington 130-grain Bronze Point (270W3)

G1 Ballistic Coefficient = 0.372

Distance • Yards	Muzzle	100	200	300	400	500	600	800	1000
Velocity • fps	3060	2802	2559	2329	2110	1904	1711	1376	1134
Energy • ft-lbs	2702	2267	1890	1565	1285	1046	845	547	371
Taylor KO Index	15.7	14.4	13.2	12.0	10.9	9.9	8.8	7.1	5.9
Path • Inches	-1.5	+1.8	0.0	-7.1	-20.6	-42.0	-71.8	-171.8	-346.1
Wind Drift • Inches	0.0	0.8	3.2	7.6	14.2	23.3	35.3	69.8	120.3

Winchester 130-grain Power-Point Plus (X2705)

G1 Ballistic Coefficient = 0.372

Distance • Yards	Muzzle	100	200	300	400	500	600	800	1000
Velocity • fps	3060	2802	2559	2323	2110	1904	1710	1375	1133
Energy • ft-lbs	2702	2267	1890	1565	1285	1046	845	546	371
Taylor KO Index	15.7	14.4	13.2	12.0	10.9	9.8	8.8	7.1	5.8
Path • Inches	-1.5	+1.8	0.0	-7.1	-20.6	-42.0	-71.8	-171.9	-346.3
Wind Drift • Inches	0.0	0.8	3.2	7.6	14.2	23.3	35.3	69.8	120.4

Winchester 130-grain Power Core 95/5 (X270WLF) LF

G1 Ballistic Coefficient = 0.360

Distance • Yards	Muzzle	100	200	300	400	500	600	800	1000
Velocity • fps	3060	2794	2544	2307	2082	1871	1675	1339	1105
Energy • ft-lbs	2702	2254	1868	1536	1252	1011	810	518	353
Taylor KO Index	15.8	14.4	13.1	11.9	10.7	9.7	8.6	6.9	5.7
Path • Inches	-1.5	+1.5	0.0	-6.9	-20.3	-41.8	-73.2	-176.6	-358.2
Wind Drift • Inches	0.0	0.8	3.4	7.9	14.7	24.2	36.8	73.0	125.9

Winchester 130-grain Power Max Bonded (S2705BP)

G1 Ballistic Coefficient = 0.341

Distance • Yards	Muzzle	100	200	300	400	500	600	800	1000
Velocity • fps	3060	2779	2515	2266	2032	1814	1613	1277	1061
Energy • ft-lbs	2702	2229	1826	1482	1191	950	752	471	325
Taylor KO Index	15.6	14.1	12.8	11.5	10.3	9.2	8.2	6.5	5.4
Path • Inches	-1.5	+1.5	0.0	-7.0	-20.8	-42.6	-75.9	-185.4	-380.9
Wind Drift • Inches	0.0	0.8	3.6	8.4	15.7	26.0	39.6	78.9	135.9

Winchester 130-grain Silvertip (X2703)

G1 Ballistic Coefficient = 0.337

Distance • Yards	Muzzle	100	200	300	400	500	600	800	1000
Velocity • fps	3060	2776	2510	2259	2022	1801	1599	1262	1051
Energy • ft-lbs	2702	2225	1818	1472	1180	936	738	460	319
Taylor KO Index	15.7	14.3	12.9	11.6	10.4	9.3	8.3	6.5	5.4
Path • Inches	-1.5	+1.5	0.0	-7.0	-20.9	-43.3	-76.6	-187.6	-386.6
Wind Drift • Inches	0.0	0.9	3.6	8.5	16.0	26.4	40.2	80.4	138.4

Winchester 130-grain E-Tip (S270WET) LF

G1 Ballistic Coefficient = 0.459

Distance • Yards	Muzzle	100	200	300	400	500	600	800	1000
Velocity • fps	3050	2841	2641	2450	2267	2092	1925	1618	1357
Energy • ft-lbs	2685	2329	2013	1732	1483	1263	1069	756	531
Taylor KO Index	15.5	14.5	13.4	12.5	11.5	10.6	9.8	8.2	6.9
Path • Inches	-1.5	+1.4	0.0	-6.4	-18.6	-37.6	-64.5	-148.4	-286.0
Wind Drift • Inches	0.0	0.6	2.6	6.0	11.1	18.1	27.1	52.3	89.1

Winchester 130-grain XP3 (SXP270WA)

G1 Ballistic Coefficient = 0.436

Distance • Yards	Muzzle	100	200	300	400	500	600	800	1000
Velocity • fps	3050	2830	2621	2420	2229	2047	1872	1557	1296
Energy • ft-lbs	2685	2311	1982	1691	1434	1209	1012	700	485
Taylor KO Index	15.7	14.6	13.5	12.4	11.5	10.5	9.6	8.0	6.7
Path • Inches	-1.5	+1.4	0.0	-6.5	-18.6	-37.7	-66.2	-153.6	-299.1
Wind Drift • Inches	0.0	0.7	2.7	6.4	11.8	19.3	28.9	56.2	96.2

Winchester 130-grain Ballistic Silvertip (SBST270)

G1 Ballistic Coefficient = 0.434

Distance • Yards	Muzzle	100	200	300	400	500	600	800	1000
Velocity • fps	3050	2828	2618	2416	2224	2040	1867	1551	1290
Energy • ft-lbs	2685	2309	1978	1685	1428	1202	1007	695	481
Taylor KO Index	15.7	14.5	13.5	12.4	11.4	10.5	9.6	8.0	6.6
Path • Inches	-1.5	+1.4	0.0	-6.5	-18.9	-38.4	-66.4	-154.1	-300.4
Wind Drift • Inches	0.0	0.7	2.8	6.4	11.9	19.4	29.1	56.6	96.8

Black Hills 130-grain Barnes Triple-Shock (1C270BHGN1) LF

G1 Ballistic Coefficient = 0.466

Distance • Yards	Muzzle	100	200	300	400	500	600	800	1000
Velocity • fps	2950	2748	2555	2371	2193	2054	1862	1567	1319
Energy • ft-lbs	2513	2181	1885	1623	1389	1183	1001	709	502
Taylor KO Index	15.2	14.1	13.1	12.2	11.3	10.6	9.6	8.1	6.8
Path • Inches	-1.5	+1.6	0.0	-6.9	-20.0	-40.3	-69.1	-159.1	-307.7
Wind Drift • Inches	0.0	0.6	2.7	6.2	11.5	18.6	28.0	54.0	91.8

Black Hills 130-grain Hornady GMX (1C270BHGN4) LF

G1 Ballistic Coefficient = 0.460

Distance • Yards	Muzzle	100	200	300	400	500	600	800	1000
Velocity • fps	2950	2746	2551	2364	2185	2014	1851	1554	1306
Energy • ft-lbs	2513	2177	1879	1613	1378	1171	989	697	492
Taylor KO Index	15.2	14.2	13.2	12.2	11.3	10.4	9.6	8.0	6.7
Path • Inches	-1.5	+1.6	0.0	-6.9	-20.1	-40.5	-69.5	-160.1	-308.8
Wind Drift • Inches	0.0	0.6	2.7	6.3	11.6	18.9	28.4	54.9	93.4

Black Hills 130-grain Barnes TSX (1C270BHGN1)

G1 Ballistic Coefficient = 0.439

Distance • Yards	Muzzle	100	200	300	400	500	600	800	1000
Velocity • fps	2950	2736	2538	2338	2151	1974	1806	1502	1256
Energy • ft-lbs	2513	2162	1852	1578	1336	1125	941	652	455
Taylor KO Index	15.2	14.1	13.1	12.1	11.1	10.2	9.3	8.6	6.5
Path • Inchcs	-1.5	+1.6	0.0	-7.0	-20.4	-41.3	-71.2	-165.1	-321.3
Wind Drift • Inches	0.0	0.7	2.8	6.6	12.3	20.0	30.1	58.5	99.8

Nosler 130-grain E-Tip (60027) LF

G1 Ballistic Coefficient = 0.467

Distance • Yards	Muzzle	100	200	300	400	500	600	800	1000
Velocity • fps	2950	2748	2556	2371	2195	2026	1865	1570	1322
Energy • ft-lbs	2512	2180	1886	1623	1391	1185	1004	712	505
Taylor KO Index	15.2	14.2	13.2	12.2	11.3	10.5	9.6	8.1	6.8
Path • Inches	-1.5	+1.7	0.0	-6.9	-20.0	-40.2	-69.0	-158.5	-305.0
Wind Drift • Inches	0.0	0.6	2.7	6.2	11.4	18.6	27.8	53.8	91.4

Remington 130-grain Copper Solid Tipped (PCS270WA) LF

G1 Ballistic Coefficient = 0.431

Distance • Yards	Muzzle	100	200	300	400	500	600	800	1000
Velocity • fps	3060	2837	2625	2422	2229	2044	1869	1551	1289
Energy • ft-lbs	2702	2323	1988	1693	1434	1206	1009	695	480
Taylor KO Index	15.6	14.4	13.4	12.3	11.3	10.4	9.5	7.9	6.6
Path • Inches	-1.5	+1.4	0.0	-6.5	-18.9	-38.3	-66.1	-153.5	-299.5
Wind Drift • Inches	0.0	0.7	2.8	6.4	11.9	19.4	29.2	56.7	97.1

Norma 130-grain Full Jacket (16907)

G1 Ballistic Coefficient = 0.365

Distance • Yards	Muzzle	100	200	300	400	500	600	800	1000
Velocity • fps	2887	2634	2395	2169	1955	1755	1571	1263	1053
Energy • ft-lbs	2407	2004	1657	1358	1104	889	712	461	326
Taylor KO Index	14.9	13.6	12.3	11.2	10.1	9.0	8.1	6.5	5.4
Path • Inches	-1.5	+1.8	0.0	-7.8	-23.1	-47.5	-83.2	-200.6	-405.6
Wind Drift • Inches	0.0	0.9	3.6	8.5	15.8	26.0	39.6	78.2	133.5

Hornady 140-grain InterLock (8556)

G1 Ballistic Coefficient = 0.487

Distance • Yards	Muzzle	100	200	300	400	500	600	800	1000
Velocity • fps	3100	2900	2709	2525	2349	2180	2019	1718	1454
Energy • ft-lbs	2987	2614	2280	1982	1715	1477	1267	917	657
Taylor KO Index	17.2	16.1	15.0	14.0	13.0	12.1	11.2	9.5	8.1
Path • Inches	-1.5	+1.3	0.0	-6.1	-17.6	-35.4	-60.6	-138.0	-263.0
Wind Drift • Inches	0.0	0.6	2.4	5.5	10.2	16.5	24.6	47.3	80.1

Federal 140-grain Nosler AccuBond (P270A1)

G1 Ballistic Coefficient = 0.500

Distance • Yards	Muzzle	100	200	300	400	500	600	800	1000
Velocity • fps	2950	2760	2580	2400	2240	2080	1927	1643	1396
Energy • ft-lbs	2705	2370	2065	1795	1555	1340	1154	839	606
Taylor KO Index	16.3	15.3	14.3	13.3	12.4	11.5	10.7	9.1	7.7
Path • Inches	-1.5	+1.5	0.0	-6.7	-19.5	-39.3	-67.0	-152.3	-289.5
Wind Drift • Inches	0.0	0.6	2.5	5.8	10.6	17.2	25.6	49.2	83.3

Winchester 140-grain AccuBond CT (S270CT)

G1 Ballistic Coefficient = 0.373

Distance • Yards	Muzzle	100	200	300	400	500	600	800	1000
Velocity • fps	2950	2751	2560	2378	2203	2035	1876	1583	1335
Energy • ft-lbs	2705	2352	2038	1757	1508	1287	1094	779	554
Taylor KO Index	16.3	15.2	14.2	13.2	12.2	11.3	10.4	8.8	7.4
Path • Inches	-1.5	+1.6	0.0	-6.9	-19.9	-40.1	-68.7	-157.4	-302.2
Wind Drift • Inches	0.0	0.6	2.6	6.1	11.3	18.3	27.5	53.0	90.0

Hornady 140-grain BTSP InterLock (8056)

G1 Ballistic Coefficient = 0.486

Distance • Yards	Muzzle	100	200	300	400	500	600	800	1000
Velocity • fps	2940	2747	2562	2385	2214	2050	1894	1606	1360
Energy • ft-lbs	2688	2346	2041	1769	1524	1307	1115	802	575
Taylor KO Index	16.3	15.2	14.2	13.2	12.3	11.4	10.5	8.9	7.6
Path • Inches	-1.5	+1.6	0.0	-6.0	-20.2	-40.3	-68.3	-156.0	-298.1
Wind Drift • Inches	0.0	0.6	2.6	6.0	11.0	17.8	26.7	51.4	87.0

Remington 140-grain Core-Lokt Ultra Bonded (PRC270WB)

G1 Ballistic Coefficient = 0.360

Distance • Yards	Muzzle	100	200	300	400	500	600	800	1000
Velocity • fps	2925	2667	2424	2193	1975	1771	1582	1268	1064
Energy • ft-lbs	2659	2211	1826	1495	1212	975	778	500	352
Taylor KO Index	16.3	14.8	13.5	12.2	11.0	9.8	8.8	7.1	5.9
Path • Inches	-1.5	+1.7	0.0	-7.6	-22.5	-46.4	-81.6	-198.3	-405.2
Wind Drift • Inches	0.0	0.8	3.6	8.4	15.8	26.0	39.5	78.2	133.7

Remington 140-grain Swift A-Frame PSP (RS270WA)

G1 Ballistic Coefficient = 0.339

Distance • Yards	Muzzle	100	200	300	400	500	600	800	1000
Velocity • fps	2925	2652	2394	2152	1923	1711	1517	1207	1023
Energy • ft-lbs	2659	2186	1782	1439	1150	910	716	453	325
Taylor KO Index	16.2	14.7	13.3	11.9	10.7	9.5	8.4	6.7	5.7
Path • Inches	-1.5	+1.7	0.0	-7.8	-23.2	-48.0	-84.9	208.1	426.9
Wind Drift • Inches	0.0	0.9	3.8	9.0	17.0	28.0	42.7	85.0	144.6

Federal 145-grain Fusion (F270FSLR1)

G1 Ballistic Coefficient = 0.415

Distance • Yards	Muzzle	100	200	300	400	500	600	800	1000
Velocity • fps	2200	2010	1830	1662	1507	1367	1247	1072	966
Energy • ft-lbs	1560	1295	1070	899	731	602	501	370	301
Taylor KO Index	12.7	11.6	10.6	9.6	8.7	7.9	7.2	6.2	5.6
Path • Inches	-1.5	+3.6	0.0	-14.0	-40.6	-82.5	-143.1	-335.4	-648.5
Wind Drift • Inches	0.0	1.1	4.6	10.9	20.3	33.1	49.6	93.5	149.7

Winchester 150-grain XP3 (SXP270W)

G1 Ballistic Coefficient = 0.504

Distance • Yards	Muzzle	100	200	300	400	500	600	800	1000
Velocity • fps	2950	2763	2583	2411	2245	2086	1934	1651	1405
Energy • ft-lbs	2898	2542	2223	1936	1679	1449	1246	908	658
Taylor KO Index	17.5	16.4	15.3	14.3	13.3	12.4	11.5	9.8	8.3
Path • Inches	-1.5	+1.5	0.0	-6.7	-19.5	-39.1	-66.8	-151.6	-287.8
Wind Drift • Inches	0.0	0.6	2.5	5.7	10.5	17.0	25.4	48.7	82.4

Fiocchi 150-grain SST Polymer Tip BT (270HSB)

G1 Ballistic Coefficient = 0.525

Distance • Yards	Muzzle	100	200	300	400	500	600	800	1000
Velocity • fps	2860	2684	2515	2352	2195	2044	1900	1632	1399
Energy • ft-lbs	2723	2399	2106	1842	1604	1391	1203	887	652
Taylor KO Index	17.0	16.0	15.0	14.0	13.1	12.2	11.3	9.7	8.3
Path • Inches	-1.5	+1.7	0.0	-7.2	-20.6	-41.3	-70.3	-158.9	-300.0
Wind Drift • Inches	0.0	0.6	2.5	5.7	10.5	16.9	25.3	48.4	81.4

Norma 150-grain Oryx (16901)

G1 Ballistic Coefficient = 0.373

Distance • Yards	Muzzle	100	200	300	400	500	600	800	1000
Velocity • fps	2854	2608	2376	2155	1947	1751	1571	1269	1070
Energy • ft-lbs	2714	2267	1880	1547	1282	1022	822	536	381
Taylor KO Index	16.9	15.5	14.1	12.8	11.6	10.4	9.3	7.5	6.4
Path • Inches	-1.5	+1.8	0.0	-8.0	-23.5	-48.2	-84.3	-202.3	-407.2
Wind Drift • Inches	0.0	0.8	3.6	8.4	15.7	25.8	39.1	77.2	131.5

Federal 150-grain Fusion (F270FS2)

G1 Ballistic Coefficient = 0.470

Distance • Yards	Muzzle	100	200	300	400	500	600	800	1000
Velocity • fps	2850	2660	2470	2290	2120	1950	1795	1511	1277
Energy • ft-lbs	2705	2345	2030	1745	1495	1270	1073	761	543
Taylor KO Index	17.0	15.8	14.7	13.6	12.6	11.6	10.7	9.0	7.6
Path • Inches	-1.5	+1.7	0.0	-7.4	-21.6	-43.3	-74.4	-171.0	-329.0
Wind Drift • Inches	0.0	0.7	2.8	6.5	12.0	19.4	29.1	56.2	95.4

Fiocchi 150-grain PSP (270SPE)

G1 Ballistic Coefficient = 0.483

Distance • Yards	Muzzle	100	200	300	400	500	600	800	1000
Velocity • fps	2850	2660	2478	2303	2135	1974	1820	1540	1305
Energy • ft-lbs	2705	2357	2045	1766	1518	1298	1103	790	567
Taylor KO Index	17.0	15.8	14.8	13.7	12.7	11.8	10.8	9.2	7.8
Path • Inches	-1.5	+1.7	0.0	-7.4	-21.3	-43.0	-73.5	-168.1	-321.9
Wind Drift • Inches	0.0	0.6	2.7	6.3	11.6	18.8	28.1	54.2	91.8

Remington 150-grain Core-Lokt Soft Point (R270W4)

G1 Ballistic Coefficient = 0.261

Distance • Yards	Muzzle	100	200	300	400	500	600	800	1000
Velocity • fps	2850	2504	2183	1886	1618	1385	1199	982	864
Energy • ft-lbs	2705	2087	1587	1185	872	639	479	321	249
Taylor KO Index	16.9	14.9	13.0	11.2	9.6	8.2	7.1	5.8	5.1
Path • Inches	-1.5	+2.0	0.0	-9.4	-28.6	-61.2	-112.1	-292.6	-617.1
Wind Drift • Inches	0.0	1.2	5.3	12.8	24.5	41.3	63.9	125.3	203.4

Federal 150-grain Sierra GameKing BTSP (P270C)

G1 Ballistic Coefficient = 0.480

Distance • Yards	Muzzle	100	200	300	400	500	600	800	1000
Velocity • fps	2830	2640	2460	2280	2110	1950	1797	1518	1286
Energy • ft-lbs	2665	2320	2010	1735	1485	1270	1076	768	551
Taylor KO Index	16.8	15.7	14.6	13.5	12.5	11.6	10.7	9.0	7.6
Path • Inches	-1.5	+1.7	0.0	-7.5	-21.8	-43.7	-75.0	-171.7	-329.2
Wind Drift • Inches	0.0	0.7	2.8	6.4	11.8	19.2	28.7	55.4	93.8

Federal 150-grain Nosler Partition (P270E)

G1 Ballistic Coefficient = 0.468

Distance • Yards	Muzzle	100	200	300	400	500	600	800	1000
Velocity • fps	2830	2630	2450	2270	2090	1930	1774	1492	1281
Energy • ft-lbs	2665	2310	1990	1710	1460	1240	1049	742	530
Taylor KO Index	16.8	15.6	14.5	13.5	12.4	11.5	10.5	8.9	7.6
Path • Inches	-1.5	+1.7	0.0	-7.6	-22.0	-44.1	-75.9	-174.4	-336.0
Wind Drift • Inches	0.0	0.7	2.8	6.6	12.2	19.8	29.6	57.2	97.2

Federal 150-grain Soft Point RN (270B)

G1 Ballistic Coefficient = 0.260

Distance • Yards	Muzzle	100	200	300	400	500	600	800	1000
Velocity • fps	2830	2490	2170	1870	1610	1370	1186	975	859
Energy • ft-lbs	2665	2055	1565	1165	860	630	468	317	246
Taylor KO Index	16.8	14.8	12.9	11.1	9.6	8.1	7.0	5.8	5.1
Path • Inches	-1.5	+2.1	0.0	-9.4	-29.2	-62.6	-114.5	-298.8	-629.2
Wind Drift • Inches	0.0	1.3	5.4	13.0	24.9	42.0	64.9	127.0	205.5

Winchester 150-grain Power Point Plus (X2704)

G1 Ballistic Coefficient = 0.345

Distance • Yards	Muzzle	100	200	300	400	500	600	800	1000
Velocity • fps	2850	2585	2336	2100	1879	1673	1487	1190	1015
Energy • ft-lbs	2705	2226	1817	1468	1175	932	737	472	343
Taylor KO Index	16.9	15.3	13.9	12.5	11.2	9.9	8.8	7.1	6.0
Path • Inches	-1.5	+2.2	0.0	-8.6	-25.0	-51.4	-89.0	-217.6	-444.4
Wind Drift • Inches	0.0	0.9	3.9	9.2	17.3	28.5	43.5	86.2	146.0

Hornady 150-grain SP InterLock (8058)

G1 Ballistic Coefficient = 0.457

Distance • Yards	Muzzle	100	200	300	400	500	600	800	1000
Velocity • fps	2840	2641	2450	2267	2092	1926	1767	1481	1249
Energy • ft-lbs	2686	2322	1999	1712	1458	1235	1040	781	520
Taylor KO Index	16.9	15.7	14.5	13.5	12.4	11.4	10.5	8.8	7.4
Path • Inches	-1.5	+2.0	0.0	-7.8	-22.5	-45.0	-75.9	-174.9	-337.9
Wind Drift • Inches	0.0	0.7	2.9	6.7	12.4	20.1	30.1	58.3	98.9

RWS 150-grain Cone Point (211 7282)

G1 Ballistic Coefficient = 0.365

Distance • Yards	Muzzle	100	200	300	400	500	600	800	1000
Velocity • fps	2799	2551	2317	2095	1886	1691	1512	1221	1039
Energy • ft-lbs	2610	2169	1788	1462	1185	952	762	497	360
Taylor KO Index	16.6	15.2	13.8	12.5	11.2	10.1	9.0	7.3	6.2
Path • Inches	-1.5	+1.9	0.0	-8.4	-24.8	-51.0	-89.6	-217.5	-441.7
Wind Drift • Inches	0.0	0.9	3.8	8.8	16.6	27.3	41.4	81.8	138.4

Sellier and Bellot 150-grain SP (V330852U)

G1 Ballistic Coefficient = 0.381

Distance • Yards	Muzzle	100	200	300	400	500	600	800	1000
Velocity • fps	2625	2387	2170	1973	1781	1602	1439	1180	1022
Energy • ft-lbs	2289	1893	1565	1294	1056	855	690	464	348
Taylor KO Index	15.6	14.2	12.9	11.8	10.6	9.5	8.6	7.0	6.1
Path • Inches	-1.5	+2.3	0.0	-9.6	-28.2	-57.8	-101.2	-243.6	-487.8
Wind Drift • Inches	0.0	0.9	3.9	9.3	17.3	28.5	43.2	84.4	140.9

Norma 156-grain Vulcan (16941)

G1 Ballistic Coefficient = 0.344

Distance • Yards	Muzzle	100	200	300	400	500	600	800	1000
Velocity • fps	2854	2586	2333	2094	1872	1667	1481	1185	1012
Energy • ft-lbs	2822	2317	1886	1520	1214	963	760	487	355
Taylor KO Index	17.6	16.0	14.4	12.9	11.6	10.3	9.1	7.3	6.2
Path • Inches	-1.5	+1.9	0.0	-8.3	-24.6	-50.8	-89.6	-219.1	-447.6
Wind Drift • Inches	0.0	0.9	3.9	9.3	17.4	28.7	43.8	86.8	146.9

The bolt of this Christensen Arms rifle in .270 WSM has two plungers rather than one to help eject the case. (Safari Press Archives)

.270 Winchester Short Magnum

When Winchester announced the .300 WSM (2001), nearly everyone knew that a .270 WSM was sure to follow, and probably sooner rather than later. After only one year the .270 WSM was in their catalogs. The idea behind all the Short Magnums is that they can be used in the "short" actions, the ones that were designed for .308-size cases. The .270 WSM follows the recent trend toward unbelted magnums. The case volume, and also the performance, falls right between the .270 Winchester and the .270 Weatherby.

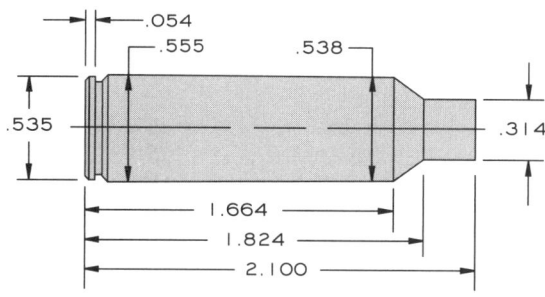

Relative Recoil Factor = 2.00

Specifications:

Controlling Agency for Standardization of this Ammunition: SAAMI

Bullet Weight Grains	Velocity fps	Maximum Average Pressure	
		Copper Crusher	Transducer

Specifications Pending

Standard barrel for velocity testing: Pending *

*Preliminary Value: 24 inches long—1 turn in 10-inch twist inches

Availability:

Cor-Bon 110-grain DPX (DPX270WSM)
G1 Ballistic Coefficient = 0.350

Distance • Yards	Muzzle	100	200	300	400	500	600	800	1000
Velocity • fps	3550	3244	2960	2693	2441	2203	1979	1579	1257
Energy • ft-lbs	3079	2572	2140	1772	1456	1186	956	607	386
Taylor KO Index	15.5	14.2	12.9	11.8	10.7	9.6	8.6	6.9	5.5
Path • Inches	-1.5	+0.9	0.0	-4.9	-14.6	-30.1	-52.9	-127.4	-258.8
Wind Drift • Inches	0.0	0.7	2.9	6.7	12.4	20.3	30.7	60.8	106.4

Federal 110-grain Barnes Tipped Triple-Shock (P270WSMF)
G1 Ballistic Coefficient = 0.378

Distance • Yards	Muzzle	100	200	300	400	500	600	800	1000
Velocity • fps	3500	3220	2960	2710	2470	2250	2040	1655	1337
Energy • ft-lbs	2990	2530	2135	1790	1495	1235	1018	669	437
Taylor KO Index	15.2	14.0	12.9	11.8	10.8	9.8	8.9	7.2	5.8
Path • Inches	-1.5	+0.9	0.0	-4.9	-14.6	-30.0	-52.2	-123.8	-247.0
Wind Drift • Inches	0.0	0.6	2.7	6.3	11.6	18.9	28.4	55.8	96.8

Remington 130-grain Core-Lokt Soft Point (R270WSM1)
G1 Ballistic Coefficient = 0.336

Distance • Yards	Muzzle	100	200	300	400	500	600	800	1000
Velocity • fps	3285	2986	2707	2444	2196	1963	1746	1373	1113
Energy • ft-lbs	3114	2573	2114	1724	1392	1112	880	544	358
Taylor KO Index	17.0	15.4	14.0	12.6	11.3	10.1	9.0	7.1	5.7
Path • Inches	-1.5	+1.2	0.0	-6.0	-17.8	-36.8	-64.9	-158.7	-327.1
Wind Drift • Inches	0.0	0.8	3.3	7.8	14.5	23.8	36.3	72.5	126.5

Federal 130-grain Nosler Partition (P270WSMP)
G1 Ballistic Coefficient = 0.415

Distance • Yards	Muzzle	100	200	300	400	500	600	800	1000
Velocity • fps	3280	3040	2810	2590	2380	2180	1990	1644	1353
Energy • ft-lbs	3105	2665	2275	1935	1635	1375	1144	780	528
Taylor KO Index	16.9	15.6	14.5	13.3	12.2	11.2	10.2	8.5	7.0
Path • Inches	-1.5	+1.1	0.0	-5.6	-16.3	-33.2	-57.5	-134.3	-263.2
Wind Drift • Inches	0.0	0.6	2.6	6.1	11.3	18.4	27.7	53.9	92.7

Winchester 130-grain XP3 (SXP270SA)
G1 Ballistic Coefficient = 0.436

Distance • Yards	Muzzle	100	200	300	400	500	600	800	1000
Velocity • fps	3275	3043	2824	2614	2415	2224	2042	1707	1417
Energy • ft-lbs	3096	2673	2301	1973	1683	1427	1205	841	580
Taylor KO Index	16.8	15.7	14.5	13.4	12.4	11.4	10.5	8.8	7.3
Path • Inches	-1.5	+1.1	0.0	-5.5	-16.1	-32.7	-56.2	-130.1	-252.4
Wind Drift • Inches	0.0	0.6	2.5	5.8	10.7	17.4	26.0	50.3	86.1

Federal 130-grain Soft Point (270WSME)

G1 Ballistic Coefficient = 0.366

Distance • Yards	Muzzle	100	200	300	400	500	600	800	1000
Velocity • fps	3250	2980	2720	2480	2250	2030	1826	1465	1189
Energy • ft-lbs	3050	2560	2140	1775	1460	1195	963	619	408
Taylor KO Index	16.8	15.4	14.0	12.8	11.6	10.5	9.4	7.6	6.1
Path • Inches	-1.5	+1.2	0.0	-5.9	-17.5	-36.0	-63.0	-150.9	-304.6
Wind Drift • Inches	0.0	0.7	3.0	7.1	13.3	21.7	32.9	65.1	113.0

Cor-Bon 130-grain DPX (DPX270WSM130/20)

G1 Ballistic Coefficient = 0.430

Distance • Yards	Muzzle	100	200	300	400	500	600	800	1000
Velocity • fps	3300	3064	2840	2627	2424	2230	2045	1704	1412
Energy • ft-lbs	3144	2710	2329	1993	1697	1436	1208	839	575
Taylor KO Index	16.8	15.6	14.5	13.4	12.3	11.3	10.4	8.7	7.2
Path • Inches	-1.5	+1.1	0.0	-5.4	-15.9	-32.3	-55.7	-129.1	-251.0
Wind Drift • Inches	0.0	0.6	2.5	5.8	10.8	17.5	26.2	50.8	87.0

Federal 130-grain Nosler Ballistic Tip (P270WSMB)

G1 Ballistic Coefficient = 0.435

Distance • Yards	Muzzle	100	200	300	400	500	600	800	1000
Velocity • fps	3300	3070	2840	2630	2430	2240	2056	1717	1425
Energy • ft-lbs	3145	2710	2335	2000	1705	1445	1221	851	586
Taylor KO Index	17.0	15.8	14.6	13.5	12.5	11.5	10.6	8.8	7.3
Path • Inches	-1.5	+1.1	0.0	-5.4	-15.8	-32.2	-55.4	-128.3	-248.9
Wind Drift • Inches	0.0	0.6	2.5	5.8	10.6	17.3	25.9	50.1	85.7

Norma 130-grain Nosler BST (16925)

G1 Ballistic Coefficient = 0.433

Distance • Yards	Muzzle	100	200	300	400	500	600	800	1000
Velocity • fps	3281	3047	2825	2614	2412	2220	2036	1698	1407
Energy • ft-lbs	3108	2681	2305	1973	1680	1423	1197	832	572
Taylor KO Index	16.9	15.7	14.6	13.5	12.5	11.5	10.5	8.8	7.3
Path • Inches	-1.5	+1.1	0.0	-5.5	-16.1	-32.6	-56.3	-130.5	-253.5
Wind Drift • Inches	0.0	0.6	2.5	5.8	10.8	17.5	26.2	50.9	87.1

Federal 130-grain Trophy Bonded Tip (P270WSMTT1)

G1 Ballistic Coefficient = 0.443

Distance • Yards	Muzzle	100	200	300	400	500	600	800	1000
Velocity • fps	3280	3050	2830	2620	2430	2240	2061	1727	1438
Energy • ft-lbs	3105	2685	2315	1990	1700	1445	1226	861	597
Taylor KO Index	16.9	15.7	14.6	13.5	12.5	11.5	10.6	8.9	7.4
Path • Inches	-1.5	+1.1	0.0	-5.5	-16.0	-32.4	-55.7	-128.5	-248.6
Wind Drift • Inches	0.0	0.6	2.4	5.7	10.5	17.0	25.3	49.3	84.2

Federal 130-grain Barnes Triple-Shock X-Bullet (P270WSMD)

G1 Ballistic Coefficient = 0.433

Distance • Yards	Muzzle	100	200	300	400	500	600	800	1000
Velocity • fps	3280	3050	2820	2610	2410	2220	2036	1698	1408
Energy • ft-lbs	3105	2675	2300	1970	1675	1420	1197	833	572
Taylor KO Index	16.9	15.7	14.5	13.4	12.4	11.4	10.5	8.7	7.2
Path • Inches	-1.5	+1.1	0.0	-5.5	-16.1	-32.7	-56.3	-130.5	-253.5
Wind Drift • Inches	0.0	0.6	2.5	5.8	10.8	17.5	26.2	50.9	87.1

Winchester 130-grain E-Tip (S270SET) LF

G1 Ballistic Coefficient = 0.459

Distance • Yards	Muzzle	100	200	300	400	500	600	800	1000
Velocity • fps	3275	3055	2845	2645	2454	2271	2095	1770	1483
Energy • ft-lbs	3096	2693	2336	2020	1738	1488	1268	904	635
Taylor KO Index	16.9	15.8	14.7	13.7	12.7	11.7	10.8	9.1	7.7
Path • Inches	-1.5	+1.1	0.0	-5.4	-15.8	-32.0	-54.9	-126.0	-242.0
Wind Drift • Inches	0.0	0.6	2.4	5.5	10.1	16.3	24.4	47.0	80.0

Winchester 130-grain Ballistic Silvertip (SBST2705)

G1 Ballistic Coefficient = 0.432

Distance • Yards	Muzzle	100	200	300	400	500	600	800	1000
Velocity • fps	3275	3041	2820	2609	2408	2215	2031	1693	1403
Energy • ft-lbs	3096	2669	2295	1964	1673	1416	1191	827	568
Taylor KO Index	16.9	15.7	14.6	13.5	12.4	11.4	10.5	8.7	7.2
Path • Inches	-1.5	+1.1	0.0	-5.5	-16.1	-32.8	-56.5	-131.3	-256.5
Wind Drift • Inches	0.0	0.6	2.5	5.9	10.8	17.6	26.4	51.1	87.5

Winchester 130-grain Power Max Bonded (X270SBP)
G1 Ballistic Coefficient = 0.340

Distance • Yards	Muzzle	100	200	300	400	500	600	800	1000
Velocity • fps	3275	2980	2704	2445	2200	1968	1754	1383	1121
Energy • ft-lbs	3096	2563	2111	1725	1396	1119	888	552	363
Taylor KO Index	16.9	15.4	14.0	12.6	11.4	10.2	9.1	7.1	5.8
Path • Inches	-1.5	+1.2	0.0	-6.0	-17.8	-36.8	-64.9	-158.1	-324.9
Wind Drift • Inches	0.0	0.8	3.3	7.7	14.3	23.6	35.9	71.6	124.8

Norma 130-grain FULL JACKET (16931)
G1 Ballistic Coefficient = 0.365

Distance • Yards	Muzzle	100	200	300	400	500	600	800	1000
Velocity • fps	3150	2882	2630	2391	2166	1953	1753	1405	1149
Energy • ft-lbs	2865	2399	1998	1652	1355	1101	888	570	381
Taylor KO Index	16.3	14.9	13.6	12.3	11.2	10.1	9.1	7.3	5.9
Path • Inches	-1.5	+1.4	0.0	-6.4	-18.9	-38.8	-67.9	-162.9	-329.3
Wind Drift • Inches	0.0	0.8	3.2	7.5	13.9	22.8	34.6	68.5	118.6

Federal 140-grain Nosler AccuBond (P270WSMA1)
G1 Ballistic Coefficient = 0.498

Distance • Yards	Muzzle	100	200	300	400	500	600	800	1000
Velocity • fps	3200	3000	2810	2620	2450	2280	2117	1813	1542
Energy • ft-lbs	3185	2785	2450	2140	1860	1615	1394	1022	739
Taylor KO Index	17.7	16.6	15.6	14.5	13.6	12.6	11.7	10.0	8.5
Path • Inches	-1.5	+1.2	0.0	-5.6	-16.2	-32.7	-55.8	-126.8	-240.3
Wind Drift • Inches	0.0	0.5	2.2	5.2	9.5	15.4	22.9	43.8	74.0

Winchester 140-grain AccuBond CT (S270WSMCT)
G1 Ballistic Coefficient = 0.473

Distance • Yards	Muzzle	100	200	300	400	500	600	800	1000
Velocity • fps	3200	2989	2789	2579	2413	2236	2067	1753	1476
Energy • ft-lbs	3184	2779	2418	2097	1810	1555	1329	955	678
Taylor KO Index	17.7	16.6	15.5	14.3	13.4	12.4	11.5	9.7	8.2
Path • Inches	-1.5	+1.2	0.0	-5.7	-16.5	-33.3	-57.1	-130.6	-249.8
Wind Drift • Inches	0.0	0.6	2.4	5.5	10.1	16.3	24.4	46.9	79.6

Nosler 140-grain AccuBond (60030 and 15210)
G1 Ballistic Coefficient = 0.496

Distance • Yards	Muzzle	100	200	300	400	500	600	800	1000
Velocity • fps	3100	2904	2715	2535	2362	2196	2037	1740	1478
Energy • ft-lbs	2987	2620	2292	1998	1734	1499	1291	941	679
Taylor KO Index	17.2	16.1	15.1	14.1	13.1	12.2	11.3	9.7	8.2
Path • Inches	-1.5	+1.3	0.0	-6.0	-17.5	-35.2	-60.1	-136.4	-259.2
Wind Drift • Inches	0.0	0.6	2.3	5.4	10.0	16.1	24.0	46.1	77.9

Winchester 140-grain XP3 (SXP279SA)
G1 Ballistic Coefficient = 0.437

Distance • Yards	Muzzle	100	200	300	400	500	600	800	1000
Velocity • fps	3275	3043	2824	2614	2415	2224	2042	1707	1417
Energy • ft-lbs	3096	2673	2301	1973	1683	1427	1205	841	580
Taylor KO Index	16.8	15.7	14.5	13.4	12.4	11.4	10.5	8.8	7.3
Path • Inches	-1.5	+1.1	0.0	-5.5	-16.1	-32.7	-56.2	-130.1	-252.4
Wind Drift • Inches	0.0	0.6	2.5	5.8	10.7	17.4	26.0	50.3	86.1

Winchester 150-grain Power Point (X270WSM)
G1 Ballistic Coefficient = 0.344

Distance • Yards	Muzzle	100	200	300	400	500	600	800	1000
Velocity • fps	3150	2867	2601	2350	2113	1890	1684	1332	1093
Energy • ft-lbs	3304	2737	2252	1839	1487	1190	945	591	398
Taylor KO Index	18.8	17.1	15.5	14.0	12.6	11.3	10.0	7.9	6.5
Path • Inches	-1.5	+1.4	0.0	-6.5	-19.4	-40.1	-70.5	-171.6	-351.9
Wind Drift • Inches	0.0	0.8	3.4	8.0	14.9	24.6	37.4	74.6	129.3

Winchester 150-grain XP3 (SXP2705)
G1 Ballistic Coefficient = 0.504

Distance • Yards	Muzzle	100	200	300	400	500	600	800	1000
Velocity • fps	3120	2926	2740	2561	2389	2224	2066	1770	1507
Energy • ft-lbs	3242	2850	2499	2184	1901	1648	1423	1044	757
Taylor KO Index	18.5	17.4	16.3	15.2	14.2	13.2	12.3	10.5	8.9
Path • Inches	-1.5	+1.3	0.0	-5.9	-17.1	-34.5	-58.8	-133.4	-252.7
Wind Drift • Inches	0.0	0.6	2.3	5.3	9.7	15.7	23.4	44.8	75.6

Norma 150-grain ORYX (16932)

G1 Ballistic Coefficient = 0.373

Distance • Yards	Muzzle	100	200	300	400	500	600	800	1000
Velocity • fps	3117	2865	2611	2378	2156	1950	1755	1412	1158
Energy • ft-lbs	3237	2718	2271	1884	1553	1267	1026	665	447
Taylor KO Index	18.6	17.0	15.6	14.2	12.9	11.6	10.5	8.4	6.9
Path • Inches	-1.5	+1.4	0.0	-6.5	-19.2	-39.3	-68.6	-164.0	-329.9
Wind Drift • Inches	0.0	0.8	3.2	7.4	13.8	22.6	34.2	67.5	116.6

Federal 150-grain Nosler Partition (P270WSMC)

G1 Ballistic Coefficient = 0.470

Distance • Yards	Muzzle	100	200	300	400	500	600	800	1000
Velocity • fps	3100	2890	2690	2500	2320	2150	1984	1677	1411
Energy • ft-lbs	3200	2785	2415	2085	1790	1540	1312	937	663
Taylor KO Index	18.4	17.2	16.0	14.8	13.8	12.8	11.8	10.0	8.4
Path • Inches	-1.5	+1.3	0.0	-6.1	-17.8	-36.0	-61.6	-141.0	-270.4
Wind Drift • Inches	0.0	0.6	2.5	5.8	10.6	17.2	25.7	49.6	84.3

Federal 150-grain Fusion (F270WSMFS1)

G1 Ballistic Coefficient = 0.500

Distance • Yards	Muzzle	100	200	300	400	500	600	800	1000
Velocity • fps	3060	2870	2680	2500	2330	2170	2014	1721	1463
Energy • ft-lbs	3120	2735	2395	2090	1815	1565	1351	987	713
Taylor KO Index	18.2	17.1	16.0	14.9	13.9	12.9	12.0	10.3	8.7
Path • Inches	-1.5	+1.4	0.0	-6.2	-17.9	-36.1	-61.6	-139.9	-265.6
Wind Drift • Inches	0.0	0.6	2.4	5.5	10.0	16.2	24.2	46.5	78.6

Federal 150-grain Nosler Partition (P270WSMC)

G1 Ballistic Coefficient = 0.470

Distance • Yards	Muzzle	100	200	300	400	500	600	800	1000
Velocity • fps	3100	2890	2690	2500	2320	2150	1984	1677	1411
Energy • ft-lbs	3200	2785	2415	2085	1790	1540	1312	937	663
Taylor KO Index	18.4	17.2	16.0	14.8	13.8	12.8	11.8	10.0	8.4
Path • Inches	-1.5	+1.3	0.0	-6.1	-17.8	-36.0	-61.6	-141.0	-270.4
Wind Drift • Inches	0.0	0.6	2.5	5.8	10.6	17.2	25.7	49.6	84.3

Nosler 150-grain Partition (15220)

G1 Ballistic Coefficient = 0.473

Distance • Yards	Muzzle	100	200	300	400	500	600	800	1000
Velocity • fps	3000	2798	2607	2422	2246	2077	1915	1618	1364
Energy • ft-lbs	2997	2607	2263	1954	1680	1437	1222	872	620
Taylor KO Index	17.6	16.4	15.3	14.2	13.2	12.2	11.2	9.5	8.0
Path • Inches	-1.5	+1.5	0.0	-6.6	-19.1	-38.6	-66.1	-151.8	-290.2
Wind Drift • Inches	0.0	0.6	2.6	6.0	11.0	17.9	26.7	51.6	87.6

Winchester 150-grain Ballistic Silvertip (SBST2705A)

G1 Ballistic Coefficient = 0.497

Distance • Yards	Muzzle	100	200	300	400	500	600	800	1000
Velocity • fps	3120	2923	2734	2554	2380	2213	2053	1755	1490
Energy • ft-lbs	3242	2845	2490	2172	1886	1631	1405	1026	740
Taylor KO Index	18.5	17.4	16.2	15.2	14.1	13.1	12.2	10.4	8.8
Path • Inches	-1.5	+1.3	0.0	-5.9	-17.2	-34.7	-59.2	-134.4	-255.2
Wind Drift • Inches	0.0	0.6	2.3	5.4	9.9	16.0	23.8	45.6	77.1

Federal 150-grain Trophy Bonded Tip (P270WSMTT2)

G1 Ballistic Coefficient = 0.469

Distance • Yards	Muzzle	100	200	300	400	500	600	800	1000
Velocity • fps	3100	2890	2700	2510	2330	2150	1984	1676	1410
Energy • ft-lbs	3200	2790	2420	2065	1800	1540	1311	936	662
Taylor KO Index	18.2	17.0	15.9	14.7	13.7	12.6	11.6	9.8	8.3
Path • Inches	-1.5	+1.3	0.0	-6.1	-17.7	-35.9	-61.6	-141.0	-270.5
Wind Drift • Inches	0.0	0.6	2.5	8.8	10.6	17.2	25.7	49.6	84.3

.270 Weatherby Magnum

The .270 Weatherby fills the space between the 7mm Weatherby Magnum and the .257 Weatherby Magnum. From a performance standpoint, it isn't different enough from the 7mm Weatherby Magnum for anyone to want both, except perhaps to be able to brag that he has a Weatherby gun in every caliber. It's a screamer: Bullet for bullet, Weatherby's .270 is 250 to 300 fps faster than the .270 Winchester, and that makes the caliber a great choice for hunting on the high plains. A good selection of bullet weights and styles is available.

Relative Recoil Factor = 2.05

Specifications:

Controlling Agency for Standardization of this Ammunition: CIP

Bullet Weight Grains	Velocity fps	Maximum Average Pressure	
		Copper Crusher	Transducer
N/S	N/S	3,800 bar	4,370 bar

Standard barrel for velocity testing: 26 inches long—1 turn in 10-inch twist

Availability:

Weatherby 100-grain Pointed-Expanding (H 270 100 SP)
G1 Ballistic Coefficient = 0.307

Distance • Yards	Muzzle	100	200	300	400	500	600	800	1000
Velocity • fps	3760	3396	3061	2751	2462	2190	1935	1490	1164
Energy • ft-lbs	3139	2560	2081	1681	1346	1065	832	493	301
Taylor KO Index	14.9	13.4	12.1	10.9	9.7	8.7	7.7	5.9	4.6
Path • Inches	-1.5	+0.8	0.0	-4.5	-13.6	-28.5	-50.6	-125.7	-264.8
Wind Drift • Inches	0.0	0.7	3.1	7.2	13.5	22.2	33.8	68.0	120.7

Nosler 130-grain Partition (15020)
G1 Ballistic Coefficient = 0.416

Distance • Yards	Muzzle	100	200	300	400	500	600	800	1000
Velocity • fps	3450	3197	2959	2734	2519	2314	2119	1759	1448
Energy • ft-lbs	3437	2952	2529	2158	1832	1546	1297	893	605
Taylor KO Index	17.6	16.3	15.1	13.9	12.8	11.8	10.8	9.4	7.4
Path • Inches	-1.5	+1.0	0.0	-4.9	-14.5	-29.6	-51.2	-119.2	-232.8
Wind Drift • Inches	0.0	0.6	2.5	5.7	10.5	17.1	25.6	49.8	85.4

Weatherby 130-grain Pointed-Expanding (H 270 130 SP)
G1 Ballistic Coefficient = 0.409

Distance • Yards	Muzzle	100	200	300	400	500	600	800	1000
Velocity • fps	3375	3123	2885	2659	2444	2240	2044	1686	1383
Energy • ft-lbs	3288	2815	2402	2041	1724	1448	1206	821	552
Taylor KO Index	17.4	16.1	14.8	13.7	12.6	11.5	10.6	8.7	7.1
Path • Inches	-1.5	+1.0	0.0	-5.2	-15.4	-31.4	-54.3	-127.0	-249.2
Wind Drift • Inches	0.0	0.6	2.6	6.0	11.1	18.0	27.0	52.7	90.7

Weatherby 130-grain Nosler Partition (N 270 130 PT)
G1 Ballistic Coefficient = 0.417

Distance • Yards	Muzzle	100	200	300	400	500	600	800	1000
Velocity • fps	3375	3127	2892	2670	2458	2256	2066	1712	1409
Energy • ft-lbs	3288	2822	2415	2058	1744	1470	1232	846	574
Taylor KO Index	17.4	16.1	14.9	13.7	12.6	11.6	10.7	8.8	7.3
Path • Inches	-1.5	+1.0	0.0	-5.2	-15.3	-31.1	-60.8	-141.9	-278.4
Wind Drift • Inches	0.0	0.6	2.5	5.9	10.8	17.6	26.4	51.3	88.1

Nosler 130-grain Ballistic Tip (15004)
G1 Ballistic Coefficient = 0.433

Distance • Yards	Muzzle	100	200	300	400	500	600	800	1000
Velocity • fps	3450	3207	2978	2760	2552	2354	2164	1813	1504
Energy • ft-lbs	3437	2970	2560	2199	1880	1600	1352	949	653
Taylor KO Index	17.6	16.3	15.2	14.0	13.0	12.0	11.0	9.2	7.7
Path • Inches	-1.5	+0.9	0.0	-4.9	-14.3	-29.1	-50.2	-116.1	-224.9
Wind Drift • Inches	0.0	0.6	2.4	5.5	10.1	16.3	24.4	47.1	80.5

Weatherby 130-grain Barnes TSX (B270 130 TSX)
G1 Ballistic Coefficient = 0.466

Distance • Yards	Muzzle	100	200	300	400	500	600	800	1000
Velocity • fps	3400	3176	2963	2761	2567	2382	2204	1872	1576
Energy • ft-lbs	3338	2912	2536	2201	1903	1638	1403	1012	717
Taylor KO Index	17.5	16.3	15.2	14.2	13.2	12.3	11.3	9.6	8.1
Path • Inches	-1.5	+1.0	0.0	-4.9	-14.4	-29.2	-50.2	-114.8	-219.5
Wind Drift • Inches	0.0	0.5	2.2	5.2	9.4	15.3	22.8	43.7	74.2

Cor-Bon 130-grain DPX (DPX270WBY130/20)
G1 Ballistic Coefficient = 0.430

Distance • Yards	Muzzle	100	200	300	400	500	600	800	1000
Velocity • fps	3360	3120	2893	2678	2472	2276	2089	1742	1442
Energy • ft-lbs	3260	2811	2417	2070	1764	1495	1259	876	601
Taylor KO Index	17.3	16.1	14.9	13.8	12.7	11.7	10.7	9.0	7.4
Path • Inches	-1.5	+1.0	0.0	-5.2	-15.3	-31.0	-53.5	-124.0	-241.0
Wind Drift • Inches	0.0	0.6	2.5	5.7	10.5	17.1	25.6	49.5	84.8

Federal 130-grain Trophy Bonded Tip (P270WBTT1)
G1 Ballistic Coefficient = 0.439

Distance • Yards	Muzzle	100	200	300	400	500	600	800	1000
Velocity • fps	3200	2970	2760	2560	2360	2170	1991	1663	1382
Energy • ft-lbs	2955	2555	2200	1885	1610	1365	1145	798	552
Taylor KO Index	16.5	15.3	14.2	13.2	12.1	11.2	10.2	8.6	7.1
Path • Inches	-1.5	+1.2	0.0	-5.8	-16.9	-34.3	-59.2	-136.9	-265.5
Wind Drift • Inches	0.0	0.6	2.6	6.0	11.0	17.8	26.7	51.8	88.6

Weatherby 140-grain AccuBond (N270 240 ACB)
G1 Ballistic Coefficient = 0.496

Distance • Yards	Muzzle	100	200	300	400	500	600	800	1000
Velocity • fps	3320	3113	2916	2727	2547	2373	2206	1893	1611
Energy • ft-lbs	3427	3014	2644	2313	2016	1751	1514	1114	807
Taylor KO Index	18.4	17.2	16.2	15.1	14.1	13.1	12.2	10.5	8.9
Path • Inches	-1.5	+1.0	0.0	-5.1	-15.0	-30.1	-51.5	-116.9	-221.4
Wind Drift • Inches	0.0	0.5	2.1	5.0	9.1	14.7	21.8	41.7	70.4

Weatherby 140-grain BST (N 270 140 BST)
G1 Ballistic Coefficient = 0.456

Distance • Yards	Muzzle	100	200	300	400	500	600	800	1000
Velocity • fps	3300	3077	2865	2663	2470	2285	2107	1777	1488
Energy • ft-lbs	3385	2943	2551	2204	1896	1622	1380	982	688
Taylor KO Index	18.3	17.0	15.9	14.8	13.7	12.7	11.7	9.9	8.3
Path • Inches	-1.5	+1.1	0.0	-5.3	-15.6	-31.5	-54.2	-124.4	-239.1
Wind Drift • Inches	0.0	0.6	2.4	5.5	10.1	16.3	24.4	47.0	80.0

Federal 140-grain Nosler AccuBond (P270WBA2)
G1 Ballistic Coefficient = 0.498

Distance • Yards	Muzzle	100	200	300	400	500	600	800	1000
Velocity • fps	3100	2900	2720	2540	2360	2200	2040	1743	1481
Energy • ft-lbs	2985	2620	2295	2000	1735	1500	1294	945	682
Taylor KO Index	17.2	16.1	15.1	14.1	13.1	12.2	11.3	9.7	8.2
Path • Inches	-1.5	+1.3	0.0	-6.0	-17.4	-35.2	-60.0	-136.2	-258.7
Wind Drift • Inches	0.0	0.6	2.3	5.4	9.9	16.0	24.0	45.9	77.7

Nosler 140-grain Ballistic Tip (15045)
G1 Ballistic Coefficient = 0.456

Distance • Yards	Muzzle	100	200	300	400	500	600	800	1000
Velocity • fps	3300	3077	2865	2663	2470	2285	2108	1778	1489
Energy • ft-lbs	3386	2944	2535	2205	1897	1623	1381	983	690
Taylor KO Index	18.1	16.9	15.7	14.6	13.5	12.5	11.6	9.7	8.2
Path • Inches	-1.5	+1.1	0.0	-5.3	-15.6	-31.5	-54.1	-124.4	-239.0
Wind Drift • Inches	0.0	0.6	2.4	5.5	10.1	16.3	24.4	46.9	79.9

Weatherby 150-grain SP (H 270 150 SP)

G1 Ballistic Coefficient = 0.462

Distance • Yards	Muzzle	100	200	300	400	500	600	800	1000
Velocity • fps	3245	3028	2821	2623	2434	2253	2078	1756	1473
Energy • ft-lbs	3507	3053	2650	2292	1973	1690	1439	1027	723
Taylor KO Index	19.3	18.0	16.7	15.6	14.4	13.4	12.4	10.5	8.8
Path • Inches	-1.5	+1.2	0.0	-5.5	-16.1	-32.6	-55.9	-128.3	-246.2
Wind Drift • Inches	0.0	0.6	2.4	5.5	10.2	16.4	24.6	47.3	80.5

Weatherby 150-grain Nosler Partition (N 270 150 PT)

G1 Ballistic Coefficient = 0.466

Distance • Yards	Muzzle	100	200	300	400	500	600	800	1000
Velocity • fps	3245	3029	2823	2627	2439	2259	2087	1786	1484
Energy • ft-lbs	3507	3055	2655	2298	1981	1699	1451	1040	734
Taylor KO Index	19.3	18.0	16.8	15.6	14.5	13.4	12.4	10.6	8.8
Path • Inches	-1.5	+1.2	0.0	-5.5	-16.1	-32.5	-55.7	-127.6	-244.5
Wind Drift • Inches	0.0	0.6	2.4	5.5	10.1	16.3	24.3	46.8	79.5

Nosler 150-grain Partition (15085)

G1 Ballistic Coefficient = 0.465

Distance • Yards	Muzzle	100	200	300	400	500	600	800	1000
Velocity • fps	3150	2939	2737	2544	2360	2183	2014	1700	1428
Energy • ft-lbs	3306	2877	2496	2157	1855	1587	1351	963	679
Taylor KO Index	18.5	17.3	16.1	14.9	13.9	12.8	11.8	10.0	8.4
Path • Inches	-1.5	+1.3	0.0	-5.9	-17.2	-34.8	-59.6	-136.7	-262.4
Wind Drift • Inches	0.0	0.6	2.4	5.7	10.5	17.0	25.4	49.0	83.3

Young boys especially like the "cool" look of the semiautomatic AR family of guns, and it is a great way to introduce them to shooting sports. (Safari Press Archives)

7-30 Waters

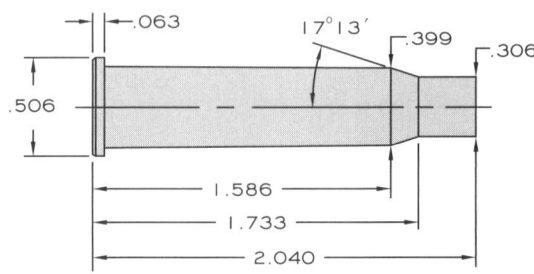

The name of the 7-30 Waters could be a bit misleading. This is not a 7x30mm cartridge that you might get fooled into expecting from the designation. Instead it is a 7mm-(.30-30), hence 7-30. Ken Waters reportedly designed this to be a flat-shooting round that could be used in 7mm conversions of Winchester Model 94 guns. The pressure limits are, like the .30-30, very mild by today's standards. A word of warning here, the only factory ammunition available today should definitely NOT be used in guns with tubular magazines. If you have a tubular magazine gun chambered for the 7-30 Waters, you are stuck with handloading with flatnose bullets.

Relative Recoil Factor = 1.46

Specifications:

Controlling Agency for Standardization of this Ammunition: SAAMI

Bullet Weight Grains	Velocity fps	Maximum Average Pressure	
		Copper Crusher	Transducer
120	2,700	40,000 cup	45,000 psi

Standard barrel for velocity testing: 24 inches long—1 turn in 9.5-inch twist

Availability:

Federal 120-grain Sierra GameKing BTSP (P730A)

G1 Ballistic Coefficient = 0.219

Distance • Yards	Muzzle	100	200	300	400	500	600	800	1000
Velocity • fps	2700	2300	1930	1600	1330	1140	1014	866	764
Energy • ft-lbs	1940	1405	990	685	470	345	274	200	156
Taylor KO Index	13.1	11.2	9.4	7.8	6.5	5.6	4.9	4.2	3.7
Path • Inches	-1.5	2.4	0.0	-11.9	-37.2	-81.9	-153.2	-403.5	-839.9
Wind Drift • Inches	0.0	1.6	7.0	17.3	33.6	56.8	86.6	160.9	252.1

Berit Aagaard with a whitetail shot with a 7mm-08. (Photo from *Guns and Hunting* by Finn Aagaard, 2012, Safari Press Archives)

7x57mm Mauser

The 7x57 Mauser is the granddaddy of all modern centerfire-rifle cartridges. It was introduced in 1892 for a bolt-action rifle that ultimately became known as the Spanish Mauser. Used in Cuba in the Spanish-American War, the rifle and cartridge so far outperformed the .30-40 Krag used by the U.S. Army at that time that it moved the War Department to expedite development of a new U.S. military rifle and cartridge. The U.S. Army wanted a .30 caliber and started by simply necking the 7x57 to .30 caliber. That's the easiest kind of designing, simply copy what works for someone else and put your own name on it. In any event, the outcome of all this was the 1903 Springfield rifle and the .30-06 cartridge, both classics. The 7x57 has been a commercial cartridge in the U.S. since the early 1900s. It has never been hugely popular but at an age in excess of 100 years, it is still a very functional number. Because some old guns that can't stand today's working pressures still exist, factory ammunition for the 7x57 Mauser is loaded to rather mild pressure levels.

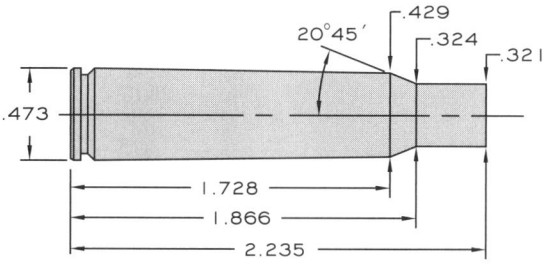

Relative Recoil Factor = 1.68

Specifications:

Controlling Agency for Standardization of this Ammunition: SAAMI

Bullet Weight Grains	Velocity fps	Maximum Average Pressure	
		Copper Crusher	Transducer
120	2,700	40,000 cup	45,000 psi
139	2,650	46,000 cup	51,000 psi
145	2,680	46,000 cup	51,000 psi
154	2,600	46,000 cup	51,000 psi
160	2,500	46,000 cup	51,000 psi
175	2,420	46,000 cup	51,000 psi

Standard barrel for velocity testing: 24 inches long—1 turn in 8.75-inch twist

Availability:

Hornady 139-grain SST (81553)
G1 Ballistic Coefficient = 0.486

Distance • Yards	Muzzle	100	200	300	400	500	600	800	1000
Velocity • fps	2760	2576	2396	2226	2061	1905	1760	1495	1275
Energy • ft-lbs	2351	2045	1772	1528	1311	1120	956	690	502
Taylor KO Index	15.6	14.5	13.5	12.6	11.6	10.7	9.9	8.4	7.2
Path • Inches	-1.5	+1.9	0.0	-7.9	-22.9	-46.1	-79.3	-180.9	-345.1
Wind Drift • Inches	0.0	0.7	2.8	6.4	11.9	19.3	28.8	55.4	93.6

Hornady 139-grain BTSP Interlock (8155)
G1 Ballistic Coefficient = 0.392

Distance • Yards	Muzzle	100	200	300	400	500	600	800	1000
Velocity • fps	2680	2455	2241	2038	1846	1667	1503	1232	1055
Energy • ft-lbs	2216	1860	1550	1282	1052	858	698	468	344
Taylor KO Index	15.1	13.8	12.6	11.5	10.4	9.4	8.5	6.9	5.8
Path • Inches	-1.5	+2.1	0.0	-9.1	-27.4	-56.9	-94.5	-224.8	-447.1
Wind Drift • Inches	0.0	0.9	3.7	8.7	16.2	26.6	40.3	78.8	132.6

Sellier & Bellot 139-grain SP (V330902U)
G1 Ballistic Coefficient = 0.250

Distance • Yards	Muzzle	100	200	300	400	500	600	800	1000
Velocity • fps	2651	2306	1987	1697	1443	1235	1088	923	819
Energy • ft-lbs	2170	1641	1219	889	642	471	365	263	207
Taylor KO Index	15.0	13.0	11.2	9.6	8.1	7.0	6.1	5.2	4.6
Path • Inches	-1.5	+2.6	0.0	-11.4	-35.0	-75.5	-138.8	-359.2	-743.1
Wind Drift • Inches	0.0	1.4	6.2	15.0	28.9	48.6	74.5	140.9	223.0

Hornady 139-grain GMX (81556) LF
G1 Ballistic Coefficient = 0.487

Distance • Yards	Muzzle	100	200	300	400	500	600	800	1000
Velocity • fps	2740	2555	2378	2208	2045	1889	1743	1475	1256
Energy • ft-lbs	2317	2015	1745	1504	1290	1101	938	672	487
Taylor KO Index	14.4	13.4	12.5	11.6	10.8	9.9	9.2	7.8	6.6
Path • Inches	-1.5	+1.9	0.0	-8.1	-23.3	-46.8	-80.1	-183.1	-350.6
Wind Drift • Inches	0.0	0.7	2.8	6.6	12.2	19.7	29.6	57.0	96.2

Federal 140-grain Soft Point (7B)
G1 Ballistic Coefficient = 0.430

Distance • Yards	Muzzle	100	200	300	400	500	600	800	1000
Velocity • fps	2660	2450	2260	2070	1890	1730	1573	1305	1113
Energy • ft-lbs	2200	1865	1585	1330	1110	930	770	530	385
Taylor KO Index	15.1	13.9	12.8	11.8	10.7	9.8	8.9	7.4	6.3
Path • Inches	-1.5	+2.1	0.0	-9.0	-26.0	-52.8	-91.3	-214.2	-421.6
Wind Drift • Inches	0.0	0.8	3.4	7.9	14.7	24.0	36.2	70.4	118.7

Federal 140-grain Nosler Partition (P7C)
G1 Ballistic Coefficient = 0.430

Distance • Yards	Muzzle	100	200	300	400	500	600	800	1000
Velocity • fps	2660	2450	2260	2070	1890	1730	1573	1305	1113
Energy • ft-lbs	2200	1865	1585	1330	1110	930	770	530	385
Taylor KO Index	15.1	13.9	12.8	11.8	10.7	9.8	8.9	7.4	6.3
Path • Inches	-1.5	+2.1	0.0	-9.0	-26.0	-52.8	-91.3	-214.2	-421.6
Wind Drift • Inches	0.0	0.8	3.4	7.9	14.7	24.0	36.2	70.4	118.7

Sellier & Bellot 140-grain FMJ (V330972U)
G1 Ballistic Coefficient = 0.382

Distance • Yards	Muzzle	100	200	300	400	500	600	800	1000
Velocity • fps	2621	2405	2206	2023	1844	1679	1520	1258	1079
Energy • ft-lbs	2119	1783	1500	1262	1058	873	719	492	362
Taylor KO Index	14.9	13.7	12.5	11.5	10.5	9.5	8.6	7.1	6.1
Path • Inches	-1.5	+2.2	0.0	-9.4	-27.2	-55.4	-96.0	-226.6	-447.7
Wind Drift • Inches	0.0	0.8	3.6	8.4	15.6	25.4	38.4	74.7	125.5

Winchester 145-grain Power-Point (X7MM1)
G1 Ballistic Coefficient = 0.355

Distance • Yards	Muzzle	100	200	300	400	500	600	800	1000
Velocity • fps	2660	2413	2180	1959	1754	1564	1396	1137	990
Energy • ft-lbs	2279	1875	1530	1236	990	788	628	417	318
Taylor KO Index	15.6	14.2	12.8	11.5	10.3	9.2	8.2	6.7	5.8
Path • Inches	-1.5	2.2	0.0	-9.6	-28.3	-58.3	-102.6	249.2	-503.1
Wind Drift • Inches	0.0	1.0	4.2	9.9	18.5	30.6	66.6	91.1	151.6

Norma 156-grain Oryx (17001)
G1 Ballistic Coefficient = 0.332

Distance • Yards	Muzzle	100	200	300	400	500	600	800	1000
Velocity • fps	2641	2377	2128	1894	1679	1484	1314	1074	944
Energy • ft-lbs	2417	1957	1569	1243	977	763	599	399	309
Taylor KO Index	16.7	15.0	13.5	12.0	10.6	9.4	8.3	6.8	6.0
Path • Inches	-1.5	+2.3	0.0	-10.0	-29.8	-62.1	-110.5	-274.6	-557.3
Wind Drift • Inches	0.0	1.1	4.6	10.9	20.5	33.9	51.8	101.4	166.9

Remington 160-grain Core-Lokt Ultra Bonded (PRC7MMRC)
G1 Ballistic Coefficient = 0.415

Distance • Yards	Muzzle	100	200	300	400	500	600	800	1000
Velocity • fps	2950	2724	2510	2305	2109	1924	1750	1439	1198
Energy • ft-lbs	3091	2636	2237	1887	1581	1315	1088	736	510
Taylor KO Index	19.1	17.7	16.3	15.0	13.7	12.5	11.4	9.3	7.8
Path • Inches	-1.5	+1.4	0.0	-7.1	-20.8	-42.4	-73.3	-171.8	-338.1
Wind Drift • Inches	0.0	0.7	3.0	7.1	13.1	21.4	32.3	62.3	108.1

Sellier & Bellot 173-grain SPCE (V330912U)
G1 Ballistic Coefficient = 0.307

Distance • Yards	Muzzle	100	200	300	400	500	600	800	1000
Velocity • fps	2379	2109	1869	1657	1459	1288	1153	985	884
Energy • ft-lbs	2171	1706	1341	1053	818	638	511	373	301
Taylor KO Index	16.7	14.8	13.1	11.6	10.2	9.0	8.1	6.9	6.2
Path • Inches	-1.5	+3.1	0.0	-13.0	-38.8	-81.1	-144.9	-355.9	-701.4
Wind Drift • Inches	0.0	1.3	5.6	13.3	25.0	41.4	62.6	118.1	187.1

Federal 175-grain Soft Point RN (7A)

G1 Ballistic Coefficient = 0.273

Distance • Yards	Muzzle	100	200	300	400	500	600	800	1000
Velocity • fps	2390	2090	1810	1560	1350	1180	1059	917	823
Energy • ft-lbs	2220	1395	1275	950	750	540	436	327	263
Taylor KO Index	17.0	14.8	12.9	11.1	9.6	8.4	7.5	6.5	5.8
Path • Inches	-1.5	+3.2	0.0	-14.1	-42.4	-90.4	-161.9	-404.8	-813.3
Wind Drift • Inches	0.0	1.5	6.6	15.8	30.1	50.0	75.3	139.0	216.8

Berit Aagaard poses with a 300-pound aoudad ram taken by a Brno Mauser 7x57mm. (Photo from *Guns and Hunting* by Finn Aagaard, 2012, Safari Press Archives)

7mm-08 Remington

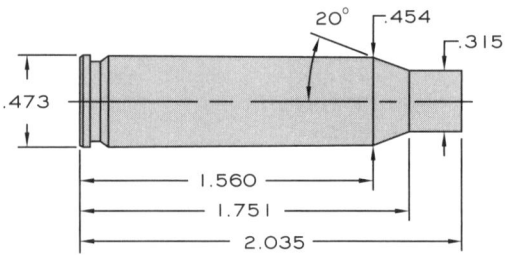

It took quite a while for the wildcat 7mm-08 to go commercial. There were 7mm wildcat versions of the .308 as early as about 1956 or '57, soon after the .308 appeared. Remington announced the 7mm-08 in 1980. All the cartridges based on the .308 Winchester (7.62 NATO) are excellent, and the 7mm-08 is certainly no exception. With a volume about 18 percent smaller than the .30-06, the .308 variants work out very nicely for calibers all the way down to the .243. The 7mm-08 and the .260 Remington (6.5mm-08) are the latest to become commercial but may very well be the best of all.

Relative Recoil Factor = 1.80

Specifications:

Controlling Agency for Standardization of this Ammunition: SAAMI

Bullet Weight Grains	Velocity fps	Maximum Average Pressure	
		Copper Crusher	Transducer
120	2,990	52,000 cup	61,000 psi
140	2,845	52,000 cup	61,000 psi

Standard barrel for velocity testing: 24 inches long—1 turn in 9.5-inch twist

Availability:

Federal 120-grain Fusion (F708FS2)
G1 Ballistic Coefficient = 0.334

Distance • Yards	Muzzle	100	200	300	400	500	600	800	1000
Velocity • fps	3000	2720	2460	2210	1970	1750	1550	1226	1031
Energy • ft-lbs	2400	1970	1605	1295	1035	820	641	400	283
Taylor KO Index	14.6	13.2	12.0	10.8	9.6	8.5	7.5	6.0	5.0
Path • Inches	-1.5	+1.6	0.0	-7.4	-22.0	-45.6	-80.7	-198.3	-408.7
Wind Drift • Inches	0.0	0.9	3.7	8.8	16.6	27.5	41.9	83.7	143.3

Cor-Bon 120-grain DPX (DPX7MM08120/20)
G1 Ballistic Coefficient = 0.344

Distance • Yards	Muzzle	100	200	300	400	500	600	800	1000
Velocity • fps	3000	2731	2477	2237	2010	1798	1603	1276	1065
Energy • ft-lbs	2399	1987	1635	1334	1077	862	685	434	302
Taylor KO Index	14.6	13.3	12.1	10.9	9.8	8.8	7.8	6.2	5.2
Path • Inches	-1.5	+1.6	0.0	-7.3	-21.5	-44.4	-78.1	-189.8	-387.4
Wind Drift • Inches	0.0	0.8	3.6	8.4	15.7	25.9	39.4	78.3	134.4

Nosler 120-grain Ballistic Tip (18200)
G1 Ballistic Coefficient = 0.424

Distance • Yards	Muzzle	100	200	300	400	500	600	800	1000
Velocity • fps	3000	2776	2563	2360	2169	1982	1808	1494	1242
Energy • ft-lbs	2399	2055	1752	1485	1252	1048	871	595	411
Taylor KO Index	14.6	13.5	12.5	11.5	10.6	9.6	8.8	7.3	6.0
Path • Inches	-1.5	+1.5	0.0	-6.8	-19.9	-40.4	-69.7	-162.6	-318.5
Wind Drift • Inches	0.0	0.7	2.9	6.8	12.5	20.4	30.7	59.8	102.3

Remington 120-grain Hollow Point (R7M082)
G1 Ballistic Coefficient = 0.344

Distance • Yards	Muzzle	100	200	300	400	500	600	800	1000
Velocity • fps	3000	2725	2467	2223	1992	1778	1582	1256	1050
Energy • ft-lbs	2398	1979	1621	1316	1058	842	667	420	294
Taylor KO Index	14.6	13.3	12.0	10.8	9.7	8.7	7.7	6.7	5.1
Path • Inches	-1.5	+1.9	0.0	-7.6	-21.7	-44.9	-79.1	-193.1	-395.8
Wind Drift • Inches	0.0	0.9	3.6	8.6	16.0	26.5	40.4	80.4	138.0

Hornady 139-grain SST (80573)

G1 Ballistic Coefficient = 0.486

Distance • Yards	Muzzle	100	200	300	400	500	600	800	1000
Velocity • fps	2950	2757	2571	2393	2222	2059	1902	1614	1367
Energy • ft-lbs	2686	2345	2040	1768	1524	1308	1117	804	577
Taylor KO Index	16.6	15.5	14.5	13.5	12.5	11.6	10.7	9.1	7.7
Path • Inches	-1.5	+1.5	0.0	-6.8	-19.7	-39.6	-67.8	-154.7	-295.5
Wind Drift • Inches	0.0	0.6	2.6	5.9	10.9	17.7	26.5	51.0	86.5

Hornady 139-grain GMX (80576) LF

G1 Ballistic Coefficient = 0.486

Distance • Yards	Muzzle	100	200	300	400	500	600	800	1000
Velocity • fps	2910	2718	2534	2358	2189	2026	1871	1587	1344
Energy • ft-lbs	2613	2280	1982	1716	1478	1267	1081	777	558
Taylor KO Index	16.4	15.3	14.3	13.3	12.3	11.4	10.6	8.9	7.6
Path • Inches	-1.5	+1.6	0.0	-7.0	-20.3	-40.9	-69.9	-159.6	-305.1
Wind Drift • Inches	0.0	0.6	2.6	6.1	11.2	18.1	27.1	52.1	88.3

Federal 140-grain Fusion (F708FS1)

G1 Ballistic Coefficient = 0.390

Distance • Yards	Muzzle	100	200	300	400	500	600	800	1000
Velocity • fps	2850	2620	2390	2180	1980	1790	1615	1315	1104
Energy • ft-lbs	2525	2125	1780	1480	1220	995	811	537	379
Taylor KO Index	16.2	14.9	13.6	12.4	11.2	10.2	9.2	7.5	6.3
Path • Inches	-1.5	+1.8	0.0	-7.9	-23.2	-47.2	-82.2	-195.4	-389.7
Wind Drift • Inches	0.0	0.8	3.4	8.0	14.9	24.4	36.9	72.5	123.7

Nosler 140-grain AccuBond (18205) + (60042)

G1 Ballistic Coefficient = 0.485

Distance • Yards	Muzzle	100	200	300	400	500	600	800	1000
Velocity • fps	2825	2636	2455	2281	2114	1955	1803	1526	1294
Energy • ft-lbs	2482	2181	1874	1618	1390	1188	1011	724	521
Taylor KO Index	16.0	15.0	13.9	13.0	12.0	11.1	10.2	8.7	7.3
Path • Inches	-1.5	+1.8	0.0	-7.5	-21.8	-43.8	-74.9	-171.3	-327.9
Wind Drift • Inches	0.0	0.7	2.7	6.3	11.7	19.0	28.4	54.8	92.7

Remington 140-grain Core-Lokt Pointed Soft Point (R7M081)

G1 Ballistic Coefficient = 0.390

Distance • Yards	Muzzle	100	200	300	400	500	600	800	1000
Velocity • fps	2860	2625	2402	2189	1988	1798	1621	1318	1106
Energy • ft-lbs	2542	2142	1793	1490	1228	1005	817	540	380
Taylor KO Index	16.2	14.9	13.6	12.4	11.3	10.2	9.2	7.5	6.3
Path • Inches	-1.5	+1.8	0.0	-9.2	-22.9	-46.8	-81.6	-194.0	-387.1
Wind Drift • Inches	0.0	0.8	3.4	8.0	14.8	24.3	36.8	72.3	123.4

Federal 140-grain Nosler Partition (P708A)

G1 Ballistic Coefficient = 0.431

Distance • Yards	Muzzle	100	200	300	400	500	600	800	1000
Velocity • fps	2800	2590	2396	2200	2020	1850	1676	1388	1169
Energy • ft-lbs	2435	2085	1775	1500	1265	1060	873	599	425
Taylor KO Index	15.9	14.7	13.6	12.5	11.5	10.5	9.5	7.9	6.6
Path • Inches	-1.5	+1.8	0.0	-8.0	-23.1	-46.6	-81.0	-189.0	-369.8
Wind Drift • Inches	0.0	0.8	3.1	7.3	13.6	22.2	33.4	65.0	110.6

Winchester 140-grain Power-Point (X708)

G1 Ballistic Coefficient = 0.360

Distance • Yards	Muzzle	100	200	300	400	500	600	800	1000
Velocity • fps	2800	2549	2312	2087	1876	1679	1499	1208	1031
Energy • ft-lbs	2437	2020	1661	1354	1094	876	699	454	330
Taylor KO Index	15.9	14.5	13.1	11.9	10.7	9.5	8.5	6.9	5.9
Path • Inches	-1.5	+1.9	0.0	-8.5	-24.9	-51.3	-90.4	-220.1	-447.9
Wind Drift • Inches	0.0	0.9	3.8	9.0	16.8	27.7	42.2	83.3	140.9

Remington 140-grain AccuTip Boat Tail (PRA7M08RB)

G1 Ballistic Coefficient = 0.486

Distance • Yards	Muzzle	100	200	300	400	500	600	800	1000
Velocity • fps	2860	2670	2489	2314	2146	1986	1832	1552	1316
Energy • ft-lbs	2542	2216	1925	1664	1432	1225	1044	749	538
Taylor KO Index	16.2	15.2	14.1	13.1	12.1	11.3	10.4	8.8	7.5
Path • Inches	-1.5	+1.7	0.0	-7.3	-21.1	-42.5	-72.4	-166.5	-319.7
Wind Drift • Inches	0.0	0.6	2.7	6.2	11.4	18.6	27.8	53.6	90.7

Remington 140-grain Core-Lokt Pointed Soft Point (RL7M081)

G1 Ballistic Coefficient = 0.390

Distance • Yards	Muzzle	100	200	300	400	500	600	800	1000
Velocity • fps	2361	2151	1951	1764	1590	1433	1294	1091	972
Energy • ft-lbs	1732	1437	1183	967	786	638	521	370	294
Taylor KO Index	13.4	12.2	11.1	10.0	9.0	8.1	7.3	6.2	5.5
Path • Inches	-1.5	+3.0	0.0	-12.2	-35.6	-72.8	-127.0	-301.9	-593.4
Wind Drift • Inches	0.0	1.1	4.5	10.6	19.8	32.4	48.9	93.6	151.9

Norma 140-grain Nosler BST (17068)

G1 Ballistic Coefficient = 0.485

Distance • Yards	Muzzle	100	200	300	400	500	600	800	1000
Velocity • fps	2822	2633	2452	2278	2112	1953	1801	1524	1293
Energy • ft-lbs	2476	2156	1870	1614	1387	1186	1009	723	520
Taylor KO Index	16.0	15.0	13.9	12.9	12.0	11.1	10.2	8.7	7.3
Path • Inches	-1.5	+1.8	0.0	-7.6	-21.7	-43.7	-74.8	-171.4	-329.5
Wind Drift • Inches	0.0	0.7	2.7	6.4	11.7	19.0	28.4	58.4	92.8

Federal 140-grain Barnes Triple-Shock X-Bullet (P708C)

G1 Ballistic Coefficient = 0.395

Distance • Yards	Muzzle	100	200	300	400	500	600	800	1000
Velocity • fps	2820	2590	2370	2160	1960	1780	1608	1310	1103
Energy • ft-lbs	2470	2085	1745	1450	1200	980	802	533	378
Taylor KO Index	16.0	14.7	13.5	12.3	11.1	10.1	9.1	7.4	6.3
Path • Inches	-1.5	+1.8	0.0	-8.1	-23.6	-48.1	-83.7	-198.5	-395.0
Wind Drift • Inches	0.0	0.8	3.4	8.0	14.9	24.4	37.0	72.5	123.5

Federal 140-grain Nosler Ballistic Tip (P708B)

G1 Ballistic Coefficient = 0.489

Distance • Yards	Muzzle	100	200	300	400	500	600	800	1000
Velocity • fps	2800	2610	2430	2260	2100	1940	1791	1518	1289
Energy • ft-lbs	2440	2135	1840	1590	1360	1165	997	716	517
Taylor KO Index	15.9	14.8	13.8	12.8	11.9	11.0	10.2	8.6	7.3
Path • Inches	-1.5	+1.6	0.0	-8.4	-25.4	-44.5	-76.2	-173.9	-332.6
Wind Drift • Inches	0.0	0.7	2.7	6.4	11.7	19.0	28.5	54.9	92.9

Federal 140-grain Trophy Bonded Tip (P708TT2)

G1 Ballistic Coefficient = 0.431

Distance • Yards	Muzzle	100	200	300	400	500	600	800	1000
Velocity • fps	2800	2590	2390	2200	2010	1840	1677	1390	1170
Energy • ft-lbs	2435	2085	1770	1500	1260	1050	875	601	426
Taylor KO Index	15.9	14.7	13.6	12.5	11.4	10.5	9.5	7.9	6.6
Path • Inches	-1.5	+1.8	0.0	-8.0	-23.2	-46.8	-81.0	-188.8	-369.4
Wind Drift • Inches	0.0	0.8	3.1	7.3	13.6	22.1	33.4	64.9	110.4

Winchester 140-grain Ballistic Silvertip (SBST708)

G1 Ballistic Coefficient = 0.455

Distance • Yards	Muzzle	100	200	300	400	500	600	800	1000
Velocity • fps	2770	2572	2382	2200	2026	1860	1704	1425	1205
Energy • ft-lbs	2386	2056	1764	1504	1276	1076	903	632	451
Taylor KO Index	15.7	14.6	13.5	12.5	11.5	10.6	9.7	8.1	6.8
Path • Inches	-1.5	+1.9	0.0	-8.0	-23.2	-46.9	-80.8	-186.7	-361.7
Wind Drift • Inches	0.0	0.7	3.0	7.0	13.0	21.1	31.7	61.4	104.2

Federal 150-grain Soft Point (708CS)

G1 Ballistic Coefficient = 0.414

Distance • Yards	Muzzle	100	200	300	400	500	600	800	1000
Velocity • fps	2650	2440	2230	2040	1860	1690	1532	1265	1082
Energy • ft-lbs	2340	1980	1660	1390	1150	950	782	533	390
Taylor KO Index	16.1	14.8	13.6	12.4	11.3	10.3	9.3	7.7	6.6
Path • Inches	-1.5	2.2	0.0	-9.2	-26.7	-54.4	-94.0	-221.4	-436.1
Wind Drift • Inches	0.0	0.8	3.5	8.3	15.5	25.3	38.2	74.5	125.5

Cor-Bon 168-grain VLD (PM7MM08168/20)

G1 Ballistic Coefficient = 0.600

Distance • Yards	Muzzle	100	200	300	400	500	600	800	1000
Velocity • fps	2700	2551	2407	2268	2133	2003	1878	1644	1436
Energy • ft-lbs	2720	2429	2162	1919	1698	1497	1316	1008	769
Taylor KO Index	18.4	17.4	16.4	15.5	14.5	13.7	12.8	11.2	9.8
Path • Inches	-1.5	+1.9	0.0	-7.9	-22.6	-45.0	-76.0	-169.0	-313.9
Wind Drift • Inches	0.0	0.6	2.3	5.4	9.8	15.8	23.5	44.5	74.2

7x64mm Brenneke

This cartridge has been around for a very long time. Wilhelm Brenneke designed the 7x64mm in 1917. It is for all practical purposes a 7mm-06, although Wilhelm undoubtedly used a Mauser cartridge for his development brass. In terms of muzzle velocity, the 7x64mm is very close to the .280 Remington, the .270 Winchester, and the .284 Winchester when you compare similar weight bullets, but certainly slower than the 7mm Remington Magnum, the 7mm Weatherby Magnum, and several others. It has never been very popular in the U.S., perhaps because there are too many similar performers in the ammunition catalogs.

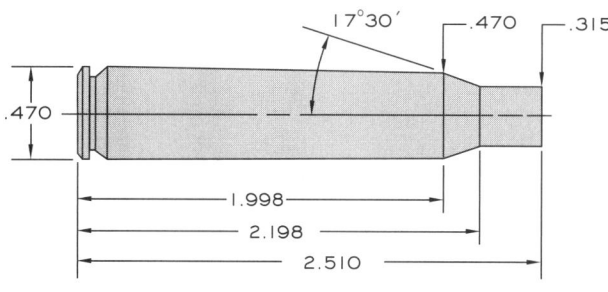

Relative Recoil Factor = 1.85

Specifications:

Controlling Agency for Standardization of this Ammunition: CIP

Bullet Weight Grains	Velocity fps	Maximum Average Pressure Copper Crusher	Transducer
N/A	N/A	---	60,000 psi (estimated)

Standard barrel for velocity testing: 24 inches long—1 turn in 8.66-inch twist

Availability:

Sellier & Bellot 139-grain SP (V331102U)
G1 Ballistic Coefficient = 0.229

Distance • Yards	Muzzle	100	200	300	400	500	600	800	1000
Velocity • fps	2808	2454	2144	1874	1615	1388	1205	988	871
Energy • ft-lbs	2432	1857	1418	1082	805	595	448	301	234
Taylor KO Index	15.8	13.8	12.1	10.6	9.1	7.8	6.8	5.6	4.9
Path • Inches	-1.5	+2.1	0.0	-9.6	-29.2	-62.4	-114.4	-297.6	-614.0
Wind Drift • Inches	0.0	1.2	5.3	12.7	24.2	40.7	62.9	123.2	199.8

Norma 140-grain Nosler AccuBond (17069)
G1 Ballistic Coefficient = 0.485

Distance • Yards	Muzzle	100	200	300	400	500	600	800	1000
Velocity • fps	2953	2759	2572	2394	2223	2059	1902	1613	1365
Energy • ft-lbs	2712	2366	2058	1782	1537	1318	1125	809	579
Taylor KO Index	16.8	15.7	14.6	13.6	12.6	11.7	10.8	9.2	7.8
Path • Inches	-1.5	+1.5	0.0	-6.8	-19.7	-39.6	-67.7	-154.8	-297.0
Wind Drift • Inches	0.0	0.6	2.6	6.0	11.0	17.8	26.6	51.1	86.7

Remington 140 Core-Lokt PSP (R7X641)
G1 Ballistic Coefficient = 0.390

Distance • Yards	Muzzle	100	200	300	400	500	600	800	1000
Velocity • fps	2950	2710	2483	2266	2061	1867	1686	1370	1139
Energy • ft-lbs	2705	2283	1916	1597	1320	1083	883	583	403
Taylor KO Index	16.8	15.4	14.1	12.9	11.7	10.6	9.6	7.8	6.5
Path • Inches	-1.5	+1.5	0.0	-7.3	-21.3	-43.6	-75.9	-180.2	-359.2
Wind Drift • Inches	0.0	0.8	3.2	7.6	14.1	23.2	35.0	68.9	118.1

Norma 156-grain Oryx (17053)
G1 Ballistic Coefficient = 0.330

Distance • Yards	Muzzle	100	200	300	400	500	600	800	1000
Velocity • fps	2789	2516	2259	2017	1793	1587	1403	1127	975
Energy • ft-lbs	2695	2193	1768	1410	1114	872	682	440	330
Taylor KO Index	17.7	15.9	14.3	12.8	11.3	10.0	8.9	7.1	6.2
Path • Inches	-1.5	+2.0	0.0	-8.8	-26.3	-54.6	-97.1	-241.6	-496.2
Wind Drift • Inches	0.0	1.0	4.2	10.0	18.8	31.2	47.7	94.4	158.0

Norma 170-grain ORYX (17020)

G1 Ballistic Coefficient = 0.324

Distance • Yards	Muzzle	100	200	300	400	500	600	800	1000
Velocity • fps	2756	2481	2222	1979	1753	1547	1366	1102	960
Energy • ft-lbs	2868	2324	1864	1479	1160	904	705	458	348
Taylor KO Index	19.0	17.1	15.3	13.6	12.1	10.7	9.4	7.6	6.6
Path • Inches	-1.5	+2.0	0.0	-9.1	-27.2	-56.7	-100.8	-249.8	-512.9
Wind Drift • Inches	0.0	1.0	4.4	10.4	19.6	32.5	49.7	98.2	163.2

Norma 170-grain Plastic Point (17019)

G1 Ballistic Coefficient = 0.378

Distance • Yards	Muzzle	100	200	300	400	500	600	800	1000
Velocity • fps	2723	2488	2265	2053	1852	1666	1496	1218	1042
Energy • ft-lbs	2800	2337	1936	1591	1296	1048	845	560	410
Taylor KO Index	18.8	17.2	15.6	14.2	12.8	11.5	10.3	8.4	7.2
Path • Inches	-1.5	+2.1	0.0	-8.9	-26.0	-53.4	-93.4	-223.5	-447.7
Wind Drift • Inches	0.0	0.9	3.8	8.9	16.6	27.2	41.3	81.1	136.7

Norma 170-grain Vulcan HP (17018)

G1 Ballistic Coefficient = 0.353

Distance • Yards	Muzzle	100	200	300	400	500	600	800	1000
Velocity • fps	2723	2472	2234	2010	1798	1604	1429	1157	1000
Energy • ft-lbs	2800	2307	1884	1525	1221	971	771	506	378
Taylor KO Index	18.8	17.0	15.4	13.9	12.4	11.1	9.9	8.0	6.9
Path • Inches	-1.5	+2.1	0.0	-9.1	-26.9	-55.4	-97.5	-237.2	-480.7
Wind Drift • Inches	0.0	1.0	4.0	10.0	18.0	29.7	45.2	89.0	149.1

Sellier & Bellot 173-grain SPCE (V331112U)

G1 Ballistic Coefficient = 0.325

Distance • Yards	Muzzle	100	200	300	400	500	600	800	1000
Velocity • fps	2526	2265	2019	1791	1581	1396	1240	1033	919
Energy • ft-lbs	2449	1971	1567	1232	961	749	590	410	325
Taylor KO Index	17.7	15.9	14.2	12.6	11.1	9.8	8.7	7.3	6.5
Path • Inches	-1.5	+2.6	0.0	-11.2	-33.3	-69.4	-123.4	-304.5	-615.3
Wind Drift • Inches	0.0	1.2	5.0	11.8	22.3	37.0	56.3	108.7	175.7

Remington 175-grain Core-Lokt PSP (R7X642)

G1 Ballistic Coefficient = 0.427

Distance • Yards	Muzzle	100	200	300	400	500	600	800	1000
Velocity • fps	2600	2397	2203	2018	1842	1678	1526	1267	1088
Energy • ft-lbs	2626	2232	1885	1582	1319	1093	905	624	461
Taylor KO Index	18.5	17.0	15.6	14.3	13.1	11.9	10.8	9.0	7.7
Path • Inches	-1.5	+2.0	0.0	-9.5	-27.5	-55.8	-96.5	-225.9	-442.2
Wind Drift • Inches	0.0	0.8	3.5	8.3	15.3	25.1	37.8	73.3	123.1

.284 Winchester

Winchester's .284 came on the scene in 1963. In a search for a cartridge case with the .30-06's volume that would work in shorter actions, Winchester went to a larger (0.5008 inch) body with the .30-06's 0.473-inch head size. This technique of a fat body and reduced head anticipated the design of the .300 WSM by 38 years. A cartridge with this form of reduced head size is known as having a "rebated head." The .284 never really set the world on fire as a 7mm, but two of its wildcat spin-offs, the 6mm-284 and the 6.5mm-284, have earned a following among long-range target shooters. Only two loadings remain in the current catalogs.

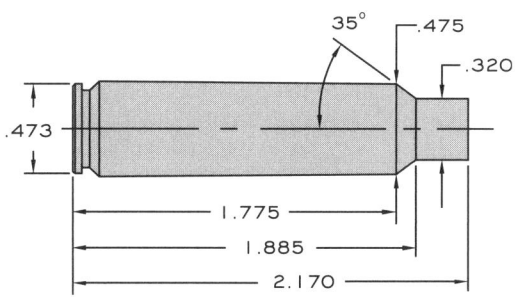

Relative Recoil Factor = 1.93

Specifications:

Controlling Agency for Standardization of this Ammunition: SAAMI

Bullet Weight Grains	Velocity fps	Maximum Average Pressure	
		Copper Crusher	Transducer
125	3,125	54,000 cup	56,000 psi
150	2,845	54,000 cup	56,000 psi

Standard barrel for velocity testing: 24 inches long—1 turn in 10-inch twist

Availability:

Cor-Bon 140-grain T- DPX (DPX284140/20)
G1 Ballistic Coefficient = 0.350

Distance • Yards	Muzzle	100	200	300	400	500	600	800	1000
Velocity • fps	2900	2636	2367	2152	1930	1724	1535	1226	1037
Energy • ft-lbs	2615	2161	1772	1440	1158	924	733	468	334
Taylor KO Index	16.5	15.0	13.6	12.2	11.0	9.8	8.7	7.0	5.9
Path • Inches	-1.5	+1.8	0.0	-7.9	-23.3	-48.1	-84.6	-206.0	-420.2
Wind Drift • Inches	0.0	0.9	3.7	8.8	16.5	27.3	41.5	82.4	140.4

Winchester 150-grain Power-Point (X2842)
G1 Ballistic Coefficient = 0.367

Distance • Yards	Muzzle	100	200	300	400	500	600	800	1000
Velocity • fps	2860	2609	2371	2145	1933	1734	1551	1248	1052
Energy • ft-lbs	2724	2243	1830	1480	1185	940	744	476	345
Taylor KO Index	17.4	15.9	14.4	13.1	11.8	10.6	9.4	7.6	6.4
Path • Inches	-1.5	+1.8	0.0	-8.0	-23.6	-48.6	-85.2	-206.9	-421.4
Wind Drift • Inches	0.0	0.9	3.6	8.6	16.0	26.4	40.2	79.4	135.3

.280 Remington

Remington introduced its .280 cartridge in 1957. This cartridge has been the source of numerous articles that compare it with the .30-06 and the .270 Winchester. There really isn't much of a story here. Both the .270 and the .280 are little more than .30-06 cartridges necked to .270 and 7mm respectively. As a result, the performance of these cartridges is so similar that it comes down to exactly which bullet you prefer to use. Both are excellent calibers, and if you have a .280 gun, there's absolutely no reason to feel you have any more, or any less, gun than your friend's .270.

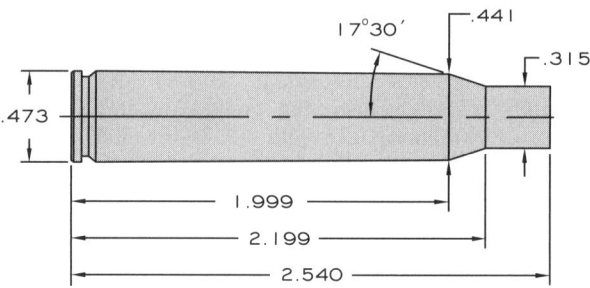

Relative Recoil Factor = 1.95

Specifications:

Controlling Agency for Standardization of this Ammunition: SAAMI

Bullet Weight Grains	Velocity fps	Maximum Average Pressure	
		Copper Crusher	Transducer
120	3,135	52,000 cup	60,000 psi
140	2,985	52,000 cup	60,000 psi
150	2,875	52,000 cup	60,000 psi
165	2,800	52,000 cup	60,000 psi

Standard barrel for velocity testing: 24 inches long—1 turn in 10-inch twist

Availability:

Hornady 139-grain SST (81583)
G1 Ballistic Coefficient = 0.486

Distance • Yards	Muzzle	100	200	300	400	500	600	800	1000
Velocity • fps	3090	2890	2699	2516	2341	2172	2010	1710	1447
Energy • ft-lbs	2946	2578	2249	1954	1691	1456	1248	903	647
Taylor KO Index	17.4	16.3	15.2	14.2	13.2	12.2	11.3	9.6	8.2
Path • Inches	-1.5	+1.3	0.0	-6.1	-17.7	-35.7	-61.0	-139.1	-265.2
Wind Drift • Inches	0.0	0.6	2.4	5.6	10.2	16.6	24.8	47.6	80.6

Hornady 139-grain GMX (81586) LF
G1 Ballistic Coefficient = 0.486

Distance • Yards	Muzzle	100	200	300	400	500	600	800	1000
Velocity • fps	3070	2871	2681	2499	2324	2156	1995	1696	1435
Energy • ft-lbs	2908	2544	2218	1927	1666	1434	1229	888	636
Taylor KO Index	17.3	16.2	15.1	14.1	13.1	12.2	11.3	9.6	8.1
Path • Inches	-1.5	+1.4	0.0	-6.2	-18.0	-36.2	-61.9	-141.2	-269.2
Wind Drift • Inches	0.0	0.6	2.4	5.6	10.3	16.7	25.0	48.0	81.4

Remington 140-grain Core-Lokt Pointed Soft Point (R280R3)
G1 Ballistic Coefficient = 0.391

Distance • Yards	Muzzle	100	200	300	400	500	600	800	1000
Velocity • fps	3000	2758	2528	2309	2102	1905	1723	1400	1159
Energy • ft-lbs	2797	2363	1986	1657	1373	1128	923	610	418
Taylor KO Index	17.0	15.7	14.4	13.1	11.9	10.8	9.8	8.0	6.6
Path • Inches	-1.5	+1.5	0.0	-7.0	-20.5	-42.0	-72.9	-172.9	-344.6
Wind Drift • Inches	0.0	0.8	3.2	7.4	13.8	22.5	34.1	67.0	115.1

Federal 140-grain Fusion (F280FS1)
G1 Ballistic Coefficient = 0.485

Distance • Yards	Muzzle	100	200	300	400	500	600	800	1000
Velocity • fps	2990	2790	2610	2430	2260	2090	1931	1639	1387
Energy • ft-lbs	2780	2425	2115	1830	1580	1355	1160	835	598
Taylor KO Index	17.0	15.8	14.8	13.8	12.8	11.9	11.0	9.3	7.9
Path • Inches	-1.5	+1.5	0.0	-6.5	-19.1	-38.5	-65.8	-150.2	-286.8
Wind Drift • Inches	0.0	0.6	2.5	5.8	10.8	17.4	26.0	50.1	85.0

Winchester 140-grain Ballistic Silvertip (SBST280)

G1 Ballistic Coefficient = 0.485

Distance • Yards	Muzzle	100	200	300	400	500	600	800	1000
Velocity • fps	3040	2842	2653	2471	2297	2130	1969	1672	1414
Energy • ft-lbs	2872	2511	2187	1898	1640	1410	1205	869	622
Taylor KO Index	17.3	16.1	15.1	14.0	13.0	12.1	11.2	9.5	5.0
Path • Inches	-1.5	+1.4	0.0	-6.3	-18.4	-37.0	-63.4	-144.6	-276.1
Wind Drift • Inches	0.0	0.6	2.5	5.7	10.5	17.0	25.4	48.9	83.0

Remington 140-grain AccuTip (PRA280RA)

G1 Ballistic Coefficient = 0.486

Distance • Yards	Muzzle	100	200	300	400	500	600	800	1000
Velocity • fps	3000	2804	2607	2437	2265	2099	1940	1647	1394
Energy • ft-lbs	2797	2444	2129	1846	1594	1369	1170	843	604
Taylor KO Index	17.0	15.9	14.8	13.8	12.9	11.9	11.0	9.4	7.9
Path • Inches	-1.5	+1.5	0.0	-6.6	-19.0	-38.1	-65.3	-149.1	-285.7
Wind Drift • Inches	0.0	0.6	2.5	5.8	10.7	17.3	25.9	49.8	84.4

Federal 140-grain Nosler Ballistic Tip (P280D)

G1 Ballistic Coefficient = 0.486

Distance • Yards	Muzzle	100	200	300	400	500	600	800	1000
Velocity • fps	2990	2790	2610	2430	2260	2090	1932	1640	1387
Energy • ft-lbs	2780	2425	2115	1830	1580	1355	1160	836	599
Taylor KO Index	17.0	15.8	14.8	13.8	12.8	11.9	11.0	9.3	7.9
Path • Inches	-1.5	+1.5	0.0	-6.5	-19.1	-38.5	-65.8	-102.5	-286.7
Wind Drift • Inches	0.0	0.6	2.5	5.8	10.7	17.4	26.0	50.1	84.9

Federal 140-grain Barnes Triple-Shock X-Bullet (P280E) LF

G1 Ballistic Coefficient = 0.393

Distance • Yards	Muzzle	100	200	300	400	500	600	800	1000
Velocity • fps	2960	2720	2500	2280	2080	1880	1699	1382	1148
Energy • ft-lbs	2725	2305	1935	1620	1340	1105	898	594	410
Taylor KO Index	16.8	15.4	14.2	13.0	11.8	10.7	9.7	7.8	6.5
Path • Inches	-1.5	+1.6	0.0	-7.2	-21.1	-42.8	-75.0	-177.8	-354.2
Wind Drift • Inches	0.0	0.8	3.2	7.5	14.0	22.8	34.6	67.9	116.5

Federal 140-grain Trophy Bonded Tip (P280TT2)

G1 Ballistic Coefficient = 0.432

Distance • Yards	Muzzle	100	200	300	400	500	600	800	1000
Velocity • fps	2950	2730	2520	2330	2140	1960	1790	1484	1239
Energy • ft-lbs	2705	2320	1980	1680	1420	1190	996	685	477
Taylor KO Index	16.8	15.4	14.3	13.2	12.2	11.1	10.2	8.4	7.0
Path • Inches	-1.5	+1.6	0.0	-7.0	-28.6	-41.5	-71.8	-166.9	-325.9
Wind Drift • Inches	0.0	0.7	2.9	6.8	12.5	20.4	30.7	59.8	102.1

Cor-Bon 140-grain DPX (DPX280140/20)

G1 Ballistic Coefficient = 0.350

Distance • Yards	Muzzle	100	200	300	400	500	600	800	1000
Velocity • fps	3000	2730	2476	2236	2009	1797	1601	1274	1068
Energy • ft-lbs	2799	2318	1906	1554	1255	1004	797	505	351
Taylor KO Index	17.0	15.5	14.1	12.7	11.4	10.2	9.1	7.2	6.0
Path • Inches	-1.5	+1.6	0.0	-7.3	-21.5	-44.4	-78.2	-190.0	-388.1
Wind Drift • Inches	0.0	0.8	3.6	8.4	15.7	25.9	39.4	78.4	134.7

Federal 150-grain Nosler Partition (P280A)

G1 Ballistic Coefficient = 0.458

Distance • Yards	Muzzle	100	200	300	400	500	600	800	1000
Velocity • fps	2890	2690	2490	2310	2130	1960	1800	1509	1269
Energy • ft-lbs	2780	2405	2070	1770	1510	1275	1080	759	537
Taylor KO Index	17.6	16.4	15.2	14.1	13.0	11.9	11.0	9.2	7.7
Path • Inches	-1.5	+1.7	0.0	-7.2	-21.1	-42.5	-73.1	-168.4	-325.5
Wind Drift • Inches	0.0	0.7	2.8	6.5	12.1	19.6	29.5	57.1	97.0

Federal 150-grain Soft Point (280B)

G1 Ballistic Coefficient = 0.417

Distance • Yards	Muzzle	100	200	300	400	500	600	800	1000
Velocity • fps	2890	2670	2460	2260	2060	1880	1710	1408	1176
Energy • ft-lbs	2780	2370	2015	1695	1420	1180	974	660	461
Taylor KO Index	17.6	16.2	15.0	13.8	12.5	11.4	10.4	8.6	7.2
Path • Inches	-1.5	+1.7	0.0	-7.5	-21.8	-44.3	-76.7	-179.8	-353.7
Wind Drift • Inches	0.0	0.8	3.1	7.3	13.5	22.0	33.2	64.8	110.7

Remington 150-grain Core-Lokt Pointed Soft Point (R280R1)

G1 Ballistic Coefficient = 0.346

Distance • Yards	Muzzle	100	200	300	400	500	600	800	1000
Velocity • fps	2890	2624	2373	2135	1912	1705	1517	1211	1027
Energy • ft-lbs	2781	2293	1875	1518	1217	968	766	489	351
Taylor KO Index	17.6	16.0	14.4	13.0	11.6	10.4	9.2	7.4	6.3
Path • Inches	-1.5	+1.8	0.0	-8.0	-23.6	-48.8	-86.0	-209.8	-428.7
Wind Drift • Inches	0.0	0.9	3.8	9.0	16.8	27.8	42.4	84.1	143.0

Norma 156-grain Oryx (17048)

G1 Ballistic Coefficient = 0.330

Distance • Yards	Muzzle	100	200	300	400	500	600	800	1000
Velocity • fps	2789	2516	2259	2017	1793	1587	1403	1127	975
Energy • ft-lbs	2695	2193	1768	1410	1114	872	682	440	330
Taylor KO Index	17.7	15.9	14.3	12.8	11.3	10.0	8.9	7.1	6.2
Path • Inches	-1.5	+2.0	0.0	-8.8	-26.3	-54.6	-97.1	-241.6	-496.2
Wind Drift • Inches	0.0	1.0	4.2	10.0	18.8	31.2	47.7	94.4	158.0

Nosler 160-grain Partition (15805)

G1 Ballistic Coefficient = 0.475

Distance • Yards	Muzzle	100	200	300	400	500	600	800	1000
Velocity • fps	2775	2584	2402	2226	2058	1898	1746	1472	1248
Energy • ft-lbs	2737	2374	2050	1762	1506	1280	1084	770	553
Taylor KO Index	18.0	16.8	15.6	14.4	13.4	12.3	11.3	9.6	8.1
Path • Inches	-1.5	+1.9	0.0	-7.9	-22.8	-46.0	-78.8	-180.8	-347.6
Wind Drift • Inches	0.0	0.7	2.9	6.7	12.3	20.0	30.0	57.9	97.9

Federal 160-grain Trophy Bonded Tip (P280TT1)

G1 Ballistic Coefficient = 0.522

Distance • Yards	Muzzle	100	200	300	400	500	600	800	1000
Velocity • fps	2800	2630	2460	2290	2140	1990	1847	1583	1356
Energy • ft-lbs	2785	2450	2145	1870	1625	1405	1212	890	653
Taylor KO Index	18.2	17.1	16.0	14.9	13.9	12.9	12.0	10.3	8.8
Path • Inches	-1.5	+1.8	0.0	-7.5	-21.8	-43.6	-74.0	-167.4	-316.8
Wind Drift • Inches	0.0	0.6	2.6	5.9	10.9	17.6	26.3	50.4	84.7

Remington 165-grain Core-Lokt Soft Point (R280R2)

G1 Ballistic Coefficient = 0.291

Distance • Yards	Muzzle	100	200	300	400	500	600	800	1000
Velocity • fps	2820	2510	2220	1950	1701	1479	1291	1042	914
Energy • ft-lbs	2913	2308	1805	1393	1060	801	611	398	306
Taylor KO Index	18.9	16.8	14.9	13.1	11.4	9.9	8.6	7.0	6.1
Path • Inches	-1.5	+2.0	0.0	-9.1	-27.4	-57.8	-104..2	-265.2	-553.6
Wind Drift • Inches	0.0	1.1	4.8	11.4	21.7	36.2	55.8	110.4	181.7

Norma 170-grain Plastic Point (17060)

G1 Ballistic Coefficient = 0.373

Distance • Yards	Muzzle	100	200	300	400	500	600	800	1000
Velocity • fps	2707	2469	2244	2031	1829	1643	1473	1198	1030
Energy • ft-lbs	2764	2300	1899	1555	1263	1019	819	542	400
Taylor KO Index	18.7	17.0	15.5	14.0	12.6	11.3	10.2	8.3	7.1
Path • Inches	-1.5	+2.1	0.0	-9.0	-26.5	-54.5	-95.3	-229.2	-460.0
Wind Drift • Inches	0.0	0.9	3.8	9.1	17.0	27.9	42.4	83.3	140.0

Norma 170-grain Vulcan (17051)

G1 Ballistic Coefficient = 0.353

Distance • Yards	Muzzle	100	200	300	400	500	600	800	1000
Velocity • fps	2592	2348	2117	1900	1696	1512	1349	1107	972
Energy • ft-lbs	2537	2081	1692	1362	1086	863	687	463	357
Taylor KO Index	17.9	16.2	14.6	13.1	11.7	10.4	9.3	7.6	6.7
Path • Inches	-1.5	+2.4	0.0	-10.2	-30.1	-62.1	-109.4	-266.0	-535.8
Wind Drift • Inches	0.0	1.0	4.4	10.3	19.4	32.0	48.6	94.9	156.6

7mm Remington Magnum

Since it became a standardized cartridge in 1962, the 7mm Remington Magnum has easily been the most popular 7mm cartridge in the inventory. That popularity applies to reloaders as well. The 7mm RM has been on the top ten list of reloading die sales for years. The overall length, slightly smaller than the full-length magnums, allows the 7mm RM to be used in standard-length actions, yet the cartridge case volume is large enough to give excellent ballistics. The caliber is versatile. With the lighter bullets the velocities are right around 3,200 fps, and that translates into a flat shooter right out to the longest practical hunting ranges. With the 175-grain bullets in the 2,850 fps class, you have a gun that easily outperforms the legendary .30-06.

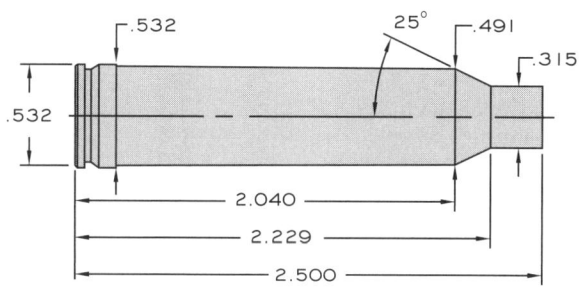

Relative Recoil Factor = 2.06

Specifications:

Controlling Agency for Standardization of this Ammunition: SAAMI

Bullet Weight Grains	Velocity fps	Maximum Average Pressure Copper Crusher	Transducer
125	3,290	52,000 cup	61,000 psi
139	3,150	52,000 cup	61,000 psi
150	3,100	52,000 cup	61,000 psi
154	3,035	52,000 cup	61,000 psi
160–162	2,940	52,000 cup	61,000 psi
175	2,850	52,000 cup	61,000 psi

Standard barrel for velocity testing: 24 inches long—1 turn in 9.5-inch twist

Availability:

Federal 110-grain Barnes Triple-Shock X-Bullet (P7RP) LF

G1 Ballistic Coefficient = 0.429

Distance • Yards	Muzzle	100	200	300	400	500	600	800	1000
Velocity • fps	3500	3200	2920	2650	2400	2170	1948	1551	1240
Energy • ft-lbs	2990	2500	2075	1720	1410	1145	927	588	375
Taylor KO Index	15.6	14.3	13.0	11.8	10.7	9.7	8.7	6.9	5.5
Path • Inches	-1.5	+0.9	0.0	-5.1	-15.0	-31.2	-54.2	-131.3	-266.9
Wind Drift • Inches	0.0	0.7	2.9	6.8	12.6	20.6	31.2	61.9	108.2

Hornady 139-grain SST (80593)

G1 Ballistic Coefficient = 0.486

Distance • Yards	Muzzle	100	200	300	400	500	600	800	1000
Velocity • fps	3240	3033	2836	2648	2467	2293	2126	1814	1537
Energy • ft-lbs	3239	2839	2482	2163	1877	1622	1396	1016	729
Taylor KO Index	18.3	17.1	16.0	14.9	13.9	12.9	12.0	10.2	8.7
Path • Inches	-1.5	+1.1	0.0	-5.5	-15.9	-32.1	-54.8	-124.8	-237.5
Wind Drift • Inches	0.0	0.6	2.2	5.2	9.6	15.5	23.1	44.3	75.0

Hornady 139-grain BTSP InterLock (8059)

G1 Ballistic Coefficient = 0.453

Distance • Yards	Muzzle	100	200	300	400	500	600	800	1000
Velocity • fps	3150	2933	2727	2530	2341	2160	1987	1668	1394
Energy • ft-lbs	3063	2656	2296	1976	1692	1440	1218	859	600
Taylor KO Index	17.8	16.5	15.4	14.3	13.2	12.2	11.2	9.4	7.9
Path • Inches	-1.5	+1.3	0.0	-6.1	-17.7	-35.5	-60.3	-139.2	-269.8
Wind Drift • Inches	0.0	0.6	2.5	5.9	10.8	17.6	26.3	50.8	86.6

PMC 139-grain BTSP (7MMHIA)

G1 Ballistic Coefficient = 0.453

Distance • Yards	Muzzle	100	200	300	400	500	600	800	1000
Velocity • fps	3150	2933	2727	2529	2341	2160	1987	1669	1396
Energy • ft-lbs	3062	2655	2294	1974	1691	1440	1219	860	601
Taylor KO Index	17.8	16.5	15.4	14.3	13.2	12.2	11.2	9.4	7.9
Path • Inches	-1.5	+1.3	0.0	-6.0	-17.4	-35.1	-60.3	-138.9	-268.0
Wind Drift • Inches	0.0	0.6	2.5	5.9	10.8	17.5	26.2	50.7	86.5

Hornady 139-grain GMX (80592) LF

G1 Ballistic Coefficient = 0.486

Distance • Yards	Muzzle	100	200	300	400	500	600	800	1000
Velocity • fps	3190	2986	2791	2604	2425	2253	2088	1780	1507
Energy • ft-lbs	3140	2751	2403	2092	1814	1566	1345	978	701
Taylor KO Index	18.0	16.8	15.7	14.7	13.7	12.7	11.8	10.0	8.5
Path • Inches	-1.5	+1.2	0.0	-5.7	-16.5	-33.2	-56.8	-129.3	-246.2
Wind Drift • Inches	0.0	0.6	2.3	5.3	9.8	15.8	23.6	45.3	76.8

Winchester 140-grain AccuBond CT (S7MMCTA)

G1 Ballistic Coefficient = 0.460

Distance • Yards	Muzzle	100	200	300	400	500	600	800	1000
Velocity • fps	3180	2965	2760	2565	2377	2197	2023	1707	1430
Energy • ft-lbs	3143	2733	2368	2044	1756	1501	1275	906	636
Taylor KO Index	18.1	16.8	15.7	14.6	13.5	12.5	11.5	9.7	8.1
Path • Inches	-1.5	+1.2	0.0	-5.8	-16.9	-34.2	-58.6	-134.9	-260.4
Wind Drift • Inches	0.0	0.6	2.4	5.7	10.5	17.0	25.4	49.0	83.5

Remington 140-grain Core-Lokt Pointed Soft Point, Boat Tail (R7MM4)

G1 Ballistic Coefficient = 0.390

Distance • Yards	Muzzle	100	200	300	400	500	600	800	1000
Velocity • fps	3175	2923	2684	2458	2243	2039	1845	1500	1225
Energy • ft-lbs	3133	2655	2240	1878	1564	1292	1059	700	469
Taylor KO Index	18.0	16.6	15.2	14.0	12.7	11.6	10.5	8.5	7.0
Path • Inches	-1.5	+1.3	0.0	-6.1	-18.0	-36.8	-64.1	-152.3	-305.3
Wind Drift • Inches	0.0	0.7	2.9	6.9	12.7	20.8	31.4	61.7	106.5

Remington 140-grain Core-Lokt Ultra Bonded (PRC7MMRA)

G1 Ballistic Coefficient = 0.409

Distance • Yards	Muzzle	100	200	300	400	500	600	800	1000
Velocity • fps	3175	2934	2707	2490	2283	2086	1898	1560	1283
Energy • ft-lbs	3133	2676	2277	1927	1620	1353	1120	756	512
Taylor KO Index	18.0	16.7	15.4	14.1	13.0	11.8	10.8	8.9	7.3
Path • Inches	-1.5	+1.3	0.0	-6.0	-17.7	-36.0	-62.4	-146.8	-290.9
Wind Drift • Inches	0.0	0.6	2.5	5.8	10.8	17.6	26.6	51.9	89.4

Federal 140-grain Nosler Partition (P7RG)

G1 Ballistic Coefficient = 0.439

Distance • Yards	Muzzle	100	200	300	400	500	600	800	1000
Velocity • fps	3150	2930	2710	2510	2320	2130	1954	1630	1356
Energy • ft-lbs	3085	2660	2290	1960	1670	1415	1188	827	572
Taylor KO Index	17.9	16.6	15.4	14.3	13.2	12.1	11.1	9.3	7.3
Path • Inches	-1.5	+1.3	0.0	-6.0	-17.5	-35.6	-61.3	-142.2	-277.3
Wind Drift • Inches	0.0	0.6	2.6	6.1	11.2	18.2	27.3	53.0	90.6

Nosler 140-grain AccuBond (18415) + (60033)

G1 Ballistic Coefficient = 0.485

Distance • Yards	Muzzle	100	200	300	400	500	600	800	1000
Velocity • fps	3150	2947	2753	2567	2389	2217	2053	1748	1478
Energy • ft-lbs	3085	2700	2356	2049	1774	1529	1311	950	680
Taylor KO Index	17.9	16.7	15.6	14.6	13.6	12.6	11.7	9.9	8.4
Path • Inches	-1.5	+1.2	0.0	-5.8	-17.0	-34.2	-58.5	-133.4	-254.2
Wind Drift • Inches	0.0	0.6	2.4	5.4	10.0	16.2	24.2	46.4	78.7

Winchester 140-grain Power Core 95/5 (X7MMRMLF) LF

G1 Ballistic Coefficient = 0.416

Distance • Yards	Muzzle	100	200	300	400	500	600	800	1000
Velocity • fps	3100	2867	2646	2435	2234	2043	1962	1535	1268
Energy • ft-lbs	2987	2555	2176	1843	1552	1297	1078	732	500
Taylor KO Index	17.6	16.3	15.0	13.8	12.7	11.6	10.6	8.7	7.2
Path • Inches	-1.5	+1.4	0.0	-6.3	-18.6	-37.8	-65.3	-152.7	-299.9
Wind Drift • Inches	0.0	0.7	2.8	6.6	12.2	19.9	29.9	58.4	100.2

Cor-Bon 140-grain DPX (DPX7MM140/20)

G1 Ballistic Coefficient = 0.350

Distance • Yards	Muzzle	100	200	300	400	500	600	800	1000
Velocity • fps	3200	2918	2654	2404	2168	1946	1738	1380	1125
Energy • ft-lbs	3184	2648	2190	1797	1462	1177	940	592	393
Taylor KO Index	18.2	16.6	15.1	13.7	12.3	11.1	9.9	7.8	6.4
Path • Inches	-1.5	+1.3	0.0	-6.2	-18.5	-38.2	-67.2	-162.7	-332.2
Wind Drift • Inches	0.0	0.8	3.3	7.7	14.3	23.5	35.7	71.1	123.4

Remington 140-grain AccuTip Boat Tail (PRA7MMRA)

G1 Ballistic Coefficient = 0.486

Distance • Yards	Muzzle	100	200	300	400	500	600	800	1000
Velocity • fps	3175	2971	2777	2591	2412	2241	2076	1769	1498
Energy • ft-lbs	3133	2744	2397	2086	1808	1560	1340	973	698
Taylor KO Index	18.0	16.9	15.8	14.7	13.7	12.7	11.8	10.0	8.5
Path • Inches	-1.5	+1.2	0.0	-5.7	-16.7	-33.6	-57.4	-130.7	-248.9
Wind Drift • Inches	0.0	0.6	2.3	5.4	9.9	15.9	23.8	45.7	77.3

Black Hills 140-grain Barnes Triple Shock (1C7MMRMBHGN1) LF

G1 Ballistic Coefficient = 0.471

Distance • Yards	Muzzle	100	200	300	400	500	600	800	1000
Velocity • fps	3150	2930	2710	2510	2320	2130	1954	1630	1356
Energy • ft-lbs	3085	2660	2290	1960	1670	1415	1188	827	572
Taylor KO Index	17.9	16.6	15.4	14.3	13.2	12.1	11.1	9.3	7.3
Path • Inches	-1.5	+1.3	0.0	-6.0	-17.5	-35.6	-61.3	-142.2	-277.3
Wind Drift • Inches	0.0	0.6	2.6	6.1	11.2	18.2	27.3	53.0	90.6

Federal 140-grain Trophy Bonded Tip (P7RTT2)

G1 Ballistic Coefficient = 0.429

Distance • Yards	Muzzle	100	200	300	400	500	600	800	1000
Velocity • fps	3150	2920	2710	2500	2300	2110	1931	1604	1329
Energy • ft-lbs	3085	2655	2275	1940	1645	1390	1160	800	549
Taylor KO Index	17.9	16.6	15.4	14.2	13.1	12.0	11.0	9.1	7.5
Path • Inches	-1.5	+1.3	0.0	-6.1	-17.6	-35.9	-62.0	-144.0	-280.8
Wind Drift • Inches	0.0	0.6	2.7	6.2	11.5	18.7	28.1	54.6	93.6

Federal 140-grain Barnes Triple-Shock X-Bullet (P7RM) LF

G1 Ballistic Coefficient = 0.396

Distance • Yards	Muzzle	100	200	300	400	500	600	800	1000
Velocity • fps	3120	2870	2640	2420	2210	2010	1822	1485	1220
Energy • ft-lbs	3025	2565	2165	1815	1515	1250	1032	686	463
Taylor KO Index	17.7	16.3	15.0	13.7	12.6	11.4	10.3	8.4	6.9
Path • Inches	-1.5	+1.4	0.0	-6.3	-18.7	-38.1	-66.2	-156.2	-310.1
Wind Drift • Inches	0.0	0.7	3.0	6.9	12.8	20.9	31.6	62.0	106.9

Norma 140-grain Barnes Triple-Shock (17054) LF

G1 Ballistic Coefficient = 0.477

Distance • Yards	Muzzle	100	200	300	400	500	600	800	1000
Velocity • fps	3117	2912	2716	2529	2350	2178	2012	1706	1439
Energy • ft-lbs	3021	2637	2294	1988	1717	1474	1259	905	644
Taylor KO Index	17.7	16.5	15.4	14.4	13.3	12.4	11.4	9.7	8.2
Path • Inches	-1.5	+1.3	0.0	-6.0	-17.5	-35.2	-60.3	-138.1	-265.0
Wind Drift • Inches	0.0	0.6	2.4	5.8	10.3	18.7	25.0	48.2	81.8

Winchester 140-grain E-Tip (P7RTT2) LF

G1 Ballistic Coefficient = 0.498

Distance • Yards	Muzzle	100	200	300	400	500	600	800	1000
Velocity • fps	3100	2905	2718	2538	2366	2200	2041	1745	1483
Energy • ft-lbs	2988	2623	2296	2003	1740	1505	1295	946	683
Taylor KO Index	17.6	16.5	15.4	14.4	13.4	12.5	11.6	9.9	8.4
Path • Inches	-1.5	+1.3	0.0	-6.0	-17.5	-35.1	-60.0	-136.1	-258.4
Wind Drift • Inches	0.0	0.6	2.3	5.4	9.9	16.0	23.9	45.3	77.5

Winchester 140-grain Power Core 95/5 (X7MMRMLF) LF

G1 Ballistic Coefficient = 0.416

Distance • Yards	Muzzle	100	200	300	400	500	600	800	1000
Velocity • fps	3100	2867	2646	2435	2234	2043	1962	1535	1268
Energy • ft-lbs	2987	2555	2176	1843	1552	1297	1078	732	500
Taylor KO Index	17.6	16.3	15.0	13.8	12.7	11.6	10.6	8.7	7.2
Path • Inches	-1.5	+1.4	0.0	-6.3	-18.6	-37.8	-65.3	-152.7	-299.9
Wind Drift • Inches	0.0	0.7	2.8	6.6	12.2	19.9	29.9	58.4	100.2

Remington 140-grain Core-Lokt Pointed Soft Point – Managed Recoil (RL7MM4)

G1 Ballistic Coefficient = 0.388

Distance • Yards	Muzzle	100	200	300	400	500	600	800	1000
Velocity • fps	2710	2482	2265	2059	1865	1683	1515	1238	1057
Energy • ft-lbs	2283	1915	1595	1318	1081	880	714	476	348
Taylor KO Index	15.4	14.1	12.9	11.7	10.6	9.6	8.6	7.0	6.0
Path • Inches	-1.5	+2.1	0.0	-8.9	-26.0	-53.1	-92.6	-220.7	-439.8
Wind Drift • Inches	0.0	0.9	3.7	8.6	16.1	26.5	40.1	78.5	132.5

Winchester 140-grain Ballistic Silvertip (SBST7A)

G1 Ballistic Coefficient = 0.460

Distance • Yards	Muzzle	100	200	300	400	500	600	800	1000
Velocity • fps	3110	2899	2697	2504	2319	2142	1971	1659	1390
Energy • ft-lbs	3008	2612	2261	1949	1671	1426	1208	856	601
Taylor KO Index	17.7	16.5	16.3	14.2	13.2	12.2	11.2	9.4	7.9
Path • Inches	-1.5	+1.3	0.0	-6.1	-17.8	-35.9	-61.7	-141.8	-273.0
Wind Drift • Inches	0.0	0.6	2.5	5.9	10.8	17.6	26.3	50.8	86.5

Winchester 140-grain E-Tip (P7RTT2) LF

G1 Ballistic Coefficient = 0.498

Distance • Yards	Muzzle	100	200	300	400	500	600	800	1000
Velocity • fps	3100	2905	2718	2538	2366	2200	2041	1745	1483
Energy • ft-lbs	2988	2623	2296	2003	1740	1505	1295	946	683
Taylor KO Index	17.6	16.5	15.4	14.4	13.4	12.5	11.6	9.9	8.4
Path • Inches	-1.5	+1.3	0.0	-6.0	-17.5	-35.1	-60.0	-136.1	-258.4
Wind Drift • Inches	0.0	0.6	2.3	5.4	9.9	16.0	23.9	45.3	77.5

Remington 140-grain Solid Copper Tipped (PCS7MMA) LF

G1 Ballistic Coefficient = 0.468

Distance • Yards	Muzzle	100	200	300	400	500	600	800	1000
Velocity • fps	3175	2964	2762	2570	2385	2208	2039	1725	1450
Energy • ft-lbs	3133	2730	2372	2053	1768	1516	1293	925	654
Taylor KO Index	18.0	16.8	15.7	14.6	13.5	12.5	11.6	9.8	8.2
Path • Inches	-1.5	+1.2	0.0	-5.8	-16.9	-34.0	-58.4	-133.7	-256.2
Wind Drift • Inches	0.0	0.6	2.4	5.6	10.3	16.7	24.9	48.0	81.6

Norma 150-grain Swift Scirocco (17062)

G1 Ballistic Coefficient = 0.537

Distance • Yards	Muzzle	100	200	300	400	500	600	800	1000
Velocity • fps	3166	2981	2804	2634	2471	2313	2161	1874	1514
Energy • ft-lbs	3339	2961	2620	2312	2033	1782	1555	1170	868
Taylor KO Index	19.3	18.1	17.1	16.0	15.0	14.1	13.2	11.4	9.8
Path • Inches	-1.5	+1.2	0.0	-5.6	-16.3	-32.6	-55.5	-124.7	-233.8
Wind Drift • Inches	0.0	0.5	2.1	4.8	8.9	14.3	21.2	40.4	67.8

Nosler 150-grain Partition (18420)

G1 Ballistic Coefficient = 0.456

Distance • Yards	Muzzle	100	200	300	400	500	600	800	1000
Velocity • fps	3100	2887	2683	2489	2303	2125	1955	1642	1374
Energy • ft-lbs	3202	2776	2399	2064	1767	1504	1273	898	629
Taylor KO Index	18.9	17.6	16.3	15.1	14.0	12.9	11.9	10.0	8.4
Path • Inches	-1.5	+1.3	0.0	-6.2	-18.0	-36.3	-62.4	-143.7	-277.1
Wind Drift • Inches	0.0	0.6	2.6	6.0	11.0	17.8	26.7	51.6	88.0

Federal 150-grain Sierra GameKing (P7RD)

G1 Ballistic Coefficient = 0.435

Distance • Yards	Muzzle	100	200	300	400	500	600	800	1000
Velocity • fps	3110	2890	2680	2470	2280	2090	1914	1593	1323
Energy • ft-lbs	3220	2775	2385	2035	1730	1460	1221	845	584
Taylor KO Index	18.9	17.6	16.3	15.0	13.9	12.7	11.6	9.7	8.1
Path • Inches	-1.5	+1.4	0.0	-6.2	-18.1	-36.7	-63.4	-147.1	-286.2
Wind Drift • Inches	0.0	0.6	2.7	6.2	11.5	18.8	28.2	54.8	93.7

Remington 150-grain AccuTip Boat Tail (PRA7MMRB)

G1 Ballistic Coefficient = 0.530

Distance • Yards	Muzzle	100	200	300	400	500	600	800	1000
Velocity • fps	3110	2926	2749	2579	2415	2258	2105	1820	1563
Energy • ft-lbs	3221	2850	2516	2215	1943	1697	1476	1103	814
Taylor KO Index	18.9	17.8	16.7	15.7	14.7	13.7	12.8	11.1	9.5
Path • Inches	-1.5	+1.3	0.0	-5.9	-17.0	-34.1	-58.0	-130.7	-248.3
Wind Drift • Inches	0.0	0.5	2.2	5.0	9.2	14.8	22.1	42.1	70.8

Remington 150-grain Core-Lokt Pointed Soft Point (R7MM2)

G1 Ballistic Coefficient = 0.346

Distance • Yards	Muzzle	100	200	300	400	500	600	800	1000
Velocity • fps	3110	2830	2568	2320	2085	1866	1662	1317	1085
Energy • ft-lbs	3221	2667	2196	1792	1448	1160	921	578	392
Taylor KO Index	18.9	17.2	15.6	14.1	12.7	11.4	10.1	8.0	6.6
Path • Inches	-1.5	+1.3	0.0	-6.6	-20.2	-43.4	-72.5	-177.6	-367.2
Wind Drift • Inches	0.0	0.8	3.4	8.1	15.1	24.9	37.9	75.5	130.6

Federal 150-grain Soft Point (7RA)

G1 Ballistic Coefficient = 0.361

Distance • Yards	Muzzle	100	200	300	400	500	600	800	1000
Velocity • fps	3110	2840	2590	2350	2120	1910	1711	1367	1122
Energy • ft-lbs	3220	2685	2230	1835	1495	1210	975	623	420
Taylor KO Index	18.9	17.3	15.8	14.3	12.9	11.6	10.4	8.3	6.8
Path • Inches	-1.5	+1.4	0.0	-6.6	-19.6	-44.6	-70.4	-169.7	-344.2
Wind Drift • Inches	0.0	0.8	3.3	7.7	14.4	23.6	35.8	71.1	123.0

Winchester 150-grain Power Point Bonded (X7MMR1BP)

G1 Ballistic Coefficient = 0.393

Distance • Yards	Muzzle	100	200	300	400	500	600	800	1000
Velocity • fps	3090	2844	2612	2381	2181	1981	1794	1460	1201
Energy • ft-lbs	3180	2694	2272	1904	1584	1307	1072	710	481
Taylor KO Index	18.8	17.3	15.9	14.5	13.3	12.1	10.9	8.9	7.3
Path • Inches	-1.5	+1.4	0.0	-6.5	-19.1	-39.1	-67.9	-160.5	-319.2
Wind Drift • Inches	0.0	0.7	3.0	7.0	13.1	21.4	32.3	63.5	109.4

Winchester 150-grain Power-Point (X7MMR1)

G1 Ballistic Coefficient = 0.373

Distance • Yards	Muzzle	100	200	300	400	500	600	800	1000
Velocity • fps	3090	2831	2587	2356	2136	1929	1735	1396	1147
Energy • ft-lbs	3100	2670	2229	1848	1520	1239	1003	649	438
Taylor KO Index	18.8	17.2	15.7	14.3	13.0	11.7	10.6	8.5	7.0
Path • Inches	-1.5	+1.4	0.0	-6.6	-19.5	-40.1	-70.1	-168.5	-342.0
Wind Drift • Inches	0.0	0.8	3.2	7.5	13.9	22.9	34.6	68.5	118.2

Federal 150-grain Fusion (F7RFS1)

G1 Ballistic Coefficient = 0.511

Distance • Yards	Muzzle	100	200	300	400	500	600	800	1000
Velocity • fps	3050	2860	2680	2510	2340	2180	2026	1738	1483
Energy • ft-lbs	3100	2725	2390	2090	1820	1580	1367	1006	733
Taylor KO Index	18.6	17.4	16.3	15.3	14.3	13.3	12.3	10.6	9.0
Path • Inches	-1.5	+1.4	0.0	-6.2	-17.9	-36.1	-61.5	-139.2	-263.4
Wind Drift • Inches	0.0	0.6	2.3	5.4	9.8	15.9	23.7	45.4	76.6

Federal 150-grain Nosler Ballistic Tip (P7RH)

G1 Ballistic Coefficient = 0.493

Distance • Yards	Muzzle	100	200	300	400	500	600	800	1000
Velocity • fps	3110	2910	2720	2540	2370	2200	2039	1739	1475
Energy • ft-lbs	3220	2825	2470	2150	1865	1610	1385	1008	725
Taylor KO Index	18.9	17.7	16.6	15.5	14.4	13.4	12.4	10.6	9.0
Path • Inches	-1.5	+1.3	0.0	-6.0	-17.4	-35.0	-59.8	-136.1	-259.6
Wind Drift • Inches	0.0	0.6	2.4	5.4	10.0	16.2	24.1	46.3	78.3

Remington 150-grain Swift Scirocco Bonded (PRSC7MMB)

G1 Ballistic Coefficient = 0.534

Distance • Yards	Muzzle	100	200	300	400	500	600	800	1000
Velocity • fps	3110	2927	2751	2582	2419	2262	2111	1827	1571
Energy • ft-lbs	3437	3044	2689	2369	2080	1819	1584	1186	877
Taylor KO Index	20.2	19.0	17.9	16.8	15.7	14.7	13.7	11.9	10.2
Path • Inches	-1.5	+1.3	0.0	-5.9	-17.0	-34.0	-57.9	-130.2	-244.5
Wind Drift • Inches	0.0	0.5	1.9	4.5	8.2	13.3	19.7	37.6	63.1

Winchester 150-grain Ballistic Silvertip (SBST7)

G1 Ballistic Coefficient = 0.493

Distance • Yards	Muzzle	100	200	300	400	500	600	800	1000
Velocity • fps	3100	2903	2714	2533	2359	2192	2030	1732	1469
Energy • ft-lbs	3200	2806	2453	2136	1853	1600	1373	999	718
Taylor KO Index	18.9	17.7	16.5	15.4	14.4	13.3	12.4	10.5	8.9
Path • Inches	-1.5	+1.3	0.0	-6.0	-17.5	-35.1	-60.2	-137.0	-260.6
Wind Drift • Inches	0.0	0.6	2.4	5.5	10.0	16.2	24.2	46.5	78.8

Norma 150-grain FULL JACKET (17026)

G1 Ballistic Coefficient = 0.443

Distance • Yards	Muzzle	100	200	300	400	500	600	800	1000
Velocity • fps	2995	2780	2576	2380	2192	2013	1844	1536	1283
Energy • ft-lbs	2988	2575	2209	1886	1601	1351	1133	786	548
Taylor KO Index	18.2	16.9	15.7	14.5	13.3	12.3	11.2	9.3	7.8
Path • Inches	-1.5	+1.5	0.0	-6.7	-19.7	-39.8	-68.6	-158.9	-308.8
Wind Drift • Inches	0.0	0.7	2.8	6.5	12.0	19.5	29.2	56.8	97.0

Hornady 154-grain SST (8061)

G1 Ballistic Coefficient = 0.525

Distance • Yards	Muzzle	100	200	300	400	500	600	800	1000
Velocity • fps	3100	2914	2736	2565	2401	2242	2089	1803	1546
Energy • ft-lbs	3286	2904	2560	2250	1970	1718	1493	1112	818
Taylor KO Index	19.4	18.2	17.1	16.0	15.0	14.0	13.1	11.3	9.7
Path • Inches	-1.5	+1.3	0.0	-5.9	-17.2	-34.5	-58.7	-132.2	-249.0
Wind Drift • Inches	0.0	0.5	2.2	5.1	9.3	15.1	22.4	42.8	72.0

Hornady 154-grain InterBond (80628)

G1 Ballistic Coefficient = 0.525

Distance • Yards	Muzzle	100	200	300	400	500	600	800	1000
Velocity • fps	3100	2914	2736	2565	2401	2242	2089	1803	1546
Energy • ft-lbs	3286	2904	2560	2250	1970	1718	1493	1112	818
Taylor KO Index	19.4	18.2	17.1	16.0	15.0	14.0	13.1	11.3	9.7
Path • Inches	-1.5	+1.3	0.0	-5.9	-17.2	-34.5	-58.7	-132.2	-249.0
Wind Drift • Inches	0.0	0.5	2.2	5.1	9.3	15.1	22.5	42.8	72.0

Hornady 154-grain SST (8061)

G1 Ballistic Coefficient = 0.525

Distance • Yards	Muzzle	100	200	300	400	500	600	800	1000
Velocity • fps	3100	2914	2736	2565	2401	2242	2089	1803	1546
Energy • ft-lbs	3286	2904	2560	2250	1970	1718	1493	1112	818
Taylor KO Index	19.4	18.2	17.1	16.0	15.0	14.0	13.1	11.3	9.7
Path • Inches	-1.5	+1.3	0.0	-5.9	-17.2	-34.5	-58.7	-132.2	-249.0
Wind Drift • Inches	0.0	0.5	2.2	5.1	9.3	15.1	22.4	42.8	72.0

Hornady 154-grain SST (8062)

G1 Ballistic Coefficient = 0.525

Distance • Yards	Muzzle	100	200	300	400	500	600	800	1000
Velocity • fps	3035	2852	2677	2508	2345	2189	2037	1755	1504
Energy • ft-lbs	3149	2781	2449	2150	1880	1638	1420	1054	774
Taylor KO Index	19.0	17.8	16.7	15.7	14.7	13.7	12.7	11.0	9.4
Path • Inches	-1.5	+1.4	0.0	-6.2	-18.0	-36.1	-61.5	-138.9	-262.4
Wind Drift • Inches	0.0	0.6	2.3	5.2	9.6	15.5	23.2	44.2	74.4

Hornady 154-grain Soft Point InterLock (8060)

G1 Ballistic Coefficient = 0.434

Distance • Yards	Muzzle	100	200	300	400	500	600	800	1000
Velocity • fps	3035	2814	2604	2404	2212	2029	1857	1542	1283
Energy • ft-lbs	3151	2708	2319	1977	1674	1408	1179	813	563
Taylor KO Index	19.0	17.6	16.3	15.0	13.8	12.7	11.6	9.6	8.0
Path • Inches	-1.5	+1.3	0.0	-6.7	-19.3	-39.3	-67.1	-156.3	-306.2
Wind Drift • Inches	0.0	0.7	2.8	6.5	12.0	19.5	29.3	57.0	97.4

Black Hills 154-grain Hornady SST (1C7MMRMBHGN4)

G1 Ballistic Coefficient = 0.525

Distance • Yards	Muzzle	100	200	300	400	500	600	800	1000
Velocity • fps	3000	2818	2644	2476	2314	2159	2009	1730	1482
Energy • ft-lbs	3078	2717	2391	2097	1832	1594	1381	1023	751
Taylor KO Index	18.7	17.6	16.5	15.5	14.5	13.5	12.6	10.8	9.3
Path • Inches	-1.5	+1.4	0.0	-6.4	-18.5	-37.1	-63.2	-142.6	-268.8
Wind Drift • Inches	0.0	0.6	2.3	5.3	9.8	15.8	23.6	45.0	75.9

Norma 156-grain ORYX (17047)

G1 Ballistic Coefficient = 0.330

Distance • Yards	Muzzle	100	200	300	400	500	600	800	1000
Velocity • fps	2953	2670	2404	2153	1918	1700	1503	1190	1010
Energy • ft-lbs	3021	2470	2002	1607	1275	1002	783	491	353
Taylor KO Index	18.7	16.9	15.2	13.6	12.1	10.8	9.5	7.5	6.4
Path • Inches	-1.5	+1.7	0.0	-7.7	-23.0	-47.8	-84.6	-208.9	-431.5
Wind Drift • Inches	0.0	0.9	3.9	9.2	17.3	28.7	43.8	87.5	148.9

Federal 160-grain Nosler Partition (P7RF)
G1 Ballistic Coefficient = 0.475

Distance • Yards	Muzzle	100	200	300	400	500	600	800	1000
Velocity • fps	2950	2750	2560	2380	2210	2040	1880	1587	1339
Energy • ft-lbs	3090	2690	2335	2015	1730	1480	1255	895	637
Taylor KO Index	19.1	17.9	16.6	15.4	14.3	13.2	12.2	10.3	8.7
Path • Inches	-1.5	+1.6	0.0	-6.8	-19.8	-40.0	-68.5	-157.0	-301.2
Wind Drift • Inches	0.0	0.6	2.6	6.1	11.2	18.2	27.3	52.7	89.5

Winchester 160-grain AccuBond CT (S7MMCT)
G1 Ballistic Coefficient = 0.511

Distance • Yards	Muzzle	100	200	300	400	500	600	800	1000
Velocity • fps	2950	2766	2590	2420	2257	2099	1947	1666	1421
Energy • ft-lbs	3091	2718	2382	2080	1809	1566	1347	987	718
Taylor KO Index	19.1	18.0	16.8	15.7	14.7	13.6	12.6	10.8	9.2
Path • Inches	-1.5	+1.5	0.0	-6.7	-19.4	-38.9	-66.3	-150.4	-285.9
Wind Drift • Inches	0.0	0.6	2.4	5.6	10.3	16.7	25.0	47.8	80.7

Nosler 160-grain AccuBond (18405)
G1 Ballistic Coefficient = 0.531

Distance • Yards	Muzzle	100	200	300	400	500	600	800	1000
Velocity • fps	2925	2748	2578	2415	2257	2105	1960	1698	1448
Energy • ft-lbs	3040	2684	2636	2072	1811	1575	1365	1013	746
Taylor KO Index	19.0	17.8	16.7	15.7	14.7	13.7	12.7	11.0	9.4
Path • Inches	-1.5	+1.6	0.0	-6.8	-19.6	-39.1	-66.6	-150.1	-282.8
Wind Drift • Inches	0.0	0.6	2.4	5.5	10.0	16.2	24.1	46.1	77.6

Federal 160-grain Nosler AccuBond (P7RA1)
G1 Ballistic Coefficient = 0.535

Distance • Yards	Muzzle	100	200	300	400	500	600	800	1000
Velocity • fps	2900	2730	2560	2390	2240	2090	1946	1378	1441
Energy • ft-lbs	2990	2635	2320	2035	1775	1545	1346	1001	738
Taylor KO Index	18.8	17.7	16.6	15.5	14.5	13.6	12.6	10.9	9.4
Path • Inches	-1.5	+1.6	0.0	-6.9	-19.9	-39.9	-67.7	-152.5	-287.0
Wind Drift • Inches	0.0	0.6	2.4	5.5	10.1	16.3	24.2	46.3	77.8

Winchester 160-grain XP3 (SXP7RM)
G1 Ballistic Coefficient = 0.512

Distance • Yards	Muzzle	100	200	300	400	500	600	800	1000
Velocity • fps	2950	2766	2590	2420	2257	2100	1948	1668	1423
Energy • ft-lbs	3091	2718	2382	2080	1809	1599	1348	989	720
Taylor KO Index	19.1	18.0	16.8	15.7	14.7	13.6	12.6	10.9	9.2
Path • Inches	-1.5	+1.5	0.0	-6.7	-19.4	-38.9	-66.3	-150.2	-284.6
Wind Drift • Inches	0.0	0.6	2.4	5.6	10.3	16.7	24.9	47.7	80.6

Remington 160-grain A-Frame PSP (RS7MMA)
G1 Ballistic Coefficient = 0.382

Distance • Yards	Muzzle	100	200	300	400	500	600	800	1000
Velocity • fps	2900	2659	2430	2212	2006	1812	1632	1322	1106
Energy • ft-lbs	2987	2511	2097	1739	1430	1166	946	621	435
Taylor KO Index	18.8	17.3	15.8	14.4	13.0	11.8	10.6	8.6	7.2
Path • Inches	-1.5	+1.5	0.0	-7.6	-22.4	-45.8	-79.8	-190.3	-381.0
Wind Drift • Inches	0.0	0.8	3.4	8.0	14.8	24.3	36.8	72.5	123.9

Federal 160-grain Barnes Triple Shock X-Bullet (P7RN) LF
G1 Ballistic Coefficient = 0.512

Distance • Yards	Muzzle	100	200	300	400	500	600	800	1000
Velocity • fps	2940	2760	2580	2410	2240	2090	1940	1661	1417
Energy • ft-lbs	3070	2695	2360	2060	1785	1545	1337	980	714
Taylor KO Index	19.1	17.9	16.7	15.6	14.5	13.6	12.6	10.8	9.2
Path • Inches	-1.5	+1.6	0.0	-6.7	-19.5	-39.3	-66.8	-151.4	-286.8
Wind Drift • Inches	0.0	0.6	2.4	5.6	10.4	16.8	25.0	48.0	81.0

Cor-Bon 160-grain DPX (DPX7MM160/20)
G1 Ballistic Coefficient = 0.400

Distance • Yards	Muzzle	100	200	300	400	500	600	800	1000
Velocity • fps	2925	2692	2471	2261	2061	1871	1694	1384	1152
Energy • ft-lbs	3040	2576	2170	1816	1509	1245	1020	680	472
Taylor KO Index	19.0	17.5	16.0	14.7	13.4	12.1	11.0	9.0	7.5
Path • Inches	-1.5	+1.6	0.0	-7.4	-21.5	-43.9	-76.3	-180.2	-357.4
Wind Drift • Inches	0.0	0.8	3.2	7.5	13.9	22.7	34.3	67.4	115.3

Federal 160-grain Trophy Bonded Tip (P7RTT1)

G1 Ballistic Coefficient = 0.520

Distance • Yards	Muzzle	100	200	300	400	500	600	800	1000
Velocity • fps	2900	2720	2550	2380	2220	2070	1923	1650	1412
Energy • ft-lbs	2990	2630	2310	2015	1755	1525	1314	968	709
Taylor KO Index	18.8	17.7	16.6	15.4	14.4	13.4	12.5	10.7	9.2
Path • Inches	-1.5	+1.6	0.0	-6.9	-20.0	-40.2	-68.4	-154.7	-292.5
Wind Drift • Inches	0.0	0.6	2.4	5.6	10.4	16.8	25.0	47.9	80.8

Hornady 162-grain SST (80633)

G1 Ballistic Coefficient = 0.549

Distance • Yards	Muzzle	100	200	300	400	500	600	800	1000
Velocity • fps	3030	2855	2688	2525	2370	2219	2073	1800	1554
Energy • ft-lbs	3302	2933	2598	2295	2020	1772	1546	1166	869
Taylor KO Index	19.9	18.8	17.7	16.6	15.6	14.6	13.6	11.8	10.2
Path • Inches	-1.5	+1.4	0.0	-6.2	-17.9	-35.7	-60.7	-136.1	-254.7
Wind Drift • Inches	0.0	0.5	2.2	5.0	9.2	14.8	22.0	41.8	70.1

Black Hills 162-grain Hornady A-MAX (1C7MMRMBHGN1)

G1 Ballistic Coefficient = 0.625

Distance • Yards	Muzzle	100	200	300	400	500	600	800	1000
Velocity • fps	2950	2799	2652	2511	2373	2240	2112	1867	1643
Energy • ft-lbs	3131	2818	2531	2268	2027	1806	1604	1255	971
Taylor KO Index	19.4	18.4	17.4	16.5	15.6	14.7	13.9	12.3	10.8
Path • Inches	-1.5	+1.5	0.0	-6.4	-18.4	-36.5	-61.5	-136.0	-250.7
Wind Drift • Inches	0.0	0.5	2.0	4.5	8.3	13.3	19.6	37.0	61.6

Hornady 162-grain BTSP InterLock (8063)

G1 Ballistic Coefficient = 0.515

Distance • Yards	Muzzle	100	200	300	400	500	600	800	1000
Velocity • fps	2940	2757	2582	2413	2251	2094	1945	1667	1424
Energy • ft-lbs	3110	2735	2399	2095	1823	1578	1361	1000	729
Taylor KO Index	19.3	18.1	17.0	15.9	14.8	13.8	12.8	11.0	9.4
Path • Inches	-1.5	+1.5	0.0	-6.7	-19.7	-39.3	-66.7	-150.9	-285.6
Wind Drift • Inches	0.0	0.6	2.4	5.6	10.3	16.7	24.9	47.6	80.3

Federal 165-grain Sierra GameKing BTSP (P7RE)

G1 Ballistic Coefficient = 0.459

Distance • Yards	Muzzle	100	200	300	400	500	600	800	1000
Velocity • fps	2950	2750	2550	2360	2180	2010	1845	1547	1299
Energy • ft-lbs	3190	2760	2380	2045	1745	1480	1251	880	621
Taylor KO Index	19.7	18.4	17.1	15.8	14.6	13.5	12.4	10.4	8.7
Path • Inches	-1.5	+1.6	0.0	-6.9	-20.1	-40.5	-69.7	-160.4	-309.7
Wind Drift • Inches	0.0	0.7	2.7	6.3	11.7	19.0	28.5	55.2	93.9

Norma 170-grain ORYX (17023)

G1 Ballistic Coefficient = 0.324

Distance • Yards	Muzzle	100	200	300	400	500	600	800	1000
Velocity • fps	2887	2604	2338	2087	1853	1638	1445	1149	986
Energy • ft-lbs	3147	2560	2063	1645	1297	1013	788	498	367
Taylor KO Index	19.9	18.0	16.1	14.4	12.8	11.3	10.0	7.9	6.8
Path • Inches	-1.5	+1.8	0.0	-8.2	-24.4	-50.8	-90.3	-223.8	-462.6
Wind Drift • Inches	0.0	1.0	4.1	9.7	18.3	30.3	46.3	92.3	155.7

Norma 170-grain Plastic Point (17027)

G1 Ballistic Coefficient = 0.378

Distance • Yards	Muzzle	100	200	300	400	500	600	800	1000
Velocity • fps	2953	2705	2470	2247	2037	1839	1654	1335	1111
Energy • ft-lbs	3293	2763	2304	1907	1567	1276	1032	673	466
Taylor KO Index	20.4	18.7	17.0	15.5	14.0	12.7	11.4	9.2	7.7
Path • Inches	-1.5	+1.6	0.0	-7.3	-21.6	-44.2	-77.3	-185.6	-375.6
Wind Drift • Inches	0.0	0.7	3.0	7.1	13.2	21.6	32.8	64.8	111.1

Norma 170-grain Vulcan HP (17024)

G1 Ballistic Coefficient = 0.353

Distance • Yards	Muzzle	100	200	300	400	500	600	800	1000
Velocity • fps	2953	2688	2438	2201	1977	1769	1577	1258	1055
Energy • ft-lbs	3293	2728	2244	1830	1476	1181	938	597	420
Taylor KO Index	20.4	18.5	16.8	15.2	13.6	12.2	10.9	8.7	7.3
Path • Inches	-1.5	+1.6	0.0	-7.5	-22.3	-46.0	-81.0	-197.9	-406.5
Wind Drift • Inches	0.0	0.8	3.2	7.7	14.4	23.7	36.1	71.6	122.6

Federal 175-grain Soft Point (7RB)

G1 Ballistic Coefficient = 0.428

Distance • Yards	Muzzle	100	200	300	400	500	600	800	1000
Velocity • fps	2860	2650	2440	2240	2060	1880	1713	1417	1187
Energy • ft-lbs	3180	2720	2310	1960	1640	1370	1141	780	548
Taylor KO Index	20.3	18.8	17.3	15.9	14.6	13.3	12.2	10.1	8.4
Path • Inches	-1.5	+1.7	0.0	-7.6	-22.1	-44.9	-77.5	-180.8	-354.1
Wind Drift • Inches	0.0	0.7	3.1	7.2	13.3	21.7	32.6	63.6	108.4

Federal 175-grain Fusion (F7RFS2)

G1 Ballistic Coefficient = 0.538

Distance • Yards	Muzzle	100	200	300	400	500	600	800	1000
Velocity • fps	2760	2290	2430	2270	2120	1980	1841	1586	1365
Energy • ft-lbs	2960	2610	2295	2010	1750	1520	1318	978	724
Taylor KO Index	19.6	18.4	17.3	16.1	15.1	14.1	13.1	11.3	9.7
Path • Inches	-1.5	+1.8	0.0	-7.8	-22.3	-44.5	-75.4	-170.0	-320.3
Wind Drift • Inches	0.0	0.6	2.5	5.8	10.7	17.4	25.9	49.5	83.1

Federal 175-grain Trophy Bonded Bear Claw (P7RT1)

G1 Ballistic Coefficient = 0.407

Distance • Yards	Muzzle	100	200	300	400	500	600	800	1000
Velocity • fps	2750	2530	2320	2120	1930	1750	1568	1435	1102
Energy • ft-lbs	2940	2485	2090	1745	1450	1195	977	659	472
Taylor KO Index	19.5	18.0	16.5	15.1	13.7	12.4	11.3	10.2	7.8
Path • Inches	-1.5	+2.0	0.0	-8.5	-24.6	-50.2	-87.2	-205.7	-406.9
Wind Drift • Inches	0.0	0.8	3.4	8.0	14.9	24.5	37.0	72.2	122.5

Remington 175-grain Core-Lokt Pointed Soft Point (R7MM3)

G1 Ballistic Coefficient = 0.428

Distance • Yards	Muzzle	100	200	300	400	500	600	800	1000
Velocity • fps	2860	2645	2442	2244	2057	1879	1713	1417	1187
Energy • ft-lbs	3178	2718	2313	1956	1644	1372	1141	780	548
Taylor KO Index	20.3	18.8	17.3	15.9	14.6	13.3	12.2	10.1	8.4
Path • Inches	-1.5	+2.0	0.0	-7.9	22.7	-45.8	-77.5	-180.8	-354.1
Wind Drift • Inches	0.0	0.7	3.1	7.2	13.3	21.7	32.6	63.6	108.4

Winchester 175-grain Power-Point (X7MMR2)

G1 Ballistic Coefficient = 0.428

Distance • Yards	Muzzle	100	200	300	400	500	600	800	1000
Velocity • fps	2860	2645	2442	2244	2057	1879	1713	1417	1187
Energy • ft-lbs	3178	2718	2313	1956	1644	1372	1141	780	548
Taylor KO Index	20.3	18.8	17.3	15.9	14.6	13.3	12.2	10.1	8.4
Path • Inches	-1.5	+2.0	0.0	-7.9	22.7	-45.8	-77.5	-180.8	-354.1
Wind Drift • Inches	0.0	0.7	3.1	7.2	13.3	21.7	32.6	63.6	108.4

Nosler 175-grain Partition (18410)

G1 Ballistic Coefficient = 0.519

Distance • Yards	Muzzle	100	200	300	400	500	600	800	1000
Velocity • fps	2800	2624	2455	2293	2136	1986	1842	1577	1350
Energy • ft-lbs	3047	2677	2343	2043	1773	1532	1319	967	708
Taylor KO Index	19.9	18.6	17.4	16.3	15.2	14.1	13.1	11.2	9.6
Path • Inches	-1.5	+1.8	0.0	-7.6	-21.7	-43.5	-74.2	-168.0	-318.1
Wind Drift • Inches	0.0	0.6	2.6	6.0	11.0	17.8	26.5	50.8	85.6

7mm Winchester Short Magnum (WSM)

When Winchester introduced the .300 WSM, most everyone present at that introduction predicted that a .270 WSM and a 7mm WSM would soon follow. *Guns & Ammo* magazine printed calculated performance figures for a series of wildcats based on the .300 WSM in June of 2001. The 7mm as a factory cartridge came into being in 2002. The short magnums as a group give improved performance when compared with cartridges based on the .308 or the .30-06 cases but can't quite equal the full-length magnum performances except in a few rare instances.

Relative Recoil Factor = 2.0

Specifications:

Controlling Agency for Standardization of this Ammunition: SAAMI

Bullet Weight Grains	Velocity fps	Maximum Average Pressure Copper Crusher	Transducer
N/A	N/A	---	65,000 psi

Standard barrel for velocity testing: 24 inches long—1 turn in 10-inch twist

Availability:

Winchester 140-grain AccuBond CT (S7MMWSMCTA)
G1 Ballistic Coefficient = 0.461

Distance • Yards	Muzzle	100	200	300	400	500	600	800	1000
Velocity • fps	3225	3008	2801	2604	2415	2233	2061	1740	1459
Energy • ft-lbs	3233	2812	2439	2107	1812	1550	1321	941	662
Taylor KO Index	18.3	17.1	15.9	14.8	13.7	12.7	11.7	9.9	8.3
Path • Inches	-1.5	+1.2	0.0	-5.6	-16.4	-33.1	-56.7	-130.4	-251.4
Wind Drift • Inches	0.0	0.6	2.4	5.6	10.3	16.6	24.8	47.9	81.5

Winchester 140-grain Ballistic Silvertip (SBST7MMS)
G1 Ballistic Coefficient = 0.461

Distance • Yards	Muzzle	100	200	300	400	500	600	800	1000
Velocity • fps	3225	3008	2801	2604	2415	2233	2061	1740	1459
Energy • ft-lbs	3233	2812	2439	2107	1812	1550	1321	941	662
Taylor KO Index	18.3	17.1	15.9	14.8	13.7	12.7	11.7	9.9	8.3
Path • Inches	-1.5	+1.2	0.0	-5.6	-16.4	-33.1	-56.7	-130.4	-251.4
Wind Drift • Inches	0.0	0.6	2.4	5.6	10.3	16.6	24.8	47.9	81.5

Federal 140-grain Nosler Ballistic Tip (P7WSMB)
G1 Ballistic Coefficient = 0.482

Distance • Yards	Muzzle	100	200	300	400	500	600	800	1000
Velocity • fps	3310	3100	2900	2700	2520	2340	2170	1852	1567
Energy • ft-lbs	3405	2985	2610	2270	1975	1705	1465	1066	763
Taylor KO Index	18.8	17.6	16.5	15.3	14.3	13.3	12.3	10.5	8.9
Path • Inches	-1.5	+1.1	0.0	-5.2	-15.2	-30.7	-52.5	-119.6	-227.6
Wind Drift • Inches	0.0	0.5	2.2	5.1	9.4	15.2	22.7	43.5	73.6

Cor-Bon 140-grain DPX (DPX7WSM140/20)
G1 Ballistic Coefficient = 0.350

Distance • Yards	Muzzle	100	200	300	400	500	600	800	1000
Velocity • fps	3200	2918	2654	2404	2168	1946	1738	1380	1125
Energy • ft-lbs	3184	2648	2190	1797	1462	1177	940	592	393
Taylor KO Index	18.2	16.6	15.1	13.7	12.3	11.1	9.9	7.8	6.4
Path • Inches	-1.5	+1.3	0.0	-6.2	-18.5	-38.2	-67.2	-162.7	-332.2
Wind Drift • Inches	0.0	0.8	3.3	7.7	14.3	23.5	35.7	71.1	123.4

Federal 140-grain Trophy Bonded Tip (P7WSMTT2)

G1 Ballistic Coefficient = 0.429

Distance • Yards	Muzzle	100	200	300	400	500	600	800	1000
Velocity • fps	3200	2970	2750	2540	2340	2150	1969	1636	1355
Energy • ft-lbs	3185	2740	2370	2010	1705	1440	1205	832	571
Taylor KO Index	18.2	16.9	15.6	14.4	13.3	12.2	11.2	9.3	7.7
Path • Inches	-1.5	+1.2	0.0	-5.8	-17.0	-34.7	-59.8	-138.8	-270.5
Wind Drift • Inches	0.0	0.6	2.6	6.1	11.2	18.3	27.5	53.3	91.4

Federal 150-grain Soft Point (7WSME)

G1 Ballistic Coefficient = 0.360

Distance • Yards	Muzzle	100	200	300	400	500	600	800	1000
Velocity • fps	3200	2830	2580	2340	2110	1900	1703	1360	1118
Energy • ft-lbs	3200	2670	2215	1820	1485	1200	966	616	417
Taylor KO Index	18.9	17.2	15.7	14.2	12.8	11.6	10.4	8.3	6.8
Path • Inches	-1.5	+1.4	0.0	-6.6	-19.7	-40.4	-71.0	-171.2	-347.3
Wind Drift • Inches	0.0	0.8	3.3	7.8	14.5	23.8	36.1	71.6	123.7

Federal 150-grain Fusion (F7WSMFS1)

G1 Ballistic Coefficient = 0.517

Distance • Yards	Muzzle	100	200	300	400	500	600	800	1000
Velocity • fps	3100	2910	2730	2560	2390	2230	2076	1786	1528
Energy • ft-lbs	3200	2820	2480	2175	1900	1650	1435	1063	778
Taylor KO Index	18.9	17.7	16.6	15.6	14.6	13.6	12.6	10.9	9.3
Path • Inches	-1.5	+1.3	0.0	-6.0	-17.2	-34.7	-59.0	-133.3	-251.6
Wind Drift • Inches	0.0	0.5	2.2	5.2	9.5	15.3	22.8	43.6	73.6

Winchester 150-grain Power Max Bonded (X7WSMBP)

G1 Ballistic Coefficient = 0.393

Distance • Yards	Muzzle	100	200	300	400	500	600	800	1000
Velocity • fps	3200	2948	2710	2484	2270	2066	1873	1526	1248
Energy • ft-lbs	3410	2894	2446	2056	1716	1421	1169	776	520
Taylor KO Index	19.5	17.9	16.5	15.1	13.9	12.6	11.4	9.3	7.6
Path • Inches	-1.5	+1.3	0.0	-6.0	-17.6	-36.1	-62.6	-147.9	-293.6
Wind Drift • Inches	0.0	0.7	2.9	6.7	12.5	20.3	30.7	60.2	104.0

Winchester 140-grain E-Tip (S7SET) LF

G1 Ballistic Coefficient = 0.498

Distance • Yards	Muzzle	100	200	300	400	500	600	800	1000
Velocity • fps	3150	2952	2763	2582	2408	2241	2080	1926	1513
Energy • ft-lbs	3085	2710	2374	2073	1803	1561	1345	985	712
Taylor KO Index	17.9	16.8	15.7	14.7	13.7	12.7	11.8	10.9	8.6
Path • Inches	-1.5	+1.2	0.0	-5.8	-16.8	-33.9	-57.8	-131.3	-249.0
Wind Drift • Inches	0.0	0.6	2.3	5.3	9.7	15.7	23.4	44.8	75.6

Winchester 150-grain PowerPoint (X7MMWSM)

G1 Ballistic Coefficient = 0.346

Distance • Yards	Muzzle	100	200	300	400	500	600	800	1000
Velocity • fps	3200	2915	2648	2396	2157	1933	1723	1364	1112
Energy • ft-lbs	3410	2830	2335	1911	1550	1245	989	620	412
Taylor KO Index	19.5	17.7	16.1	14.6	13.1	11.8	10.5	8.3	6.8
Path • Inches	-1.5	+1.3	0.0	-6.3	-18.6	-38.5	-67.8	-164.6	-337.2
Wind Drift • Inches	0.0	0.8	3.3	7.8	14.5	23.9	36.3	72.4	125.8

Winchester 160-grain AccuBond CT (S7MMWSMCT)

G1 Ballistic Coefficient = 0.229

Distance • Yards	Muzzle	100	200	300	400	500	600	800	1000
Velocity • fps	3050	2862	2682	2509	2342	2182	2029	1741	1486
Energy • ft-lbs	3306	2911	2556	2237	1950	1692	1462	1077	785
Taylor KO Index	19.8	18.6	17.4	16.3	15.2	14.2	13.2	11.3	9.6
Path • Inches	-1.5	+1.4	0.0	-6.2	-17.9	-36.0	-61.4	-139.1	-263.7
Wind Drift • Inches	0.0	0.6	2.3	5.4	9.8	15.8	23.6	45.2	76.3

Winchester 160-grain XP3 (SXSP7WSM)

G1 Ballistic Coefficient = 0.229

Distance • Yards	Muzzle	100	200	300	400	500	600	800	1000
Velocity • fps	3050	2862	2682	2509	2342	2182	2029	1741	1486
Energy • ft-lbs	3306	2911	2556	2237	1950	1692	1462	1077	785
Taylor KO Index	19.8	18.6	17.4	16.3	15.2	14.2	13.2	11.3	9.6
Path • Inches	-1.5	+1.4	0.0	-6.2	-17.9	-36.0	-61.4	-139.1	-263.7
Wind Drift • Inches	0.0	0.6	2.3	5.4	9.8	15.8	23.6	45.2	76.3

Federal 160-grain Trophy Bonded Tip (P7WSMTT1)

G1 Ballistic Coefficient = 0.520

Distance • Yards	Muzzle	100	200	300	400	500	600	800	1000
Velocity • fps	3000	2830	2640	2470	2310	2150	2001	1720	1471
Energy • ft-lbs	3195	2620	2480	2170	1895	1645	1422	1051	769
Taylor KO Index	19.5	18.4	17.1	16.0	15.0	14.0	13.0	11.2	9.5
Path • Inches	-1.5	+1.4	0.0	-6.4	-18.6	-37.2	-63.4	-143.3	-270.6
Wind Drift • Inches	0.0	0.6	2.3	5.4	9.9	16.0	23.8	45.6	76.9

Federal 160-grain Barnes Triple-Shock X-Bullet (P7WSMD) LF

G1 Ballistic Coefficient = 0.511

Distance • Yards	Muzzle	100	200	300	400	500	600	800	1000
Velocity • fps	2990	2800	2620	2450	2290	2130	1977	1694	1445
Energy • ft-lbs	3175	2790	2445	2135	1855	1605	1389	1020	742
Taylor KO Index	19.4	18.2	17.0	15.9	14.9	13.8	12.8	11.0	9.4
Path • Inches	-1.5	+1.5	0.0	-6.5	-18.8	-37.8	-64.4	-145.8	-276.2
Wind Drift • Inches	0.0	0.6	2.4	5.5	10.2	16.4	24.5	46.9	79.2

Cor-Bon 180-grain VLD (PM7WSM180/20)

G1 Ballistic Coefficient = 0.650

Distance • Yards	Muzzle	100	200	300	400	500	600	800	1000
Velocity • fps	2850	2708	2570	2436	2307	2181	2059	1828	1615
Energy • ft-lbs	3247	2931	2641	2373	2127	1902	1695	1336	1043
Taylor KO Index	20.8	19.8	18.8	17.8	16.8	15.9	15.0	13.3	11.8
Path • Inches	-1.5	+1.6	0.0	-6.9	-19.6	-38.9	-65.5	-144.4	-265.1
Wind Drift • Inches	0.0	0.5	2.0	4.6	8.3	13.3	19.7	37.1	61.5

Kinuno Mbogo and a kongoni *or Coke hartebeest he shot with Finn Aagaard's 7x64. Obviously the hunter is happy!* (Photo from *Gu[..] and Hunting* by Finn Aagaard, 2012, Safari Press Archives)

7mm Remington Short Action Ultra Mag

The flavor of the month at the 2002 SHOT show was the announcement of a whole collection of cartridges based on short, fat cases. For the most part, these cartridges are designed with an overall length the same as the .308 and its necked down versions. Remington decided that it should not be upstaged, so it introduced both this cartridge and also a longer version.

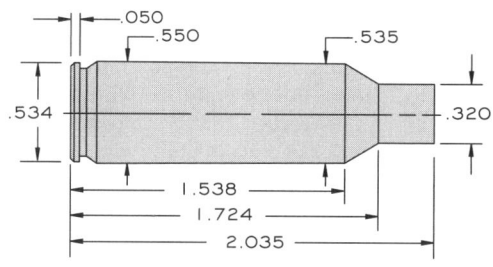

Relative Recoil Factor = 2.07

Specifications:

Controlling Agency for Standardization of this Ammunition: SAAMI

Bullet Weight Grains	Velocity fps	Maximum Average Pressure	
		Copper Crusher	Transducer
*	*	---	65,000 psi

Standard barrel for velocity testing: 24 inches long—1 turn in 10-inch twist

Availability:

Nosler 140-grain AccuBond (16019)

G1 Ballistic Coefficient = 0.485

Distance • Yards	Muzzle	100	200	300	400	500	600	800	1000
Velocity • fps	3150	2947	2753	2567	2389	2217	2053	1748	1478
Energy • ft-lbs	3085	2700	2356	2049	1774	1529	1311	950	680
Taylor KO Index	17.9	16.7	15.6	14.6	13.6	12.6	11.7	9.9	8.4
Path • Inches	-1.5	+1.2	0.0	-5.8	-17.0	-34.2	-58.5	-133.4	-254.2
Wind Drift • Inches	0.0	0.6	2.4	5.4	10.0	16.2	24.2	46.4	78.7

Remington 150-grain Core-Lokt Pointed Soft Point (PR7SM2)

G1 Ballistic Coefficient = 0.346

Distance • Yards	Muzzle	100	200	300	400	500	600	800	1000
Velocity • fps	3110	2831	2568	2321	2087	1867	1662	1317	1085
Energy • ft-lbs	3221	2669	2196	1793	1449	1160	920	577	392
Taylor KO Index	19.9	17.2	15.6	14.1	12.7	11.4	10.1	8.0	6.6
Path • Inches	-1.5	+1.4	0.0	-6.7	-19.9	-41.2	-72.4	-176.2	-361.0
Wind Drift • Inches	0.0	0.8	3.4	8.1	15.1	24.9	37.9	75.6	130.7

Nosler 150-grain Partition (16030)

G1 Ballistic Coefficient = 0.456

Distance • Yards	Muzzle	100	200	300	400	500	600	800	1000
Velocity • fps	3000	2791	2592	2402	2219	2046	1880	1577	1322
Energy • ft-lbs	2997	2595	2238	1921	1641	1394	1178	829	583
Taylor KO Index	18.3	17.0	15.8	14.6	13.5	12.5	11.4	9.6	8.0
Path • Inches	-1.5	+1.5	0.0	-6.7	-19.4	-39.1	-67.2	-154.8	-299.0
Wind Drift • Inches	0.0	0.6	2.7	6.2	11.5	18.7	28.0	54.2	92.2

Remington 160-grain Core-Lokt Ultra Bonded (PR7SM4)

G1 Ballistic Coefficient = 0.414

Distance • Yards	Muzzle	100	200	300	400	500	600	800	1000
Velocity • fps	2960	2733	2518	2313	2117	1929	1753	1441	1198
Energy • ft-lbs	3112	2654	2252	1900	1592	1322	1093	738	510
Taylor KO Index	19.2	17.7	16.3	15.0	13.7	12.5	11.4	9.4	7.8
Path • Inches	-1.5	+1.6	0.0	-7.1	-20.7	-42.9	-72.9	-171.5	-340.0
Wind Drift • Inches	0.0	0.7	3.0	7.1	13.1	21.4	32.3	68.1	108.1

Nosler 160-grain AccuBond (16042)

G1 Ballistic Coefficient = 0.531

Distance • Yards	Muzzle	100	200	300	400	500	600	800	1000
Velocity • fps	2850	2676	2509	2348	2193	2044	1900	1635	1403
Energy • ft-lbs	2886	2545	2237	1959	1709	1484	1283	950	700
Taylor KO Index	18.5	17.4	16.3	15.2	14.2	13.3	12.3	10.6	9.1
Path • Inches	-1.5	+1.7	0.0	-7.2	-20.7	-44.5	-70.6	-159.3	-300.4
Wind Drift • Inches	0.0	0.6	2.4	5.7	10.4	16.8	25.1	48.0	80.8

Nosler 160-grain Partition (16058)

G1 Ballistic Coefficient = 0.475

Distance • Yards	Muzzle	100	200	300	400	500	600	800	1000
Velocity • fps	2850	2658	2471	2292	2122	1958	1803	1521	1286
Energy • ft-lbs	2886	2507	2169	1868	1600	1363	1155	822	588
Taylor KO Index	18.5	17.3	16.0	14.9	13.8	12.7	11.7	9.9	8.3
Path • Inches	-1.5	+1.7	0.0	-7.4	-21.5	-43.3	-74.1	-170.0	-326.6
Wind Drift • Inches	0.0	0.7	2.8	6.4	11.3	19.2	28.8	55.6	94.2

7mm Remington Ultra Magnum

Remington introduced the 7mm Ultra Magnum in 2001. This cartridge should not be confused with their 7mm Short Action Ultra Magnum. As you can see from the drawings, it is a larger and more potent family member. For a given bullet weight, this larger cartridge case can produce over 200 fps more velocity. If you own one of these guns, you will have to be careful which ammo you ask for at the gun store.

Since 2008 Remington has offered this ammunition loaded to three power levels. That translates into three recoil levels. It doesn't seem to make a whole lot of sense to buy too much gun and then get wimpy ammo because you can't stand the recoil.

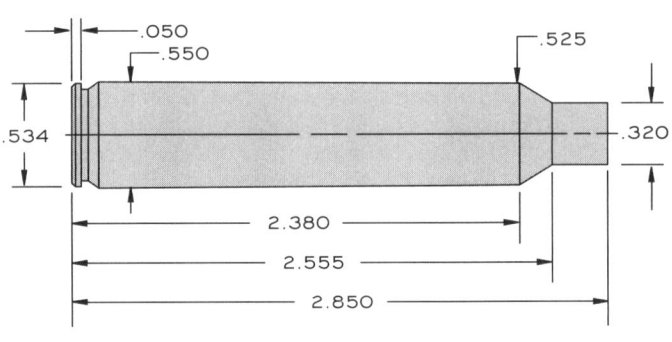

Relative Recoil Factor = 2.13

Specifications:

Controlling Agency for Standardization of this Ammunition: SAAMI

Bullet Weight Grains	Velocity fps	Maximum Average Pressure Copper Crusher	Transducer
N/A	N/A	---	65,000 psi

Standard barrel for velocity testing: 24 inches long—1 turn in 10-inch twist

Availability:

Remington 140-grain Core-Lokt Ultra Bonded – Power Level 3 (PR7UM1)   G1 Ballistic Coefficient = 0.409

Distance • Yards	Muzzle	100	200	300	400	500	600	800	1000
Velocity • fps	3425	3170	2929	2701	2485	2278	2081	1720	1411
Energy • ft-lbs	3646	3123	2667	2268	1919	1613	1347	920	619
Taylor KO Index	19.5	18.0	16.6	15.3	14.1	12.9	11.8	9.8	8.0
Path • Inches	-1.5	+1.0	0.0	-5.0	-14.9	-30.3	-52.5	-122.7	-240.5
Wind Drift • Inches	0.0	0.6	2.5	5.9	10.9	17.9	26.5	51.5	86.6

Remington 140-grain Core-Lokt Pointed Soft Point – Power Level 1 (R7UM1-P1)   G1 Ballistic Coefficient = 0.410

Distance • Yards	Muzzle	100	200	300	400	500	600	800	1000
Velocity • fps	3000	2768	2549	2339	2139	1950	1772	1454	1205
Energy • ft-lbs	3797	2384	2019	1700	1422	1181	976	657	452
Taylor KO Index	17.0	15.7	14.5	13.3	12.1	11.1	10.1	8.3	6.8
Path • Inches	-1.5	+1.5	0.0	-6.9	-20.1	-41.0	-71.0	-166.8	-329.1
Wind Drift • Inches	0.0	0.7	3.0	7.0	13.0	21.3	32.1	62.8	107.6

Remington 150-grain Swift Scirocco Bonded – Power Level 3 (PRSC7UM1)   G1 Ballistic Coefficient = 0.534

Distance • Yards	Muzzle	100	200	300	400	500	600	800	1000
Velocity • fps	3325	3132	2945	2771	2602	2438	2281	1983	1711
Energy • ft-lbs	3682	3267	2894	2558	2254	1979	1733	1310	975
Taylor KO Index	20.2	19.1	17.9	16.9	15.8	14.8	13.9	12.1	10.4
Path • Inches	-1.5	+1.0	0.0	-5.0	-14.6	-29.3	-49.8	-112.0	-209.8
Wind Drift • Inches	0.0	0.5	2.0	4.6	8.4	13.4	20.0	37.8	63.4

Nosler 160-grain AccuBond (60048)   G1 Ballistic Coefficient = 0.540

Distance • Yards	Muzzle	100	200	300	400	500	600	800	1000
Velocity • fps	3225	3039	2861	2690	2525	2366	2214	1925	1661
Energy • ft-lbs	3695	3281	2908	2571	2265	1989	1741	1316	981
Taylor KO Index	20.9	19.7	18.6	17.5	16.4	15.4	14.4	12.5	10.8
Path • Inches	-1.5	+1.1	0.0	-5.4	-15.6	-31.2	-53.1	-119.2	-223.1
Wind Drift • Inches	0.0	0.5	2.0	4.7	8.6	13.8	20.5	38.9	65.3

Nosler 160-grain Partition (16705)

G1 Ballistic Coefficient = 0.456

Distance • Yards	Muzzle	100	200	300	400	500	600	800	1000
Velocity • fps	3225	3015	2813	2621	2437	2260	2092	1777	1499
Energy • ft-lbs	3695	3228	2812	2441	2110	1815	1555	1122	798
Taylor KO Index	20.9	19.6	18.3	17.0	15.8	14.7	13.6	11.5	9.7
Path • Inches	-1.5	+1.2	0.0	-5.6	-16.2	-32.7	-56.0	-127.9	-244.2
Wind Drift • Inches	0.0	0.6	2.3	5.4	9.9	16.0	23.9	46.0	78.0

Cor-Bon 160-grain DPX (DPX7MM)

G1 Ballistic Coefficient = 0.400

Distance • Yards	Muzzle	100	200	300	400	500	600	800	1000
Velocity • fps	3200	2952	2718	2496	2284	2083	1893	1549	1270
Energy • ft-lbs	3639	3098	2626	2214	1855	1542	1273	852	573
Taylor KO Index	20.8	19.2	17.6	16.2	14.8	13.5	12.3	10.1	8.2
Path • Inches	-1.5	+1.2	0.0	-6.0	-17.5	-35.8	-62.0	-145.9	-288.6
Wind Drift • Inches	0.0	0.7	2.8	6.6	12.2	19.9	30.0	58.8	101.3

Remington 160-grain Core-Lokt Ultra Bonded – Power Level 2 (PR7UM2-P2)

G1 Ballistic Coefficient = 0.415

Distance • Yards	Muzzle	100	200	300	400	500	600	800	1000
Velocity • fps	2950	2724	2508	2303	2108	1922	1748	1438	1196
Energy • ft-lbs	3091	2635	2235	1884	1578	1312	1086	734	508
Taylor KO Index	19.1	17.7	16.3	14.9	13.7	12.5	11.3	9.3	7.8
Path • Inches	-1.5	+1.6	0.0	-7.1	-20.8	-42.4	-73.4	-171.9	-338.6
Wind Drift • Inches	0.0	0.7	3.0	7.1	13.2	21.5	32.4	63.3	108.4

Remington 175-grain Swift A-Frame – Power Level 3 (PR7UM5)

G1 Ballistic Coefficient = 0.548

Distance • Yards	Muzzle	100	200	300	400	500	600	800	1000
Velocity • fps	3025	2850	2681	2519	2363	2212	2066	1793	1547
Energy • ft-lbs	3555	3157	2795	2467	2170	1901	1659	1443	930
Taylor KO Index	21.5	20.2	19.0	17.9	16.8	15.7	14.7	12.7	11.0
Path • Inches	-1.5	+1.4	0.0	-6.2	-18.0	-35.9	-61.0	-136.8	-256.2
Wind Drift • Inches	0.0	0.5	2.2	5.0	9.2	14.9	22.1	42.1	70.6

The first animal Craig Boddington ever took with a 7mm Remington Magnum was this ostrich, shot in 1979 in the Namib Desert. He borrowed the rifle from PH Ben Nolte, and it performed wonderfully on a tough running shot. (Photo from Safari Rifles II by Craig Boddington, 2009, Safari Press Archives)

7mm Weatherby Magnum

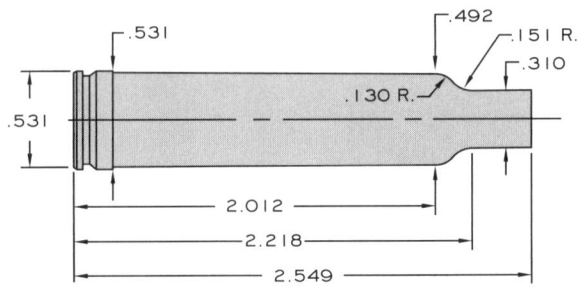

This cartridge was part of the early family of Weatherby magnums. The case is considerably shorter than the full-length magnums (2.549 inches vs: 2.850) but slightly longer than the 7mm Remington Magnum (2.500 inches). The 7mm Weatherby is loaded to a higher pressure standard than the 7mm Remington Magnum. The combination of greater volume and higher pressure gives this cartridge about another 150–200 fps velocity, but it's a push when compared with the 7mm Ultra Magnum. Using a 26-inch barrel for data collection doesn't hurt either.

Relative Recoil Factor = 2.20

Specifications:

Controlling Agency for Standardization of this Ammunition: CIP

Bullet Weight Grains	Velocity fps	Maximum Average Pressure	
		Copper Crusher	Transducer
N/S	N/S	3,800 bar	4,370 bar

Standard barrel for velocity testing: 26 inches long—1 turn in 10-inch twist

Availability:

Weatherby 120-grain Barnes TTSX (B7MM120TTSX) LF
G1 Ballistic Coefficient = 0.370

Distance • Yards	Muzzle	100	200	300	400	500	600	800	1000
Velocity • fps	3430	3148	2884	2635	2399	2178	1966	1585	1278
Energy • ft-lbs	3136	2841	2216	1850	1534	1262	1030	670	435
Taylor KO Index	16.7	15.3	14.0	12.8	11.7	10.6	9.6	7.7	6.2
Path • Inches	-1.5	+1.0	0.0	-5.2	-15.4	-31.7	-55.3	-131.8	-264.5
Wind Drift • Inches	0.0	0.7	2.8	6.6	12.2	19.9	30.0	59.1	102.8

Weatherby 139-grain Soft Point (H 7MM 139 SP)
G1 Ballistic Coefficient = 0.392

Distance • Yards	Muzzle	100	200	300	400	500	600	800	1000
Velocity • fps	3340	3079	2834	2601	2380	2170	1970	1608	1310
Energy • ft-lbs	3443	2926	2478	2088	1748	1453	1198	798	530
Taylor KO Index	18.8	17.4	16.0	14.7	13.4	12.2	11.1	9.1	7.4
Path • Inches	-1.5	+0.8	0.0	-5.1	-16.0	-32.8	-56.9	-134.2	-266.0
Wind Drift • Inches	0.0	0.7	2.7	6.4	11.8	19.2	29.0	56.7	98.1

Nosler 140-grain AccuBond (18022)
G1 Ballistic Coefficient = 0.485

Distance • Yards	Muzzle	100	200	300	400	500	600	800	1000
Velocity • fps	3340	3128	2926	2732	2548	2370	2200	1880	1593
Energy • ft-lbs	3469	3042	2661	2322	2018	1748	1504	1099	789
Taylor KO Index	19.0	17.8	16.6	15.5	14.5	13.5	12.5	10.7	9.0
Path • Inches	-1.5	+1.0	0.0	-5.1	-14.8	-30.0	-51.3	-116.8	-221.9
Wind Drift • Inches	0.0	0.5	2.2	5.0	9.2	14.9	22.2	42.6	72.0

Weatherby 140-grain Nosler Partition (N 7mm 140 PT)
G1 Ballistic Coefficient = 0.434

Distance • Yards	Muzzle	100	200	300	400	500	600	800	1000
Velocity • fps	3303	3069	2847	2636	2434	2241	2057	1717	1424
Energy • ft-lbs	3391	2927	2519	2159	1841	1562	1315	917	631
Taylor KO Index	18.8	17.4	16.2	15.0	13.8	12.7	11.7	9.8	8.1
Path • Inches	-1.5	+0.8	0.0	-5.0	-15.8	-32.1	-55.4	-128.2	-248.7
Wind Drift • Inches	0.0	0.6	2.5	5.8	10.6	17.3	25.9	50.1	85.8

Weatherby 140-grain Barnes TSX (B 7MM 140 TSX) LF
G1 Ballistic Coefficient = 0.47

Distance • Yards	Muzzle	100	200	300	400	500	600	800	1000
Velocity • fps	3250	3039	2837	2645	2460	2283	2114	1797	1517
Energy • ft-lbs	3284	2871	2503	2175	1882	1621	1389	1005	716
Taylor KO Index	18.5	17.3	16.1	15.0	14.0	13.0	12.0	10.2	8.6
Path • Inches	-1.5	+1.1	0.0	-5.5	-15.9	-32.1	-55.0	-125.4	-239.4
Wind Drift • Inches	0.0	0.6	2.3	5.3	9.8	15.8	23.6	45.3	76.8

Weatherby 150-grain BST (N 7MM 150 BST)

G1 Ballistic Coefficient = 0.494

Distance • Yards	Muzzle	100	200	300	400	500	600	800	1000
Velocity • fps	3300	3093	2896	2708	2527	2353	2187	1875	1594
Energy • ft-lbs	3627	3187	2793	2442	2127	1844	1594	1171	849
Taylor KO Index	20.1	18.8	17.6	16.5	15.4	14.3	13.3	11.4	9.7
Path • Inches	-1.5	+1.1	0.0	-5.2	-15.2	-30.6	-52.3	-118.7	-225.1
Wind Drift • Inches	0.0	0.5	2.2	5.0	9.2	14.8	22.1	42.3	71.4

Weatherby 154-grain SP (H 7MM 154 SP)

G1 Ballistic Coefficient = 0.433

Distance • Yards	Muzzle	100	200	300	400	500	600	800	1000
Velocity • fps	3260	3027	2807	2597	2396	2206	2025	1689	1401
Encrgy • ft-lbs	3625	3135	2694	2306	1963	1662	1402	975	671
Taylor KO Index	20.4	18.9	17.5	16.2	15.0	13.8	12.7	10.6	8.8
Path • Inches	-1.5	1.2	0.0	-5.6	-16.3	-33.1	-57.0	-132.1	-256.5
Wind Drift • Inches	0.0	0.6	2.5	5.9	10.9	17.6	26.7	51.1	87.5

Hornady 154-grain InterBond (80689)

G1 Ballistic Coefficient = 0.530

Distance • Yards	Muzzle	100	200	300	400	500	600	800	1000
Velocity • fps	3200	3012	2832	2659	2492	2331	2176	1884	1620
Energy • ft-lbs	3501	3101	2741	2416	2123	1858	1619	1214	897
Taylor KO Index	20.0	18.8	17.7	16.6	15.6	14.6	13.6	11.8	10.1
Path • Inches	-1.5	1.2	0.0	-5.5	-15.9	-32.0	-54.4	-122.6	-230.2
Wind Drift • Inches	0.0	0.5	2.1	4.8	8.9	14.3	21.2	40.4	67.9

Weatherby 160-grain Nosler Partition (N 7MM 160 PT)

G1 Ballistic Coefficient = 0.475

Distance • Yards	Muzzle	100	200	300	400	500	600	800	1000
Velocity • fps	3200	2991	2791	2600	2417	2241	2072	1758	1483
Energy • ft-lbs	3688	3177	2767	2401	2075	1781	1526	1099	781
Taylor KO Index	20.8	19.4	18.1	16.9	15.7	14.5	13.5	11.4	9.6
Path • Inches	-1.5	1.2	0.0	-5.7	-16.5	-33.3	-57.0	-130.2	-248.9
Wind Drift • Inches	0.0	0.6	2.4	5.5	10.0	16.2	24.2	46.6	79.0

Nosler 160-grain AccuBond (18063)

G1 Ballistic Coefficient = 0.531

Distance • Yards	Muzzle	100	200	300	400	500	600	800	1000
Velocity • fps	3050	2868	2694	2525	2364	2209	2059	1778	1526
Energy • ft-lbs	3306	2924	2579	2268	1987	1733	1506	1123	828
Taylor KO Index	19.8	18.6	17.5	16.4	15.3	14.3	13.4	11.5	9.9
Path • Inches	-1.5	1.4	0.0	-6.2	-17.8	-35.6	-60.6	-136.5	-256.8
Wind Drift • Inches	0.0	0.5	2.2	5.2	9.4	15.2	22.7	43.3	72.8

Weatherby 160-grain Nosler AccuBond (N 7MM 160 ACB)

G1 Ballistic Coefficient = 0.529

Distance • Yards	Muzzle	100	200	300	400	500	600	800	1000
Velocity • fps	3050	2868	2694	2596	2364	2206	2055	1774	1522
Energy • ft-lbs	3306	2923	2578	2267	1986	1733	1501	1118	823
Taylor KO Index	19.8	18.6	17.5	16.9	15.3	14.3	13.3	11.5	9.9
Path • Inches	-1.5	1.4	0.0	-6.2	-17.8	-35.7	-60.7	-136.8	-257.4
Wind Drift • Inches	0.0	0.5	2.2	5.2	9.5	15.3	22.8	43.5	73.2

Federal 160-grain Trophy Bonded Tip (P7WBTT1)

G1 Ballistic Coefficient = 0.525

Distance • Yards	Muzzle	100	200	300	400	500	600	800	1000
Velocity • fps	3100	2910	2730	2560	2390	2240	2088	1801	1544
Energy • ft-lbs	3415	3015	2655	2330	2035	1775	1549	1153	848
Taylor KO Index	20.1	16.9	17.7	16.6	15.5	14.5	13.6	11.7	10.0
Path • Inches	-1.5	1.3	0.0	-6.0	-17.2	-34.5	-58.7	-132.4	-249.2
Wind Drift • Inches	0.0	0.5	2.2	5.1	9.4	15.1	22.5	42.9	72.2

Weatherby 175-grain SP (H 7MM 175 SP)

G1 Ballistic Coefficient = 0.462

Distance • Yards	Muzzle	100	200	300	400	500	600	800	1000
Velocity • fps	3070	2861	2662	2471	2288	2113	1945	1637	1373
Energy • ft-lbs	3662	3181	2753	2373	2034	1735	1470	1041	733
Taylor KO Index	21.8	20.3	18.9	17.5	16.2	15.0	13.8	11.6	9.7
Path • Inches	-1.5	1.4	0.0	-6.3	-18.3	-37.0	-63.4	-145.7	-280.5
Wind Drift • Inches	0.0	0.6	2.3	5.4	9.9	16.0	24.0	46.3	78.9

7mm Dakota

The Dakota series of cartridges are (except for the .450 Dakota) based on the .404 Jeffery case. Since a good big case will outperform a good little case any day, the 7mm Dakota is no wimp. It's right up there with the other top-end 7mm performers. Ammunition is only available through Dakota Arms.

Relative Recoil Factor = 2.27

Specifications:

Controlling Agency for Standardization of this Ammunition: Factory

Availability:

Dakota 140-grain Swift (7MM-140FAF)

G1 Ballistic Coefficient = 0.355

Distance • Yards	Muzzle	100	200	300	400	500	600	800	1000
Velocity • fps	3400	3109	2836	2579	2337	2107	1892	1508	1211
Energy • ft-lbs	3595	3005	2501	2069	1698	1381	1113	707	456
Taylor KO Index	19.3	17.7	16.1	14.6	13.3	12.0	10.7	8.6	6.9
Path • Inches	-1.5	+1.0	0.0	-5.4	-16.0	-33.0	-57.9	-139.4	-283.1
Wind Drift • Inches	0.0	0.6	2.7	6.3	11.6	19.1	28.9	57.3	100.0

Dakota 140 Nosler Partition (7MM-140NPT)

G1 Ballistic Coefficient = 0.434

Distance • Yards	Muzzle	100	200	300	400	500	600	800	1000
Velocity • fps	3400	3160	2933	2718	2512	2316	2128	1780	1477
Energy • ft-lbs	3595	3105	2675	2296	1962	1667	1408	986	678
Taylor KO Index	19.3	17.9	16.7	15.4	14.3	13.2	12.1	10.1	8.4
Path • Inches	-1.5	+1.0	0.0	-5.0	-14.8	-30.1	-51.8	-118.0	-232.5
Wind Drift • Inches	0.0	0.5	2.2	5.0	9.2	15.0	22.4	43.3	74.0

Dakota 140-grain Nosler Ballistic Tip (7MM-140NBT)

G1 Ballistic Coefficient = 0.485

Distance • Yards	Muzzle	100	200	300	400	500	600	800	1000
Velocity • fps	3400	3185	2980	2784	2597	2418	2245	1922	1630
Energy • ft-lbs	3595	3154	2761	2411	2098	1818	1568	1149	827
Taylor KO Index	19.3	18.1	16.9	15.8	14.8	13.7	12.8	10.9	9.3
Path • Inches	-1.5	+1.0	0.0	-4.9	-14.3	-28.8	-49.3	-112.2	-213.0
Wind Drift • Inches	0.0	0.5	1.9	4.4	8.1	13.1	19.5	27.6	63.1

Dakota 160-grain Nosler Partition (7MM-160NPT)

G1 Ballistic Coefficient = 0.47

Distance • Yards	Muzzle	100	200	300	400	500	600	800	1000
Velocity • fps	3200	2990	2791	2600	2416	2241	2072	1759	1482
Energy • ft-lbs	3639	3178	2767	2401	2075	1784	1526	1099	781
Taylor KO Index	20.8	19.4	18.1	16.9	15.7	14.5	13.5	11.4	9.6
Path • Inches	-1.5	+1.2	0.0	-5.7	-16.5	-33.2	-57.0	-130.3	-249.9
Wind Drift • Inches	0.0	0.5	2.1	4.9	9.0	14.6	21.8	41.9	71.1

Dakota 160-grain Swift (7MM-160FAF)

G1 Ballistic Coefficient = 0.45

Distance • Yards	Muzzle	100	200	300	400	500	600	800	1000
Velocity • fps	3200	2979	2769	2568	2377	2193	2018	1694	1414
Energy • ft-lbs	3639	3154	2754	2344	2007	1709	1447	1019	710
Taylor KO Index	20.8	19.3	18.0	16.7	15.4	14.2	13.1	11.0	9.2
Path • Inches	-1.5	+1.2	0.0	-5.8	-16.8	-34.0	-58.4	-134.8	-261.4
Wind Drift • Inches	0.0	0.5	2.2	5.2	9.6	15.6	23.3	45.0	76.8

7mm STW (Shooting Times Westerner)

The 7mm STW began when Layne Simpson decided that he wanted a high-performance 7mm. His wildcat cartridge started as a 8mm Remington Magnum case necked to 7mm. This provided a case with a little more volume than the 7mm Weatherby and, therefore, a little more performance. The idea was that rechambering 7mm Remington Magnums for the 7mm STW cartridge would provide an easy conversion. That was a completely feasible plan if the original chambers were exactly the nominal size, but the real world isn't that way. To make these conversions practical, the 7mm STW shape was "improved" by straightening out the body just a little. The 7mm STW received SAAMI standardization in late 1996.

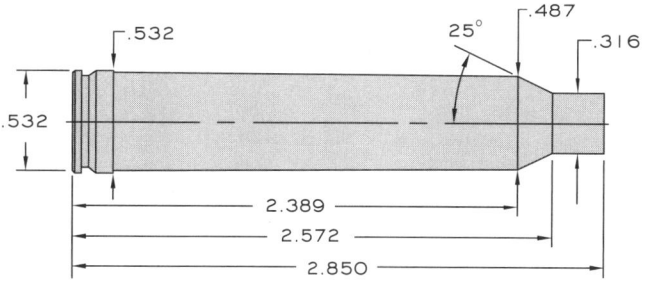

Relative Recoil Factor = 2.31

Specifications:

Controlling Agency for Standardization of this Ammunition: SAAMI

Bullet Weight Grains	Velocity fps	Maximum Average Pressure	
		Copper Crusher	Transducer
140	3,325	54,000 cup	65,000 psi
160	3,250	54,000 cup	65,000 psi

Standard barrel for velocity testing: 24 inches long—1 turn in 10-inch twist

Availability:

Nosler 140-grain Partition (60046)

G1 Ballistic Coefficient = 0.442

Distance • Yards	Muzzle	100	200	300	400	500	600	800	1000
Velocity • fps	3300	3069	2851	2643	2445	2255	2075	1739	1448
Energy • ft-lbs	3385	2928	2527	2171	1858	1581	1339	941	652
Taylor KO Index	18.7	17.4	16.2	15.0	13.9	12.8	11.8	9.9	8.2
Path • Inches	-1.5	+1.1	0.0	-5.4	-15.7	-31.9	-55.0	-126.8	-245.2
Wind Drift • Inches	0.0	0.6	2.4	5.7	10.4	16.9	25.3	48.9	83.6

Nosler 140-grain AccuBond (17036)

G1 Ballistic Coefficient = 0.434

Distance • Yards	Muzzle	100	200	300	400	500	600	800	1000
Velocity • fps	3300	3066	2844	2632	2430	2238	2054	1714	1422
Energy • ft-lbs	3386	2922	2514	2154	1837	1557	1312	914	629
Taylor KO Index	18.7	17.4	16.2	14.9	13.8	12.7	11.7	9.7	8.1
Path • Inches	-1.5	+1.1	0.0	-5.4	-15.8	-32.2	-55.5	-128.5	-249.4
Wind Drift • Inches	0.0	0.6	2.5	5.8	10.7	17.3	25.9	50.2	86.0

A² 140-grain Nosler Ballistic Tip (None)

G1 Ballistic Coefficient = 0.540

Distance • Yards	Muzzle	100	200	300	400	500	600	800	1000
Velocity • fps	3450	3254	3067	2888	2715	2550	2389	2086	1808
Energy • ft-lbs	3700	3291	2924	2592	2292	2021	1775	1354	1016
Taylor KO Index	19.6	18.5	17.4	16.4	15.4	14.5	13.6	11.8	10.3
Path • Inches	-1.5	+0.9	0.0	-4.6	-13.4	-26.8	-45.7	-102.6	-191.6
Wind Drift • Inches	0.0	0.5	1.9	4.3	7.9	12.6	18.7	35.4	59.2

Nosler 140-grain Ballistic Tip (17013)

G1 Ballistic Coefficient = 0.485

Distance • Yards	Muzzle	100	200	300	400	500	600	800	1000
Velocity • fps	3300	3090	2889	2698	2514	2338	2169	1852	1569
Energy • ft-lbs	3386	2968	2596	2263	1966	1700	1463	1067	765
Taylor KO Index	18.7	17.6	16.4	15.3	14.3	13.3	12.3	10.5	8.9
Path • Inches	-1.5	+1.1	0.0	-5.2	-15.3	-30.8	-52.7	-120.0	-228.2
Wind Drift • Inches	0.0	0.5	2.2	5.1	9.4	15.2	22.6	43.3	73.3

Nosler 150-grain Partition (17050)

G1 Ballistic Coefficient = 0.456

Distance • Yards	Muzzle	100	200	300	400	500	600	800	1000
Velocity • fps	3175	2958	2751	2554	2365	2184	2011	1692	1418
Energy • ft-lbs	3358	2915	2522	2173	1664	1590	1348	954	668
Taylor KO Index	19.3	18.0	16.7	15.5	14.4	13.3	12.2	10.3	8.6
Path • Inches	-1.5	+1.2	0.0	-5.8	-17.0	-34.4	-59.1	-136.0	-262.0
Wind Drift • Inches	0.0	0.6	2.5	5.8	10.6	17.2	25.8	49.8	84.9

A² 160-grain Nosler Partition (None)

G1 Ballistic Coefficient = 0.565

Distance • Yards	Muzzle	100	200	300	400	500	600	800	1000
Velocity • fps	3250	3071	2900	2735	2576	2422	2272	1991	1732
Energy • ft-lbs	3752	3351	2987	2657	2357	2084	1835	1409	1066
Taylor KO Index	21.0	19.9	18.8	17.8	16.7	15.7	14.7	12.9	11.2
Path • Inches	-1.5	+1.1	0.0	-5.2	-15.1	-30.3	-51.3	-114.7	-213.3
Wind Drift • Inches	0.0	0.5	1.9	4.4	8.1	13.0	19.3	36.4	60.8

Federal 160-grain Nosler AccuBond (P7STWA1)

G1 Ballistic Coefficient = 0.530

Distance • Yards	Muzzle	100	200	300	400	500	600	800	1000
Velocity • fps	3100	2920	2740	2570	2410	2250	2097	1812	1556
Energy • ft-lbs	3415	3025	2670	2350	2055	1805	1563	1167	861
Taylor KO Index	20.1	19.0	17.8	16.7	15.6	14.6	13.6	11.8	10.1
Path • Inches	-1.5	+1.3	0.0	-5.9	-17.0	-34.3	-58.4	-131.7	-248.1
Wind Drift • Inches	0.0	0.5	2.2	5.0	9.2	14.9	22.2	42.3	71.2

Nosler 160-grain AccuBond (17068) + (60047)

G1 Ballistic Coefficient = 0.531

Distance • Yards	Muzzle	100	200	300	400	500	600	800	1000
Velocity • fps	3075	2892	2717	2548	2386	2292	2078	1796	1542
Energy • ft-lbs	3360	2973	2623	2308	2023	1766	1535	1146	845
Taylor KO Index	20.0	18.8	17.6	16.5	15.5	14.5	13.5	11.7	10.0
Path • Inches	-1.5	+1.3	0.0	-6.0	-17.4	-35.0	-59.5	-134.0	-252.0
Wind Drift • Inches	0.0	0.5	2.2	5.1	9.3	15.1	22.4	42.8	71.9

Nosler 160-grain Partition (17071)

G1 Ballistic Coefficient = 0.475

Distance • Yards	Muzzle	100	200	300	400	500	600	800	1000
Velocity • fps	3075	2871	2677	2490	2311	2140	1975	1672	1409
Energy • ft-lbs	3360	2930	2546	2203	1898	1627	1387	994	706
Taylor KO Index	20.0	18.6	17.4	16.2	15.0	13.9	12.8	10.9	9.1
Path • Inches	-1.5	+1.4	0.0	-6.2	-18.1	-36.4	-62.4	-142.7	-273.3
Wind Drift • Inches	0.0	0.6	2.5	5.8	10.6	17.2	25.7	49.5	84.0

Federal 160-grain Trophy Bonded Tip (P7STWTT1)

G1 Ballistic Coefficient = 0.524

Distance • Yards	Muzzle	100	200	300	400	500	600	800	1000
Velocity • fps	3100	2910	2730	2560	2390	2240	2088	1801	1544
Energy • ft-lbs	3415	3015	2655	2330	2035	1775	1549	1153	847
Taylor KO Index	20.1	18.9	17.7	16.6	15.5	14.5	13.6	11.7	10.0
Path • Inches	-1.5	+1.3	0.0	-6.0	-17.2	-34.6	-58.7	-132.4	-249.3
Wind Drift • Inches	0.0	0.5	2.2	5.1	9.4	15.1	22.5	42.9	72.2

7.21mm (.284) Lazzeroni Firebird

Here's the 7mm Lazzeroni Magnum cartridge. This big boomer is in the same general class as the 7mm Dakota and the 7mm STW. The 140-grain bullet loading has more than a 400 fps velocity edge over the 7mm Remington Magnum.

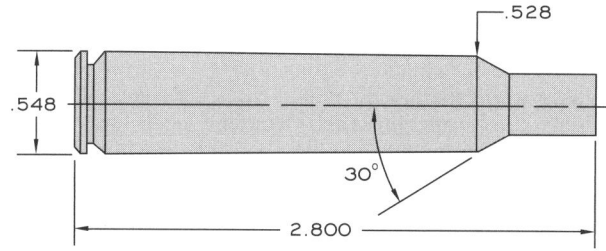

Relative Recoil Factor = 2.33

Specifications:

Controlling Agency for Standardization of this Ammunition: Factory

Availability:

Lazzeroni 120-grain (721FH120)

G1 Ballistic Coefficient = 0.472

Distance • Yards	Muzzle	100	200	300	400	500	600	800	1000
Velocity • fps	3950	3698	3461	3237	3024	2821	2628	2265	1931
Energy • ft-lbs	4158	3645	3193	2792	2437	2121	1841	1367	994
Taylor KO Index	19.2	18.0	16.9	15.8	14.7	13.7	12.8	11.0	9.4
Path • Inches	-1.5	+0.5	0.0	-3.4	-10.2	-20.7	-35.6	-81.4	-154.7
Wind Drift • Inches	0.0	0.4	1.8	4.2	7.8	12.5	18.5	35.0	58.8

Lazzeroni 140-grain Nosler Partition (721FH140P)

G1 Ballistic Coefficient = 0.560

Distance • Yards	Muzzle	100	200	300	400	500	600	800	1000
Velocity • fps	3750	3522	3306	3101	2905	2718	2539	2200	1889
Energy • ft-lbs	4372	3857	3399	2990	2625	2297	2004	1506	1110
Taylor KO Index	22.4	21.0	19.7	18.4	17.2	16.0	14.9	12.9	11.0
Path • Inches	-1.5	0.6	0.0	-3.8	-11.3	-22.8	-39.1	-88.7	-167.5
Wind Drift • Inches	0.0	0.4	1.8	4.3	7.8	12.5	18.5	35.0	58.7

The first animal Craig Boddington took with his new custom-made 7x57 was this Cape hartebeest. It was also the longest shot he has ever made with the 7x57, a bit over 300 yards. (Photo from Safari Rifles II by Craig Boddington, 2009, Safari Press Archives)

.30 M1 Carbine

Developed shortly before WWII, the .30 M1 Carbine represented the U.S. Army's idea of what was needed in an "assault-rifle" cartridge. Today's largely uninformed, but nevertheless, highly opinionated press and electronic media have succeeded in making the term "assault rifle" stand for high power, thus implying the guns must be extremely dangerous. The fact is just the reverse. Assault rifles are (and need to be) very low-power rifles. They are designed to be controllable by the average soldier in fully automatic fire. Look at the relative recoil figure. The .30 M1 Carbine is not suitable for use on deer-size game.

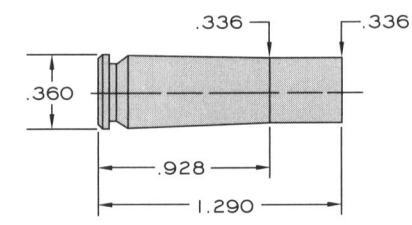

Relative Recoil Factor = 0.99

Specifications:

Controlling Agency for Standardization of this Ammunition: SAAMI

Bullet Weight Grains	Velocity fps	Maximum Average Pressure	
		Copper Crusher	Transducer
110	1,965	40,000 cup	40,000 psi

Standard barrel for velocity testing: 20 inches long—1 turn in 20-inch twist

Availability:

Cor-Bon 100-grain DPX (DPX30100/20)
G1 Ballistic Coefficient = 0.170

Distance • Yards	Muzzle	100	200	300	400	500	600	800	1000
Velocity • fps	2025	1490	1123	945	838	756	686	568	472
Energy • ft-lbs	911	493	280	198	156	127	104	72	50
Taylor KO Index	8.9	6.6	4.9	4.2	3.7	3.3	3.0	2.5	2.1
Path • Inches	-1.5	+7.3	0.0	-34.8	-108.4	-231.6	-415.8	-1023	-2072
Wind Drift • Inches	0.0	4.4	19.6	45.2	78.6	119.0	166.4	284.4	438.8

Federal 110-grain Soft Point RN (30CA)
G1 Ballistic Coefficient = 0.166

Distance • Yards	Muzzle	100	200	300	400	500	600	800	1000
Velocity • fps	1990	1570	1240	1040	920	840	774	665	575
Energy • ft-lbs	965	600	375	260	210	175	146	108	81
Taylor KO Index	9.6	7.6	6.0	5.0	4.5	4.1	3.7	3.2	2.8
Path • Inches	-1.5	+6.4	0.0	-27.7	-81.8	-167.8	-346.2	-835.0	-1643
Wind Drift • Inches	0.0	3.4	15.1	35.7	63.5	97.2	136.1	230.9	349.9

Magtech 110-grain SP (30B)
G1 Ballistic Coefficient = 0.168

Distance • Yards	Muzzle	100	200	300	400	500	600	800	1000
Velocity • fps	1990	1570	1241	1039	926	845	779	670	581
Energy • ft-lbs	968	602	376	264	209	174	148	110	82
Taylor KO Index	9.6	7.6	6.0	5.0	4.5	4.1	3.8	3.2	2.8
Path • Inches	-1.5	+6.4	0.0	-29.1	-90.2	-191.9	-342.3	-825.2	-1622
Wind Drift • Inches	0.0	3.4	14.8	35.2	62.8	95.9	134.6	228.2	345.6

Remington 110-grain Soft Point (R30CAR) + (L30R1)
G1 Ballistic Coefficient = 0.167

Distance • Yards	Muzzle	100	200	300	400	500	600	800	1000
Velocity • fps	1990	1567	1236	1035	923	842	776	667	578
Energy • ft-lbs	967	600	373	262	208	173	147	109	81
Taylor KO Index	9.6	7.6	6.0	5.0	4.5	4.1	3.8	3.2	2.8
Path • Inches	-1.5	+6.4	0.0	-27.7	-81.8	-167.8	-334.5	-830.7	-1634
Wind Drift • Inches	0.0	3.4	15.0	35.5	63.2	96.6	135.4	229.7	348.0

Speer 110-grain GDSP (24465)
G1 Ballistic Coefficient = 0.168

Distance • Yards	Muzzle	100	200	300	400	500	600	800	1000
Velocity • fps	1990	1571	1241	1039	926	845	779	670	581
Energy • ft-lbs	967	602	376	264	209	174	148	110	82
Taylor KO Index	9.6	7.6	6.0	5.0	4.5	4.1	3.8	3.2	2.8
Path • Inches	-1.5	+6.4	0.0	-29.1	-90.2	-191.9	-342.3	-825.0	-1622
Wind Drift • Inches	0.0	3.4	14.8	35.2	62.6	95.9	134.6	228.2	345.6

Winchester 110-grain Hollow Soft Point (X30M1)

G1 Ballistic Coefficient = 0.167

Distance • Yards	Muzzle	100	200	300	400	500	600	800	1000
Velocity • fps	1990	1567	1236	1035	923	842	776	667	578
Energy • ft-lbs	967	600	373	262	208	173	147	109	81
Taylor KO Index	9.6	7.6	6.0	5.0	4.5	4.1	3.8	3.2	2.8
Path • Inches	-1.5	+6.4	0.0	-27.7	-81.8	-167.8	-334.5	-830.7	-1634
Wind Drift • Inches	0.0	3.4	15.0	35.5	63.2	96.6	135.4	229.7	348.0

Sellier & Bellot 110-grain FMJ (V332652U)

G1 Ballistic Coefficient = 0.155

Distance • Yards	Muzzle	100	200	300	400	500	600	800	1000
Velocity • fps	2024	1567	1217	1015	903	821	754	642	549
Energy • ft-lbs	1001	600	362	252	199	165	139	101	74
Taylor KO Index	9.8	7.6	5.9	4.9	4.4	4.0	3.6	3.1	2.7
Path • Inches	-1.5	+6.5	0.0	-30.4	-93.7	-200.0	-358.0	-869.4	-1724
Wind Drift • Inches	0.0	3.6	16.0	37.8	67.1	102.4	143.6	244.0	371.2

Aguila 110-grain FMJ (1E302110)

G1 Ballistic Coefficient = 0.165

Distance • Yards	Muzzle	100	200	300	400	500	600	800	1000
Velocity • fps	1990	1563	1232	1032	919	839	772	663	573
Energy • ft-lbs	968	597	371	260	206	172	146	107	80
Taylor KO Index	9.6	7.6	6.0	5.0	4.4	4.1	3.7	3.2	2.8
Path • Inches	-1.5	+6.5	0.0	-29.5	-91.5	-194.7	-347.4	-838.1	-1650
Wind Drift • Inches	0.0	3.4	15.1	35.8	63.7	97.4	136.6	231.7	351.2s

American Eagle (Federal) 110-grain Full Metal Jacket (AE30CB)

G1 Ballistic Coefficient = 0.166

Distance • Yards	Muzzle	100	200	300	400	500	600	800	1000
Velocity • fps	1990	1570	1240	1040	920	840	774	665	575
Energy • ft-lbs	965	600	375	260	210	175	146	108	81
Taylor KO Index	9.6	7.6	6.0	5.0	4.5	4.1	3.7	3.2	2.8
Path • Inches	-1.5	6.4	0.0	-27.7	-81.8	-167.8	-346.2	-835.0	-1643
Wind Drift • Inches	0.0	3.4	15.1	35.7	63.5	97.2	136.1	230.9	349.9

Magtech 110-grain FMJ (30A)

G1 Ballistic Coefficient = 0.184

Distance • Yards	Muzzle	100	200	300	400	500	600	800	1000
Velocity • fps	1990	1604	1290	1080	960	878	812	706	619
Energy • ft-lbs	968	628	407	285	225	188	161	122	94
Taylor KO Index	9.6	7.8	6.2	5.2	4.6	4.2	3.9	3.4	3.0
Path • Inches	-1.5	+6.1	0.0	-27.1	-84.0	-178.4	-318.1	-764.1	-1492
Wind Drift • Inches	0.0	3.0	13.3	31.8	57.4	88.5	124.6	211.5	319.3

UMC (Remington) 110-grain Metal Case (L30CR1)

G1 Ballistic Coefficient = 0.167

Distance • Yards	Muzzle	100	200	300	400	500	600	800	1000
Velocity • fps	1990	1567	1236	1035	923	842	776	667	578
Energy • ft-lbs	967	600	373	262	208	173	147	109	81
Taylor KO Index	9.6	7.6	6.0	5.0	4.5	4.1	3.8	3.2	2.8
Path • Inches	-1.5	+6.4	0.0	-27.7	-81.8	-167.8	-334.5	-830.7	-1634
Wind Drift • Inches	0.0	3.4	15.0	35.5	63.2	96.6	135.4	229.7	348.0

USA (Winchester) 110-grain Full Metal Jacket (Q3132)

G1 Ballistic Coefficient = 0.180

Distance • Yards	Muzzle	100	200	300	400	500	600	800	1000
Velocity • fps	1990	1596	1279	1070	952	870	804	697	610
Energy • ft-lbs	967	622	399	280	221	185	158	119	91
Taylor KO Index	9.6	7.7	6.2	5.2	4.6	4.2	3.9	3.4	3.0
Path • Inches	-1.5	6.5	0.0	-27.9	-86.0	-181.6	-323.8	-778.3	-1522
Wind Drift • Inches	0.0	3.1	13.7	32.6	58.6	90.2	126.9	215.4	325.4

7.62x39mm Russian

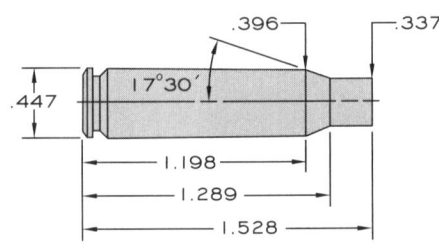

Assault rifles came into general usage during WWII when military authorities recognized that an all-up infantry rifle (the Garand for instance) was not the best weapon for street fighting. The 7.62x39mm was the Soviet answer to this need. Introduced in 1943, it has become the standard infantry cartridge for many of the world's armies. Because some guns chambered for this cartridge use a 0.310-inch groove-diameter barrel and some others use a 0.308-inch one, it's a good idea to know the dimensions of both your gun and your ammunition before you start shooting. The SAAMI standard for this round uses bullets with a diameter tolerance of 0.309 to 0.311 inch.

Relative Recoil Factor = 1.29

Specifications:

Controlling Agency for Standardization of this Ammunition: SAAMI

Bullet Weight Grains	Velocity fps	Maximum Average Pressure	
		Copper Crusher	Transducer
123	2,350	50,000 cup	45,000 psi

Standard barrel for velocity testing: 20 inches long—1 turn in 9.45-inch twist

Availability:

Cor-Bon 108-grain MPG (MPG76239108/20) LF
G1 Ballistic Coefficient = 0.165

Distance • Yards	Muzzle	100	200	300	400	500	600	800	1000
Velocity • fps	2500	2004	1575	1240	1036	922	841	717	618
Energy • ft-lbs	1499	964	595	369	258	204	170	123	92
Taylor KO Index	12.0	9.6	7.5	5.9	5.0	4.4	4.0	3.4	3.0
Path • Inches	-1.5	+3.7	0.0	-18.0	-58.6	-131.5	-245.3	-629.6	-1288
Wind Drift • Inches	0.0	2.5	11.1	27.9	53.8	86.9	125.8	220.2	337.6

Sellier & Bellot 123-grain SP (V332252U)
G1 Ballistic Coefficient = 0.273

Distance • Yards	Muzzle	100	200	300	400	500	600	800	1000
Velocity • fps	2437	2135	1855	1601	1381	1203	1076	927	830
Energy • ft-lbs	1622	1245	940	701	521	395	316	235	188
Taylor KO Index	13.2	11.6	10.0	8.7	7.5	6.5	5.8	5.0	4.5
Path • Inches	-1.5	+3.1	0.0	-13.2	-40.2	-85.5	-154.8	-388.8	-784.8
Wind Drift • Inches	0.0	1.5	6.4	15.4	29.2	48.6	73.6	136.7	214.2

Winchester 123-grain Soft Point (X76239)
G1 Ballistic Coefficient = 0.243

Distance • Yards	Muzzle	100	200	300	400	500	600	800	1000
Velocity • fps	2365	2030	1731	1465	1248	1093	992	864	772
Energy • ft-lbs	1527	1129	818	586	425	327	269	204	163
Taylor KO Index	12.8	11.0	9.4	7.9	6.8	5.9	5.4	4.7	4.2
Path • Inches	-1.5	+3.8	0.0	-15.4	-46.3	-98.4	-183.9	-461.6	-926.2
Wind Drift • Inches	0.0	1.8	7.6	18.5	35.3	58.5	87.0	157.0	242.1

USA (Winchester) 123-grain Soft Point (Q3174)
G1 Ballistic Coefficient = 0.245

Distance • Yards	Muzzle	100	200	300	400	500	600	800	1000
Velocity • fps	2355	2026	1726	1463	1247	1093	994	866	775
Energy • ft-lbs	1515	1121	814	584	425	326	270	205	164
Taylor KO Index	12.7	11.0	9.3	7.9	6.7	5.9	5.4	4.7	4.2
Path • Inches	-1.5	+3.8	0.0	-15.6	-47.6	-102.0	-184.1	-461.5	-924.8
Wind Drift • Inches	0.0	1.8	7.6	18.4	35.2	58.2	87.0	156.1	240.7

Federal 123-grain Fusion (F76239FS1)
G1 Ballistic Coefficient = 0.296

Distance • Yards	Muzzle	100	200	300	400	500	600	800	1000
Velocity • fps	2350	2080	1820	1590	1390	1220	1096	947	852
Energy • ft-lbs	1510	1180	905	695	525	405	328	245	198
Taylor KO Index	12.8	11.3	9.9	8.7	7.6	6.6	6.0	5.2	4.6
Path • Inches	-1.5	+3.3	0.0	-14.0	-41.6	-87.7	-156.6	-387.0	-772.5
Wind Drift • Inches	0.0	1.4	6.1	14.7	27.8	46.0	69.4	128.8	201.9

Federal 123-grain Soft Point (76239B)
G1 Ballistic Coefficient = 0.273

Distance • Yards	Muzzle	100	200	300	400	500	600	800	1000
Velocity • fps	2350	2060	1780	1540	1330	1160	1047	910	817
Energy • ft-lbs	1510	1155	870	645	480	370	300	226	182
Taylor KO Index	12.7	11.1	9.6	8.3	7.2	6.3	5.7	4.9	4.4
Path • Inches	-1.5	+3.4	0.0	-14.5	-43.8	-93.0	-168.0	-418.3	-836.7
Wind Drift • Inches	0.0	1.6	6.7	16.2	30.8	51.1	76.7	140.6	218.5

Hornady 123-grain A-MAX (8078)
G1 Ballistic Coefficient = 0.270

Distance • Yards	Muzzle	100	200	300	400	500	600	800	1000
Velocity • fps	2360	2060	1784	1535	1323	1158	1046	908	815
Energy • ft-lbs	1521	1159	869	644	488	366	299	225	182
Taylor KO Index	12.8	11.1	9.7	8.3	7.2	6.3	5.7	4.9	4.4
Path • Inches	-1.5	+3.4	0.0	-14.4	-43.6	-92.8	-167.8	-418.5	-838.0
Wind Drift • Inches	0.0	1.6	6.8	16.3	31.0	51.4	77.2	141.4	219.9

Cor-Bon 123-grain DPX (DPX762x39123/20)
G1 Ballistic Coefficient = 0.270

Distance • Yards	Muzzle	100	200	300	400	500	600	800	1000
Velocity • fps	2300	2204	1733	1491	1288	1133	1028	898	807
Energy • ft-lbs	1445	1097	821	608	453	351	289	220	178
Taylor KO Index	12.4	11.9	9.4	8.1	7.0	6.1	5.6	4.9	4.4
Path • Inches	-1.5	+3.6	0.0	-15.3	-46.3	-98.3	-177.4	-439.4	-874.2
Wind Drift • Inches	0.0	1.6	7.0	16.9	32.1	53.0	79.2	143.7	222.3

Sellier & Bellot 123-grain FMJ (V322242U)
G1 Ballistic Coefficient = 0.192

Distance • Yards	Muzzle	100	200	300	400	500	600	800	1000
Velocity • fps	2421	1999	1627	1319	1104	979	895	773	678
Energy • ft-lbs	1601	1092	723	475	333	262	219	163	126
Taylor KO Index	13.1	10.8	8.8	7.1	6.0	5.8	4.8	4.2	3.7
Path • Inches	-1.5	+3.7	0.0	-17.0	-54.1	-119.7	-221.5	-562.7	-1138
Wind Drift • Inches	0.0	2.2	9.7	24.0	46.2	75.4	110.2	194.0	296.9

UMC (Remington) 123-grain MC (L762391)
G1 Ballistic Coefficient = 0.266

Distance • Yards	Muzzle	100	200	300	400	500	600	800	1000
Velocity • fps	2365	2060	1780	1528	1314	1149	1038	902	809
Energy • ft-lbs	1527	1159	865	638	472	371	294	222	179
Taylor KO Index	12.9	11.3	9.7	8.4	7.2	6.3	5.6	4.9	4.4
Path • Inches	-1.5	+3.4	0.0	-14.4	-43.9	-93.6	-169.4	-423.3	-848.5
Wind Drift • Inches	0.0	1.6	6.9	16.6	31.6	52.4	78.5	143.6	223.1

PMC 123-grain Full Metal Jacket (7.62A)
G1 Ballistic Coefficient = 0.277

Distance • Yards	Muzzle	100	200	300	400	500	600	800	1000
Velocity • fps	2350	2072	1817	1583	1368	1171	1056	917	824
Energy • ft-lbs	1495	1162	894	678	507	371	305	230	185
Taylor KO Index	12.7	11.2	9.8	8.6	7.4	6.3	5.7	5.0	4.5
Path • Inches	-1.5	+3.4	0.0	-14.3	-43.4	-91.9	-165.8	-412.3	-215.4
Wind Drift • Inches	0.0	1.6	6.6	15.9	30.3	50.2	75.3	138.4	215.4

Lapua 123-grain FMJ (4317235)
G1 Ballistic Coefficient = 0.247

Distance • Yards	Muzzle	100	200	300	400	500	600	800	1000
Velocity • fps	2345	2031	1735	1463	1248	1095	995	868	777
Energy • ft-lbs	1507	1131	825	586	425	327	271	206	165
Taylor KO Index	12.8	11.1	9.5	8.0	6.8	6.0	5.4	4.7	4.2
Path • Inches	-1.5	+3.6	0.0	-15.4	-47.1	-101.3	-184.4	-461.4	-923.5
Wind Drift • Inches	0.0	1.8	7.6	18.4	35.0	57.9	86.1	155.2	239.3

Fiocchi 124-grain FMJ (762SOVA)
G1 Ballistic Coefficient = 0.297

Distance • Yards	Muzzle	100	200	300	400	500	600	800	1000
Velocity • fps	2375	2100	1844	1611	1405	1234	1107	953	857
Energy • ft-lbs	1552	1214	936	715	544	419	337	250	202
Taylor KO Index	13.0	11.5	10.1	8.8	7.7	6.8	6.1	5.2	4.7
Path • Inches	-1.5	+3.2	0.0	-13.5	-40.4	-85.1	-152.5	-377.6	-755.6
Wind Drift • Inches	0.0	1.4	6.0	14.4	27.3	45.2	68.3	127.3	200.2

American Eagle (Federal) 124-grain Full Metal Jacket (AE76239A)

G1 Ballistic Coefficient = 0.297

Distance • Yards	Muzzle	100	200	300	400	500	600	800	1000
Velocity • fps	2350	2080	1820	1600	1390	1220	1096	947	852
Energy • ft-lbs	1520	1190	915	700	535	410	331	247	220
Taylor KO Index	12.8	11.3	9.9	8.7	7.6	6.7	6.0	5.2	4.6
Path • Inches	-1.5	+3.3	0.0	-13.8	-41.5	-87.2	-156.4	-386.4	-771.5
Wind Drift • Inches	0.0	1.4	6.1	14.7	27.8	45.9	69.2	128.7	201.7

Remington 125-grain Pointed Soft Point (R762391)

G1 Ballistic Coefficient = 0.267

Distance • Yards	Muzzle	100	200	300	400	500	600	800	1000
Velocity • fps	2365	2062	1783	1533	1320	1154	1041	904	811
Energy • ft-lbs	1552	1180	882	652	483	370	301	227	183
Taylor KO Index	13.0	11.3	9.8	8.4	7.3	6.3	5.7	5.0	4.5
Path • Inches	-1.5	+3.4	0.0	-14.4	-43.7	-93.2	-168.7	-421.3	-844.3
Wind Drift • Inches	0.0	1.6	6.8	16.5	31.4	52.0	78.0	142.9	222.0

Cor-Bon 125-grain Jacketed Hollow Point (SD762x39125/20)

G1 Ballistic Coefficient = 0.275

Distance • Yards	Muzzle	100	200	300	400	500	600	800	1000
Velocity • fps	2400	2102	1826	1577	1362	1189	1067	922	826
Energy • ft-lbs	1600	1227	926	691	515	392	316	236	189
Taylor KO Index	13.3	11.7	10.1	8.8	7.6	6.6	5.9	5.1	4.6
Path • Inches	-1.5	+3.2	0.0	-13.7	-41.6	-88.6	-160.7	-398.6	-785.9
Wind Drift • Inches	0.0	1.4	5.8	14.0	26.7	44.3	66.9	123.7	193.2

Cor-Bon 150-grain Jacketed Soft Point (HT762x39150/20)

G1 Ballistic Coefficient = 0.325

Distance • Yards	Muzzle	100	200	300	400	500	600	800	1000
Velocity • fps	2300	2052	1821	1609	1420	1259	1134	978	881
Energy • ft-lbs	1762	1403	1105	863	672	528	428	318	258
Taylor KO Index	15.3	13.7	12.1	10.7	9.5	8.4	7.6	6.5	5.9
Path • Inches	-1.5	+3.4	0.0	-13.9	-41.4	-86.3	-153.6	-373.7	-730.1
Wind Drift • Inches	0.0	1.2	5.1	12.2	23.0	38.0	57.2	106.7	168.1

Cor-Bon 150-grain Glaser Blue (05600/na)

G1 Ballistic Coefficient = 0.225

Distance • Yards	Muzzle	100	200	300	400	500	600	800	1000
Velocity • fps	2300	1948	1632	1363	1158	1027	941	822	732
Energy • ft-lbs	1527	1095	768	536	387	305	256	195	155
Taylor KO Index	13.2	11.1	9.3	7.8	6.6	5.9	5.4	4.7	4.2
Path • Inches	-1.5	+3.9	0.0	-17.1	-53.0	-114.9	-209.6	-522.8	-1043
Wind Drift • Inches	0.0	2.0	8.7	21.2	40.4	66.1	97.0	171.6	262.4

.30-30 Winchester

For over 100 years the .30-30 Winchester has been what most hunters would call your basic deer rifle. Despite the black-powder type of designation, the .30-30 was the first cartridge of its class loaded with smokeless powder. There are lots of .30 caliber cartridges with better performance but few combine with a carbine length, lever-action rifle to offer a better combination of enough power and a super handy rifle for woods-type hunting. There is a little something to remember when looking at the performance data for the .30-30 Winchester: The velocity specifications are measured in a 24-inch barrel and a lot of guns that are chambered for the .30-30 have 20-inch barrels. This will cost about 100–150 fps in the short barrel. Time has proved that this doesn't make much difference because .30-30s are seldom used where long shots are required.

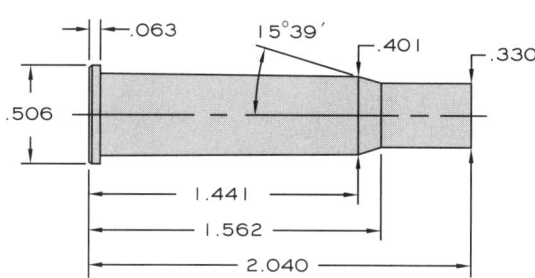

Relative Recoil Factor = 1.60

Specifications:

Controlling Agency for Standardization of this Ammunition: SAAMI

Bullet Weight Grains	Velocity fps	Maximum Average Pressure	
		Copper Crusher	Transducer
55 (Saboted)	3,365	N/S	38,000 psi
125	2,550	38,000 cup	42,000 psi
150	2,370	38,000 cup	42,000 psi
170	2,180	38,000 cup	42,000 psi

Standard barrel for velocity testing: 24 inches long—1 turn in 12-inch twist

Availability:

Remington 125-grain Core-Lokt Pointed Soft Point – Managed Recoil (RL30301)
G1 Ballistic Coefficient = 0.215

Distance • Yards	Muzzle	100	200	300	400	500	600	800	1000
Velocity • fps	2175	1820	1508	1255	1082	975	899	787	700
Energy • ft-lbs	1313	919	631	437	325	264	224	172	136
Taylor KO Index	12.0	10.0	8.5	6.9	6.0	5.4	4.9	4.3	3.9
Path • Inches	-1.5	+4.6	0.0	-20.0	-62.2	-134.1	-242.6	-595.1	-1174
Wind Drift • Inches	0.0	2.3	9.9	24.2	45.5	72.8	105.1	182.5	276.9

Federal 125-grain Hi-Shok Hollow Point (3030C)
G1 Ballistic Coefficient = 0.174

Distance • Yards	Muzzle	100	200	300	400	500	600	800	1000
Velocity • fps	2570	2090	1660	1320	1080	960	874	747	647
Energy • ft-lbs	1830	1210	770	480	320	260	212	155	116
Taylor KO Index	14.1	11.5	9.1	7.3	5.9	5.3	4.8	4.1	3.6
Path • Inches	-1.5	+3.3	0.0	-16.0	-50.9	-109.5	-220.2	-570.9	-117.3
Wind Drift • Inches	0.0	2.2	10.0	25.1	48.9	80.2	117.4	207.6	319.2

Hornady 140-grain MonoFlex (82731)
G1 Ballistic Coefficient = 0.280

Distance • Yards	Muzzle	100	200	300	400	500	600	800	1000
Velocity • fps	2500	2201	1922	1668	1443	1255	1114	950	850
Energy • ft-lbs	1943	1505	1149	865	647	490	386	281	224
Taylor KO Index	15.4	13.6	11.8	10.3	8.9	7.7	6.9	5.9	5.2
Path • Inches	-1.5	+2.9	0.0	-12.3	-37.3	-79.0	-142.8	-359.8	-730.8
Wind Drift • Inches	0.0	1.4	6.0	14.3	27.3	45.5	69.1	130.3	206.6

Federal 150-grain Fusion (F3030FS1)
G1 Ballistic Coefficient = 0.269

Distance • Yards	Muzzle	100	200	300	400	500	600	800	1000
Velocity • fps	2390	2090	1810	1550	1340	1170	1052	912	818
Energy • ft-lbs	1900	1450	1085	805	595	455	369	277	223
Taylor KO Index	15.8	13.8	11.9	10.2	8.8	7.7	6.9	6.0	5.4
Path • Inches	-1.5	+3.2	0.0	-14.2	-42.8	-91.5	-163.8	-409.9	-823.5
Wind Drift • Inches	0.0	1.6	6.7	16.1	30.6	50.8	76.5	140.8	219.4

Federal 150-grain Soft Point FN (3030A)

G1 Ballistic Coefficient = 0.220

Distance • Yards	Muzzle	100	200	300	400	500	600	800	1000
Velocity • fps	2390	2020	1680	1400	1180	1040	950	826	733
Energy • ft-lbs	1900	1355	945	650	460	355	300	227	179
Taylor KO Index	15.8	13.3	11.1	9.2	7.8	6.9	6.3	5.5	4.8
Path • Inches	-1.5	+3.6	0.0	-15.9	-49.1	-104.5	-198.2	-500.9	-1009
Wind Drift • Inches	0.0	1.9	8.4	20.6	39.6	65.3	96.4	172.1	264.2

Fiocchi 150-grain PSP (303B)

G1 Ballistic Coefficient = 0.186

Distance • Yards	Muzzle	100	200	300	400	500	600	800	1000
Velocity • fps	2390	1960	1582	1277	1075	958	976	756	662
Energy • ft-lbs	1902	1279	834	543	385	305	256	190	146
Taylor KO Index	15.8	12.9	10.4	8.4	7.1	6.3	5.8	5.0	4.4
Path • Inches	-1.5	+3.9	0.0	-18.0	-57.4	-127.0	-234.4	-592.8	-1197
Wind Drift • Inches	0.0	2.3	10.3	25.5	48.8	79.0	114.7	200.8	306.6

Hornady 150-grain Round Nose InterLock (8080)

G1 Ballistic Coefficient = 0.186

Distance • Yards	Muzzle	100	200	300	400	500	600	800	1000
Velocity • fps	2390	1959	1581	1276	1074	957	876	756	662
Energy • ft-lbs	1902	1278	832	542	384	305	226	190	146
Taylor KO Index	15.8	12.9	10.4	8.4	7.1	6.3	5.8	5.0	4.4
Path • Inches	-1.5	+3.7	0.0	-18.0	-57.4	-127.0	-234.4	-592.8	-1197
Wind Drift • Inches	0.0	2.3	10.3	25.5	48.8	79.0	114.7	200.8	306.6

PMC 150-grain RNSP (3030HIA)

G1 Ballistic Coefficient = 0.186

Distance • Yards	Muzzle	100	200	300	400	500	600	800	1000
Velocity • fps	2390	1959	1581	1276	1074	957	876	756	662
Energy • ft-lbs	1902	1278	832	542	384	305	256	190	146
Taylor KO Index	15.8	12.9	10.4	8.4	7.1	6.3	5.8	5.0	4.4
Path • Inches	-1.5	+3.9	0.0	-18.0	-57.4	-127.0	-234.4	-592.8	-1197
Wind Drift • Inches	0.0	2.3	10.3	25.5	48.8	79.0	114.7	200.8	306.7

Remington 150-grain Core-Lokt Soft Point (R30301)

G1 Ballistic Coefficient = 0.193

Distance • Yards	Muzzle	100	200	300	400	500	600	800	1000
Velocity • fps	2390	1973	1605	1303	1095	974	891	771	677
Energy • ft-lbs	1902	1296	858	565	399	316	265	198	153
Taylor KO Index	15.8	13.0	10.6	8.6	7.2	6.4	5.9	5.1	4.5
Path • Inches	-1.5	3.8	0.0	-17.5	-55.6	-122.5	-226.0	-571.8	-1153
Wind Drift • Inches	0.0	2.2	9.8	24.3	46.7	76.0	110.7	194.4	297.1

Winchester 150-grain Power-Point (X30306)

G1 Ballistic Coefficient = 0.218

Distance • Yards	Muzzle	100	200	300	400	500	600	800	1000
Velocity • fps	2390	2018	1684	1398	1177	1036	945	822	729
Energy • ft-lbs	1902	1356	944	651	461	357	298	225	177
Taylor KO Index	15.8	13.3	11.1	9.2	7.8	6.8	6.2	5.4	4.8
Path • Inches	-1.5	+3.6	0.0	-16.0	-49.9	-108.8	-200.0	-505.5	-1018
Wind Drift • Inches	0.0	2.0	8.5	20.9	40.1	66.1	97.4	173.6	266.4

Winchester 150-grain Silvertip (X30302)

G1 Ballistic Coefficient = 0.218

Distance • Yards	Muzzle	100	200	300	400	500	600	800	1000
Velocity • fps	2390	2018	1684	1398	1177	1036	945	822	729
Energy • ft-lbs	1902	1356	944	651	461	357	298	225	177
Taylor KO Index	15.8	13.3	11.1	9.2	7.8	6.8	6.2	5.4	4.8
Path • Inches	-1.5	+3.6	0.0	-16.0	-49.9	-108.8	-200.0	-505.5	-1018
Wind Drift • Inches	0.0	2.0	8.5	20.9	40.1	66.1	97.4	173.6	266.4

Sellier & Bellot 150-grain SP (V330352U)

G1 Ballistic Coefficient = 0.190

Distance • Yards	Muzzle	100	200	300	400	500	600	800	1000
Velocity • fps	2388	1966	1594	1291	1086	967	885	765	671
Energy • ft-lbs	1900	1287	846	556	393	311	261	195	150
Taylor KO Index	15.8	13.0	10.5	8.5	7.2	6.4	5.8	5.0	4.4
Path • Inches	-1.5	+3.8	0.0	-17.7	-56.4	-124.5	-229.7	-580.8	-1172
Wind Drift • Inches	0.0	2.3	10.0	24.8	47.6	77.2	112.3	197.0	300.9

Winchester 150-grain Ballistic Silvertip (SBST3030)

G1 Ballistic Coefficient = 0.233

Distance • Yards	Muzzle	100	200	300	400	500	600	800	1000
Velocity • fps	2390	2040	1723	1447	1225	1072	976	849	757
Energy • ft-lbs	1902	1386	989	697	499	383	317	240	191
Taylor KO Index	15.0	13.5	11.4	9.6	8.1	7.1	6.4	5.6	5.0
Path • Inches	-1.5	+3.8	0.0	-15.6	-47.9	-102.3	-187.5	-473.2	-952.6
Wind Drift • Inches	0.0	1.8	7.9	19.2	36.8	61.0	90.6	163.0	251.1

Winchester 150-grain Power Max Bonded (X30306BP)

G1 Ballistic Coefficient = 0.224

Distance • Yards	Muzzle	100	200	300	400	500	600	800	1000
Velocity • fps	2390	2028	1702	1420	1198	1052	959	834	741
Energy • ft-lbs	1902	1370	965	672	478	368	306	232	183
Taylor KO Index	15.8	13.4	11.2	9.4	7.9	6.9	6.3	5.5	4.9
Path • Inches	-1.5	+3.5	0.0	-15.7	-48.7	-105.9	-194.5	-491.3	-988.9
Wind Drift • Inches	0.0	1.9	8.2	20.1	38.6	63.8	94.4	168.9	259.6

Winchester 150-grain Hollow Point (X30301)

G1 Ballistic Coefficient = 0.218

Distance • Yards	Muzzle	100	200	300	400	500	600	800	1000
Velocity • fps	2390	2018	1684	1398	1177	1036	945	822	729
Energy • ft-lbs	1902	1356	944	651	461	357	298	225	177
Taylor KO Index	15.8	13.3	11.1	9.2	7.8	6.8	6.2	5.4	4.8
Path • Inches	-1.5	+3.6	0.0	-16.0	-49.9	-108.8	-200.0	-505.5	-1018
Wind Drift • Inches	0.0	2.0	8.5	20.9	40.1	66.1	97.4	173.6	266.4

Cor-Bon 150-grain DPX (DPX3030150/20)

G1 Ballistic Coefficient = 0.225

Distance • Yards	Muzzle	100	200	300	400	500	600	800	1000
Velocity • fps	2300	1948	1632	1364	1159	1028	943	823	733
Energy • ft-lbs	1762	1264	888	620	448	352	296	226	179
Taylor KO Index	15.2	12.9	10.8	9.0	7.6	6.8	6.2	5.4	4.8
Path • Inches	-1.5	+3.9	0.0	-17.1	-53.0	-114.7	-209.2	-521.8	-1040
Wind Drift • Inches	0.0	2.0	8.7	21.2	40.4	66.0	96.8	171.2	261.8

Federal 150-grain Barnes Triple-Shock X-Bullet (PSC3030WA) LF

G1 Ballistic Coefficient = 0.185

Distance • Yards	Muzzle	100	200	300	400	500	600	800	1000
Velocity • fps	2220	1800	1450	1180	1020	920	848	734	643
Energy • ft-lbs	1640	1085	695	460	345	280	240	180	138
Taylor KO Index	14.7	11.9	9.6	7.8	6.7	6.1	5.6	4.8	4.2
Path • Inches	-1.5	+4.7	0.0	-21.7	-68.6	-147.8	-267.8	-662.1	-1317
Wind Drift • Inches	0.0	2.6	11.5	28.2	52.3	83.7	119.7	206.5	313.4

Remington 150-grain Copper Solid Tip (PSC3030WA) LF

G1 Ballistic Coefficient = 0.186

Distance • Yards	Muzzle	100	200	300	400	500	600	800	1000
Velocity • fps	2200	1806	1448	1185	1022	924	852	739	648
Energy • ft-lbs	1641	1066	698	468	348	284	242	182	140
Taylor KO Index	14.5	11.9	9.6	7.8	6.7	6.1	5.6	4.9	4.3
Path • Inches	-1.5	+5.2	0.0	-21.8	-68.6	-149.0	-267.9	-660.3	-1310
Wind Drift • Inches	0.0	2.6	11.4	28.0	52.3	82.8	118.4	204.1	309.6

Hornady 170-grain FTX (82730)

G1 Ballistic Coefficient = 0.330

Distance • Yards	Muzzle	100	200	300	400	500	600	800	1000
Velocity • fps	2400	2150	1916	1699	1501	1329	1189	1009	904
Energy • ft-lbs	2046	1643	1304	1025	801	628	502	362	291
Taylor KO Index	16.9	15.1	13.5	12.0	10.6	9.4	8.4	7.1	6.4
Path • Inches	-1.5	+3.0	0.0	-12.6	-37.2	-77.4	-137.2	-335.8	-670.6
Wind Drift • Inches	0.0	1.2	5.3	12.6	23.6	39.0	59.1	112.4	179.4

Federal 170-grain Fusion (F3030FS2)

G1 Ballistic Coefficient = 0.312

Distance • Yards	Muzzle	100	200	300	400	500	600	800	1000
Velocity • fps	2200	1950	1720	1510	1330	1180	1073	940	851
Energy • ft-lbs	1825	1435	1115	860	665	525	435	333	273
Taylor KO Index	16.5	14.6	12.9	11.3	9.9	8.8	8.0	7.0	6.4
Path • Inches	-1.5	+3.0	0.0	-15.6	-46.8	-97.8	-173.5	-420.4	-824.9
Wind Drift • Inches	0.0	1.5	6.4	15.2	28.6	46.9	69.9	127.7	198.2

Federal 170-grain Soft Point RN (3030B)

G1 Ballistic Coefficient = 0.255

Distance • Yards	Muzzle	100	200	300	400	500	600	800	1000
Velocity • fps	2200	1900	1620	1380	1190	1060	975	858	771
Energy • ft-lbs	1830	1355	990	720	535	425	359	278	225
Taylor KO Index	16.5	14.2	12.1	10.3	8.9	7.9	7.3	6.4	5.8
Path • Inches	-1.5	+4.1	0.0	-17.4	-52.4	-109.4	-204.5	-501.5	-988.1
Wind Drift • Inches	0.0	1.9	8.0	19.3	36.6	59.7	87.7	155.6	238.0

Fiocchi 170-grain FSP (3030C)

G1 Ballistic Coefficient = 0.205

Distance • Yards	Muzzle	100	200	300	400	500	600	800	1000
Velocity • fps	2200	1827	1501	1240	1066	961	886	773	684
Energy • ft-lbs	1827	1259	850	580	429	349	296	226	177
Taylor KO Index	16.5	13.7	11.2	9.3	8.0	7.2	6.6	5.8	5.1
Path • Inches	-1.5	+4.5	0.0	-20.2	-63.1	-136.5	-247.6	-609.7	-1203
Wind Drift • Inches	0.0	2.4	10.3	25.2	47.4	75.8	109.1	189.2	287.2

Hornady 170-grain Flat Point InterLock (8085)

G1 Ballistic Coefficient = 0.186

Distance • Yards	Muzzle	100	200	300	400	500	600	800	1000
Velocity • fps	2200	1796	1450	1185	1022	924	850	737	645
Energy • ft-lbs	1827	1218	793	530	395	322	273	205	157
Taylor KO Index	16.5	13.4	10.8	8.9	7.6	6.9	6.4	5.5	4.8
Path • Inches	-1.5	+4.7	0.0	-21.3	-67.4	-146.6	-266.5	-658.7	-1310
Wind Drift • Inches	0.0	2.6	11.4	28.0	52.5	83.2	119.4	205.5	312.0

Remington 170-grain Soft Point Core-Lokt (R30302)

G1 Ballistic Coefficient = 0.255

Distance • Yards	Muzzle	100	200	300	400	500	600	800	1000
Velocity • fps	2200	1895	1619	1381	1191	1061	975	858	771
Energy • ft-lbs	1827	1355	989	720	535	425	359	278	225
Taylor KO Index	16.5	15.0	13.6	11.6	10.0	8.9	7.3	6.4	5.8
Path • Inches	-1.5	+4.1	0.0	-17.5	-53.3	-113.6	-204.5	-501.5	-988.1
Wind Drift • Inches	0.0	1.9	8.0	19.3	36.6	59.7	87.7	155.6	238.0

Winchester 170-grain Silvertip (X30304)

G1 Ballistic Coefficient = 0.278

Distance • Yards	Muzzle	100	200	300	400	500	600	800	1000
Velocity • fps	2200	1920	1665	1439	1251	1110	1015	893	806
Energy • ft-lbs	1827	1392	1046	781	590	465	389	301	245
Taylor KO Index	16.5	14.4	12.5	10.8	9.4	8.3	7.6	6.7	6.0
Path • Inches	-1.5	+4.0	0.0	-16.6	-50.2	-106.0	-190.0	-464.2	-913.3
Wind Drift • Inches	0.0	1.7	7.2	17.4	32.9	53.8	79.7	143.2	220.1

Winchester 170-grain Power-Point (X30303)

G1 Ballistic Coefficient = 0.241

Distance • Yards	Muzzle	100	200	300	400	500	600	800	1000
Velocity • fps	2200	1879	1591	1346	1157	1034	952	837	749
Energy • ft-lbs	1827	1332	955	783	506	405	342	264	212
Taylor KO Index	16.5	14.1	11.9	10.1	8.7	7.7	7.1	6.3	5.6
Path • Inches	-1.5	+4.2	0.0	-18.1	-55.5	-118.7	-214.5	-526.7	-1039
Wind Drift • Inches	0.0	2.0	8.5	20.7	39.1	63.6	92.9	163.7	249.6

Winchester 170-grain Power Core 95/5 (X33030WLF) LF

G1 Ballistic Coefficient = 0.237

Distance • Yards	Muzzle	100	200	300	400	500	600	800	1000
Velocity • fps	2390	2047	1736	1463	1241	1085	986	857	765
Energy • ft-lbs	2157	1583	1138	808	582	445	367	278	221
Taylor KO Index	17.9	15.3	13.0	10.9	9.3	8.1	7.4	6.4	5.7
Path • Inches	-1.5	+3.7	0.0	-15.4	-47.2	-100.6	-184.2	-464.3	-934.3
Wind Drift • Inches	0.0	1.8	7.7	18.8	36.0	59.6	88.7	160.0	246.7

Remington 170-grain Core-Lokt Hollow Point (R30303)

G1 Ballistic Coefficient = 0.255

Distance • Yards	Muzzle	100	200	300	400	500	600	800	1000
Velocity • fps	2200	1895	1619	1381	1191	1061	975	858	771
Energy • ft-lbs	1827	1355	989	720	535	425	359	278	225
Taylor KO Index	18.5	16.0	13.6	11.6	10.0	8.9	7.3	6.4	5.8
Path • Inches	-1.5	+4.1	0.0	-17.5	-53.3	-113.6	-204.5	-501.5	-988.1
Wind Drift • Inches	0.0	1.9	8.0	19.3	36.6	59.7	87.7	155.6	238.0

.300 Savage

The .300 Savage cartridge was introduced in 1921 for the Savage Model 99 lever-action rifle. On the basis of case volume, it should fall just a little bit short of the .308's performance. But because the .300 Savage was to be used in guns that were not quite as strong as the current standards for bolt-action guns, the pressure specifications are about 10 percent lower than those for the .308. Even with this handicap, the .300 Savage is an entirely adequate hunting cartridge, outperforming the venerable .30-30 by about 250 fps. In spite of the performance advantage, the cartridge has never achieved the popularity of the .30-30.

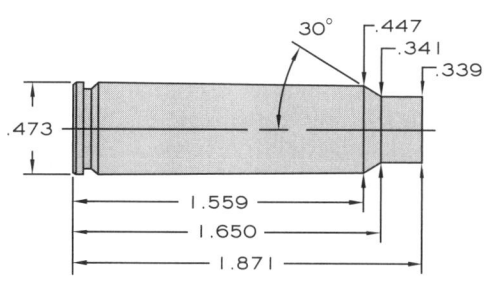

Relative Recoil Factor = 1.78

Specifications:

Controlling Agency for Standardization of this Ammunition: SAAMI

Bullet Weight Grains	Velocity fps	Maximum Average Pressure	
		Copper Crusher	Transducer
150	2,615	46,000 cup	47,000 psi
180	2,340	46,000 cup	47,000 psi

Standard barrel for velocity testing: 24 inches long—1 turn in 12-inch twist

Availability:

Federal 150-grain Soft Point (300A)
G1 Ballistic Coefficient = 0.313

Distance • Yards	Muzzle	100	200	300	400	500	600	800	1000
Velocity • fps	2630	2350	2100	1850	1630	1430	1262	1039	919
Energy • ft-lbs	2305	1845	1460	1145	885	685	530	360	282
Taylor KO Index	17.4	15.5	13.9	12.2	10.8	9.4	8.3	6.9	6.1
Path • Inches	-1.5	+2.4	0.0	-10.4	-30.9	-64.4	-115.6	-288.2	-589.1
Wind Drift • Inches	0.0	1.2	4.9	11.6	22.0	36.6	55.8	108.8	177.2

Remington 150-grain Core-Lokt Pointed Soft Point (R30SV2)
G1 Ballistic Coefficient = 0.314

Distance • Yards	Muzzle	100	200	300	400	500	600	800	1000
Velocity • fps	2630	2354	2095	1853	1631	1432	1265	1041	921
Energy • ft-lbs	2303	1845	1462	1143	806	685	533	361	283
Taylor KO Index	17.4	15.5	13.8	12.2	10.8	9.5	8.3	6.9	6.1
Path • Inches	-1.5	+2.4	0.0	-10.4	-30.9	-64.6	-115.3	-287.3	-587.1
Wind Drift • Inches	0.0	1.2	4.8	11.6	21.9	36.4	55.6	108.4	176.6

Winchester 150-grain Power-Point (X3001)
G1 Ballistic Coefficient = 0.294

Distance • Yards	Muzzle	100	200	300	400	500	600	800	1000
Velocity • fps	2630	2336	2061	1810	1575	1372	1206	1002	889
Energy • ft-lbs	2303	1817	1415	1091	826	627	484	334	263
Taylor KO Index	18.5	16.4	14.5	12.7	11.1	9.7	8.5	7.1	6.3
Path • Inches	-1.5	+2.5	0.0	-10.0	-32.1	-67.6	-121.8	-307.4	-631.1
Wind Drift • Inches	0.0	1.2	5.2	12.5	23.8	39.7	60.8	117.7	189.9

Federal 180-grain Soft Point (300B)
G1 Ballistic Coefficient = 0.383

Distance • Yards	Muzzle	100	200	300	400	500	600	800	1000
Velocity • fps	2350	2140	1940	1750	1570	1410	1275	1077	962
Energy • ft-lbs	2205	1825	1495	1215	985	800	650	464	370
Taylor KO Index	18.6	16.9	15.4	13.9	12.4	11.2	10.1	8.5	7.6
Path • Inches	-1.5	+3.1	0.0	-12.4	-36.1	-73.8	-129.8	-309.4	-608.5
Wind Drift • Inches	0.0	1.1	4.6	10.9	20.4	33.4	50.3	96.1	155.4

.307 Winchester

This is an interesting cartridge. If you look at the dimensions, you will find that the .307 Winchester is externally identical to the .308 except that the .307 has a rim. Introduced in 1982, the .307 was designed to be used in U.S. Repeating Arms Company's new M94 Angle Eject lever-action rifles. Ballistics are similar to the .308, but the .307 has thicker case walls (smaller internal volume) and pays a small performance penalty for that reason.

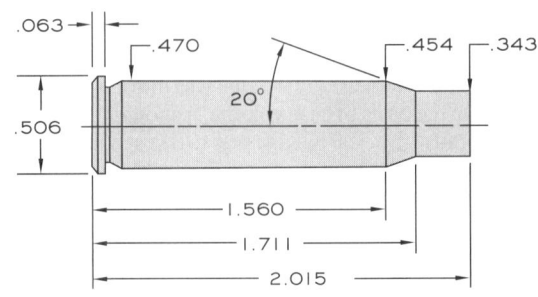

Relative Recoil Factor = 1.93

Specifications:

Controlling Agency for Standardization of this Ammunition: SAAMI

Bullet Weight Grains	Velocity fps	Maximum Average Pressure	
		Copper Crusher	Transducer
150	2,705	52,000 cup	N/S
180	2,450	52,000 cup	N/S

Standard barrel for velocity testing: 24 inches long—1 turn in 12-inch twist

Availability:

Winchester 180-grain Power-Point (X3076)

G1 Ballistic Coefficient = 0.253

Distance • Yards	Muzzle	100	200	300	400	500	600	800	1000
Velocity • fps	2510	2179	1874	1599	1362	1177	1051	904	306
Energy • ft-lbs	2538	1898	1404	1022	742	554	442	327	260
Taylor KO Index	19.9	17.3	14.8	12.7	10.8	9.3	8.3	7.2	6.4
Path • Inches	-1.5	2.9	0.0	-12.9	-39.6	-85.1	-155.6	-396.6	-808.2
Wind Drift • Inches	0.0	1.6	6.6	16.1	30.9	51.7	78.3	145.3	227.4

.308 Marlin Express

The .308 Marlin Express is a 2008 development resulting from a joint effort of Marlin and Hornady. Marlin wanted a modern cartridge for its lever-action rifle. The result is a spin-off of the .307 Winchester. Since the only current .307 loading uses the 180-grain Winchester Power Point bullet, it is not suitable for use in tubular magazines, hence a new cartridge designation. Ballistics are said to be similar to the .308 but fall just a little short of that. Still, it is a very powerful lever-action cartridge.

Relative Recoil Factor = 1.95

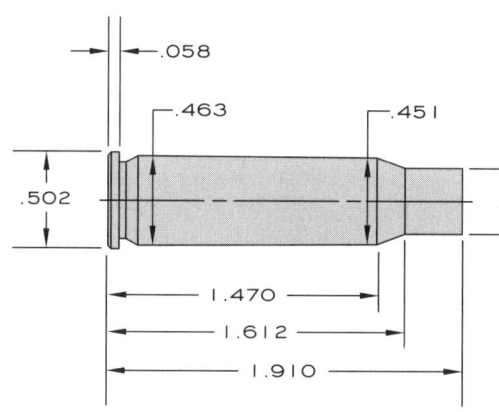

Specifications:

Controlling Agency for Standardization of this Ammunition: SAAMI

Bullet Weight Grains	Velocity fps	Maximum Average Pressure	
		Copper Crusher	Transducer
160	2,660	N/S	47,500 psi

Standard barrel for velocity testing: 24 inches long—1 turn in 12-inch twist

Availability:

Hornady 140-grain MonoFlex (82734)
G1 Ballistic Coefficient = 0.432

Distance • Yards	Muzzle	100	200	300	400	500	600	800	1000
Velocity • fps	2800	2531	2278	2039	1817	1610	1426	1145	988
Energy • ft-lbs	2437	1991	1612	1292	1026	806	633	408	303
Taylor KO Index	17.2	15.6	14.0	12.6	11.2	9.9	8.8	7.1	6.1
Path • Inches	-1.5	+2.0	0.0	-8.7	-25.8	-53.5	-94.7	-233.0	-477.1
Wind Drift • Inches	0.0	1.0	4.1	9.8	18.4	30.4	46.4	91.9	154.2

Hornady 160-grain FTX (82733)
G1 Ballistic Coefficient = 0.432

Distance • Yards	Muzzle	100	200	300	400	500	600	800	1000
Velocity • fps	2660	2430	2226	2026	1836	1659	1497	1229	1055
Energy • ft-lbs	2513	2111	1761	1457	1197	978	796	536	395
Taylor KO Index	18.7	17.1	15.7	14.3	12.9	11.7	10.5	8.7	7.4
Path • Inches	-1.5	+2.2	0.0	-9.2	-26.9	-55.0	-95.7	-227.4	-451.4
Wind Drift • Inches	0.0	0.9	3.7	8.7	16.2	26.7	40.3	78.8	132.5

.308 Winchester (7.62mm NATO)

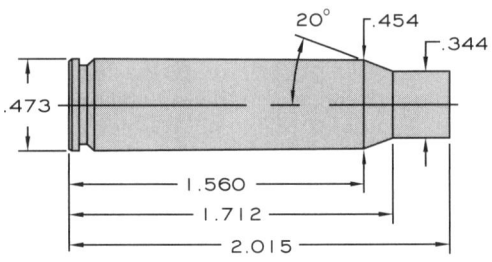

The U.S. Army was working on potential replacement for the .30-06 cartridge during WWII. The best candidate was called the T65, and by the early 1950s it was in the final stages of a serious testing process. While the T65 was not adopted as a standardized military cartridge until 1955, Winchester, who had participated in the development process, jumped the gun (no pun intended) and introduced the .308 Winchester in 1952. Any standardized U.S. military cartridge is almost certain to be "popular," and the .308 has lived up to that expectation. With just a little less recoil than the .30-06 (and a little less performance), the .308 has spawned a series of X-08 neck-downs. The number of different loadings available testify to the popularity of this caliber.

Relative Recoil Factor = 1.95

Specifications:

Controlling Agency for Standardization of this Ammunition: SAAMI

Bullet Weight Grains	Velocity fps	Maximum Average Pressure	
		Copper Crusher	Transducer
55 (Saboted)	3,750	N/S	52,000 psi
110	3,150	52,000 cup	62,000 psi
125	3,030	52,000 cup	62,000 psi
150	2,800	52,000 cup	62,000 psi
165–168	2,670	52,000 cup	62,000 psi
180	2,600	52,000 cup	62,000 psi
200	2,440	52,000 cup	62,000 psi

Standard barrel for velocity testing: 24 inches long—1 turn in 12-inch twist

Availability:

Hornady 110-grain TAP-FPD (80898)
G1 Ballistic Coefficient = 0.290

Distance • Yards	Muzzle	100	200	300	400	500	600	800	1000
Velocity • fps	3165	2830	2519	2228	1957	1708	1483	1145	970
Energy • ft-lbs	2446	1956	1549	1212	935	712	537	320	230
Taylor KO Index	15.3	13.7	12.2	10.8	9.5	8.3	7.2	5.5	4.7
Path • Inches	-1.5	+1.4	0.0	-6.9	-20.9	-44.1	-79.3	-202.3	-431.1
Wind Drift • Inches	0.0	1.0	4.1	9.7	18.3	30.6	47.1	95.4	163.2

Ultramax 110-grain Sierra HP (308R1)
G1 Ballistic Coefficient = 0.160

Distance • Yards	Muzzle	100	200	300	400	500	600	800	1000
Velocity • fps	2875	2321	1833	1424	1133	976	879	741	635
Energy • ft-lbs	2019	1316	821	496	314	233	189	134	98
Taylor KO Index	13.9	11.2	8.9	6.9	5.5	4.7	4.3	3.6	3.1
Path • Inches	-1.5	+2.5	0.0	-13.1	-43.7	-101.2	-195.2	-527.1	-1115
Wind Drift • Inches	0.0	2.1	9.3	23.7	47.2	74.4	118.2	213.0	331.0

Lapua 123-grain Full Metal Jacket (4317527)
G1 Ballistic Coefficient = 0.274

Distance • Yards	Muzzle	100	200	300	400	500	600	800	1000
Velocity • fps	2936	2599	2286	1995	1728	1490	1289	1031	901
Energy • ft-lbs	2355	1845	1428	1087	816	606	454	290	222
Taylor KO Index	15.9	14.1	12.4	10.8	9.4	8.1	7.0	5.6	4.9
Path • Inches	-1.5	+1.8	0.0	-8.5	-25.9	-54.9	-99.8	-258.4	-547.2
Wind Drift • Inches	0.0	1.1	4.8	11.6	22.0	37.0	57.2	114.0	188.2

Remington 125-grain Core-Lokt Pointed Soft Point-Managed Recoil (RL308W1)
G1 Ballistic Coefficient = 0.314

Distance • Yards	Muzzle	100	200	300	400	500	600	800	1000
Velocity • fps	2660	2382	2121	1878	1653	1453	1280	1049	926
Energy • ft-lbs	1964	1575	1249	976	759	585	455	306	238
Taylor KO Index	14.6	13.1	11.7	10.3	9.1	8.0	7.0	5.8	5.1
Path • Inches	-1.5	+2.3	0.0	-10.1	-30.1	-63.0	-112.4	-280.2	-573.8
Wind Drift • Inches	0.0	1.1	4.8	11.4	21.5	35.8	54.8	107.1	175.1

Nosler 125-grain Ballistic Tip (23230S308ETA)

G1 Ballistic Coefficient = 0.366

Distance • Yards	Muzzle	100	200	300	400	500	600	800	1000
Velocity • fps	3175	2907	2653	2414	2188	1974	1774	1422	1160
Energy • ft-lbs	2798	2345	1954	1617	1328	1082	874	561	374
Taylor KO Index	17.5	16.0	14.6	13.3	12.0	10.9	9.8	7.8	6.4
Path • Inches	-1.5	+1.3	0.0	-6.3	-18.5	-38.0	-66.5	-159.5	-322.1
Wind Drift • Inches	0.0	0.8	3.1	7.4	13.7	22.5	34.1	67.4	116.8

Cor-Bon 125-grain JHP (SD308125/20)

G1 Ballistic Coefficient = 0.250

Distance • Yards	Muzzle	100	200	300	400	500	600	800	1000
Velocity • fps	3100	2718	2367	2043	1747	1485	1268	1006	875
Energy • ft-lbs	2668	2052	1556	1159	848	613	446	281	213
Taylor KO Index	17.1	14.9	13.0	11.2	9.6	8.2	7.0	5.5	4.8
Path • Inches	-1.5	+1.6	0.0	-7.9	-24.1	-51.9	-95.5	-254.4	-550.1
Wind Drift • Inches	0.0	1.2	4.9	11.9	22.8	38.6	60.1	120.9	200.0

Cor-Bon 130-grain Glaser Blue (05800/na)

G1 Ballistic Coefficient = 0.200

Distance • Yards	Muzzle	100	200	300	400	500	600	800	1000
Velocity • fps	3000	2539	2122	1748	1427	1182	1029	863	754
Energy • ft-lbs	2597	1861	1300	882	588	403	306	215	164
Taylor KO Index	17.2	14.5	12.1	10.0	8.2	6.8	5.9	4.9	4.3
Path • Inches	-1.5	+2.0	0.0	-9.8	-31.2	-69.8	-133.2	-365.6	-785.1
Wind Drift • Inches	0.0	1.5	6.7	16.5	32.4	55.6	86.2	163.9	260.1

Cor-Bon 130-grain T-DPX (DPX308130/20)

G1 Ballistic Coefficient = 0.300

Distance • Yards	Muzzle	100	200	300	400	500	600	800	1000
Velocity • fps	3000	2687	2394	2121	1866	1633	1426	1120	961
Energy • ft-lbs	2599	2084	1655	1299	1005	770	587	362	267
Taylor KO Index	17.2	16.4	13.7	12.1	10.7	9.3	8.2	6.4	5.5
Path • Inches	-1.5	+1.7	0.0	-7.8	-23.3	-48.9	-87.6	-221.6	-466.2
Wind Drift • Inches	0.0	1.0	4.2	10.0	19.0	31.7	48.7	97.9	165.2

Cor-Bon 140-grain MPG (MPG308140/20) LF

G1 Ballistic Coefficient = 0.300

Distance • Yards	Muzzle	100	200	300	400	500	600	800	1000
Velocity • fps	2875	2591	2286	2020	1773	1550	1355	1081	940
Energy • ft-lbs	2570	2055	1625	1269	978	747	571	363	275
Taylor KO Index	17.7	15.8	14.1	12.4	10.9	9.5	8.3	6.7	5.8
Path • Inches	-1.5	+1.9	0.0	-8.6	-25.7	-54.0	-96.8	-244.8	-511.2
Wind Drift • Inches	0.0	1.1	4.5	10.7	20.2	33.7	51.8	103.2	171.9

USA (Winchester) 147-grain Full Metal Jacket (USA3081)

G1 Ballistic Coefficient = 0.417

Distance • Yards	Muzzle	100	200	300	400	500	600	800	1000
Velocity • fps	2800	2582	2374	2176	1987	1812	1646	1356	1141
Energy • ft-lbs	2559	2176	1840	1545	1289	1072	884	600	425
Taylor KO Index	18.1	16.7	15.4	14.1	12.9	11.7	10.6	8.8	7.4
Path • Inches	-1.5	+1.9	0.0	-8.0	-23.5	-46.6	-82.4	-193.4	-380.7
Wind Drift • Inches	0.0	0.8	3.2	7.6	14.1	23.1	34.8	67.9	115.5

Sellier & Bellot 147-grain FMJ (V331402U)

G1 Ballistic Coefficient = 0.409

Distance • Yards	Muzzle	100	200	300	400	500	600	800	1000
Velocity • fps	2788	2567	2357	2156	1966	1786	1619	1329	1121
Energy • ft-lbs	2538	2152	1813	1518	1262	1041	855	577	410
Taylor KO Index	18.0	16.6	15.2	13.9	12.7	11.6	10.5	8.6	7.3
Path • Inches	-1.5	+1.9	0.0	-8.2	-23.8	-48.5	-84.2	-198.3	-391.6
Wind Drift • Inches	0.0	0.8	3.3	7.8	14.5	23.8	35.9	70.2	119.3

PMC 147-grain FMJ-BT (308B)

G1 Ballistic Coefficient = 0.445

Distance • Yards	Muzzle	100	200	300	400	500	600	800	1000
Velocity • fps	2780	2575	2380	2194	2018	1851	1692	1410	1190
Energy • ft-lbs	2522	2164	1848	1571	1329	1118	935	650	463
Taylor KO Index	18.0	16.7	15.4	14.2	13.1	12.0	10.8	9.1	7.7
Path • Inches	-1.5	+1.9	0.0	-8.0	-23.3	-47.1	-81.0	-187.8	-365.1
Wind Drift • Inches	0.0	0.7	3.0	7.1	13.2	21.5	32.4	62.8	106.7

Nosler 150-grain AccuBond (23201) + (60056)

G1 Ballistic Coefficient = 0.435

Distance • Yards	Muzzle	100	200	300	400	500	600	800	1000
Velocity • fps	3025	2805	2596	2397	2206	2025	1852	1540	1282
Energy • ft-lbs	3047	2621	2245	1913	1620	1365	1143	790	548
Taylor KO Index	20.0	18.5	17.1	15.8	14.6	13.4	12.2	10.2	8.5
Path • Inches	-1.5	+1.5	0.0	-6.6	-19.3	-39.2	-67.5	-156.7	-305.2
Wind Drift • Inches	0.0	0.7	2.8	6.5	12.0	19.5	29.3	57.0	97.4

Hornady 150-grain InterBond (80938)

G1 Ballistic Coefficient = 0.415

Distance • Yards	Muzzle	100	200	300	400	500	600	800	1000
Velocity • fps	3000	2772	2555	2348	2151	1963	1786	1470	1220
Energy • ft-lbs	2997	2558	2173	1836	1540	1282	1063	720	496
Taylor KO Index	19.8	18.3	16.9	15.5	14.2	13.0	11.8	9.7	8.1
Path • Inches	-1.5	+1.5	0.0	-6.9	-20.0	-40.7	-70.5	-165.2	-324.9
Wind Drift • Inches	0.0	0.7	3.0	6.9	12.8	20.9	31.5	61.6	105.5

Hornady 150-grain SST (80933)

G1 Ballistic Coefficient = 0.332

Distance • Yards	Muzzle	100	200	300	400	500	600	800	1000
Velocity • fps	3000	2772	2555	2348	1963	1744	1543	1220	1026
Energy • ft-lbs	2997	2458	2000	1611	1284	1013	794	495	351
Taylor KO Index	19.6	18.3	16.9	15.5	13.0	11.5	10.2	8.1	6.8
Path • Inches	-1.5	+1.6	0.0	-7.4	-22.1	-45.8	-81.0	-199.4	-411.6
Wind Drift • Inches	0.0	0.9	3.8	8.9	16.7	27.7	42.3	84.4	144.4

Fiocchi 150-grain SST Polymer Tip BT (308HSA)

G1 Ballistic Coefficient = 0.415

Distance • Yards	Muzzle	100	200	300	400	500	600	800	1000
Velocity • fps	2860	2638	2428	2227	2035	1854	1684	1386	1161
Energy • ft-lbs	2723	2318	1963	1651	1397	1144	945	640	449
Taylor KO Index	18.9	17.4	16.0	14.7	13.4	12.2	11.1	9.1	7.7
Path • Inches	-1.5	+1.8	0.0	-7.7	-22.4	-45.5	-78.7	-184.7	-363.8
Wind Drift • Inches	0.0	0.8	3.2	7.4	13.4	22.5	33.9	66.2	113.0

Federal 150-grain Nosler Partition (P308S)

G1 Ballistic Coefficient = 0.389

Distance • Yards	Muzzle	100	200	300	400	500	600	800	1000
Velocity • fps	2840	2600	2380	2170	1970	1780	1604	1305	1097
Energy • ft-lbs	2685	2260	1885	1565	1285	1050	857	567	401
Taylor KO Index	18.7	17.2	15.7	14.3	13.0	11.7	10.6	8.6	7.2
Path • Inches	-1.5	+1.8	0.0	-8.0	-23.3	-47.7	-83.1	-197.7	-394.5
Wind Drift • Inches	0.0	0.8	3.4	8.1	15.0	24.6	37.3	73.3	125.0

Winchester 150-grain XP3 (SXP308)

G1 Ballistic Coefficient = 0.437

Distance • Yards	Muzzle	100	200	300	400	500	600	800	1000
Velocity • fps	2825	2616	2417	2226	2044	1871	1707	1418	1192
Energy • ft-lbs	2658	2279	1945	1650	1392	1166	971	670	473
Taylor KO Index	18.6	17.3	16.0	14.7	13.5	12.3	11.3	9.4	7.9
Path • Inches	-1.5	+1.8	0.0	-7.8	-22.6	-45.7	-78.8	-183.4	-357.8
Wind Drift • Inches	0.0	0.7	3.0	7.1	13.2	21.5	32.4	63.0	107.2

Federal 150-grain Fusion (F308FS1)

G1 Ballistic Coefficient = 0.414

Distance • Yards	Muzzle	100	200	300	400	500	600	800	1000
Velocity • fps	2820	2600	2390	2190	2000	1820	1653	1360	1143
Energy • ft-lbs	2650	2250	1905	1600	1335	1105	911	616	435
Taylor KO Index	18.6	17.2	15.8	14.5	13.2	12.0	10.9	9.0	7.5
Path • Inches	-1.5	+1.8	0.0	-8.0	-23.1	-46.9	-81.4	-191.3	-377.0
Wind Drift • Inches	0.0	0.8	3.2	7.6	14.1	23.0	34.7	67.8	115.5

Federal 150-grain Soft Point (308A)

G1 Ballistic Coefficient = 0.315

Distance • Yards	Muzzle	100	200	300	400	500	600	800	1000
Velocity • fps	2820	2530	2260	2010	1770	1560	1373	1100	956
Energy • ft-lbs	2650	2140	1705	1345	1050	810	628	403	304
Taylor KO Index	18.6	16.7	14.9	13.3	11.7	10.3	9.1	7.3	6.3
Path • Inches	-1.5	+2.0	0.0	-8.8	-26.3	-54.8	-97.6	+243.8	-504.6
Wind Drift • Inches	0.0	1.0	4.4	10.4	19.6	32.6	50.0	99.2	165.5

Hornady 150-grain InterBond (80939)

G1 Ballistic Coefficient = 0.416

Distance • Yards	Muzzle	100	200	300	400	500	600	800	1000
Velocity • fps	2820	2601	2392	2197	2003	1823	1656	1363	1145
Energy • ft-lbs	2648	2252	1905	1601	1336	1107	914	619	437
Taylor KO Index	18.6	17.2	15.8	14.5	13.2	12.0	10.9	9.0	7.6
Path • Inches	-1.5	+1.3	0.0	-7.9	-23.1	-47.0	-81.3	-190.8	-375.8
Wind Drift • Inches	0.0	0.8	3.2	7.6	14.0	22.9	34.6	67.6	115.0

Hornady 150-grain BTSP InterLock (8091)

G1 Ballistic Coefficient = 0.349

Distance • Yards	Muzzle	100	200	300	400	500	600	800	1000
Velocity • fps	2820	2560	2315	2084	1866	1664	1481	1190	1017
Energy • ft-lbs	2648	2183	1785	1446	1159	992	731	471	345
Taylor KO Index	18.6	16.9	15.3	13.8	12.3	11.0	9.8	7.9	6.7
Path • Inches	-1.5	+1.9	0.0	-8.4	-24.9	-51.4	-90.5	-220.4	-448.8
Wind Drift • Inches	0.0	0.9	3.9	9.2	17.3	28.5	43.4	86.0	145.3

PMC 150-grain BTSP (308HIA)

G1 Ballistic Coefficient = 0.349

Distance • Yards	Muzzle	100	200	300	400	500	600	800	1000
Velocity • fps	2820	2560	2315	2084	1866	1664	1481	1190	1017
Energy • ft-lbs	2648	2183	1785	1446	1159	922	731	471	345
Taylor KO Index	18.6	16.9	15.3	13.8	12.3	11.0	9.8	7.9	6.7
Path • Inches	-1.5	+1.9	0.0	-8.4	-24.9	-51.4	-90.5	-220.4	-448.8
Wind Drift • Inches	0.0	0.9	3.9	9.2	17.3	28.5	43.4	86.0	145.3

Remington 150-grain Core-Lokt Ultra Bonded (PRC308WA)

G1 Ballistic Coefficient = 0.331

Distance • Yards	Muzzle	100	200	300	400	500	600	800	1000
Velocity • fps	2820	2546	2288	2048	1819	1611	1423	1140	983
Energy • ft-lbs	2648	2159	1744	1394	1102	864	674	433	322
Taylor KO Index	18.0	16.8	15.1	13.5	12.0	10.6	9.4	7.5	6.5
Path • Inches	-1.5	+1.9	0.0	-8.6	-25.3	-53.1	-94.1	-232.4	-477.6
Wind Drift • Inches	0.0	1.0	4.1	9.8	18.5	30.6	46.8	92.9	155.9

Remington 150-grain Core-Lokt Pointed Soft Point (R308W1)

G1 Ballistic Coefficient = 0.315

Distance • Yards	Muzzle	100	200	300	400	500	600	800	1000
Velocity • fps	2820	2533	2263	2009	1774	1560	1373	1100	955
Energy • ft-lbs	2649	2137	1705	1344	1048	810	628	403	304
Taylor KO Index	18.6	16.7	14.9	13.3	11.7	10.3	9.1	7.3	6.3
Path • Inches	-1.5	+2.0	0.0	-8.8	-26.2	-54.8	-98.1	-247.0	-510.1
Wind Drift • Inches	0.0	1.0	4.4	10.4	19.6	32.6	50.0	99.2	165.4

Winchester 150-grain Power-Point (X3085)

G1 Ballistic Coefficient = 0.294

Distance • Yards	Muzzle	100	200	300	400	500	600	800	1000
Velocity • fps	2820	2513	2227	1960	1713	1492	1302	1049	919
Energy • ft-lbs	2648	2104	1651	1279	977	742	565	367	281
Taylor KO Index	18.6	16.6	14.7	12.9	11.3	9.8	8.6	6.9	6.1
Path • Inches	-1.5	+2.0	0.0	-9.1	-27.2	-57.3	-103.2	-262.2	-546.9
Wind Drift • Inches	0.0	1.1	4.7	11.3	21.4	35.8	55.0	108.9	179.6

Sellier and Bellot 150-grain SPCE (V331412U)

G1 Ballistic Coefficient = 0.326

Distance • Yards	Muzzle	100	200	300	400	500	600	800	1000
Velocity • fps	2756	2470	2213	1983	1757	1552	1371	1105	962
Energy • ft-lbs	2524	2027	1627	1307	1029	802	626	407	308
Taylor KO Index	18.2	16.3	14.6	13.1	11.6	10.2	9.0	7.3	6.3
Path • Inches	-1.5	+2.1	0.0	-9.1	-27.2	-56.6	-100.8	-251.6	-515.6
Wind Drift • Inches	0.0	1.0	4.3	10.3	19.5	32.3	49.4	97.5	162.2

Fiocchi 150 PSP (308B)

G1 Ballistic Coefficient = 0.348

Distance • Yards	Muzzle	100	200	300	400	500	600	800	1000
Velocity • fps	2750	2494	2252	2024	1811	1613	1434	1158	999
Energy • ft-lbs	2518	2072	1690	1365	1092	864	685	446	333
Taylor KO Index	18.2	16.5	14.9	13.4	12.0	10.6	9.5	7.6	6.6
Path • Inches	-1.5	+2.1	0.0	-8.9	-26.4	-54.6	-96.2	-234.5	-476.5
Wind Drift • Inches	0.0	1.0	4.0	9.6	18.0	29.7	45.3	89.4	149.8

Hornady 150-grain SST InterLock (8093)

G1 Ballistic Coefficient = 0.415

Distance • Yards	Muzzle	100	200	300	400	500	600	800	1000
Velocity • fps	2820	2601	2392	2192	2003	1823	1656	1362	1145
Energy • ft-lbs	2648	2252	1905	1601	1336	1107	913	618	437
Taylor KO Index	18.6	17.2	15.8	14.5	13.2	12.0	10.9	9.0	7.6
Path • Inches	-1.5	+1.8	0.0	-7.9	-23.1	-47.1	-81.3	-190.9	-376.1
Wind Drift • Inches	0.0	0.8	3.2	7.6	14.0	22.9	34.6	67.6	115.2

Lapua 150-grain Mega Soft Point (4317498)

G1 Ballistic Coefficient = 0.323

Distance • Yards	Muzzle	100	200	300	400	500	600	800	1000
Velocity • fps	2789	2511	2249	2003	1774	1565	1381	1109	963
Energy • ft-lbs	2591	2100	1685	1337	1049	816	635	410	309
Taylor KO Index	18.4	16.6	14.8	13.2	11.7	10.3	9.1	7.3	6.4
Path • Inches	-1.5	+2.0	0.0	-8.9	-26.6	-55.3	-98.7	-247.0	-507.9
Wind Drift • Inches	0.0	1.0	4.3	10.3	19.3	32.1	49.1	97.2	162.2

Hornady 150-grain GMX (8094) LF

G1 Ballistic Coefficient = 0.416

Distance • Yards	Muzzle	100	200	300	400	500	600	800	1000
Velocity • fps	2940	2715	2500	2296	2101	1919	1745	1436	1196
1Energy • ft-lbs	2878	2454	2082	1756	1470	1236	1014	687	476
Taylor KO Index	19.4	17.9	16.5	15.2	13.9	12.7	11.5	9.5	7.9
Path • Inches	-1.5	+1.6	0.0	-7.2	-21.0	-42.6	-73.8	-172.9	-340.1
Wind Drift • Inches	0.0	0.7	3.0	7.1	13.2	21.5	32.4	63.3	108.3

Norma 150-grain Nosler BST (17625)

G1 Ballistic Coefficient = 0.435

Distance • Yards	Muzzle	100	200	300	400	500	600	800	1000
Velocity • fps	2822	2612	2412	2221	2039	1865	1702	1413	1188
Energy • ft-lbs	2653	2274	1939	1644	1384	1159	965	665	470
Taylor KO Index	18.6	17.2	15.9	14.7	13.5	12.3	11.2	9.3	7.8
Path • Inches	-1.5	+1.8	0.0	-7.8	-22.6	-45.9	-79.1	-184.2	-369.6
Wind Drift • Inches	0.0	0.7	3.1	7.2	13.3	21.6	32.6	63.4	107.8

Federal 150-grain Nosler Ballistic Tip (P308F)

G1 Ballistic Coefficient = 0.433

Distance • Yards	Muzzle	100	200	300	400	500	600	800	1000
Velocity • fps	2820	2610	2410	2220	2040	1860	1695	1405	1181
Energy • ft-lbs	2650	2270	1935	1640	1380	1155	957	658	465
Taylor KO Index	18.6	17.2	15.9	14.7	13.5	12.3	11.2	9.3	7.8
Path • Inches	-1.5	+1.8	0.0	-7.8	-22.7	-46.0	-79.5	-185.3	-362.2
Wind Drift • Inches	0.0	0.7	3.1	7.2	13.4	21.8	32.8	64.0	108.8

Federal 150-grain Barnes Triple Shock X-Bullet (P308V) LF

G1 Ballistic Coefficient = 0.371

Distance • Yards	Muzzle	100	200	300	400	500	600	800	1000
Velocity • fps	2820	2570	2340	2120	1910	1720	1543	1247	1056
Energy • ft-lbs	2650	2205	1825	1495	1215	980	793	518	372
Taylor KO Index	18.8	17.0	15.4	14.0	12.6	11.4	10.2	8.2	7.0
Path • Inches	-1.5	+1.9	0.0	-8.3	-24.2	-49.9	-87.0	-209.1	-421.1
Wind Drift • Inches	0.0	0.9	3.6	8.6	16.1	26.4	40.1	79.2	134.4

Remington 150-grain Swift Scirocco Bonded (PRSC308WA)

G1 Ballistic Coefficient = 0.435

Distance • Yards	Muzzle	100	200	300	400	500	600	800	1000
Velocity • fps	2820	2611	2410	2219	2037	1863	1699	1410	1185
Energy • ft-lbs	2648	2269	1935	1640	1381	1156	962	662	468
Taylor KO Index	18.6	17.2	15.9	14.6	13.4	12.3	11.2	9.3	7.8
Path • Inches	-1.5	+1.8	0.0	-7.8	-22.7	-46.0	-79.3	-184.7	-360.7
Wind Drift • Inches	0.0	0.7	3.1	7.2	13.3	21.7	32.7	63.6	108.1

Winchester 150-grain Power Max Bonded (X3085BP)

G1 Ballistic Coefficient = 0.325

Distance • Yards	Muzzle	100	200	300	400	500	600	800	1000
Velocity • fps	2820	2542	2280	2034	1808	1594	1407	1127	974
Energy • ft-lbs	2648	2152	1731	1378	1089	846	660	423	316
Taylor KO Index	18.6	16.8	15.0	13.4	11.9	10.5	9.3	7.4	6.4
Path • Inches	-1.5	+2.0	0.0	-8.7	-25.1	-53.6	-95.2	-235.9	-485.9
Wind Drift • Inches	0.0	1.0	4.2	10.0	18.8	31.3	47.8	94.9	158.9

Winchester 150-grain E-Tip (S308ETA) LF

G1 Ballistic Coefficient = 0.469

Distance • Yards	Muzzle	100	200	300	400	500	600	800	1000
Velocity • fps	2810	2616	2430	2251	2080	1917	1763	1483	1255
Energy • ft-lbs	2629	2279	1966	1688	1441	1224	1035	733	524
Taylor KO Index	18.5	17.3	16.0	14.9	13.7	12.7	11.6	9.8	8.3
Path • Inches	-1.5	+1.8	0.0	-7.7	-22.3	-44.9	-77.0	-176.9	-340.6
Wind Drift • Inches	0.0	0.7	2.8	6.6	12.2	19.9	29.8	57.6	97.7

Winchester 150-grain Ballistic Silvertip (SBST308)

G1 Ballistic Coefficient = 0.435

Distance • Yards	Muzzle	100	200	300	400	500	600	800	1000
Velocity • fps	2810	2601	2401	2211	2028	1856	1692	1404	1181
Energy • ft-lbs	2629	2253	1920	1627	1370	1147	954	657	465
Taylor KO Index	18.5	17.2	15.8	14.6	13.4	12.2	11.2	9.3	7.8
Path • Inches	-1.5	+1.8	0.0	-7.8	-22.8	-46.2	-80.0	-186.9	-367.0
Wind Drift • Inches	0.0	0.7	3.1	7.2	13.4	21.8	32.8	63.8	108.6

Ultramax 150-grain Nosler Ballistic Tip (308R2)

G1 Ballistic Coefficient = 0.433

Distance • Yards	Muzzle	100	200	300	400	500	600	800	1000
Velocity • fps	2700	2495	2299	2112	1934	1766	1609	1336	1134
Energy • ft-lbs	2429	2074	1761	1486	1246	1039	862	594	429
Taylor KO Index	17.8	16.5	15.2	13.9	12.8	11.7	10.6	8.8	7.5
Path • Inches	-1.5	+2.1	0.0	-8.6	-25.1	-50.9	-87.8	-204.9	-400.4
Wind Drift • Inches	0.0	0.8	3.3	7.7	14.2	23.3	35.0	68.1	115.2

Fiocchi 150-grain FMJ-BT (308A)

G1 Ballistic Coefficient = 0.398

Distance • Yards	Muzzle	100	200	300	400	500	600	800	1000
Velocity • fps	2890	2659	2438	2228	2028	1839	1664	1358	1135
Energy • ft-lbs	2781	2356	1980	1653	1370	1126	922	615	429
Taylor KO Index	19.1	17.5	16.1	14.7	13.4	12.1	11.0	9.0	7.5
Path • Inches	-1.5	+1.7	0.0	-7.6	-22.2	-45.3	-78.7	-186.1	-369.6
Wind Drift • Inches	0.0	0.8	3.3	7.6	14.2	23.3	35.2	69.0	118.0

Lapua 150-grain Lock Base (4317538)

G1 Ballistic Coefficient = 0.488

Distance • Yards	Muzzle	100	200	300	400	500	600	800	1000
Velocity • fps	2789	2603	2424	2253	2088	1930	1781	1508	1282
Energy • ft-lbs	2586	2253	1954	1687	1449	1239	1054	756	546
Taylor KO Index	18.4	17.2	16.0	14.9	13.8	12.7	11.8	10.0	8.5
Path • Inches	-1.5	+1.8	0.0	-7.7	-22.4	-45.0	-76.9	-175.7	-336.2
Wind Drift • Inches	0.0	0.6	2.5	5.8	10.6	17.3	25.9	49.9	84.3

Norma 150-grain Full Metal Jacket (17523)

G1 Ballistic Coefficient = 0.438

Distance • Yards	Muzzle	100	200	300	400	500	600	800	1000
Velocity • fps	2723	2519	2325	2139	1961	1793	1636	1360	1153
Energy • ft-lbs	2470	2115	1800	1524	1281	1071	892	617	443
Taylor KO Index	18.0	16.6	15.3	14.1	12.9	11.8	10.8	9.0	7.6
Path • Inches	-1.5	+2.0	0.0	-8.4	-24.5	-49.6	-85.6	-199.2	-388.6
Wind Drift • Inches	0.0	0.8	3.2	7.5	13.9	22.6	34.0	66.2	112.2

Norma 150-grain FULL JACKET (17622)

G1 Ballistic Coefficient = 0.423

Distance • Yards	Muzzle	100	200	300	400	500	600	800	1000
Velocity • fps	2657	2449	2251	2062	1883	1714	1557	1290	1101
Energy • ft-lbs	2352	1999	1688	1417	1181	979	808	554	404
Taylor KO Index	17.5	16.2	14.9	13.6	12.4	11.3	10.3	8.5	7.3
Path • Inches	-1.5	+1.8	0.0	-9.0	-26.3	-53.3	-92.3	-216.3	-424.4
Wind Drift • Inches	0.0	0.8	3.4	8.1	15.0	24.5	37.0	72.0	121.3

American Eagle (Federal) 150-grain Full Metal jacket Boat-Tail (AE308D)

G1 Ballistic Coefficient = 0.409

Distance • Yards	Muzzle	100	200	300	400	500	600	800	1000
Velocity • fps	2820	2600	2390	2180	1990	1810	1640	1346	1131
Energy • ft-lbs	2650	2245	1895	1585	1320	1090	896	603	428
Taylor KO Index	18.6	17.2	15.8	14.4	13.1	11.9	10.8	8.9	7.5
Path • Inches	-1.5	+1.8	0.0	-8.0	-23.3	-47.2	-82.0	-193.2	-381.9
Wind Drift • Inches	0.0	0.8	3.3	7.7	14.3	23.4	35.3	69.1	117.8

Magtech 150-grain FMJ (308A)

G1 Ballistic Coefficient = 0.456

Distance • Yards	Muzzle	100	200	300	400	500	600	800	1000
Velocity • fps	2820	2620	2429	2245	2070	1903	1745	1460	1232
Energy • ft-lbs	2649	2287	1965	1680	1427	1206	1014	711	506
Taylor KO Index	18.6	17.3	16.0	14.8	13.7	12.6	11.5	9.6	8.1
Path • Inches	-1.5	+1.8	0.0	-7.7	-22.3	-45.0	-77.4	-178.7	-345.8
Wind Drift • Inches	0.0	0.7	2.9	6.8	12.6	20.5	30.7	59.5	101.0

Remington 150-grain Copper Solid Tipped (PCS308WA) LF

G1 Ballistic Coefficient = 0.400

Distance • Yards	Muzzle	100	200	300	400	500	600	800	1000
Velocity • fps	2820	2593	2376	2171	1975	1791	1620	1324	1114
Energy • ft-lbs	2648	2238	1881	1569	1299	1068	874	584	414
Taylor KO Index	18.6	17.1	15.7	14.3	13.0	11.8	10.7	8.7	7.4
Path • Inches	-1.5	+2.0	0.0	-8.0	-23.4	-47.8	-83.0	-196.4	-389.6
Wind Drift • Inches	0.0	0.8	3.4	7.9	14.7	24.0	36.3	71.1	121.1

UMC (Remington) 150-grain MC (L308W4)

G1 Ballistic Coefficient = 0.315

Distance • Yards	Muzzle	100	200	300	400	500	600	800	1000
Velocity • fps	2820	2533	2263	2010	1776	1561	1373	1100	955
Energy • ft-lbs	2649	2137	1707	1347	1050	812	628	403	304
Taylor KO Index	18.6	16.7	14.9	13.3	11.7	10.3	9.1	7.3	6.3
Path • Inches	-1.5	+2.0	0.0	-8.8	-26.2	-54.8	-98.1	-247.0	-510.1
Wind Drift • Inches	0.0	1.0	4.4	10.4	19.6	32.6	50.0	99.2	165.4

Cor-Bon 155-grain Scenar (PM300155/20)

G1 Ballistic Coefficient = 0.254

Distance • Yards	Muzzle	100	200	300	400	500	600	800	1000
Velocity • fps	2900	2540	2208	1902	1625	1385	1194	975	857
Energy • ft-lbs	2895	2220	1678	1245	909	660	491	327	253
Taylor KO Index	19.8	17.3	15.1	13.0	11.1	9.4	8.1	6.6	5.8
Path • Inches	-1.5	+2.0	0.0	-9.2	-28.0	-60.1	-110.5	-290.8	-617.0
Wind Drift • Inches	0.0	1.2	5.3	12.9	24.8	41.8	64.8	127.4	207.1

Hornady 155-grain A-MAX (8095PM)

G1 Ballistic Coefficient = 0.435

Distance • Yards	Muzzle	100	200	300	400	500	600	800	1000
Velocity • fps	2850	2639	2438	2245	2062	1887	1723	1430	1200
Energy • ft-lbs	2795	2397	2045	1735	1463	1225	1022	704	496
Taylor KO Index	19.4	18.0	16.6	15.3	14.1	12.9	11.8	9.8	8.2
Path • Inches	-1.5	+1.8	0.0	-7.6	-22.1	-44.9	-77.4	-180.0	-351.4
Wind Drift • Inches	0.0	0.7	3.0	7.1	13.1	21.3	32.1	62.4	106.3

Hornady 155 Palma Match (8095PM)

G1 Ballistic Coefficient = 0.435

Distance • Yards	Muzzle	100	200	300	400	500	600	800	1000
Velocity • fps	2850	2639	2438	2245	2062	1887	1723	1430	1200
Energy • ft-lbs	2795	2397	2045	1735	1463	1225	1022	704	496
Taylor KO Index	19.4	18.0	16.6	15.3	14.1	12.9	11.8	9.8	8.2
Path • Inches	-1.5	+1.8	0.0	-7.6	-22.1	-44.9	-77.4	-180.0	-351.4
Wind Drift • Inches	0.0	0.7	3.0	7.1	13.1	21.3	32.1	62.4	106.3

Lapua 155-grain Scenar (4317073)

G1 Ballistic Coefficient = 0.452

Distance • Yards	Muzzle	100	200	300	400	500	600	800	1000
Velocity • fps	2820	2621	2428	2243	2064	1895	1735	1451	1182
Energy • ft-lbs	2725	2354	2020	1723	1466	1237	1031	725	478
Taylor KO Index	19.2	17.9	16.6	15.3	14.1	12.9	11.8	9.9	8.1
Path • Inches	-1.5	+1.8	0.0	-7.7	-22.4	-45.2	-77.7	-179.8	-348.4
Wind Drift • Inches	0.0	0.7	2.9	6.9	12.7	20.7	31.1	60.2	102.3

Hornady 155-grain TAP-FPD (80928)

G1 Ballistic Coefficient = 0.435

Distance • Yards	Muzzle	100	200	300	400	500	600	800	1000
Velocity • fps	2785	2577	2379	2189	2008	1836	1674	1389	1171
Energy • ft-lbs	2669	2285	1947	1649	1387	1160	964	664	472
Taylor KO Index	19.0	17.6	16.2	14.9	13.7	12.5	11.4	9.5	8.0
Path • Inches	-1.5	+1.9	0.0	-8.0	-23.3	-47.3	-81.6	-190.2	-371.4
Wind Drift • Inches	0.0	0.8	3.1	7.3	13.6	22.1	33.3	64.8	110.0

Black Hills 155-grain Hornady A-MAX (1C308BHGN3)

G1 Ballistic Coefficient = 0.435

Distance • Yards	Muzzle	100	200	300	400	500	600	800	1000
Velocity • fps	2750	2543	2346	2158	1978	1808	1668	1368	1158
Energy • ft-lbs	2603	2227	1895	1603	1347	1125	935	644	460
Taylor KO Index	18.8	17.3	16.0	14.7	13.5	12.3	11.2	9.3	7.9
Path • Inches	-1.5	+1.9	0.0	-8.3	-24.0	-48.7	-84.0	-195.9	-382.6
Wind Drift • Inches	0.0	0.8	3.2	7.4	13.8	22.5	33.9	66.0	112.0

Hornady 155-grain BTHP (80926)

G1 Ballistic Coefficient = 0.405

Distance • Yards	Muzzle	100	200	300	400	500	600	800	1000
Velocity • fps	2610	2396	2191	1997	1816	1642	1486	1226	1057
Energy • ft-lbs	2344	1975	1652	1372	1134	928	760	518	384
Taylor KO Index	17.6	16.3	14.9	13.6	12.4	11.2	10.1	8.4	7.2
Path • Inches	-1.5	+2.3	0.0	-9.5	-27.2	-56.7	-98.6	-233.0	-460.2
Wind Drift • Inches	0.0	0.9	3.7	8.7	16.2	26.6	40.2	78.2	131.2

Hornady 165-grain SST (80983)

G1 Ballistic Coefficient = 0.447

Distance • Yards	Muzzle	100	200	300	400	500	600	800	1000
Velocity • fps	2840	2635	2439	2252	2079	1902	1741	1452	1222
Energy • ft-lbs	2955	2544	2180	1858	1574	1325	1111	773	547
Taylor KO Index	20.6	19.1	18.1	16.3	15.1	13.8	12.6	10.5	8.9
Path • Inches	-1.5	+1.8	0.0	-7.6	-22.1	-44.7	-76.9	-178.2	-345.9
Wind Drift • Inches	0.0	0.7	3.0	6.9	12.7	20.7	31.2	60.5	102.8

Nosler 165-grain AccuBond (60049)

G1 Ballistic Coefficient = 0.475

Distance • Yards	Muzzle	100	200	300	400	500	600	800	1000
Velocity • fps	2800	2611	2431	2258	2091	1932	1781	1507	1279
Energy • ft-lbs	2872	2497	2165	1868	1602	1367	1163	832	599
Taylor KO Index	20.3	19.0	17.6	16.4	15.2	14.0	12.9	10.9	9.3
Path • Inches	-1.5	+1.8	0.0	-7.7	-22.2	-44.7	-76.5	-175.1	-335.5
Wind Drift • Inches	0.0	0.7	2.8	6.4	11.9	19.3	28.9	55.7	94.3

Nosler 165-grain Partition (23205) + (60053)

G1 Ballistic Coefficient = 0.410

Distance • Yards	Muzzle	100	200	300	400	500	600	800	1000
Velocity • fps	2800	2579	2368	2168	1978	1798	1629	1339	1127
Energy • ft-lbs	2872	2436	2054	1721	1433	1185	973	657	466
Taylor KO Index	20.3	18.7	17.2	15.7	14.4	13.1	11.8	9.7	8.2
Path • Inches	-1.5	+1.9	0.0	-8.1	-23.6	-48.0	-83.2	-195.9	-386.8
Wind Drift • Inches	0.0	0.8	3.3	7.7	14.4	23.5	35.6	69.5	118.2

Federal 165-grain Fusion (F308FS2)

G1 Ballistic Coefficient = 0.446

Distance • Yards	Muzzle	100	200	300	400	500	600	800	1000
Velocity • fps	2700	2500	2310	2130	1950	1790	1636	1364	1159
Energy • ft-lbs	2670	2290	1955	1660	1400	1175	980	682	492
Taylor KO Index	19.6	18.2	16.8	15.5	14.2	13.0	11.9	9.9	8.4
Path • Inches	-1.5	+2.0	0.0	-8.6	-24.8	-50.1	-86.4	-200.6	-390.1
Wind Drift • Inches	0.0	0.8	3.2	7.4	13.8	22.4	33.8	65.5	110.7

Federal 165-grain Sierra GameKing BTSP (P308C)

G1 Ballistic Coefficient = 0.405

Distance • Yards	Muzzle	100	200	300	400	500	600	800	1000
Velocity • fps	2700	2480	2270	2070	1880	1710	1546	1270	1082
Energy • ft-lbs	2670	2285	1890	1575	1300	1070	876	591	429
Taylor KO Index	19.6	18.0	16.5	15.0	13.6	12.4	11.2	9.2	7.9
Path • Inches	-1.5	+2.1	0.0	-8.9	-25.6	-52.7	-91.2	-215.6	-426.5
Wind Drift • Inches	0.0	0.8	3.5	8.3	15.4	25.3	38.2	74.7	126.2

Hornady 165-grain BTSP InterLock (8098)

G1 Ballistic Coefficient = 0.435

Distance • Yards	Muzzle	100	200	300	400	500	600	800	1000
Velocity • fps	2700	2496	2301	2115	1937	1770	1612	1339	1137
Energy • ft-lbs	2670	2283	1940	1639	1375	1148	953	657	473
Taylor KO Index	19.6	18.1	16.7	15.4	14.1	12.9	11.7	9.7	8.3
Path • Inches	-1.5	+2.0	0.0	-8.7	-25.2	-51.0	-87.7	-205.2	-402.9
Wind Drift • Inches	0.0	0.8	3.3	7.6	14.2	23.2	34.9	67.8	114.6

Norma 165-grain Swift A-Frame (17612)

G1 Ballistic Coefficient = 0.367

Distance • Yards	Muzzle	100	200	300	400	500	600	800	1000
Velocity • fps	2700	2459	2231	2015	1811	1623	1453	1181	1018
Energy • ft-lbs	2672	2216	1824	1488	1202	965	774	571	380
Taylor KO Index	19.6	17.9	16.2	14.6	13.1	11.8	10.5	8.6	7.4
Path • Inches	-1.5	+2.1	0.0	-9.1	-26.9	-55.3	-97.1	-235.7	-476.0
Wind Drift • Inches	0.0	0.9	3.9	9.3	17.4	28.6	43.5	85.4	143.2

Federal 165-grain Trophy Bonded Tip (P308TT4)

G1 Ballistic Coefficient = 0.451

Distance • Yards	Muzzle	100	200	300	400	500	600	800	1000
Velocity • fps	2880	2680	2480	2290	2110	1940	1779	1487	1250
Energy • ft-lbs	3040	2620	2250	1920	1635	1375	1160	810	572
Taylor KO Index	20.9	19.5	18.0	16.6	15.3	14.1	12.9	10.8	9.1
Path • Inches	-1.5	+1.7	0.0	-7.3	-21.4	-43.0	-74.2	-171.4	-332.2
Wind Drift • Inches	0.0	0.7	2.9	6.7	12.3	20.1	30.2	58.5	99.5

Federal 165-grain Trophy Bonded Tip (P308TT2)

G1 Ballistic Coefficient = 0.453

Distance • Yards	Muzzle	100	200	300	400	500	600	800	1000
Velocity • fps	2700	2500	2310	2130	1960	1800	1648	1506	1170
Energy • ft-lbs	2670	2295	1960	1685	1410	1180	832	696	502
Taylor KO Index	19.6	18.2	16.8	15.5	14.2	13.1	12.0	10.9	8.5
Path • Inches	-1.5	+2.0	0.0	-8.6	-24.7	-49.9	-85.8	-198.7	-385.4
Wind Drift • Inches	0.0	0.8	3.1	7.3	13.5	22.1	33.2	64.2	108.6

Remington 165-grain AccuTip Boat Tail (PRA308WB)

G1 Ballistic Coefficient = 0.447

Distance • Yards	Muzzle	100	200	300	400	500	600	800	1000
Velocity • fps	2700	2501	2311	2129	1956	1792	1638	1367	1160
Energy • ft-lbs	2620	2292	1957	1661	1401	1176	983	684	493
Taylor KO Index	19.6	18.2	16.8	15.5	14.2	13.0	11.9	9.9	8.1
Path • Inches	-1.5	+2.1	0.0	-8.9	-24.8	-54.6	-97.8	-244.3	-505.4
Wind Drift • Inches	0.0	1.0	4.4	10.4	19.7	32.7	50.1	99.4	165.8

Nosler 165-grain Ballistic Tip (60050)

G1 Ballistic Coefficient = 0.483

Distance • Yards	Muzzle	100	200	300	400	500	600	800	1000
Velocity • fps	2800	2611	2431	2258	2091	1932	1781	1507	1279
Energy • ft-lbs	2872	2497	2165	1868	1602	1367	1163	832	599
Taylor KO Index	20.3	19.0	17.6	16.4	15.2	14.0	12.9	10.9	9.3
Path • Inches	-1.5	+1.8	0.0	-7.7	-22.2	-44.7	-76.5	-175.1	-335.5
Wind Drift • Inches	0.0	0.7	2.8	6.4	11.9	19.3	28.9	55.7	94.3

Hornady 165-grain GMX (8099) LF

G1 Ballistic Coefficient = 0.447

Distance • Yards	Muzzle	100	200	300	400	500	600	800	1000
Velocity • fps	2750	2549	2357	2173	1998	1831	1674	1397	1181
Energy • ft-lbs	2770	2380	2036	1730	1462	1228	1027	715	512
Taylor KO Index	20.0	18.5	17.1	15.8	14.5	13.3	12.2	10.1	8.6
Path • Inches	-1.5	+1.9	0.0	-8.2	-23.8	-48.1	-82.8	-192.0	-373.0
Wind Drift • Inches	0.0	0.7	3.1	7.2	13.4	21.8	32.7	63.5	107.6

Ultramax 165-grain Nosler Ballistic Tip (308R3)

G1 Ballistic Coefficient = 0.475

Distance • Yards	Muzzle	100	200	300	400	500	600	800	1000
Velocity • fps	2680	2494	2315	2143	1979	1823	1676	1413	1204
Energy • ft-lbs	2632	2279	1964	1684	1436	1218	1029	732	531
Taylor KO Index	19.5	18.1	16.1	15.6	14.4	13.2	12.2	10.3	8.7
Path • Inches	-1.5	+2.0	0.0	-8.6	-24.7	-49.7	-85.2	-195.8	-376.5
Wind Drift • Inches	0.0	0.7	3.0	7.0	12.9	21.0	31.5	60.9	102.7

Ultramax 165-grain Speer Boat Tail (308R4)

G1 Ballistic Coefficient = 0.471

Distance • Yards	Muzzle	100	200	300	400	500	600	800	1000
Velocity • fps	2680	2492	2312	2139	1974	1817	1669	1405	1197
Energy • ft-lbs	2632	2279	1959	1677	1428	1210	1020	723	525
Taylor KO Index	19.5	18.1	16.8	15.5	14.3	13.2	12.1	10.2	8.7
Path • Inches	-1.5	+2.0	0.0	-8.6	-24.8	-49.9	-85.6	-196.8	-379.1
Wind Drift • Inches	0.0	0.7	3.0	7.1	13.1	21.2	31.9	61.6	103.9

Fiocchi 165-grain Sierra GameKing HPBT (308GKB)
G1 Ballistic Coefficient = 0.446

Distance • Yards	Muzzle	100	200	300	400	500	600	800	1000
Velocity • fps	2675	2433	2205	1988	1785	1598	1430	1165	1008
Energy • ft-lbs	2622	2169	1781	1488	1168	936	749	497	372
Taylor KO Index	19.4	17.7	16.0	14.4	13.0	11.6	10.4	8.5	7.3
Path • Inches	-1.5	+2.2	0.0	-9.4	-27.6	-56.7	-99.5	-240.3	-483.6
Wind Drift • Inches	0.0	1.0	4.0	9.5	17.8	29.3	44.5	87.3	145.9

Federal 165-grain Barnes Triple Shock X-Bullet (P308H)
G1 Ballistic Coefficient = 0.380

Distance • Yards	Muzzle	100	200	300	400	500	600	800	1000
Velocity • fps	2650	2420	2200	1990	1800	1620	1453	1189	1026
Energy • ft-lbs	2575	2145	1775	1455	1185	960	774	518	386
Taylor KO Index	19.2	17.6	16.0	14.4	13.1	11.8	10.5	8.6	7.4
Path • Inches	-1.5	+2.2	0.0	-9.5	-27.5	-56.6	-99.0	-237.1	-473.7
Wind Drift • Inches	0.0	0.9	3.9	9.2	17.2	28.2	42.7	83.6	140.0

Lapua 167-grain Scenar (C317515)
G1 Ballistic Coefficient = 0.470

Distance • Yards	Muzzle	100	200	300	400	500	600	800	1000
Velocity • fps	2756	2564	2381	2205	2036	1875	1723	1450	1229
Energy • ft-lbs	2817	2489	2102	1803	1537	1304	1101	758	560
Taylor KO Index	20.3	18.8	17.5	16.2	15.0	13.8	12.7	10.7	9.0
Path • Inches	-1.5	+1.9	0.0	-8.0	-23.2	-46.8	-80.4	-185.3	-358.6
Wind Drift • Inches	0.0	0.7	2.9	6.8	12.6	20.4	30.7	59.2	100.3

Lapua 167-grain Scenar (4317515)
G1 Ballistic Coefficient = 0.460

Distance • Yards	Muzzle	100	200	300	400	500	600	800	1000
Velocity • fps	2690	2497	2312	2135	1966	1805	1655	1387	1179
Energy • ft-lbs	2684	2313	1983	1691	1433	1209	1016	714	516
Taylor KO Index	19.8	18.3	17.0	15.7	14.4	13.3	12.2	10.2	8.7
Path • Inches	-1.5	+2.0	0.0	-8.6	-24.8	-50.0	-85.9	-198.9	-386.6
Wind Drift • Inches	0.0	0.8	3.1	7.2	13.4	21.8	32.7	63.2	106.8

American Eagle (Federal) 168-grain Open Tip Match (A76251M1A)
G1 Ballistic Coefficient = 0.457

Distance • Yards	Muzzle	100	200	300	400	500	600	800	1000
Velocity • fps	2650	2460	2280	2100	1930	1770	1620	1358	1158
Energy • ft-lbs	2620	2255	1930	1645	1395	1175	980	688	500
Taylor KO Index	19.6	18.2	16.9	15.5	14.3	13.1	12.0	10.0	8.6
Path • Inches	-1.5	+2.1	0.0	-8.9	-25.5	-51.6	-89.1	-206.1	-399.0
Wind Drift • Inches	0.0	0.8	3.2	7.4	13.8	22.4	33.7	65.2	109.9

Nosler 168-grain E-Tip (60051) LF
G1 Ballistic Coefficient = 0.510

Distance • Yards	Muzzle	100	200	300	400	500	600	800	1000
Velocity • fps	2750	2574	2404	2241	2085	1934	1790	1527	1306
Energy • ft-lbs	2821	2471	2156	1873	1622	1395	1195	870	636
Taylor KO Index	20.3	19.0	17.8	16.6	15.4	14.3	13.2	11.3	9.7
Path • Inches	-1.5	+1.9	0.0	-7.9	-22.7	-45.6	-77.8	-176.7	-335.6
Wind Drift • Inches	0.0	0.6	2.7	6.2	11.5	18.6	27.8	53.3	89.8

Nosler 168-grain Custom Competition (60054)
G1 Ballistic Coefficient = 0.469

Distance • Yards	Muzzle	100	200	300	400	500	600	800	1000
Velocity • fps	2750	2558	2375	2199	2031	1869	1717	1445	1225
Energy • ft-lbs	2821	2441	2104	1804	1539	1303	1100	779	580
Taylor KO Index	20.3	18.9	17.6	16.3	15.0	13.8	12.7	10.7	9.1
Path • Inches	-1.5	+1.9	0.0	-8.1	-23.4	-47.1	-80.8	-186.0	-358.3
Wind Drift • Inches	0.0	0.7	2.9	6.8	12.6	20.5	30.8	59.5	100.8

Cor-Bon 168-grain HPBT (PM308168/20)
G1 Ballistic Coefficient = 0.460

Distance • Yards	Muzzle	100	200	300	400	500	600	800	1000
Velocity • fps	2700	2507	2321	2144	1974	1813	1662	1393	1183
Energy • ft-lbs	2720	2344	2011	1715	1454	1227	1030	724	522
Taylor KO Index	20.0	18.5	17.2	15.8	14.6	13.4	12.3	10.3	8.7
Path • Inches	-1.5	+2.0	0.0	-8.5	-24.6	-49.6	-85.2	-196.7	-380.4
Wind Drift • Inches	0.0	0.7	3.1	7.2	13.3	21.6	32.5	62.9	106.4

Hornady 168-grain A-MAX Match (8096)

G1 Ballistic Coefficient = 0.475

Distance • Yards	Muzzle	100	200	300	400	500	600	800	1000
Velocity • fps	2700	2513	2333	2161	1996	1839	1691	1425	1213
Energy • ft-lbs	2719	2355	2030	1742	1486	1261	1067	758	549
Taylor KO Index	20.0	18.6	17.2	16.0	14.8	13.6	12.5	10.5	9.0
Path • Inches	-1.5	+2.0	0.0	-8.4	-24.3	-48.9	-83.8	-192.4	-370.1
Wind Drift • Inches	0.0	0.7	3.0	6.9	12.8	20.8	31.2	60.2	101.6

Hornady 168-grain TAP-FPD (80968)

G1 Ballistic Coefficient = 0.475

Distance • Yards	Muzzle	100	200	300	400	500	600	800	1000
Velocity • fps	2700	2513	2333	2161	1996	1839	1690	1424	1212
Energy • ft-lbs	2719	2355	2030	1742	1486	1261	1065	757	548
Taylor KO Index	20.0	18.6	17.2	16.0	14.8	13.6	12.5	10.5	9.0
Path • Inches	-1.5	+2.0	0.0	-8.4	-24.3	-48.9	-83.8	-192.8	-370.5
Wind Drift • Inches	0.0	0.7	3.0	6.9	12.8	20.8	31.2	60.3	101.8

Hornady 168-grain BTHP Match (8097) + Moly (80973)

G1 Ballistic Coefficient = 0.450

Distance • Yards	Muzzle	100	200	300	400	500	600	800	1000
Velocity • fps	2700	2503	2314	2133	1960	1797	1644	1373	1166
Energy • ft-lbs	2719	2336	1997	1697	1433	1204	1008	703	507
Taylor KO Index	20.0	18.5	17.1	15.8	14.5	13.3	12.2	10.1	8.6
Path • Inches	-1.5	+2.0	0.0	-8.5	-24.7	-50.0	-86.0	-199.4	-387.1
Wind Drift • Inches	0.0	0.8	3.1	7.4	13.6	22.2	33.4	64.7	109.4

PMC 168-grain BTHP (308HMA)

G1 Ballistic Coefficient = 0.450

Distance • Yards	Muzzle	100	200	300	400	500	600	800	1000
Velocity • fps	2700	2503	2314	2133	1960	1797	1644	1373	1166
Energy • ft-lbs	2719	2336	1997	1697	1433	1204	1008	703	507
Taylor KO Index	20.0	18.5	17.1	15.8	14.5	13.3	12.2	10.2	8.6
Path • Inches	-1.5	+2.0	0.0	-8.5	-24.7	-50.0	-86.0	-199.4	-387.1
Wind Drift • Inches	0.0	0.8	3.1	7.4	13.6	22.2	33.4	64.7	109.4

Remington 168-grain Boat Tail HP Match (R308W7)

G1 Ballistic Coefficient = 0.476

Distance • Yards	Muzzle	100	200	300	400	500	600	800	1000
Velocity • fps	2680	2496	2314	2143	1979	1823	1677	1414	1205
Energy • ft-lbs	2678	2318	1998	1713	1460	1239	1049	746	541
Taylor KO Index	19.8	18.4	17.1	15.8	14.6	13.5	12.4	10.5	8.9
Path • Inches	-1.5	+2.1	0.0	-8.6	-24.7	-49.9	-85.2	-196.1	-378.8
Wind Drift • Inches	0.0	0.7	3.0	7.0	12.9	21.0	35.9	69.9	118.3

Ultramax 168-grain Sierra HP Match (308R5)

G1 Ballistic Coefficient = 0.430

Distance • Yards	Muzzle	100	200	300	400	500	600	800	1000
Velocity • fps	2680	2475	2278	2091	1913	1745	1588	1318	1121
Energy • ft-lbs	2680	2285	1937	1631	1365	1136	941	648	469
Taylor KO Index	19.8	18.3	16.8	15.5	14.1	12.9	11.7	9.7	8.3
Path • Inches	-1.5	+2.1	0.0	-8.8	-25.6	-51.9	-89.6	-209.5	-409.9
Wind Drift • Inches	0.0	0.8	3.3	7.2	14.5	23.7	35.8	69.5	117.4

Winchester 168-grain Sierra MatchKing BTHP (S308M)

G1 Ballistic Coefficient = 0.453

Distance • Yards	Muzzle	100	200	300	400	500	600	800	1000
Velocity • fps	2680	2485	2297	2118	1948	1786	1635	1367	1163
Energy • ft-lbs	2680	2303	1970	1674	1415	1190	997	698	505
Taylor KO Index	19.8	18.4	17.0	15.7	14.4	13.2	12.1	10.1	8.6
Path • Inches	-1.5	+2.1	0.0	-8.7	-25.1	-50.7	-87.2	-202.0	-391.6
Wind Drift • Inches	0.0	0.7	3.2	7.4	13.7	22.3	33.5	64.8	109.5

Winchester 168-grain E-Tip (S308ETB) LF

G1 Ballistic Coefficient = 0.503

Distance • Yards	Muzzle	100	200	300	400	500	600	800	1000
Velocity • fps	2670	2494	2325	2163	2007	1858	1716	1481	1251
Energy • ft-lbs	2859	2320	2016	1744	1502	1287	1099	796	584
Taylor KO Index	19.7	18.4	17.2	16.0	14.8	13.7	12.7	10.9	9.2
Path • Inches	-1.5	+2.0	0.0	-8.5	-24.4	-49.0	-83.7	-190.6	-363.1
Wind Drift • Inches	0.0	0.7	2.8	6.6	12.2	19.8	29.6	56.8	95.5

Winchester 168-grain Ballistic Silvertip (SBST308A)

G1 Ballistic Coefficient = 0.476

Distance • Yards	Muzzle	100	200	300	400	500	600	800	1000
Velocity • fps	2670	2484	2306	2134	1971	1815	1669	1408	1200
Energy • ft-lbs	2659	2301	1983	1699	1449	1229	1040	739	537
Taylor KO Index	19.7	18.4	17.0	15.8	14.6	13.4	12.3	10.4	8.9
Path • Inches	-1.5	+2.1	0.0	-8.6	-24.8	-50.0	-85.6	-197.8	-382.1
Wind Drift • Inches	0.0	0.7	3.0	7.0	13.0	21.1	31.7	61.1	103.0

Black Hills 168-grain Barnes Triple Shock (1C308BHGN1) LF

G1 Ballistic Coefficient = 0.510

Distance • Yards	Muzzle	100	200	300	400	500	600	800	1000
Velocity • fps	2650	2477	2311	2151	1998	1851	1712	1460	1252
Energy • ft-lbs	2620	2290	1993	1727	1489	1278	1093	795	585
Taylor KO Index	19.6	18.3	17.1	15.9	14.8	13.7	12.7	10.8	9.3
Path • Inches	-1.5	+2.1	0.0	-8.6	-24.7	-49.6	-84.6	-192.7	-367.9
Wind Drift • Inches	0.0	0.6	2.5	5.9	10.9	17.7	26.5	50.8	85.4

Black Hills 168-grain Hornady A-MAX (1C308BHGN2)

G1 Ballistic Coefficient = 0.475

Distance • Yards	Muzzle	100	200	300	400	500	600	800	1000
Velocity • fps	2650	2465	2287	2116	1953	1798	1652	1393	1189
Energy • ft-lbs	2620	2267	1951	1671	1423	1207	1019	724	527
Taylor KO Index	19.6	18.2	16.9	15.6	14.4	13.3	12.2	10.3	8.8
Path • Inches	-1.5	+2.1	0.0	-8.8	-25.3	-51.0	-87.5	-201.1	-386.9
Wind Drift • Inches	0.0	0.7	3.0	7.1	13.2	21.4	32.2	62.0	104.5

Black Hills 168-grain Match Hollow Point (D308N1)

G1 Ballistic Coefficient = 0.450

Distance • Yards	Muzzle	100	200	300	400	500	600	800	1000
Velocity • fps	2650	2447	2272	2095	1926	1766	1617	1354	1154
Energy • ft-lbs	2620	2252	1926	1638	1384	1164	975	684	497
Taylor KO Index	19.6	18.1	16.8	15.5	14.2	13.1	12.0	10.0	8.5
Path • Inches	-1.5	+2.1	0.0	-8.9	-25.7	-51.9	-89.3	-206.7	-400.5
Wind Drift • Inches	0.0	0.8	3.2	7.5	13.8	22.5	33.9	65.5	110.5

Federal 168-grain Sierra MatchKing BTHP (GM308M)

G1 Ballistic Coefficient = 0.464

Distance • Yards	Muzzle	100	200	300	400	500	600	800	1000
Velocity • fps	2650	2460	2280	2100	1940	1780	1632	1371	1169
Energy • ft-lbs	2620	2265	1935	1650	1400	1180	994	701	510
Taylor KO Index	19.6	18.2	16.9	15.5	14.3	13.2	12.1	10.1	8.6
Path • Inches	-1.5	+2.1	0.0	-8.9	-25.5	-51.5	-88.5	-204.2	-394.4
Wind Drift • Inches	0.0	0.8	3.1	7.3	13.5	22.0	33.1	64.0	107.9

Fiocchi 168-grain Sierra MatchKing HPBT (308MKB)

G1 Ballistic Coefficient = 0.438

Distance • Yards	Muzzle	100	200	300	400	500	600	800	1000
Velocity • fps	2650	2455	2267	2086	1906	1736	1583	1318	1124
Energy • ft-lbs	2619	2248	1916	1623	1355	1124	935	648	472
Taylor KO Index	19.6	18.1	16.8	15.4	14.1	12.8	11.7	9.7	8.3
Path • Inches	-1.5	+2.2	0.0	-9.0	-26.1	-52.8	-91.1	-212.2	-413.9
Wind Drift • Inches	0.0	0.8	3.3	7.8	14.5	23.6	35.5	69.0	116.3

Sellier and Bellot 168-grain HPBT (V331452U)

G1 Ballistic Coefficient = 0.493

Distance • Yards	Muzzle	100	200	300	400	500	600	800	1000
Velocity • fps	2628	2444	2274	2115	1958	1809	1667	1414	1211
Energy • ft-lbs	2579	2231	1930	1670	1431	1221	1037	746	547
Taylor KO Index	19.4	18.1	16.8	15.6	14.5	13.4	12.3	10.5	9.0
Path • Inches	-1.5	+2.2	0.0	-8.8	-25.5	-51.2	-87.6	-200.6	-385.2
Wind Drift • Inches	0.0	0.7	3.0	6.9	12.8	20.7	31.1	59.7	100.5

Cor-Bon 168-grain DPX (DPX308168/20)

G1 Ballistic Coefficient = 0.425

Distance • Yards	Muzzle	100	200	300	400	500	600	800	1000
Velocity • fps	2600	2396	2200	2014	1839	1673	1520	1262	1084
Energy • ft-lbs	2522	2141	1807	1514	1261	1044	862	594	438
Taylor KO Index	19.2	17.7	16.3	14.9	13.6	12.4	11.2	9.3	8.0
Path • Inches	-1.5	+2.3	0.0	-9.5	-27.6	-56.0	-96.8	-226.9	-444.7
Wind Drift • Inches	0.0	0.8	3.5	8.3	15.5	25.2	38.1	73.9	124.1

Norma 168-grain Coated Hollow Point (17615)

G1 Ballistic Coefficient = 0.470

Distance • Yards	Muzzle	100	200	300	400	500	600	800	1000
Velocity • fps	2549	2366	2191	2023	1863	1712	1570	1324	1139
Energy • ft-lbs	2424	2089	1791	1527	1295	1093	920	654	484
Taylor KO Index	18.8	17.5	16.2	15.0	13.8	12.7	11.6	9.8	8.4
Path • Inches	-1.5	+2.4	0.0	-9.6	-27.8	-55.9	-96.0	-221.2	-426.2
Wind Drift • Inches	0.0	0.8	3.3	7.6	14.1	23.0	34.5	66.5	111.4

Federal 170-grain Fusion (F308FSLR1)

G1 Ballistic Coefficient = 0.284

Distance • Yards	Muzzle	100	200	300	400	500	600	800	1000
Velocity • fps	2000	1740	1510	1312	1157	1049	974	867	787
Energy • ft-lbs	1510	1145	860	650	505	415	358	284	234
Taylor KO Index	15.0	13.0	11.3	9.8	8.7	7.8	7.3	6.5	5.9
Path • Inches	-1.5	+5.0	0.0	-20.4	-61.0	-127.5	-225.6	-536.7	-1033
Wind Drift • Inches	0.0	1.9	8.1	19.2	35.8	57.5	83.4	146.0	221.5

Lapua 170-grain Naturalis (N317201)

G1 Ballistic Coefficient = 0.316

Distance • Yards	Muzzle	100	200	300	400	500	600	800	1000
Velocity • fps	2625	2352	2094	1855	1633	1436	1269	1045	924
Energy • ft-lbs	2597	2085	1652	1297	1007	779	608	412	322
Taylor KO Index	19.6	17.6	15.7	13.9	12.2	10.7	9.5	7.8	6.9
Path • Inches	-1.5	+2.4	0.0	-10.4	-30.9	-64.6	-115.2	-286.5	-584.7
Wind Drift • Inches	0.0	1.1	4.8	11.5	21.8	36.2	55.2	107.6	175.4

Lapua 170-grain Naturalis (N317201)

G1 Ballistic Coefficient = 0.316

Distance • Yards	Muzzle	100	200	300	400	500	600	800	1000
Velocity • fps	2625	2352	2094	1855	1633	1436	1269	1045	924
Energy • ft-lbs	2597	2085	1652	1297	1007	779	608	412	322
Taylor KO Index	19.6	17.6	15.7	13.9	12.2	10.7	9.5	7.8	6.9
Path • Inches	-1.5	+2.4	0.0	-10.4	-30.9	-64.6	-115.2	-286.5	-584.7
Wind Drift • Inches	0.0	1.1	4.8	11.5	21.8	36.2	55.2	107.6	175.4

Cor-Bon 175-grain HPBT (PM308175/20)

G1 Ballistic Coefficient = 0.490

Distance • Yards	Muzzle	100	200	300	400	500	600	800	1000
Velocity • fps	2650	2470	2297	2132	1973	1821	1678	1422	1218
Energy • ft-lbs	2730	2372	2052	1766	1513	1289	1095	786	574
Taylor KO Index	20.4	19.0	17.7	16.4	15.2	14.0	12.9	10.9	9.4
Path • Inches	-1.5	+2.1	0.0	-8.7	-25.1	-50.4	-86.2	-197.2	-377.5
Wind Drift • Inches	0.0	0.7	3.0	6.9	12.7	20.6	30.9	59.5	100.2

Remington 175-grain MatchKing BTHP (RM308WA)

G1 Ballistic Coefficient = 0.496

Distance • Yards	Muzzle	100	200	300	400	500	600	800	1000
Velocity • fps	2609	2433	2264	2102	1946	1798	1657	1407	1207
Energy • ft-lbs	2644	2300	1992	1716	1472	1256	1066	769	566
Taylor KO Index	20.1	18.7	17.4	16.2	15.0	13.8	12.8	10.8	9.3
Path • Inches	-1.5	+2.2	0.0	-9.0	-25.9	-51.9	-88.8	-202.8	-387.6
Wind Drift • Inches	0.0	0.7	3.0	7.0	12.8	20.8	31.2	60.0	100.8

Federal 175-grain Sierra MatchKing BTHP (GM308M2)

G1 Ballistic Coefficient = 0.503

Distance • Yards	Muzzle	100	200	300	400	500	600	800	1000
Velocity • fps	2600	2430	2260	2100	1950	1800	1662	1414	1214
Energy • ft-lbs	2625	2290	1985	1715	1475	1260	1073	777	573
Taylor KO Index	20.0	18.7	17.4	16.2	15.0	13.9	12.8	10.9	9.3
Path • Inches	-1.5	+2.2	0.0	-9.1	-25.8	-51.8	-88.9	-202.7	-386.5
Wind Drift • Inches	0.0	0.7	3.0	6.9	12.7	20.6	30.8	59.2	99.4

Black Hills 175-grain Match Hollow Point (D308N5)

G1 Ballistic Coefficient = 0.496

Distance • Yards	Muzzle	100	200	300	400	500	600	800	1000
Velocity • fps	2600	2420	2260	2090	1940	1790	1651	1402	1203
Energy • ft-lbs	2627	2284	1977	1703	1460	1245	1060	764	562
Taylor KO Index	20.0	18.6	17.4	16.1	14.9	13.8	12.7	10.8	9.3
Path • Inches	-1.5	+2.2	0.0	-9.1	-26.1	-52.4	-89.5	-204.8	-393.0
Wind Drift • Inches	0.0	0.7	3.0	7.0	12.9	21.0	31.3	60.2	111.2

Fiocchi 175-grain Sierra MatchKing HPBT (308MKD)

G1 Ballistic Coefficient = 0.496

Distance • Yards	Muzzle	100	200	300	400	500	600	800	1000
Velocity • fps	2595	2419	2251	2089	1934	1786	1648	1400	1202
Energy • ft-lbs	2816	2274	1968	1695	1453	1240	1056	761	561
Taylor KO Index	20.0	18.6	17.3	16.1	14.9	13.8	12.7	10.8	9.3
Path • Inches	-1.5	+2.2	0.0	-9.1	-26.2	-52.6	-89.8	-205.2	-392.0
Wind Drift • Inches	0.0	0.7	3.0	7.0	12.9	21.0	31.4	60.3	101.3

Hornady 178-grain BTHP (8077)

G1 Ballistic Coefficient = 0.530

Distance • Yards	Muzzle	100	200	300	400	500	600	800	1000
Velocity • fps	2780	2609	2444	2285	2132	1985	1845	1585	1361
Energy • ft-lbs	3054	2690	2361	2064	1797	1558	1346	994	733
Taylor KO Index	21.8	20.4	19.1	17.9	16.7	15.5	14.5	12.4	10.7
Path • Inches	-1.5	+1.8	0.0	-7.6	-21.9	-43.9	-74.7	-168.6	-318.3
Wind Drift • Inches	0.0	0.6	2.5	5.9	10.8	17.5	26.1	49.9	83.9

Federal 180-grain Fusion (F308FS3)

G1 Ballistic Coefficient = 0.502

Distance • Yards	Muzzle	100	200	300	400	500	600	800	1000
Velocity • fps	2600	2430	2260	2100	1950	1800	1661	1414	1214
Energy • ft-lbs	2700	2365	2040	1765	1515	1295	1104	799	589
Taylor KO Index	20.6	19.2	17.9	16.6	15.4	14.3	13.2	11.2	9.6
Path • Inches	-1.5	+2.2	0.0	-9.1	-25.8	-51.9	-89.0	-202.8	-386.7
Wind Drift • Inches	0.0	0.7	3.0	6.9	12.7	20.6	30.8	59.2	99.4

Federal 180-grain Soft Point (308B)

G1 Ballistic Coefficient = 0.385

Distance • Yards	Muzzle	100	200	300	400	500	600	800	1000
Velocity • fps	2570	2350	2130	1930	1740	1570	1413	1164	1014
Energy • ft-lbs	2640	2195	1815	1485	1210	980	798	542	411
Taylor KO Index	20.4	18.6	16.9	15.3	13.8	12.4	11.2	9.2	8.0
Path • Inches	-1.5	+2.4	0.0	-10.0	-29.5	-60.7	-105.4	-251.8	-500.8
Wind Drift • Inches	0.0	1.0	4.0	9.5	17.7	29.0	44.0	85.7	143.9

Federal 180-grain Nosler Partition (P308E)

G1 Ballistic Coefficient = 0.474

Distance • Yards	Muzzle	100	200	300	400	500	600	800	1000
Velocity • fps	2570	2390	2210	2050	1890	1730	1591	1342	1152
Energy • ft-lbs	2640	2280	1955	1670	1420	1200	1012	720	531
Taylor KO Index	20.4	18.9	17.5	16.2	15.0	13.7	12.6	10.6	9.1
Path • Inches	-1.5	+2.3	0.0	-9.4	-26.9	-54.7	-93.9	-216.1	-416.0
Wind Drift • Inches	0.0	0.8	3.2	7.5	13.8	22.5	33.8	65.1	109.2

Federal 180-grain Trophy Bonded Tip (P308TT1)

G1 Ballistic Coefficient = 0.499

Distance • Yards	Muzzle	100	200	300	400	500	600	800	1000
Velocity • fps	2620	2450	2280	2120	1960	1810	1671	1419	1217
Energy • ft-lbs	2745	2390	2070	1790	1535	1315	1116	806	592
Taylor KO Index	20.8	19.4	18.1	16.8	15.5	14.3	13.2	11.2	9.6
Path • Inches	-1.5	+2.2	0.0	-8.9	-25.5	-51.1	-87.7	-200.2	-382.1
Wind Drift • Inches	0.0	0.7	2.9	6.9	12.7	20.6	30.8	59.1	99.4

Remington 180-grain Core-Lokt Ultra Bonded (PRC308WC)

G1 Ballistic Coefficient = 0.384

Distance • Yards	Muzzle	100	200	300	400	500	600	800	1000
Velocity • fps	2620	2404	2198	2002	1818	1644	1487	1225	1055
Energy • ft-lbs	2743	2309	1930	1601	1320	1080	884	600	445
Taylor KO Index	20.8	19.0	17.4	15.9	14.4	13.0	11.8	9.7	8.4
Path • Inches	-1.5	+2.3	0.0	-9.5	-27.7	-56.4	-98.2	-233.4	-463.9
Wind Drift • Inches	0.0	0.9	3.7	8.7	16.2	26.6	40.2	78.4	131.6

Remington 180-grain Core-Lokt Pointed Soft Point (R308W3)

G1 Ballistic Coefficient = 0.384

Distance • Yards	Muzzle	100	200	300	400	500	600	800	1000
Velocity • fps	2620	2393	2178	1974	1782	1604	1443	1184	1025
Energy • ft-lbs	2743	2288	1896	1557	1269	1028	833	561	420
Taylor KO Index	21.7	19.0	17.2	15.6	14.1	12.7	11.4	9.4	8.1
Path • Inches	-1.5	+2.3	0.0	-9.7	-28.3	-57.8	-101.2	-243.0	-486.1
Wind Drift • Inches	0.0	0.9	3.9	9.2	17.2	28.3	42.8	83.7	138.8

Remington 180-grain Core-Lokt Soft Point (R308W2)

G1 Ballistic Coefficient = 0.248

Distance • Yards	Muzzle	100	200	300	400	500	600	800	1000
Velocity • fps	2620	2274	1955	1666	1414	1212	1071	912	810
Energy • ft-lbs	2743	2066	1527	1109	799	587	459	333	262
Taylor KO Index	21.7	18.0	15.5	13.2	11.2	9.6	8.5	7.2	6.4
Path • Inches	-1.5	+2.6	0.0	-11.8	-36.3	-78.2	-145.6	-372.7	-751.2
Wind Drift • Inches	0.0	1.5	6.4	15.5	29.8	50.1	76.5	143.8	226.6

Winchester 180-grain Silvertip (X3083)

G1 Ballistic Coefficient = 0.384

Distance • Yards	Muzzle	100	200	300	400	500	600	800	1000
Velocity • fps	2620	2393	2178	1974	1782	1604	1443	1184	1025
Energy • ft-lbs	2743	2280	1896	1557	1269	1028	833	561	420
Taylor KO Index	21.7	19.0	17.2	15.6	14.1	12.7	11.4	9.4	8.1
Path • Inches	-1.5	+2.6	0.0	-9.9	-28.9	-58.8	-101.2	-243.0	-486.1
Wind Drift • Inches	0.0	0.9	3.9	9.2	17.2	28.3	42.8	83.7	138.8

Winchester 180-grain Power-Point (X3086)

G1 Ballistic Coefficient = 0.381

Distance • Yards	Muzzle	100	200	300	400	500	600	800	1000
Velocity • fps	2620	2392	2176	1971	1779	1600	1437	1179	1022
Energy • ft-lbs	2743	2287	1892	1553	1264	1023	826	556	417
Taylor KO Index	20.8	18.9	17.2	15.6	14.1	12.7	11.4	9.3	8.1
Path • Inches	-1.5	+2.3	0.0	-9.7	-28.3	-58.0	-101.4	-242.6	-483.9
Wind Drift • Inches	0.0	0.9	3.9	9.3	17.4	28.5	43.2	84.5	141.0

Norma 180-grain Nosler Partition (17635)

G1 Ballistic Coefficient = 0.474

Distance • Yards	Muzzle	100	200	300	400	500	600	800	1000
Velocity • fps	2612	2428	2252	2082	1920	1766	1622	1367	1170
Energy • ft-lbs	2728	2357	2057	1734	1474	1247	1052	747	547
Taylor KO Index	20.7	19.2	17.8	16.5	15.2	14.0	12.8	10.8	9.3
Path • Inches	-1.5	+2.2	0.0	-9.1	-26.2	-52.8	-90.5	-208.2	-400.7
Wind Drift • Inches	0.0	0.8	3.1	7.3	13.5	22.0	33.0	63.5	106.9

Norma 180-grain Plastic Point (17628)

G1 Ballistic Coefficient = 0.366

Distance • Yards	Muzzle	100	200	300	400	500	600	800	1000
Velocity • fps	2612	2375	2151	1939	1741	1559	1395	1148	997
Energy • ft-lbs	2728	2255	1849	1503	1211	971	778	523	397
Taylor KO Index	20.7	18.8	17.0	15.4	13.8	12.3	11.0	9.1	7.9
Path • Inches	-1.5	+2.3	0.0	-9.9	-29.0	-59.7	-104.8	-252.8	-507.0
Wind Drift • Inches	0.0	1.0	4.1	9.8	18.3	30.2	45.8	89.5	148.6

Norma 180-grain Oryx (17675)

G1 Ballistic Coefficient = 0.288

Distance • Yards	Muzzle	100	200	300	400	500	600	800	1000
Velocity • fps	2612	2305	2019	1775	1543	1341	1180	986	876
Energy • ft-lbs	2728	2124	1629	1232	952	719	557	389	307
Taylor KO Index	20.7	18.3	16.0	14.1	12.2	10.6	9.3	7.8	6.9
Path • Inches	-1.5	+2.5	0.0	-11.1	-33.1	-69.8	-127.0	-322.1	-651.0
Wind Drift • Inches	0.0	1.3	5.4	13.0	24.7	41.2	63.1	121.4	195.0

Norma 180-grain Alaska (17636)

G1 Ballistic Coefficient = 0.257

Distance • Yards	Muzzle	100	200	300	400	500	600	800	1000
Velocity • fps	2612	2269	1953	1687	1420	1215	1074	915	812
Energy • ft-lbs	2728	2059	1526	1111	802	590	461	335	264
Taylor KO Index	20.7	18.0	15.5	13.2	11.2	9.6	8.5	7.2	6.4
Path • Inches	-1.5	+2.7	0.0	-11.9	-36.3	-78.3	-145.5	-371.9	-749.0
Wind Drift • Inches	0.0	1.5	6.4	15.4	29.6	49.8	76.0	142.8	225.2

Black Hills 180-grain Nosler AccuBond (1C308BHGN4)

G1 Ballistic Coefficient = 0.507

Distance • Yards	Muzzle	100	200	300	400	500	600	800	1000
Velocity • fps	2550	2380	2216	2059	1909	1765	1630	1389	1197
Energy • ft-lbs	2600	2264	1964	1695	1456	1246	1063	772	573
Taylor KO Index	20.2	18.8	17.6	16.3	15.1	14.0	12.9	11.0	9.5
Path • Inches	-1.5	+2.3	0.0	-9.4	-27.0	-54.2	-92.6	-210.9	-401.8
Wind Drift • Inches	0.0	0.7	3.0	7.0	13.0	21.0	31.4	60.3	101.0

Sellier and Bellot 180-grain Soft Point (V331432U)

G1 Ballistic Coefficient = 0.284

Distance • Yards	Muzzle	100	200	300	400	500	600	800	1000
Velocity • fps	2454	2146	1877	1642	1422	1241	1106	947	847
Energy • ft-lbs	2414	1846	1412	1080	809	615	489	358	287
Taylor KO Index	19.4	17.0	14.9	13.0	11.3	9.8	8.8	7.5	6.7
Path • Inches	-1.5	+3.0	0.0	-12.8	-38.7	-82.1	-148.6	-370.8	-735.9
Wind Drift • Inches	0.0	1.4	6.0	14.5	27.6	45.9	69.6	130.5	205.7

Norma 180-grain Vulcan (17660)

G1 Ballistic Coefficient = 0.315

Distance • Yards	Muzzle	100	200	300	400	500	600	800	1000
Velocity • fps	2612	2338	2081	1841	1619	1423	1257	1038	919
Energy • ft-lbs	2728	2185	1731	1355	1049	810	632	431	338
Taylor KO Index	20.7	18.5	16.5	14.6	12.8	11.3	10.0	8.2	7.3
Path • Inches	-1.5	+2.4	0.0	-10.5	-31.4	-65.6	-117.0	-291.1	-593.8
Wind Drift • Inches	0.0	1.2	4.9	11.7	22.0	36.6	56.0	108.9	177.1

Federal 180-grain Barnes MRX Bullet (P308W) LF

G1 Ballistic Coefficient = 0.549

Distance • Yards	Muzzle	100	200	300	400	500	600	800	1000
Velocity • fps	2600	2440	2290	2140	2000	1860	1730	1492	1291
Energy • ft-lbs	2700	2385	2095	1835	1600	1390	1197	890	667
Taylor KO Index	20.6	19.3	18.1	16.9	15.8	14.7	13.7	11.6	10.2
Path • Inches	-1.5	+2.2	0.0	-8.9	-25.2	-50.0	-85.6	-192.8	-362.8
Wind Drift • Inches	0.0	0.6	2.7	6.2	11.5	18.6	27.7	52.9	88.5

Sellier and Bellot 180-grain FMJ (V331422U)

G1 Ballistic Coefficient = 0.502

Distance • Yards	Muzzle	100	200	300	400	500	600	800	1000
Velocity • fps	2411	2240	2081	1932	1787	1649	1521	1298	1131
Energy • ft-lbs	2331	2011	1735	1497	1276	1087	985	673	511
Taylor KO Index	19.1	17.7	16.5	15.3	14.2	13.1	12.0	10.3	9.0
Path • Inches	-1.5	+2.7	0.0	-10.7	-30.8	-61.7	-105.5	-240.8	-459.0
Wind Drift • Inches	0.0	0.7	3.0	6.9	12.8	20.8	31.1	59.5	98.8

Lapua 185-grain Scenar (C317523)

G1 Ballistic Coefficient = 0.521

Distance • Yards	Muzzle	100	200	300	400	500	600	800	1000
Velocity • fps	2608	2440	2279	2124	1974	1832	1696	1451	1249
Energy • ft-lbs	2795	2447	2134	1853	1602	1379	1183	866	641
Taylor KO Index	21.2	19.9	18.6	17.3	16.1	14.9	13.8	11.8	10.2
Path • Inches	-1.5	+2.2	0.0	-8.9	-25.5	-51.0	-86.9	-197.4	-375.6
Wind Drift • Inches	0.0	0.7	2.8	6.6	12.1	19.6	29.4	56.2	94.3

Lapua 185-grain Mega SP (4317189)

G1 Ballistic Coefficient = 0.310

Distance • Yards	Muzzle	100	200	300	400	500	600	800	1000
Velocity • fps	2510	2238	1983	1746	1532	1345	1193	1003	894
Energy • ft-lbs	2589	2057	1615	1253	964	744	585	413	329
Taylor KO Index	20.4	18.2	16.1	14.2	12.5	10.9	9.7	8.2	7.3
Path • Inches	-1.5	+2.7	0.0	-11.6	-34.7	-72.7	-130.8	-326.0	-651.9
Wind Drift • Inches	0.0	1.2	5.3	12.6	23.9	39.7	60.4	115.7	185.5

Lapua 185-grain Scenar (4317523)

G1 Ballistic Coefficient = 0.427

Distance • Yards	Muzzle	100	200	300	400	500	600	800	1000
Velocity • fps	2477	2279	2090	1911	1742	1584	1440	1203	1049
Energy • ft-lbs	2521	2134	1795	1500	1246	1031	852	595	452
Taylor KO Index	20.5	18.6	17.0	15.6	14.2	12.9	11.7	9.8	8.5
Path • Inches	-1.5	+2.6	0.0	-10.6	-30.7	-62.4	-108.0	-254.0	-498.0
Wind Drift • Inches	0.0	0.9	3.8	8.9	16.5	27.0	40.6	78.5	130.4

Cor-Bon 185-grain HPBT Subsonic (PM308S190/20)

G1 Ballistic Coefficient = 0.470

Distance • Yards	Muzzle	100	200	300	400	500	600	800	1000
Velocity • fps	1000	959	924	893	865	839	815	771	732
Energy • ft-lbs	422	388	361	337	316	297	280	251	226
Taylor KO Index	8.4	8.0	7.7	7.5	7.2	7.0	6.8	6.4	6.1
Path • Inches	-1.5	+18.1	0.0	-58.8	-161.1	-310.0	-508.3	-1066	-1859
Wind Drift • Inches	0.0	1.1	4.4	9.7	17.0	26.3	37.4	65.5	101.3

Lapua 185-grain FMJ – Boat Tail (4317590)

G1 Ballistic Coefficient = 0.490

Distance • Yards	Muzzle	100	200	300	400	500	600	800	1000
Velocity • fps	2495	2335	2181	2033	1864	1703	1570	1333	1152
Energy • ft-lbs	2558	2240	1954	1698	1427	1192	1012	730	546
Taylor KO Index	20.3	19.0	17.8	16.5	15.2	13.9	12.8	10.9	9.4
Path • Inches	-1.5	+2.4	0.0	-9.8	-27.2	-55.1	-98.4	-225.7	-433.3
Wind Drift • Inches	0.0	0.8	3.4	7.9	14.7	24.0	33.5	64.3	107.5

Cor-Bon 190-grain HPBT (PM308190/20)

G1 Ballistic Coefficient = 0.520

Distance • Yards	Muzzle	100	200	300	400	500	600	800	1000
Velocity • fps	2600	2432	2271	2115	1966	1823	1688	1444	1243
Energy • ft-lbs	2853	2496	2176	1888	1631	1403	1203	880	652
Taylor KO Index	21.7	20.3	19.0	17.7	16.4	15.2	14.1	12.1	10.4
Path • Inches	-1.5	+2.2	0.0	-8.9	-25.7	-51.4	-87.6	-198.8	-377.1
Wind Drift • Inches	0.0	0.7	2.8	6.6	12.2	19.8	29.6	56.7	95.1

Lapua 200-grain Subsonic (4317340)

G1 Ballistic Coefficient = 0.470

Distance • Yards	Muzzle	100	200	300	400	500	600	800	1000
Velocity • fps	1050	1001	961	926	894	866	839	791	748
Energy • ft-lbs	490	445	410	381	355	333	313	278	249
Taylor KO Index	9.2	8.8	8.5	8.1	7.9	7.6	7.4	7.0	6.6
Path • Inches	-1.5	+16.6	0.0	-54.2	-148.7	-286.5	-470.4	-987.5	-1724
Wind Drift • Inches	0.0	1.2	4.8	10.5	18.3	28.0	39.7	68.7	105.4

Hunter and guide with a magnificent gemsbok. Note the tape holding the gunstock together. Sometimes in the field emergency repairs are needed, and a roll of duct tape comes in very handy. (Photo from *Out of Bullets* by Johnny Chilton, 2011, Safari Press Archives)

.30-06 Springfield

It's well over 100 years old, but who's counting? The .30-06 is the standard by which every other U.S. cartridge is judged. Conceived in response to the embarrassment inflicted by the 7x57mm Mausers in Cuba in the Spanish-American War, the first cut at this cartridge was little more than a 7x57 necked to .30 caliber. In its original form, the 1903 cartridge for the Model 1903 rifle used a 220-grain roundnose bullet at 2,300 fps. Three years later the 1906 version (hence the .30-06 name) adopted a 150-grain pointed bullet at 2,700 fps.

The basic design of U.S. military rifle ammunition changed very little until the 7.62 NATO cartridge (.308 Winchester) was adopted in 1952. The .30-06 remains the most popular caliber in the inventory in this country. There are over 100 loadings available today. At this rate, we might see .30-06s around in the year 2100.

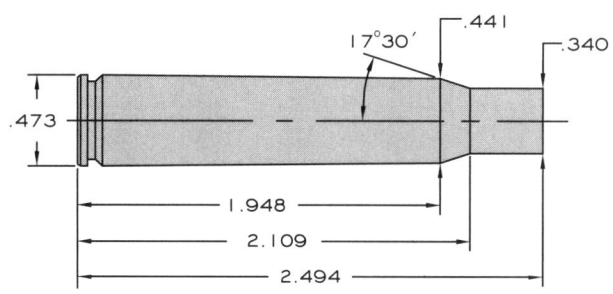

Relative Recoil Factor = 2.19

Specifications:

Controlling Agency for Standardization of this Ammunition: SAAMI

Bullet Weight Grains	Velocity fps	Maximum Average Pressure	
		Copper Crusher	Transducer
110	3,300	50,000 cup	60,000 psi
125	3,125	50,000 cup	60,000 psi
150	2,900	50,000 cup	60,000 psi
165–168	2,790	50,000 cup	60,000 psi
180	2,690	50,000 cup	60,000 psi
200	2,450	50,000 cup	60,000 psi
220	2,400	50,000 cup	60,000 psi

Standard barrel for velocity testing: 24 inches long—1 turn in 10-inch twist

Availability:

Remington 55-grain Accelerator Pointed Soft Point (R30069)

G1 Ballistic Coefficient = 0.197

Distance • Yards	Muzzle	100	200	300	400	500	600	800	1000
Velocity • fps	4080	3484	2964	2499	2080	1706	1388	1009	850
Energy • ft-lbs	2033	1482	1073	763	528	355	235	124	88
Taylor KO Index	9.9	8.4	7.2	6.0	5.0	4.1	3.4	2.4	2.1
Path • Inches	-1.5	+0.7	0.0	-4.7	-15.0	-33.6	-65.1	-194.4	-542.0
Wind Drift • Inches	0.0	1.1	4.6	11.1	21.3	36.4	57.8	123.2	212.0

Federal 110-grain Barnes Triple-Shock X-Bullet (13063) LF

G1 Ballistic Coefficient = 0.331

Distance • Yards	Muzzle	100	200	300	400	500	600	800	1000
Velocity • fps	3400	3090	2800	2520	2270	2030	1804	1413	1134
Energy • ft-lbs	2825	2330	1910	1555	1255	1000	735	488	311
Taylor KO Index	16.5	15.0	13.6	12.2	11.0	9.8	8.7	6.8	5.5
Path • Inches	-1.5	+1.1	0.0	-5.5	-16.5	-34.4	-60.6	-148.2	-306.6
Wind Drift • Inches	0.0	0.8	3.2	7.5	14.0	23.1	35.2	70.4	123.4

Lapua 123-grain Full Metal Jacket (4317577)

G1 Ballistic Coefficient = 0.274

Distance • Yards	Muzzle	100	200	300	400	500	600	800	1000
Velocity • fps	2936	2599	2286	1995	1728	490	1289	1031	900
Energy • ft-lbs	2364	1853	1433	1092	819	609	454	290	221
Taylor KO Index	15.9	14.1	12.4	10.8	9.4	8.1	7.0	5.6	4.9
Path • Inches	-1.5	+1.8	0.0	-8.5	-25.9	-54.9	-100.5	-262.6	-549.9
Wind Drift • Inches	0.0	1.1	4.8	11.6	22.0	37.0	57.1	113.9	188.0

Federal 125-grain Soft Point (3006CS)

G1 Ballistic Coefficient = 0.267

Distance • Yards	Muzzle	100	200	300	400	500	600	800	1000
Velocity • fps	3140	2780	2450	2140	1850	1590	1368	1065	918
Energy • ft-lbs	2735	2145	1680	1265	950	705	520	315	234
Taylor KO Index	17.3	15.3	13.5	11.8	10.2	8.7	7.5	5.9	5.0
Path • Inches	-1.5	+1.5	0.0	-7.4	-22.3	-48.0	-86.8	-226.8	-488.8
Wind Drift • Inches	0.0	1.1	4.5	10.8	20.5	34.5	53.5	108.4	182.3

Remington 125-grain Pointed Soft Point (R30061)

G1 Ballistic Coefficient = 0.268

Distance • Yards	Muzzle	100	200	300	400	500	600	800	1000
Velocity • fps	3140	2780	2447	2138	1853	1595	1371	1067	919
Energy • ft-lbs	2736	2145	1662	1269	953	706	522	316	234
Taylor KO Index	17.3	15.3	13.5	11.8	10.2	8.8	7.5	5.9	5.1
Path • Inches	-1.5	+1.5	0.0	-7.4	-22.4	-47.6	-87.2	-230.5	-493.1
Wind Drift • Inches	0.0	1.1	4.5	10.8	20.5	34.4	53.3	108.0	181.8

Winchester 125-grain Pointed Soft Point (X30062)

G1 Ballistic Coefficient = 0.268

Distance • Yards	Muzzle	100	200	300	400	500	600	800	1000
Velocity • fps	3140	2780	2447	2138	1853	1595	1371	1067	919
Energy • ft-lbs	2736	2145	1662	1269	953	706	522	316	234
Taylor KO Index	17.3	15.3	13.5	11.8	10.2	8.8	7.5	5.9	5.1
Path • Inches	-1.5	+1.5	0.0	-7.4	-22.4	-47.6	-87.2	-230.5	-493.1
Wind Drift • Inches	0.0	1.1	4.5	10.8	20.5	34.4	53.3	108.0	181.8

Remington 125-grain Core-Lokt Pointed Soft Point – Managed Recoil (RL30062)

G1 Ballistic Coefficient = 0.267

Distance • Yards	Muzzle	100	200	300	400	500	600	800	1000
Velocity • fps	2660	2335	2034	1757	1509	1300	1139	956	846
Energy • ft-lbs	1964	1513	1148	856	632	469	360	254	200
Taylor KO Index	14.6	12.8	11.2	9.7	8.3	7.2	6.3	5.3	4.7
Path • Inches	-1.5	+2.5	0.0	-10.9	-33.3	-71.3	-130.6	-335.2	-680.0
Wind Drift • Inches	0.0	1.3	5.7	13.8	26.5	44.4	68.1	130.6	208.4

Cor-Bon 130-grain Glaser Blue (06000/na)

G1 Ballistic Coefficient = 0.200

Distance • Yards	Muzzle	100	200	300	400	500	600	800	1000
Velocity • fps	3100	2623	2203	1819	1436	1223	1052	815	762
Energy • ft-lbs	2775	1995	1401	955	637	432	320	221	168
Taylor KO Index	17.7	15.0	12.6	10.4	8.5	7.0	6.0	4.7	4.4
Path • Inches	-1.5	+1.8	0.0	-9.1	-28.9	-65.3	-125.9	-343.5	-722.5
Wind Drift • Inches	0.0	1.5	6.4	15.7	30.9	53.2	82.9	159.8	255.4

USA (Winchester) 147-grain Full Metal Jacket (USA3006)

G1 Ballistic Coefficient = 0.421

Distance • Yards	Muzzle	100	200	300	400	500	600	800	1000
Velocity • fps	3020	2794	2579	2374	2178	1992	1815	1497	1243
Energy • ft-lbs	2976	2548	2171	1839	1548	1295	1075	732	504
Taylor KO Index	19.5	18.1	16.7	15.4	14.1	12.9	11.7	9.7	8.0
Path • Inches	-1.5	+1.5	0.0	-6.7	-19.6	-39.9	-69.0	-161.6	-318.9
Wind Drift • Inches	0.0	0.7	2.9	6.8	12.5	20.4	30.7	59.8	102.5

Hornady 150-grain InterBond (81098)

G1 Ballistic Coefficient = 0.415

Distance • Yards	Muzzle	100	200	300	400	500	600	800	1000
Velocity • fps	3080	2848	2627	2417	2216	2025	1845	1519	1256
Energy • ft-lbs	3159	2700	2298	1945	1636	1366	1134	769	526
Taylor KO Index	20.3	18.8	17.3	16.0	14.6	13.4	12.2	10.0	8.3
Path • Inches	-1.5	+1.4	0.0	-6.4	-18.9	-38.4	-66.4	-155.3	-305.2
Wind Drift • Inches	0.0	0.7	2.8	6.7	12.3	20.1	30.3	59.2	101.6

Hornady 150-grain SST (81093)

G1 Ballistic Coefficient = 0.415

Distance • Yards	Muzzle	100	200	300	400	500	600	800	1000
Velocity • fps	3080	2848	2627	2417	2216	2025	1845	1519	1256
Energy • ft-lbs	3159	2700	2298	1945	1636	1366	1134	739	526
Taylor KO Index	20.3	18.8	17.3	16.0	14.6	13.4	12.2	10.0	8.3
Path • Inches	-1.5	+1.4	0.0	-6.4	-18.9	-38.4	-66.4	-155.3	-305.2
Wind Drift • Inches	0.0	0.7	2.8	6.7	12.3	20.1	30.3	59.2	101.6

Nosler 150-grain Partition (60055)

G1 Ballistic Coefficient = 0.393

Distance • Yards	Muzzle	100	200	300	400	500	600	800	1000
Velocity • fps	3000	2759	2531	2314	2109	1913	1729	1407	1165
Energy • ft-lbs	2997	2535	2133	1783	1481	1219	966	660	452
Taylor KO Index	19.8	18.2	16.7	15.3	13.9	12.6	11.4	9.3	7.7
Path • Inches	-1.5	+1.5	0.0	-7.0	-20.5	-41.8	-72.7	-172.1	-342.5
Wind Drift • Inches	0.0	0.8	3.1	7.4	13.7	22.4	33.8	66.4	114.1

Fiocchi 150-grain SST Polymer Tip BT (3006HSA)

G1 Ballistic Coefficient = 0.415

Distance • Yards	Muzzle	100	200	300	400	500	600	800	1000
Velocity • fps	2925	2700	2486	2283	2088	1904	1731	1424	1187
Energy • ft-lbs	2849	2428	2059	1735	1453	1208	999	676	470
Taylor KO Index	19.3	17.8	16.6	15.1	13.8	12.6	11.4	9.4	7.8
Path • Inches	-1.5	+1.6	0.0	-7.3	-21.2	-43.2	-74.7	-175.3	-345.0
Wind Drift • Inches	0.0	0.7	3.1	7.2	13.3	21.7	32.7	64.0	109.4

Winchester 150-grain XP3 (SXP3006A)

G1 Ballistic Coefficient = 0.439

Distance • Yards	Muzzle	100	200	300	400	500	600	800	1000
Velocity • fps	2925	2712	2508	2313	2127	1950	1783	1481	1239
Energy • ft-lbs	2849	2448	2095	1782	1507	1266	1059	731	511
Taylor KO Index	19.3	17.9	16.6	15.3	14.0	12.9	11.8	9.8	8.2
Path • Inches	-1.5	+1.6	0.0	-7.2	-20.8	-42.2	-72.7	-168.9	-329.2
Wind Drift • Inches	0.0	0.7	2.9	6.8	12.5	20.4	30.7	59.6	101.7

Winchester 150-grain Power-Point (X30061)

G1 Ballistic Coefficient = 0.294

Distance • Yards	Muzzle	100	200	300	400	500	600	800	1000
Velocity • fps	2920	2607	2314	2041	1788	1558	1358	1079	937
Energy • ft-lbs	2839	2263	1783	1387	1064	808	615	388	292
Taylor KO Index	19.3	17.2	15.3	13.5	11.8	10.3	9.0	7.1	6.2
Path • Inches	-1.5	+1.8	0.0	-8.3	-25.1	-52.8	-94.9	-241.2	-506.4
Wind Drift • Inches	0.0	1.1	4.5	10.7	20.3	33.8	52.1	104.0	173.6

Federal 150-grain Nosler AccuBond (P3006A2)

G1 Ballistic Coefficient = 0.433

Distance • Yards	Muzzle	100	200	300	400	500	600	800	1000
Velocity • fps	2910	2700	2490	2300	2110	1930	1762	1462	1223
Energy • ft-lbs	2820	2420	2070	1760	1485	1245	1035	712	498
Taylor KO Index	19.2	17.8	16.4	15.2	13.9	12.7	11.6	9.6	8.1
Path • Inches	-1.5	+1.6	0.0	-7.2	-21.1	-42.6	-73.9	-172.1	-335.9
Wind Drift • Inches	0.0	0.7	3.0	6.9	12.8	20.8	31.3	60.8	103.8

Federal 150-grain Sierra GameKing BTSP (P3006G)

G1 Ballistic Coefficient = 0.379

Distance • Yards	Muzzle	100	200	300	400	500	600	800	1000
Velocity • fps	2910	2670	2440	2220	2010	1810	1628	1316	1101
Energy • ft-lbs	2820	2370	1975	1635	1345	1095	883	577	404
Taylor KO Index	19.2	17.6	16.1	14.7	13.3	11.9	10.7	8.7	7.3
Path • Inches	-1.5	+1.7	0.0	-7.6	-22.3	-45.5	-79.7	-190.4	-382.0
Wind Drift • Inches	0.0	0.8	3.4	8.0	14.9	24.5	37.2	73.2	125.2

Federal 150-grain Soft Point (3006A)

G1 Ballistic Coefficient = 0.314

Distance • Yards	Muzzle	100	200	300	400	500	600	800	1000
Velocity • fps	2910	2620	2340	2080	1840	1620	1423	1128	971
Energy • ft-lbs	2820	2280	1825	1445	1130	875	675	424	314
Taylor KO Index	19.2	17.3	15.4	13.7	12.1	10.7	9.4	7.4	6.4
Path • Inches	-1.5	+1.8	0.0	-8.2	-24.4	-50.8	-91.2	-230.1	-478.8
Wind Drift • Inches	0.0	1.0	4.2	10.0	18.8	31.2	47.9	95.6	160.9

Fiocchi 150-grain PSP (3006B)

G1 Ballistic Coefficient = 0.380

Distance • Yards	Muzzle	100	200	300	400	500	600	800	1000
Velocity • fps	2910	2667	2438	2217	2009	1814	1633	1322	1105
Energy • ft-lbs	2820	2368	1976	1637	1345	1096	889	582	407
Taylor KO Index	19.1	17.6	16.1	14.6	13.3	12.0	10.8	8.7	7.3
Path • Inches	-1.5	+1.7	0.0	-7.6	-22.2	-45.6	-79.4	-189.6	-379.9
Wind Drift • Inches	0.0	0.8	3.4	8.0	14.8	24.3	36.9	72.6	124.3

Hornady 150-grain SST InterLock (8109)

G1 Ballistic Coefficient = 0.416

Distance • Yards	Muzzle	100	200	300	400	500	600	800	1000
Velocity • fps	2910	2686	2473	2270	2077	1893	1722	1417	1182
Energy • ft-lbs	2820	2431	2037	1716	1436	1193	988	669	465
Taylor KO Index	19.2	17.7	16.3	15.0	13.7	12.5	11.4	9.4	7.8
Path • Inches	-1.5	+1.7	0.0	-7.4	-21.5	-43.7	-75.6	-177.9	-352.4
Wind Drift • Inches	0.0	0.7	3.1	7.2	13.4	21.8	32.9	64.4	110.0

Hornady 150-grain InterBond (81099)

G1 Ballistic Coefficient = 0.416

Distance • Yards	Muzzle	100	200	300	400	500	600	800	1000
Velocity • fps	2910	2686	2473	2270	2077	1893	1722	1417	1182
Energy • ft-lbs	2820	24.3	2037	1716	1436	1193	988	669	465
Taylor KO Index	19.2	17.7	16.3	15.0	13.7	12.5	11.4	9.4	7.8
Path • Inches	-1.5	+1.7	0.0	-7.4	-21.5	-43.6	-75.6	-177.9	-352.4
Wind Drift • Inches	0.0	1.0	4.2	10.0	18.8	31.2	47.9	95.6	160.9

Hornady 150-grain SST (8109)

G1 Ballistic Coefficient = 0.415

Distance • Yards	Muzzle	100	200	300	400	500	600	800	1000
Velocity • fps	2910	2686	2473	2270	2077	1893	1721	1415	1181
Energy • ft-lbs	2820	2403	2037	1716	1436	1193	986	667	465
Taylor KO Index	19.2	17.7	16.3	15.0	13.7	12.5	11.4	9.3	7.8
Path • Inches	-1.5	+1.7	0.0	-7.4	-21.5	-43.7	-75.6	-177.4	-349.2
Wind Drift • Inches	0.0	0.7	3.1	7.2	13.4	21.9	33.0	64.5	110.2

Hornady 150-grain BTSP InterLock (8111)

G1 Ballistic Coefficient = 0.349

Distance • Yards	Muzzle	100	200	300	400	500	600	800	1000
Velocity • fps	2910	2645	2395	2159	1937	1729	1539	1228	1037
Energy • ft-lbs	2820	2330	1911	1553	1249	996	789	503	359
Taylor KO Index	19.2	17.4	15.7	14.1	12.7	11.3	10.1	8.0	6.8
Path • Inches	-1.5	+1.7	0.0	-7.8	-23.1	-46.7	-84.3	-206.7	-424.6
Wind Drift • Inches	0.0	0.9	3.7	8.8	16.5	27.2	41.5	82.3	140.3

Hornady 150-grain SP InterLock (8110)

G1 Ballistic Coefficient = 0.338

Distance • Yards	Muzzle	100	200	300	400	500	600	800	1000
Velocity • fps	2910	2637	2380	2137	1909	1697	1504	1196	1016
Energy • ft-lbs	2820	2315	1886	1521	1213	959	754	477	344
Taylor KO Index	19.2	17.4	15.0	14.1	12.6	11.2	9.9	7.9	6.7
Path • Inches	-1.5	+1.8	0.0	-7.9	-23.5	-48.7	-86.2	-213.2	-440.2
Wind Drift • Inches	0.0	0.9	3.9	9.1	17.2	28.4	43.3	86.1	146.4

PMC 150-grain BTSP (3006HIA)

G1 Ballistic Coefficient = 0.349

Distance • Yards	Muzzle	100	200	300	400	500	600	800	1000
Velocity • fps	2910	2645	2395	2159	1937	1729	1540	1230	1039
Energy • ft-lbs	2820	2330	1911	1553	1249	996	790	504	360
Taylor KO Index	19.2	17.5	15.8	14.2	12.8	11.4	10.2	8.1	6.9
Path • Inches	-1.5	+1.7	0.0	-7.8	-23.1	-47.8	-84.0	-204.6	-417.4
Wind Drift • Inches	0.0	0.9	3.7	8.8	16.5	27.2	41.4	82.2	140.1

Remington 150-grain Core-Lokt Ultra Bonded (PRC3006A)

G1 Ballistic Coefficient = 0.331

Distance • Yards	Muzzle	100	200	300	400	500	600	800	1000
Velocity • fps	2910	2631	2368	2121	1889	1674	1481	1176	1003
Energy • ft-lbs	2820	2331	1868	1498	1188	933	731	461	335
Taylor KO Index	19.2	17.4	15.6	14.0	12.5	11.0	9.5	7.8	6.6
Path • Inches	-1.5	+1.8	0.0	-8.0	-23.8	-49.3	-87.7	-217.8	-450.8
Wind Drift • Inches	0.0	0.9	4.0	9.4	17.6	29.1	44.5	88.7	150.4

Remington 150-grain Core-Lokt Pointed Soft Point (R30062)

G1 Ballistic Coefficient = 0.315

Distance • Yards	Muzzle	100	200	300	400	500	600	800	1000
Velocity • fps	2910	2617	2342	2083	1843	1622	1427	1131	972
Energy • ft-lbs	2820	2281	1827	1445	1131	876	678	426	315
Taylor KO Index	19.2	17.3	15.5	13.7	12.2	10.7	9.4	7.5	6.4
Path • Inches	-1.5	+1.8	0.0	-8.2	-24.4	-50.9	-91.0	-229.3	-477.1
Wind Drift • Inches	0.0	1.0	4.2	9.9	18.7	31.1	47.7	95.1	160.2

Winchester 150-grain Silvertip (X30063)
G1 Ballistic Coefficient = 0.314

Distance • Yards	Muzzle	100	200	300	400	500	600	800	1000
Velocity • fps	2910	2617	2342	2083	1843	1622	1423	1128	971
Energy • ft-lbs	2820	2281	1827	1445	1131	876	675	424	314
Taylor KO Index	19.2	17.3	15.5	13.7	12.2	10.7	9.4	7.5	6.4
Path • Inches	-1.5	+2.1	0.0	-8.5	-25.0	-51.8	-92.1	-230.1	-478.8
Wind Drift • Inches	0.0	1.0	4.2	10.0	18.8	31.2	47.9	95.6	160.9

Federal 150-grain Fusion (F3006FS1)
G1 Ballistic Coefficient = 0.408

Distance • Yards	Muzzle	100	200	300	400	500	600	800	1000
Velocity • fps	2900	2670	2460	2250	2060	1870	1696	1391	1161
Energy • ft-lbs	2800	2380	2015	1695	1410	1170	959	644	449
Taylor KO Index	19.1	17.6	16.2	14.9	13.6	12.3	11.2	9.2	7.7
Path • Inches	-1.5	+1.7	0.0	-7.5	-21.8	-44.1	-77.0	-181.1	-357.8
Wind Drift • Inches	0.0	0.8	3.2	7.4	13.7	22.5	33.9	66.4	113.5

Sellier & Bellot 150-grain SPCE (V331602U)
G1 Ballistic Coefficient = 0.324

Distance • Yards	Muzzle	100	200	300	400	500	600	800	1000
Velocity • fps	2887	2604	2338	2087	1853	1638	1445	1149	986
Energy • ft-lbs	2777	2259	1821	1452	1145	894	696	440	324
Taylor KO Index	19.1	17.2	15.4	13.8	12.2	10.8	9.5	7.6	6.5
Path • Inches	-1.5	+1.8	0.0	-8.2	-24.4	-50.8	-90.3	-223.8	-462.6
Wind Drift • Inches	0.0	1.0	4.1	9.7	18.3	30.3	46.3	92.4	155.7

Hornady 150-grain GMX (8112) LF
G1 Ballistic Coefficient = 0.415

Distance • Yards	Muzzle	100	200	300	400	500	600	800	1000
Velocity • fps	3080	2848	2627	2417	2216	2025	1845	1519	1256
Energy • ft-lbs	3159	2700	2298	1945	1636	1366	1134	739	526
Taylor KO Index	20.3	18.8	17.3	16.0	14.6	13.4	12.2	10.0	8.3
Path • Inches	-1.5	+1.4	0.0	-6.4	-18.9	-38.4	-66.4	-155.3	-305.2
Wind Drift • Inches	0.0	0.7	2.8	6.7	12.3	20.1	30.3	59.2	101.6

Norma 150-grain Nosler Ballistic Tip (17654)
G1 Ballistic Coefficient = 0.435

Distance • Yards	Muzzle	100	200	300	400	500	600	800	1000
Velocity • fps	2936	2721	2516	2320	2133	1955	1786	1483	1239
Energy • ft-lbs	2872	2467	2109	1794	1516	1273	1063	733	512
Taylor KO Index	19.4	17.9	16.6	15.3	14.1	12.9	11.8	9.8	8.2
Path • Inches	-1.5	+1.6	0.0	-7.1	-20.7	-41.9	-72.3	-168.0	-327.6
Wind Drift • Inches	0.0	0.7	2.9	6.8	12.5	20.4	30.7	59.7	101.8

Winchester 150-grain Power Core 95/5 (X30066LF) LF
G1 Ballistic Coefficient = 0.344

Distance • Yards	Muzzle	100	200	300	400	500	600	800	1000
Velocity • fps	2920	2651	2397	2158	1932	1722	1531	1220	1032
Energy • ft-lbs	2839	2340	1914	1550	1243	988	781	496	354
Taylor KO Index	19.3	17.5	15.8	14.2	12.8	11.4	10.1	8.1	6.8
Path • Inches	-1.5	+1.7	0.0	-7.8	-23.1	-47.8	-84.2	-205.7	-420.8
Wind Drift • Inches	0.0	0.9	3.8	8.9	16.7	27.5	42.0	83.5	142.2

Winchester 150-grain Power Max Bonded (X30061BP)
G1 Ballistic Coefficient = 0.325

Distance • Yards	Muzzle	100	200	300	400	500	600	800	1000
Velocity • fps	2920	2636	2368	2117	1882	1664	1469	1165	995
Energy • ft-lbs	2836	2313	1868	1492	1179	922	719	452	330
Taylor KO Index	19.3	17.4	15.6	14.0	12.4	11.0	9.7	7.7	6.6
Path • Inches	-1.5	+1.8	0.0	-8.0	-23.8	-49.4	-87.7	-217.2	-449.2
Wind Drift • Inches	0.0	1.0	4.0	9.5	17.9	29.6	45.4	90.4	153.2

Federal 150-grain Nosler Ballistic Tip (P3006P)
G1 Ballistic Coefficient = 0.433

Distance • Yards	Muzzle	100	200	300	400	500	600	800	1000
Velocity • fps	2910	2700	2490	2300	2110	1930	1762	1462	1223
Energy • ft-lbs	2820	2420	2070	1760	1485	1245	1035	712	498
Taylor KO Index	19.2	17.8	16.4	15.2	13.9	12.7	11.6	9.6	8.1
Path • Inches	-1.5	+1.6	0.0	-7.2	-21.1	-42.6	-73.9	-172.1	-335.9
Wind Drift • Inches	0.0	0.7	3.0	6.9	12.8	20.8	31.3	60.8	103.8

Remington 150-grain Swift Scirocco Bonded (PRSC3006C)
G1 Ballistic Coefficient = 0.43

Distance • Yards	Muzzle	100	200	300	400	500	600	800	1000
Velocity • fps	2910	2696	2492	2298	2111	1934	1766	1466	1226
Energy • ft-lbs	2820	2421	2069	1758	1485	1246	1039	716	501
Taylor KO Index	19.2	17.8	16.4	15.2	13.9	12.8	11.7	9.7	8.1
Path • Inches	-1.5	+1.6	0.0	-7.3	-21.1	-42.3	-73.8	-172.2	-337.8
Wind Drift • Inches	0.0	0.7	2.9	6.9	12.7	20.7	31.1	60.5	103.3

Remington 150-grain AccuTip Boat Tail (PRA3006A)
G1 Ballistic Coefficient = 0.41

Distance • Yards	Muzzle	100	200	300	400	500	600	800	1000
Velocity • fps	2910	2686	2473	2270	2077	1893	1720	1414	1179
Energy • ft-lbs	2820	2403	2037	1716	1436	1193	985	666	463
Taylor KO Index	19.2	17.7	15.9	16.3	15.0	13.7	12.5	11.4	9.3
Path • Inches	-1.5	+1.7	0.0	-7.4	-21.5	-43.7	-75.7	-178.2	-353.3
Wind Drift • Inches	0.0	0.7	3.1	7.2	13.4	21.9	33.0	64.6	110.4

Remington 150-grain Bronze Point (R30063)
G1 Ballistic Coefficient = 0.36

Distance • Yards	Muzzle	100	200	300	400	500	600	800	1000
Velocity • fps	2910	2656	2416	2189	1974	1773	1587	1275	1070
Energy • ft-lbs	2820	2349	1944	1596	1298	1047	839	542	381
Taylor KO Index	19.2	17.5	15.9	14.4	13.0	11.7	10.5	8.4	7.1
Path • Inches	-1.5	+1.7	0.0	-7.7	-22.7	-46.6	-81.8	-198.3	-403.9
Wind Drift • Inches	0.0	0.8	3.6	8.4	15.6	25.7	39.1	77.3	132.0

Winchester 150-grain E-Tip (81093) LF
G1 Ballistic Coefficient = 0.46

Distance • Yards	Muzzle	100	200	300	400	500	600	800	1000
Velocity • fps	2900	2702	2512	2330	2156	1989	1831	1542	1301
Energy • ft-lbs	2801	2431	2102	1809	1548	1318	1117	792	564
Taylor KO Index	19.1	17.8	16.6	15.4	14.2	13.1	12.1	10.2	8.6
Path • Inches	-1.5	+1.6	0.0	-7.2	-20.7	-41.8	-71.6	-164.5	-316.4
Wind Drift • Inches	0.0	0.7	2.7	6.3	11.7	19.0	28.4	54.9	93.2

Winchester 150-grain Ballistic Silvertip (SBST3006)
G1 Ballistic Coefficient = 0.43

Distance • Yards	Muzzle	100	200	300	400	500	600	800	1000
Velocity • fps	2900	2687	2483	2289	2103	1926	1759	1459	1221
Energy • ft-lbs	2801	2404	2054	1745	1473	1236	1030	710	497
Taylor KO Index	19.1	17.7	16.4	15.1	13.9	12.7	11.6	9.6	8.1
Path • Inches	-1.5	+1.7	0.0	-7.3	-21.2	-43.0	-74.4	-173.5	-340.6
Wind Drift • Inches	0.0	0.7	3.0	6.9	12.8	20.8	31.3	60.9	103.8

Fiocchi 150-grain FMJ BT (3006A)
G1 Ballistic Coefficient = 0.39

Distance • Yards	Muzzle	100	200	300	400	500	600	800	1000
Velocity • fps	3000	2762	2536	2322	2117	1924	1743	1422	1177
Energy • ft-lbs	2970	2540	2142	1795	1493	1233	1012	673	462
Taylor KO Index	19.8	18.2	16.7	15.3	14.0	12.7	11.5	9.4	7.8
Path • Inches	-1.5	+1.5	0.0	-7.0	-20.4	-41.6	-72.2	-170.4	-338.2
Wind Drift • Inches	0.0	0.7	3.1	7.2	13.5	22.0	33.2	65.3	112.1

American Eagle (Federal) 150-grain Full Metal Jacket Boat-Tail (AE3006N)
G1 Ballistic Coefficient = 0.40

Distance • Yards	Muzzle	100	200	300	400	500	600	800	1000
Velocity • fps	2910	2680	2470	2260	2060	1880	1706	1399	1167
Energy • ft-lbs	2820	2395	2025	1700	1420	1175	970	652	454
Taylor KO Index	19.2	17.7	16.3	14.9	13.6	12.4	11.3	9.2	7.7
Path • Inches	-1.5	+1.7	0.0	-7.4	-21.7	-43.8	-76.2	-179.3	-354.1
Wind Drift • Inches	0.0	0.8	3.1	7.3	13.6	22.3	33.6	65.8	112.5

Remington 150-grain Copper Solid Tipped (PSC3006A) LF
G1 Ballistic Coefficient = 0.40

Distance • Yards	Muzzle	100	200	300	400	500	600	800	1000
Velocity • fps	2910	2678	2458	2248	2048	1860	1683	1375	1147
Energy • ft-lbs	2820	2388	2012	1683	1397	1152	944	630	438
Taylor KO Index	19.2	17.7	16.2	14.8	13.5	12.3	11.1	9.1	7.6
Path • Inches	-1.5	+1.7	0.0	-7.4	-21.8	-44.5	-77.2	-182.4	-361.8
Wind Drift • Inches	0.0	0.8	3.2	7.8	14.0	22.9	34.6	67.9	116.1

UMC (Remington) 150-grain Metal Case (L30062)

G1 Ballistic Coefficient = 0.315

Distance • Yards	Muzzle	100	200	300	400	500	600	800	1000
Velocity • fps	2910	2617	2342	2085	1842	1623	1427	1131	972
Energy • ft-lbs	2820	2281	1827	1448	1133	878	678	426	315
Taylor KO Index	19.2	17.2	15.4	13.8	12.2	10.7	9.4	7.5	6.5
Path • Inches	-1.5	+1.8	0.0	-8.2	-24.4	-50.9	-91.0	-229.3	-477.1
Wind Drift • Inches	0.0	1.0	4.2	9.9	18.7	31.1	47.7	95.1	160.2

Norma 150-grain Full Metal Jacket (17563)

G1 Ballistic Coefficient = 0.483

Distance • Yards	Muzzle	100	200	300	400	500	600	800	1000
Velocity • fps	2789	2601	2421	2248	2082	1924	1773	1499	1273
Energy • ft-lbs	2591	2254	1953	1684	1444	1233	1047	749	540
Taylor KO Index	18.4	17.2	16.0	14.8	13.7	12.7	11.7	9.9	8.4
Path • Inches	-1.5	+1.8	0.0	-7.8	-22.4	-45.1	-77.2	-176.7	-338.6
Wind Drift • Inches	0.0	0.7	2.8	6.5	12.0	19.4	29.1	56.0	94.8

Norma 150-grain Full Jacket (17651)

G1 Ballistic Coefficient = 0.423

Distance • Yards	Muzzle	100	200	300	400	500	600	800	1000
Velocity • fps	2772	2557	2353	2158	1973	1797	1636	1351	1139
Energy • ft-lbs	2492	2121	1796	1511	1262	1048	892	608	432
Taylor KO Index	18.3	16.9	15.5	14.2	13.0	11.9	10.8	8.9	7.5
Path • Inches	-1.5	1.9	0.0	-8.2	-23.9	-48.5	-83.9	-197.3	-389.9
Wind Drift • Inches	0.0	0.8	3.3	7.7	14.2	23.2	34.8	67.9	115.3

Federal 150-grain Full Metal Jacket (AE3006M1)

G1 Ballistic Coefficient = 0.410

Distance • Yards	Muzzle	100	200	300	400	500	600	800	1000
Velocity • fps	2740	2520	2310	2120	1930	1750	1687	1305	1106
Energy • ft-lbs	2500	2120	1785	1490	1240	1020	839	568	408
Taylor KO Index	18.1	16.6	15.2	14.0	12.7	11.6	11.1	8.6	7.3
Path • Inches	-1.5	+2.0	0.0	-8.6	-24.7	-50.5	-87.5	-206.1	-406.9
Wind Drift • Inches	0.0	0.8	3.4	8.0	14.9	24.3	36.7	71.8	121.8

Black Hills 155-grain Hornady A-MAX (1C3006BHGN4)

G1 Ballistic Coefficient = 0.435

Distance • Yards	Muzzle	100	200	300	400	500	600	800	1000
Velocity • fps	2900	2687	2483	2289	2103	1927	1760	1461	1223
Energy • ft-lbs	2895	2485	2123	1804	1523	1278	1066	734	515
Taylor KO Index	19.8	18.3	16.9	15.6	14.3	13.1	12.0	10.0	8.3
Path • Inches	-1.5	+1.7	0.0	-7.3	-21.3	-43.1	-74.4	-172.9	-337.3
Wind Drift • Inches	0.0	0.7	2.9	6.9	12.7	20.8	31.2	60.8	103.7

Hornady 165-grain InterBond (81158)

G1 Ballistic Coefficient = 0.447

Distance • Yards	Muzzle	100	200	300	400	500	600	800	1000
Velocity • fps	2960	2750	2599	2357	2173	1997	1831	1529	1280
Energy • ft-lbs	3209	2769	2380	2034	1729	1461	1228	856	600
Taylor KO Index	21.5	20.0	18.9	17.1	15.8	14.5	13.3	11.1	9.3
Path • Inches	-1.5	+1.6	0.0	-6.9	-20.1	-40.7	-70.0	-161.8	-313.8
Wind Drift • Inches	0.0	0.7	2.8	6.5	12.0	19.5	29.3	56.8	96.8

Hornady 165-grain SST (81153)

G1 Ballistic Coefficient = 0.447

Distance • Yards	Muzzle	100	200	300	400	500	600	800	1000
Velocity • fps	2960	2750	2549	2357	2173	1997	1831	1529	1280
Energy • ft-lbs	3209	2769	2380	2034	1729	1461	1228	856	600
Taylor KO Index	21.5	20.0	18.5	17.1	15.8	14.5	13.3	11.1	9.3
Path • Inches	-1.5	+1.6	0.0	-6.9	-20.1	-40.7	-70.0	-161.8	-313.8
Wind Drift • Inches	0.0	0.7	2.8	6.5	12.0	19.5	29.3	56.8	96.8

Federal 165-grain Nosler Partition (P3006AD)

G1 Ballistic Coefficient = 0.410

Distance • Yards	Muzzle	100	200	300	400	500	600	800	1000
Velocity • fps	2830	2310	2400	2190	2000	1820	1651	1356	1138
Energy • ft-lbs	2935	2490	2100	1760	1465	1210	999	559	475
Taylor KO Index	20.5	18.9	17.4	15.9	14.5	13.2	12.0	9.8	8.3
Path • Inches	-1.5	+1.8	0.0	-7.9	-23.0	-46.8	-81.2	-191.1	-377.3
Wind Drift • Inches	0.0	0.8	3.2	7.6	14.2	23.2	35.0	68.4	116.5

Federal 165-grain Sierra GameKing BTSP (P3006D)

G1 Ballistic Coefficient = 0.402

Distance • Yards	Muzzle	100	200	300	400	500	600	800	1000
Velocity • fps	2800	2580	2360	2160	1970	1780	1610	1318	1111
Energy • ft-lbs	2870	2430	2045	1710	1415	1165	950	637	453
Taylor KO Index	20.3	18.7	17.1	15.7	14.3	12.9	11.7	9.6	8.1
Path • Inches	-1.5	+1.9	0.0	-8.2	-23.7	-48.3	-84.2	-198.9	-394.2
Wind Drift • Inches	0.0	0.8	3.4	7.9	14.7	24.1	36.5	71.4	121.4

Fiocchi 165-grain PSP (3006C)

G1 Ballistic Coefficient = 0.404

Distance • Yards	Muzzle	100	200	300	400	500	600	800	1000
Velocity • fps	2800	2576	2363	2161	1968	1786	1615	1323	1115
Energy • ft-lbs	2872	2432	2046	1710	1419	1169	956	642	456
Taylor KO Index	20.3	18.7	17.2	15.7	14.3	13.0	11.7	9.6	8.1
Path • Inches	-1.5	+1.9	0.0	-8.1	-23.7	-48.3	-83.9	-198.1	-392.3
Wind Drift • Inches	0.0	0.8	3.4	7.9	14.6	24.0	36.2	70.9	120.6

Hornady 165-grain SST InterLock (81154

G1 Ballistic Coefficient = 0.447

Distance • Yards	Muzzle	100	200	300	400	500	600	800	1000
Velocity • fps	2800	2597	2403	2217	2039	1870	1711	1427	1204
Energy • ft-lbs	2872	2470	2115	1800	1523	1281	1073	747	531
Taylor KO Index	20.3	18.9	17.4	16.1	14.8	13.6	12.4	10.4	8.7
Path • Inches	-1.5	+1.8	0.0	-7.9	-22.8	-46.2	-79.4	-184.1	-357.6
Wind Drift • Inches	0.0	0.7	3.0	7.0	13.0	21.2	31.8	61.8	104.9

Hornady 165-grain InterBond (81159)

G1 Ballistic Coefficient = 0.446

Distance • Yards	Muzzle	100	200	300	400	500	600	800	1000
Velocity • fps	2800	2597	2403	2217	2039	1870	1709	1425	1201
Energy • ft-lbs	2872	2470	2115	1800	1523	1281	1071	744	529
Taylor KO Index	20.3	18.9	17.4	16.1	14.8	13.6	12.4	10.3	8.7
Path • Inches	-1.5	+1.8	0.0	-7.9	-22.8	-46.2	-79.5	-184.4	-358.3
Wind Drift • Inches	0.0	0.7	3.0	7.0	13.0	21.2	31.9	62.0	105.3

Hornady 165-grain BTSP Interlock (8115)

G1 Ballistic Coefficient = 0.435

Distance • Yards	Muzzle	100	200	300	400	500	600	800	1000
Velocity • fps	2800	2591	2392	2202	2020	1848	1685	1398	1177
Energy • ft-lbs	2873	2460	2097	1777	1495	1252	1041	716	508
Taylor KO Index	20.3	18.8	17.4	16.0	14.7	13.4	12.2	10.1	8.5
Path • Inches	-1.5	+1.8	0.0	-8.0	-23.3	-47.0	-80.6	-188.4	-370.0
Wind Drift • Inches	0.0	0.8	3.1	7.3	13.4	21.9	33.0	64.2	109.1

Nosler 165-grain AccuBond (23422) + (60057)

G1 Ballistic Coefficient = 0.475

Distance • Yards	Muzzle	100	200	300	400	500	600	800	1000
Velocity • fps	2800	2609	2425	2249	2080	1919	1766	1490	1262
Energy • ft-lbs	2873	2494	2155	1854	1586	1350	1143	813	584
Taylor KO Index	20.3	18.9	17.6	16.3	15.1	13.9	12.8	10.8	9.2
Path • Inches	-1.5	+1.8	0.0	-7.7	-22.3	-45.0	-77.1	-176.9	-340.0
Wind Drift • Inches	0.0	0.7	2.8	6.6	12.1	19.7	29.5	57.0	96.5

Nosler 165-grain Partition (23420)

G1 Ballistic Coefficient = 0.410

Distance • Yards	Muzzle	100	200	300	400	500	600	800	1000
Velocity • fps	2800	2579	2368	2168	1977	1797	1629	1339	1127
Energy • ft-lbs	2873	2438	2056	1723	1433	1184	973	657	461
Taylor KO Index	20.3	18.7	17.2	15.7	14.4	13.0	11.8	9.7	8.2
Path • Inches	-1.5	+1.9	0.0	-8.1	-23.6	-48.0	-83.2	-195.9	-386.8
Wind Drift • Inches	0.0	0.8	3.3	7.7	14.4	23.5	35.6	69.5	118.2

Remington 165-grain Core-Lokt Pointed Soft Point (R3006B)

G1 Ballistic Coefficient = 0.33

Distance • Yards	Muzzle	100	200	300	400	500	600	800	1000
Velocity • fps	2800	2534	2283	2047	1825	1621	1437	1154	993
Energy • ft-lbs	2872	2352	1909	1534	1220	963	757	488	362
Taylor KO Index	20.3	18.4	16.6	14.9	13.2	11.8	10.4	8.4	7.2
Path • Inches	-1.5	+2.0	0.0	-8.7	-25.9	-53.2	-94.3	-233.1	-477.9
Wind Drift • Inches	0.0	1.0	4.1	9.6	18.1	30.0	45.7	90.5	152.1

Winchester 165-grain Pointed Soft Point (X30065)

G1 Ballistic Coefficient = 0.341

Distance • Yards	Muzzle	100	200	300	400	500	600	800	1000
Velocity • fps	2800	2536	2286	2051	1831	1627	1445	1160	998
Energy • ft-lbs	2867	2355	1915	1514	1228	970	765	493	365
Taylor KO Index	20.3	18.4	16.6	14.9	13.3	11.8	10.5	8.4	7.2
Path • Inches	-1.5	+2.0	0.0	-8.6	-25.6	-53.0	-93.5	-229.1	-467.8
Wind Drift • Inches	0.0	1.0	4.0	9.6	18.0	29.7	45.3	89.6	150.8

Federal 165-grain Fusion (F3006FS2)

G1 Ballistic Coefficient = 0.456

Distance • Yards	Muzzle	100	200	300	400	500	600	800	1000
Velocity • fps	2790	2590	2400	2200	2040	1880	1722	1442	1218
Energy • ft-lbs	2850	2460	2110	1800	1525	1285	1087	762	543
Taylor KO Index	20.3	18.8	17.4	16.0	14.8	13.6	12.5	10.5	8.8
Path • Inches	-1.5	+1.8	0.0	-7.9	-22.9	-46.0	-79.3	-183.2	-354.6
Wind Drift • Inches	0.0	0.7	3.0	6.9	12.8	20.8	31.2	60.5	102.6

Hornady 165-grain GMX (8116) LF

G1 Ballistic Coefficient = 0.447

Distance • Yards	Muzzle	100	200	300	400	500	600	800	1000
Velocity • fps	2940	2730	2531	2339	2156	1981	1816	1516	1270
Energy • ft-lbs	3166	2731	2346	2004	1703	1438	1203	842	591
Taylor KO Index	21.3	19.8	18.4	17.0	15.7	14.4	13.2	11.0	9.2
Path • Inches	-1.5	+1.6	0.0	-7.0	-20.4	-41.3	-71.1	-164.4	-318.9
Wind Drift • Inches	0.0	0.7	2.8	6.6	12.1	19.7	29.6	57.4	97.8

Fiocchi 165-grain Sierra GameKing HPBT (3006GKB)

G1 Ballistic Coefficient = 0.357

Distance • Yards	Muzzle	100	200	300	400	500	600	800	1000
Velocity • fps	2850	2598	2355	2125	1908	1708	1524	1225	1039
Energy • ft-lbs	2976	2473	2032	1654	1334	1069	852	550	396
Taylor KO Index	20.7	18.9	17.1	15.4	13.9	12.4	11.1	8.9	7.5
Path • Inches	-1.5	+1.8	0.0	-8.1	-24.0	-49.5	-86.9	-210.6	-427.3
Wind Drift • Inches	0.0	0.9	3.7	8.8	16.5	27.3	41.5	82.0	139.3

Federal 165-grain Nosler Ballistic Tip (P3006Q)

G1 Ballistic Coefficient = 0.476

Distance • Yards	Muzzle	100	200	300	400	500	600	800	1000
Velocity • fps	2800	2610	2430	2250	2080	1920	1768	1491	1263
Energy • ft-lbs	2870	2495	2155	1855	1585	1350	1145	814	584
Taylor KO Index	20.3	18.9	17.6	16.3	15.1	13.9	12.8	10.8	9.2
Path • Inches	-1.5	+1.8	0.0	-7.7	-22.3	-45.0	-77.1	-177.2	-341.8
Wind Drift • Inches	0.0	0.7	2.8	6.6	12.1	19.7	29.5	56.9	96.3

Federal 165-grain Trophy Bonded Tip (P3006TT2)

G1 Ballistic Coefficient = 0.452

Distance • Yards	Muzzle	100	200	300	400	500	600	800	1000
Velocity • fps	2800	2600	2410	2220	2040	1880	1722	1439	1214
Energy • ft-lbs	2870	2475	2120	1805	1530	1290	1086	759	540
Taylor KO Index	20.3	18.9	17.5	16.1	14.8	13.6	12.5	10.4	8.8
Path • Inches	-1.5	+1.8	0.0	-7.9	-22.8	-45.8	-79.0	-182.7	-354.2
Wind Drift • Inches	0.0	0.7	3.0	6.9	12.8	20.9	31.4	60.9	103.3

Federal 165-grain Barnes Triple-Shock X-Bullet (P3006AF) LF

G1 Ballistic Coefficient = 0.380

Distance • Yards	Muzzle	100	200	300	400	500	600	800	1000
Velocity • fps	2800	2560	2340	2120	1920	1730	1555	1262	1069
Energy • ft-lbs	2870	2405	2000	1650	1350	1095	886	584	419
Taylor KO Index	20.3	18.6	17.0	15.4	13.9	12.6	11.3	9.2	7.8
Path • Inches	-1.5	+1.9	0.0	-8.4	-24.2	-49.9	-87.0	-208.1	-416.9
Wind Drift • Inches	0.0	0.9	3.6	8.4	15.8	25.9	39.2	77.2	131.0

Remington 165-grain AccuTip Boat Tail (PRA3006B)

G1 Ballistic Coefficient = 0.447

Distance • Yards	Muzzle	100	200	300	400	500	600	800	1000
Velocity • fps	2800	2597	2403	2217	2039	1870	1711	1427	1204
Energy • ft-lbs	2872	2470	2115	1800	1523	1281	1073	747	531
Taylor KO Index	20.3	18.9	17.4	16.1	14.8	13.6	12.4	10.4	8.7
Path • Inches	-1.5	+1.8	0.0	-7.9	-22.8	-46.2	-79.4	-184.1	-357.6
Wind Drift • Inches	0.0	0.7	3.0	7.0	13.0	21.2	31.8	61.8	104.9

Remington 165-grain Solid Copper Tipped (PSC3006B) LF

G1 Ballistic Coefficient = 0.447

Distance • Yards	Muzzle	100	200	300	400	500	600	800	1000
Velocity • fps	2800	2597	2403	2217	2039	1870	1711	1427	1204
Energy • ft-lbs	2872	2470	2115	1800	1523	1281	1073	747	531
Taylor KO Index	20.3	18.9	17.4	16.1	14.8	13.6	12.4	10.4	8.7
Path • Inches	-1.5	+2.1	0.0	-8.2	-23.4	-47.1	-79.4	-184.1	-357.6
Wind Drift • Inches	0.0	0.7	3.0	7.0	13.0	21.2	31.8	61.8	104.9

Cor-Bon 168-grain T- DPX (DPX3006168/20)

G1 Ballistic Coefficient = 0.450

Distance • Yards	Muzzle	100	200	300	400	500	600	800	1000
Velocity • fps	2900	2694	2497	2309	2128	1955	1792	1497	1257
Energy • ft-lbs	3138	2708	2326	1988	1689	1427	1198	837	590
Taylor KO Index	21.4	19.9	18.5	17.1	15.7	14.5	13.2	11.1	9.3
Path • Inches	-1.5	+1.3	0.0	-7.2	-21.0	-42.5	-73.1	-168.9	-327.4
Wind Drift • Inches	0.0	0.7	2.8	6.6	12.2	19.9	29.9	58.1	98.8

Nosler 168-grain E-Tip (60084) LF

G1 Ballistic Coefficient = 0.510

Distance • Yards	Muzzle	100	200	300	400	500	600	800	1000
Velocity • fps	2800	2622	2451	2286	2128	1975	1828	1561	1334
Energy • ft-lbs	2924	2564	2241	1949	1689	1455	1247	910	664
Taylor KO Index	20.7	19.4	18.1	16.9	15.7	14.6	13.5	11.5	9.9
Path • Inches	-1.5	+1.8	0.0	-7.6	-21.8	-43.8	-74.7	-169.5	-321.8
Wind Drift • Inches	0.0	0.6	2.6	6.1	11.2	18.1	27.4	51.9	87.5

Winchester 168-grain Ballistic Silvertip (SBST3006A)

G1 Ballistic Coefficient = 0.475

Distance • Yards	Muzzle	100	200	300	400	500	600	800	1000
Velocity • fps	2790	2599	2416	2240	2072	1911	1758	1482	1256
Energy • ft-lbs	2093	2520	2177	1872	1601	1362	1153	820	588
Taylor KO Index	20.6	19.2	17.9	16.6	15.3	14.1	13.0	11.0	9.3
Path • Inches	-1.5	+1.8	0.0	-7.8	-22.5	-45.2	-77.8	-178.9	-345.4
Wind Drift • Inches	0.0	0.7	2.8	6.6	12.2	19.8	29.7	57.3	97.1

Sellier & Bellot 168-grain HPBT (V331642U)

G1 Ballistic Coefficient = 0.462

Distance • Yards	Muzzle	100	200	300	400	500	600	800	1000
Velocity • fps	2786	2590	2402	2222	2050	1886	1731	1452	1228
Energy • ft-lbs	2896	2503	2158	1843	1568	1327	1118	787	563
Taylor KO Index	18.4	17.1	15.9	14.7	13.5	12.4	11.4	9.6	8.1
Path • Inches	-1.5	+1.8	0.0	-7.9	-22.8	-46.0	-79.1	-182.2	-351.9
Wind Drift • Inches	0.0	0.7	2.9	6.8	12.6	20.5	30.8	59.6	101.0

Black Hills 168-grain Hornady A-MAX (1C3006BHGN3)

G1 Ballistic Coefficient = 0.475

Distance • Yards	Muzzle	100	200	300	400	500	600	800	1000
Velocity • fps	2750	2561	2379	2205	2038	1879	1729	1457	1237
Energy • ft-lbs	2822	2447	2112	1814	1550	1318	1115	792	571
Taylor KO Index	20.3	18.9	17.6	16.3	15.1	13.9	12.8	10.8	9.1
Path • Inches	-1.5	+1.9	0.0	-8.1	-23.3	-46.9	-80.4	-184.4	-354.6
Wind Drift • Inches	0.0	0.7	2.9	6.7	12.4	20.2	30.3	58.5	99.0

Hornady 168-grain A-MAX Match (M1 Garand) (81170)

G1 Ballistic Coefficient = 0.475

Distance • Yards	Muzzle	100	200	300	400	500	600	800	1000
Velocity • fps	2710	2522	2342	2170	2004	1847	1698	1432	1218
Energy • ft-lbs	2739	2373	2046	1756	1499	1272	1076	765	554
Taylor KO Index	20.0	18.9	17.3	16.0	14.8	13.7	12.6	10.6	9.0
Path • Inches	-1.5	+2.3	0.0	-8.6	-24.7	-49.4	-83.1	-190.8	-366.9
Wind Drift • Inches	0.0	0.7	3.0	6.9	12.7	20.7	31.0	59.9	101.1

Black Hills 168-grain Hornady Match Hollow Point (1C3006BHGN2)

G1 Ballistic Coefficient = 0.450

Distance • Yards	Muzzle	100	200	300	400	500	600	800	1000
Velocity • fps	2700	2503	2313	2133	1960	1796	1643	1372	1165
Energy • ft-lbs	2720	2337	1997	1697	1433	1204	1077	702	506
Taylor KO Index	20.0	18.5	17.1	15.8	14.5	13.3	12.1	10.1	8.6
Path • Inches	-1.5	+2.0	0.0	-8.5	-24.7	-50.0	-86.1	-200.2	-390.8
Wind Drift • Inches	0.0	0.8	3.2	7.4	13.6	22.2	33.4	64.8	109.5

Federal 168-grain Sierra MatchKing BTHP (GM3006M)

G1 Ballistic Coefficient = 0.464

Distance • Yards	Muzzle	100	200	300	400	500	600	800	1000
Velocity • fps	2700	2510	2320	2150	1980	1820	1670	1402	1191
Energy • ft-lbs	2720	2350	2010	1720	1460	1230	1040	733	529
Taylor KO Index	20.0	18.6	17.1	15.9	14.6	13.5	12.3	10.4	8.8
Path • Inches	-1.5	+2.0	0.0	-8.5	-24.5	-49.4	-84.8	-196.0	-380.4
Wind Drift • Inches	0.0	0.7	3.0	7.1	13.2	21.4	32.2	62.2	105.0

Fiocchi 168-grain Sierra MatchKing HPBT (3006MKB)

G1 Ballistic Coefficient = 0.439

Distance • Yards	Muzzle	100	200	300	400	500	600	800	1000
Velocity • fps	2695	2499	2309	2127	1946	1774	1618	1346	1143
Energy • ft-lbs	2708	2330	1989	1688	1413	1174	977	676	489
Taylor KO Index	19.9	18.5	17.1	15.7	14.4	13.1	12.0	9.9	8.4
Path • Inches	-1.5	+2.0	0.0	-8.6	-25.1	-50.8	-87.5	-203.8	-397.2
Wind Drift • Inches	0.0	0.8	3.2	7.6	14.1	22.9	34.5	67.0	113.4

Federal 170-grain Fusion (F3006FSLR1)

G1 Ballistic Coefficient = 0.284

Distance • Yards	Muzzle	100	200	300	400	500	600	800	1000
Velocity • fps	2000	1740	1510	1312	1157	1049	974	867	787
Energy • ft-lbs	1510	1145	860	650	505	415	358	284	234
Taylor KO Index	15.0	13.0	11.3	9.8	8.7	7.8	7.3	6.5	5.9
Path • Inches	-1.5	+5.0	0.0	-20.4	-61.0	-127.5	-225.6	-536.7	-1033
Wind Drift • Inches	0.0	1.9	8.1	19.2	35.8	57.5	83.4	146.0	221.5

Lapua 170-grain Naturalis (N317202)

G1 Ballistic Coefficient = 0.314

Distance • Yards	Muzzle	100	200	300	400	500	600	800	1000
Velocity • fps	2723	2441	2177	1929	1701	1494	1316	1069	938
Energy • ft-lbs	2800	2257	1789	1406	1092	843	654	431	332
Taylor KO Index	20.7	18.3	16.3	14.4	12.7	11.2	9.8	8.0	7.0
Path • Inches	-1.5	+2.2	0.0	-9.6	-28.5	-59.5	-106.2	-265.1	-545.2
Wind Drift • Inches	0.0	1.1	4.6	11.0	20.7	34.5	52.8	103.9	171.2

Hornady 180-grain InterBond (81188)

G1 Ballistic Coefficient = 0.480

Distance • Yards	Muzzle	100	200	300	400	500	600	800	1000
Velocity • fps	2820	2630	2447	2272	2104	1944	1791	1513	1283
Energy • ft-lbs	3178	2764	2393	2063	1769	1509	1282	916	658
Taylor KO Index	22.3	20.8	19.4	18.0	16.7	15.4	14.2	12.0	10.2
Path • Inches	-1.5	+1.8	0.0	-7.6	-21.9	-44.1	-75.5	-172.9	-331.6
Wind Drift • Inches	0.0	0.7	2.8	6.4	11.8	19.2	28.8	55.6	94.1

Hornady 180-grain SST (81183)

G1 Ballistic Coefficient = 0.480

Distance • Yards	Muzzle	100	200	300	400	500	600	800	1000
Velocity • fps	2820	2630	2447	2272	2104	1944	1791	1513	1283
Energy • ft-lbs	3178	2764	2393	2063	1769	1509	1282	916	658
Taylor KO Index	22.3	20.8	19.4	18.0	16.7	15.4	14.2	12.0	10.2
Path • Inches	-1.5	+1.8	0.0	-7.6	-21.9	-44.1	-75.5	-172.9	-331.6
Wind Drift • Inches	0.0	0.7	2.8	6.4	11.8	19.2	28.8	55.6	94.1

Winchester 180-grain XP3 (SPX3006)

G1 Ballistic Coefficient = 0.527

Distance • Yards	Muzzle	100	200	300	400	500	600	800	1000
Velocity • fps	2750	2579	2414	2280	2103	1957	1817	1559	1338
Energy • ft-lbs	3022	2658	2330	2034	1768	1530	1319	972	716
Taylor KO Index	21.8	20.4	19.1	18.1	16.7	15.5	14.4	12.3	10.6
Path • Inches	-1.5	+1.9	0.0	-7.8	-22.5	-45.1	-76.7	-173.4	-327.7
Wind Drift • Inches	0.0	0.6	2.6	6.0	11.0	17.9	26.7	51.1	85.9

Winchester 180-grain AccuBond CT (S3006CT)

G1 Ballistic Coefficient = 0.436

Distance • Yards	Muzzle	100	200	300	400	500	600	800	1000
Velocity • fps	2750	2573	2403	2239	2082	1931	1788	1525	1304
Energy • ft-lbs	3022	2646	2303	2004	1732	1491	1278	930	680
Taylor KO Index	21.8	20.4	19.0	17.7	16.5	15.3	14.2	12.1	10.3
Path • Inches	-1.5	+1.9	0.0	-7.9	-22.8	-45.6	-77.9	-176.9	-336.1
Wind Drift • Inches	0.0	0.6	2.7	6.2	11.5	18.6	17.8	53.4	90.1

Sellier and Bellot 180-grain Soft Point (V330352U)

G1 Ballistic Coefficient = 0.361

Distance • Yards	Muzzle	100	200	300	400	500	600	800	1000
Velocity • fps	2707	2451	2219	2009	1804	1613	1441	1170	1010
Energy • ft-lbs	2937	2407	1973	1617	1301	1041	831	547	408
Taylor KO Index	21.4	19.4	17.6	15.9	14.3	12.8	11.4	9.3	8.0
Path • Inches	-1.5	+2.1	0.0	-9.1	-26.9	-55.5	-97.6	-237.8	-481.4
Wind Drift • Inches	0.0	1.0	4.0	9.4	17.6	29.1	44.2	87.0	145.8

Black Hills 180-grain Nosler AccuBond (1C3006BHGN5)

G1 Ballistic Coefficient = 0.475

Distance • Yards	Muzzle	100	200	300	400	500	600	800	1000
Velocity • fps	2700	2510	2330	2160	2000	1840	1691	1425	1213
Energy • ft-lbs	2670	2315	1995	1710	1460	1240	1048	745	539
Taylor KO Index	19.6	18.2	16.6	15.7	14.5	13.4	12.3	10.3	8.8
Path • Inches	-1.5	+2.0	0.0	-8.5	-24.4	-48.7	-83.8	-192.4	-370.1
Wind Drift • Inches	0.0	0.7	3.0	6.9	12.8	20.8	31.2	60.2	101.6

Federal 180-grain Fusion (F3006FS3)

G1 Ballistic Coefficient = 0.495

Distance • Yards	Muzzle	100	200	300	400	500	600	800	1000
Velocity • fps	2700	2520	2350	2190	2030	1870	1726	1465	1251
Energy • ft-lbs	2915	2540	2205	1905	1640	1403	1191	858	626
Taylor KO Index	21.4	20.0	18.6	17.3	16.1	14.8	13.7	11.6	9.9
Path • Inches	-1.5	+2.0	0.0	-8.4	-23.9	-47.8	-82.2	-187.6	-358.2
Wind Drift • Inches	0.0	0.7	2.8	6.6	12.2	19.8	29.6	57.0	96.1

Federal 180-grain Nosler AccuBond (P3006A1)

G1 Ballistic Coefficient = 0.475

Distance • Yards	Muzzle	100	200	300	400	500	600	800	1000
Velocity • fps	2700	2510	2330	2160	2000	1840	1691	1425	1213
Energy • ft-lbs	2670	2315	1995	1710	1460	1240	1048	745	539
Taylor KO Index	19.6	18.2	16.6	15.7	14.5	13.4	12.3	10.3	8.8
Path • Inches	-1.5	+2.0	0.0	-8.5	-24.4	-48.7	-83.8	-192.4	-370.1
Wind Drift • Inches	0.0	0.7	3.0	6.9	12.8	20.8	31.2	60.2	101.6

Federal 180-grain Nosler Partition (P3006F)

G1 Ballistic Coefficient = 0.475

Distance • Yards	Muzzle	100	200	300	400	500	600	800	1000
Velocity • fps	2700	2510	2330	2160	2000	1840	1691	1425	1213
Energy • ft-lbs	2670	2315	1995	1710	1460	1240	1048	745	539
Taylor KO Index	19.6	18.2	16.6	15.7	14.5	13.4	12.3	10.3	8.8
Path • Inches	-1.5	+2.0	0.0	-8.5	-24.4	-48.7	-83.8	-192.4	-370.1
Wind Drift • Inches	0.0	0.7	3.0	6.9	12.8	20.8	31.2	60.2	101.6

Federal 180-grain Soft Point (3006B)

G1 Ballistic Coefficient = 0.38?

Distance • Yards	Muzzle	100	200	300	400	500	600	800	1000
Velocity • fps	2700	2470	2250	2050	1850	1670	1505	1230	1053
Energy • ft-lbs	2915	2440	2025	1670	1365	1110	905	605	443
Taylor KO Index	21.4	19.6	17.8	16.2	14.7	13.2	11.9	9.7	8.3
Path • Inches	-1.5	+2.1	0.0	-9.0	-26.2	-54.0	-93.6	-223.3	-445.1
Wind Drift • Inches	0.0	0.9	3.7	8.7	16.3	26.7	40.5	79.3	133.6

Hornady 180-grain SST InterLock (81184)

G1 Ballistic Coefficient = 0.48?

Distance • Yards	Muzzle	100	200	300	400	500	600	800	1000
Velocity • fps	2700	2515	2337	2165	2003	1847	1700	1436	1223
Energy • ft-lbs	2913	2527	2182	1875	1603	1363	1155	824	598
Taylor KO Index	21.4	19.9	18.5	17.1	15.9	14.6	13.5	11.4	9.7
Path • Inches	-1.5	+2.0	0.0	-8.4	-24.2	-48.7	-83.4	-191.2	-367.0
Wind Drift • Inches	0.0	0.7	2.9	6.8	12.6	20.5	30.8	59.4	100.2

Hornady 180-grain SP InterLock (8118)

G1 Ballistic Coefficient = 0.42?

Distance • Yards	Muzzle	100	200	300	400	500	600	800	1000
Velocity • fps	2700	2491	2292	2102	1921	1751	1592	1318	1120
Energy • ft-lbs	2913	2480	2099	1765	1475	1225	1013	694	501
Taylor KO Index	21.4	19.7	18.2	16.6	15.2	13.9	12.6	10.4	8.9
Path • Inches	-1.5	2.1	0.0	-8.7	25.3	-51.3	-88.7	-207.6	-407.2
Wind Drift • Inches	0.0	0.8	3.3	7.8	14.6	23.8	35.9	69.8	118.1

Norma 180-grain Nosler Partition (17649)

G1 Ballistic Coefficient = 0.474

Distance • Yards	Muzzle	100	200	300	400	500	600	800	1000
Velocity • fps	2700	2512	2333	2160	1995	1838	1689	1423	1211
Energy • ft-lbs	2914	2524	2175	1865	1591	1350	1141	810	587
Taylor KO Index	21.4	19.9	18.5	17.1	15.8	14.6	13.4	11.3	9.6
Path • Inches	-1.5	+2.0	0.0	-8.4	-24.3	-48.9	-83.9	-192.7	-370.7
Wind Drift • Inches	0.0	0.7	3.0	6.9	12.8	20.8	31.3	60.4	101.9

Norma 180-grain Swift A-Frame (17518)

G1 Ballistic Coefficient = 0.400

Distance • Yards	Muzzle	100	200	300	400	500	600	800	1000
Velocity • fps	2700	2479	2268	2067	1877	1699	1535	1259	1074
Energy • ft-lbs	2914	2456	2056	1708	1408	1158	942	634	461
Taylor KO Index	21.4	19.6	18.0	16.4	14.9	13.5	12.2	10.0	8.5
Path • Inches	-1.5	+2.0	0.0	-8.8	-25.9	-52.8	-92.0	-218.9	-436.5
Wind Drift • Inches	0.0	0.9	3.6	8.4	15.7	25.7	38.8	75.9	128.2

Norma 180-grain Oryx (17674)

G1 Ballistic Coefficient = 0.288

Distance • Yards	Muzzle	100	200	300	400	500	600	800	1000
Velocity • fps	2700	2395	2110	1846	1606	1394	1221	1006	890
Energy • ft-lbs	2914	2301	1786	1367	1031	777	596	405	316
Taylor KO Index	21.4	19.0	16.7	14.6	12.7	11.0	9.7	8.0	7.0
Path • Inches	-1.5	+2.3	0.0	-10.1	-30.6	-64.8	-117.4	-299.9	-612.3
Wind Drift • Inches	0.0	1.2	5.1	12.3	23.5	39.2	60.2	117.4	190.3

Norma 180-grain Alaska (17648)

G1 Ballistic Coefficient = 0.257

Distance • Yards	Muzzle	100	200	300	400	500	600	800	1000
Velocity • fps	2700	2359	2044	1755	1500	1285	1124	944	836
Energy • ft-lbs	2914	2225	1670	1231	899	660	505	356	280
Taylor KO Index	21.4	18.7	16.2	13.9	11.9	10.2	8.9	7.5	6.6
Path • Inches	-1.5	+2.4	0.0	-10.7	-33.0	-71.0	-131.0	-338.4	-688.8
Wind Drift • Inches	0.0	1.4	5.8	14.2	27.2	45.7	70.2	134.6	214.7

Remington 180-grain Core-Lokt Ultra Bonded (PRC3006C)

G1 Ballistic Coefficient = 0.402

Distance • Yards	Muzzle	100	200	300	400	500	600	800	1000
Velocity • fps	2700	2480	2270	2070	1882	1704	1540	1264	1077
Energy • ft-lbs	2913	2457	2059	1713	1415	1161	948	638	464
Taylor KO Index	21.4	19.6	18.0	16.4	14.9	13.5	12.2	10.0	8.5
Path • Inches	-1.5	+2.1	0.0	-8.9	-25.8	-52.7	-91.7	-218.0	-434.4
Wind Drift • Inches	0.0	0.8	3.6	8.4	15.6	25.5	38.6	75.4	127.3

Remington 180-grain Core-Lokt Pointed Soft Point (R30065)

G1 Ballistic Coefficient = 0.384

Distance • Yards	Muzzle	100	200	300	400	500	600	800	1000
Velocity • fps	2700	2469	2250	2042	1846	1663	1496	1221	1046
Energy • ft-lbs	2913	2436	2023	1666	1362	1105	895	596	438
Taylor KO Index	21.4	19.6	17.8	16.2	14.6	13.2	11.8	9.7	8.3
Path • Inches	-1.5	+2.1	0.0	-9.0	26.3	-54.0	-94.3	-226.5	-454.6
Wind Drift • Inches	0.0	0.9	3.7	8.8	16.5	27.0	41.0	80.3	135.2

Remington 180-grain Swift A-Frame PSP (RS3006A)

G1 Ballistic Coefficient = 0.377

Distance • Yards	Muzzle	100	200	300	400	500	600	800	1000
Velocity • fps	2700	2465	2243	2032	1833	1648	1479	1205	1034
Energy • ft-lbs	2913	2429	2010	1650	1343	1085	874	580	428
Taylor KO Index	21.4	19.5	17.8	16.1	14.5	13.1	11.7	9.5	8.2
Path • Inches	-1.5	+2.1	0.0	-9.1	-26.6	-54.4	-95.4	-230.1	-463.2
Wind Drift • Inches	0.0	0.9	3.8	9.0	16.8	27.7	42.0	82.3	138.4

Remington 180-grain Core-Lokt Soft Point (R30064)

G1 Ballistic Coefficient = 0.248

Distance • Yards	Muzzle	100	200	300	400	500	600	800	1000
Velocity • fps	2700	2348	2023	1727	1466	1251	1097	926	820
Energy • ft-lbs	2913	2203	1635	1192	859	625	481	343	269
Taylor KO Index	21.4	18.6	16.0	13.7	11.6	9.9	8.7	7.3	6.5
Path • Inches	-1.5	+2.4	0.0	-11.0	-33.8	-72.8	-135.9	-351.5	-714.6
Wind Drift • Inches	0.0	1.4	6.1	14.8	28.5	48.0	73.7	140.2	222.7

Winchester 180-grain Silvertip (X30066)

G1 Ballistic Coefficient = 0.384

Distance • Yards	Muzzle	100	200	300	400	500	600	800	1000
Velocity • fps	2700	2469	2250	2042	1846	1663	1496	1221	1046
Energy • ft-lbs	2913	2436	2023	1666	1362	1105	895	596	438
Taylor KO Index	21.4	19.6	17.8	16.2	14.6	13.2	11.8	9.7	8.3
Path • Inches	-1.5	+2.1	0.0	-9.0	26.3	-54.0	-94.3	-226.5	-454.6
Wind Drift • Inches	0.0	0.9	3.7	8.8	16.5	27.0	41.0	80.3	135.2

Winchester 180-grain Power-Point (X30064)

G1 Ballistic Coefficient = 0.381

Distance • Yards	Muzzle	100	200	300	400	500	600	800	1000
Velocity • fps	2700	2468	2247	2038	1840	1657	1490	1215	1043
Energy • ft-lbs	2913	2433	2018	1659	1354	1097	887	591	435
Taylor KO Index	21.4	19.5	17.8	16.1	14.6	13.1	11.8	9.6	8.3
Path • Inches	-1.5	2.1	0.0	-9.0	-26.4	-54.2	-94.6	-226.2	-452.1
Wind Drift • Inches	0.0	0.9	3.8	8.9	16.6	27.3	41.3	81.0	136.4

Fiocchi 180-grain SST Polymer Tip BT (3006HSC)

G1 Ballistic Coefficient = 0.480

Distance • Yards	Muzzle	100	200	300	400	500	600	800	1000
Velocity • fps	2675	2490	2313	2144	1981	1827	1681	1420	1211
Energy • ft-lbs	2859	2479	2139	1837	1569	1333	1130	806	586
Taylor KO Index	21.2	19.7	18.3	17.0	15.7	14.5	13.3	11.2	9.6
Path • Inches	-1.5	+2.1	0.0	-8.6	-24.7	-49.7	-85.2	-195.3	-375.0
Wind Drift • Inches	0.0	0.7	3.0	6.9	12.8	20.8	31.2	60.2	101.5

Norma 180-grain Nosler AccuBond (17563)

G1 Ballistic Coefficient = 0.507

Distance • Yards	Muzzle	100	200	300	400	500	600	800	1000
Velocity • fps	2674	2499	2331	2169	2015	1867	1726	1471	1260
Energy • ft-lbs	2859	2497	2172	1881	1623	1393	1191	865	635
Taylor KO Index	21.2	19.8	18.5	17.2	16.0	14.8	13.7	11.7	10.0
Path • Inches	-1.5	+2.0	0.0	-8.4	24.3	-48.7	-83.1	-189.0	-359.7
Wind Drift • Inches	0.0	0.7	2.8	6.5	12.0	19.5	29.2	56.0	94.3

Cor-Bon 180-grain T- DPX (DPX3006180/20)

G1 Ballistic Coefficient = 0.475

Distance • Yards	Muzzle	100	200	300	400	500	600	800	1000
Velocity • fps	2750	2581	2379	2205	2038	1879	1729	1457	1237
Energy • ft-lbs	3023	2622	2263	1944	1661	1412	1195	849	612
Taylor KO Index	21.8	20.4	18.8	17.5	16.1	14.9	13.7	11.5	9.8
Path • Inches	-1.5	+1.9	0.0	-8.1	23.3	-46.9	-80.4	-184.4	-354.6
Wind Drift • Inches	0.0	0.7	2.9	6.7	12.4	20.2	30.3	58.5	99.0

Nosler 180-grain Ballistic Tip (23400) + (60058)

G1 Ballistic Coefficient = 0.507

Distance • Yards	Muzzle	100	200	300	400	500	600	800	1000
Velocity • fps	2750	2572	2402	2238	2080	1929	1785	1522	1300
Energy • ft-lbs	3023	2646	2306	2002	1729	1487	1273	926	676
Taylor KO Index	21.8	20.4	19.0	17.7	16.5	15.3	14.1	12.1	10.3
Path • Inches	-1.5	+1.9	0.0	-7.9	22.8	-45.7	-78.0	-177.3	-337.1
Wind Drift • Inches	0.0	0.6	2.7	6.3	11.6	18.7	28.0	53.7	90.6

Winchester 180-grain E-Tip (S3006ET) LF

G1 Ballistic Coefficient = 0.52

Distance • Yards	Muzzle	100	200	300	400	500	600	800	1000
Velocity • fps	2750	2578	2412	2252	2099	1951	1810	1552	1331
Energy • ft-lbs	3022	2655	2325	2027	1760	1521	1310	963	708
Taylor KO Index	21.8	20.4	19.1	17.8	16.6	15.5	14.3	12.3	10.3
Path • Inches	-1.5	+1.9	0.0	-7.9	-22.6	-45.2	-77.0	-174.1	-329.5
Wind Drift • Inches	0.0	0.6	2.6	6.1	11.2	18.0	26.9	51.6	86.8

Black Hills 180-grain Barnes Triple Shock (1C3006BHGN1) LF

G1 Ballistic Coefficient = 0.45

Distance • Yards	Muzzle	100	200	300	400	500	600	800	1000
Velocity • fps	2700	2504	2315	2135	1964	1801	1647	1377	1170
Energy • ft-lbs	2913	2506	2143	1823	1541	1296	1085	758	547
Taylor KO Index	21.4	19.8	18.3	16.9	15.6	14.3	13.0	10.9	9.3
Path • Inches	-1.5	+2.0	0.0	-8.5	-24.7	-49.9	-85.9	-199.8	-385.6
Wind Drift • Inches	0.0	0.8	3.1	7.3	13.6	22.1	33.2	64.3	108.7

ederal 180-grain Barnes MRX Bullet (P3006AH) LF
G1 Ballistic Coefficient = 0.554

Distance • Yards	Muzzle	100	200	300	400	500	600	800	1000
Velocity • fps	2700	2540	2380	2230	2090	1950	1817	1571	1358
Energy • ft-lbs	2915	2575	2270	1990	1740	1515	1320	987	738
Taylor KO Index	21.4	20.1	18.8	17.7	16.6	15.4	14.4	12.4	10.8
Path • Inches	-1.5	+1.9	0.0	-8.1	-23.3	-46.1	-78.3	-176.0	-330.3
Wind Drift • Inches	0.0	0.6	2.5	5.8	10.7	17.3	25.8	49.2	82.6

ederal 180-grain Trophy Bonded Tip (P3006TT1)
G1 Ballistic Coefficient = 0.502

Distance • Yards	Muzzle	100	200	300	400	500	600	800	1000
Velocity • fps	2700	2520	2350	2190	2030	1880	1738	1479	1264
Energy • ft-lbs	2915	2540	2210	1910	1645	1410	1207	874	639
Taylor KO Index	21.4	20.0	18.6	17.3	16.1	14.9	13.8	11.7	10.0
Path • Inches	-1.5	+2.0	0.0	-8.4	-23.9	-47.7	-81.7	-186.0	-354.4
Wind Drift • Inches	0.0	0.7	2.8	6.5	12.0	19.5	29.1	56.0	94.3

ederal 180-grain Triple-Shock X-Bullet (P3006AE) LF
G1 Ballistic Coefficient = 0.452

Distance • Yards	Muzzle	100	200	300	400	500	600	800	1000
Velocity • fps	2700	2504	2315	2135	1964	1801	1647	1377	1170
Energy • ft-lbs	2913	2506	2143	1823	1541	1296	1085	758	547
Taylor KO Index	21.4	19.8	18.3	16.9	15.6	14.3	13.0	10.9	9.3
Path • Inches	-1.5	+2.0	0.0	-8.5	-24.7	-49.9	-85.9	-199.8	-385.6
Wind Drift • Inches	0.0	0.8	3.1	7.3	13.6	22.1	33.2	64.3	108.7

orma 180-grain Plastic Point (17653)
G1 Ballistic Coefficient = 0.366

Distance • Yards	Muzzle	100	200	300	400	500	600	800	1000
Velocity • fps	2700	2458	2229	2013	1809	1621	1450	1179	1016
Energy • ft-lbs	2914	2416	1987	1620	1309	1050	841	555	413
Taylor KO Index	21.4	19.5	17.7	15.9	14.3	12.8	11.5	9.3	8.0
Path • Inches	-1.5	+2.1	0.0	-9.2	-26.9	-55.4	-97.3	-236.3	477.3
Wind Drift • Inches	0.0	0.9	3.9	9.3	17.4	28.7	43.6	85.7	143.7

orma 180-grain Vulcan (17659)
G1 Ballistic Coefficient = 0.315

Distance • Yards	Muzzle	100	200	300	400	500	600	800	1000
Velocity • fps	2700	2420	2158	1912	1685	1481	1305	1063	934
Energy • ft-lbs	2914	2350	1868	1466	1135	877	681	451	349
Taylor KO Index	21.4	19.2	17.1	15.1	13.3	11.7	10.3	8.4	7.4
Path • Inches	-1.5	+2.2	0.0	-9.7	-29.0	-60.8	-108.8	-273.2	-557.7
Wind Drift • Inches	0.0	1.1	4.6	11.1	21.0	34.8	53.3	104.7	172.1

emington 180-grain Swift Scirocco Bonded (PRSC3006B)
G1 Ballistic Coefficient = 0.500

Distance • Yards	Muzzle	100	200	300	400	500	600	800	1000
Velocity • fps	2700	2522	2351	2186	2028	1878	1734	1474	1259
Energy • ft-lbs	2913	2542	2208	1910	1644	1409	1202	868	634
Taylor KO Index	21.4	20.0	18.6	17.3	16.1	14.9	13.7	11.7	10.0
Path • Inches	-1.5	+2.0	0.0	-8.3	-23.9	-47.9	-81.8	-186.8	-357.7
Wind Drift • Inches	0.0	0.7	2.8	6.6	12.1	19.6	29.3	56.3	94.9

Vinchester 180-grain Power Max Bonded (X30064BP)
G1 Ballistic Coefficient = 0.394

Distance • Yards	Muzzle	100	200	300	400	500	600	800	1000
Velocity • fps	2700	2475	2262	2058	1866	1687	1522	1374	1065
Energy • ft-lbs	2913	2448	2044	1693	1392	1137	926	621	453
Taylor KO Index	21.4	19.6	17.9	16.3	14.8	13.4	12.1	10.9	8.4
Path • Inches	-1.5	+2.1	0.0	-8.9	-26.1	-53.2	-92.6	-220.0	-437.4
Wind Drift • Inches	0.0	0.9	3.6	8.6	15.9	26.1	39.6	77.4	130.6

emington 180-grain Bronze Point (R30066)
G1 Ballistic Coefficient = 0.412

Distance • Yards	Muzzle	100	200	300	400	500	600	800	1000
Velocity • fps	2700	2485	2280	2084	1899	1725	1562	1287	1095
Energy • ft-lbs	2913	2468	2077	1736	1441	1189	976	662	479
Taylor KO Index	21.4	19.7	18.1	16.5	15.0	13.7	12.4	10.2	8.7
Path • Inches	-1.5	+2.1	0.0	-8.8	25.5	-52.0	-90.4	-213.8	-424.1
Wind Drift • Inches	0.0	0.8	3.5	8.1	15.1	24.8	37.4	72.9	123.3

Remington 180-grain AccuTip Boat Tail (PRA3006C)

G1 Ballistic Coefficient = 0.480

Distance • Yards	Muzzle	100	200	300	400	500	600	800	1000
Velocity • fps	2725	2539	2360	2188	2024	1867	1719	1452	1235
Energy • ft-lbs	2967	2576	2226	1914	1637	1393	1181	842	610
Taylor KO Index	21.6	20.1	18.7	17.3	16.0	14.8	13.6	11.5	9.8
Path • Inches	-1.5	+2.0	0.0	-8.2	23.7	-47.7	-81.6	-187.1	-359.2
Wind Drift • Inches	0.0	0.7	2.9	6.8	12.5	20.2	30.4	58.5	98.9

Fiocchi 180-grain Sierra MatchKing HPBT (3006MKD)

G1 Ballistic Coefficient = 0.496

Distance • Yards	Muzzle	100	200	300	400	500	600	800	1000
Velocity • fps	2665	2487	2315	2151	1992	1841	1699	1442	1234
Energy • ft-lbs	2838	2471	2143	1849	1587	1355	1154	832	608
Taylor KO Index	21.1	19.7	18.3	17.0	15.8	14.6	13.5	11.4	9.8
Path • Inches	-1.5	+2.1	0.0	-8.6	-24.6	-49.5	-84.6	-193.3	-369.2
Wind Drift • Inches	0.0	0.7	2.9	6.7	12.4	20.2	30.2	58.1	97.9

Federal 180-grain Trophy Bonded Tip (P3006TT4)

G1 Ballistic Coefficient = 0.498

Distance • Yards	Muzzle	100	200	300	400	500	600	800	1000
Velocity • fps	2880	2700	2520	2350	2180	2020	1870	1592	1354
Energy • ft-lbs	3316	2902	2530	2197	1898	1633	1398	1013	733
Taylor KO Index	22.8	21.4	20.0	18.6	17.3	16.0	14.8	12.6	10.7
Path • Inches	-1.5	+1.6	0.0	-7.2	-20.6	-41.5	-70.8	-161.1	-306.8
Wind Drift • Inches	0.0	0.6	2.6	6.0	11.0	17.8	26.7	51.3	86.6

Winchester 180-grain Ballistic Silvertip (SBST3006B)

G1 Ballistic Coefficient = 0.507

Distance • Yards	Muzzle	100	200	300	400	500	600	800	1000
Velocity • fps	2750	2572	2402	2237	2080	1928	1785	1522	1300
Energy • ft-lbs	3022	2644	2305	2001	1728	1486	1273	926	676
Taylor KO Index	21.8	20.4	19.0	17.7	16.5	15.3	14.1	12.1	10.3
Path • Inches	-1.5	+1.9	0.0	-7.9	-22.8	-45.7	-78.0	-177.3	-337.1
Wind Drift • Inches	0.0	0.6	2.7	6.3	11.6	18.7	28.0	53.7	90.6

A² 180-grain Mono; Dead Tough (None)

G1 Ballistic Coefficient = 0.264

Distance • Yards	Muzzle	100	200	300	400	500	600	800	1000
Velocity • fps	2700	2365	2054	1769	1524	1310	1146	958	849
Energy • ft-lbs	2913	2235	1687	1251	928	686	525	367	288
Taylor KO Index	21.4	18.7	16.3	14.0	12.1	10.4	9.1	7.6	6.7
Path • Inches	-1.5	+2.4	0.0	-10.6	-32.4	-69.1	-127.5	-328.8	-669.9
Wind Drift • Inches	0.0	1.3	5.7	13.7	26.2	44.1	67.8	130.4	208.8

Sellier and Bellot 180-grain FMJ (V331612U)

G1 Ballistic Coefficient = 0.502

Distance • Yards	Muzzle	100	200	300	400	500	600	800	1000
Velocity • fps	2674	2494	2326	2169	2014	1866	1725	1469	1258
Energy • ft-lbs	2883	2507	2182	1897	1622	1392	1189	863	633
Taylor KO Index	21.2	19.8	18.4	17.2	16.0	14.8	13.7	11.6	10.0
Path • Inches	-1.5	+2.0	0.0	-8.4	-24.3	-48.7	-83.2	-189.2	-360.1
Wind Drift • Inches	0.0	0.6	2.5	5.9	10.9	17.6	26.3	50.6	85.1

Lapua 185-grain Mega (4317563)

G1 Ballistic Coefficient = 0.31

Distance • Yards	Muzzle	100	200	300	400	500	600	800	1000
Velocity • fps	2625	2346	2083	1839	1616	1417	1250	1032	913
Energy • ft-lbs	2831	2261	1784	1390	1072	825	642	437	343
Taylor KO Index	21.4	19.1	17.0	15.0	13.2	11.5	10.2	8.4	7.4
Path • Inches	-1.5	+2.4	0.0	-10.5	-31.3	-65.5	-117.8	-295.8	-599.2
Wind Drift • Inches	0.0	1.2	4.9	11.8	22.3	37.1	56.8	110.4	179.4

Norma 200-grain Oryx (17677)

G1 Ballistic Coefficient = 0.33

Distance • Yards	Muzzle	100	200	300	400	500	600	800	1000
Velocity • fps	2625	2368	2125	1897	1687	1496	1329	1088	956
Energy • ft-lbs	3061	2490	2006	1599	1265	994	784	526	406
Taylor KO Index	23.1	20.8	18.7	16.7	14.8	13.2	11.7	9.6	8.4
Path • Inches	-1.5	+2.3	0.0	-10.0	-29.9	-62.0	-110.0	-271.7	-549.4
Wind Drift • Inches	0.0	1.1	4.5	10.6	20.0	33.2	50.6	98.8	162.7

Lapua 200-grain Mega SP (4317567)

G1 Ballistic Coefficient = 0.329

Distance • Yards	Muzzle	100	200	300	400	500	600	800	1000
Velocity • fps	2543	2284	2040	1813	1604	1417	1259	1045	927
Energy • ft-lbs	2873	2317	1849	1460	1143	892	704	485	382
Taylor KO Index	22.4	20.1	18.0	16.0	14.1	12.5	11.1	9.2	8.2
Path • Inches	-1.5	+2.6	0.0	-11.0	-32.6	-67.8	-121.0	-299.5	-601.0
Wind Drift • Inches	0.0	1.2	4.8	11.5	21.8	36.0	54.8	106.2	172.3

Remington 220-grain Core-Lokt Soft Point (R30067)

G1 Ballistic Coefficient = 0.294

Distance • Yards	Muzzle	100	200	300	400	500	600	800	1000
Velocity • fps	2410	2130	1870	1632	1422	1246	1112	954	856
Energy • ft-lbs	2837	2216	1708	1301	988	758	604	445	358
Taylor KO Index	23.3	20.6	18.1	15.8	13.7	12.1	10.8	9.3	8.3
Path • Inches	-1.5	+3.1	0.0	-13.1	-39.4	-83.0	-150.2	-372.1	-735.3
Wind Drift • Inches	0.0	1.4	6.0	14.3	27.1	45.0	68.1	127.5	200.8

Federal 220-grain Soft Point (3006HS)

G1 Ballistic Coefficient = 0.294

Distance • Yards	Muzzle	100	200	300	400	500	600	800	1000
Velocity • fps	2400	2120	1860	1620	1410	1240	1110	954	856
Energy • ft-lbs	2815	2195	1690	1285	975	750	602	444	358
Taylor KO Index	23.2	20.5	18.0	15.7	13.6	12.0	10.7	9.2	8.3
Path • Inches	-1.5	+3.1	0.0	-13.5	-40.0	-84.5	-150.2	-373.2	-748.9
Wind Drift • Inches	0.0	1.4	6.0	14.4	27.2	45.2	68.3	127.7	201.0

good solid bench, a good rest for the rifle, and a decent shooter will soon tell if the rifle and ammo combination chosen is what is desirable for accuracy. (Safari
ess Archives)

7.82mm (.308) Lazzeroni Patriot

Lazzeroni has two 30-caliber cartridges. The Patriot is the shorter of the two. A short fireplug of a case, the Patriot uses an unbelted case with a 0.548-inch head diameter to drive a 130-grain bullet to .300 Winchester Magnum velocity.

Relative Recoil Factor = 2.23

Specifications:

Controlling Agency for Standardization of this Ammunition: Factory

Availability:

Lazzeroni 130-grain Nosler

G1 Ballistic Coefficient = 0.429

Distance • Yards	Muzzle	100	200	300	400	500	600	800	1000
Velocity • fps	3571	3318	3080	2855	2640	2436	2242	1879	1558
Energy • ft-lbs	3681	3180	2740	2354	2013	1713	1451	1019	701
Taylor KO Index	20.4	19.0	17.6	16.3	15.1	13.9	12.8	10.7	8.9
Path • Inches	-1.5	+0.8	0.0	-4.5	-13.3	-27.0	-46.6	-108.1	-210.4
Wind Drift • Inches	0.0	0.6	2.3	5.3	9.7	15.8	23.6	45.5	77.7

Lazzeroni 180-grain Nosler Partition (782PT180P)

G1 Ballistic Coefficient = 0.585

Distance • Yards	Muzzle	100	200	300	400	500	600	800	1000
Velocity • fps	3184	3000	2825	2656	2493	2336	2185	1900	1640
Energy • ft-lbs	4052	3600	3191	2821	2485	2182	1909	1443	1075
Taylor KO Index	25.2	23.8	22.4	21.0	19.7	18.5	17.3	15.0	13.0
Path • Inches	-1.5	+1.2	0.0	-5.5	-16.0	-32.1	-54.5	-122.4	-229.5
Wind Drift • Inches	0.0	0.5	2.0	4.7	8.7	14.0	20.8	39.4	66.1

*Finn Aagaard fires on a target gong while a proctor scores his shots during a contest. (*Photo from *Guns and Hunting* by Finn Aagaard, 2012, Safari Press Archives)

.300 H&H Magnum

The .300 H&H Magnum (introduced about 1920) wasn't the first belted magnum; the .375 H&H actually arrived about ten years earlier. Still the .300 H&H can be called the father of all the .300 magnums. Starting with the H&H case, wildcatters improved, reshaped, necked, and generally reformed the case into nearly every configuration they, or anyone else, could imagine. The .300 H&H has never been hugely popular in the U.S., but several of its offspring are near the top of the charts.

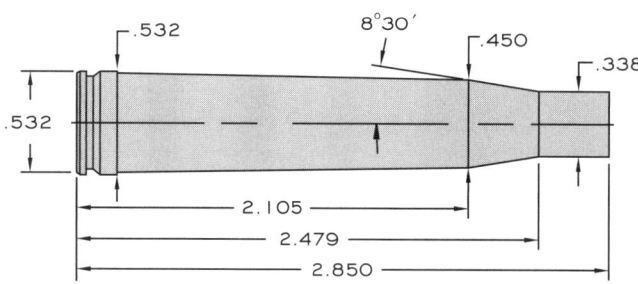

Relative Recoil Factor = 2.34

Specifications:

Controlling Agency for Standardization of this Ammunition: SAAMI

Bullet Weight Grains	Velocity fps	Maximum Average Pressure	
		Copper Crusher	Transducer
150	3,110	54,000 cup	N/S
180	2,780	54,000 cup	N/S
220	2,565	54,000 cup	N/S

Standard barrel for velocity testing: 24 inches long—1 turn in 10-inch twist

Availability:

Nosler 150-grain AccuBond (20002)

G1 Ballistic Coefficient = 0.435

Distance • Yards	Muzzle	100	200	300	400	500	600	800	1000
Velocity • fps	3200	2972	2755	2548	2350	2162	1983	1653	1373
Energy • ft-lbs	3410	2941	2527	2162	1840	1557	1310	911	628
Taylor KO Index	21.1	19.6	18.2	16.8	15.5	14.3	13.1	10.9	9.1
Path • Inches	-1.5	+1.2	0.0	-5.8	-17.0	-34.5	-59.4	-137.6	-267.2
Wind Drift • Inches	0.0	0.6	2.6	6.0	11.1	18.0	27.0	52.3	59.6

Nosler 150-grain AccuBond (20002)

G1 Ballistic Coefficient = 0.435

Distance • Yards	Muzzle	100	200	300	400	500	600	800	1000
Velocity • fps	3200	2972	2755	2548	2350	2162	1983	1653	1373
Energy • ft-lbs	3410	2941	2527	2162	1840	1557	1310	911	628
Taylor KO Index	21.1	19.6	18.2	16.8	15.5	14.3	13.1	10.9	9.1
Path • Inches	-1.5	+1.2	0.0	-5.8	-17.0	-34.5	-59.4	-137.6	-267.2
Wind Drift • Inches	0.0	0.6	2.6	6.0	11.1	18.0	27.0	52.3	59.6

Nosler 180-grain Partition (20063)

G1 Ballistic Coefficient = 0.474

Distance • Yards	Muzzle	100	200	300	400	500	600	800	1000
Velocity • fps	2950	2751	2561	2379	2205	2038	1879	1587	1339
Energy • ft-lbs	3478	3025	2622	2262	1942	1660	1411	1006	717
Taylor KO Index	23.4	21.8	20.3	18.8	17.5	16.1	14.9	12.6	10.6
Path • Inches	-1.5	+1.6	0.0	-6.8	-19.8	-40.0	-68.6	-157.1	-301.3
Wind Drift • Inches	0.0	0.6	2.6	6.1	11.2	18.3	27.3	52.8	89.6

Federal 180-grain Nosler Partition (P300HA)

G1 Ballistic Coefficient = 0.358

Distance • Yards	Muzzle	100	200	300	400	500	600	800	1000
Velocity • fps	2880	2620	2380	2150	1930	1730	1546	1240	1047
Energy • ft-lbs	3315	2750	2260	1840	1480	1190	956	615	439
Taylor KO Index	22.8	20.8	18.8	17.0	15.3	13.7	12.2	9.8	8.3
Path • Inches	-1.5	+1.8	0.0	-8.0	-23.4	-48.6	-84.9	-206.9	-422.6
Wind Drift • Inches	0.0	0.9	3.7	8.7	16.2	26.8	40.7	80.6	137.3

Nosler 180-grain Partition (20084)

G1 Ballistic Coefficient = 0.482

Distance • Yards	Muzzle	100	200	300	400	500	600	800	1000
Velocity • fps	2750	2563	2383	2211	2046	1890	1741	1492	1251
Energy • ft-lbs	3358	2917	2522	2171	1860	1586	1347	962	695
Taylor KO Index	24.2	22.6	21.0	19.5	18.0	16.6	15.3	13.0	11.0
Path • Inches	-1.5	+1.9	0.0	-8.0	-23.2	-46.6	-79.8	-182.7	-350.5
Wind Drift • Inches	0.0	0.7	2.8	6.6	12.2	19.9	29.8	57.4	97.1

Hornady 180-grain Interbond (8210)

G1 Ballistic Coefficient = 0.480

Distance • Yards	Muzzle	100	200	300	400	500	600	800	1000
Velocity • fps	2870	2678	2493	2316	2146	1984	1829	1547	1309
Energy • ft-lbs	3292	2865	2484	2144	1841	1573	1338	956	685
Taylor KO Index	22.7	21.2	19.7	18.3	17.0	15.7	14.5	12.3	10.4
Path • Inches	-1.5	+1.7	0.0	-7.3	-21.0	-42.4	-72.6	-166.0	-318.2
Wind Drift • Inches	0.0	0.6	2.7	6.3	11.6	18.7	28.1	54.1	91.7

Nosler 180-grain Ballistic Tip (20051)

G1 Ballistic Coefficient = 0.506

Distance • Yards	Muzzle	100	200	300	400	500	600	800	1000
Velocity • fps	2950	2764	2586	2414	2249	2091	1939	1857	1412
Energy • ft-lbs	3478	3053	2672	2329	2022	1748	1502	1098	787
Taylor KO Index	23.4	21.9	20.5	19.1	17.8	16.6	15.4	14.7	11.2
Path • Inches	-1.5	+1.5	0.0	-6.7	-19.4	-39.0	-66.6	-151.1	-286.7
Wind Drift • Inches	0.0	0.6	2.4	5.7	10.4	16.9	25.2	48.4	81.7

Federal 180-grain Barnes Triple-Shock X-Bullet (P300HB) LF

G1 Ballistic Coefficient = 0.456

Distance • Yards	Muzzle	100	200	300	400	500	600	800	1000
Velocity • fps	2880	2680	2480	2290	2120	1950	1790	1499	1261
Energy • ft-lbs	3315	2860	2460	2105	1790	1510	1280	898	636
Taylor KO Index	22.8	21.2	19.6	18.1	16.8	15.4	14.2	11.9	10.0
Path • Inches	-1.5	+1.7	0.0	-7.3	-21.3	-42.9	-73.8	-170.2	-329.1
Wind Drift • Inches	0.0	0.7	2.8	6.6	12.2	19.8	29.8	57.6	98.0

Federal 180-grain Trophy Bonded Tip (P300HTT1)

G1 Ballistic Coefficient = 0.498

Distance • Yards	Muzzle	100	200	300	400	500	600	800	1000
Velocity • fps	2880	2700	2520	2350	2180	2020	1870	1592	1354
Energy • ft-lbs	3315	2900	2530	2200	1900	1635	1398	1013	733
Taylor KO Index	22.8	21.4	20.0	18.6	17.3	16.0	14.8	12.6	10.7
Path • Inches	-1.5	+1.6	0.0	-7.1	-20.6	-41.4	-70.8	+161.1	-306.8
Wind Drift • Inches	0.0	0.6	2.6	3.0	11.0	17.8	26.7	51.3	86.6

Nosler 200-grain AccuBond (20070 & 60060)

G1 Ballistic Coefficient = 0.588

Distance • Yards	Muzzle	100	200	300	400	500	600	800	1000
Velocity • fps	2750	2596	2448	2304	2165	2032	1903	1662	1448
Energy • ft-lbs	3358	2993	2661	2357	2082	1834	1609	1227	931
Taylor KO Index	24.2	22.8	21.5	20.3	19.1	17.9	16.7	14.6	12.7
Path • Inches	-1.5	+1.8	0.0	-7.6	-21.8	-43.4	-73.5	-163.8	-304.7
Wind Drift • Inches	0.0	0.6	2.3	5.3	9.8	15.4	23.4	44.4	74.1

.300 Remington Short Action Ultra Magnum

In 2002 Remington introduced their version of a "short magnum." Like the .300 WSM, this cartridge is designed to be fired from .308-length actions. There are claims that the short, fat cases somehow provide unusual "efficiency" and that these cartridges can exceed the performance of the larger magnums. While these short magnums are useful cartridges with excellent performance, the evidence of higher efficiency is tenuous at best.

Relative Recoil Factor = 2.36

Specifications:

Controlling Agency for Standardization of this Ammunition: SAAMI

Bullet Weight Grains	Velocity fps	Maximum Average Pressure Copper Crusher	Transducer
150	3,200	N/A	65,000 psi
180	2,960	N/A	65,000 psi

Standard barrel for velocity testing: 24 inches long—1 turn in 10-inch twist

Availability:

Nosler 150-grain AccuBond (19045)
G1 Ballistic Coefficient = 0.435

Distance • Yards	Muzzle	100	200	300	400	500	600	800	1000
Velocity • fps	3200	2972	2755	2548	2350	2162	1983	1653	1373
Energy • ft-lbs	3410	2941	2527	2162	1840	1557	1310	911	628
Taylor KO Index	21.1	19.6	18.2	16.8	15.5	14.3	13.1	10.9	9.1
Path • Inches	-1.5	+1.2	0.0	-5.8	-17.0	-34.5	-59.4	-137.6	-267.2
Wind Drift • Inches	0.0	0.6	2.6	6.0	11.1	18.0	27.0	52.3	89.6

Nosler 150-grain Partition (19015)
G1 Ballistic Coefficient = 0.387

Distance • Yards	Muzzle	100	200	300	400	500	600	800	1000
Velocity • fps	3200	2944	2702	2474	2256	2050	1855	1507	1231
Energy • ft-lbs	3410	2887	2432	2038	1695	1400	1147	758	505
Taylor KO Index	21.1	19.4	17.8	16.3	14.9	13.5	12.2	9.9	8.1
Path • Inches	-1.5	+1.3	0.0	-6.0	-17.8	-36.3	-63.2	-149.7	-298.2
Wind Drift • Inches	0.0	0.7	2.9	6.8	12.7	20.7	31.3	61.5	106.4

Remington 150-grain Core-Lokt Ultra Bonded (PR300SM1)
G1 Ballistic Coefficient = 0.330

Distance • Yards	Muzzle	100	200	300	400	500	600	800	1000
Velocity • fps	3200	2901	2622	2359	2112	1880	1666	1305	1071
Energy • ft-lbs	3410	2803	2290	1854	1485	1177	924	567	382
Taylor KO Index	21.1	19.1	17.3	15.6	13.9	12.4	11.0	8.6	7.1
Path • Inches	-1.5	+1.3	0.0	-6.4	-19.1	-39.6	-69.9	-171.8	-356.0
Wind Drift • Inches	0.0	0.8	3.1	7.4	13.8	22.8	34.8	69.7	121.0

Nosler 150-grain Ballistic Tip (19020)
G1 Ballistic Coefficient = 0.435

Distance • Yards	Muzzle	100	200	300	400	500	600	800	1000
Velocity • fps	3200	2972	2755	2548	2350	2162	1983	1653	1373
Energy • ft-lbs	3410	2941	2527	2162	1840	1557	1310	911	628
Taylor KO Index	21.1	19.6	18.2	16.8	15.5	14.3	13.1	10.9	9.1
Path • Inches	-1.5	+1.2	0.0	-5.8	-17.0	-34.5	-59.4	-137.6	-267.2
Wind Drift • Inches	0.0	0.6	2.6	6.0	11.1	18.0	27.0	52.3	89.6

Remington 165-grain Core-Lokt Pointed Soft Point (PR300SM2)
G1 Ballistic Coefficient = 0.338

Distance • Yards	Muzzle	100	200	300	400	500	600	800	1000
Velocity • fps	3075	2792	2527	2276	2040	1819	1612	1272	1056
Energy • ft-lbs	3464	2856	2339	1898	1525	1213	952	593	409
Taylor KO Index	22.3	20.3	18.3	16.5	14.8	13.2	11.7	9.2	7.7
Path • Inches	-1.5	+1.5	0.0	-7.0	-20.7	-42.1	-75.7	-186.7	-387.7
Wind Drift • Inches	0.0	0.8	3.6	8.4	15.8	26.1	39.8	79.5	137.0

Nosler 165-grain Partition (60061)

G1 Ballistic Coefficient = 0.417

Distance • Yards	Muzzle	100	200	300	400	500	600	800	1000
Velocity • fps	3000	2772	2556	2350	2154	1967	1791	1475	1225
Energy • ft-lbs	3297	2815	2393	2023	1700	1417	1167	798	550
Taylor KO Index	20.8	20.1	18.6	17.1	15.6	14.3	13.0	10.7	8.9
Path • Inches	-1.5	+1.5	0.0	-6.8	-20.0	-40.7	-70.3	-164.6	-323.5
Wind Drift • Inches	0.0	0.7	2.9	6.9	12.7	20.8	31.3	61.2	104.8

Nosler 165-grain Ballistic Tip (19030)

G1 Ballistic Coefficient = 0.475

Distance • Yards	Muzzle	100	200	300	400	500	600	800	1000
Velocity • fps	3000	2799	2608	2424	2248	2080	1919	1623	1369
Energy • ft-lbs	3297	2871	2491	2153	1851	1585	1350	965	687
Taylor KO Index	21.8	20.3	18.9	17.6	16.3	15.1	13.9	11.8	9.9
Path • Inches	-1.5	+1.5	0.0	-6.6	-19.1	-38.5	-65.9	-150.9	-289.2
Wind Drift • Inches	0.0	0.6	2.6	6.0	11.0	17.8	26.6	51.3	87.0

Remington 180-grain Core-Lokt Ultra Bonded (PR300SM4)

G1 Ballistic Coefficient = 0.403

Distance • Yards	Muzzle	100	200	300	400	500	600	800	1000
Velocity • fps	2960	2727	2506	2295	2094	1904	1727	1412	1173
Energy • ft-lbs	3501	2972	2509	2105	1753	1449	1192	797	550
Taylor KO Index	23.4	21.6	19.8	18.2	16.6	15.1	13.7	11.2	9.3
Path • Inches	-1.5	+1.6	0.0	-7.1	-20.9	-42.6	-73.9	-174.2	-345.0
Wind Drift • Inches	0.0	0.7	2.8	6.6	12.2	19.9	30.0	58.9	101.0

Nosler 180-grain AccuBond (19005)

G1 Ballistic Coefficient = 0.507

Distance • Yards	Muzzle	100	200	300	400	500	600	800	1000
Velocity • fps	2900	2716	2540	2370	2207	2051	1901	1624	1385
Energy • ft-lbs	3361	2948	2578	2245	1946	1681	1445	1051	766
Taylor KO Index	23.0	21.5	20.1	18.8	17.5	16.2	15.1	12.9	11.0
Path • Inches	-1.5	+1.6	0.0	-7.0	-20.2	-40.6	-69.2	-157.0	-297.8
Wind Drift • Inches	0.0	0.6	2.5	5.8	10.7	17.3	25.8	49.5	83.6

Nosler 180-grain Partition (60062)

G1 Ballistic Coefficient = 0.481

Distance • Yards	Muzzle	100	200	300	400	500	600	800	1000
Velocity • fps	2900	2707	2522	2344	2175	2011	1854	1589	1328
Energy • ft-lbs	3361	2929	2542	2196	1891	1616	1374	984	705
Taylor KO Index	23.0	21.4	20.0	18.6	17.2	15.9	14.7	12.6	10.5
Path • Inches	-1.5	+1.6	0.0	-7.1	-20.5	-41.4	-70.8	-161.9	-310.0
Wind Drift • Inches	0.0	0.6	2.6	6.2	11.3	18.4	27.6	53.1	90.0

Nosler 180-grain E-Tip (19012) LF

G1 Ballistic Coefficient = 0.532

Distance • Yards	Muzzle	100	200	300	400	500	600	800	1000
Velocity • fps	2900	2724	2556	2394	2237	2087	1943	1674	1437
Energy • ft-lbs	3361	2965	2611	2290	2000	1741	1509	1120	826
Taylor KO Index	23.0	21.6	20.2	19.0	17.7	16.5	15.4	13.3	11.4
Path • Inches	-1.5	+1.6	0.0	-6.9	-19.9	-39.8	-67.8	-152.8	-287.9
Wind Drift • Inches	0.0	0.6	2.4	5.5	10.1	16.3	24.4	46.5	78.3

Nosler 180-grain Ballistic Tip (19010)

G1 Ballistic Coefficient = 0.507

Distance • Yards	Muzzle	100	200	300	400	500	600	800	1000
Velocity • fps	2900	2716	2540	2370	2207	2051	1901	1624	1385
Energy • ft-lbs	3361	2948	2578	2245	1946	1681	1445	1051	766
Taylor KO Index	23.0	21.5	20.1	18.8	17.5	16.2	15.1	12.9	11.0
Path • Inches	-1.5	+1.6	0.0	-7.0	-20.2	-40.6	-69.2	-157.0	-297.8
Wind Drift • Inches	0.0	0.6	2.5	5.8	10.7	17.3	25.8	49.5	83.6

Remington 190-grain MatchKing BTHP (RM300SM7)

G1 Ballistic Coefficient = 0.533

Distance • Yards	Muzzle	100	200	300	400	500	600	800	1000
Velocity • fps	2900	2725	2557	2395	2239	2089	1944	1676	1439
Energy • ft-lbs	3547	3133	2758	2420	2115	1840	1595	1185	874
Taylor KO Index	24.2	22.8	21.4	20.0	18.7	17.5	16.3	14.0	12.0
Path • Inches	-1.5	+1.6	0.0	-6.9	-19.9	-39.8	-67.7	-152.6	-275.5
Wind Drift • Inches	0.0	0.5	2.1	5.0	9.1	14.7	21.9	41.8	70.3

Nosler 200-grain AccuBond (19040)

G1 Ballistic Coefficient = 0.588

Distance • Yards	Muzzle	100	200	300	400	500	600	800	1000
Velocity • fps	2800	2644	2495	2249	2209	2074	1944	1699	1480
Energy • ft-lbs	3481	3105	2764	2451	2167	1910	1678	1282	973
Taylor KO Index	24.6	23.3	22.0	20.7	19.4	18.3	17.1	15.0	13.0
Path • Inches	-1.5	+1.7	0.0	-7.3	-21.0	-41.7	-70.6	-157.2	-292.4
Wind Drift • Inches	0.0	0.6	2.2	5.2	9.5	15.3	22.8	43.2	71.1

Nosler 200-grain Partition (19035)

G1 Ballistic Coefficient = 0.482

Distance • Yards	Muzzle	100	200	300	400	500	600	800	1000
Velocity • fps	2800	2611	2429	2255	2088	1930	1779	1504	1277
Energy • ft-lbs	3481	3026	2621	2258	1937	1653	1406	1005	724
Taylor KO Index	24.6	23.0	21.4	19.8	18.4	17.0	15.7	13.2	11.2
Path • Inches	-1.5	+1.8	0.0	-7.7	-22.3	-44.8	-76.6	-175.3	-336.0
Wind Drift • Inches	0.0	0.7	2.8	6.5	11.9	19.3	29.0	55.9	94.5

These are actual groups shot at 300, 400, and 500 yards under field conditions. Nothing can prepare you for shots in the field unless you go out and shoot under the circumstances you expect to encounter. (Safari Press Archives)

.300 Winchester Short Magnum

The first cartridge fad in the twenty-first century was the short magnum. Winchester's .30-caliber version was introduced in 2001. Like most of the other short magnums, the .300 WSM is designed for actions that will handle cartridges with an overall length of 2.810 inches, the same length as the .308 Winchester. This fat, little cartridge is based on the .404 Jeffery case, much modified. As could be expected, this cartridge has already spawned a host of "necked to" spin-offs.

Relative Recoil Factor = 2.36

Specifications:

Controlling Agency for Standardization of this Ammunition: SAAMI

Bullet Weight Grains	Velocity fps	Maximum Average Pressure Copper Crusher	Transducer
150	3,300	N/A	65,000 psi
180	2,970	N/A	65,000 psi

Standard barrel for velocity testing: 24 inches long—1 turn in 10-inch twist

Availability:

Federal 130-grain Barnes Tipped Triple-Shock (P300WSMK) LF
G1 Ballistic Coefficient = 0.35

Distance • Yards	Muzzle	100	200	300	400	500	600	800	1000
Velocity • fps	3500	3200	2930	2670	2420	2190	1970	1576	1262
Energy • ft-lbs	3535	2965	2475	2055	1690	1385	1120	717	460
Taylor KO Index	20.0	18.3	16.8	15.3	13.8	12.5	11.3	9.0	7.2
Path • Inches	-1.5	+0.9	0.0	-5.0	-14.9	-30.8	-54.0	-129.5	-261.9
Wind Drift • Inches	0.0	0.7	2.8	6.7	12.4	20.2	30.5	60.4	105.4

Remington 150-grain Core-Lokt Pointed Soft Point (R300WSM1)
G1 Ballistic Coefficient = 0.29

Distance • Yards	Muzzle	100	200	300	400	500	600	800	1000
Velocity • fps	3320	2977	2600	2364	2087	1830	1597	1220	1010
Energy • ft-lbs	3671	2952	2356	1861	1451	1116	849	497	340
Taylor KO Index	21.9	19.6	17.2	15.6	13.8	12.1	10.5	8.1	6.7
Path • Inches	-1.5	+1.2	0.0	-6.2	-18.6	-39.0	-69.9	-177.1	-377.9
Wind Drift • Inches	0.0	0.9	3.8	8.9	16.8	27.9	42.9	87.1	151.4

Winchester 150-grain XP3 (SXP300SA)
G1 Ballistic Coefficient = 0.43

Distance • Yards	Muzzle	100	200	300	400	500	600	800	1000
Velocity • fps	3300	3068	2847	2637	2437	2246	2063	1725	1433
Energy • ft-lbs	3626	3134	2699	2316	1979	1679	1418	991	684
Taylor KO Index	21.8	20.2	18.8	17.4	16.1	14.8	13.6	11.4	9.5
Path • Inches	-1.5	+1.1	0.0	-5.4	-16.0	-32.0	-55.2	-127.8	-247.6
Wind Drift • Inches	0.0	0.6	2.5	5.7	10.6	17.1	25.7	49.7	84.9

Winchester 150-grain Power-Point (X300WSM1)
G1 Ballistic Coefficient = 0.29

Distance • Yards	Muzzle	100	200	300	400	500	600	800	1000
Velocity • fps	3270	2931	2617	2324	2050	1796	1566	1201	1000
Energy • ft-lbs	3561	2861	2281	1798	1399	1074	817	481	333
Taylor KO Index	21.6	19.3	17.3	15.3	13.5	11.9	10.3	7.9	6.6
Path • Inches	-1.5	+1.3	0.0	-6.4	-19.2	-40.4	-72.4	-183.7	-391.7
Wind Drift • Inches	0.0	0.9	3.8	9.1	17.1	28.5	43.8	89.1	154.1

Federal 150-grain Fusion (F300WSMFS1)
G1 Ballistic Coefficient = 0.41

Distance • Yards	Muzzle	100	200	300	400	500	600	800	1000
Velocity • fps	3250	3010	2770	2560	2350	2150	1962	1617	1330
Energy • ft-lbs	3520	3010	2565	2175	1830	1535	1282	872	590
Taylor KO Index	21.5	19.9	18.3	16.9	15.5	14.2	12.9	10.7	8.8
Path • Inches	-1.5	+1.2	0.0	-5.7	-16.7	-34.2	-58.9	-137.8	-270.5
Wind Drift • Inches	0.0	0.6	2.7	6.2	11.5	18.8	28.2	55.1	94.8

Winchester 150-grain E-Tip (S300SETA) LF

G1 Ballistic Coefficient = 0.469

Distance • Yards	Muzzle	100	200	300	400	500	600	800	1000
Velocity • fps	3300	3083	2877	2679	2491	2310	2137	1813	1527
Energy • ft-lbs	3626	3165	2756	2391	2066	1777	1521	1095	776
Taylor KO Index	21.8	20.3	19.0	17.7	16.4	15.2	14.1	12.0	10.1
Path • Inches	-1.5	+1.1	0.0	-5.3	-15.4	-31.2	-53.5	-122.5	-233.8
Wind Drift • Inches	0.0	0.6	2.3	5.3	9.7	15.8	23.5	45.2	76.7

Winchester 150-grain Ballistic Silvertip (SBST300S)

G1 Ballistic Coefficient = 0.425

Distance • Yards	Muzzle	100	200	300	400	500	600	800	1000
Velocity • fps	3300	3061	2834	2619	2414	2218	2032	1688	1395
Energy • ft-lbs	3628	3121	2676	2285	1941	1638	1375	950	648
Taylor KO Index	21.8	20.2	18.7	17.3	15.9	14.6	13.4	11.1	9.2
Path • Inches	-1.5	+1.1	0.0	-5.4	-15.9	-32.4	-56.0	-130.5	-255.7
Wind Drift • Inches	0.0	0.6	2.5	5.9	10.9	17.7	26.6	51.6	88.6

Winchester 150-grain Power Max Bonded (X300SBP)

G1 Ballistic Coefficient = 0.325

Distance • Yards	Muzzle	100	200	300	400	500	600	800	1000
Velocity • fps	3270	2962	2675	2406	2152	1914	1694	1322	1078
Energy • ft-lbs	3561	2922	2384	1927	1542	1220	956	583	387
Taylor KO Index	21.6	19.5	17.7	15.9	14.2	12.6	11.2	8.7	7.1
Path • Inches	-1.5	+1.2	0.0	-6.1	-18.3	-37.9	-67.2	-165.6	-344.4
Wind Drift • Inches	0.0	0.8	3.4	8.1	15.2	25.0	38.2	76.7	133.6

Federal 150-grain Nosler Ballistic Tip (P300WSMD)

G1 Ballistic Coefficient = 0.434

Distance • Yards	Muzzle	100	200	300	400	500	600	800	1000
Velocity • fps	3250	3020	2800	2590	2390	2200	2018	1683	1397
Energy • ft-lbs	3520	3035	2610	2235	1905	1615	1357	944	650
Taylor KO Index	21.5	19.9	18.5	17.1	15.8	14.5	13.3	11.4	9.2
Path • Inches	-1.5	+1.2	0.0	-5.6	-16.3	-33.3	-57.4	-132.9	-258.2
Wind Drift • Inches	0.0	0.6	2.5	5.9	10.9	17.6	26.5	51.3	87.8

Norma 150-grain Nosler BST (17570)

G1 Ballistic Coefficient = 0.435

Distance • Yards	Muzzle	100	200	300	400	500	600	800	1000
Velocity • fps	3215	2985	2767	2560	2364	2175	1994	1663	1381
Energy • ft-lbs	3437	2963	2547	2179	1861	1576	1325	921	635
Taylor KO Index	21.2	19.7	18.3	16.9	15.6	14.4	13.2	11.0	9.1
Path • Inches	-1.5	+1.2	0.0	-5.7	-16.8	-34.1	-58.8	-136.1	-264.3
Wind Drift • Inches	0.0	0.7	2.8	6.3	11.0	17.9	26.8	52.0	88.9

Cor-Bon 150-grain DPX (DPX300WSM150/20)

G1 Ballistic Coefficient = 0.370

Distance • Yards	Muzzle	100	200	300	400	500	600	800	1000
Velocity • fps	3100	2825	2565	2321	2089	1872	1670	1327	1093
Energy • ft-lbs	3202	2658	2193	1794	1454	1168	930	587	398
Taylor KO Index	20.5	18.6	16.9	15.3	13.8	12.4	11.0	8.8	7.2
Path • Inches	-1.5	+1.4	0.0	-6.7	-20.0	-41.2	-72.3	-175.5	-358.4
Wind Drift • Inches	0.0	0.8	3.4	8.0	15.0	24.6	37.5	74.5	128.8

Norma 150-grain FULL JACKET (17573)

G1 Ballistic Coefficient = 0.423

Distance • Yards	Muzzle	100	200	300	400	500	600	800	1000
Velocity • fps	2953	2731	2519	2318	2126	1944	1771	1463	1218
Energy • ft-lbs	2905	2485	2116	1791	1507	1259	1045	713	495
Taylor KO Index	19.5	18.0	16.6	15.3	14.0	12.8	11.7	9.7	8.0
Path • Inches	-1.5	+1.6	0.0	-7.1	-20.6	-41.9	-72.4	-169.0	-331.4
Wind Drift • Inches	0.0	0.7	3.0	6.9	12.8	20.9	31.5	61.4	105.1

Federal 165-grain Nosler Partition (P300WSME)

G1 Ballistic Coefficient = 0.412

Distance • Yards	Muzzle	100	200	300	400	500	600	800	1000
Velocity • fps	3120	2880	2660	2450	2240	2050	1866	1535	1267
Energy • ft-lbs	3565	3045	2590	2190	1840	1535	1277	864	588
Taylor KO Index	22.7	20.9	19.3	17.8	16.3	14.9	13.5	11.1	9.2
Path • Inches	-1.5	+1.4	0.0	-6.2	-18.3	-37.4	-64.7	-151.5	-298.0
Wind Drift • Inches	0.0	0.7	2.8	6.6	12.2	19.9	30.0	58.6	100.7

Federal 165-grain Fusion (F300WSMFS1)

G1 Ballistic Coefficient = 0.453

Distance • Yards	Muzzle	100	200	300	400	500	600	800	1000
Velocity • fps	3100	2890	2680	2480	2300	2120	1950	1636	1368
Energy • ft-lbs	3520	3050	2630	2260	1930	1640	1393	981	686
Taylor KO Index	22.5	21.0	19.5	18.0	16.7	15.4	14.2	11.9	9.9
Path • Inches	-1.5	+1.4	0.0	-6.2	-18.0	-36.5	-62.6	-144.2	-278.0
Wind Drift • Inches	0.0	0.6	2.6	6.0	11.0	17.9	26.9	52.0	88.6

Federal 165-grain Trophy Bonded Tip (P300WSMTT2)

G1 Ballistic Coefficient = 0.451

Distance • Yards	Muzzle	100	200	300	400	500	600	800	1000
Velocity • fps	3130	2910	2710	2510	2320	2140	1968	1651	1379
Energy • ft-lbs	3590	3110	2680	2305	1970	1675	1419	988	697
Taylor KO Index	22.7	21.1	19.7	18.2	16.8	15.5	14.3	12.0	10.0
Path • Inches	-1.5	+1.3	0.0	-6.0	-17.6	-35.7	-61.4	-141.4	-273.0
Wind Drift • Inches	0.0	0.6	2.6	5.9	11.0	17.8	26.6	51.3	87.9

Federal 165-grain Barnes Triple-Shock X-Bullet (P300WSMG) LF

G1 Ballistic Coefficient = 0.381

Distance • Yards	Muzzle	100	200	300	400	500	600	800	1000
Velocity • fps	3130	2870	2630	2400	2190	1980	1788	1446	1185
Energy • ft-lbs	3590	3025	2540	2115	1750	1435	1172	766	514
Taylor KO Index	22.7	20.8	19.1	17.4	15.9	14.4	13.0	10.5	8.6
Path • Inches	-1.5	+1.4	0.0	-6.3	-18.8	-38.5	-67.1	-159.6	-319.5
Wind Drift • Inches	0.0	0.7	3.1	7.2	13.3	21.8	33.0	65.1	112.4

Black Hills 175-grain Sierra MatchKing (1C300WSMBHGN2)

G1 Ballistic Coefficient = 0.496

Distance • Yards	Muzzle	100	200	300	400	500	600	800	1000
Velocity • fps	2950	2760	2578	2404	2236	2074	1920	1635	1388
Energy • ft-lbs	3381	2961	2584	2245	1943	1672	1433	1039	749
Taylor KO Index	22.7	21.3	19.9	18.5	17.2	16.0	14.8	12.6	10.7
Path • Inches	-1.5	+1.5	0.0	-6.8	-19.6	-39.3	-67.2	-153.1	-292.5
Wind Drift • Inches	0.0	0.6	2.5	5.8	10.7	17.3	25.9	49.7	84.1

Remington 180-grain AccuTip Boat Tail (PRA300WSMB)

G1 Ballistic Coefficient = 0.480

Distance • Yards	Muzzle	100	200	300	400	500	600	800	1000
Velocity • fps	3010	2812	2622	2440	2265	2097	1937	1641	1386
Energy • ft-lbs	3621	3159	2746	2378	2050	1757	1500	1077	768
Taylor KO Index	23.8	22.3	20.8	19.3	17.9	16.6	15.3	13.0	11.0
Path • Inches	-1.5	+1.4	0.0	-6.5	-18.9	-38.0	-65.1	-148.8	-284.6
Wind Drift • Inches	0.0	0.6	2.5	5.8	10.8	17.5	26.1	50.3	85.3

Winchester 180-grain XP3 (SXP300S)

G1 Ballistic Coefficient = 0.527

Distance • Yards	Muzzle	100	200	300	400	500	600	800	1000
Velocity • fps	3010	2829	2655	2488	2326	2171	2022	1743	1494
Energy • ft-lbs	3621	3198	2817	2473	2162	1883	1634	1214	893
Taylor KO Index	23.8	22.4	21.0	19.7	18.4	17.2	16.0	13.8	11.8
Path • Inches	-1.5	+1.4	0.0	-6.4	-18.3	-36.8	-62.6	-141.1	-265.8
Wind Drift • Inches	0.0	0.6	2.3	5.3	9.7	15.6	23.3	44.5	74.9

Winchester 180-grain AccuBond CT (S300WSMCT)

G1 Ballistic Coefficient = 0.509

Distance • Yards	Muzzle	100	200	300	400	500	600	800	1000
Velocity • fps	3010	2822	2643	2470	2304	2144	1991	1705	1454
Energy • ft-lbs	3622	3185	2795	2439	2121	1837	1584	1163	845
Taylor KO Index	23.8	22.4	20.9	19.6	18.2	17.0	15.8	13.5	11.5
Path • Inches	-1.5	+1.4	0.0	-6.4	-18.5	-37.2	-63.5	-143.8	-272.4
Wind Drift • Inches	0.0	0.6	2.4	5.5	10.1	16.3	24.3	46.6	78.6

Federal 180-grain Nosler Partition (P300WSMB)

G1 Ballistic Coefficient = 0.473

Distance • Yards	Muzzle	100	200	300	400	500	600	800	1000
Velocity • fps	2980	2780	2590	2410	2230	2060	1900	1605	1353
Energy • ft-lbs	3550	3090	2680	2315	1990	1700	1443	1029	732
Taylor KO Index	23.6	22.0	20.5	19.1	17.7	16.3	15.0	12.7	10.7
Path • Inches	-1.5	+1.5	0.0	-6.6	-18.4	-39.1	-67.0	-153.6	-294.8
Wind Drift • Inches	0.0	0.6	2.6	6.0	11.1	18.0	27.0	52.1	88.5

Federal 180-grain Soft Point (300WSMC)

G1 Ballistic Coefficient = 0.386

Distance • Yards	Muzzle	100	200	300	400	500	600	800	1000
Velocity • fps	2980	2740	2500	2280	2080	1880	1696	1375	1140
Energy • ft-lbs	3550	2990	2505	2085	1720	1410	1150	756	520
Taylor KO Index	23.6	21.7	19.8	18.1	16.5	14.9	13.4	10.9	9.0
Path • Inches	-1.5	+1.6	0.0	-7.1	-21.0	-42.7	-74.6	-177.4	-354.4
Wind Drift • Inches	0.0	0.8	3.2	7.6	14.1	23.1	35.0	68.8	118.2

Winchester 180-grain Power-Point (X300WSM)

G1 Ballistic Coefficient = 0.438

Distance • Yards	Muzzle	100	200	300	400	500	600	800	1000
Velocity • fps	2970	2755	2549	2353	2166	1987	1818	1512	1263
Energy • ft-lbs	3526	3034	2598	2214	1827	1578	1322	912	638
Taylor KO Index	23.5	21.8	20.2	18.6	17.2	15.7	14.4	12.0	10.0
Path • Inches	-1.5	+1.6	0.0	-7.0	-20.3	-41.0	-70.2	-162.8	-316.9
Wind Drift • Inches	0.0	0.7	2.8	6.6	12.2	19.9	29.9	58.1	99.2

Federal 180-grain Nosler AccuBond (P300WSMA1)

G1 Ballistic Coefficient = 0.507

Distance • Yards	Muzzle	100	200	300	400	500	600	800	1000
Velocity • fps	2960	2770	2600	2420	2260	2100	1948	1666	1420
Energy • ft-lbs	3500	3075	2690	2350	2040	1765	1517	1110	806
Taylor KO Index	23.4	21.9	20.6	19.2	17.9	16.6	15.4	13.2	11.2
Path • Inches	-1.5	+1.5	0.0	-6.6	-19.2	-38.7	-66.0	-149.8	-284.0
Wind Drift • Inches	0.0	0.6	2.4	5.6	10.4	16.8	25.0	48.0	81.1

Black Hills 180-grain Nosler AccuBond (1C300WSMBHGN1)

G1 Ballistic Coefficient = 0.509

Distance • Yards	Muzzle	100	200	300	400	500	600	800	1000
Velocity • fps	2950	2765	2587	2417	2252	2094	1943	1662	1417
Energy • ft-lbs	3478	3056	2676	2335	2028	1754	1510	1105	803
Taylor KO Index	23.4	21.9	20.5	19.1	17.8	16.6	15.4	13.2	11.2
Path • Inches	-1.5	+1.5	0.0	-6.7	-19.4	-38.9	-66.4	-150.8	-286.8
Wind Drift • Inches	0.0	0.6	2.5	5.8	10.7	17.3	25.9	49.7	84.1

Federal 180-grain Fusion (F300WSM1)

G1 Ballistic Coefficient = 0.487

Distance • Yards	Muzzle	100	200	300	400	500	600	800	1000
Velocity • fps	2950	2760	2570	2390	2220	2060	1904	1616	1369
Energy • ft-lbs	3480	3035	2640	2285	1970	1690	1449	1044	749
Taylor KO Index	23.3	21.9	20.4	18.9	17.6	16.3	15.1	12.8	10.8
Path • Inches	-1.5	+1.5	0.0	-6.8	-19.7	-39.7	-67.7	-154.5	-295.0
Wind Drift • Inches	0.0	0.6	2.6	5.9	10.9	17.7	26.4	50.9	86.2

Norma 180-grain ORYX (17571)

G1 Ballistic Coefficient = 0.288

Distance • Yards	Muzzle	100	200	300	400	500	600	800	1000
Velocity • fps	2904	2568	2288	2011	1756	1525	1327	1059	923
Energy • ft-lbs	3371	2673	2094	1617	1232	930	704	448	341
Taylor KO Index	23.0	20.5	18.1	15.9	13.9	12.1	10.5	8.4	7.3
Path • Inches	-1.5	+1.8	0.0	-8.5	-25.7	-54.3	-97.8	-250.0	-525.5
Wind Drift • Inches	0.0	1.1	4.6	11.1	21.0	35.1	54.0	107.7	178.3

Nosler 180-grain AccuBond (22600) + (60063)

G1 Ballistic Coefficient = 0.507

Distance • Yards	Muzzle	100	200	300	400	500	600	800	1000
Velocity • fps	2900	2716	2540	2371	2208	2051	1901	1624	1385
Energy • ft-lbs	3362	2950	2580	2247	1949	1682	1445	1055	766
Taylor KO Index	23.0	21.5	20.1	18.8	17.5	16.2	15.1	12.9	11.0
Path • Inches	-1.5	+1.6	0.0	-7.0	-20.2	-40.6	-69.2	-157.0	-297.9
Wind Drift • Inches	0.0	0.6	2.5	5.8	10.7	17.3	25.8	49.5	83.6

Cor-Bon 180-grain DPX (DPX300WSM180/20)

G1 Ballistic Coefficient = 0.450

Distance • Yards	Muzzle	100	200	300	400	500	600	800	1000
Velocity • fps	3000	2789	2588	2395	2211	2035	1867	1562	1308
Energy • ft-lbs	3598	3110	2677	2294	1954	1655	1394	976	684
Taylor KO Index	23.8	22.1	20.5	19.0	17.5	16.1	14.8	12.4	10.4
Path • Inches	-1.5	+1.5	0.0	-8.7	-19.4	-39.3	-67.6	-156.2	-302.3
Wind Drift • Inches	0.0	0.6	2.7	6.3	11.7	19.0	28.4	55.1	94.0

Federal 180-grain Barnes MRX-Bullet (P300WSMJ) LF

G1 Ballistic Coefficient = 0.552

Distance • Yards	Muzzle	100	200	300	400	500	600	800	1000
Velocity • fps	2980	2810	2640	2480	2330	2180	2038	1769	1528
Energy • ft-lbs	3550	3150	2790	2465	2170	1900	1660	1251	933
Taylor KO Index	23.6	22.3	20.9	19.6	18.5	17.3	16.1	14.0	12.1
Path • Inches	-1.5	+1.5	0.0	-6.4	-18.4	-37.0	-62.8	-140.9	-262.7
Wind Drift • Inches	0.0	0.5	2.2	5.1	9.3	15.0	22.4	42.6	71.4

Winchester 180-grain Ballistic Silvertip (SBST300SA)

G1 Ballistic Coefficient = 0.507

Distance • Yards	Muzzle	100	200	300	400	500	600	800	1000
Velocity • fps	3010	2822	2641	2468	2301	2141	1987	1701	1450
Energy • ft-lbs	3621	3182	2788	2434	2116	1832	1579	1157	840
Taylor KO Index	23.8	22.4	20.9	19.5	18.2	17.0	15.7	13.5	11.5
Path • Inches	-1.5	+1.4	0.0	-6.4	-18.6	-37.3	-63.6	-144.1	-273.2
Wind Drift • Inches	0.0	0.6	2.4	5.5	10.1	16.4	24.4	46.8	79.0

Winchester 180-grain E-Tip (S300SET)

G1 Ballistic Coefficient = 0.523

Distance • Yards	Muzzle	100	200	300	400	500	600	800	1000
Velocity • fps	3010	2827	2652	2484	2321	2165	2015	1734	1483
Energy • ft-lbs	3621	3195	2811	2465	2153	1873	1623	1203	882
Taylor KO Index	23.8	22.4	21.0	19.7	18.4	17.1	16.0	13.7	11.7
Path • Inches	-1.5	+1.4	0.0	-6.4	-18.3	-36.8	-62.8	-141.7	-267.2
Wind Drift • Inches	0.0	0.6	2.3	5.3	9.8	15.8	23.5	45.0	75.7

Federal 180-grain Barnes Triple Shock X-Bullet (P300WSMF) LF

G1 Ballistic Coefficient = 0.456

Distance • Yards	Muzzle	100	200	300	400	500	600	800	1000
Velocity • fps	2980	2770	2570	2380	2200	2030	1865	1564	1312
Energy • ft-lbs	3550	3070	2645	2265	1935	1640	1391	978	688
Taylor KO Index	23.6	21.9	20.4	18.8	17.4	16.1	14.8	12.4	10.4
Path • Inches	-1.5	+1.5	0.0	-6.7	-19.7	-39.8	-68.2	-157.2	-303.7
Wind Drift • Inches	0.0	0.6	2.7	6.3	11.6	18.8	28.3	54.7	93.2

Remington 180-grain Swift Scirocco Bonded (PRSC300WSMB)

G1 Ballistic Coefficient = 0.507

Distance • Yards	Muzzle	100	200	300	400	500	600	800	1000
Velocity • fps	2980	2793	2614	2442	2276	2116	1964	1680	1432
Energy • ft-lbs	3549	3118	2730	2382	2070	1790	1542	1128	819
Taylor KO Index	23.6	22.1	20.7	19.3	18.0	16.8	15.6	13.3	11.3
Path • Inches	-1.5	+1.5	0.0	-6.6	-19.0	-38.1	-65.0	-147.5	-279.6
Wind Drift • Inches	0.0	0.6	2.4	5.6	10.3	16.6	24.8	47.5	80.3

Winchester 180-grain Power Max Bonded (X300WSMBP)

G1 Ballistic Coefficient = 0.394

Distance • Yards	Muzzle	100	200	300	400	500	600	800	1000
Velocity • fps	2970	2731	2505	2290	2085	1892	1711	1393	1156
Energy • ft-lbs	3525	2982	2508	2096	1738	1430	1170	775	534
Taylor KO Index	23.5	21.6	19.8	18.1	16.5	15.0	13.6	11.0	9.2
Path • Inches	-1.5	+1.6	0.0	-7.1	-20.9	-42.7	-74.3	-175.9	-349.9
Wind Drift • Inches	0.0	0.8	3.2	7.4	13.8	22.6	34.2	67.2	115.3

Federal 180-grain Trophy Bonded Tip (P300WSMJ)

G1 Ballistic Coefficient = 0.500

Distance • Yards	Muzzle	100	200	300	400	500	600	800	1000
Velocity • fps	2960	2770	2590	2420	2250	2090	1936	1651	1404
Energy • ft-lbs	3500	3070	2680	2335	2025	1745	1499	1090	788
Taylor KO Index	23.4	21.9	20.5	19.2	17.8	16.6	15.3	13.1	11.1
Path • Inches	-1.5	+1.5	0.0	-6.6	-19.3	-38.9	-64.4	-151.0	-286.4
Wind Drift • Inches	0.0	0.6	2.5	5.7	10.5	17.0	25.5	48.9	62.7

Cor-Bon 190-grain HPBT (PM300WSM190/20)

G1 Ballistic Coefficient = 0.530

Distance • Yards	Muzzle	100	200	300	400	500	600	800	1000
Velocity • fps	2900	2724	2555	2392	2236	2085	1940	1670	1433
Energy • ft-lbs	3549	3132	2755	2415	2109	1834	1588	1177	866
Taylor KO Index	24.2	22.6	21.4	20.0	18.7	17.4	16.2	14.0	12.0
Path • Inches	-1.5	+1.6	0.0	-6.9	-19.9	-39.9	-67.9	-153.1	-288.6
Wind Drift • Inches	0.0	0.6	2.4	5.5	10.2	16.4	24.5	46.8	78.7

.300 Winchester Magnum

The full-length magnums are too long to chamber in standard-length (read .30-06 length) actions. When Winchester introduced the .300 Winchester Magnum in 1963, the idea was to obtain a high-performance cartridge that could be chambered in their Model 70 actions. The effort was highly successful. The .300 Winchester Magnum sits comfortably in the top ten of reloading die sales, one place ahead of the 7mm Remington Magnum. Reloading-die sales are a pretty good indication of a cartridge's popularity. The .300 Winchester Magnum's popularity is well deserved. It provides performance virtually identical to the .300 H&H in a shorter action. Some factory loads drive 180-grain bullets in excess of 3,000 fps.

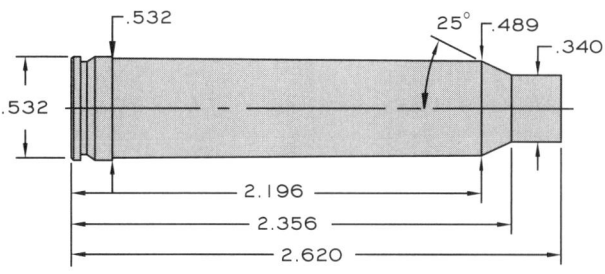

Relative Recoil Factor = 2.39

Specifications:

Controlling Agency for Standardization of this Ammunition: SAAMI

Bullet Weight Grains	Velocity fps	Maximum Average Pressure Copper Crusher	Transducer
150	3,275	54,000 cup	64,000 psi
180	2,950	54,000 cup	64,000 psi
190	2,875	54,000 cup	64,000 psi
200	2,800	54,000 cup	64,000 psi
220	2,665	54,000 cup	64,000 psi

Standard barrel for velocity testing: 24 inches long—1 turn in 10-inch twist

Availability:

Federal 130-grain Barnes Tipped Triple-Shock (P300WL)
G1 Ballistic Coefficient = 0.358

Distance • Yards	Muzzle	100	200	300	400	500	600	800	1000
Velocity • fps	3500	3200	2930	2670	2420	2190	1971	1578	1263
Energy • ft-lbs	3535	2965	2475	2055	1690	1385	1122	719	461
Taylor KO Index	20.0	18.3	16.8	15.3	13.8	12.5	11.3	9.0	7.2
Path • Inches	-1.5	+0.9	0.0	-5.0	-14.9	-30.8	-53.9	-129.3	-261.6
Wind Drift • Inches	0.0	0.7	2.8	6.6	12.3	20.2	30.5	60.3	105.2

Remington 150-grain Core-Lokt Ultra Bonded (PRC300WA)
G1 Ballistic Coefficient = 0.310

Distance • Yards	Muzzle	100	200	300	400	500	600	800	1000
Velocity • fps	3290	2967	2660	2384	2120	1873	1645	1271	1042
Energy • ft-lbs	3605	2931	2366	1893	1496	1168	902	538	362
Taylor KO Index	21.7	19.6	17.6	15.7	14.0	12.4	10.9	8.4	6.9
Path • Inches	-1.5	+1.2	0.0	-6.1	-18.4	-38.5	-68.6	-171.4	-360.9
Wind Drift • Inches	0.0	0.9	3.6	8.5	16.0	26.4	40.5	81.7	142.8

PMC 150-grain BTSP (300HIA)
G1 Ballistic Coefficient = 0.349

Distance • Yards	Muzzle	100	200	300	400	500	600	800	1000
Velocity • fps	3290	3275	2988	2718	2464	2224	1998	1786	1417
Energy • ft-lbs	3572	2972	2460	2022	1648	1329	1063	669	439
Taylor KO Index	21.6	19.7	17.9	16.3	14.7	13.2	11.8	9.4	7.6
Path • Inches	-1.5	+1.2	0.0	-5.9	-17.6	-36.3	-63.8	-154.4	-315.2
Wind Drift • Inches	0.0	0.8	3.2	7.4	13.9	22.8	34.6	68.9	120.0

Winchester 150-grain Power-Point (X300WM1)
G1 Ballistic Coefficient = 0.295

Distance • Yards	Muzzle	100	200	300	400	500	600	800	1000
Velocity • fps	3290	2951	2636	2342	2068	1813	1581	1212	1005
Energy • ft-lbs	3605	2900	2314	1827	1424	1095	833	489	337
Taylor KO Index	21.7	19.5	17.4	15.5	13.6	12.0	10.4	8.0	6.6
Path • Inches	-1.5	+1.3	0.0	-6.3	-19.0	-39.8	-71.5	-183.2	-393.3
Wind Drift • Inches	0.0	0.9	3.8	9.0	17.0	28.2	39.0	79.2	137.2

Hornady 150-grain SST (82014)

G1 Ballistic Coefficient = 0.415

Distance • Yards	Muzzle	100	200	300	400	500	600	800	1000
Velocity • fps	3275	3032	2802	2584	2375	2177	1989	1644	1353
Energy • ft-lbs	3572	3061	2615	2223	1879	1578	1318	900	610
Taylor KO Index	21.6	20.0	18.5	17.1	15.7	14.4	13.1	10.9	8.9
Path • Inches	-1.5	+1.1	0.0	-5.6	-16.4	-33.4	-57.7	-134.6	-263.6
Wind Drift • Inches	0.0	0.6	2.6	6.1	11.3	18.4	27.6	53.6	92.6

Hornady 150-grain BTSP InterLock (8201)

G1 Ballistic Coefficient = 0.350

Distance • Yards	Muzzle	100	200	300	400	500	600	800	1000
Velocity • fps	3275	2988	2718	2469	2224	1998	1789	1419	1149
Energy • ft-lbs	3573	2974	2461	2023	1648	1330	1066	671	440
Taylor KO Index	21.6	19.7	17.9	16.3	14.7	13.2	11.8	9.4	7.6
Path • Inches	-1.5	+1.2	0.0	-6.0	-17.8	-36.5	-63.8	-155.2	-319.8
Wind Drift • Inches	0.0	0.8	3.2	7.4	13.9	22.8	34.6	68.7	119.7

Winchester 150-grain XP3 (SXP300WMA)

G1 Ballistic Coefficient = 0.438

Distance • Yards	Muzzle	100	200	300	400	500	600	800	1000
Velocity • fps	3260	3030	2811	2603	2404	2214	2034	1699	1412
Energy • ft-lbs	3539	3057	2632	2256	1925	1633	1378	962	664
Taylor KO Index	21.5	20.0	18.6	17.2	15.9	14.6	13.4	11.2	9.3
Path • Inches	-1.5	+1.1	0.0	-5.6	-16.2	-33.0	-56.8	-131.3	-254.6
Wind Drift • Inches	0.0	0.6	2.5	5.8	10.7	17.4	26.1	50.5	86.4

Federal 150-grain Fusion (F300WFS1)

G1 Ballistic Coefficient = 0.411

Distance • Yards	Muzzle	100	200	300	400	500	600	800	1000
Velocity • fps	3200	2960	2730	2510	2300	2110	1922	1583	1302
Energy • ft-lbs	3410	2915	2480	2100	1770	1480	1231	835	565
Taylor KO Index	21.1	19.5	18.0	16.6	15.2	13.9	12.7	10.4	8.6
Path • Inches	-1.5	+1.2	0.0	-5.9	-17.3	-35.4	-61.1	-143.1	-281.2
Wind Drift • Inches	0.0	0.7	2.7	6.4	11.8	19.3	29.0	56.6	97.3

Federal 150-grain Soft Point (300WGS)

G1 Ballistic Coefficient = 0.391

Distance • Yards	Muzzle	100	200	300	400	500	600	800	1000
Velocity • fps	3150	2900	2660	2440	2220	2020	1830	1488	1218
Energy • ft-lbs	3305	2800	2360	1975	1645	1355	1115	737	495
Taylor KO Index	20.8	19.1	17.6	16.1	14.7	13.3	12.1	9.8	8.0
Path • Inches	-1.5	+1.3	0.0	-6.2	-18.3	-37.5	-65.2	-154.2	-306.9
Wind Drift • Inches	0.0	0.7	3.0	6.9	12.8	21.0	31.7	62.2	107.4

Remington 150-grain Core-Lokt Pointed Soft Point – Managed Recoil (RL300W1)

G1 Ballistic Coefficient = 0.314

Distance • Yards	Muzzle	100	200	300	400	500	600	800	1000
Velocity • fps	2650	2373	2113	1870	1646	1446	1275	1046	924
Energy • ft-lbs	2339	1875	1486	1164	902	696	541	365	284
Taylor KO Index	17.5	15.7	13.9	12.3	10.9	9.5	8.4	6.9	6.1
Path • Inches	-1.5	+2.4	0.0	-10.2	-30.4	-63.5	-113.4	-282.6	-578.4
Wind Drift • Inches	0.0	1.1	4.8	11.5	21.7	36.0	55.0	107.5	-175.4

Hornady 150-grain GMX (82012) LF

G1 Ballistic Coefficient = 0.415

Distance • Yards	Muzzle	100	200	300	400	500	600	800	1000
Velocity • fps	3400	3150	2914	2690	2477	2273	2080	1723	1418
Energy • ft-lbs	3850	3304	2827	2409	2043	1721	1441	989	669
Taylor KO Index	22.4	20.8	19.2	17.8	16.3	15.0	13.7	11.4	9.4
Path • Inches	-1.5	+1.0	0.0	-5.1	-15.0	-30.6	-53.0	-123.5	-241.5
Wind Drift • Inches	0.0	0.6	2.5	5.8	10.8	17.5	26.2	51.0	87.6

Winchester 150-grain Power Max Bonded (X30WM1BP)

G1 Ballistic Coefficient = 0.325

Distance • Yards	Muzzle	100	200	300	400	500	600	800	1000
Velocity • fps	3290	2981	2693	2422	2168	1929	1708	1332	1084
Energy • ft-lbs	3605	2959	2415	1954	1565	1239	971	591	391
Taylor KO Index	21.7	19.7	17.8	16.0	14.3	12.7	11.3	8.8	7.2
Path • Inches	-1.5	+1.2	0.0	-6.0	-19.0	-37.4	-66.2	-163.2	-399.4
Wind Drift • Inches	0.0	0.8	3.4	8.0	15.0	24.8	37.8	76.0	132.6

Winchester 150-grain E-Tip (S300WMETA) LF   G1 Ballistic Coefficient = 0.486

Distance • Yards	Muzzle	100	200	300	400	500	600	800	1000
Velocity • fps	3260	3083	2877	2679	2491	2310	2142	1828	1549
Energy • ft-lbs	3539	3088	2687	2329	2011	1728	1528	1114	800
Taylor KO Index	21.5	20.3	19.0	17.7	16.4	15.2	14.1	12.1	10.2
Path • Inches	-1.5	+1.1	0.0	-5.4	-15.9	-32.1	-54.1	-123.1	-234.4
Wind Drift • Inches	0.0	0.5	2.2	5.2	9.5	15.4	22.9	43.9	74.3

Winchester 150-grain PowerCore 95/5 (X300WMLF) LF   G1 Ballistic Coefficient = 0.344

Distance • Yards	Muzzle	100	200	300	400	500	600	800	1000
Velocity • fps	3260	2969	2697	2441	2199	1971	1758	1390	1128
Energy • ft-lbs	3539	2936	2423	1984	1610	1294	1030	644	424
Taylor KO Index	21.5	19.6	17.8	16.1	14.5	13.0	11.6	9.2	7.4
Path • Inches	-1.5	+1.2	0.0	-6.0	-17.9	-37.0	-65.1	-158.1	-324.1
Wind Drift • Inches	0.0	0.8	3.2	7.6	14.2	23.4	35.6	70.9	123.4

Norma 150-grain Nosler BST (17551)   G1 Ballistic Coefficient = 0.435

Distance • Yards	Muzzle	100	200	300	400	500	600	800	1000
Velocity • fps	3248	3017	2798	2590	2390	2199	2017	1683	1396
Energy • ft-lbs	3515	3033	2609	2235	1903	1611	1356	944	650
Taylor KO Index	21.4	19.9	18.5	17.1	15.8	14.5	13.3	11.1	9.2
Path • Inches	-1.5	+1.2	0.0	-5.6	-16.4	-33.3	-57.4	-133.0	-258.4
Wind Drift • Inches	0.0	0.6	2.5	5.9	10.9	17.6	26.3	51.3	87.8

Norma 150-grain Barnes Triple-Shock (17546) LF   G1 Ballistic Coefficient = 0.428

Distance • Yards	Muzzle	100	200	300	400	500	600	800	1000
Velocity • fps	3215	2932	2761	2550	2350	2158	1976	1641	1358
Energy • ft-lbs	3444	2962	2539	2167	1840	1552	1300	897	615
Taylor KO Index	21.2	19.4	18.2	16.8	15.5	14.2	13.0	10.8	9.0
Path • Inches	-1.5	+1.2	0.0	-5.8	-16.9	-34.4	-59.3	-137.7	-268.4
Wind Drift • Inches	0.0	0.6	2.6	6.1	11.2	18.2	27.4	53.2	91.3

Remington 150-grain Copper Solid Tipped (PCS300WA) LF   G1 Ballistic Coefficient = 0.440

Distance • Yards	Muzzle	100	200	300	400	500	600	800	1000
Velocity • fps	3290	3037	2799	2572	2357	2152	1958	1605	1390
Energy • ft-lbs	3605	3072	2608	2204	1850	1542	1277	858	574
Taylor KO Index	21.7	20.0	18.5	17.0	15.6	14.2	12.9	10.6	8.7
Path • Inches	-1.5	+1.1	0.0	-5.6	-16.4	-33.6	-58.2	-136.8	-270.2
Wind Drift • Inches	0.0	0.7	2.7	6.4	11.7	19.1	28.8	56.4	97.2

Norma 150-grain FULL JACKET (17545)   G1 Ballistic Coefficient = 0.423

Distance • Yards	Muzzle	100	200	300	400	500	600	800	1000
Velocity • fps	3067	2839	2622	2416	2220	2033	1855	1534	1272
Energy • ft-lbs	3134	2686	2293	1947	1643	1377	1147	784	539
Taylor KO Index	20.2	18.7	17.3	15.9	14.7	13.4	12.2	10.1	8.4
Path • Inches	-1.5	+1.4	0.0	-6.4	-18.9	-38.4	-66.4	-154.7	-302.9
Wind Drift • Inches	0.0	0.7	2.8	6.6	12.1	19.8	29.8	58.0	99.4

Hornady 165-grain SST (82024)   G1 Ballistic Coefficient = 0.452

Distance • Yards	Muzzle	100	200	300	400	500	600	800	1000
Velocity • fps	3100	2885	2680	2483	2296	2117	1948	1632	1364
Energy • ft-lbs	3620	3049	2630	2259	1930	1641	1388	976	682
Taylor KO Index	22.5	20.9	19.5	18.0	16.7	15.4	14.1	11.8	9.9
Path • Inches	-1.5	+1.4	0.0	-6.4	-18.6	-37.2	-62.6	-144.7	-280.8
Wind Drift • Inches	0.0	0.6	2.6	6.0	11.1	18.0	27.0	52.2	89.0

Hornady 165-grain InterBond (82028)   G1 Ballistic Coefficient = 0.447

Distance • Yards	Muzzle	100	200	300	400	500	600	800	1000
Velocity • fps	3260	3035	2821	2617	2422	2235	2057	1727	1440
Energy • ft-lbs	3893	3373	2915	2505	2148	1830	1550	1093	760
Taylor KO Index	23.7	22.0	20.5	19.0	17.6	16.2	14.9	12.5	10.5
Path • Inches	-1.5	+1.1	0.0	-5.5	-16.1	-32.7	-56.2	-129.5	-250.0
Wind Drift • Inches	0.0	0.6	2.4	5.7	10.5	17.0	25.4	49.1	83.7

Hornady 165-grain SST (82024)

G1 Ballistic Coefficient = 0.452

Distance • Yards	Muzzle	100	200	300	400	500	600	800	1000
Velocity • fps	3100	2885	2680	2483	2296	2117	1948	1632	1364
Energy • ft-lbs	3620	3049	2630	2259	1930	1641	1388	976	682
Taylor KO Index	22.5	20.9	19.5	18.0	16.7	15.4	14.1	11.8	9.9
Path • Inches	-1.5	+1.4	0.0	-6.4	-18.6	-37.2	-62.6	-144.7	-280.8
Wind Drift • Inches	0.0	0.6	2.6	6.0	11.1	18.0	27.0	52.2	89.0

Hornady 165-grain BTSP InterLock (8202)

G1 Ballistic Coefficient = 0.435

Distance • Yards	Muzzle	100	200	300	400	500	600	800	1000
Velocity • fps	3100	2877	2665	2462	2269	2084	1908	1587	1319
Energy • ft-lbs	3522	3033	2603	2221	1887	1592	1334	923	637
Taylor KO Index	22.5	20.9	19.3	17.8	16.5	15.1	13.9	11.5	9.6
Path • Inches	-1.5	+1.3	0.0	-6.5	-18.5	-37.3	-63.8	-148.5	-290.5
Wind Drift • Inches	0.0	0.6	2.7	6.3	11.6	18.7	28.3	55.0	94.1

Nosler 165-grain Partition (23320)

G1 Ballistic Coefficient = 0.417

Distance • Yards	Muzzle	100	200	300	400	500	600	800	1000
Velocity • fps	3100	2867	2647	2436	2236	2045	1864	1537	1271
Energy • ft-lbs	3521	3011	2567	2174	1832	1532	1274	866	592
Taylor KO Index	22.5	20.8	19.2	17.7	16.2	14.8	13.5	11.2	9.2
Path • Inches	-1.5	+1.4	0.0	-6.3	-18.6	-37.7	-65.2	-152.5	-299.2
Wind Drift • Inches	0.0	0.7	2.8	6.6	12.2	19.8	29.8	58.2	99.9

Federal 165-grain Fusion (F300WFS2)

G1 Ballistic Coefficient = 0.451

Distance • Yards	Muzzle	100	200	300	400	500	600	800	1000
Velocity • fps	3080	2870	2660	2460	2280	2100	1930	1617	1352
Energy • ft-lbs	3475	3005	2590	2225	1900	1610	1365	959	670
Taylor KO Index	22.4	20.8	19.3	17.9	16.6	15.2	14.0	11.7	9.8
Path • Inches	-1.5	+1.4	0.0	-6.3	-18.3	-37.1	-63.6	-146.7	-283.6
Wind Drift • Inches	0.0	0.6	2.6	6.1	11.2	18.2	27.3	52.8	90.1

Federal 165-grain Nosler Partition (P300WK)

G1 Ballistic Coefficient = 0.410

Distance • Yards	Muzzle	100	200	300	400	500	600	800	1000
Velocity • fps	3050	2820	2590	2380	2180	1990	1809	1484	1227
Energy • ft-lbs	3410	2905	2465	2080	1740	1450	1199	808	552
Taylor KO Index	22.1	20.5	18.8	17.3	15.8	14.4	13.1	10.4	8.9
Path • Inches	-1.5	+1.5	0.0	-6.6	-19.4	-39.5	-68.4	-160.5	-316.4
Wind Drift • Inches	0.0	0.7	2.9	6.9	12.7	20.8	31.3	61.2	105.1

Hornady 165-grain GMX (82026) LF

G1 Ballistic Coefficient = 0.447

Distance • Yards	Muzzle	100	200	300	400	500	600	800	1000
Velocity • fps	3260	3035	2821	2617	2422	2235	2057	1727	1440
Energy • ft-lbs	3893	3373	2915	2505	2148	1830	1550	1093	760
Taylor KO Index	23.7	22.0	20.5	19.0	17.6	16.2	14.9	12.5	10.5
Path • Inches	-1.5	+1.1	0.0	-5.5	-16.1	-32.7	-56.2	-129.5	-250.0
Wind Drift • Inches	0.0	0.6	2.4	5.7	10.5	17.0	25.4	49.1	83.7

Black Hills 165-grain Hornady GMX (1C300WMBHGN4) LF

G1 Ballistic Coefficient = 0.447

Distance • Yards	Muzzle	100	200	300	400	500	600	800	1000
Velocity • fps	3050	2835	2631	2435	2248	2069	1898	1597	1326
Energy • ft-lbs	3408	2946	2536	2173	1852	1568	1321	923	644
Taylor KO Index	22.1	20.6	19.1	17.7	16.3	15.0	13.8	11.6	9.6
Path • Inches	-1.5	+1.4	0.0	-6.4	-18.8	-38.0	-65.4	-151.0	-292.4
Wind Drift • Inches	0.0	0.6	2.7	6.2	11.5	18.6	28.0	54.2	92.6

Federal 165-grain Trophy Bonded Tip (P300WTT2)

G1 Ballistic Coefficient = 0.453

Distance • Yards	Muzzle	100	200	300	400	500	600	800	1000
Velocity • fps	3050	2840	2630	2440	2250	2080	1912	1603	1341
Energy • ft-lbs	3410	2950	2540	2180	1860	1575	1339	941	659
Taylor KO Index	22.1	20.6	19.1	17.7	16.3	15.1	13.9	11.6	9.7
Path • Inches	-1.5	+1.4	0.0	-6.4	-18.7	-37.9	-64.9	-149.7	-289.1
Wind Drift • Inches	0.0	0.6	2.6	6.1	11.3	18.4	27.5	53.3	90.8

Federal 165-grain Barnes Triple-Shock X-Bullet (P300WR) LF

G1 Ballistic Coefficient = 0.381

Distance • Yards	Muzzle	100	200	300	400	500	600	800	1000
Velocity • fps	3050	2800	2560	2340	2120	1920	1730	1398	1152
Energy • ft-lbs	3410	2870	2400	1995	1650	1345	1096	716	487
Taylor KO Index	22.1	20.3	18.6	17.0	15.4	13.9	12.6	10.1	8.4
Path • Inches	-1.5	+1.5	0.0	-6.7	-20.0	-40.8	-71.3	-169.7	-340.4
Wind Drift • Inches	0.0	0.8	3.2	7.4	13.9	22.7	34.4	68.7	116.8

Cor-Bon 168-grain T-DPX (DPX300WM168/20) L

G1 Ballistic Coefficient = 0.450

Distance • Yards	Muzzle	100	200	300	400	500	600	800	1000
Velocity • fps	3125	2908	2701	2504	2315	2134	1962	1645	1374
Energy • ft-lbs	3644	3155	2723	2340	2000	1700	1436	1009	704
Taylor KO Index	23.1	21.5	20.0	18.5	17.1	15.8	14.5	12.2	10.2
Path • Inches	-1.5	+1.3	0.0	-6.1	-17.7	-35.8	-61.6	-142.1	-274.6
Wind Drift • Inches	0.0	0.6	2.6	6.0	11.0	17.9	26.8	51.8	88.4

Black Hills 178-grain Hornady A-MAX (1C300WMBHGN2)

G1 Ballistic Coefficient = 0.495

Distance • Yards	Muzzle	100	200	300	400	500	600	800	1000
Velocity • fps	3000	2808	2623	2446	2276	2113	1956	1666	1414
Energy • ft-lbs	3558	3116	2720	2366	2048	1765	1513	1098	791
Taylor KO Index	23.5	22.0	20.5	19.2	17.8	16.5	15.3	13.0	11.1
Path • Inches	-1.5	+1.5	0.0	-6.5	-18.8	-37.9	-64.8	-147.4	-280.5
Wind Drift • Inches	0.0	0.6	2.4	5.7	10.5	17.0	25.3	48.6	82.3

Hornady 178-grain A-MAX (8203)

G1 Ballistic Coefficient = 0.495

Distance • Yards	Muzzle	100	200	300	400	500	600	800	1000
Velocity • fps	2960	2770	2587	2412	2243	2081	1927	1640	1393
Energy • ft-lbs	3462	3031	2645	2298	1988	1712	1467	1064	767
Taylor KO Index	23.2	21.7	20.3	18.9	17.6	16.3	15.1	12.8	10.9
Path • Inches	-1.5	+1.5	0.0	-6.7	-19.4	-39.1	-66.7	-151.8	-289.1
Wind Drift • Inches	0.0	0.6	2.5	5.8	10.7	17.3	25.8	50.0	83.8

Hornady 180-grain InterBond (82198)

G1 Ballistic Coefficient = 0.480

Distance • Yards	Muzzle	100	200	300	400	500	600	800	1000
Velocity • fps	3130	2927	2732	2546	2366	2195	2031	1726	1458
Energy • ft-lbs	3917	3424	2953	2589	2238	1925	1650	1191	850
Taylor KO Index	24.8	23.2	21.6	20.2	18.7	17.4	16.1	13.7	11.5
Path • Inches	-1.5	+1.3	0.0	-5.9	-17.3	-34.8	-59.6	-135.9	-259.4
Wind Drift • Inches	0.0	0.6	2.4	5.5	10.4	16.5	24.6	47.3	80.2

Hornady 180-grain SST (82193)

G1 Ballistic Coefficient = 0.480

Distance • Yards	Muzzle	100	200	300	400	500	600	800	1000
Velocity • fps	3130	2927	2732	2546	2366	2195	2029	1724	1456
Energy • ft-lbs	3917	3424	2953	2589	2238	1925	1646	1188	847
Taylor KO Index	24.8	23.2	21.6	20.2	18.7	17.4	16.1	13.7	11.5
Path • Inches	-1.5	+1.3	0.0	-5.9	-17.3	-34.8	-59.6	-136.0	-259.8
Wind Drift • Inches	0.0	0.6	2.4	5.6	10.2	16.5	24.7	47.4	80.4

Fiocchi 180-grain SST Polymer Tip BT (300WMHSA)

G1 Ballistic Coefficient = 0.480

Distance • Yards	Muzzle	100	200	300	400	500	600	800	1000
Velocity • fps	3000	2802	2613	2432	2258	2091	1931	1637	1383
Energy • ft-lbs	3597	3139	2729	2363	2037	1748	1491	1071	765
Taylor KO Index	24.8	23.2	21.6	20.2	18.7	17.4	16.1	13.7	11.5
Path • Inches	-1.5	+1.5	0.0	-6.6	-19.0	-38.3	-65.6	-149.8	-286.4
Wind Drift • Inches	0.0	0.6	2.5	5.9	10.8	17.5	26.2	50.4	85.5

Fiocchi 180-grain PSP InterLock BT (300WMB)

G1 Ballistic Coefficient = 0.452

Distance • Yards	Muzzle	100	200	300	400	500	600	800	1000
Velocity • fps	3000	2790	2590	2399	2216	2040	1872	1714	1313
Energy • ft-lbs	3597	3112	2682	2300	1962	1664	1401	982	689
Taylor KO Index	23.8	22.1	20.5	19.0	17.6	16.2	14.8	13.6	10.4
Path • Inches	-1.5	+1.5	0.0	-6.7	-19.4	-39.3	-67.5	-155.7	-301.2
Wind Drift • Inches	0.0	0.6	2.7	6.3	11.6	18.8	28.3	54.8	93.4

Winchester 180-grain XP3 (SXP300WM)

G1 Ballistic Coefficient = 0.527

Distance • Yards	Muzzle	100	200	300	400	500	600	800	1000
Velocity • fps	3000	2819	2646	2479	2318	2163	2012	1734	1486
Energy • ft-lbs	3597	3176	2797	2455	2147	1869	1619	1202	883
Taylor KO Index	23.8	22.3	21.0	19.6	18.4	17.1	15.9	13.7	11.8
Path • Inches	-1.5	+1.4	0.0	-6.4	-18.5	-37.0	-63.1	-142.3	-268.1
Wind Drift • Inches	0.0	0.6	2.3	5.3	9.8	15.8	23.5	44.8	75.5

Federal 180-grain Nosler AccuBond (P00WA1)

G1 Ballistic Coefficient = 0.507

Distance • Yards	Muzzle	100	200	300	400	500	600	800	1000
Velocity • fps	2960	2770	2600	2420	2260	2100	1947	1664	1418
Energy • ft-lbs	3500	3075	2690	2350	2040	1765	1515	1108	804
Taylor KO Index	23.4	21.9	20.6	19.2	17.9	16.6	15.4	13.2	11.2
Path • Inches	-1.5	+1.5	0.0	-6.6	-19.2	-38.7	-66.1	-150.0	-284.3
Wind Drift • Inches	0.0	0.6	2.4	5.6	10.4	16.8	25.1	48.1	81.3

Federal 180-grain Fusion (F300WFS3)

G1 Ballistic Coefficient = 0.482

Distance • Yards	Muzzle	100	200	300	400	500	600	800	1000
Velocity • fps	2960	2760	2570	2390	2220	2060	1902	1612	1363
Energy • ft-lbs	3480	3035	2640	2285	1970	1690	1447	1038	743
Taylor KO Index	23.4	21.9	20.4	18.9	17.6	16.3	15.1	12.8	10.8
Path • Inches	-1.5	+1.5	0.0	-6.8	-19.7	-39.7	-67.5	-154.3	-295.0
Wind Drift • Inches	0.0	0.6	2.6	6.0	11.0	17.8	26.6	51.3	87.0

Federal 180-grain Soft Point (300WBS)

G1 Ballistic Coefficient = 0.439

Distance • Yards	Muzzle	100	200	300	400	500	600	800	1000
Velocity • fps	2960	2750	2540	2340	2160	1980	1812	1507	1259
Energy • ft-lbs	3500	3010	2580	2195	1860	1565	1313	908	634
Taylor KO Index	23.4	21.8	20.1	18.5	17.1	15.7	14.4	11.9	10.0
Path • Inches	-1.5	+1.6	0.0	-7.0	-20.3	-41.1	-70.6	-164.4	-321.6
Wind Drift • Inches	0.0	0.7	2.8	6.6	12.3	20.0	30.0	58.2	99.4

Federal 180-grain Nosler Partition (P300WD2)

G1 Ballistic Coefficient = 0.362

Distance • Yards	Muzzle	100	200	300	400	500	600	800	1000
Velocity • fps	2960	2700	2460	2200	2010	1800	1611	1291	1078
Energy • ft-lbs	3500	2915	2410	1980	1605	1295	1038	666	464
Taylor KO Index	23.4	21.4	19.5	17.4	15.9	14.3	12.8	10.2	8.5
Path • Inches	-1.5	+1.6	0.0	-7.4	-21.9	-45.0	-78.9	-190.4	-385.9
Wind Drift • Inches	0.0	0.8	3.5	8.2	15.4	25.3	38.5	76.3	130.8

Hornady 180-grain SST InterLock (82194)

G1 Ballistic Coefficient = 0.481

Distance • Yards	Muzzle	100	200	300	400	500	600	800	1000
Velocity • fps	2960	2764	2575	2395	2222	2057	1900	1608	1359
Energy • ft-lbs	3501	3052	2650	2292	1974	1691	1443	1034	739
Taylor KO Index	23.4	21.9	20.4	19.0	17.6	16.3	15.0	12.7	10.8
Path • Inches	-1.5	+1.6	0.0	-7.0	-20.1	-39.9	-67.6	-154.8	-297.3
Wind Drift • Inches	0.0	0.6	2.6	6.0	11.0	17.9	26.7	51.5	87.4

Hornady 180-grain SP InterLock (8200)

G1 Ballistic Coefficient = 0.426

Distance • Yards	Muzzle	100	200	300	400	500	600	800	1000
Velocity • fps	2960	2739	2528	2328	2136	1953	1782	1473	1227
Energy • ft-lbs	3501	2998	2555	2165	1823	1523	1269	868	602
Taylor KO Index	23.4	21.7	20.0	18.4	16.9	15.5	14.1	11.7	9.7
Path • Inches	-1.5	+1.6	0.0	-7.0	-20.5	-41.6	-71.8	-167.4	-327.9
Wind Drift • Inches	0.0	0.7	2.9	6.9	12.7	20.7	31.2	60.8	103.9

Remington 180-grain Core-Lokt Ultra Bonded (PRC300WC)

G1 Ballistic Coefficient = 0.402

Distance • Yards	Muzzle	100	200	300	400	500	600	800	1000
Velocity • fps	2960	2727	2505	2294	2093	1903	1723	1408	1169
Energy • ft-lbs	3501	2971	2508	2103	1751	1448	1187	793	547
Taylor KO Index	23.4	21.6	19.8	18.2	16.6	15.1	13.6	11.2	9.3
Path • Inches	-1.5	1.6	0.0	-7.2	-20.9	-42.7	-74.1	-175.5	-350.2
Wind Drift • Inches	0.0	0.8	3.1	7.3	13.6	22.2	33.5	65.8	112.8

Remington 180-grain Core-Lokt Pointed Soft Point (R300W2)

G1 Ballistic Coefficient = 0.383

Distance • Yards	Muzzle	100	200	300	400	500	600	800	1000
Velocity • fps	2960	2715	2482	2262	2052	1856	1674	1355	1126
Energy • ft-lbs	3501	2945	2463	2044	1683	1375	1120	734	507
Taylor KO Index	23.4	21.5	19.7	17.9	16.3	14.7	13.3	10.7	8.9
Path • Inches	-1.5	+1.7	0.0	-7.4	-21.3	-43.7	-76.1	-181.3	-363.0
Wind Drift • Inches	0.0	0.8	3.3	7.7	14.4	23.6	35.7	70.3	120.6

Remington 180-grain Swift A-Frame PSP (RS300WMB)

G1 Ballistic Coefficient = 0.377

Distance • Yards	Muzzle	100	200	300	400	500	600	800	1000
Velocity • fps	2960	2712	2476	2253	2042	1842	1657	1338	1113
Energy • ft-lbs	3501	2938	2451	2029	1666	1356	1097	715	495
Taylor KO Index	23.4	21.5	19.6	17.8	16.2	14.6	13.1	10.6	8.8
Path • Inches	-1.5	+1.5	0.0	-7.3	-21.5	-44.0	-76.9	-183.7	-369.0
Wind Drift • Inches	0.0	0.8	3.3	7.8	14.6	24.0	36.4	71.9	123.3

Winchester 180-grain Power-Point (X30WM2)

G1 Ballistic Coefficient = 0.438

Distance • Yards	Muzzle	100	200	300	400	500	600	800	1000
Velocity • fps	2960	2745	2540	2340	2157	1979	1811	1506	1258
Energy • ft-lbs	3501	3011	2578	2196	1859	1565	1311	907	633
Taylor KO Index	23.4	21.7	20.1	18.5	17.1	15.7	14.3	11.9	10.0
Path • Inches	-1.5	+1.6	0.0	-7.0	-20.3	-41.0	-70.7	-164.1	-319.4
Wind Drift • Inches	0.0	0.7	2.8	6.6	12.3	20.0	30.0	58.4	99.6

Norma 180-grain Nosler AccuBond (17548)

G1 Ballistic Coefficient = 0.507

Distance • Yards	Muzzle	100	200	300	400	500	600	800	1000
Velocity • fps	2953	2767	2588	2417	2252	2093	1941	1660	1414
Energy • ft-lbs	3485	3061	2678	2335	2027	1752	1507	1101	799
Taylor KO Index	23.4	21.9	20.5	19.1	17.8	16.6	15.4	13.1	11.2
Path • Inches	-1.5	+1.5	0.0	-6.7	-19.4	-38.9	-66.4	-150.7	-285.9
Wind Drift • Inches	0.0	0.6	2.4	5.7	10.4	16.9	25.2	48.3	81.6

Black Hills 180-grain Nosler AccuBond (1C300WMBHGN3)

G1 Ballistic Coefficient = 0.507

Distance • Yards	Muzzle	100	200	300	400	500	600	800	1000
Velocity • fps	2950	2764	2586	2414	2249	2091	1939	1657	1412
Energy • ft-lbs	3479	3054	2673	2330	2023	1748	1503	1098	797
Taylor KO Index	23.4	21.9	20.5	19.1	17.8	16.6	15.4	13.1	11.2
Path • Inches	-1.5	+1.5	0.0	-6.7	-19.4	-39.0	-66.6	-151.0	-286.6
Wind Drift • Inches	0.0	0.6	2.4	5.7	10.4	16.9	25.2	48.4	81.7

Nosler 180-grain AccuBond (23305) + (60059)

G1 Ballistic Coefficient = 0.507

Distance • Yards	Muzzle	100	200	300	400	500	600	800	1000
Velocity • fps	2950	2764	2586	2414	2249	2091	1939	1657	1412
Energy • ft-lbs	3479	3054	2673	2330	2023	1748	1503	1098	797
Taylor KO Index	23.4	21.9	20.5	19.1	17.8	16.6	15.4	13.1	11.2
Path • Inches	-1.5	+1.5	0.0	-6.7	-19.4	-39.0	-66.6	-151.0	-286.6
Wind Drift • Inches	0.0	0.6	2.4	5.7	10.4	16.9	25.2	48.4	81.7

Winchester 180-grain AccuBond CT (S300WMCT)

G1 Ballistic Coefficient = 0.510

Distance • Yards	Muzzle	100	200	300	400	500	600	800	1000
Velocity • fps	2950	2765	2588	2417	2253	2095	1944	1664	1419
Energy • ft-lbs	3478	3055	2676	2334	2028	1754	1511	1109	805
Taylor KO Index	23.4	21.9	20.5	19.1	17.8	16.6	15.4	13.2	11.2
Path • Inches	-1.5	+1.5	0.0	-6.7	-19.4	-39.0	-66.4	-150.5	-285.4
Wind Drift • Inches	0.0	0.6	2.4	5.6	10.4	16.8	25.0	48.0	81.0

Norma 180-grain Oryx (17547)

G1 Ballistic Coefficient = 0.288

Distance • Yards	Muzzle	100	200	300	400	500	600	800	1000
Velocity • fps	2920	2600	2301	2023	1766	1534	1334	1062	924
Energy • ft-lbs	3409	2702	2117	1636	1247	941	712	451	342
Taylor KO Index	23.1	20.6	18.2	16.0	14.0	12.1	10.6	8.4	7.3
Path • Inches	-1.5	+1.8	0.0	-8.4	-25.4	-53.6	-96.7	-247.2	-520.4
Wind Drift • Inches	0.0	1.1	4.6	11.0	20.8	34.8	53.7	107.2	178.3

Norma 180-grain Swift A-Frame (17519)

G1 Ballistic Coefficient = 0.400

Distance • Yards	Muzzle	100	200	300	400	500	600	800	1000
Velocity • fps	2920	2687	2466	2256	2056	1867	1689	1379	1149
Energy • ft-lbs	3409	2887	2432	2035	1690	1393	1141	761	528
Taylor KO Index	23.1	21.3	19.5	17.9	16.3	14.8	13.4	10.9	9.1
Path • Inches	-1.5	+1.7	0.0	-7.4	-21.6	-44.1	-76.7	-181.9	-363.1
Wind Drift • Inches	0.0	0.8	3.2	7.5	13.9	22.8	34.5	67.6	115.7

Norma 180-grain Plastic Point (17687)

G1 Ballistic Coefficient = 0.366

Distance • Yards	Muzzle	100	200	300	400	500	600	800	1000
Velocity • fps	3018	2758	2513	2281	2063	1856	1664	1334	1105
Energy • ft-lbs	3641	3042	2525	2080	1701	1377	1106	712	488
Taylor KO Index	23.9	21.8	19.9	18.1	16.3	14.7	13.2	10.6	8.8
Path • Inches	-1.5	+1.5	0.0	-7.0	-20.8	-42.8	-75.0	-181.3	-369.6
Wind Drift • Inches	0.0	0.8	3.4	8.0	14.8	24.2	36.8	72.9	125.4

Cor-Bon 180-grain DPX (DPX300WM180/20)

G1 Ballistic Coefficient = 0.450

Distance • Yards	Muzzle	100	200	300	400	500	600	800	1000
Velocity • fps	3000	2789	2588	2395	2211	2035	1867	1562	1308
Energy • ft-lbs	3598	3110	2677	2294	1954	1655	1394	976	684
Taylor KO Index	23.8	22.1	20.5	19.0	17.5	16.1	14.8	12.4	10.4
Path • Inches	-1.5	+1.5	0.0	-6.7	-19.4	-39.3	-67.6	-156.2	-302.3
Wind Drift • Inches	0.0	0.6	2.7	6.3	11.7	19.0	28.4	55.1	94.0

Sellier & Bellot 180-grain PTS (V332552U)

G1 Ballistic Coefficient = 0.480

Distance • Yards	Muzzle	100	200	300	400	500	600	800	1000
Velocity • fps	2936	2741	2554	2374	2202	2037	1880	1591	1345
Energy • ft-lbs	3446	3003	2607	2254	1939	1659	1413	1012	723
Taylor KO Index	23.3	21.7	20.2	18.8	17.4	16.1	14.9	12.6	10.7
Path • Inches	-1.5	+1.6	0.0	-6.9	-20.0	-40.2	-68.9	-157.6	-301.7
Wind Drift • Inches	0.0	0.6	2.6	6.1	11.2	18.1	27.1	52.2	88.6

Federal 180-grain Barnes MRX-Bullet (P300WX) LF

G1 Ballistic Coefficient = 0.550

Distance • Yards	Muzzle	100	200	300	400	500	600	800	1000
Velocity • fps	2960	2790	2620	2470	2310	2160	2017	1749	1509
Energy • ft-lbs	3500	3110	2750	2430	2135	1870	1627	1223	911
Taylor KO Index	23.4	22.1	20.8	19.6	18.3	17.1	16.0	13.9	12.0
Path • Inches	-1.5	+1.5	0.0	-6.5	-18.7	-37.5	-63.9	-143.4	-268.7
Wind Drift • Inches	0.0	0.5	2.2	5.2	9.5	15.3	22.8	43.3	72.7

Federal 180-grain Trophy Bonded Tip (P300WTT1)

G1 Ballistic Coefficient = 0.501

Distance • Yards	Muzzle	100	200	300	400	500	600	800	1000
Velocity • fps	2960	2770	2590	2420	2250	2090	1936	1652	1405
Energy • ft-lbs	3500	3070	2680	2335	2025	1745	1499	1091	789
Taylor KO Index	23.4	20.9	20.5	19.2	17.8	16.6	15.3	13.1	11.1
Path • Inches	-1.5	+1.5	0.0	-6.6	-19.3	-38.9	-66.4	-150.9	-286.8
Wind Drift • Inches	0.0	0.6	2.5	5.7	10.5	17.0	25.5	48.9	82.6

Federal 180-grain Barnes Triple-Shock X-Bullet (P300WP) LF

G1 Ballistic Coefficient = 0.455

Distance • Yards	Muzzle	100	200	300	400	500	600	800	1000
Velocity • fps	2960	2750	2550	2360	2180	2010	1846	1547	1297
Energy • ft-lbs	3500	3025	2605	2235	1905	1610	1363	956	673
Taylor KO Index	23.4	21.8	20.2	18.7	17.3	15.9	14.6	12.3	10.3
Path • Inches	-1.5	+1.3	0.0	-6.8	-20.0	-40.4	-69.4	-160.1	-309.5
Wind Drift • Inches	0.0	0.7	2.7	6.4	11.8	19.1	28.7	55.6	94.6

Remington 180-grain Swift Scirocco (PRSC300WB)

G1 Ballistic Coefficient = 0.508

Distance • Yards	Muzzle	100	200	300	400	500	600	800	1000
Velocity • fps	2960	2774	2595	2424	2259	2100	1949	1667	1421
Energy • ft-lbs	3501	3075	2692	2348	2039	1762	1519	1111	807
Taylor KO Index	23.4	22.0	20.6	19.2	17.9	16.6	15.4	13.2	11.3
Path • Inches	-1.5	+1.5	0.0	-6.7	-19.3	-38.7	-66.0	-149.8	-284.9
Wind Drift • Inches	0.0	0.6	2.4	5.6	10.4	16.8	25.0	47.9	81.0

Remington 180-grain AccuTip Boat Tail (PRA300WC)

G1 Ballistic Coefficient = 0.481

Distance • Yards	Muzzle	100	200	300	400	500	600	800	1000
Velocity • fps	2960	2764	2577	2397	2224	2058	1899	1608	1359
Energy • ft-lbs	3501	3053	2653	2295	1976	1693	1442	1033	738
Taylor KO Index	23.4	21.9	20.4	19.0	17.6	16.3	15.0	12.7	10.8
Path • Inches	-1.5	+1.5	0.0	-6.8	-19.6	-39.5	-67.6	-154.6	-295.9
Wind Drift • Inches	0.0	0.6	2.6	6.0	11.0	17.9	26.8	51.6	87.5

Winchester 180-grain Power Max Bonded (X30WM2)

G1 Ballistic Coefficient = 0.394

Distance • Yards	Muzzle	100	200	300	400	500	600	800	1000
Velocity • fps	2960	2722	2496	2281	2077	1884	1703	1387	1152
Energy • ft-lbs	3501	2961	2490	2080	1724	1419	1160	769	530
Taylor KO Index	23.4	21.6	19.8	18.1	16.4	14.9	13.5	11.0	9.1
Path • Inches	-1.5	+1.6	0.0	-7.2	-21.1	-43.1	-74.9	-177.3	-352.7
Wind Drift • Inches	0.0	0.8	3.2	7.5	13.9	22.7	34.4	67.6	115.9

Black Hills 180-grain Barnes TSX (1C300WMBHGN1) LF

G1 Ballistic Coefficient = 0.453

Distance • Yards	Muzzle	100	200	300	400	500	600	800	1000
Velocity • fps	2950	2743	2545	2355	2174	2001	1836	1537	1289
Energy • ft-lbs	3478	3007	2589	2218	1890	1600	1348	944	664
Taylor KO Index	23.4	21.7	20.2	18.7	17.2	15.8	14.5	12.2	10.2
Path • Inches	-1.5	+1.6	0.0	-6.9	-20.2	-40.8	-70.1	-161.7	-312.8
Wind Drift • Inches	0.0	0.7	2.8	6.4	11.9	19.3	28.9	56.0	95.5

Winchester 180-grain Ballistic Silvertip (SBST300)

G1 Ballistic Coefficient = 0.508

Distance • Yards	Muzzle	100	200	300	400	500	600	800	1000
Velocity • fps	2950	2764	2586	2415	2250	2092	1941	1660	1415
Energy • ft-lbs	3478	3054	2673	2333	2026	1751	1507	1102	800
Taylor KO Index	23.4	21.9	20.5	19.1	17.8	16.6	15.4	13.1	11.2
Path • Inches	-1.5	+1.5	0.0	-6.7	-19.4	-38.9	-66.5	-151.0	-287.2
Wind Drift • Inches	0.0	0.6	2.4	5.7	10.4	16.8	25.1	48.2	81.4

Winchester 180-grain E-Tip (S300WMET)

G1 Ballistic Coefficient = 0.523

Distance • Yards	Muzzle	100	200	300	400	500	600	800	1000
Velocity • fps	2950	2770	2597	2430	2270	2116	1966	1670	1447
Energy • ft-lbs	3478	3066	2695	2361	2059	1789	1546	1142	837
Taylor KO Index	23.4	21.9	20.6	19.2	18.0	16.8	15.6	13.4	11.5
Path • Inches	-1.5	+1.5	0.0	-6.7	-19.2	-36.8	-65.7	-148.4	-280.3
Wind Drift • Inches	0.0	0.6	2.4	5.5	10.1	16.3	24.3	46.4	78.2

A² 180-grain Monolithic; Dead Tough

G1 Ballistic Coefficient = 0.263

Distance • Yards	Muzzle	100	200	300	400	500	600	800	1000
Velocity • fps	3120	2756	2420	2108	1820	1559	1338	1047	905
Energy • ft-lbs	3890	3035	2340	1776	1324	972	716	438	328
Taylor KO Index	24.7	21.8	19.2	16.7	14.4	12.3	10.6	8.3	7.2
Path • Inches	-1.5	+1.6	0.0	-7.6	-22.9	-49.0	-90.2	-239.6	-511.5
Wind Drift • Inches	0.0	1.1	4.6	11.1	21.1	35.6	55.2	111.7	186.8

Black Hills 190-grain Match Hollow Point (2C300WMN1)

G1 Ballistic Coefficient = 0.560

Distance • Yards	Muzzle	100	200	300	400	500	600	800	1000
Velocity • fps	2950	2781	2619	2462	2311	2165	2025	1761	1524
Energy • ft-lbs	3672	3265	2894	2559	2254	1978	1731	1309	980
Taylor KO Index	24.7	23.2	21.9	20.6	19.3	18.1	16.9	14.7	12.7
Path • Inches	-1.5	+1.5	0.0	-6.6	-18.9	-37.7	-63.9	-143.2	-268.2
Wind Drift • Inches	0.0	0.5	2.2	5.1	9.3	15.0	22.3	42.5	71.2

Cor-Bon 190-grain HPBT (PM300WM190/20)

G1 Ballistic Coefficient = 0.530

Distance • Yards	Muzzle	100	200	300	400	500	600	800	1000
Velocity • fps	2900	2724	2555	2392	2236	2085	1940	1670	1433
Energy • ft-lbs	3549	3132	2755	2415	2109	1834	1588	1177	866
Taylor KO Index	24.2	22.6	21.4	20.0	18.7	17.4	16.2	14.0	12.0
Path • Inches	-1.5	+1.6	0.0	-6.9	-19.9	-39.9	-67.9	-153.1	-288.6
Wind Drift • Inches	0.0	0.6	2.4	5.5	10.2	16.4	24.5	46.8	78.7

Federal 190-grain Sierra MatchKing BTHP (GM300WM)

G1 Ballistic Coefficient = 0.534

Distance • Yards	Muzzle	100	200	300	400	500	600	800	1000
Velocity • fps	2900	2730	2560	2400	2240	2090	1946	1678	1441
Energy • ft-lbs	3500	3125	2760	2420	2115	1845	1598	1188	876
Taylor KO Index	24.2	22.8	21.4	20.1	18.7	17.5	16.3	14.0	12.0
Path • Inches	-1.5	+1.6	0.0	-6.9	-19.9	-39.8	-67.7	-152.5	-287.1
Wind Drift • Inches	0.0	0.6	2.4	5.5	10.1	16.3	24.2	46.3	77.9

Remington 200-grain Swift A-Frame PSP (RS300WA)

G1 Ballistic Coefficient = 0.395

Distance • Yards	Muzzle	100	200	300	400	500	600	800	1000
Velocity • fps	2825	2595	2377	2169	1971	1786	1612	1315	1106
Energy • ft-lbs	3544	2990	2508	2088	1726	1416	1154	769	544
Taylor KO Index	24.9	22.8	20.9	19.1	17.3	15.7	14.2	11.6	9.7
Path • Inches	-1.5	+1.8	0.0	-8.0	-23.4	-47.8	-83.3	-198.4	-397.1
Wind Drift • Inches	0.0	0.8	3.4	8.0	14.8	24.3	36.7	72.0	122.7

Nosler 200-grain AccuBond (23300)

G1 Ballistic Coefficient = 0.588

Distance • Yards	Muzzle	100	200	300	400	500	600	800	1000
Velocity • fps	2800	2645	2494	2349	2209	2073	1942	1697	1478
Energy • ft-lbs	3483	3107	2764	2451	2167	1909	1676	1280	971
Taylor KO Index	24.6	23.3	21.9	20.7	19.4	18.2	17.1	14.9	13.0
Path • Inches	-1.5	+1.7	0.0	-7.3	-21.0	-41.7	-70.6	-157.3	-292.7
Wind Drift • Inches	0.0	0.6	2.2	5.2	9.5	15.3	22.8	43.3	72.3

Norma 200-grain Oryx (17676)

G1 Ballistic Coefficient = 0.338

Distance • Yards	Muzzle	100	200	300	400	500	600	800	1000
Velocity • fps	2789	2523	2272	2036	1813	1610	1427	1148	989
Energy • ft-lbs	3455	2828	2293	1841	1461	1151	905	584	435
Taylor KO Index	24.5	22.2	20.0	17.9	16.0	14.2	12.6	10.1	8.7
Path • Inches	-1.5	+2.0	0.0	-8.8	-26.0	-53.8	-95.1	-233.5	-477.5
Wind Drift • Inches	0.0	1.0	4.1	9.7	18.3	30.3	46.2	91.4	153.4

Federal 200-grain Trophy Bonded Bear Claw (P300WT1)

G1 Ballistic Coefficient = 0.396

Distance • Yards	Muzzle	100	200	300	400	500	600	800	1000
Velocity • fps	2700	2480	2260	2060	1870	1690	1525	1249	1067
Energy • ft-lbs	3235	2720	2275	1885	1550	1265	1033	693	505
Taylor KO Index	23.8	21.8	19.9	18.1	16.5	14.9	13.4	11.0	9.4
Path • Inches	-1.5	+2.1	0.0	-8.9	-25.9	-53.3	-92.4	-219.4	-435.9
Wind Drift • Inches	0.0	0.9	3.6	8.5	15.9	26.0	39.4	77.0	130.0

.300 Dakota

This is another of Dakota Arms' line of proprietary cartridges based on the .404 Jeffery case. Performance-wise, it is somewhat more potent than the .300 Winchester Magnum and a push with the .300 Weatherby Magnum, depending on which loads you use for comparison. Proprietary cartridges are not new to the ammunition business. Many of the British cartridges started life as proprietary numbers. Sometimes, like the .375 H&H (and numerous others), they become standardized.

Relative Recoil Factor = 2.60

Specifications:

Controlling Agency for Standardization of this Ammunition: Factory

Availability:

Dakota 180-grain Nosler Partition (300-180NPT)

G1 Ballistic Coefficient = 0.474

Distance • Yards	Muzzle	100	200	300	400	500	600	800	1000
Velocity • fps	3250	3038	2835	2642	2457	2279	2108	1791	1509
Energy • ft-lbs	4223	3689	3214	2790	2413	2076	1777	1282	911
Taylor KO Index	25.7	24.1	22.5	20.9	19.5	18.0	16.7	14.2	12.0
Path • Inches	-1.5	+1.1	0.0	-5.5	-15.9	-32.1	-55.0	-125.9	-241.4
Wind Drift • Inches	0.0	0.6	2.3	5.4	9.8	15.9	23.8	45.6	77.4

Dakota 180-grain Fail-Safe (300-180WFS)

G1 Ballistic Coefficient = 0.412

Distance • Yards	Muzzle	100	200	300	400	500	600	800	1000
Velocity • fps	3250	3006	2776	2557	2349	2150	1961	1616	1329
Energy • ft-lbs	4223	3614	3081	2614	2205	1848	1537	1044	706
Taylor KO Index	25.7	23.8	22.0	20.3	18.6	17.0	15.5	12.8	10.5
Path • Inches	-1.5	+1.2	0.0	-5.7	-16.7	-34.0	-58.9	-138.2	-273.1
Wind Drift • Inches	0.0	0.6	2.7	6.2	11.6	18.8	28.3	55.2	94.9

Dakota 180-grain Swift A-Frame (300-180FAF)

G1 Ballistic Coefficient = 0.400

Distance • Yards	Muzzle	100	200	300	400	500	600	800	1000
Velocity • fps	3250	2999	2763	2538	2324	2121	1928	1578	1292
Energy • ft-lbs	4223	3597	2051	2575	2160	1798	1486	996	667
Taylor KO Index	2537	23.8	21.9	20.1	18.4	16.8	15.3	12.5	10.2
Path • Inches	-1.5	+1.2	0.0	-5.8	-16.9	-34.5	-59.9	-141.3	-281.2
Wind Drift • Inches	0.0	0.7	2.8	6.5	12.0	19.5	29.4	57.5	99.1

Dakota 180-grain Nosler Ballistic Tip (300-180NBT)

G1 Ballistic Coefficient = 0.507

Distance • Yards	Muzzle	100	200	300	400	500	600	800	1000
Velocity • fps	3250	3051	2861	2679	2504	2336	2174	1870	1596
Energy • ft-lbs	4223	3722	3273	2869	2507	2181	1890	1398	1019
Taylor KO Index	25.7	24.2	22.7	21.2	19.8	18.5	17.2	14.8	12.6
Path • Inches	-1.5	+1.1	0.0	-5.4	-15.6	-31.4	-53.5	-121.1	-229.4
Wind Drift • Inches	0.0	0.5	2.2	5.0	9.1	14.7	21.9	41.8	70.4

Dakota 200-grain Nosler Partition (300-200NPT)

G1 Ballistic Coefficient = 0.481

Distance • Yards	Muzzle	100	200	300	400	500	600	800	1000
Velocity • fps	3050	2850	2659	2475	2299	2130	1969	1670	1410
Energy • ft-lbs	4132	3609	3140	2722	2349	2016	1722	1239	884
Taylor KO Index	26.8	25.1	23.4	21.8	20.2	18.7	17.3	14.7	12.4
Path • Inches	-1.5	+1.4	0.0	-6.3	-18.3	-36.9	-63.2	-144.4	-277.0
Wind Drift • Inches	0.0	0.6	2.5	5.7	10.6	17.1	25.6	49.2	83.5

Dakota 200-grain Swift A-Frame (300-200FAF)

G1 Ballistic Coefficient = 0.444

Distance • Yards	Muzzle	100	200	300	400	500	600	800	1000
Velocity • fps	3050	2834	2628	2431	2242	2062	1891	1578	1317
Energy • ft-lbs	4132	3567	3067	2624	2233	1889	1588	1107	770
Taylor KO Index	26.8	24.9	23.1	21.4	19.7	18.1	16.6	13.9	11.6
Path • Inches	-1.5	+1.4	0.0	-6.5	-18.8	-38.1	-65.6	-152.0	-296.4
Wind Drift • Inches	0.0	0.6	2.7	6.3	11.6	18.8	28.2	54.8	93.6

.300 RCM (Ruger Compact Magnum)

This is another example of the latest trend in magnum cartridge design, the nonbelted magnum. These cartridges generally have base diameters the same as the belt diameters found on the classic belted magnums. It is very similar in both dimensions and proportions to the Winchester Short Magnum, but they are definitely NOT Interchangeable.

Relative Recoil Factor = 2.45

Cartridge dimensions: 0.052, 0.532, 0.516, 0.532, 0.330, 1.609, 1.748, 2.100

Specifications:

Controlling Agency for Standardization of this Ammunition: SAAMI*

Bullet Weight Grains	Velocity fps	Maximum Average Pressure	
		Copper Crusher	Transducer
150	N/A	N/S	65,000 psi
165	N/A	N/S	65,000 psi
180	N/A	N/S	65,000 psi

Standard barrel for velocity testing: 24 inches long—1 turn in 10-inch twist

Availability:

Hornady 150-grain SST (82231)

G1 Ballistic Coefficient = 0.415

Distance • Yards	Muzzle	100	200	300	400	500	600	800	1000
Velocity • fps	3310	3065	2833	2613	2404	2204	2014	1665	1370
Energy • ft-lbs	3648	3128	2673	2274	1924	1618	1351	924	626
Taylor KO Index	21.8	20.2	18.7	17.2	15.9	14.5	13.3	11.0	9.0
Path • Inches	-1.5	+1.1	0.0	-5.4	-16.0	-32.6	-56.3	-131.4	257.2
Wind Drift • Inches	0.0	0.6	2.6	6.0	11.2	18.2	27.3	53.1	91.2

Hornady 150-grain GMX (82230) LF

G1 Ballistic Coefficient = 0.415

Distance • Yards	Muzzle	100	200	300	400	500	600	800	1000
Velocity • fps	3265	3023	2793	2575	2367	2169	1981	1636	1347
Energy • ft-lbs	3550	3042	2598	2208	1866	1567	1307	892	605
Taylor KO Index	21.5	20.0	18.4	17.0	15.6	14.3	13.1	10.8	8.9
Path • Inches	-1.5	+1.2	0.0	-5.6	-16.5	-33.6	-58.1	-135.6	-265.8
Wind Drift • Inches	0.0	0.6	2.6	6.2	11.4	18.5	27.8	54.2	93.1

Hornady 165-grain SST (82232)

G1 Ballistic Coefficient = 0.447

Distance • Yards	Muzzle	100	200	300	400	500	600	800	1000
Velocity • fps	3185	2964	2753	2552	2360	2176	2000	1677	1399
Energy • ft-lbs	3716	3217	2776	2386	2040	1734	1466	1030	717
Taylor KO Index	23.1	21.5	20.0	18.5	17.1	15.8	14.5	12.2	10.2
Path • Inches	-1.5	+1.2	0.0	-5.8	-17.0	-34.4	-59.2	-136.6	-264.0
Wind Drift • Inches	0.0	0.6	2.5	5.9	10.8	17.5	26.3	50.8	86.7

Hornady 180-grain SST (82235)

G1 Ballistic Coefficient = 0.480

Distance • Yards	Muzzle	100	200	300	400	500	600	800	1000
Velocity • fps	3040	2840	2649	2466	2290	2121	1960	1662	1403
Energy • ft-lbs	3693	3223	2804	2430	2096	1799	1536	1104	787
Taylor KO Index	24.1	22.5	21.0	19.5	18.1	16.8	15.5	13.2	11.1
Path • Inches	-1.5	+1.4	0.0	-6.4	-18.5	-37.2	-63.7	-145.4	-278.1
Wind Drift • Inches	0.0	0.6	2.5	5.8	10.6	17.2	25.7	49.6	84.1

.300 Weatherby Magnum

If you ask most any group of shooters to name one caliber that best describes the term "high-powered rifle," chances are the answer will be the .300 Weatherby Magnum. Until it was recently passed by the .30-.378 Weatherby (and approximately equaled by the .300 RUM), Roy Weatherby's .300 Magnum was the performance leader of the .30 caliber rifles. The design has been around since the WWII years. Perhaps the most obvious identifying feature about Weatherby cartridges is the venturi shoulder. Whether or not the shoulder adds anything to velocity, the appearance is enough to ensure that most shooters will instantly recognize it as a Weatherby.

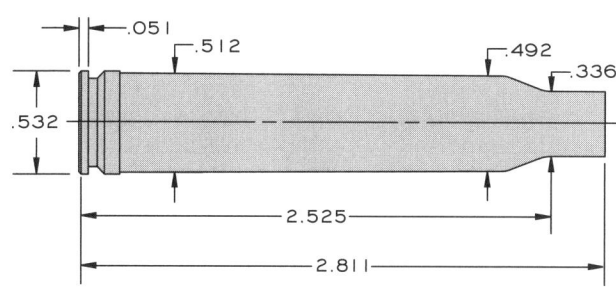

Relative Recoil Factor = 2.63

Specifications:

Controlling Agency for Standardization of this Ammunition: SAAMI*

Bullet Weight Grains	Velocity fps	Maximum Average Pressure	
		Copper Crusher	Transducer
180	3,185	N/S	65,000 psi
190	3,015	N/S	65,000 psi
220	2,835	N/S	65,000 psi

Standard barrel for velocity testing: 24 inches long—1 turn in 10-inch twist

* Ammunition for the .300 Weatherby Magnum is also manufactured in Europe under CIP standards. Ammunition manufactured to CIP specifications will exhibit somewhat different performance values, but can be safely fired in rifles in good condition.

Availability:

Weatherby 130-grain Barnes TTSX (B300130TTSX) LF

G1 Ballistic Coefficient = 0.350

Distance • Yards	Muzzle	100	200	300	400	500	600	800	1000
Velocity • fps	3650	3334	3041	2766	2508	2265	2036	1621	1287
Energy • ft-lbs	3847	3209	2670	2210	1817	1481	1196	759	478
Taylor KO Index	20.9	19.1	17.4	15.8	14.3	13.0	11.8	9.3	7.4
Path • Inches	-1.5	+0.8	0.0	-4.6	-13.7	-28.4	-49.9	-120.2	-244.5
Wind Drift • Inches	0.0	0.7	2.8	6.5	12.1	19.8	29.9	59.2	103.6

Weatherby 150-grain Nosler Partition (N 300 150 PT)

G1 Ballistic Coefficient = 0.387

Distance • Yards	Muzzle	100	200	300	400	500	600	800	1000
Velocity • fps	3540	3263	3004	2759	2528	2307	2097	1714	1390
Energy • ft-lbs	4173	3547	3004	2536	2128	1773	1466	979	644
Taylor KO Index	23.4	21.5	19.8	18.2	16.7	15.2	13.8	11.3	9.2
Path • Inches	-1.5	+1.0	0.0	-4.9	-14.6	-29.7	-50.2	-118.7	-236.8
Wind Drift • Inches	0.0	0.6	2.5	6.0	11.1	18.0	27.1	53.0	91.7

Weatherby 150-grain SP (H 300 150 SP)

G1 Ballistic Coefficient = 0.338

Distance • Yards	Muzzle	100	200	300	400	500	600	800	1000
Velocity • fps	3540	3225	2932	2657	2399	2155	1925	1518	1206
Energy • ft-lbs	4173	3462	2862	2351	1916	1547	1234	768	485
Taylor KO Index	23.4	21.3	19.4	17.5	15.8	14.2	12.7	10.0	8.0
Path • Inches	-1.5	+1.0	0.0	-5.2	-15.4	-31.8	-54.5	-133.1	-276.1
Wind Drift • Inches	0.0	0.7	2.9	6.8	12.6	20.6	37.3	64.4	113.0

Nosler 150-grain AccuBond (22000)

G1 Ballistic Coefficient = 0.434

Distance • Yards	Muzzle	100	200	300	400	500	600	800	1000
Velocity • fps	3475	3231	3002	2783	2575	2376	2185	1832	1522
Energy • ft-lbs	4022	3478	3001	2580	2209	1881	1591	1118	771
Taylor KO Index	22.9	21.3	19.8	18.4	17.0	15.7	14.4	12.1	10.0
Path • Inches	-1.5	+0.9	0.0	-4.8	-14.1	-28.6	-49.3	-114.0	-220.7
Wind Drift • Inches	0.0	0.6	2.3	5.4	10.0	16.1	24.1	46.5	79.4

Hornady 150-grain InterBond (82219)

G1 Ballistic Coefficient = 0.41?

Distance • Yards	Muzzle	100	200	300	400	500	600	800	1000
Velocity • fps	3375	3126	2891	2669	2456	2254	2060	1705	1402
Energy • ft-lbs	3793	3255	2784	2371	2009	1692	1413	969	655
Taylor KO Index	22.3	20.6	19.1	17.6	16.2	14.9	13.6	11.3	9.3
Path • Inches	-1.5	+1.0	0.0	-5.2	-15.3	-31.2	-53.9	-125.8	-246.1
Wind Drift • Inches	0.0	0.6	2.5	5.9	10.9	17.7	26.6	51.7	88.8

Weatherby 165-grain SP (H 300 165 SP)

G1 Ballistic Coefficient = 0.38?

Distance • Yards	Muzzle	100	200	300	400	500	600	800	1000
Velocity • fps	3390	3123	2872	2634	2409	2195	1990	1621	1317
Energy • ft-lbs	4210	3573	3021	2542	2126	1765	1452	963	635
Taylor KO Index	24.6	22.7	20.9	19.1	17.5	15.9	14.4	11.8	9.6
Path • Inches	-1.5	+1.0	0.0	-5.3	-15.5	-31.8	-55.4	-131.3	-262.8
Wind Drift • Inches	0.0	0.6	2.7	6.3	11.7	19.1	28.8	56.5	97.8

Nosler 165-grain Partition (22049)

G1 Ballistic Coefficient = 0.41(

Distance • Yards	Muzzle	100	200	300	400	500	600	800	1000
Velocity • fps	3250	3006	2774	2554	2344	2145	1956	1611	1324
Energy • ft-lbs	3870	3310	2819	2390	2013	1686	1403	951	643
Taylor KO Index	23.6	21.8	20.1	18.5	17.0	15.6	14.2	11.7	9.6
Path • Inches	-1.5	+1.2	0.0	-5.7	-16.8	-34.1	-59.1	-138.2	-271.7
Wind Drift • Inches	0.0	0.7	2.7	6.3	11.6	18.9	28.4	55.5	95.4

Weatherby 165-grain BST (N 300 165 BST)

G1 Ballistic Coefficient = 0.47?

Distance • Yards	Muzzle	100	200	300	400	500	600	800	1000
Velocity • fps	3350	3133	2927	2730	2542	2361	2187	1862	1572
Energy • ft-lbs	4111	3596	3138	2730	2367	2042	1753	1271	906
Taylor KO Index	24.3	22.7	21.3	19.8	18.5	17.1	15.9	13.5	11.4
Path • Inches	-1.5	+1.0	0.0	-5.1	-14.8	-30.0	-51.4	-117.4	-224.5
Wind Drift • Inches	0.0	0.5	2.2	5.1	9.4	15.2	22.7	43.5	73.8

Weatherby 165-grain Barnes TSX (B300165TSX) LF

G1 Ballistic Coefficient = 0.39?

Distance • Yards	Muzzle	100	200	300	400	500	600	800	1000
Velocity • fps	3330	3073	2831	2601	2383	2175	1978	1620	1323
Energy • ft-lbs	4064	3460	2936	2479	2080	1734	1434	961	641
Taylor KO Index	24.2	22.3	20.6	18.9	17.3	15.8	14.4	11.8	9.6
Path • Inches	-1.5	+1.1	0.0	-5.4	-16.0	-32.8	-56.9	-133.7	-264.4
Wind Drift • Inches	0.0	0.6	2.7	6.3	11.7	19.0	29.6	55.9	96.6

Weatherby 180-grain AccuBond (N 300 180 ACB)

G1 Ballistic Coefficient = 0.50?

Distance • Yards	Muzzle	100	200	300	400	500	600	800	1000
Velocity • fps	3250	3051	2806	2676	2503	2334	2173	1869	1596
Energy • ft-lbs	4223	3721	3271	2867	2504	2178	1888	1397	1018
Taylor KO Index	25.7	24.2	22.2	21.2	19.8	18.5	17.2	14.8	12.6
Path • Inches	-1.5	+1.1	0.0	-5.4	-15.6	-31.4	-53.5	-121.2	-223.0
Wind Drift • Inches	0.0	0.5	1.9	4.5	8.2	13.2	19.7	37.6	63.5

Weatherby 180-grain Nosler Partition (N 300 180 PT)

G1 Ballistic Coefficient = 0.47?

Distance • Yards	Muzzle	100	200	300	400	500	600	800	1000
Velocity • fps	3240	3028	2826	2634	2449	2271	2102	1784	1503
Energy • ft-lbs	4195	3665	3194	2772	2396	2062	1764	1272	904
Taylor KO Index	25.7	24.0	22.4	20.9	19.4	18.0	16.6	14.1	11.9
Path • Inches	-1.5	+1.2	0.0	-5.5	-16.0	-32.4	-55.4	-126.8	-243.2
Wind Drift • Inches	0.0	0.6	2.3	5.4	9.9	16.0	23.8	45.8	77.8

Weatherby 180-grain SP (H 300 180 SP)

G1 Ballistic Coefficient = 0.42?

Distance • Yards	Muzzle	100	200	300	400	500	600	800	1000
Velocity • fps	3240	3004	2781	2569	2366	2173	1987	1649	1363
Energy • ft-lbs	4195	3607	3091	2637	2237	1886	1579	1087	743
Taylor KO Index	25.7	23.8	22.0	20.3	18.7	17.2	15.7	13.1	10.8
Path • Inches	-1.5	+1.2	0.0	-5.7	-16.6	-33.8	-58.4	-136.1	-267.1
Wind Drift • Inches	0.0	0.6	2.6	6.0	11.2	18.2	27.3	53.1	91.1

A² 180-grain Monolithic; Dead Tough (None)

G1 Ballistic Coefficient = 0.264

Distance • Yards	Muzzle	100	200	300	400	500	600	800	1000
Velocity • fps	3180	2811	2471	2155	1863	1602	1375	1065	916
Energy • ft-lbs	4041	3158	2440	1856	1387	1026	755	454	336
Taylor KO Index	25.2	22.3	19.6	17.1	14.8	12.7	10.9	8.4	7.3
Path • Inches	-1.5	+1.5	0.0	-7.2	-21.8	-46.7	-85.8	-227.9	-489.5
Wind Drift • Inches	0.0	1.1	4.5	10.8	20.5	34.4	53.4	108.6	182.9

Nosler 180-grain AccuBond (22010)

G1 Ballistic Coefficient = 0.507

Distance • Yards	Muzzle	100	200	300	400	500	600	800	1000
Velocity • fps	3175	2980	2792	2613	2441	2275	2116	1818	1551
Energy • ft-lbs	4026	3548	3116	2728	2380	2068	1790	1321	961
Taylor KO Index	25.1	23.6	22.1	20.7	19.3	18.0	16.8	14.4	12.3
Path • Inches	-1.5	+1.2	0.0	-5.7	-16.4	-33.0	-56.4	-127.8	-241.4
Wind Drift • Inches	0.0	0.5	2.2	5.1	9.4	15.2	22.6	43.2	72.9

Nosler 180-grain Partition (22005)

G1 Ballistic Coefficient = 0.471

Distance • Yards	Muzzle	100	200	300	400	500	600	800	1000
Velocity • fps	3175	2966	2767	2576	2394	2218	2052	1740	1466
Energy • ft-lbs	4029	3517	3059	2653	2290	1967	1683	1210	860
Taylor KO Index	25.1	23.5	21.9	20.4	19.0	17.6	16.3	13.8	11.6
Path • Inches	-1.5	+1.2	0.0	-5.8	-16.8	-33.9	-58.0	-132.7	-253.7
Wind Drift • Inches	0.0	0.6	2.4	5.5	10.1	16.4	24.5	47.2	80.1

Weatherby 180-grain Spitzer (G300180SR)

G1 Ballistic Coefficient = 0.331

Distance • Yards	Muzzle	100	200	300	400	500	600	800	1000
Velocity • fps	3150	2855	2579	2319	2074	1845	1634	1282	1059
Energy • ft-lbs	3967	3259	2659	2150	1720	1361	1068	657	448
Taylor KO Index	24.9	22.6	20.4	18.4	16.4	14.6	12.9	10.2	8.4
Path • Inches	-1.5	+1.4	0.0	-6.6	-19.8	-41.0	-72.5	-178.4	-369.6
Wind Drift • Inches	0.0	0.8	3.6	8.4	15.7	25.9	39.6	79.3	137.2

Hornady 180-grain SP InterLock (8222)

G1 Ballistic Coefficient = 0.426

Distance • Yards	Muzzle	100	200	300	400	500	600	800	1000
Velocity • fps	3120	2891	2673	2466	2268	2079	1901	1575	1304
Energy • ft-lbs	3890	3340	2856	2430	2055	1727	1444	991	680
Taylor KO Index	24.7	22.9	21.2	19.5	18.0	16.5	15.1	12.5	10.3
Path • Inches	-1.5	+1.3	0.0	-6.2	-18.1	-36.8	-63.6	-148.4	-291.7
Wind Drift • Inches	0.0	0.7	2.7	6.4	11.8	19.2	28.8	56.0	96.0

Remington 180-grain Core-Lokt Pointed Soft Point (R300WB1)

G1 Ballistic Coefficient = 0.383

Distance • Yards	Muzzle	100	200	300	400	500	600	800	1000
Velocity • fps	3120	2866	2627	2400	2184	1976	1786	1445	1185
Energy • ft-lbs	3890	3284	2758	2301	1905	1565	1275	835	561
Taylor KO Index	24.7	22.7	20.8	19.0	17.3	15.7	14.1	11.4	9.4
Path • Inches	-1.5	+1.4	0.0	-6.4	-18.9	-38.7	-67.5	-161.0	-324.7
Wind Drift • Inches	0.0	0.7	3.1	7.2	13.3	21.8	33.0	65.0	112.2

Federal 180-grain Nosler Partition (P300WBA)

G1 Ballistic Coefficient = 0.473

Distance • Yards	Muzzle	100	200	300	400	500	600	800	1000
Velocity • fps	3080	2880	2680	2490	2320	2140	1975	1671	1407
Energy • ft-lbs	3790	3305	2870	2485	2140	1835	1580	1116	791
Taylor KO Index	24.4	22.8	21.2	19.7	18.4	16.9	15.6	13.2	11.1
Path • Inches	-1.5	+1.4	0.0	-6.2	-17.9	-36.3	62.3	-142.5	-273.1
Wind Drift • Inches	0.0	0.6	2.5	5.8	10.6	17.2	25.8	49.6	84.3

Weatherby 180-grain BST (N 300 180 BST)

G1 Ballistic Coefficient = 0.507

Distance • Yards	Muzzle	100	200	300	400	500	600	800	1000
Velocity • fps	3250	3051	2806	2676	2503	2334	2173	1869	1596
Energy • ft-lbs	4223	3721	3271	2867	2504	2178	1888	1397	1018
Taylor KO Index	25.7	24.2	22.2	21.2	19.8	18.5	17.2	14.8	12.6
Path • Inches	-1.5	+1.1	0.0	-5.4	-15.6	-31.4	-53.5	-121.2	-223.0
Wind Drift • Inches	0.0	0.5	1.9	4.5	8.2	13.2	19.7	37.6	63.5

Weatherby 180-grain Barnes TSX (B 300 180 BST) LF

G1 Ballistic Coefficient = 0.453

Distance • Yards	Muzzle	100	200	300	400	500	600	800	1000
Velocity • fps	3240	3018	2807	2609	2414	2230	2054	1728	1444
Energy • ft-lbs	4197	3642	3151	2715	2319	1987	1687	1194	834
Taylor KO Index	25.7	23.9	22.2	20.7	19.1	17.7	16.3	13.7	11.4
Path • Inches	-1.5	+1.2	0.0	-5.6	-16.3	-33.0	-56.6	-130.3	-251.1
Wind Drift • Inches	0.0	0.6	2.4	5.7	10.4	16.9	25.3	48.7	83.1

Cor-Bon 180-grain DPX (DPX300WBY180/20)

G1 Ballistic Coefficient = 0.450

Distance • Yards	Muzzle	100	200	300	400	500	600	800	1000
Velocity • fps	3200	2979	2769	2569	2377	2194	2018	1695	1415
Energy • ft-lbs	4094	3548	3066	2638	2259	1924	1628	1148	801
Taylor KO Index	25.3	23.6	21.9	20.3	18.8	17.4	16.0	13.4	11.2
Path • Inches	-1.5	+1.2	0.0	-5.8	-16.8	-34.0	-58.4	-134.6	-259.7
Wind Drift • Inches	0.0	0.6	2.5	5.8	10.6	17.3	25.9	50.0	85.3

Federal 180-grain Barnes Triple-Shock X-Bullet (P300WBE) LF

G1 Ballistic Coefficient = 0.455

Distance • Yards	Muzzle	100	200	300	400	500	600	800	1000
Velocity • fps	3110	2900	2690	2500	2310	2130	1960	1646	1377
Energy • ft-lbs	3865	3350	2895	2485	2130	1810	1536	1083	758
Taylor KO Index	24.6	23.0	21.3	19.8	18.3	16.9	15.5	13.0	10.9
Path • Inches	-1.5	+1.3	0.0	-6.1	-17.8	-36.2	-62.0	-142.8	-275.5
Wind Drift • Inches	0.0	0.6	2.6	6.0	11.0	17.8	26.6	51.5	87.6

Federal 180-grain Trophy Bonded Tip (P300WBTT1)

G1 Ballistic Coefficient = 0.499

Distance • Yards	Muzzle	100	200	300	400	500	600	800	1000
Velocity • fps	3100	2910	2721	2540	2370	2200	2042	1745	1483
Energy • ft-lbs	3840	3375	2955	2580	2240	1940	1666	1218	879
Taylor KO Index	24.6	23.0	21.5	20.1	18.8	17.4	16.2	13.8	11.7
Path • Inches	-1.5	+1.3	0.0	-6.0	-17.4	-35.1	-59.9	-136.1	-258.3
Wind Drift • Inches	0.0	0.6	2.3	5.4	9.9	16.0	23.9	45.8	77.4

Nosler 180-grain E-Tip (20058 & 60068) LF

G1 Ballistic Coefficient = 0.532

Distance • Yards	Muzzle	100	200	300	400	500	600	800	1000
Velocity • fps	3050	2868	2695	2527	2366	2211	2062	1781	1530
Energy • ft-lbs	3718	3287	2903	2552	2237	1954	1699	1269	936
Taylor KO Index	24.2	22.7	21.3	20.0	18.7	17.5	16.3	14.1	12.1
Path • Inches	-1.5	+1.4	0.0	-6.2	-17.8	-35.6	-60.5	-136.2	-256.2
Wind Drift • Inches	0.0	0.5	2.2	5.1	9.4	15.2	22.6	43.1	72.5

Weatherby 200-grain Nosler Partition (N 300 200 PT)

G1 Ballistic Coefficient = 0.481

Distance • Yards	Muzzle	100	200	300	400	500	600	800	1000
Velocity • fps	3060	2860	2668	2485	2308	2139	1977	1677	1416
Energy • ft-lbs	4158	3631	3161	2741	2366	2032	1735	1249	891
Taylor KO Index	26.9	25.2	23.5	21.9	20.3	18.8	17.4	14.8	12.5
Path • Inches	-1.5	+1.4	0.0	-6.3	-18.2	-36.6	-62.7	-143.3	-274.9
Wind Drift • Inches	0.0	0.6	2.5	5.7	10.5	17.0	25.4	49.0	83.1

Nosler 200-grain AccuBond (22064)

G1 Ballistic Coefficient = 0.587

Distance • Yards	Muzzle	100	200	300	400	500	600	800	1000
Velocity • fps	2950	2789	2634	2484	2339	2199	2064	1808	1577
Energy • ft-lbs	3864	3454	3081	2741	2430	2148	1892	1453	1104
Taylor KO Index	26.0	24.5	23.2	21.9	20.6	19.4	18.2	15.9	13.9
Path • Inches	-1.5	+1.5	0.0	-6.5	-18.6	-37.1	-62.8	-139.9	-259.7
Wind Drift • Inches	0.0	0.5	2.1	4.8	8.8	14.2	21.1	40.0	66.8

Nosler 200-grain Partition (22001)

G1 Ballistic Coefficient = 0.481

Distance • Yards	Muzzle	100	200	300	400	500	600	800	1000
Velocity • fps	2950	2754	2567	2387	2215	2050	1893	1603	1355
Energy • ft-lbs	3864	3369	2926	2530	2178	1866	1591	1141	816
Taylor KO Index	26.0	24.2	22.6	21.0	19.5	18.0	16.7	14.1	11.9
Path • Inches	-1.5	+1.5	0.0	-6.8	-19.8	-39.8	-68.1	-155.7	-297.9
Wind Drift • Inches	0.0	0.6	2.6	6.0	11.1	17.9	26.8	51.7	87.7

Weatherby 220-grain Hornady Round Nose-Expanding (H 300 220 RN)   G1 Ballistic Coefficient = 0.300

Distance • Yards	Muzzle	100	200	300	400	500	600	800	1000
Velocity • fps	2845	2543	2260	1996	1751	1530	1337	1071	933
Energy • ft-lbs	3954	3158	2495	1946	1497	1143	873	560	426
Taylor KO Index	27.5	24.6	21.9	19.3	16.9	14.8	12.9	10.4	9.0
Path • Inches	-1.5	+2.0	0.0	-8.8	-26.4	-55.2	-100.0	-255.0	-528.7
Wind Drift • Inches	0.0	1.1	4.5	10.8	20.5	34.2	52.8	104.8	173.9

PH Russell Lovemore and hunter Derek McDonald with a fine Cape kudu taken with a .300 Winchester Magnum and 180-grain Hornady bullets. Today the .300 Winchester Magnum rivals the .30-06 in popularity. (Photo from *Safari Rifles II* by Craig Boddington, 2009, Safari Press Archives)

.300 Remington Ultra Magnum

It's clear that just plain old "magnum" isn't good enough any more. Now we have the .300 ULTRA Magnum, introduced in 1999. This is Remington's entry into the .30-caliber class, where they have been absent for many years. The .300 RUM (now that's an unfortunate abbreviation) is something a little different from the standard belted magnum. For openers, it's beltless. It is actually based on a .404 Jeffery case necked to .308 and with a rebated head. It follows a recent trend to do away with the belt on magnum cartridges. While the data below are based on the SAAMI standard barrel length of 24 inches, you may find 26-inch data quoted by various sources. That extra two inches makes about a 50 fps difference in muzzle velocity.

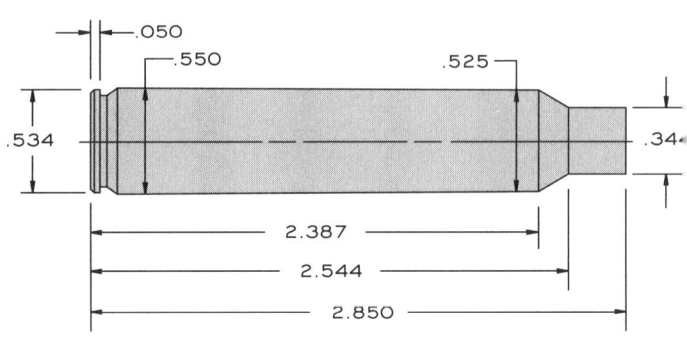

Relative Recoil Factor = 2.64

Specifications:

Controlling Agency for Standardization of this Ammunition: SAAMI

Bullet Weight Grains	Velocity fps	Maximum Average Pressure Copper Crusher	Transducer
180	3,250	N/A	65,000 psi

Standard barrel for velocity testing: 24 inches long—1 turn in 10-inch twist

Availability:

Remington 150-grain Swift Scirocco Bonded – Power Level 3 (PR300UM5)
G1 Ballistic Coefficient = 0.430

Distance • Yards	Muzzle	100	200	300	400	500	600	800	1000
Velocity • fps	3450	3208	2980	2762	2556	2358	2170	1820	1512
Energy • ft-lbs	3964	3427	2956	2541	2175	1852	1569	1103	761
Taylor KO Index	22.8	21.2	19.7	18.2	16.9	15.6	14.3	12.0	10.0
Path • Inches	-1.5	+0.9	0.0	-4.9	-14.3	-29.0	-50.1	-115.7	-223.8
Wind Drift • Inches	0.0	0.5	2.1	4.9	9.0	14.6	21.8	42.1	71.9

Nosler 150-grain AccuBond (21088)
G1 Ballistic Coefficient = 0.435

Distance • Yards	Muzzle	100	200	300	400	500	600	800	1000
Velocity • fps	3525	3278	3046	2826	2615	2415	2225	1869	1553
Energy • ft-lbs	4138	3580	3091	2658	2278	1943	1649	1163	804
Taylor KO Index	23.3	21.7	20.1	18.7	17.3	15.9	14.7	12.3	10.2
Path • Inches	-1.5	+0.9	0.0	-4.6	-13.6	-27.7	-47.7	-110.2	-213.1
Wind Drift • Inches	0.0	0.6	2.3	5.3	9.7	15.8	23.6	45.4	77.5

Remington 150-grain Core-Lokt Pointed Soft Point – Power Level 1 (PR300UM5)
G1 Ballistic Coefficient = 0.314

Distance • Yards	Muzzle	100	200	300	400	500	600	800	1000
Velocity • fps	2910	2617	2342	2083	1843	1622	1423	1128	971
Energy • ft-lbs	2820	2281	1827	1445	1131	876	675	424	314
Taylor KO Index	19.2	17.3	15.5	13.7	12.2	10.7	9.4	7.4	6.4
Path • Inches	-1.5	+1.6	0.0	-8.2	-24.4	-50.9	-90.8	-227.1	-472.3
Wind Drift • Inches	0.0	1.0	4.2	10.0	18.8	31.2	47.9	95.6	161.0

Remington 150-grain Core-Lokt PSP – Power Level 1 – Managed Recoil (RL300M1)
G1 Ballistic Coefficient = 0.314

Distance • Yards	Muzzle	100	200	300	400	500	600	800	1000
Velocity • fps	2815	2528	2258	2005	1770	1556	1367	1098	953
Energy • ft-lbs	2639	2127	1698	1339	1043	806	622	400	303
Taylor KO Index	18.6	16.7	14.9	13.2	11.7	10.3	9.0	7.2	6.3
Path • Inches	-1.5	+1.8	0.0	-8.8	-26.4	-55.1	-98.3	-245.7	-508.4
Wind Drift • Inches	0.0	1.0	4.4	10.5	19.8	32.8	50.4	99.9	166.4

Remington 150-grain AccuTip Boat Tail – Power Level 1 (PRA300UM1-P1)   G1 Ballistic Coefficient = 0.415

Distance • Yards	Muzzle	100	200	300	400	500	600	800	1000
Velocity • fps	2910	2686	2473	2270	2077	1893	1719	1414	1179
Energy • ft-lbs	2820	2403	2037	1716	1436	1193	985	666	463
Taylor KO Index	19.2	17.7	16.3	15.0	13.7	12.5	11.3	9.3	7.8
Path • Inches	-1.5	+1.5	0.0	-7.4	-21.5	-43.7	-75.7	-177.6	-349.8
Wind Drift • Inches	0.0	0.7	3.1	7.2	13.4	21.9	33.1	64.6	110.5

Nosler 165-grain Partition (60064)   G1 Ballistic Coefficient = 0.417

Distance • Yards	Muzzle	100	200	300	400	500	600	800	1000
Velocity • fps	3350	3103	2871	2650	2440	2239	2046	1694	1394
Energy • ft-lbs	4111	3537	3020	2573	2181	1837	1538	1056	716
Taylor KO Index	24.3	22.5	20.8	19.2	17.7	16.3	14.9	10.3	10.1
Path • Inches	-1.5	+1.1	0.0	-5.3	-15.5	-31.6	-54.7	-127.4	-249.0
Wind Drift • Inches	0.0	0.6	2.5	5.9	10.9	17.8	26.6	51.8	89.0

Remington 165-grain Solid Copper Tipped (PCS300UMB)   G1 Ballistic Coefficient = 0.447

Distance • Yards	Muzzle	100	200	300	400	500	600	800	1000
Velocity • fps	3200	2978	2767	2565	2372	2188	2012	1687	1407
Energy • ft-lbs	3751	3248	2804	2410	2061	1753	1483	1042	725
Taylor KO Index	23.2	21.6	20.1	18.6	17.2	15.9	14.6	12.2	10.2
Path • Inches	-1.5	+1.3	0.0	-6.1	-17.4	-35.0	-58.6	-135.1	-261.1
Wind Drift • Inches	0.0	0.6	2.5	5.8	10.7	17.4	26.1	50.4	86.1

Nosler 180-grain AccuBond (21005 & 60065)   G1 Ballistic Coefficient = 0.507

Distance • Yards	Muzzle	100	200	300	400	500	600	800	1000
Velocity • fps	3250	3051	2861	2678	2504	2335	2175	1871	1599
Energy • ft-lbs	4221	3721	3272	2868	2506	2180	1891	1400	1020
Taylor KO Index	25.7	24.2	22.7	21.2	19.8	18.5	17.2	14.8	12.7
Path • Inches	-1.5	+1.1	0.0	-5.4	-15.6	-31.4	-53.5	-121.1	-228.7
Wind Drift • Inches	0.0	0.5	2.2	5.0	9.1	14.7	21.9	41.7	70.4

Nosler 180-grain Partition (21001 & 60066)   G1 Ballistic Coefficient = 0.473

Distance • Yards	Muzzle	100	200	300	400	500	600	800	1000
Velocity • fps	3250	3038	2835	2642	2456	2278	2107	1789	1508
Energy • ft-lbs	4221	3688	3213	2789	2412	2075	1775	1280	909
Taylor KO Index	25.7	24.1	22.5	20.9	19.5	18.0	16.7	14.2	11.9
Path • Inches	-1.5	+1.1	0.0	-5.5	-15.9	-32.2	-55.1	-125.9	-240.7
Wind Drift • Inches	0.0	0.6	2.3	5.4	9.8	15.9	23.8	45.7	77.6

Remington 180-grain Swift A-Frame PSP (RS300UM1)   G1 Ballistic Coefficient = 0.377

Distance • Yards	Muzzle	100	200	300	400	500	600	800	1000
Velocity • fps	3250	2985	2735	2499	2274	2061	1861	1503	1223
Energy • ft-lbs	4221	3560	2989	2495	2067	1698	1385	903	597
Taylor KO Index	25.7	23.6	21.7	19.8	18.0	16.3	14.7	11.9	9.7
Path • Inches	-1.5	+1.2	0.0	-5.9	-17.3	-35.5	-61.9	-147.3	-295.2
Wind Drift • Inches	0.0	0.7	2.9	6.9	12.8	20.9	31.6	62.4	108.1

Cor-Bon 180-grain DPX (DPX300RUM180/20)   G1 Ballistic Coefficient = 0.450

Distance • Yards	Muzzle	100	200	300	400	500	600	800	1000
Velocity • fps	3300	3074	2860	2655	2460	2273	2094	1762	1472
Energy • ft-lbs	4354	3778	3269	2818	2419	2065	1753	1241	868
Taylor KO Index	26.1	24.3	22.7	21.0	19.5	18.0	16.6	14.0	11.7
Path • Inches	-1.5	+1.1	0.0	-5.4	-15.6	-31.7	-54.5	-125.4	-241.6
Wind Drift • Inches	0.0	0.6	2.4	5.6	10.2	16.6	24.8	47.8	81.4

Nosler 180-grain Ballistic Tip (21048)   G1 Ballistic Coefficient = 0.507

Distance • Yards	Muzzle	100	200	300	400	500	600	800	1000
Velocity • fps	3250	3051	2861	2678	2504	2335	2175	1871	1599
Energy • ft-lbs	4221	3721	3272	2868	2506	2180	1891	1400	1020
Taylor KO Index	25.7	24.2	22.7	21.2	19.8	18.5	17.2	14.8	12.7
Path • Inches	-1.5	+1.1	0.0	-5.4	-15.6	-31.4	-53.5	-121.1	-228.7
Wind Drift • Inches	0.0	0.5	2.2	5.0	9.1	14.7	21.9	41.7	70.4

Nosler 180-grain E-Tip (21007) LF

G1 Ballistic Coefficient = 0.532

Distance • Yards	Muzzle	100	200	300	400	500	600	800	1000
Velocity • fps	3150	2964	2787	2616	2451	2293	2141	1854	1594
Energy • ft-lbs	3965	3511	3104	2735	2401	2101	1832	1374	1016
Taylor KO Index	24.9	23.5	22.1	20.7	19.4	18.2	17.0	14.7	12.6
Path • Inches	-1.5	+1.2	0.0	-5.7	-16.5	-33.1	-56.3	-126.6	-237.8
Wind Drift • Inches	0.0	0.5	2.1	4.9	9.0	14.5	21.6	41.1	69.0

Remington 180-grain Swift Scirocco Bonded (PR300UM3)

G1 Ballistic Coefficient = 0.500

Distance • Yards	Muzzle	100	200	300	400	500	600	800	1000
Velocity • fps	3250	3048	2856	2672	2495	2325	2161	1855	1580
Energy • ft-lbs	4221	3714	3260	2853	2487	2160	1868	1376	998
Taylor KO Index	25.7	24.1	22.6	21.2	19.8	18.4	17.1	14.7	12.5
Path • Inches	-1.5	+1.1	0.0	-5.4	-15.7	-31.5	-53.8	-122.0	-231.0
Wind Drift • Inches	0.0	0.5	2.2	5.0	9.2	14.9	22.2	42.5	71.4

Remington 180-grain Core-Lokt Ultra Bonded – Power Level 3 (PR300UM4)

G1 Ballistic Coefficient = 0.389

Distance • Yards	Muzzle	100	200	300	400	500	600	800	1000
Velocity • fps	3230	2988	2742	2508	2287	2076	1881	1530	1249
Energy • ft-lbs	4171	3535	2983	2503	2086	1725	1415	935	623
Taylor KO Index	25.6	23.7	21.7	19.9	18.1	16.4	14.9	12.1	9.9
Path • Inches	-1.5	+1.2	0.0	-5.9	-17.4	-35.5	-61.7	-146.0	-290.5
Wind Drift • Inches	0.0	0.6	2.6	6.0	11.2	18.3	27.7	54.3	93.9

Remington 180-grain Core-Lokt Ultra Bonded – Power Level 2 (PR300UM2-P2)

G1 Ballistic Coefficient = 0.395

Distance • Yards	Muzzle	100	200	300	400	500	600	800	1000
Velocity • fps	2980	2742	2515	2300	2096	1902	1719	1400	1160
Energy • ft-lbs	3549	3004	2528	2114	1765	1445	1181	783	538
Taylor KO Index	23.6	21.7	19.9	18.2	16.6	15.1	13.6	11.1	9.2
Path • Inches	-1.5	+1.4	0.0	-7.1	-20.7	-42.4	-73.7	-174.4	-346.8
Wind Drift • Inches	0.0	0.8	3.2	7.4	13.7	22.5	34.0	66.8	114.6

Federal 180-grain Trophy Bonded Tip (P300RUMTT1)

G1 Ballistic Coefficient = 0.498

Distance • Yards	Muzzle	100	200	300	400	500	600	800	1000
Velocity • fps	3200	3000	2810	2630	2450	2280	2119	1815	1544
Energy • ft-lbs	4090	3600	3155	2760	2405	2085	1795	1317	953
Taylor KO Index	25.3	23.8	22.3	20.8	19.4	18.1	16.8	14.4	12.2
Path • Inches	-1.5	+1.2	0.0	-5.6	-16.2	-32.6	-55.8	-126.6	-240.1
Wind Drift • Inches	0.0	0.5	2.2	5.2	9.5	15.3	22.8	43.7	73.8

Federal 180-grain Barnes Triple-Shock X-Bullet (P300RUMD) LF

G1 Ballistic Coefficient = 0.454

Distance • Yards	Muzzle	100	200	300	400	500	600	800	1000
Velocity • fps	3150	2930	2730	2530	2340	2160	1988	1670	1397
Energy • ft-lbs	3965	3440	2970	2555	2190	1865	1580	1115	780
Taylor KO Index	24.9	23.2	21.6	20.0	18.5	17.1	15.7	13.2	11.1
Path • Inches	-1.5	+1.3	0.0	-6.0	-17.3	-35.1	-60.3	-138.9	-267.8
Wind Drift • Inches	0.0	0.6	2.5	5.9	10.8	17.5	26.3	50.7	86.4

Remington 180-grain Swift Scirocco Bonded – Power Level 2 (PR300UM3-P2)

G1 Ballistic Coefficient = 0.507

Distance • Yards	Muzzle	100	200	300	400	500	600	800	1000
Velocity • fps	2980	2793	2614	2442	2225	2116	1962	1678	1430
Energy • ft-lbs	3549	3118	2730	2382	2070	1790	1540	1126	817
Taylor KO Index	23.6	22.1	20.7	19.3	17.6	16.8	15.5	13.3	11.3
Path • Inches	-1.5	+1.3	0.0	-6.6	-19.0	-38.1	-65.1	-147.6	-279.9
Wind Drift • Inches	0.0	0.6	2.4	5.6	10.3	16.1	24.8	47.6	80.4

Nosler 200-grain AccuBond (21073 & 60067)

G1 Ballistic Coefficient = 0.587

Distance • Yards	Muzzle	100	200	300	400	500	600	800	1000
Velocity • fps	3050	2885	2727	2574	2426	2283	2145	1863	1644
Energy • ft-lbs	4131	3697	3302	2942	2614	2315	2044	1576	1200
Taylor KO Index	26.8	25.4	24.0	22.7	21.3	20.1	18.9	16.4	14.5
Path • Inches	-1.5	+1.3	0.0	-6.0	-17.3	-34.5	-58.3	-129.7	-240.7
Wind Drift • Inches	0.0	0.5	2.0	4.6	8.4	13.6	20.1	38.0	63.5

Federal 200-grain Nosler Partition (P300RUMC)

G1 Ballistic Coefficient = 0.484

Distance • Yards	Muzzle	100	200	300	400	500	600	800	1000
Velocity • fps	3070	2870	2680	2490	2320	2150	1990	1690	1429
Energy • ft-lbs	4185	3655	3180	2760	2380	2045	1758	1269	907
Taylor KO Index	27.0	25.3	23.6	21.9	20.4	18.9	17.5	14.9	12.6
Path • Inches	-1.5	+1.4	0.0	-6.2	-18.0	-36.4	-62.1	-141.6	-270.4
Wind Drift • Inches	0.0	0.6	2.4	5.6	10.4	16.8	25.2	48.4	82.0

Remington 200-grain Swift A-Frame – Power Level 3 (RS300UM2)

G1 Ballistic Coefficient = 0.395

Distance • Yards	Muzzle	100	200	300	400	500	600	800	1000
Velocity • fps	3032	2791	2562	2345	2138	1942	1756	1430	1181
Energy • ft-lbs	4083	3459	2916	2442	2030	1675	1370	908	619
Taylor KO Index	26.7	24.6	22.5	20.6	18.8	17.1	15.5	12.6	10.4
Path • Inches	-1.5	+1.4	0.0	-6.8	-19.9	-40.7	-70.8	-167.4	-332.7
Wind Drift • Inches	0.0	0.7	3.1	7.2	13.4	21.9	33.1	65.0	111.8

Remington's Eddie Stevenson (right) with Craig Boddington and a gemsbok taken with the .300 Remington Ultra Mag. The fastest .30 calibers mated with tough bullets are wonderful for plains game in open country. (Photo from *Safari Rifles II* by Craig Boddington, 2009, Safari Press Archives)

7.82mm (.308) Lazzeroni Warbird

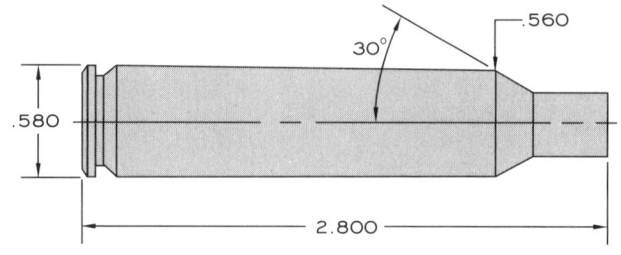

If the 7.82mm Patriot is Lazzeroni's "short" magnum, the Warbird is certainly a "long" magnum. This cartridge shares the lead in the .30-caliber velocity derby with the .30-.378 Weatherby. It is an extremely potent cartridge, and one with rather specialized applications.

Relative Recoil Factor = 2.80

Specifications:

Controlling Agency for Standardization of this Ammunition: Factory

Availability:

Lazzeroni 130-grain (782WB130)

G1 Ballistic Coefficient = 0.429

Distance • Yards	Muzzle	100	200	300	400	500	600	800	1000
Velocity • fps	3975	3697	3438	3193	2962	2742	2533	2143	1791
Energy • ft-lbs	4562	3948	3412	2944	2533	2172	1853	1327	926
Taylor KO Index	22.7	21.1	19.7	18.3	16.9	15.7	14.5	12.3	10.2
Path • Inches	-1.5	+0.5	0.0	-3.5	-10.4	-21.2	-36.7	-84.9	-164.2
Wind Drift • Inches	0.0	0.5	2.0	4.7	8.6	13.8	20.6	39.3	66.7

Lazzeroni 150-grain Nosler Partition (782WB150P)

G1 Ballistic Coefficient = 0.491

Distance • Yards	Muzzle	100	200	300	400	500	600	800	1000
Velocity • fps	3775	3542	3323	3114	2915	2724	2542	2199	1884
Energy • ft-lbs	4747	4181	3679	3231	2831	2473	2154	1611	1182
Taylor KO Index	24.9	23.4	21.9	20.6	19.2	18.1	16.8	14.5	12.4
Path • Inches	-1.5	+0.6	0.0	-3.8	-11.2	-22.6	38.8	-88.0	-166.6
Wind Drift • Inches	0.0	0.4	1.9	4.3	7.8	12.6	18.6	35.3	59.3

Lazzeroni 180-grain Nosler Partition (782WB180P)

G1 Ballistic Coefficient = 0.549

Distance • Yards	Muzzle	100	200	300	400	500	600	800	1000
Velocity • fps	3550	3352	3163	2983	2810	2643	2483	2179	1897
Energy • ft-lbs	5038	4493	4001	3558	3157	2794	2466	1899	1439
Taylor KO Index	29.1	26.5	25.1	23.6	22.3	20.9	19.7	17.3	15.0
Path • Inches	-1.5	+0.9	0.0	-4.3	-12.4	-25.0	-42.6	-95.4	-177.8
Wind Drift • Inches	0.0	0.4	1.8	4.1	7.4	11.9	17.7	33.3	55.5

Lazzeroni 200-grain Swift A-Frame (782WB200A)

G1 Ballistic Coefficient = 0.55

Distance • Yards	Muzzle	100	200	300	400	500	600	800	1000
Velocity • fps	3350	3162	2983	2810	2644	2484	2330	2038	1768
Energy • ft-lbs	4985	4442	3952	3509	3106	2742	2412	1844	1389
Taylor KO Index	29.5	27.8	26.3	24.7	23.3	21.9	20.5	17.9	15.6
Path • Inches	-1.5	+1.0	0.0	-4.9	-14.2	-28.5	-48.4	-108.4	-202.3
Wind Drift • Inches	0.0	0.5	1.9	4.4	8.0	12.8	19.0	36.0	60.1

.30-378 Weatherby

Whenever a new cartridge case is introduced, the wildcat builders have a field day. The basic case is necked down, necked up, improved, and otherwise modified to conform to the whims of the wildcat "designer." The .378 Weatherby offered a tremendous opportunity. This big case was necked to both .338 and .30 calibers soon after its introduction. The wildcat versions have been around for a long time. I fired a .30-378 in 1965, and that certainly wasn't the first gun in this caliber. Recently Weatherby has begun offering guns and ammunition in both .30-378 and .333-378 calibers. The .30-378 is the hottest .30 caliber in the standard inventory.

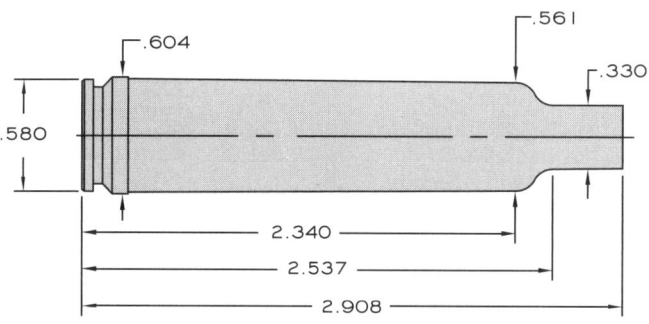

Relative Recoil Factor = 2.55

Specifications:

Controlling Agency for Standardization of this Ammunition: Factory

Standard barrel for velocity testing: 26 inches long—1 turn in 10-inch twist

Availability:

Weatherby 130-grain Barnes TTSX (B303130TTSX) LF

G1 Ballistic Coefficient = 0.350

Distance • Yards	Muzzle	100	200	300	400	500	600	800	1000
Velocity • fps	3740	3417	3118	2889	2577	2329	2097	1673	1326
Energy • ft-lbs	4039	3372	2807	2327	1917	1567	1269	808	508
Taylor KO Index	21.4	19.5	17.8	16.2	14.7	13.3	12.0	9.6	7.6
Path • Inches	-1.5	+0.7	0.0	-4.3	-13.0	-26.9	-47.2	-113.6	-230.7
Wind Drift • Inches	0.0	0.6	2.7	6.3	11.7	19.2	28.9	57.1	100.0

Nosler 165-grain Partition (23028)

G1 Ballistic Coefficient = 0.410

Distance • Yards	Muzzle	100	200	300	400	500	600	800	1000
Velocity • fps	3450	3194	2952	2723	2506	2299	2102	1739	1427
Energy • ft-lbs	4360	3737	3193	2717	2301	1935	1620	1108	747
Taylor KO Index	25.0	23.2	21.4	19.8	18.2	16.7	15.3	12.6	10.4
Path • Inches	-1.5	+1.0	0.0	-5.0	-14.6	-29.8	-51.6	-120.4	-235.9
Wind Drift • Inches	0.0	0.6	2.5	5.8	10.7	17.4	26.1	50.8	87.3

Weatherby 165-grain BST (N 303 165 BST)

G1 Ballistic Coefficient = 0.475

Distance Yards	Muzzle	100	200	300	400	500	600	800	1000
Velocity • fps	3500	3275	3062	2859	2665	2480	2301	1966	1664
Energy • ft-lbs	4488	3930	3435	2995	2603	2253	1940	1417	1014
Taylor KO Index	25.4	23.8	22.2	20.8	19.3	18.0	16.7	14.3	11.9
Path • Inches	-1.5	+0.9	0.0	-4.6	-13.4	-27.2	46.6	106.3	202.8
Wind Drift Inches	0.0	0.5	2.1	4.7	8.9	14.3	21.4	40.9	69.1

Weatherby 165-grain Barnes TSX (B303165TSX) LF

G1 Ballistic Coefficient = 0.398

Distance • Yards	Muzzle	100	200	300	400	500	600	800	1000
Velocity • fps	3450	3186	2937	2702	2479	2265	2068	1699	1396
Energy • ft-lbs	4362	3719	3161	2675	2251	1882	1567	1057	704
Taylor KO Index	25.0	23.1	21.3	19.6	18.0	16.4	15.0	12.3	10.1
Path • Inches	-1.5	+1.0	0.0	-5.0	-14.8	-30.2	-52.4	-123.0	-242.5
Wind Drift • Inches	0.0	0.6	2.6	6.0	11.1	18.0	27.1	52.9	91.2

Weatherby 180-grain AccuBond (N303180ACB)

G1 Ballistic Coefficient = 0.507

Distance • Yards	Muzzle	100	200	300	400	500	600	800	1000
Velocity • fps	3420	3213	3015	2826	2645	2471	2307	1992	1705
Energy • ft-lbs	4676	4126	3634	3193	2797	2441	2127	1586	1163
Taylor KO Index	27.1	25.4	23.9	22.4	20.9	19.6	18.3	15.8	13.5
Path • Inches	-1.5	+0.9	0.0	-4.8	-13.9	-28.0	-47.7	-107.9	-203.5
Wind Drift • Inches	0.0	0.5	2.0	4.7	8.5	13.7	20.4	37.8	65.2

Nosler 180-grain AccuBond (23010)

G1 Ballistic Coefficient = 0.507

Distance • Yards	Muzzle	100	200	300	400	500	600	800	1000
Velocity • fps	3400	3194	2998	2810	2630	2457	2291	1978	1693
Energy • ft-lbs	4620	4077	3592	3155	2763	2412	2099	1564	1145
Taylor KO Index	26.9	25.3	23.7	22.3	20.8	19.5	18.1	15.7	13.4
Path • Inches	-1.5	+1.0	0.0	-4.8	-14.1	-28.3	-48.4	-109.4	-206.2
Wind Drift • Inches	0.0	0.5	2.0	4.7	8.6	13.8	20.6	39.1	65.8

Nosler 180-grain Partition (23001)

G1 Ballistic Coefficient = 0.474

Distance • Yards	Muzzle	100	200	300	400	500	600	800	1000
Velocity • fps	3400	3180	2971	2772	2581	2398	2224	1895	1601
Energy • ft-lbs	4620	4042	3528	3070	2662	2298	1977	1436	1025
Taylor KO Index	26.9	25.2	23.5	22.0	20.4	19.0	17.6	15.0	12.7
Path • Inches	-1.5	+1.0	0.0	-4.9	-14.4	-29.0	-49.8	-113.5	-216.4
Wind Drift • Inches	0.0	0.5	2.2	5.0	9.3	15.0	22.3	42.7	72.3

Weatherby 180-grain BST (N 303 180 BST)

G1 Ballistic Coefficient = 0.507

Distance • Yards	Muzzle	100	200	300	400	500	600	800	1000
Velocity • fps	3420	3213	3015	2826	2645	2471	2305	1990	1703
Energy • ft-lbs	4676	4126	3634	3194	2797	2441	2124	1583	1160
Taylor KO Index	27.1	25.4	23.9	22.4	20.9	19.6	19.3	15.8	13.5
Path • Inches	-1.5	+0.9	0.0	-4.8	-13.9	-28.0	-47.8	-108.0	-203.7
Wind Drift • Inches	0.0	0.4	1.8	4.2	7.7	12.4	18.4	35.0	58.8

Weatherby 180-grain Barnes TSX (B 303 180 TSX) LF

G1 Ballistic Coefficient = 0.453

Distance • Yards	Muzzle	100	200	300	400	500	600	800	1000
Velocity • fps	3360	3132	2916	2709	2513	2324	2144	1809	1513
Energy • ft-lbs	4513	3921	3398	2935	2524	2160	1838	1308	915
Taylor KO Index	26.6	24.8	23.1	21.5	19.9	18.4	17.0	14.3	12.0
Path • Inches	-1.5	+1.0	0.0	-5.1	-15.0	-30.4	-52.2	-120.0	-230.7
Wind Drift • Inches	0.0	0.6	2.3	5.4	9.9	16.0	24.0	46.2	78.6

Weatherby 200-grain Partition (N 303 200 PT)

G1 Ballistic Coefficient = 0.481

Distance Yards	Muzzle	100	200	300	400	500	600	800	1000
Velocity • fps	3160	2955	2759	2572	2392	2220	2054	1746	1475
Energy • ft-lbs	4434	3877	3381	2938	2541	2188	1873	1354	966
Taylor KO Index	27.8	26.0	24.3	22.6	21.0	19.5	18.1	15.4	13.0
Path • Inches	-1.5	+1.2	0.0	-5.8	-16.9	-34.1	-58.3	-133.1	-254.9
Wind Drift • Inches	0.0	0.6	2.4	5.5	10.1	16.3	24.3	46.7	79.1

Nosler 200-grain AccuBond (23044)

G1 Ballistic Coefficient = 0.587

Distance • Yards	Muzzle	100	200	300	400	500	600	800	1000
Velocity • fps	3150	2982	2820	2664	2513	2367	2226	1958	1712
Energy • ft-lbs	4406	3949	3531	3151	2805	2488	2201	1703	1302
Taylor KO Index	27.7	26.2	24.8	23.4	22.1	20.8	19.6	17.2	15.1
Path • Inches	-1.5	+1.2	0.0	-5.6	-16.1	-32.1	-54.3	-120.7	-223.6
Wind Drift • Inches	0.0	0.5	0.9	4.4	8.1	13.0	19.0	36.3	60.4

Nosler 200-grain Partition (23005)

G1 Ballistic Coefficient = 0.481

Distance • Yards	Muzzle	100	200	300	400	500	600	800	1000
Velocity • fps	3150	2945	2750	2563	2383	2211	2047	1740	1470
Energy • ft-lbs	4406	3852	3357	2916	2522	2170	1861	1345	960
Taylor KO Index	27.7	25.9	24.2	22.6	21.0	19.5	18.0	15.3	12.9
Path • Inches	-1.5	+1.3	0.0	-5.9	-17.0	-34.3	-58.7	-133.9	-255.5
Wind Drift • Inches	0.0	0.6	2.4	5.5	10.1	16.3	24.4	46.8	79.4

.303 British

The .303 British was Britain's service-rifle cartridge from 1888 until it was replaced in, about, 1957 by the 7.62 NATO round (.308). It started life as a black-powder number firing a 215-grain bullet. When smokeless powder came into service use, the black powder was replaced with cordite. Cordite is a tubular "powder" that was manufactured for this cartridge in sticks about 0.040 inches in diameter and 1.625 inches long, like pencil leads. Since it would have been almost impossible to load the cordite sticks into the 0.310-inch-diameter case mouth, a flock of little old ladies in tennis shoes loaded the sticks into the cases by hand before the neck and shoulder were formed. After the propellant was inserted, the case forming was completed and the bullet seated. Talk about hard ways to do easy things.

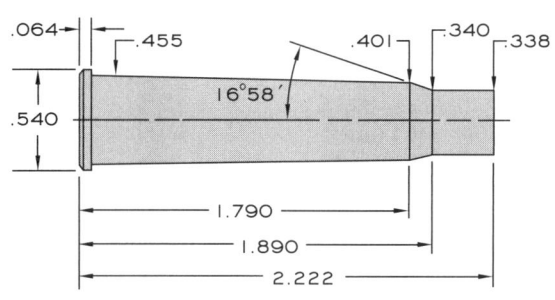

Relative Recoil Factor = 1.98

Specifications:

Controlling Agency for Standardization of this Ammunition: SAAMI

Bullet Weight Grains	Velocity fps	Maximum Average Pressure	
		Copper Crusher	Transducer
150	2,685	45,000 cup	49,000 psi
180	2,450	45,000 cup	49,000 psi
215	2,155	45,000 cup	49,000 psi

Standard barrel for velocity testing: 24 inches long—1 turn in 10-inch twist

Availability:

Federal 150-grain Soft Point (303B)

G1 Ballistic Coefficient = 0.357

Distance • Yards	Muzzle	100	200	300	400	500	600	800	1000
Velocity • fps	2690	2440	2210	1980	1780	1590	1420	1153	999
Energy • ft-lbs	2400	1980	1620	1310	1055	840	672	443	333
Taylor KO Index	18.1	16.4	14.9	13.3	12.0	10.7	9.5	7.7	6.7
Path • Inches	-1.5	+2.2	0.0	-9.4	-27.6	-56.8	99.8	243.8	493.7
Wind Drift • Inches	0.0	1.0	4.1	9.6	18.1	29.8	40.8	80.2	132.9

Hornady 150-grain SP InterLock (8225)

G1 Ballistic Coefficient = 0.361

Distance • Yards	Muzzle	100	200	300	400	500	600	800	1000
Velocity • fps	2685	2441	2210	1992	1767	1598	1427	1161	1004
Energy • ft-lbs	2401	1984	1627	1321	1064	850	679	449	336
Taylor KO Index	18.1	16.4	14.9	13.4	11.9	10.8	9.6	7.8	6.7
Path • Inches	-1.5	+1.0	0.0	-9.3	-27.4	-56.5	-99.5	-242.4	-490.0
Wind Drift • Inches	0.0	1.0	4.0	9.5	17.9	29.5	40.3	79.2	132.4

Sellier and Bellot 150-grain Soft Point (V331312U)

G1 Ballistic Coefficient = 0.276

Distance • Yards	Muzzle	100	200	300	400	500	600	800	1000
Velocity • fps	2654	2322	2032	1778	1534	1326	1163	972	863
Energy • ft-lbs	2341	1792	1372	1050	784	586	451	315	248
Taylor KO Index	17.7	15.5	13.5	11.8	10.2	8.8	7.8	6.5	5.8
Path • Inches	-1.5	+2.4	0.0	-10.8	-32.8	-69.8	-127.2	-325.4	-660.2
Wind Drift • Inches	0.0	1.3	5.5	13.3	25.4	42.6	65.3	125.8	201.7

UMC (Remington) 174-grain Metal Case (L303B1)

G1 Ballistic Coefficient = 0.315

Distance • Yards	Muzzle	100	200	300	400	500	600	800	1000
Velocity • fps	2475	2209	1960	1729	1520	1337	1189	1003	896
Energy • ft-lbs	2366	1885	1484	1155	892	691	547	389	310
Taylor KO Index	19.1	17.1	15.2	13.4	11.8	10.3	9.2	7.8	7.0
Path • Inches	-1.5	+2.8	0.0	-11.9	-35.6	-74.3	-133.4	-330.8	-659.0
Wind Drift • Inches	0.0	1.2	5.3	1.2.7	23.9	39.7	60.3	115.1	184.2

Federal 180-grain Soft Point (303AS)

G1 Ballistic Coefficient = 0.327

Distance • Yards	Muzzle	100	200	300	400	500	600	800	1000
Velocity • fps	2460	2210	1970	1740	1540	1360	1213	1021	912
Energy • ft-lbs	2420	1945	1545	1215	950	740	588	416	332
Taylor KO Index	19.7	17.7	15.8	13.9	12.3	10.9	9.7	8.2	7.3
Path • Inches	-1.5	+2.8	0.0	-11.9	-35.3	-73.6	-130.4	-320.4	-643.6
Wind Drift • Inches	0.0	1.2	5.1	12.2	23.0	38.1	57.8	110.6	177.6

Remington 180-grain Core-Lokt Soft Point (R303B1)

G1 Ballistic Coefficient = 0.247

Distance • Yards	Muzzle	100	200	300	400	500	600	800	1000
Vclocity • fps	2460	2124	1817	1542	1311	1137	1023	884	788
Energy • ft-lbs	2418	1803	1319	950	690	517	418	313	248
Taylor KO Index	19.9	17.1	14.7	12.5	10.6	9.2	8.2	7.1	6.3
Path • Inches	-1.5	+2.9	0.0	-13.6	-42.2	-91.0	-168.2	-420.7	833.5
Wind Drift • Inches	0.0	1.6	7.0	17.1	32.8	54.8	82.4	151.1	234.9

Winchester 180-grain Power-Point (X303B1)

G1 Ballistic Coefficient = 0.370

Distance • Yards	Muzzle	100	200	300	400	500	600	800	1000
Velocity • fps	2460	2233	2018	1816	1629	1459	1311	1093	968
Energy • ft-lbs	2418	1993	1627	1318	1060	851	687	477	375
Taylor KO Index	19.9	18.0	16.3	14.7	13.2	11.8	10.5	8.8	7.8
Path • Inches	-1.5	+2.8	0.0	-11.3	-33.2	-68.2	-120.0	-289.9	-574.9
Wind Drift • Inches	0.0	1.1	4.5	10.6	19.8	32.6	49.3	95.2	155.4

Sellier & Bellot 180-grain FMJ (V331302U)

G1 Ballistic Coefficient = 0.505

Distance • Yards	Muzzle	100	200	300	400	500	600	800	1000
Velocity • fps	2438	2272	2112	1959	1813	1674	1545	1318	1146
Energy • ft-lbs	2378	2063	1783	1534	1314	1121	954	695	525
Taylor KO Index	19.5	18.2	16.9	15.7	14.5	13.4	12.4	10.5	9.2
Path • Inches	-1.5	+2.6	0.0	-10.4	-29.9	-60.0	-102.6	-234.0	-445.7
Wind Drift • Inches	0.0	0.8	3.2	7.5	13.9	22.6	33.7	64.5	107.5

In the late 1950s, an article in Outdoor Life *magazine about calling predators led Bill Quimby to order a call from a mail-order supplier. This big coyote came to his .300 Savage the first morning he tried it outside Tucson, Arizona.* (Photo from *Sixty Years a Hunter* by Bill Quimby, 2010, Safari Press Archives)

.32-20 Winchester

The .32-20 dates back to 1882, which is about 130 years ago. Winchester designed the .32-20 to be a midpower rifle cartridge, but it was also useful in revolvers. As is true with many of these old designations, the .32-20 isn't a .32 at all. The bore diameter is 0.305 inch. Today two companies make ammunition for this old-timer. Velocities and pressures are very low. The .32-20 isn't the caliber you would choose if you wanted a flat-shooting varmint rifle.

Relative Recoil Factor = 0.55

Specifications:

Controlling Agency for Standardization of this Ammunition: SAAMI

Bullet Weight Grains	Velocity fps	Maximum Average Pressure	
		Copper Crusher	Transducer
100	1,200	16,000 cup	N/S

Standard barrel for velocity testing: 24 inches long—1 turn in 20-inch twist

Availability:

 Remington 100-grain Lead (R32201)

G1 Ballistic Coefficient = 0.166

Distance • Yards	Muzzle	100	200	300	400	500	600	800	1000
Velocity • fps	1210	1021	913	834	769	712	660	569	489
Energy • ft-lbs	325	231	185	154	131	113	97	72	53
Taylor KO Index	5.4	4.5	4.1	3.7	3.4	3.2	2.9	2.5	2.1
Path • Inches	-1.5	+15.8	0.0	-57.5	-165.1	-331.5	-552.1	-1252	-2373
Wind Drift • Inches	0.0	4.2	15.5	32.4	54.8	82.6	116.0	201.2	314.4

Winchester 100-grain Lead (X32201)

G1 Ballistic Coefficient = 0.166

Distance • Yards	Muzzle	100	200	300	400	500	600	800	1000
Velocity • fps	1210	1021	913	834	769	712	660	569	489
Energy • ft-lbs	325	231	185	154	131	113	97	72	53
Taylor KO Index	5.4	4.5	4.1	3.7	3.4	3.2	2.9	2.5	2.1
Path • Inches	-1.5	+15.8	0.0	-57.5	-165.1	-331.5	-552.1	-1252	-2373
Wind Drift • Inches	0.0	4.2	15.5	32.4	54.8	82.6	116.0	201.2	314.4

.32 Winchester Special

Winchester's .32 Winchester Special cartridge is another very old number. This cartridge is a real .32, using bullets that are 0.321 inch in diameter. It can be thought of as the .30-30 necked up to .32 caliber. Starting in 1895, Winchester's Model 94 rifles were available chambered for the .32 Winchester Special. There have not been any guns built in this caliber since about 1960, but the ammo companies continue to offer this caliber. Hornady has recently started producing this caliber.

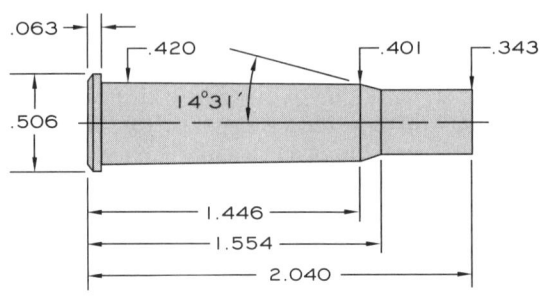

Relative Recoil Factor = 1.95

Specifications:

Controlling Agency for Standardization of this Ammunition: SAAMI

Bullet Weight Grains	Velocity fps	Maximum Average Pressure	
		Copper Crusher	Transducer
170	2,235	38,000 cup	42,000 psi

Standard barrel for velocity testing: 24 inches long—1 turn in 16-inch twist

Availability:

Hornady 165-grain FTX (82732)
G1 Ballistic Coefficient = 0.31

Distance • Yards	Muzzle	100	200	300	400	500	600	800	1000
Velocity • fps	2410	2145	1897	1669	1465	1290	1152	983	882
Energy • ft-lbs	2128	1685	1318	1020	786	610	486	354	285
Taylor KO Index	18.3	16.3	14.4	12.7	11.1	9.8	8.7	7.5	6.7
Path • Inches	-1.5	+3.0	0.0	-12.8	-38.1	-79.6	-142.2	-350.9	-703.4
Wind Drift • Inches	0.0	1.3	5.6	13.4	25.2	41.8	63.3	119.5	189.6

Federal 170-grain Soft Point FN (32A)
G1 Ballistic Coefficient = 0.24

Distance • Yards	Muzzle	100	200	300	400	500	600	800	1000
Velocity • fps	2250	1920	1630	1380	1180	1050	964	845	757
Energy • ft-lbs	1910	1395	1000	715	525	415	315	270	216
Taylor KO Index	17.6	15.0	12.7	10.8	9.2	8.2	7.5	6.6	5.9
Path • Inches	-1.5	+4.0	0.0	-17.2	-52.3	-109.8	-204.4	-505.5	-1002
Wind Drift • Inches	0.0	1.9	8.2	20.0	37.9	61.9	91.0	161.6	247.2

Remington 170-grain Core-Lokt Soft Point (R32WS2)
G1 Ballistic Coefficient = 0.23

Distance • Yards	Muzzle	100	200	300	400	500	600	800	1000
Velocity • fps	2250	1921	1626	1372	1175	1044	957	839	749
Energy • ft-lbs	1910	1393	998	710	521	411	346	266	212
Taylor KO Index	17.5	14.9	12.6	10.7	9.1	8.1	7.5	6.5	5.7
Path • Inches	-1.5	+4.0	0.0	-17.3	-53.2	-114.2	-207.5	-502.7	-976.0
Wind Drift • Inches	0.0	1.9	8.4	20.3	38.6	63.0	92.4	163.8	250.2

Winchester 170-grain Power-Point (X32WS2
G1 Ballistic Coefficient = 0.20

Distance • Yards	Muzzle	100	200	300	400	500	600	800	1000
Velocity • fps	2250	1870	1537	1267	1082	971	893	777	686
Energy • ft-lbs	1911	1320	892	606	442	356	301	228	178
Taylor KO Index	17.5	14.5	11.9	9.8	8.4	7.5	7.0	6.1	5.3
Path • Inches	-1.5	+4.3	0.0	-19.2	-60.2	-130.8	-236.4	-571.2	-1113
Wind Drift • Inches	0.0	2.3	10.0	24.5	46.4	74.7	108.0	188.1	286.1

8x57mm Mauser JS

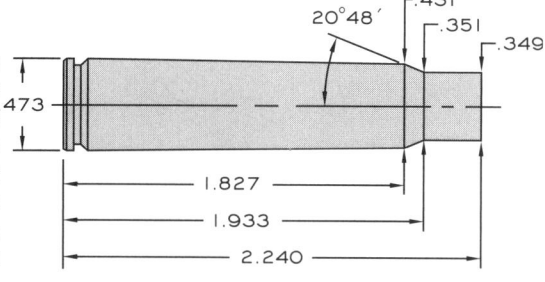

The 8x57mm Mauser was adopted as the standard German military cartridge in 1888. It has been around for a long time, and so have some of the guns that use this caliber. Exactly because some of these old guns still exist, the SAAMI pressure specifications are very modest (see below). Interestingly, the CIP (European) specifications allow the European manufacturers to load to considerably better performance. There are now some American producers who are loading for "modern rifles in good condition." They are loading to pressures similar to the .30-06. When they say "modern in good condition," they really mean it. If you have a very old 8mm Mauser gun, be very careful what ammo you put through it. If you have any doubts, have a good talk with your gunsmith.

You may see three versions of this cartridge: JS, JR, and JRS. The JR and JRS versions are rimmed and are mainly used in single-shot and multibarrel guns.

Relative Recoil Factor = 1.81

Specifications:

Controlling Agency for Standardization of this Ammunition: SAAMI (see above)

Bullet Weight Grains	Velocity fps	Maximum Average Pressure Copper Crusher	Transducer
170	2,340	37,000 cup	35,000 psi

Standard barrel for velocity testing: 24 inches long—1 turn in 9.5-inch twist

Availability:

Norma 123-grain FULL JACKET (18009)
G1 Ballistic Coefficient = 0.191

Distance • Yards	Muzzle	100	200	300	400	500	600	800	1000
Velocity • fps	2822	2357	1939	1572	1278	1079	963	819	715
Energy • ft-lbs	2176	1519	1029	678	466	318	253	183	140
Taylor KO Index	16.0	13.4	11.0	8.9	7.3	6.1	5.5	4.6	4.1
Path • Inches	-1.5	+2.4	0.0	-11.8	-37.8	-85.2	-162.4	-436.3	-919.0
Wind Drift • Inches	0.0	1.8	7.8	19.3	37.9	64.4	97.7	179.9	281.2

Federal 170-grain Soft Point (8A)
G1 Ballistic Coefficient = 0.310

Distance • Yards	Muzzle	100	200	300	400	500	600	800	1000
Velocity • fps	2250	2030	1810	1620	1440	1200	1037	947	856
Energy • ft-lbs	1910	1550	1240	990	785	630	446	339	277
Taylor KO Index	17.6	15.9	14.2	12.7	11.3	9.4	8.5	7.4	6.7
Path • Inches	-1.5	+3.5	0.0	-14.3	-41.6	-86.1	-166.1	-404.7	-798.4
Wind Drift • Inches	0.0	1.5	6.2	14.8	27.9	46.0	68.8	126.6	197.4

Remington 170-grain Core-Lokt Soft Point (R8MSR)
G1 Ballistic Coefficient = 0.205

Distance • Yards	Muzzle	100	200	300	400	500	600	800	1000
Velocity • fps	2360	1969	1622	1333	1123	997	912	792	699
Energy • ft-lbs	2102	1463	993	671	476	375	314	237	185
Taylor KO Index	18.5	15.4	12.7	10.5	8.8	7.8	7.2	6.2	5.5
Path • Inches	-1.5	+3.8	0.0	-17.2	-54.1	-118.6	-218.1	-549.3	-1103
Wind Drift • Inches	0.0	1.9	8.4	20.7	39.6	64.6	94.5	116.5	254.5

Winchester 170-grain Power-Point (X8MM)
G1 Ballistic Coefficient = 0.205

Distance • Yards	Muzzle	100	200	300	400	500	600	800	1000
Velocity • fps	2360	1969	1622	1333	1123	997	912	791	697
Energy • ft-lbs	2102	1463	993	671	476	375	314	236	184
Taylor KO Index	18.5	15.4	12.7	10.5	8.8	7.8	7.2	6.2	5.5
Path • Inches	-1.5	+3.8	0.0	-17.2	-54.1	-118.6	-217.5	-533.2	-1048
Wind Drift • Inches	0.0	2.1	9.3	23.0	44.0	71.8	105.0	184.9	282.5

Nosler 180-grain Ballistic Tip (24605)

G1 Ballistic Coefficient = 0.39

Distance • Yards	Muzzle	100	200	300	400	500	600	800	1000
Velocity • fps	2600	2380	2171	1973	1787	1613	1454	1198	1037
Energy • ft-lbs	2702	2264	1883	1556	1276	1040	846	574	430
Taylor KO Index	21.6	19.8	18.0	16.4	14.6	13.4	12.1	10.0	8.6
Path • Inches	-1.5	+2.3	0.0	-9.7	-28.4	-58.0	-101.1	-240.3	-476.5
Wind Drift • Inches	0.0	0.9	3.8	9.0	16.9	27.7	41.9	81.6	136.4

Sellier & Bellot 196-grain SPCE (V331812U)

G1 Ballistic Coefficient = 0.28

Distance • Yards	Muzzle	100	200	300	400	500	600	800	1000
Velocity • fps	2591	2292	2012	1755	1523	1324	1167	980	872
Energy • ft-lbs	2932	2286	1763	1341	1010	763	593	418	331
Taylor KO Index	23.4	20.7	18.2	15.9	13.8	12.0	10.6	8.9	7.9
Path • Inches	-1.5	+2.6	0.0	-11.2	-33.8	-71.5	-129.0	-326.3	-668.3
Wind Drift • Inches	0.0	1.3	5.5	13.2	25.2	42.0	64.2	123.2	197.2

Norma 196-grain Oryx (18004)

G1 Ballistic Coefficient = 0.33

Distance • Yards	Muzzle	100	200	300	400	500	600	800	1000
Velocity • fps	2526	2270	2028	1803	1596	1412	1255	1044	927
Energy • ft-lbs	2778	2242	1791	1415	1109	868	686	475	374
Taylor KO Index	22.8	20.5	18.3	16.3	14.4	12.8	11.4	9.4	8.4
Path • Inches	-1.5	+2.6	0.0	-11.1	-33.0	-68.8	-122.3	-302.2	-605.3
Wind Drift • Inches	0.0	1.2	4.9	11.6	21.8	36.1	54.9	106.2	172.0

Norma 196-grain Alaska (18003)

G1 Ballistic Coefficient = 0.30

Distance • Yards	Muzzle	100	200	300	400	500	600	800	1000
Velocity • fps	2526	2248	1988	1747	1530	1341	1188	998	889
Energy • ft-lbs	2778	2200	1720	1328	1020	783	614	433	344
Taylor KO Index	22.8	20.3	18.0	15.8	13.8	12.1	10.7	9.0	8.0
Path • Inches	-1.5	+1.6	0.0	-11.5	-34.6	-72.7	-130.6	-326.7	-654.4
Wind Drift • Inches	0.0	1.2	5.3	12.8	24.1	40.1	61.2	117.2	187.9

Norma 196-grain Vulcan (18020)

G1 Ballistic Coefficient = 0.34

Distance • Yards	Muzzle	100	200	300	400	500	600	800	1000
Velocity • fps	2526	2281	2049	1832	1634	1453	1297	1075	951
Energy • ft-lbs	2778	2264	1828	1461	1162	919	732	503	394
Taylor KO Index	22.8	20.6	18.5	16.6	14.8	13.1	11.7	9.7	8.6
Path • Inches	-1.5	+2.6	0.0	-10.9	-32.2	-66.6	-117.5	-286.2	-574.8
Wind Drift • Inches	0.0	1.1	4.2	9.8	18.5	30.5	46.4	89.9	146.8

Sellier & Bellot 196-grain FMJ (V331802U)

G1 Ballistic Coefficient = 0.56

Distance • Yards	Muzzle	100	200	300	400	500	600	800	1000
Velocity • fps	2558	2406	2258	2116	1979	1848	1723	1493	1298
Energy • ft-lbs	2848	2519	2220	1949	1705	1486	1292	970	733
Taylor KO Index	23.1	21.8	20.4	19.1	17.9	16.7	15.6	13.5	11.7
Path • Inches	-1.5	+2.3	0.0	-9.1	-25.9	-51.7	-87.6	-196.5	-368.2
Wind Drift • Inches	0.0	0.6	2.7	6.2	11.4	18.3	27.3	51.9	86.6

Nosler 200-grain AccuBond (24610)

G1 Ballistic Coefficient = 0.45

Distance • Yards	Muzzle	100	200	300	400	500	600	800	1000
Velocity • fps	2475	2287	2108	1937	1775	1623	1482	1246	1082
Energy • ft-lbs	2720	2322	1972	1660	1399	1169	976	689	520
Taylor KO Index	22.8	21.1	19.5	17.9	16.4	15.0	13.7	11.5	10.0
Path • Inches	-1.5	+2.6	0.0	-11.2	-33.8	-71.5	-129.0	-326.3	-668.3
Wind Drift • Inches	0.0	0.9	3.6	8.4	15.5	25.3	38.1	73.3	112.1

Nosler 200-grain Partition (24615)

G1 Ballistic Coefficient = 0.42

Distance • Yards	Muzzle	100	200	300	400	500	600	800	1000
Velocity • fps	2475	2276	2088	1909	1740	1582	1438	1202	1048
Energy • ft-lbs	2720	2301	1936	1618	1344	1112	918	642	488
Taylor KO Index	22.8	21.0	19.3	17.6	16.1	14.6	13.3	11.1	9.7
Path • Inches	-1.5	+2.6	0.0	-10.6	-30.8	-62.5	-108.1	-253.4	-495.1
Wind Drift • Inches	0.0	0.9	3.8	8.6	16.6	27.1	40.8	78.8	130.7

.325 Winchester Short Magnum

This is the large end of the WSM line, introduced in late 2004. Like many other cartridges, the .325 designation is slightly inaccurate. This cartridge actually uses 8mm bullets. That's good news because reloaders don't have to worry about a whole new caliber. Performance is very similar to the 8mm Remington Magnum, but that performance is produced with a shorter case.

Relative Recoil Factor = 2.78

Specifications:

Controlling Agency for Standardization of this Ammunition: SAAMI

Bullet Weight Grains	Velocity fps	Maximum Average Pressure Copper Crusher	Transducer
200	2,950	N/S	65,000 psi

Standard barrel for velocity testing: 24 inches long—1 turn in 10-inch twist

Availability:

Winchester 180-grain Ballistic Silvertip (SBST325S)
G1 Ballistic Coefficient = 0.439

Distance • Yards	Muzzle	100	200	300	400	500	600	800	1000
Velocity • fps	3060	2841	2632	2432	2242	2060	1888	1573	1310
Energy • ft-lbs	3743	3226	2769	2365	2009	1696	1425	989	686
Taylor KO Index	25.6	23.7	22.0	20.3	18.7	17.2	15.8	13.1	10.9
Path • Inches	-1.5	+1.4	0.0	-6.4	-18.7	-38.0	-65.4	-151.6	-294.7
Wind Drift • Inches	0.0	0.6	2.4	5.7	10.5	17.1	25.6	49.8	85.1

Winchester 200-grain XP3 (SXP325S)
G1 Ballistic Coefficient = 0.500

Distance • Yards	Muzzle	100	200	300	400	500	600	800	1000
Velocity • fps	2950	2762	2581	2408	2241	2081	1928	1644	1398
Energy • ft-lbs	3864	3387	2959	2574	2230	1923	1651	1201	868
Taylor KO Index	27.8	26.0	24.3	22.9	21.1	19.6	18.2	15.5	13.2
Path • Inches	-1.5	+1.5	0.0	-6.8	-19.5	-39.2	-66.9	-152.1	-289.2
Wind Drift • Inches	0.0	0.6	2.5	5.8	10.6	17.1	25.6	49.2	83.1

Winchester 200-grain AccuBond (S325WSMCT)
G1 Ballistic Coefficient = 0.477

Distance • Yards	Muzzle	100	200	300	400	500	600	800	1000
Velocity • fps	2950	2753	2565	2384	2210	2044	1885	1594	1346
Energy • ft-lbs	3866	3367	2922	2524	2170	1856	1578	1128	804
Taylor KO Index	27.4	25.6	23.8	22.1	20.5	19.0	17.5	14.8	12.5
Path • Inches	-1.5	+1.5	0.0	-6.8	-19.8	-39.9	-68.4	-156.5	-299.8
Wind Drift • Inches	0.0	0.6	2.3	5.5	10.1	16.3	24.4	47.1	79.9

Nosler 200-grain AccuBond (23065 & 60077)
G1 Ballistic Coefficient = 0.450

Distance • Yards	Muzzle	100	200	300	400	500	600	800	1000
Velocity • fps	2900	2693	2496	2307	2127	1955	1792	1497	1257
Energy • ft-lbs	3734	3221	2767	2364	2009	1698	1427	996	702
Taylor KO Index	26.8	24.9	23.0	21.3	19.6	18.0	16.5	13.8	11.6
Path • Inches	-1.5	+1.6	0.0	-7.2	-21.0	-42.5	-73.1	-168.9	-327.4
Wind Drift • Inches	0.0	0.7	2.8	6.6	12.2	19.9	29.9	58.1	98.8

Winchester 220-grain Power-Point (X325WSM)
G1 Ballistic Coefficient = 0.388

Distance • Yards	Muzzle	100	200	300	400	500	600	800	1000
Velocity • fps	2840	2605	2382	2169	1968	1779	1603	1304	1097
Energy • ft-lbs	3941	3316	2772	2300	1893	1547	1019	831	588
Taylor KO Index	29.0	26.6	24.3	22.2	20.1	18.2	16.4	13.3	11.2
Path • Inches	-1.5	+1.8	0.0	-8.0	-23.3	-47.6	-83.2	-197.8	-394.9
Wind Drift • Inches	0.0	0.7	3.1	7.3	13.5	22.2	33.6	66.1	112.6

Winchester 220-grain Power Max Bonded (XP325WSMBP)

G1 Ballistic Coefficient = 0.403

Distance • Yards	Muzzle	100	200	300	400	500	600	800	1000
Velocity • fps	2840	2613	2398	2192	1997	1815	1643	1346	1129
Energy • ft-lbs	3939	3336	2808	2347	1948	1609	1320	885	623
Taylor KO Index	28.8	26.5	24.3	22.3	20.3	18.4	16.7	13.7	11.5
Path • Inches	-1.5	+1.8	0.0	-7.9	-23.0	-46.2	-81.2	-191.6	-379.5
Wind Drift • Inches	0.0	0.8	3.3	7.7	14.4	23.5	35.4	69.4	118.4

Craig Boddington took a nice gemsbok bull when shooting a 8mm Remington Magnum with 220-grain Sierra bullets. (Photo from *Safari Rifles II* by Craig Boddington, 2009, Safari Press Archives)

8mm Remington Magnum

The 8mm Remington Magnum is a "full-length" magnum cartridge. That is, it is the same length as the .375 H&H, the grandfather of all belted magnums. This gives the 8mm enough case volume to produce some impressive ballistics. Nosler has added this caliber to its loaded ammunition line. This cartridge is at least enough for all North American game.

Relative Recoil Factor = 2.77

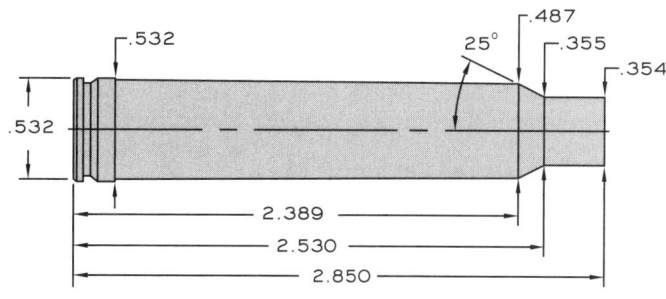

Specifications:

Controlling Agency for Standardization of this Ammunition: SAAMI

Bullet Weight Grains	Velocity fps	Maximum Average Pressure	
		Copper Crusher	Transducer
185	3,065	54,000 cup	65,000 psi
220	2,815	54,000 cup	65,000 psi

Standard barrel for velocity testing: 24 inches long—1 turn in 9.5-inch twist

Availability:

Nosler 180-grain Ballistic Tip (24410)

G1 Ballistic Coefficient = 0.394

Distance • Yards	Muzzle	100	200	300	400	500	600	800	1000
Velocity • fps	3200	2949	2711	2486	2271	2068	1876	1529	1252
Energy • ft-lbs	4092	3475	2937	2469	2061	1709	1407	935	627
Taylor KO Index	26.6	24.5	22.5	20.6	18.9	17.2	15.6	12.7	10.4
Path • Inches	-1.5	+1.2	0.0	-6.0	-17.6	-36.0	-62.6	-147.6	-292.9
Wind Drift • Inches	0.0	0.7	2.9	6.7	12.4	20.3	30.6	60.0	103.6

Nosler 200-grain AccuBond (24400)

G1 Ballistic Coefficient = 0.450

Distance • Yards	Muzzle	100	200	300	400	500	600	800	1000
Velocity • fps	3000	2789	2587	2394	2210	2034	1867	1562	1308
Energy • ft-lbs	3996	3453	2972	2546	2169	1838	1549	1084	759
Taylor KO Index	27.7	25.7	23.9	22.1	20.4	18.8	17.2	14.4	12.1
Path • Inches	-1.5	+1.5	0.0	-6.7	-19.4	-39.3	-67.6	-156.2	-302.3
Wind Drift • Inches	0.0	0.6	2.7	6.3	11.7	19.0	28.4	55.1	94.0

Remington 200-grain Swift A-Frame PSP (RS8MMRA)

G1 Ballistic Coefficient = 0.332

Distance • Yards	Muzzle	100	200	300	400	500	600	800	1000
Velocity • fps	2900	2623	2361	2115	1884	1671	1478	1175	1002
Energy • ft-lbs	3734	3054	2476	1987	1577	1240	971	613	446
Taylor KO Index	26.8	24.2	21.8	19.5	17.4	15.4	13.6	10.8	9.2
Path • Inches	-1.5	+1.8	0.0	-8.0	-23.9	-49.6	-88.1	-218.9	-452.7
Wind Drift • Inches	0.0	0.9	4.0	9.4	17.6	29.2	44.6	88.8	150.4

Nosler 200-grain Partition (24405)

G1 Ballistic Coefficient = 0.426

Distance • Yards	Muzzle	100	200	300	400	500	600	800	1000
Velocity • fps	3000	2777	2565	2363	2170	1987	1813	1500	1247
Energy • ft-lbs	3996	3424	2921	2478	2090	1752	1460	999	691
Taylor KO Index	27.7	25.6	23.7	21.8	20.0	18.3	16.7	13.8	11.5
Path • Inches	-1.5	+1.5	0.0	-6.8	-19.8	-40.3	-69.5	-162.1	-317.2
Wind Drift • Inches	0.0	0.7	2.9	6.7	12.4	20.2	30.5	59.4	101.6

.338 RCM (Ruger Compact Magnum)

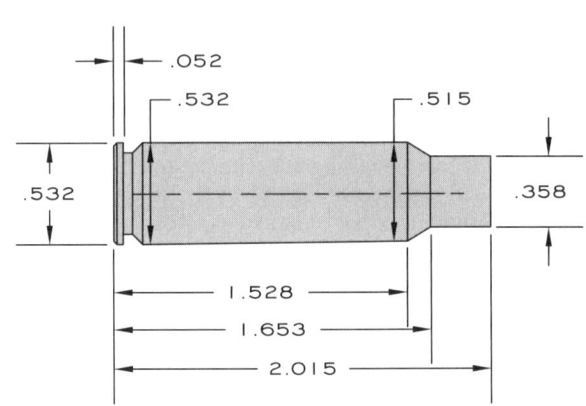

The .338 RCM is the .338-caliber version of this shortened and nonbelted magnum. The case length is identical to the .308 Winchester, and that allows the .338 RCM to be used in short actions. The performance is, for all practical purposes, identical to the .338 Winchester Magnum.

Relative Recoil Factor = 2.93

Specifications:

Controlling Agency for Standardization of this Ammunition: SAAMI

Bullet Weight Grains	Velocity fps	Maximum Average Pressure	
		Copper Crusher	Transducer
200	2,940	54,000 cup	64,000 psi
210	2,855	54,000 cup	64,000 psi
225	2,770	54,000 cup	64,000 psi
250	2,645	54,000 cup	64,000 psi
300	2,415	54,000 cup	64,000 psi

Standard barrel for velocity testing: 24 inches long—1 turn in 10-inch twist

Availability:

Hornady 185-grain GMX (82238) LF
G1 Ballistic Coefficient = 0.420

Distance • Yards	Muzzle	100	200	300	400	500	600	800	1000
Velocity • fps	2980	2755	2542	2338	2143	1958	1784	1471	1223
Energy • ft-lbs	3647	3118	2653	2242	1887	1575	1307	889	615
Taylor KO Index	26.6	24.6	22.7	20.9	19.1	17.5	15.9	13.1	10.9
Path • Inches	-1.5	+1.5	0.0	-6.9	-20.2	-41.2	-71.2	-166.3	-326.5
Wind Drift • Inches	0.0	0.7	2.9	6.9	12.8	20.8	31.4	61.2	104.8

Hornady 200-grain SST (82237)
G1 Ballistic Coefficient = 0.45

Distance • Yards	Muzzle	100	200	300	400	500	600	800	1000
Velocity • fps	2950	2744	2547	2358	2177	2004	1840	1542	1294
Energy • ft-lbs	3846	3342	2879	2468	2104	1784	1504	1056	744
Taylor KO Index	28.5	26.5	24.6	22.8	21.0	19.4	17.8	14.9	12.5
Path • Inches	-1.5	+1.6	0.0	-6.9	-20.1	-40.7	-69.9	-161.2	-311.6
Wind Drift • Inches	0.0	0.7	2.7	6.4	11.8	19.2	28.8	55.7	94.9

Hornady 225-grain InterBond (82138)
G1 Ballistic Coefficient = 0.51

Distance • Yards	Muzzle	100	200	300	400	500	600	800	1000
Velocity • fps	2750	2575	2404	2245	2089	1940	1798	1537	1316
Energy • ft-lbs	3778	3313	2894	2518	2180	1880	1615	1180	865
Taylor KO Index	29.9	28.0	26.1	24.4	22.7	21.1	19.5	16.7	14.3
Path • Inches	-1.5	+1.8	0.0	-7.9	-22.7	-45.4	-77.5	-175.7	-333.2
Wind Drift • Inches	0.0	0.6	2.6	6.2	11.3	18.4	27.4	52.6	88.6

Hornady 225-grain SST (82236)
G1 Ballistic Coefficient = 0.51

Distance • Yards	Muzzle	100	200	300	400	500	600	800	1000
Velocity • fps	2750	2575	2407	2245	2089	1940	1798	1537	1316
Energy • ft-lbs	3778	3313	2894	2518	2180	1880	1615	1180	865
Taylor KO Index	29.9	28.0	26.2	24.4	22.7	21.1	19.5	16.7	14.3
Path • Inches	-1.5	+1.9	0.0	-7.9	-22.7	-45.4	-77.5	-175.7	-333.2
Wind Drift • Inches	0.0	0.6	2.6	6.2	11.3	18.4	27.4	52.6	88.6

.338 Winchester Magnum

The .338 Winchester Magnum and the .264 Winchester Magnum were both introduced in 1958. The .338 is still going great while the .264 is on its last legs. That may well be because the .338 Winchester Magnum filled a big void in the power spectrum of cartridges available at that time. The .338 is enough cartridge for most any North American game, though possibly marginal for the largest Alaskan bears. At the same time, it can't realistically be called an "African" caliber. The .338 Winchester Magnum is one of the family of shortened magnums designed to fit into a standard length Model 70 action. The nearest big factory competition comes from the 8mm Remington Magnum, which never achieved the popularity of the .338 Winchester Magnum.

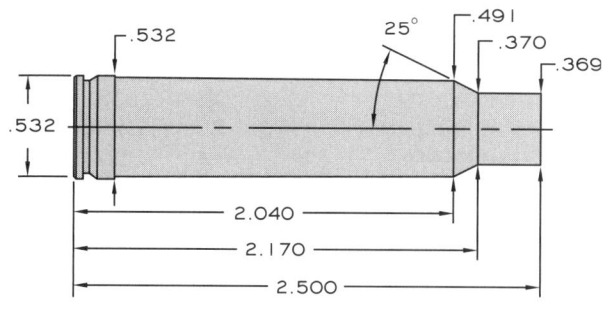

Relative Recoil Factor = 2.93

Specifications:

Controlling Agency for Standardization of this Ammunition: SAAMI

Bullet Weight Grains	Velocity fps	Maximum Average Pressure	
		Copper Crusher	Transducer
200	2,940	54,000 cup	64,000 psi
210	2,855	54,000 cup	64,000 psi
225	2,770	54,000 cup	64,000 psi
250	2,645	54,000 cup	64,000 psi
300	2,415	54,000 cup	64,000 psi

Standard barrel for velocity testing: 24 inches long—1 turn in 10-inch twist

Availability:

Federal 180-grain Nosler AccuBond (P338A3)

G1 Ballistic Coefficient = 0.372

Distance • Yards	Muzzle	100	200	300	400	500	600	800	1000
Velocity • fps	3120	2860	2610	2380	2160	1950	1754	1411	1157
Energy • ft-lbs	3890	3270	2730	2265	1865	1525	1230	796	535
Taylor KO Index	27.1	24.9	22.7	20.7	18.8	16.9	15.2	12.3	10.1
Path • Inches	-1.5	+1.4	0.0	-6.4	-19.1	-39.1	-68.6	-164.0	-330.1
Wind Drift • Inches	0.0	0.7	2.8	6.7	12.4	20.3	30.8	60.9	105.2

Nosler 180-grain AccuBond (24042)

G1 Ballistic Coefficient = 0.372

Distance • Yards	Muzzle	100	200	300	400	500	600	800	1000
Velocity • fps	3100	2840	2595	2363	2143	1935	1740	1400	1149
Energy • ft-lbs	3842	3225	2693	2232	1836	1497	1210	783	528
Taylor KO Index	26.9	24.7	22.6	20.5	18.6	16.8	15.1	12.2	10.0
Path • Inches	-1.5	+1.4	0.0	-6.6	-19.4	-39.8	-69.6	-166.5	-335.2
Wind Drift • Inches	0.0	0.8	3.2	7.5	13.9	22.8	34.6	68.3	118.0

Hornady 185-grain GMX (82226) LF

G1 Ballistic Coefficient = 0.420

Distance • Yards	Muzzle	100	200	300	400	500	600	800	1000
Velocity • fps	3080	2850	2632	2424	2226	2036	1857	1533	1270
Energy • ft-lbs	3896	3337	2845	2413	2034	1703	1417	966	662
Taylor KO Index	27.5	25.5	23.5	21.7	19.9	18.2	16.6	13.7	11.3
Path • Inches	-1.5	+1.4	0.0	-6.4	-18.8	-38.2	-66.0	-154.0	-301.9
Wind Drift • Inches	0.0	0.7	2.8	6.6	12.2	19.8	29.8	58.2	99.8

Hornady 200-grain SST (82223)

G1 Ballistic Coefficient = 0.420

Distance • Yards	Muzzle	100	200	300	400	500	600	800	1000
Velocity • fps	3030	2820	2619	2428	2244	2068	1901	1594	1336
Energy • ft-lbs	4076	3531	3047	2617	2236	1899	1605	1129	793
Taylor KO Index	29.3	27.2	25.3	23.4	21.7	20.0	18.4	15.4	12.9
Path • Inches	-1.5	+1.4	0.0	-6.5	-18.9	-38.3	-65.8	-151.5	-292.6
Wind Drift • Inches	0.0	0.6	2.6	6.2	11.4	18.4	27.6	53.5	91.2

Winchester 200-grain Power-Point (X3381)

G1 Ballistic Coefficient = 0.308

Distance • Yards	Muzzle	100	200	300	400	500	600	800	1000
Velocity • fps	2960	2658	2375	2110	1862	1635	1432	1129	969
Energy • ft-lbs	3890	3137	2505	1977	1539	1187	911	566	417
Taylor KO Index	28.6	25.7	22.9	20.4	18.0	15.8	13.9	10.9	9.4
Path • Inches	-1.5	+2.0	0.0	-8.2	-24.3	-50.4	-88.9	-225.5	-471.9
Wind Drift • Inches	0.0	1.0	4.2	9.9	18.8	31.2	47.9	95.4	162.0

Nosler 200-grain AccuBond (25050)

G1 Ballistic Coefficient = 0.414

Distance • Yards	Muzzle	100	200	300	400	500	600	800	1000
Velocity • fps	2950	2724	2509	2303	2108	1922	1747	1437	1195
Energy • ft-lbs	3866	3295	2795	2357	1973	1641	1356	917	635
Taylor KO Index	28.5	26.3	24.2	22.2	20.4	18.6	16.9	13.9	11.5
Path • Inches	-1.5	+1.6	0.0	-7.1	-20.8	-42.4	-73.4	-172.1	-338.9
Wind Drift • Inches	0.0	0.7	3.0	7.1	13.2	21.5	32.4	63.4	108.5

Winchester 200-grain Power Max Bonded (X3381BP)

G1 Ballistic Coefficient = 0.323

Distance • Yards	Muzzle	100	200	300	400	500	600	800	1000
Velocity • fps	2960	2671	2400	2146	1907	1686	1487	1175	999
Energy • ft-lbs	3890	3189	2559	2044	1614	1262	982	613	444
Taylor KO Index	28.6	25.8	23.2	20.7	18.4	16.3	14.4	11.3	9.6
Path • Inches	-1.5	+1.7	0.0	-7.8	-23.1	-48.1	-85.3	-211.6	-438.6
Wind Drift • Inches	0.0	0.9	4.0	9.4	17.8	29.3	44.8	89.5	152.1

Winchester 200-grain E-Tip (S338ET) LF

G1 Ballistic Coefficient = 0.425

Distance • Yards	Muzzle	100	200	300	400	500	600	800	1000
Velocity • fps	2950	2729	2519	2319	2127	1945	1774	1466	1222
Energy • ft-lbs	3864	3292	2805	2376	1999	1671	1397	955	663
Taylor KO Index	28.5	26.4	24.3	22.4	20.5	18.8	17.1	14.2	11.8
Path • Inches	-1.5	+1.6	0.0	-7.1	-20.6	-41.9	-72.4	-168.9	-330.8
Wind Drift • Inches	0.0	0.7	3.0	6.9	12.8	21.8	31.4	61.1	104.5

Winchester 200-grain Ballistic Silvertip (SBST338)

G1 Ballistic Coefficient = 0.414

Distance • Yards	Muzzle	100	200	300	400	500	600	800	1000
Velocity • fps	2950	2724	2509	2303	2108	1922	1749	1438	1196
Energy • ft-lbs	3866	3295	2795	2357	1973	1641	1358	919	636
Taylor KO Index	28.5	26.3	24.2	22.2	20.4	18.6	16.9	13.9	11.6
Path • Inches	-1.5	+1.6	0.0	-7.1	-20.8	-42.4	-73.4	-172.6	-342.0
Wind Drift • Inches	0.0	0.7	3.0	7.1	13.2	21.5	32.4	63.2	108.3

Nosler 210-grain Partition (25019)

G1 Ballistic Coefficient = 0.400

Distance • Yards	Muzzle	100	200	300	400	500	600	800	1000
Velocity • fps	2950	2716	2494	2282	2081	1891	1712	1398	1162
Energy • ft-lbs	4059	3441	2901	2430	2020	1667	1367	912	630
Taylor KO Index	29.9	27.5	25.3	23.1	21.1	19.2	17.4	14.2	11.8
Path • Inches	-1.5	+1.6	0.0	-7.2	-21.1	-43.1	-74.8	-176.6	-350.3
Wind Drift • Inches	0.0	0.8	3.2	7.4	13.7	22.4	33.9	66.5	113.9

Federal 210-grain Nosler Partition (P338A2)

G1 Ballistic Coefficient = 0.39

Distance • Yards	Muzzle	100	200	300	400	500	600	800	1000
Velocity • fps	2830	2600	2380	2180	1980	1790	1617	1319	1110
Energy • ft-lbs	3735	3155	2650	2210	1825	1500	1219	812	574
Taylor KO Index	28.7	26.4	24.1	22.1	20.1	18.2	16.4	13.4	11.3
Path • Inches	-1.5	+1.8	0.0	-8.0	-23.4	-47.5	-82.8	-196.3	-390.2
Wind Drift • Inches	0.0	0.7	3.0	7.1	13.3	21.8	32.9	64.6	110.0

Federal 225-grain Fusion (F338FS1)

G1 Ballistic Coefficient = 0.55

Distance • Yards	Muzzle	100	200	300	400	500	600	800	1000
Velocity • fps	2850	2690	2530	2370	2220	2080	1938	1680	1452
Energy • ft-lbs	4060	3600	3190	2810	2470	2160	1877	1411	1053
Taylor KO Index	31.0	29.2	27.5	25.7	24.1	22.6	21.1	18.3	15.8
Path • Inches	-1.5	+1.7	0.0	-7.1	-20.5	-40.8	-69.3	-155.5	-291.2
Wind Drift • Inches	0.0	0.6	2.3	5.4	9.9	16.0	23.8	45.2	75.9

Hornady 225-grain InterBond (82338)

G1 Ballistic Coefficient = 0.576

Distance • Yards	Muzzle	100	200	300	400	500	600	800	1000
Velocity • fps	2840	2680	2526	2377	2233	2096	1960	1709	1484
Energy • ft-lbs	4318	3590	3189	2823	2492	2194	1919	1459	1101
Taylor KO Index	30.9	29.1	27.4	25.8	24.3	22.8	21.3	18.6	16.1
Path • Inches	-1.5	+1.7	0.0	-7.1	-20.4	-40.7	-68.9	-153.8	-286.7
Wind Drift • Inches	0.0	0.6	2.2	5.2	9.5	15.4	22.8	43.4	72.6

Hornady 225-grain SST (82233)

G1 Ballistic Coefficient = 0.515

Distance • Yards	Muzzle	100	200	300	400	500	600	800	1000
Velocity • fps	2840	2662	2490	2325	2166	2014	1868	1599	1367
Energy • ft-lbs	4031	3540	3099	2702	2345	2027	1744	1277	933
Taylor KO Index	30.9	28.9	27.1	25.3	23.5	21.9	20.3	17.4	14.9
Path • Inches	-1.5	+1.7	0.0	-7.3	-21.1	-42.3	-72.0	-163.2	-309.0
Wind Drift • Inches	0.0	0.6	2.5	5.9	10.8	17.5	26.1	50.1	84.5

Winchester 225-grain AccuBond CT (S338CT)

G1 Ballistic Coefficient = 0.548

Distance • Yards	Muzzle	100	200	300	400	500	600	800	1000
Velocity • fps	2800	2634	2474	2319	2170	2026	1888	1632	1408
Energy • ft-lbs	3918	3467	3058	2688	2353	2052	1781	1331	990
Taylor KO Index	30.4	28.6	26.9	25.2	23.6	22.0	20.5	17.7	15.3
Path • Inches	-1.5	+1.8	0.0	-7.4	-21.3	-42.6	-72.5	-162.9	-306.0
Wind Drift • Inches	0.0	0.5	2.2	5.0	9.3	15.0	22.3	42.3	71.4

Hornady 225-grain SST (82234)

G1 Ballistic Coefficient = 0.431

Distance • Yards	Muzzle	100	200	300	400	500	600	800	1000
Velocity • fps	2785	2575	2375	2184	2001	1828	1666	1381	1164
Energy • ft-lbs	3875	3313	2818	2382	2000	1670	1388	953	677
Taylor KO Index	30.3	28.0	25.8	23.7	21.7	19.9	18.1	15.0	12.6
Path • Inches	-1.5	+1.9	0.0	-8.0	-23.4	-47.4	-82.0	-191.2	-374.0
Wind Drift • Inches	0.0	0.8	3.2	7.4	13.7	22.3	33.8	65.5	111.2

Remington 225-grain Swift A-Frame PSP (RS338WA)

G1 Ballistic Coefficient = 0.337

Distance • Yards	Muzzle	100	200	300	400	500	600	800	1000
Velocity • fps	2785	2517	2266	2029	1808	1605	1422	1143	987
Energy • ft-lbs	3871	3165	2565	2057	1633	1286	1010	652	486
Taylor KO Index	30.3	27.3	24.6	22.0	19.6	17.4	15.5	12.5	10.8
Path • Inches	-1.5	+2.0	0.0	-8.8	-25.2	-54.1	-95.9	-237.4	-486.6
Wind Drift • Inches	0.0	1.0	4.1	9.8	18.4	30.5	46.5	91.9	154.1

Remington 225-grain Core-Lokt Ultra Bonded (PRC338WA)

G1 Ballistic Coefficient = 0.456

Distance • Yards	Muzzle	100	200	300	400	500	600	800	1000
Velocity • fps	2780	2582	2392	2210	2036	1871	1715	1435	1213
Energy • ft-lbs	3860	3329	2858	2440	2071	1748	1469	1029	735
Taylor KO Index	30.2	28.1	26.0	24.0	22.1	20.3	18.6	15.6	13.2
Path • Inches	-1.5	+1.9	0.0	-7.9	-23.0	-46.5	-80.0	-184.7	-357.6
Wind Drift • Inches	0.0	0.6	2.7	6.2	11.6	18.8	28.2	54.7	92.8

Remington 225-grain Core-Lokt Pointed Soft Point (R338W1)

G1 Ballistic Coefficient = 0.435

Distance • Yards	Muzzle	100	200	300	400	500	600	800	1000
Velocity • fps	2780	2572	2374	2184	2003	1832	1670	1386	1169
Energy • ft-lbs	3860	3305	2837	2389	1999	1663	1394	960	683
Taylor KO Index	30.2	27.9	25.8	23.7	21.8	19.9	18.2	15.1	12.7
Path • Inches	-1.5	+1.9	0.0	-8.1	-23.4	-47.5	-82.0	-191.6	-376.4
Wind Drift • Inches	0.0	0.8	3.1	7.3	13.6	22.2	33.4	64.9	110.2

Nosler 225-grain AccuBond (25027)

G1 Ballistic Coefficient = 0.550

Distance • Yards	Muzzle	100	200	300	400	500	600	800	1000
Velocity • fps	2750	2586	2428	2275	2128	1987	1851	1600	1381
Energy • ft-lbs	3779	3342	2946	2588	2264	1973	1713	1278	953
Taylor KO Index	29.9	28.1	26.4	24.7	23.1	21.6	20.1	17.4	15.0
Path • Inches	-1.5	+1.8	0.0	-7.8	-22.2	-44.4	-75.4	-169.4	-318.2
Wind Drift • Inches	0.0	0.6	2.5	5.7	10.5	17.0	25.3	48.3	81.1

Nosler 225-grain Partition (25036)

G1 Ballistic Coefficient = 0.454

Distance • Yards	Muzzle	100	200	300	400	500	600	800	1000
Velocity • fps	2750	2552	2363	2182	2008	1844	1688	1412	1195
Energy • ft-lbs	3779	3255	2790	2378	2015	1698	1425	997	714
Taylor KO Index	29.9	27.7	25.7	23.7	21.8	20.0	18.3	15.3	13.0
Path • Inches	-1.5	+1.9	0.0	-8.2	-23.6	-47.8	-82.2	-190.0	-368.1
Wind Drift • Inches	0.0	0.7	3.0	7.1	13.1	21.4	32.1	62.2	105.4

Cor-Bon 225-grain DPX (DPX338225/20)

G1 Ballistic Coefficient = 0.375

Distance • Yards	Muzzle	100	200	300	400	500	600	800	1000
Velocity • fps	2800	2559	2330	2114	1909	1717	1541	1249	1059
Energy • ft-lbs	3918	3272	2714	2183	1780	1441	1161	762	548
Taylor KO Index	30.4	27.8	25.3	23.0	20.7	18.7	16.7	13.6	11.5
Path • Inches	-1.5	+1.9	0.0	-8.3	-24.5	-50.2	-87.8	-210.4	-422.6
Wind Drift • Inches	0.0	0.9	3.6	8.6	16.0	26.3	39.9	78.6	133.4

Federal 225-grain Barnes Triple-Shock X-Bullet (P338K) LF

G1 Ballistic Coefficient = 0.385

Distance • Yards	Muzzle	100	200	300	400	500	600	800	1000
Velocity • fps	2800	2570	2340	2130	1930	1740	1568	1275	1078
Energy • ft-lbs	3915	3290	2740	2270	1865	1520	1228	812	581
Taylor KO Index	30.4	27.9	25.4	23.1	21.0	18.9	17.0	13.9	11.7
Path • Inches	-1.5	+1.9	0.0	-8.3	-24.1	-49.5	-86.3	-205.9	-411.4
Wind Drift • Inches	0.0	0.8	3.5	8.3	15.5	25.5	38.6	75.8	128.8

Federal 225-grain Trophy Bonded Bear Claw (P338T1)

G1 Ballistic Coefficient = 0.373

Distance • Yards	Muzzle	100	200	300	400	500	600	800	1000
Velocity • fps	2730	2490	2260	2050	1840	1660	1489	1210	1037
Energy • ft-lbs	3725	3100	2560	2095	1700	1376	1108	732	537
Taylor KO Index	29.7	27.1	24.6	22.3	20.0	18.0	16.2	13.1	11.3
Path • Inches	-1.5	+2.1	0.0	-8.9	-25.9	-53.7	-93.4	-224.4	-450.4
Wind Drift • Inches	0.0	0.9	3.8	9.0	16.7	27.5	41.8	82.1	138.4

Norma 225-grain FULL JACKET (18510)

G1 Ballistic Coefficient = 0.49

Distance • Yards	Muzzle	100	200	300	400	500	600	800	1000
Velocity • fps	2657	2480	2310	2146	1996	1841	1700	1445	1237
Energy • ft-lbs	3528	3074	2668	2304	1980	1694	1444	1043	764
Taylor KO Index	28.9	26.9	25.1	23.3	21.6	20.0	18.5	15.7	13.4
Path • Inches	-1.5	+2.0	0.0	-8.6	-24.8	-49.7	-84.9	-193.6	-369.5
Wind Drift • Inches	0.0	0.7	2.9	6.7	12.4	20.1	30.1	57.8	97.2

Norma 230-grain Oryx (18511)

G1 Ballistic Coefficient = 0.37

Distance • Yards	Muzzle	100	200	300	400	500	600	800	1000
Velocity • fps	2756	2514	2284	2066	1863	1673	1499	1215	1039
Energy • ft-lbs	3880	3228	2665	2181	1773	1429	1147	754	551
Taylor KO Index	30.6	27.6	25.4	22.9	20.7	18.6	16.6	13.5	11.5
Path • Inches	-1.5	+2.0	0.0	-8.7	-25.5	-52.4	-91.8	-220.9	-444.5
Wind Drift • Inches	0.0	0.8	3.4	8.0	15.0	24.7	37.5	73.7	124.5

A² 250-grain Triad (None)

G1 Ballistic Coefficient = 0.30

Distance • Yards	Muzzle	100	200	300	400	500	600	800	1000
Velocity • fps	2700	2407	2133	1877	1653	1447	1271	1039	916
Energy • ft-lbs	4046	3216	2526	1956	1516	1162	897	600	466
Taylor KO Index	32.6	29.1	25.7	22.7	20.0	17.5	15.4	12.6	11.1
Path • Inches	-1.5	+2.3	0.0	-9.8	-29.8	-62.1	-112.0	-283.4	-578.8
Wind Drift • Inches	0.0	1.1	4.8	11.6	21.9	36.5	56.0	109.6	179.2

A² 250-grain Nosler Partition (None)

G1 Ballistic Coefficient = 0.6

Distance • Yards	Muzzle	100	200	300	400	500	600	800	1000
Velocity • fps	2700	2568	2439	2314	2193	2075	1962	1747	1550
Energy • ft-lbs	4046	3659	3302	2972	2669	2390	2138	1694	1333
Taylor KO Index	32.6	31.0	29.4	27.9	26.5	25.0	23.8	21.2	18.8
Path • Inches	-1.5	+1.9	0.0	-7.7	-22.0	-43.4	-72.9	-160.2	-293.6
Wind Drift • Inches	0.0	0.5	2.0	4.7	8.6	13.8	20.4	38.3	63.4

Federal 250-grain Nosler Partition (P338B2)

G1 Ballistic Coefficient = 0.471

Distance • Yards	Muzzle	100	200	300	400	500	600	800	1000
Velocity • fps	2660	2470	2290	2120	1960	1800	1653	1391	1186
Energy • ft-lbs	3925	3395	2920	2495	2120	1785	1517	1075	781
Taylor KO Index	32.1	29.8	27.6	25.6	23.7	21.7	20.0	16.8	14.4
Path • Inches	-1.5	+2.1	0.0	-8.8	-25.2	-50.7	-87.1	-200.9	-388.9
Wind Drift • Inches	0.0	0.7	3.1	7.2	13.2	21.5	32.3	62.3	105.1

Remington 250-grain Core-Lokt Pointed Soft Point (R338W2)

G1 Ballistic Coefficient = 0.432

Distance • Yards	Muzzle	100	200	300	400	500	600	800	1000
Velocity • fps	2660	2456	2261	2075	1898	1731	1575	1308	1115
Energy • ft-lbs	3927	3348	2837	2389	1999	1663	1378	950	1115
Taylor KO Index	32.1	29.6	27.3	25.0	22.9	20.9	19.1	15.8	13.5
Path • Inches	-1.5	+2.1	0.0	-8.9	-26.0	-52.7	-91.3	-214.1	-420.9
Wind Drift • Inches	0.0	0.8	3.4	7.9	14.6	23.9	36.0	70.0	118.2

Nosler 250-grain Partition (24040)

G1 Ballistic Coefficient = 0.473

Distance • Yards	Muzzle	100	200	300	400	500	600	800	1000
Velocity • fps	2600	2416	2240	2071	1909	1756	1613	1360	1165
Energy • ft-lbs	3754	3341	2786	2381	2024	1713	1444	1026	753
Taylor KO Index	31.4	29.2	27.0	25.0	23.0	21.2	19.5	16.4	14.1
Path • Inches	-1.5	+2.2	0.0	-9.2	-26.5	-53.3	-91.5	-210.5	-405.2
Wind Drift • Inches	0.0	0.8	3.2	7.4	13.6	22.1	33.2	64.0	107.6

![Hartmann zebra photograph]

Hartmann zebra, taken with a .338 Winchester Magnum. No matter what you run into, there are few things in Africa a .338 can't handle. (Photo from *Safari Rifles II* by Craig Boddington, 2009, Safari Press Archives)

.330 Dakota

The .330 Dakota is a .338 caliber cartridge based on the .404 Jeffery case. As is true with several other relatively new rounds based on this same case—the original .404 Jeffery design dates to 1910—the performance is impressive. Dakota currently loads six bullet styles for this caliber.

Relative Recoil Factor = 3.11

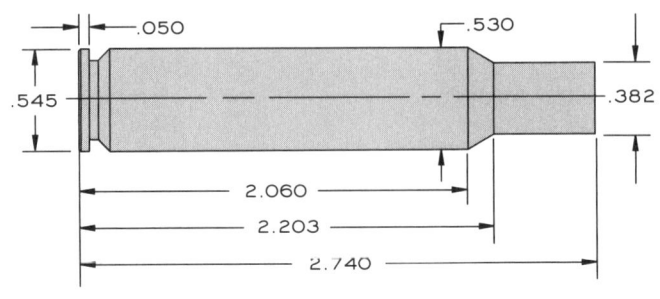

Specifications:

Controlling Agency for Standardization of this Ammunition: Factory

Standard barrel for velocity testing: 24 inches long

Availability:

Dakota 200-grain Nosler Ballistic Tip (330-200NBT)

G1 Ballistic Coefficient = 0.414

Distance • Yards	Muzzle	100	200	300	400	500	600	800	1000
Velocity • fps	3100	2866	2644	2432	2230	2038	1856	1528	1261
Energy • ft-lbs	4269	3648	3104	2627	2209	1845	1530	1036	707
Taylor KO Index	30.0	27.8	25.6	23.6	21.6	19.7	18.0	14.8	12.2
Path • Inches	-1.5	+1.4	0.0	-6.4	-18.6	-37.8	-65.5	-153.8	-304.4
Wind Drift • Inches	0.0	0.7	2.8	6.6	12.3	20.0	30.1	58.8	101.1

Dakota 225-grain Nosler Partition (330-225NPT)

G1 Ballistic Coefficient = 0.454

Distance • Yards	Muzzle	100	200	300	400	500	600	800	1000
Velocity • fps	2950	2743	2545	2356	2175	2002	1837	1538	1290
Energy • ft-lbs	4349	3760	3238	2774	2364	2003	1687	1183	832
Taylor KO Index	32.1	29.9	27.7	25.7	23.7	21.8	20.0	16.8	14.1
Path • Inches	-1.5	+1.6	0.0	-6.9	-20.1	-40.7	-70.0	-161.9	-314.6
Wind Drift • Inches	0.0	0.7	2.8	6.4	11.8	19.2	28.9	56.0	95.3

Dakota 225-grain Swift A-Frame (330-225FAF)

G1 Ballistic Coefficient = 0.384

Distance • Yards	Muzzle	100	200	300	400	500	600	800	1000
Velocity • fps	2950	2706	2475	2256	2048	1851	1668	1351	1124
Energy • ft-lbs	4349	3660	3062	2544	2096	1713	1391	912	631
Taylor KO Index	32.1	29.5	27.0	24.6	22.3	20.2	18.2	14.7	12.2
Path • Inches	-1.5	+1.6	0.0	-7.3	-21.5	-44.0	-76.7	-183.6	-370.2
Wind Drift • Inches	0.0	0.8	3.3	7.7	14.4	23.6	35.8	70.5	120.9

Dakota 230-grain Fail-Safe (330-230WFS)

G1 Ballistic Coefficient = 0.436

Distance • Yards	Muzzle	100	200	300	400	500	600	800	1000
Velocity • fps	2950	2735	2529	2333	2146	1967	1798	1493	1247
Energy • ft-lbs	4446	3820	3268	2781	2352	1977	1651	1139	794
Taylor KO Index	32.9	30.5	28.2	26.0	23.9	21.9	20.0	16.6	13.9
Path • Inches	-1.5	+1.6	0.0	-7.0	-20.4	-41.4	-71.4	-166.5	-326.3
Wind Drift • Inches	0.0	0.7	2.9	6.7	12.4	20.2	30.4	59.1	100.9

Dakota 250-grain Nosler Partition (330-250NPT)

G1 Ballistic Coefficient = 0.47

Distance • Yards	Muzzle	100	200	300	400	500	600	800	1000
Velocity • fps	2800	2608	2423	2247	2077	1915	1762	1484	1287
Energy • ft-lbs	4353	3776	3261	2803	2396	2037	1724	1223	877
Taylor KO Index	33.9	31.6	29.4	27.2	25.1	23.2	21.3	18.0	15.6
Path • Inches	-1.5	+1.8	0.0	-7.7	-22.4	-45.1	-77.3	-177.9	-343.7
Wind Drift • Inches	0.0	0.7	2.8	6.6	12.2	19.8	29.7	57.3	97.1

Dakota 250-grain Swift A-Frame (330-250FAF)

G1 Ballistic Coefficient = 0.427

Distance • Yards	Muzzle	100	200	300	400	500	600	800	1000
Velocity • fps	2800	2587	2385	2191	2006	1832	1667	1379	1161
Energy • ft-lbs	4353	3717	3158	2666	2235	1863	1544	1056	748
Taylor KO Index	33.9	31.3	28.9	26.5	24.3	22.2	20.2	16.7	14.1
Path • Inches	-1.5	+1.8	0.0	-8.0	-23.2	-47.1	-81.4	-191.0	-376.5
Wind Drift • Inches	0.0	0.8	3.2	7.4	13.7	22.4	33.8	65.8	111.9

The then new Remington Ultra Mag cartridges were tested in the Selous Reserve in Tanzania in 2000. It included the fast, powerful .338 Remington Ultra Mag. It worked perfectly on a longish shot on this Lichtenstein hartebeest. (Photo from *Safari Rifles II* by Craig Boddington, 2009, Safari Press Archives)

.338 A-Square Excaliber

Developed in 1978, the .338 A^2 is a .338 caliber variation on the .378 Weatherby cartridge. Performance is similar to the .338-378, muzzle velocities being a little higher for two of the three bullet weights and a bit slower for the third. This cartridge was designed to fill the need for a flat-shooting round for long-range hunting of large game.

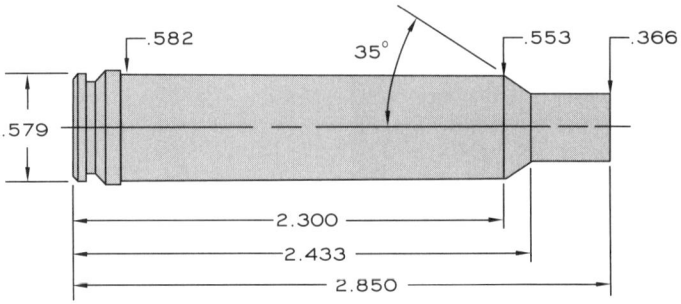

Relative Recoil Factor = 3.15

Specifications:

Controlling Agency for Standardization of this Ammunition: Factory

Standard barrel for velocity testing: 26 inches long—1 turn in 10-inch twist

Availability:

A^2 200-grain Nosler Ballistic Tip (None)

)G1 Ballistic Coefficient = 0.457

Distance • Yards	Muzzle	100	200	300	400	500	600	800	1000
Velocity • fps	3500	3266	3045	2835	2634	2442	2260	1916	1609
Energy • ft-lbs	5440	4737	4117	3568	3081	2648	2268	1631	1150
Taylor KO Index	33.8	31.5	29.4	27.4	25.4	23.6	21.9	18.6	15.6
Path • Inches	-1.5	+0.9	0.0	-4.6	-13.6	-27.6	-47.4	-108.7	-209.1
Wind Drift • Inches	0.0	0.5	2.2	5.1	9.3	15.0	22.4	43.0	73.0

A^2 250-grain Nosler Partition (None)

G1 Ballistic Coefficient = 0.676

Distance • Yards	Muzzle	100	200	300	400	500	600	800	1000
Velocity • fps	3120	2974	2834	2697	2565	2436	2313	2073	1849
Energy • ft-lbs	5403	4911	4457	4038	3652	3295	2967	2386	1898
Taylor KO Index	37.7	35.9	34.2	32.6	31.0	29.4	28.0	25.1	22.4
Path • Inches	-1.5	+1.2	0.0	-5.6	-15.9	-31.5	52.9	116.1	211.9
Wind Drift • Inches	0.0	0.4	1.7	3.8	7.0	11.2	16.5	30.9	51.0

A^2 250-grain Triad (None)

G1 Ballistic Coefficient = 0.30

Distance • Yards	Muzzle	100	200	300	400	500	600	800	1000
Velocity • fps	3120	2799	2500	2220	1938	1715	1498	1163	983
Energy • ft-lbs	5403	4348	3469	2736	2128	1634	1246	751	537
Taylor KO Index	37.7	33.8	30.2	26.8	23.6	20.7	18.1	14.1	11.9
Path • Inches	-1.5	+1.5	0.0	-7.1	-21.2	-44.6	-80.1	-204.5	-433.9
Wind Drift • Inches	0.0	1.0	4.0	9.5	17.9	29.8	45.8	92.6	158.3

.338 Remington Ultra Magnum

When the .300 Remington Ultra Magnum was introduced in 1999, it was clear that either Remington was going to follow with a couple more calibers based on the same case or the wildcatters would do the job for them. The .338 RUM is good big cartridge, just slightly milder than the .338-378 Weatherby. The trend in most new large magnums is to do away with the belt. The .338 RUM headspaces off the shoulder.

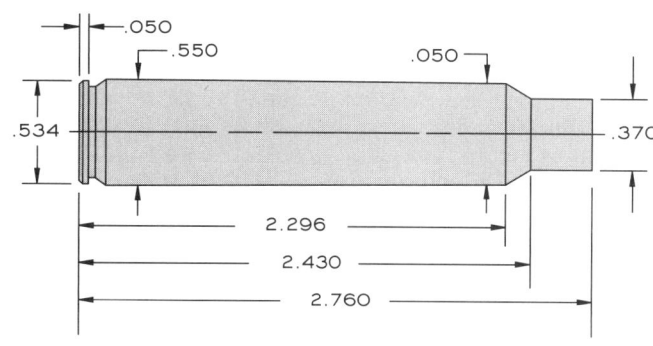

Relative Recoil Factor = 3.18

Specifications:

Controlling Agency for Standardization of this Ammunition: SAAMI

Bullet Weight Grains	Velocity fps	Maximum Average Pressure	
		Copper Crusher	Transducer
25	4,000	52,000 cup	65,000 psi

Standard barrel for velocity testing: 24 inches long—1 turn in 9-inch twist

Availability:

Nosler 200-grain AccuBond (24001)
G1 Ballistic Coefficient = 0.414

Distance • Yards	Muzzle	100	200	300	400	500	600	800	1000
Velocity • fps	3150	2913	2699	2475	2272	2078	1894	1560	1287
Energy • ft-lbs	4408	3770	3212	2722	2292	1917	1593	1081	736
Taylor KO Index	31.9	29.5	27.4	25.1	23.0	21.1	19.2	15.8	13.1
Path • Inches	-1.5	+1.3	0.0	-6.1	-17.9	-36.5	-63.1	-147.6	-290.0
Wind Drift • Inches	0.0	0.7	2.8	6.5	12.0	19.5	29.4	57.4	98.6

Federal 210-grain Nosler Partition (P338RUMA)
G1 Ballistic Coefficient = 0.401

Distance • Yards	Muzzle	100	200	300	400	500	600	800	1000
Velocity • fps	3050	2810	2580	2370	2160	1970	1787	1460	1206
Energy • ft-lbs	4335	3685	3115	2615	2180	1805	1489	994	678
Taylor KO Index	30.9	28.5	26.2	24.0	21.9	20.0	18.1	14.8	12.2
Path • Inches	-1.5	+1.4	0.0	-6.7	-19.6	-39.9	-69.2	-163.0	-322.8
Wind Drift • Inches	0.0	0.7	3.0	7.0	13.0	21.3	32.1	63.0	108.3

Federal 225-grain Nosler AccuBond (P338RUMA1)
G1 Ballistic Coefficient = 0.549

Distance • Yards	Muzzle	100	200	300	400	500	600	800	1000
Velocity • fps	3020	2850	2680	2520	2260	2210	2065	1793	1548
Energy • ft-lbs	4555	4045	3580	3165	2785	2440	2131	1606	1197
Taylor KO Index	32.8	31.0	29.1	27.4	25.6	24.0	22.4	19.5	16.8
Path • Inches	-1.5	+1.4	0.0	-6.2	-17.9	-36.0	-61.1	-137.1	-256.7
Wind Drift • Inches	0.0	0.5	2.0	4.5	8.3	13.4	19.9	37.8	63.4

Nosler 225-grain AccuBond (24032) + (60083)
G1 Ballistic Coefficient = 0.550

Distance • Yards	Muzzle	100	200	300	400	500	600	800	1000
Velocity • fps	2975	2803	2637	2477	2323	2174	2031	1762	1521
Energy • ft-lbs	4423	3925	3475	3066	2696	2362	2061	1551	1158
Taylor KO Index	32.3	30.5	28.6	26.9	25.2	23.6	22.1	19.1	16.5
Path • Inches	-1.5	+1.5	0.0	-6.5	-18.6	-37.2	-63.2	-141.7	-265.4
Wind Drift • Inches	0.0	0.5	2.2	5.1	9.4	15.1	22.5	42.9	72.0

Nosler 225-grain Partition (24044)

G1 Ballistic Coefficient = 0.454

Distance • Yards	Muzzle	100	200	300	400	500	600	800	1000
Velocity • fps	2975	2767	2569	2378	2196	2023	1857	1556	1304
Energy • ft-lbs	4423	3826	3297	2827	2411	2044	1723	1209	850
Taylor KO Index	32.3	30.1	27.9	25.8	23.9	22.0	20.2	16.9	14.2
Path • Inches	-1.5	+1.5	0.0	-6.8	-19.8	-40.0	-68.6	-158.3	-306.0
Wind Drift • Inches	0.0	0.7	2.7	6.3	11.7	19.0	28.5	55.2	94.0

Cor-Bon 225-grain DPX (DPX338RUM225/20)

G1 Ballistic Coefficient = 0.375

Distance • Yards	Muzzle	100	200	300	400	500	600	800	1000
Velocity • fps	3000	2748	2510	2284	2069	1867	1679	1353	1121
Energy • ft-lbs	4498	3774	3148	2606	2140	1743	1409	915	628
Taylor KO Index	32.6	29.9	27.3	24.8	22.5	20.3	18.2	14.7	12.2
Path • Inches	-1.5	+1.6	0.0	-7.1	-20.9	-42.8	-74.7	-178.7	-359.4
Wind Drift • Inches	0.0	0.8	3.3	7.8	14.4	23.7	35.9	71.0	122.0

Remington 250-grain Core-Lokt Pointed Soft Point (PR338UM2)

G1 Ballistic Coefficient = 0.432

Distance • Yards	Muzzle	100	200	300	400	500	600	800	1000
Velocity • fps	2860	2647	2443	2249	2064	1887	1722	1427	1196
Energy • ft-lbs	4540	3888	3314	2807	2363	1977	1646	1130	795
Taylor KO Index	34.5	32.0	29.5	27.1	24.9	22.8	20.8	17.2	14.4
Path • Inches	-1.5	+1.7	0.0	-7.6	-22.0	-44.7	-77.1	-179.6	-351.1
Wind Drift • Inches	0.0	0.7	2.7	6.4	11.8	19.3	29.0	56.5	96.3

Remington 250-grain Swift A-Frame PSP (PR338UM1)

G1 Ballistic Coefficient = 0.428

Distance • Yards	Muzzle	100	200	300	400	500	600	800	1000
Velocity • fps	2860	2645	2440	2244	2057	1879	1713	1417	1187
Energy • ft-lbs	4540	3882	3303	2794	2347	1960	1629	1115	783
Taylor KO Index	34.5	31.9	29.5	27.1	24.8	22.7	20.7	17.1	14.3
Path • Inches	-1.5	+1.7	0.0	-7.6	-22.1	-44.9	-77.5	-180.8	-354.1
Wind Drift • Inches	0.0	0.7	2.8	6.4	12.0	20.0	29.4	57.2	97.6

Nosler 250-grain Partition (24056)

G1 Ballistic Coefficient = 0.473

Distance • Yards	Muzzle	100	200	300	400	500	600	800	1000
Velocity • fps	2850	2656	2470	2291	2120	1956	1801	1518	1283
Energy • ft-lbs	4510	3916	3386	2914	2495	2125	1800	1279	915
Taylor KO Index	34.4	32.1	29.8	27.7	25.6	23.6	21.7	18.3	15.5
Path • Inches	-1.5	+1.7	0.0	-7.4	-21.5	-43.6	-74.2	-170.3	-327.2
Wind Drift • Inches	0.0	0.7	2.8	6.4	11.9	19.3	28.9	55.8	94.6

8.59mm (.338) Lazzeroni Titan

Lazzeroni's 8.59mm Titan is a very large capacity .338 with performance to match the case volume. This cartridge significantly outperforms the .338 Winchester Magnum. The gun is certainly adequate for any North American game, including large bears.

Relative Recoil Factor = 3.30

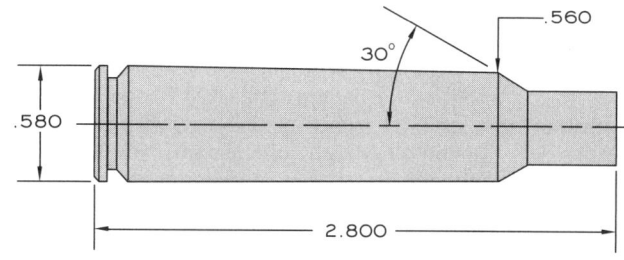

Specifications:

Controlling Agency for Standardization of this Ammunition: Factory

Availability:

Lazzeroni 185-grain
G1 Ballistic Coefficient = 0.501

Distance • Yards	Muzzle	100	200	300	400	500	600	800	1000
Velocity • fps	3550	3334	3129	2933	2746	2566	2394	2068	1771
Energy • ft-lbs	5178	4568	4023	3535	3098	2706	2354	1758	1288
Taylor KO Index	31.8	29.9	28.0	26.3	24.6	23.0	21.4	18.5	15.9
Path • Inches	-1.5	+0.8	0.0	-4.4	-12.8	-25.8	-44.1	-99.9	-188.8
Wind Drift • Inches	0.0	0.5	2.0	4.5	8.2	13.3	19.7	37.4	62.9

Lazzeroni 225-grain Nosler Partition (859TN225P)
G1 Ballistic Coefficient = 0.536

Distance • Yards	Muzzle	100	200	300	400	500	600	800	1000
Velocity • fps	3300	3110	2927	2752	2584	2421	2265	1970	1700
Energy • ft-lbs	5422	4832	4282	3785	3336	2929	2564	1939	1444
Taylor KO Index	35.9	33.8	31.8	29.9	28.1	26.3	24.2	21.5	18.5
Path • Inches	-1.5	+1.0	0.0	-5.1	-14.8	-29.7	-50.6	-113.6	-213.2
Wind Drift • Inches	0.0	0.5	2.0	4.6	8.4	13.5	20.0	38.0	63.8

Lazzeroni 250-grain Swift A-Frame (859TN250A)
G1 Ballistic Coefficient = 0.570

Distance • Yards	Muzzle	100	200	300	400	500	600	800	1000
Velocity • fps	3150	2977	2810	2649	2494	2344	2201	1927	1676
Energy • ft-lbs	5510	4920	4384	3896	3453	3050	2689	2062	1560
Taylor KO Index	38.0	35.9	33.9	32.0	30.1	28.3	26.6	23.3	20.3
Path • Inches	-1.5	+1.2	0.0	-5.6	-16.2	-32.4	-54.8	-122.4	-228.0
Wind Drift • Inches	0.0	0.5	2.0	4.6	8.3	13.4	19.9	37.6	62.9

.340 Weatherby Magnum

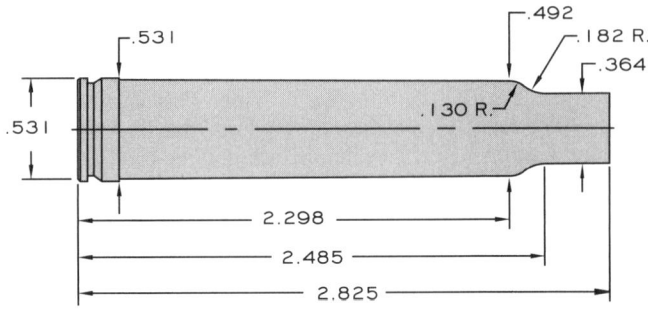

The .340 Weatherby Magnum was Weatherby's 1962 entry in the .338 derby. It follows the pattern of the .300 Weatherby Magnum, being little more than the .300 necked up to .340. Until the recent introduction of the .338-378 and other high-volume cartridges, the .340 Weatherby was the king of the .338s in terms of performance. As with many other cartridge names, the .340 number has no significance. This is a .338 caliber cartridge, and a potent one, too.

Relative Recoil Factor = 3.30

Specifications:

Controlling Agency for Standardization of this Ammunition: CIP

Bullet Weight Grains	Velocity fps	Maximum Average Pressure	
		Copper Crusher	Transducer
N/S	N/S	3,800 bar	4,370 bar

Standard barrel for velocity testing: 26 inches long—1 turn in 10-inch twist

Availability:

Weatherby 200-grain AccuBond (N 340 200 ACB)
G1 Ballistic Coefficient = 0.413

Distance • Yards	Muzzle	100	200	300	400	500	600	800	1000
Velocity • fps	3225	2984	2755	2538	2331	2134	1946	1605	1321
Energy • ft-lbs	4820	3955	3372	2862	2414	2023	1682	1144	775
Taylor KO Index	31.1	28.3	26.6	24.5	22.5	20.6	18.8	15.5	12.8
Path • Inches	-1.5	+1.2	0.0	-5.8	-17.0	-34.6	-59.9	-140.0	-274.8
Wind Drift • Inches	0.0	0.6	2.7	6.3	11.6	18.9	28.5	55.5	95.5

Weatherby 200-grain SP (H 340 200 SP)
G1 Ballistic Coefficient = 0.361

Distance • Yards	Muzzle	100	200	300	400	500	600	800	1000
Velocity • fps	3221	2946	2688	2444	2213	1995	1789	1429	1162
Energy • ft-lbs	4607	3854	3208	2652	2174	1767	1421	907	599
Taylor KO Index	31.1	28.4	26.0	23.6	21.4	19.3	17.3	13.8	11.3
Path • Inches	-1.5	+2.3	0.0	-6.1	-18.0	-37.0	-65.0	-157.0	-321.1
Wind Drift • Inches	0.0	0.8	3.1	7.3	13.6	22.4	34.0	67.4	117.0

Nosler 200-grain AccuBond (29022)
G1 Ballistic Coefficient = 0.41

Distance • Yards	Muzzle	100	200	300	400	500	600	800	1000
Velocity • fps	3200	2961	2734	2518	2312	2116	1930	1592	1311
Energy • ft-lbs	4549	3894	3320	2817	2375	1990	1655	1126	764
Taylor KO Index	30.9	28.6	26.4	24.3	22.3	20.4	18.6	15.4	12.7
Path • Inches	-1.5	+1.2	0.0	-5.9	-17.3	-35.2	-60.9	-142.3	-279.4
Wind Drift • Inches	0.0	0.7	2.7	6.3	11.7	19.1	27.8	56.0	96.3

Nosler 210-grain Partition (29033)
G1 Ballistic Coefficient = 0.40

Distance • Yards	Muzzle	100	200	300	400	500	600	800	1000
Velocity • fps	3225	2882	2651	2432	2224	2025	1838	1503	1236
Energy • ft-lbs	4555	3873	3278	2759	2306	1913	1576	1053	712
Taylor KO Index	31.7	29.2	26.9	24.7	22.6	20.5	18.6	15.2	12.5
Path • Inches	-1.5	+1.4	0.0	-6.3	-18.5	-37.8	-65.5	-154.3	-305.4
Wind Drift • Inches	0.0	0.7	2.9	6.8	12.6	20.6	31.1	60.9	104.9

Weatherby 210-grain Nosler Partition (H 340 210 PT)
G1 Ballistic Coefficient = 0.40

Distance • Yards	Muzzle	100	200	300	400	500	600	800	1000
Velocity • fps	3211	2963	2728	2505	2293	2092	1899	1554	1273
Energy • ft-lbs	4807	4093	3470	2927	2452	2040	1683	1126	756
Taylor KO Index	32.6	30.0	27.7	25.4	23.3	21.2	19.3	15.8	12.9
Path • Inches	-1.5	+1.2	0.0	-5.9	-17.4	-35.5	-61.6	-145.4	-289.5
Wind Drift • Inches	0.0	0.7	2.8	6.6	12.1	19.8	29.9	58.5	100.9

Weatherby 225-grain SP (H 340 225 SP)

G1 Ballistic Coefficient = 0.397

Distance • Yards	Muzzle	100	200	300	400	500	600	800	1000
Velocity • fps	3066	2824	2595	2377	2170	1973	1786	1456	1200
Energy • ft-lbs	4696	3984	3364	2822	2352	1944	1595	1060	720
Taylor KO Index	33.3	30.7	28.2	25.8	23.6	21.4	19.5	15.9	13.1
Path • Inches	-1.5	+1.4	0.0	-6.6	-19.4	-39.6	-68.8	-163.1	-326.0
Wind Drift • Inches	0.0	1.4	3.0	7.0	13.1	21.4	32.3	63.4	109.2

Nosler 225-grain AccuBond (29040)

G1 Ballistic Coefficient = 0.550

Distance • Yards	Muzzle	100	200	300	400	500	600	800	1000
Velocity • fps	3000	2827	2660	2499	2344	2195	2051	1780	1537
Energy • ft-lbs	4498	3993	3536	3122	2746	2407	2101	1583	1180
Taylor KO Index	32.6	30.7	28.9	27.1	25.5	23.8	22.3	19.3	16.7
Path • Inches	-1.5	+1.4	0.0	-6.4	-18.3	-36.5	-62.0	-139.0	-260.3
Wind Drift • Inches	0.0	0.5	2.2	5.1	9.3	15.0	22.2	42.3	71.4

Nosler 225-grain Partition (29059)

G1 Ballistic Coefficient = 0.454

Distance • Yards	Muzzle	100	200	300	400	500	600	800	1000
Velocity • fps	3000	2791	2591	2400	2217	2042	1876	1572	1317
Energy • ft-lbs	4498	3892	3356	2879	2457	2085	1759	1235	867
Taylor KO Index	32.6	30.3	28.1	26.1	24.1	22.2	20.4	17.1	14.3
Path • Inches	-1.5	+1.5	0.0	-6.7	-19.4	-39.2	-67.4	-155.3	-300.0
Wind Drift • Inches	0.0	0.6	2.7	6.3	11.5	18.8	28.3	54.5	92.8

Cor-Bon 225-grain DPX (DPX340WBY225/20)

G1 Ballistic Coefficient = 0.375

Distance • Yards	Muzzle	100	200	300	400	500	600	800	1000
Velocity • fps	3100	2842	2599	2368	2150	1943	1749	1409	1157
Energy • ft-lbs	4802	4037	3376	2803	2309	1886	1529	993	669
Taylor KO Index	33.7	30.9	28.2	25.7	23.4	21.1	19.0	15.3	12.6
Path • Inches	-1.5	+1.4	0.0	-6.6	-19.4	-39.7	-69.3	-165.4	-332.4
Wind Drift • Inches	0.0	0.8	3.2	7.4	13.8	22.6	34.2	67.5	116.6

Weatherby 225-grain Barnes TSX (B 340 225 TSX) LF

G1 Ballistic Coefficient = 0.385

Distance • Yards	Muzzle	100	200	300	400	500	600	800	1000
Velocity • fps	2970	2726	2495	2276	2067	1871	1686	1376	1134
Energy • ft-lbs	4408	3714	3111	2588	2136	1749	1421	933	643
Taylor KO Index	32.3	29.6	27.1	24.7	22.5	20.3	18.3	14.9	12.3
Path • Inches	-1.5	+1.6	0.0	-7.2	-21.1	-43.2	-75.3	-179.1	-358.2
Wind Drift • Inches	0.0	0.7	3.2	7.6	14.2	23.3	35.2	69.4	119.2

Weatherby 250-grain Nosler Partition (N 340 250 PT)

G1 Ballistic Coefficient = 0.473

Distance • Yards	Muzzle	100	200	300	400	500	600	800	1000
Velocity • fps	2941	2743	2553	2371	2197	2029	1869	1577	1330
Energy • ft-lbs	4801	4176	3618	3120	2678	2286	1940	1381	983
Taylor KO Index	35.5	33.1	30.8	28.6	26.5	24.5	22.6	19.1	16.1
Path • Inches	-1.5	+1.6	0.0	-6.9	-20.0	-40.4	-69.1	-158.7	-306.0
Wind Drift • Inches	0.0	0.6	2.6	6.2	11.4	18.4	27.6	53.2	90.3

Weatherby 250-grain SP (H 340 250 SP)

G1 Ballistic Coefficient = 0.431

Distance • Yards	Muzzle	100	200	300	400	500	600	800	1000
Velocity • fps	2963	2745	2537	2338	2149	1968	1796	1489	1241
Energy • ft-lbs	4873	4182	3572	3035	2563	2150	1792	1230	855
Taylor KO Index	35.8	33.1	30.6	28.2	25.9	23.8	21.7	18.0	15.0
Path • Inches	-1.5	+1.6	0.0	-7.0	-20.3	-41.2	-71.2	-166.1	-326.4
Wind Drift • Inches	0.0	0.7	2.9	6.8	12.5	20.4	30.6	59.6	101.9

² 250-grain Nosler Partition (None)

G1 Ballistic Coefficient = 0.579

Distance • Yards	Muzzle	100	200	300	400	500	600	800	1000
Velocity • fps	2820	2684	2552	2424	2299	2179	2063	1841	1635
Energy • ft-lbs	4414	3999	3615	3261	2935	2635	2364	1882	1485
Taylor KO Index	34.0	32.4	30.8	29.3	27.6	26.3	25.0	22.3	19.8
Path • Inches	-1.5	+1.7	0.0	-7.2	-20.7	-41.3	-66.2	-145.3	-265.8
Wind Drift • Inches	0.0	0.5	1.9	4.4	8.0	12.9	19.1	35.8	59.2

Nosler 250-grain Partition (29099)

G1 Ballistic Coefficient = 0.473

Distance • Yards	Muzzle	100	200	300	400	500	600	800	1000
Velocity • fps	2800	2608	2424	2247	2078	1916	1763	1485	1258
Energy • ft-lbs	4353	3776	3262	2804	2397	2038	1725	1225	879
Taylor KO Index	33.8	31.5	29.3	27.1	25.1	23.1	21.3	17.9	15.2
Path • Inches	-1.5	+1.8	0.0	-7.7	-22.4	-45.1	-77.3	-177.4	-341.1
Wind Drift • Inches	0.0	0.7	2.8	6.6	12.2	19.8	29.7	57.3	97.0

A² 250-grain Triad (None)

G1 Ballistic Coefficient = 0.303

Distance • Yards	Muzzle	100	200	300	400	500	600	800	1000
Velocity • fps	2820	2520	2238	1976	1741	1522	1333	1071	935
Energy • ft-lbs	4414	3524	2781	2166	1683	1286	987	636	485
Taylor KO Index	34.0	30.4	27.0	23.9	21.0	18.4	16.1	13.0	11.3
Path • Inches	-1.5	+2.0	0.0	-9.0	-26.8	-56.2	-101.2	-257.4	-532.4
Wind Drift • Inches	0.0	1.1	4.6	10.9	20.6	34.3	52.7	104.5	173.2

Finn Aagaard personally favored the .338 only when really heavy bullets were used; he took this 6x6 elk in British Columbia. (Photo from *Guns and Hunting* by Finn Aagaard, 2012, Safari Press Archives)

.338 Lapua Magnum

This is yet another example of a big, beltless magnum case. It has found some favor with long-range shooters, especially when using the 300-grain match bullets. The use of heavy bullets raises a small problem because the more or less standard .338 twist rate of one turn in 10 inches isn't fast enough to stabilize the long bullets. It never gets to be a serious problem since nearly all competition long-range shooting is done with custom-built guns and the barrel twist can be anything the shooter prefers. So far this caliber is a special purpose number in the U.S.

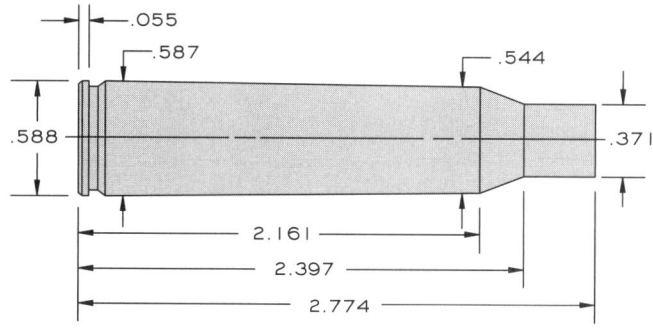

Relative Recoil Factor = 3.33

Specifications:

Controlling Agency for Standardization of this Ammunition: SAAMI

Bullet Weight Grains	Velocity fps	Maximum Average Pressure	
		Copper Crusher	Transducer
N/A	N/A	N/A	65,000 psi

Standard barrel for velocity testing: N/A

Availability:

Nosler 225-grain AccuBond (28039)
G1 Ballistic Coefficient = 0.549

Distance • Yards	Muzzle	100	200	300	400	500	600	800	1000
Velocity • fps	3025	2850	2682	2521	2365	2214	2069	1796	1551
Energy • ft-lbs	4571	4058	3594	3175	2798	2449	2139	1613	1202
Taylor KO Index	32.9	31.0	29.1	27.4	25.7	24.1	22.5	19.5	16.9
Path • Inches	-1.5	+1.4	0.0	-6.2	-17.9	-35.9	-60.9	-136.6	-255.7
Wind Drift • Inches	0.0	0.5	2.2	5.0	9.2	14.8	22.0	41.9	70.8

Cor-Bon 225-grain DPX (DPX338L225/20)
G1 Ballistic Coefficient = 0.375

Distance • Yards	Muzzle	100	200	300	400	500	600	800	1000
Velocity • fps	3100	2842	2599	2368	2150	1943	1749	1409	1157
Energy • ft-lbs	4802	4037	3376	2803	2309	1886	1529	993	669
Taylor KO Index	33.7	30.9	28.2	25.7	23.4	21.1	19.0	15.3	12.6
Path • Inches	-1.5	+1.4	0.0	-6.6	-19.4	-39.7	-69.3	-165.4	-332.4
Wind Drift • Inches	0.0	0.8	3.2	7.4	13.8	22.6	34.2	67.5	116.6

Black Hills 225-grain Sierra MatchKing (2C338LAPN2)
G1 Ballistic Coefficient = 0.675

Distance • Yards	Muzzle	100	200	300	400	500	600	800	1000
Velocity • fps	2950	2810	2674	2542	2414	2290	2169	1959	1725
Energy • ft-lbs	4832	4384	3970	3589	3237	2912	2613	2088	1652
Taylor KO Index	35.6	33.9	32.3	30.7	29.1	27.6	26.2	23.4	20.8
Path • Inches	-1.5	+1.4	0.0	-6.3	-18.0	-35.7	-60.0	-91.7	-240.4
Wind Drift • Inches	0.0	0.4	1.8	4.2	7.6	12.7	17.9	33.6	55.6

Lapua 231-grain Naturalis (N318020)
G1 Ballistic Coefficient = 0.380

Distance • Yards	Muzzle	100	200	300	400	500	600	800	1000
Velocity • fps	3018	2768	2532	2308	2095	1894	1706	1379	1140
Energy • ft-lbs	4673	3932	3289	2732	2252	1840	1493	975	666
Taylor KO Index	33.7	30.9	28.2	25.7	23.4	21.1	19.0	15.4	12.7
Path • Inches	-1.5	+1.5	0.0	-7.0	-20.5	-41.9	-73.1	-174.3	-349.4
Wind Drift • Inches	0.0	0.8	3.2	7.6	14.1	23.1	35.0	69.0	118.8

Nosler 250-grain AccuBond (28075)

G1 Ballistic Coefficient = 0.584

Distance • Yards	Muzzle	100	200	300	400	500	600	800	1000
Velocity • fps	2850	2692	2539	2392	2250	2112	1979	1729	1505
Energy • ft-lbs	4508	4022	3578	3176	2810	2476	2174	1660	1258
Taylor KO Index	34.4	32.5	30.6	28.9	27.2	25.5	23.9	20.9	18.2
Path • Inches	-1.5	+1.6	0.0	-7.0	-20.2	-40.2	-68.0	-151.6	-282.0
Wind Drift • Inches	0.0	0.5	2.2	5.1	9.3	15.0	22.4	42.0	70.9

Cor-Bon 250-grain HPBT (PM338250/20)

G1 Ballistic Coefficient = 0.500

Distance • Yards	Muzzle	100	200	300	400	500	600	800	1000
Velocity • fps	3000	2810	2627	2452	2284	2122	1967	1679	1428
Energy • ft-lbs	4997	4393	3833	3339	2896	2500	2148	1565	1132
Taylor KO Index	36.2	33.9	31.7	29.6	27.6	25.6	23.9	20.3	17.2
Path • Inches	-1.5	+1.4	0.0	-6.5	-18.8	-37.8	-64.4	-146.4	-278.1
Wind Drift • Inches	0.0	0.6	2.4	5.6	10.3	16.7	25.0	47.9	81.0

Lapua 250-grain Scenar (4318017)

G1 Ballistic Coefficient = 0.675

Distance • Yards	Muzzle	100	200	300	400	500	600	800	1000
Velocity • fps	2969	2828	2692	2559	2431	2306	2185	1953	1740
Energy • ft-lbs	4895	4441	4023	3637	3281	2953	2650	2119	1677
Taylor KO Index	35.9	34.2	32.6	31.0	29.4	27.9	26.5	23.6	21.1
Path • Inches	-1.5	+1.4	0.0	-6.2	-17.8	-35.2	-59.1	-129.8	-237.1
Wind Drift • Inches	0.0	0.4	1.6	3.7	6.8	10.8	16.0	30.0	49.5

Remington 250-grain Scenar (RM338LMR1)

G1 Ballistic Coefficient = 0.675

Distance • Yards	Muzzle	100	200	300	400	500	600	800	1000
Velocity • fps	2960	2820	2683	2551	2423	2299	2178	1947	1732
Energy • ft-lbs	4863	4412	3996	3613	3259	2932	2633	2105	1666
Taylor KO Index	35.7	34.0	32.4	30.8	29.2	27.8	26.3	23.5	20.9
Path • Inches	-1.5	+1.4	0.0	-6.3	-17.9	-35.4	-59.5	-130.6	-238.6
Wind Drift • Inches	0.0	0.4	1.8	4.1	7.5	12.1	17.8	33.4	55.3

Federal 250-grain Sierra MatchKing BTHP (GM338LM)

G1 Ballistic Coefficient = 0.588

Distance • Yards	Muzzle	100	200	300	400	500	600	800	1000
Velocity • fps	2950	2790	2630	2480	2340	2200	2065	1810	1578
Energy • ft-lbs	4830	4320	3850	3425	3035	2685	2368	1819	1363
Taylor KO Index	35.6	33.7	31.7	29.9	28.2	26.6	24.9	21.8	19.0
Path • Inches	-1.5	+1.5	0.0	-6.5	-18.5	-37.1	-62.8	-139.8	-259.5
Wind Drift • Inches	0.0	0.5	2.1	4.8	8.8	14.2	21.1	39.9	66.6

Hornady 250-grain BTHP (8230)

G1 Ballistic Coefficient = 0.670

Distance • Yards	Muzzle	100	200	300	400	500	600	800	1000
Velocity • fps	2900	2760	2625	2494	2366	2242	2122	1893	1681
Energy • ft-lbs	4668	4229	3825	3452	3108	2791	2501	1991	1570
Taylor KO Index	36.0	33.3	31.7	30.1	28.6	27.1	25.6	22.9	20.3
Path • Inches	-1.5	+1.5	0.0	-6.6	-18.8	-37.1	-62.4	-137.2	-250.6
Wind Drift • Inches	0.0	0.5	1.9	4.3	7.8	12.5	18.5	34.8	57.6

Lapua 250-grain Lock Base (4318033)

G1 Ballistic Coefficient = 0.66?

Distance • Yards	Muzzle	100	200	300	400	500	600	800	1000
Velocity • fps	2953	2810	2671	2537	2406	2280	2157	1923	1705
Energy • ft-lbs	4842	4384	3962	3573	3215	2886	2583	2052	1615
Taylor KO Index	35.6	33.9	32.2	30.6	29.0	27.5	26.0	23.2	20.6
Path • Inches	-1.5	+1.4	0.0	-6.3	-18.1	-35.8	-60.2	-132.4	-242.5
Wind Drift • Inches	0.0	0.4	1.7	3.8	7.0	11.2	16.5	31.0	51.3

Cor-Bon 265-grain DPX (DPX338L265/20)

G1 Ballistic Coefficient = 0.44?

Distance • Yards	Muzzle	100	200	300	400	500	600	800	1000
Velocity • fps	2800	2596	2401	2214	2036	1867	1701	1423	1199
Energy • ft-lbs	4614	3967	3393	2886	2440	2051	1715	1191	847
Taylor KO Index	36.8	33.2	30.7	28.3	26.1	23.9	21.8	18.2	15.3
Path • Inches	-1.5	+1.8	0.0	-7.9	-22.9	-46.2	-79.6	-184.7	-359.0
Wind Drift • Inches	0.0	0.7	3.0	7.1	13.1	21.3	32.0	62.2	105.6

Hornady 285-grain BTHP (82306)

G1 Ballistic Coefficient = 0.700

Distance • Yards	Muzzle	100	200	300	400	500	600	800	1000
Velocity • fps	2745	2623	2504	2388	2275	2165	2058	1854	1665
Energy • ft-lbs	4768	4352	3966	3608	3275	2966	2682	2177	1754
Taylor KO Index	37.8	36.1	34.5	32.9	31.3	29.8	28.3	25.5	22.9
Path • Inches	-1.5	+1.8	0.0	-7.3	-20.8	-40.9	-68.5	-149.1	-270.4
Wind Drift • Inches	0.0	0.4	1.8	4.2	7.6	12.2	17.9	33.5	55.2

Black Hills 300-grain Sierra MatchKing (2C338LAPN1)

G1 Ballistic Coefficient = 0.768

Distance • Yards	Muzzle	100	200	300	400	500	600	800	1000
Velocity • fps	2800	2681	2564	2451	2341	2233	2129	1928	1739
Energy • ft-lbs	5224	4788	4382	4004	3651	3323	3019	2476	2013
Taylor KO Index	40.6	38.8	37.1	35.5	33.9	32.3	30.8	27.9	25.2
Path • Inches	-1.5	+1.7	0.0	-7.0	-19.7	-38.8	-64.9	-140.8	-254.3
Wind Drift • Inches	0.0	0.4	1.7	3.9	7.1	11.3	16.7	31.1	51.1

Cor-Bon 300-grain HPBT (PM338300/20)

G1 Ballistic Coefficient = 0.500

Distance • Yards	Muzzle	100	200	300	400	500	600	800	1000
Velocity • fps	2800	2618	2443	2275	2114	1959	1811	1542	1314
Energy • ft-lbs	5224	4567	3977	3449	2977	2557	2196	1583	1150
Taylor KO Index	40.6	37.9	35.4	33.0	30.6	28.4	26.2	22.3	19.0
Path • Inches	-1.5	+1.8	0.0	-7.6	-22.0	-44.1	-75.3	-171.5	-326.6
Wind Drift • Inches	0.0	0.6	2.7	6.2	11.4	18.5	27.7	53.2	89.9

Dakota 300-grain MatchKing (338-300MKG)

G1 Ballistic Coefficient = 0.768

Distance • Yards	Muzzle	100	200	300	400	500	600	800	1000
Velocity • fps	2800	2681	2564	2451	2341	2233	2128	1927	1739
Energy • ft-lbs	5224	4788	4382	4003	3651	3323	3018	2475	2014
Taylor KO Index	33.9	32.5	31.0	29.7	28.3	27.0	25.8	23.3	21.1
Path • Inches	-1.5	+1.7	0.0	-7.0	-19.7	-38.8	-64.8	-140.8	-254.4
Wind Drift • Inches	0.0	0.4	1.7	3.9	7.1	11.3	16.7	31.1	51.1

Lapua 300-grain Scenar (4318013)

G1 Ballistic Coefficient = 0.739

Distance • Yards	Muzzle	100	200	300	400	500	600	800	1000
Velocity • fps	2723	2600	2482	2367	2255	2145	2042	1842	1653
Energy • ft-lbs	4938	4504	4102	3731	3387	3066	2778	2259	1819
Taylor KO Index	39.4	37.7	36.0	34.3	32.7	31.1	29.6	26.7	23.9
Path • Inches	-1.5	+1.8	0.0	-7.5	-21.1	-41.6	-106.2	-151.9	-275.5
Wind Drift • Inches	0.0	0.4	1.8	4.2	7.7	12.3	18.2	34.0	55.9

Cor-Bon 300-grain HPBT Subsonic (PM338S300/20)

G1 Ballistic Coefficient = 0.500

Distance • Yards	Muzzle	100	200	300	400	500	600	800	1000
Velocity • fps	1000	962	929	899	872	847	824	782	745
Energy • ft-lbs	666	616	575	539	507	479	453	408	369
Taylor KO Index	14.5	13.9	13.5	13.0	12.6	12.3	11.9	11.3	10.8
Path • Inches	-1.5	+18.0	0.0	-58.3	-159.6	-306.7	-502.3	-1051	-1828
Wind Drift • Inches	0.0	1.1	4.1	9.2	16.0	27.4	35.2	61.6	95.2

.338 Norma Magnum

Introduced in 2008, the .338 Norma Magnum is touted as a cartridge especially developed for long-range shooting competition. The case has somewhat less capacity than the .338 Lapua Magnum and with a very slightly reduced performance level. At the time of this writing in 2011, Norma does not list this round in their catalog. Black Hills offers the only factory load available.

Relative Recoil Factor = 3.75

Specifications:

Controlling Agency for Standardization of this Ammunition: CIP

Bullet Weight Grains	Velocity fps	Maximum Average Pressure	
		Copper Crusher	Transducer
N/A	N/A	55,150 cup	63,850 psi

Standard barrel for velocity testing: 26 inches long

Availability:

Black Hills 300-grain Sierra MatchKing (2C338NORMAGN1)

G1 Ballistic Coefficient = 0.768

Distance • Yards	Muzzle	100	200	300	400	500	600	800	1000
Velocity • fps	2725	2608	2493	2382	2273	2167	2064	1867	1683
Energy • ft-lbs	4948	4531	4142	3780	3443	3130	2840	2323	1888
Taylor KO Index	39.5	37.8	36.1	34.5	32.9	31.4	29.9	27.0	24.4
Path • Inches	-1.5	+1.8	0.0	-7.4	-20.9	-41.2	-68.9	-149.3	-270.2
Wind Drift • Inches	0.0	0.4	1.8	4.0	7.4	11.8	17.4	32.4	53.2

Dunlop Farren with Finn Aagaard and Kinuno, the tracker, after taking an oryx with a .338. (Photo from *Guns and Hunting* by Finn Aagaard, 2012, Safari Press Archives)

.338-378 Weatherby

As with the .30-378, Weatherby has responded to what wildcatters have been doing for years and "formalized" the .338-378. This is a very potent cartridge, one that is suitable for all but the most dangerous African game. Driving a 250-grain bullet in excess of 3,000 fps produces a recoil that certainly gets your attention. This is hardly a caliber for plinking tin cans.

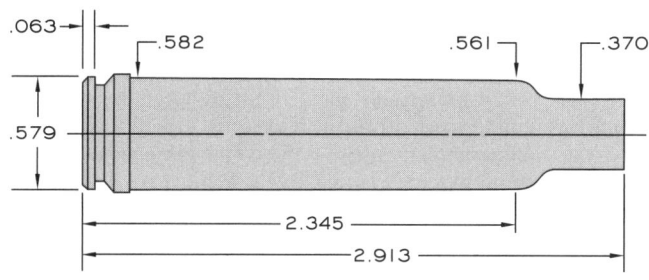

Relative Recoil Factor = 3.40

Specifications

Controlling Agency for Standardization of this Ammunition: Factory

Standard barrel for velocity testing: 26 inches long—1 turn in 10-inch twist

Availability:

Weatherby 200-grain AccuBond (N 333 200 ACB)
G1 Ballistic Coefficient = 0.414

Distance • Yards	Muzzle	100	200	300	400	500	600	800	1000
Velocity • fps	3380	3130	2894	2670	2457	2254	2061	1705	1402
Energy • ft-lbs	5075	4371	3720	3160	2681	2256	1886	1292	873
Taylor KO Index	32.6	30.2	27.9	25.8	23.7	21.8	19.9	16.5	13.5
Path • Inches	-1.5	+1.0	0.0	-5.2	-15.3	-31.1	-53.8	-125.5	-245.7
Wind Drift • Inches	0.0	0.6	2.5	5.9	10.9	17.7	26.6	51.7	88.9

Nosler 225-grain AccuBond (27031)
G1 Ballistic Coefficient = 0.550

Distance • Yards	Muzzle	100	200	300	400	500	600	800	1000
Velocity • fps	3250	3066	2890	2721	2558	2400	2248	1961	1699
Energy • ft-lbs	5278	4698	4174	3699	3269	2879	2526	1923	1442
Taylor KO Index	35.3	33.3	31.4	29.6	27.8	26.1	24.4	21.3	18.5
Path • Inches	-1.5	+1.1	0.0	-5.3	-15.2	-30.5	-51.8	-116.2	-216.9
Wind Drift • Inches	0.0	0.5	2.0	4.6	8.3	13.4	19.9	37.7	63.1

Weatherby 225-grain Barnes TSX (B 333 225 TSX) LF
G1 Ballistic Coefficient = 0.389

Distance • Yards	Muzzle	100	200	300	400	500	600	800	1000
Velocity • fps	3180	2794	2778	2591	2410	2238	1845	1499	1226
Energy • ft-lbs	5052	4280	3608	3024	2516	2077	1702	1123	752
Taylor KO Index	34.5	32.3	30.2	28.1	26.2	24.3	20.0	16.3	13.3
Path • Inches	-1.5	+1.3	0.0	-6.1	-18.0	-36.8	-64.0	-151.4	-301.6
Wind Drift • Inches	0.0	0.7	2.9	6.9	12.7	20.8	31.4	61.8	106.8

Weatherby 250-grain Nosler Partition (N 333 250 PT)
G1 Ballistic Coefficient = 0.473

Distance • Yards	Muzzle	100	200	300	400	500	600	800	1000
Velocity • fps	3060	2856	2662	2478	2297	2125	1961	1658	1396
Energy • ft-lbs	5197	4528	3933	3401	2927	2507	2134	1526	1082
Taylor KO Index	36.9	34.5	32.1	29.9	27.7	25.7	23.7	20.1	16.9
Path • Inches	-1.5	+1.4	0.0	-6.3	-18.3	-36.9	63.2	-144.8	-278.7
Wind Drift • Inches	0.0	0.6	2.5	5.8	10.7	17.4	26.0	50.1	85.1

Nosler 250-grain Partition (27068)
G1 Ballistic Coefficient = 0.473

Distance • Yards	Muzzle	100	200	300	400	500	600	800	1000
Velocity • fps	3050	2847	2652	2466	2287	2116	1952	1650	1390
Energy • ft-lbs	5165	4499	3905	3376	2905	2486	2116	1512	1073
Taylor KO Index	36.8	34.4	32.0	29.8	27.6	25.5	23.6	19.9	16.8
Path • Inches	-1.5	+1.4	0.0	-6.4	-18.4	-37.2	63.7	-145.8	-279.4
Wind Drift • Inches	0.0	0.6	2.5	5.8	10.8	17.5	26.1	50.4	85.6

.357 Magnum (Rifle Data)

The high-performance pistol cartridges, especially the .357 Magnum and the .44 Remington Magnum, can be effectively shot from rifle-length barrels. They are especially useful in carbine-size rifles with light recoil and provide good effectiveness at short ranges. The loads shown below have the same product numbers as the comparable pistol loads produced by these same companies. They represent identical loading levels. While all pistol ammunition in this caliber is suitable for use in rifles chambered for this cartridge, you can occasionally find some ammunition marked for "Rifles Only." See the .357 Magnum pistol listing for more information.

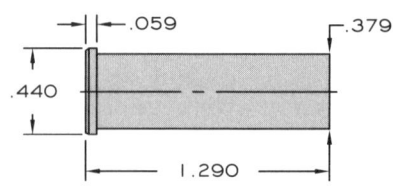

Relative Recoil Factor = 1.30

Specifications:

Controlling Agency for Standardization of this Ammunition: SAAMI

See pistol data section for detailed specifications.
The data below were taken in the barrel lengths listed with each loading.

Availability:

Winchester 158-grain Jacketed Soft Point (X3575P) [20-inch barrel data]   G1 Ballistic Coefficient = 0.163

Distance • Yards	Muzzle	100	200	300	400	500	600	800	1000
Velocity • fps	1830	1427	1138	980	883	809	745	638	548
Energy • ft-lbs	1175	715	454	337	274	229	195	143	105
Taylor KO Index	14.8	11.5	9.2	7.9	7.1	6.5	6.0	5.1	4.4
Path • Inches	-1.5	+7.9	0.0	-34.6	-105.4	-220.7	-372.4	-871.7	-1689
Wind Drift • Inches	0.0	3.9	16.7	38.2	66.2	100.0	139.2	234.8	355.9

Federal 180-grain Hollow Point (C357G) [18-inch barrel data]   G1 Ballistic Coefficient = 0.210

Distance • Yards	Muzzle	100	200	300	400	500	600	800	1000
Velocity • fps	1550	1260	1100	980	900	840	787	698	662
Energy • ft-lbs	960	655	480	385	325	285	248	195	155
Taylor KO Index	14.2	11.8	10.8	9.0	8.3	7.7	7.2	6.4	5.7
Path • Inches	-1.5	+9.9	0.0	-38.5	-111.9	-225.7	-398.0	-908.5	-1708
Wind Drift • Inches	0.0	3.5	14.2	31.4	53.5	80.2	111.1	186.1	279.4

Winchester 158-grain Jacketed Hollow Point (X3574P) [20-inch barrel data]   G1 Ballistic Coefficient = 0.163

Distance • Yards	Muzzle	100	200	300	400	500	600	800	1000
Velocity • fps	1830	1427	1138	980	883	809	745	638	548
Energy • ft-lbs	1175	715	454	337	274	229	195	143	105
Taylor KO Index	14.8	11.5	9.2	7.9	7.1	6.5	6.0	5.1	4.4
Path • Inches	-1.5	+7.9	0.0	-34.6	-105.4	-220.7	-372.4	-871.7	-1689
Wind Drift • Inches	0.0	3.9	16.7	38.2	66.2	100.0	139.2	234.8	355.9

Winchester 180-grain Partition Gold (S357P) [20-inch barrel data]   G1 Ballistic Coefficient = 0.13

Distance • Yards	Muzzle	100	200	300	400	500	600	800	1000
Velocity • fps	1550	1160	965	854	769	698	637	530	442
Energy • ft-lbs	960	538	372	291	237	195	162	112	78
Taylor KO Index	14.2	1.06	8.9	7.8	7.1	6.4	5.8	4.9	4.1
Path • Inches	-1.5	+12.1	0.0	-49.3	-146.3	-302.2	-529.5	-1262	-2505
Wind Drift • Inches	0.0	5.6	22.0	46.2	77.4	115.4	160.7	275.7	429.7

.35 Remington

When Remington introduced their new .35 caliber in 1908, it was intended as a mild round for the Model 6 semiautomatic rifle. Later it was offered in a number of slide-action and lever-action guns. Even with the pressure levels very low by modern standards, this cartridge offers more punch than the .30-30, but falls considerably short of what can be achieved by the .35 Whelen. While this cartridge is adequate for deer-size game under brush-hunting conditions, it's popularity is slowly declining except for the single-shot-pistol (one-hand rifle) market.

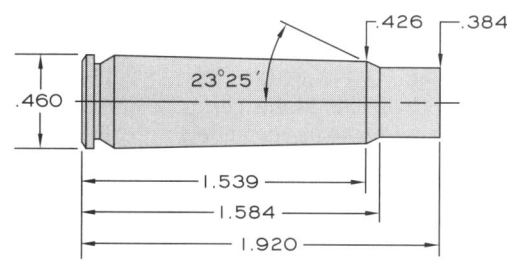

Relative Recoil Factor = 1.87

Specifications:

Controlling Agency for Standardization of this Ammunition: SAAMI

Bullet Weight Grains	Velocity fps	Maximum Average Pressure	
		Copper Crusher	Transducer
150	2,275	35,000 cup	33,500 psi
200	2,055	35,000 cup	33,500 psi

Standard barrel for velocity testing: 24 inches long—1 turn in 16-inch twist

Availability:

Remington 150-grain Core-Lokt Pointed Soft Point (R35R1)

G1 Ballistic Coefficient = 0.184

Distance • Yards	Muzzle	100	200	300	400	500	600	800	1000
Velocity • fps	2300	1874	1506	1218	1039	934	857	739	645
Energy • ft-lbs	1762	1169	755	494	359	291	245	182	138
Taylor KO Index	17.7	14.4	11.6	9.4	8.0	7.2	6.6	5.7	5.0
Path • Inches	-1.5	+4.3	0.0	-19.9	-63.2	-138.7	-250.3	-608.0	-1195
Wind Drift • Inches	0.0	2.5	11.0	27.2	51.6	82.4	118.6	205.7	312.9

Federal 200-grain Soft Point RN (35A)

G1 Ballistic Coefficient = 0.193

Distance • Yards	Muzzle	100	200	300	400	500	600	800	1000
Velocity • fps	2080	1700	1380	1140	1000	910	842	733	643
Energy • ft-lbs	1920	1280	840	575	445	368	315	238	184
Taylor KO Index	21.3	17.4	14.2	11.7	10.3	9.3	8.6	7.5	6.6
Path • Inches	-1.5	+5.4	0.0	-23.3	-70.0	-144.0	280.8	-665.4	-1285
Wind Drift • Inches	0.0	2.7	12.0	29.0	53.3	83.4	118.3	202.3	305.6

Hornady 200-grain FTX (82735)

G1 Ballistic Coefficient = 0.300

Distance • Yards	Muzzle	100	200	300	400	500	600	800	1000
Velocity • fps	2225	1963	1721	1503	1361	1167	1061	928	839
Energy • ft-lbs	2198	1711	1315	1003	769	605	500	383	313
Taylor KO Index	22.8	20.1	17.6	15.4	13.5	11.9	10.9	9.5	8.6
Path • Inches	-1.5	+3.8	0.0	-15.6	-46.7	-97.8	-174.6	-425.5	-838.0
Wind Drift • Inches	0.0	1.5	6.5	15.7	29.5	48.5	72.4	131.9	204.6

Remington 200-grain Core-Lokt Soft Point (R35R2)

G1 Ballistic Coefficient = 0.193

Distance • Yards	Muzzle	100	200	300	400	500	600	800	1000
Velocity • fps	2080	1700	1380	1140	1000	910	842	733	643
Energy • ft-lbs	1920	1280	840	575	445	368	315	238	184
Taylor KO Index	21.3	17.4	14.2	11.7	10.3	9.3	8.6	7.5	6.6
Path • Inches	-1.5	+5.4	0.0	-23.3	-70.0	-144.0	280.8	-665.4	-1285
Wind Drift • Inches	0.0	2.7	12.0	29.0	53.3	83.4	118.3	202.3	305.6

Winchester 200-grain Power-Point (X35R1)

G1 Ballistic Coefficient = 0.193

Distance • Yards	Muzzle	100	200	300	400	500	600	800	1000
Velocity • fps	2020	1646	1335	1114	985	901	833	725	637
Energy • ft-lbs	1812	1203	791	551	431	366	308	234	180
Taylor KO Index	20.7	16.9	13.7	11.4	10.1	9.2	8.5	7.4	6.5
Path • Inches	-1.5	+5.8	0.0	-25.4	-78.8	-168.0	-292.8	-689.6	-1327
Wind Drift • Inches	0.0	2.8	12.4	29.8	54.3	84.4	119.4	203.2	306.5

.356 Winchester

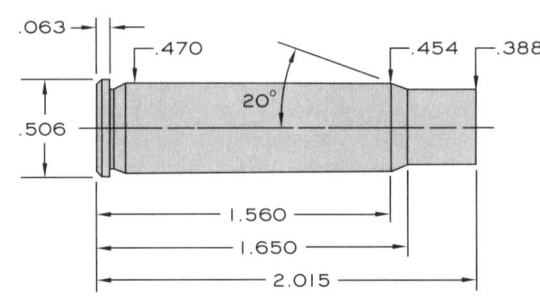

Designed as a rimmed version of the .358 Winchester for use in lever-action rifles, the .356 Winchester is a .35-caliber version of the .307 Winchester. The internal ballistics of this cartridge suffer a little because of the lower pressures necessary for it to be compatible with the lever-action guns. The exterior ballistics also suffer from the flatnose bullet that is needed for use in tubular magazines. Still, the .356 is an excellent choice for deer-size game, especially in Eastern-type conditions where game is nearly always taken at very modest ranges. In very light rifles, the .356's recoil gets to be a problem for recoil-sensitive shooters. Like the .307, short-barrel guns will yield lower muzzle velocities.

Relative Recoil Factor = 2.22

Specifications:

Controlling Agency for Standardization of this Ammunition: SAAMI

Bullet Weight Grains	Velocity fps	Maximum Average Pressure	
		Copper Crusher	Transducer
200	2,370	52,000 cup	N/S
250	2,075	52,000 cup	N/S

Standard barrel for velocity testing: 24 inches long—1 turn in 12-inch twist

Availability:

Winchester 200-grain Power-Point (X3561)

G1 Ballistic Coefficient = 0.239

Distance • Yards	Muzzle	100	200	300	400	500	600	800	1000
Velocity • fps	2460	2114	1797	1517	1284	1113	1005	870	774
Energy • ft-lbs	2688	1985	1434	1022	732	550	449	336	266
Taylor KO Index	25.2	21.6	18.4	15.5	13.1	11.4	10.3	8.9	7.9
Path • Inches	-1.5	+3.2	0.0	-14.0	-43.3	-93.7	173.6	433.7	859.0
Wind Drift • Inches	0.0	1.7	7.3	17.8	34.3	57.1	85.7	156.2	242.2

A .325 Winchester Short Magnum was used to drop this gemsbo bull in its tracks. The .325 is essentially an 8mm cartridge, faste than the 8x57, slower than the 8mm Remington Magnum, an very close to the .338 Winchester Magnum in performance o game. (Photo from *Safari Rifles II* by Craig Boddingtor 2009, Safari Press Archives)

256

.35 Whelen

Named for Col. Townsend Whelen, this cartridge is nothing but a .30-06 necked to .35 caliber. There is some disagreement whether Colonel Whelen himself or gunmaker James Howe developed the cartridge in the early 1920s, but that matters little since they, in all probability, worked together. Besides, necking an existing cartridge to another caliber is hardly the epitome of the cartridge designer's art. The .35 Whelen can handle bullets up to 250 grains, and that produces some impressive Taylor KO Index numbers, especially for a cartridge based on the .30-06 case.

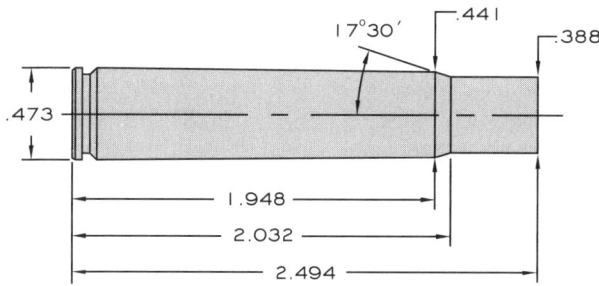

Relative Recoil Factor = 2.64

Specifications:

Controlling Agency for Standardization of this Ammunition: SAAMI

Bullet Weight Grains	Velocity fps	Maximum Average Pressure	
		Copper Crusher	Transducer
200	2,660	52,000 cup	N/S

Standard barrel for velocity testing: 24 inches long—1 turn in 12-inch twist

Availability:

Federal 180-grain Fusion (F35FS1)

G1 Ballistic Coefficient = 0.315

Distance • Yards	Muzzle	100	200	300	400	500	600	800	1000
Velocity • fps	2700	2420	2160	1910	1690	1480	1306	1064	935
Energy • ft-lbs	2915	2340	1860	1460	1135	880	682	453	350
Taylor KO Index	24.9	22.3	19.9	17.6	15.6	13.6	12.0	9.8	8.6
Path • Inches	-1.5	+2.2	0.0	-9.6	-29.1	-60.7	-108.1	-269.4	-553.1
Wind Drift • Inches	0.0	1.1	4.6	11.1	20.9	34.8	53.2	104.5	171.9

Hornady 200-grain SP (81193)

G1 Ballistic Coefficient = 0.280

Distance • Yards	Muzzle	100	200	300	400	500	600	800	1000
Velocity • fps	2910	2582	2277	1993	1732	1499	1300	1041	909
Energy • ft-lbs	3760	2961	2303	1764	1332	997	751	481	367
Taylor KO Index	29.8	26.4	23.3	20.4	17.7	15.3	13.3	10.6	9.3
Path • Inches	-1.5	+1.9	0.0	-8.6	-26.0	-55.1	-99.8	-256.8	-541.6
Wind Drift • Inches	0.0	1.1	4.8	11.4	21.7	36.3	56.0	111.6	184.5

Remington 200-grain Core-Lokt Pointed Soft Point (R35WH1)

G1 Ballistic Coefficient = 0.294

Distance • Yards	Muzzle	100	200	300	400	500	600	800	1000
Velocity • fps	2675	2378	2100	1842	1606	1399	1227	1012	896
Energy • ft-lbs	3177	2510	1968	1506	1146	869	669	455	356
Taylor KO Index	27.3	24.3	21.4	18.8	16.4	14.3	12.6	10.4	9.2
Path • Inches	-1.5	+2.3	0.0	-10.3	-30.9	-65.0	-117.9	-299.6	-610.0
Wind Drift • Inches	0.0	1.2	5.1	12.2	23.2	38.7	59.3	115.6	187.4

Nosler 225-grain AccuBond (30004) + (60081)

G1 Ballistic Coefficient = 0.421

Distance • Yards	Muzzle	100	200	300	400	500	600	800	1000
Velocity • fps	2800	2584	2379	2183	1996	1819	1654	1364	1148
Energy • ft-lbs	3918	3338	2828	2381	1990	1654	1366	930	659
Taylor KO Index	32.1	29.7	27.3	25.0	22.9	20.9	19.0	15.7	13.2
Path • Inches	-1.5	+1.9	0.0	-8.0	-23.4	-47.4	-82.1	-192.3	-377.9
Wind Drift • Inches	0.0	0.8	3.2	7.5	14.0	22.8	34.4	67.2	114.2

Federal 225-grain Trophy Bonded Bear Claw (P35WT1)

G1 Ballistic Coefficient = 0.346

Distance • Yards	Muzzle	100	200	300	400	500	600	800	1000
Velocity • fps	2600	2350	2120	1900	1690	1500	1335	1096	963
Energy • ft-lbs	3375	2760	2240	1795	1430	1130	891	600	464
Taylor KO Index	29.8	27.0	24.3	21.8	19.4	17.2	15.3	12.6	11.1
Path • Inches	-1.5	+2.4	0.0	-10.1	-30.2	-62.2	-110.2	-269.0	-543.2
Wind Drift • Inches	0.0	1.0	4.4	10.5	19.8	32.7	49.7	97.1	159.9

Nosler 250-grain Partition (30020)

G1 Ballistic Coefficient = 0.447

Distance • Yards	Muzzle	100	200	300	400	500	600	800	1000
Velocity • fps	2550	2357	2173	1998	1831	1674	1529	1280	1103
Energy • ft-lbs	3609	3084	2621	2215	1861	1555	1299	910	676
Taylor KO Index	32.6	30.1	27.8	25.5	23.4	21.4	19.5	16.4	14.1
Path • Inches	-1.5	+2.4	0.0	-9.8	-28.2	-57.1	-98.4	-228.6	-444.1
Wind Drift • Inches	0.0	0.8	3.4	8.1	15.0	24.4	36.7	71.0	118.8

Norma 250-grain ORYX (19009)

G1 Ballistic Coefficient = 0.340

Distance • Yards	Muzzle	100	200	300	400	500	600	800	1000
Velocity • fps	2428	2184	1954	1740	1545	1372	1227	1034	924
Energy • ft-lbs	3273	2648	2121	1682	1325	1045	836	593	474
Taylor KO Index	31.0	27.9	25.0	22.2	19.8	17.5	15.7	13.2	11.8
Path • Inches	-1.5	+2.9	0.0	-12.0	-35.6	-73.8	-130.6	-318.3	-635.8
Wind Drift • Inches	0.0	1.2	5.0	11.9	22.4	36.9	55.9	107.0	172.1

Remington 250-grain Pointed Soft Point (R35WH3)

G1 Ballistic Coefficient = 0.410

Distance • Yards	Muzzle	100	200	300	400	500	600	800	1000
Velocity • fps	2400	2196	2005	1823	1653	1497	1357	1139	1006
Energy • ft-lbs	3197	2680	2230	1844	1570	1244	1023	720	562
Taylor KO Index	30.6	28.0	25.6	23.2	21.1	19.1	17.4	14.6	12.9
Path • Inches	-1.5	+2.9	0.0	-11.5	-33.6	-68.3	-118.9	-281.6	-552.0
Wind Drift • Inches	0.0	1.0	4.2	9.8	18.2	29.8	44.8	86.2	114.3

The European metric calibers are not seen nearly as often in Africa as in the old days, not even with European hunters. PH Justin Seymour-Smith of Zimbabwe is pictured with a Franz Sodia 9.3x74R that was used to take this eland. (Safari Press Archives)

.350 Remington Magnum

The .350 Remington Magnum was introduced in 1965. It anticipated the current craze for short magnums by nearly 40 years. At its introduction, it was intended for the Model 660 Magnum carbine that sported an 18-inch barrel. I remember it as a savage kicker in those very light guns. Since the case volume was virtually identical to the .30-06, it has slightly better performance than the .35 Whelen (because it uses a somewhat higher working pressure.) Along about 1990, the .350 RM dropped off the radar only to reappear in 2005.

Relative Recoil Factor = 2.65

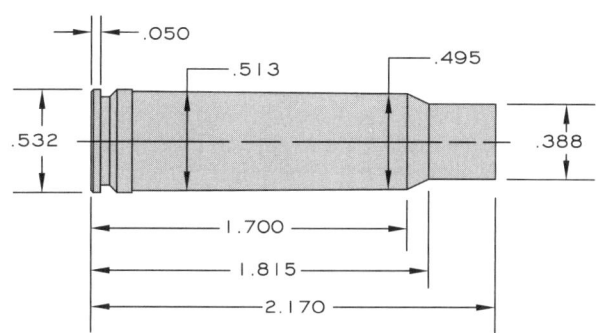

Specifications:

Controlling Agency for Standardization of this Ammunition: SAAMI

Bullet Weight Grains	Velocity fps	Maximum Average Pressure	
		Copper Crusher	Transducer
200	2,690	53,000 cup	N/S

Standard barrel for velocity testing: 20 inches long—1 turn in 16-inch twist

Availability:

Remington 200-grain Core-Lokt Pointed Soft Point (R350M1)

G1 Ballistic Coefficient = 0.293

Distance • Yards	Muzzle	100	200	300	400	500	600	800	1000
Velocity • fps	2775	2471	2186	1921	1678	1461	1276	1037	911
Energy • ft-lbs	3419	2711	2127	1639	1250	947	724	477	369
Taylor KO Index	28.2	25.1	22.2	19.5	17.1	14.9	13.0	10.5	9.3
Path • Inches	-1.5	+2.1	0.0	-9.4	-28.3	-59.7	-107.5	-272.7	-566.8
Wind Drift • Inches	0.0	1.1	4.8	11.6	22.0	36.7	56.4	111.2	182.4

Nosler 225-grain Partition (31010)

G1 Ballistic Coefficient = 0.430

Distance • Yards	Muzzle	100	200	300	400	500	600	800	1000
Velocity • fps	2550	2350	2159	1977	1805	1644	1495	1246	1075
Energy • ft-lbs	3250	2760	2330	1954	1629	1351	1118	775	578
Taylor KO Index	29.3	27.0	24.8	22.7	20.7	18.9	17.2	14.3	12.3
Path • Inches	-1.5	+2.4	0.0	-9.9	-28.7	-58.2	-100.6	-235.3	-460.0
Wind Drift • Inches	0.0	0.9	3.6	8.4	15.7	25.6	38.6	74.8	125.2

9.3x74R

Introduced at about the same time (or perhaps a little before) as the .375 H&H, the 9.3x74R was intended primarily for the combination guns that are so popular in continental Europe. Being rimmed, it also found a good home in double rifles and single-shot guns. It is still popular in Europe but seldom seen in the U.S. Reloading is limited. One reason for this is that there are very few bullets available for this caliber (0.366 inches) here in the U.S., although they are available from European sources. In addition, dies are hard to find, and many of the cases utilize Berdan priming. This is an effective caliber, very nearly the same performance as the redoubtable .375 H&H.

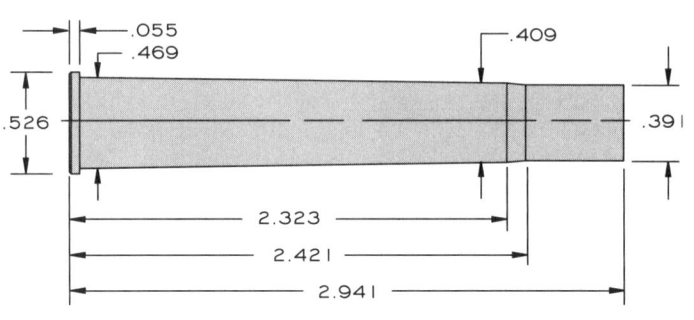

Relative Recoil Factor = 3.45

Specifications:

Controlling Agency for Standardization of this Ammunition: CIP

Bullet Weight Grains	Velocity fps	Maximum Average Pressure	
		Copper Crusher	Transducer
All	N/A	43,541 cup	49,347 psi

Standard barrel for velocity testing: N/A

Availability:

Norma 232-grain Vulcan (19321)
G1 Ballistic Coefficient = 0.278

Distance • Yards	Muzzle	100	200	300	400	500	600	800	1000
Velocity • fps	2560	2254	1970	1710	1477	1281	1132	959	855
Energy • ft-lbs	3377	2619	2000	1506	1125	846	660	473	376
Taylor KO Index	31.1	27.3	23.9	20.7	17.9	15.5	13.7	11.6	10.4
Path • Inches	-1.5	+2.7	0.0	-11.7	-35.4	-75.1	-136.1	-344.8	-704.8
Wind Drift • Inches	0.0	1.4	5.8	13.9	26.6	44.4	67.7	128.8	204.6

Nosler 250-grain AccuBond (33033)
G1 Ballistic Coefficient = 0.495

Distance • Yards	Muzzle	100	200	300	400	500	600	800	1000
Velocity • fps	2550	2375	2208	2048	1895	1749	1613	1370	1179
Energy • ft-lbs	3609	3132	2706	2328	1993	1698	1444	1042	772
Taylor KO Index	33.5	31.2	29.0	26.9	24.9	23.0	21.2	18.0	15.5
Path • Inches	-1.5	+2.3	0.0	-9.5	-27.2	-54.7	-93.6	-213.8	-408.7
Wind Drift • Inches	0.0	0.7	3.1	7.2	13.3	21.6	32.3	62.1	104.0

Nosler 250-grain Partition (33048)
G1 Ballistic Coefficient = 0.483

Distance • Yards	Muzzle	100	200	300	400	500	600	800	1000
Velocity • fps	2300	2132	1971	1818	1673	1538	1414	1208	1067
Energy • ft-lbs	3359	2886	2467	2099	1777	1502	1271	927	723
Taylor KO Index	34.6	32.1	29.6	27.3	25.2	23.1	21.3	18.2	16.0
Path • Inches	-1.5	+3.1	0.0	-12.0	-34.6	-69.7	-119.4	-274.2	-524.5
Wind Drift • Inches	0.0	0.9	3.7	8.6	16.0	25.9	38.8	73.9	121.4

Norma 285-grain Plastic Point (19325)
G1 Ballistic Coefficient = 0.32

Distance • Yards	Muzzle	100	200	300	400	500	600	800	1000
Velocity • fps	2362	2137	1925	1727	1546	1384	1245	1053	942
Energy • ft-lbs	3532	2891	2345	1887	1513	1212	901	701	561
Taylor KO Index	35.2	31.8	28.7	25.7	23.0	20.6	18.6	15.7	14.0
Path • Inches	-1.5	+3.1	0.0	-12.5	-36.7	-75.6	-132.9	-320.7	-630.3
Wind Drift • Inches	0.0	1.2	4.8	11.4	21.4	35.2	53.1	101.3	163.0

Norma 285-grain Alaska (19320)

G1 Ballistic Coefficient = 0.365

Distance • Yards	Muzzle	100	200	300	400	500	600	800	1000
Velocity • fps	2329	2106	1897	1701	1522	1363	1228	1044	937
Energy • ft-lbs	3440	2814	2281	1835	1466	1176	955	690	556
Taylor KO Index	34.7	31.4	28.3	25.3	22.7	20.3	18.3	15.6	14.0
Path • Inches	-1.5	+3.2	0.0	-12.9	-37.8	-77.8	-136.6	-327.8	-646.3
Wind Drift • Inches	0.0	1.2	4.9	11.6	21.8	35.8	54.0	102.6	164.5

Norma 285-grain Oryx (19332)

G1 Ballistic Coefficient = 0.330

Distance • Yards	Muzzle	100	200	300	400	500	600	800	1000
Velocity • fps	2329	2084	1854	1642	1452	1288	1157	993	894
Energy • ft-lbs	3440	2753	2180	1710	1334	1050	848	624	506
Taylor KO Index	34.7	31.1	27.6	24.5	21.6	19.2	17.2	14.8	13.3
Path • Inches	-1.5	+3.3	0.0	-13.4	-39.8	-82.8	-146.7	-357.3	-708.7
Wind Drift • Inches	0.0	1.3	5.5	13.1	24.6	40.6	61.3	115.2	182.4

Federal 286-grain Swift A-Frame (P93745A)

G1 Ballistic Coefficient = 0.384

Distance • Yards	Muzzle	100	200	300	400	500	600	800	1000
Velocity • fps	2360	2150	1950	1760	1580	1420	1283	1083	966
Energy • ft-lbs	3535	2930	2405	1960	1590	1285	1046	745	593
Taylor KO Index	35.5	32.3	29.3	26.5	23.8	21.4	19.3	16.3	14.5
Path • Inches	-1.5	+3.0	0.0	-12.3	-35.8	-73.4	-128.2	-305.6	-601.2
Wind Drift • Inches	0.0	1.1	4.6	10.8	20.1	33.0	49.8	95.2	154.2

Hornady 286-grain SP-RP (82304)

G1 Ballistic Coefficient = 0.365

Distance • Yards	Muzzle	100	200	300	400	500	600	800	1000
Velocity • fps	2360	2136	1924	1727	1545	1383	1245	1053	943
Energy • ft-lbs	3536	2896	2351	1893	1516	1214	965	705	565
Taylor KO Index	35.7	32.3	29.1	26.1	23.4	20.9	18.8	15.9	14.3
Path • Inches	-1.5	+3.1	0.0	-12.5	-36.7	75.5	-132.6	-318.5	-629.5
Wind Drift • Inches	0.0	1.2	4.8	11.4	21.4	35.2	53.1	101.2	162.9

Federal 286-grain Barnes Triple-Shock X-Bullet (P9374C) LF

G1 Ballistic Coefficient = 0.411

Distance • Yards	Muzzle	100	200	300	400	500	600	800	1000
Velocity • fps	2360	2160	1970	1770	1630	1470	1335	1125	990
Energy • ft-lbs	3535	2965	2465	2035	1675	1375	1132	804	634
Taylor KO Index	35.7	32.7	29.8	27.1	24.6	22.2	20.2	17.0	15.1
Path • Inches	-1.5	+3.0	0.0	-12.0	-34.8	-70.8	-123.0	-289.8	-566.2
Wind Drift • Inches	0.0	1.0	4.2	10.0	18.6	30.4	45.7	87.6	142.9

Federal 286-grain Barnes Banded Solid (P9374D) LF

G1 Ballistic Coefficient = 0.259

Distance • Yards	Muzzle	100	200	300	400	500	600	800	1000
Velocity • fps	2360	2050	1760	1500	1290	1130	1023	890	798
Energy • ft-lbs	3535	2660	1965	1435	1055	805	665	505	405
Taylor KO Index	35.5	30.8	26.5	22.6	19.4	17.0	15.4	13.4	12.0
Path • Inches	-1.5	+3.4	0.0	-14.7	-44.9	-96.0	-174.2	-435.6	-872.8
Wind Drift • Inches	0.0	1.6	7.1	17.2	32.7	54.2	81.0	147.6	228.6

.375 Winchester

In 1978 Winchester introduced the .375 Winchester. This cartridge is certainly not a .375 magnum. In terms of both energy and Taylor KO Index, the .375 is inferior to the .356 Winchester. This cartridge can be thought of as a .30-30 necked to .375 caliber. The only bullet that is factory loaded is a 200-grain Power-Point. It is intended for the Model 94 Big Bore lever action that tolerates high pressures. No full-size rifles are currently being chambered for this cartridge, but it is chambered in some single-shot pistols.

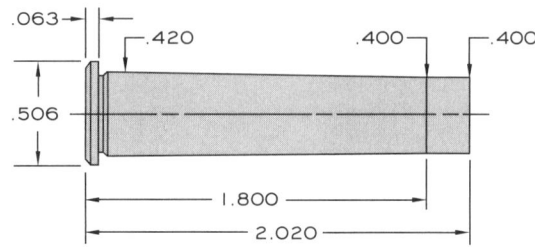

Relative Recoil Factor = 1.98

Specifications:

Controlling Agency for Standardization of this Ammunition: SAAMI

Bullet Weight Grains	Velocity fps	Maximum Average Pressure	
		Copper Crusher	Transducer
200	2,180	52,000 cup	N/S
250	1,885	52,000 cup	N/S

Standard barrel for velocity testing: 24 inches long—1 turn in 12-inch twist

Availability:

Winchester 200-grain Power-Point (X375W

G1 Ballistic Coefficient = 0.215

Distance • Yards	Muzzle	100	200	300	400	500	600	800	1000
Velocity • fps	2200	1841	1526	1268	1089	980	904	790	700
Energy • ft-lbs	2150	1506	1034	714	527	427	363	277	218
Taylor KO Index	23.6	19.8	16.4	13.6	11.7	10.5	9.7	8.5	7.5
Path • Inches	-1.5	+4.4	0.0	-19.5	-60.7	-131.1	-236.0	-566.8	-1098
Wind Drift • Inches	0.0	2.2	9.8	23.8	44.9	72.2	104.4	181.7	275.9

A nice Zambezi Valley dagga *bull, taken in 2007 with the .375 Ruger.* (Photo from *Safari Rifles II* by Craig Boddington, 2009, Safari Press Archives)

.375 Ruger

Sometime about 2007 Ruger decided it wanted to get into the .375-caliber business. The rifle they had in mind was the company's Model 77 Hawkeye, which has a standard length (same length as .30-06) action. Now, Ruger is not an ammunition developer but knew where to go for that work—Hornady. The Hornady folks developed a cartridge that used the standard 3.280-inch overall length and the 0.532-inch rim diameter, the same as a .375 H&H Magnum. The new cartridge follows the recent trend of not bothering with a belt, so the body diameter at rear of the case is the same as the belt diameter of the magnums but slightly larger than the body diameter of the .375 H&H. Clever! They won't interchange. The volume of the .375 Ruger is slightly larger than the .375 H&H, but some of the published data for the Ruger are for 20-inch barrels. Under those conditions, the performance is just a tad below that old standard, but when compared in the same length barrels the situation gets reversed. This is a potent round.

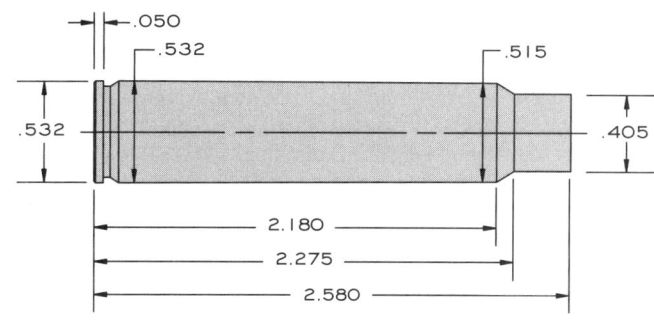

Relative Recoil Factor = 3.30

Specifications:

Controlling Agency for Standardization of this Ammunition: SAAMI

Bullet Weight Grains	Velocity fps	Maximum Average Pressure	
		Copper Crusher	Transducer
–	-	N/S	62,000 psi

Standard barrel for velocity testing: 24 inches long—1 turn in 12-inch twist

Availability:

Hornady 270-grain SP-RP (8231)

G1 Ballistic Coefficient = 0.380

Distance • Yards	Muzzle	100	200	300	400	500	600	800	1000
Velocity • fps	2840	2600	2372	2156	1951	1759	1582	1282	1081
Energy • ft-lbs	4835	4052	3373	2786	2283	1855	1501	986	700
Taylor KO Index	41.2	37.7	34.4	31.3	28.3	25.5	2.98	18.6	15.7
Path • Inches	-1.5	+1.8	0.0	-8.0	-23.6	-48.2	-84.2	-201.3	-403.4
Wind Drift • Inches	0.0	0.8	3.5	8.3	15.4	25.3	38.4	75.6	128.7

Hornady 300-grain DGX (82333)

G1 Ballistic Coefficient = 0.275

Distance • Yards	Muzzle	100	200	300	400	500	600	800	1000
Velocity • fps	2660	2344	2050	1780	1536	1328	1164	973	864
Energy • ft-lbs	4713	3660	2800	2110	1572	1174	903	631	497
Taylor KO Index	42.9	37.8	33.0	28.7	24.8	21.4	18.8	15.7	13.9
Path • Inches	-1.5	+2.4	0.0	-10.8	-32.6	-69.2	-125.6	-321.2	-663.4
Wind Drift • Inches	0.0	1.3	5.5	13.3	25.4	42.6	65.3	125.8	201.8

Hornady 300-grain DGS (8232)

G1 Ballistic Coefficient = 0.275

Distance • Yards	Muzzle	100	200	300	400	500	600	800	1000
Velocity • fps	2660	2344	2050	1780	1536	1328	1164	973	864
Energy • ft-lbs	4713	3660	2800	2110	1572	1174	903	631	497
Taylor KO Index	42.9	37.8	33.0	28.7	24.8	21.4	18.8	15.7	13.9
Path • Inches	-1.5	+2.4	0.0	-10.8	-32.6	-69.2	-125.6	-321.2	-663.4
Wind Drift • Inches	0.0	1.3	5.5	13.3	25.4	42.6	65.3	125.8	201.8

.375 Holland & Holland Magnum

The British riflemaking firm of Holland and Holland wanted something new and different in 1912 when it introduced its .375 H&H. It was common for gunmakers in those days to have their own proprietary cartridges. Of course, we don't do that any more—not much we don't. The names Weatherby, Lazzeroni, A[2], and Dakota all come to mind as carrying on that practice, at least to a degree. The .375 H&H is the smallest "African" cartridge allowed in many countries. It has proved its worth in over 90 years of exemplary field history. A few years after its introduction, the .375 H&H led to the .300 H&H, and that number begat nearly all the other belted magnums we use today. The .375 H&H has a well-earned place in cartridge history.

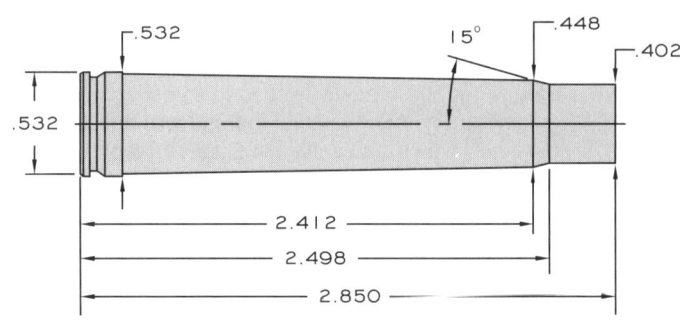

Relative Recoil Factor = 3.42

Specifications:

Controlling Agency for Standardization of this Ammunition: SAAMI*

Bullet Weight Grains	Velocity fps	Maximum Average Pressure	
		Copper Crusher	Transducer
270	2,680	53,000 cup	62,000 psi
300	2,515	53,000 cup	62,000 psi

Standard barrel for velocity testing: 24 inches long—1 turn in 12-inch twist

* Some ammunition that is available in this caliber is manufactured to CIP (European) specifications.

Availability:

Cor-Bon 235-grain DPX (DPX375HH235/20)
G1 Ballistic Coefficient = 0.300

Distance • Yards	Muzzle	100	200	300	400	500	600	800	1000
Velocity • fps	3000	2687	2395	2122	1867	1634	1427	1121	962
Energy • ft-lbs	4697	3769	2995	2350	1820	1394	1063	656	483
Taylor KO Index	37.8	33.8	30.2	26.7	23.5	20.6	18.0	14.1	12.1
Path • Inches	-1.5	+1.7	0.0	-7.8	-23.3	-48.8	-87.5	-221.2	-465.2
Wind Drift • Inches	0.0	1.0	4.2	10.0	19.0	31.6	48.6	97.6	164.9

Kynoch 235-grain Soft Nose or Solid
G1 Ballistic Coefficient = 0.340

Distance • Yards	Muzzle	100	200	300	400	500	600	800	1000
Velocity • fps	2800	2535	2284	2048	1827	1623	1440	1156	995
Energy • ft-lbs	4090	3360	2723	2189	1742	1375	1083	698	517
Taylor KO Index	35.3	32.0	28.8	25.9	23.1	20.5	18.2	14.6	12.6
Path • Inches	-1.5	+2.0	0.0	-8.7	-25.6	-53.1	-93.8	-230.0	-470.2
Wind Drift • Inches	0.0	0.9	3.6	8.6	16.2	26.9	59.0	81.2	136.5

Federal 250-grain Trophy Bonded Bear Claw (P375T4)
G1 Ballistic Coefficient = 0.339

Distance • Yards	Muzzle	100	200	300	400	500	600	800	1000
Velocity • fps	2670	2410	2170	1940	1730	1530	1359	1107	968
Energy • ft-lbs	3955	3230	2610	2090	1655	1305	1026	680	520
Taylor KO Index	35.9	32.4	29.1	26.1	23.2	20.5	18.2	14.9	13.0
Path • Inches	-1.5	+2.3	0.0	-9.6	-28.6	-59.4	-105.0	-257.5	-523.1
Wind Drift • Inches	0.0	1.0	4.4	10.3	19.4	32.2	49.0	92.6	159.4

Nosler 260-grain AccuBond (34010) + (60090)
G1 Ballistic Coefficient = 0.473

Distance • Yards	Muzzle	100	200	300	400	500	600	800	1000
Velocity • fps	2750	2560	2378	2203	2036	1876	1725	1453	1233
Energy • ft-lbs	4367	3784	3265	2802	2393	2032	1718	1219	878
Taylor KO Index	38.4	35.8	33.2	30.8	28.4	26.2	24.1	20.3	17.2
Path • Inches	-1.5	+1.9	0.0	-8.1	-23.3	-47.0	-80.5	-185.0	-355.8
Wind Drift • Inches	0.0	0.7	2.9	6.8	12.5	20.3	30.5	58.9	99.6

Nosler 260-grain Partition (34001) + (60090)
G1 Ballistic Coefficient = 0.314

Distance • Yards	Muzzle	100	200	300	400	500	600	800	1000
Velocity • fps	2750	2467	2201	1952	1721	1513	1331	1077	943
Energy • ft-lbs	4367	3514	2797	2200	1710	1321	1023	670	513
Taylor KO Index	38.4	34.5	30.7	27.3	24.0	21.1	18.6	15.0	13.2
Path • Inches	-1.5	+2.1	0.0	-9.3	-27.8	-58.2	-103.8	-159.1	-533.8
Wind Drift • Inches	0.0	1.1	4.5	10.8	20.4	34.0	52.0	102.6	169.7

Federal 260-grain Nosler AccuBond (P375A1)
G1 Ballistic Coefficient = 0.470

Distance • Yards	Muzzle	100	200	300	400	500	600	800	1000
Velocity • fps	2700	2510	2330	2160	1990	1830	1682	1415	1204
Energy • ft-lbs	4210	3640	3130	2685	2235	1935	1633	1157	837
Taylor KO Index	37.6	35.0	32.5	30.1	27.7	25.5	23.4	19.7	16.8
Path • Inches	-1.5	+2.0	0.0	-8.5	-24.5	-48.9	-84.2	-193.8	-373.3
Wind Drift • Inches	0.0	0.6	2.7	6.3	11.7	19.0	28.4	54.9	92.8

Hornady 270-grain SP-RP SPF (8503)
G1 Ballistic Coefficient = 0.380

Distance • Yards	Muzzle	100	200	300	400	500	600	800	1000
Velocity • fps	2800	2562	2336	2122	1919	1729	1555	1262	1069
Energy • ft-lbs	4699	3935	3272	2699	2208	1793	1449	955	695
Taylor KO Index	40.5	37.1	33.8	30.7	27.8	25.0	22.5	18.3	15.5
Path • Inches	-1.5	+2.0	0.0	-8.3	-24.3	-49.8	-87.0	-208.1	-416.9
Wind Drift • Inches	0.0	0.9	3.6	8.4	15.8	25.9	39.2	56.2	131.0

Federal 270-grain Soft Point (375A)
G1 Ballistic Coefficient = 0.324

Distance • Yards	Muzzle	100	200	300	400	500	600	800	1000
Velocity • fps	2690	2420	2170	1920	1700	1500	1326	1079	946
Energy • ft-lbs	4340	3510	2810	2220	1740	1351	1055	698	537
Taylor KO Index	39.0	35.1	31.5	27.8	24.7	21.8	19.2	15.6	13.7
Path • Inches	-1.5	+2.2	0.0	-9.7	-28.8	-60.2	-107.3	-267.8	-545.7
Wind Drift • Inches	0.0	1.1	4.5	10.8	20.4	33.8	51.6	101.4	167.2

Remington 270-grain Soft Point (R375M1)
G1 Ballistic Coefficient = 0.267

Distance • Yards	Muzzle	100	200	300	400	500	600	800	1000
Velocity • fps	2690	2363	2060	1780	1530	1317	1152	963	854
Energy • ft-lbs	4338	3347	2543	1900	1403	1039	796	556	438
Taylor KO Index	39.0	34.3	29.9	25.8	22.2	26.4	16.7	14.0	12.4
Path • Inches	-1.5	+2.4	0.0	-10.6	-32.3	-69.0	-125.8	-323.7	-671.3
Wind Drift • Inches	0.0	1.3	5.6	13.6	26.0	43.6	67.0	129.0	206.6

Kynoch 270-grain Soft Nose
G1 Ballistic Coefficient = 0.370

Distance • Yards	Muzzle	100	200	300	400	500	600	800	1000
Velocity • fps	2650	2415	2189	1977	1778	1594	1428	1166	1011
Energy • ft-lbs	4210	3496	2874	2344	1895	1524	1223	816	612
Taylor KO Index	38.4	35.0	31.7	28.7	25.8	23.1	20.7	16.9	14.7
Path • Inches	-1.5	+2.2	0.0	-9.5	-28.0	-57.5	-100.9	-244.4	-491.8
Wind Drift • Inches	0.0	1.0	4.0	9.5	17.7	29.2	44.2	86.6	144.6

Cor-Bon 270-grain DPX (DPX375HH270/20)
G1 Ballistic Coefficient = 0.325

Distance • Yards	Muzzle	100	200	300	400	500	600	800	1000
Velocity • fps	2625	2358	2108	1873	1656	1462	1294	1063	938
Energy • ft-lbs	4132	3336	2664	2104	1645	1281	1004	677	527
Taylor KO Index	38.0	34.1	30.5	27.1	24.0	21.1	18.7	15.4	13.6
Path • Inches	-1.5	+2.4	0.0	-10.2	-30.5	-63.4	-112.7	-278.7	-567.4
Wind Drift • Inches	0.0	1.1	4.7	11.1	21.0	34.9	53.2	103.8	169.9

Norma 270-grain Barnes TSX (19530) LF
G1 Ballistic Coefficient = 0.326

Distance • Yards	Muzzle	100	200	300	400	500	600	800	1000
Velocity • fps	2625	2539	2109	1875	1659	1464	1297	1065	939
Energy • ft-lbs	4132	3338	2668	2108	1650	1286	1009	680	529
Taylor KO Index	38.0	36.7	30.5	27.1	24.0	21.2	18.8	15.4	13.6
Path • Inches	-1.5	+2.3	0.0	-10.2	-30.4	-63.3	-112.4	-277.9	-565.6
Wind Drift • Inches	0.0	1.1	4.7	11.1	20.9	34.7	53.0	103.4	169.4

Hornady 300-grain DGS (8509)

G1 Ballistic Coefficient = 0.276

Distance • Yards	Muzzle	100	200	300	400	500	600	800	1000
Velocity • fps	2705	2376	2072	1804	1560	1356	1185	983	870
Energy • ft-lbs	4873	3760	2861	2167	1621	1222	936	644	504
Taylor KO Index	43.6	38.3	33.4	29.1	25.1	21.9	19.1	15.8	14.0
Path • Inches	-1.5	+2.7	0.0	-10.8	-32.1	-68.4	-121.6	-312.5	-637.8
Wind Drift • Inches	0.0	1.3	5.4	13.0	24.7	41.4	63.6	123.5	199.0

Hornady 300-grain DGX-SPF (82332)

G1 Ballistic Coefficient = 0.275

Distance • Yards	Muzzle	100	200	300	400	500	600	800	1000
Velocity • fps	2685	2368	2072	1800	1554	1342	1175	978	867
Energy • ft-lbs	4801	3733	2860	2157	1608	1200	920	638	501
Taylor KO Index	43.2	38.1	33.3	28.9	25.0	21.6	18.9	15.7	13.9
Path • Inches	-1.5	+2.4	0.0	-10.5	-31.9	-67.7	-122.9	-314.7	-657.5
Wind Drift • Inches	0.0	1.3	5.4	13.1	25.1	42.0	64.5	124.7	200.5

Norma 300-grain Swift A-Frame (19503)

G1 Ballistic Coefficient = 0.325

Distance • Yards	Muzzle	100	200	300	400	500	600	800	1000
Velocity • fps	2559	2296	2049	1819	1607	1418	1258	1044	926
Energy • ft-lbs	4363	3514	2798	2204	1721	1341	1055	726	571
Taylor KO Index	41.2	37.0	33.0	29.3	25.9	22.9	20.3	16.8	14.9
Path • Inches	-1.5	+2.6	0.0	-10.9	-32.3	-67.3	-119.6	-295.3	-598.2
Wind Drift • Inches	0.0	1.2	4.9	11.6	21.8	36.2	55.2	106.9	173.6

Norma 300-grain Oryx (19520)

G1 Ballistic Coefficient = 0.320

Distance • Yards	Muzzle	100	200	300	400	500	600	800	1000
Velocity • fps	2559	2292	2041	1807	1595	1405	1245	1034	919
Energy • ft-lbs	4363	3500	2775	2176	1694	1315	1033	713	562
Taylor KO Index	41.2	36.9	32.9	29.1	25.7	22.6	20.1	16.7	14.8
Path • Inches	-1.5	+2.6	0.0	-11.0	-32.6	-68.0	-121.1	-299.7	-607.9
Wind Drift • Inches	0.0	1.2	4.9	11.8	22.3	36.9	56.3	109.0	176.5

Federal 300-grain Soft Point (375B)

G1 Ballistic Coefficient = 0.32

Distance • Yards	Muzzle	100	200	300	400	500	600	800	1000
Velocity • fps	2530	2270	2020	1790	1580	1390	1233	1029	915
Energy • ft-lbs	4265	3425	2720	2135	1665	1295	1013	705	558
Taylor KO Index	40.8	36.6	32.6	28.8	25.5	22.4	19.9	16.6	14.7
Path • Inches	-1.5	+2.6	0.0	-11.1	-33.5	-69.5	-124.0	-306.5	-620.0
Wind Drift • Inches	0.0	1.2	5.0	12.0	22.6	37.4	57.0	109.9	177.5

Hornady 300-grain DGX (82332)

G1 Ballistic Coefficient = 0.27

Distance • Yards	Muzzle	100	200	300	400	500	600	800	1000
Velocity • fps	2530	2223	1938	1678	1448	1256	1113	947	846
Energy • ft-lbs	4263	3282	2503	1875	1396	1050	825	598	477
Taylor KO Index	40.8	35.8	31.2	27.0	23.3	20.2	17.9	15.3	13.6
Path • Inches	-1.5	+2.8	0.0	-12.1	-36.7	77.9	-141.2	-357.5	-728.7
Wind Drift • Inches	0.0	1.4	6.0	14.4	27.4	45.8	69.7	131.7	208.4

Winchester 300-grain Nosler Partition (S375SLSP)

G1 Ballistic Coefficient = 0.39

Distance • Yards	Muzzle	100	200	300	400	500	600	800	1000
Velocity • fps	2530	2376	2112	1919	1737	1569	1417	1174	1024
Energy • ft-lbs	4263	3572	2970	2452	2010	1640	1338	919	699
Taylor KO Index	40.8	37.3	34.0	30.9	28.0	25.3	22.8	18.9	16.5
Path • Inches	-1.5	+2.5	0.0	-10.3	-30.1	-61.5	-107.0	-253.8	-501.5
Wind Drift • Inches	0.0	1.0	4.0	9.3	17.4	28.5	43.0	83.6	138.7

A² 300-grain Triad (None)

G1 Ballistic Coefficient = 0.28

Distance • Yards	Muzzle	100	200	300	400	500	600	800	1000
Velocity • fps	2550	2251	1973	1717	1496	1302	1151	971	866
Energy • ft-lbs	4331	3375	2592	1964	1491	1130	882	629	499
Taylor KO Index	41.1	36.3	31.8	27.7	24.1	21.0	18.5	15.6	14.0
Path • Inches	-1.5	+2.7	0.0	-11.7	-35.1	-75.1	-134.8	-339.8	-682.0
Wind Drift • Inches	0.0	1.3	5.6	13.5	25.7	42.9	65.4	124.8	198.8

Federal 300-grain Nosler Partition (P375F)

G1 Ballistic Coefficient = 0.399

Distance • Yards	Muzzle	100	200	300	400	500	600	800	1000
Velocity • fps	2530	2320	2110	1920	1740	1570	1419	1176	1026
Energy • ft-lbs	4265	3570	2970	2450	2010	1640	1342	922	701
Taylor KO Index	40.8	37.4	34.0	30.9	28.0	25.3	22.9	19.0	16.5
Path • Inches	-1.5	+2.5	0.0	-10.2	-30.1	-61.6	-106.8	-253.3	-500.4
Wind Drift • Inches	0.0	0.9	3.9	9.3	17.3	28.4	42.9	83.3	138.3

Remington 300-grain Swift A-Frame PSP (RS375MA)

G1 Ballistic Coefficient = 0.350

Distance • Yards	Muzzle	100	200	300	400	500	600	800	1000
Velocity • fps	2522	2280	2051	1836	1637	1458	1302	1079	955
Energy • ft-lbs	4238	3462	2802	2245	1786	1415	1129	776	608
Taylor KO Index	40.6	36.7	33.1	29.6	26.4	23.5	21.0	17.4	15.4
Path • Inches	-1.5	+2.6	0.0	-10.9	-32.2	-66.5	-117.2	-285.0	-571.8
Wind Drift • Inches	0.0	1.1	4.6	10.8	20.4	33.6	51.1	99.0	161.7

Federal 300-grain Swift A-Frame (P375SA)

G1 Ballistic Coefficient = 0.325

Distance • Yards	Muzzle	100	200	300	400	500	600	800	1000
Velocity • fps	2450	2190	1950	1730	1530	1350	1203	1015	907
Energy • ft-lbs	4000	3205	2540	1995	1555	1215	964	686	549
Taylor KO Index	39.5	35.3	31.4	27.9	24.7	21.8	19.4	16.4	14.3
Path • Inches	-1.5	+2.9	0.0	-12.0	-35.8	-74.4	-132.3	-325.1	-652.7
Wind Drift • Inches	0.0	1.2	5.2	12.4	23.3	38.6	58.6	111.9	179.3

Nosler 300-grain Partition (34005)

G1 Ballistic Coefficient = 0.398

Distance • Yards	Muzzle	100	200	300	400	500	600	800	1000
Velocity • fps	2450	2240	2040	1851	1674	1511	1366	1140	1004
Energy • ft-lbs	4000	3342	2772	2282	1867	1522	1243	866	672
Taylor KO Index	39.5	36.1	32.9	29.8	27.0	24.3	22.0	18.4	16.2
Path • Inches	-1.5	+2.7	0.0	-11.1	-32.4	-66.2	-115.2	-273.1	-538.0
Wind Drift • Inches	0.0	1.0	4.2	9.8	18.2	29.9	45.1	87.1	143.3

Federal 300-grain Fusion Safari (F375FS1)

G1 Ballistic Coefficient = 0.351

Distance • Yards	Muzzle	100	200	300	400	500	600	800	1000
Velocity • fps	2400	2200	1970	1750	1550	1380	1237	1043	933
Energy • ft-lbs	3965	3210	2570	2040	1610	1265	1019	725	580
Taylor KO Index	38.6	35.4	31.7	28.1	24.9	22.2	19.9	16.8	15.0
Path • Inches	-1.5	+2.9	0.0	-11.9	-35.4	-73.2	-131.1	-317.4	-631.2
Wind Drift • Inches	0.0	1.2	4.9	11.6	21.9	36.0	54.5	104.2	167.8

Cor-Bon 300-grain DPX (EH375HH300DPX/10)

G1 Ballistic Coefficient = 0.330

Distance • Yards	Muzzle	100	200	300	400	500	600	800	1000
Velocity • fps	2600	2339	2093	1863	1650	1459	1294	1065	941
Energy • ft-lbs	4504	3645	2918	2312	1814	1418	1116	756	590
Taylor KO Index	41.8	37.6	33.6	29.9	26.5	23.4	20.8	17.1	15.1
Path • Inches	-1.5	+2.4	0.0	-10.4	-30.9	-64.2	-114.0	-280.7	-569.6
Wind Drift • Inches	0.0	1.1	4.7	11.1	20.9	34.7	52.8	103.0	168.4

Federal 300-grain Trophy Bonded Bear Claw (P375T3)

G1 Ballistic Coefficient = 0.342

Distance • Yards	Muzzle	100	200	300	400	500	600	800	1000
Velocity • fps	2600	2350	2110	1890	1680	1490	1326	1089	959
Energy • ft-lbs	4505	3670	2965	2305	1880	1480	1172	790	612
Taylor KO Index	41.9	37.9	34.0	30.5	27.1	24.0	21.4	17.5	15.5
Path • Inches	-1.5	+2.4	0.0	-10.1	-30.4	-62.9	-111.0	-271.5	-548.8
Wind Drift • Inches	0.0	1.1	4.5	10.6	20.0	33.1	50.4	98.3	161.7

Federal 300-grain Barnes Triple-Shock X-Bullet (P375H) LF

G1 Ballistic Coefficient = 0.358

Distance • Yards	Muzzle	100	200	300	400	500	600	800	1000
Velocity • fps	2470	2240	2010	1800	1610	1440	1291	1076	956
Energy • ft-lbs	4065	3325	2700	2170	1735	1380	1110	722	608
Taylor KO Index	39.6	36.1	32.4	29.0	25.9	23.2	20.8	17.3	15.4
Path • Inches	-1.5	+2.7	0.0	-11.3	-33.6	-69.0	-121.0	-292.8	-584.1
Wind Drift • Inches	0.0	1.1	4.6	10.9	20.5	33.7	51.1	98.6	160.5

Federal 300-grain Trophy Bonded Bear Claw (P375T1)
G1 Ballistic Coefficient = 0.343

Distance • Yards	Muzzle	100	200	300	400	500	600	800	1000
Velocity • fps	2400	2160	1930	1720	1530	1360	1219	1031	923
Energy • ft-lbs	3835	3105	2485	1975	1555	1230	990	708	567
Taylor KO Index	38.7	34.8	31.1	27.7	24.7	21.9	19.6	16.6	14.9
Path • Inches	-1.5	+3.0	0.0	-12.4	-36.6	-75.9	-133.2	-323.9	-645.1
Wind Drift • Inches	0.0	1.2	5.0	12.0	22.5	37.1	56.2	107.1	171.9

Hornady 300-grain DGS-SPF (8509)
G1 Ballistic Coefficient = 0.275

Distance • Yards	Muzzle	100	200	300	400	500	600	800	1000
Velocity • fps	2660	2344	2050	1780	1536	1328	1164	973	864
Energy • ft-lbs	4713	3660	2800	2110	1572	1174	903	631	497
Taylor KO Index	42.3	37.7	32.9	28.6	24.7	21.3	18.7	15.6	13.9
Path • Inches	-1.5	+2.4	0.0	-10.8	-32.6	-69.2	-125.6	-321.2	-663.4
Wind Drift • Inches	0.0	1.3	5.5	13.3	25.4	42.6	65.3	125.8	201.8

Norma 300-grain SOLID (19355)
G1 Ballistic Coefficient = 0.229

Distance • Yards	Muzzle	100	200	300	400	500	600	800	1000
Velocity • fps	2550	2183	1847	1550	1301	1120	1005	866	769
Energy • ft-lbs	4333	3174	2274	1600	1128	836	673	500	394
Taylor KO Index	41.0	35.1	29.7	24.9	20.9	18.0	16.2	13.9	12.4
Path • Inches	-1.5	+2.9	0.0	-13.2	-41.1	-89.6	-165.7	-427.2	-874.7
Wind Drift • Inches	0.0	1.7	7.3	17.0	34.4	57.6	86.8	159.3	247.8

Hornady 300-grain DGS (82322)
G1 Ballistic Coefficient = 0.275

Distance • Yards	Muzzle	100	200	300	400	500	600	800	1000
Velocity • fps	2530	2223	1938	1678	1448	1256	1113	947	846
Energy • ft-lbs	4263	3282	2503	1875	1396	1050	825	598	477
Taylor KO Index	40.8	35.8	31.2	27.0	23.3	20.2	17.9	15.3	13.6
Path • Inches	-1.5	+2.8	0.0	-12.1	-36.7	77.9	-141.2	-357.5	-728.7
Wind Drift • Inches	0.0	1.4	6.0	14.4	27.4	45.8	69.7	131.7	208.4

Winchester 300-grain Nosler Solid (S375SLS) LF
G1 Ballistic Coefficient = 0.198

Distance • Yards	Muzzle	100	200	300	400	500	600	800	1000
Velocity • fps	2530	2110	1734	1413	1170	1021	928	799	703
Energy • ft-lbs	4263	2965	2002	1329	911	694	573	426	329
Taylor KO Index	40.8	34.0	27.9	22.8	18.9	16.5	15.0	12.9	11.3
Path • Inches	-1.5	+3.2	0.0	-15.0	-47.5	-105.4	-196.6	-507.2	-1037
Wind Drift • Inches	0.0	2.0	8.7	21.7	42.0	69.8	103.3	184.8	284.6

Federal 300-grain Barnes Banded Solid (P375J)
G1 Ballistic Coefficient = 0.306

Distance • Yards	Muzzle	100	200	300	400	500	600	800	1000
Velocity • fps	2500	2230	1970	1730	1520	1330	1180	995	889
Energy • ft-lbs	4165	3300	2585	2000	1535	1185	927	660	526
Taylor KO Index	40.2	35.8	31.7	27.8	24.4	21.4	19.0	16.0	14.3
Path • Inches	-1.5	+2.8	0.0	-11.8	-35.3	-74.0	-132.4	-329.5	-666.7
Wind Drift • Inches	0.0	1.3	5.4	12.9	24.4	40.5	61.6	117.8	188.3

Norma 300-grain Barnes Banded Solid (19505) LF
G1 Ballistic Coefficient = 0.30?

Distance • Yards	Muzzle	100	200	300	400	500	600	800	1000
Velocity • fps	2493	2220	1964	1727	1513	1328	1179	995	889
Energy • ft-lbs	4141	3282	2569	1987	1526	1175	926	660	527
Taylor KO Index	40.2	35.8	31.6	27.8	24.4	21.4	19.0	16.0	14.3
Path • Inches	-1.5	+2.8	0.0	-11.9	-35.7	-74.3	-132.9	-330.4	-668.1
Wind Drift • Inches	0.0	1.3	5.4	12.9	24.4	40.5	61.6	117.6	188.0

Federal 300-grain Trophy Bonded Sledgehammer Solid (P375T2)
G1 Ballistic Coefficient = 0.30?

Distance • Yards	Muzzle	100	200	300	400	500	600	800	1000
Velocity • fps	2440	2120	1820	1550	1320	1150	1034	895	800
Energy • ft-lbs	3965	2980	2195	1595	1165	875	713	534	426
Taylor KO Index	39.3	34.2	29.3	25.0	21.3	18.5	16.7	14.4	12.9
Path • Inches	-1.5	+3.2	0.0	-13.8	-42.1	-90.4	-164.9	-416.9	-842.8
Wind Drift • Inches	0.0	1.6	6.9	16.7	32.0	53.3	80.2	147.4	229.3

Norma 350-grain Soft Nose (19525)

G1 Ballistic Coefficient = 0.339

Distance • Yards	Muzzle	100	200	300	400	500	600	800	1000
Velocity • fps	2300	2062	1840	1635	1450	1291	1162	999	900
Energy • ft-lbs	4112	3306	2633	2078	1635	1295	1050	775	630
Taylor KO Index	43.2	38.8	34.6	30.7	27.3	24.3	21.8	18.8	16.9
Path • Inches	-1.5	+3.4	0.0	-13.7	-40.4	-83.7	-148.0	-358.3	-707.8
Wind Drift • Inches	0.0	1.3	5.4	12.9	24.3	40.0	60.2	113.0	178.9

Norma 350-grain Full Metal Jacket (19526)

G1 Ballistic Coefficient = 0.324

Distance • Yards	Muzzle	100	200	300	400	500	600	800	1000
Velocity • fps	2300	2052	1821	1608	1419	1258	1133	977	881
Energy • ft-lbs	4112	3273	2578	2010	1565	1231	998	743	604
Taylor KO Index	43.2	38.6	34.2	30.2	26.7	23.7	21.3	18.4	16.6
Path • Inches	-1.5	+3.4	0.0	-13.9	-41.4	-86.2	-153.0	-372.6	-737.7
Wind Drift • Inches	0.0	1.4	5.7	13.6	25.7	42.3	63.6	118.8	187.1

Finn Aagaard used the .375 H&H for most of his own buffalo hunting—because he felt that he could place his shots more accurately with it than with his iron-sighted .458, and that counts for more than 200 grains of extra lead. He almost invariably carried it when guiding hunters on nondangerous game—the antelope, zebra, warthog, and such that are illogically lumped together as "plains game." (Photo from Guns and Hunting by Finn Aagaard, 2012, Safari Press Archives)

.375 Dakota

This is the .375 version of Dakota's line of proprietary cartridges. It produces 2,700 fps with a 300-grain bullet. That level of performance puts it a little ahead of the .375 H&H but also a little behind the .378 Weatherby. Dakota is currently loading four bullets in this caliber.

Relative Recoil Factor = 3.60

Specifications:

Controlling Agency for Standardization of this Ammunition: Factory

Standard barrel for velocity testing: 26 inches long

Availability:

Dakota 250-grain Swift A-Frame (375-250FAF)

G1 Ballistic Coefficient = 0.271

Distance • Yards	Muzzle	100	200	300	400	500	600	800	1000
Velocity • fps	2900	2562	2249	1958	1691	1456	1259	1015	889
Energy • ft-lbs	4670	3645	2807	2128	1588	1177	881	572	439
Taylor KO Index	38.9	34.4	30.2	26.3	22.7	19.6	16.9	13.6	11.9
Path • Inches	-1.5	+1.9	0.0	-8.8	-26.8	-57.3	-104.7	-273.6	-570.5
Wind Drift • Inches	0.0	1.2	5.0	11.9	22.7	38.2	59.1	117.2	192.4

Dakota 300-grain Woodleigh Soft Nose (375-300WSN)

G1 Ballistic Coefficient = 0.340

Distance • Yards	Muzzle	100	200	300	400	500	600	800	1000
Velocity • fps	2700	2440	2195	1965	1750	1553	1379	1118	974
Energy • ft-lbs	4837	3968	3211	2572	2040	1607	1267	833	632
Taylor KO Index	43.5	39.3	35.4	31.7	28.2	25.0	22.2	18.0	15.7
Path • Inches	-1.5	+2.2	0.0	-9.4	27.9	-57.8	-102.5	-253.2	-515.1
Wind Drift • Inches	0.0	1.0	4.3	10.1	19.1	31.6	48.1	94.6	157.3

Dakota 300-grain Swift A-Frame (375-300FAF)

G1 Ballistic Coefficient = 0.325

Distance • Yards	Muzzle	100	200	300	400	500	600	800	1000
Velocity • fps	2700	2429	2173	1934	1712	1511	1335	1084	950
Energy • ft-lbs	4857	3930	3148	2492	1953	1521	1188	784	601
Taylor KO Index	43.5	39.1	35.0	31.2	27.6	24.3	21.5	17.5	15.3
Path • Inches	-1.5	+2.2	0.0	-9.6	-28.6	-59.5	-106.1	-264.7	-539.8
Wind Drift • Inches	0.0	1.1	4.5	10.7	20.2	33.4	51.1	100.5	166.0

Dakota 300-grain Woodleigh Solid (375-300WSO)

G1 Ballistic Coefficient = 0.307

Distance • Yards	Muzzle	100	200	300	400	500	600	800	1000
Velocity • fps	2700	2413	2145	1894	1663	1456	1280	1046	921
Energy • ft-lbs	4857	3881	3064	2389	1842	1413	1092	728	585
Taylor KO Index	43.5	38.9	34.6	30.5	26.8	23.5	20.6	16.9	14.8
Path • Inches	-1.5	+2.2	0.0	-9.8	-29.5	-61.9	-111.1	-280.5	-572.9
Wind Drift • Inches	0.0	1.1	4.8	11.4	21.6	36.0	55.2	108.3	177.3

.375 Remington Ultra Magnum

Introduced in the year 2002, the .375 RUM is currently the largest in the ultra magnum series. It continues what seems to be the recent factory preference for "unbelted" magnums. Since the belt on the classic magnums like the .375 H&H is little more than a headspace feature (no significant strength is added), the unbelted design probably simplifies manufacturing without degrading performance.

Relative Recoil Factor = 3.70

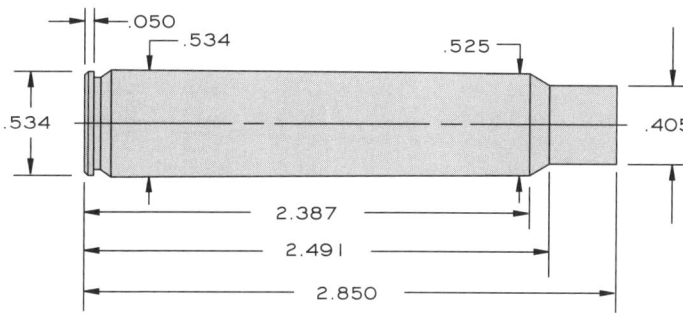

Specifications:

Controlling Agency for Standardization of this Ammunition: SAAMI*

Bullet Weight Grains	Velocity fps	Maximum Average Pressure Copper Crusher	Transducer
N/A	N/A	N/S	65,000 psi

Standard barrel for velocity testing: xxx inches long

Availability:

Cor-Bon 235-grain DPX (DPX375RUM235/20)
G1 Ballistic Coefficient = 0.300

Distance • Yards	Muzzle	100	200	300	400	500	600	800	1000
Velocity • fps	3200	2873	2569	2284	2018	1772	1548	1196	1000
Energy • ft-lbs	5345	4308	3444	2724	2126	1639	1252	746	522
Taylor KO Index	40.3	36.2	32.3	28.8	25.4	22.3	19.5	15.1	12.6
Path • Inches	-1.5	+1.4	0.0	-6.7	-20.0	-42.0	-75.0	-189.4	-401.2
Wind Drift • Inches	0.0	0.9	3.8	9.2	17.2	28.7	44.1	89.2	153.7

Nosler 260-grain AccuBond (35016)
G1 Ballistic Coefficient = 0.473

Distance • Yards	Muzzle	100	200	300	400	500	600	800	1000
Velocity • fps	2950	2751	2561	2379	2204	2037	1877	1584	1336
Energy • ft-lbs	5025	4371	3788	3268	2805	2395	2035	1450	1031
Taylor KO Index	41.2	38.4	35.8	33.2	30.8	28.4	26.2	22.1	18.7
Path • Inches	-1.5	+1.6	0.0	-6.9	-19.9	-40.0	-98.6	-157.3	-301.8
Wind Drift • Inches	0.0	0.6	2.6	6.1	11.3	18.3	27.4	52.9	89.8

Nosler 260-grain Partition (35032)
G1 Ballistic Coefficient = 0.314

Distance • Yards	Muzzle	100	200	300	400	500	600	800	1000
Velocity • fps	2950	2654	2377	2116	1873	1649	1449	1144	980
Energy • ft-lbs	5025	4068	3262	2586	2026	1571	1213	756	555
Taylor KO Index	41.2	37.1	33.2	29.6	26.2	23.0	20.2	16.0	13.7
Path • Inches	-1.5	+1.7	0.0	-7.9	-23.6	-49.3	-87.8	-219.5	-457.1
Wind Drift • Inches	0.0	1.0	4.1	9.8	18.4	30.5	46.8	93.6	158.3

Remington 270-grain Soft Point (PR375UM2)
G1 Ballistic Coefficient = 0.267

Distance • Yards	Muzzle	100	200	300	400	500	600	800	1000
Velocity • fps	2900	2558	2241	1947	1678	1442	1244	1006	882
Energy • ft-lbs	5041	3922	3010	2272	1689	1246	928	606	466
Taylor KO Index	42.1	37.1	32.5	28.2	24.3	20.9	18.0	14.6	12.8
Path • Inches	-1.5	+1.9	0.0	-8.9	-27.1	-58.0	-106.2	-278.3	-580.2
Wind Drift • Inches	0.0	1.2	5.0	12.1	23.2	39.0	60.3	119.5	195.7

Remington 300-grain Swift A-Frame PSP (PR375UM3)
G1 Ballistic Coefficient = 0.350

Distance • Yards	Muzzle	100	200	300	400	500	600	800	1000
Velocity • fps	2760	2505	2263	2035	1822	1624	1445	1165	1003
Energy • ft-lbs	5073	4178	3412	2759	2210	1757	1391	905	671
Taylor KO Index	44.5	40.4	36.5	32.8	29.4	26.2	23.3	18.8	16.2
Path • Inches	-1.5	+2.0	0.0	-8.8	-26.1	-54.0	-95.3	-233.8	-476.5
Wind Drift • Inches	0.0	1.0	4.0	9.5	17.6	29.4	44.8	88.4	148.4

Nosler 300-grain Partition (35055)

G1 Ballistic Coefficient = 0.398

Distance • Yards	Muzzle	100	200	300	400	500	600	800	1000
Velocity • fps	2750	2525	2311	2107	1914	1733	1566	1282	1087
Energy • ft-lbs	5039	4284	3559	2959	2442	2002	1633	1094	787
Taylor KO Index	44.3	40.7	37.2	34.0	30.8	27.9	25.2	20.7	17.5
Path • Inches	-1.5	+2.0	0.0	-8.5	-24.9	-50.8	-88.2	-209.1	-415.1
Wind Drift • Inches	0.0	0.8	3.5	8.2	15.3	25.1	38.0	74.3	126.0

Cor-Bon 300-grain DPX (DPX375RUM300/20)

G1 Ballistic Coefficient = 0.325

Distance • Yards	Muzzle	100	200	300	400	500	600	800	1000
Velocity • fps	2750	2475	2218	1976	1751	1546	1366	1102	961
Energy • ft-lbs	5039	4082	3279	2602	2043	1593	1243	810	615
Taylor KO Index	44.2	39.8	35.6	30.2	28.1	24.8	22.0	17.7	15.4
Path • Inches	-1.5	+2.1	0.0	-9.2	-27.3	-56.9	-101.0	-250.3	-513.6
Wind Drift • Inches	0.0	1.0	4.4	10.4	19.6	32.5	49.7	98.0	162.9

A stormy day on the Alaska Peninsula produced this big bear for Jim Clifton (left) and Finn Aagaard. Clifton took the trophy, a ten-footer, with a Whitfor in .375 H&H. (Photo from *Guns and Hunting* by Finn Aagaard, 2012, Safari Press Archives)

.375 A-Square

The .375 A² is Art Alphin's high-performance vision for this caliber. Driving a 300-grain bullet at better than 2,900 fps, the .375 A² stands between the .375 H&H and the .378 Weatherby. This is generally more rifle than is needed for North American hunting use, but easily fills the African medium-rifle requirement.

Relative Recoil Factor = 3.94

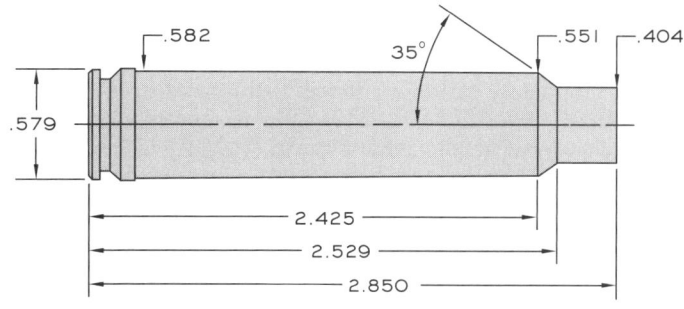

Specifications:

Controlling Agency for Standardization of this Ammunition: Factory

Availability:

A² 300-grain Triad (None)

G1 Ballistic Coefficient = 0.287

Distance • Yards	Muzzle	100	200	300	400	500	600	800	1000
Velocity • fps	2920	2598	2294	2012	1762	1531	1331	1060	922
Energy • ft-lbs	5679	4488	3505	2698	2068	1582	1180	748	567
Taylor KO Index	47.1	41.9	37.0	32.4	28.4	24.7	21.4	17.1	14.9
Path • Inches	-1.5	+1.8	0.0	-8.4	-25.5	-53.7	-97.6	-252.1	-527.0
Wind Drift • Inches	0.0	1.1	4.6	11.0	20.9	35.0	53.9	107.7	178.9

This elephant with 70-plus-pound tusks was taken with Finn's .375 after a long stalk. Facing are Finn and Kinuno, Finn's loyal tracker. (Photo from *Guns and Hunting* by Finn Aagaard, 2012, Safari Press Archives)

.375 Weatherby Magnum

This is probably the second oldest in Weatherby's line of proprietary cartridges. It was first offered in 1945, went out of production around 1965, but has recently been reintroduced. It is nothing more than a blown-out .375 H&H, but that said, its increased volume gives it a lot more punch. This cartridge has been largely superseded by the .378 Weatherby Magnum.

Relative Recoil Factor = 3.75

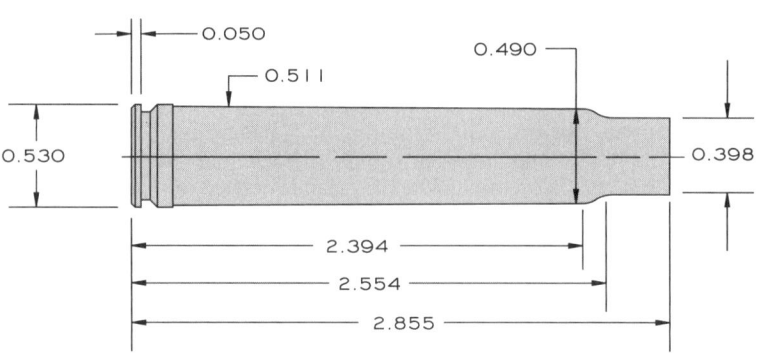

Specifications:

Controlling Agency for Standardization of this Ammunition: Factory
Typical barrel for velocity testing: 26 inches long—1 turn in 12-inch twist

Availability:

Nosler 260-grain AccuBond (36008)

G1 Ballistic Coefficient = 0.472

Distance • Yards	Muzzle	100	200	300	400	500	600	800	1000
Velocity • fps	3000	2799	2606	2422	2245	2076	1913	1616	1361
Energy • ft-lbs	5195	4521	3921	3386	2910	2489	2114	1507	1070
Taylor KO Index	41.8	39.0	36.3	33.7	31.3	28.9	26.6	22.5	19.0
Path • Inches	-1.5	+1.5	0.0	-6.6	-19.1	-38.6	-66.1	-151.5	-290.7
Wind Drift • Inches	0.0	0.6	2.6	6.0	11.0	17.9	26.8	51.7	87.6

Weatherby 300-grain Nosler Partition (N375300PT)

G1 Ballistic Coefficient = 0.398

Distance • Yards	Muzzle	100	200	300	400	500	600	800	1000
Velocity • fps	2800	2572	2366	2140	1963	1760	1588	1296	1094
Energy • ft-lbs	5224	4480	3696	3076	2541	2084	1680	1118	797
Taylor KO Index	45.0	41.3	38.0	34.4	31.5	28.3	25.5	20.8	17.6
Path • Inches	-1.5	+1.9	0.0	-8.2	-24.0	-49.0	-85.3	-202.5	-403.3
Wind Drift • Inches	0.0	0.8	3.5	8.1	15.1	24.8	37.6	73.7	125.2

A lion dead looks innocent but one alive that charges brings a whole new dimension to the game.
(Photo from *Out of Bullets* by Johnny Chilton, 2011, Safari Press Archives)

.378 Weatherby Magnum

The .378 Weatherby Magnum goes back quite a long way. Its story starts about 1953. Roy Weatherby wanted something that was significantly more potent than the .375 H&H. His .375 Weatherby Magnum was better than the H&H but not enough better to satisfy Roy. Being very well aware that in cartridge design there's no substitute for case volume for producing velocity and bullet energy, the Weatherby company developed the .378. This cartridge takes a larger action than the standard full-length magnums like the .375. It's a big boomer and kills at both ends. If you are recoil sensitive, you would be well advised to pick a different caliber. Since its introduction, the .378 Weatherby Magnum has become the basic cartridge case for the .30-378, the .338-378, and numerous wildcats.

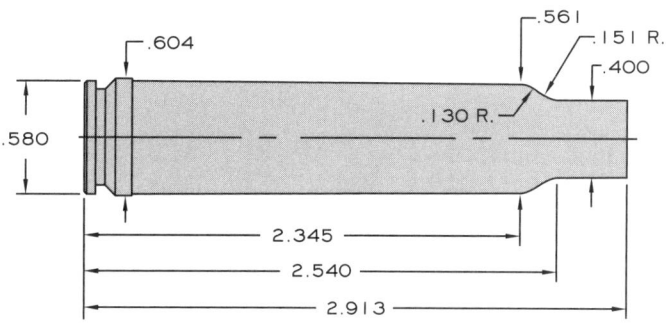

Relative Recoil Factor = 3.95

Specifications:

Controlling Agency for Standardization of this Ammunition: CIP

Bullet Weight Grains	Velocity fps	Maximum Average Pressure	
		Copper Crusher	Transducer
N/S	N/S	3,800 bar	4,370 bar

Typical barrel for velocity testing: 26 inches long—1 turn in 12-inch twist

Availability:

Nosler 260-grain AccuBond (37042)

G1 Ballistic Coefficient = 0.473

Distance • Yards	Muzzle	100	200	300	400	500	600	800	1000
Velocity • fps	3100	2894	2698	2510	2330	2157	1992	1686	1420
Energy • ft-lbs	5549	4838	4204	3639	3136	2688	2291	1641	1165
Taylor KO Index	43.3	40.4	37.7	35.1	32.5	30.1	27.8	23.5	19.8
Path • Inches	-1.5	+1.3	0.0	-6.1	-17.8	-35.8	-61.3	-140.3	-268.7
Wind Drift • Inches	0.0	0.6	2.5	5.7	10.5	17.0	25.5	49.1	83.3

Weatherby 270-grain SP (H 378 270 SP)

G1 Ballistic Coefficient = 0.380

Distance • Yards	Muzzle	100	200	300	400	500	600	800	1000
Velocity • fps	3180	2921	2677	2445	2225	2017	1819	1470	1201
Energy • ft-lbs	6062	5115	4295	3583	2968	2438	1985	1297	865
Taylor KO Index	46.4	42.6	39.0	35.6	32.4	29.4	26.4	21.3	17.4
Path • Inches	-1.5	+1.3	0.0	-6.1	-18.1	-37.1	-64.8	-154.9	-312.7
Wind Drift • Inches	0.0	0.7	3.0	7.0	13.1	21.4	32.4	63.8	110.4

Weatherby 270-grain Barnes TSX (B 378 270 TSX) LF

G1 Ballistic Coefficient = 0.502

Distance • Yards	Muzzle	100	200	300	400	500	600	800	1000
Velocity • fps	3060	2868	2684	2507	2337	2174	2017	1725	1468
Energy • ft-lbs	5615	4932	4319	3770	3276	2834	2440	1785	1292
Taylor KO Index	44.4	41.6	38.9	36.4	33.9	31.5	29.3	25.0	21.3
Path • Inches	-1.5	+1.4	0.0	-6.2	-17.9	-36.0	-61.5	-139.6	-264.8
Wind Drift • Inches	0.0	0.6	2.4	5.4	10.1	16.2	24.1	46.2	78.2

Weatherby 300-grain Round Nose-Expanding (H 378 300 RN)

G1 Ballistic Coefficient = 0.250

Distance • Yards	Muzzle	100	200	300	400	500	600	800	1000
Velocity • fps	2925	2558	2220	1908	1627	1383	1189	971	852
Energy • ft-lbs	5699	4360	3283	2424	1764	1274	942	628	484
Taylor KO Index	47.4	41.4	36.0	30.9	26.4	22.4	19.2	15.6	13.7
Path • Inches	-1.5	+1.9	0.0	-9.0	-27.8	-60.0	-111.3	-295.0	-615.4
Wind Drift • Inches	0.0	1.3	5.4	13.0	24.9	42.2	65.4	128.8	209.2

A² 300-grain Triad (None)

G1 Ballistic Coefficient = 0.287

Distance • Yards	Muzzle	100	200	300	400	500	600	800	1000
Velocity • fps	2900	2577	2276	1997	1747	1518	1320	1054	919
Energy • ft-lbs	5602	4424	3452	2656	2034	1535	1161	740	563
Taylor KO Index	47.0	41.7	36.9	32.4	28.3	24.6	21.3	17.0	14.8
Path • Inches	-1.5	+1.9	0.0	-8.6	-25.9	-54.6	-99.2	-256.1	-534.4
Wind Drift • Inches	0.0	1.1	4.6	11.1	21.1	35.4	54.5	108.6	180.0

Weatherby 300-grain Full Metal Jacket (H 378 300 FJ)

G1 Ballistic Coefficient = 0.275

Distance • Yards	Muzzle	100	200	300	400	500	600	800	1000
Velocity • fps	2925	2591	2280	1991	1725	1489	1287	1031	901
Energy • ft-lbs	5699	4470	3461	2640	1983	1476	1104	708	540
Taylor KO Index	47.4	42.0	36.9	32.3	27.9	24.1	20.7	16.6	14.5
Path • Inches	-1.5	+1.8	0.0	-8.6	-26.1	-55.4	-101.1	-263.7	-551.7
Wind Drift • Inches	0.0	1.1	4.8	11.5	21.9	36.8	57.1	113.8	187.8

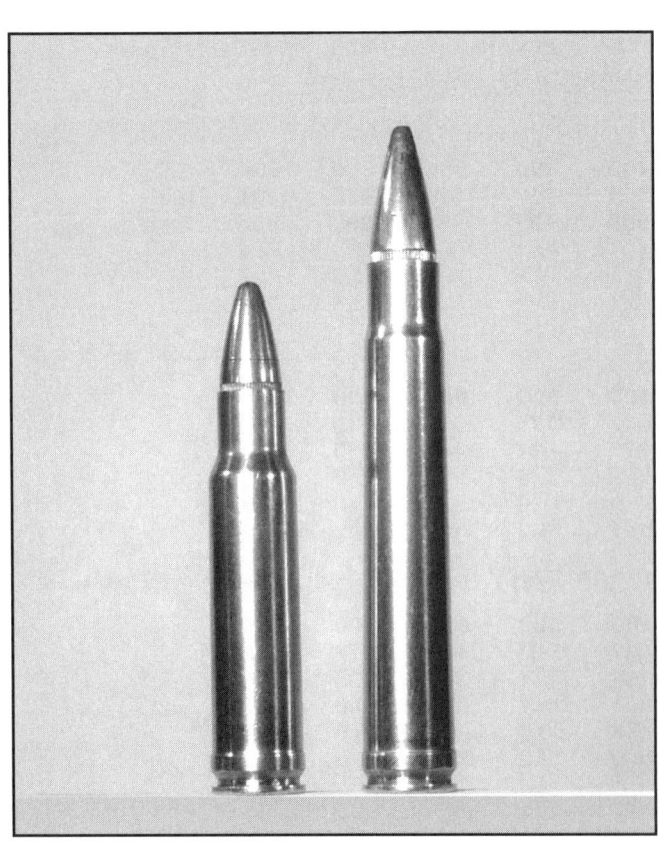

(Left to right) .350 Remington Magnum, .375 Weatherby Magnum.

.38-40 Winchester

Perhaps the first thing you should know about the .38-40 is that it isn't. It isn't a .38-caliber round, that is. The .38-40 uses bullets 0.4005 inches in diameter and is really a .40 caliber. I suppose that doesn't matter too much because the .38 Special isn't a .38 either. The .38-40 goes back to the beginning of centerfire cartridges. It was introduced in 1874 when the idea that using the same cartridge in both rifles and pistols had lots of support.

Relative Recoil Factor = 0.95

Specifications:

Controlling Agency for Standardization of this Ammunition: SAAMI

Bullet Weight Grains	Velocity fps	Maximum Average Pressure	
		Copper Crusher	Transducer
180	1,150	14,000 cup	N/S

Standard barrel for velocity testing: 24 inches long—1 turn in 36-inch twist

Availability:

Winchester 180-grain Soft Point (X3840)
G1 Ballistic Coefficient = 0.173

Distance • Yards	Muzzle	100	200	300	400	500	600	800	1000
Velocity • fps	1160	999	901	827	764	710	660	572	494
Energy • ft-lbs	538	399	324	273	233	201	174	131	98
Taylor KO Index	11.9	10.3	9.3	8.5	7.9	7.3	6.8	5.9	5.1
Path • Inches	-1.5	+16.6	0.0	-59.4	-169.8	-339.6	-566.4	-1277	-2405
Wind Drift • Inches	0.0	3.9	14.2	29.9	50.8	77.0	108.6	189.7	297.5

Cowboy Action Ammunition:

Black Hills 180-grain FPL (DCB3840N1)
G1 Ballistic Coefficient = 0.175

Distance • Yards	Muzzle	100	200	300	400	500	600	800	1000
Velocity • fps	800	742	690	643	599	559	522	457	403
Energy • ft-lbs	256	220	190	165	144	125	109	83	65
Taylor KO Index	7.7	7.2	6.7	6.2	5.8	5.4	5.0	4.4	3.9
Path • Inches	-1.5	+30.9	0.0	-104.0	-292.4	-578.0	-975.9	-2181	-4088
Wind Drift • Inches	0.0	2.6	10.4	23.8	43.0	68.5	100.8	188.4	311.6

...is leopard was shot in Masailand. (From left) Tracker Nyamaiya, PH ...y Carr-Hartley, and Johnny Chilton. (Photo from Out of Bullets ... Johnny Chilton, 2011, Safari Press Archives)

.38-55 Winchester

This is another example of an old, old cartridge, originally designed for use with black powder. Unlike the .38-40, this was a rifle cartridge from its introduction. The cases (not complete rounds) have been used for many years in Schutzen rifle competition where they have established a fine record for accuracy. The cartridge is nearing the end of its useful life. After all, it has only been around since about 1884.

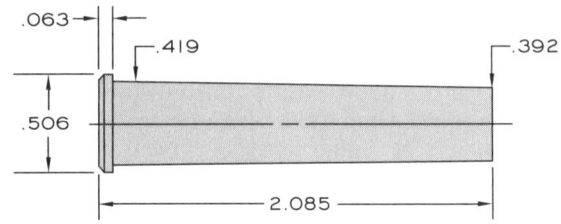

Relative Recoil Factor = 1.51

Specifications:

Controlling Agency for Standardization of this Ammunition: SAAMI

Bullet Weight Grains	Velocity fps	Maximum Average Pressure Copper Crusher	Transducer
2,550	1,320	30,000 cup	N/S

Standard barrel for velocity testing: 24 inches long—1 turn in 18-inch twist

Availability:

Winchester 255-grain Soft Point (X3855)
G1 Ballistic Coefficient = 0.355

Distance • Yards	Muzzle	100	200	300	400	500	600	800	1000
Velocity • fps	1320	1190	1091	1018	963	917	877	809	751
Energy • ft-lbs	987	802	674	587	525	476	436	371	319
Taylor KO Index	18.1	16.3	15.0	14.0	13.2	12.6	12.0	11.1	10.3
Path • Inches	-1.5	+11.5	0.0	-40.7	-114.8	-226.5	-374.8	-809.7	-1448
Wind Drift • Inches	0.0	2.2	8.6	18.8	32.2	48.4	67.3	112.7	168.3

Cowboy Action Ammunition:

Black Hills 255-grain FNL (2CCB3855N1)
G1 Ballistic Coefficient = 0.35

Distance • Yards	Muzzle	100	200	300	400	500	600	800	1000
Velocity • fps	1250	1136	1052	989	939	897	860	796	741
Energy • ft-lbs	885	731	627	554	500	456	419	359	311
Taylor KO Index	17.2	15.6	14.4	13.6	12.9	12.3	11.8	10.9	10.2
Path • Inches	-1.5	+12.7	0.0	-44.1	-123.6	-242.6	-404.9	-874.7	-1566
Wind Drift • Inches	0.0	2.2	8.3	17.9	30.5	45.8	63.8	107.4	161.1

.405 Winchester

The .405 Winchester is an oldie that suddenly was revived. It was introduced in 1904 for Winchester's 1895 rifle. Despite its age, this cartridge was never a black-powder number. At the time of its introduction, it was the most powerful lever-action cartridge made. Hornady brought it back to production in 2002. Potential users are advised to consult the factory before firing this ammo in any guns except brand-new models that have been factory chambered specifically for this cartridge.

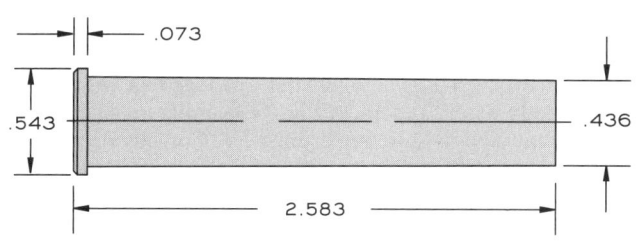

Relative Recoil Factor = 2.75

Specifications:

Controlling Agency for Standardization of this Ammunition: SAAMI

Bullet Weight Grains	Velocity fps	Maximum Average Pressure	
		Copper Crusher	Transducer
300	2,200	N/S	46,000 psi

Standard barrel for velocity testing: N/A—1 turn in 14-inch twist

Availability:

Hornady 300-grain FP InterLock (8240)

G1 Ballistic Coefficient = 0.225

Distance • Yards	Muzzle	100	200	300	400	500	600	800	1000
Velocity • fps	2200	1857	1553	1300	1117	1002	923	810	722
Energy • ft-lbs	3224	2297	1607	1126	831	669	568	437	347
Taylor KO Index	38.8	32.7	27.4	22.9	19.7	17.6	16.3	14.3	12.7
Path • Inches	-1.5	+4.4	0.0	-18.9	-58.5	-125.8	-227.8	-560.3	-1106
Wind Drift • Inches	0.0	2.3	9.2	22.5	42.5	68.6	99.6	174.2	265.0

Hornady 300-grain SP InterLock (8241)

G1 Ballistic Coefficient = 0.250

Distance • Yards	Muzzle	100	200	300	400	500	600	800	1000
Velocity • fps	2200	1890	1610	1370	1181	1053	967	851	764
Energy • ft-lbs	3224	2379	1727	1250	929	738	624	482	389
Taylor KO Index	38.8	33.3	28.4	24.1	20.8	18.5	17.0	15.0	13.5
Path • Inches	-1.5	+4.2	0.0	-17.7	-54.0	-115.2	-207.8	-509.8	-1005
Wind Drift • Inches	0.0	1.9	8.2	19.8	37.4	61.0	89.4	158.3	241.8

.44-40 Winchester

The .44-40 was developed for Winchester's Model 1873 rifle. It is yet another example of an early centerfire black-powder cartridge that has been used in both rifles and pistols. While it is greatly outperformed by the .44 Remington Magnum, this caliber is making a comeback in Cowboy Action Shooting events. The Cowboy Action ammunition is loaded to a very much milder standard than the current SAAMI specification.

Relative Recoil Factor = 1.07

Specifications:

Controlling Agency for Standardization of this Ammunition: SAAMI

Bullet Weight Grains	Velocity fps	Maximum Average Pressure	
		Copper Crusher	Transducer
200	1,175	13,000 cup	N/S

Standard barrel for velocity testing: 24 inches long—1 turn in 36-inch twist

Availability:

Winchester 200-grain Soft Point (X4440)

G1 Ballistic Coefficient = 0.16▶

Distance • Yards	Muzzle	100	200	300	400	500	600	800	1000
Velocity • fps	1190	1006	900	822	756	699	645	553	472
Energy • ft-lbs	629	449	360	300	254	217	185	136	99
Taylor KO Index	14.5	12.3	11.0	10.0	9.2	8.5	7.9	6.7	5.8
Path • Inches	-1.5	+16.4	0.0	-59.3	-170.3	-342.3	-571.4	-1301	-2478
Wind Drift • Inches	0.0	4.3	15.6	32.8	55.5	84.0	118.3	206.6	324.7

Cowboy Action Ammunition:

Black Hills 200-grain RNFP (DCB4440N1) (24-inch barrel)

G1 Ballistic Coefficient = 0.16◀

Distance • Yards	Muzzle	100	200	300	400	500	600	800	1000
Velocity • fps	800	737	681	630	584	541	502	473	378
Energy • ft-lbs	284	241	206	176	151	130	112	83	63
Taylor KO Index	9.8	9.0	8.3	7.7	7.1	6.6	6.1	5.3	4.6
Path • Inches	-1.5	+31.3	0.0	-106.5	-300.8	-597.7	-1015	-2296	-4364
Wind Drift • Inches	0.0	2.8	11.4	26.1	47.3	75.6	111.7	210.2	350.6

.44 Remington Magnum (Rifle Data)

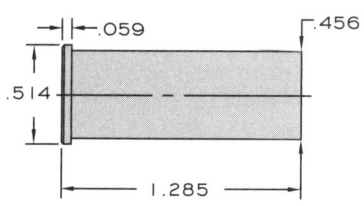

The .44 Remington Magnum defined the high-powered pistol from its introduction in 1956 until the .454 Casull came into being a few years ago. Both have now been overtaken by some even bigger boomers. As a pistol, there are few shooters who can get more than a few shots out of the .44 Remington Magnum without beginning to flinch or concentrating on not flinching. Pistol shooting is one thing, and rifle shooting is another. In carbine-style guns, the .44 Remington Magnum becomes a good choice for deer-size game. The comparison with the .30-30 is interesting. On an energy basis, the .30-30 is a little better. But when Taylor KO indices are compared, the .44 Remington Magnum is the clear choice. It comes down to whether you believe energy (lighter bullet, higher velocity) is "everything" or if you like heavy bullets at modest velocity. It becomes the "shooter's choice." All the cartridges listed here can be used in pistols and any pistol cartridge in this caliber can be used in properly chambered modern rifles in good condition. See the pistol listing for more information.

Relative Recoil Factor = 1.90

Specifications:

Controlling Agency for Standardization of this Ammunition: SAAMI

See pistol data section for detailed specifications.
All data in this section taken in 20-inch barrels.

Availability:

Remington 240-grain Soft Point (R44MG2)
G1 Ballistic Coefficient = 0.167

Distance • Yards	Muzzle	100	200	300	400	500	600	800	1000
Velocity • fps	1760	1380	1114	970	878	806	744	640	551
Energy • ft-lbs	1650	1015	661	501	411	346	295	218	162
Taylor KO Index	25.9	20.3	16.4	14.3	12.9	11.9	11.0	9.5	8.2
Path • Inches	-1.5	+8.5	0.0	-36.3	-109.5	-227.8	-382.6	-889.8	-1714
Wind Drift • Inches	0.0	4.0	16.8	37.9	65.2	98.1	136.3	229.5	347.5

Federal 240-grain Hollow Point (C44A)
G1 Ballistic Coefficient = 0.169

Distance • Yards	Muzzle	100	200	300	400	500	600	800	1000
Velocity • fps	1760	1390	1120	980	890	810	749	646	561
Energy • ft-lbs	1650	1025	670	510	415	350	299	223	168
Taylor KO Index	26.3	20.0	16.1	14.1	12.8	11.7	10.8	9.3	8.1
Path • Inches	-1.5	+8.4	0.0	-35.6	-106.0	-217.7	-396.4	-934.2	-1810
Wind Drift • Inches	0.0	3.9	16.6	37.4	64.5	97.1	135.0	227.5	344.4

Remington 240-grain Semi-Jacketed Hollow Point (R44MG3)
G1 Ballistic Coefficient = 0.167

Distance • Yards	Muzzle	100	200	300	400	500	600	800	1000
Velocity • fps	1760	1380	1114	970	878	806	744	640	551
Energy • ft-lbs	1650	1015	661	501	411	346	295	218	162
Taylor KO Index	25.9	20.3	16.4	14.3	12.9	11.9	11.0	9.5	8.2
Path • Inches	-1.5	+8.5	0.0	-36.3	-109.5	-227.8	-382.6	-889.8	-1714
Wind Drift • Inches	0.0	4.0	16.8	37.9	65.2	98.1	136.3	229.5	347.5

Winchester 240-grain Hollow Soft Point (X44MHSP2)
G1 Ballistic Coefficient = 0.159

Distance • Yards	Muzzle	100	200	300	400	500	600	800	1000
Velocity • fps	1760	1362	1094	953	861	789	727	621	531
Energy • ft-lbs	1651	988	638	484	395	332	282	205	150
Taylor KO Index	26.1	20.4	16.1	14.1	12.7	11.7	10.8	9.2	7.9
Path • Inches	-1.5	+8.7	0.0	-37.1	-110.4	-227.0	-394.8	-922.4	-1788
Wind Drift • Inches	0.0	4.2	17.7	39.6	68.0	102.0	141.8	239.2	363.3

Winchester 250-grain Platinum Tip (S44PTHP)

G1 Ballistic Coefficient = 0.188

Distance • Yards	Muzzle	100	200	300	400	500	600	800	1000
Velocity • fps	1830	1475	1201	1032	931	857	796	695	611
Energy • ft-lbs	1859	1208	801	591	481	408	352	268	208
Taylor KO Index	26.9	21.7	17.6	15.1	13.7	12.6	11.7	10.2	9.0
Path • Inches	-1.5	+7.3	0.0	-31.4	-95.2	-199.1	-350.2	-825.8	-1592
Wind Drift • Inches	0.0	3.3	14.3	33.1	58.2	88.5	123.6	208.4	313.8

Winchester 250-grain Partition Gold (S44MP)

G1 Ballistic Coefficient = 0.188

Distance • Yards	Muzzle	100	200	300	400	500	600	800	1000
Velocity • fps	1810	1455	1188	1025	926	853	794	693	609
Energy • ft-lbs	1818	1175	738	583	476	404	350	267	206
Taylor KO Index	26.6	21.4	17.4	15.0	13.6	12.5	11.7	10.2	8.9
Path • Inches	-1.5	+7.5	0.0	-32.0	-96.9	-202.2	-355.1	-835.6	-1609
Wind Drift • Inches	0.0	3.3	14.4	33.3	58.4	88.6	123.7	208.3	313.6

The big bears of North America call for big guns. (Photo courtesy of Dennis Harms, from *Ask the Grizzly/Brown Bear Guides,* by J.Y. Jones, 20 Safari Press Archives)

.444 Marlin

In 1964 Marlin reopened its production line for its Model 1895 lever action. To make the new offering more attractive, Marlin chambered the gun for a new cartridge that had been developed by Remington, the .444. The .444 and the .45-70 are near enough in physical size to make you wonder, "Why bother?" The answer is that the .444 and the new Model 1895 were designed for a pressure level of 42,000 psi compared to the .45-70's 28,000 psi limit. While the .444 makes an excellent brush gun, the caliber never caught on with the American hunter.

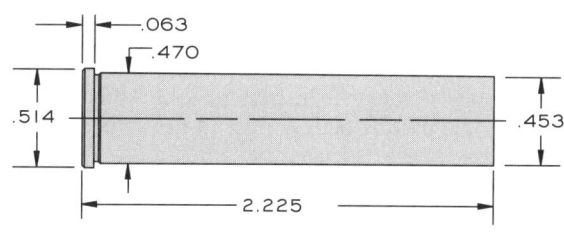

Relative Recoil Factor = 2.54

Specifications:

Controlling Agency for Standardization of this Ammunition: SAAMI

Bullet Weight Grains	Velocity fps	Maximum Average Pressure	
		Copper Crusher	Transducer
240	2,320	44,000 cup	42,000 psi
265	2,100	44,000 cup	42,000 psi

Standard barrel for velocity testing: 24 inches long—1 turn in 38-inch twist

Availability:

Remington 240-grain Soft Point (R444M)
G1 Ballistic Coefficient = 0.146

Distance • Yards	Muzzle	100	200	300	400	500	600	800	1000
Velocity • fps	2350	1815	1377	1087	941	846	769	646	544
Energy • ft-lbs	2942	1755	1010	630	472	381	315	222	158
Taylor KO Index	34.6	26.8	20.3	16.0	13.9	12.5	11.4	9.5	8.0
Path • Inches	-1.5	+4.7	0.0	-23.4	-76.0	-168.1	-297.5	-730.8	-1469
Wind Drift • Inches	0.0	3.1	14.2	35.4	65.5	102.5	145.6	250.8	384.2

Hornady 265-grain FTX (82744)
G1 Ballistic Coefficient = 0.225

Distance • Yards	Muzzle	100	200	300	400	500	600	800	1000
Velocity • fps	2325	1971	1652	1380	1171	1035	948	827	736
Energy • ft-lbs	3180	2285	1606	1120	807	631	529	402	319
Taylor KO Index	37.8	32.1	26.9	22.5	19.1	16.8	15.4	13.5	12.0
Path • Inches	-1.5	+3.8	0.0	-16.6	-51.7	-112.0	-204.8	-512.5	-1025
Wind Drift • Inches	0.0	2.0	8.5	20.8	39.8	65.3	90.6	170.4	261.0

Cor-Bon 280-grain BondedCore SP (HT444M280BC/20)
G1 Ballistic Coefficient = 0.250

Distance • Yards	Muzzle	100	200	300	400	500	600	800	1000
Velocity • fps	2200	1890	1610	1369	1180	1051	966	850	763
Energy • ft-lbs	3009	2200	1612	1165	865	687	581	449	362
Taylor KO Index	37.9	32.6	27.8	23.6	20.3	18.1	16.7	14.7	13.2
Path • Inches	-1.5	+4.2	0.0	-17.7	-54.0	-115.4	-208.1	-510.8	-1007
Wind Drift • Inches	0.0	1.9	8.2	19.8	37.5	61.1	89.1	158.6	242.3

Cor-Bon 305-grain FP Penetrator (HT444M305FPN/20)
G1 Ballistic Coefficient = 0.250

Distance • Yards	Muzzle	100	200	300	400	500	600	800	1000
Velocity • fps	2100	1799	1530	1303	1133	1021	945	835	751
Energy • ft-lbs	2988	2191	1585	1151	870	707	605	472	382
Taylor KO Index	39.4	33.8	28.7	24.5	21.3	19.2	17.7	15.7	14.1
Path • Inches	-1.5	+4.7	0.0	-19.6	-59.8	-127.1	-227.6	-551.0	-1075
Wind Drift • Inches	0.0	2.0	8.8	21.1	39.5	63.6	92.4	161.5	245.1

.45-70 Government

The .45-70 Government was originally manufactured for the Model 1873 Springfield rifle. In 1866 Erskine Allin developed a way to convert 1865 Springfield muzzleloading rifles to fire a .58 caliber, rimfire cartridge. In 1873, an Allin-style rifle won a competition for a new breechloading rifle. Those rifles became known as "Trapdoors" because of their breeching mechanism. Today's factory loadings are held to very mild levels because of these old guns, but handloaders have been known to push pressures up to as high as 50,000 psi in modern guns.

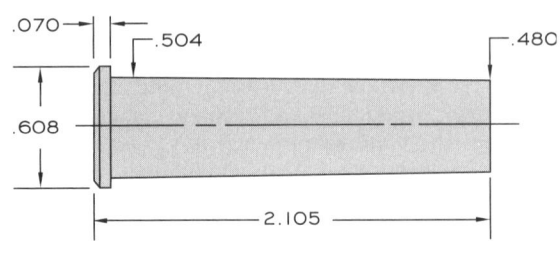

Relative Recoil Factor = 2.43

Specifications:

Controlling Agency for Standardization of this Ammunition: SAAMI

Bullet Weight Grains	Velocity fps	Maximum Average Pressure	
		Copper Crusher	Transducer
300	1,830	28,000 cup	28,000 psi
405	1,320	28,000 cup	28,000 psi

Standard barrel for velocity testing: 24 inches long—1 turn in 20-inch twist

Availability:

Federal 300-grain Hollow Point (4570AS)
G1 Ballistic Coefficient = 0.29

Distance • Yards	Muzzle	100	200	300	400	500	600	800	1000
Velocity • fps	1850	1610	1400	1230	1100	1010	945	848	773
Energy • ft-lbs	2280	1730	1305	1000	800	680	595	480	398
Taylor KO Index	36.3	31.6	27.5	24.1	21.6	19.8	18.5	16.6	15.2
Path • Inches	-1.5	+5.9	0.0	-23.9	-71.0	-146.8	-256.3	-597.5	-1133
Wind Drift • Inches	0.0	2.1	8.7	20.6	37.6	59.4	85.0	146.2	220.1

Winchester 300-grain Jacketed Hollow Point (X4570H)
G1 Ballistic Coefficient = 0.28

Distance • Yards	Muzzle	100	200	300	400	500	600	800	1000
Velocity • fps	1880	1650	1425	1235	1105	1010	948	849	772
Energy • ft-lbs	2355	1815	1355	1015	810	680	599	480	397
Taylor KO Index	36.9	32.4	28.0	24.2	21.7	19.8	18.6	16.7	15.2
Path • Inches	-1.5	+5.8	0.0	-23.3	-69.4	-144.1	-249.8	-575.7	-1078
Wind Drift • Inches	0.0	2.1	8.8	20.7	38.0	60.1	85.2	147.0	221.5

Winchester 300-grain Partition Gold (SPGX4570)
G1 Ballistic Coefficient = 0.2

Distance • Yards	Muzzle	100	200	300	400	500	600	800	1000
Velocity • fps	1880	1558	1292	1103	988	910	849	748	665
Energy • ft-lbs	2355	1616	1112	811	651	551	480	373	294
Taylor KO Index	36.9	30.6	25.4	21.7	19.4	17.9	16.7	14.7	13.1
Path • Inches	-1.5	+6.5	0.0	-27.3	-83.0	-174.2	-300.6	-695.7	-1318
Wind Drift • Inches	0.0	2.8	12.0	28.2	50.8	78.4	110.5	187.1	280.8

Remington 300-grain Semi-Jacketed Hollow Point (R4570L)
G1 Ballistic Coefficient = 0.2

Distance • Yards	Muzzle	100	200	300	400	500	600	800	1000
Velocity • fps	1810	1497	1244	1073	969	895	834	735	653
Energy • ft-lbs	2182	1492	1031	767	625	533	463	360	284
Taylor KO Index	35.5	29.4	24.4	21.1	19.0	17.6	16.4	14.4	12.8
Path • Inches	-1.5	+7.1	0.0	-29.6	-89.3	-186.0	-318.2	-731.7	-1381
Wind Drift • Inches	0.0	2.9	12.6	29.3	52.1	79.8	111.8	-188.6	-282.7

Hornady 325-grain FTX (13063)

G1 Ballistic Coefficient = 0.230

Distance • Yards	Muzzle	100	200	300	400	500	600	800	1000
Velocity • fps	2050	1729	1450	1225	1072	974	847	798	714
Energy • ft-lbs	3032	2158	1516	1083	829	685	590	459	368
Taylor KO Index	43.6	36.8	30.8	26.0	22.8	20.7	18.0	17.0	15.2
Path • Inches	-1.5	+5.1	0.0	-21.8	-66.9	-142.0	-253.8	-611.2	-1188
Wind Drift • Inches	0.0	2.3	9.9	23.9	44.4	70.5	101.1	174.4	263.5

Winchester 375-grain Dual Bond (S4570B)

G1 Ballistic Coefficient = 0.267

Distance • Yards	Muzzle	100	200	300	400	500	600	800	1000
Velocity • fps	1500	1292	1135	1029	955	897	848	767	699
Energy • ft-lbs	1873	1391	1072	881	759	670	599	490	406
Taylor KO Index	36.8	31.7	27.8	25.2	23.4	22.8	20.8	18.8	17.2
Path • Inches	-1.5	+9.7	0.0	-36.6	-106.1	-213.7	-364.5	-816.9	-1508
Wind Drift • Inches	0.0	2.8	11.3	25.1	43.3	65.3	90.7	151.7	226.4

Remington 405-grain Soft Point (R4570G)

G1 Ballistic Coefficient = 0.281

Distance • Yards	Muzzle	100	200	300	400	500	600	800	1000
Velocity • fps	1330	1168	1055	977	918	869	824	749	685
Energy • ft-lbs	1590	1227	1001	858	758	679	611	505	421
Taylor KO Index	35.2	30.9	28.0	25.9	24.3	23.0	21.8	19.8	18.2
Path • Inches	-1.5	+12.0	0.0	-43.1	-122.7	-243.5	-402.8	-880.3	-1594
Wind Drift • Inches	0.0	2.8	10.8	23.2	39.3	58.8	81.5	136.6	204.7

Cor-Bon 405-grain FP Penetrator (HT4570405FPN/20)

G1 Ballistic Coefficient = 0.280

Distance • Yards	Muzzle	100	200	300	400	500	600	800	1000
Velocity • fps	1650	1427	1242	1105	1012	945	891	804	734
Energy • ft-lbs	2449	1832	1389	1100	923	804	716	584	486
Taylor KO Index	43.7	37.8	32.9	29.3	26.8	25.1	23.6	21.3	19.5
Path • Inches	-1.5	+7.8	0.0	-30.4	-89.1	-181.7	-313.2	-712.4	-1327
Wind Drift • Inches	0.0	2.4	10.2	23.4	41.4	63.4	89.0	150.0	223.9

Cor-Bon 460-grain HC (HT4570460HC/20)

G1 Ballistic Coefficient = 0.300

Distance • Yards	Muzzle	100	200	300	400	500	600	800	1000
Velocity • fps	1650	1441	1264	1128	1033	965	911	826	756
Energy • ft-lbs	2782	2120	1632	1301	1091	952	848	697	585
Taylor KO Index	49.7	43.4	38.0	33.9	31.1	29.0	27.4	24.9	22.8
Path • Inches	-1.5	+7.7	0.0	-29.5	-86.2	-175.4	-302.0	-695.4	-1274
Wind Drift • Inches	0.0	2.3	9.5	21.8	38.8	59.8	84.2	142.3	212.5

Cowboy Action Ammunition:

PMC 405-grain LFP (45-70CA)

G1 Ballistic Coefficient = 0.301

Distance • Yards	Muzzle	100	200	300	400	500	600	800	1000
Velocity • fps	1350	1193	1078	999	938	889	846	772	709
Energy • ft-lbs	1639	1280	1046	897	792	711	643	536	452
Taylor KO Index	35.8	31.6	28.6	26.5	24.9	23.6	22.4	20.5	18.8
Path • Inches	-1.5	+11.5	0.0	-41.1	-116.5	-230.1	-386.1	-842.0	-1520
Wind Drift • Inches	0.0	2.6	10.2	22.0	37.5	56.2	78.1	130.6	195.3

Black Hills 405-grain FPL (2CCB4570N1)

G1 Ballistic Coefficient = 0.280

Distance • Yards	Muzzle	100	200	300	400	500	600	800	1000
Velocity • fps	1250	1111	1016	948	894	848	807	736	675
Energy • ft-lbs	1406	1110	926	809	719	646	586	487	409
Taylor KO Index	33.1	29.4	26.9	25.1	23.7	22.5	21.4	19.5	17.9
Path • Inches	-1.5	+13.3	0.0	-46.9	-132.5	-261.6	-439.1	-960.3	-1741
Wind Drift • Inches	0.0	2.7	10.3	21.9	37.1	55.6	77.3	130.3	196.7

.450 Marlin

Marlin and Hornady teamed to introduce this cartridge in 2001. Marlin wanted a high-performance cartridge to pick up in modern guns where the ancient .45-70 left off. While the cartridge volume is about the same as the .45-70, the working pressure is nearly double, producing a significant improvement in performance. Hornady's designers put an extra-wide belt on this cartridge, thereby reducing the number of possible wrong chambers that can be found. At the present time, the only loading uses flatpoint bullets for use in tubular magazine guns.

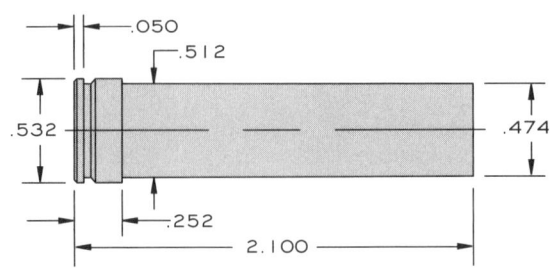

Relative Recoil Factor = 3.27

Specifications:

Controlling Agency for Standardization of this Ammunition: SAAMI

Bullet Weight Grains	Velocity fps	Maximum Average Pressure Copper Crusher	Transducer
300	N/A	N/A	44,000 psi

Standard barrel for velocity testing: 18.5 inches long—1 turn in 20-inch twist (preliminary)

Availability:

Hornady 325-grain FTX (82750)

G1 Ballistic Coefficient = 0.23

Distance • Yards	Muzzle	100	200	300	400	500	600	800	1000
Velocity • fps	2225	1887	1585	1331	1139	1017	936	821	732
Energy • ft-lbs	3572	2569	1813	1278	936	747	632	486	387
Taylor KO Index	47.3	40.1	33.7	28.3	24.2	21.6	19.9	17.5	15.6
Path • Inches	-1.5	+4.2	0.0	-18.2	-56.1	-120.8	-219.0	-540.6	-1070
Wind Drift • Inches	0.0	2.0	8.9	21.6	40.9	66.5	97.0	170.5	259.8

Hornady 350-grain FP InterLock (8250)

G1 Ballistic Coefficient = 0.19

Distance • Yards	Muzzle	100	200	300	400	500	600	800	1000
Velocity • fps	2100	1720	1397	1156	1011	920	850	741	653
Energy • ft-lbs	3427	2298	1516	1039	795	658	562	427	332
Taylor KO Index	48.1	39.4	32.0	26.5	23.2	21.1	19.5	17.0	15.0
Path • Inches	-1.5	+5.2	0.0	-23.2	-72.4	-155.5	-279.7	-680.0	-1336
Wind Drift • Inches	0.0	2.7	11.6	28.2	52.2	82.0	116.6	199.9	302.2

Donna Boddington took her first buffalo with the Ruger No. 1 .450/.400-3". (Photo from *Safari Rifles II* by Craig Boddington 2009, Safari Press Archives)

.50 BMG (Browning Machine Gun)

The .50 BMG's history dates from about the beginning of WWI. It was, as the name implies, designed for John Browning's heavy machine gun. In addition to it being mounted on a tripod, the gun found applications in ground vehicles, and during most of WWII it was the primary machine gun on aircraft. Sometime in the 1980s a group of shooters formed the .50 Caliber Shooters Association to use guns in .50 BMG caliber for long- range competition. Unfortunately, the surplus military ammunition is not nearly accurate enough for this purpose, and various companies have, from time to time, manufactured high-quality target ammo in this caliber.

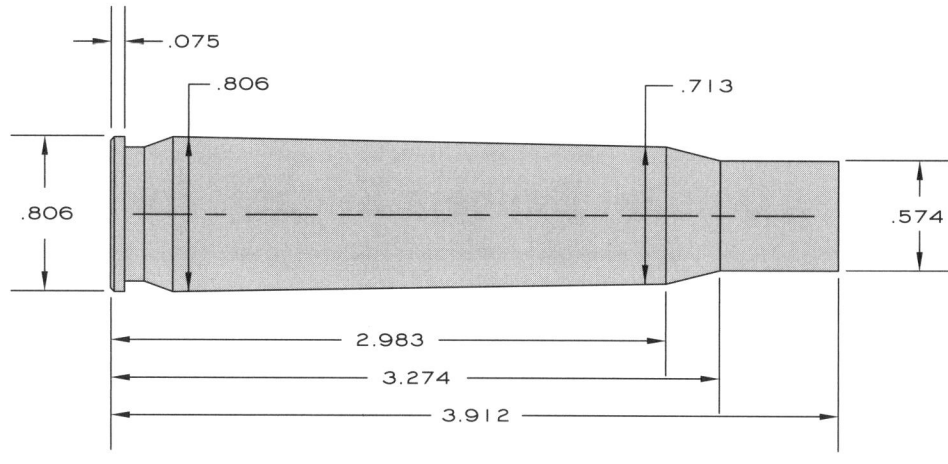

Relative Recoil Factor = 9.16

Specifications:

Controlling Agency for Standardization of this Ammunition: Factory

Availability:

Magtech 624-grain FMJ (50BMGA)

G1 Ballistic Coefficient = 0.715

Distance • Yards	Muzzle	100	200	300	400	500	600	800	1000
Velocity • fps	2952	2820	2691	2566	2444	2326	2211	1991	1785
Energy • ft-lbs	12077	11018	10036	9125	8281	7499	6776	5493	4414
Taylor KO Index	133.7	127.1	121.9	116.2	110.7	105.3	100.1	90.2	80.8
Path • Inches	-1.5	+1.4	0.0	-6.2	-17.8	-35.1	-58.9	-128.6	-233.4
Wind Drift • Inches	0.0	0.4	1.7	3.9	7.1	11.4	16.8	31.3	51.6

PMC 700-grain Sinterfire/Frangible (50F)

G1 Ballistic Coefficient = 0.500

Distance • Yards	Muzzle	100	200	300	400	500	600	800	1000
Velocity • fps	2800	2618	2443	2275	2114	1959	1811	1542	1314
Energy • ft-lbs	12189	10656	9281	8048	6946	5966	5101	3694	2682
Taylor KO Index	142.2	133.0	124.1	115.6	107.4	99.5	92.0	78.3	66.8
Path • Inches	-1.5	+1.8	0.0	-7.6	-22.0	-44.1	-75.3	-171.5	-326.6
Wind Drift • Inches	0.0	0.6	2.7	6.2	11.4	18.5	27.7	53.2	89.9

Hornady 750-grain A-MAX (8270)

G1 Ballistic Coefficient = 0.500

Distance • Yards	Muzzle	100	200	300	400	500	600	800	1000
Velocity • fps	2815	2632	2457	2288	2126	1971	1823	1552	1322
Energy • ft-lbs	13197	11538	10052	8719	7528	6468	5536	4011	2910
Taylor KO Index	41.9	39.1	36.4	33.8	31.4	29.0	26.7	22.6	19.0
Path • Inches	-1.5	+1.8	0.0	-7.5	-21.7	-43.6	-74.4	-169.4	-322.6
Wind Drift • Inches	0.0	0.6	2.6	6.2	11.3	18.4	27.5	52.8	89.2

Data for Calibers Suitable for Dangerous Game

(Sometimes known as "African" Calibers)

For most hunters, the ultimate experience is an African hunt. Somehow, bagging a white-tailed deer doesn't seem to come up to the same standard as a lion or a Cape buffalo. Professional hunters in Africa learned early on that if they wanted to stay in the business (read alive) they wanted guns with reliable stopping power. That led to large calibers, and generally to rather heavy rifles. Be sure to read what Craig Boddington has to say about rifles for this class of game. The cartridges listed here represent a good selection of the cartridges that are the most suitable for someone planning (or just dreaming about) a hunt for dangerous game.

Because these guns are very seldom fired at great distances, the yardage distances listed are in 50-yard increments out to 300 yards, with data for 400 yards and 500 yards included mostly for comparison purposes. The path information is based on iron sights (a 0.9-inch sight height).

In the heavy-barrel Ruger No. 1, recoil from the .450/.400-3" is quite mild. (Photo from *Safari Rifles II* by Craig Boddington, 2009, Safari Press Archives)

All elephants are potentially dangerous, but unless a bull is in musth he c̲ generally be counted on to demonstrate, threaten, and be a bully, but i̲ extremely rare for an unwounded bull to press home a serious charge. (Ph̲ from *Elephant!* by Craig Boddington, 2012, Safari Press Archives)

.450/.400 Nitro Express 3"

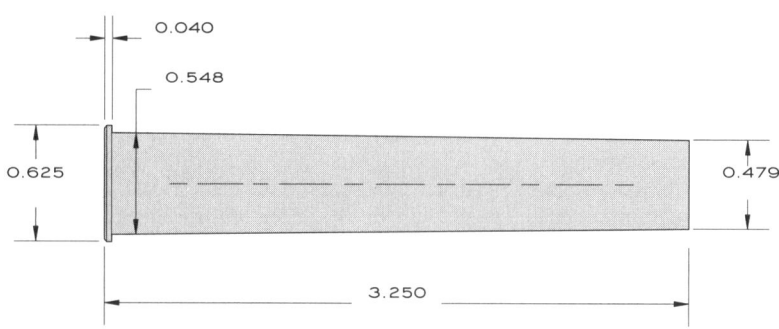

Hornady has opted to support this classic caliber. It follows the established British convention in naming cartridges—formed by necking down (or up) another cartridge—to put the parent cartridge first and the "necked-to" size second. So, the .450/.400 is a .450 Nitro Express necked to .400 caliber. This is an oldie and has been largely made obsolete by the .375 H&H.

Relative Recoil Factor = 4.27

Specifications:

Controlling Agency for Standardization of this Ammunition: CIP

Bullet Weight Grains	Velocity fps	Maximum Average Pressure Copper Crusher	Transducer
N/S	N/S	36,300 cup	40,600 psi

Standard barrel for velocity testing: 26 inches long—1 turn in 10-inch twist

Availability:

Hornady 400-grain DGX (82433)
G1 Ballistic Coefficient = 0.326

Distance • Yards	Muzzle	50	100	150	200	250	300	400	500
Velocity • fps	2050	1933	1820	1712	1608	1512	1420	1260	1135
Energy • ft-lbs	3732	3319	2941	2604	2298	2030	1791	1412	1145
Taylor KO Index	47.8	45.1	42.4	39.9	37.5	35.3	31.3	29.4	26.5
Path • Inches	-0.9	+1.8	+2.2	0.0	-5.2	-13.8	-26.1	-64.3	-124.4
Wind Drift • Inches	0.0	0.4	1.6	3.7	6.7	10.8	15.9	29.6	48.1

Hornady 400-grain DGS (8242)
G1 Ballistic Coefficient = 0.326

Distance • Yards	Muzzle	50	100	150	200	250	300	400	500
Velocity • fps	2050	1933	1820	1712	1608	1512	1420	1260	1135
Energy • ft-lbs	3732	3319	2941	2604	2298	2030	1791	1412	1145
Taylor KO Index	47.8	45.1	42.4	39.9	37.5	35.3	31.3	29.4	26.5
Path • Inches	-0.9	+1.8	+2.2	0.0	-5.2	-13.8	-26.1	-64.3	-124.4
Wind Drift • Inches	0.0	0.4	1.6	3.7	6.7	10.8	15.9	29.6	48.1

This great grizzly was taken near the Noatak River in the Alaskan Arctic. (Photo from Ask the Grizzly/Brown Bear Guides, by J.Y. Jones, 2011, Safari Press Archives)

.400 A-Square Dual Purpose Magnum (DPM)

In 1992 the A-Square Company was asked to design a .40-caliber cartridge (it's actually closer to a .41) that would be legal for dangerous game in Africa and that used the basic H&H belted magnum case. The result was the .400 Pondoro. In 2005 A-Square announced the .400 DPM, which is identical to the .400 Pondoro. The dual-purpose designation comes from using pistol bullets for light loads while keeping A-Square's Triad bullet loadings for major hunting applications. The pistol bullets are no longer offered as an option.

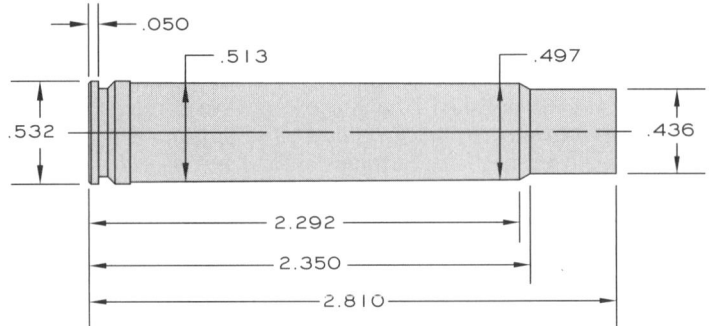

Relative Recoil Factor = 4.27

Specifications:

Controlling Agency for Standardization of this Ammunition: Factory

Standard barrel for velocity testing: 26 inches long—1 turn in 10-inch twist

Availability:

A² 400-grain Triad (None)

G1 Ballistic Coefficient = 0.325

Distance • Yards	Muzzle	50	100	150	200	250	300	400	500
Velocity • fps	2400	2271	2146	2026	1908	1797	1689	1491	1319
Energy • ft-lbs	5116	4584	4092	3646	3236	2869	2533	1975	1545
Taylor KO Index	56.2	53.2	50.3	47.5	44.7	42.1	39.6	34.9	30.9
Path • Inches	-0.9	+1.2	+1.5	0.0	-3.6	-9.7	-18.4	-45.4	-88.1
Wind Drift • Inches	0.0	0.3	1.3	2.9	5.4	8.6	12.8	24.0	39.7

Modern softpoints are so good today that under most circumstances you could probably hunt rhino with the very best modern bullets. But what if you get a charge? Better load up with solids! (Photo from *Safari Rifles II* by Craig Boddington, 2009, Safari Press Archives)

.404 Dakota

With their .404 Dakota, Dakota Arms enters into the African class of cartridges. By conservative standards, the .404 is a medium African rifle. While perhaps not quite enough for the toughest of the dangerous game animals in Africa, it is certainly plenty for anything on the North American continent.

Relative Recoil Factor = 4.27

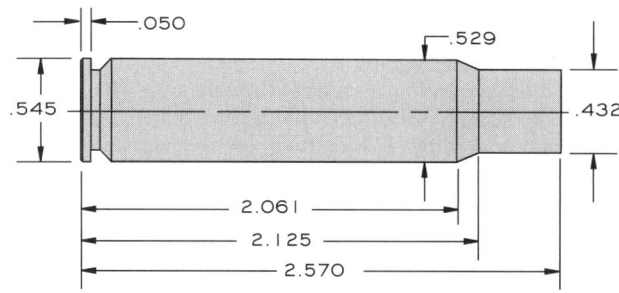

Specifications:

Controlling Agency for Standardization of this Ammunition: Factory

Standard barrel for velocity testing: 26 inches long

Availability:

Dakota 350-grain Woodleigh Soft Nose (404-350WSN)

G1 Ballistic Coefficient = 0.357

Distance • Yards	Muzzle	50	100	150	200	250	300	400	500
Velocity • fps	2550	2429	2310	2196	2084	1976	1871	1673	1493
Energy • ft-lbs	5055	4585	4150	3747	3376	3034	2721	2175	1732
Taylor KO Index	53.9	51.4	48.9	46.4	44.1	41.8	39.6	35.4	31.6
Path • Inches	-0.9	+0.9	+1.3	0.0	-3.1	-8.2	-15.5	-37.8	-72.8
Wind Drift • Inches	0.0	0.3	1.0	2.4	4.4	7.1	10.4	19.6	32.3

Dakota 400-grain Woodleigh Soft Nose (404-400WSN)

G1 Ballistic Coefficient = 0.354

Distance • Yards	Muzzle	50	100	150	200	250	300	400	500
Velocity • fps	2400	2282	2167	2055	1947	1842	1741	1553	1385
Energy • ft-lbs	5117	4625	4170	3751	3366	3014	2693	2142	1704
Taylor KO Index	58.0	55.2	52.4	49.7	47.1	44.5	42.1	37.5	33.5
Path • Inches	-0.9	+1.1	+1.5	0.0	-3.6	-9.4	-29.1	-43.6	-84.5
Wind Drift • Inches	0.0	0.3	1.2	2.7	4.9	7.8	11.6	21.7	35.7

Dakota 400-grain Woodleigh Solid (404-400WSO)

G1 Ballistic Coefficient = 0.358

Distance • Yards	Muzzle	50	100	150	200	250	300	400	500
Velocity • fps	2400	2283	2169	2059	1951	1848	1748	1561	1394
Energy • ft-lbs	5117	4630	4180	3765	3383	3033	2714	2164	1725
Taylor KO Index	58.0	55.2	52.4	49.8	47.2	44.7	42.3	37.7	33.7
Path • Inches	-0.9	+1.1	+1.5	0.0	-3.5	-9.4	-29.0	-43.4	-83.4
Wind Drift • Inches	0.0	0.3	1.1	2.6	4.8	7.7	11.4	21.4	35.2

This great coastal grizzly was taken in British Columbia. As Ruark said, "Use Enough Gun," but you do not need to overdo it. (Photo from Ask the Grizzly/ Brown Bear Guides, by J.Y. Jones, 2011, Safari Press Archives)

.404 Jeffery

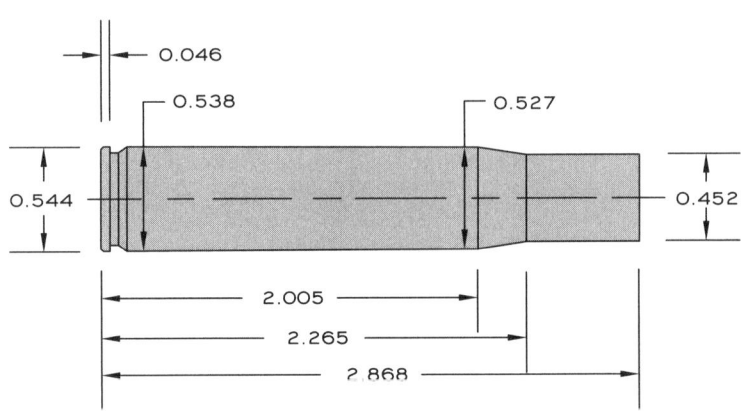

This is a classic example of why you can't always tell the exact bullet size from the name of the cartridge. This cartridge (also called the .404 Rimless Nitro Express) actually uses .423 inch diameter bullets. It was designed by Jeffery along about 1910. There is nothing .404 about it. The cartridge, being rimless, is most suited to bolt-action guns. It is only slightly milder than the .416 Rigby and is well respected as a medium-caliber African cartridge.

Relative Recoil Factor = 4.21

Specifications:

Controlling Agency for Standardization of this Ammunition: SAAMI

Bullet Weight Grains	Velocity fps	Maximum Average Pressure	
		Copper Crusher	Transducer
		N/S	N/S

Typical barrel for velocity testing: 26 inches long—1 turn in 16.5-inch twist

Availability:

Hornady 400-grain DGX (8239)

G1 Ballistic Coefficient = 0.315

Distance • Yards	Muzzle	50	100	150	200	250	300	400	500
Velocity • fps	2300	2170	2046	1924	1809	1697	1592	1399	1238
Energy • ft-lbs	4698	4185	3716	3289	2906	2557	2251	1738	1362
Taylor KO Index	53.9	50.8	47.9	45.1	42.4	39.8	37.3	32.8	29.0
Path • Inches	-0.9	+1.3	+1.7	0.0	-4.0	-10.8	-20.5	-50.8	-98.9
Wind Drift • Inches	0.0	0.3	1.4	3.2	5.9	9.5	14.1	26.6	43.8

Hornady 400-grain DGS (8238)

G1 Ballistic Coefficient = 0.31

Distance • Yards	Muzzle	50	100	150	200	250	300	400	500
Velocity • fps	2300	2170	2046	1924	1809	1697	1592	1399	1238
Energy • ft-lbs	4698	4185	3716	3289	2906	2557	2251	1738	1362
Taylor KO Index	53.9	50.8	47.9	45.1	42.4	39.8	37.3	32.8	29.0
Path • Inches	-0.9	+1.3	+1.7	0.0	-4.0	-10.8	-20.5	-50.8	-98.9
Wind Drift • Inches	0.0	0.3	1.4	3.2	5.9	9.5	14.1	26.6	43.8

.416 Rigby

When it was introduced in 1911, this cartridge represented what the British thought to be a minimum African cartridge. The .416 soon showed that it really wasn't minimum at all. It's a large case, roughly the same volume as the .378 and the .460 Weatherby Magnums. The .416 Rigby has always been loaded to a much more modest pressure level than the more modern cartridges of similar caliber. Today, there are other cartridges that offer similar performance in slightly smaller cases, but none of these is really a whole lot "better" in the field—where performance really counts—than the .416 Rigby.

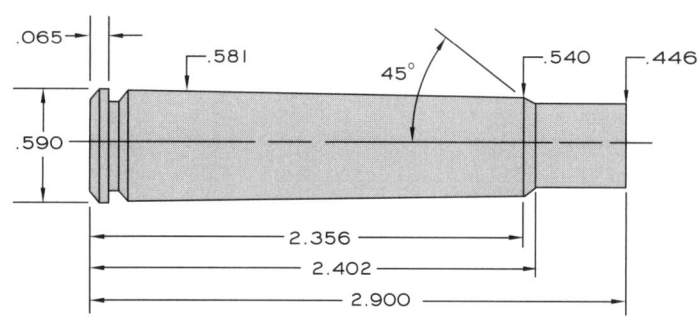

Relative Recoil Factor = 4.27

Specifications:

Controlling Agency for Standardization of this Ammunition: CIP

Bullet Weight Grains	Velocity fps	Maximum Average Pressure Copper Crusher	Transducer
N/S	N/S	N/S	47,137 psi

Typical barrel for velocity testing: 26 inches long—1 turn in 16.5-inch twist

Availability:

Cor-Bon 350-grain DPX (DPX416350/20)
G1 Ballistic Coefficient = 0.325

Distance • Yards	Muzzle	50	100	150	200	250	300	400	500
Velocity • fps	2750	2611	2476	2345	2218	2095	1976	1751	1546
Energy • ft-lbs	5879	5299	4766	4275	3825	3412	3036	2384	1858
Taylor KO Index	57.2	54.3	51.5	48.8	46.1	43.6	41.1	36.4	32.2
Path • Inches	-0.9	+0.7	+1.1	0.0	-2.6	-7.1	-13.5	-33.2	-64.4
Wind Drift • Inches	0.0	0.2	1.0	2.4	4.4	7.0	10.4	19.6	32.5

Hornady 400-grain DGX (82663)
G1 Ballistic Coefficient = 0.320

Distance • Yards	Muzzle	50	100	150	200	250	300	400	500
Velocity • fps	2415	2284	2157	2034	1915	1801	1691	1490	1316
Energy • ft-lbs	5180	4634	4131	3675	3256	2882	2540	1973	1538
Taylor KO Index	57.4	54.3	51.3	48.4	45.5	42.8	40.2	35.4	31.3
Path • Inches	-0.9	+1.1	+1.5	0.0	-3.6	-9.6	-18.2	-45.1	-87.7
Wind Drift • Inches	0.0	0.3	1.3	2.9	5.4	8.7	12.9	24.3	40.2

A² 400-grain Dead Tough SP, Monolithic Solid, Lion Load SP (None)
G1 Ballistic Coefficient = 0.317

Distance • Yards	Muzzle	50	100	150	200	250	300	400	500
Velocity • fps	2400	2268	2140	2016	1897	1782	1673	1471	1298
Energy • ft-lbs	5115	4570	4069	3612	3194	2823	2487	1923	1497
Taylor KO Index	57.1	53.9	50.9	47.9	45.1	42.4	39.8	35.0	30.9
Path • Inches	-0.9	+1.2	+1.5	0.0	-3.7	-9.8	-18.6	-46.0	-89.8
Wind Drift • Inches	0.0	0.3	1.3	3.0	5.5	8.6	13.2	24.8	41.1

Winchester 400-grain Nosler Partition (S416RSLP)
G1 Ballistic Coefficient = 0.390

Distance • Yards	Muzzle	50	100	150	200	250	300	400	500
Velocity • fps	2370	2263	2159	2058	1960	1864	1772	1598	1439
Energy • ft-lbs	4988	4551	4141	3763	3410	3087	2788	2266	1840
Taylor KO Index	56.3	53.8	51.3	48.9	46.6	44.3	42.1	38.0	34.2
Path • Inches	-0.9	+1.2	+1.5	0.0	-3.6	-9.4	-17.7	-43.0	-81.9
Wind Drift • Inches	0.0	0.3	1.1	2.4	4.4	7.1	10.5	19.6	32.2

Federal 400-grain Fusion Safari (F416FS1)
G1 Ballistic Coefficient = 0.387

Distance • Yards	Muzzle	50	100	150	200	250	300	400	500
Velocity • fps	2350	2243	2180	2097	1970	1843	1770	1590	1420
Energy • ft-lbs	4906	4470	4064	3687	3339	3018	2775	2230	1790
Taylor KO Index	55.9	53.3	51.8	49.8	46.8	43.8	42.1	37.8	33.8
Path • Inches	-0.9	+1.2	+1.5	0.0	-3.6	-9.6	-18.1	-44.0	-83.8
Wind Drift • Inches	0.0	0.3	1.1	2.5	4.6	7.3	10.7	20.1	32.9

Federal 400-grain Swift A-Frame (P416SA)
G1 Ballistic Coefficient = 0.366

Distance • Yards	Muzzle	50	100	150	200	250	300	400	500
Velocity • fps	2350	2237	2130	2020	1920	1816	1720	1540	1380
Energy • ft-lbs	4905	4446	4020	3625	3265	2931	2635	2110	1695
Taylor KO Index	55.9	53.2	50.6	48.0	45.6	43.2	40.9	36.6	32.8
Path • Inches	-0.9	+1.2	+1.6	0.0	-3.7	-9.8	-18.4	-30.1	-86.3
Wind Drift • Inches	0.0	0.3	1.2	2.6	4.8	7.8	11.5	21.4	35.2

Norma 400-grain Swift A-Frame (11069)
G1 Ballistic Coefficient = 0.367

Distance • Yards	Muzzle	50	100	150	200	250	300	400	500
Velocity • fps	2350	2237	2127	2021	1917	1817	1721	1541	1380
Energy • ft-lbs	4906	4446	4021	3627	3265	2934	2631	2109	1691
Taylor KO Index	55.9	53.2	50.6	48.0	45.6	43.2	40.9	36.6	32.8
Path • Inches	-0.9	+1.2	+1.6	0.0	-3.7	-9.7	-18.4	-45.0	-86.3
Wind Drift • Inches	0.0	0.3	1.2	2.6	4.8	7.7	11.4	21.4	35.2

Federal 400-grain Trophy Bonded Bear Claw (P416T1)
G1 Ballistic Coefficient = 0.373

Distance • Yards	Muzzle	50	100	150	200	250	300	400	500
Velocity • fps	2300	2190	2080	1980	1880	1783	1690	1520	1360
Energy • ft-lbs	4700	4262	3860	3483	3140	2823	2540	2045	1645
Taylor KO Index	54.7	52.1	49.4	47.1	44.7	42.4	40.2	36.1	32.3
Path • Inches	-0.9	+1.3	+1.6	0.0	-3.9	-10.2	-19.2	-46.8	-89.6
Wind Drift • Inches	0.0	0.3	1.2	2.7	4.9	7.8	11.6	21.6	35.5

Federal 400-grain Barnes Triple-Shock X-Bullet (P416D)
G1 Ballistic Coefficient = 0.39

Distance • Yards	Muzzle	50	100	150	200	250	300	400	500
Velocity • fps	2400	2293	2190	2090	1990	1890	1800	1630	1460
Energy • ft-lbs	4700	4669	4255	3865	3515	3174	2875	2345	1905
Taylor KO Index	57.1	54.5	52.1	49.7	47.3	44.9	42.8	38.7	34.7
Path • Inches	-0.9	+1.1	+1.5	0.0	-3.5	-9.1	-17.2	-41.8	-79.6
Wind Drift • Inches	0.0	0.2	1.0	2.4	4.4	7.0	10.3	19.3	31.6

Cor-Bon 400-grain DPX (EH416R400DPX/10)
G1 Ballistic Coefficient = 0.35

Distance • Yards	Muzzle	50	100	150	200	250	300	400	500
Velocity • fps	2375	2256	2141	2029	1920	1815	1715	1527	1361
Energy • ft-lbs	5011	4522	4071	3656	3275	2928	2612	2071	1646
Taylor KO Index	56.5	53.6	50.9	48.2	45.6	43.1	40.8	36.3	32.4
Path • Inches	-0.9	+1.2	+1.5	0.0	-3.6	-9.7	-18.3	-44.8	-86.3
Wind Drift • Inches	0.0	0.3	1.2	2.7	5.0	8.0	11.9	22.3	36.7

Hornady 400-grain DGS (8265)
G1 Ballistic Coefficient = 0.32

Distance • Yards	Muzzle	50	100	150	200	250	300	400	500
Velocity • fps	2415	2284	2157	2034	1915	1801	1691	1490	1316
Energy • ft-lbs	5180	4634	4131	3675	3256	2882	2540	1973	1538
Taylor KO Index	57.4	54.3	51.3	48.4	45.5	42.8	40.2	35.4	31.3
Path • Inches	-0.9	+1.1	+1.5	0.0	-3.6	-9.6	-18.2	-45.1	-87.7
Wind Drift • Inches	0.0	0.3	1.3	2.9	5.4	8.7	12.9	24.3	40.2

Federal 400-grain Barnes Banded Solid (P416E)
G1 Ballistic Coefficient = 0.3

Distance • Yards	Muzzle	50	100	150	200	250	300	400	500
Velocity • fps	2400	2293	2190	2090	1990	1890	1800	1630	1460
Energy • ft-lbs	4700	4669	4255	3865	3515	3174	2875	2345	1905
Taylor KO Index	57.1	54.5	52.1	49.7	47.3	44.9	42.8	38.7	34.7
Path • Inches	-0.9	+1.1	+1.5	0.0	-3.5	-9.1	-17.2	-41.8	-79.6
Wind Drift • Inches	0.0	0.2	1.0	2.4	4.4	7.0	10.3	19.3	31.6

Cor-Bon 400-grain FPBS (EH416R400FPBS/10)

G1 Ballistic Coefficient = 0.350

Distance • Yards	Muzzle	50	100	150	200	250	300	400	500
Velocity • fps	2375	2256	2141	2029	1920	1815	1715	1527	1361
Energy • ft-lbs	5011	4522	4071	3656	3275	2928	2612	2071	1646
Taylor KO Index	56.5	53.6	50.9	48.2	45.6	43.1	40.8	36.3	32.4
Path • Inches	-0.9	+1.2	+1.5	0.0	-3.6	-9.7	-18.3	-44.8	-86.3
Wind Drift • Inches	0.0	0.3	1.2	2.7	5.0	8.0	11.9	22.3	36.7

Norma 400-grain SOLID (11076)

G1 Ballistic Coefficient = 0.247

Distance • Yards	Muzzle	50	100	150	200	250	300	400	500
Velocity • fps	2375	2208	2047	1893	1748	1610	1483	1264	1106
Energy • ft-lbs	5011	4330	3728	3185	2713	2304	1954	1420	1086
Taylor KO Index	56.5	52.5	48.7	45.0	41.6	38.3	35.3	30.0	26.3
Path • Inches	-0.9	+1.3	+1.7	0.0	-4.1	-11.1	-21.4	-54.6	-109.6
Wind Drift • Inches	0.0	0.4	1.7	4.0	7.4	12.0	18.0	34.4	57.0

Federal 400-grain Trophy Bonded Sledgehammer Solid (P416T2)

G1 Ballistic Coefficient = 0.273

Distance • Yards	Muzzle	50	100	150	200	250	300	400	500
Velocity • fps	2370	2218	2070	1932	1800	1671	1550	1340	1170
Energy • ft-lbs	4990	4372	3815	3316	2870	2479	2135	1590	1215
Taylor KO Index	56.3	52.7	49.2	45.9	42.8	39.7	36.8	31.9	27.8
Path • Inches	-0.9	+1.2	+1.6	0.0	-4.0	-10.6	-20.4	-51.4	-101.9
Wind Drift • Inches	0.0	0.4	1.6	3.6	6.6	10.7	16.0	30.4	50.4

Winchester 400-grain Nosler Solid (S416RSLS)

G1 Ballistic Coefficient = 0.263

Distance • Yards	Muzzle	50	100	150	200	250	300	400	500
Velocity • fps	2370	2213	2061	1916	1778	1647	1525	1310	1145
Energy • ft-lbs	4988	4350	3774	3263	2807	2410	2064	1522	1163
Taylor KO Index	56.3	52.6	49.0	45.5	42.3	39.2	36.2	31.1	27.2
Path • Inches	-0.9	+1.3	+1.7	0.0	-4.0	-10.8	-20.9	-52.6	-104.8
Wind Drift • Inches	0.0	0.4	1.6	3.8	6.9	11.2	16.7	31.9	52.9

Norma 400-grain Barnes Banded Solid (11050) LF

G1 Ballistic Coefficient = 0.388

Distance • Yards	Muzzle	50	100	150	200	250	300	400	500
Velocity • fps	2097	1997	1900	1806	1715	1628	1545	1392	1260
Energy • ft-lbs	3907	3543	3207	2897	2613	2355	2121	1721	1410
Taylor KO Index	49.8	47.5	45.2	42.9	40.8	38.7	36.7	33.1	30.0
Path • Inches	-0.9	+1.7	+2.0	0.0	-4.7	-12.4	-23.3	-56.5	-107.7
Wind Drift • Inches	0.0	0.3	1.3	2.9	5.4	8.6	12.6	23.5	38.2

Kynoch 410-grain Solid; Soft Nose

G1 Ballistic Coefficient = 0.426

Distance • Yards	Muzzle	50	100	150	200	250	300	400	500
Velocity • fps	2300	2204	2110	2018	1929	1842	1758	1599	1453
Energy • ft-lbs	4817	4422	4053	3708	3387	3090	2815	2328	1923
Taylor KO Index	56.0	53.7	51.4	49.2	47.0	44.9	42.8	39.0	35.4
Path • Inches	-0.9	+1.2	+1.6	0.0	-3.7	-9.8	-18.4	-44.4	-84.1
Wind Drift • Inches	0.0	0.2	1.0	2.3	4.2	6.8	10.0	18.5	30.2

Norma 450-grain Soft Nose (11070)

G1 Ballistic Coefficient = 0.357

Distance • Yards	Muzzle	50	100	150	200	250	300	400	500
Velocity • fps	2150	2040	1933	1830	1731	1635	1545	1379	1239
Energy • ft-lbs	4620	4158	3735	3347	2996	2673	2385	1902	1534
Taylor KO Index	57.5	54.6	51.7	48.9	46.3	43.7	41.3	36.9	33.1
Path • Inches	-0.9	+1.6	+2.0	0.0	-4.6	-12.0	-22.7	-55.5	-106.6
Wind Drift • Inches	0.0	0.3	1.3	3.1	5.7	9.1	13.4	25.0	40.9

.416 Remington Magnum

When Remington wanted to get onto the .416 bandwagon, the company simply necked up its 8mm magnum to make the .416 Remington Magnum (introduced in 1988). If you have to pick a single gun for your next African safari, this caliber should be on your short list. When the major factories began offering premium hunting bullets, obtained from outside suppliers, the terminal performance of factory ammo took a huge jump. All the loadings available today feature premium-style bullets. For all practical purposes, the .416 Remington Magnum offers the same performance as the legendary .416 Rigby.

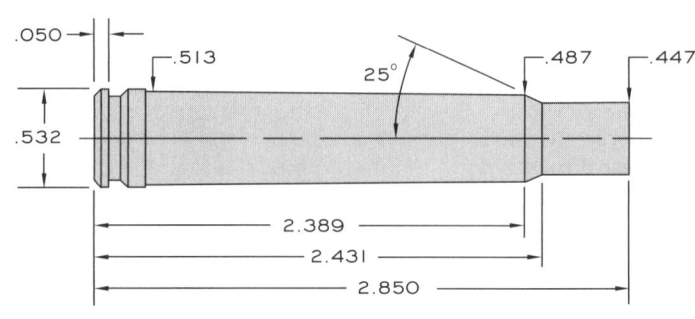

Relative Recoil Factor = 4.32

Specifications:

Controlling Agency for Standardization of this Ammunition: SAAMI

Bullet Weight Grains	Velocity fps	Maximum Average Pressure Copper Crusher	Transducer
350	2,525	54,000 cup	65,000 psi
400	2,400	54,000 cup	65,000 psi

Standard barrel for velocity testing: 24 inches long—1 turn in 14-inch twist

Availability:

Federal 400-grain Trophy Bonded Bear Claw (P416RT1)
G1 Ballistic Coefficient = 0.375

Distance • Yards	Muzzle	50	100	150	200	250	300	400	500
Velocity • fps	2400	2288	2180	2073	1970	1871	1770	1590	1430
Energy • ft-lbs	5115	4652	4215	3819	3445	3110	2790	2250	1805
Taylor KO Index	57.1	54.4	51.8	49.3	46.8	44.5	42.1	37.8	34.0
Path • Inches	-0.9	+1.1	+1.5	0.0	-3.5	-9.2	-17.4	-42.5	-81.4
Wind Drift • Inches	0.0	0.3	1.1	2.5	4.6	7.3	10.8	20.2	33.2

Federal 400-grain Swift A-Frame (P416RSA)
G1 Ballistic Coefficient = 0.365

Distance • Yards	Muzzle	50	100	150	200	250	300	400	500
Velocity • fps	2400	2285	2180	2065	1960	1858	1760	1580	1410
Energy • ft-lbs	5115	4640	4200	3789	3420	3087	2760	2153	1775
Taylor KO Index	57.1	54.3	51.8	49.1	46.6	44.2	41.8	37.6	33.5
Path • Inches	-0.9	+1.1	+1.5	0.0	-3.5	-9.3	-17.6	-43.0	-82.4
Wind Drift • Inches	0.0	0.3	1.1	2.6	4.7	7.5	11.1	20.9	34.3

Federal 400-grain Fusion Safari (F416RFS1)
G1 Ballistic Coefficient = 0.35

Distance • Yards	Muzzle	50	100	150	200	250	300	400	500
Velocity • fps	2400	2282	2130	2057	1920	1845	1730	1550	1390
Energy • ft-lbs	5117	4628	4176	3759	3376	3025	2645	2125	1705
Taylor KO Index	57.1	54.2	50.6	48.9	45.6	43.9	41.1	36.8	33.0
Path • Inches	-0.9	+1.1	+1.5	0.0	-3.5	-9.4	-17.8	-43.4	-83.5
Wind Drift • Inches	0.0	0.3	1.2	2.6	4.8	7.8	11.5	21.5	35.4

Hornady 400-grain DGX (82673)
G1 Ballistic Coefficient = 0.31

Distance • Yards	Muzzle	50	100	150	200	250	300	400	500
Velocity • fps	2400	2269	2142	2019	1901	1787	1678	1477	1304
Energy • ft-lbs	5115	4574	4075	3622	3208	2836	2500	1937	1510
Taylor KO Index	57.1	53.9	50.9	48.0	45.2	42.5	39.9	35.1	31.0
Path • Inches	-0.9	+1.2	+1.5	0.0	-3.7	-9.7	-18.5	-45.8	-89.1
Wind Drift • Inches	0.0	0.3	1.3	3.0	5.5	8.8	13.0	24.6	40.7

Remington 400-grain Swift A-Frame Pointed Soft Point (R416RA)

G1 Ballistic Coefficient = 0.368

Distance • Yards	Muzzle	50	100	150	200	250	300	400	500
Velocity • fps	2400	2286	2175	2067	1962	1861	1764	1580	1415
Energy • ft-lbs	5115	4643	4201	3797	3419	3078	2764	2218	1779
Taylor KO Index	57.1	54.3	51.7	49.1	46.6	44.2	41.9	37.6	33.6
Path • Inches	-0.9	+1.1	+1.5	0.0	-3..5	-9.3	17.6	-42.8	-82.2
Wind Drift • Inches	0.0	0.3	1.1	2.6	4.7	7.5	11.0	20.7	34.0

Winchester 400-grain Nosler Partition (S416SLSP)

G1 Ballistic Coefficient = 0.390

Distance • Yards	Muzzle	50	100	150	200	250	300	400	500
Velocity • fps	2400	2293	2188	2086	1987	1890	1797	1621	1460
Energy • ft-lbs	5115	4669	4251	3865	3504	3174	2868	2333	1894
Taylor KO Index	57.1	54.5	52.0	49.6	47.2	44.9	42.7	38.5	34.7
Path • Inches	-0.9	+1.1	+1.5	0.0	-3.5	-9.1	-17.2	-41.8	-79.6
Wind Drift • Inches	0.0	0.2	1.0	2.4	4.4	7.0	10.3	19.3	31.6

Federal 400-grain Barnes Triple-Shock X-Bullet (P416RB) LF

G1 Ballistic Coefficient = 0.390

Distance • Yards	Muzzle	50	100	150	200	250	300	400	500
Velocity • fps	2370	2263	2160	2058	1960	1864	1770	1600	1440
Energy • ft-lbs	4990	4551	4145	3763	3415	3087	2795	2280	1850
Taylor KO Index	56.3	53.8	51.3	48.9	46.6	44.3	42.1	38.0	34.2
Path • Inches	-0.9	+1.2	+1.5	0.0	-3.6	-9.4	-17.7	-43.0	-81.9
Wind Drift • Inches	0.0	0.3	1.1	2.4	4.4	7.1	10.5	19.6	32.2

Cor-Bon 400-grain DPX (EH416400DPX/10)

G1 Ballistic Coefficient = 0.350

Distance • Yards	Muzzle	50	100	150	200	250	300	400	500
Velocity • fps	2350	2232	2117	2006	1898	1794	1694	1509	1345
Energy • ft-lbs	4906	4425	3982	3574	3200	2859	2550	2022	1607
Taylor KO Index	55.9	53.3	51.8	49.8	46.8	43.8	42.1	37.8	33.8
Path • Inches	-0.9	+1.2	+1.6	0.0	-3.7	-9.9	-18.7	-45.9	-88.4
Wind Drift • Inches	0.0	0.3	1.2	2.8	5.1	8.2	12.1	22.6	37.3

A² 400-grain Dead Tough SP, Monolithic Solid, Lion Load SP (None)

G1 Ballistic Coefficient = 0.317

Distance • Yards	Muzzle	50	100	150	200	250	300	400	500
Velocity • fps	2380	2249	2121	1998	1880	1766	1657	1457	1286
Energy • ft-lbs	5031	4492	3998	3548	3139	2770	2439	1887	1470
Taylor KO Index	56.6	53.5	50.4	47.5	44.7	42.0	39.4	34.6	30.6
Path • Inches	-0.9	+1.2	+1.6	0.0	-3.7	-9.9	-18.9	-46.9	-91.5
Wind Drift • Inches	0.0	0.3	1.3	3.0	5.6	9.0	13.3	25.1	41.6

Federal 400-grain Barnes Banded Solid (P416RB) LF

G1 Ballistic Coefficient = 0.390

Distance • Yards	Muzzle	50	100	150	200	250	300	400	500
Velocity • fps	2400	2293	2190	2086	1990	1890	1790	1620	1460
Energy • ft-lbs	5115	4669	4250	3865	3500	3174	2860	2325	1885
Taylor KO Index	57.1	54.5	52.1	49.6	47.3	44.9	42.6	38.5	37.4
Path • Inches	-0.9	+1.1	+1.5	0.0	-3.5	-9.1	-17.2	-41.8	-79.6
Wind Drift • Inches	0.0	0.3	1.0	2.4	4.4	7.0	10.3	19.3	31.6

Federal 400-grain Trophy Bonded Sledgehammer Solid (P416T2)

G1 Ballistic Coefficient = 0.275

Distance • Yards	Muzzle	50	100	150	200	250	300	400	500
Velocity • fps	2400	2248	2100	1962	1820	1699	1570	1360	1190
Energy • ft-lbs	5115	4491	3920	3419	2950	2565	2200	1635	1245
Taylor KO Index	57.1	53.4	49.9	46.6	43.3	40.4	37.3	32.3	28.3
Path • Inches	-0.9	+1.2	+1.6	0.0	-3.8	-10.3	-19.8	-49.8	-98.5
Wind Drift • Inches	0.0	0.4	1.5	3.5	6.4	10.4	15.6	29.6	49.2

Hornady 400-grain DGS (82674)

G1 Ballistic Coefficient = 0.319

Distance • Yards	Muzzle	50	100	150	200	250	300	400	500
Velocity • fps	2400	2269	2142	2019	1901	1787	1678	1477	1304
Energy • ft-lbs	5115	4574	4075	3622	3208	2836	2500	1937	1510
Taylor KO Index	57.1	53.9	50.9	48.0	45.2	42.5	39.9	35.1	31.0
Path • Inches	-0.9	+1.2	+1.5	0.0	-3.7	-9.7	-18.5	-45.8	-89.1
Wind Drift • Inches	0.0	0.3	1.3	3.0	5.5	8.8	13.0	24.6	40.7

Norma 400-grain SOLID (11075)

G1 Ballistic Coefficient = 0.247

Distance • Yards	Muzzle	50	100	150	200	250	300	400	500
Velocity • fps	2400	2232	2070	1915	1768	1630	1501	1279	1115
Energy • ft-lbs	5117	4424	3807	3259	2778	2360	2002	1452	1105
Taylor KO Index	57.1	53.1	49.2	45.5	42.0	38.7	35.7	30.4	28.5
Path • Inches	-0.9	+1.2	+1.6	0.0	-4.0	-10.8	-20.9	-53.3	-107.1
Wind Drift • Inches	0.0	0.4	1.7	4.0	7.3	11.8	17.7	33.9	56.3

Winchester 400-grain Nosler Solid (S416SLS)

G1 Ballistic Coefficient = 0.263

Distance • Yards	Muzzle	50	100	150	200	250	300	400	500
Velocity • fps	2400	2242	2089	1943	1806	1671	1547	1328	1157
Energy • ft-lbs	5115	4464	3876	3354	2896	2481	2124	1565	1190
Taylor KO Index	57.1	53.3	49.7	46.2	42.9	39.7	36.8	31.6	27.5
Path • Inches	-0.9	+1.2	+1.6	0.0	-3.9	-10.5	-20.2	-51.2	-101.9
Wind Drift • Inches	0.0	0.4	1.6	3.7	6.8	11.0	16.4	31.3	52.0

Cor-Bon 400-grain FPBS (EH416400FPBS/10)

G1 Ballistic Coefficient = 0.350

Distance • Yards	Muzzle	50	100	150	200	250	300	400	500
Velocity • fps	2350	2232	2117	2006	1898	1794	1694	1509	1345
Energy • ft-lbs	4906	4425	3982	3574	3200	2859	2550	2022	1607
Taylor KO Index	55.9	53.3	51.8	49.8	46.8	43.8	42.1	37.8	33.8
Path • Inches	-0.9	+1.2	+1.6	0.0	-3.7	-9.9	-18.7	-45.9	-88.4
Wind Drift • Inches	0.0	0.3	1.2	2.8	5.1	8.2	12.1	22.6	37.3

Norma 450-grain Soft Nose (11072)

G1 Ballistic Coefficient = 0.358

Distance • Yards	Muzzle	50	100	150	200	250	300	400	500
Velocity • fps	2150	2040	1933	1830	1731	1637	1546	1381	1241
Energy • ft-lbs	4620	4158	3735	3347	2996	2677	2390	1907	1709
Taylor KO Index	57.5	54.6	51.7	48.9	46.3	43.8	41.3	36.9	33.2
Path • Inches	-0.9	+1.6	+2.0	0.0	-4.5	-12.0	-22.7	-55.4	-106.4
Wind Drift • Inches	0.0	0.3	1.3	3.1	5.6	9.0	13.4	25.0	40.8

Norma 450-grain Full Metal Jacket (11073)

G1 Ballistic Coefficient = 0.350

Distance • Yards	Muzzle	50	100	150	200	250	300	400	500
Velocity • fps	2150	2037	1928	1823	1722	1626	1534	1337	1227
Energy • ft-lbs	4620	4148	3713	3321	2963	2643	2353	1868	1504
Taylor KO Index	57.5	54.5	51.6	48.8	46.1	43.5	41.0	36.6	32.8
Path • Inches	-0.9	+1.6	+2.0	0.0	-4.6	-12.1	-22.9	-56.0	-107.8
Wind Drift • Inches	0.0	0.3	1.4	3.2	5.8	9.3	13.7	25.6	41.9

Botswana PH Christo Spykerman and tracker Rubin with a [Botswana tusker shot with a .416 Remington loaded with 45 grain Norma solids. (Safari Press Archives)

.500/.416 Nitro Express 3¼"

Even though it is based on the .500 Nitro Express case that started life about 1880, this caliber is a relatively new number. Norma provides both a softnose bullet as well as a full-metal-jacket bullet in its PH series.

Rimmed cartridges work best in double rifles and other single-load guns. They are generally avoided in magazine rifles because the rims sometimes cause feeding problems.

Relative Recoil Factor = 4.35

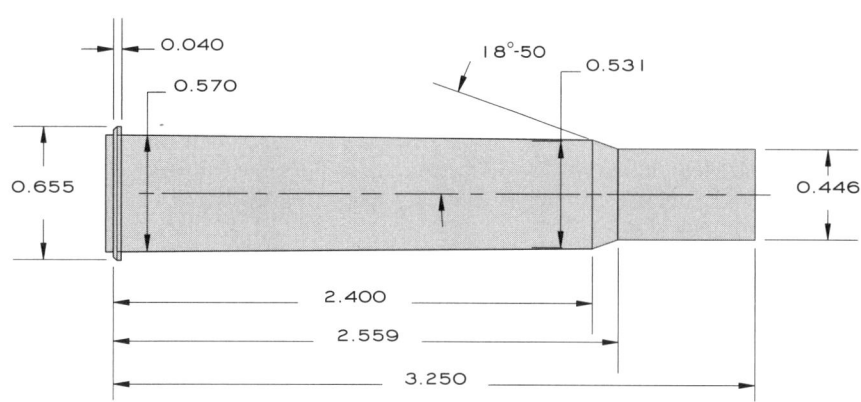

Specifications:

Controlling Agency for Standardization of this Ammunition: SAAMI

Bullet Weight Grains	Velocity fps	Maximum Average Pressure	
		Copper Crusher	Transducer
350	2,525	54,000 cup	65,000 psi
400	2,400	54,000 cup	65,000 psi

Standard barrel for velocity testing: 24 inches long—1 turn in 14-inch twist

Availability:

Norma 410-grain SOFT NOSE (11053)

G1 Ballistic Coefficient = 0.307

Distance • Yards	Muzzle	50	100	150	200	250	300	400	500
Velocity • fps	2325	2191	2062	1937	1817	1703	1594	1397	1233
Energy • ft-lbs	4922	4373	3872	3417	3007	2641	2315	1778	1384
Taylor KO Index	56.7	53.4	50.2	47.2	44.3	41.5	38.8	34.0	30.0
Path • Inches	-0.9	+1.3	+1.7	0.0	-4.0	-10.6	-20.2	-50.2	-98.1
Wind Drift • Inches	0.0	0.3	1.4	3.3	6.0	9.6	14.3	27.0	44.6

Norma 410-grain FULL METAL JACKET (11052)

G1 Ballistic Coefficient = 0.307

Distance • Yards	Muzzle	50	100	150	200	250	300	400	500
Velocity • fps	2325	2191	2062	1937	1817	1703	1594	1397	1233
Energy • ft-lbs	4922	4373	3872	3417	3007	2641	2315	1778	1384
Taylor KO Index	56.7	53.4	50.2	47.2	44.3	41.5	38.8	34.0	30.0
Path • Inches	-0.9	+1.3	+1.7	0.0	-4.0	-10.6	-20.2	-50.2	-98.1
Wind Drift • Inches	0.0	0.3	1.4	3.3	6.0	9.6	14.3	27.0	44.6

.416 Ruger

The world needs another .416 cartridge like the return of the plague. The .416 Ruger is a nonbelted magnum with a cartridge-case length of 2.580 inches. That's only very slightly longer than the .458 Winchester cartridge; therefore, this cartridge will undoubtedly fit into standard length actions. The performance is in the same ballpark as the .416 Remington Magnum. Hornady worked with Ruger to develop this cartridge.

Relative Recoil Factor = 4.40

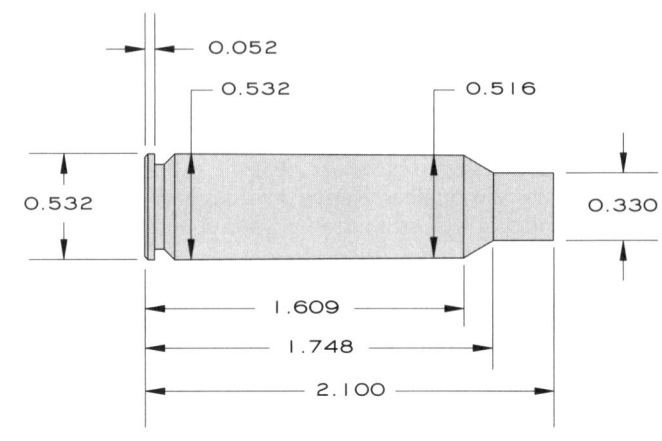

Specifications:

Controlling Agency for Standardization of this Ammunition: SAAMI

Bullet Weight Grains	Velocity fps	Maximum Average Pressure	
		Copper Crusher	Transducer
350	2,525	54,000 cup	65,000 psi
400	2,400	54,000 cup	65,000 psi

Standard barrel for velocity testing: 24 inches long—1 turn in 14-inch twist

Availability:

Hornady 400-grain DGX (82665)

G1 Ballistic Coefficient = 0.330

Distance • Yards	Muzzle	50	100	150	200	250	300	400	500
Velocity • fps	2400	2273	2151	2031	1917	1805	1700	1502	1331
Energy • ft-lbs	5116	4591	4109	3666	3264	2896	2568	2006	1573
Taylor KO Index	57.1	54.0	51.1	48.3	45.6	42.9	40.6	35.7	31.6
Path • Inches	-1.5	+1.2	+1.5	0.0	-3.6	-9.6	-18.3	-45.0	-87.2
Wind Drift • Inches	0.0	0.3	1.2	2.9	5.3	8.4	12.5	23.6	39.0

Hornady 400-grain DGS (82666)

G1 Ballistic Coefficient = 0.33

Distance • Yards	Muzzle	50	100	150	200	250	300	400	500
Velocity • fps	2400	2273	2151	2031	1917	1805	1700	1502	1331
Energy • ft-lbs	5116	4591	4109	3666	3264	2896	2568	2006	1573
Taylor KO Index	57.1	54.0	51.1	48.3	45.6	42.9	40.6	35.7	31.6
Path • Inches	-1.5	+1.2	+1.5	0.0	-3.6	-9.6	-18.3	-45.0	-87.2
Wind Drift • Inches	0.0	0.3	1.2	2.9	5.3	8.4	12.5	23.6	39.0

.416 Dakota

The .416 Dakota is the largest round that Dakota bases on the .404 Jeffery. Power-wise, it is right in the middle of the .416s, being a bit more potent than the .416 Rigby and Remington and a slight bit milder than the .416 Weatherby and Lazzeroni. I think it would take a lifetime of hunting to establish whether any practical difference is to be seen in the terminal performance of the whole collection of .416s.

Relative Recoil Factor = 4.56

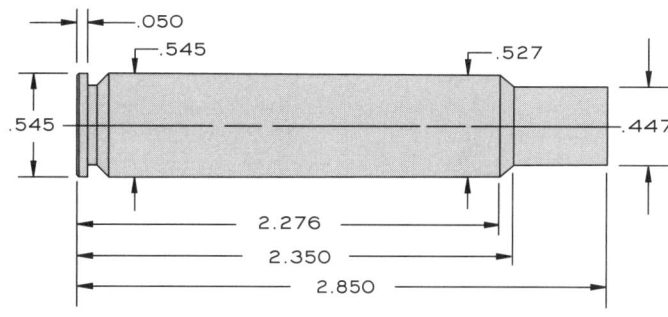

Specifications:

Controlling Agency for Standardization of this Ammunition: Factory

Standard barrel for velocity testing: 24 inches long

Availability:

Dakota 400-grain Swift A-Frame (416-400FAF)

G1 Ballistic Coefficient = 0.370

Distance • Yards	Muzzle	50	100	150	200	250	300	400	500
Velocity • fps	2500	2384	2271	2161	2055	1951	1851	1661	1488
Energy • ft-lbs	5553	5050	4583	4150	3750	3381	3042	2450	1966
Taylor KO Index	60.9	58.1	55.3	52.7	50.1	47.5	45.1	40.5	36.3
Path • Inches	-0.9	+1.0	+1.3	0.0	-3.2	-8.4	-16.0	-39.0	-74.7
Wind Drift • Inches	0.0	0.2	1.0	2.4	4.4	7.0	10.3	19.3	31.8

Dakota 350-grain Woodleigh Soft Nose (416-410WSN)

G1 Ballistic Coefficient = 0.357

Distance • Yards	Muzzle	50	100	150	200	250	300	-400	500
Velocity • fps	2500	2386	2274	2166	2060	1958	1858	1670	1499
Energy • ft-lbs	5691	5183	4710	4271	3865	3490	3145	2541	2045
Taylor KO Index	60.9	58.1	55.4	52.8	50.2	47.7	45.3	40.7	36.5
Path • Inches	-0.9	+1.0	+1.3	0.0	-3.2	-8.4	-15.9	38.8	-74.2
Wind Drift • Inches	0.0	0.2	1.0	2.4	4.3	6.9	10.2	19.0	31.3

Dakota 410-grain Woodleigh Solid (416-410WSO)

G1 Ballistic Coefficient = 0.357

Distance • Yards	Muzzle	50	100	150	200	250	300	400	500
Velocity • fps	2500	2374	2252	2134	2019	1908	1801	1600	1420
Energy • ft-lbs	5691	5134	4620	4146	3712	3314	2953	2331	1836
Taylor KO Index	60.9	57.8	54.9	52.0	49.2	46.5	43.9	39.0	34.6
Path • Inches	-0.9	+1.0	+1.4	0.0	-3.3	-8.7	-16.4	-40.4	-78.2
Wind Drift • Inches	0.0	0.3	1.1	2.6	4.8	7.7	11.4	21.4	35.3

.416 Weatherby Magnum

The .416 Weatherby was Weatherby's answer to the popularity of the .416 caliber generally. Introduced in 1989, it is really a .416-378, that is, a .378 Weatherby case necked to .416. That puts this cartridge's performance between the .378 Weatherby and the .460 Weatherby. That's pretty awesome company. The .416 Weatherby Magnum is a truly "African" caliber, suitable for any game anywhere.

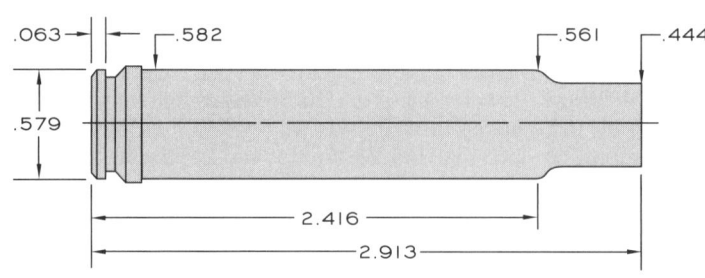

Relative Recoil Factor = 4.77

Specifications:

Controlling Agency for Standardization of this Ammunition: CIP

Bullet Weight Grains	Velocity fps	Maximum Average Pressure Copper Crusher	Transducer
N/S	N/S	3,800 bar	4,370 bar

Standard barrel for velocity testing: 26 inches long—1 turn in 14-inch twist

Availability:

Weatherby 350-grain Barnes-X (B 416 350 XS) LF
G1 Ballistic Coefficient = 0.520

Distance • Yards	Muzzle	50	100	150	200	250	300	400	500
Velocity • fps	2880	2790	2702	2615	2530	2447	2365	2207	2054
Energy • ft-lbs	6448	6051	5674	5317	4977	4655	4349	3785	3279
Taylor KO Index	59.9	58.0	56.2	54.4	52.6	50.9	49.2	45.9	42.7
Path • Inches	-0.9	+0.5	+0.9	0.0	-2.1	-5.6	-10.6	-25.2	-47.0
Wind Drift • Inches	0.0	0.2	0.6	1.4	2.5	3.9	5.7	10.5	17.0

Weatherby 400-grain Hornady Soft Point (H 416 400 RN)
G1 Ballistic Coefficient = 0.311

Distance • Yards	Muzzle	50	100	150	200	250	300	400	500
Velocity • fps	2700	2556	2417	2282	2152	2025	1903	1676	1470
Energy • ft-lbs	6474	5805	5189	4626	4113	3642	3216	2493	1918
Taylor KO Index	64.2	60.8	57.5	54.2	51.2	48.1	45.2	39.8	34.9
Path • Inches	-0.9	+0.8	+1.1	0.0	-2.8	-7.5	-14.3	-35.5	-69.2
Wind Drift • Inches	0.0	0.3	1.1	2.6	4.7	7.6	11.3	21.3	35.4

A² 400-grain Dead Tough SP, Monolithic Solid, Lion Load SP (None)
G1 Ballistic Coefficient = 0.32

Distance • Yards	Muzzle	50	100	150	200	250	300	400	500
Velocity • fps	2600	2463	2328	2202	2073	1957	1841	1624	1430
Energy • ft-lbs	6004	5390	4813	4307	3834	3402	3011	2343	1817
Taylor KO Index	61.8	58.5	55.3	52.3	49.3	46.5	43.8	38.6	34.0
Path • Inches	-0.9	+0.9	+1.2	0.0	-3.0	-8.1	-15.4	-38.1	-74.2
Wind Drift • Inches	0.0	0.3	1.1	2.6	4.8	7.8	11.5	21.8	36.1

10.57mm (.416) Lazzeroni Meteor

The 10.57mm Meteor is the largest caliber in the Lazzeroni line. As is true of all the Lazzeroni line, the 10.57 has performance considerably better than most of the older, more traditional cartridges in the same caliber. There's nothing mild about the Meteor. In fact, it drives a 400-grain bullet 100 fps faster than any other .416 cartridge. This is a cartridge for serious African hunters.

Relative Recoil Factor = 4.95

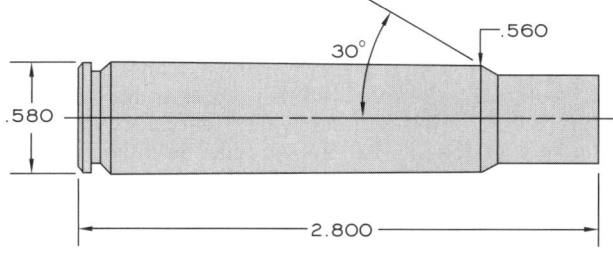

Specifications:

Controlling Agency for Standardization of this Ammunition: Factory

Availability:

Lazzeroni 400-grain (None)

G1 Ballistic Coefficient = 0.550

Distance Yards	Muzzle	50	100	150	200	250	300	400	500
Velocity • fps	2800	2716	2634	2554	2474	2396	2320	2171	2028
Energy • ft-lbs	6965	6555	6165	5793	5440	5102	4784	4190	3656
Taylor KO Index	66.6	64.6	62.6	60.7	58.8	57.0	55.1	51.6	48.2
Path • Inches	-0.9	+0.6	+0.9	0.0	-2.3	-5.9	-11.1	-26.5	-48.2
Wind Drift • Inches	0.0	0.1	0.6	1.3	2.4	3.8	5.6	10.3	16.6

Hunting with Rory Muil, Craig Boddington used a .416 Rigby to take this "PAC" (Problem Animal Control) elephant in Zimbabwe's Zambezi Valley. The various .416s are an ideal combination of penetration, power, and versatility, and are absolutely wonderful when hunting elephants. (Photo from *Elephant!* by Craig Boddington, 2012, Safari Press Archives)

.450 Nitro Express 3¼"

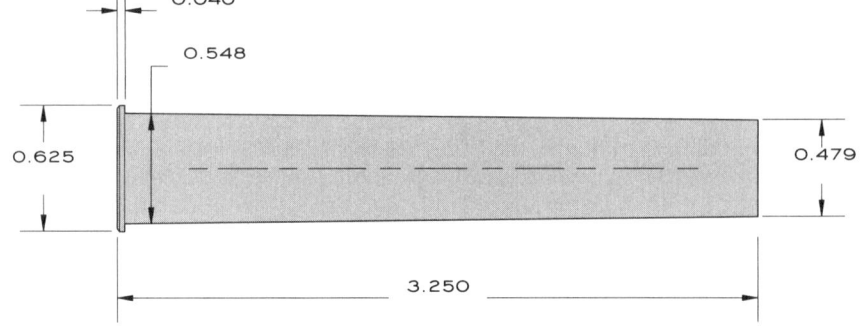

The .450 Nitro Express is a smokless number derived from an earlier black-powder cartridge of very similar dimensions. Rigby introduced this version about 1897. Hornady has recently introduced it into their lineup of "African"-type cartridges.

Relative Recoil Factor = 4.75

Specifications:

Controlling Agency for Standardization of this Ammunition: CIP

Bullet Weight Grains	Velocity fps	Maximum Average Pressure	
		Copper Crusher	Transducer
N/S	N/S	39,200 cup	44,300 psi

Standard barrel for velocity testing: 24 inches long

Availability:

Hornady 480-grain DGX (8255)

G1 Ballistic Coefficient = 0.550

Distance Yards	Muzzle	50	100	150	200	250	300	400	500
Velocity • fps	2150	2008	1872	1742	1618	1504	1397	1218	1087
Energy • ft-lbs	4927	4297	3733	3230	2792	2406	2080	1574	1268
Taylor KO Index	67.5	63.1	58.8	54.7	50.8	47.2	43.9	38.1	34.1
Path • Inches	-0.9	+1.7	+2.1	0.0	-5.0	-13.2	-25.4	-63.7	-125.6
Wind Drift • Inches	0.0	0.4	1.8	4.1	7.6	12.2	18.2	34.3	55.8

Hornady 480-grain DGS (8256)

G1 Ballistic Coefficient = 0.550

Distance Yards	Muzzle	50	100	150	200	250	300	400	500
Velocity • fps	2150	2008	1872	1742	1618	1504	1397	1218	1087
Energy • ft-lbs	4927	4297	3733	3230	2792	2406	2080	1574	1268
Taylor KO Index	67.5	63.1	58.8	54.7	50.8	47.2	43.9	38.1	34.1
Path • Inches	-0.9	+1.7	+2.1	0.0	-5.0	-13.2	-25.4	-63.7	-125.6
Wind Drift • Inches	0.0	0.4	1.8	4.1	7.6	12.2	18.2	34.3	55.8

Tim Danklef (left) and PH *Andrew Dawson prep traditional doubles. Danklef used his W. J. Jeffery .475 N 2 to take this fine Zimbabwe tusker. As always, Daws carried his William Evans .470. (Photo from* Elephant! Craig Boddington, 2012, Safari Press Archives)

.458 Winchester Magnum

The .458 Winchester Magnum is the largest of Winchester's slightly shortened, belted-magnum calibers. It was introduced in 1956, and its adaptability to standard-length actions made it popular in the U.S. The question of relative popularity between this cartridge and the British "African" cartridges is that the British cartridges are nearly always utilized in custom-made (read very expensive) rifles, while the .458 Winchester Magnum is available in the more or less over-the-counter Model 70 African. Performance-wise, the .458 Winchester Magnum is bullet-for-bullet about 450 fps slower than the .460 Weatherby Magnum, which is based on the much larger volume .378 Weatherby Magnum case.

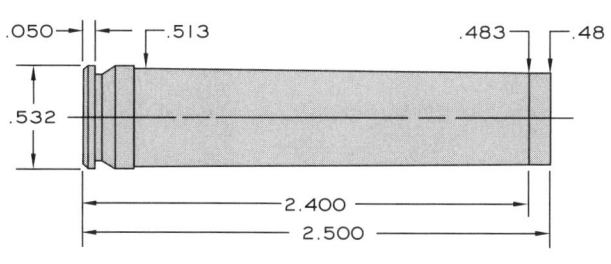

Relative Recoil Factor = 4.36

Specifications:

Controlling Agency for Standardization of this Ammunition: SAAMI

Bullet Weight Grains	Velocity fps	Maximum Average Pressure Copper Crusher	Transducer
500	2,025	53,000 cup	N/S
510	2,025	53,000 cup	N/S

Standard barrel for velocity testing: 24 inches long—1 turn in 14-inch twist

Availability:

Federal 400-grain Trophy Bonded Bear Claw (P458T1)

G1 Ballistic Coefficient = 0.353

Distance • Yards	Muzzle	50	100	150	200	250	300	400	500
Velocity • fps	2250	2136	2030	1917	1810	1714	1620	1440	1290
Energy • ft-lbs	4495	4052	3640	3265	2920	2609	2325	1845	1480
Taylor KO Index	58.9	55.9	53.1	50.2	47.4	44.9	42.4	37.7	32.8
Path • Inches	-0.9	+1.4	+1.8	0.0	-4.1	-10.9	-20.6	-50.4	-97.0
Wind Drift • Inches	0.0	0.3	1.3	2.9	5.4	8.6	12.7	23.8	39.1

A² 465-grain Dead Tough SP, Monolithic Solid, Lion Load SP (None)

G1 Ballistic Coefficient = 0.358

Distance • Yards	Muzzle	50	100	150	200	250	300	400	500
Velocity • fps	2220	2108	1999	1894	1791	1694	1601	1429	1280
Energy • ft-lbs	5088	4589	4127	3704	3312	2965	2563	2042	1639
Taylor KO Index	67.5	64.1	60.8	57.6	54.5	51.5	48.7	43.5	38.9
Path • Inches	-0.9	+1.4	+1.8	0.0	-4.2	-11.2	-21.1	-51.7	-99.4
Wind Drift • Inches	0.0	0.3	1.3	3.0	5.4	8.7	12.8	23.9	39.2

Hornady 500-grain DGX (85833)

G1 Ballistic Coefficient = 0.360

Distance • Yards	Muzzle	50	100	150	200	250	300	400	500
Velocity • fps	2140	2007	1880	1758	1643	1532	1432	1254	1119
Energy • ft-lbs	5084	4475	3926	3422	2996	2608	2276	1745	1392
Taylor KO Index	70.0	65.7	61.5	57.5	53.7	50.1	46.8	41.0	36.6
Path • Inches	-0.9	+1.7	+2.1	0.0	-4.9	-13.0	-24.8	-61.8	-120.9
Wind Drift • Inches	0.0	0.4	1.7	3.8	7.0	11.4	16.9	31.7	51.7

Norma 500-grain Swift A-Frame (11120)

G1 Ballistic Coefficient = 0.361

Distance • Yards	Muzzle	50	100	150	200	250	300	400	500
Velocity • fps	2116	2008	1903	1802	1705	1612	1523	1362	1226
Energy • ft-lbs	4972	4477	4022	3606	3227	2885	2577	2060	1669
Taylor KO Index	69.2	65.7	62.3	59.0	55.8	52.7	49.8	44.6	40.1
Path • Inches	-0.9	+1.6	+2.0	0.0	-4.7	-12.4	-23.4	-57.2	-109.9
Wind Drift • Inches	0.0	0.3	1.4	3.1	5.7	9.2	13.6	25.3	41.2

Federal 500-grain Swift A-Frame (P458SA)

G1 Ballistic Coefficient = 0.360

Distance • Yards	Muzzle	50	100	150	200	250	300	400	500
Velocity • fps	2090	1983	1880	1778	1680	1590	1500	1350	1210
Energy • ft-lbs	4850	4365	3915	3511	3145	2806	2510	2010	1630
Taylor KO Index	68.4	64.9	61.5	58.2	55.0	52.0	49.1	44.2	39.6
Path • Inches	-0.9	+1.7	+2.1	0.0	-4.8	-12.7	-24.1	-58.8	-112.8
Wind Drift • Inches	0.0	0.3	1.4	3.2	5.8	9.4	13.8	25.8	41.9

Federal 500-grain Trophy Bonded Bear Claw (P458T2)

G1 Ballistic Coefficient = 0.284

Distance • Yards	Muzzle	50	100	150	200	250	300	400	500
Velocity • fps	2090	1954	1820	1700	1580	1474	1370	1200	1080
Energy • ft-lbs	4850	4241	3685	3211	2775	2412	2080	1595	1285
Taylor KO Index	68.4	63.9	59.5	55.6	49.1	48.2	44.8	39.3	35.3
Path • Inches	-0.9	+1.8	+2.2	0.0	-5.2	-13.9	-26.6	-66.5	-130.4
Wind Drift • Inches	0.0	0.4	1.8	4.1	7.6	12.3	18.2	34.1	55.3

Winchester 500-grain Nosler Partition (S458WSLSP)

G1 Ballistic Coefficient = 0.330

Distance • Yards	Muzzle	50	100	150	200	250	300	400	500
Velocity • fps	2010	1896	1786	1680	1580	1486	1397	1243	1124
Energy • ft-lbs	4485	3991	3538	3135	2771	2451	2168	1715	1401
Taylor KO Index	65.8	62.0	58.4	55.0	51.7	48.6	45.7	40.7	36.8
Path • Inches	-0.9	+1.9	+2.3	0.0	-5.4	-14.3	-27.1	-66.7	-128.8
Wind Drift • Inches	0.0	0.4	1.6	3.7	6.8	10.9	16.1	29.9	48.1

Hornady 500-grain DGS (8585)

G1 Ballistic Coefficient = 0.360

Distance • Yards	Muzzle	50	100	150	200	250	300	400	500
Velocity • fps	2140	2007	1880	1758	1643	1532	1432	1254	1119
Energy • ft-lbs	5084	4475	3926	3422	2996	2608	2276	1745	1392
Taylor KO Index	70.0	65.7	61.5	57.5	53.7	50.1	46.8	41.0	36.6
Path • Inches	-0.9	+1.7	+2.1	0.0	-4.9	-13.0	-24.8	-61.8	-120.9
Wind Drift • Inches	0.0	0.4	1.7	3.8	7.0	11.4	16.9	31.7	51.7

Federal 500-grain Barnes Triple-Shock X-Bullet (P458D) LF

G1 Ballistic Coefficient = 0.41

Distance • Yards	Muzzle	50	100	150	200	250	300	400	500
Velocity • fps	2050	1957	1870	1780	1700	1615	1540	1390	1270
Energy • ft-lbs	4665	4255	3865	3520	3190	2897	2620	2150	1780
Taylor KO Index	67.1	64.0	61.2	58.2	55.6	52.8	50.4	45.5	41.5
Path • Inches	-0.9	+1.7	+2.1	0.0	-4.9	-12.8	-24.0	-57.8	-109.6
Wind Drift • Inches	0.0	0.3	1.2	2.8	5.2	8.2	12.1	22.4	36.4

Norma 500-grain Barnes Solid (11110) LF

G1 Ballistic Coefficient = 0.39

Distance • Yards	Muzzle	50	100	150	200	250	300	400	500
Velocity • fps	2067	1969	1874	1783	1694	1609	1528	1380	1251
Energy • ft-lbs	4745	4307	3902	3529	3188	2876	2594	2114	1738
Taylor KO Index	67.6	64.4	61.3	58.3	55.4	52.6	50.0	45.1	40.9
Path • Inches	-0.9	+1.7	+2.1	0.0	-4.8	-12.7	-23.9	-57.9	-110.2
Wind Drift • Inches	0.0	0.3	1.3	3.0	5.4	8.6	12.6	23.5	38.2

Federal 500-grain Barnes Banded Solid (P458E) LF

G1 Ballistic Coefficient = 0.39

Distance • Yards	Muzzle	50	100	150	200	250	300	400	500
Velocity • fps	2050	1953	1860	1767	1680	1594	1520	1370	1240
Energy • ft-lbs	4665	4234	3830	3466	3130	2822	2545	2075	1710
Taylor KO Index	67.1	63.9	60.8	57.8	55.0	52.1	49.7	44.8	40.6
Path • Inches	-0.9	+1.8	+2.1	0.0	-4.9	-12.9	-24.4	-59.0	-112.4
Wind Drift • Inches	0.0	0.3	1.3	3.0	5.4	8.7	12.8	17.8	38.7

Winchester 500-grain Nosler Solid (S458WSLS) LF

G1 Ballistic Coefficient = 0.24

Distance • Yards	Muzzle	50	100	150	200	250	300	400	500
Velocity • fps	2010	1856	1710	1573	1447	1333	1233	1083	986
Energy • ft-lbs	4485	3824	3246	2747	2323	1973	1688	1302	1080
Taylor KO Index	65.8	60.7	55.9	51.5	47.3	43.6	40.3	35.4	32.3
Path • Inches	-0.9	+2.0	+2.5	0.0	-6.0	-16.3	-31.4	-79.6	-157.3
Wind Drift • Inches	0.0	0.5	2.2	5.2	9.6	15.4	22.9	42.5	67.5

Federal 500-grain Trophy Bonded Sledgehammer Solid (P458T3)

G1 Ballistic Coefficient = 0.336

Distance • Yards	Muzzle	50	100	150	200	250	300	400	500
Velocity • fps	1950	1838	1730	1628	1530	1440	1350	1210	1100
Energy • ft-lbs	4220	3752	3320	2945	2595	2304	2030	1615	1335
Taylor KO Index	63.8	60.1	56.6	53.3	50.1	47.1	44.2	39.6	36.0
Path • Inches	-0.9	+2.1	+2.5	0.0	-5.8	-15.3	-28.9	-71.1	-136.9
Wind Drift • Inches	0.0	0.4	1.7	3.9	7.0	11.3	16.6	30.8	49.6

he Good Old Days! The elephant is a ninety-eight-pounder; the rifle is an early Winchester Model 70 African in .458 Winchester Magnum. Despite ll the bashing, the .458 Winchester has probably accounted for more sport-hunted elephants than any other cartridge since WW II! (Photo from lephant! by Craig Boddington, 2012, Safari Press Archives)

.458 Lott

Jack Lott decided in the late 1960s that the .458 Winchester needed "improvement." His answer was the .458 Lott, which was introduced in 1971 as a wildcat. The A-Square company picked up on the caliber and for a number of years has furnished brass and loaded ammunition to its specifications. In 2002, Hornady took the lead to obtain SAAMI standardization for this caliber. Now several companies are producing ammo for the .458 Lott. The .458 Lott uses a case that's 0.300 inches longer than the .458 Winchester (it can be thought of as a full-length magnum), increasing both the case volume and the ballistic performance.

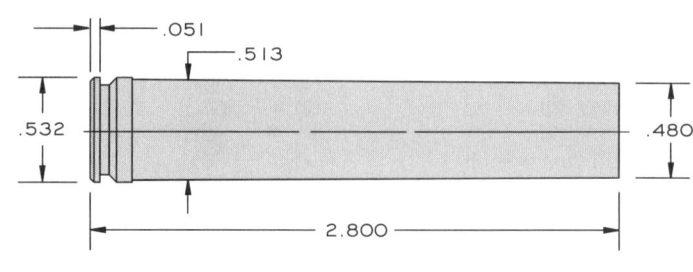

Relative Recoil Factor = 5.05

Specifications:

Controlling Agency for Standardization of this Ammunition: SAAMI

Bullet Weight Grains	Velocity fps	Maximum Average Pressure	
		Copper Crusher	Transducer
500	N/S	53,700 cup	62,400 psi

Standard barrel for velocity testing: 24 inches long—1 turn in 10-inch twist

Availability:

A² 465-grain Triad

G1 Ballistic Coefficient = 0.357

Distance • Yards	Muzzle	50	100	150	200	250	300	400	500
Velocity • fps	2380	2263	2150	2039	1932	1829	1730	1551	1378
Energy • ft-lbs	5848	5290	4773	4295	3855	3455	3091	2485	1962
Taylor KO Index	72.4	68.9	65.4	62.0	58.8	55.6	52.6	47.2	41.9
Path • Inches	-0.9	+1.2	+1.9	0.0	-3.6	-9.6	-18.1	-44.2	-85.2
Wind Drift • Inches	0.0	0.3	1.2	2.7	4.9	7.8	11.6	21.7	35.8

Federal 500-grain Trophy Bonded Bear Claw (P458LT1)

G1 Ballistic Coefficient = 0.282

Distance • Yards	Muzzle	50	100	150	200	250	300	400	500
Velocity • fps	2300	2156	2020	1883	1760	1634	1520	1320	1160
Energy • ft-lbs	5875	5180	4515	3937	3421	2965	2567	1933	1498
Taylor KO Index	75.2	70.5	66.1	61.6	57.6	53.5	49.7	43.2	37.9
Path • Inches	-0.9	+1.4	+1.8	0.0	-4.2	-11.2	-21.5	-54.0	-106.5
Wind Drift • Inches	0.0	0.4	1.6	3.6	6.7	10.8	16.1	30.4	50.3

Federal 500-grain Fusion Safari (F458LFS1)

G1 Ballistic Coefficient = 0.28

Distance • Yards	Muzzle	50	100	150	200	250	300	400	500
Velocity • fps	2300	2155	2020	1881	1750	1632	1520	1310	1160
Energy • ft-lbs	5875	5158	4505	3931	3405	2958	2550	1920	1485
Taylor KO Index	75.2	70.5	66.1	61.5	57.3	53.4	49.7	42.9	37.9
Path • Inches	-0.9	+1.4	+1.8	0.0	-4.2	-11.2	-21.6	-54.1	-106.8
Wind Drift • Inches	0.0	0.4	1.6	3.6	6.7	10.8	16.1	22.7	50.5

Hornady 500-grain DGX (82613)

G1 Ballistic Coefficient = 0.29

Distance • Yards	Muzzle	50	100	150	200	250	300	400	500
Velocity • fps	2300	2162	2028	1900	1777	1660	1549	1352	1192
Energy • ft-lbs	5872	5190	4567	4010	3506	3060	2665	2029	1577
Taylor KO Index	75.2	70.7	66.3	62.2	58.1	54.3	50.7	44.2	39.0
Path • Inches	-0.9	+1.4	+1.7	0.0	-4.1	-11.0	-21.1	-52.6	-103.2
Wind Drift • Inches	0.0	0.4	1.5	3.5	6.4	10.2	15.2	28.8	47.5

Federal 500-grain Barnes Triple-Shock X Bullet (P458LA) LF

G1 Ballistic Coefficient = 0.41

Distance • Yards	Muzzle	50	100	150	200	250	300	400	500
Velocity • fps	2280	2182	2090	1992	1900	1813	1730	1560	1420
Energy • ft-lbs	5770	5285	4825	4406	4000	3649	3305	2715	2225
Taylor KO Index	74.6	71.4	68.4	65.2	62.2	59.3	56.6	51.0	46.5
Path • Inches	-0.9	+1.3	+1.6	0.0	-3.8	-10.1	-19.0	-45.8	-86.8
Wind Drift • Inches	0.0	0.3	1.1	2.4	4.4	7.1	10.4	19.4	31.6

Cor-Bon 500-grain DPX (EH458L500DPX/10)

G1 Ballistic Coefficient = 0.280

Distance • Yards	Muzzle	50	100	150	200	250	300	400	500
Velocity • fps	2100	1962	1830	1704	1585	1474	1372	1200	1076
Energy • ft-lbs	4897	4275	3718	3224	2790	2412	2090	1598	1287
Taylor KO Index	68.7	64.2	59.9	55.7	51.9	48.2	44.9	39.3	35.2
Path • Inches	-0.9	+1.8	+2.2	0.0	-5.2	-13.8	-26.5	-66.4	-130.4
Wind Drift • Inches	0.0	0.4	1.8	4.2	7.7	12.4	18.4	34.5	56.0

Federal 500-grain Barnes Banded Solid (P458LG) LF

G1 Ballistic Coefficient = 0.395

Distance • Yards	Muzzle	50	100	150	200	250	300	400	500
Velocity • fps	2300	2196	2100	1997	1900	1809	1720	1550	1400
Energy • ft-lbs	5875	5357	4875	4429	4010	3635	3280	2670	2170
Taylor KO Index	75.2	71.8	68.7	65.3	62.2	59.2	56.3	50.7	45.8
Path • Inches	-0.9	+1.3	+1.6	0.0	-3.8	-10.0	-18.9	-45.7	-87.0
Wind Drift • Inches	0.0	0.3	1.1	2.5	4.6	7.4	10.8	20.2	33.1

Federal 500-grain Trophy Bonded Sledgehammer Solid (P458LT2)

G1 Ballistic Coefficient = 0.330

Distance • Yards	Muzzle	50	100	150	200	250	300	400	500
Velocity • fps	2300	2176	2060	1940	1830	1721	1620	1430	1270
Energy • ft-lbs	5875	5260	4690	4181	3700	3291	2900	2260	1780
Taylor KO Index	75.2	71.2	67.4	63.5	59.9	56.3	53.0	46.8	41.5
Path • Inches	-0.9	+1.3	+1.7	0.0	-4.0	-10.6	-20.1	-49.6	-96.1
Wind Drift • Inches	0.0	0.3	1.3	3.1	5.6	9.0	13.3	25.1	41.3

Hornady 500-grain DGS (8262)

G1 Ballistic Coefficient = 0.295

Distance • Yards	Muzzle	50	100	150	200	250	300	400	500
Velocity • fps	2300	2162	2028	1900	1777	1660	1549	1352	1192
Energy • ft-lbs	5872	5190	4567	4010	3506	3060	2665	2029	1577
Taylor KO Index	75.2	70.7	66.3	62.2	58.1	54.3	50.7	44.2	39.0
Path • Inches	-0.9	+1.4	+1.7	0.0	-4.1	-11.0	-21.1	-52.6	-103.2
Wind Drift • Inches	0.0	0.4	1.5	3.5	6.4	10.2	15.2	28.8	47.5

Norma 500-grain SOLID (11117)

G1 Ballistic Coefficient = 0.269

Distance • Yards	Muzzle	50	100	150	200	250	300	400	500
Velocity • fps	2300	2149	2003	1864	1731	1606	1485	1285	1130
Energy • ft-lbs	5875	5127	4457	3858	3328	2864	2462	1833	1419
Taylor KO Index	75.2	70.3	65.5	61.0	56.6	52.5	48.7	42.0	37.0
Path • Inches	-0.9	+1.4	+1.8	0.0	-4.3	-11.5	-22.0	-55.6	-110.2
Wind Drift • Inches	0.0	0.4	1.6	3.8	7.0	11.4	17.0	32.3	53.3

Cor-Bon 500-grain FPBS (EH458L500FPBS/10)

G1 Ballistic Coefficient = 0.280

Distance • Yards	Muzzle	50	100	150	200	250	300	400	500
Velocity • fps	2100	1962	1830	1704	1585	1474	1372	1200	1076
Energy • ft-lbs	4897	4275	3718	3224	2790	2412	2090	1598	1287
Taylor KO Index	68.7	64.2	59.9	55.7	51.9	48.2	44.9	39.3	35.2
Path • Inches	-0.9	+1.8	+2.2	0.0	-5.2	-13.8	-26.5	-66.4	-130.4
Wind Drift • Inches	0.0	0.4	1.8	4.2	7.7	12.4	18.4	34.5	56.0

Norma 550-grain Soft Nose (11113)

G1 Ballistic Coefficient = 0.328

Distance • Yards	Muzzle	50	100	150	200	250	300	400	500
Velocity • fps	2100	1982	1868	1758	1654	1554	1461	1295	1162
Energy • ft-lbs	4897	4361	3874	3434	3039	2683	2370	1862	1499
Taylor KO Index	68.7	64.8	62.1	57.5	54.1	50.8	47.8	42.4	38.0
Path • Inches	-0.9	+1.7	+2.1	0.0	-4.9	-13.0	-24.7	-60.8	-117.8
Wind Drift • Inches	0.0	0.4	1.5	3.5	6.4	10.3	15.3	28.6	46.6

Norma 550-grain Full Metal Jacket (11114)

G1 Ballistic Coefficient = 0.328

Distance • Yards	Muzzle	50	100	150	200	250	300	400	500
Velocity • fps	2100	1982	1868	1758	1654	1554	1461	1295	1162
Energy • ft-lbs	4897	4361	3874	3434	3039	2683	2370	1862	1499
Taylor KO Index	68.7	64.8	62.1	57.5	54.1	50.8	47.8	42.4	38.0
Path • Inches	-0.9	+1.7	+2.1	0.0	-4.9	-13.0	-24.7	-60.8	-117.8
Wind Drift • Inches	0.0	0.4	1.5	3.5	6.4	10.3	15.3	28.6	46.6

.450 Dakota

To complete its line of large-volume cartridges, Dakota changed to an even larger case than the .404 Jeffery. The basic case for the .450 Dakota is the .416 Rigby, but for comparison it is very much like a .378 (or .460) Weatherby without the belt. No matter what its origin, the actual case is huge and develops the performance you would expect from such a big boomer. You are not going to see many of these guns in benchrest matches.

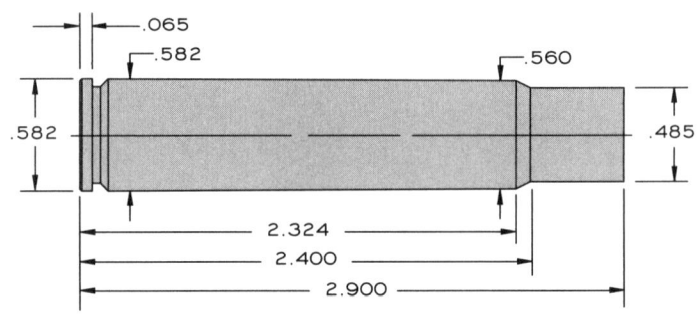

Relative Recoil Factor = 5.80

Specifications:

Controlling Agency for Standardization of this Ammunition: Factory

Standard barrel for velocity testing: 24 inches long

Availability:

Dakota 400-grain Swift A-Frame (450-400FAF)

G1 Ballistic Coefficient = 0.320

Distance • Yards	Muzzle	50	100	150	200	250	300	400	500
Velocity • fps	2700	2560	2425	2293	2166	2042	1923	1699	1496
Energy • ft-lbs	6477	5823	5223	4672	4167	3706	3266	2564	1988
Taylor KO Index	70.7	67.0	63.5	60.0	56.7	53.4	50.3	44.5	39.2
Path • Inches	-0.9	+0.8	+1.1	0.0	-2.8	-7.4	-14.1	-35.0	-68.0
Wind Drift • Inches	0.0	0.3	1.1	2.5	4.6	7.3	10.9	20.6	34.1

Dakota 500-grain Swift A-Frame (450-500FAF)

G1 Ballistic Coefficient = 0.361

Distance • Yards	Muzzle	50	100	150	200	250	300	400	500
Velocity • fps	2550	2430	2313	2199	2089	1982	1878	1681	1502
Energy • ft-lbs	7221	6577	5942	5372	4845	4360	3915	3139	2505
Taylor KO Index	83.4	79.5	75.7	71.9	68.3	64.8	61.4	55.0	49.1
Path • Inches	-0.9	+0.9	+1.3	0.0	-3.1	-8.1	-15.4	-37.7	-72.3
Wind Drift • Inches	0.0	0.2	1.0	2.4	4.4	7.0	10.3	19.3	31.9

Dakota 550-grain Woodleigh Soft Nose (450-550WSN)

G1 Ballistic Coefficient = 0.48

Distance • Yards	Muzzle	50	100	150	200	250	300	400	500
Velocity • fps	2450	2361	2275	2190	2106	2025	1946	1793	1649
Energy • ft-lbs	7332	6812	6321	5857	5420	5010	4624	3926	3320
Taylor KO Index	88.2	85.0	81.9	78.8	75.8	66.2	63.7	58.7	53.9
Path • Inches	-0.9	+1.0	+1.3	0.0	-3.2	-8.3	-15.5	-37.0	-69.4
Wind Drift • Inches	0.0	0.2	0.8	1.9	3.4	5.4	7.9	14.6	23.8

Dakota 550-grain Woodleigh Solid (450-550WSO)

G1 Ballistic Coefficient = 0.42

Distance • Yards	Muzzle	50	100	150	200	250	300	400	500
Velocity • fps	2450	2350	2253	2158	2065	1974	1886	1718	1562
Energy • ft-lbs	7332	6748	6200	5688	5209	4762	4347	3607	2981
Taylor KO Index	88.2	84.6	81.1	77.7	74.3	71.0	67.9	61.8	56.2
Path • Inches	-0.9	+1.0	+1.4	0.0	-3.2	-8.5	-16.0	-38.5	-73.0
Wind Drift • Inches	0.0	0.2	0.9	2.1	3.8	6.2	9.1	16.8	27.5

Dakota 600-grain Barnes Solid (450-600BSO) LF

G1 Ballistic Coefficient = 0.45

Distance • Yards	Muzzle	50	100	150	200	250	300	400	500
Velocity • fps	2350	2259	2169	2082	1996	1913	1832	1677	1534
Energy • ft-lbs	7359	6798	6270	5774	5311	4877	4473	3749	3134
Taylor KO Index	92.3	88.7	85.1	81.7	78.4	75.1	71.9	65.8	60.2
Path • Inches	-0.9	+1.2	+1.5	0.0	-3.5	-9.2	-17.2	-41.4	-77.9
Wind Drift • Inches	0.0	0.2	0.9	2.1	3.8	6.1	9.0	16.6	27.1

.460 Weatherby Magnum

In terms of muzzle energy, the .460 Weatherby until recently was the most "powerful" American cartridge available to the serious hunter. The cartridge case is from the same basic head size as the .378. With factory loadings that produce in excess of 7,500 ft-lbs of muzzle energy, this is enough gun for any game anywhere. The .460 Weatherby Magnum is a caliber that very few people shoot for "fun." The recoil is fearsome even with a heavy gun. Weatherby introduced the .460 in 1958.

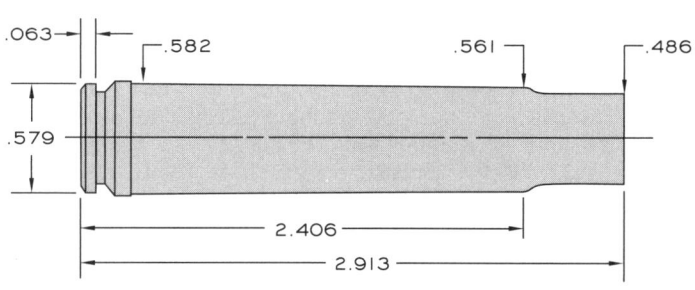

Relative Recoil Factor = 5.86

Specifications:

Controlling Agency for Standardization of this Ammunition: CIP

Bullet Weight Grains	Velocity fps	Maximum Average Pressure	
		Copper Crusher	Transducer
N/S	N/S	3,800 bar	4,370 bar

Typical barrel for velocity testing: 26 inches long—1 turn in 16-inch twist

Availability:

Weatherby 450-grain Barnes-X (B 460 450 TSX)

G1 Ballistic Coefficient = 0.368

Distance • Yards	Muzzle	500	100	150	200	250	300	400	500
Velocity • fps	2660	2539	2422	2308	2197	2088	1983	1783	1598
Energy • ft-lbs	7072	6445	5864	5322	4823	4357	3932	3178	2553
Taylor KO Index	78.3	74.8	71.3	68.0	64.7	61.5	58.4	52.5	47.0
Path • Inches	-0.9	+0.8	+1.1	0.0	-2.8	-7.3	-13.9	-33.9	-64.9
Wind Drift • Inches	0.0	0.2	1.0	2.2	4.0	6.4	9.4	17.7	29.1

Weatherby 500-grain Round Nose-Expanding (H 460 500 RN)

G1 Ballistic Coefficient = 0.287

Distance • Yards	Muzzle	50	100	150	200	250	300	400	500
Velocity • fps	2600	2448	2301	2158	2022	1890	1764	1533	1333
Energy • ft-lbs	7504	6654	5877	5174	4539	3965	3456	2608	1972
Taylor KO Index	85.1	80.1	75.3	70.6	66.1	61.8	57.7	50.2	43.6
Path • Inches	-0.9	+0.9	+1.3	0.0	-3.2	-8.4	-16.1	-40.4	-79.3
Wind Drift • Inches	0.0	0.3	1.3	3.0	5.4	8.8	13.1	25.0	41.7

A² 500-grain Dead Tough SP, Monolithic Solid, Lion Load SP (None)

G1 Ballistic Coefficient = 0.377

Distance • Yards	Muzzle	50	100	150	200	250	300	400	500
Velocity • fps	2580	2464	2349	2241	2131	2030	1929	1737	1560
Energy • ft-lbs	7389	6743	6126	5578	5040	4576	4133	3351	2702
Taylor KO Index	84.4	80.6	76.8	73.3	69.7	66.4	63.1	56.8	51.0
Path • Inches	-0.9	+0.9	+1.2	0.0	-3.0	-7.8	-14.8	-36.0	-68.8
Wind Drift • Inches	0.0	0.2	1.0	2.2	4.1	6.5	9.6	18.0	29.6

Weatherby 500-grain Full Metal Jacket (B 460 500 FJ)

G1 Ballistic Coefficient = 0.295

Distance • Yards	Muzzle	50	100	150	200	250	300	400	500
Velocity • fps	2600	2452	2309	2170	2037	1907	1784	1557	1357
Energy • ft-lbs	7504	6676	5917	5229	4605	4039	3534	2690	2046
Taylor KO Index	85.1	80.2	75.5	71.0	66.6	62.4	58.4	50.9	44.4
Path • Inches	0.9	+0.9	+1.3	0.0	-3.1	-8.3	-15.9	-39.8	-78.3
Wind Drift • Inches	0.0	0.3	1.2	2.9	5.3	8.5	12.7	24.1	40.2

.500/.465 Nitro Express

The most common practice in the U.S. when naming "necked to" cartridges is to put the caliber first and the cartridge of origin second. An example would be the .30-.378 Weatherby which is a .378 WM case necked to .30 caliber. The Brits seem to do just the opposite. The .500/.465 Nitro Express is basically a .500 Nitro Express (it's actually closer to a .470 NE case) necked to .465 caliber. This cartridge was introduced in 1906 or 1907 by Holland & Holland when the British government outlawed .450 caliber guns in India.

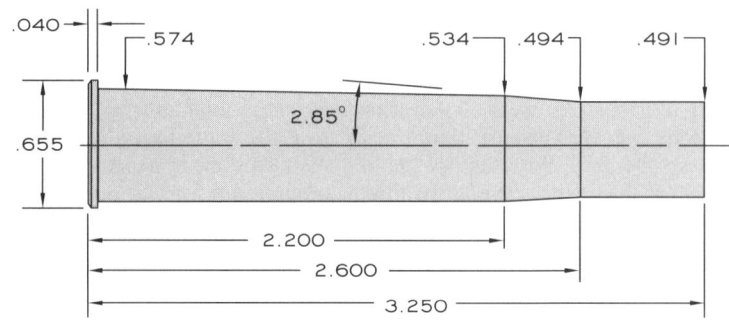

Relative Recoil Factor = 4.59

Specifications:

Controlling Agency for Standardization of this Ammunition: CIP

Bullet Weight Grains	Velocity fps	Maximum Average Pressure Copper Crusher	Transducer
N/S	N/S	2,200 bar	2,450 bar

Standard barrel for velocity testing: 26 inches long—1 turn in 30-inch twist

Availability:

A² 480-grain Dead Tough SP, Monolithic Solid, Lion Load SP (None)

G1 Ballistic Coefficient = 0.350

Distance • Yards	Muzzle	50	100	150	200	250	300	400	500
Velocity • fps	2150	2038	1926	1823	1722	1626	1534	1366	1226
Energy • ft-lbs	4926	4426	3960	3545	3160	2817	2507	1990	1601
Taylor KO Index	69.7	66.1	62.5	59.1	55.9	52.7	49.8	44.3	39.8
Path • Inches	-0.9	+1.6	+2.0	0.0	-4.6	-12.1	-22.8	-56.1	-108.1
Wind Drift • Inches	0.0	0.3	1.4	3.2	5.8	9.3	13.7	25.7	42.0

Kynoch 480-grain Soft Point

G1 Ballistic Coefficient = 0.41

Distance • Yards	Muzzle	50	100	150	200	250	300	400	500
Velocity • fps	2150	2054	1962	1872	1784	1700	1619	1467	1332
Energy • ft-lbs	4930	4490	4100	3735	3394	3091	2793	2295	1891
Taylor KO Index	69.7	66.6	63.6	60.7	57.9	55.1	52.5	47.6	43.2
Path • Inches	-0.9	+1.5	+1.9	0.0	-4.4	-11.5	-21.6	-52.1	-98.9
Wind Drift • Inches	0.0	0.3	1.2	2.6	4.8	7.7	11.3	21.0	34.3

.470 Nitro Express

If there is one cartridge that illustrates the "African" tradition, it would be the .470 Nitro Express. It is a large rimmed cartridge in the classic British tradition. Being rimmed, its use is largely confined to double rifles. Most of the traditional British African cartridges have one thing in common: They drive a large caliber, heavy bullet at velocities in the 2,000 to 2,300 fps range. When you have a 500-grain bullet humping along at 2,150 fps, you have tremendous stopping power, and the double rifle provides the shooter with a reserve shot without any manipulating of the gun. While the recoil of these cartridges is formidable, the weight of a double rifle tames down the worst of the jolt. Besides, who feels recoil in the face of a charging buffalo?

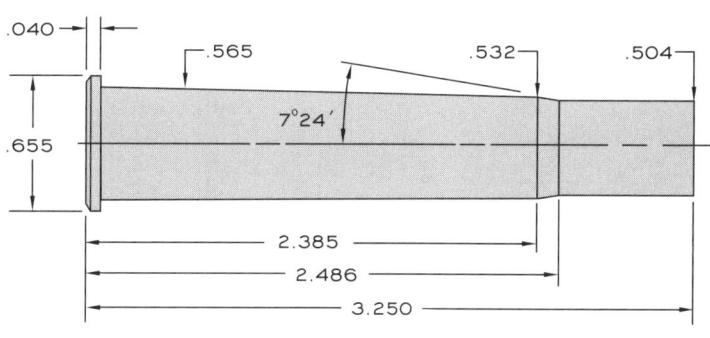

Relative Recoil Factor = 4.84

Specifications:

Controlling Agency for Standardization of this Ammunition: CIP

Bullet Weight Grains	Velocity fps	Maximum Average Pressure	
		Copper Crusher	Transducer
N/S	N/S	2,400 bar	2,700 bar

Standard barrel for velocity testing: 26 inches long—1 turn in 21-inch twist

Availability:

A² 500-grain Dead Tough SP, Monolithic Solid, Lion Load SP (None)
G1 Ballistic Coefficient = 0.325

Distance • Yards	Muzzle	50	100	150	200	250	300	400	500
Velocity • fps	2150	2029	1912	1800	1693	1590	1493	1320	1180
Energy • ft-lbs	5132	4572	4058	3597	3182	2806	2475	1935	1545
Taylor KO Index	72.9	68.8	64.8	61.0	57.4	53.9	50.6	44.7	40.0
Path • Inches	-0.9	+1.6	+2.0	0.0	-4.7	-12.4	-23.6	58.2	-113.0
Wind Drift • Inches	0.0	0.4	1.5	3.4	6.3	10.1	15.0	28.1	38.3

Federal 500-grain Swift A-Frame (P470SA)
G1 Ballistic Coefficient = 0.363

Distance • Yards	Muzzle	50	100	150	200	250	300	400	500
Velocity • fps	2150	2042	1940	1835	1740	1643	1550	1390	1250
Energy • ft-lbs	5130	4629	4160	3739	3350	2998	2680	2140	1730
Taylor KO Index	72.8	69.1	65.7	62.1	58.9	55.6	52.5	47.1	42.3
Path • Inches	-0.9	+1.6	+1.9	0.0	-4.5	-11.9	-22.6	-55.1	-105.6
Wind Drift • Inches	0.0	0.3	1.3	3.0	5.6	8.9	13.2	24.6	40.1

Federal 500-grain Trophy Bonded Bear Claw (P470T1)
G1 Ballistic Coefficient = 0.299

Distance • Yards	Muzzle	50	100	150	200	250	300	400	500
Velocity • fps	2150	2019	1890	1771	1660	1547	1450	1270	1130
Energy • ft-lbs	5130	4526	3975	3485	3045	2659	2320	1785	1420
Taylor KO Index	72.8	68.4	64.0	60.0	56.2	52.4	49.1	43.0	38.3
Path • Inches	-0.9	+1.6	+2.0	0.0	-4.8	-12.8	-24.4	-60.7	-118.6
Wind Drift • Inches	0.0	0.4	1.6	3.8	6.9	11.1	16.5	31.0	50.6

Hornady 500-grain DGX (8263)
G1 Ballistic Coefficient = 0.290

Distance • Yards	Muzzle	50	100	150	200	250	300	400	500
Velocity • fps	2150	2015	1885	1760	1643	1531	1429	1249	1114
Energy • ft-lbs	5132	4508	3946	3442	2998	2604	2267	1732	1378
Taylor KO Index	72.8	68.2	63.8	59.6	55.6	51.8	48.4	42.3	37.7
Path • Inches	-0.9	+1.6	+2.0	0.0	-4.9	-12.9	-24.7	-61.7	-121.1
Wind Drift • Inches	0.0	0.4	1.7	3.9	7.1	11.5	17.1	32.2	52.5

Norma 500-grain Soft Nose (11205)
G1 Ballistic Coefficient = 0.396

Distance • Yards	Muzzle	50	100	150	200	250	300	400	500
Velocity • fps	2100	2002	1906	1814	1725	1639	1575	1406	1274
Energy • ft-lbs	4897	4449	4035	3654	3304	2984	2693	2195	1802
Taylor KO Index	71.1	67.8	64.5	61.4	58.4	55.5	52.7	47.6	43.1
Path • Inches	-0.9	+1.6	+2.0	0.0	-4.7	-12.2	-23.1	-55.8	-106.2
Wind Drift • Inches	0.0	0.3	1.2	2.9	5.2	8.4	12.3	22.9	37.2

Federal 500-grain Barnes Triple-Shock X-Bullet (P470C) LF
G1 Ballistic Coefficient = 0.363

Distance • Yards	Muzzle	50	100	150	200	250	300	400	500
Velocity • fps	2150	2042	1940	1835	1740	1643	1550	1390	1250
Energy • ft-lbs	5130	4629	4160	3739	3350	2998	2680	2140	1730
Taylor KO Index	72.8	69.1	65.7	62.1	58.9	55.6	52.5	47.1	42.3
Path • Inches	-0.9	+1.6	+1.9	0.0	-4.5	-11.9	-22.6	-55.1	-105.6
Wind Drift • Inches	0.0	0.3	1.3	3.0	5.6	8.9	13.2	24.6	40.1

Hornady 500-grain DGS (8264)
G1 Ballistic Coefficient = 0.290

Distance • Yards	Muzzle	50	100	150	200	250	300	400	500
Velocity • fps	2150	2015	1885	1760	1643	1531	1429	1249	1114
Energy • ft-lbs	5132	4508	3946	3442	2998	2604	2267	1732	1378
Taylor KO Index	72.8	68.2	63.8	59.6	55.6	51.8	48.4	42.3	37.7
Path • Inches	-0.9	+1.6	+2.0	0.0	-4.9	-12.9	-24.7	-61.7	-121.1
Wind Drift • Inches	0.0	0.4	1.7	3.9	7.1	11.5	17.1	32.2	52.5

Kynoch 500-grain Solid or Soft Nose
G1 Ballistic Coefficient = 0.382

Distance • Yards	Muzzle	50	100	150	200	250	300	400	500
Velocity • fps	2125	2023	1923	1827	1734	1644	1559	1402	1266
Energy • ft-lbs	5015	4543	4107	3706	3338	3003	2699	2183	1780
Taylor KO Index	72.0	68.6	65.2	61.9	58.8	55.7	52.8	47.5	42.9
Path • Inches	-0.9	+1.6	+2.0	0.0	-4.6	-12.0	-22.8	-55.3	-105.6
Wind Drift • Inches	0.0	0.3	1.3	2.9	5.4	8.6	12.6	23.5	38.3

Cor-Bon 500-grain DPX (EH470500DPX/10)
G1 Ballistic Coefficient = 0.300

Distance • Yards	Muzzle	50	100	150	200	250	300	400	500
Velocity • fps	2100	1971	1847	1728	1616	1510	1411	1241	1112
Energy • ft-lbs	4897	4314	3789	3318	2900	2532	2212	1710	1374
Taylor KO Index	71.1	66.7	62.5	58.5	54.7	51.1	47.8	42.0	37.6
Path • Inches	-0.9	+1.7	+2.2	0.0	-5.1	-13.4	-26.7	-63.8	-124.5
Wind Drift • Inches	0.0	0.4	1.7	4.0	7.1	11.4	17.0	31.8	51.7

Federal 500-grain Trophy Bonded Sledgehammer Solid (P470T2)
G1 Ballistic Coefficient = 0.278

Distance • Yards	Muzzle	50	100	150	200	250	300	400	500
Velocity • fps	2150	2009	1880	1745	1630	1508	1410	1230	1090
Energy • ft-lbs	5130	4483	3900	3381	2935	2526	2195	1670	1330
Taylor KO Index	72.8	68.0	63.7	59.1	55.2	51.1	47.7	41.6	36.9
Path • Inches	-0.9	+1.7	+2.1	0.0	-4.9	-13.2	-25.2	-63.3	-124.7
Wind Drift • Inches	0.0	0.4	1.8	4.1	7.5	12.1	18.0	33.8	55.2

Federal 500-grain Barnes Banded Solid (P470D) LF
G1 Ballistic Coefficient = 0.24

Distance • Yards	Muzzle	50	100	150	200	250	300	400	500
Velocity • fps	2150	1988	1840	1688	1560	1426	1320	1140	1020
Energy • ft-lbs	5130	4389	3740	3136	2690	2260	1930	1440	1165
Taylor KO Index	72.8	67.3	62.3	57.2	52.8	48.3	44.7	38.6	34.5
Path • Inches	-0.9	+1.7	+2.2	0.0	-5.2	-14.1	-27.2	-69.4	-138.5
Wind Drift • Inches	0.0	0.5	2.0	4.8	8.8	14.3	21.3	40.2	64.8

Cor-Bon 500-grain FPBS (EH470500FPBS/10)
G1 Ballistic Coefficient = 0.30

Distance • Yards	Muzzle	50	100	150	200	250	300	400	500
Velocity • fps	2100	1971	1847	1728	1616	1510	1411	1241	1112
Energy • ft-lbs	4897	4314	3789	3318	2900	2532	2212	1710	1374
Taylor KO Index	71.1	66.7	62.5	58.5	54.7	51.1	47.8	42.0	37.6
Path • Inches	-0.9	+1.7	+2.2	0.0	-5.1	-13.4	-26.7	-63.8	-124.5
Wind Drift • Inches	0.0	0.4	1.7	4.0	7.1	11.4	17.0	31.8	51.7

Norma 500-grain Full Metal Jacket (11206)

G1 Ballistic Coefficient = 0.391

Distance • Yards	Muzzle	50	100	150	200	250	300	400	500
Velocity • fps	2100	2001	1904	1811	1721	1634	1551	1399	1266
Energy • ft-lbs	4897	4445	4027	3642	3290	2965	2672	2172	1781
Taylor KO Index	71.1	67.7	64.5	61.3	58.3	55.3	52.5	47.4	42.9
Path • Inches	-0.9	+1.3	+2.0	0.0	-4.7	-12.3	-23.2	-56.1	-106.9
Wind Drift • Inches	0.0	0.3	1.3	2.9	5.3	8.5	12.5	23.2	37.8

...irty years ago it was rare to see a double rifle on safari, but in 2009 on one safari there were a 500/.465, .600, .577, and .700! A double rifle is not ...ential for elephant hunting, and it is not for everyone, but the double has made a dramatic comeback. (Photo from *Elephant!* by Craig Boddington, ...12, Safari Press Archives)

.475 Nitro Express Number 2

The .475 Nitro Express Number 2 was, like many other British cartridges, introduced in 1907. This cartridge has a case head somewhat larger than the .470 and .500 NE cartridges and has a case length of 3.500 inches. It's a big one. Over the years there have been a variety of loadings (some very mild and some pretty hot) for this cartridge, so care must be used when selecting ammunition for an old gun.

Relative Recoil Factor = 4.95

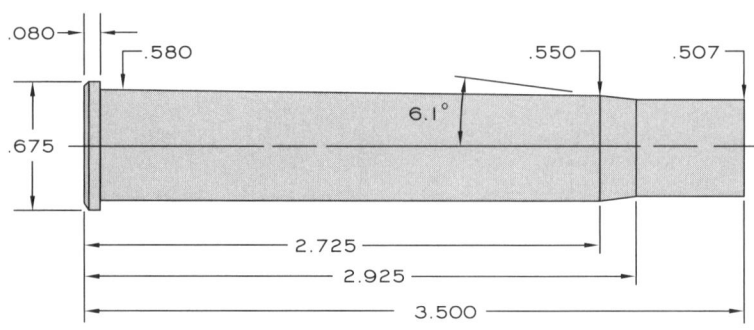

Specifications:

Controlling Agency for Standardization of this Ammunition: CIP

Bullet Weight Grains	Velocity fps	Maximum Average Pressure	
		Copper Crusher	Transducer
N/S	N/S	2,450 bar	2,750 bar

Standard barrel for velocity testing: 26 inches long—1 turn in 18-inch twist

Availability:

Kynoch 480-grain Solid or Soft Nose

G1 Ballistic Coefficient = 0.348

Distance • Yards	Muzzle	50	100	150	200	250	300	400	500
Velocity • fps	2200	2084	1974	1867	1763	1664	1570	1397	1250
Energy • ft-lbs	5160	4636	4155	3716	3315	2953	2627	2081	1665
Taylor KO Index	71.8	68.0	64.0	60.9	57.4	54.3	51.2	45.6	40.8
Path • Inches	-0.9	+1.5	+1.8	0.0	-4.4	-11.5	-21.8	-53.4	-103.1
Wind Drift • Inches	0.0	0.3	1.3	3.1	5.6	9.0	13.4	25.1	41.0

A² 500-grain Dead Tough SP, Monolithic Solid, Lion Load SP (None)

G1 Ballistic Coefficient = 0.33

Distance • Yards	Muzzle	50	100	150	200	250	300	400	500
Velocity • fps	2200	2080	1964	1852	1744	1641	1548	1366	1219
Energy • ft-lbs	5375	4804	4283	3808	3378	2991	2645	2073	1650
Taylor KO Index	74.8	70.7	66.8	63.0	59.3	55.8	52.6	46.4	41.4
Path • Inches	-0.9	+1.5	+1.9	0.0	-4.4	-11.7	-22.2	-54.7	-106.0
Wind Drift • Inches	0.0	0.3	1.4	3.2	5.9	9.6	14.1	26.5	43.5

.495 A-Square

Art Alphin's .495 A² cartridge is basically a slightly shortened .460 Weatherby case necked to hold a 0.510-diameter bullet. When it was designed in 1977, it was intended to provide a cartridge firing .50-caliber bullets that could be adapted to existing bolt actions. Performance-wise, this cartridge is more powerful than either the .500 Nitro Express or the .505 Gibbs.

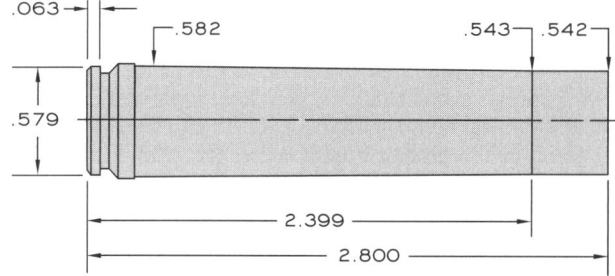

Relative Recoil Factor = 6.03

Specifications:

Controlling Agency for Standardization of this Ammunition: Factory

Standard barrel for velocity testing: 26 inches long—1 turn in 10-inch twist

Availability:

A² 570-grain Dead Tough SP, Monolithic Solid, Lion Load SP (None)

G1 Ballistic Coefficient = 0.348

Distance • Yards	Muzzle	50	100	150	200	250	300	400	500
Velocity • fps	2350	2231	2117	2003	1896	1790	1690	1504	1340
Energy • ft-lbs	6989	6302	5671	5081	4552	4058	3616	2863	2272
Taylor KO Index	97.6	92.7	87.9	83.2	78.7	74.3	70.2	62.5	55.6
Path • Inches	-0.9	+1.2	+1.6	0.0	-3.8	-9.9	-18.8	-46.1	-88.9
Wind Drift • Inches	0.0	0.3	1.2	2.8	5.1	8.2	12.2	22.8	37.6

Many times it's the cover and the elephant that determines the shot you get, so while you might prefer a body shot to a brain shot—or vice versa—you should study all the vital shots and be ready for what's available. (Photo from *Elephant!* by Craig Boddington, 2012, Safari Press Archives)

.500 Nitro Express 3"

This cartridge was first introduced as a black-powder version (the .500 Express) about 1880. The smokeless-powder version came along about 1890. While it is an excellent cartridge and was highly respected by some of the professional hunters, the .500 NE never reached the popularity of the .470 NE.

Relative Recoil Factor = 5.52

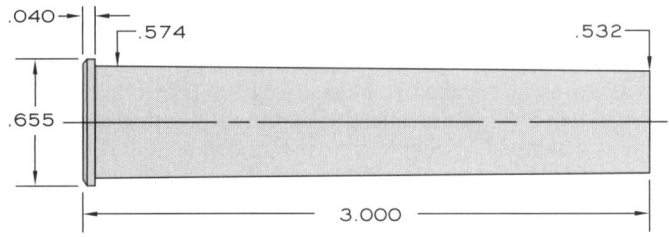

Specifications:

Controlling Agency for Standardization of this Ammunition: CIP

Bullet Weight Grains	Velocity fps	Maximum Average Pressure Copper Crusher	Transducer
N/S	N/S	2,500 bar	2,800 bar

Standard barrel for velocity testing: 26 inches long—1 turn in 15-inch twist

Availability:

A² 570-grain Dead Tough SP, Monolithic Solid, Lion Load SP (None)
G1 Ballistic Coefficient = 0.350

Distance • Yards	Muzzle	50	100	150	200	250	300	400	500
Velocity • fps	2150	2038	1928	1823	1722	1626	1534	1366	1226
Energy • ft-lbs	5850	5256	4703	4209	3752	3345	2977	2364	1901
Taylor KO Index	89.3	84.6	80.1	75.7	71.5	67.5	63.7	56.7	50.9
Path • Inches	-0.9	+1.6	+2.0	0.0	-4.6	-12.1	-22.9	-56.0	-108.1
Wind Drift • Inches	0.0	0.3	1.4	3.2	5.8	9.3	13.7	25.7	42.0

Hornady 570-grain DGX (8268)
G1 Ballistic Coefficient = 0.280

Distance • Yards	Muzzle	50	100	150	200	250	300	400	500
Velocity • fps	2150	2013	1881	1755	1635	1524	1419	1240	1106
Energy • ft-lbs	5850	5130	4477	3901	3384	2939	2547	1946	1549
Taylor KO Index	88.9	83.6	77.8	72.6	67.6	63.0	58.7	51.3	45.8
Path • Inches	-0.9	+1.6	+2.1	0.0	-4.9	-13.0	-24.9	-62.2	-122.2
Wind Drift • Inches	0.0	0.4	1.7	4.0	7.2	11.7	17.4	32.7	53.4

Federal 570-grain Swift A-Frame (P500NSA)
G1 Ballistic Coefficient = 0.300

Distance • Yards	Muzzle	50	100	150	200	250	300	400	500
Velocity • fps	2100	1973	1850	1734	1630	1518	1420	1250	1120
Energy • ft-lbs	5580	4929	4335	3807	3340	2919	2560	1985	1595
Taylor KO Index	86.9	81.6	76.5	71.7	67.4	62.8	58.7	51.7	46.3
Path • Inches	-0.9	+1.7	+2.1	0.0	-5.0	-13.4	-25.5	-63.2	-123.2
Wind Drift • Inches	0.0	0.4	1.6	3.8	7.0	11.2	16.6	31.2	50.7

Norma 570-grain Soft Nose (11301)
G1 Ballistic Coefficient = 0.38

Distance • Yards	Muzzle	50	100	150	200	250	300	400	500
Velocity • fps	2100	2000	1903	1809	1719	1632	1548	1396	1263
Energy • ft-lbs	5583	5064	4585	4146	3742	3371	3036	2466	2020
Taylor KO Index	86.9	82.7	78.7	74.8	71.1	67.5	64.1	57.7	52.0
Path • Inches	-0.9	+1.6	+2.0	0.0	-4.7	-12.3	-23.2	-56.3	-107.2
Wind Drift • Inches	0.0	0.3	1.3	2.9	5.3	8.5	12.6	23.4	38.0

Kynoch 570-grain Solid or Soft Nose
G1 Ballistic Coefficient = 0.38

Distance • Yards	Muzzle	50	100	150	200	250	300	400	500
Velocity • fps	2150	2048	1948	1851	1758	1688	1532	1423	1284
Energy • ft-lbs	5850	5300	4800	4337	3911	3521	3168	2563	2088
Taylor KO Index	89.3	85.0	80.9	76.9	73.0	70.1	65.7	59.1	53.3
Path • Inches	-0.9	+1.6	+1.9	0.0	-4.5	-11.7	-22.1	-53.8	-102.7
Wind Drift • Inches	0.0	0.3	1.2	2.9	5.2	8.4	12.4	23.0	37.5

Federal 570-grain Barnes Triple-Shock X Bullet (P500NC) LF

G1 Ballistic Coefficient = 0.369

Distance • Yards	Muzzle	50	100	150	200	250	300	400	500
Velocity • fps	2100	1995	1890	1794	1700	1609	1520	1370	1230
Energy • ft-lbs	5580	5038	4530	4076	3655	3277	2935	2355	1915
Taylor KO Index	86.9	82.5	78.2	74.2	70.3	66.6	62.9	56.7	50.9
Path • Inches	-0.9	+1.7	+2.0	0.0	-4.8	-12.5	-23.6	-57.6	-110.2
Wind Drift • Inches	0.0	0.3	1.3	3.1	5.6	9.0	13.3	24.8	40.5

Hornady 570-grain DGS (8269)

G1 Ballistic Coefficient = 0.286

Distance • Yards	Muzzle	50	100	150	200	250	300	400	500
Velocity • fps	2150	2013	1881	1755	1635	1524	1419	1240	1106
Energy • ft-lbs	5850	5130	4477	3901	3384	2939	2547	1946	1549
Taylor KO Index	88.9	83.6	77.8	72.6	67.6	63.0	58.7	51.3	45.8
Path • Inches	-0.9	+1.6	+2.1	0.0	-4.9	-13.0	-24.9	-62.2	-122.2
Wind Drift • Inches	0.0	0.4	1.7	4.0	7.2	11.7	17.4	32.7	53.4

Federal 570-grain Barnes Banded Solid (P500ND) LF

G1 Ballistic Coefficient = 0.243

Distance • Yards	Muzzle	50	100	150	200	250	300	400	500
Velocity • fps	2100	1942	1790	1649	1520	1395	1190	1120	1010
Energy • ft-lbs	5580	4772	4055	3441	2910	2465	2100	1585	1290
Taylor KO Index	86.9	80.3	74.0	68.2	62.9	57.7	53.4	46.3	41.8
Path • Inches	-0.9	+1.8	+2.3	0.0	-5.5	-14.8	-28.5	-72.6	-144.4
Wind Drift • Inches	0.0	0.5	2.1	4.9	9.0	14.6	21.7	40.8	65.5

Norma 570-grain Full Metal Jacket (11302)

G1 Ballistic Coefficient = 0.370

Distance • Yards	Muzzle	50	100	150	200	250	300	400	500
Velocity • fps	2100	1995	1893	1795	1701	1610	1524	1367	1233
Energy • ft-lbs	5583	5039	4538	4081	3664	3282	2940	2364	1924
Taylor KO Index	86.9	82.5	78.3	74.3	70.4	66.6	63.0	56.5	51.0
Path • Inches	-0.9	+1.7	+2.0	0.0	-4.7	-12.5	-23.6	-57.5	-110.0
Wind Drift • Inches	0.0	0.3	1.3	3.1	5.6	9.0	13.3	24.8	40.4

Cartridges for double rifles, and to a slightly lesser extent single shots, should have rimmed cases. The extraction system is relatively weak compared to the camming power of a turnbolt, and the big surface of the rim is essential to ensure positive extraction/ejection. (Photo from *Safari Rifles II* by Craig Boddington, 2009, Safari Press Archives)

.500 Jeffery

While the British hunters tended to prefer the big double rifles which used rimmed cartridges, German hunters preferred bolt-action rifles. This cartridge originated as the 12.7 x 70 mm Schuler. Jeffery picked it up and offered it in his line. For many years it was the most powerful cartridge available for bolt-action rifles.

Relative Recoil Factor = 5.78

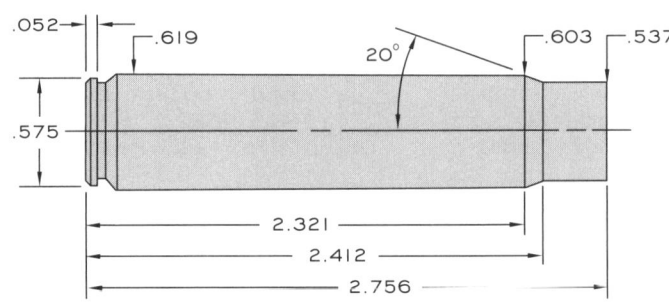

Specifications:

Controlling Agency for Standardization of this Ammunition: CIP

Bullet Weight Grains	Velocity fps	Maximum Average Pressure	
		Copper Crusher	Transducer
N/S	N/S	2,850 bar	3,250 bar

Standard barrel for velocity testing: 26 inches long—1 turn in 10-inch twist

Availability:

Kynoch 535-grain Solid or Soft Nose
G1 Ballistic Coefficient = 0.350

Distance • Yards	Muzzle	50	100	150	200	250	300	400	500
Velocity • fps	2400	2280	2164	2051	1942	1836	1734	1545	1376
Energy • ft-lbs	6844	6179	5565	4999	4480	4006	3574	2835	2250
Taylor KO Index	93.5	88.9	84.3	79.9	75.7	71.6	67.6	60.2	53.6
Path • Inches	-0.9	+1.1	+1.5	0.0	-3.6	-9.4	-17.9	-43.8	+84.5
Wind Drift • Inches	0.0	0.3	1.2	2.7	4.9	7.9	11.7	22.0	36.2

Cor-Bon 535-grain FPBS (EH500J535FPBS/10)
G1 Ballistic Coefficient = 0.350

Distance • Yards	Muzzle	50	100	150	200	250	300	400	500
Velocity • fps	2300	2183	2070	1960	1854	1752	1654	1472	1314
Energy • ft-lbs	6286	5664	5091	4565	4084	3645	3249	2576	2052
Taylor KO Index	89.3	84.8	80.4	76.1	72.0	68.0	64.2	57.2	51.0
Path • Inches	-0.9	+1.3	+1.7	0.0	-3.9	-10.4	-19.7	-48.2	-92.8
Wind Drift • Inches	0.0	0.3	1.2	2.9	5.2	8.4	12.4	23.4	38.4

Norma 540-grain SOLID (11318)
G1 Ballistic Coefficient = 0.23

Distance • Yards	Muzzle	50	100	150	200	250	300	400	500
Velocity • fps	2400	2226	2059	1900	1749	1608	1476	1253	1094
Energy • ft-lbs	6908	5943	5086	4331	3670	3099	2614	1883	1435
Taylor KO Index	94.1	87.2	80.7	74.5	68.5	63.0	57.8	49.1	42.9
Path • Inches	-0.9	+1.3	+1.6	0.0	-4.1	-11.0	-21.3	-54.5	-110.0
Wind Drift • Inches	0.0	0.4	1.8	4.1	7.6	12.3	18.5	35.4	58.7

Norma 570-grain SOFT NOSE (11316)
G1 Ballistic Coefficient = 0.38

Distance • Yards	Muzzle	50	100	150	200	250	300	400	500
Velocity • fps	2200	2097	1997	1901	1807	1716	1630	1468	1325
Energy • ft-lbs	6127	5568	5050	4573	4134	3730	3362	2729	2222
Taylor KO Index	91.0	86.7	82.6	78.6	74.7	71.0	67.4	60.7	54.8
Path • Inches	-0.9	+1.4	+1.8	0.0	-4.2	-11.1	-20.9	-50.8	-96.8
Wind Drift • Inches	0.0	0.3	1.2	2.7	5.0	8.0	11.8	21.9	35.8

Cor-Bon 570-grain DPX (EH500J570DPX/10)
G1 Ballistic Coefficient = 0.37

Distance • Yards	Muzzle	50	100	150	200	250	300	400	500
Velocity • fps	2300	2192	2088	1986	1887	1791	1699	1527	1373
Energy • ft-lbs	6697	6085	5517	4992	4508	4063	3656	2953	2388
Taylor KO Index	95.1	90.7	86.4	82.2	78.1	74.1	70.3	63.2	56.8
Path • Inches	-0.9	+1.3	+1.6	0.0	-3.8	-10.1	-19.1	-46.5	-88.8
Wind Drift • Inches	0.0	0.3	1.1	2.6	4.8	7.7	11.3	21.2	34.7

Norma 570-grain FULL METAL JACKET (11315)

G1 Ballistic Coefficient = 0.370

Distance • Yards	Muzzle	50	100	150	200	250	300	400	500
Velocity • fps	2200	2092	1987	1886	1788	1694	1604	1437	1292
Energy • ft-lbs	6127	5541	4999	4502	4048	3632	3255	2614	2112
Taylor KO Index	91.0	86.5	82.2	77.2	74.0	70.1	66.4	59.4	53.4
Path • Inches	-0.9	+1.5	+1.8	0.0	-4.3	-11.3	-21.3	-51.9	-99.4
Wind Drift • Inches	0.0	0.3	1.2	2.9	5.3	8.4	12.5	23.3	38.1

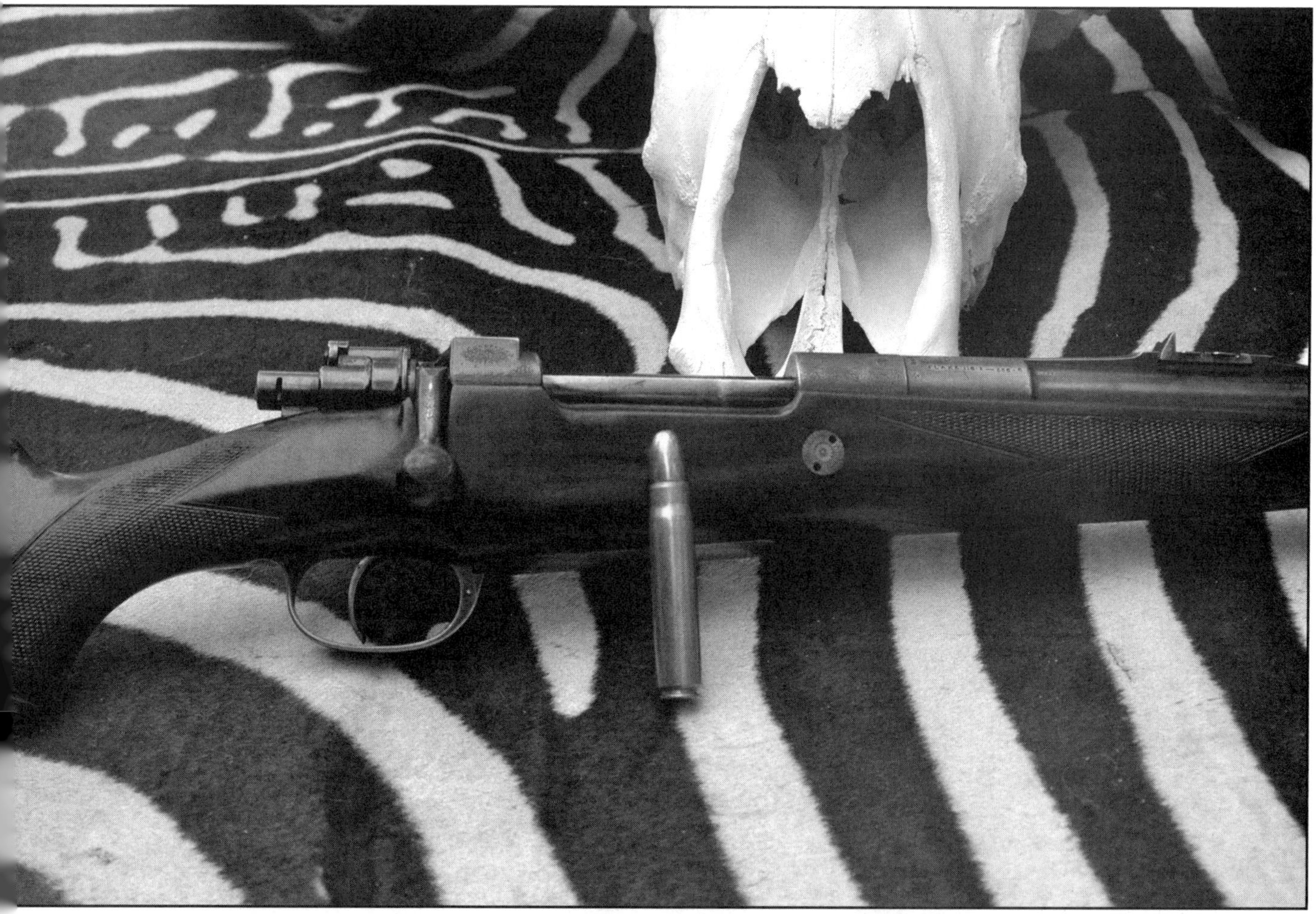

Fletcher Jamieson's .500 Jeffery is one of the most famous big-bore bolt actions in the world. It is one of twenty-odd original Jeffery rifles in this ambering, and the only one made to special order. Jamieson was instrumental in helping John Taylor write *African Rifles and Cartridges.* (Photo ·m *Elephant!* by Craig Boddington, 2012, Safari Press Archives)

.500 A-Square

Companion cartridge to the .495 A², the .500 can be thought of as an "improved" .495. The case is 0.1 inch longer and the body is blown out to 0.568 just at the shoulder. These changes allow the .500 A² to produce well over 100 ft-lbs more muzzle energy than the .495. It is a potent round for bolt-action rifles.

Relative Recoil Factor = 6.68

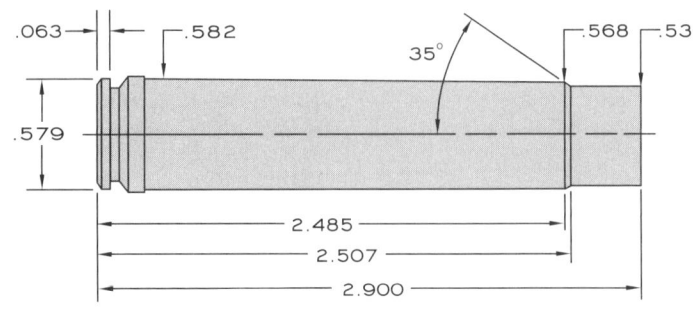

Specifications:

Controlling Agency for Standardization of this Ammunition: Factory

Standard barrel for velocity testing: 26 inches long—1 turn in 10-inch twist

Availability:

A² 600-grain Dead Tough SP, Monolithic Solid, Lion Load SP (None)

G1 Ballistic Coefficient = 0.357

Distance • Yards	Muzzle	50	100	150	200	250	300	400	500
Velocity • fps	2470	2351	2235	2122	2031	1907	1804	1612	1438
Energy • ft-lbs	8127	7364	6654	6001	5397	4844	4337	3461	2755
Taylor KO Index	108.0	102.8	97.7	92.8	88.0	83.4	78.9	70.5	62.9
Path • Inches	-0.9	+1.0	+1.4	0.0	-3.3	-8.8	-16.6	-40.7	-78.3
Wind Drift • Inches	0.0	0.3	1.1	2.5	4.6	7.4	11.0	20.6	33.9

The Mauser extractor is unquestionably the most positive ever used on a bolt gun. (Photo from *Guns and Hunting* by Finn Aagaard, 2012, Safari Press Archives)

.505 Rimless Magnum (Gibbs)

Introduced about 1913, the .505 Gibbs was a proprietary cartridge designed for use in bolt-action rifles. The ballistics of this cartridge are such that it's doubtful that anyone would chamber a new rifle for it today, but the cartridge was certainly well respected in the pre-WWII period.

Relative Recoil Factor = 5.44

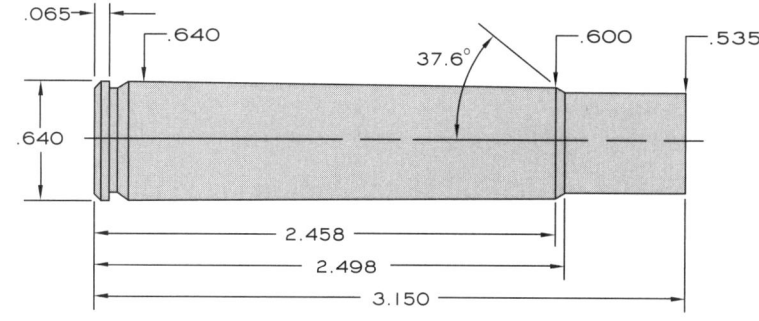

Specifications:

Agency for Standardization of this Ammunition: CIP

Bullet Weight Grains	Velocity fps	Maximum Average Pressure	
		Copper Crusher	Transducer
N/S	N/S	2,400 bar	2,700 bar

Standard barrel for velocity testing: 26 inches long—1 turn in 16-inch twist

Availability:

A² 525-grain Dead Tough SP, Monolithic Solid, Lion Load SP (None)

G1 Ballistic Coefficient = 0.339

Distance • Yards	Muzzle	50	100	150	200	250	300	400	500
Velocity • fps	2300	2179	2063	1949	1840	1735	1634	1449	1290
Energy • ft-lbs	6166	5539	4962	4430	3948	3510	3115	2450	1940
Taylor KO Index	87.1	82.5	78.1	73.8	69.7	65.7	61.9	54.9	48.9
Path • Inches	-0.9	+1.3	+1.7	0.0	-4.0	-10.5	-19.9	-49.0	-94.8
Wind Drift • Inches	0.0	0.3	1.3	3.0	5.4	8.7	12.9	24.3	40.0

Cor-Bon 525-grain DPX (EH505G525DPX/10)

G1 Ballistic Coefficient = 0.340

Distance • Yards	Muzzle	50	100	150	200	250	300	400	500
Velocity • fps	2300	2180	2063	1950	1842	1737	1637	1452	1293
Energy • ft-lbs	6168	5541	4964	4436	3954	3518	3124	2460	1948
Taylor KO Index	87.1	82.6	78.1	73.9	69.8	65.8	62.0	55.0	49.0
Path • Inches	-0.9	+1.3	+1.7	0.0	-4.0	-10.5	-19.9	-48.8	-94.4
Wind Drift • Inches	0.0	0.3	1.3	3.0	5.4	8.7	12.9	24.2	39.8

Kynoch 525-grain Solid or Soft Nose

G1 Ballistic Coefficient = 0.350

Distance • Yards	Muzzle	50	100	150	200	250	300	400	500
Velocity • fps	2300	2183	2070	1960	1853	1751	1653	1471	1313
Energy • ft-lbs	6168	5558	4995	4478	4005	3575	3186	2525	2010
Taylor KO Index	87.1	82.7	78.4	74.2	70.2	66.3	62.6	55.7	49.7
Path • Inches	-0.9	+1.3	+1.7	0.0	-3.9	-10.4	-19.7	-48.2	-93.0
Wind Drift • Inches	0.0	0.3	1.2	2.9	5.2	8.4	12.5	23.4	38.5

Cor-Bon 525-grain FPBS (EH505G525FPBS/10)

G1 Ballistic Coefficient = 0.340

Distance • Yards	Muzzle	50	100	150	200	250	300	400	500
Velocity • fps	2300	2180	2063	1950	1842	1737	1637	1452	1293
Energy • ft-lbs	6168	5541	4964	4436	3954	3518	3124	2460	1948
Taylor KO Index	87.1	82.6	78.1	73.9	69.8	65.8	62.0	55.0	49.0
Path • Inches	-0.9	+1.3	+1.7	0.0	-4.0	-10.5	-19.9	-48.8	-94.4
Wind Drift • Inches	0.0	0.3	1.3	3.0	5.4	8.7	12.9	24.2	39.8

Norma 540-grain SOLID (11312)

G1 Ballistic Coefficient = 0.242

Distance • Yards	Muzzle	50	100	150	200	250	300	400	500
Velocity • fps	2300	2132	1972	1819	1674	1539	1416	1209	1067
Energy • ft-lbs	6345	5453	4462	3967	3362	2842	2404	1754	1366
Taylor KO Index	89.6	83.1	76.8	70.9	65.2	60.0	55.2	47.1	41.6
Path • Inches	-0.9	+1.4	+1.8	0.0	-4.5	-12.0	-23.3	-59.6	-119.6
Wind Drift • Inches	0.0	0.4	1.8	4.3	8.0	12.9	19.3	36.8	60.6

Norma 600-grain Protected Point (11310)

G1 Ballistic Coefficient = 0.381

Distance • Yards	Muzzle	50	100	150	200	250	300	400	500
Velocity • fps	2100	1998	1899	1803	1711	1623	1538	1384	1251
Energy • ft-lbs	5877	5319	4805	4334	3904	3509	3154	2551	2084
Taylor KO Index	91.4	87.0	82.7	78.5	74.5	70.7	67.0	60.3	54.5
Path • Inches	-0.9	+1.7	+2.0	0.0	-4.7	-12.4	-23.4	-56.8	-108.4
Wind Drift • Inches	0.0	0.3	1.3	3.0	5.4	8.7	12.9	23.9	39.0

Norma 600-grain Full Metal Jacket (11311)

G1 Ballistic Coefficient = 0.381

Distance • Yards	Muzzle	50	100	150	200	250	300	400	500
Velocity • fps	2100	1998	1899	1803	1711	1623	1538	1384	1251
Energy • ft-lbs	5877	5319	4805	4334	3904	3509	3154	2551	2084
Taylor KO Index	91.4	87.0	82.7	78.5	74.5	70.7	67.0	60.3	54.5
Path • Inches	-0.9	+1.7	+2.0	0.0	-4.7	-12.4	-23.4	-56.8	-108.4
Wind Drift • Inches	0.0	0.3	1.3	3.0	5.4	8.7	12.9	23.9	39.0

Legendary Zimbabwe hunter Richard Harland is a staunch bolt-action man. He used a .458 during most of his career with Rhodesian Parks a Wildlife, but today his pride and joy is an original George Gibbs .505, one of just seventy-five original .505s made by Gibbs. (Photo from Elepha by Craig Boddington, 2012, Safari Press Archives)

.577 Nitro Express 3"

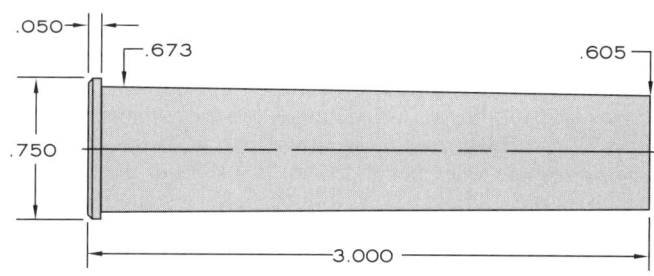

The .577 Nitro Express started life in about 1880 as a black-powder cartridge (obviously not called "Nitro" then). Even before 1900, in the very dawning of the smokeless powder era, the .577 was introduced as a Nitro version. There was also a 2¾-inch version. With all these variations, any owner of a .577-caliber gun would be well advised to be very sure he knows just exactly which cartridges are suitable for use in his gun. The consequences of getting the wrong ammo are just too severe.

Relative Recoil Factor = 6.93

Specifications:

Controlling Agency for Standardization of this Ammunition: CIP

Bullet Weight Grains	Velocity fps	Maximum Average Pressure	
		Copper Crusher	Transducer
N/S	N/S	2,200 bar	2,450 bar

Standard barrel for velocity testing: 26 inches long—1 turn in 30-inch twist

Availability:

A² 750-grain Dead Tough SP, Monolithic Solid, Lion Load SP (None)
G1 Ballistic Coefficient = 0.315

Distance • Yards	Muzzle	50	100	150	200	250	300	400	500
Velocity • fps	2050	1929	1811	1701	1595	1495	1401	1240	1116
Energy • ft-lbs	6998	6197	5463	4817	4234	3721	3272	2560	2075
Taylor KO Index	128.5	120.8	113.5	106.6	100.0	93.7	87.8	77.7	69.9
Path • Inches	-0.9	+1.8	+2.2	0.0	-5.2	-13.9	-26.5	-65.6	-127.6
Wind Drift • Inches	0.0	0.4	1.6	3.8	7.0	11.2	16.6	30.9	50.2

Kynoch 750-grain Solid or Soft Nose
G1 Ballistic Coefficient = 0.430

Distance • Yards	Muzzle	50	100	150	200	250	300	400	500
Velocity • fps	2050	1960	1874	1790	1708	1630	1554	1414	1291
Energy • ft-lbs	7010	6400	5860	5335	4860	4424	4024	3332	2774
Taylor KO Index	128.5	122.8	117.5	112.2	107.1	102.2	97.4	86.6	80.9
Path • Inches	-0.9	+1.7	+2.1	0.0	-4.8	-12.6	-23.7	-57.0	-107.8
Wind Drift • Inches	0.0	0.3	1.2	2.7	5.0	7.9	11.6	21.5	34.8

Cor-Bon 750-grain DPX (EH577N750DPX/10)
G1 Ballistic Coefficient = 0.350

Distance • Yards	Muzzle	50	100	150	200	250	300	400	500
Velocity • fps	2050	1941	1835	1734	1637	1544	1457	1301	1174
Energy • ft-lbs	7000	6274	5611	5007	4462	3972	3537	2820	2296
Taylor KO Index	126.7	120.0	113.4	107.2	101.2	95.5	90.1	80.4	72.6
Path • Inches	-0.9	+1.8	+2.2	0.0	-5.1	-13.4	-25.4	-62.1	-119.4
Wind Drift • Inches	0.0	0.4	1.5	3.4	6.2	9.9	14.6	27.3	44.3

Cor-Bon 750-grain FPBS (EH577N750FPBS/10)
G1 Ballistic Coefficient = 0.350

Distance • Yards	Muzzle	50	100	150	200	250	300	400	500
Velocity • fps	2050	1941	1835	1734	1637	1544	1457	1301	1174
Energy • ft-lbs	7000	6274	5611	5007	4462	3972	3537	2820	2296
Taylor KO Index	126.7	120.0	113.4	107.2	101.2	95.5	90.1	80.4	72.6
Path • Inches	-0.9	+1.8	+2.2	0.0	-5.1	-13.4	-25.4	-62.1	-119.4
Wind Drift • Inches	0.0	0.4	1.5	3.4	6.2	9.9	14.6	27.3	44.3

.577 Tyrannosaur

The .577 Tyrannosaur is an A[2] development, designed (about 1993) to produce a "big stopper" cartridge for a bolt-action rifle. The cartridge certainly achieves that goal, falling only about one Taylor KO index point (154 to 155) short of the legendary .600 NE. The case has a huge volume, holding something on the order of 150 grains of propellant. I think Art Alphin wins the naming award for this cartridge.

Relative Recoil Factor = 8.31

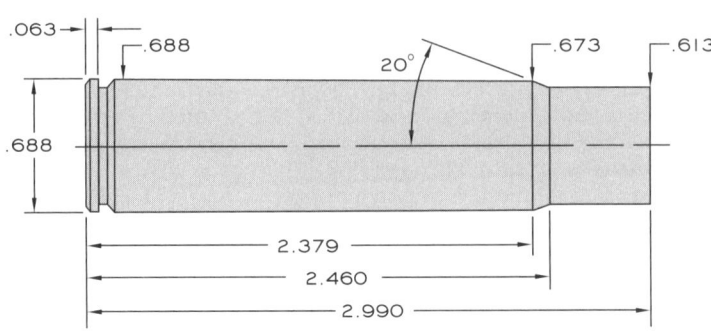

Specifications:

Controlling Agency for Standardization of this Ammunition: Factory

Bullet Weight Grains	Velocity fps	Maximum Average Pressure	
		Copper Crusher	Transducer
N/S	N/S	53,000 cup	65,000 psi

Standard barrel for velocity testing: 26 inches long—1 turn in 12-inch twist

Availability:

A[2] 750 Dead Tough SP, Monolithic Solid, Lion Load SP (None)

G1 Ballistic Coefficient = 0.318

Distance • Yards	Muzzle	50	100	150	200	250	300	400	500
Velocity • fps	2460	2327	2197	2072	1950	1835	1617	1576	1336
Energy • ft-lbs	10,077	9018	8039	7153	6335	5609	4906	3831	2975
Taylor KO Index	154.2	145.9	137.7	129.9	122.2	115.0	101.4	98.8	83.7
Path • Inches	-0.9	+1.1	+1.4	0.0	-3.5	-9.2	-17.5	-43.4	-84.6
Wind Drift • Inches	0.0	0.3	1.2	2.9	5.2	8.5	12.6	23.8	39.5

*This is not a staged photo. The shooter
at maximum recoil from a full-power l[
from a .600 Nitro Express double. The m[
behind him with the fish net (which is a j[
played on the shooter) is the owner of
gun, English double-rifle expert Cal Pa[
from Alaska.* (Safari Press Archives)

.600 Nitro Express

Developed by Jeffery in 1903, the .600 Nitro Express has long held the position of the ultimate "elephant gun". If John "Pondoro" Taylor's KO index has any meaning, the .600 Nitro is more potent as a stopper than the .50 BMG. That comparison puts us right into the center of the energy vs. momentum argument, and the .600 NE is one of the foremost examples for the momentum advocates. Guns are still being built for this cartridge. Expensive guns.

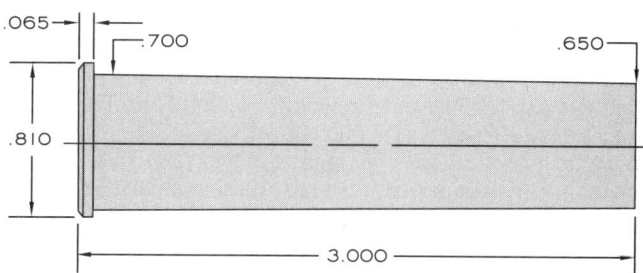

Relative Recoil Factor = 7.91

Specifications:

Controlling Agency for Standardization of this Ammunition: CIP

Bullet Weight Grains	Velocity fps	Maximum Average Pressure	
		Copper Crusher	Transducer
N/S	N/S	2,200 bar	2,450 bar

Standard barrel for velocity testing: 26 inches long—1 turn in 30-inch twist

Availability:

A² 900-grain Dead Tough SP, Monolithic Solid, Lion Load SP (None)

G1 Ballistic Coefficient = 0.272

Distance • Yards	Muzzle	50	100	150	200	250	300	400	500
Velocity • fps	1950	1814	1680	1564	1452	1349	1256	1111	1014
Energy • ft-lbs	7596	6581	5634	4891	4212	3635	3155	2470	2056
Taylor KO Index	155.9	145.1	134.4	125.1	116.1	107.9	100.4	88.8	81.1
Path • Inches	-0.9	+2.2	+2.6	0.0	-6.2	-16.5	-31.6	-79.3	-154.9
Wind Drift • Inches	0.0	0.5	2.1	4.8	8.8	14.1	20.9	38.6	61.4

Kynoch 900-grain Solid or Soft Nose

G1 Ballistic Coefficient = 0.262

Distance • Yards	Muzzle	50	100	150	200	250	300	400	500
Velocity • fps	1950	1807	1676	1551	1435	1330	1237	1094	999
Energy • ft-lbs	7600	6330	5620	4808	4117	3534	3059	2392	1996
Taylor KO Index	155.9	144.5	134.0	124.0	114.8	106.4	98.9	87.5	79.9
Path • Inches	-0.9	+2.2	+2.6	0.0	-6.3	-16.8	-32.2	-81.1	-158.6
Wind Drift • Inches	0.0	0.5	2.1	5.0	9.1	14.7	21.8	40.3	63.8

l Papas of Alaska is the man who wrote the book on the .600 Nitro Express; holding one in this picture. (Safari Press Archives)

.700 Nitro Express

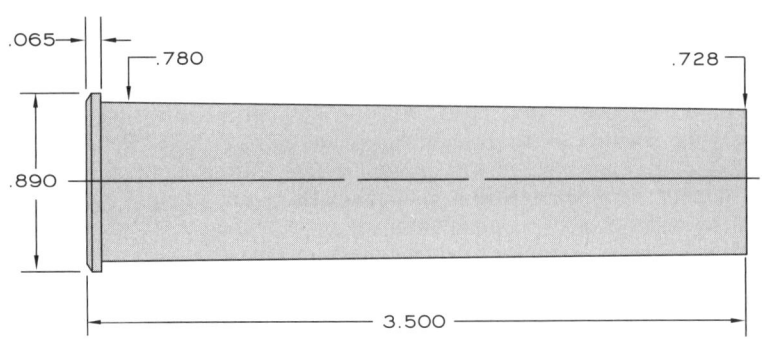

The .700 Nitro Express first appeared about 1988. As near as I can tell, the main reason for a .700 Nitro Express is that there already was a .600 Nitro Express. The .700 NE gives its owner a great "My gun is bigger than your gun" position. One wonders if it is really all that much "better." Because there are very few rifles chambered for this caliber, they are all custom made and very, very, expensive.

Relative Recoil Factor = 9.00

Specifications:

Controlling Agency for Standardization of this Ammunition: CIP

Bullet Weight Grains	Velocity fps	Maximum Average Pressure	
		Copper Crusher	Transducer
N/S	N/S	2,200 bar*	2,450 bar*

Standard barrel for velocity testing: 26 inches long*

* Estimated

Availability:

Kynoch 1000-grain Solid
G1 Ballistic Coefficient = 0.31

Distance • Yards	Muzzle	50	100	150	200	250	300	400	500
Velocity • fps	2000	1879	1762	1651	1547	1448	1357	1203	1088
Energy • ft-lbs	8900	7842	6902	6061	5316	4663	4096	3216	2632
Taylor KO Index	206.3	193.8	181.7	170.3	159.6	149.4	140.0	124.1	112.2
Path • Inches	-0.9	+2.0	+2.4	0.0	-5.6	-14.8	28.2	-69.8	-135.6
Wind Drift • Inches	0.0	0.4	1.7	4.0	7.3	11.8	24.3	32.4	52.3

A² 1000-grain Monolithic (None)
G1 Ballistic Coefficient = 0.30

Distance • Yards	Muzzle	50	100	150	200	250	300	400	500
Velocity • fps	1900	1782	1669	1562	1461	1369	1285	1146	1048
Energy • ft-lbs	8015	7950	6188	5419	4740	4166	3665	2917	2438
Taylor KO Index	196.0	183.8	172.1	161.1	150.7	141.1	132.5	118.2	108.1
Path • Inches	-0.9	+2.3	+2.7	0.0	-6.3	-16.6	-31.6	-78.1	-151.4
Wind Drift • Inches	0.0	0.5	1.9	4.3	7.9	12.7	18.7	34.6	55.1

Rifle Prep for Safari

How To Prepare Your Rifle, and Yourself, for an African Hunt

Story and photos by Craig Boddington

These days I generally use a standard two-gun hard case like this Pelican case, using the extra space for binoculars and other accessories. Just make sure the rifles cannot rub together and won't shift, and use good, sturdy locks.

It sure is fun to be a gunwriter, trying out all the cool new stuff. It's just marvelous to go on a hunt with a rifle received just a few days earlier, untested, untried, often just barely sighted in. For better or worse, that's what the job entails. So this is very much a "do as I say, not as I do" article, for which I apologize.

On the other hand, since I have used a wide variety of rifles under a tremendous range of conditions, I have pretty much seen it all, which has given me some idea what to look out for. Six or even broken stocks. Multiple failures to feed, in multiple ways. Ejectors that didn't eject, magazine boxes that went "bombs away," safeties that didn't function, iron sights that fell off, scope mounts that sheared. And a whole lot more.

Anything made by man can fail, so some of this falls under Murphy's Law and is unavoidable.

However, it is also my experience that anything that doesn't work properly is unlikely to start working; and anything that works well during a reasonable break-in period is likely to keep working. Similarly, any incipient failure is likely to show itself fairly quickly. There is no precise rule of thumb, but I think it's unwise to take a rifle on safari without running a minimum of fifty rounds down the barrel. A hundred is much better. Regardless of the number of rounds fired, a half-dozen range sessions is minimal, and a dozen much better. This establishes consistency, and also familiarity—because our interest here is in preparing you as well as your rifle.

You can do the math. If we're talking a hard-kicking rifle, then five or six cartridges per range session might be optimum (along with plenty of shooting of lighter calibers). This means ten or a dozen range sessions are necessary. The message

329

is: Plan well ahead. Most professional hunters will tell you that they have a big problem with clients who bring new, unfamiliar rifles that they don't know very well and, all too often, are scared to death of. So start early—or take your old favorite. But, for the sake of argument, let's assume you don't have a suitable rifle for your safari. So you make your choice, and in the months (not weeks or days) before your safari, these are things you should think about. These considerations are not in any essential order, and, in fact, most are both simultaneous and continuous.

Check the Rifle

This is ongoing during your range sessions. The action should be completely checked for functioning: Extraction, ejection, functioning of the

During your preparations, periodically inspect the rifle from muzzle to butt, including the sling, sling swivels, and sling-swivel studs. Having a sling come undone while the rifle is slung can be catastrophic.

safety, smooth feeding. This last isn't as simple as it sounds. A lot of bolt actions feed well with one or two rounds in the magazine but stumble a bit when the magazine is full. Others are exactly opposite. Check the feeding with all combinations of a partially full or full magazine. Shoot the rifle under the same conditions to make sure the floorplate doesn't come undone during recoil.

Make sure the action screws are tight and stay tight. Don't stop your continuous inspection at the action. I've seen recoil pads come loose—and of course wooden stocks can warp and split. As shooting sessions progress, examine the tang area and any recoil reinforcing lugs for incipient cracks. If they start, they can only get worse, but it's amazing what a good hand with epoxy can accomplish. Some years ago I arrived in Africa to find the stock of my .30-06 broken at the wrist. My PH and I worked some magic with epoxy and a dowel rod. I used it on that safari with no problems.

Pay attention to little things like sling swivel studs. Most common are the screw-in studs. I hate them because they invariably start to unscrew, and as this progresses they become loose. One of the worst things that can happen in the field is for a sling swivel stud, especially the forward one, to pop out while the rifle is slung over the shoulder. The rifle is catapulted backward and invariably lands on the scope. Two-screw studs are superior, though rarely supplied; barrel band swivels are the sturdiest.

Again, all these inspections should be continuous. Mechanical things like safeties and functioning are generally either sound or not, but it may take some time for loose screws and recoil cracks to appear.

Check the Sights

This one is obvious. On a scoped rifle, the mount and ring screws should be checked periodically. Tight is good, over-tightening is not good, and broken screws are very bad. Scopes can fail, and recoil is the great enemy of both scopes and mounts. I have seen virtually every brand of scope fail, and I've seen hard-recoiling guns literally eat mounts. However, under the theory that things that work are likely to keep on working, almost every scope failure I have seen has taken place on the range, not in the field. If, as you should, you shoot many times more on the range than you do on safari, that's where problems should show up.

Not so obvious is that we tend to consider iron sights as sturdy fixtures. Often this isn't the case. Many iron sights, especially as provided by American

There are many good rifle slings and multiple sound ways to carry ammo. Shown are a traditional culling belt, an ammo wallet, and belt slides from Texas Hunt Company and Murray Custom Leather. The slings are, right to left, Murray, Texas Hunt Co., and Trader Keith's. Whatever you choose, practice with them!

manufacturers, are flimsy affairs that seem to be mostly supplied for ornamentation, not for actual use. If there is any intent whatsoever to use iron sights, it's important to check them very carefully and make sure they are indeed bulletproof. If there is serious intent to use iron sights, aftermarket sights installed by a good gunsmith are often the best choice.

Check the Ammo

During your practice and preparation, you will, I hope, experiment with a variety of loads to see what your rifle likes best. At some point you must decide on the ammo you will actually use on your safari. There are a couple of cautions here. American shooters tend to be enamored of two things: Accuracy and velocity. When choosing ammunition for African hunting, it is essential to pick bullets that will provide the penetration you

need on the largest game you intend to hunt. These may not be the most accurate or fastest loads in your rifle, but bullet performance is far more important. There are exceptions, but African shooting is rarely done past two hundred yards. So you can sacrifice a bit of accuracy to get the penetration you need on the largest game you intend to hunt with any rifle in your battery.

Another word on velocity is needed. You don't necessarily need the fastest ammo out there. No game animal at African shooting distances will know the difference between one hundred feet per second more or less. However, especially on the big bores, it is a highly recommended chore to check the velocity of your ammo on a chronograph. Two years ago in Botswana, my buddy Bill Jones was hunting elephant with his .577 double, obviously enough gun and legendary for penetration. Except, on both his elephant bulls, we had failure to penetrate with well-

While practicing from sticks and other positions is essential, there are times when there is no choice but to raise the rifle and shoot—especially on dangerou[s] game at close range. Spend a lot of time shooting offhand out to at least 50 yards!

One of the final steps before packing your duffel should be both a visual inspection of all the ammo you intend to take, plus running each cartridge in and out of the chamber to ensure smooth feeding.

placed frontal brain shots, and both animals wer[e] taken with follow-up shots as they turned away.

This went against everything we thought w[e] knew. Smelling a rat, we chronographed the amm[o] when we got home. Actual velocity was just 1,80[0] fps, not the 2,050 fps the .577 is supposed to delive[r]. My theory is that 1,800 fps wasn't enough veloci[ty] to overcome resistance against the big .577-in[ch] bullet. No one knows exactly how much veloci[ty] is required for a given weight and style of bullet [to] overcome resistance and penetrate, but you shou[ld] check your loads and be certain they are close [to] the specified velocity.

A final item on ammo is that you should ma[ke] a real, actual check: The last time you're on t[he] range, when you've selected the perfect load a[nd] you're doing your final zeroing and practicing befo[re] departure, run every cartridge you plan to ta[ke] through the magazine and/or into and out of t[he] chamber. Visual inspection is great, but minor der[...]

in cases or slightly malformed bullet tips mean little. A case that won't fit your chamber means a whole bunch! Twice when I've failed to do this, once with a .300 H&H and another time with a .340 Weatherby Magnum, I had cartridges with the bullets seated a bit long, and I wound up with a bullet stuck in the lands and an action full of powder.

Essential Accessories

There really aren't very many accessories in the firearms line. An extra scope set in rings isn't a bad idea, especially if you're taking just one rifle. Of course, the one time I needed a spare scope I didn't have one. Zambian PH Pete Fisher came to my rescue, stripping a Leupold off his son's rifle and loaning it to me for the duration. Don't count on extra scopes lying around—but whether you need one depends somewhat on your luck!

Lens covers are optional, at least in the dry season. Absent the chance of rain I generally don't use them—but I do concede that they also protect your scope lens from the ever-present African dust. I leave that one up to you. Items you absolutely do need include: Rifle sling, ammo carrier, soft case, and minimal cleaning kit.

Rifle slings are a matter of personal preference. I tend to like simple carrying straps that are somewhat wider to better distribute the weight—but not so wide that you can't readily wrap into the hasty sling. The "stretchy" slings like the excellent Vero Vellini line are very good, and I also like the "modified military" padded sling from Texas Hunt Company—but there is also nothing like a simple sling of good leather (Dick Murray makes some of the best). Canvas webbing also makes a great sling, which has the tremendous advantage of minimal slipping on your shoulder. The sling from Trader Keith's is a perfect example. Whatever sling you choose, inspect both the sling and detachable swivels—and don't get so attached to your sling that you use it until it breaks. (Yes, I did that once!)

You should have a soft gun case that will protect your rifle from scratching—and from dust—while bumping along in the Land Cruiser. With baggage weight requirements dropping and overweight charges escalating, I recommend as simple and light a "gun bag" as possible. Come to think of it, you may not have to bring one. Many outfitters supply them as a matter of course—but if you aren't certain there will be one, you should bring your own, especially if you are taking really nice rifles on your safari.

Most camps will have at least a basic cleaning rod and some oil. This is probably all you really need, but it's good to bring at least some minimal

me rifles feed well with just one or two cartridges in the magazine, but hiccup when the magazine is full—and vice versa. Check your rifle's feeding with variations between fully loaded and empty, and if there's a problem get it fixed.

All screws, but perhaps especially the scope mount and ring screws, should be checked periodically to make sure nothing is working loose.

gear. For the last couple of seasons I've stuck an Otis palm-sized cleaning kit in my gun case. It weighs almost nothing and has just about everything you really need. On normal safaris you'll need to wipe off the fingerprints daily and swab the grit out of the action. During the rainy season or safaris in a forest, even a daily thorough cleaning won't stay ahead of the rust!

How you carry your ammo in the field is also a matter of personal preference. There are many options, but you should make a decision and then practice using it. It is not a good idea to have cartridges rattling around in your pockets (most PHs will quickly cure this problem!)—and it's a much worse idea to walk away from the vehicle without plenty of extra ammo. There is no set number, but I usually carry at least a dozen cartridges. A running gun battle with a buffalo, or even a kudu in the brush, can eat up ammo at an embarrassing rate—and there is nothing worse than to run out of ammo with a job not quite finished!

There are cartridge wallets in both leathe and cloth, cartridge belts with the same option and jackets and vests with cartridge loops. Som guys with two-rifle batteries carry some of eac cartridge all the time, a good idea as long as yo don't get mixed up. Whatever you choose, practi how you will carry your ammo and how you w reload. If you're using a cartridge carrier wi exposed loops, make certain the loops are tig enough to prevent the cartridges from falling o while running or crawling.

Plan for Travel

Physical preparations for traveling to Afri with firearms are simple. You need a good, sturd lockable hard gun case with strong locks, eith key or combination. These days I prefer a two-ri hard case wide enough to accommodate two scop rifles. If there's space left over, I put a binocul rangefinder, cleaning kit, knives, and other sm items in the case. I travel a lot with firearms, a

I've destroyed a lot of cases. I don't believe the metal cases are stronger than hard plastic—or vice versa—but I'm certain that, across the board, hinges, latches, and square corners are the breaking points! Cases I've had good luck with include SKB, Pelican, and Americase, in no particular order. Cases with wheels are great. When packing I remove the bolt and wrap it in a silicone rag or a clean pair of socks, not just for ease of inspection but because the bolt creates a projection that will cause problems if the baggage smashers drop the case from two or three stories.

In recent years, rules for carrying ammunition have swung back and forth: with the gun, separate from the gun, in a lockable case, in your baggage. You need to check with your airline, but understand that the rules may change from day to day (and supervisor to supervisor). Fairly constant is the following: must have original factory containers and be no more than (maximum) of five kilograms (eleven pounds). This is plenty for even a three-rifle battery, while adequate shotgun ammo for a good bird shoot is hopeless. These days I put my ammo in a small lockable metal ammo carrier, and I put it in my duffel bag, prepared to shift as the wind blows. I started out locking this case, but after TSA cut several locks to look inside I put the lock and a key inside the case, available if requested.

The second element is the paperwork, which varies radically with carrier and country. Always tell your travel agent you are traveling with firearms, and demand that he or she finds out if any advance notification is required. Sometimes yes, sometimes no . . . so start early, and make certain all the boxes have been checked. Similarly, some countries require gun permits in advance. Others, notably Namibia, South Africa, and Zimbabwe, do not; you can get them on arrival, although you'll save time and stress by filling out the forms in advance. Mandatory is the U.S. Customs Form 4457, which serves as a "gun license" for U.S. citizens worldwide. These little details drive you to making your firearms selection well ahead of your departure date—and once you have started the paperwork, God help you if you change your mind! Be certain of your choice before you declare.

Practice, Practice

Start early, finish late! We can distill all this "safari rifle" stuff into just three key elements: absolute reliability; a bullet that will ensure penetration on the largest game you will hunt with a given rifle; and

practice! This last is the most important. Within broad parameters, how well you shoot is more important than what you are shooting.

Too many of us are fixated on shooting tight little groups off the bench. Accuracy builds confidence, and precise shooting does have field applications in situations where you can "bench up" with a pack over a rock or log. Most African field shooting is more fluid. The most common complaints from PHs are not only that their clients don't shoot very well but also that they don't get their shots off quickly enough—usually because they can't find a steady rest. These ills can be cured through practice—but you have to practice smart. Get yourself a set of good shooting sticks. Learn how to use them, and then put them in your gun case and take them with you. I've used several varieties and all are good, but the ones Donna and I are using now are the takedown sticks from African Sporting Creations with screw-in lower legs. There are many tricks, but every shooter must learn the proper height for utmost stability, and how to place the supporting hand.

Three-legged shooting sticks are almost universal in Africa because they are fast and get you above low brush and thornbushes. On many safaris, the vast majority of shots will be taken from sticks—but African shooting isn't just about sticks. Prone is occasionally used when terrain and vegetation allow, as are sitting and kneeling. And for close encounters in thick brush, especially with dangerous game, good old offhand shooting is a critical skill.

Your practice regimen should include the full range of shooting positions, plus sticks, and a bit off the bench to verify zero and accuracy. However, you must practice smart. This is not immersion therapy, and your first day on safari is the test. Budget time for multiple range sessions for months before your safari. Shooting is shooting, so you can work out with sticks and shooting positions with a .22 rimfire or a light-recoiling varmint rifle, consciously limiting your exposure to heavy recoil.

It is necessary to fire your actual safari rifles, and you should fire them a lot—but not all at once. Concentrate not only on becoming thoroughly familiar with your rifles and not only on getting steady and making that first shot count, but also on working the action or finding the second trigger of your double and backing up your own shot. Take this a step further and practice reloading from your chosen ammo carrier, enough until you know where the cartridges are and can get them into the rifle without looking down.

The three-legged, shooting-stick system is almost universal in Africa. Usually carried by a tracker or your PH, they're fast and they get you above the lo[...] brush—but they take some getting used to. Get a set and make it part of your range gear.

For many of us, access to ranges is a problem. You or a handloading buddy can make up dummy cartridges (always inspect them carefully and keep them away from any loaded rounds!), and just like we did in the Marines, tremendous training can be achieved by dry-firing and reloading drills. An air gun in a basement range is great practice.

Another option is to consider a shooting school where you can indeed indulge in some immersion therapy under the guidance of a skilled instructor. There are several around, and I'm sure most are very good. The one I am personally most familiar with, and had curriculum input to, is the SAAM Safari course, held at Tim Fallon's FTW Ranch in the Texas Hill Country. During that four-day course, students fire several hundred rounds on static ranges, at lifelike plains-game targets from field positions, and dangerous-game targets that may be static, may be charging, or may appear out of nowhere. They have even included brain shots at crocodile and hippo, and lion and leopard from a

blind. To avoid a permanent flinch, much trainin[...] is done with "ranch" .223s that have express sight[...] and scopes, and with light loads for some of th[...] big calibers.

At some point on your first safari day you wi[...] repair to a range to "check zero" on your rifle. Th[...] is important after a long journey, but your PH an[...] his trackers will work tirelessly to get you as clos[...] as they can, and they really don't much care if you[...] rifle is an inch or two off at a hundred yards. Th[...] range day is primarily their opportunity to evalua[...] if you handle your gun safely, how your equipmen[...] functions and your familiarity with it, and ho[...] quickly and well you can set up on the sticks an[...] take a shot. This is not an exam you can cram fo[...] so whether you practice at a local range, in yo[...] back forty, in your basement with a BB gun and d[...] fire drills, or at a formal school (preferably, son[...] combination of all!), the shooting you do before yo[...] safari is, ultimately, far more important than t[...] specific rifles you choose to take.

Data for Pistol and Revolver Cartridges

The listings in this section include only data for pistol-length barrels. Where the velocity data from longer barrels are available, these calibers have also been shown in the rifle section. To allow for easier comparisons, the data in this section extend to 200 yards. This increased distance is useful for some of the high-powered, large-caliber cartridges but is a totally impractical range for the small-caliber numbers. Nevertheless, we have included the data for all calibers so that the reader can make comparisons if and when he wishes. Some rifle manufacturers have chambered guns for pistol calibers such as the .44 Remington Magnum and the .357 Magnum. Many of the larger-caliber pistol rounds would also make excellent cartridges for guns in the lever-action carbine class.

.17 Remington Fireball

When Remington introduced its XP-100 Pistol in 1964, they also introduced a new cartridge designed especially for the XP-100. The .221 Remington Fireball was a shortened .222 Remington with a little less performance. Shooters grabbed the XP-100 and rebarreled it in numerous calibers—anything but the .221. Remington finally stopped making the .221 but recently restarted production and also introduced this .17 Remington Fireball cartridge.

Relative Recoil Factor = 0.43

Specifications:

Controlling Agency for Standardization of this Ammunition: SAAMI

Bullet Weight Grains	Velocity fps	Maximum Average Pressure	
		Copper Crusher	Transducer
--	--	52,000 cup	N/S

Standard barrel for velocity testing: 10.50 inches long—1 turn in 12-inch twist

Availability:

Remington 20-grain Accutip-V (PRA17FB)
G1 Ballistic Coefficient = 0.18

Distance • Yards	Muzzle	100	200	300	400	500	600	800	1000
Velocity • fps	4000	3380	2840	2360	1930	1555	1255	949	806
Energy • ft-lbs	710	507	358	247	165	107	70	40	29
Taylor KO Index	2.0	1.7	1.4	1.2	1.0	0.8	0.6	0.5	0.4
Path • Inches	-1.5	+0.8	0.0	-5.2	-16.6	-37.6	-73.2	-219.4	-519.7
Wind Drift • Inches	0.0	1.2	5.0	12.2	23.7	41.1	65.8	138.9	233.4

Remington 25-grain UMC Jacketed Hollow Point (L17FBV)
G1 Ballistic Coefficient = 0.19

Distance • Yards	Muzzle	100	200	300	400	500	600	800	1000
Velocity • fps	3850	3280	2780	2330	1925	1569	1276	966	823
Energy • ft-lbs	823	597	429	301	206	137	90	52	38
Taylor KO Index	2.4	2.1	1.8	1.5	1.2	1.0	0.8	0.6	0.5
Path • Inches	-1.5	+0.9	0.0	-5.4	-17.4	-38.9	-74.8	-219.3	-512.3
Wind Drift • Inches	0.0	1.2	5.0	12.0	23.3	40.0	63.8	133.6	225.3

The Thomson Contender is a favorite for handgun hunters becaus *can be ordered in so many different calibers and configurations. (Sa* Press Archives)

.25 Auto (ACP) – 6.35 Browning

The smallest of the current factory production centerfire pistol rounds, the great virtue of the .25 Auto is that it fits very small pistols. The power is so low, much lower than a .22 WMR, that its use for self-defense is limited to the deterrent effect that the mere showing of any gun provides. This cartridge has been around since 1908. That's a long time. There are still many different loadings available, so someone must be buying this ammunition.

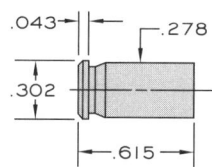

Relative Recoil Factor = 0.17

Specifications:

Controlling Agency for Standardization of this Ammunition: SAAMI

Bullet Weight Grains	Velocity fps	Maximum Average Pressure	
		Copper Crusher	Transducer
35 XTP-HP	900	18,000 cup	25,000 psi
45 XP	805	18,000 cup	25,000 psi
50 MC-FMC	755	18,000 cup	25,000 psi

Standard barrel for velocity testing: 2 inches long—1 turn in a 16-inch twist

Availability:

Cor-Bon 35-grain Glaser Safety Slug (00200/na)

G1 Ballistic Coefficient = 0.075

Distance • Yards	Muzzle	25	50	75	100	125	150	175	200
Velocity • fps	1100	1014	950	899	855	815	779	746	715
Energy • ft-lbs	94	80	70	63	57	52	47	43	40
Taylor KO Index	1.4	1.3	1.2	1.2	1.1	1.0	1.0	1.0	0.9
Mid-Range Trajectory Height • Inches	0.0	0.2	1.0	2.6	4.8	7.8	12.1	17.3	23.9
Drop • Inches	0.0	-1.0	-4.0	-9.5	-17.6	-28.8	-43.2	-61.1	-83.0

Fiocchi 35-grain XTP (25XTP)

G1 Ballistic Coefficient = 0.072

Distance • Yards	Muzzle	25	50	75	100	125	150	175	200
Velocity • fps	900	854	813	776	742	709	679	650	597
Energy • ft-lbs	63	57	51	47	43	39	36	33	30
Taylor KO Index	1.1	1.1	1.0	1.0	1.0	0.9	0.9	0.8	0.8
Mid-Range Trajectory Height • Inches	0.0	0.3	1.5	3.5	6.6	10.5	16.3	23.1	31.9
Drop • Inches	0.0	-1.4	-5.8	-13.4	-24.7	-39.9	-59.4	-83.7	-113.1

Hornady 35-grain JHP/XTP (90012)

G1 Ballistic Coefficient = 0.073

Distance • Yards	Muzzle	25	50	75	100	125	150	175	200
Velocity • fps	900	854	813	777	742	711	681	653	625
Energy • ft-lbs	63	57	51	47	43	39	36	33	30
Taylor KO Index	1.1	1.1	1.0	1.0	0.9	0.9	0.9	0.8	0.8
Mid-Range Trajectory Height • Inches	0.0	0.3	1.5	3.5	6.6	10.5	16.2	23.0	31.6
Drop • Inches	0.0	-1.4	-5.8	-13.4	-24.6	-39.8	-59.2	-83.3	-112.4

Speer 35-grain Gold Dot Hollow Point (23602)

G1 Ballistic Coefficient = 0.091

Distance • Yards	Muzzle	25	50	75	100	125	150	175	200
Velocity • fps	900	863	830	799	770	743	717	693	669
Energy • ft-lbs	63	58	53	50	46	43	40	37	35
Taylor KO Index	1.1	1.1	1.1	1.0	1.0	0.9	0.9	0.9	0.8
Mid-Range Trajectory Height • Inches	0.0	0.3	1.5	3.4	6.3	10.2	15.4	21.8	29.4
Drop • Inches	0.0	-1.4	-5.7	-13.1	-24.0	-38.5	-57.0	-79.6	-106.9

Winchester 45-grain Expanding Point (X25AXP)

G1 Ballistic Coefficient = 0.059

Distance • Yards	Muzzle	25	50	75	100	125	150	175	200
Velocity • fps	815	770	729	690	655	621	589	559	530
Energy • ft-lbs	66	59	53	48	42	39	35	31	28
Taylor KO Index	1.3	1.2	1.2	1.1	1.1	1.0	1.0	0.9	0.9
Mid-Range Trajectory Height • Inches	0.0	0.4	1.8	4.4	8.2	13.6	20.6	29.3	41.0
Drop • Inches	0.0	-1.7	-7.1	-16.5	-30.6	-49.6	-74.3	-105.2	-143.0

Fiocchi 50-grain FMJ (25AP)

G1 Ballistic Coefficient = 0.156

Distance • Yards	Muzzle	25	50	75	100	125	150	175	200
Velocity • fps	800	783	766	751	735	721	706	692	678
Energy • ft-lbs	71	68	65	63	60	58	55	53	51
Taylor KO Index	1.5	1.4	1.4	1.4	1.3	1.3	1.3	1.3	1.2
Mid-Range Trajectory Height • Inches	0.0	0.4	1.7	3.9	7.4	11.3	17.4	23.8	32.2
Drop • Inches	0.0	-1.7	-7.0	-16.0	-28.8	-45.6	-66.6	-92.0	-121.9

Sellier & Bellot 50-grain FMJ (V320022U)

G1 Ballistic Coefficient = 0.084

Distance • Yards	Muzzle	25	50	75	100	125	150	175	200
Velocity • fps	781	751	723	696	671	646	623	600	579
Energy • ft-lbs	69	64	59	54	50	46	43	40	37
Taylor KO Index	1.4	1.4	1.3	1.3	1.2	1.2	1.1	1.1	1.1
Mid-Range Trajectory Height • Inches	0.0	0.4	1.9	4.5	8.3	13.5	20.3	28.7	38.9
Drop • Inches	0.0	-1.8	-7.5	-17.3	-31.7	-50.8	-75.2	-105.1	-141.1

American Eagle (Federal) 50-grain Full Metal Jacket (AE25AP)

G1 Ballistic Coefficient = 0.11

Distance • Yards	Muzzle	25	50	75	100	125	150	175	200
Velocity • fps	760	740	720	700	680	662	644	627	610
Energy • ft-lbs	65	62	60	57	55	52	50	48	46
Taylor KO Index	1.4	1.4	1.3	1.3	1.2	1.2	1.2	1.2	1.1
Mid-Range Trajectory Height • Inches	0.0	0.5	1.9	4.6	8.4	13.1	20.1	27.7	37.7
Drop • Inches	0.0	-1.9	-7.8	-17.9	-32.5	-51.7	-75.9	-105.4	-140.3

Magtech 50-grain FMJ (25A)

G1 Ballistic Coefficient = 0.08

Distance • Yards	Muzzle	25	50	75	100	125	150	175	200
Velocity • fps	760	733	707	682	659	636	614	593	573
Energy • ft-lbs	64	60	56	52	48	45	42	39	36
Taylor KO Index	1.4	1.3	1.3	1.2	1.2	1.1	1.1	1.1	1.0
Mid-Range Trajectory Height • Inches	0.0	0.5	2.0	4.7	8.7	14.0	21.0	29.7	40.2
Drop • Inches	0.0	-1.9	-7.9	-18.2	-33.2	-53.2	-78.5	-109.6	-146.8

Remington 50-grain Metal Case (R25AP)

G1 Ballistic Coefficient = 0.08

Distance • Yards	Muzzle	25	50	75	100	125	150	175	200
Velocity • fps	760	733	707	682	659	636	614	593	573
Energy • ft-lbs	64	60	56	52	48	45	42	39	36
Taylor KO Index	1.4	1.3	1.3	1.2	1.2	1.1	1.1	1.1	1.0
Mid-Range Trajectory Height • Inches	0.0	0.5	2.0	4.7	8.7	14.0	21.0	29.7	40.2
Drop • Inches	0.0	-1.9	-7.9	-18.2	-33.2	-53.2	-78.5	-109.6	-146.8

UMC (Remington) 50-grain Metal Case (L25AP)

G1 Ballistic Coefficient = 0.0

Distance • Yards	Muzzle	25	50	75	100	125	150	175	200
Velocity • fps	760	733	707	682	659	636	614	593	573
Energy • ft-lbs	64	60	56	52	48	45	42	39	36
Taylor KO Index	1.4	1.3	1.3	1.2	1.2	1.1	1.1	1.1	1.0
Mid-Range Trajectory Height • Inches	0.0	0.5	2.0	4.7	8.7	14.0	21.0	29.7	40.2
Drop • Inches	0.0	-1.9	-7.9	-18.2	-33.2	-53.2	-78.5	-109.6	-146.8

Speer 50-grain TMJ RN Lawman (53607)

G1 Ballistic Coefficient = 0.1

Distance • Yards	Muzzle	25	50	75	100	125	150	175	200
Velocity • fps	760	738	717	696	677	658	639	622	604
Energy • ft-lbs	64	60	57	54	51	48	45	43	41
Taylor KO Index	1.4	1.4	1.3	1.3	1.2	1.2	1.2	1.1	1.1
Mid-Range Trajectory Height • Inches	0.0	0.5	1.9	4.2	8.5	13.2	20.2	28.0	38.1
Drop • Inches	0.0	-1.9	-7.8	-18.0	-32.6	-51.9	-76.3	-106.0	-141.3

USA (Winchester) 50-grain Full Metal Jacket (Q4203)

G1 Ballistic Coefficient = 0.100

Distance • Yards	Muzzle	25	50	75	100	125	150	175	200
Velocity • fps	760	736	707	690	669	649	629	609	591
Energy • ft-lbs	64	60	56	53	50	47	44	41	39
Taylor KO Index	1.4	1.3	1.3	1.2	1.2	1.2	1.1	1.1	1.1
Mid-Range Trajectory Height • Inches	0.0	0.5	2.0	4.6	8.6	13.8	20.5	28.9	39.0
Drop • Inches	0.0	-1.9	-7.9	-18.1	-32.8	-52.4	-77.2	-107.5	-143.6

Aguila 50-grain FMJ (1E252110)

G1 Ballistic Coefficient = 0.090

Distance • Yards	Muzzle	25	50	75	100	125	150	175	200
Velocity • fps	755	728	703	679	656	633	612	591	571
Energy • ft-lbs	63	59	55	51	48	45	42	39	36
Taylor KO Index	1.4	1.3	1.3	1.2	1.2	1.1	1.1	1.1	1.0
Mid-Range Trajectory Height • Inches	0.0	0.3	1.6	4.3	8.4	13.8	20.8	29.6	40.3
Drop • Inches	0.0	-1.9	-8.0	-18.4	-33.6	-53.8	-79.4	-110.8	-148.4

Blazer (CCI) 50-grain TMJ (3501)

G1 Ballistic Coefficient = 0.111

Distance • Yards	Muzzle	25	50	75	100	125	150	175	200
Velocity • fps	755	733	712	692	673	654	636	619	602
Energy • ft-lbs	63	60	56	53	50	48	45	42	40
Taylor KO Index	1.4	1.3	1.3	1.3	1.2	1.2	1.2	1.1	1.1
Mid-Range Trajectory Height • Inches	0.0	0.5	2.0	4.7	8.6	13.7	20.4	28.6	38.5
Drop • Inches	0.0	-1.9	-7.9	-18.2	-33.0	-52.6	-77.2	-107.2	-142.9

PMC 50-grain Full Metal Jacket (25A)

G1 Ballistic Coefficient = 0.084

Distance • Yards	Muzzle	25	50	75	100	125	150	175	200
Velocity • fps	750	722	695	670	645	622	599	578	557
Energy • ft-lbs	63	58	54	50	46	43	40	37	34
Taylor KO Index	1.4	1.3	1.3	1.2	1.2	1.1	1.1	1.0	1.0
Mid-Range Trajectory Height • Inches	0.0	0.5	2.1	4.9	9.0	14.6	21.9	31.0	42.0
Drop • Inches	0.0	-2.0	-8.1	-18.8	-34.3	-55.0	-81.3	-113.7	-152.6

The 1911 Colt remains as one of the world's most popular semiautomatic pistols fully 100 years after its introduction. (Safari Press Archives)

.30 Luger (7.65mm)

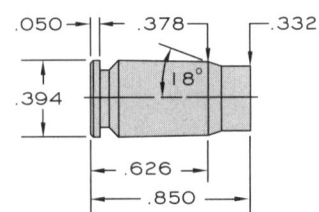

The .30 Luger actually predates the 9mm Luger, having been introduced in 1900. It is very similar to the .30 Mauser but is not interchangeable. The design is a rimless bottlenecked case that anticipated the .357 Sig design by nearly 100 years. Winchester has recently started making this ammunition. Ammo companies must think that they can sell enough to pay for tooling up the line . . . or they wouldn't bother.

Relative Recoil Factor = 0.52

Specifications:

Controlling Agency for Standardization of this Ammunition: SAAMI

Bullet Weight Grains	Velocity fps	Maximum Average Pressure Copper Crusher	Transducer
93	1,190	28,000 cup	N/S

Standard barrel for velocity testing: 4.5 inches long—1 turn in 11-inch twist

Availability:

Fiocchi 88-grain Full Metal Jacket (765A)

G1 Ballistic Coefficient = 0.18

Distance • Yards	Muzzle	25	50	75	100	125	150	175	200
Velocity • fps	1200	1147	1102	1063	1029	999	973	949	927
Energy • ft-lbs	297	272	251	233	219	206	195	186	177
Taylor KO Index	4.9	4.7	4.5	4.4	4.2	4.1	4.0	3.9	3.8
Mid-Range Trajectory Height • Inches	0.0	0.2	0.8	1.9	3.6	5.8	8.6	12.1	16.3
Drop • Inches	0.0	-0.8	-3.2	-7.4	-13.6	-21.7	-32.1	-44.7	-59.7

Winchester 93-grain Full Metal Jacket (X30LP)

G1 Ballistic Coefficient = 0.18

Distance • Yards	Muzzle	25	50	75	100	125	150	175	200
Velocity • fps	1220	1165	1110	1075	1040	1009	982	957	934
Energy • ft-lbs	305	280	255	239	225	210	199	189	180
Taylor KO Index	5.0	4.8	4.6	4.4	4.3	4.1	4.0	3.9	3.8
Mid-Range Trajectory Height • Inches	0.0	0.1	0.9	2.0	3.5	5.6	8.4	11.8	15.9
Drop • Inches	0.0	-0.8	-3.1	-7.2	-13.2	-21.1	-31.2	-43.5	-58.2

Fiocchi 93-grain Jacketed Soft Point (765B)

G1 Ballistic Coefficient = 0.18

Distance • Yards	Muzzle	25	50	75	100	125	150	175	200
Velocity • fps	1210	1156	1109	1069	1035	1004	978	953	931
Energy • ft-lbs	305	276	254	236	221	208	197	188	179
Taylor KO Index	5.0	4.7	4.5	4.4	4.2	4.1	4.0	3.9	3.8
Mid-Range Trajectory Height • Inches	0.0	0.2	0.8	1.9	3.5	5.6	8.5	11.9	16.2
Drop • Inches	0.0	-0.8	-3.2	-7.3	-13.4	-21.4	-31.7	-44.2	-59.1

.32 Short Colt

Here's another oldie (vintage of about 1875) that's been recently reintroduced into manufacturing. As with the .30 Luger, it's hard to see where the sales potential comes from, but since I know practically nothing about that part of the business, I hope the factory folks are smarter than I am and know what they are doing.

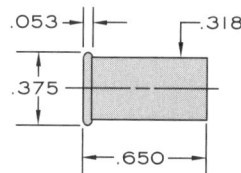

Relative Recoil Factor = 0.27

Specifications:

Controlling Agency for Standardization of this Ammunition: SAAMI

Bullet Weight Grains	Velocity fps	Maximum Average Pressure	
		Copper Crusher	Transducer
80 L	700	13,000 cup	N/S

Standard barrel for velocity testing: 4 inches long, vented—1 turn in 16-inch twist

Availability:

Winchester 80-grain Lead-Round Nose (X32SCP)

G1 Ballistic Coefficient = 0.054

Distance • Yards	Muzzle	25	50	75	100	125	150	175	200
Velocity • fps	745	702	665	625	590	557	526	496	467
Energy • ft-lbs	100	88	79	69	62	55	49	44	39
Taylor KO Index	2.6	2.5	2.4	2.2	2.1	2.0	1.9	1.8	1.7
Mid-Range Trajectory Height • Inches	0.0	0.5	2.2	5.0	9.9	16.4	25.2	36.6	50.7
Drop • Inches	0.0	-2.0	-8.5	-19.9	-36.8	-60.0	-90.2	-128.2	-175.0

Author Bob Forker in action with a six-shooter—just another day at the range for him.

.32 Smith & Wesson (Short Version)

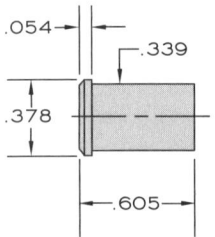

The little .32 Smith & Wesson cartridge dates clear back to 1878. It was designed when "pocket" pistols were all the rage. This is a meek and mild cartridge and can't be considered for any serious self-defense role. Both the muzzle energy and the Taylor KO indices are far too low to be effective. Check out the .32 Smith & Wesson Long.

Relative Recoil Factor = 0.27

Specifications:

Controlling Agency for Standardization of this Ammunition: SAAMI

Bullet Weight Grains	Velocity fps	Maximum Average Pressure	
		Copper Crusher	Transducer
85–88 L	700	12,000 cup	N/S

Standard barrel for velocity testing: 4 inches long, vented—1 turn in 18.75-inch twist

Availability:

Magtech 85-grain LRN (32SWA)
G1 Ballistic Coefficient = 0.115

Distance • Yards	Muzzle	25	50	75	100	125	150	175	200
Velocity • fps	680	662	645	627	610	594	578	562	547
Energy • ft-lbs	87	83	78	74	70	67	63	60	57
Taylor KO Index	2.6	2.5	2.5	2.4	2.3	2.3	2.2	2.1	2.1
Mid-Range Trajectory Height • Inches	0.0	0.2	2.1	5.3	10.1	16.8	24.9	34.9	46.8
Drop • Inches	0.0	-2.4	-9.8	-22.3	-40.5	-64.4	-94.5	-131.1	-174.6

Winchester 85-grain Lead Round Nose (X32SWP)
G1 Ballistic Coefficient = 0.11

Distance • Yards	Muzzle	25	50	75	100	125	150	175	200
Velocity • fps	680	662	645	627	610	594	578	562	547
Energy • ft-lbs	87	83	78	74	70	67	63	60	57
Taylor KO Index	2.6	2.5	2.5	2.4	2.3	2.3	2.2	2.1	2.1
Mid-Range Trajectory Height • Inches	0.0	0.2	2.1	5.3	10.1	16.8	24.9	34.9	46.8
Drop • Inches	0.0	-2.4	-9.8	-22.3	-40.5	-64.4	-94.5	-131.1	-174.6

Remington 88-grain Lead Round Nose (R32SW)
G1 Ballistic Coefficient = 0.11

Distance • Yards	Muzzle	25	50	75	100	125	150	175	200
Velocity • fps	680	662	645	627	610	594	578	562	547
Energy • ft-lbs	87	83	78	74	70	67	63	60	57
Taylor KO Index	2.6	2.5	2.5	2.4	2.3	2.3	2.2	2.1	2.1
Mid-Range Trajectory Height • Inches	0.0	0.2	2.1	5.3	10.1	16.8	24.9	34.9	46.8
Drop • Inches	0.0	-2.4	-9.8	-22.3	-40.5	-64.4	-94.5	-131.1	-174.6

.32 Smith & Wesson Long

This is another very old cartridge (introduced in 1896). While it originally was considered a self-defense round for pocket pistols, today's use of this caliber is almost exclusively for target pistols. In some competitions, .32 caliber pistols meet the minimum power rules and have the virtue of very light recoil and fine accuracy. The use of this caliber in competition seems more common in Europe than in the U.S.

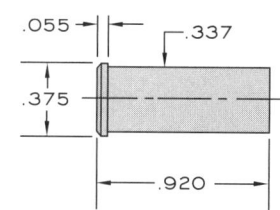

Relative Recoil Factor = 0.35

Specifications:

Controlling Agency for Standardization of this Ammunition: SAAMI

Bullet Weight Grains	Velocity fps	Maximum Average Pressure	
		Copper Crusher	Transducer
98 L	775	12,000 cup	15,000 psi

Standard barrel for velocity testing: 4 inches long—1 turn in 18.75-inch twist

Availability:

Lapua 83-grain LWC (4318023)

G1 Ballistic Coefficient = 0.029

Distance • Yards	Muzzle	25	50	75	100	125	150	175	200
Velocity • fps	787	704	632	567	509	456	408	363	323
Energy • ft-lbs	114	91	74	59	48	38	31	24	19
Taylor KO Index	2.9	2.6	2.4	2.1	1.9	1.7	1.5	1.4	1.2
Mid-Range Trajectory Height • Inches	0.0	0.5	2.2	5.6	11.0	19.5	31.4	49.2	71.9
Drop • Inches	0.0	-1.9	-8.2	-19.9	-38.4	-65.3	-102.7	-153.3	-220.7

Fiocchi 97-grain LRN (32SWLL)

G1 Ballistic Coefficient = 0.121

Distance • Yards	Muzzle	25	50	75	100	125	150	175	200
Velocity • fps	800	778	757	737	718	700	682	664	648
Energy • ft-lbs	138	130	124	117	111	105	100	95	90
Taylor KO Index	3.5	3.4	3.3	3.2	3.1	3.0	2.9	2.9	2.8
Mid-Range Trajectory Height • Inches	0.0	0.4	1.8	4.1	7.6	11.8	18.0	24.9	38.8
Drop • Inches	0.0	-1.7	-7.0	-16.2	-29.2	-46.6	-68.3	-94.7	-126.1

Fiocchi 97-grain FMJ (32SWLA

G1 Ballistic Coefficient = 0.121

Distance • Yards	Muzzle	25	50	75	100	125	150	175	200
Velocity • fps	800	778	757	737	718	700	682	664	648
Energy • ft-lbs	138	130	124	117	111	105	100	95	90
Taylor KO Index	3.5	3.4	3.3	3.2	3.1	3.0	2.9	2.9	2.8
Mid-Range Trajectory Height • Inches	0.0	0.4	1.8	4.1	7.6	11.8	18.0	24.9	38.8
Drop • Inches	0.0	-1.7	-7.0	-16.2	-29.2	-46.6	-68.3	-94.7	-126.1

Lapua 98-grain LWC (4318026)

G1 Ballistic Coefficient = 0.061

Distance • Yards	Muzzle	25	50	75	100	125	150	175	200
Velocity • fps	787	727	673	623	578	535	496	459	424
Energy • ft-lbs	135	115	98	85	73	62	53	46	39
Taylor KO Index	3.4	3.2	2.9	2.7	2.5	2.4	2.2	2.0	1.9
Mid-Range Trajectory Height • Inches	0.0	0.5	2.1	5.0	9.7	16.3	25.5	37.9	53.6
Drop • Inches	0.0	-1.8	-7.8	-18.6	-35.0	-57.8	-88.2	-127.4	-177.0

Federal 98-grain Lead Wadcutter (C32LA)

G1 Ballistic Coefficient = 0.041

Distance • Yards	Muzzle	25	50	75	100	125	150	175	200
Velocity • fps	780	720	670	620	570	531	492	455	421
Energy • ft-lbs	130	115	95	80	70	61	52	45	38
Taylor KO Index	3.4	3.1	2.9	2.7	2.5	2.3	2.1	2.0	1.8
Mid-Range Trajectory Height • Inches	0.0	0.5	2.1	5.1	9.8	16.4	26.0	38.6	54.6
Drop • Inches	0.0	-1.9	-8.0	-18.9	-35.6	-58.8	-89.8	-129.8	-180.3

Magtech 98-grain LWC (32SWLB)

G1 Ballistic Coefficient = 0.030

Distance • Yards	Muzzle	25	50	75	100	125	150	175	200
Velocity • fps	682	615	554	499	449	402	360	322	288
Energy • ft-lbs	101	82	67	54	44	36	29	23	18
Taylor KO Index	3.0	2.7	2.4	2.2	2.0	1.8	1.6	1.4	1.3
Mid-Range Trajectory Height • Inches	0.0	0.3	2.6	6.8	14.0	24.7	40.3	63.1	92.9
Drop • Inches	0.0	-2.5	-10.8	-26.2	-50.3	-85.3	-133.8	-199.1	-285.6

Aguila 98-grain Lead (1E322340)

G1 Ballistic Coefficient = 0.119

Distance • Yards	Muzzle	25	50	75	100	125	150	175	200
Velocity • fps	705	687	670	651	635	618	602	587	572
Energy • ft-lbs	108	103	98	92	88	83	79	75	71
Taylor KO Index	3.1	3.0	2.9	2.8	2.8	2.7	2.7	2.6	2.5
Mid-Range Trajectory Height • Inches	0.0	0.5	2.3	5.3	10.5	15.6	23.1	32.3	43.3
Drop • Inches	0.0	-2.2	-9.1	-20.8	-37.6	-59.8	-87.6	-121.5	-161.7

Magtech 98-grain LRN (32SWLA)

G1 Ballistic Coefficient = 0.119

Distance • Yards	Muzzle	25	50	75	100	125	150	175	200
Velocity • fps	705	687	670	651	635	618	602	587	572
Energy • ft-lbs	108	103	98	92	88	83	79	75	71
Taylor KO Index	3.1	3.0	2.9	2.8	2.8	2.7	2.7	2.6	2.5
Mid-Range Trajectory Height • Inches	0.0	0.5	2.3	5.3	10.5	15.6	23.1	32.3	43.3
Drop • Inches	0.0	-2.2	-9.1	-20.8	-37.6	-59.8	-87.6	-121.5	-161.7

Remington 98-grain LRN (R32SWL)

G1 Ballistic Coefficient = 0.119

Distance • Yards	Muzzle	25	50	75	100	125	150	175	200
Velocity • fps	705	687	670	651	635	618	602	587	572
Energy • ft-lbs	108	103	98	92	88	83	79	75	71
Taylor KO Index	3.1	3.0	2.9	2.8	2.8	2.7	2.7	2.6	2.5
Mid-Range Trajectory Height • Inches	0.0	0.5	2.3	5.3	10.5	15.6	23.1	32.3	43.3
Drop • Inches	0.0	-2.2	-9.1	-20.8	-37.6	-59.8	-87.6	-121.5	-161.7

Winchester 98-grain Lead Round Nose (X32SWLP)

G1 Ballistic Coefficient = 0.11

Distance • Yards	Muzzle	25	50	75	100	125	150	175	200
Velocity • fps	705	687	670	651	635	618	602	587	572
Energy • ft-lbs	108	103	98	92	88	83	79	75	71
Taylor KO Index	3.1	3.0	2.9	2.8	2.8	2.7	2.7	2.6	2.5
Mid-Range Trajectory Height • Inches	0.0	0.5	2.3	5.3	10.5	15.6	23.1	32.3	43.3
Drop • Inches	0.0	-2.2	-9.1	-20.8	-37.6	-59.8	-87.6	-121.5	-161.7

Magtech 98-grain SJHP (32SWLC)

G1 Ballistic Coefficient = 0.0

Distance • Yards	Muzzle	25	50	75	100	125	150	175	200
Velocity • fps	778	610	479	373	287	219	166	127	66
Energy • ft-lbs	132	81	50	30	18	10	6.0	3.5	0.9
Taylor KO Index	3.4	2.7	2.1	1.6	1.3	1.0	0.7	0.5	0.3
Mid-Range Trajectory Height • Inches	0.0	1.8	3.0	8.5	18.5	36.1	65.1	112.2	183.3
Drop • Inches	0.0	-2.1	-10.2	-28.0	-61.8	-111.0	-196.6	-303.5	-492.4

Sellier & Bellot 100-grain WC (V311302U)

G1 Ballistic Coefficient = 0.0

Distance • Yards	Muzzle	25	50	75	100	125	150	175	200
Velocity • fps	735	704	675	647	620	595	570	547	524
Energy • ft-lbs	120	110	101	93	85	79	72	66	61
Taylor KO Index	3.3	3.2	3.0	2.9	2.8	2.7	2.6	2.5	2.4
Mid-Range Trajectory Height • Inches	0.0	0.2	1.8	4.7	9.1	15.0	23.0	32.8	45.1
Drop • Inches	0.0	-2.1	-8.5	-19.8	-36.2	-58.2	-86.4	-121.4	-163.5

Fiocchi 100-grain LWC (32LA)

G1 Ballistic Coefficient = 0.0

Distance • Yards	Muzzle	25	50	75	100	125	150	175	200
Velocity • fps	730	652	583	522	466	415	368	327	289
Energy • ft-lbs	116	95	74	60	47	38	30	24	19
Taylor KO Index	3.3	2.9	2.6	2.3	2.1	1.9	1.7	1.5	1.3
Mid-Range Trajectory Height • Inches	0.0	0.2	2.2	6.1	12.5	22.5	37.0	58.6	87.2
Drop • Inches	0.0	-2.2	-9.5	-23.3	-45.0	-76.8	-121.3	-181.9	-263.0

Sellier & Bellot 100-grain LRN (V311312U)

G1 Ballistic Coefficient = 0.220

Distance • Yards	Muzzle	25	50	75	100	125	150	175	200
Velocity • fps	886	871	856	842	829	816	803	791	779
Energy • ft-lbs	174	168	163	158	153	148	143	139	135
Taylor KO Index	3.9	3.9	3.8	3.8	3.7	3.6	3.6	3.5	3.5
Mid-Range Trajectory Height • Inches	0.0	0.3	1.4	3.3	5.8	9.4	13.8	18.8	25.3
Drop • Inches	0.0	-1.4	-5.7	-12.9	-23.2	-36.6	-53.3	-73.4	-97.0

ith hard-kicking handguns, you need not only hearing and eye protection, like you do for all shooting, but also a good grip and well-made shooting ves. This is the author and creator of this book, Bob Forker, in action. (Safari Press Archives)

.32 H&R Magnum

Designed in 1983, the .32 H&R (Harrington and Richardson) Magnum brings modern technology and high velocity to the .32-caliber pistol market. This cartridge is sometimes touted for hunting use but can't be taken seriously for that purpose. The .32-caliber guns have long been more popular in Europe as target pistols. This is still a new cartridge, and it will take more time before its true worth has been established by active shooters. The velocity data for this pistol round may be misleading since the standard velocity test barrel is a rifle length of 24 inches.

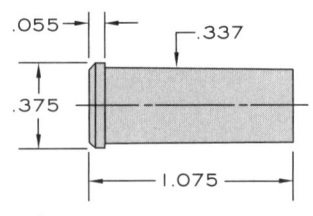

Relative Recoil Factor = 0.42

Specifications:

Controlling Agency for Standardization of this Ammunition: SAAMI

Bullet Weight Grains	Velocity fps	Maximum Average Pressure	
		Copper Crusher	Transducer
85 JHP	1,120	21,000 cup	N/S
95 LSWC	1,020	21,000 cup	N/S

Standard barrel for velocity testing: 24 inches long—1 turn in 12-inch twist

Availability:

Federal 85-grain Hi-Shok JHP (C32HRB)
G1 Ballistic Coefficient = 0.137

Distance • Yards	Muzzle	25	50	75	100	125	150	175	200
Velocity • fps	1120	1070	1020	990	950	921	894	870	847
Energy • ft-lbs	235	215	195	185	170	160	151	143	135
Taylor KO Index	4.2	4.1	3.9	3.8	3.6	3.5	3.4	3.3	3.2
Mid-Range Trajectory Height • Inches	0.0	0.2	1.0	2.3	4.2	6.7	10.1	14.1	19.2
Drop • Inches	0.0	-0.9	-3.7	-8.6	-15.7	-25.3	-37.4	-52.2	-69.9

Black Hills 85-grain JHP (D32H&RN1)
G1 Ballistic Coefficient = 0.12

Distance • Yards	Muzzle	25	50	75	100	125	150	175	200
Velocity • fps	1100	1050	1020	970	930	901	874	849	825
Energy • ft-lbs	230	210	195	175	165	153	144	136	129
Taylor KO Index	4.2	4.0	3.8	3.7	3.5	3.4	3.3	3.2	3.2
Mid-Range Trajectory Height • Inches	0.0	0.2	1.0	2.3	4.3	7.0	10.5	14.8	19.9
Drop • Inches	0.0	-0.9	-3.8	-8.9	-16.4	-26.2	-38.8	-54.2	-72.5

Federal 95-grain Lead Semi-Wadcutter (C32HRA)
G1 Ballistic Coefficient = 0.10

Distance • Yards	Muzzle	25	50	75	100	125	150	175	200
Velocity • fps	1020	970	930	890	860	831	803	777	753
Energy • ft-lbs	220	200	180	170	155	146	136	127	120
Taylor KO Index	4.3	4.1	3.9	3.8	3.6	3.5	3.4	3.3	3.2
Mid-Range Trajectory Height • Inches	0.0	0.3	1.2	2.7	5.1	8.2	12.3	17.2	23.5
Drop • Inches	0.0	-1.1	-4.5	-10.4	-19.0	-30.6	-45.2	-63.3	-85.0

Cowboy Action Loads:

Black Hills 90-grain FPL (DCB32H&RN2)
G1 Ballistic Coefficient = 0.1

Distance • Yards	Muzzle	25	50	75	100	125	150	175	200
Velocity • fps	750	733	716	700	685	670	655	640	626
Energy • ft-lbs	112	107	103	98	94	90	86	82	78
Taylor KO Index	3.0	3.0	2.9	2.8	2.8	2.7	2.7	2.6	2.5
Mid-Range Trajectory Height • Inches	0.0	0.5	2.0	4.6	8.5	13.5	20.0	27.8	37.1
Drop • Inches	0.0	-2.0	-8.0	-18.2	-32.9	-52.2	-76.3	-105.5	-140.0

.32 Auto (7.65mm Browning)

John Browning designed this little cartridge in 1899. The fact that it is still around proves that some shooters find it useful. There is some good news and some bad news about the .32 Auto. The good news is that the cartridge design accepts very lightweight and compact guns. Since concealed carry has been legalized in many states, the .32 Auto finds many advocates. The bad news is that the .32 Auto isn't really enough cartridge to provide much serious self-defense. It's a difficult trade-off to get just right, and it's one that's made only slightly easier with the availability of high-performance bullets.

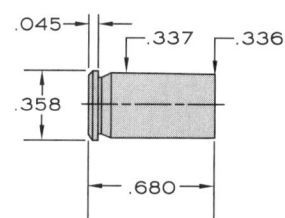

Relative Recoil Factor = 0.29

Specifications:

Controlling Agency for Standardization of this Ammunition: SAAMI

Bullet Weight Grains	Velocity fps	Maximum Average Pressure Copper Crusher	Transducer
60 STHP	960	15,000 cup	20,500 psi
71 MC	900	15,000 cup	20,500 psi

Standard barrel for velocity testing: 4 inches long—1 turn in 16-inch twist

Availability:

Cor-Bon 55-grain Glaser Safety Slug (00400/na)
G1 Ballistic Coefficient = 0.100

Distance • Yards	Muzzle	25	50	75	100	125	150	175	200
Velocity • fps	1100	1033	980	936	899	865	834	806	779
Energy • ft-lbs	148	130	117	107	99	91	85	79	74
Taylor KO Index	2.7	2.5	2.4	2.3	2.2	2.1	2.0	2.0	1.9
Mid-Range Trajectory Height • Inches	0.0	0.2	1.0	2.4	4.5	7.4	11.2	15.7	21.5
Drop • Inches	0.0	-0.9	-3.9	-9.2	-16.8	-27.2	-40.6	-57.0	-76.7

Cor-Bon 55-grain Pow'RBall (PB3255/20)
G1 Ballistic Coefficient = 0.100

Distance • Yards	Muzzle	25	50	75	100	125	150	175	200
Velocity • fps	1100	1033	980	936	899	865	834	806	779
Energy • ft-lbs	148	130	117	107	99	91	85	79	74
Taylor KO Index	2.7	2.5	2.4	2.3	2.2	2.1	2.0	2.0	1.9
Mid-Range Trajectory Height • Inches	0.0	0.2	1.0	2.4	4.5	7.4	11.2	15.7	21.5
Drop • Inches	0.0	-0.9	-3.9	-9.2	-16.8	-27.2	-40.6	-57.0	-76.7

Fiocchi 60-grain Jacketed Hollowpoint (32APHP)
G1 Ballistic Coefficient = 0.120

Distance • Yards	Muzzle	25	50	75	100	125	150	175	200
Velocity • fps	1100	1043	997	957	924	893	865	839	815
Energy • ft-lbs	161	145	132	122	113	106	100	94	88
Taylor KO Index	2.9	2.8	2.7	2.6	2.5	2.4	2.3	2.2	2.2
Mid-Range Trajectory Height • Inches	0.0	0.2	1.0	2.4	4.4	7.1	10.7	14.9	20.4
Drop • Inches	0.0	-0.9	-3.9	-9.0	-16.5	-26.5	-39.2	-54.9	-73.6

Cor-Bon 60-grain DPX (DPX3260/20)
G1 Ballistic Coefficient = 0.100

Distance • Yards	Muzzle	25	50	75	100	125	150	175	200
Velocity • fps	1050	978	934	897	863	833	804	778	753
Energy • ft-lbs	141	127	116	107	99	92	86	81	75
Taylor KO Index	2.8	2.6	2.5	2.4	2.3	2.2	2.2	2.1	2.0
Mid-Range Trajectory Height • Inches	0.0	0.3	1.1	2.7	5.0	8.1	12.1	17.2	23.2
Drop • Inches	0.0	-1.1	-4.4	-10.2	-18.7	-30.1	-44.6	-62.5	-83.8

Cor-Bon 60-grain JHP (SD3260/20)
G1 Ballistic Coefficient = 0.100

Distance • Yards	Muzzle	25	50	75	100	125	150	175	200
Velocity • fps	1050	978	934	897	863	833	804	778	753
Energy • ft-lbs	141	127	116	107	99	92	86	81	75
Taylor KO Index	2.8	2.6	2.5	2.4	2.3	2.2	2.2	2.1	2.0
Mid-Range Trajectory Height • Inches	0.0	0.3	1.1	2.7	5.0	8.1	12.1	17.2	23.2
Drop • Inches	0.0	-1.1	-4.4	-10.2	-18.7	-30.1	-44.6	-62.5	-83.8

Fiocchi 60-grain XTPHP (32XTP)

G1 Ballistic Coefficient = 0.120

Distance • Yards	Muzzle	25	50	75	100	125	150	175	200
Velocity • fps	1000	960	926	895	867	841	817	794	772
Energy • ft-lbs	133	123	114	107	100	94	89	84	80
Taylor KO Index	2.7	2.6	2.5	2.4	2.3	2.2	2.2	2.1	2.1
Mid-Range Trajectory Height • Inches	0.0	0.3	1.2	2.8	5.1	8.2	12.3	17.0	23.1
Drop • Inches	0.0	-1.1	-4.6	-10.6	-19.3	-30.9	-45.6	-63.5	-84.9

Hornady 60-grain JHP/XTP (90062)

G1 Ballistic Coefficient = 0.091

Distance • Yards	Muzzle	25	50	75	100	125	150	175	200
Velocity • fps	1000	949	906	868	834	803	774	747	721
Energy • ft-lbs	133	120	109	100	93	86	80	74	69
Taylor KO Index	2.7	2.5	2.4	2.3	2.2	2.1	2.1	2.0	1.9
Mid-Range Trajectory Height • Inches	0.0	0.3	1.2	2.8	5.3	8.6	13.0	18.4	25.0
Drop • Inches	0.0	-1.1	-4.7	-10.8	-19.9	-32.1	-47.6	-66.8	-89.9

PMC 60-grain Jacketed Hollow Point (32B)

G1 Ballistic Coefficient = 0.060

Distance • Yards	Muzzle	25	50	75	100	125	150	175	200
Velocity • fps	980	849	820	791	763	722	685	650	617
Energy • ft-lbs	117	111	98	87	78	69	63	56	51
Taylor KO Index	2.6	2.3	2.2	2.1	2.1	1.9	1.8	1.8	1.7
ctory Height Inches	0.0	0.2	0.9	2.7	5.5	9.6	14.9	21.5	29.9
Drop • Inches	0.0	-1.2	-5.0	-11.8	-21.9	-35.7	-53.6	-76.2	-103.8

Winchester 60-grain Silvertip Hollow Point (X32ASHP)

G1 Ballistic Coefficient = 0.101

Distance • Yards	Muzzle	25	50	75	100	125	150	175	200
Velocity • fps	970	930	895	864	835	809	784	760	737
Energy • ft-lbs	125	115	107	99	93	87	82	77	72
Taylor KO Index	2.6	2.5	2.4	2.3	2.2	2.2	2.1	2.1	2.0
Mid-Range Trajectory Height • Inches	0.0	0.3	1.3	2.9	5.4	8.7	13.0	18.4	24.6
Drop • Inches	0.0	-1.2	-4.9	-11.3	-20.6	-33.0	-48.7	-67.9	-90.9

Speer 60-grain Gold Dot (23604)

G1 Ballistic Coefficient = 0.11

Distance • Yards	Muzzle	25	50	75	100	125	150	175	200
Velocity • fps	960	925	894	866	840	816	793	771	750
Energy • ft-lbs	123	114	107	100	94	89	84	79	75
Taylor KO Index	2.6	2.5	2.4	2.3	2.2	2.2	2.1	2.1	2.0
Mid-Range Trajectory Height • Inches	0.0	0.2	1.2	3.0	5.4	8.7	13.0	18.3	24.7
Drop • Inches	0.0	-1.2	-5.0	-11.4	-20.8	-33.2	-48.9	-68.0	-90.8

Federal 65-grain Hydra-Shok JHP (P32HS1)

G1 Ballistic Coefficient = 0.1

Distance • Yards	Muzzle	25	50	75	100	125	150	175	200
Velocity • fps	925	890	860	830	810	786	763	741	720
Energy • ft-lbs	125	115	105	100	95	89	84	79	75
Taylor KO Index	2.7	2.6	2.5	2.4	2.3	2.3	2.2	2.1	2.1
Mid-Range Trajectory Height • Inches	0.0	0.3	1.4	3.2	5.9	9.4	14.1	19.5	26.6
Drop • Inches	0.0	-1.3	-5.3	-12.3	-22.3	-35.7	-52.6	-73.2	-97.8

Magtech 65-grain JHP (GG32A)

G1 Ballistic Coefficient = 0.1

Distance • Yards	Muzzle	25	50	75	100	125	150	175	200
Velocity • fps	922	897	874	853	833	813	795	777	761
Energy • ft-lbs	123	116	110	105	100	96	91	87	84
Taylor KO Index	2.7	2.6	2.5	2.5	2.4	2.4	2.3	2.3	2.2
Mid-Range Trajectory Height • Inches	0.0	0.1	0.8	2.7	5.3	8.5	13.0	18.2	24.7
Drop • Inches	0.0	-1.3	-5.3	-12.1	-22.0	-34.9	-51.2	-70.8	-94.1

Magtech 71-grain LRN (32C)

G1 Ballistic Coefficient = 0.1

Distance • Yards	Muzzle	25	50	75	100	125	150	175	200
Velocity • fps	905	879	855	831	810	789	770	751	733
Energy • ft-lbs	129	122	115	109	103	98	93	89	85
Taylor KO Index	2.9	2.8	2.7	2.6	2.6	2.5	2.5	2.4	2.3
Mid-Range Trajectory Height • Inches	0.0	0.3	1.4	3.2	5.8	9.5	14.1	19.8	26.5
Drop • Inches	0.0	-1.4	-5.5	-12.7	-23.0	-36.5	-53.6	-74.3	-98.9

Magtech 71-grain JHP (32B)

G1 Ballistic Coefficient = 0.132

Distance • Yards	Muzzle	25	50	75	100	125	150	175	200
Velocity • fps	905	879	855	831	810	789	770	751	733
Energy • ft-lbs	129	122	115	109	103	98	93	89	85
Taylor KO Index	2.9	2.8	2.7	2.6	2.6	2.5	2.5	2.4	2.3
Mid-Range Trajectory Height • Inches	0.0	0.3	1.4	3.2	5.8	9.5	14.1	19.8	26.5
Drop • Inches	0.0	-1.4	-5.5	-12.7	-23.0	-36.5	-53.6	-74.3	-98.9

Speer 71-grain TMJ RN Lawman Training (53632)

G1 Ballistic Coefficient = 0.105

Distance • Yards	Muzzle	25	50	75	100	125	150	175	200
Velocity • fps	900	868	838	811	785	761	738	716	694
Energy • ft-lbs	128	119	111	104	97	91	86	81	76
Taylor KO Index	2.8	2.7	2.7	2.6	2.5	2.4	2.3	2.3	2.2
Mid-Range Trajectory Height • Inches	0.0	0.3	1.5	3.4	6.2	10.0	15.0	20.8	28.3
Drop • Inches	0.0	-1.4	-5.6	-13.0	-23.6	-37.8	-55.8	-77.7	-103.8

Aguila 71-grain FMJ (1E322110)

G1 Ballistic Coefficient = 0.132

Distance • Yards	Muzzle	25	50	75	100	125	150	175	200
Velocity • fps	905	879	854	831	810	789	770	751	733
Energy • ft-lbs	129	122	115	109	103	98	93	89	85
Taylor KO Index	2.9	2.8	2.7	2.6	2.6	2.5	2.5	2.4	2.3
Mid-Range Trajectory Height • Inches	0.0	0.3	1.4	3.2	5.8	9.5	14.1	19.8	26.5
Drop • Inches	0.0	-1.4	-5.5	-12.7	-23.0	-36.5	-53.6	-74.3	-98.9

Magtech 71-grain FMC (32A)

G1 Ballistic Coefficient = 0.132

Distance • Yards	Muzzle	25	50	75	100	125	150	175	200
Velocity • fps	905	879	855	831	810	789	770	751	733
Energy • ft-lbs	129	122	115	109	103	98	93	89	85
Taylor KO Index	2.9	2.8	2.7	2.6	2.6	2.5	2.5	2.4	2.3
Mid-Range Trajectory Height • Inches	0.0	0.3	1.4	3.2	5.8	9.5	14.1	19.8	26.5
Drop • Inches	0.0	-1.4	-5.5	-12.7	-23.0	-36.5	-53.6	-74.3	-98.9

Remington 71-grain Metal Case (R32AP)

G1 Ballistic Coefficient = 0.132

Distance • Yards	Muzzle	25	50	75	100	125	150	175	200
Velocity • fps	905	879	855	831	810	789	770	751	733
Energy • ft-lbs	129	122	115	109	103	98	93	89	85
Taylor KO Index	2.9	2.8	2.7	2.6	2.6	2.5	2.5	2.4	2.3
Mid-Range Trajectory Height • Inches	0.0	0.3	1.4	3.2	5.8	9.5	14.1	19.8	26.5
Drop • Inches	0.0	-1.4	-5.5	-12.7	-23.0	-36.5	-53.6	-74.3	-98.9

UMC (Remington) 71-grain Metal Case (L32AP)

G1 Ballistic Coefficient = 0.132

Distance • Yards	Muzzle	25	50	75	100	125	150	175	200
Velocity • fps	905	879	855	831	810	789	770	751	733
Energy • ft-lbs	129	122	115	109	103	98	93	89	85
Taylor KO Index	2.9	2.8	2.7	2.6	2.6	2.5	2.5	2.4	2.3
Mid-Range Trajectory Height • Inches	0.0	0.3	1.4	3.2	5.8	9.5	14.1	19.8	26.5
Drop • Inches	0.0	-1.4	-5.5	-12.7	-23.0	-36.5	-53.6	-74.3	-98.9

USA (Winchester) 71-grain Full Metal Jacket (Q4255)

G1 Ballistic Coefficient = 0.132

Distance • Yards	Muzzle	25	50	75	100	125	150	175	200
Velocity • fps	905	879	855	831	810	789	770	751	733
Energy • ft-lbs	129	122	115	109	103	98	93	89	85
Taylor KO Index	2.9	2.8	2.7	2.6	2.6	2.5	2.5	2.4	2.3
Mid-Range Trajectory Height • Inches	0.0	0.3	1.4	3.2	5.8	9.5	14.1	19.8	26.5
Drop • Inches	0.0	-1.4	-5.5	-12.7	-23.0	-36.5	-53.6	-74.3	-98.9

American Eagle (Federal) 71-grain Full Metal Jacket (AE32AP)

G1 Ballistic Coefficient = 0.123

Distance • Yards	Muzzle	25	50	75	100	125	150	175	200
Velocity • fps	900	870	850	820	800	778	758	738	719
Energy • ft-lbs	130	120	115	105	100	96	91	86	82
Taylor KO Index	2.8	2.8	2.7	2.6	2.5	2.5	2.4	2.3	2.3
Mid-Range Trajectory Height • Inches	0.0	0.3	1.4	3.3	6.1	9.7	14.5	20.1	27.2
Drop • Inches	0.0	-1.4	-5.6	-12.8	-23.8	-37.2	-54.6	-75.8	-101.1

Blazer (CCI) 71-grain TMJ (3503)

G1 Ballistic Coefficient = 0.112

Distance • Yards	Muzzle	25	50	75	100	125	150	175	200
Velocity • fps	900	870	843	816	791	768	746	725	705
Energy • ft-lbs	128	119	112	105	99	93	88	83	78
Taylor KO Index	2.8	2.8	2.7	2.6	2.5	2.4	2.4	2.3	2.2
Mid-Range Trajectory Height • Inches	0.0	0.3	1.4	3.3	6.1	9.9	14.7	20.7	27.8
Drop • Inches	0.0	-1.4	-5.6	-12.9	-23.5	-37.6	-55.3	-76.9	-102.6

PMC 71-grain Full Metal Jacket (32A)

G1 Ballistic Coefficient = 0.101

Distance • Yards	Muzzle	25	50	75	100	125	150	175	200
Velocity • fps	900	869	838	809	781	756	732	710	688
Energy • ft-lbs	128	118	110	103	96	90	85	79	75
Taylor KO Index	2.8	2.8	2.7	2.6	2.5	2.4	2.3	2.2	2.2
Mid-Range Trajectory Height • Inches	0.0	0.3	1.5	3.4	6.2	10.0	15.0	20.9	28.6
Drop • Inches	0.0	-1.4	-5.6	-13.0	-23.7	-38.0	-56.1	-78.2	-104.6

Sellier and Bellot 73-grain FMJ (SBA03201)

G1 Ballistic Coefficient = 0.116

Distance • Yards	Muzzle	25	50	75	100	125	150	175	200
Velocity • fps	1043	998	955	920	889	861	834	810	786
Energy • ft-lbs	177	162	148	137	128	120	113	106	100
Taylor KO Index	3.4	3.2	3.1	3.0	2.9	2.8	2.7	2.6	2.6
Mid-Range Trajectory Height • Inches	0.0	0.3	1.1	2.6	4.8	7.7	11.5	16.3	22.0
Drop • Inches	0.0	-1.0	-4.3	-9.9	-18.0	-29.0	-42.8	-59.8	-80.0

Fiocchi 73-grain FMJ (32AP)

G1 Ballistic Coefficient = 0.110

Distance • Yards	Muzzle	25	50	75	100	125	150	175	200
Velocity • fps	1000	962	929	899	873	847	824	802	781
Energy • ft-lbs	160	150	137	131	120	116	110	104	99
Taylor KO Index	3.3	3.1	3.0	2.9	2.8	2.8	2.7	2.6	2.5
Mid-Range Trajectory Height • Inches	0.0	0.3	1.2	2.7	5.0	8.1	12.1	16.8	22.9
Drop • Inches	0.0	-1.1	-4.6	-10.6	-19.2	-30.7	-45.3	-63.0	-84.1

In thick, dense jungles in South America, handguns were often carried on hunts in days gone by when jaguar hunting was still legal. (Photo from *Jaguar Hunting* by Tony de Almeida, 2012, Safari Press Archives)

.327 Federal Magnum

This is the newest .32-caliber round, introduced in 2008. Like several other caliber designations in this book, the .327 number is misleading. This is a conventional .32 caliber that uses .312-diameter bullets; there's nothing .327 about it. The case is 0.125 inches longer than the .32 H&R Magnum's case, and it uses somewhat higher working pressures so that the performance gains over the .32 H&R Magnum are substantial. This cartridge puts .38 Special performance (or even a little better) into a .32-caliber package. It will be interesting to see if it will gain favor as an ankle gun.

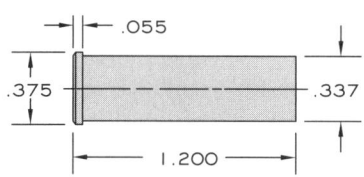

Relative Recoil Factor = 0.67

Specifications:

Controlling Agency for Standardization of this Ammunition: SAAMI

Bullet Weight Grains	Velocity fps	Maximum Average Pressure Copper Crusher	Transducer
85	1,600*	N/S	45,000 psi
100	1,500*	N/S	45,000 psi

Standard barrel for velocity testing: 5 inches long (1 Piece)—1 turn in 12-inch twist *

* Estimated preliminary numbers

Availability:

Federal 85-grain Hydra-Shok JHP (PD327HS1 H)

G1 Ballistic Coefficient = 0.140

Distance • Yards	Muzzle	25	50	75	100	125	150	175	200
Velocity • fps	1400	1310	1220	1150	1090	1043	1003	968	938
Energy • ft-lbs	370	320	280	250	225	205	190	177	166
Taylor KO Index	5.3	5.0	4.6	4.4	4.1	4.0	3.8	3.7	3.6
Mid-Range Trajectory Height • Inches	0.0	0.1	0.6	1.5	2.9	4.8	7.4	10.6	14.5
Drop • Inches	0.0	-0.6	-2.4	-5.8	-10.7	-17.5	-26.3	-37.2	-50.5

American Eagle (Federal) 85-grain Jacketed Soft Point (AE327A)

G1 Ballistic Coefficient = 0.140

Distance • Yards	Muzzle	25	50	75	100	125	150	175	200
Velocity • fps	1400	1310	1220	1150	1090	1043	1003	968	938
Energy • ft-lbs	588	511	447	396	357	326	301	281	264
Taylor KO Index	5.3	5.0	4.6	4.4	4.1	4.0	3.8	3.7	3.6
Mid-Range Trajectory Height • Inches	0.0	0.1	0.6	1.5	2.9	4.8	7.4	10.6	14.5
Drop • Inches	0.0	-0.6	-2.4	-5.8	-10.7	-17.5	-26.3	-37.2	-50.5

American Eagle (Federal) 100-grain Jacketed Soft Point (AE327)

G1 Ballistic Coefficient = 0.160

Distance • Yards	Muzzle	25	50	75	100	125	150	175	200
Velocity • fps	1500	1410	1320	1250	1180	1124	1076	1036	1001
Energy • ft-lbs	500	440	390	345	310	280	257	238	223
Taylor KO Index	6.7	6.3	5.9	5.6	5.3	5.0	4.8	4.6	4.5
Mid-Range Trajectory Height • Inches	0.0	0.1	0.6	1.3	2.5	4.1	6.3	9.1	12.5
Drop • Inches	0.0	-0.5	-2.1	-4.9	-9.2	-15.0	-22.5	-31.9	-43.3

Speer 100-grain GDHP (23913)

G1 Ballistic Coefficient = 0.160

Distance • Yards	Muzzle	25	50	75	100	125	150	175	200
Velocity • fps	1500	1408	1324	1248	1181	1124	1076	1036	1001
Energy • ft-lbs	500	440	389	346	310	280	257	238	223
Taylor KO Index	6.7	6.3	5.9	5.6	5.3	5.0	4.8	4.6	4.5
Mid-Range Trajectory Height • Inches	0.0	0.1	0.5	1.3	2.5	4.1	6.3	9.1	12.5
Drop • Inches	0.0	-0.5	-2.1	-4.9	-9.2	-15.0	-22.5	-31.9	-43.3

Speer 115-grain GDHP (23914)

G1 Ballistic Coefficient = 0.180

Distance • Yards	Muzzle	25	50	75	100	125	150	175	200
Velocity • fps	1380	1307	1240	1181	1130	1086	1049	1016	987
Energy • ft-lbs	486	436	393	357	326	301	281	264	249
Taylor KO Index	7.1	6.7	6.4	6.1	5.8	5.6	5.4	5.2	5.1
Mid-Range Trajectory Height • Inches	0.0	0.2	0.6	1.5	2.8	4.6	7.0	9.9	13.6
Drop • Inches	0.0	-0.6	-2.5	-5.7	-10.6	-17.1	-25.5	-35.8	-48.3

.357 Magnum

At the time of its introduction in 1935, the .357 Magnum (sometimes called the .357 Smith and Wesson Magnum) was the world's most powerful revolver cartridge. Credit for its development is usually given to Col. D. B. Wesson of Smith & Wesson and to Phil Sharpe. The .357 design is essentially a lengthened .38 Special, but the working pressure level has been pushed up to 35,000 psi, which more than doubles the .38 Special's 17,000 psi level and produces much better performance. Before the .357 was introduced, the .38 Special was the standard police sidearm in the U.S., almost the universal choice. It wasn't very long until the increased effectiveness of the .357 Magnum became known. Today, factories offer over 50 loadings in this caliber. For law enforcement application today, revolvers have largely been replaced by semiautomatic pistols, not necessarily resulting in an improvement in reliability.

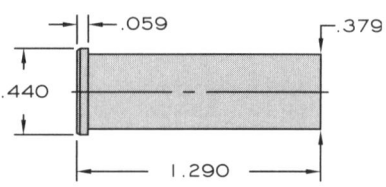

Relative Recoil Factor = 0.89

Specifications:

Controlling Agency for Standardization of this Ammunition: SAAMI

Bullet Weight Grains	Velocity fps	Maximum Average Pressure	
		Copper Crusher	Transducer
110 JHP	1,270	45,000 cup	35,000 psi
125 SJHP	1,500*	45,000 cup	35,000 psi
125 JSP	1,400	45,000 cup	35,000 psi
140 SJHP	1,330	45,000 cup	35,000 psi
145 STHP	1,270	45,000 cup	35,000 psi
158 MP-L	1,220	45,000 cup	35,000 psi
158 SWC	1,220	45,000 cup	35,000 psi
158 SJHP	1,220	45,000 cup	35,000 psi
180 STHP	1,400*	45,000 cup	35,000 psi

Standard barrel for velocity testing: 4 inches long [vented];
(* 12.493 inch, alternate one-piece barrel)—1 turn in 18.75-inch twist

Availability:

Cor-Bon 80-grain Glaser Safety Slug (02600/02800)
G1 Ballistic Coefficient = 0.08

Distance • Yards	Muzzle	25	50	75	100	125	150	175	200
Velocity • fps	1600	1409	1249	1125	1036	971	919	875	836
Energy • ft-lbs	455	353	277	225	191	168	150	136	124
Taylor KO Index	6.5	5.7	5.1	4.6	4.2	4.0	3.8	3.6	3.4
Mid-Range Trajectory Height • Inches	0.0	0.1	0.6	1.4	2.8	4.9	7.7	11.5	16.0
Drop • Inches	0.0	-0.5	-2.0	-5.0	-9.6	-16.3	-25.3	-36.9	-51.3

PMC 90-grain EMJ NT (357EMA)
G1 Ballistic Coefficient = 0.14

Distance • Yards	Muzzle	25	50	75	100	125	150	175	200
Velocity • fps	1200	1132	1078	1033	994	961	932	905	881
Energy • ft-lbs	288	257	232	213	198	185	174	164	155
Taylor KO Index	5.5	5.2	4.9	4.7	4.6	4.4	4.3	4.2	4.0
Mid-Range Trajectory Height • Inches	0.0	0.2	0.9	2.0	3.7	6.1	9.2	12.8	17.5
Drop • Inches	0.0	-0.8	-3.3	-7.6	-14.0	-22.6	-33.5	-46.9	-63.0

Magtech 95-grain SCHP (FD357A)
G1 Ballistic Coefficient = 0.14

Distance • Yards	Muzzle	25	50	75	100	125	150	175	200
Velocity • fps	1411	1315	1230	1157	1097	1048	1007	972	941
Energy • ft-lbs	420	365	319	283	254	232	214	199	187
Taylor KO Index	6.9	6.4	6.0	5.6	5.3	5.1	4.9	4.7	4.6
Mid-Range Trajectory Height • Inches	0.0	0.1	0.6	1.5	2.9	4.7	7.3	10.4	14.3
Drop • Inches	0.0	-0.6	-2.4	-5.7	-10.6	-17.2	-25.9	-36.7	-49.8

Cor-Bon 100-grain Pow'RBall (PB357100/20)

G1 Ballistic Coefficient = 0.090

Distance • Yards	Muzzle	25	50	75	100	125	150	175	200
Velocity • fps	1450	1299	1175	1082	1013	959	914	875	840
Energy • ft-lbs	467	375	307	260	228	204	186	170	157
Taylor KO Index	7.4	6.6	6.0	5.5	5.2	4.9	4.7	4.5	4.3
Mid-Range Trajectory Height • Inches	0.0	0.1	0.6	1.6	3.1	5.3	8.3	12.1	16.8
Drop • Inches	0.0	-0.6	-2.4	-5.8	-11.1	-18.5	-28.2	-40.6	-55.8

Federal 100-grain Barnes Expander (P357XB1) LF

G1 Ballistic Coefficient = 0.090

Distance • Yards	Muzzle	25	50	75	100	125	150	175	200
Velocity • fps	1400	1330	1260	1200	1140	1095	1055	1022	992
Energy • ft-lbs	610	545	490	445	405	373	346	325	306
Taylor KO Index	10.0	9.5	9.0	8.6	8.2	7.8	7.6	7.3	7.0
Mid-Range Trajectory Height • Inches	0.0	0.1	0.6	1.4	2.7	4.5	6.8	9.7	13.3
Drop • Inches	0.0	-0.6	-2.4	-5.6	-10.3	-16.7	-24.9	-35.0	-47.2

CCI 109-grain Shotshell (3709)

Muzzle Velocity = 1000 fps -- Loaded with #9 Shot

Winchester 110-grain Jacketed Flat Point (SC357NT) LF

G1 Ballistic Coefficient = 0.090

Distance • Yards	Muzzle	25	50	75	100	125	150	175	200
Velocity • fps	1275	1182	1105	1047	998	959	924	893	865
Energy • ft-lbs	397	341	298	268	243	225	209	195	183
Taylor KO Index	7.2	6.6	6.2	5.9	5.6	5.4	5.2	5.0	4.9
Mid-Range Trajectory Height • Inches	0.0	0.1	0.8	1.8	3.5	5.8	8.8	12.5	17.0
Drop • Inches	0.0	-0.7	-3.0	-7.0	-13.0	-21.1	-31.6	-44.6	-60.2

Cor-Bon 110-grain Jacketed Hollow Point (SD357110/20)

G1 Ballistic Coefficient = 0.090

Distance • Yards	Muzzle	25	50	75	100	125	150	175	200
Velocity • fps	1500	1356	1233	1134	1058	1000	953	914	878
Energy • ft-lbs	550	449	372	314	274	245	222	204	188
Taylor KO Index	8.4	7.6	6.9	6.4	5.9	5.6	5.3	5.1	4.9
Mid-Range Trajectory Height • Inches	0.0	0.1	0.5	1.4	2.8	4.8	7.5	10.9	15.0
Drop • Inches	0.0	-0.5	-2.2	-5.4	-10.2	-16.9	-25.7	-36.9	-50.5

Remington 110-grain Semi-Jacketed Hollow Point (R357M7)

G1 Ballistic Coefficient = 0.090

Distance • Yards	Muzzle	25	50	75	100	125	150	175	200
Velocity • fps	1295	1182	1094	1027	975	932	894	861	830
Energy • ft-lbs	410	341	292	258	232	212	195	181	168
Taylor KO Index	7.2	6.6	6.1	5.7	5.5	5.2	5.0	4.8	4.7
Mid-Range Trajectory Height • Inches	0.0	0.2	0.8	1.9	3.5	5.9	9.1	13.0	17.8
Drop • Inches	0.0	-0.7	-2.9	-7.0	-13.1	-21.4	-32.2	-45.6	-61.8

USA (Winchester) 110-grain Jacketed Hollow Point (Q4204)

G1 Ballistic Coefficient = 0.090

Distance • Yards	Muzzle	25	50	75	100	125	150	175	200
Velocity • fps	1295	1183	1095	1029	977	934	896	863	832
Energy • ft-lbs	410	342	292	259	233	213	196	182	169
Taylor KO Index	7.2	6.6	6.1	5.7	5.5	5.2	5.0	4.8	4.7
Mid-Range Trajectory Height • Inches	0.0	0.2	0.8	1.9	3.5	5.9	9.0	13.0	17.7
Drop • Inches	0.0	-0.7	-2.9	-7.0	-13.1	-21.4	-32.1	-45.5	-61.6

Fiocchi 125-grain SJSP (357C)

G1 Ballistic Coefficient = 0.090

Distance • Yards	Muzzle	25	50	75	100	125	150	175	200
Velocity • fps	1500	1426	1357	1293	1235	1182	1136	1095	1060
Energy • ft-lbs	625	565	511	464	423	388	358	333	312
Taylor KO Index	9.6	9.1	8.7	8.3	7.9	7.6	7.3	7.0	6.8
Mid-Range Trajectory Height • Inches	0.0	0.1	0.5	1.2	2.3	3.8	5.8	8.3	11.4
Drop • Inches	0.0	-0.5	-2.1	-4.8	-8.9	-14.3	-21.4	-30.1	-40.6

Hornady 125-grain FTX (90500)

G1 Ballistic Coefficient = 0.090

Distance • Yards	Muzzle	25	50	75	100	125	150	175	200
Velocity • fps	1500	1398	1301	1224	1153	1096	1048	1008	974
Energy • ft-lbs	624	543	470	416	370	334	305	282	263
Taylor KO Index	9.6	8.9	8.3	7.8	7.4	7.0	6.7	6.4	6.2
Mid-Range Trajectory Height • Inches	0.0	0.1	0.5	1.3	2.5	4.2	6.5	9.4	13.0
Drop • Inches	0.0	-0.5	-2.1	-5.0	-9.4	-15.3	-23.1	-32.9	-44.8

Fiocchi 125-grain SJHP (357D)

G1 Ballistic Coefficient = 0.190

Distance • Yards	Muzzle	25	50	75	100	125	150	175	200
Velocity • fps	1450	1375	1306	1243	1187	1137	1095	1058	1026
Energy • ft-lbs	584	525	474	429	391	359	333	311	292
Taylor KO Index	9.3	8.8	8.3	7.9	7.6	7.3	7.0	6.8	6.6
Mid-Range Trajectory Height • Inches	0.0	0.1	0.5	1.3	2.5	4.1	6.3	9.0	12.3
Drop • Inches	0.0	-0.5	-2.2	-5.2	-9.5	-15.4	-23.0	-32.4	-43.8

PMC 125-grain JSP (357B)

G1 Ballistic Coefficient = 0.125

Distance • Yards	Muzzle	25	50	75	100	125	150	175	200
Velocity • fps	1450	1339	1240	1158	1090	1037	994	956	924
Energy • ft-lbs	584	497	427	372	330	299	274	254	237
Taylor KO Index	9.2	8.5	7.9	7.4	6.9	6.6	6.3	6.1	5.9
Mid-Range Trajectory Height • Inches	0.0	0.1	0.6	1.5	2.8	4.7	7.2	10.4	14.3
Drop • Inches	0.0	-0.6	-2.3	-5.5	-10.3	-16.9	-25.5	-36.2	-49.2

UMC (Remington) 125-grain JSP (L357M12)

G1 Ballistic Coefficient = 0.125

Distance • Yards	Muzzle	25	50	75	100	125	150	175	200
Velocity • fps	1450	1339	1240	1158	1090	1037	994	956	924
Energy • ft-lbs	584	497	427	372	330	299	274	254	237
Taylor KO Index	9.2	8.5	7.9	7.4	6.9	6.6	6.3	6.1	5.9
Mid-Range Trajectory Height • Inches	0.0	0.1	0.6	1.5	2.8	4.7	7.2	10.4	14.3
Drop • Inches	0.0	-0.6	-2.3	-5.5	-10.3	-16.9	-25.5	-36.2	-49.2

Winchester 125-grain Jacketed Flat Point (WC3571)

G1 Ballistic Coefficient = 0.12?

Distance • Yards	Muzzle	25	50	75	100	125	150	175	200
Velocity • fps	1370	1269	1183	1112	1055	1009	970	936	906
Energy • ft-lbs	521	447	389	343	309	283	261	243	228
Taylor KO Index	8.7	8.1	7.5	7.1	6.7	6.4	6.2	6.0	5.8
Mid-Range Trajectory Height • Inches	0.0	0.1	0.6	1.6	3.0	5.1	7.8	11.2	15.3
Drop • Inches	0.0	-0.6	-2.6	-6.1	-11.3	-18.5	-27.8	-39.4	-53.3

Black Hills 125-grain Jacketed Hollow Point-Gold Dot (M357N2)

G1 Ballistic Coefficient = 0.15?

Distance • Yards	Muzzle	25	50	75	100	125	150	175	200
Velocity • fps	1500	1402	1313	1233	1164	1106	1058	1018	984
Energy • ft-lbs	625	546	478	422	376	340	311	288	269
Taylor KO Index	9.5	8.9	8.3	7.8	7.4	7.1	6.7	6.5	6.3
Mid-Range Trajectory Height • Inches	0.0	0.1	0.6	1.3	2.5	4.2	6.4	9.3	12.7
Drop • Inches	0.0	-0.5	-2.1	-5.0	-9.3	-15.2	-22.9	-32.4	-44.0

Hornady 125-grain JHP/XTP (90502)

G1 Ballistic Coefficient = 0.15

Distance • Yards	Muzzle	25	50	75	100	125	150	175	200
Velocity • fps	1500	1403	1314	1234	1166	1108	1060	1020	985
Energy • ft-lbs	625	546	476	423	377	341	312	289	270
Taylor KO Index	9.5	8.9	8.4	7.8	7.4	7.1	6.7	6.5	6.3
Mid-Range Trajectory Height • Inches	0.0	0.1	0.6	1.3	2.5	4.2	6.4	9.3	12.7
Drop • Inches	0.0	-0.5	-2.1	-5.0	-9.3	-15.2	-22.8	-32.4	-44.0

Fiocchi 125-grain SJHP (357D)

G1 Ballistic Coefficient = 0.19

Distance • Yards	Muzzle	25	50	75	100	125	150	175	200
Velocity • fps	1450	1375	1306	1243	1187	1137	1095	1058	1026
Energy • ft-lbs	584	525	474	429	391	359	333	311	292
Taylor KO Index	9.3	8.8	8.3	7.9	7.6	7.3	7.0	6.8	6.6
Mid-Range Trajectory Height • Inches	0.0	0.1	0.5	1.3	2.5	4.1	6.3	9.0	12.3
Drop • Inches	0.0	-0.5	-2.2	-5.2	-9.5	-15.4	-23.0	-32.4	-43.8

PMC 125-grain Jacketed Hollow Point (357B)

G1 Ballistic Coefficient = 0.1

Distance • Yards	Muzzle	25	50	75	100	125	150	175	200
Velocity • fps	1450	1337	1235	1155	1093	1039	995	958	926
Energy • ft-lbs	584	498	428	373	332	300	275	255	238
Taylor KO Index	9.2	8.5	7.9	7.4	7.0	6.6	6.3	6.1	5.9
Mid-Range Trajectory Height • Inches	0.0	0.1	0.6	1.4	2.8	4.7	7.2	10.4	14.2
Drop • Inches	0.0	-0.6	-2.3	-5.5	-10.2	-16.8	-25.4	-36.1	-49.1

Remington 125-grain Semi-Jacketed Hollow Point (R357M1)

G1 Ballistic Coefficient = 0.125

Distance • Yards	Muzzle	25	50	75	100	125	150	175	200
Velocity • fps	1450	1339	1240	1158	1090	1037	994	956	924
Energy • ft-lbs	583	497	427	372	330	299	274	254	237
Taylor KO Index	9.2	8.5	7.9	7.4	6.9	6.6	6.3	6.1	5.9
Mid-Range Trajectory Height • Inches	0.0	0.1	0.6	1.5	2.8	4.6	7.2	10.4	14.3
Drop • Inches	0.0	-0.6	-2.3	-5.5	-10.3	-16.9	-25.5	-36.2	-49.2

Speer 125-grain Gold Dot Hollow Point (23920)

G1 Ballistic Coefficient = 0.140

Distance • Yards	Muzzle	25	50	75	100	125	150	175	200
Velocity • fps	1450	1350	1261	1183	1118	1065	1021	984	952
Energy • ft-lbs	583	506	441	389	347	315	290	269	252
Taylor KO Index	9.2	8.6	8.0	7.5	7.1	6.8	6.5	6.3	6.1
Mid-Range Trajectory Height • Inches	0.0	0.1	0.6	1.4	2.7	4.6	7.0	10.1	13.8
Drop • Inches	0.0	-0.5	-2.3	-5.4	-10.0	-16.4	-24.7	-35.1	-47.7

Winchester 125-grain Jacketed HP (X3576P)

G1 Ballistic Coefficient = 0.124

Distance • Yards	Muzzle	25	50	75	100	125	150	175	200
Velocity • fps	1450	1338	1240	1157	1090	1036	992	955	922
Energy • ft-lbs	583	497	427	371	380	298	273	253	236
Taylor KO Index	9.2	8.5	7.9	7.4	6.9	6.6	6.3	6.1	5.9
Mid-Range Trajectory Height • Inches	0.0	0.1	0.6	1.5	2.8	4.7	7.3	10.6	14.5
Drop • Inches	0.0	-0.6	-2.3	-5.5	-10.3	-16.9	-25.6	-36.4	-49.6

Federal 125-grain Hi-Shok JHP (C357B)

G1 Ballistic Coefficient = 0.133

Distance • Yards	Muzzle	25	50	75	100	125	150	175	200
Velocity • fps	1440	1340	1240	1160	1100	1049	1005	969	937
Energy • ft-lbs	575	495	425	373	333	305	281	261	244
Taylor KO Index	9.2	8.5	7.9	7.4	7.0	6.7	6.4	6.2	3.0
Mid-Range Trajectory Height • Inches	0.0	0.1	0.6	1.5	2.8	4.7	7.2	10.4	14.3
Drop • Inches	0.0	-0.6	-2.3	-5.5	-10.3	-16.8	-25.4	-36.0	-49.0

Cor-Bon 125-grain Jacketed Hollow Point (SD357125/20)

G1 Ballistic Coefficient = 0.150

Distance • Yards	Muzzle	25	50	75	100	125	150	175	200
Velocity • fps	1400	1311	1232	1163	1106	1058	1018	984	953
Energy • ft-lbs	544	477	421	376	339	311	288	269	252
Taylor KO Index	8.9	8.4	7.9	7.4	7.1	6.7	6.5	6.3	6.1
Mid-Range Trajectory Height • Inches	0.0	0.1	0.6	1.5	2.9	4.7	7.2	10.3	14.2
Drop • Inches	0.0	-0.6	-2.4	-5.7	-10.6	-17.2	-25.8	-36.5	-49.5

Magtech 125-grain GDHP (GG357A)

G1 Ballistic Coefficient = 0.163

Distance • Yards	Muzzle	25	50	75	100	125	150	175	200
Velocity • fps	1378	1298	1221	1163	1110	1065	1027	994	965
Energy • ft-lbs	527	468	417	376	342	315	293	274	259
Taylor KO Index	8.8	8.3	7.8	7.4	7.1	6.8	6.6	6.4	6.2
Mid-Range Trajectory Height • Inches	0.0	0.1	0.6	1.5	2.9	4.7	7.2	10.2	14.0
Drop • Inches	0.0	-0.6	-2.5	-5.8	-10.8	-17.4	-26.1	-36.7	-49.6

Winchester 125-grain Bonded PDX1 (S357MPDB)

G1 Ballistic Coefficient = 0.129

Distance • Yards	Muzzle	25	50	75	100	125	150	175	200
Velocity • fps	1325	1232	1153	1089	1038	995	959	927	898
Energy • ft-lbs	487	421	369	329	299	275	255	239	224
Taylor KO Index	8.5	7.9	7.4	7.0	6.6	6.4	6.1	5.9	5.7
Mid-Range Trajectory Height • Inches	0.0	0.2	0.7	1.7	3.3	5.4	8.2	11.8	16.1
Drop • Inches	0.0	-0.6	-2.7	-6.4	-12.0	-19.6	-29.3	-41.4	-56.1

Cor-Bon 125-grain DPX (DPX357125/20) + (TR357125/20)

G1 Ballistic Coefficient = 0.110

Distance • Yards	Muzzle	25	50	75	100	125	150	175	200
Velocity • fps	1300	1196	1113	1048	997	955	918	886	856
Energy • ft-lbs	469	397	344	305	276	253	234	218	203
Taylor KO Index	8.3	7.6	7.1	6.7	6.4	6.1	5.9	5.6	5.5
Mid-Range Trajectory Height • Inches	0.0	0.2	0.8	1.9	3.5	5.8	8.9	12.7	17.4
Drop • Inches	0.0	-0.7	-2.9	-6.8	-12.8	-20.9	-31.4	-44.4	-60.2

Remington 125-grain Golden Saber (GS357MA)

G1 Ballistic Coefficient = 0.148

Distance • Yards	Muzzle	25	50	75	100	125	150	175	200
Velocity • fps	1220	1152	1095	1049	1009	976	946	919	895
Energy • ft-lbs	413	369	333	305	283	264	249	235	222
Taylor KO Index	7.8	7.3	7.0	6.7	6.4	6.2	6.0	5.9	5.7
Mid-Range Trajectory Height • Inches	0.0	0.2	0.8	1.9	3.5	5.8	8.8	12.4	16.8
Drop • Inches	0.0	-0.8	-3.2	-7.4	-13.5	-21.8	-32.3	-45.3	-60.7

Magtech 125-grain FMJ Flat (357Q)

G1 Ballistic Coefficient = 0.171

Distance • Yards	Muzzle	25	50	75	100	125	150	175	200
Velocity • fps	1405	1326	1254	1191	1136	1089	1049	1015	985
Energy • ft-lbs	548	488	437	394	358	329	305	286	269
Taylor KO Index	9.0	8.5	8.0	7.6	7.2	6.9	6.7	6.5	6.3
Mid-Range Trajectory Height • Inches	0.0	0.1	0.6	1.5	2.8	4.5	6.9	9.8	13.5
Drop • Inches	0.0	-0.6	-2.4	-5.6	-10.3	-16.7	-24.9	-35.1	-47.5

Federal 130-grain Hydra-Shok JHP (PD357HS2 H)

G1 Ballistic Coefficient = 0.172

Distance • Yards	Muzzle	25	50	75	100	125	150	175	200
Velocity • fps	1410	1330	1260	1190	1140	1093	1052	1018	988
Energy • ft-lbs	575	510	455	410	375	345	320	299	282
Taylor KO Index	9.3	8.8	8.4	7.9	7.6	7.2	7.0	6.7	6.6
Mid-Range Trajectory Height • Inches	0.0	0.1	0.6	1.5	2.8	4.5	6.9	9.7	13.3
Drop • Inches	0.0	-0.6	-2.4	-5.5	-10.2	-16.6	-24.7	-34.9	-47.1

Speer 135-grain GDHP Short Barrel (23917)

G1 Ballistic Coefficient = 0.141

Distance • Yards	Muzzle	25	50	75	100	125	150	175	200
Velocity • fps	990	957	928	901	877	854	833	812	793
Energy • ft-lbs	294	275	258	244	231	219	208	198	189
Taylor KO Index	6.8	6.6	6.4	6.2	6.0	5.9	5.7	5.6	5.5
Mid-Range Trajectory Height • Inches	0.0	0.3	1.2	2.8	5.1	8.1	12.0	16.7	22.6
Drop • Inches	0.0	-1.1	-4.6	-10.7	-19.4	-30.9	-45.4	-66.0	-83.9

Hornady 140-grain JHP/XTP (90552)

G1 Ballistic Coefficient = 0.169

Distance • Yards	Muzzle	25	50	75	100	125	150	175	200
Velocity • fps	1350	1275	1208	1150	1100	1058	1023	991	964
Energy • ft-lbs	566	506	454	411	376	348	325	306	289
Taylor KO Index	9.6	9.1	8.6	8.2	7.9	7.6	7.3	7.1	6.9
Mid-Range Trajectory Height • Inches	0.0	0.2	0.7	1.6	3.0	4.9	7.4	10.5	14.3
Drop • Inches	0.0	-0.6	-2.6	-6.0	-11.1	-18.0	-26.8	-37.7	-50.8

Hornady 140-grain FTX (92755)

G1 Ballistic Coefficient = 0.16

Distance • Yards	Muzzle	25	50	75	100	125	150	175	200
Velocity • fps	1440	1353	1274	1203	1143	1092	1049	1013	981
Energy • ft-lbs	644	569	504	450	406	371	342	319	299
Taylor KO Index	10.3	9.7	9.1	8.6	8.2	7.8	7.5	7.2	7.0
Mid-Range Trajectory Height • Inches	0.0	0.1	0.6	1.4	2.7	4.5	6.8	9.7	13.3
Drop • Inches	0.0	-0.6	-2.3	-5.4	-9.9	-16.2	-24.2	-34.3	-46.4

Cor-Bon 140-grain Jacketed Hollow Point (SD357140/20)

G1 Ballistic Coefficient = 0.16

Distance • Yards	Muzzle	25	50	75	100	125	150	175	200
Velocity • fps	1300	1227	1163	1108	1063	1025	991	962	936
Energy • ft-lbs	525	468	420	382	351	326	306	288	272
Taylor KO Index	9.3	8.8	8.3	7.9	7.6	7.3	7.1	6.9	6.7
Mid-Range Trajectory Height • Inches	0.0	0.2	0.7	1.7	3.2	5.3	8.0	11.2	15.3
Drop • Inches	0.0	-0.7	-2.8	-6.5	-12.0	-19.4	-28.9	-40.6	-54.6

Fiocchi 142-grain Truncated Cone Point (357F)

G1 Ballistic Coefficient = 0.14

Distance • Yards	Muzzle	25	50	75	100	125	150	175	200
Velocity • fps	1420	1326	1245	1170	1110	1060	1018	983	952
Energy • ft-lbs	636	555	487	432	388	354	327	305	286
Taylor KO Index	10.3	9.6	9.0	8.5	8.0	7.7	7.4	7.1	6.9
Mid-Range Trajectory Height • Inches	0.0	0.1	0.6	1.5	2.8	4.7	7.2	10.2	14.1
Drop • Inches	0.0	-0.6	-2.4	-5.6	-10.4	-16.9	-25.4	-36.0	-48.8

Winchester 145-grain Silvertip Hollow Point (X357SHP)

G1 Ballistic Coefficient = 0.163

Distance • Yards	Muzzle	25	50	75	100	125	150	175	200
Velocity • fps	1290	1219	1155	1104	1060	1023	990	962	936
Energy • ft-lbs	535	478	428	393	361	337	316	298	282
Taylor KO Index	9.5	9.0	8.5	8.1	7.8	7.6	7.3	7.1	6.9
Mid-Range Trajectory Height • Inches	0.0	0.2	0.8	1.7	3.5	5.2	7.9	11.3	15.3
Drop • Inches	0.0	-0.7	-2.8	-6.6	-12.1	-19.6	-29.1	-40.9	-55.0

Fiocchi 148-grain Jacketed Hollowpoint (357E)

G1 Ballistic Coefficient = 0.132

Distance • Yards	Muzzle	25	50	75	100	125	150	175	200
Velocity • fps	1310	1221	1145	1084	1035	994	958	927	899
Energy • ft-lbs	564	490	431	387	352	324	302	282	266
Taylor KO Index	9.9	9.2	8.6	8.2	7.8	7.5	7.2	7.0	6.9
Mid-Range Trajectory Height • Inches	0.0	0.2	0.7	1.8	3.3	5.5	8.3	11.8	16.2
Drop • Inches	0.0	-0.7	-2.8	-6.6	-12.2	-19.8	-29.7	-41.9	-56.6

PMC 150-grain Starfire Hollow Point (357SFA)

G1 Ballistic Coefficient = 0.141

Distance • Yards	Muzzle	25	50	75	100	125	150	175	200
Velocity • fps	1200	1131	1076	1031	992	960	930	904	879
Energy • ft-lbs	480	427	386	354	329	307	288	272	257
Taylor KO Index	9.2	8.7	8.2	7.9	7.6	7.3	7.1	6.9	6.7
Mid-Range Trajectory Height • Inches	0.0	0.1	0.8	2.0	3.7	6.0	9.1	12.8	17.4
Drop • Inches	0.0	-0.8	-3.3	-7.6	-14.0	-22.6	-33.4	-46.8	-62.8

PMC 158-grain Jacketed Soft Point (357A)

G1 Ballistic Coefficient = 0.142

Distance • Yards	Muzzle	25	50	75	100	125	150	175	200
Velocity • fps	1200	1132	1078	1033	994	961	931	905	880
Energy • ft-lbs	505	450	408	374	347	324	304	287	272
Taylor KO Index	9.7	9.1	8.7	8.3	8.0	7.7	7.5	7.3	7.1
Mid-Range Trajectory Height • Inches	0.0	0.1	0.8	2.0	3.7	6.0	9.0	12.8	17.3
Drop • Inches	0.0	-0.8	-3.3	-7.6	-14.0	-22.5	-33.4	-46.8	-62.7

Sellier and Bellot 158-grain SP (V311112U)

G1 Ballistic Coefficient = 0.154

Distance • Yards	Muzzle	25	50	75	100	125	150	175	200
Velocity • fps	1263	1193	1131	1080	1098	1001	970	942	916
Energy • ft-lbs	561	489	449	409	378	352	330	311	295
Taylor KO Index	10.2	9.6	9.1	8.7	8.4	8.1	7.8	7.6	7.4
Mid-Range Trajectory Height • Inches	0.0	0.2	0.8	1.8	3.4	5.6	8.4	11.8	16.1
Drop • Inches	0.0	-0.7	-3.0	-6.9	-12.7	-20.5	-30.5	-42.8	-57.5

American Eagle (Federal) 158-grain Jacketed Soft Point (AE357A)

G1 Ballistic Coefficient = 0.196

Distance • Yards	Muzzle	25	50	75	100	125	150	175	200
Velocity • fps	1240	1190	1140	1100	1060	1029	1001	976	953
Energy • ft-lbs	540	495	455	425	395	371	351	334	319
Taylor KO Index	10.0	9.6	9.2	8.9	8.5	8.3	8.1	7.9	7.7
Mid-Range Trajectory Height • Inches	0.0	0.2	0.8	1.8	3.4	5.4	8.1	11.3	15.4
Drop • Inches	0.0	-0.7	-3.0	-7.0	-12.7	-20.4	-30.2	-42.1	-56.2

Magtech 158-grain LSWC (357C)

G1 Ballistic Coefficient = 0.146

Distance • Yards	Muzzle	25	50	75	100	125	150	175	200
Velocity • fps	1235	1164	1104	1056	1015	980	949	922	897
Energy • ft-lbs	535	475	428	391	361	337	316	298	282
Taylor KO Index	9.9	9.4	8.9	8.5	8.2	7.9	7.6	7.4	7.2
Mid-Range Trajectory Height • Inches	0.0	0.2	0.8	1.9	3.5	5.8	8.7	12.3	16.6
Drop • Inches	0.0	-0.7	-3.1	-7.2	-13.3	-21.4	-31.8	-44.6	-59.8

Magtech 158-grain SJSP (MP357A) (357E)

G1 Ballistic Coefficient = 0.146

Distance • Yards	Muzzle	25	50	75	100	125	150	175	200
Velocity • fps	1235	1164	1104	1056	1015	980	949	922	897
Energy • ft-lbs	535	475	428	391	361	337	316	298	282
Taylor KO Index	9.9	9.4	8.9	8.5	8.2	7.9	7.6	7.4	7.2
Mid-Range Trajectory Height • Inches	0.0	0.2	0.8	1.9	3.5	5.8	8.7	12.3	16.6
Drop • Inches	0.0	-0.7	-3.1	-7.2	-13.3	-21.4	-31.8	-44.6	-59.8

Remington 158-grain Soft Point (R357M3)

G1 Ballistic Coefficient = 0.146

Distance • Yards	Muzzle	25	50	75	100	125	150	175	200
Velocity • fps	1235	1164	1104	1056	1015	980	949	922	897
Energy • ft-lbs	535	475	428	391	361	337	316	298	282
Taylor KO Index	9.9	9.4	8.9	8.5	8.2	7.9	7.6	7.4	7.2
Mid-Range Trajectory Height • Inches	0.0	0.2	0.8	1.9	3.5	5.8	8.7	12.3	16.6
Drop • Inches	0.0	-0.7	-3.1	-7.2	-13.3	-21.4	-31.8	-44.6	-59.8

Remington 158-grain LSW (R357M5)

G1 Ballistic Coefficient = 0.146

Distance • Yards	Muzzle	25	50	75	100	125	150	175	200
Velocity • fps	1235	1164	1104	1056	1015	980	949	922	897
Energy • ft-lbs	535	475	428	391	361	337	316	298	282
Taylor KO Index	9.9	9.4	8.9	8.5	8.2	7.9	7.6	7.4	7.2
Mid-Range Trajectory Height • Inches	0.0	0.2	0.8	1.9	3.5	5.8	8.7	12.3	16.6
Drop • Inches	0.0	-0.7	-3.1	-7.2	-13.3	-21.4	-31.8	-44.6	-59.8

Winchester 158-grain Jacketed Soft Point (X3575P)

G1 Ballistic Coefficient = 0.146

Distance • Yards	Muzzle	25	50	75	100	125	150	175	200
Velocity • fps	1235	1164	1104	1056	1015	980	949	922	897
Energy • ft-lbs	535	475	428	391	361	337	316	298	282
Taylor KO Index	9.9	9.4	8.9	8.5	8.2	7.9	7.6	7.4	7.2
Mid-Range Trajectory Height • Inches	0.0	0.2	0.8	1.9	3.5	5.8	8.7	12.3	16.6
Drop • Inches	0.0	-0.7	-3.1	-7.2	-13.3	-21.4	-31.8	-44.6	-59.8

Magtech 158-grain LFN (357L)

G1 Ballistic Coefficient = 0.138

Distance • Yards	Muzzle	25	50	75	100	125	150	175	200
Velocity • fps	1085	1037	997	963	933	905	880	856	834
Energy • ft-lbs	413	378	349	325	305	287	272	257	244
Taylor KO Index	8.8	8.4	8.1	7.8	7.5	7.3	7.1	6.9	6.7
Mid-Range Trajectory Height • Inches	0.0	0.2	1.0	2.3	4.3	7.0	10.5	14.6	20.0
Drop • Inches	0.0	-1.0	-3.9	-9.1	-16.6	-26.6	-39.2	-54.6	-73.1

Aguila 158-grain Semi-Jacketed Hollow Point (1E572821)

G1 Ballistic Coefficient = 0.079

Distance • Yards	Muzzle	25	50	75	100	125	150	175	200
Velocity • fps	1545	1360	1207	1093	1014	952	903	860	822
Energy • ft-lbs	835	649	512	420	361	318	286	260	237
Taylor KO Index	12.4	11.0	9.7	8.8	8.2	7.7	7.3	6.9	6.6
Mid-Range Trajectory Height • Inches	0.0	0.1	0.6	1.5	3.0	5.2	8.1	12.0	16.8
Drop • Inches	0.0	-0.5	-2.2	-5.3	-10.3	-17.4	-26.9	-39.0	-54.1

Black Hills 158-grain Jacketed Hollow Point-Gold Dot (M357N3)

G1 Ballistic Coefficient = 0.17

Distance • Yards	Muzzle	25	50	75	100	125	150	175	200
Velocity • fps	1250	1188	1134	1088	1049	1016	986	960	936
Energy • ft-lbs	548	495	451	416	386	362	341	323	307
Taylor KO Index	10.0	9.5	9.1	8.7	8.4	8.2	7.9	7.7	7.5
Mid-Range Trajectory Height • Inches	0.0	0.2	0.8	1.8	3.4	5.4	8.2	11.6	14.3
Drop • Inches	0.0	-0.7	-3.0	-6.9	-12.7	-20.4	-30.2	-42.3	-56.6

Blazer (CCI) 158-grain JHP Brass (5207)

G1 Ballistic Coefficient = 0.16

Distance • Yards	Muzzle	25	50	75	100	125	150	175	200
Velocity • fps	1250	1186	1130	1084	1044	1010	980	953	929
Energy • ft-lbs	548	493	448	412	382	358	337	319	303
Taylor KO Index	10.1	9.6	9.1	8.8	8.4	8.2	7.9	7.7	7.5
Mid-Range Trajectory Height • Inches	0.0	0.2	0.7	1.8	3.4	5.4	8.3	11.6	15.8
Drop • Inches	0.0	-0.7	-3.0	-7.0	-12.8	-20.6	-30.5	-42.7	-57.3

Fiocchi 158-grain Hornady XTP/JHP (357XTP)

G1 Ballistic Coefficient = 0.20

Distance • Yards	Muzzle	25	50	75	100	125	150	175	200
Velocity • fps	1250	1197	1149	1109	1072	1041	1013	988	966
Energy • ft-lbs	548	503	463	431	403	380	360	343	327
Taylor KO Index	10.1	9.7	9.3	9.0	8.7	8.4	8.2	8.0	7.8
Mid-Range Trajectory Height • Inches	0.0	0.1	0.7	1.7	3.3	5.3	7.9	11.0	15.0
Drop • Inches	0.0	-0.7	-3.0	-6.8	-12.5	-20.0	-29.5	-41.2	-55.0

Hornady 158-grain JHP/XTP (90562)

G1 Ballistic Coefficient = 0.207

Distance • Yards	Muzzle	25	50	75	100	125	150	175	200
Velocity • fps	1250	1197	1150	1109	1073	1042	1014	989	966
Energy • ft-lbs	548	503	464	431	404	381	361	343	327
Taylor KO Index	10.0	9.6	9.2	8.9	8.6	8.4	8.2	8.0	7.8
Mid-Range Trajectory Height • Inches	0.0	0.2	0.8	1.8	3.3	5.2	7.9	11.1	14.9
Drop • Inches	0.0	-0.7	-3.0	-6.8	-12.5	-20.0	-29.5	-41.1	-54.9

Federal 158-grain JHP (P357HS1)

G1 Ballistic Coefficient = 0.196

Distance • Yards	Muzzle	25	50	75	100	125	150	175	200
Velocity • fps	1240	1190	1140	1100	1060	1029	1001	976	953
Energy • ft-lbs	540	495	455	425	395	371	351	334	319
Taylor KO Index	10.0	9.6	9.2	8.9	8.5	8.3	8.1	7.9	7.7
Mid-Range Trajectory Height • Inches	0.0	0.2	0.8	1.8	3.4	5.4	8.1	11.3	15.4
Drop • Inches	0.0	-0.7	-3.0	-7.0	-12.7	-20.4	-30.2	-42.1	-56.2

Federal 158-grain JHP (C357E)

G1 Ballistic Coefficient = 0.196

Distance • Yards	Muzzle	25	50	75	100	125	150	175	200
Velocity • fps	1240	1190	1140	1100	1060	1029	1001	976	953
Energy • ft-lbs	540	495	455	425	395	371	351	334	319
Taylor KO Index	10.0	9.6	9.2	8.9	8.5	8.3	8.1	7.9	7.7
Mid-Range Trajectory Height • Inches	0.0	0.2	0.8	1.8	3.4	5.4	8.1	11.3	15.4
Drop • Inches	0.0	-0.7	-3.0	-7.0	-12.7	-20.4	-30.2	-42.1	-56.2

Magtech 158-grain SJHP (357B)

G1 Ballistic Coefficient = 0.146

Distance • Yards	Muzzle	25	50	75	100	125	150	175	200
Velocity • fps	1235	1164	1104	1056	1015	980	949	922	897
Energy • ft-lbs	535	475	428	391	361	337	316	298	282
Taylor KO Index	9.9	9.4	8.9	8.5	8.2	7.9	7.6	7.4	7.2
Mid-Range Trajectory Height • Inches	0.0	0.2	0.8	1.9	3.5	5.8	8.7	12.3	16.6
Drop • Inches	0.0	-0.7	-3.1	-7.2	-13.3	-21.4	-31.8	-44.6	-59.8

Remington 158-grain Semi-Jacket Hollow Point (R357M2)

G1 Ballistic Coefficient = 0.146

Distance • Yards	Muzzle	25	50	75	100	125	150	175	200
Velocity • fps	1235	1164	1104	1056	1015	980	949	922	897
Energy • ft-lbs	535	475	428	391	361	337	316	298	282
Taylor KO Index	9.9	9.4	8.9	8.5	8.2	7.9	7.6	7.4	7.2
Mid-Range Trajectory Height • Inches	0.0	0.2	0.8	1.9	3.5	5.8	8.7	12.3	16.6
Drop • Inches	0.0	-0.7	-3.1	-7.2	-13.3	-21.4	-31.8	-44.6	-59.8

Speer 158-grain GDHP (23960)

G1 Ballistic Coefficient = 0.168

Distance • Yards	Muzzle	25	50	75	100	125	150	175	200
Velocity • fps	1235	1173	1119	1074	1036	1003	974	948	924
Energy • ft-lbs	535	483	440	405	377	353	333	315	299
Taylor KO Index	10.0	9.5	9.0	8.7	8.4	8.1	7.9	7.7	7.5
Mid-Range Trajectory Height • Inches	0.0	0.1	0.7	1.8	3.4	5.6	8.4	11.8	16.1
Drop • Inches	0.0	-0.7	-3.0	-7.1	-13.0	-31.0	-31.1	-43.5	-58.3

Winchester 158-grain Jacketed Hollow Point (X3574P)

G1 Ballistic Coefficient = 0.146

Distance • Yards	Muzzle	25	50	75	100	125	150	175	200
Velocity • fps	1235	1164	1104	1056	1015	980	949	922	897
Energy • ft-lbs	535	475	428	391	361	337	316	298	282
Taylor KO Index	9.9	9.4	8.9	8.5	8.2	7.9	7.6	7.4	7.2
Mid-Range Trajectory Height • Inches	0.0	0.2	0.8	1.9	3.5	5.8	8.7	12.3	16.6
Drop • Inches	0.0	-0.7	-3.1	-7.2	-13.3	-21.4	-31.8	-44.6	-59.8

Blazer (CCI) 158-grain JHP (3542)

G1 Ballistic Coefficient = 0.163

Distance • Yards	Muzzle	25	50	75	100	125	150	175	200
Velocity • fps	1150	1099	1056	1019	987	959	933	910	888
Energy • ft-lbs	464	424	391	364	342	323	306	290	277
Taylor KO Index	9.3	8.9	8.5	8.2	8.0	7.7	7.5	7.3	7.2
Mid-Range Trajectory Height • Inches	0.0	0.2	0.9	2.1	4.0	6.3	9.4	13.1	17.8
Drop • Inches	0.0	-0.9	-3.5	-8.1	-14.8	-23.7	-35.0	-48.7	-65.1

Sellier and Bellot 158-grain Non-Tox TFMJ (SBA03574)

G1 Ballistic Coefficient = 0.157

Distance • Yards	Muzzle	25	50	75	100	125	150	175	200
Velocity • fps	1263	1193	1133	1083	1041	1005	973	946	920
Energy • ft-lbs	561	499	450	411	380	354	333	314	297
Taylor KO Index	10.2	9.6	9.1	8.7	8.4	8.1	7.8	7.6	7.4
Mid-Range Trajectory Height • Inches	0.0	0.2	0.8	1.8	3.4	5.6	8.3	11.7	16.0
Drop • Inches	0.0	-0.7	-2.9	-6.9	-12.6	-20.4	-30.4	-42.6	-57.2

Sellier and Bellot 158-grain FMJ (SBA03571)

G1 Ballistic Coefficient = 0.154

Distance • Yards	Muzzle	25	50	75	100	125	150	175	200
Velocity • fps	1263	1193	1131	1080	1038	1001	970	942	916
Energy • ft-lbs	561	499	449	409	378	352	330	311	295
Taylor KO Index	10.2	9.6	9.1	8.7	8.4	8.1	7.8	7.6	7.4
Mid-Range Trajectory Height • Inches	0.0	0.2	0.8	1.8	3.4	5.6	8.4	11.8	16.1
Drop • Inches	0.0	-0.7	-3.0	-6.9	-12.7	-20.5	-30.5	-42.8	-57.5

Magtech 158-grain FMJ-Flat (357D)

G1 Ballistic Coefficient = 0.180

Distance • Yards	Muzzle	25	50	75	100	125	150	175	200
Velocity • fps	1235	1177	1125	1082	1045	1013	985	959	936
Energy • ft-lbs	535	468	444	411	383	360	340	323	307
Taylor KO Index	10.0	9.5	9.1	8.7	8.4	8.2	7.9	7.7	7.5
Mid-Range Trajectory Height • Inches	0.0	0.1	0.7	1.8	3.4	5.5	8.2	11.6	15.7
Drop • Inches	0.0	-0.7	-3.0	-7.1	-12.9	-20.8	-30.7	-42.9	-57.4

PMC 158-grain EMJ (357EMA) LF

G1 Ballistic Coefficient = 0.142

Distance • Yards	Muzzle	25	50	75	100	125	150	175	200
Velocity • fps	1200	1132	1078	1033	994	961	932	905	881
Energy • ft-lbs	505	451	408	374	347	324	305	288	272
Taylor KO Index	9.7	9.1	8.7	8.3	8.0	7.8	7.5	7.3	7.2
Mid-Range Trajectory Height • Inches	0.0	0.2	0.9	2.0	3.7	6.1	9.2	13.0	17.5
Drop • Inches	0.0	-0.8	-3.3	-7.6	-14.0	-22.6	-33.5	-46.9	-63.0

Speer 170-grain Gold Dot Soft Point (23959)

G1 Ballistic Coefficient = 0.18

Distance • Yards	Muzzle	25	50	75	100	125	150	175	200
Velocity • fps	1180	1130	1087	1051	1019	990	965	942	920
Energy • ft-lbs	526	482	446	417	392	370	352	335	320
Taylor KO Index	10.2	9.8	9.4	9.1	8.8	8.6	8.4	8.2	8.0
Mid-Range Trajectory Height • Inches	0.0	0.2	0.9	2.0	3.7	5.9	8.8	12.4	16.7
Drop • Inches	0.0	-0.8	-3.3	-7.6	-14.0	-22.4	-33.0	-45.9	-61.3

Cor-Bon 180-grain Bonded-Core SP (HT357180BC/20)

G1 Ballistic Coefficient = 0.18

Distance • Yards	Muzzle	25	50	75	100	125	150	175	200
Velocity • fps	1200	1148	1104	1065	1032	1002	976	953	931
Energy • ft-lbs	576	527	487	453	426	402	381	363	346
Taylor KO Index	11.0	10.5	10.1	9.8	9.5	9.2	9.0	8.7	8.5
Mid-Range Trajectory Height • Inches	0.0	0.2	0.8	1.9	3.6	5.8	8.6	12.0	16.3
Drop • Inches	0.0	-0.8	-3.2	-7.4	-13.5	-21.7	-32.0	-44.6	-59.6

Winchester 180-grain Partition Gold (S357P)

G1 Ballistic Coefficient = 0.18

Distance • Yards	Muzzle	25	50	75	100	125	150	175	200
Velocity • fps	1180	1131	1088	1052	1020	992	967	944	923
Energy • ft-lbs	557	511	473	442	416	394	374	356	340
Taylor KO Index	10.8	10.4	10.0	9.6	9.3	9.1	8.9	8.7	8.5
Mid-Range Trajectory Height • Inches	0.0	0.2	0.8	2.0	3.6	5.9	8.7	12.3	16.5
Drop • Inches	0.0	-0.8	-3.3	-7.6	-13.9	-22.3	-32.9	-45.7	-61.0

Federal 180-grain Cast Core (P357J)

G1 Ballistic Coefficient = 0.2

Distance • Yards	Muzzle	25	50	75	100	125	150	175	200
Velocity • fps	1130	1090	1060	1030	1000	976	954	933	914
Energy • ft-lbs	510	475	450	425	400	381	364	348	334
Taylor KO Index	10.4	10.0	9.7	9.5	9.2	9.0	8.8	8.6	8.4
Mid-Range Trajectory Height • Inches	0.0	0.2	0.9	2.1	3.9	6.3	9.3	12.8	17.4
Drop • Inches	0.0	-0.9	-3.6	-8.2	-14.9	-23.8	-34.9	-48.5	-64.5

Federal 180-grain Swift A-Frame (P357SA)

G1 Ballistic Coefficient = 0.185

Distance • Yards	Muzzle	25	50	75	100	125	150	175	200
Velocity • fps	1130	1090	1050	1020	990	965	942	920	900
Energy • ft-lbs	510	470	440	415	390	372	355	339	324
Taylor KO Index	10.4	10.0	9.7	9.4	9.1	8.9	8.7	8.5	8.3
Mid-Range Trajectory Height • Inches	0.0	0.2	0.9	2.1	3.9	6.3	9.4	13.0	17.7
Drop • Inches	0.0	-0.9	-3.6	-8.3	-15.0	-24.0	-35.3	-49.1	-65.4

Remington 180-grain Semi-Jacketed Hollow Point (R357M10)

G1 Ballistic Coefficient = 0.165

Distance • Yards	Muzzle	25	50	75	100	125	150	175	200
Velocity • fps	1145	1095	1053	1017	985	958	932	909	888
Energy • ft-lbs	524	479	443	413	388	367	348	330	315
Taylor KO Index	10.5	10.0	9.6	9.3	9.0	8.8	8.6	8.3	8.2
Mid-Range Trajectory Height • Inches	0.0	0.2	0.9	2.1	3.9	6.3	9.4	13.2	17.7
Drop • Inches	0.0	-0.8	-3.5	-8.1	-14.9	-23.8	-35.1	-48.9	-65.2

Federal 180-grain JHP (C357G)

G1 Ballistic Coefficient = 0.205

Distance • Yards	Muzzle	25	50	75	100	125	150	175	200
Velocity • fps	1080	1050	1020	1000	970	948	929	910	893
Energy • ft-lbs	465	440	415	395	380	360	345	331	319
Taylor KO Index	9.9	9.7	9.4	9.2	8.9	8.7	8.6	8.4	8.2
Mid-Range Trajectory Height • Inches	0.0	0.2	0.9	2.3	4.2	6.7	9.9	13.6	18.4
Drop • Inches	0.0	-1.0	-3.9	-8.9	-16.1	-25.7	-37.6	-52.1	-69.2

Cor-Bon 200-grain Hard Cast   (HT357200HC/20)

G1 Ballistic Coefficient = 0.150

Distance • Yards	Muzzle	25	50	75	100	125	150	175	200
Velocity • fps	1150	1095	1049	1010	977	947	921	896	874
Energy • ft-lbs	587	532	489	453	424	399	377	357	339
Taylor KO Index	11.7	11.2	10.7	10.3	10.0	9.7	9.4	9.1	8.9
Mid-Range Trajectory Height • Inches	0.0	0.2	0.9	2.1	4.0	6.4	9.6	13.4	18.2
Drop • Inches	0.0	-0.8	-3.5	-8.2	-14.9	-24.0	-35.4	-49.4	-66.2

Cowboy Action Loads:

Black Hills 158-grain CNL (DCB357N1)

G1 Ballistic Coefficient = 0.150

Distance • Yards	Muzzle	25	50	75	100	125	150	175	200
Velocity • fps	800	782	765	749	733	717	702	688	674
Energy • ft-lbs	225	215	205	197	188	181	173	166	159
Taylor KO Index	6.4	6.3	6.1	6.0	5.9	5.8	5.7	5.5	5.4
Mid-Range Trajectory Height • Inches	0.0	0.4	1.8	4.1	7.4	11.8	17.4	24.2	32.4
Drop • Inches	0.0	-1.7	-7.0	-16.0	-28.8	-45.2	-66.8	-92.3	-124.4

Fiocchi 158-grain LRNFP (357CA)

G1 Ballistic Coefficient = 0.139

Distance • Yards	Muzzle	25	50	75	100	125	150	175	200
Velocity • fps	700	684	669	654	640	625	612	598	585
Energy • ft-lbs	172	164	157	150	144	137	131	126	120
Taylor KO Index	5.7	5.5	5.4	5.3	5.2	5.1	4.9	4.8	4.7
Mid-Range Trajectory Height • Inches	0.0	0.6	2.3	5.3	9.7	15.5	22.9	31.9	42.6
Drop • Inches	0.0	-2.3	-9.1	-20.9	-37.7	-59.8	-87.5	-121.0	-160.6

.357 SIG

This cartridge was developed in 1994 especially for the Sig P229 pistol. Design-wise, the .357 SIG comes pretty close to being a 10mm Auto or .40 S&W cartridge necked to .357 caliber. The design follows the concept of the .30 Luger and the wildcat .38-45 cartridge that received some interest in the early 1960s. The case volume enables the .357 SIG's performance to approach that of the .357 Magnum.

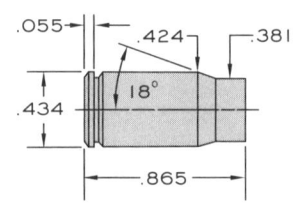

Relative Recoil Factor = 0.76

Specifications:

Controlling Agency for Standardization of this Ammunition: SAAMI

Bullet Weight Grains	Velocity fps	Maximum Average Pressure Copper Crusher	Transducer
125 FMJ	1,350	N/S	40,000 psi

Standard barrel for velocity testing: 4 inches long—1 turn in 16-inch twist

Availability:

Cor-Bon 80-grain Glaser Safety Slug (02500/-2700)
G1 Ballistic Coefficient = 0.080

Distance • Yards	Muzzle	25	50	75	100	125	150	175	200
Velocity • fps	1650	1453	1285	1151	1055	985	931	885	845
Energy • ft-lbs	484	375	293	235	198	173	154	139	127
Taylor KO Index	6.7	5.9	5.2	4.7	4.3	4.0	3.8	3.6	3.4
Mid-Range Trajectory Height • Inches	0.0	0.1	0.5	1.3	2.7	4.7	7.4	11.0	15.5
Drop • Inches	0.0	-0.4	-1.9	-4.7	-9.1	-15.5	-24.1	-35.2	-49.1

Cor-Bon 100-grain Pow'RBall (PB357SIG100/20)
G1 Ballistic Coefficient = 0.10

Distance • Yards	Muzzle	25	50	75	100	125	150	175	200
Velocity • fps	1600	1445	1309	1194	1104	1036	983	939	901
Energy • ft-lbs	569	464	381	317	217	238	215	196	180
Taylor KO Index	8.2	7.4	6.7	6.1	5.6	5.3	5.0	4.8	4.6
Mid-Range Trajectory Height • Inches	0.0	0.1	0.5	1.3	2.6	4.4	6.9	10.1	14.2
Drop • Inches	0.0	-0.5	-2.0	-4.7	-9.0	-15.1	-23.2	-33.5	-46.3

Winchester 105-grain Jacketed Flat Point NT (SC357SNT)
G1 Ballistic Coefficient = 0.12

Distance • Yards	Muzzle	25	50	75	100	125	150	175	200
Velocity • fps	1370	1267	1179	1108	1050	1004	965	932	901
Energy • ft-lbs	438	375	324	286	257	235	217	202	189
Taylor KO Index	7.3	6.8	6.3	5.9	5.6	5.4	5.2	5.0	4.8
Mid-Range Trajectory Height • Inches	0.0	0.2	0.7	1.7	3.1	5.2	8.0	11.4	15.7
Drop • Inches	0.0	-0.6	-2.6	-6.1	-11.4	-18.7	-28.1	-39.8	-54.0

Cor-Bon 115-grain JHP (SD357SIG115/20)
G1 Ballistic Coefficient = 0.14

Distance • Yards	Muzzle	25	50	75	100	125	150	175	200
Velocity • fps	1500	1395	1301	1217	1146	1088	1040	1000	966
Energy • ft-lbs	575	497	432	378	336	302	276	255	211
Taylor KO Index	8.8	8.2	7.6	7.1	6.7	6.4	6.1	5.8	5.6
Mid-Range Trajectory Height • Inches	0.0	0.1	0.5	1.3	2.5	4.3	6.6	9.5	13.1
Drop • Inches	0.0	-0.5	-2.1	-5.0	-9.4	-15.4	-23.3	-33.2	-45.2

Fiocchi 124-grain XTP JHP (357SGXTP)
G1 Ballistic Coefficient = 0.1

Distance • Yards	Muzzle	25	50	75	100	125	150	175	200
Velocity • fps	1350	1267	1194	1131	1080	1036	1000	968	940
Energy • ft-lbs	505	442	392	353	321	296	275	258	243
Taylor KO Index	8.6	8.0	7.6	7.2	6.8	6.6	6.3	6.1	6.0
Mid-Range Trajectory Height • Inches	0.0	0.2	0.7	1.6	3.1	5.0	7.6	10.9	14.9
Drop • Inches	0.0	-0.6	-2.6	-6.1	-11.3	-16.4	-27.4	-38.7	-52.3

Hornady 124-grain JHP/XTP (9130)

G1 Ballistic Coefficient = 0.177

Distance • Yards	Muzzle	25	50	75	100	125	150	175	200
Velocity • fps	1350	1278	1208	1157	1108	1067	1031	1000	973
Energy • ft-lbs	502	450	405	369	338	313	293	276	261
Taylor KO Index	8.4	7.9	7.5	7.2	6.9	6.6	6.3	6.1	6.0
Mid-Range Trajectory Height • Inches	0.0	0.1	0.5	1.4	2.8	4.8	7.2	10.3	14.0
Drop • Inches	0.0	-0.6	-2.6	-6.2	-11.5	-17.8	-26.5	-37.3	-50.1

Fiocchi 124-grain FMJ (357SIGAP)

G1 Ballistic Coefficient = 0.152

Distance • Yards	Muzzle	25	50	75	100	125	150	175	200
Velocity • fps	1350	1267	1194	1132	1080	1038	1001	969	941
Energy • ft-lbs	505	442	395	353	324	297	276	259	244
Taylor KO Index	8.6	8.0	7.6	7.2	6.8	6.6	6.3	6.1	6.0
Mid-Range Trajectory Height • Inches	0.0	0.1	0.6	1.6	3.0	5.0	7.6	10.8	14.8
Drop • Inches	0.0	-0.6	-2.6	-6.1	-11.3	-18.3	-27.4	-38.6	-52.2

Cor-Bon 125-grain JHP (SD357SIG125/20)

G1 Ballistic Coefficient = 0.150

Distance • Yards	Muzzle	25	50	75	100	125	150	175	200
Velocity • fps	1425	1333	1251	1180	1119	1069	1027	991	960
Energy • ft-lbs	564	494	435	386	348	317	293	273	256
Taylor KO Index	9.1	8.5	8.0	7.5	7.1	6.8	6.5	6.3	6.1
Mid-Range Trajectory Height • Inches	0.0	0.1	0.6	1.4	2.7	4.6	7.0	10.1	13.8
Drop • Inches	0.0	-0.6	-2.3	-5.5	-10.2	-16.7	-25.1	-35.5	-48.1

Speer 125-grain Gold Dot (23918)

G1 Ballistic Coefficient = 0.141

Distance • Yards	Muzzle	25	50	75	100	125	150	175	200
Velocity • fps	1375	1283	1203	1135	1079	1033	995	961	932
Energy • ft-lbs	525	457	402	358	323	297	275	257	241
Taylor KO Index	8.8	8.2	7.7	7.2	6.9	6.6	6.3	6.1	5.9
Mid-Range Trajectory Height • Inches	0.0	0.2	0.7	1.6	3.0	5.0	7.6	10.8	14.8
Drop • Inches	0.0	-0.6	-2.5	-6.0	-11.0	-18.0	-27.0	-38.2	-51.8

Winchester 125-grain Bonded PDX1 (S357SPDB)

G1 Ballistic Coefficient = 0.143

Distance • Yards	Muzzle	25	50	75	100	125	150	175	200
Velocity • fps	1350	1262	1186	1122	1069	1026	989	956	928
Energy • ft-lbs	506	442	391	349	317	292	271	254	239
Taylor KO Index	8.6	8.1	7.6	7.2	6.8	6.6	6.3	6.1	5.9
Mid-Range Trajectory Height • Inches	0.0	0.2	0.7	1.6	3.1	5.1	7.8	11.1	15.1
Drop • Inches	0.0	-0.6	-2.6	-6.1	-11.4	-18.6	-27.8	-39.2	-53.0

American Eagle (Federal) 125-grain Full Metal Jacket (AE357S2)

G1 Ballistic Coefficient = 0.153

Distance • Yards	Muzzle	25	50	75	100	125	150	175	200
Velocity • fps	1350	1270	1190	1130	1080	1039	1002	970	942
Energy • ft-lbs	510	445	395	355	325	299	279	261	246
Taylor KO Index	8.6	8.1	7.6	7.2	6.9	6.6	6.4	6.2	6.0
Mid-Range Trajectory Height • Inches	0.0	0.1	0.5	1.4	2.9	5.0	7.5	10.3	14.6
Drop • Inches	0.0	-0.6	-2.6	-6.1	-11.3	-18.3	-27.3	-38.5	-51.9

Winchester 125-grain Brass Enclosed Base (WC357SIG)

G1 Ballistic Coefficient = 0.140

Distance • Yards	Muzzle	25	50	75	100	125	150	175	200
Velocity • fps	1350	1261	1183	1118	1065	1021	984	952	923
Energy • ft-lbs	506	441	389	347	315	290	269	252	237
Taylor KO Index	8.6	8.1	7.6	7.1	6.8	6.5	6.3	6.1	5.9
Mid-Range Trajectory Height • Inches	0.0	0.2	0.7	1.6	3.1	5.1	7.8	11.2	15.3
Drop • Inches	0.0	-0.6	-2.6	-6.2	-11.4	-18.6	-27.9	-39.4	-53.3

Speer 125-grain TMJ FN (53919)

G1 Ballistic Coefficient = 0.147

Distance • Yards	Muzzle	25	50	75	100	125	150	175	200
Velocity • fps	1325	1242	1171	1111	1061	1020	985	954	926
Energy • ft-lbs	487	429	380	343	313	289	269	253	239
Taylor KO Index	8.5	7.9	7.5	7.1	6.8	6.5	6.3	6.1	5.9
Mid-Range Trajectory Height • Inches	0.0	0.2	0.7	1.7	3.2	5.2	7.9	11.3	15.4
Drop • Inches	0.0	-0.6	-2.7	-6.3	-11.7	-19.1	-28.5	-40.1	-54.2

Cor-Bon 125-grain DPX (DPX357SIG125/20)

G1 Ballistic Coefficient = 0.110

Distance • Yards	Muzzle	25	50	75	100	125	150	175	200
Velocity • fps	1350	1238	1146	1074	1018	972	933	899	868
Energy • ft-lbs	505	426	365	320	288	262	242	224	209
Taylor KO Index	8.6	7.9	7.3	6.8	6.5	6.2	5.9	5.7	5.5
Mid-Range Trajectory Height • Inches	0.0	0.2	0.7	1.7	3.3	5.5	8.5	12.1	16.7
Drop • Inches	0.0	-0.6	-2.7	-6.4	-12.0	-19.6	-29.6	-42.1	-57.3

Federal 125-grain Jacketed Hollow Point (P357S1)

G1 Ballistic Coefficient = 0.153

Distance • Yards	Muzzle	25	50	75	100	125	150	175	200
Velocity • fps	1350	1270	1190	1130	1080	1039	1002	970	942
Energy • ft-lbs	510	445	395	355	325	299	279	261	246
Taylor KO Index	8.6	8.1	7.6	7.2	6.9	6.6	6.4	6.2	6.0
Mid-Range Trajectory Height • Inches	0.0	0.1	0.5	1.4	2.9	5.0	7.5	10.3	14.6
Drop • Inches	0.0	-0.6	-2.6	-6.1	-11.3	-18.3	-27.3	-38.5	-51.9

UMC (Remington) 125-grain Jacketed Hollow Point (L357S2)

G1 Ballistic Coefficient = 0.119

Distance • Yards	Muzzle	25	50	75	100	125	150	175	200
Velocity • fps	1350	1246	1157	1088	1032	988	950	916	886
Energy • ft-lbs	506	431	372	329	296	271	250	233	218
Taylor KO Index	8.6	7.9	7.4	6.9	6.6	6.3	6.1	5.8	5.6
Mid-Range Trajectory Height • Inches	0.0	0.2	0.5	1.6	2.9	5.3	8.1	11.7	15.9
Drop • Inches	0.0	-0.6	-2.7	-6.3	-11.8	-19.3	-28.9	-40.9	-55.5

USA (Winchester) 125-grain Jacketed Hollow Point (USA357SJHP)

G1 Ballistic Coefficient = 0.142

Distance • Yards	Muzzle	25	50	75	100	125	150	175	200
Velocity • fps	1350	1262	1185	1120	1067	1024	987	954	926
Energy • ft-lbs	506	442	390	348	316	291	270	253	238
Taylor KO Index	8.6	8.0	7.6	7.1	6.8	6.5	6.3	6.1	5.9
Mid-Range Trajectory Height • Inches	0.0	0.1	0.5	1.5	2.9	5.0	7.7	11.0	15.0
Drop • Inches	0.0	-0.6	-2.6	-6.2	-11.4	-18.6	-27.8	-39.2	-52.9

Speer 125-grain TMJFN Lawman Clean-Fire (53232)

G1 Ballistic Coefficient = 0.14

Distance • Yards	Muzzle	25	50	75	100	125	150	175	200
Velocity • fps	1350	1265	1190	1127	1074	1031	994	962	934
Energy • ft-lbs	506	444	393	352	320	295	274	257	242
Taylor KO Index	8.6	8.1	7.6	7.2	6.8	6.6	6.3	6.1	6.0
Mid-Range Trajectory Height • Inches	0.0	0.2	0.7	1.7	3.1	5.1	7.7	11.0	15.0
Drop • Inches	0.0	-0.6	-2.6	-6.1	-11.4	-18.5	-27.6	-39.0	-52.6

UMC (Remington) 125-grain MC (L357S1)

G1 Ballistic Coefficient = 0.11

Distance • Yards	Muzzle	25	50	75	100	125	150	175	200
Velocity • fps	1350	1245	1157	1087	1032	987	948	915	885
Energy • ft-lbs	506	430	372	328	296	270	250	232	217
Taylor KO Index	8.6	7.9	7.4	6.9	6.6	6.3	6.0	5.8	5.6
Mid-Range Trajectory Height • Inches	0.0	0.2	0.7	1.7	3.2	5.4	8.2	11.9	16.2
Drop • Inches	0.0	-0.6	-2.7	-6.3	-11.8	-19.3	-29.1	-41.3	-56.0

USA (Winchester) 125-grain Full Metal Jacket-Flat Nose (Q4209)

G1 Ballistic Coefficient = 0.14

Distance • Yards	Muzzle	25	50	75	100	125	150	175	200
Velocity • fps	1350	1262	1185	1120	1067	1024	987	954	926
Energy • ft-lbs	506	442	390	348	316	291	270	253	238
Taylor KO Index	8.6	8.0	7.6	7.1	6.8	6.5	6.3	6.1	5.9
Mid-Range Trajectory Height • Inches	0.0	0.1	0.5	1.5	2.9	5.0	7.7	11.0	15.0
Drop • Inches	0.0	-0.6	-2.6	-6.2	-11.4	-18.6	-27.8	-39.2	-52.9

Speer 125-grain TMJ FN (53919)

G1 Ballistic Coefficient = 0.1

Distance • Yards	Muzzle	25	50	75	100	125	150	175	200
Velocity • fps	1325	1242	1171	1111	1061	1020	985	954	926
Energy • ft-lbs	487	429	380	343	313	289	269	253	239
Taylor KO Index	8.5	7.9	7.5	7.1	6.8	6.5	6.3	6.1	5.9
Mid-Range Trajectory Height • Inches	0.0	0.2	0.7	1.7	3.2	5.2	7.9	11.3	15.4
Drop • Inches	0.0	-0.6	-2.7	-6.3	-11.7	-19.1	-28.5	-40.1	-54.2

Sellier and Bellot 140 FMJ (V311162U)

G1 Ballistic Coefficient = 0.112

Distance • Yards	Muzzle	25	50	75	100	125	150	175	200
Velocity • fps	1352	1247	1150	1078	1022	976	938	904	873
Energy • ft-lbs	563	479	408	362	325	296	273	254	237
Taylor KO Index	9.7	8.9	8.2	7.7	7.3	7.0	6.7	6.5	6.2
Mid-Range Trajectory Height • Inches	0.0	0.2	0.7	1.7	3.3	5.5	8.4	12.1	16.5
Drop • Inches	0.0	-0.6	-2.7	-6.4	-11.9	-19.5	-29.4	-41.8	-56.8

Hornady 147-grain JHP/XTP (9131)

G1 Ballistic Coefficient = 0.233

Distance • Yards	Muzzle	25	50	75	100	125	150	175	200
Velocity • fps	1225	1180	1138	1104	1072	1045	1019	997	976
Energy • ft-lbs	490	455	422	398	375	356	339	324	311
Taylor KO Index	9.2	8.8	8.5	8.3	8.0	7.8	7.6	7.5	7.3
Mid-Range Trajectory Height • Inches	0.0	0.2	0.6	1.6	3.1	5.3	8.0	11.2	15.0
Drop • Inches	0.0	-0.7	-3.0	-7.0	-12.8	-20.4	-30.0	-41.7	-55.6

.380 Auto (9mm Browning Short)

Another of the John Browning designs, the .380 Auto was first introduced as the 9mm Browning Short in 1908. Most decisions involving ballistics involve trade-offs. The .380 Auto makes an interesting example. The cartridge is far better than the .32 Auto in terms of energy and stopping power, but still falls considerably short of either the 9mm Luger or the .38 Special. At the same time, the modest power lends itself to simple and very compact guns that are easier to carry than the more powerful calibers. The trade does not have an obvious "perfect" choice. The .380 Auto is generally accepted as "enough" gun by persons wanting a gun that's comfortable to carry (for occasional use or, perhaps, never), but not "enough" gun by the law enforcement types who want all the power they can handle.

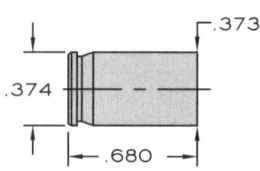

Relative Recoil Factor = 0.41

Specifications:

Controlling Agency for Standardization of this Ammunition: SAAMI

Bullet Weight Grains	Velocity fps	Maximum Average Pressure	
		Copper Crusher	Transducer
85 STHP	990	17,000 cup	21,500 psi
88–90 JHP	980	17,000 cup	21,500 psi
95 MC	945	17,000 cup	21,500 psi
100 FMJ	910	17,000 cup	21,500 psi

Standard barrel for velocity testing: 3.75 inches long—1 turn in 16-inch twist

Availability:

Cor-Bon 70-grain Glaser Safety Slug (00600/00800)
G1 Ballistic Coefficient = 0.07

Distance • Yards	Muzzle	25	50	75	100	125	150	175	200
Velocity • fps	1200	1077	992	929	878	833	793	757	723
Energy • ft-lbs	224	180	153	134	120	108	98	89	81
Taylor KO Index	4.3	3.8	3.5	3.3	3.1	3.0	2.8	2.7	2.6
Mid-Range Trajectory Height • Inches	0.0	0.2	1.0	2.3	4.4	7.4	11.3	16.3	22.6
Drop • Inches	0.0	-0.8	-3.5	-8.4	-15.8	-26.0	-39.4	-56.2	-78.8

Cor-Bon 70-grain Pow'RBall (PB380070/20)
G1 Ballistic Coefficient = 0.07

Distance • Yards	Muzzle	25	50	75	100	125	150	175	200
Velocity • fps	1100	1009	942	889	843	802	765	730	698
Energy • ft-lbs	188	158	138	123	110	100	91	83	76
Taylor KO Index	3.9	3.6	3.4	3.2	3.0	2.9	2.7	2.6	2.5
Mid-Range Trajectory Height • Inches	0.0	0.2	1.1	2.6	4.9	8.1	12.4	17.8	24.6
Drop • Inches	0.0	-1.0	-4.0	-9.6	-17.9	-29.2	-43.9	-62.4	-84.9

Magtech 77-grain SCHP (FD380A)
G1 Ballistic Coefficient = 0.1

Distance • Yards	Muzzle	25	50	75	100	125	150	175	200
Velocity • fps	1099	1039	991	951	916	885	856	830	805
Energy • ft-lbs	207	185	168	155	144	134	125	118	111
Taylor KO Index	4.3	4.1	3.9	3.7	3.6	3.5	3.4	3.3	3.2
Mid-Range Trajectory Height • Inches	0.0	0.2	1.0	2.3	4.4	7.2	10.8	15.3	20.7
Drop • Inches	0.0	-0.9	-3.9	-9.0	-16.6	-26.7	-39.6	-55.5	-76.6

Cor-Bon 80-grain DPX (DPX38080/20)
G1 Ballistic Coefficient = 0.0

Distance • Yards	Muzzle	25	50	75	100	125	150	175	200
Velocity • fps	1050	982	928	883	843	807	774	743	714
Energy • ft-lbs	196	171	153	138	126	116	106	98	84
Taylor KO Index	4.3	4.0	3.8	3.6	3.4	3.3	3.1	3.0	2.9
Mid-Range Trajectory Height • Inches	0.0	0.3	1.1	2.7	5.1	8.3	12.6	17.9	24.6
Drop • Inches	0.0	-1.0	-4.3	-10.1	-18.7	-30.4	-45.4	-64.0	-86.6

Magtech 85-grain GGJHP (GG380A)

G1 Ballistic Coefficient = 0.111

Distance • Yards	Muzzle	25	50	75	100	125	150	175	200
Velocity • fps	1082	1025	978	939	904	873	845	818	794
Energy • ft-lbs	221	198	181	166	154	144	135	126	119
Taylor KO Index	4.7	4.4	4.2	4.1	3.9	3.8	3.7	3.5	3.4
Mid-Range Trajectory Height • Inches	0.0	0.2	1.0	2.4	4.5	7.4	11.1	15.7	21.3
Drop • Inches	0.0	-1.0	-4.0	-9.3	-17.1	-27.5	-40.8	-57.1	-76.6

Winchester 85-grain Silvertip Hollow Point (X380ASHP)

G1 Ballistic Coefficient = 0.113

Distance • Yards	Muzzle	25	50	75	100	125	150	175	200
Velocity • fps	1000	958	921	890	860	833	808	784	762
Energy • ft-lbs	189	173	160	149	140	131	123	116	110
Taylor KO Index	4.3	4.1	4.0	3.8	3.7	3.6	3.5	3.4	3.3
Mid-Range Trajectory Height • Inches	0.0	0.3	1.2	2.8	5.1	8.2	12.3	17.3	23.4
Drop • Inches	0.0	-1.1	-4.6	-10.6	-19.4	-31.1	-45.9	-64.0	-85.6

Remington 88-grain Jacketed Hollow Point (R380A1)

G1 Ballistic Coefficient = 0.129

Distance • Yards	Muzzle	25	50	75	100	125	150	175	200
Velocity • fps	990	954	920	894	868	844	821	800	779
Energy • ft-lbs	191	178	165	156	146	139	132	125	119
Taylor KO Index	4.4	4.2	4.1	4.0	3.9	3.8	3.7	3.6	3.5
Mid-Range Trajectory Height • Inches	0.0	0.3	1.2	2.8	5.1	8.2	12.2	17.1	23.0
Drop • Inches	0.0	-1.1	-4.7	-10.7	-19.5	-31.2	-45.8	-63.8	-85.0

Cor-Bon 90-grain JHP (SD38090/20)

G1 Ballistic Coefficient = 0.100

Distance • Yards	Muzzle	25	50	75	100	125	150	175	200
Velocity • fps	1050	994	948	908	874	842	813	786	760
Energy • ft-lbs	220	197	180	165	153	142	132	123	116
Taylor KO Index	4.8	4.5	4.3	4.2	4.0	3.9	3.7	3.6	3.5
Mid-Range Trajectory Height • Inches	0.0	0.2	1.1	2.6	4.8	7.8	11.8	16.8	22.7
Drop • Inches	0.0	-1.0	-4.3	-9.9	-18.2	-29.3	-43.5	-60.9	-81.9

Hornady 90-grain FTX (90080)

G1 Ballistic Coefficient = 0.100

Distance • Yards	Muzzle	25	50	75	100	125	150	175	200
Velocity • fps	1000	953	913	878	846	817	790	764	740
Energy • ft-lbs	200	182	167	154	143	133	125	117	109
Taylor KO Index	4.6	4.4	4.2	4.0	3.9	3.8	3.6	3.5	3.4
Mid-Range Trajectory Height • Inches	0.0	0.2	1.2	2.8	5.2	8.5	12.7	17.9	24.3
Drop • Inches	0.0	-1.1	-4.6	-10.8	-19.7	-31.6	-46.9	-65.6	-88.0

Black Hills 90-grain JHP (D380N1)

G1 Ballistic Coefficient = 0.100

Distance • Yards	Muzzle	25	50	75	100	125	150	175	200
Velocity • fps	1000	953	913	878	846	817	789	764	739
Energy • ft-lbs	200	182	167	154	143	133	125	117	109
Taylor KO Index	4.6	4.3	4.2	4.0	3.9	3.7	3.6	3.5	3.4
Mid-Range Trajectory Height • Inches	0.0	0.3	1.2	2.8	5.2	8.4	12.6	17.9	24.2
Drop • Inches	0.0	-1.1	-4.6	-10.8	-19.7	-31.6	-46.8	-65.4	-87.8

Federal 90-grain Hydra-Shok JHP (PD380HS1 H)

G1 Ballistic Coefficient = 0.103

Distance • Yards	Muzzle	25	50	75	100	125	150	175	200
Velocity • fps	1000	950	910	880	850	821	794	769	745
Energy • ft-lbs	200	180	165	155	145	135	126	118	111
Taylor KO Index	4.6	4.4	4.2	4.0	3.9	3.8	3.6	3.5	3.4
Mid-Range Trajectory Height • Inches	0.0	0.3	1.2	2.8	5.2	8.5	12.6	17.6	24.1
Drop • Inches	0.0	-1.1	-4.6	-10.7	-19.6	-31.5	-46.6	-65.2	-87.4

Hornady 90-grain JHP/XTP (90102)

G1 Ballistic Coefficient = 0.100

Distance • Yards	Muzzle	25	50	75	100	125	150	175	200
Velocity • fps	1000	953	913	878	846	817	790	764	740
Energy • ft-lbs	200	182	167	154	143	133	125	117	109
Taylor KO Index	4.6	4.4	4.2	4.0	3.9	3.7	3.6	3.5	3.4
Mid-Range Trajectory Height • Inches	0.0	0.3	1.2	2.8	5.2	8.4	12.7	18.0	24.3
Drop • Inches	0.0	-1.1	-4.6	-10.8	-19.7	-31.6	-46.9	-65.6	-88.0

Speer 90-grain GDHP (23606)

G1 Ballistic Coefficient = 0.101

Distance • Yards	Muzzle	25	50	75	100	125	150	175	200
Velocity • fps	990	945	907	873	842	813	786	761	737
Energy • ft-lbs	196	179	164	152	142	132	124	116	109
Taylor KO Index	4.5	4.3	4.2	4.0	3.9	3.7	3.6	3.5	3.4
Mid-Range Trajectory Height • Inches	0.0	0.3	1.2	2.8	5.3	8.6	12.8	18.2	24.6
Drop • Inches	0.0	-1.1	-4.7	-10.9	-20.0	-32.1	-47.6	-66.5	-89.2

Fiocchi 90-grain XTP JHP (380XTP)

G1 Ballistic Coefficient = 0.099

Distance • Yards	Muzzle	25	50	75	100	125	150	175	200
Velocity • fps	975	932	894	861	831	802	775	750	726
Energy • ft-lbs	190	174	160	148	138	128	120	112	105
Taylor KO Index	4.5	4.3	4.1	3.9	3.8	3.7	3.5	3.4	3.3
Mid-Range Trajectory Height • Inches	0.0	0.3	1.3	2.9	5.4	8.8	13.2	18.7	25.3
Drop • Inches	0.0	-1.2	-4.9	-11.3	-20.6	-33.1	-48.9	-68.4	-91.8

Fiocchi 90-grain FMJHP (380APHP)

G1 Ballistic Coefficient = 0.099

Distance • Yards	Muzzle	25	50	75	100	125	150	175	200
Velocity • fps	975	932	894	861	831	802	775	750	726
Energy • ft-lbs	190	174	160	148	138	128	120	112	105
Taylor KO Index	4.5	4.3	4.1	3.9	3.8	3.7	3.5	3.4	3.3
Mid-Range Trajectory Height • Inches	0.0	0.3	1.3	2.9	5.4	8.8	13.2	18.7	25.3
Drop • Inches	0.0	-1.2	-4.9	-11.3	-20.6	-33.1	-48.9	-68.4	-91.8

Aguila 90-grain JHP (1E802112)

G1 Ballistic Coefficient = 0.10

Distance • Yards	Muzzle	25	50	75	100	125	150	175	200
Velocity • fps	945	906	872	840	812	785	759	735	712
Energy • ft-lbs	179	164	152	141	132	123	115	108	101
Taylor KO Index	4.3	4.2	4.0	3.9	3.7	3.6	3.5	3.4	3.3
Mid-Range Trajectory Height • Inches	0.0	0.3	1.3	3.1	5.7	9.3	13.8	18.3	26.5
Drop • Inches	0.0	-1.2	-5.2	-11.9	-21.7	-34.9	-51.8	-71.9	-96.4

PMC 90-grain Full Metal Jacket (380A)

G1 Ballistic Coefficient = 0.09

Distance • Yards	Muzzle	25	50	75	100	125	150	175	200
Velocity • fps	920	896	853	822	791	764	739	715	693
Energy • ft-lbs	169	156	144	134	125	117	109	102	96
Taylor KO Index	4.2	4.1	3.9	3.8	3.6	3.5	3.4	3.3	3.2
Mid-Range Trajectory Height • Inches	0.0	0.3	1.4	3.3	6.0	9.7	14.6	20.6	27.9
Drop • Inches	0.0	-1.3	-5.4	-12.5	-22.9	-36.7	-54.2	-75.8	-101.5

PMC 90-grain EMJ NT (380EMA)

G1 Ballistic Coefficient = 0.09

Distance • Yards	Muzzle	25	50	75	100	125	150	175	200
Velocity • fps	920	896	853	822	791	764	739	715	693
Energy • ft-lbs	169	156	144	134	125	117	109	102	96
Taylor KO Index	4.2	4.1	3.9	3.8	3.6	3.5	3.4	3.3	3.2
Mid-Range Trajectory Height • Inches	0.0	0.3	1.4	3.3	6.0	9.7	14.6	20.6	27.9
Drop • Inches	0.0	-1.3	-5.4	-12.5	-22.9	-36.7	-54.2	-75.8	-101.5

Sellier and Bellot 92-grain FMJ (V310332U)

G1 Ballistic Coefficient = 0.1

Distance • Yards	Muzzle	25	50	75	100	125	150	175	200
Velocity • fps	955	919	884	854	826	800	775	752	729
Energy • ft-lbs	187	173	161	149	139	131	123	115	109
Taylor KO Index	4.5	4.3	4.1	4.0	3.9	3.7	3.6	3.5	3.4
Mid-Range Trajectory Height • Inches	0.0	0.3	1.3	3.0	5.6	9.0	13.4	18.9	25.6
Drop • Inches	0.0	-1.2	-5.0	-11.6	-21.2	-33.9	-50.1	-69.8	-93.4

Magtech 95-grain LRN (380A)(380D) (MP380A)

G1 Ballistic Coefficient = 0.0

Distance • Yards	Muzzle	25	50	75	100	125	150	175	200
Velocity • fps	951	900	861	817	781	752	722	693	666
Energy • ft-lbs	190	171	156	141	128	119	110	101	94
Taylor KO Index	4.6	4.3	4.1	3.9	3.7	3.6	3.5	3.3	3.2
Mid-Range Trajectory Height • Inches	0.0	0.3	1.4	3.2	5.9	9.6	14.5	20.7	28.2
Drop • Inches	0.0	-1.2	-5.2	-12.1	-22.2	-35.7	-53.2	-74.7	-100.7

Winchester 95-grain Bonded PDX1 (S380PDB)

G1 Ballistic Coefficient = 0.117

Distance • Yards	Muzzle	25	50	75	100	125	150	175	200
Velocity • fps	1000	960	924	893	865	838	814	790	768
Energy • ft-lbs	211	194	180	168	158	148	140	132	125
Taylor KO Index	4.8	4.6	4.5	4.3	4.2	4.0	3.9	3.8	3.7
Mid-Range Trajectory Height • Inches	0.0	0.3	1.2	2.8	5.1	8.2	12.3	17.3	23.3
Drop • Inches	0.0	-1.1	-4.6	-10.6	-19.4	-31.0	-45.7	-63.8	-85.2

Magtech 95-grain JHP (380B)

G1 Ballistic Coefficient = 0.116

Distance • Yards	Muzzle	25	50	75	100	125	150	175	200
Velocity • fps	951	917	886	858	832	807	784	762	741
Energy • ft-lbs	191	177	166	155	146	137	130	122	116
Taylor KO Index	4.6	4.4	4.3	4.2	4.0	3.9	3.8	3.7	3.6
Mid-Range Trajectory Height • Inches	0.0	0.3	1.3	3.0	5.5	8.9	13.3	18.7	25.2
Drop • Inches	0.0	-1.2	-5.0	-11.6	-21.2	-33.8	-49.9	-69.4	-92.7

PMC 95-grain Starfire Hollow Point (380SFA)

G1 Ballistic Coefficient = 0.088

Distance • Yards	Muzzle	25	50	75	100	125	150	175	200
Velocity • fps	925	884	847	813	783	755	727	702	677
Energy • ft-lbs	180	165	151	140	129	120	112	104	97
Taylor KO Index	4.4	4.2	4.1	3.9	3.8	3.6	3.5	3.4	3.3
Mid-Range Trajectory Height • Inches	0.0	0.3	1.4	3.3	6.1	9.8	14.7	20.9	28.3
Drop • Inches	0.0	-1.3	-5.4	-12.5	-22.9	-36.8	-54.5	-76.3	-102.5

American Eagle (Federal) 95-grain Full Metal Jacket (AE380AP)

G1 Ballistic Coefficient = 0.105

Distance • Yards	Muzzle	25	50	75	100	125	150	175	200
Velocity • fps	980	940	900	870	840	812	787	762	739
Energy • ft-lbs	205	185	170	160	145	139	131	123	115
Taylor KO Index	4.7	4.6	4.4	4.2	4.1	3.9	3.8	3.7	3.6
Mid-Range Trajectory Height • Inches	0.0	0.3	1.2	2.9	5.3	8.7	13.0	18.1	24.7
Drop • Inches	0.0	-1.2	-4.8	-11.1	-20.3	-32.5	-48.1	-67.1	-89.9

Fiocchi 95-grain FMJ (380AP)

G1 Ballistic Coefficient = 0.077

Distance • Yards	Muzzle	25	50	75	100	125	150	175	200
Velocity • fps	960	908	864	824	789	756	725	695	668
Energy • ft-lbs	194	174	157	143	131	120	111	102	94
Taylor KO Index	4.7	4.4	4.2	4.0	3.8	3.7	3.5	3.4	3.2
Mid-Range Trajectory Height • Inches	0.0	0.3	1.3	3.1	5.8	9.5	14.5	20.4	28.1
Drop • Inches	0.0	-1.2	-5.1	-11.8	-21.8	-35.3	-52.5	-73.9	-99.8

Remington 95-grain Metal Case (R380AP)

G1 Ballistic Coefficient = 0.077

Distance • Yards	Muzzle	25	50	75	100	125	150	175	200
Velocity • fps	955	904	865	821	785	753	722	693	665
Energy • ft-lbs	190	172	160	142	130	120	110	101	93
Taylor KO Index	4.6	4.4	4.2	4.0	3.8	3.6	3.5	3.3	3.2
Mid-Range Trajectory Height • Inches	0.0	0.3	1.4	3.1	5.9	9.5	14.4	20.6	28.1
Drop • Inches	0.0	-1.2	-5.1	-12.0	-22.0	-35.5	-52.9	-74.4	-100.4

MC (Remington) 95-grain Metal Case (L380AP)

G1 Ballistic Coefficient = 0.077

Distance • Yards	Muzzle	25	50	75	100	125	150	175	200
Velocity • fps	955	904	865	821	785	753	722	693	665
Energy • ft-lbs	190	172	160	142	130	120	110	101	93
Taylor KO Index	4.6	4.4	4.2	4.0	3.8	3.6	3.5	3.3	3.2
Mid-Range Trajectory Height • Inches	0.0	0.3	1.4	3.1	5.9	9.5	14.4	20.6	28.1
Drop • Inches	0.0	-1.2	-5.1	-12.0	-22.0	-35.5	-52.9	-74.4	-100.4

MC (Remington) 95-grain FNEB (LL380AP)

G1 Ballistic Coefficient = 0.077

Distance • Yards	Muzzle	25	50	75	100	125	150	175	200
Velocity • fps	955	904	865	821	785	753	722	693	665
Energy • ft-lbs	190	172	160	142	130	120	110	101	93
Taylor KO Index	4.6	4.4	4.2	4.0	3.8	3.6	3.5	3.3	3.2
Mid-Range Trajectory Height • Inches	0.0	0.3	1.4	3.1	5.9	9.5	14.4	20.6	28.1
Drop • Inches	0.0	-1.2	-5.1	-12.0	-22.0	-35.5	-52.9	-74.4	-100.4

USA (Winchester) 95-grain Full Metal Jacket (Q4206)

G1 Ballistic Coefficient = 0.082

Distance • Yards	Muzzle	25	50	75	100	125	150	175	200
Velocity • fps	955	907	865	828	794	762	733	705	678
Energy • ft-lbs	190	173	160	145	133	123	113	105	97
Taylor KO Index	4.6	4.4	4.2	4.0	3.8	3.5	3.4	3.3	3.2
Mid-Range Trajectory Height • Inches	0.0	0.3	1.3	3.1	5.8	9.4	14.2	20.2	27.5
Drop • Inches	0.0	-1.2	-5.1	-11.9	-21.8	-35.2	-52.3	-73.4	-98.8

Winchester 95-grain Win Clean Brass Enclosed Base (WC3801)

G1 Ballistic Coefficient = 0.102

Distance • Yards	Muzzle	25	50	75	100	125	150	175	200
Velocity • fps	955	916	881	849	820	793	768	744	721
Energy • ft-lbs	192	177	164	152	142	133	124	117	110
Taylor KO Index	4.6	4.4	4.3	4.1	4.0	3.8	3.7	3.6	3.5
Mid-Range Trajectory Height • Inches	0.0	0.3	1.3	3.0	5.6	9.1	13.6	19.0	25.9
Drop • Inches	0.0	-1.2	-5.0	-11.7	-21.3	-34.1	-50.4	-70.4	-94.3

Magtech 95-grain FMJ (380B)

G1 Ballistic Coefficient = 0.078

Distance • Yards	Muzzle	25	50	75	100	125	150	175	200
Velocity • fps	951	900	861	817	781	752	722	693	666
Energy • ft-lbs	190	171	156	141	128	119	110	101	94
Taylor KO Index	4.6	4.3	4.1	3.9	3.7	3.6	3.5	3.3	3.2
Mid-Range Trajectory Height • Inches	0.0	0.3	1,4	3.2	5.9	9.6	14.5	20.7	28.2
Drop • Inches	0.0	-1.2	-5.2	-12.1	-22.2	-35.7	-53.2	-74.7	-100.7

Magtech 95-grain FEB (CR380A)

G1 Ballistic Coefficient = 0.07

Distance • Yards	Muzzle	25	50	75	100	125	150	175	200
Velocity • fps	951	900	861	817	781	748	718	688	661
Energy • ft-lbs	190	171	157	141	128	118	109	100	92
Taylor KO Index	4.6	4.4	4.2	4.0	3.8	3.6	3.5	3.3	3.2
Mid-Range Trajectory Height • Inches	0.0	0.3	1,4	3.2	6.0	9.7	14.7	20.8	28.6
Drop • Inches	0.0	-1.2	-5.2	-12.1	-22.2	-35.9	-53.5	-75.3	-101.7

Speer 95-grain TMJ RN Lawman (53608)

G1 Ballistic Coefficient = 0.13

Distance • Yards	Muzzle	25	50	75	100	125	150	175	200
Velocity • fps	950	920	892	867	843	821	800	780	761
Energy • ft-lbs	190	178	168	159	150	142	135	128	122
Taylor KO Index	4.6	4.5	4.3	4.2	4.1	4.0	3.9	3.8	3.7
Mid-Range Trajectory Height • Inches	0.0	0.3	1.3	3.0	5.5	8.8	13.1	18.1	24.5
Drop • Inches	0.0	-1.2	-5.0	-11.6	-21.0	-33.4	-49.1	-68.2	-90.9

Aguila 95-grain FMJ (1E802110)

G1 Ballistic Coefficient = 0.08

Distance • Yards	Muzzle	25	50	75	100	125	150	175	200
Velocity • fps	945	897	856	819	784	753	723	695	669
Energy • ft-lbs	188	170	155	141	130	120	110	102	94
Taylor KO Index	4.6	4.3	4.1	4.0	3.8	3.6	3.5	3.4	3.2
Mid-Range Trajectory Height • Inches	0.0	0.3	1.4	3.2	5.9	9.7	14.6	20.8	28.4
Drop • Inches	0.0	-1.3	-5.2	-12.2	-22.3	-36.1	-53.5	-75.2	-101.4

Blazer (CCI) 95-grain FMJ RN (5202)

G1 Ballistic Coefficient = 0.1

Distance • Yards	Muzzle	25	50	75	100	125	150	175	200
Velocity • fps	945	915	888	862	839	817	796	776	757
Energy • ft-lbs	188	177	166	157	149	141	134	127	121
Taylor KO Index	4.6	4.4	4.3	4.2	4.1	3.9	3.8	3.7	3.7
Mid-Range Trajectory Height • Inches	0.0	0.3	1.3	3.0	5.5	8.4	13.2	18.4	24.7
Drop • Inches	0.0	-1.2	-5.1	-11.7	-21.2	-33.8	-49.6	-68.9	-91.8

Blazer (CCI) 95-grain TMJ (3505)

G1 Ballistic Coefficient = 0.1

Distance • Yards	Muzzle	25	50	75	100	125	150	175	200
Velocity • fps	945	915	888	862	839	817	796	776	757
Energy • ft-lbs	188	177	166	157	149	141	134	127	121
Taylor KO Index	4.6	4.4	4.3	4.2	4.1	3.9	3.8	3.7	3.7
Mid-Range Trajectory Height • Inches	0.0	0.3	1.3	3.0	5.5	8.4	13.2	18.4	24.7
Drop • Inches	0.0	-1.2	-5.1	-11.7	-21.2	-33.8	-49.6	-68.9	-91.8

Black Hills 100-grain Full Metal Jacket (D380N3)

G1 Ballistic Coefficient = 0.125

Distance • Yards	Muzzle	25	50	75	100	125	150	175	200
Velocity • fps	950	918	889	863	838	815	793	772	752
Energy • ft-lbs	200	187	176	165	156	147	140	132	126
Taylor KO Index	4.8	4.7	4.5	4.4	4.3	4.2	4.0	3.9	3.8
Mid-Range Trajectory Height • Inches	0.0	0.3	1.3	3.0	5.5	8.8	13.1	18.4	24.8
Drop • Inches	0.0	-1.2	-5.0	-11.6	-21.1	-33.6	-49.5	-68.8	-91.7

ahead, make my day and show yourself! This Ruger Blackhawk is being aimed at a ground squirrel. (Safari Press Archives)

9mm Makarov (9x18mm)

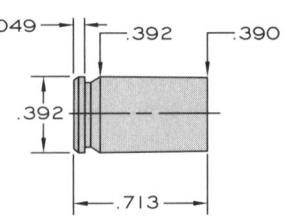

During the Cold War days, the 9mm Makarov cartridge was the standard military and police cartridge for the USSR and several Eastern Block countries. Since the collapse of the USSR, Makarov pistols have begun to appear in the U.S., both new and surplus. Power-wise, this cartridge falls in the gap between the .380 Auto and the 9mm Luger. The dimensions are such that these rounds won't fit Luger chambers, and it isn't a good idea to try to use .380 or other 9mm cartridges in the Makarov.

Relative Recoil Factor = 0.41

Specifications:

Controlling Agency for Standardization of this Ammunition: Factory
Barrel for velocity testing: 3.75 inches long—1 turn in 9.45-inch twist

Availability:

Cor-Bon 70-grain Pow'R Ball (PB09MAK)

G1 Ballistic Coefficient = 0.070

Distance • Yards	Muzzle	25	50	75	100	125	150	175	200
Velocity • fps	1250	1111	1016	948	894	847	806	768	734
Energy • ft-lbs	242	192	161	140	124	112	101	92	84
Taylor KO Index	4.4	3.9	3.6	3.4	3.2	3.0	2.9	2.7	2.6
Mid-Range Trajectory Height • Inches	0.0	0.2	0.9	2.2	4.2	7.1	10.9	15.8	21.8
Drop • Inches	0.0	-0.8	-3.3	-7.9	-14.9	-24.7	-37.4	-53.6	-73.4

Cor-Bon 75-grain Glaser Safety Slug (00600/00800)

G1 Ballistic Coefficient = 0.07

Distance • Yards	Muzzle	25	50	75	100	125	150	175	200
Velocity • fps	1150	1049	977	921	874	832	795	761	729
Energy • ft-lbs	210	183	159	141	127	115	105	96	88
Taylor KO Index	4.4	4.0	3.7	3.5	3.3	3.2	3.0	2.9	2.8
Mid-Range Trajectory Height • Inches	0.0	0.2	1.0	2.4	4.6	7.6	11.6	16.5	22.8
Drop • Inches	0.0	-0.9	-3.7	-8.8	-16.5	-27.1	-40.7	-57.8	-78.7

Hornady 95-grain XTP (91002)

G1 Ballistic Coefficient = 0.12

Distance • Yards	Muzzle	25	50	75	100	125	150	175	200
Velocity • fps	1000	963	930	901	874	849	826	804	783
Energy • ft-lbs	211	196	182	171	161	152	144	138	129
Taylor KO Index	4.8	4.6	4.5	4.3	4.2	4.1	4.0	3.9	3.8
Mid-Range Trajectory Height • Inches	0.0	0.3	1.2	2.7	5.0	8.1	12.1	17.0	22.8
Drop • Inches	0.0	-1.1	-4.6	-10.5	-19.2	-30.7	-45.2	-62.8	-83.9

Fiocchi 95-grain FMJ (9MAK)

G1 Ballistic Coefficient = 0.0

Distance • Yards	Muzzle	25	50	75	100	125	150	175	200
Velocity • fps	1020	956	905	861	823	787	754	723	694
Energy • ft-lbs	219	193	173	156	142	131	120	110	102
Taylor KO Index	4.9	4.6	4.4	4.2	4.0	3.8	3.7	3.5	3.4
Mid-Range Trajectory Height • Inches	0.0	0.3	1.2	2.8	5.3	8.7	13.2	18.8	25.8
Drop • Inches	0.0	-1.1	-4.6	-10.7	-19.7	-32.0	-49.8	-67.3	-91.1

Blazer (CCI) 95-grain TMJ (3506)

G1 Ballistic Coefficient = 0.1

Distance • Yards	Muzzle	25	50	75	100	125	150	175	200
Velocity • fps	1050	1006	969	936	907	880	856	833	811
Energy • ft-lbs	220	214	188	185	174	164	154	146	139
Taylor KO Index	5.1	4.9	4.7	4.5	4.4	4.3	4.1	4.0	3.9
Mid-Range Trajectory Height • Inches	0.0	0.3	1.1	2.5	4.6	7.5	11.1	15.7	21.1
Drop • Inches	0.0	-1.0	-4.2	-9.6	-17.6	-28.2	-41.6	-58.0	-77.4

Sellier & Bellot 95-grain FMJ (V310912U)

G1 Ballistic Coefficient = 0.108

Distance • Yards	Muzzle	25	50	75	100	125	150	175	200
Velocity • fps	1017	971	931	897	866	837	810	786	762
Energy • ft-lbs	218	199	183	170	158	148	139	130	122
Taylor KO Index	4.9	4.7	4.5	4.3	4.2	4.1	3.9	3.8	3.7
Mid-Range Trajectory Height • Inches	0.0	0.2	1.1	2.7	5.0	8.1	12.1	17.2	23.2
Drop • Inches	0.0	-1.1	-4.5	-10.4	-19.0	-30.4	-45.0	-62.9	-84.4

Winchester 95-grain FMJ (MC918M)

G1 Ballistic Coefficient = 0.112

Distance • Yards	Muzzle	25	50	75	100	125	150	175	200
Velocity • fps	1015	970	933	899	869	841	815	791	768
Energy • ft-lbs	216	199	182	171	159	149	140	132	124
Taylor KO Index	4.9	4.7	4.5	4.3	4.2	4.1	3.9	3.8	3.7
Mid-Range Trajectory Height • Inches	0.0	0.3	1.2	2.7	5.0	8.1	12.1	17.1	23.1
Drop • Inches	0.0	-1.1	-4.5	-10.4	-19.0	-30.4	-44.9	-62.7	-84.0

American Eagle (Federal) 95-grain Full Metal Jacket (AE9MK)

G1 Ballistic Coefficient = 0.112

Distance • Yards	Muzzle	25	50	75	100	125	150	175	200
Velocity • fps	1000	960	920	890	860	832	807	783	760
Energy • ft-lbs	210	195	180	165	155	146	137	129	122
Taylor KO Index	4.8	4.7	4.5	4.3	4.2	4.0	3.9	3.8	3.7
Mid-Range Trajectory Height • Inches	0.0	0.3	1.2	2.8	5.1	8.3	12.4	17.3	23.6
Drop • Inches	0.0	-1.1	-4.6	-10.6	-19.4	-31.2	-46.0	-64.2	-86.0

9mm Luger (9mm Parabellum) (9x19mm)

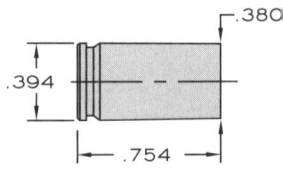

When he first introduced his 9mm pistol cartridge in 1902, Georgi Luger couldn't have even imagined what was going to happen to his brainchild. The German navy adopted the cartridge in 1904 and the German army in 1908. That in itself would be a pretty glowing resume for any cartridge design, but its history was just starting. By WWII the 9mm Luger had gone on to become the standard military and police pistol cartridge in most of Western Europe. Metric cartridges weren't very popular in the U.S. before WWII, and the cartridge didn't get a lot of use here. With the increased popularity of semiautomatic pistols in the U.S., the 9mm caliber ended up on more and more shooters' ammo shelves. Then in 1985, the U.S. army adopted the M-9 Beretta pistol as a "replacement" for the 1911A1 .45 Auto and guaranteed the 9mm Luger at least another 50 years of useful life. Performance-wise, the 9mm Luger packs a little more punch than the .38 Special but falls well short of the .357 Magnum.

Relative Recoil Factor = 0.65

Specifications:

Controlling Agency for Standardization of this Ammunition: SAAMI

Bullet Weight Grains	Velocity fps	Maximum Average Pressure	
		Copper Crusher	Transducer
88 JHP	1,500	33,000 cup	35,000 psi
95 JSP	1,330	33,000 cup	35,000 psi
100 JHP	1,210	33,000 cup	35,000 psi
115 MC	1,125	33,000 cup	35,000 psi
115 JHP	1,145	33,000 cup	35,000 psi
115 STHP	1,210	33,000 cup	35,000 psi
124 NC	1,090	33,000 cup	35,000 psi
147 MC	985	33,000 cup	35,000 psi

Standard barrel for velocity testing: 4 inches long—1 turn in 10-inch twist

Availability:

CCI 64-grain Shotshell (3706)
Shot cartridge using # 11 Shot @ 1450 fps muzzle velocity.

Cor-Bon 80-grain Glaser Safety Slug +P (01000/01200)
G1 Ballistic Coefficient = 0.0

Distance • Yards	Muzzle	25	50	75	100	125	150	175	200
Velocity • fps	1500	1324	1181	1076	1001	944	896	855	818
Energy • ft-lbs	399	311	248	206	178	158	143	130	119
Taylor KO Index	6.1	5.4	4.8	4.4	4.1	3.8	3.6	3.5	3.3
Mid-Range Trajectory Height • Inches	0.0	0.1	0.6	1.6	3.2	5.4	8.4	12.8	17.2
Drop • Inches	0.0	-0.5	-2.3	-5.6	-10.8	-18.2	-28.0	-40.5	-56.0

Cor-Bon 90-grain JHP +P (SD0990/20)
G1 Ballistic Coefficient = 0.0

Distance • Yards	Muzzle	25	50	75	100	125	150	175	200
Velocity • fps	1500	1333	1195	1091	1016	958	911	870	834
Energy • ft-lbs	450	355	285	238	206	184	166	151	139
Taylor KO Index	6.9	6.1	5.5	5.0	4.7	4.4	4.2	4.0	3.8
Mid-Range Trajectory Height • Inches	0.0	0.1	0.6	1.5	3.0	5.2	8.1	11.9	16.6
Drop • Inches	0.0	-0.5	-2.3	-5.5	-10.6	-17.8	-27.4	-39.6	-54.6

Magetech 92.6-grain SCHP (FD9A)
G1 Ballistic Coefficient = 0.1

Distance • Yards	Muzzle	25	50	75	100	125	150	175	200
Velocity • fps	1330	1246	1173	1112	1062	1020	984	953	925
Energy • ft-lbs	364	319	283	254	232	214	199	187	176
Taylor KO Index	6.3	5.9	5.5	5.2	5.0	4.8	4.6	4.5	4.4
Mid-Range Trajectory Height • Inches	0.0	0.2	0.7	1.7	3.2	5.2	7.9	11.2	15.4
Drop • Inches	0.0	-0.6	-2.7	-6.3	-11.7	-19.0	-28.4	-40.0	-54.0

Magtech 95-grain JSP – FLAT (9D)

G1 Ballistic Coefficient = 0.146

Distance • Yards	Muzzle	25	50	75	100	125	150	175	200
Velocity • fps	1345	1260	1185	1122	1071	1028	991	959	931
Energy • ft-lbs	380	335	295	266	242	223	207	194	183
Taylor KO Index	6.5	6.1	5.7	5.4	5.2	5.0	4.8	4.6	4.5
Mid-Range Trajectory Height • Inches	0.0	0.2	0.7	1.7	3.1	5.1	7.8	11.0	15.1
Drop • Inches	0.0	-0.6	-2.6	-6.2	-11.4	-18.6	-27.8	-39.2	-53.0

Cor-Bon 95-grain DPX (DPX0995/20)

G1 Ballistic Coefficient = 0.135

Distance • Yards	Muzzle	25	50	75	100	125	150	175	200
Velocity • fps	1300	1214	1142	1082	1034	994	959	929	901
Energy • ft-lbs	357	311	275	247	226	208	194	182	171
Taylor KO Index	6.3	5.8	5.5	5.2	5.0	4.8	4.6	4.5	4.3
Mid-Range Trajectory Height • Inches	0.0	0.2	0.8	1.8	3.4	5.5	8.3	11.9	16.2
Drop • Inches	0.0	-0.7	-2.8	-6.6	-12.3	-20.0	-29.9	-42.2	-56.9

Cor-Bon 100-grain Pow'RBall +P (PB91000/20)

G1 Ballistic Coefficient = 0.100

Distance • Yards	Muzzle	25	50	75	100	125	150	175	200
Velocity • fps	1475	1335	1215	1120	1048	993	947	908	873
Energy • ft-lbs	483	396	328	279	244	219	199	183	169
Taylor KO Index	7.5	6.8	6.2	5.7	5.3	5.1	4.8	4.6	4.4
Mid-Range Trajectory Height • Inches	0.0	0.1	0.6	1.5	3.0	5.0	7.8	11.3	15.7
Drop • Inches	0.0	-0.5	-2.3	-5.5	-10.5	-17.4	-26.5	-38.1	-52.3

PMC 100-grain SinterFire/Frangible (9SF)

G1 Ballistic Coefficient = 0.146

Distance • Yards	Muzzle	25	50	75	100	125	150	175	200
Velocity • fps	1250	1177	1115	1065	1023	987	956	927	902
Energy • ft-lbs	346	308	276	252	232	216	203	191	181
Taylor KO Index	6.3	6.0	5.7	5.4	5.2	5.0	4.8	4.7	4.6
Mid-Range Trajectory Height • Inches	0.0	0.2	0.8	1.9	3.5	5.7	8.6	12.2	16.6
Drop • Inches	0.0	-0.7	-3.0	-7.0	-13.0	-21.0	-31.3	-43.9	-59.1

Federal 105-grain EFMJ – Low Recoil (PDCSP2 H)

G1 Ballistic Coefficient = 0.111

Distance • Yards	Muzzle	25	50	75	100	125	150	175	200
Velocity • fps	1230	1140	1070	1010	970	932	898	868	840
Energy • ft-lbs	355	305	265	240	220	203	188	176	164
Taylor KO Index	6.6	6.1	5.7	5.4	5.2	5.0	4.8	4.6	4.5
Mid-Range Trajectory Height • Inches	0.0	0.2	0.8	2.0	3.8	6.3	9.5	13.4	18.4
Drop • Inches	0.0	-0.8	-3.2	-7.5	-13.9	-22.7	-33.9	-47.8	-64.7

Winchester 105-grain Jacketed Soft Point NT (SC9NT)

G1 Ballistic Coefficient = 0.137

Distance • Yards	Muzzle	25	50	75	100	125	150	175	200
Velocity • fps	1200	1131	1074	1028	989	955	925	898	873
Energy • ft-lbs	336	298	269	246	228	213	200	188	178
Taylor KO Index	6.4	6.0	5.7	5.5	5.3	5.1	4.9	4.8	4.7
Mid-Range Trajectory Height • Inches	0.0	0.2	0.8	2.0	3.7	6.1	9.1	13.0	17.5
Drop • Inches	0.0	-0.8	-3.3	-7.6	-14.1	-22.6	-33.6	-47.1	-63.2

Hornady 115-grain FTX (90250)

G1 Ballistic Coefficient = 0.130

Distance • Yards	Muzzle	25	50	75	100	125	150	175	200
Velocity • fps	1135	1075	1026	986	950	920	892	866	843
Energy • ft-lbs	329	295	269	248	231	216	203	192	181
Taylor KO Index	6.6	6.3	6.0	5.8	5.6	5.4	5.2	5.1	4.9
Mid-Range Trajectory Height • Inches	0.0	0.2	0.9	2.2	4.1	6.7	10.0	14.2	19.2
Drop • Inches	0.0	-0.9	-3.6	-8.4	-15.5	-24.9	-37.0	-51.7	-69.3

Cor-Bon 115-grain JHP +P (SD09115/20)

G1 Ballistic Coefficient = 0.130

Distance • Yards	Muzzle	25	50	75	100	125	150	175	200
Velocity • fps	1350	1254	1172	1104	1050	1006	968	935	906
Energy • ft-lbs	460	402	351	312	282	258	239	223	210
Taylor KO Index	7.9	7.3	6.9	6.5	6.1	5.9	5.7	5.5	5.3
Mid-Range Trajectory Height • Inches	0.0	0.1	0.6	1.6	3.1	5.2	7.9	11.4	15.6
Drop • Inches	0.0	-0.6	-2.6	-6.2	-11.6	-18.9	-28.4	-40.2	-54.5

Magtech 115-grain JHP +P + (9H)

G1 Ballistic Coefficient = 0.165

Distance • Yards	Muzzle	25	50	75	100	125	150	175	200
Velocity • fps	1328	1254	1187	1131	1084	1043	1009	978	951
Energy • ft-lbs	451	401	360	327	300	278	260	244	231
Taylor KO Index	7.8	7.3	6.9	6.6	6.3	6.1	5.9	5.7	5.6
Mid-Range Trajectory Height • Inches	0.0	0.2	0.7	1.6	3.1	5.0	7.6	10.9	14.7
Drop • Inches	0.0	-0.6	-2.7	-6.2	-11.5	-18.6	-27.7	-38.9	-52.4

Black Hills 115-grain Barnes Tac-XP +P (D9N10)

G1 Ballistic Coefficient = 0.150

Distance • Yards	Muzzle	25	50	75	100	125	150	175	200
Velocity • fps	1250	1179	1119	1069	1027	991	960	933	907
Energy • ft-lbs	400	355	320	292	269	251	236	222	210
Taylor KO Index	7.3	6.9	6.5	6.3	6.0	5.8	5.6	5.5	5.3
Mid-Range Trajectory Height • Inches	0.0	0.2	0.7	1.8	3.4	5.6	8.5	12.1	16.4
Drop • Inches	0.0	-0.7	-3.0	-7.0	-13.0	-20.9	-31.1	-43.7	-58.7

Black Hills 115-grain EXP (Extra Power) HP (D9M6)

G1 Ballistic Coefficient = 0.150

Distance • Yards	Muzzle	25	50	75	100	125	150	175	200
Velocity • fps	1250	1179	1118	1068	1026	991	960	932	907
Energy • ft-lbs	400	355	319	291	269	251	235	222	210
Taylor KO Index	7.3	6.9	6.5	6.2	6.0	5.8	5.6	5.5	5.3
Mid-Range Trajectory Height • Inches	0.0	0.2	0.8	1.8	3.5	6.0	8.5	12.0	16.2
Drop • Inches	0.0	-0.7	-3.0	-7.0	-13.0	-20.9	-31.1	-43.5	-58.4

Cor-Bon 115-grain DPX +P (DPX09115/20) + (TR09115/20)

G1 Ballistic Coefficient = 0.14

Distance • Yards	Muzzle	25	50	75	100	125	150	175	200
Velocity • fps	1250	1176	1115	1064	1022	986	954	926	901
Energy • ft-lbs	399	354	317	286	267	248	233	219	207
Taylor KO Index	7.3	6.9	6.5	6.2	6.0	5.8	5.6	5.4	5.3
Mid-Range Trajectory Height • Inches	0.0	0.2	0.8	1.9	3.5	5.7	8.6	12.2	16.6
Drop • Inches	0.0	-0.7	-3.0	-7.1	-13.0	-21.1	-31.3	-44.0	-59.2

Fiocchi 115-grain Jacketed Hollowpoint (9APHP)

G1 Ballistic Coefficient = 0.14

Distance • Yards	Muzzle	25	50	75	100	125	150	175	200
Velocity • fps	1250	1176	1114	1063	1021	985	954	926	900
Energy • ft-lbs	400	353	317	289	266	248	232	219	207
Taylor KO Index	7.3	6.9	6.5	6.2	6.0	5.8	5.6	5.4	5.3
Mid-Range Trajectory Height • Inches	0.0	0.2	0.8	1.8	3.5	5.6	8.5	12.1	16.4
Drop • Inches	0.0	-0.7	-3.0	-7.1	-13.0	-21.0	-31.3	-43.8	-58.9

Remington 115-grain Jacketed Hollow Point +P (R9MM6)

G1 Ballistic Coefficient = 0.1

Distance • Yards	Muzzle	25	50	75	100	125	150	175	200
Velocity • fps	1250	1175	1113	1061	1019	983	951	923	897
Energy • ft-lbs	399	353	315	288	265	247	231	218	206
Taylor KO Index	7.3	6.9	6.5	6.2	6.0	5.7	5.6	5.4	5.2
Mid-Range Trajectory Height • Inches	0.0	0.2	0.8	1.9	3.5	5.6	8.5	12.1	16.4
Drop • Inches	0.0	-0.7	-3.0	-7.1	-13.0	-21.1	-31.3	-44.0	-59.1

Magtech 115-grain JHP +P (GG9A)

G1 Ballistic Coefficient = 0.1

Distance • Yards	Muzzle	25	50	75	100	125	150	175	200
Velocity • fps	1246	1188	1137	1093	1056	1023	994	968	945
Energy • ft-lbs	397	360	330	305	285	267	252	240	228
Taylor KO Index	7.3	6.9	6.6	6.4	6.2	6.0	5.8	5.7	5.5
Mid-Range Trajectory Height • Inches	0.0	0.2	0.8	1.8	3.4	5.5	8.2	11.5	15.5
Drop • Inches	0.0	-0.7	-3.0	-6.9	-12.7	-20.4	-30.2	-42.2	-56.4

Sellier and Bellot 115-grain JHP (V310422U)

G1 Ballistic Coefficient = 0.0

Distance • Yards	Muzzle	25	50	75	100	125	150	175	200
Velocity • fps	1237	1129	1027	964	912	869	830	794	962
Energy • ft-lbs	395	329	272	237	213	193	176	161	148
Taylor KO Index	7.2	6.6	6.0	5.6	5.3	5.1	4.9	4.6	4.5
Mid-Range Trajectory Height • Inches	0.0	0.2	0.8	2.1	4.0	6.7	10.3	14.8	20.3
Drop • Inches	0.0	-0.8	-3.3	-7.8	-14.7	-24.1	-36.3	-51.6	-70.1

Winchester 115-grain Silvertip Hollow Point (X9MMSHP)
G1 Ballistic Coefficient = 0.143

Distance • Yards	Muzzle	25	50	75	100	125	150	175	200
Velocity • fps	1225	1154	1095	1047	1007	973	942	915	890
Energy • ft-lbs	383	340	306	280	259	242	227	214	202
Taylor KO Index	7.1	6.7	6.4	6.1	5.9	5.7	5.5	5.4	5.2
Mid-Range Trajectory Height • Inches	0.0	0.2	0.8	1.9	3.6	5.8	8.8	12.5	16.9
Drop • Inches	0.0	-0.8	-3.1	-7.3	-13.5	-21.8	-32.3	-45.3	-60.8

USA (Winchester) 115-grain Jacketed Hollow Point (USA9JHP)
G1 Ballistic Coefficient = 0.142

Distance • Yards	Muzzle	25	50	75	100	125	150	175	200
Velocity • fps	1225	1154	1095	1047	1006	972	941	914	865
Energy • ft-lbs	383	340	306	280	259	241	226	213	202
Taylor KO Index	7.2	6.7	6.4	6.1	5.9	5.7	5.5	5.3	5.1
Mid-Range Trajectory Height • Inches	0.0	0.2	0.8	1.9	3.6	5.9	8.9	12.6	17.1
Drop • Inches	0.0	-0.8	-3.1	-7.3	-13.5	-21.8	-32.5	-45.5	-61.2

Speer 115-grain GDHP (23614)
G1 Ballistic Coefficient = 0.125

Distance • Yards	Muzzle	25	50	75	100	125	150	175	200
Velocity • fps	1210	1133	1071	1022	981	945	913	885	859
Energy • ft-lbs	374	328	293	267	246	228	213	200	188
Taylor KO Index	7.1	6.6	6.2	6.0	5.7	5.5	5.3	5.2	5.0
Mid-Range Trajectory Height • Inches	0.0	0.2	0.8	2.0	3.8	6.2	9.3	13.3	18.0
Drop • Inches	0.0	-0.8	-3.2	-7.6	-14.0	-22.7	-33.9	-47.6	-64.1

Federal 115-grain JHP (C9BP)
G1 Ballistic Coefficient = 0.119

Distance • Yards	Muzzle	25	50	75	100	125	150	175	200
Velocity • fps	1180	1110	1050	1000	960	925	894	866	840
Energy • ft-lbs	355	310	280	255	235	219	204	192	180
Taylor KO Index	6.9	6.5	6.1	5.8	5.6	5.4	5.2	5.1	4.9
Mid-Range Trajectory Height • Inches	0.0	0.2	0.9	2.1	4.0	6.6	9.8	13.8	18.8
Drop • Inches	0.0	-0.8	-3.4	-8.0	-14.7	-23.8	-35.5	-49.8	-67.1

PMC 115-grain Jacketed Hollow Point (9B)
G1 Ballistic Coefficient = 0.138

Distance • Yards	Muzzle	25	50	75	100	125	150	175	200
Velocity • fps	1160	1099	1049	1007	971	940	912	886	862
Energy • ft-lbs	344	308	281	259	241	226	212	201	190
Taylor KO Index	6.8	6.4	6.1	5.9	5.7	5.5	5.3	5.2	5.0
Mid-Range Trajectory Height • Inches	0.0	0.2	0.9	2.1	4.0	6.4	9.6	13.6	18.4
Drop • Inches	0.0	-0.8	-3.5	-8.1	-14.8	-23.9	-35.4	-49.6	-66.4

Hornady 115-grain JHP/XTP (90252)
G1 Ballistic Coefficient = 0.141

Distance • Yards	Muzzle	25	50	75	100	125	150	175	200
Velocity • fps	1155	1095	1047	1006	971	940	913	887	864
Energy • ft-lbs	341	306	280	258	241	226	213	201	191
Taylor KO Index	6.7	6.4	6.1	5.9	5.7	5.5	5.3	5.2	5.1
Mid-Range Trajectory Height • Inches	0.0	0.2	0.9	2.1	4.0	6.3	9.5	13.5	18.2
Drop • Inches	0.0	-0.8	-3.5	-8.1	-14.9	-24.0	-35.4	-49.5	-66.3

Magtech 115-grain JHP (9C)
G1 Ballistic Coefficient = 0.141

Distance • Yards	Muzzle	25	50	75	100	125	150	175	200
Velocity • fps	1155	1095	1047	1006	971	940	913	887	864
Energy • ft-lbs	341	306	280	258	241	226	213	201	191
Taylor KO Index	6.7	6.4	6.1	5.9	5.7	5.5	5.3	5.2	5.1
Mid-Range Trajectory Height • Inches	0.0	0.2	0.9	2.1	4.0	6.3	9.5	13.5	18.2
Drop • Inches	0.0	-0.8	-3.5	-8.1	-14.9	-24.0	-35.4	-49.5	-66.3

Remington 115-grain Jacketed Hollow Point (R9MM1)
G1 Ballistic Coefficient = 0.141

Distance • Yards	Muzzle	25	50	75	100	125	150	175	200
Velocity • fps	1155	1095	1047	1006	971	940	913	887	864
Energy • ft-lbs	341	306	280	258	241	226	213	201	191
Taylor KO Index	6.7	6.4	6.1	5.9	5.7	5.5	5.3	5.2	5.1
Mid-Range Trajectory Height • Inches	0.0	0.2	0.9	2.1	4.0	6.3	9.5	13.5	18.2
Drop • Inches	0.0	-0.8	-3.5	-8.1	-14.9	-24.0	-35.4	-49.5	-66.3

Black Hills 115-grain JHP (D9SCN2)

G1 Ballistic Coefficient = 0.150

Distance • Yards	Muzzle	25	50	75	100	125	150	175	200
Velocity • fps	1150	1095	1049	1010	977	947	921	896	872
Energy • ft-lbs	338	306	281	261	244	229	217	205	195
Taylor KO Index	6.7	6.4	6.1	5.9	5.7	5.5	5.4	5.2	5.1
Mid-Range Trajectory Height • Inches	0.0	0.2	0.9	2.2	3.9	6.4	9.6	13.5	18.2
Drop • Inches	0.0	-0.8	-3.5	-8.2	-14.9	-24.0	-35.4	-49.4	-66.2

UMC (Remington) 115-grain Jacketed Hollow Point (L9MM1)

G1 Ballistic Coefficient = 0.155

Distance • Yards	Muzzle	25	50	75	100	125	150	175	200
Velocity • fps	1135	1084	1041	1005	973	945	919	896	874
Energy • ft-lbs	329	300	277	258	242	228	216	205	195
Taylor KO Index	6.6	6.3	6.1	5.9	5.7	5.5	5.4	5.2	5.1
Mid-Range Trajectory Height • Inches	0.0	0.2	0.9	2.2	4.0	6.5	9.7	13.5	18.3
Drop • Inches	0.0	-0.9	-3.6	-8.3	-15.2	-24.4	-36.0	-50.1	-67.0

Sellier and Bellot 115-grain FMJ (V311522U)

G1 Ballistic Coefficient = 0.10?

Distance • Yards	Muzzle	25	50	75	100	125	150	175	200
Velocity • fps	1280	1180	1089	1026	975	933	897	864	834
Energy • ft-lbs	421	358	304	269	243	222	205	191	178
Taylor KO Index	7.5	6.9	6.4	6.0	5.7	5.5	5.2	5.1	4.9
Mid-Range Trajectory Height • Inches	0.0	0.2	0.8	1.9	3.7	6.1	9.2	13.3	18.2
Drop • Inches	0.0	-0.7	-3.0	-7.1	-13.3	-21.7	-32.7	-46.3	-62.9

Sellier and Bellot 115-grain Non-Tox FMJ (V310452U) LF

G1 Ballistic Coefficient = 0.10?

Distance • Yards	Muzzle	25	50	75	100	125	150	175	200
Velocity • fps	1280	1173	1089	1026	975	933	897	864	834
Energy • ft-lbs	418	351	303	269	243	222	205	191	178
Taylor KO Index	7.5	6.9	6.4	6.0	5.7	5.5	5.3	5.1	4.9
Mid-Range Trajectory Height • Inches	0.0	0.1	0.4	1.5	3.2	5.7	8.8	12.9	17.8
Drop • Inches	0.0	-0.7	-3.0	-7.1	-13.3	-21.7	-32.7	-46.3	-62.9

Sellier and Bellot 115-grain FMJ (V211552U)

G1 Ballistic Coefficient = 0.1?

Distance • Yards	Muzzle	25	50	75	100	125	150	175	200
Velocity • fps	1280	1173	1089	1026	975	933	897	864	834
Energy • ft-lbs	421	358	304	269	243	222	205	191	178
Taylor KO Index	7.5	6.9	6.4	6.0	5.7	5.5	5.2	5.1	4.9
Mid-Range Trajectory Height • Inches	0.0	0.2	0.8	1.9	3.7	6.1	9.2	13.3	18.2
Drop • Inches	0.0	-0.7	-3.0	-7.1	-13.3	-21.7	-32.7	-46.3	-62.9

Fiocchi 115-grain FMJ (9AP)

G1 Ballistic Coefficient = 0.1?

Distance • Yards	Muzzle	25	50	75	100	125	150	175	200
Velocity • fps	1260	1191	1131	1082	1040	1004	973	946	920
Energy • ft-lbs	405	362	326	299	276	258	242	228	216
Taylor KO Index	7.4	7.0	6.6	6.3	6.1	5.9	5.7	5.5	5.4
Mid-Range Trajectory Height • Inches	0.0	0.2	0.8	1.8	3.4	5.5	8.4	11.7	16.0
Drop • Inches	0.0	-0.7	-3.0	-6.9	-12.7	-20.5	-30.4	-42.7	-57.4

Aguila 115-grain FMJ (1E07704)

G1 Ballistic Coefficient = 0.1

Distance • Yards	Muzzle	25	50	75	100	125	150	175	200
Velocity • fps	1250	1176	1113	1062	1020	983	952	924	898
Energy • ft-lbs	399	353	317	288	266	247	231	218	206
Taylor KO Index	7.3	6.9	6.5	6.2	6.0	5.7	5.6	5.4	5.3
Mid-Range Trajectory Height • Inches	0.0	0.2	0.8	1.9	3.5	5.7	8.6	12.3	16.6
Drop • Inches	0.0	-0.7	-3.0	-7.1	-13.0	-21.1	-31.4	-44.1	-59.4

Speer 115-grain TMJ RN (53615)

G1 Ballistic Coefficient = 0.1?

Distance • Yards	Muzzle	25	50	75	100	125	150	175	200
Velocity • fps	1200	1145	1098	1058	1024	994	967	943	920
Energy • ft-lbs	368	335	308	286	268	252	239	227	218
Taylor KO Index	7.0	6.7	6.4	6.2	6.0	5.8	5.6	5.5	5.4
Mid-Range Trajectory Height • Inches	0.0	0.2	0.8	2.0	3.6	5.8	8.7	12.3	16.5
Drop • Inches	0.0	-0.8	-3.2	-7.4	-13.6	-21.9	-32.3	-45.1	-60.3

eer 115-grain TMJ RN Lawman (53615)

G1 Ballistic Coefficient = 0.151

Distance • Yards	Muzzle	25	50	75	100	125	150	175	200
Velocity • fps	1200	1137	1084	1040	1003	971	942	916	892
Energy • ft-lbs	368	330	300	276	257	241	227	214	203
Taylor KO Index	7.0	6.6	6.3	6.1	5.9	5.7	5.5	5.4	5.2
Mid-Range Trajectory Height • Inches	0.0	0.2	0.8	2.0	3.7	6.0	9.0	12.8	17.2
Drop • Inches	0.0	-0.8	-3.3	-7.6	-13.9	-22.4	-33.1	-46.4	-62.2

A (Winchester) 115-grain Brass Enclosed Base (WC91)

G1 Ballistic Coefficient = 0.171

Distance • Yards	Muzzle	25	50	75	100	125	150	175	200
Velocity • fps	1190	1135	1088	1048	1014	984	957	933	910
Energy • ft-lbs	362	329	302	281	262	247	234	222	212
Taylor KO Index	6.9	6.6	6.3	6.1	5.9	5.8	5.6	5.5	5.3
Mid-Range Trajectory Height • Inches	0.0	0.2	0.9	2.1	3.7	5.9	8.8	12.4	16.7
Drop • Inches	0.0	-0.8	-3.3	-7.6	-13.9	-22.2	-32.8	-45.8	-61.2

A (Winchester) 115-grain Full Metal Jacket (Q4172) + (USA9MMVP)

G1 Ballistic Coefficient = 0.143

Distance • Yards	Muzzle	25	50	75	100	125	150	175	200
Velocity • fps	1190	1125	1071	1027	990	958	929	903	878
Energy • ft-lbs	362	323	293	270	250	234	220	208	197
Taylor KO Index	6.9	6.6	6.2	6.0	5.8	5.6	5.4	5.3	5.1
Mid-Range Trajectory Height • Inches	0.0	0.2	0.9	2.0	3.8	6.1	9.1	12.9	17.5
Drop • Inches	0.0	-0.8	-3.3	-7.7	-14.2	-22.8	-33.8	-47.3	-63.4

nerican Eagle (Federal) 115-grain Full Metal Jacket (AE9DP)

G1 Ballistic Coefficient = 0.119

Distance • Yards	Muzzle	25	50	75	100	125	150	175	200
Velocity • fps	1180	1110	1050	1000	960	925	894	866	840
Energy • ft-lbs	355	310	280	255	235	219	204	192	180
Taylor KO Index	6.9	6.5	6.1	5.8	5.6	5.4	5.2	5.1	4.9
Mid-Range Trajectory Height • Inches	0.0	0.2	0.9	2.1	4.0	6.5	9.8	13.8	18.8
Drop • Inches	0.0	-0.8	-3.4	-8.0	-14.7	-23.8	-35.5	-49.8	-67.1

ick Hills 115-grain FMJ (D9NSCN1)

G1 Ballistic Coefficient = 0.140

Distance • Yards	Muzzle	25	50	75	100	125	150	175	200
Velocity • fps	1150	1091	1043	1003	968	938	910	885	861
Energy • ft-lbs	338	304	278	257	239	225	211	200	189
Taylor KO Index	6.7	6.4	6.1	5.8	5.6	5.5	5.3	5.2	5.0
Mid-Range Trajectory Height • Inches	0.0	0.2	0.9	2.2	4.0	6.5	9.7	13.7	18.5
Drop • Inches	0.0	-0.8	-3.5	-8.2	-15.0	-24.2	-35.8	-50.0	-67.1

ick Hills 115-grain Full Metal Jacket (D9N1)

G1 Ballistic Coefficient = 0.140

Distance • Yards	Muzzle	25	50	75	100	125	150	175	200
Velocity • fps	1150	1091	1042	1002	967	937	910	884	861
Energy • ft-lbs	336	304	278	256	239	224	211	200	189
Taylor KO Index	6.7	6.4	6.1	5.8	5.7	5.5	5.3	5.2	5.0
Mid-Range Trajectory Height • Inches	0.0	0.2	0.9	2.1	4.0	6.4	9.6	13.6	18.3
Drop • Inches	0.0	-0.8	-3.5	-8.2	-15.0	-24.1	-35.7	-49.9	-66.8

IC 115-grain Full Metal Jacket (9A)

G1 Ballistic Coefficient = 0.165

Distance • Yards	Muzzle	25	50	75	100	125	150	175	200
Velocity • fps	1150	1099	1057	1020	988	960	935	912	890
Energy • ft-lbs	338	309	285	266	250	236	223	212	202
Taylor KO Index	6.7	6.4	6.2	6.0	5.8	5.6	5.5	5.3	5.2
Mid-Range Trajectory Height • Inches	0.0	0.2	0.9	2.1	3.9	6.3	9.4	13.2	17.7
Drop • Inches	0.0	-0.9	-3.5	-8.1	-14.8	-23.6	-34.9	-48.6	-65.0

IC 115-grain EMJ NT (9EMA)

G1 Ballistic Coefficient = 0.165

Distance • Yards	Muzzle	25	50	75	100	125	150	175	200
Velocity • fps	1150	1099	1057	1020	988	960	935	912	890
Energy • ft-lbs	338	309	285	266	250	236	223	212	202
Taylor KO Index	6.7	6.4	6.2	6.0	5.8	5.6	5.5	5.3	5.2
Mid-Range Trajectory Height • Inches	0.0	0.2	0.9	2.1	3.9	6.3	9.4	13.2	17.7
Drop • Inches	0.0	-0.9	-3.5	-8.1	-14.8	-23.6	-34.9	-48.6	-65.0

Blazer (CCI) 115-grain FMJ RN (5200)

G1 Ballistic Coefficient = 0.

Distance • Yards	Muzzle	25	50	75	100	125	150	175	200
Velocity • fps	1145	1091	1046	1008	975	946	920	895	873
Energy • ft-lbs	335	304	278	259	247	229	216	205	195
Taylor KO Index	6.7	6.4	6.1	5.9	5.7	5.5	5.4	5.2	5.1
Mid-Range Trajectory Height • Inches	0.0	0.2	0.9	2.1	3.9	6.4	9.6	13.5	18.2
Drop • Inches	0.0	-0.9	-3.5	-8.2	-15.0	-24.1	-35.6	-49.7	-66.5

Blazer (CCI) 115-grain FMJ (3509)

G1 Ballistic Coefficient = 0.

Distance • Yards	Muzzle	25	50	75	100	125	150	175	200
Velocity • fps	1145	1089	1047	1005	971	942	915	891	868
Energy • ft-lbs	341	303	280	258	241	227	214	203	192
Taylor KO Index	6.7	6.4	6.1	5.9	5.7	5.5	5.4	5.2	5.1
Mid-Range Trajectory Height • Inches	0.0	0.2	0.9	2.1	3.9	6.4	9.6	13.5	18.2
Drop • Inches	0.0	-0.9	-3.5	-8.2	-15.1	-24.2	-35.7	-49.8	-66.6

Magtech 115-grain FEB (CR9A)

G1 Ballistic Coefficient = 0

Distance • Yards	Muzzle	25	50	75	100	125	150	175	200
Velocity • fps	1135	1084	1041	1005	973	945	919	896	874
Energy • ft-lbs	330	300	277	258	242	228	216	205	195
Taylor KO Index	6.6	6.3	6.1	5.9	5.7	5.5	5.4	5.2	5.1
Mid-Range Trajectory Height • Inches	0.0	0.2	0.9	2.2	4.0	6.5	9.7	13.5	18.3
Drop • Inches	0.0	-0.9	-3.6	-8.3	-15.2	-24.4	-36.0	-50.1	-67.0

Magtech 115-grain FMJ (9A –MP9A)

G1 Ballistic Coefficient = 0

Distance • Yards	Muzzle	25	50	75	100	125	150	175	200
Velocity • fps	1135	1079	1027	994	961	931	905	880	857
Energy • ft-lbs	330	297	270	253	235	222	209	198	188
Taylor KO Index	6.6	6.3	6.0	5.8	5.6	5.4	5.3	5.1	5.0
Mid-Range Trajectory Height • Inches	0.0	0.2	0.9	2.1	4.0	6.5	9.8	13.8	18.6
Drop • Inches	0.0	-0.9	-3.6	-8.4	-15.4	-24.6	-36.4	-50.8	-68.0

Remington 115-grain Metal Case (R9MM3)

G1 Ballistic Coefficient = 0

Distance • Yards	Muzzle	25	50	75	100	125	150	175	200
Velocity • fps	1135	1084	1041	1005	973	945	920	896	874
Energy • ft-lbs	329	300	277	258	242	228	216	205	195
Taylor KO Index	6.6	6.3	6.1	5.9	5.7	5.5	5.4	5.2	5.1
Mid-Range Trajectory Height • Inches	0.0	0.2	0.9	2.1	4.0	6.4	9.6	13.5	17.4
Drop • Inches	0.0	-0.9	-3.6	-8.3	-15.2	-24.3	-35.9	-50.0	-66.7

UMC (Remington) 115-grain Metal Case (L9MM3)

G1 Ballistic Coefficient = 0

Distance • Yards	Muzzle	25	50	75	100	125	150	175	200
Velocity • fps	1135	1084	1041	1005	973	945	920	896	874
Energy • ft-lbs	329	300	277	258	242	228	216	205	195
Taylor KO Index	6.6	6.3	6.1	5.9	5.7	5.5	5.4	5.2	5.1
Mid-Range Trajectory Height • Inches	0.0	0.2	0.9	2.1	4.0	6.4	9.6	13.5	17.4
Drop • Inches	0.0	-0.9	-3.6	-8.3	-15.2	-24.3	-35.9	-50.0	-66.7

UMC (Remington) 115-grain Flat Nose Enclosed Base (Leadless)Metal Case (LL9MM11)

G1 Ballistic Coefficient = 0

Distance • Yards	Muzzle	25	50	75	100	125	150	175	200
Velocity • fps	1135	1084	1041	1004	973	944	919	895	873
Energy • ft-lbs	329	300	277	258	242	228	216	205	195
Taylor KO Index	6.6	6.3	6.1	5.9	5.7	5.5	5.4	5.2	5.1
Mid-Range Trajectory Height • Inches	0.0	0.2	0.9	2.1	4.0	6.4	9.6	13.5	18.2
Drop • Inches	0.0	-0.9	-3.6	-8.3	-15.2	-24.3	-35.9	-50.0	-66.8

Lapua 116-grain FMJ (4319200)

G1 Ballistic Coefficient = 0

Distance • Yards	Muzzle	25	50	75	100	125	150	175	200
Velocity • fps	1198	1118	1056	1006	964	929	897	868	841
Energy • ft-lbs	300	322	287	261	240	222	207	194	182
Taylor KO Index	7.0	6.6	6.2	5.9	5.7	5.5	5.3	5.1	5.0
Mid-Range Trajectory Height • Inches	0.0	0.2	0.9	2.1	3.8	6.3	9.5	13.5	18.3
Drop • Inches	0.0	-0.8	-3.3	-7.8	-14.4	-23.3	-34.7	-48.7	-65.6

Aguila 117-grain JHP (1E092112)

G1 Ballistic Coefficient = 0.145

Distance • Yards	Muzzle	25	50	75	100	125	150	175	200
Velocity • fps	1250	1176	1115	1064	1022	986	954	926	901
Energy • ft-lbs	406	360	323	294	271	253	237	223	211
Taylor KO Index	7.4	7.0	6.6	6.3	6.1	5.9	5.7	5.5	5.4
Mid-Range Trajectory Height • Inches	0.0	0.2	0.7	1.8	3.5	5.7	8.6	12.2	16.6
Drop • Inches	0.0	-0.7	-3.0	-7.1	-13.0	-21.1	-31.3	-44.0	-59.2

Lapua 120-grain CEPP SUPER (4319175)

G1 Ballistic Coefficient = 0.129

Distance • Yards	Muzzle	25	50	75	100	125	150	175	200
Velocity • fps	1181	1111	1055	1010	971	938	909	881	856
Energy • ft-lbs	372	329	297	272	251	234	220	207	195
Taylor KO Index	7.2	6.8	6.4	6.1	5.9	5.7	5.5	5.4	5.2
Mid-Range Trajectory Height • Inches	0.0	0.2	0.9	2.1	3.9	6.3	9.5	13.4	18.1
Drop • Inches	0.0	-0.8	-3.4	-7.9	-14.6	-23.4	-34.7	-48.7	-65.4

Lapua 120-grain FMJ CEPP EXTRA Lead Free (4319178) LF

G1 Ballistic Coefficient = 0.129

Distance • Yards	Muzzle	25	50	75	100	125	150	175	200
Velocity • fps	1181	1111	1055	1010	971	938	909	881	856
Energy • ft-lbs	372	329	297	272	251	234	220	207	195
Taylor KO Index	7.2	6.8	6.4	6.1	5.9	5.7	5.5	5.4	5.2
Mid-Range Trajectory Height • Inches	0.0	0.2	0.9	2.1	3.9	6.3	9.5	13.4	18.1
Drop • Inches	0.0	-0.8	-3.4	-7.9	-14.6	-23.4	-34.7	-48.7	-65.4

Fiocchi 123-grain TC Enclosed Base, Leadless Primer (9TCEB123) LF

G1 Ballistic Coefficient = 0.148

Distance • Yards	Muzzle	25	50	75	100	125	150	175	200
Velocity • fps	1180	1119	1068	1026	990	959	931	905	882
Energy • ft-lbs	383	345	314	290	270	253	239	226	214
Taylor KO Index	7.4	7.1	6.7	6.5	6.2	6.0	5.9	5.7	5.6
Mid-Range Trajectory Height • Inches	0.0	0.2	0.9	2.1	3.8	6.2	9.3	13.0	17.7
Drop • Inches	0.0	-0.8	-3.4	-7.8	-14.3	-23.0	-34.1	-47.7	-63.9

Lapua 123-grain FMJ Combat (4319163)

G1 Ballistic Coefficient = 0.136

Distance • Yards	Muzzle	25	50	75	100	125	150	175	200
Velocity • fps	1165	1102	1050	1007	971	939	911	885	861
Energy • ft-lbs	371	331	301	277	258	241	227	214	202
Taylor KO Index	7.3	6.9	6.5	6.3	6.1	5.9	5.7	5.5	5.4
Mid-Range Trajectory Height • Inches	0.0	0.2	0.9	2.1	3.9	6.3	9.5	13.5	18.2
Drop • Inches	0.0	-0.8	-3.4	-8.0	-14.8	-23.8	-35.2	-49.2	-66.0

Lapua 123-grain FMJ (4319177)

G1 Ballistic Coefficient = 0.134

Distance • Yards	Muzzle	25	50	75	100	125	150	175	200
Velocity • fps	1050	1007	970	938	909	883	859	836	814
Energy • ft-lbs	301	277	257	240	226	213	201	191	181
Taylor KO Index	6.5	6.3	6.1	5.9	5.7	5.5	5.4	5.2	5.1
Mid-Range Trajectory Height • Inches	0.0	0.2	1.1	2.5	4.6	7.4	11.0	15.5	20.9
Drop • Inches	0.0	-1.0	-4.2	-9.6	-17.6	-28.1	-41.4	-57.6	-76.9

Sellier and Bellot 124-grain SP (V311592U)

G1 Ballistic Coefficient = 0.079

Distance • Yards	Muzzle	25	50	75	100	125	150	175	200
Velocity • fps	1165	1074	991	935	888	847	810	777	745
Energy • ft-lbs	372	316	269	241	217	198	181	166	153
Taylor KO Index	7.3	6.8	6.2	5.9	5.6	5.3	5.1	4.9	4.7
Mid-Range Trajectory Height • Inches	0.0	0.2	1.0	2.3	4.4	7.3	11.2	16.1	22.1
Drop • Inches	0.0	-0.9	-3.6	-8.6	-16.1	-26.3	-39.5	-56.1	-76.3

Magtech 124-grain JSP (9S)

G1 Ballistic Coefficient = 0.173

Distance • Yards	Muzzle	25	50	75	100	125	150	175	200
Velocity • fps	1109	1067	1030	999	971	946	923	901	881
Energy • ft-lbs	339	313	292	275	259	246	234	224	214
Taylor KO Index	7.0	6.7	6.5	6.3	6.1	6.0	5.8	5.7	5.6
Mid-Range Trajectory Height • Inches	0.0	0.2	1.0	2.2	4.1	6.6	9.8	13.7	18.4
Drop • Inches	0.0	-0.9	-3.7	-8.6	-15.6	-25.0	-36.7	-51.0	-68.0

Magtech 124-grain LRN (9E)

G1 Ballistic Coefficient = 0.173

Distance • Yards	Muzzle	25	50	75	100	125	150	175	200
Velocity • fps	1109	1067	1030	999	971	946	923	901	881
Energy • ft-lbs	339	313	292	275	259	246	234	224	214
Taylor KO Index	7.0	6.7	6.5	6.3	6.1	6.0	5.8	5.7	5.6
Mid-Range Trajectory Height • Inches	0.0	0.2	1.0	2.2	4.1	6.6	9.8	13.7	18.4
Drop • Inches	0.0	-0.9	-3.7	-8.6	-15.6	-25.0	-36.7	-51.0	-68.0

Fiocchi 124-grain Hornady XTP JHP (9XTPC)

G1 Ballistic Coefficient = 0.165

Distance • Yards	Muzzle	25	50	75	100	125	150	175	200
Velocity • fps	1290	1220	1159	1107	1063	1025	993	964	939
Energy • ft-lbs	458	410	370	337	311	290	272	256	243
Taylor KO Index	8.1	7.7	7.3	7.0	6.7	6.5	6.3	6.1	5.9
Mid-Range Trajectory Height • Inches	0.0	0.1	0.7	1.7	3.2	5.3	7.9	11.3	15.3
Drop • Inches	0.0	-0.7	-2.8	-6.6	-12.1	-19.6	-29.1	-40.8	-54.8

Black Hills 124-grain Jacketed Hollow Point +P (D9N9)

G1 Ballistic Coefficient = 0.180

Distance • Yards	Muzzle	25	50	75	100	125	150	175	200
Velocity • fps	1250	1190	1137	1092	1054	1020	991	965	941
Energy • ft-lbs	430	390	356	328	306	287	271	256	244
Taylor KO Index	7.9	7.5	7.2	6.9	6.6	6.4	6.2	6.1	5.9
Mid-Range Trajectory Height • Inches	0.0	0.2	0.8	1.8	3.4	5.4	8.2	11.5	15.5
Drop • Inches	0.0	-0.7	-3.0	-6.9	-12.6	-20.4	-30.1	-42.1	-56.3

Speer 124-grain GDHP +P (23617)

G1 Ballistic Coefficient = 0.135

Distance • Yards	Muzzle	25	50	75	100	125	150	175	200
Velocity • fps	1220	1146	1085	1037	996	962	931	903	877
Energy • ft-lbs	410	362	324	296	273	255	239	224	212
Taylor KO Index	7.7	7.2	6.8	6.5	6.3	6.1	5.9	5.7	5.5
Mid-Range Trajectory Height • Inches	0.0	0.2	0.8	2.0	3.7	6.0	9.1	12.9	17.4
Drop • Inches	0.0	-0.8	-3.2	-7.4	-13.7	-22.2	-33.0	-46.3	-62.3

Winchester 124-grain Bonded PDX1 +P (S9MMPDB)

G1 Ballistic Coefficient = 0.17

Distance • Yards	Muzzle	25	50	75	100	125	150	175	200
Velocity • fps	1200	1143	1095	1054	1019	988	961	936	913
Energy • ft-lbs	396	360	330	306	286	269	254	241	230
Taylor KO Index	7.6	7.2	6.9	6.6	6.4	6.2	6.1	5.9	5.8
Mid-Range Trajectory Height • Inches	0.0	0.2	0.8	1.9	3.6	5.9	8.8	12.4	16.7
Drop • Inches	0.0	-0.8	-3.2	-7.5	-13.7	-22.0	-32.5	-45.4	-60.7

Fiocchi 124-grain JHP (9APBHP)

G1 Ballistic Coefficient = 0.16

Distance • Yards	Muzzle	25	50	75	100	125	150	175	200
Velocity • fps	1180	1125	1077	1038	1005	975	948	924	901
Energy • ft-lbs	383	348	319	297	278	262	247	235	224
Taylor KO Index	7.4	7.1	6.8	6.5	6.3	6.1	6.0	5.8	5.7
Mid-Range Trajectory Height • Inches	0.0	0.2	0.8	2.0	3.7	6.0	9.0	12.7	17.2
Drop • Inches	0.0	-0.8	-3.3	-7.7	-14.1	-22.7	-33.5	-46.8	-62.6

Remington 124-grain Brass-Jacketed Hollow Point +P (GS9MMD)

G1 Ballistic Coefficient = 0.19

Distance • Yards	Muzzle	25	50	75	100	125	150	175	200
Velocity • fps	1180	1131	1089	1053	1021	994	968	946	924
Energy • ft-lbs	384	352	327	305	287	272	258	246	235
Taylor KO Index	7.4	7.1	6.8	6.6	6.4	6.3	6.1	6.0	5.8
Mid-Range Trajectory Height • Inches	0.0	0.2	0.8	2.0	3.8	5.8	8.7	12.3	16.5
Drop • Inches	0.0	-0.8	-3.3	-7.6	-13.9	-22.3	-32.8	-45.6	-60.9

Black Hills 124-grain Jacketed Hollow Point (D9N3)

G1 Ballistic Coefficient = 0.1

Distance • Yards	Muzzle	25	50	75	100	125	150	175	200
Velocity • fps	1150	1103	1063	1028	998	971	947	925	904
Energy • ft-lbs	363	335	311	291	274	260	247	235	225
Taylor KO Index	7.2	6.9	6.7	6.5	6.3	6.1	5.9	5.8	5.7
Mid-Range Trajectory Height • Inches	0.0	0.2	0.9	2.1	3.8	6.1	9.2	12.9	17.3
Drop • Inches	0.0	-0.8	-3.5	-8.0	-14.6	-23.4	-34.4	-47.9	-63.9

Speer 124-grain GDHP (23618)

G1 Ballistic Coefficient = 0.135

Distance • Yards	Muzzle	25	50	75	100	125	150	175	200
Velocity • fps	1150	1089	1039	999	963	932	904	879	855
Energy • ft-lbs	364	327	297	275	255	239	225	213	201
Taylor KO Index	7.3	6.9	6.6	6.3	6.1	5.9	5.7	5.5	5.4
Mid-Range Trajectory Height • Inches	0.0	0.2	0.9	2.2	4.0	6.5	9.8	13.8	18.7
Drop • Inches	0.0	-0.8	-3.5	-8.2	-15.1	-24.3	-36.0	-50.4	-67.6

Speer 124-grain GDHP-SB +P (23611) (Short Barrel)

G1 Ballistic Coefficient = 0.135

Distance • Yards	Muzzle	25	50	75	100	125	150	175	200
Velocity • fps	1150	1089	1039	999	963	932	904	879	855
Energy • ft-lbs	364	327	297	275	255	239	225	213	201
Taylor KO Index	7.3	6.9	6.6	6.3	6.1	5.9	5.7	5.5	5.4
Mid-Range Trajectory Height • Inches	0.0	0.2	0.9	2.2	4.0	6.5	9.8	13.8	18.7
Drop • Inches	0.0	-0.8	-3.5	-8.2	-15.1	-24.3	-36.0	-50.4	-67.6

Remington 124-grain Brass Jacketed Hollow Point (GS9MMB)

G1 Ballistic Coefficient = 0.150

Distance • Yards	Muzzle	25	50	75	100	125	150	175	200
Velocity • fps	1125	1074	1031	995	963	935	910	886	864
Energy • ft-lbs	349	318	293	273	255	241	228	216	206
Taylor KO Index	7.1	6.8	6.5	6.3	6.1	5.9	5.7	5.6	5.4
Mid-Range Trajectory Height • Inches	0.0	0.2	1.0	2.2	4.0	6.5	9.8	13.8	18.5
Drop • Inches	0.0	-0.9	-3.7	-8.5	-15.5	-24.8	-36.6	-50.9	-68.0

Remington 124-grain Brass Jacketed Hollow Point (GS9MMB)

G1 Ballistic Coefficient = 0.150

Distance • Yards	Muzzle	25	50	75	100	125	150	175	200
Velocity • fps	1125	1074	1031	995	963	935	910	886	864
Energy • ft-lbs	349	318	293	273	255	241	228	216	206
Taylor KO Index	7.1	6.8	6.5	6.3	6.1	5.9	5.7	5.6	5.4
Mid-Range Trajectory Height • Inches	0.0	0.2	1.0	2.2	4.0	6.5	9.8	13.8	18.5
Drop • Inches	0.0	-0.9	-3.7	-8.5	-15.5	-24.8	-36.6	-50.9	-68.0

Federal 124-grain Hydra-Shok JHP (P9HS1)

G1 Ballistic Coefficient = 0.149

Distance • Yards	Muzzle	25	50	75	100	125	150	175	200
Velocity • fps	1120	1070	1030	990	960	932	906	883	861
Energy • ft-lbs	345	315	290	270	255	239	226	215	204
Taylor KO Index	7.0	6.7	6.5	6.2	6.0	5.9	5.7	5.6	5.4
Mid-Range Trajectory Height • Inches	0.0	0.2	0.9	2.2	4.1	6.6	9.8	13.9	18.7
Drop • Inches	0.0	-0.9	-3.7	-8.5	-15.6	-25.0	-36.8	-51.3	-68.6

Hornady 124-grain JHP/XTP (90242)

G1 Ballistic Coefficient = 0.173

Distance • Yards	Muzzle	25	50	75	100	125	150	175	200
Velocity • fps	1110	1067	1030	999	971	946	923	901	881
Energy • ft-lbs	339	314	292	275	260	246	235	224	214
Taylor KO Index	7.0	6.7	6.5	6.3	6.1	6.0	5.8	5.7	5.6
Mid-Range Trajectory Height • Inches	0.0	0.2	1.0	2.2	4.1	6.5	9.7	13.6	18.3
Drop • Inches	0.0	-0.9	-3.7	-8.6	-15.6	-24.9	-36.6	-50.9	-67.8

Magtech 124-grain JHP (GG9B)

G1 Ballistic Coefficient = 0.165

Distance • Yards	Muzzle	25	50	75	100	125	150	175	200
Velocity • fps	1096	1054	1017	986	958	933	910	888	868
Energy • ft-lbs	331	306	285	268	253	240	228	217	208
Taylor KO Index	6.9	6.6	6.4	6.2	6.0	5.9	5.7	5.6	5.5
Mid-Range Trajectory Height • Inches	0.0	0.2	1.0	2.3	4.2	6.8	10.0	14.1	18.9
Drop • Inches	0.0	-0.9	-3.9	-8.8	-16.0	-25.6	-37.7	-52.4	-69.8

PMC 124-grain Starfire Hollow Point (9SFB)

G1 Ballistic Coefficient = 0.143

Distance • Yards	Muzzle	25	50	75	100	125	150	175	200
Velocity • fps	1090	1043	1003	969	939	912	887	864	842
Energy • ft-lbs	327	299	277	259	243	229	217	206	195
Taylor KO Index	6.9	6.6	6.3	6.1	5.9	5.8	5.6	5.4	5.3
Mid-Range Trajectory Height • Inches	0.0	0.2	1.0	2.3	4.3	6.9	10.3	14.5	19.5
Drop • Inches	0.0	-0.9	-3.9	-9.0	-16.4	-26.2	-38.7	-53.8	-71.9

Fiocchi 124-grain FMJ Truncated Cone (9APC)

G1 Ballistic Coefficient = 0.148

Distance • Yards	Muzzle	25	50	75	100	125	150	175	200
Velocity • fps	1200	1136	1082	1038	1000	968	939	913	888
Energy • ft-lbs	396	355	322	297	275	258	243	229	217
Taylor KO Index	7.6	7.2	6.8	6.5	6.3	6.1	5.9	5.8	5.6
Mid-Range Trajectory Height • Inches	0.0	0.2	0.9	2.0	3.7	6.0	9.1	12.7	17.3
Drop • Inches	0.0	-0.8	-3.3	-7.6	-13.9	-22.4	-33.2	-46.5	-62.4

Fiocchi 124-grain Metal Case (9APB)

G1 Ballistic Coefficient = 0.148

Distance • Yards	Muzzle	25	50	75	100	125	150	175	200
Velocity • fps	1180	1119	1068	1026	990	959	931	905	882
Energy • ft-lbs	383	345	314	290	270	253	239	226	214
Taylor KO Index	7.4	7.1	6.7	6.5	6.2	6.0	5.9	5.7	5.6
Mid-Range Trajectory Height • Inches	0.0	0.2	0.9	2.1	3.8	6.2	9.3	13.0	17.7
Drop • Inches	0.0	-0.8	-3.4	-7.8	-14.3	-23.0	-34.1	-47.7	-63.9

American Eagle (Federal) 124-grain Full Metal Jacket (AE9AP)

G1 Ballistic Coefficient = 0.165

Distance • Yards	Muzzle	25	50	75	100	125	150	175	200
Velocity • fps	1150	1100	1050	1010	980	981	925	901	878
Energy • ft-lbs	365	330	305	280	265	249	235	223	212
Taylor KO Index	7.3	6.9	6.6	6.4	6.2	6.0	5.8	5.7	5.5
Mid-Range Trajectory Height • Inches	0.0	0.2	0.9	2.1	3.9	6.4	9.5	13.4	18.1
Drop • Inches	0.0	-0.8	-3.5	-8.1	-14.9	-23.9	-35.3	-49.2	-65.8

USA (Winchester) 124-grain Full Metal Jacket (USA9MM) + (Q4318)

G1 Ballistic Coefficient = 0.20?

Distance • Yards	Muzzle	25	50	75	100	125	150	175	200
Velocity • fps	1140	1091	1050	1015	984	957	932	909	888
Energy • ft-lbs	358	328	303	284	267	252	239	228	217
Taylor KO Index	7.2	6.9	6.6	6.4	6.2	6.0	5.9	5.7	5.6
Mid-Range Trajectory Height • Inches	0.0	0.2	0.9	2.1	3.9	6.3	9.4	13.3	17.9
Drop • Inches	0.0	-0.9	-3.6	-8.2	-15.0	-24.0	-35.3	-49.2	-65.7

USA (Winchester) 124-grain FMJ (USA9MM) + (Q4318)

G1 Ballistic Coefficient = 0.16?

Distance • Yards	Muzzle	25	50	75	100	125	150	175	200
Velocity • fps	1140	1091	1050	1015	984	957	932	821	802
Energy • ft-lbs	311	291	275	258	244	232	220	210	200
Taylor KO Index	7.2	6.9	6.6	6.4	6.2	6.0	5.9	5.7	5.6
Mid-Range Trajectory Height • Inches	0.0	0.2	0.9	2.1	3.9	6.3	9.4	13.2	17.8
Drop • Inches	0.0	-0.9	-3.5	-8.2	-15.0	-23.9	-35.3	-49.1	-65.5

USA (Winchester) 124-grain Brass Enclosed Base (WC92)

G1 Ballistic Coefficient = 0.18?

Distance • Yards	Muzzle	25	50	75	100	125	150	175	200
Velocity • fps	1130	1087	1049	1017	988	963	939	918	898
Energy • ft-lbs	352	325	303	285	269	255	243	232	222
Taylor KO Index	7.1	6.9	6.6	6.4	6.2	6.1	5.9	5.8	5.7
Mid-Range Trajectory Height • Inches	0.0	0.2	0.9	2.1	3.9	6.4	9.4	13.3	17.8
Drop • Inches	0.0	-0.9	-3.6	-8.3	-15.1	-24.1	-35.4	-49.2	-65.6

American Eagle (Federal) 124-grain TMJ (53615)

G1 Ballistic Coefficient = 0.1?

Distance • Yards	Muzzle	25	50	75	100	125	150	175	200
Velocity • fps	1120	1070	1030	990	960	932	907	883	861
Energy • ft-lbs	345	315	290	270	255	239	226	215	204
Taylor KO Index	7.0	6.7	6.5	6.2	6.0	5.9	5.7	5.6	5.4
Mid-Range Trajectory Height • Inches	0.0	0.2	1.0	2.2	4.1	6.7	9.9	14.0	18.8
Drop • Inches	0.0	-0.9	-3.7	-8.5	-15.6	-25.0	-36.9	-51.5	-68.8

American Eagle (Federal) 124-grain Full Metal Jacket (AE9N1)

G1 Ballistic Coefficient = 0.1?

Distance • Yards	Muzzle	25	50	75	100	125	150	175	200
Velocity • fps	1120	1070	1030	990	960	932	906	883	961
Energy • ft-lbs	345	315	290	270	255	239	226	215	204
Taylor KO Index	7.0	6.7	6.5	6.2	6.0	5.9	5.7	5.6	5.4
Mid-Range Trajectory Height • Inches	0.0	0.2	0.9	2.2	4.1	6.6	9.8	13.9	18.7
Drop • Inches	0.0	-0.9	-3.7	-8.5	-15.6	-25.0	-36.8	-51.3	-68.6

Aguila 124-grain FMJ (1E092110)

G1 Ballistic Coefficient = 0.168

Distance • Yards	Muzzle	25	50	75	100	125	150	175	200
Velocity • fps	1115	1071	1033	1000	971	945	922	900	879
Energy • ft-lbs	342	316	294	275	260	246	234	223	213
Taylor KO Index	7.0	6.8	6.5	6.3	6.1	6.0	5.8	5.7	5.5
Mid-Range Trajectory Height • Inches	0.0	0.2	1.0	2.2	4.1	6.6	9.8	13.7	18.4
Drop • Inches	0.0	-0.9	-3.7	-8.5	-15.5	-24.8	-36.6	-50.8	-67.8

Hornady 124-grain TAP-FPD (90248)

G1 Ballistic Coefficient = 0.173

Distance • Yards	Muzzle	25	50	75	100	125	150	175	200
Velocity • fps	1110	1067	1030	999	971	946	923	901	881
Energy • ft-lbs	339	314	292	275	260	246	235	224	214
Taylor KO Index	7.0	6.7	6.5	6.3	6.1	6.0	5.8	5.7	5.6
Mid-Range Trajectory Height • Inches	0.0	0.2	1.0	2.2	4.1	6.5	9.7	13.6	18.3
Drop • Inches	0.0	-0.9	-3.7	-8.6	-15.6	-24.9	-36.6	-50.9	67.8

PMC 124-grain EMJ NT (9EMG)

G1 Ballistic Coefficient = 0.142

Distance • Yards	Muzzle	25	50	75	100	125	150	175	200
Velocity • fps	1110	1059	1017	980	949	921	895	871	849
Energy • ft-lbs	339	309	285	265	248	234	221	209	199
Taylor KO Index	7.0	6.7	6.4	6.2	6.0	5.8	5.6	5.5	5.4
Mid-Range Trajectory Height • Inches	0.0	0.2	1.0	2.3	4.2	6.8	10.1	14.3	19.3
Drop • Inches	0.0	-0.9	-3.8	-8.7	-15.9	-25.5	-37.7	-52.6	-70.3

PMC 124-grain Full Metal Jacket (9G)

G1 Ballistic Coefficient = 0.142

Distance • Yards	Muzzle	25	50	75	100	125	150	175	200
Velocity • fps	1110	1059	1017	980	949	921	895	871	849
Energy • ft-lbs	339	309	285	265	248	234	221	209	199
Taylor KO Index	7.0	6.7	6.4	6.2	6.0	5.8	5.6	5.5	5.4
Mid-Range Trajectory Height • Inches	0.0	0.2	1.0	2.3	4.2	6.8	10.1	14.3	19.3
Drop • Inches	0.0	-0.9	-3.8	-8.7	-15.9	-25.5	-37.7	-52.6	-70.3

Remington 124-grain Metal Case (R9MM2)

G1 Ballistic Coefficient = 0.173

Distance • Yards	Muzzle	25	50	75	100	125	150	175	200
Velocity • fps	1110	1067	1030	999	971	946	923	901	881
Energy • ft-lbs	339	314	292	275	260	246	235	224	214
Taylor KO Index	7.0	6.7	6.5	6.3	6.1	6.0	5.8	5.7	5.6
Mid-Range Trajectory Height • Inches	0.0	0.2	1.0	2.2	4.1	6.5	9.7	13.6	18.3
Drop • Inches	0.0	-0.9	-3.7	-8.6	-15.6	-24.9	-36.6	-50.9	-67.8

Magtech 124-grain FMC (9B)

G1 Ballistic Coefficient = 0.174

Distance • Yards	Muzzle	25	50	75	100	125	150	175	200
Velocity • fps	1109	1067	1030	999	971	946	923	901	881
Energy • ft-lbs	339	314	292	275	260	246	235	224	214
Taylor KO Index	7.0	6.7	6.5	6.3	6.1	6.0	5.8	5.7	5.6
Mid-Range Trajectory Height • Inches	0.0	0.2	1.0	2.2	4.1	6.5	9.7	13.6	18.3
Drop • Inches	0.0	-0.9	-3.7	-8.6	-15.6	-24.9	-36.6	-50.9	-67.8

Magtech 124-grain FEB (CR9B)

G1 Ballistic Coefficient = 0.173

Distance • Yards	Muzzle	25	50	75	100	125	150	175	200
Velocity • fps	1109	1067	1030	999	971	946	923	901	881
Energy • ft-lbs	339	313	292	275	260	246	234	224	214
Taylor KO Index	7.0	6.7	6.5	6.3	6.1	6.0	5.8	5.7	5.6
Mid-Range Trajectory Height • Inches	0.0	0.2	1.0	2.2	4.1	6.6	9.8	13.6	18.4
Drop • Inches	0.0	-0.9	-3.7	-8.6	-15.6	-25.0	-36.7	-51.0	-68.0

MC (Remington) 124-grain Metal Case (L9MM2)

G1 Ballistic Coefficient = 0.183

Distance • Yards	Muzzle	25	50	75	100	125	150	175	200
Velocity • fps	1100	1061	1030	998	971	947	925	905	886
Energy • ft-lbs	339	310	292	274	259	247	236	225	216
Taylor KO Index	6.9	6.7	6.5	6.3	6.1	6.0	5.8	5.7	5.6
Mid-Range Trajectory Height • Inches	0.0	0.2	1.0	2.2	4.1	6.6	9.7	13.7	18.3
Drop • Inches	0.0	-0.9	-3.8	-8.7	-15.8	25.1	36.9	-51.2	-68.2

387

UMC (Remington) 124-grain FMJ (LL9MM2)

G1 Ballistic Coefficient = 0.182

Distance • Yards	Muzzle	25	50	75	100	125	150	175	200
Velocity • fps	1100	1061	1030	997	971	947	925	904	885
Energy • ft-lbs	333	310	290	274	260	247	236	225	216
Taylor KO Index	6.9	6.7	6.5	6.3	6.1	6.0	5.8	5.7	5.6
Mid-Range Trajectory Height • Inches	0.0	0.2	0.9	2.2	4.1	6.6	9.8	13.6	18.4
Drop • Inches	0.0	-0.9	-3.8	-8.7	-15.8	-25.2	-37.0	-51.3	-68.4

Blazer (CCI) 124-grain FMJ RN (5201)

G1 Ballistic Coefficient = 0.165

Distance • Yards	Muzzle	25	50	75	100	125	150	175	200
Velocity • fps	1090	1049	1014	982	955	930	907	886	866
Energy • ft-lbs	327	303	283	266	251	238	227	216	206
Taylor KO Index	6.9	6.6	6.4	6.2	6.0	5.9	5.7	5.6	5.5
Mid-Range Trajectory Height • Inches	0.0	0.2	1.0	2.3	4.2	6.8	10.1	14.2	19.0
Drop • Inches	0.0	-0.9	-3.8	-8.9	-16.2	-25.8	-38.0	-52.8	-70.4

Blazer (CCI) 124-grain TMJ (3560)

G1 Ballistic Coefficient = 0.165

Distance • Yards	Muzzle	25	50	75	100	125	150	175	200
Velocity • fps	1090	1049	1014	982	955	930	907	886	866
Energy • ft-lbs	327	303	283	266	251	238	227	216	206
Taylor KO Index	6.9	6.6	6.4	6.2	6.0	5.9	5.7	5.6	5.5
Mid-Range Trajectory Height • Inches	0.0	0.2	1.0	2.3	4.2	6.8	10.1	14.2	19.0
Drop • Inches	0.0	-0.9	-3.8	-8.9	-16.2	-25.8	-38.0	-52.8	-70.4

Speer Clean-Fire 124-grain – TMJ RN Lawman (53824) LF

G1 Ballistic Coefficient = 0.18?

Distance • Yards	Muzzle	25	50	75	100	125	150	175	200
Velocity • fps	1090	1052	1019	991	965	941	920	900	881
Energy • ft-lbs	327	305	286	270	256	244	233	223	214
Taylor KO Index	6.9	6.6	6.4	6.2	6.1	5.9	5.8	5.7	5.6
Mid-Range Trajectory Height • Inches	0.0	0.2	1.0	2.3	4.2	7.0	10.0	13.7	18.6
Drop • Inches	0.0	-0.9	-3.8	-8.8	-16.0	-25.6	-37.5	-52.1	-69.3

Speer 124-grain – TMJ RN Lawman (53616)

G1 Ballistic Coefficient = 0.18?

Distance • Yards	Muzzle	25	50	75	100	125	150	175	200
Velocity • fps	1090	1052	1019	991	965	941	920	900	881
Energy • ft-lbs	327	305	286	270	256	244	233	223	214
Taylor KO Index	6.9	6.6	6.4	6.2	6.1	5.9	5.8	5.7	5.6
Mid-Range Trajectory Height • Inches	0.0	0.2	1.0	2.3	4.2	7.0	10.0	13.7	18.6
Drop • Inches	0.0	-0.9	-3.8	-8.8	-16.0	-25.6	-37.5	-52.1	-69.3

Blazer (CCI) 124-grain FMJ (3578)

G1 Ballistic Coefficient = 0.16?

Distance • Yards	Muzzle	25	50	75	100	125	150	175	200
Velocity • fps	1090	1049	1014	982	955	930	907	886	866
Energy • ft-lbs	327	303	283	266	251	238	227	216	206
Taylor KO Index	6.9	6.6	6.4	6.2	6.0	5.9	5.7	5.6	5.5
Mid-Range Trajectory Height • Inches	0.0	0.2	1.0	2.3	4.2	6.8	10.1	14.2	19.0
Drop • Inches	0.0	-0.9	-3.8	-8.9	-16.2	-25.8	-38.0	-52.8	-70.4

Sellier and Bellot 124-grain FMJ (V310492U)

G1 Ballistic Coefficient = 0.0?

Distance • Yards	Muzzle	25	50	75	100	125	150	175	200
Velocity • fps	1181	1087	1001	944	896	855	818	784	752
Energy • ft-lbs	382	324	275	245	221	201	184	169	156
Taylor KO Index	7.4	6.9	6.3	6.0	5.7	5.4	5.2	4.9	4.7
Mid-Range Trajectory Height • Inches	0.0	0.2	0.9	2.3	4.3	7.2	11.0	15.8	21.7
Drop • Inches	0.0	-0.8	-3.5	-8.4	-15.7	-25.7	-38.7	-55.0	-74.7

Cor-Bon 125-grain JHP +P (SD09125/20)

G1 Ballistic Coefficient = 0.1?

Distance • Yards	Muzzle	25	50	75	100	125	150	175	200
Velocity • fps	1250	1174	1110	1058	1016	979	947	919	893
Energy • ft-lbs	434	383	342	311	286	266	249	234	221
Taylor KO Index	7.9	7.5	7.1	6.7	6.5	6.2	6.0	5.8	5.7
Mid-Range Trajectory Height • Inches	0.0	0.1	0.7	1.8	3.5	5.7	8.6	12.3	16.7
Drop • Inches	0.0	-0.7	-3.0	-7.1	-13.1	-21.2	-31.6	-44.4	-59.7

Hornady 125-grain HAP (90275)

G1 Ballistic Coefficient = 0.182

Distance • Yards	Muzzle	25	50	75	100	125	150	175	200
Velocity • fps	1100	1061	1030	997	971	947	925	904	885
Energy • ft-lbs	339	312	292	276	259	249	237	227	217
Taylor KO Index	7.5	7.2	6.9	6.6	6.4	6.2	6.0	5.9	5.7
Mid-Range Trajectory Height • Inches	0.0	0.2	1.0	2.3	4.1	6.6	9.8	13.8	18.4
Drop • Inches	0.0	-0.9	-3.8	-8.7	-15.8	-25.2	-37.0	-51.3	-68.4

Federal 135-grain Hydra-Shok JHP (PD9HS5 H)

G1 Ballistic Coefficient = 0.192

Distance • Yards	Muzzle	25	50	75	100	125	150	175	200
Velocity • fps	1060	1030	1000	970	950	929	909	890	872
Energy • ft-lbs	335	315	300	280	270	259	248	238	228
Taylor KO Index	7.3	7.1	6.9	6.7	6.5	6.4	6.2	6.1	6.0
Mid-Range Trajectory Height • Inches	0.0	0.2	1.0	2.4	4.4	7.0	10.3	14.2	19.2
Drop • Inches	0.0	-1.0	-4.0	-9.2	-16.8	-26.7	-39.1	-54.2	-72.0

Winchester 147-grain Silvertip Hollow Point (X9MMST147)

G1 Ballistic Coefficient = 0.201

Distance • Yards	Muzzle	25	50	75	100	125	150	175	200
Velocity • fps	1010	985	962	940	921	902	885	869	853
Energy • ft-lbs	333	317	302	289	277	266	256	246	237
Taylor KO Index	7.5	7.3	7.2	7.0	6.9	6.7	6.6	6.5	6.4
Mid-Range Trajectory Height • Inches	0.0	0.3	1.1	2.6	4.7	7.4	11.0	15.3	20.4
Drop • Inches	0.0	-1.1	-4.4	-10.1	-18.2	-28.9	-42.2	-58.3	-77.3

Fiocchi 147-grain Jacketed Hollowpoint (9APDHP)

G1 Ballistic Coefficient = 0.158

Distance • Yards	Muzzle	25	50	75	100	125	150	175	200
Velocity • fps	1005	974	946	920	897	875	855	836	817
Energy • ft-lbs	329	310	292	277	262	250	239	228	218
Taylor KO Index	7.5	7.3	7.1	6.9	6.7	6.5	6.4	6.2	6.1
Mid-Range Trajectory Height • Inches	0.0	0.3	1.2	2.7	4.9	7.8	11.6	15.9	21.6
Drop • Inches	0.0	-1.2	-4.5	-10.3	-18.7	-29.8	-43.7	-60.6	-80.6

Federal 147-grain Hydra-Shok JHP (P9HS2)

G1 Ballistic Coefficient = 0.192

Distance • Yards	Muzzle	25	50	75	100	125	150	175	200
Velocity • fps	1000	980	950	930	910	892	874	857	841
Energy • ft-lbs	325	310	295	285	275	260	249	240	231
Taylor KO Index	7.5	7.3	7.1	7.0	6.8	6.7	6.5	6.4	6.3
Mid-Range Trajectory Height • Inches	0.0	0.3	1.2	2.6	4.8	7.7	11.3	15.5	21.0
Drop • Inches	0.0	-1.1	-4.5	-10.3	-18.6	-29.5	-43.2	-59.7	-79.1

Winchester 147-grain Bonded PDX1 (S9MMPDB1)

G1 Ballistic Coefficient = 0.205

Distance • Yards	Muzzle	25	50	75	100	125	150	175	200
Velocity • fps	1000	976	954	934	915	897	881	865	849
Energy • ft-lbs	326	311	297	285	273	263	253	244	236
Taylor KO Index	7.5	7.3	7.1	7.0	6.8	6.7	6.6	6.4	6.3
Mid-Range Trajectory Height • Inches	0.0	0.3	1.1	2.6	4.8	7.6	11.2	15.6	20.8
Drop • Inches	0.0	-1.1	-4.5	-10.3	-18.5	-29.4	-42.9	-59.3	-78.5

Magtech 147-grain JHP Subsonic (9K)

G1 Ballistic Coefficient = 0.204

Distance • Yards	Muzzle	25	50	75	100	125	150	175	200
Velocity • fps	990	967	945	926	907	890	873	858	843
Energy • ft-lbs	320	305	292	280	268	259	249	240	232
Taylor KO Index	7.4	7.2	7.0	6.9	6.8	6.7	6.5	6.4	6.3
Mid-Range Trajectory Height • Inches	0.0	0.3	1.2	2.7	4.8	7.7	11.3	15.8	21.0
Drop • Inches	0.0	-1.1	-4.6	-10.5	-18.9	-29.9	-43.7	-60.3	-79.9

Remington 147-grain Jacketed Hollow Point (Subsonic) (R9MM8)

G1 Ballistic Coefficient = 0.184

Distance • Yards	Muzzle	25	50	75	100	125	150	175	200
Velocity • fps	990	964	941	920	900	881	863	846	830
Energy • ft-lbs	320	304	289	276	264	253	243	234	225
Taylor KO Index	7.4	7.2	7.0	6.9	6.7	6.6	6.5	6.3	6.2
Mid-Range Trajectory Height • Inches	0.0	0.3	1.1	2.7	4.9	7.8	11.5	16.0	21.9
Drop • Inches	0.0	-1.1	-4.6	-10.5	-19.0	-30.1	-44.1	-61.0	-80.8

Remington 147-grain Brass-Jacketed Hollow Point (GS9MMC)

G1 Ballistic Coefficient = 0.184

Distance • Yards	Muzzle	25	50	75	100	125	150	175	200
Velocity • fps	990	964	941	920	900	881	863	846	830
Energy • ft-lbs	320	304	289	276	264	253	243	234	225
Taylor KO Index	7.4	7.2	7.0	6.9	6.7	6.6	6.5	6.3	6.2
Mid-Range Trajectory Height • Inches	0.0	0.3	1.1	2.7	4.9	7.8	11.5	16.0	21.9
Drop • Inches	0.0	-1.1	-4.6	-10.5	-19.0	-30.1	-44.1	-61.0	-80.8

Winchester 147-grain SXT (S9)

G1 Ballistic Coefficient = 0.210

Distance • Yards	Muzzle	25	50	75	100	125	150	175	200
Velocity • fps	990	967	947	927	909	892	876	861	846
Energy • ft-lbs	320	306	293	281	270	260	251	242	234
Taylor KO Index	7.4	7.2	7.1	6.9	6.8	6.7	6.5	6.4	6.3
Mid-Range Trajectory Height • Inches	0.0	0.3	1.2	2.7	4.8	7.7	11.3	15.7	20.9
Drop • Inches	0.0	-1.1	-4.6	-10.4	-18.8	-29.8	-43.6	-60.1	-79.6

USA (Winchester) 147-grain Jacketed Hollow Point (USA9JHP2)

G1 Ballistic Coefficient = 0.200

Distance • Yards	Muzzle	25	50	75	100	125	150	175	200
Velocity • fps	990	966	945	925	906	888	872	856	840
Energy • ft-lbs	320	305	291	279	268	258	248	239	231
Taylor KO Index	7.4	7.2	7.1	6.9	6.8	6.6	6.5	6.4	6.3
Mid-Range Trajectory Height • Inches	0.0	0.3	1.2	2.7	4.9	7.8	11.4	15.9	21.2
Drop • Inches	0.0	-1.1	-4.6	-10.5	-18.9	-30.0	-43.8	-60.5	-80.1

USA (Winchester) 147-grain Brass Enclosed Base (WC93)

G1 Ballistic Coefficient = 0.200

Distance • Yards	Muzzle	25	50	75	100	125	150	175	200
Velocity • fps	990	966	945	925	906	888	872	856	840
Energy • ft-lbs	320	305	291	279	268	258	248	239	231
Taylor KO Index	7.4	7.2	7.1	6.9	6.8	6.6	6.5	6.4	6.3
Mid-Range Trajectory Height • Inches	0.0	0.3	1.2	2.7	4.9	7.8	11.4	15.9	21.2
Drop • Inches	0.0	-1.1	-4.6	-10.5	-18.9	-30.0	-43.8	-60.5	-80.1

Speer 147-grain GDHP (23619)

G1 Ballistic Coefficient = 0.16

Distance • Yards	Muzzle	25	50	75	100	125	150	175	200
Velocity • fps	985	957	932	909	887	867	848	830	813
Energy • ft-lbs	317	299	283	270	257	246	235	225	216
Taylor KO Index	7.4	7.2	7.0	6.8	6.6	6.5	6.3	6.2	6.1
Mid-Range Trajectory Height • Inches	0.0	0.3	1.2	2.7	5.0	8.0	11.8	16.5	22.1
Drop • Inches	0.0	-1.1	-4.6	-10.7	-19.3	-30.7	-45.0	-62.3	-82.7

Fiocchi 147-grain XTPHP (9XTPB)

G1 Ballistic Coefficient = 0.21

Distance • Yards	Muzzle	25	50	75	100	125	150	175	200
Velocity • fps	980	959	938	920	903	886	870	855	841
Energy • ft-lbs	313	300	287	276	266	256	247	239	231
Taylor KO Index	7.3	7.2	7.0	6.9	6.8	6.6	6.5	6.4	6.3
Mid-Range Trajectory Height • Inches	0.0	0.3	1.2	2.7	5.0	7.9	11.6	15.8	21.3
Drop • Inches	0.0	-1.2	-4.7	-10.6	-19.2	-30.4	-44.4	-61.2	-81.0

Hornady 147-grain XTP (90282)

G1 Ballistic Coefficient = 0.21

Distance • Yards	Muzzle	25	50	75	100	125	150	175	200
Velocity • fps	975	954	935	916	899	883	867	853	838
Energy • ft-lbs	310	297	285	274	264	255	246	237	229
Taylor KO Index	7.3	7.1	7.0	6.8	6.7	6.6	6.5	6.4	6.2
Mid-Range Trajectory Height • Inches	0.0	0.3	1.2	2.8	5.0	7.9	11.7	16.2	21.5
Drop • Inches	0.0	-1.2	-4.7	-10.7	-19.4	-30.7	-44.8	-61.7	-81.7

American Eagle (Federal) 147-grain Full Metal Jacket Flat Point (AE9FP)

G1 Ballistic Coefficient = 0.1

Distance • Yards	Muzzle	25	50	75	100	125	150	175	200
Velocity • fps	1000	980	950	930	910	892	874	857	841
Energy • ft-lbs	325	310	295	285	275	260	249	240	231
Taylor KO Index	7.5	7.3	7.1	7.0	6.8	6.7	6.5	6.4	6.3
Mid-Range Trajectory Height • Inches	0.0	0.3	1.2	2.6	4.8	7.7	11.3	15.5	21.0
Drop • Inches	0.0	-1.1	-4.5	-10.3	-18.6	-29.5	-43.2	-59.7	-79.1

American Eagle (Federal) 147-grain TMJ (AE9N2)

G1 Ballistic Coefficient = 0.192

Distance • Yards	Muzzle	25	50	75	100	125	150	175	200
Velocity • fps	1000	980	950	930	910	892	874	857	841
Energy • ft-lbs	325	310	295	285	275	260	249	240	231
Taylor KO Index	7.5	7.3	7.1	7.0	6.8	6.7	6.5	6.4	6.3
Mid-Range Trajectory Height • Inches	0.0	0.3	1.2	2.6	4.8	7.7	11.3	15.5	21.0
Drop • Inches	0.0	-1.1	-4.5	-10.3	-18.6	-29.5	-43.2	-59.7	-79.1

Fiocchi 147-grain Metal Case (9APD)

G1 Ballistic Coefficient = 0.165

Distance • Yards	Muzzle	25	50	75	100	125	150	175	200
Velocity • fps	1000	971	944	920	898	877	858	839	821
Energy • ft-lbs	326	308	297	276	263	251	240	230	220
Taylor KO Index	7.5	7.3	7.1	6.9	6.7	6.6	6.4	6.3	6.1
Mid-Range Trajectory Height • Inches	0.0	0.3	1.2	2.7	4.9	7.8	11.6	15.9	21.6
Drop • Inches	0.0	-1.1	-4.5	-10.4	-18.8	-29.9	-43.8	-60.7	-80.7

Magtech 147-grain FMC-FLAT (Subsonic) (9G)

G1 Ballistic Coefficient = 0.204

Distance • Yards	Muzzle	25	50	75	100	125	150	175	200
Velocity • fps	990	967	945	926	907	890	873	858	843
Energy • ft-lbs	320	305	292	280	268	259	249	240	232
Taylor KO Index	7.4	7.2	7.0	6.9	6.8	6.7	6.5	6.4	6.3
Mid-Range Trajectory Height • Inches	0.0	0.3	1.2	2.7	4.8	7.7	11.3	15.8	21.0
Drop • Inches	0.0	-1.1	-4.6	-10.5	-18.9	-29.9	-43.7	-60.3	-79.9

UMC (Remington) 147-grain FNEB (LL9MM9)

G1 Ballistic Coefficient = 0.185

Distance • Yards	Muzzle	25	50	75	100	125	150	175	200
Velocity • fps	990	965	941	920	900	881	864	847	831
Energy • ft-lbs	320	304	289	276	264	254	244	234	225
Taylor KO Index	7.4	7.2	7.0	6.9	6.7	6.6	6.5	6.3	6.2
Mid-Range Trajectory Height • Inches	0.0	0.3	1.2	2.7	4.9	7.8	11.6	15.9	21.5
Drop • Inches	0.0	-1.1	-4.6	-10.5	-19.0	-30.1	-44.1	-61.0	-80.8

MC (Remington) 147-grain Metal Case (Match) (L9MM9)

G1 Ballistic Coefficient = 0.184

Distance • Yards	Muzzle	25	50	75	100	125	150	175	200
Velocity • fps	990	964	941	920	900	881	863	846	830
Energy • ft-lbs	320	304	289	276	264	253	243	234	225
Taylor KO Index	7.4	7.2	7.0	6.9	6.7	6.6	6.5	6.3	6.2
Mid-Range Trajectory Height • Inches	0.0	0.3	1.1	2.7	4.9	7.8	11.5	16.0	21.4
Drop • Inches	0.0	-1.1	-4.6	-10.5	-19.0	-30.1	-44.1	-61.0	-80.8

SA (Winchester) 147-grain Full Metal Jacket – Flat Nose (USA9MM1)

G1 Ballistic Coefficient = 0.199

Distance • Yards	Muzzle	25	50	75	100	125	150	175	200
Velocity • fps	990	966	945	924	906	888	871	855	840
Energy • ft-lbs	320	305	292	279	268	257	248	239	230
Taylor KO Index	7.4	7.2	7.1	6.9	6.8	6.6	6.5	6.4	6.3
Mid-Range Trajectory Height • Inches	0.0	0.2	1.1	2.6	4.8	7.7	11.4	15.8	21.1
Drop • Inches	0.0	-1.1	-4.6	-10.5	-18.9	-30.0	-43.8	-60.4	-80.1

peer 147-grain TMJ FN Lawman (53620)

G1 Ballistic Coefficient = 0.188

Distance • Yards	Muzzle	25	50	75	100	125	150	175	200
Velocity • fps	985	957	932	909	888	868	849	831	814
Energy • ft-lbs	317	299	284	270	257	246	235	225	216
Taylor KO Index	7.4	7.2	7.0	6.8	6.6	6.5	6.3	6.2	6.1
Mid-Range Trajectory Height • Inches	0.0	0.3	1.2	2.8	5.0	8.0	11.8	16.3	22.1
Drop • Inches	0.0	-1.1	-4.6	-10.7	-19.3	-30.7	-45.0	-62.2	-82.7

peer 147-grain Clean-Fire TMJ FN Lawman (53826)

G1 Ballistic Coefficient = 0.188

Distance • Yards	Muzzle	25	50	75	100	125	150	175	200
Velocity • fps	985	957	932	909	888	868	849	831	814
Energy • ft-lbs	317	299	284	270	257	246	235	225	216
Taylor KO Index	7.4	7.2	7.0	6.8	6.6	6.5	6.3	6.2	6.1
Mid-Range Trajectory Height • Inches	0.0	0.3	1.2	2.8	5.0	8.0	11.8	16.3	22.1
Drop • Inches	0.0	-1.1	-4.6	-10.7	-19.3	-30.7	-45.0	-62.2	-82.7

Hornady 147-grain TAP-FPD (90288)

G1 Ballistic Coefficient = 0.212

Distance • Yards	Muzzle	25	50	75	100	125	150	175	200
Velocity • fps	975	954	935	916	899	883	867	852	838
Energy • ft-lbs	310	297	285	274	264	254	245	237	229
Taylor KO Index	7.3	7.1	7.0	6.8	6.7	6.6	6.5	6.4	6.3
Mid-Range Trajectory Height • Inches	0.0	0.3	1.2	2.7	5.0	7.8	11.6	15.9	21.5
Drop • Inches	0.0	-1.2	-4.7	-10.7	-19.4	-30.7	-44.8	-61.8	-81.7

Blazer (CCI) 147-grain TMJ (3582)

G1 Ballistic Coefficient = 0.190

Distance • Yards	Muzzle	25	50	75	100	125	150	175	200
Velocity • fps	950	929	908	890	872	855	839	824	809
Energy • ft-lbs	295	282	269	259	248	239	230	222	214
Taylor KO Index	7.1	6.9	6.8	6.6	6.5	6.4	6.3	6.1	6.0
Mid-Range Trajectory Height • Inches	0.0	0.3	1.3	2.9	5.3	8.4	12.4	17.2	22.9
Drop • Inches	0.0	-1.2	-5.0	-11.3	-20.4	-32.4	-47.4	-65.4	-86.6

Blazer (CCI) 147-grain Clean-Fire – TMJ (3462)

G1 Ballistic Coefficient = 0.189

Distance • Yards	Muzzle	25	50	75	100	125	150	175	200
sVelocity • fps	950	928	908	890	872	855	839	823	808
Energy • ft-lbs	295	281	269	258	248	239	230	221	213
Taylor KO Index	7.1	6.9	6.8	6.7	6.5	6.4	6.3	6.2	6.0
Mid-Range Trajectory Height • Inches	0.0	0.3	1.3	2.9	5.3	8.4	12.4	17.2	22.9
Drop • Inches	0.0	-1.2	-5.0	-11.3	-20.4	-32.4	-47.4	-65.4	-86.6

Cor-Bon 147-grain FMJ (PM09147/50)

G1 Ballistic Coefficient = 0.200

Distance • Yards	Muzzle	25	50	75	100	125	150	175	200
Velocity • fps	900	883	866	850	835	821	807	793	780
Energy • ft-lbs	264	254	245	236	228	220	213	205	199
Taylor KO Index	6.7	6.6	6.5	6.4	6.2	6.1	6.0	5.9	5.8
Mid-Range Trajectory Height • Inches	0.0	0.3	1.4	3.2	5.8	9.2	13.5	18.5	24.9
Drop • Inches	0.0	-1.4	-5.5	-12.6	-22.6	-35.8	-52.1	-71.8	-95.0

Fiocchi 158-grain FMJ (Subsonic) (9APE)

G1 Ballistic Coefficient = 0.15

Distance • Yards	Muzzle	25	50	75	100	125	150	175	200
Velocity • fps	940	916	893	872	852	833	815	798	781
Energy • ft-lbs	309	294	280	267	254	244	233	223	214
Taylor KO Index	7.6	7.4	7.2	7.0	6.8	6.7	6.5	6.4	6.3
Mid-Range Trajectory Height • Inches	0.0	0.3	1.3	3.0	5.5	8.7	12.9	17.7	24.0
Drop • Inches	0.0	-1.2	-5.1	-11.7	-21.1	-33.5	-49.0	-67.9	-90.1

9mm Largo (9x23mm)*

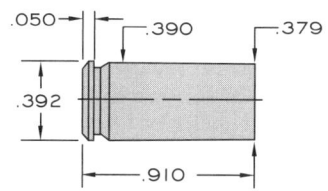

The 9mm Largo is a lengthened version of the 9mm Luger. It originated in Spain about 1913 as a variation on the earlier Bergmann-Bayard. The CCI ammunition loaded for this caliber was originally manufactured to support surplus Spanish military pistols. This cartridge has a potential for mixing with the Winchester 9x23mm.

Multiple cautions apply here. While several European 9mms are sometimes called 9x23mm, they are loaded to a much lower pressure specification than the current Winchester 9x23mm loadings. The Winchester ammunition should never be used in any 9mm Largo guns. The Largo is very similar to the .380 Super Auto, but again, the two are not interchangeable.

Relative Recoil Factor = 0.82

Specifications:

Controlling Agency for Standardization of this Ammunition: Factory

Barrel used for velocity testing: 4 inches long—1 turn in 10-inch twist

Availability:

CCI 124-grain HP (3513) (Until Supply Exhausted)

G1 Ballistic Coefficient = 0.121

Distance • Yards	Muzzle	25	50	75	100	125	150	175	200
Velocity • fps	1190	1114	1055	1006	966	931	900	872	846
Energy • ft-lbs	390	342	306	279	257	239	223	209	197
Taylor KO Index	7.5	7.0	6.6	6.3	6.1	5.9	5.7	5.5	5.3
Mid-Range Trajectory Height • Inches	0.0	0.2	0.7	2.1	3.7	6.3	9.5	13.5	18.3
Drop • Inches	0.0	-0.8	-3.4	-7.9	-14.5	-23.4	-34.7	-48.8	-65.7

CAUTION!

The source for this ammunition is probably dried up. See the cautions above. Do not use Winchester 9x23 ammo as a substitute for Largo ammunition. With 124 or 125-grain bullets, if the velocity of the ammunition is over 1,200 fps maximum, it should not be used in Largo pistols. If you have a pistol that needs Largo ammunition, reloading may be your only way to keep shooting. The same cautions apply to reloading. Be very sure you are not using Winchester 9x23 data.

9x23mm Winchester

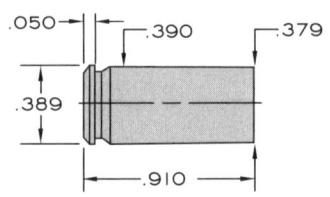

Please review the comments for the 9mm Largo. This is high-pressure ammunition that meets the performance specifications of the USPSA/IPSC competition rules. It should NOT be used in pistols that might also be marked 9x23mm but were not designed for the high pressures. If your 9x23mm gun is not a competition gun, it probably was not designed for this ammunition. **USE CAUTION!** There is a real potential for a mixup here!

Relative Recoil Factor = 0.82

Specifications:

Controlling Agency for Standardization of this Ammunition: Factory

Barrel used for velocity testing: 5 inches long—The twist rate is not available

Availability:

Cor-Bon 100-grain Pow'RBall (PB9X23100/20)

G1 Ballistic Coefficient = 0.100

Distance • Yards	Muzzle	25	50	75	100	125	150	175	200
Velocity • fps	1600	1445	1309	1194	1104	1036	983	936	901
Energy • ft-lbs	569	464	381	317	271	238	215	196	180
Taylor KO Index	8.2	7.4	6.7	6.1	5.6	5.3	5.0	4.8	4.6
Mid-Range Trajectory Height • Inches	0.0	0.1	0.5	1.3	2.5	4.4	6.9	10.2	14.2
Drop • Inches	0.0	-0.4	-2.0	-4.7	-9.0	-15.1	-23.2	-33.5	-46.3

USA (Winchester) 124-grain Jacketed Flat Point (Q4304)

G1 Ballistic Coefficient = 0.18

Distance • Yards	Muzzle	25	50	75	100	125	150	175	200
Velocity • fps	1460	1381	1308	1242	1183	1131	1087	1050	1017
Energy • ft-lbs	587	525	471	425	385	353	326	304	285
Taylor KO Index	9.2	8.7	8.2	7.8	7.4	7.1	6.9	6.6	6.4
Mid-Range Trajectory Height • Inches	0.0	0.1	0.6	1.3	2.5	4.2	6.3	9.1	12.3
Drop • Inches	0.0	-0.5	-2.2	-5.1	-9.5	-15.4	-23.0	-32.4	-43.8

Cor-Bon 125-grain JHP (SD9X23125/20)

G1 Ballistic Coefficient = 0.18

Distance • Yards	Muzzle	25	50	75	100	125	150	175	200
Velocity • fps	1450	1371	1299	1233	1175	1125	1082	1045	1013
Energy • ft-lbs	584	522	468	422	383	351	325	303	285
Taylor KO Index	9.2	8.7	8.3	7.8	7.5	7.2	6.9	6.6	6.4
Mid-Range Trajectory Height • Inches	0.0	0.1	0.6	1.4	2.6	4.2	6.5	9.2	12.6
Drop • Inches	0.0	-0.5	-2.2	-5.2	-9.6	-15.6	-23.3	-32.8	-44.4

Winchester 125-grain Silvertip Hollow Point (X923W)

G1 Ballistic Coefficient = 0.13

Distance • Yards	Muzzle	25	50	75	100	125	150	175	200
Velocity • fps	1450	1344	1249	1170	1103	1051	1007	970	937
Energy • ft-lbs	583	502	433	380	338	306	281	261	244
Taylor KO Index	9.2	8.5	7.9	7.4	7.0	6.7	6.4	6.2	6.0
Mid-Range Trajectory Height • Inches	0.0	0.1	0.6	1.5	2.8	4.6	7.1	10.2	14.0
Drop • Inches	0.0	-0.5	-2.3	-5.4	-10.2	-16.6	-25.1	-35.6	-48.4

Cor-Bon 125-grain DPX (DPX9X23125/20)

G1 Ballistic Coefficient = 0.1

Distance • Yards	Muzzle	25	50	75	100	125	150	175	200
Velocity • fps	1350	1261	1183	1118	1065	1021	984	952	923
Energy • ft-lbs	506	441	389	347	315	290	269	252	237
Taylor KO Index	8.6	8.0	7.5	7.1	6.8	6.5	6.3	6.1	5.9
Mid-Range Trajectory Height • Inches	0.0	0.2	0.7	1.7	3.1	5.1	7.8	11.1	15.3
Drop • Inches	0.0	-0.6	-2.6	-6.2	-11.4	-18.6	-27.9	-39.4	-53.3

.38 Short Colt

The history of the .38 Short Colt is rather cloudy. This cartridge seems to have appeared on the scene sometime along about the 1880s. It has the same dimensions as the .38 Long Colt except for the case length and the overall length. Why it is still in the inventory is a mystery. Perhaps the best excuse for this cartridge is: If there hadn't been a .38 SHORT Colt, there couldn't have been a .38 LONG Colt.

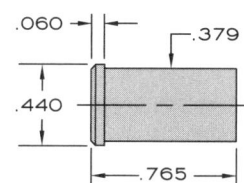

Relative Recoil Factor = 0.17

Specifications:

Controlling Agency for Standardization of this Ammunition: SAAMI

Bullet Weight Grains	Velocity fps	Maximum Average Pressure	
		Copper Crusher	Transducer
125 L	775	12,000 cup	N/S

Standard barrel for velocity testing: 4 inches long (vented)—1 turn in 16-inch twist

Availability:

Remington 125-grain Lead (R38SC)

G1 Ballistic Coefficient = 0.102

Distance • Yards	Muzzle	25	50	75	100	125	150	175	200
Velocity • fps	730	707	685	665	645	626	607	589	571
Energy • ft-lbs	150	140	130	123	115	109	102	96	91
Taylor KO Index	4.7	4.5	4.4	4.2	4.1	4.0	3.9	3.8	3.6
Mid-Range Trajectory Height • Inches	0.0	0.5	2.1	5.0	9.2	14.8	22.1	31.1	41.9
Drop • Inches	0.0	-2.1	-8.5	-19.5	-35.5	-56.6	-83.3	-115.9	-154.8

Paper targets are fine and well but try to shoot at critters that fidget and will not sit still whenever the opportunity arises. (Safari Press Archives)

.38 Long Colt

The .38 Long Colt was the U.S. Army's pistol cartridge just before the adoption of the .45 Auto. It is easy to see why the .45 was welcomed as a huge improvement in stopping power. It dates back to 1875, and at that time it was certainly loaded with black powder. Ammo from a .38 Long Colt can be fired in .38 Special revolvers, but the .38 Special ammo is much too long to be fired in a .38 Long Colt revolver. The .38 Special pressures are also way higher than those of the .38 Long Colt. No one had loaded for this cartridge for quite a while until Black Hills began to produce it for the Cowboy Action shooters. It is interesting that the new ammo has a higher velocity than the old, old factory loadings.

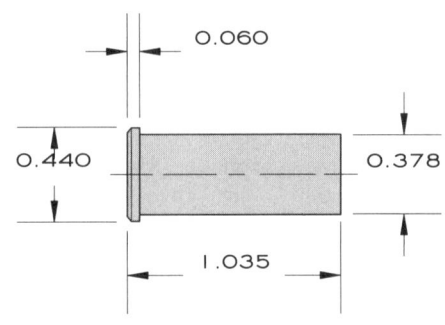

Relative Recoil Factor = 0.69

Specifications:

Controlling Agency for Standardization of this Ammunition: SAAMI

Bullet Weight Grains	Velocity fps	Maximum Average Pressure	
		Copper Crusher	Transducer
130 L	725	12,000 cup	N/S
150 L	725	12.000 cup	N/S

Standard barrel for velocity testing: 4 inches long (vented)—1 turn in 16-inch twist

Availability:
Cowboy Action Loads:

Black Hills 158-grain RNL (DCB3840N1)

G1 Ballistic Coefficient = 0.145

Distance • Yards	Muzzle	25	50	75	100	125	150	175	200
Velocity • fps	950	922	897	873	852	831	811	792	775
Energy • ft-lbs	317	298	282	268	254	242	231	220	211
Taylor KO Index	7.7	7.5	7.3	7.1	6.9	6.7	6.6	6.4	6.3
Mid-Range Trajectory Height • Inches	0.0	0.3	1.3	3.0	5.4	8.7	12.8	17.9	24.0
Drop • Inches	0.0	-1.2	-5.0	-11.5	-20.8	-33.1	-48.6	-67.4	-89.6

A Ruger .22 LR and a California ground squirrel. A match made for people who want to learn how to shoot a handgun well. (Safari Press Archives)

.38 Smith & Wesson

The .38 Smith & Wesson might be thought of as a forerunner of the .38 Special. Designed in 1877, it was once very popular as a police cartridge. The very modest performance (145–146 grains at 685 fps) is about like the mildest wadcutter target loads for the .38 Special.

Relative Recoil Factor = 0.45

Specifications:

Controlling Agency for Standardization of this Ammunition: SAAMI

Bullet Weight Grains	Velocity fps	Maximum Average Pressure	
		Copper Crusher	Transducer
145–146	680	13,000 cup	14,500 psi

Standard barrel for velocity testing: 4 inches long—1 turn in 18.75-inch twist

Availability:

Winchester 145-grain Lead Round Nose (X38SWP)
G1 Ballistic Coefficient = 0.125

Distance • Yards	Muzzle	25	50	75	100	125	150	175	200
Velocity • fps	685	668	650	635	620	605	590	575	561
Energy • ft-lbs	150	145	138	131	125	119	113	107	102
Taylor KO Index	5.1	5.0	4.9	4.8	4.6	4.5	4.4	4.3	4.2
Mid-Range Trajectory Height • Inches	0.0	0.6	2.4	5.6	10.2	16.3	24.2	33.8	45.3
Drop • Inches	0.0	-2.4	-9.6	-21.9	-39.6	-63.0	-92.3	-127.8	-169.9

Magtech 146-grain LRN (38SWA)
G1 Ballistic Coefficient = 0.135

Distance • Yards	Muzzle	25	50	75	100	125	150	175	200
Velocity • fps	686	670	655	640	625	611	597	584	570
Energy • ft-lbs	153	146	139	133	127	121	116	110	105
Taylor KO Index	5.1	5.0	4.9	4.8	4.7	4.6	4.5	4.4	4.3
Mid-Range Trajectory Height • Inches	0.0	0.6	2.4	5.6	10.2	16.2	23.9	33.4	44.6
Drop • Inches	0.0	-2.3	-9.5	-21.8	-39.3	-62.4	-91.4	-126.4	-167.8

Remington 146-grain Lead (R38SW)
G1 Ballistic Coefficient = 0.125

Distance • Yards	Muzzle	25	50	75	100	125	150	175	200
Velocity • fps	685	668	650	635	620	605	590	575	561
Energy • ft-lbs	150	145	138	131	125	119	113	107	102
Taylor KO Index	5.1	5.0	4.9	4.8	4.6	4.5	4.4	4.3	4.2
Mid-Range Trajectory Height • Inches	0.0	0.6	2.4	5.6	10.2	16.3	24.2	33.8	45.3
Drop • Inches	0.0	-2.4	-9.6	-21.9	-39.6	-63.0	-92.3	-127.8	-169.9

.38 Super Auto Colt

The .38 Super Auto Colt was introduced clear back in 1929 as an improved version of the even older (1900) .38 Auto. The very minimal rim feeds much better from automatic pistol magazines than ammunition with a more standard rim size, like the .38 Special. This cartridge should NOT be used in pistols chambered for .38 ACP. Performance is good, better than the .38 Special but nowhere near the .357 Magnum with the same bullet weights. The performance of the .38 Super Auto Colt is approximately equal to the 9mm, which may explain why its popularity seems to be in decline.

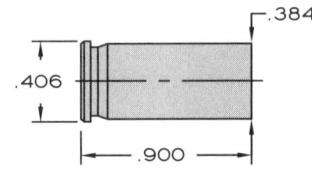

Relative Recoil Factor = 0.71

Specifications:

Controlling Agency for Standardization of this Ammunition: SAAMI

Bullet Weight Grains	Velocity fps	Maximum Average Pressure	
		Copper Crusher	Transducer
115 JHP	1,280	33,000 cup	36,500 psi
125 JHP	1,230	33,000 cup	36,500 psi
130 FMC	1,200	33,000 cup	36,500 psi

Standard barrel for velocity testing: 5 inches long—1 turn in 16-inch twist

Availability:

Cor-Bon 80-grain Glaser Safety Slug (01400/01600)

G1 Ballistic Coefficient = 0.08(

Distance • Yards	Muzzle	25	50	75	100	125	150	175	200
Velocity • fps	1650	1453	1285	1151	1055	985	931	885	845
Energy • ft-lbs	483	375	293	235	198	173	154	139	127
Taylor KO Index	6.7	5.9	5.2	4.7	4.3	4.0	3.8	3.6	3.4
Mid-Range Trajectory Height • Inches	0.0	0.1	0.5	1.3	2.7	4.7	7.4	11.0	15.5
Drop • Inches	0.0	-0.4	-1.9	-4.7	-9.1	-15.5	-24.1	-35.2	-49.1

Cor-Bon 100-grain Pow'RBall (PB38X100/20)

G1 Ballistic Coefficient = 0.10

Distance • Yards	Muzzle	25	50	75	100	125	150	175	200
Velocity • fps	1525	1379	1252	1149	1070	1009	961	920	884
Energy • ft-lbs	516	422	348	293	254	226	205	188	174
Taylor KO Index	7.8	7.1	6.4	5.9	5.5	5.2	4.9	4.7	4.5
Mid-Range Trajectory Height • Inches	0.0	0.1	0.5	1.4	2.7	4.7	7.4	10.8	15.1
Drop • Inches	0.0	-0.5	-2.2	-5.2	-9.9	-16.4	-25.1	-36.2	-49.8

Cor-Bon 115-grain JHP +P (SD38X115/20)

G1 Ballistic Coefficient = 0.12

Distance • Yards	Muzzle	25	50	75	100	125	150	175	200
Velocity • fps	1425	1316	1222	1143	1079	1026	986	949	918
Energy • ft-lbs	519	443	381	333	297	270	248	230	215
Taylor KO Index	8.3	7.7	7.1	6.7	6.3	6.0	5.8	5.6	5.4
Mid-Range Trajectory Height • Inches	0.0	0.1	0.6	1.5	2.9	4.9	7.5	10.8	14.8
Drop • Inches	0.0	-0.6	-2.4	-5.7	-10.6	-17.4	-26.3	-37.4	-50.8

Cor-Bon 125-grain DPX +P (DPX38X125/20)

G1 Ballistic Coefficient = 0.1

Distance • Yards	Muzzle	25	50	75	100	125	150	175	200
Velocity • fps	1350	1261	1183	1118	1065	1021	984	952	923
Energy • ft-lbs	506	441	389	347	315	290	269	252	237
Taylor KO Index	8.6	8.0	7.5	7.1	6.8	6.5	6.3	6.1	5.9
Mid-Range Trajectory Height • Inches	0.0	0.2	0.7	1.7	3.1	5.1	7.8	11.1	15.3
Drop • Inches	0.0	-0.6	-2.6	-6.2	-11.4	-18.6	-27.9	-39.4	-53.3

Cor-Bon 125-grain JHP +P (SD38X125/20)

G1 Ballistic Coefficient = 0.1

Distance • Yards	Muzzle	25	50	75	100	125	150	175	200
Velocity • fps	1325	1255	1193	1138	1092	1053	1019	989	962
Energy • ft-lbs	487	437	395	360	331	308	288	271	257
Taylor KO Index	8.4	8.0	7.6	7.2	6.9	6.7	6.5	6.3	6.1
Mid-Range Trajectory Height • Inches	0.0	0.2	0.7	1.6	3.1	5.0	7.5	10.6	14.5
Drop • Inches	0.0	-0.6	-2.7	-6.2	-11.4	-18.5	-27.5	-38.6	-51.9

Winchester 125-grain Silvertip Hollow Point [+P Load] (X38ASHP)

G1 Ballistic Coefficient = 0.183

Distance • Yards	Muzzle	25	50	75	100	125	150	175	200
Velocity • fps	1240	1182	1130	1087	1050	1018	989	964	941
Energy • ft-lbs	427	388	354	328	306	288	272	258	246
Taylor KO Index	7.9	7.5	7.2	6.9	6.7	6.5	6.3	6.1	6.0
Mid-Range Trajectory Height • Inches	0.0	0.2	0.8	1.8	3.4	5.5	8.2	11.6	15.6
Drop • Inches	0.0	-0.7	-3.0	-7.0	-12.8	-20.6	-30.4	-42.5	-56.8

Fiocchi 129-grain FMJ (38SA)

G1 Ballistic Coefficient = 0.121

Distance • Yards	Muzzle	25	50	75	100	125	150	175	200
Velocity • fps	1150	1083	1029	986	949	916	886	859	834
Energy • ft-lbs	381	336	306	278	258	240	225	211	199
Taylor KO Index	7.5	7.1	6.8	6.5	6.2	6.0	5.8	5.6	5.5
Mid-Range Trajectory Height • Inches	0.0	0.2	0.9	2.2	4.1	6.7	10.1	14.1	19.3
Drop • Inches	0.0	-0.9	-3.6	-8.3	-15.3	-24.7	-36.7	-51.5	-69.2

Aguila 130-grain FMJ +P (1E382112)

G1 Ballistic Coefficient = 0.087

Distance • Yards	Muzzle	25	50	75	100	125	150	175	200
Velocity • fps	1220	1112	1033	973	925	883	846	813	782
Energy • ft-lbs	426	357	308	273	248	225	207	191	177
Taylor KO Index	8.1	7.4	6.8	6.5	6.1	5.9	5.6	5.4	5.2
Mid-Range Trajectory Height • Inches	0.0	0.2	0.9	2.1	4.1	6.8	10.3	14.8	20.3
Drop • Inches	0.0	-0.8	-3.3	-7.9	-14.8	-24.2	-36.3	-51.6	-70.1

USA (Winchester) 130-grain FMJ (Q4205)

G1 Ballistic Coefficient = 0.158

Distance • Yards	Muzzle	25	50	75	100	125	150	175	200
Velocity • fps	1215	1152	1099	1054	1017	984	955	929	906
Energy • ft-lbs	426	383	348	321	299	280	264	249	237
Taylor KO Index	8.1	7.7	7.3	7.0	6.8	6.5	6.3	6.2	6.0
Mid-Range Trajectory Height • Inches	0.0	0.2	0.8	1.9	3.6	5.8	8.7	12.4	16.9
Drop • Inches	0.0	-0.8	-3.2	-7.4	-13.5	-21.8	-32.2	-45.1	-60.5

MC 130-grain Full Metal Jacket (38SA)

G1 Ballistic Coefficient = 0.137

Distance • Yards	Muzzle	25	50	75	100	125	150	175	200
Velocity • fps	1090	1041	1000	965	934	907	881	857	835
Energy • ft-lbs	343	331	289	269	252	237	224	212	201
Taylor KO Index	7.2	6.9	6.6	6.4	6.2	6.0	5.8	5.7	5.5
Mid-Range Trajectory Height • Inches	0.0	0.2	1.0	2.4	4.3	7.0	10.5	14.8	19.9
Drop • Inches	0.0	-0.9	-3.9	-9.0	-16.5	-26.4	-39.0	-54.4	-72.7

.38 Special (.38 Smith & Wesson Special)

Like the .30-06 cartridge in rifles, the .38 Special is the standard by which pistol cartridges are compared. The .38 Special was introduced by Smith & Wesson in 1902, and at that time it was loaded with black powder. For many years this was the cartridge used by most police officers. That began to change in the 1950s as other, more potent cartridges came into general use. The +P loads were introduced in an attempt to "soup-up" the .38 Special, but even the +Ps fall well short of the performance obtainable with the .357 Magnum. Just a glance at the SAAMI pressure levels shows why this is true. Still, with a 148-grain wadcutter bullet, the .38 Special continues to be the most popular caliber for target revolvers. There are more different factory loadings offered for this cartridge than for any other pistol ammunition.

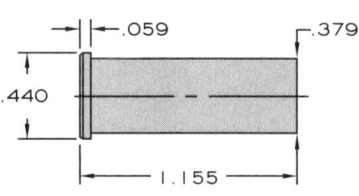

Relative Recoil Factor = 0.53

Specifications:

Controlling Agency for Standardization of this Ammunition: SAAMI

Bullet Weight Grains	Velocity fps	Maximum Average Pressure Copper Crusher	Transducer
110 STHP	945	17,000 cup	17,000 psi
158 LSWC	750	17,000 cup	17,000 psi
200 L	630	17,000 cup	17,000 psi
+P Loads			
95 STHP	1080	20,000 cup	18,500 psi
110 JHP	980	20,000 cup	18,500 psi
125 JHP	940	20,000 cup	18,500 psi
147 JHP	855	20,000 cup	18,500 psi
150 L	840	20,000 cup	18,500 psi
158 LSWC	880	20,000 cup	18,500 psi

Standard barrel for velocity testing: 4 inches long (vented)—1 turn in 18.75-inch twist.

Alternate one piece-barrel—7.710 inches.

Availability:

Cor-Bon 80-grain Glaser Safety Slug +P (02200/02400)

G1 Ballistic Coefficient = 0.08

Distance • Yards	Muzzle	25	50	75	100	125	150	175	200
Velocity • fps	1250	1125	1037	972	920	875	836	801	768
Energy • ft-lbs	278	225	191	168	150	136	124	114	105
Taylor KO Index	5.1	4.6	4.2	4.0	3.8	3.6	3.4	3.3	3.1
Mid-Range Trajectory Height • Inches	0.0	0.2	0.9	2.1	4.0	6.7	10.3	14.9	20.5
Drop • Inches	0.0	-0.8	-3.2	-7.7	-14.5	-23.8	-36.0	-51.3	-70.0

Cor-Bon 80-grain Glaser Safety Slug (01800/02000)

G1 Ballistic Coefficient = 0.08

Distance • Yards	Muzzle	25	50	75	100	125	150	175	200
Velocity • fps	1200	1089	1011	951	903	861	823	789	757
Energy • ft-lbs	256	211	182	161	145	132	120	111	102
Taylor KO Index	4.9	4.4	4.1	3.9	3.7	3.5	3.4	3.2	3.1
Mid-Range Trajectory Height • Inches	0.0	0.2	0.9	2.2	4.2	7.1	10.8	15.5	21.3
Drop • Inches	0.0	-0.8	-3.4	-8.2	-15.4	-25.2	-37.9	-53.9	-73.4

Magtech 95-grain SCHP +P (FD38A)

G1 Ballistic Coefficient = 0.0

Distance • Yards	Muzzle	25	50	75	100	125	150	175	200
Velocity • fps	1083	1017	965	921	884	850	819	790	763
Energy • ft-lbs	247	218	197	179	165	152	141	132	123
Taylor KO Index	5.3	4.9	4.7	4.5	4.3	4.1	4.0	3.8	3.7
Mid-Range Trajectory Height • Inches	0.0	0.2	1.1	2.5	4.7	7.6	11.5	16.4	22.3
Drop • Inches	0.0	-1.0	-4.0	-9.4	-17.4	-28.1	-41.9	-58.9	-79.4

Cor-Bon 100-grain Pow'RBall +P (PB38100/20)

G1 Ballistic Coefficient = 0.100

Distance • Yards	Muzzle	25	50	75	100	125	150	175	200
Velocity • fps	1150	1071	1010	962	921	885	852	823	795
Energy • ft-lbs	294	255	227	205	188	174	161	150	140
Taylor KO Index	5.9	5.5	5.2	4.9	4.7	4.5	4.3	4.2	4.1
Mid-Range Trajectory Height • Inches	0.0	0.2	1.0	2.3	4.3	7.0	10.6	15.0	20.5
Drop • Inches	0.0	-0.9	-3.6	-8.5	-15.7	-25.5	-38.1	-53.7	-72.4

CCI 109-grain – Shotshell (3709)
Number 9 Shot at 1000 fps

Hornady 110-grain FTX +P (90311)

G1 Ballistic Coefficient = 0.131

Distance • Yards	Muzzle	25	50	75	100	125	150	175	200
Velocity • fps	1090	1039	997	961	930	901	874	850	827
Energy • ft-lbs	290	264	243	226	211	198	187	177	167
Taylor KO Index	6.1	5.8	5.6	5.4	5.2	5.1	4.9	4.8	4.6
Mid-Range Trajectory Height • Inches	0.0	0.2	1.0	2.3	4.3	7.1	10.6	15.9	20.1
Drop • Inches	0.0	-1.0	-3.9	-9.0	-16.5	-26.6	-39.2	-54.8	-73.3

Hornady 110-grain FTX (90310)

G1 Ballistic Coefficient = 0.131

Distance • Yards	Muzzle	25	50	75	100	125	150	175	200
Velocity • fps	1010	972	940	910	883	858	835	813	792
Energy • ft-lbs	249	231	216	202	191	180	170	161	153
Taylor KO Index	5.7	5.5	5.3	5.1	5.0	4.8	4.7	4.6	4.4
Mid-Range Trajectory Height • Inches	0.0	0.2	1.1	2.6	4.9	7.9	11.8	16.6	22.3
Drop • Inches	0.0	-1.1	-4.5	-10.3	-18.8	-30.1	-44.3	-61.6	-82.2

Winchester 110-grain Jacketed Flat Point (SC38NT) LF

G1 Ballistic Coefficient = 0.118

Distance • Yards	Muzzle	25	50	75	100	125	150	175	200
Velocity • fps	975	938	906	876	849	824	800	778	756
Energy • ft-lbs	232	215	200	188	176	166	156	148	140
Taylor KO Index	5.5	5.3	5.1	4.9	4.8	4.6	4.5	4.4	4.2
Mid-Range Trajectory Height • Inches	0.0	0.3	1.2	2.9	5.3	8.5	12.7	17.9	24.1
Drop • Inches	0.0	-1.2	-4.8	-11.1	-20.2	-32.4	-47.7	-66.4	-88.7

Cor-Bon 110-grain DPX +P (DP38110/20) + (TR38110/20) LF

G1 Ballistic Coefficient = 0.135

Distance • Yards	Muzzle	25	50	75	100	125	150	175	200
Velocity • fps	1050	1007	971	939	910	884	860	837	816
Energy • ft-lbs	269	248	230	215	202	191	181	171	163
Taylor KO Index	5.9	5.6	5.4	5.3	5.1	5.0	4.8	4.7	4.6
Mid-Range Trajectory Height • Inches	0.0	0.3	1.0	2.5	4.6	7.4	11.1	15.4	21.0
Drop • Inches	0.0	-1.0	-4.2	-9.6	-17.6	-28.1	-41.4	-57.7	-77.0

Cor-Bon 110-grain JHP+ P (SD38110/20)

G1 Ballistic Coefficient = 0.135

Distance • Yards	Muzzle	25	50	75	100	125	150	175	200
Velocity • fps	1050	1007	971	939	910	884	860	837	816
Energy • ft-lbs	269	248	230	215	202	191	181	171	163
Taylor KO Index	5.9	5.6	5.4	5.3	5.1	5.0	4.8	4.7	4.6
Mid-Range Trajectory Height • Inches	0.0	0.3	1.0	2.5	4.6	7.4	11.1	15.4	21.0
Drop • Inches	0.0	-1.0	-4.2	-9.6	-17.6	-28.1	-41.4	-57.7	-77.0

Fiocchi 110-grain Hornady XTP JHP (38XTPB)

G1 Ballistic Coefficient = 0.131

Distance • Yards	Muzzle	25	50	75	100	125	150	175	200
Velocity • fps	1000	964	932	903	876	852	829	807	787
Energy • ft-lbs	244	227	212	199	188	177	168	159	151
Taylor KO Index	5.6	5.4	5.2	5.1	4.9	4.8	4.7	4.5	4.2
Mid-Range Trajectory Height • Inches	0.0	0.2	1.1	2.7	5.0	8.1	12.0	16.8	22.6
Drop • Inches	0.0	-1.1	-4.6	-10.5	-19.1	-30.6	-45.0	-62.6	-83.6

Remington 110-grain Semi-Jacketed Hollow Point +P (R38S10)

G1 Ballistic Coefficient = 0.129

Distance • Yards	Muzzle	25	50	75	100	125	150	175	200
Velocity • fps	995	959	926	898	871	847	824	802	782
Energy • ft-lbs	242	224	210	197	185	175	166	157	149
Taylor KO Index	5.6	5.4	5.2	5.0	4.9	4.8	4.6	4.5	4.4
Mid-Range Trajectory Height • Inches	0.0	0.3	1.2	2.7	5.1	8.1	12.1	17.0	22.8
Drop • Inches	0.0	-1.1	-4.6	-10.6	-19.4	-30.9	-45.5	-63.3	-84.5

Federal 110-grain Hydra-Shok JHP (PD38HS3H)

G1 Ballistic Coefficient = 0.127

Distance • Yards	Muzzle	25	50	75	100	125	150	175	200
Velocity • fps	980	940	910	880	860	836	813	792	772
Energy • ft-lbs	235	215	205	190	180	171	162	153	145
Taylor KO Index	5.5	5.3	5.1	4.9	4.8	4.7	4.6	4.4	4.3
Mid-Range Trajectory Height • Inches	0.0	0.3	1.2	2.8	5.2	8.4	12.5	17.3	23.5
Drop • Inches	0.0	-1.2	-4.8	-10.9	-19.9	-31.8	-46.8	-65.1	-86.8

Remington 110-grain Semi-Jacketed Hollow Point (R38S16)

G1 Ballistic Coefficient = 0.128

Distance • Yards	Muzzle	25	50	75	100	125	150	175	200
Velocity • fps	950	919	890	864	840	817	796	775	756
Energy • ft-lbs	220	206	194	182	172	163	155	147	140
Taylor KO Index	5.3	5.2	5.0	4.8	4.7	4.6	4.5	4.3	4.2
Mid-Range Trajectory Height • Inches	0.0	0.3	1.4	3.0	5.4	8.8	13.0	18.3	24.6
Drop • Inches	0.0	-1.2	-5.0	-11.6	-21.0	-33.6	-49.3	-68.6	-91.4

Winchester 110-grain Silvertip Hollow Point (X38S9HP)

G1 Ballistic Coefficient = 0.149

Distance • Yards	Muzzle	25	50	75	100	125	150	175	200
Velocity • fps	945	918	894	871	850	830	811	792	775
Energy • ft-lbs	218	206	195	185	176	168	161	153	147
Taylor KO Index	5.3	5.1	5.0	4.9	4.8	4.7	4.5	4.4	4.3
Mid-Range Trajectory Height • Inches	0.0	0.3	1.3	3.0	5.4	8.7	12.8	18.0	24.0
Drop • Inches	0.0	-1.2	-5.1	-11.6	-21.0	-33.4	-48.9	-67.8	-90.1

Fiocchi 110-grain FMJ FN (38H)

G1 Ballistic Coefficient = 0.13

Distance • Yards	Muzzle	25	50	75	100	125	150	175	200
Velocity • fps	1080	1032	992	958	928	900	874	851	829
Energy • ft-lbs	284	260	240	224	210	198	187	177	168
Taylor KO Index	6.1	5.8	5.6	5.4	5.2	5.0	4.9	4.8	4.7
Mid-Range Trajectory Height • Inches	0.0	0.2	1.0	2.4	4.4	7.1	10.7	14.9	20.2
Drop • Inches	0.0	-1.0	-4.0	-9.2	-16.7	-26.8	-39.8	-55.2	-73.9

Fiocchi 125-grain SJSP (38D)

G1 Ballistic Coefficient = 0.15

Distance • Yards	Muzzle	25	50	75	100	125	150	175	200
Velocity • fps	970	942	916	893	869	851	832	813	796
Energy • ft-lbs	261	246	233	222	209	201	192	184	176
Taylor KO Index	6.2	6.0	5.8	5.7	5.5	5.4	5.3	5.2	5.1
Mid-Range Trajectory Height • Inches	0.0	0.3	1.2	2.8	5.2	8.3	12.3	16.9	22.9
Drop • Inches	0.0	-1.2	-4.8	-11.0	-19.9	-31.7	-46.5	-64.4	-85.6

Magtech 125-grain LFN (38U)

G1 Ballistic Coefficient = 0.1

Distance • Yards	Muzzle	25	50	75	100	125	150	175	200
Velocity • fps	919	898	879	860	843	826	810	794	779
Energy • ft-lbs	234	224	214	205	197	189	182	175	169
Taylor KO Index	5.9	5.7	5.6	5.5	5.4	5.3	5.2	5.1	5.0
Mid-Range Trajectory Height • Inches	0.0	0.3	1.4	3.1	5.7	9.0	13.2	18.1	25.0
Drop • Inches	0.0	-1.3	-5.2	-12.1	-21.9	-34.7	-50.7	-70.0	-92.7

Winchester 125-grain Jacketed Flat Point (WC381)

G1 Ballistic Coefficient = 0.1

Distance • Yards	Muzzle	25	50	75	100	125	150	175	200
Velocity • fps	775	758	742	727	712	697	683	669	655
Energy • ft-lbs	167	160	153	147	141	135	129	124	119
Taylor KO Index	4.9	4.8	4.7	4.6	4.5	4.4	4.4	4.3	4.2
Mid-Range Trajectory Height • Inches	0.0	0.5	1.5	5.3	7.8	12.5	18.5	25.7	34.3
Drop • Inches	0.0	-1.8	-7.4	-17.0	-30.7	-48.6	-71.0	-98.1	-130.1

Magtech 125-grain LRN (Short) (38G)

G1 Ballistic Coefficient = 0.142

Distance • Yards	Muzzle	25	50	75	100	125	150	175	200
Velocity • fps	686	671	659	642	628	614	601	588	575
Energy • ft-lbs	130	125	120	114	109	105	100	96	92
Taylor KO Index	4.4	4.3	4.2	4.1	4.0	3.9	3.8	3.7	3.7
Mid-Range Trajectory Height • Inches	0.0	0.6	2.0	5.5	9.7	16.0	23.7	33.0	44.1
Drop • Inches	0.0	-2.3	-9.5	-21.7	-39.2	-62.2	-90.9	-125.7	-166.7

Black Hills 125-grain Jacketed Hollow Point +P (D38N2)

G1 Ballistic Coefficient = 0.140

Distance • Yards	Muzzle	25	50	75	100	125	150	175	200
Velocity • fps	1050	1008	973	942	914	888	865	842	821
Energy • ft-lbs	306	282	263	246	232	219	208	197	187
Taylor KO Indcx	6.7	6.4	6.2	6.0	5.8	5.7	5.5	5.4	5.2
Mid-Range Trajectory Height • Inches	0.0	0.2	1.1	2.5	4.6	7.3	11.0	15.4	20.8
Drop • Inches	0.0	-1.0	-4.2	-9.6	-17.5	-28.0	-41.2	-57.4	-76.6

Magtech 125-grain JHP +P (GG38A)

G1 Ballistic Coefficient = 0.230

Distance • Yards	Muzzle	25	50	75	100	125	150	175	200
Velocity • fps	1017	993	971	950	931	913	896	880	865
Energy • ft-lbs	287	274	262	250	241	231	223	215	208
Taylor KO Index	6.5	6.3	6.2	6.1	5.9	5.8	5.7	5.6	5.5
Mid-Range Trajectory Height • Inches	0.0	0.2	1.1	2.5	4.6	7.3	10.8	15.0	20.0
Drop • Inches	0.0	-1.1	-4.3	-9.9	-17.9	-28.4	-41.5	-57.3	-75.9

Remington 125-grain Brass-Jacketed Hollow Point +P (GS38SB)

G1 Ballistic Coefficient = 0.175

Distance • Yards	Muzzle	25	50	75	100	125	150	175	200
Velocity • fps	975	950	929	905	885	886	848	831	814
Energy • ft-lbs	264	250	238	227	218	208	200	192	184
Taylor KO Index	6.2	6.0	5.9	5.8	5.6	5.5	5.4	5.3	5.2
Mid-Range Trajectory Height • Inches	0.0	0.3	1.0	2.8	5.2	8.0	11.9	16.6	22.2
Drop • Inches	0.0	-1.2	-4.7	-10.8	-19.6	-31.1	-45.5	-63.0	-83.6

Cor-Bon 125-grain JHP +P (SD38125/20)

G1 Ballistic Coefficient = 0.165

Distance • Yards	Muzzle	25	50	75	100	125	150	175	200
Velocity • fps	950	925	903	882	862	843	825	808	792
Energy • ft-lbs	251	238	226	216	206	197	189	181	174
Taylor KO Index	6.1	5.9	5.8	5.6	5.5	5.4	5.3	5.2	5.0
Mid-Range Trajectory Height • Inches	0.0	0.3	1.3	2.9	5.4	8.5	12.6	17.3	23.4
Drop • Inches	0.0	-1.2	-5.0	-11.4	-20.6	-32.8	-48.0	-66.4	-88.1

MC 125-grain Starfire Hollow Point +P (38SFA)

G1 Ballistic Coefficient = 0.125

Distance • Yards	Muzzle	25	50	75	100	125	150	175	200
Velocity • fps	950	918	889	863	838	815	793	772	752
Energy • ft-lbs	251	234	219	206	195	184	174	165	157
Taylor KO Index	6.1	5.9	5.7	5.5	5.3	5.2	5.1	4.9	4.8
Mid-Range Trajectory Height • Inches	0.0	0.3	1.3	3.0	5.5	8.8	13.1	18.4	24.7
Drop • Inches	0.0	-1.2	-5.0	-11.6	-21.1	-33.6	-49.5	-68.8	-91.7

Blazer (CCI) 125-grain Jacketed Hollow +P (3514)

G1 Ballistic Coefficient = 0.130

Distance • Yards	Muzzle	25	50	75	100	125	150	175	200
Velocity • fps	945	915	887	862	838	816	795	775	755
Energy • ft-lbs	248	232	218	206	195	185	175	167	158
Taylor KO Indcx	6.0	5.8	5.7	5.5	5.4	5.2	5.1	5.0	4.8
Mid-Range Trajectory Height • Inches	0.0	0.3	1.3	3.0	5.5	8.9	13.2	18.5	24.8
Drop • Inches	0.0	-1.2	-5.1	-11.7	-21.2	-33.8	-49.7	-69.0	-92.0

Speer 125-grain Gold Dot +P (23720)

G1 Ballistic Coefficient = 0.140

Distance • Yards	Muzzle	25	50	75	100	125	150	175	200
Velocity • fps	945	917	891	867	845	824	804	785	766
Energy • ft-lbs	248	233	220	209	198	188	179	171	163
Taylor KO Index	6.0	5.9	5.7	5.5	5.4	5.3	5.1	5.0	4.9
Mid-Range Trajectory Height • Inches	0.0	0.3	1.3	3.0	5.5	8.8	13.0	18.2	24.4
Drop • Inches	0.0	-1.2	-5.1	-11.8	-21.1	-33.6	-49.3	-68.3	-90.9

Remington 125-grain Semi-Jacketed Hollow Point +P (R38S2)

G1 Ballistic Coefficient = 0.165

Distance • Yards	Muzzle	25	50	75	100	125	150	175	200
Velocity • fps	945	921	898	878	858	839	822	805	788
Energy • ft-lbs	248	235	224	214	204	196	187	180	173
Taylor KO Index	6.0	5.9	5.7	5.6	5.5	5.3	5.2	5.1	5.0
Mid-Range Trajectory Height • Inches	0.0	0.3	1.3	3.0	5.4	8.5	12.6	17.6	23.6
Drop • Inches	0.0	-1.2	-5.0	-11.5	-20.8	-33.1	-48.4	-67.0	-88.9

UMC (Remington) 125-grain Jacketed Hollow Point +P (L38S2)

G1 Ballistic Coefficient = 0.164

Distance • Yards	Muzzle	25	50	75	100	125	150	175	200
Velocity • fps	945	921	898	878	858	839	822	805	788
Energy • ft-lbs	248	235	224	214	204	196	187	180	173
Taylor KO Index	6.0	5.9	5.7	5.6	5.5	5.3	5.2	5.1	5.0
Mid-Range Trajectory Height • Inches	0.0	0.3	1.3	3.0	5.4	8.5	12.6	17.6	23.6
Drop • Inches	0.0	-1.2	-5.0	-11.5	-20.8	-33.1	-48.5	-67.0	-89.0

Winchester 125-grain Jacketed Hollow Point +P (X38S7PH)

G1 Ballistic Coefficient = 0.165

Distance • Yards	Muzzle	25	50	75	100	125	150	175	200
Velocity • fps	945	921	898	878	858	839	822	805	788
Energy • ft-lbs	248	235	224	214	204	196	187	180	173
Taylor KO Index	6.0	5.9	5.7	5.6	5.5	5.3	5.2	5.1	5.0
Mid-Range Trajectory Height • Inches	0.0	0.3	1.3	3.0	5.4	8.5	12.6	17.6	23.6
Drop • Inches	0.0	-1.2	-5.0	-11.5	-20.8	-33.1	-18.4	-67.0	-88.9

Winchester 125-grain Silvertip Hollow Point +P (X38S8HP)

G1 Ballistic Coefficient = 0.165

Distance • Yards	Muzzle	25	50	75	100	125	150	175	200
Velocity • fps	945	921	898	878	858	839	822	805	788
Energy • ft-lbs	248	235	224	214	204	196	187	180	173
Taylor KO Index	6.0	5.9	5.7	5.6	5.5	5.3	5.2	5.1	5.0
Mid-Range Trajectory Height • Inches	0.0	0.3	1.3	3.0	5.4	8.5	12.6	17.6	23.6
Drop • Inches	0.0	-1.2	-5.0	-11.5	-20.8	-33.1	-18.4	-67.0	-88.9

USA (Winchester) 125-grain Jacketed Hollow Point +P (USA38JHP)

G1 Ballistic Coefficient = 0.16

Distance • Yards	Muzzle	25	50	75	100	125	150	175	200
Velocity • fps	945	920	898	877	857	838	820	803	787
Energy • ft-lbs	248	235	224	214	204	195	187	179	172
Taylor KO Index	6.0	5.9	5.7	5.6	5.5	5.3	5.2	5.1	5.0
Mid-Range Trajectory Height • Inches	0.0	0.3	1.2	2.9	5.3	8.6	12.7	17.7	23.6
Drop • Inches	0.0	-1.2	-5.0	-11.5	-20.8	-33.1	-48.5	-67.1	-89.0

Fiocchi 125-grain SJHP (38F)

G1 Ballistic Coefficient = 0.20

Distance • Yards	Muzzle	25	50	75	100	125	150	175	200
Velocity • fps	940	921	902	885	869	853	839	824	810
Energy • ft-lbs	245	235	226	218	210	202	195	189	182
Taylor KO Index	6.0	5.9	5.8	5.6	5.5	5.4	5.3	5.3	5.2
Mid-Range Trajectory Height • Inches	0.0	0.3	1.3	2.9	5.3	8.5	12.4	17.3	23.0
Drop • Inches	0.0	-1.2	-5.1	-11.5	-20.8	-32.9	-48.0	-66.2	-87.6

Magtech 125-grain SJHP Flat (38D)

G1 Ballistic Coefficient = 0.16

Distance • Yards	Muzzle	25	50	75	100	125	150	175	200
Velocity • fps	938	914	892	871	852	833	816	799	783
Energy • ft-lbs	244	232	221	211	202	193	185	177	170
Taylor KO Index	6.0	5.8	5.7	5.6	5.4	5.3	5.2	5.1	5.0
Mid-Range Trajectory Height • Inches	0.0	0.3	1.3	3.0	5.4	8.7	12.9	17.9	24.0
Drop • Inches	0.0	-1.3	-5.1	-11.7	-21.1	-33.6	-49.1	-68.0	-90.2

Magtech 125-grain SJHP +P (38F)

G1 Ballistic Coefficient = 0.1

Distance • Yards	Muzzle	25	50	75	100	125	150	175	200
Velocity • fps	938	914	891	870	851	832	814	797	781
Energy • ft-lbs	245	232	220	210	200	192	184	176	169
Taylor KO Index	6.0	5.8	5.7	5.5	5.4	5.3	5.2	5.1	5.0
Mid-Range Trajectory Height • Inches	0.0	0.3	1.3	3.0	5.4	8.7	12.9	17.9	24.0
Drop • Inches	0.0	-1.3	-5.1	-11.7	-21.2	-33.6	-49.2	-68.1	-90.4

Hornady 125-grain JHP / XTP (90322)

G1 Ballistic Coefficient = 0.153

Distance • Yards	Muzzle	25	50	75	100	125	150	175	200
Velocity • fps	900	877	856	836	817	799	782	765	749
Energy • ft-lbs	225	214	203	194	185	177	170	163	156
Taylor KO Index	5.7	5.6	5.5	5.3	5.2	5.1	5.0	4.9	4.8
Mid-Range Trajectory Height • Inches	0.0	0.3	1.4	3.2	5.9	9.4	13.9	19.5	26.0
Drop • Inches	0.0	-1.4	-5.6	-12.7	-23.0	-36.4	-53.4	-73.8	-98.0

Speer 125-grain GDHP-SB (23921)

G1 Ballistic Coefficient = 0.142

Distance • Yards	Muzzle	25	50	75	100	125	150	175	200
Velocity • fps	860	838	818	798	780	762	745	728	712
Energy • ft-lbs	205	195	186	177	169	161	154	147	141
Taylor KO Index	5.5	5.3	5.2	5.1	5.0	4.9	4.7	4.6	4.5
Mid-Range Trajectory Height • Inches	0.0	0.4	1.6	3.6	6.5	10.4	15.4	21.1	28.6
Drop • Inches	0.0	-1.5	-6.1	-13.9	-25.2	-40.0	-58.5	-81.0	-107.5

Fiocchi 125-grain XTPHP (38XTP)

G1 Ballistic Coefficient = 0.151

Distance • Yards	Muzzle	25	50	75	100	125	150	175	200
Velocity • fps	850	830	811	793	776	759	743	727	712
Energy • ft-lbs	200	191	182	175	167	160	153	147	141
Taylor KO Index	5.4	5.3	5.2	5.1	4.9	4.8	4.7	4.6	4.5
Mid-Range Trajectory Height • Inches	0.0	0.4	1.6	3.6	6.6	10.5	15.6	21.4	28.9
Drop • Inches	0.0	-1.5	-6.2	-14.2	-25.6	-40.7	-59.5	-82.2	-109.0

Federal 125-grain NyClad HP (P38MA)

G1 Ballistic Coefficient = 0.050

Distance • Yards	Muzzle	25	50	75	100	125	150	175	200
Velocity • fps	830	780	730	680	640	602	566	531	499
Energy • ft-lbs	190	165	145	130	115	101	89	78	69
Taylor KO Index	5.3	5.0	4.7	4.3	4.1	3.8	3.6	3.4	3.2
Mid-Range Trajectory Height • Inches	0.0	0.4	1.8	4.3	8.2	13.8	21.3	31.0	43.4
Drop • Inches	0.0	-1.6	-6.9	-16.3	-30.4	-49.7	-75.1	-107.2	-147.1

UMC (Remington) 125-grain FNEB +P (LL38S2)

G1 Ballistic Coefficient = 0.211

Distance • Yards	Muzzle	25	50	75	100	125	150	175	200
Velocity • fps	975	954	935	916	899	882	867	852	837
Energy • ft-lbs	264	253	242	233	224	216	208	201	195
Taylor KO Index	6.2	6.1	6.0	5.8	5.7	5.6	5.5	5.4	5.3
Mid-Range Trajectory Height • Inches	0.0	0.3	1.2	2.8	5.0	7.9	11.7	16.0	21.5
Drop • Inches	0.0	-1.2	-4.7	-10.7	-19.4	-30.7	-44.8	-61.8	-81.8

UMC 125-grain EMJ NT +P (38EMD) LF

G1 Ballistic Coefficient = 0.124

Distance • Yards	Muzzle	25	50	75	100	125	150	175	200
Velocity • fps	950	920	892	864	837	814	792	771	751
Energy • ft-lbs	250	234	219	206	195	184	174	165	157
Taylor KO Index	6.1	5.9	5.7	5.5	5.3	5.2	5.0	4.9	4.8
Mid-Range Trajectory Height • Inches	0.0	0.3	1.3	3.0	5.5	8.9	13.2	18.3	24.8
Drop • Inches	0.0	-1.2	-5.0	-11.6	-21.1	-33.7	-49.5	-68.8	-91.8

Magtech 125-grain FMJ Flat (38Q)

G1 Ballistic Coefficient = 0.161

Distance • Yards	Muzzle	25	50	75	100	125	150	175	200
Velocity • fps	938	914	891	870	851	832	814	797	781
Energy • ft-lbs	245	232	220	210	200	192	184	176	169
Taylor KO Index	6.0	5.8	5.7	5.5	5.4	5.3	5.2	5.1	5.0
Mid-Range Trajectory Height • Inches	0.0	0.3	1.3	3.0	5.4	8.7	12.9	17.9	24.0
Drop • Inches	0.0	-1.3	-5.1	-11.7	-21.2	-33.6	-49.2	-68.1	-90.4

American Eagle (Federal) 125-grain FMJ (AE38K)

G1 Ballistic Coefficient = 0.185

Distance • Yards	Muzzle	25	50	75	100	125	150	175	200
Velocity • fps	890	870	850	830	820	807	792	778	765
Energy • ft-lbs	230	220	210	200	195	181	174	168	162
Taylor KO Index	5.7	5.5	5.4	5.3	5.2	5.1	5.0	5.0	4.9
Mid-Range Trajectory Height • Inches	0.0	0.3	1.4	3.2	5.9	9.5	13.9	19.3	25.7
Drop • Inches	0.0	-1.4	-5.6	-12.9	-23.2	-36.7	-53.6	-73.9	-97.8

UMC (Remington) 125-grain FNEB (LL38S17)

G1 Ballistic Coefficient = 0.211

Distance • Yards	Muzzle	25	50	75	100	125	150	175	200
Velocity • fps	850	836	822	808	796	783	771	759	747
Energy • ft-lbs	201	194	188	181	176	170	165	160	155
Taylor KO Index	5.4	5.3	5.2	5.2	5.1	5.0	4.9	4.8	4.8
Mid-Range Trajectory Height • Inches	0.0	0.4	1.6	3.6	6.5	10.2	15.0	20.4	27.5
Drop • Inches	0.0	-1.5	-6.2	-14.0	-25.2	-39.8	-57.9	-79.7	-105.3

Blazer (CCI) 125-grain FMJ FN Brass (5204)

G1 Ballistic Coefficient = 0.147

Distance • Yards	Muzzle	25	50	75	100	125	150	175	200
Velocity • fps	865	844	824	805	786	769	752	736	720
Energy • ft-lbs	208	198	188	180	172	164	157	150	144
Taylor KO Index	5.5	5.4	5.3	5.1	5.0	4.9	4.8	4.7	4.6
Mid-Range Trajectory Height • Inches	0.0	0.3	1.5	3.5	6.4	10.2	15.1	21.0	28.1
Drop • Inches	0.0	-1.5	-6.0	-13.7	-24.8	-39.4	-57.7	-79.8	-105.9

USA (Winchester) 125-grain Jacketed Flat Point (USA38SP)

G1 Ballistic Coefficient = 0.127

Distance • Yards	Muzzle	25	50	75	100	125	150	175	200
Velocity • fps	850	826	804	783	763	744	725	707	690
Energy • ft-lbs	201	190	179	170	162	154	146	139	132
Taylor KO Index	5.4	5.3	5.1	5.0	4.9	4.7	4.6	4.5	4.4
Mid-Range Trajectory Height • Inches	0.0	0.3	1.5	3.6	6.7	10.7	15.9	22.3	29.9
Drop • Inches	0.0	-1.5	-6.2	-14.3	-25.9	-41.3	-60.5	-83.9	-111.7

Federal 129-grain Hydra-Shok JHP +P (P38HS1)

G1 Ballistic Coefficient = 0.18

Distance • Yards	Muzzle	25	50	75	100	125	150	175	200
Velocity • fps	950	930	910	890	870	852	836	820	805
Energy • ft-lbs	255	245	235	225	215	202	194	187	180
Taylor KO Index	6.3	6.1	6.0	5.9	5.7	5.6	5.5	5.4	5.3
Mid-Range Trajectory Height • Inches	0.0	0.3	1.3	2.9	5.3	8.4	12.4	17.2	22.9
Drop • Inches	0.0	-1.2	-5.0	-11.4	-20.5	-32.5	-47.5	-65.6	-86.9

Winchester 130-grain Bonded PDX1 +P (S38PDB)

G1 Ballistic Coefficient = 0.18

Distance • Yards	Muzzle	25	50	75	100	125	150	175	200
Velocity • fps	950	928	908	888	870	853	837	821	806
Energy • ft-lbs	260	249	238	228	219	210	202	195	188
Taylor KO Index	6.3	6.2	6.0	5.9	5.8	5.7	5.5	5.4	5.3
Mid-Range Trajectory Height • Inches	0.0	0.3	1.3	2.9	5.3	8.4	12.4	17.2	23.0
Drop • Inches	0.0	-1.2	-5.0	-11.4	-20.5	-32.5	-47.5	-65.6	-86.8

Aguila 130-grain FMJ (1E38211521)

G1 Ballistic Coefficient = 0.19

Distance • Yards	Muzzle	25	50	75	100	125	150	175	200
Velocity • fps	950	929	909	890	872	855	839	824	809
Energy • ft-lbs	261	249	238	229	220	211	203	196	189
Taylor KO Index	6.3	6.2	6.0	5.9	5.8	5.7	5.6	5.5	5.4
Mid-Range Trajectory Height • Inches	0.0	0.3	1.2	2.9	5.2	8.4	12.3	17.1	22.8
Drop • Inches	0.0	-1.2	-5.0	-11.3	-20.4	-32.4	-47.4	-65.4	-86.6

Fiocchi 130-grain FMJ (38A)

G1 Ballistic Coefficient = 0.1

Distance • Yards	Muzzle	25	50	75	100	125	150	175	200
Velocity • fps	950	918	890	864	840	817	795	774	755
Energy • ft-lbs	260	244	228	215	203	193	183	173	164
Taylor KO Index	6.3	6.1	5.9	5.7	5.6	5.4	5.3	5.1	5.0
Mid-Range Trajectory Height • Inches	0.0	0.3	1.3	3.0	5.5	8.8	13.1	18.2	24.7
Drop • Inches	0.0	-1.2	-5.0	-11.6	-21.0	-33.6	-49.4	-68.6	-91.5

American Eagle (Federal) 130-grain Full Metal Jacket (AE38K)

G1 Ballistic Coefficient = 0.1

Distance • Yards	Muzzle	25	50	75	100	125	150	175	200
Velocity • fps	890	870	850	830	820	804	789	774	760
Energy • ft-lbs	230	220	210	200	195	187	180	173	167
Taylor KO Index	5.9	5.8	5.6	5.5	5.4	5.3	5.2	5.1	5.0
Mid-Range Trajectory Height • Inches	0.0	0.3	1.4	3.3	6.0	9.5	14.0	19.5	25.9
Drop • Inches	0.0	-1.4	-5.6	-12.9	-23.2	-36.8	-53.8	-74.2	-98.2

Magtech 130-grain FMJ (38T)

G1 Ballistic Coefficient = 0.190

Distance • Yards	Muzzle	25	50	75	100	125	150	175	200
Velocity • fps	800	786	773	759	746	734	721	709	698
Energy • ft-lbs	185	178	172	166	161	155	150	145	141
Taylor KO Index	5.3	5.2	5.1	5.0	5.0	4.9	4.8	4.7	4.6
Mid-Range Trajectory Height • Inches	0.0	0.4	1.8	4.0	7.3	11.6	17.0	23.5	31.2
Drop • Inches	0.0	-1.7	-7.0	-15.8	-28.5	-45.0	-65.6	-90.4	-119.4

USA (Winchester) 130-grain Full Metal Jacket (Q4171)

G1 Ballistic Coefficient = 0.150

Distance • Yards	Muzzle	25	50	75	100	125	150	175	200
Velocity • fps	800	782	765	749	733	717	702	688	673
Energy • ft-lbs	185	177	169	162	155	149	142	137	131
Taylor KO Index	5.3	5.2	5.1	5.0	4.9	4.8	4.7	4.6	4.5
Mid-Range Trajectory Height • Inches	0.0	0.4	1.8	4.1	7.4	11.8	17.4	24.3	32.4
Drop • Inches	0.0	-1.7	-7.0	-16.0	-28.8	-45.7	-66.9	-92.4	-121.5

UMC (Remington) 130-grain Metal Case (L38S11)

G1 Ballistic Coefficient = 0.217

Distance • Yards	Muzzle	25	50	75	100	125	150	175	200
Velocity • fps	790	778	766	755	743	733	722	711	701
Energy • ft-lbs	180	175	169	164	160	155	150	146	142
Taylor KO Index	5.3	5.2	5.1	5.0	4.9	4.9	4.8	4.7	4.7
Mid-Range Trajectory Height • Inches	0.0	0.4	1.8	4.1	7.4	11.7	17.2	23.8	31.5
Drop • Inches	0.0	-1.8	-7.1	-16.2	-29.0	-45.8	-66.6	-91.6	-120.9

PMC 132-grain EMJ NT (38EMG) + (38G) LF

G1 Ballistic Coefficient = 0.123

Distance • Yards	Muzzle	25	50	75	100	125	150	175	200
Velocity • fps	840	817	795	773	752	733	714	696	679
Energy • ft-lbs	207	195	185	175	166	158	150	142	135
Taylor KO Index	5.7	5.5	5.4	5.2	5.1	4.9	4.8	4.7	4.6
Mid-Range Trajectory Height • Inches	0.0	0.4	1.6	3.8	6.9	11.1	16.4	23.0	30.8
Drop • Inches	0.0	-1.6	-6.4	-14.7	-26.6	-42.3	-62.1	-86.2	-114.7

Speer 135-grain GDHP-SB +P (23921)

G1 Ballistic Coefficient = 0.142

Distance • Yards	Muzzle	25	50	75	100	125	150	175	200
Velocity • fps	860	838	818	798	780	762	745	728	712
Energy • ft-lbs	222	211	200	191	182	174	166	159	152
Taylor KO Index	5.9	5.8	5.6	5.5	5.4	5.2	5.1	5.0	4.9
Mid-Range Trajectory Height • Inches	0.0	0.4	1.5	3.6	6.5	10.4	15.3	21.4	28.6
Drop • Inches	0.0	-1.5	-6.1	-13.9	-25.2	-40.0	-58.5	-81.0	-107.5

Cor-Bon 147-grain FMJ (PM38147/50)

G1 Ballistic Coefficient = 0.140

Distance • Yards	Muzzle	25	50	75	100	125	150	175	200
Velocity • fps	900	876	853	831	811	791	773	755	738
Energy • ft-lbs	264	250	237	226	215	205	195	186	178
Taylor KO Index	6.7	6.6	6.4	6.2	6.1	5.9	5.8	5.7	5.5
Mid-Range Trajectory Height • Inches	0.0	0.3	1.4	3.3	6.0	9.6	14.2	19.8	26.5
Drop • Inches	0.0	-1.4	-5.6	-12.8	-23.1	-36.7	-53.8	-74.6	-99.1

Fiocchi 148-grain LWC (38LA)

G1 Ballistic Coefficient = 0.066

Distance • Yards	Muzzle	25	50	75	100	125	150	175	200
Velocity • fps	730	696	663	633	604	576	549	524	499
Energy • ft-lbs	175	159	144	132	119	109	99	90	82
Taylor KO Index	5.5	5.3	5.0	4.8	4.6	4.3	4.1	4.0	3.8
Mid-Range Trajectory Height • Inches	0.0	0.5	2.3	5.3	9.9	16.2	24.6	35.0	48.3
Drop • Inches	0.0	-2.1	-8.7	-20.2	-37.2	-60.1	-89.6	-126.3	-170.9

Black Hills 148-grain Match HBWC (D38N3)

G1 Ballistic Coefficient = 0.060

Distance • Yards	Muzzle	25	50	75	100	125	150	175	200
Velocity • fps	700	664	631	599	568	539	512	486	461
Energy • ft-lbs	161	145	131	118	106	96	86	78	70
Taylor KO Index	5.3	5.0	4.8	4.5	4.3	4.1	3.9	3.7	3.5
Mid-Range Trajectory Height • Inches	0.0	0.5	2.4	5.8	10.9	18.1	27.5	39.6	54.8
Drop • Inches	0.0	-2.3	-9.5	-22.2	-41.0	-66.4	-99.4	-140.6	-191.1

Federal 148-grain Lead Wadcutter Match (GM38A)

G1 Ballistic Coefficient = 0.061

Distance • Yards	Muzzle	25	50	75	100	125	150	175	200
Velocity • fps	690	650	610	570	540	508	477	448	420
Energy • ft-lbs	155	140	120	110	95	85	75	66	58
Taylor KO Index	5.2	4.9	4.6	4.3	4.1	3.8	3.6	3.4	3.2
Mid-Range Trajectory Height • Inches	0.0	0.6	2.5	6.2	11.7	19.5	30.0	43.6	60.9
Drop • Inches	0.0	-2.4	-9.9	-23.3	-43.3	-70.7	-106.6	-152.0	-208.2

Magtech 148-grain LWC (38B)

G1 Ballistic Coefficient = 0.055

Distance • Yards	Muzzle	25	50	75	100	125	150	175	200
Velocity • fps	710	670	634	599	566	536	507	479	452
Energy • ft-lbs	166	148	132	118	105	94	84	75	67
Taylor KO Index	5.4	5.1	4.8	4.5	4.3	4.0	3.8	3.6	3.4
Mid-Range Trajectory Height • Inches	0.0	0.6	2.4	5.7	10.8	17.9	27.4	39.7	54.9
Drop • Inches	0.0	-2.2	-9.3	-21.8	-40.4	-65.6	-98.4	-139.8	-190.6

Remington 148-grain Targetmaster Lead WC Match (R38S3)

G1 Ballistic Coefficient = 0.055

Distance • Yards	Muzzle	25	50	75	100	125	150	175	200
Velocity • fps	710	670	634	599	566	536	507	479	452
Energy • ft-lbs	166	148	132	118	105	94	84	75	67
Taylor KO Index	5.4	5.1	4.8	4.5	4.3	4.0	3.8	3.6	3.4
Mid-Range Trajectory Height • Inches	0.0	0.6	2.4	5.7	10.8	17.9	27.4	39.7	54.9
Drop • Inches	0.0	-2.2	-9.3	-21.8	-40.4	-65.6	-98.4	-139.8	-190.6

Winchester 148-grain Lead-Wad Cutter (X38SMRP)

G1 Ballistic Coefficient = 0.055

Distance • Yards	Muzzle	25	50	75	100	125	150	175	200
Velocity • fps	710	670	634	599	566	536	507	479	452
Energy • ft-lbs	166	148	132	118	105	94	84	75	67
Taylor KO Index	5.4	5.1	4.8	4.5	4.3	4.0	3.8	3.6	3.4
Mid-Range Trajectory Height • Inches	0.0	0.6	2.4	5.7	10.8	17.9	27.4	39.7	54.9
Drop • Inches	0.0	-2.2	-9.3	-21.8	-40.4	-65.6	-98.4	-139.8	-190.6

Sellier & Bellot 148-grain WC (V311002U)

G1 Ballistic Coefficient = 0.12

Distance • Yards	Muzzle	25	50	75	100	125	150	175	200
Velocity • fps	699	679	663	646	629	613	598	583	568
Energy • ft-lbs	162	153	145	137	130	124	117	112	106
Taylor KO Index	5.3	5.1	5.0	4.9	4.7	4.6	4.5	4.4	4.3
Mid-Range Trajectory Height • Inches	0.0	0.5	2.3	5.4	9.8	15.8	23.4	32.8	43.9
Drop • Inches	0.0	-2.3	-9.2	-21.1	-38.2	-60.8	-89.1	-123.5	-164.4

Fiocchi 148-grain SJHP (38E)

G1 Ballistic Coefficient = 0.15

Distance • Yards	Muzzle	25	50	75	100	125	150	175	200
Velocity • fps	750	734	719	704	690	676	662	649	636
Energy • ft-lbs	183	177	168	163	155	150	144	138	133
Taylor KO Index	5.7	5.5	5.4	5.3	5.2	5.1	5.0	4.9	4.8
Mid-Range Trajectory Height • Inches	0.0	0.5	2.0	4.6	8.4	13.4	19.7	27.1	36.6
Drop • Inches	0.0	-2.0	-8.0	-18.1	-32.7	-51.8	-75.7	-104.5	-138.5

USA (Winchester) 150-grain Lead Round Nose (Q4196)

G1 Ballistic Coefficient = 0.17

Distance • Yards	Muzzle	25	50	75	100	125	150	175	200
Velocity • fps	845	828	812	796	781	766	752	738	725
Energy • ft-lbs	238	228	219	211	203	196	188	182	175
Taylor KO Index	6.5	6.3	6.2	6.1	6.0	5.9	5.8	5.6	5.5
Mid-Range Trajectory Height • Inches	0.0	0.3	1.5	3.6	6.6	10.4	15.4	21.4	28.5
Drop • Inches	0.0	-1.5	-6.2	-14.3	-25.7	-40.7	-59.4	-82.0	-108.5

Sellier & Bellot 158-grain LRN (V311012U)

G1 Ballistic Coefficient = 0.1

Distance • Yards	Muzzle	25	50	75	100	125	150	175	200
Velocity • fps	997	971	946	924	903	884	865	848	831
Energy • ft-lbs	349	331	314	299	286	274	263	252	243
Taylor KO Index	8.1	7.8	7.6	7.5	7.3	7.1	7.0	6.9	6.7
Mid-Range Trajectory Height • Inches	0.0	0.3	1.2	2.7	4.9	7.8	11.5	16.0	21.3
Drop • Inches	0.0	-1.1	-4.5	-10.4	-18.8	-29.8	-43.7	-60.4	-80.2

Sellier & Bellot 158-grain SP (V311082U) + (V311082U)

G1 Ballistic Coefficient = 0.191

Distance • Yards	Muzzle	25	50	75	100	125	150	175	200
Velocity • fps	889	871	855	839	824	809	794	781	767
Energy • ft-lbs	277	266	256	247	238	229	221	214	207
Taylor KO Index	7.2	7.0	6.9	6.8	6.6	6.5	6.4	6.3	6.2
Mid-Range Trajectory Height • Inches	0.0	0.3	1.4	3.2	5.9	9.4	13.9	19.3	25.6
Drop • Inches	0.0	-1.4	-5.6	-12.9	-23.2	-36.7	-53.5	-73.8	-97.7

Fiocchi 158-grain LRN (38C)

G1 Ballistic Coefficient = 0.178

Distance • Yards	Muzzle	25	50	75	100	125	150	175	200
Velocity • fps	880	862	844	828	812	796	781	767	753
Energy • ft-lbs	271	261	250	240	231	222	214	206	199
Taylor KO Index	7.1	6.9	6.8	6.7	6.5	6.4	6.3	6.2	6.1
Mid-Range Trajectory Height • Inches	0.0	0.3	1.4	3.3	6.1	9.7	14.3	19.8	26.4
Drop • Inches	0.0	-1.4	-5.8	-13.2	-23.7	-37.6	-54.9	-75.7	-100.3

Magtech 158-grain LFN (38L)

G1 Ballistic Coefficient = 0.138

Distance • Yards	Muzzle	25	50	75	100	125	150	175	200
Velocity • fps	800	781	762	745	718	711	695	679	664
Energy • ft-lbs	225	214	204	195	186	177	169	162	155
Taylor KO Index	6.4	6.3	6.1	6.0	5.9	5.7	5.6	5.5	5.4
Mid-Range Trajectory Height • Inches	0.0	0.4	1.8	4.1	7.5	11.9	17.6	24.6	32.9
Drop • Inches	0.0	-1.7	-7.0	-16.0	-29.0	-46.0	-67.4	-93.2	-123.8

American Eagle (Federal) 158-grain Lead Round Nose (AE38B)

G1 Ballistic Coefficient = 0.192

Distance • Yards	Muzzle	25	50	75	100	125	150	175	200
Velocity • fps	770	760	750	730	720	708	696	685	643
Energy • ft-lbs	210	200	195	190	185	176	170	165	159
Taylor KO Index	6.2	6.1	6.0	5.9	5.8	5.7	5.6	5.5	5.4
Mid-Range Trajectory Height • Inches	0.0	0.5	1.9	4.3	7.9	12.5	18.3	24.9	33.6
Drop • Inches	0.0	-1.8	-7.5	-17.1	-30.7	-48.5	-70.6	-97.3	-128.5

Blazer (CCI) 158-grain LRN (3522)

G1 Ballistic Coefficient = 0.147

Distance • Yards	Muzzle	25	50	75	100	125	150	175	200
Velocity • fps	755	741	728	715	702	690	678	666	655
Energy • ft-lbs	200	193	186	179	173	167	161	156	150
Taylor KO Index	6.1	6.0	5.9	5.8	5.7	5.6	5.5	5.4	5.3
Mid-Range Trajectory Height • Inches	0.0	0.5	2.0	4.5	8.2	13.0	19.1	26.5	35.3
Drop • Inches	0.0	-1.9	-7.8	-17.8	-32.0	-50.6	-73.8	-101.7	-134.5

Magtech 158-grain LRN (38A) + (MP33A)

G1 Ballistic Coefficient = 0.147

Distance • Yards	Muzzle	25	50	75	100	125	150	175	200
Velocity • fps	755	739	723	707	692	678	663	649	636
Energy • ft-lbs	200	191	183	175	168	161	154	148	142
Taylor KO Index	6.1	6.0	5.8	5.7	5.6	5.5	5.3	5.2	5.1
Mid-Range Trajectory Height • Inches	0.0	0.5	2.0	4.6	8.3	13.2	19.5	27.2	36.3
Drop • Inches	0.0	-1.9	-7.8	-17.8	-32.4	-51.3	-75.0	-103.6	-137.3

Remington 158-grain Lead (Round Nose) (R38S5)

G1 Ballistic Coefficient = 0.147

Distance • Yards	Muzzle	25	50	75	100	125	150	175	200
Velocity • fps	755	739	723	707	692	678	663	649	636
Energy • ft-lbs	200	191	183	175	168	161	154	148	142
Taylor KO Index	6.1	6.0	5.8	5.7	5.6	5.5	5.3	5.2	5.1
Mid-Range Trajectory Height • Inches	0.0	0.5	2.0	4.6	8.3	13.2	19.5	27.2	36.3
Drop • Inches	0.0	-1.9	-7.8	-17.8	-32.4	-51.3	-75.0	-103.6	-137.3

UMC (Remington) 158-grain Lead Round Nose (L38S5)

G1 Ballistic Coefficient = 0.146

Distance • Yards	Muzzle	25	50	75	100	125	150	175	200
Velocity • fps	755	739	723	707	692	678	663	649	636
Energy • ft-lbs	200	191	183	175	168	161	154	148	142
Taylor KO Index	6.1	6.0	5.8	5.7	5.6	5.5	5.3	5.2	5.1
Mid-Range Trajectory Height • Inches	0.0	0.5	2.0	4.6	8.3	13.2	19.5	27.2	36.3
Drop • Inches	0.0	-1.9	-7.8	-17.8	-32.4	-51.3	-75.0	-103.6	-137.3

Winchester 158-grain Lead – Round Nose (X38S1P)
G1 Ballistic Coefficient = 0.147

Distance • Yards	Muzzle	25	50	75	100	125	150	175	200
Velocity • fps	755	739	723	707	692	678	663	649	636
Energy • ft-lbs	200	191	183	175	168	161	154	148	142
Taylor KO Index	6.1	6.0	5.8	5.7	5.6	5.5	5.3	5.2	5.1
Mid-Range Trajectory Height • Inches	0.0	0.5	2.0	4.6	8.3	13.2	19.5	27.2	36.3
Drop • Inches	0.0	-1.9	-7.8	-17.8	-32.4	-51.3	-75.0	-103.6	-137.3

Aguila 158-grain SJHP (1E382520)
G1 Ballistic Coefficient = 0.180

Distance • Yards	Muzzle	25	50	75	100	125	150	175	200
Velocity • fps	900	881	863	845	829	813	798	783	769
Energy • ft-lbs	284	272	261	251	241	232	223	215	207
Taylor KO Index	7.3	7.1	7.0	6.8	6.7	6.6	6.4	6.3	6.2
Mid-Range Trajectory Height • Inches	0.0	0.3	1.4	3.2	5.9	9.3	13.7	18.8	25.3
Drop • Inches	0.0	-1.4	-5.5	-12.6	-22.7	-36.0	-52.6	-72.5	-96.1

Remington 158-grain Lead Hollow Point +P (R38S12)
G1 Ballistic Coefficient = 0.188

Distance • Yards	Muzzle	25	50	75	100	125	150	175	200
Velocity • fps	890	872	855	839	823	808	794	780	766
Energy • ft-lbs	278	267	257	247	238	229	221	213	206
Taylor KO Index	7.2	7.0	6.9	6.8	6.6	6.5	6.4	6.3	6.2
Mid-Range Trajectory Height • Inches	0.0	0.3	1.4	3.3	6.6	9.4	13.8	19.2	25.6
Drop • Inches	0.0	-1.4	-5.6	-12.8	-23.2	-36.8	-53.5	-73.7	-97.6

Winchester 158-grain Lead Wad Cutter Hollow Point +P (X38SPD)
G1 Ballistic Coefficient = 0.188

Distance • Yards	Muzzle	25	50	75	100	125	150	175	200
Velocity • fps	890	872	855	839	823	808	794	780	766
Energy • ft-lbs	278	267	257	247	238	229	221	213	206
Taylor KO Index	7.2	7.0	6.9	6.8	6.6	6.5	6.4	6.3	6.2
Mid-Range Trajectory Height • Inches	0.0	0.3	1.4	3.3	6.6	9.4	13.8	19.2	25.6
Drop • Inches	0.0	-1.4	-5.6	-12.8	-23.2	-36.8	-53.5	-73.7	-97.6

Magtech 158-grain LFN (38L)
G1 Ballistic Coefficient = 0.13

Distance • Yards	Muzzle	25	50	75	100	125	150	175	200
Velocity • fps	800	781	762	745	718	711	695	679	664
Energy • ft-lbs	225	214	204	195	186	177	169	162	155
Taylor KO Index	6.4	6.3	6.1	6.0	5.9	5.7	5.6	5.5	5.4
Mid-Range Trajectory Height • Inches	0.0	0.4	1.8	4.1	7.5	11.9	17.6	24.6	32.9
Drop • Inches	0.0	-1.7	-7.0	-16.0	-29.0	-46.0	-67.4	-93.2	-123.8

Magtech 158-grain LSWC (38J)
G1 Ballistic Coefficient = 0.14

Distance • Yards	Muzzle	25	50	75	100	125	150	175	200
Velocity • fps	755	738	721	705	689	674	659	645	630
Energy • ft-lbs	200	191	182	174	167	159	152	146	139
Taylor KO Index	6.1	5.9	5.8	5.7	5.6	5.4	5.3	5.2	5.1
Mid-Range Trajectory Height • Inches	0.0	0.5	2.0	4.6	8.4	13.3	19.6	27.0	36.2
Drop • Inches	0.0	-1.9	-7.9	-18.0	-32.4	-51.5	-75.3	-104.1	-138.2

Remington 158-grain Semi-Wadcutter (R38S6)
G1 Ballistic Coefficient = 0.14

Distance • Yards	Muzzle	25	50	75	100	125	150	175	200
Velocity • fps	755	739	723	707	692	678	663	649	636
Energy • ft-lbs	200	191	183	175	168	161	154	148	142
Taylor KO Index	6.1	6.0	5.8	5.7	5.6	5.5	5.3	5.2	5.1
Mid-Range Trajectory Height • Inches	0.0	0.5	2.0	4.6	8.3	13.2	19.5	27.2	36.3
Drop • Inches	0.0	-1.9	-7.8	-17.8	-32.4	-51.3	-75.0	-103.6	-137.3

Winchester 158-grain Lead – Semi Wad Cutter (X38WCPSV)
G1 Ballistic Coefficient = 0.1

Distance • Yards	Muzzle	25	50	75	100	125	150	175	200
Velocity • fps	755	738	721	705	689	674	659	645	630
Energy • ft-lbs	200	191	182	174	167	159	152	146	139
Taylor KO Index	6.1	5.9	5.8	5.7	5.6	5.4	5.3	5.2	5.1
Mid-Range Trajectory Height • Inches	0.0	0.5	2.0	4.6	8.4	13.3	19.6	27.0	36.2
Drop • Inches	0.0	-1.9	-7.9	-18.0	-32.4	-51.5	-75.3	-104.1	-138.2

Magtech 158-grain SJFP +P (38N)

G1 Ballistic Coefficient = 0.079

Distance • Yards	Muzzle	25	50	75	100	125	150	175	200
Velocity • fps	890	849	812	778	746	717	689	662	636
Energy • ft-lbs	278	253	231	212	196	180	166	154	142
Taylor KO Index	7.2	6.9	6.6	6.3	6.0	5.8	5.6	5.3	5.1
Mid-Range Trajectory Height • Inches	0.0	0.3	1.5	3.6	6.6	10.8	16.2	23.1	31.4
Drop • Inches	0.0	-1.4	-5.9	-13.6	-24.9	-40.1	-59.6	-83.6	-112.6

Sellier & Bellot 158-grain SP (V311022U)

G1 Ballistic Coefficient = 0.057

Distance • Yards	Muzzle	25	50	75	100	125	150	175	200
Velocity • fps	889	835	784	741	701	663	628	594	563
Energy • ft-lbs	278	245	216	193	172	154	138	124	111
Taylor KO Index	7.2	6.7	6.3	6.0	5.7	5.4	5.1	4.8	4.5
Mid-Range Trajectory Height • Inches	0.0	0.4	1.6	3.8	7.1	11.7	17.9	26.0	35.9
Drop • Inches	0.0	-1.4	-6.0	-14.1	-26.2	-42.6	-64.1	-91.0	-124.1

Magtech 158-grain SJSP Flat (38C)

G1 Ballistic Coefficient = 0.175

Distance • Yards	Muzzle	25	50	75	100	125	150	175	200
Velocity • fps	744	730	723	704	692	679	667	656	644
Energy • ft-lbs	200	187	183	174	168	162	156	151	146
Taylor KO Index	6.0	5.9	5.8	5.7	5.6	5.5	5.4	5.3	5.2
Mid-Range Trajectory Height • Inches	0.0	0.5	2.0	4.7	8.5	13.4	19.7	27.4	36.4
Drop • Inches	0.0	-2.0	-8.0	-18.3	-33.0	-52.2	-76.1	-104.9	-138.7

Magtech 158-grain SJHP +P (38H)

G1 Ballistic Coefficient = 0.149

Distance • Yards	Muzzle	25	50	75	100	125	150	175	200
Velocity • fps	890	868	847	827	808	789	772	755	739
Energy • ft-lbs	278	264	251	240	229	219	209	200	192
Taylor KO Index	7.2	7.0	6.8	6.7	6.5	6.4	6.2	6.1	6.0
Mid-Range Trajectory Height • Inches	0.0	0.3	1.4	3.3	6.0	9.7	14.3	19.9	26.7
Drop • Inches	0.0	-1.4	-5.7	-13.0	-23.5	-37.3	-54.6	-75.6	-100.3

Magtech 158-grain SJHP (38E)

G1 Ballistic Coefficient = 0.192

Distance • Yards	Muzzle	25	50	75	100	125	150	175	200
Velocity • fps	807	793	779	766	753	740	728	716	704
Energy • ft-lbs	230	221	213	206	199	192	186	180	174
Taylor KO Index	6.5	6.4	6.3	6.2	6.1	6.0	5.9	5.8	5.7
Mid-Range Trajectory Height • Inches	0.0	0.4	1.7	3.9	7.2	11.3	16.6	23.1	30.6
Drop • Inches	0.0	-1.7	-6.8	-15.6	-28.0	-44.2	-64.4	-88.7	-117.3

Hornady 158-grain JHP/XTP (90362)

G1 Ballistic Coefficient = 0.145

Distance • Yards	Muzzle	25	50	75	100	125	150	175	200
Velocity • fps	800	782	765	747	731	715	699	684	670
Energy • ft-lbs	225	214	205	196	188	179	172	164	157
Taylor KO Index	6.4	6.3	6.2	6.0	5.9	5.8	5.6	5.5	5.4
Mid-Range Trajectory Height • Inches	0.0	0.4	1.8	4.1	7.4	11.8	17.5	24.4	32.5
Drop • Inches	0.0	-1.7	-7.0	-16.0	-28.9	-45.8	-67.0	-92.7	-123.0

Magtech 158-grain FMJ Flat (38Q)

G1 Ballistic Coefficient = 0.162

Distance • Yards	Muzzle	25	50	75	100	125	150	175	200
Velocity • fps	938	914	891	871	851	833	815	798	782
Energy • ft-lbs	309	293	279	266	254	243	233	224	214
Taylor KO Index	7.6	7.4	7.2	7.0	6.9	6.7	6.6	6.4	6.3
Mid-Range Trajectory Height • Inches	0.0	0.3	1.3	3.0	5.5	8.7	12.9	18.0	24.0
Drop • Inches	0.0	-1.3	-5.1	-11.7	-21.2	-33.6	-49.2	-68.0	-90.3

Speer 158-grain TMJ FN +P (53750)

G1 Ballistic Coefficient = 0.172

Distance • Yards	Muzzle	25	50	75	100	125	150	175	200
Velocity • fps	900	880	861	843	826	809	794	778	764
Energy • ft-lbs	284	272	260	249	240	230	221	213	205
Taylor KO Index	7.3	7.1	6.9	6.8	6.7	6.5	6.4	6.3	6.2
Mid-Range Trajectory Height • Inches	0.0	0.3	1.4	3.2	5.9	9.4	13.8	18.9	25.5
Drop • Inches	0.0	-1.4	-5.5	-12.6	-22.8	-36.1	-52.8	-72.9	-96.6

Speer 158-grain Clean-Fire – TMJ FN +P Lawman (53833)

G1 Ballistic Coefficient = 0.172

Distance • Yards	Muzzle	25	50	75	100	125	150	175	200
Velocity • fps	900	880	861	843	826	809	794	778	764
Energy • ft-lbs	284	272	260	249	240	230	221	213	205
Taylor KO Index	7.3	7.1	6.9	6.8	6.7	6.5	6.4	6.3	6.2
Mid-Range Trajectory Height • Inches	0.0	0.3	1.4	3.2	5.9	9.4	13.8	18.9	25.5
Drop • Inches	0.0	-1.4	-5.5	-12.6	-22.8	-36.1	-52.8	-72.9	-96.6

Blazer (CCI) 158-grain TMJ +P (3519)

G1 Ballistic Coefficient = 0.175

Distance • Yards	Muzzle	25	50	75	100	125	150	175	200
Velocity • fps	850	833	816	800	785	771	756	742	729
Energy • ft-lbs	253	243	234	225	216	208	201	193	186
Taylor KO Index	6.9	6.7	6.6	6.5	6.3	6.2	6.1	6.0	5.9
Mid-Range Trajectory Height • Inches	0.0	0.4	1.6	3.6	6.5	10.3	15.3	21.2	28.2
Drop • Inches	0.0	-1.5	-6.2	-14.1	-25.4	-40.2	-58.7	-81.0	-107.2

Blazer (CCI) 158-grain Clean-Fire – TMJ +P (3475)

G1 Ballistic Coefficient = 0.175

Distance • Yards	Muzzle	25	50	75	100	125	150	175	200
Velocity • fps	850	833	816	800	785	771	756	742	729
Energy • ft-lbs	253	243	234	225	216	208	201	193	186
Taylor KO Index	6.9	6.7	6.6	6.5	6.3	6.2	6.1	6.0	5.9
Mid-Range Trajectory Height • Inches	0.0	0.4	1.6	3.6	6.5	10.3	15.3	21.2	28.2
Drop • Inches	0.0	-1.5	-6.2	-14.1	-25.4	-40.2	-58.7	-81.0	-107.2

PMC 158-grain EMJ NT (38EMM) LF

G1 Ballistic Coefficient = 0.12

Distance • Yards	Muzzle	25	50	75	100	125	150	175	200
Velocity • fps	800	780	760	741	722	704	687	671	655
Energy • ft-lbs	225	213	202	192	183	174	166	158	150
Taylor KO Index	6.4	6.3	6.1	6.0	5.8	5.7	5.5	5.4	5.3
Mid-Range Trajectory Height • Inches	0.0	0.4	1.8	4.1	7.5	12.1	17.9	24.6	33.4
Drop • Inches	0.0	-1.7	-7.0	-16.1	-29.1	-46.3	-67.9	-94.1	-125.1

Magtech 158-grain FMJ Flat (38P) & (CR38A)

G1 Ballistic Coefficient = 0.14

Distance • Yards	Muzzle	25	50	75	100	125	150	175	200
Velocity • fps	755	739	723	707	692	678	664	650	637
Energy • ft-lbs	200	191	183	176	168	161	155	148	142
Taylor KO Index	6.1	6.0	5.8	5.7	5.6	5.5	5.4	5.3	5.1
Mid-Range Trajectory Height • Inches	0.0	0.5	2.0	4.6	8.3	13.3	19.6	27.2	36.3
Drop • Inches	0.0	-1.9	-7.9	-17.9	-32.3	-51.3	-74.9	-103.5	-137.3

Fiocchi 158-grain Full Metal Jacket (38G)

G1 Ballistic Coefficient = 0.14

Distance • Yards	Muzzle	25	50	75	100	125	150	175	200
Velocity • fps	730	714	699	684	670	656	642	629	616
Energy • ft-lbs	186	179	171	164	157	151	145	139	133
Taylor KO Index	5.9	5.8	5.6	5.5	5.4	5.3	5.2	5.1	5.0
Mid-Range Trajectory Height • Inches	0.0	0.5	2.2	4.9	8.9	14.2	20.9	28.7	38.8
Drop • Inches	0.0	-2.1	-8.4	-19.2	-34.6	-54.8	-80.1	-110.6	-146.6

Cowboy Action Loads:

Black Hills 158-grain CNL (DCB38N1)

G1 Ballistic Coefficient = 0.1

Distance • Yards	Muzzle	25	50	75	100	125	150	175	200
Velocity • fps	800	781	763	745	728	712	696	681	666
Energy • ft-lbs	225	214	204	195	186	178	170	163	155
Taylor KO Index	6.4	6.3	6.1	6.0	5.9	5.7	5.6	5.5	5.4
Mid-Range Trajectory Height • Inches	0.0	0.4	1.8	4.1	7.5	11.2	16.9	24.5	37.0
Drop • Inches	0.0	-1.7	-7.0	-16.0	-29.0	-46.0	-67.3	-93.0	-123.5

Magtech 158-grain Lead Flat Nose (38L)

G1 Ballistic Coefficient = 0.1

Distance • Yards	Muzzle	25	50	75	100	125	150	175	200
Velocity • fps	800	788	776	764	753	742	731	720	710
Energy • ft-lbs	225	218	211	205	199	193	187	182	177
Taylor KO Index	6.4	6.3	6.3	6.2	6.1	6.0	5.9	5.8	5.7
Mid-Range Trajectory Height • Inches	0.0	0.3	1.7	3.9	7.2	11.4	16.7	23.1	30.6
Drop • Inches	0.0	-1.7	-6.9	-15.8	-28.3	-44.7	-65.0	-89.4	-118.0

Winchester 158-grain Cast Lead (USA38CB)

G1 Ballistic Coefficient = 0.133

Distance • Yards	Muzzle	25	50	75	100	125	150	175	200
Velocity • fps	800	780	761	743	725	708	691	675	659
Energy • ft-lbs	225	214	203	193	185	176	168	160	153
Taylor KO Index	6.4	6.3	6.1	6.0	5.8	5.7	5.6	5.4	5.3
Mid-Range Trajectory Height • Inches	0.0	0.4	1.8	4.1	7.5	11.6	17.7	24.7	33.1
Drop • Inches	0.0	-1.7	-7.0	-16.1	-29.1	-46.2	-67.6	-93.6	-124.3

Fiocchi 158-grain LRNFP (38CA)

G1 Ballistic Coefficient = 0.139

Distance • Yards	Muzzle	25	50	75	100	125	150	175	200
Velocity • fps	640	626	612	599	585	572	560	548	536
Energy • ft-lbs	144	137	131	126	120	115	110	105	101
Taylor KO Index	5.2	5.0	4.9	4.8	4.7	4.6	4.5	4.4	4.3
Mid-Range Trajectory Height • Inches	0.0	0.7	2.8	6.4	11.6	18.5	27.4	38.1	50.9
Drop • Inches	0.0	-2.7	-10.9	-25.0	-45.1	-71.5	-104.6	-144.6	-191.9

...good ergonomic grip on any handgun makes a big difference in the comfort of shooting it, in this case a .357 Magnum. (Safari Press Archives)

.40 Smith & Wesson

This cartridge came on the scene (1990) just after the 10mm Auto was announced. It's easy to guess that some bright lad recognized that if the 10mm cartridge was simply shortened by 0.140 inches, the resulting ammunition would fit in many 9mm pistol frames and still retain enough volume to have plenty of power for personal defense. This conversion resulted in an effective round in a more compact pistol than what was needed for the 10mm. The .40 S&W has become the cartridge of choice for many law enforcement groups in the U.S.

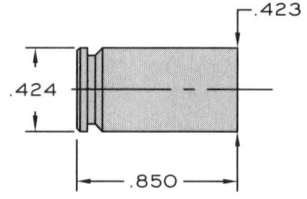

Relative Recoil Factor = 0.74

Specifications:

Controlling Agency for Standardization of this Ammunition: SAAMI

Bullet Weight Grains	Velocity fps	Maximum Average Pressure Copper Crusher	Transducer
155 STHP	1,195	N/S	35,000 psi
155 FMJ	1,115	N/S	35,000 psi
180 JHP	985	N/S	35,000 psi

Standard barrel for velocity testing: 4 inches long—1 turn in 16-inch twist

Availability:

CCI 88-grain Shotshell (3970)
Number 9 shot at 1250 fps.

Cor-Bon 115-grain Glaser Safety Slug (03000/03200)
G1 Ballistic Coefficient = 0.11

Distance • Yards	Muzzle	25	50	75	100	125	150	175	200
Velocity • fps	1400	1281	1181	1101	1039	990	948	912	880
Energy • ft-lbs	501	419	356	310	276	250	230	213	198
Taylor KO Index	9.2	8.4	7.8	7.2	6.8	6.5	6.2	6.0	5.8
Mid-Range Trajectory Height • Inches	0.0	0.2	0.7	1.6	3.1	5.2	8.1	11.6	16.0
Drop • Inches	0.0	-0.6	-2.5	-6.0	-11.2	-18.5	-28.0	-39.9	-44.4

Fiocchi 125-grain Frangible Sinterfire (40SFNT) LF
G1 Ballistic Coefficient = 0.1?

Distance • Yards	Muzzle	25	50	75	100	125	150	175	200
Velocity • fps	1265	1175	1102	1045	998	959	924	894	866
Energy • ft-lbs	444	383	337	303	278	255	237	222	208
Taylor KO Index	9.3	8.6	8.1	7.7	7.3	7.0	6.8	6.5	6.3
Mid-Range Trajectory Height • Inches	0.0	0.2	0.8	1.9	3.5	5.9	8.9	12.7	17.4
Drop • Inches	0.0	-0.9	-3.6	-8.4	-15.3	-24.5	-36.2	-50.5	-67.2

PMC 125-grain SinterFire/Frangible (40F)
G1 Ballistic Coefficient = 0.1

Distance • Yards	Muzzle	25	50	75	100	125	150	175	200
Velocity • fps	1250	1172	1107	1055	1012	975	943	914	888
Energy • ft-lbs	433	381	341	309	284	264	247	232	219
Taylor KO Index	8.9	8.4	7.9	7.5	7.2	7.0	6.7	6.5	6.3
Mid-Range Trajectory Height • Inches	0.0	0.2	0.8	1.9	3.5	5.8	8.7	12.4	16.9
Drop • Inches	0.0	-0.7	-3.0	-7.1	-13.1	-21.3	-31.7	-44.6	-60.1

Magtech 130-grain SCHP (FD40A)
G1 Ballistic Coefficient = 0.1

Distance • Yards	Muzzle	25	50	75	100	125	150	175	200
Velocity • fps	1190	1132	1082	1041	1006	975	947	922	899
Energy • ft-lbs	409	370	338	313	292	274	259	246	233
Taylor KO Index	9.1	8.6	8.2	7.9	7.7	7.4	7.2	7.0	6.8
Mid-Range Trajectory Height • Inches	0.0	0.2	0.8	2.0	3.7	6.0	9.0	12.7	17.1
Drop • Inches	0.0	-0.8	-3.3	-7.6	-14.0	-22.5	-33.2	-46.4	-62.0

Cor-Bon 135-grain JHP (SD40135/20)
G1 Ballistic Coefficient = 0.125

Distance • Yards	Muzzle	25	50	75	100	125	150	175	200
Velocity • fps	1325	1229	1148	1083	1031	988	951	919	890
Energy • ft-lbs	526	453	395	352	319	293	271	253	238
Taylor KO Index	10.2	9.5	8.9	8.4	8.0	7.6	7.3	7.1	6.9
Mid-Range Trajectory Height • Inches	0.0	0.1	0.7	1.7	3.2	5.4	8.3	11.9	16.2
Drop • Inches	0.0	-0.7	-2.7	-6.5	-12.1	-19.7	-29.6	-41.8	-56.6

Cor-Bon 135-grain Pow'RBall (PB40135/20)
G1 Ballistic Coefficient = 0.120

Distance • Yards	Muzzle	25	50	75	100	125	150	175	200
Velocity • fps	1325	1225	1143	1077	1024	981	944	911	882
Energy • ft-lbs	526	450	391	347	314	288	267	249	233
Taylor KO Index	10.2	9.5	8.8	8.3	7.9	7.6	7.3	7.0	6.8
Mid-Range Trajectory Height • Inches	0.0	0.2	0.8	1.9	3.5	5.8	8.7	12.5	17.0
Drop • Inches	0.0	-0.6	-2.8	-6.5	-12.2	-19.9	-29.8	-42.2	-57.2

Federal 135-grain Hydra-Shok JHP (PD40HS4 H)
G1 Ballistic Coefficient = 0.113

Distance • Yards	Muzzle	25	50	75	100	125	150	175	200
Velocity • fps	1200	1120	1050	1000	960	924	892	862	835
Energy • ft-lbs	430	375	330	300	275	256	238	223	209
Taylor KO Index	9.3	8.6	8.1	7.7	7.4	7.1	6.9	6.6	6.4
Mid-Range Trajectory Height • Inches	0.0	0.2	0.9	2.1	3.9	6.4	9.7	13.7	18.8
Drop • Inches	0.0	-0.8	-3.3	-7.8	-14.4	-23.4	-35.0	-49.3	-66.4

Federal 135-grain EFMJ (PD40CSP2H)
G1 Ballistic Coefficient = 0.100

Distance • Yards	Muzzle	25	50	75	100	125	150	175	200
Velocity • fps	1200	1110	1040	990	940	903	869	838	809
Energy • ft-lbs	430	370	325	290	265	244	226	210	196
Taylor KO Index	9.3	8.6	8.0	7.6	7.3	7.0	6.7	6.5	6.2
Mid-Range Trajectory Height • Inches	0.0	0.2	0.9	2.1	4.0	6.6	10.1	14.3	19.6
Drop • Inches	0.0	-0.8	-3.4	-7.9	-14.7	-24.0	-35.9	-50.7	-68.6

Winchester 140-grain Jacketed Flat Point (SC40NT) LF
G1 Ballistic Coefficient = 0.130

Distance • Yards	Muzzle	25	50	75	100	125	150	175	200
Velocity • fps	1155	1091	1039	996	960	928	900	873	849
Energy • ft-lbs	415	370	336	309	286	268	252	237	224
Taylor KO Index	9.2	8.7	8.3	8.0	7.7	7.4	7.2	7.0	6.8
Mid-Range Trajectory Height • Inches	0.0	0.2	0.9	2.1	4.0	6.5	9.7	13.8	18.6
Drop • Inches	0.0	-0.8	-3.5	-8.2	-15.1	-24.2	-35.9	-50.2	-67.4

Cor-Bon 140-grain DPX (DPX40140/20) + (TR40140/20)
G1 Ballistic Coefficient = 0.135

Distance • Yards	Muzzle	25	50	75	100	125	150	175	200
Velocity • fps	1200	1130	1073	1026	987	953	923	896	871
Energy • ft-lbs	448	397	358	327	303	283	265	250	236
Taylor KO Index	9.6	9.1	8.6	8.2	7.9	7.6	7.4	7.2	7.0
Mid-Range Trajectory Height • Inches	0.0	0.2	0.9	2.0	3.8	6.2	9.3	13.1	17.8
Drop • Inches	0.0	-0.8	-3.3	-7.7	-14.1	-22.7	-33.8	-47.4	-63.7

Black Hills 140-grain Barnes Tac-XP (D40N6) LF
G1 Ballistic Coefficient = 0.150

Distance • Yards	Muzzle	25	50	75	100	125	150	175	200
Velocity • fps	1150	1095	1049	1010	977	947	921	896	874
Energy • ft-lbs	411	373	342	317	297	279	264	250	237
Taylor KO Index	9.4	9.0	8.6	8.3	8.0	7.8	7.6	7.3	7.2
Mid-Range Trajectory Height • Inches	0.0	0.2	0.9	2.1	3.9	6.4	9.5	13.5	18.2
Drop • Inches	0.0	-0.8	-3.5	-8.2	-14.9	-24.0	-35.4	-49.4	-66.2

Cor-Bon 150-grain JHP (SD40150/20)
G1 Ballistic Coefficient = 0.140

Distance • Yards	Muzzle	25	50	75	100	125	150	175	200
Velocity • fps	1200	1132	1076	1030	992	958	929	902	877
Energy • ft-lbs	480	427	386	354	328	306	287	271	256
Taylor KO Index	10.3	9.7	9.2	8.8	8.5	8.2	8.0	7.7	7.5
Mid-Range Trajectory Height • Inches	0.0	0.2	0.8	2.0	3.7	6.1	9.1	13.0	17.6
Drop • Inches	0.0	-0.8	-3.3	-7.6	-14.0	-22.6	-33.6	-47.1	-63.2

Magtech 155-grain JHP (40D)
G1 Ballistic Coefficient = 0.166

Distance • Yards	Muzzle	25	50	75	100	125	150	175	200
Velocity • fps	1205	1146	1096	1051	1018	987	959	934	911
Energy • ft-lbs	500	452	414	383	357	335	317	300	286
Taylor KO Index	10.9	10.4	10.0	9.6	9.2	9.0	8.7	8.5	8.3
Mid-Range Trajectory Height • Inches	0.0	0.2	0.8	1.9	3.6	5.9	8.8	12.4	16.7
Drop • Inches	0.0	-0.8	-3.2	-7.4	-13.6	-21.9	-32.4	-45.3	-60.6

Magtech 155-grain JHP (GG40A)
G1 Ballistic Coefficient = 0.217

Distance • Yards	Muzzle	25	50	75	100	125	150	175	200
Velocity • fps	1205	1159	1118	1083	1052	1024	999	977	956
Energy • ft-lbs	500	463	430	404	381	361	344	329	315
Taylor KO Index	10.7	10.3	9.9	9.6	9.3	9.1	8.8	8.7	8.5
Mid-Range Trajectory Height • Inches	0.0	0.2	0.8	1.9	3.5	5.6	8.3	11.6	15.7
Drop • Inches	0.0	-0.8	-3.2	-7.3	-13.2	-21.2	-31.2	-43.4	-57.8

Remington 155-grain Jacketed Hollow Point (R40SW1)
G1 Ballistic Coefficient = 0.165

Distance • Yards	Muzzle	25	50	75	100	125	150	175	200
Velocity • fps	1205	1146	1095	1053	1017	986	956	933	910
Energy • ft-lbs	499	452	413	382	356	335	316	300	285
Taylor KO Index	10.7	10.2	9.7	9.3	9.0	8.7	8.5	8.3	8.1
Mid-Range Trajectory Height • Inches	0.0	0.2	0.8	1.9	3.6	5.8	8.7	12.3	16.6
Drop • Inches	0.0	-0.8	-3.2	-7.4	-13.6	-21.9	-32.4	-45.2	-60.5

Winchester 155-grain Silvertip Hollow Point (X40SWSTHP)
G1 Ballistic Coefficient = 0.166

Distance • Yards	Muzzle	25	50	75	100	125	150	175	200
Velocity • fps	1205	1146	1096	1054	1018	987	959	934	911
Energy • ft-lbs	500	452	414	382	357	335	317	300	285
Taylor KO Index	10.7	10.2	9.7	9.3	9.0	8.7	8.5	8.3	8.1
Mid-Range Trajectory Height • Inches	0.0	0.2	0.8	1.9	3.6	5.8	8.7	12.3	16.6
Drop • Inches	0.0	-0.8	-3.2	-7.4	-13.6	-21.9	-32.4	-45.2	-60.4

Speer 155-grain GDHP (23961)
G1 Ballistic Coefficient = 0.12

Distance • Yards	Muzzle	25	50	75	100	125	150	175	200
Velocity • fps	1200	1124	1063	1015	974	939	909	880	854
Energy • ft-lbs	496	435	389	355	326	304	284	266	251
Taylor KO Index	10.6	10.0	9.4	9.0	8.6	8.3	8.1	7.8	7.6
Mid-Range Trajectory Height • Inches	0.0	0.2	0.8	2.0	3.8	6.2	9.4	13.3	18.0
Drop • Inches	0.0	-0.8	-3.3	-7.7	-14.2	-23.0	-34.2	-48.0	-64.6

Hornady 155-grain JHP / XTP (9132)
G1 Ballistic Coefficient = 0.13

Distance • Yards	Muzzle	25	50	75	100	125	150	175	200
Velocity • fps	1180	1115	1061	1017	980	948	919	893	868
Energy • ft-lbs	479	428	388	356	331	309	291	274	260
Taylor KO Index	10.5	9.9	9.4	9.0	8.7	8.4	8.1	7.9	7.7
Mid-Range Trajectory Height • Inches	0.0	0.2	0.9	2.1	3.8	6.2	9.3	13.2	17.8
Drop • Inches	0.0	-0.8	-3.4	-7.9	-14.4	-23.2	-34.4	-48.2	+64.6

Hornady 155-grain TAP-FPD (91328)
G1 Ballistic Coefficient = 0.13

Distance • Yards	Muzzle	25	50	75	100	125	150	175	200
Velocity • fps	1180	1115	1061	1017	980	948	919	893	868
Energy • ft-lbs	479	428	388	356	331	309	291	274	260
Taylor KO Index	10.5	9.9	9.4	9.0	8.7	8.4	8.1	7.9	7.7
Mid-Range Trajectory Height • Inches	0.0	0.2	0.9	2.1	3.8	6.2	9.3	13.2	17.8
Drop • Inches	0.0	-0.8	-3.4	-7.9	-14.4	-23.2	-34.4	-48.2	+64.6

Fiocchi 155-grain XTPHP (40XTP)
G1 Ballistic Coefficient = 0.1

Distance • Yards	Muzzle	25	50	75	100	125	150	175	200
Velocity • fps	1160	1098	1047	1006	970	939	910	885	861
Energy • ft-lbs	463	415	377	348	324	303	285	269	255
Taylor KO Index	10.3	9.7	9.3	8.9	8.6	8.3	8.1	7.8	7.6
Mid-Range Trajectory Height • Inches	0.0	0.2	0.9	2.1	4.0	6.4	9.7	13.5	18.4
Drop • Inches	0.0	-0.8	-3.5	-8.1	-14.9	-23.9	-35.5	-49.6	-66.5

Black Hills 155-grain Jacketed Hollow Point (D40N1)

G1 Ballistic Coefficient = 0.125

Distance • Yards	Muzzle	25	50	75	100	125	150	175	200
Velocity • fps	1150	1085	1032	989	952	920	891	865	840
Energy • ft-lbs	450	405	367	337	312	292	273	257	243
Taylor KO Index	10.2	9.6	9.1	8.8	8.4	8.1	7.9	7.7	7.4
Mid-Range Trajectory Height • Inches	0.0	0.2	0.9	2.2	4.1	6.6	9.9	14.0	18.9
Drop • Inches	0.0	-0.9	-3.6	-8.3	-15.2	-24.5	-36.4	-50.9	-68.3

Federal 155-grain Hydra-Shok JHP (P40HS2)

G1 Ballistic Coefficient = 0.067

Distance • Yards	Muzzle	25	50	75	100	125	150	175	200
Velocity • fps	1140	1080	1030	990	950	918	890	864	839
Energy • ft-lbs	445	400	365	335	315	290	273	257	243
Taylor KO Index	10.1	9.6	9.1	8.8	8.4	8.1	7.9	7.7	7.4
Mid-Range Trajectory Height • Inches	0.0	0.2	0.9	2.2	4.1	6.6	9.9	14.1	19.0
Drop • Inches	0.0	-0.9	-3.8	-9.1	-17.1	-24.8	-36.8	-51.4	-68.9

Blazer (CCI) 155-grain TMJ (3587)

G1 Ballistic Coefficient = 0.125

Distance • Yards	Muzzle	25	50	75	100	125	150	175	200
Velocity • fps	1175	1104	1047	1001	963	929	899	872	846
Energy • ft-lbs	475	420	377	345	319	297	278	262	246
Taylor KO Index	10.4	9.8	9.3	8.9	8.5	8.2	8.0	7.7	7.5
Mid-Range Trajectory Height • Inches	0.0	0.2	0.9	2.1	4.0	6.4	9.6	13.6	18.4
Drop • Inches	0.0	-0.8	-3.4	-8.0	-14.8	-23.8	-35.3	-49.5	-66.5

Speer 155-grain TMJ FN (53957)

G1 Ballistic Coefficient = 0.125

Distance • Yards	Muzzle	25	50	75	100	125	150	175	200
Velocity • fps	1175	1105	1049	1003	965	931	901	874	848
Energy • ft-lbs	475	420	378	346	320	298	279	263	248
Taylor KO Index	10.4	9.8	9.3	8.9	8.5	8.2	8.0	7.7	7.5
Mid-Range Trajectory Height • Inches	0.0	0.2	0.9	2.1	4.0	6.4	9.7	13.8	18.7
Drop • Inches	0.0	-0.8	-3.4	-8.0	-14.7	-23.8	-35.4	-49.6	-66.7

American Eagle (Federal) 155-grain Full Metal Jacket Ball (AE40R2)

G1 Ballistic Coefficient = 0.127

Distance • Yards	Muzzle	25	50	75	100	125	150	175	200
Velocity • fps	1160	1100	1040	1000	960	927	898	871	847
Energy • ft-lbs	465	415	375	345	320	296	278	261	247
Taylor KO Index	10.3	9.7	9.2	8.9	8.5	8.2	8.0	7.7	7.5
Mid-Range Trajectory Height • Inches	0.0	0.2	0.9	2.2	4.0	6.5	9.8	13.8	18.8
Drop • Inches	0.0	-0.9	-3.5	-8.2	-15.0	-24.2	-35.9	-50.4	-67.6

Magtech 160-grain LSWC (40C)

G1 Ballistic Coefficient = 0.150

Distance • Yards	Muzzle	25	50	75	100	125	150	175	200
Velocity • fps	1165	1107	1059	1018	984	954	927	902	879
Energy • ft-lbs	484	435	398	369	343	323	305	289	274
Taylor KO Index	10.7	10.1	9.7	9.3	9.0	8.7	8.5	8.2	8.0
Mid-Range Trajectory Height • Inches	0.0	0.2	0.9	2.1	3.9	6.2	9.3	13.1	17.7
Drop • Inches	0.0	-0.8	-3.4	-8.0	-14.6	-23.4	-34.6	-48.4	-64.7

Hornady 165-grain FTX (91340)

G1 Ballistic Coefficient = 0.145

Distance • Yards	Muzzle	25	50	75	100	125	150	175	200
Velocity • fps	1045	1006	972	942	915	890	867	846	825
Energy • ft-lbs	400	371	346	325	307	290	276	262	250
Taylor KO Index	9.9	9.5	9.2	8.9	8.6	8.4	8.2	8.0	7.8
Mid-Range Trajectory Height • Inches	0.0	0.3	1.1	2.5	4.6	7.4	11.0	15.5	20.8
Drop • Inches	0.0	-1.0	-4.2	-9.7	-17.6	-28.1	-41.3	-57.5	-76.6

Cor-Bon 165-grain JHP (SD40165/20)

G1 Ballistic Coefficient = 0.150

Distance • Yards	Muzzle	25	50	75	100	125	150	175	200
Velocity • fps	1150	1094	1048	1010	976	947	920	896	873
Energy • ft-lbs	484	439	403	374	349	328	310	294	279
Taylor KO Index	10.8	10.3	9.9	9.5	9.2	8.9	8.7	8.4	8.2
Mid-Range Trajectory Height • Inches	0.0	0.2	0.9	2.1	3.9	6.3	9.5	13.5	18.1
Drop • Inches	0.0	-0.8	-3.5	-8.2	-14.9	-24.0	-35.4	-49.5	-66.2

Remington 165-grain Golden Saber (GS40SWA)

G1 Ballistic Coefficient = 0.136

Distance • Yards	Muzzle	25	50	75	100	125	150	175	200
Velocity • fps	1150	1089	1040	999	964	933	905	879	856
Energy • ft-lbs	485	435	396	366	340	319	300	283	268
Taylor KO Index	10.8	10.3	9.8	9.4	9.1	8.8	8.5	8.3	8.1
Mid-Range Trajectory Height • Inches	0.0	0.2	0.9	2.1	4.0	6.4	9.7	13.7	18.5
Drop • Inches	0.0	-0.8	-3.5	-8.2	-15.1	-24.2	-35.9	-50.1	-67.2

Speer 165-grain Gold Dot (23970)

G1 Ballistic Coefficient = 0.138

Distance • Yards	Muzzle	25	50	75	100	125	150	175	200
Velocity • fps	1150	1090	1043	1001	966	935	907	882	858
Energy • ft-lbs	485	436	399	367	342	320	302	285	270
Taylor KO Index	10.8	10.3	9.8	9.4	9.1	8.8	8.6	8.3	8.1
Mid-Range Trajectory Height • Inches	0.0	0.2	0.9	2.1	3.9	6.4	9.1	13.7	18.4
Drop • Inches	0.0	-0.8	-3.5	-8.2	-15.0	-24.2	-35.8	-50.0	-67.0

Winchester 165-grain Bonded PDX1 (S40SWPDB)

G1 Ballistic Coefficient = 0.165

Distance • Yards	Muzzle	25	50	75	100	125	150	175	200
Velocity • fps	1140	1091	1049	1014	983	955	930	907	886
Energy • ft-lbs	476	436	404	377	354	335	317	302	288
Taylor KO Index	10.7	10.3	9.9	9.6	9.3	9.0	8.8	8.6	8.4
Mid-Range Trajectory Height • Inches	0.0	0.2	0.9	2.2	4.0	6.4	9.5	13.4	18.0
Drop • Inches	0.0	-0.9	-3.6	-8.2	-15.0	-24.0	-35.4	-49.3	-65.8

Fiocchi 165-grain Jacketed Hollowpoint (40SWC)

G1 Ballistic Coefficient = 0.14

Distance • Yards	Muzzle	25	50	75	100	125	150	175	200
Velocity • fps	1100	1050	1009	973	942	914	889	865	843
Energy • ft-lbs	450	404	373	347	325	306	289	274	260
Taylor KO Index	10.4	9.9	9.5	9.2	8.9	8.6	8.4	8.2	8.0
Mid-Range Trajectory Height • Inches	0.0	0.2	1.0	2.3	4.3	6.9	10.3	14.5	19.6
Drop • Inches	0.0	-0.9	-3.8	-8.9	-16.2	-25.9	-38.3	-53.4	-71.4

PMC 165-grain JHP (40B)

G1 Ballistic Coefficient = 0.15

Distance • Yards	Muzzle	25	50	75	100	125	150	175	200
Velocity • fps	1040	1002	970	941	915	891	869	848	828
Energy • ft-lbs	396	368	345	325	307	291	277	265	251
Taylor KO Index	9.8	9.5	9.2	8.9	8.6	8.4	8.2	8.0	7.8
Mid-Range Trajectory Height • Inches	0.0	0.3	1.1	2.5	4.6	7.4	11.0	15.5	20.7
Drop • Inches	0.0	-1.0	-4.2	-9.7	-17.7	-28.2	-41.5	-57.6	-76.8

Federal 165-grain Hydra-Shok JHP (P40HS3)

G1 Ballistic Coefficient = 0.1

Distance • Yards	Muzzle	25	50	75	100	125	150	175	200
Velocity • fps	980	950	930	910	890	871	853	836	820
Energy • ft-lbs	350	330	315	300	290	278	267	256	246
Taylor KO Index	9.2	9.0	8.8	8.6	8.4	8.2	8.0	7.9	7.7
Mid-Range Trajectory Height • Inches	0.0	0.3	1.2	2.7	5.1	7.9	11.7	16.4	21.9
Drop • Inches	0.0	-1.2	-4.7	-10.7	-19.4	-30.8	-45.0	-62.3	-82.6

Speer 165-grain TMJ FN Lawman (53955)

G1 Ballistic Coefficient = 0.1

Distance • Yards	Muzzle	25	50	75	100	125	150	175	200
Velocity • fps	1150	1089	1040	999	964	932	904	879	855
Energy • ft-lbs	484	435	396	365	340	318	300	283	268
Taylor KO Index	11.1	10.5	10.1	9.7	9.3	9.0	8.7	8.5	8.3
Mid-Range Trajectory Height • Inches	0.0	0.2	0.9	2.1	4.0	6.5	9.8	13.8	18.7
Drop • Inches	0.0	-0.8	-3.5	-8.2	-15.1	-24.3	-36.0	-50.4	-67.6

UMC (Remington) 165-grain Metal Case (L40SW4)

G1 Ballistic Coefficient = 0.

Distance • Yards	Muzzle	25	50	75	100	125	150	175	200
Velocity • fps	1150	1089	1040	999	964	933	905	879	856
Energy • ft-lbs	484	435	396	366	340	319	300	283	268
Taylor KO Index	10.8	10.2	9.8	9.4	9.1	8.8	8.5	8.3	8.1
Mid-Range Trajectory Height • Inches	0.0	0.2	1.0	2.2	4.1	6.5	9.7	13.7	18.5
Drop • Inches	0.0	-0.8	-3.5	-8.2	-15.1	-24.2	-35.9	-50.1	-67.2

American Eagle (Federal) 165-grain Full Metal Jacket (AE40R3)

G1 Ballistic Coefficient = 0.155

Distance • Yards	Muzzle	25	50	75	100	125	150	175	200
Velocity • fps	1130	1080	1040	1000	970	942	917	894	872
Energy • ft-lbs	470	425	390	365	340	325	308	293	279
Taylor KO Index	10.9	10.4	10.1	9.7	9.4	9.1	8.9	8.6	8.4
Mid-Range Trajectory Height • Inches	0.0	0.2	0.9	2.2	4.0	6.5	9.7	13.7	18.4
Drop • Inches	0.0	-0.9	-3.6	-8.4	-15.3	-24.5	-36.2	-50.5	-67.2

Winchester 165-grain Brass Enclosed Base (WC401)

G1 Ballistic Coefficient = 0.197

Distance • Yards	Muzzle	25	50	75	100	125	150	175	200
Velocity • fps	1130	1089	1054	1024	996	972	950	929	910
Energy • ft-lbs	468	435	407	384	364	346	331	316	303
Taylor KO Index	10.7	10.3	9.9	9.7	9.4	9.2	9.0	8.8	8.6
Mid-Range Trajectory Height • Inches	0.0	0.2	0.9	2.1	3.9	6.2	9.2	13.0	17.4
Drop • Inches	0.0	-0.9	-3.6	-8.2	-15.0	-23.8	-35.0	-48.6	-64.7

Blazer (CCI) 165-grain TMJ (3589)

G1 Ballistic Coefficient = 0.140

Distance • Yards	Muzzle	25	50	75	100	125	150	175	200
Velocity • fps	1100	1048	1006	969	938	909	883	859	836
Energy • ft-lbs	443	403	371	344	321	303	286	270	256
Taylor KO Index	10.4	9.9	9.5	9.1	8.8	8.6	8.3	8.1	7.9
Mid-Range Trajectory Height • Inches	0.0	0.2	0.9	2.3	4.2	6.9	10.3	14.5	19.6
Drop • Inches	0.0	-0.9	-3.9	-8.9	-16.2	-26.0	-38.4	-53.6	-71.7

USA (Winchester) 165-grain Full Metal Jacket – Flat Nose (USA40SW) (USA40SWVP)

G1 Ballistic Coefficient = 0.197

Distance • Yards	Muzzle	25	50	75	100	125	150	175	200
Velocity • fps	1060	1029	1001	976	953	932	913	895	877
Energy • ft-lbs	412	388	367	349	333	319	305	293	282
Taylor KO Index	10.0	9.7	9.4	9.2	9.0	8.8	8.6	8.4	8.3
Mid-Range Trajectory Height • Inches	0.0	0.2	1.0	2.3	4.3	6.9	10.2	14.2	19.0
Drop • Inches	0.0	-1.0	-4.0	-9.2	-16.7	-26.6	-38.9	-53.9	-71.5

Blazer 165-grain FMJ FN (5210)

G1 Ballistic Coefficient = 0.136

Distance • Yards	Muzzle	25	50	75	100	125	150	175	200
Velocity • fps	1050	1007	971	940	911	885	861	838	817
Energy • ft-lbs	404	372	345	324	304	287	272	258	245
Taylor KO Index	9.8	9.5	9.2	8.9	8.6	8.3	8.1	7.9	7.7
Mid-Range Trajectory Height • Inches	0.0	0.3	1.1	2.5	4.6	7.4	11.1	15.6	21.0
Drop • Inches	0.0	-1.0	-4.2	-9.6	-17.5	-29.1	-41.4	-57.6	-76.9

Magtech 165-grain FMJ Flat (40G)

G1 Ballistic Coefficient = 0.092

Distance • Yards	Muzzle	25	50	75	100	125	150	175	200
Velocity • fps	1050	990	941	899	863	830	799	771	744
Energy • ft-lbs	404	359	325	296	273	252	234	218	203
Taylor KO Index	9.9	9.3	8.9	8.5	8.1	7.8	7.5	7.3	7.0
Mid-Range Trajectory Height • Inches	0.0	0.2	1.1	2.6	4.9	8.0	12.1	17.2	23.4
Drop • Inches	0.0	-1.0	-4.3	-10.0	-18.4	-29.7	-44.1	-62.0	-83.5

Speer 165-grain TMJ FN Lawman – Clean Fire (53982)

G1 Ballistic Coefficient = 0.135

Distance • Yards	Muzzle	25	50	75	100	125	150	175	200
Velocity • fps	1050	1007	971	939	911	884	860	837	816
Energy • ft-lbs	404	372	345	323	304	286	271	257	244
Taylor KO Index	10.1	9.7	9.4	9.1	8.8	8.5	8.3	8.1	7.9
Mid-Range Trajectory Height • Inches	0.0	0.2	1.0	2.6	4.6	7.4	11.1	15.6	21.0
Drop • Inches	0.0	-1.0	-4.2	-9.6	-17.6	-28.1	-41.4	-57.7	-77.0

Blazer (CCI) 165-grain FMJ FN – Blazer Brass (5210)

G1 Ballistic Coefficient = 0.136

Distance • Yards	Muzzle	25	50	75	100	125	150	175	200
Velocity • fps	1050	1007	971	940	911	885	861	838	817
Energy • ft-lbs	404	372	345	324	304	287	272	258	245
Taylor KO Index	9.9	9.5	9.2	8.9	8.6	8.3	8.1	7.9	7.7
Mid-Range Trajectory Height • Inches	0.0	0.3	1.1	2.5	4.6	7.4	11.1	15.4	20.9
Drop • Inches	0.0	-1.0	-4.2	-9.6	-17.5	-28.1	-41.4	-57.6	-76.9

Fiocchi 165-grain FMJ TC (40SWF)
G1 Ballistic Coefficient = 0.148

Distance • Yards	Muzzle	25	50	75	100	125	150	175	200
Velocity • fps	1020	985	954	927	901	878	856	836	816
Energy • ft-lbs	381	356	333	315	298	283	269	256	244
Taylor KO Index	9.6	9.3	9.0	8.7	8.5	8.3	8.1	7.9	7.7
Mid-Range Trajectory Height • Inches	0.0	0.3	1.1	2.6	4.8	7.7	11.4	15.8	21.4
Drop • Inches	0.0	-1.1	-4.4	-10.1	-18.3	-29.2	-42.9	-59.6	-79.4

Fiocchi 165-grain Truncated Cone Encapsulated Base (40TCEB) LF
G1 Ballistic Coefficient = 0.148

Distance • Yards	Muzzle	25	50	75	100	125	150	175	200
Velocity • fps	1020	985	954	927	901	878	856	836	816
Energy • ft-lbs	381	356	333	315	298	283	269	256	244
Taylor KO Index	9.6	9.3	9.0	8.7	8.5	8.3	8.1	7.9	7.7
Mid-Range Trajectory Height • Inches	0.0	0.3	1.1	2.6	4.8	7.7	11.4	15.8	21.4
Drop • Inches	0.0	-1.1	-4.4	-10.1	-18.3	-29.2	-42.9	-59.6	-79.4

PMC 165-grain Full Metal Jacket (40D)
G1 Ballistic Coefficient = 0.160

Distance • Yards	Muzzle	25	50	75	100	125	150	175	200
Velocity • fps	985	957	931	908	885	864	845	826	809
Energy • ft-lbs	356	335	317	301	287	274	262	250	240
Taylor KO Index	9.3	9.0	8.8	8.6	8.4	8.2	8.0	7.8	7.6
Mid-Range Trajectory Height • Inches	0.0	0.3	1.2	2.8	5.0	8.0	11.9	16.6	22.0
Drop • Inches	0.0	-1.1	-4.7	-10.7	-19.4	-30.8	-45.1	-62.5	-83.1

PMC 165-grain EMJ NT (40EMD) LF
G1 Ballistic Coefficient = 0.15

Distance • Yards	Muzzle	25	50	75	100	125	150	175	200
Velocity • fps	985	955	928	903	879	858	837	818	800
Energy • ft-lbs	355	334	315	298	283	270	257	245	234
Taylor KO Index	9.3	9.0	8.7	8.5	8.3	8.1	7.9	7.7	7.5
Mid-Range Trajectory Height • Inches	0.0	0.3	1.2	2.8	5.1	8.1	12.0	16.6	22.5
Drop • Inches	0.0	-1.1	-4.7	-10.7	-19.4	-31.0	-45.5	-63.0	-83.9

Cor-Bon 165-grain FMJ (PM40165/50)
G1 Ballistic Coefficient = 0.14

Distance • Yards	Muzzle	25	50	75	100	125	150	175	200
Velocity • fps	850	829	909	789	771	753	738	719	703
Energy • ft-lbs	265	252	239	228	218	208	198	190	181
Taylor KO Index	8.0	7.8	7.6	7.4	7.3	7.1	7.0	6.8	6.6
Mid-Range Trajectory Height • Inches	0.0	0.4	1.6	3.7	6.7	10.6	15.7	22.0	29.3
Drop • Inches	0.0	-1.5	-6.2	-14.2	-25.7	-40.9	-59.9	-82.9	-110.1

Fiocchi 170-grain FMJ Truncated Cone (40SWA)
G1 Ballistic Coefficient = 0.14

Distance • Yards	Muzzle	25	50	75	100	125	150	175	200
Velocity • fps	1020	985	954	927	901	878	856	836	816
Energy • ft-lbs	393	366	344	324	307	291	277	264	252
Taylor KO Index	9.9	9.6	9.3	9.0	8.8	8.5	8.3	8.1	7.9
Mid-Range Trajectory Height • Inches	0.0	0.3	1.1	2.6	4.8	7.7	11.4	15.8	21.4
Drop • Inches	0.0	-1.1	-4.4	-10.1	-18.3	-29.2	-42.9	-59.6	-79.4

Speer 180-grain Gold Dot (23962)
G1 Ballistic Coefficient = 0.1

Distance • Yards	Muzzle	25	50	75	100	125	150	175	200
Velocity • fps	1025	989	957	928	902	879	856	835	815
Energy • ft-lbs	420	391	366	344	325	309	293	279	266
Taylor KO Index	10.5	10.2	9.8	9.5	9.3	9.0	8.8	8.6	8.4
Mid-Range Trajectory Height • Inches	0.0	0.2	1.1	2.5	4.7	7.6	11.3	15.8	21.2
Drop • Inches	0.0	-1.1	-4.5	-10.2	-18.4	-29.4	-43.0	-59.7	-79.4

Winchester 180-grain Bonded PDX1 (S40PDB1)
G1 Ballistic Coefficient = 0.1

Distance • Yards	Muzzle	25	50	75	100	125	150	175	200
Velocity • fps	1025	995	969	945	922	902	883	864	847
Energy • ft-lbs	420	396	375	357	340	325	311	299	287
Taylor KO Index	10.5	10.2	10.0	9.7	9.5	9.3	9.1	8.9	8.7
Mid-Range Trajectory Height • Inches	0.0	0.3	1.1	2.6	4.7	7.4	11.0	15.3	20.4
Drop • Inches	0.0	-1.1	-4.3	-9.9	-17.9	-28.4	-41.6	-57.6	-76.5

Remington 180-grain Golden Saber (GS40SWB)

G1 Ballistic Coefficient = 0.178

Distance • Yards	Muzzle	25	50	75	100	125	150	175	200
Velocity • fps	1015	986	960	936	914	894	875	857	840
Energy • ft-lbs	412	389	368	350	334	320	308	294	282
Taylor KO Index	10.4	10.1	9.9	9.6	9.4	9.2	9.0	8.8	8.6
Mid-Range Trajectory Height • Inches	0.0	0.3	1.3	2.6	4.5	7.5	11.1	15.5	20.7
Drop • Inches	0.0	-1.1	-4.4	-10.0	-18.2	-28.9	-42.4	-58.6	-77.8

Remington 180-grain Jacketed Hollow Point (R40SW2)

G1 Ballistic Coefficient = 0.178

Distance • Yards	Muzzle	25	50	75	100	125	150	175	200
Velocity • fps	1015	986	960	936	914	894	875	857	840
Energy • ft-lbs	412	389	368	350	334	320	308	294	282
Taylor KO Index	10.4	10.1	9.9	9.6	9.4	9.2	9.0	8.8	8.6
Mid-Range Trajectory Height • Inches	0.0	0.3	1.3	2.6	4.5	7.5	11.1	15.5	20.7
Drop • Inches	0.0	-1.1	-4.4	-10.0	-18.2	-28.9	-42.4	-58.6	-77.8

UMC (Remington) 180-grain JHP (L40SW2)

G1 Ballistic Coefficient = 0.178

Distance • Yards	Muzzle	25	50	75	100	125	150	175	200
Velocity • fps	1015	986	960	936	914	894	875	857	840
Energy • ft-lbs	412	389	368	350	334	320	308	294	282
Taylor KO Index	10.4	10.1	9.9	9.6	9.4	9.2	9.0	8.8	8.6
Mid-Range Trajectory Height • Inches	0.0	0.3	1.3	2.6	4.5	7.5	11.1	15.5	20.7
Drop • Inches	0.0	-1.1	-4.4	-10.0	-18.2	-28.9	-42.4	-58.6	-77.8

USA (Winchester) 180-grain Jacketed Hollow Point (USA40JHP)

G1 Ballistic Coefficient = 0.170

Distance • Yards	Muzzle	25	50	75	100	125	150	175	200
Velocity • fps	1010	980	954	930	907	887	867	848	831
Energy • ft-lbs	408	384	364	345	329	314	300	288	276
Taylor KO Index	10.4	10.1	9.8	9.6	9.3	9.1	8.9	8.7	8.5
Mid-Range Trajectory Height • Inches	0.0	0.2	1.1	2.6	4.7	7.6	11.2	15.7	21.1
Drop • Inches	0.0	-1.1	-4.4	-10.2	-18.4	-29.2	-42.9	-59.4	-78.9

Black Hills 180-grain Jacketed Hollow Point (D40N2)

G1 Ballistic Coefficient = 0.175

Distance • Yards	Muzzle	25	50	75	100	125	150	175	200
Velocity • fps	1000	972	947	924	903	883	864	846	829
Energy • ft-lbs	400	378	359	341	326	312	298	286	275
Taylor KO Index	10.3	10.0	9.7	9.5	9.3	9.1	8.9	8.7	8.5
Mid-Range Trajectory Height • Inches	0.0	0.3	1.2	2.6	4.8	7.7	11.4	15.9	21.2
Drop • Inches	0.0	-1.1	-4.5	-10.3	-18.7	-29.7	-43.5	-60.2	-80.0

Federal 180-grain Hi-Shok JHP (C40SWA)

G1 Ballistic Coefficient = 0.169

Distance • Yards	Muzzle	25	50	75	100	125	150	175	200
Velocity • fps	1000	970	950	920	900	880	860	842	825
Energy • ft-lbs	400	375	360	340	325	309	296	284	272
Taylor KO Index	10.3	10.0	9.8	9.5	9.3	9.1	8.8	8.7	8.5
Mid-Range Trajectory Height • Inches	0.0	0.3	1.2	2.7	4.9	7.8	11.5	15.9	21.5
Drop • Inches	0.0	-1.1	-4.5	-10.4	-18.8	-29.8	-43.7	-60.5	-80.4

Federal 180-grain Hydra-Shok JHP (P40HS1)

G1 Ballistic Coefficient = 0.169

Distance • Yards	Muzzle	25	50	75	100	125	150	175	200
Velocity • fps	1000	970	950	920	900	880	860	842	825
Energy • ft-lbs	400	375	360	340	325	309	296	284	272
Taylor KO Index	10.3	10.0	9.8	9.5	9.3	9.1	8.8	8.7	8.5
Mid-Range Trajectory Height • Inches	0.0	0.3	1.2	2.7	4.9	7.8	11.5	15.9	21.5
Drop • Inches	0.0	-1.1	-4.5	-10.4	-18.8	-29.8	-43.7	-60.5	-80.4

Fiocchi 180-grain JHP (40SWE)

G1 Ballistic Coefficient = 0.164

Distance • Yards	Muzzle	25	50	75	100	125	150	175	200
Velocity • fps	1000	970	943	919	896	875	855	837	819
Energy • ft-lbs	399	376	355	338	321	306	293	280	268
Taylor KO Index	10.3	10.0	9.7	9.5	9.2	9.0	8.8	8.6	8.4
Mid-Range Trajectory Height • Inches	0.0	0.3	1.2	2.7	4.9	7.8	11.6	16.2	21.7
Drop • Inches	0.0	-1.1	-4.5	-10.4	-18.8	-30.0	-43.9	-60.8	-80.9

Fiocchi 180-grain XTP JHP (40XTPB)

G1 Ballistic Coefficient = 0.164

Distance • Yards	Muzzle	25	50	75	100	125	150	175	200
Velocity • fps	1000	970	943	919	896	875	855	837	819
Energy • ft-lbs	399	376	355	338	321	306	293	280	268
Taylor KO Index	10.3	10.0	9.7	9.5	9.2	9.0	8.8	8.6	8.4
Mid-Range Trajectory Height • Inches	0.0	0.3	1.2	2.7	4.9	7.8	11.6	16.2	21.7
Drop • Inches	0.0	-1.1	-4.5	-10.4	-18.8	-30.0	-43.9	-60.8	-80.9

Magtech 180-grain JHP (GG40B)

G1 Ballistic Coefficient = 0.166

Distance • Yards	Muzzle	25	50	75	100	125	150	175	200
Velocity • fps	990	962	938	913	891	871	852	834	817
Energy • ft-lbs	392	370	352	333	318	303	290	278	267
Taylor KO Index	10.2	9.9	9.6	9.4	9.2	9.0	8.8	8.6	8.4
Mid-Range Trajectory Height • Inches	0.0	0.3	1.2	2.7	5.0	7.9	11.7	16.2	21.9
Drop • Inches	0.0	-1.1	-4.6	-10.6	-19.1	-30.4	-44.6	-61.7	-82.0

Magtech 180-grain JHP (40A)

G1 Ballistic Coefficient = 0.156

Distance • Yards	Muzzle	25	50	75	100	125	150	175	200
Velocity • fps	990	960	933	909	886	865	845	826	808
Energy • ft-lbs	390	368	348	330	314	299	285	273	261
Taylor KO Index	10.2	9.9	9.6	9.3	9.1	8.9	8.7	8.5	8.3
Mid-Range Trajectory Height • Inches	0.0	0.3	1.2	2.7	5.0	7.9	11.8	16.5	22.1
Drop • Inches	0.0	-1.1	-4.6	-10.6	-19.2	-30.6	-44.8	-62.1	-82.6

PMC 180-grain Starfire Hollow Point (40SFA)

G1 Ballistic Coefficient = 0.16

Distance • Yards	Muzzle	25	50	75	100	125	150	175	200
Velocity • fps	985	958	933	910	889	869	850	832	815
Energy • ft-lbs	388	367	348	331	316	302	289	277	266
Taylor KO Index	10.1	9.9	9.6	9.4	9.1	8.9	8.7	8.6	8.4
Mid-Range Trajectory Height • Inches	0.0	0.3	1.2	2.7	5.0	7.9	11.7	16.4	21.9
Drop • Inches	0.0	-1.1	-4.6	-10.7	-19.3	-30.6	-44.9	-62.1	-82.5

Hornady 180-grain JHP / XTP (9136)

G1 Ballistic Coefficient = 0.16

Distance • Yards	Muzzle	25	50	75	100	125	150	175	200
Velocity • fps	950	926	903	882	862	844	826	809	792
Energy • ft-lbs	361	342	326	311	297	285	273	262	251
Taylor KO Index	9.8	9.5	9.3	9.1	8.9	8.7	8.5	8.3	8.1
Mid-Range Trajectory Height • Inches	0.0	0.3	1.3	2.9	5.3	8.5	12.5	17.5	23.3
Drop • Inches	0.0	-1.2	-5.0	-11.4	-20.6	-32.7	-47.9	-66.3	-87.9

Black Hills 180-grain JHP (D40SCN2)

G1 Ballistic Coefficient = 0.17

Distance • Yards	Muzzle	25	50	75	100	125	150	175	200
Velocity • fps	975	950	927	905	885	866	848	831	815
Energy • ft-lbs	380	361	343	327	313	300	288	276	266
Taylor KO Index	10.0	9.8	9.5	9.3	9.1	8.9	8.7	8.5	8.4
Mid-Range Trajectory Height • Inches	0.0	0.3	1.2	2.8	5.1	8.1	12.0	16.7	22.2
Drop • Inches	0.0	-1.2	-4.7	-10.8	-19.6	-31.1	-45.5	-63.0	-83.6

Hornady 180-grain TAP-FPD (91368)

G1 Ballistic Coefficient = 0.1

Distance • Yards	Muzzle	25	50	75	100	125	150	175	200
Velocity • fps	950	926	903	882	862	844	826	809	792
Energy • ft-lbs	361	342	326	311	297	285	273	262	251
Taylor KO Index	9.8	9.5	9.3	9.1	8.9	8.7	8.5	8.3	8.1
Mid-Range Trajectory Height • Inches	0.0	0.3	1.3	2.9	5.3	8.5	12.5	17.5	23.3
Drop • Inches	0.0	-1.2	-5.0	-11.4	-20.6	-32.7	-47.9	-66.3	-87.9

Speer 180-grain GDHP-SB Short Barrel (23974)

G1 Ballistic Coefficient = 0.1

Distance • Yards	Muzzle	25	50	75	100	125	150	175	200
Velocity • fps	950	923	898	876	854	834	815	797	779
Energy • ft-lbs	361	341	322	307	291	278	265	254	243
Taylor KO Index	9.8	9.5	9.2	9.0	8.8	8.6	8.4	8.2	8.0
Mid-Range Trajectory Height • Inches	0.0	0.3	1.3	3.0	5.4	8.6	12.8	17.6	23.9
Drop • Inches	0.0	-1.2	-5.0	-11.5	-20.8	-33.0	-48.4	-67.1	-89.2

American Eagle (Federal) 180-grain Full Metal Jacket (AE40R1)

G1 Ballistic Coefficient = 0.170

Distance • Yards	Muzzle	25	50	75	100	125	150	175	200
Velocity • fps	1000	970	950	920	900	880	861	843	826
Energy • ft-lbs	400	375	360	340	325	310	296	284	272
Taylor KO Index	10.3	10.0	9.8	9.5	9.3	9.1	8.9	8.7	8.5
Mid-Range Trajectory Height • Inches	0.0	0.3	1.2	2.7	4.9	7.8	11.5	16.1	21.5
Drop • Inches	0.0	-1.1	-4.5	-10.4	-18.7	-29.8	-43.7	-60.5	-80.3

Aguila 180-grain FMJ (1E402110)

G1 Ballistic Coefficient = 0.097

Distance • Yards	Muzzle	25	50	75	100	125	150	175	200
Velocity • fps	1100	1031	977	933	893	860	829	800	773
Energy • ft-lbs	484	425	382	348	320	296	275	256	239
Taylor KO Index	11.3	10.6	10.1	9.6	9.2	8.9	8.5	8.2	8.0
Mid-Range Trajectory Height • Inches	0.0	0.2	1.0	2.4	4.5	7.4	11.2	16.0	21.7
Drop • Inches	0.0	-0.9	-3.9	-9.2	-16.9	-27.4	-40.8	-57.4	-77.3

Magtech 180-grain FMJ Flat (40PS)

G1 Ballistic Coefficient = 0.179

Distance • Yards	Muzzle	25	50	75	100	125	150	175	200
Velocity • fps	1050	1017	988	962	938	917	896	877	859
Energy • ft-lbs	441	414	387	370	352	336	321	308	295
Taylor KO Index	10.8	10.5	10.2	9.9	9.6	9.4	9.2	9.0	8.8
Mid-Range Trajectory Height • Inches	0.0	0.2	1.1	2.4	4.6	7.1	10.6	14.6	19.7
Drop • Inches	0.0	-1.0	-4.1	-9.4	-17.1	-27.3	-40.0	-55.4	-73.7

USA (Winchester) 180-grain Full Metal Jacket (Q4238)

G1 Ballistic Coefficient = 0.193

Distance • Yards	Muzzle	25	50	75	100	125	150	175	200
Velocity • fps	1020	993	968	946	925	905	887	870	853
Energy • ft-lbs	416	394	374	357	342	328	315	303	291
Taylor KO Index	10.5	10.2	10.0	9.7	9.5	9.3	9.1	8.9	8.8
Mid-Range Trajectory Height • Inches	0.0	0.3	1.1	2.6	4.6	7.4	11.0	15.0	20.3
Drop • Inches	0.0	-1.1	-4.3	-9.9	-17.9	-28.5	-41.7	-57.6	-76.5

American Eagle (Federal) 180-grain Full Metal Jacket (AE40R1) + (AE40N1)

G1 Ballistic Coefficient = 0.164

Distance • Yards	Muzzle	25	50	75	100	125	150	175	200
Velocity • fps	1000	970	950	920	900	880	861	843	826
Energy • ft-lbs	400	375	360	340	325	310	296	284	272
Taylor KO Index	10.3	10.0	9.8	9.5	9.3	9.1	8.9	8.7	8.5
Mid-Range Trajectory Height • Inches	0.0	0.3	1.2	2.7	4.9	7.8	11.5	15.8	21.5
Drop • Inches	0.0	-1.1	-4.5	-10.4	-18.7	-29.8	-43.7	-60.5	-80.3

Blazer (CCI) 180-grain FMJ (3591)

G1 Ballistic Coefficient = 0.144

Distance • Yards	Muzzle	25	50	75	100	125	150	175	200
Velocity • fps	1000	967	937	910	886	863	841	821	802
Energy • ft-lbs	400	374	351	331	313	298	283	269	257
Taylor KO Index	10.3	10.0	9.6	9.4	9.1	8.9	8.7	8.5	8.3
Mid-Range Trajectory Height • Inches	0.0	0.3	1.2	2.7	5.0	7.9	11.8	16.5	22.2
Drop • Inches	0.0	-1.1	-4.6	-10.5	-19.0	-30.3	-44.5	-61.8	-82.3

CCI 180-grain Clean-Fire – TMJ Blazer (3477)

G1 Ballistic Coefficient = 0.144

Distance • Yards	Muzzle	25	50	75	100	125	150	175	200
Velocity • fps	1000	967	937	910	886	863	841	821	802
Energy • ft-lbs	400	374	351	331	313	298	283	269	257
Taylor KO Index	10.3	10.0	9.6	9.4	9.1	8.9	8.7	8.5	8.3
Mid-Range Trajectory Height • Inches	0.0	0.3	1.2	2.7	5.0	7.9	11.8	16.5	22.2
Drop • Inches	0.0	-1.1	-4.6	-10.5	-19.0	-30.3	-44.5	-61.8	-82.3

Fiocchi 180-grain FMJ TC (40SWD)

G1 Ballistic Coefficient = 0.175

Distance • Yards	Muzzle	25	50	75	100	125	150	175	200
Velocity • fps	1000	972	947	924	903	883	864	846	829
Energy • ft-lbs	400	378	359	341	326	312	298	286	275
Taylor KO Index	10.3	10.0	9.7	9.5	9.3	9.1	8.9	8.7	8.5
Mid-Range Trajectory Height • Inches	0.0	0.3	1.2	2.6	4.8	7.7	11.4	15.9	21.2
Drop • Inches	0.0	-1.1	-4.5	-10.3	-18.7	-28.7	-43.5	-60.2	-80.0

Speer 180-grain TMJ FN Lawman (53958)

G1 Ballistic Coefficient = 0.144

Distance • Yards	Muzzle	25	50	75	100	125	150	175	200
Velocity • fps	1000	967	937	910	886	863	841	821	802
Energy • ft-lbs	400	374	351	331	313	298	283	269	257
Taylor KO Index	10.3	10.0	9.6	9.4	9.1	8.9	8.7	8.5	8.3
Mid-Range Trajectory Height • Inches	0.0	0.3	1.2	2.7	5.0	7.9	11.8	16.5	22.2
Drop • Inches	0.0	-1.1	-4.6	-10.5	-19.0	-30.3	-44.5	-61.8	-82.3

Speer 180-grain Clean-Fire – TMJ FN Lawman (53880)

G1 Ballistic Coefficient = 0.144

Distance • Yards	Muzzle	25	50	75	100	125	150	175	200
Velocity • fps	1000	967	937	910	886	863	841	821	802
Energy • ft-lbs	400	374	351	331	313	298	283	269	257
Taylor KO Index	10.3	10.0	9.6	9.4	9.1	8.9	8.7	8.5	8.3
Mid-Range Trajectory Height • Inches	0.0	0.3	1.2	2.7	5.0	7.9	11.8	16.5	22.2
Drop • Inches	0.0	-1.1	-4.6	-10.5	-19.0	-30.3	-44.5	-61.8	-82.3

Magtech 180-grain FMJ Flat (MP40B) + (40B)

G1 Ballistic Coefficient = 0.15

Distance • Yards	Muzzle	25	50	75	100	125	150	175	200
Velocity • fps	990	960	933	909	886	865	845	826	808
Energy • ft-lbs	390	368	348	330	314	299	285	273	261
Taylor KO Index	10.2	9.9	9.6	9.3	9.1	8.9	8.7	8.5	8.3
Mid-Range Trajectory Height • Inches	0.0	0.3	1.2	2.7	5.0	7.9	11.8	16.5	22.1
Drop • Inches	0.0	-1.1	-4.6	-10.6	-19.2	-30.6	-44.8	-62.1	-82.6

Magtech 180-grain FMJ Flat (CR40A)

G1 Ballistic Coefficient = 0.15

Distance • Yards	Muzzle	25	50	75	100	125	150	175	200
Velocity • fps	990	960	933	909	886	865	845	826	808
Energy • ft-lbs	390	369	348	330	314	299	285	273	261
Taylor KO Index	10.2	9.9	9.6	9.3	9.1	8.9	8.7	8.5	8.3
Mid-Range Trajectory Height • Inches	0.0	0.3	1.2	2.7	5.0	8.0	11.9	16.3	22.1
Drop • Inches	0.0	-1.1	-4.6	-10.6	-19.2	-30.6	-44.9	-62.2	-82.7

UMC (Remington) 180-grain FNEB (LL40SW5)

G1 Ballistic Coefficient = 0.18

Distance • Yards	Muzzle	25	50	75	100	125	150	175	200
Velocity • fps	990	964	940	918	898	879	861	844	827
Energy • ft-lbs	392	371	353	337	322	309	296	285	274
Taylor KO Index	10.4	10.2	9.9	9.7	9.5	9.3	9.1	8.9	8.7
Mid-Range Trajectory Height • Inches	0.0	0.3	1.1	2.7	4.9	7.8	11.6	16.1	21.5
Drop • Inches	0.0	-1.1	-4.6	-10.5	-19.0	-30.2	-44.2	-61.1	-81.1

Winchester 180-grain Brass Enclosed Base (WC402)

G1 Ballistic Coefficient = 0.1

Distance • Yards	Muzzle	25	50	75	100	125	150	175	200
Velocity • fps	990	965	943	922	902	884	866	850	834
Energy • ft-lbs	392	372	356	340	325	312	330	289	278
Taylor KO Index	10.2	9.9	9.7	9.5	9.3	9.1	8.9	8.7	8.6
Mid-Range Trajectory Height • Inches	0.0	0.3	1.2	2.7	5.0	7.8	11.4	16.0	21.3
Drop • Inches	0.0	-1.1	-4.6	-10.6	-19.2	-30.1	-44.0	-60.7	-80.5

Blazer (CCI) 180-grain FMJ FN – Blazer Brass (5220)

G1 Ballistic Coefficient = 0.1

Distance • Yards	Muzzle	25	50	75	100	125	150	175	200
Velocity • fps	985	953	925	899	875	853	832	812	793
Energy • ft-lbs	388	363	342	323	306	291	276	263	251
Taylor KO Index	10.1	9.8	9.5	9.2	9.0	8.8	8.6	8.4	8.2
Mid-Range Trajectory Height • Inches	0.0	0.3	1.2	2.8	5.1	8.2	12.1	16.7	22.7
Drop • Inches	0.0	-1.2	-4.7	-10.8	-19.5	-31.1	-45.7	-63.4	-84.5

PMC 180-grain FMJ / FP (40E)

G1 Ballistic Coefficient = 0.

Distance • Yards	Muzzle	25	50	75	100	125	150	175	200
Velocity • fps	985	957	931	908	885	864	845	826	808
Energy • ft-lbs	388	366	346	329	313	299	285	273	261
Taylor KO Index	10.1	9.8	9.6	9.3	9.1	8.9	8.7	8.5	8.3
Mid-Range Trajectory Height • Inches	0.0	0.2	1.1	2.7	5.0	8.0	11.8	16.5	22.1
Drop • Inches	0.0	-1.1	-4.7	-10.7	-19.3	-30.8	-45.1	-62.5	-83.0

UMC (Remington) 180-grain MC (L40SW3)

G1 Ballistic Coefficient = 0.177

Distance • Yards	Muzzle	25	50	75	100	125	150	175	200
Velocity • fps	985	959	936	913	893	874	856	839	822
Energy • ft-lbs	388	368	350	333	319	305	293	281	270
Taylor KO Index	10.1	9.9	9.6	9.4	9.2	9.0	8.8	8.6	8.5
Mid-Range Trajectory Height • Inches	0.0	0.3	1.3	2.9	5.1	7.9	11.7	16.3	21.7
Drop • Inches	0.0	-1.1	-4.6	-10.6	-19.2	-30.5	-44.6	-61.7	-81.9

Black Hills 180-grain FMJ (D40SCN1)

G1 Ballistic Coefficient = 0.175

Distance • Yards	Muzzle	25	50	75	100	125	150	175	200
Velocity • fps	975	950	927	905	885	866	848	831	815
Energy • ft-lbs	380	361	343	327	313	300	288	276	266
Taylor KO Index	10.0	9.8	9.5	9.3	9.1	8.9	8.7	8.5	8.4
Mid-Range Trajectory Height • Inches	0.0	0.3	1.2	2.8	5.1	8.1	12.0	16.7	22.2
Drop • Inches	0.0	-1.2	-4.7	-10.8	-19.6	-31.1	-45.5	-63.0	-83.6

Sellier & Bellot 180-grain FMJ (V311202U)

G1 Ballistic Coefficient = 0.155

Distance • Yards	Muzzle	25	50	75	100	125	150	175	200
Velocity • fps	968	952	915	892	870	850	830	812	794
Energy • ft-lbs	375	355	335	318	303	289	276	264	252
Taylor KO Index	10.0	9.7	9.4	9.2	9.0	8.8	8.5	8.4	8.2
Mid-Range Trajectory Height • Inches	0.0	0.3	1.2	2.8	5.2	8.3	12.3	17.2	23.0
Drop • Inches	0.0	-1.2	-4.8	-11.1	-20.0	-31.9	-46.7	-64.7	-86.0

Sellier & Bellot 180-grain TFMJ (V311232U)

G1 Ballistic Coefficient = 0.152

Distance • Yards	Muzzle	25	50	75	100	125	150	175	200
Velocity • fps	968	940	914	891	869	848	818	810	792
Energy • ft-lbs	375	353	334	317	302	287	274	262	251
Taylor KO Index	10.2	9.9	9.6	9.4	9.2	8.9	8.7	8.5	8.3
Mid-Range Trajectory Height • Inches	0.0	0.3	1.2	2.8	5.2	6.3	12.3	17.2	23.0
Drop • Inches	0.0	-1.2	-4.8	-11.1	-20.0	-31.9	-46.8	-64.8	-86.2

Speer 200-grain TMJ FN Lawman – Clean Fire (53882)

G1 Ballistic Coefficient = 0.167

Distance • Yards	Muzzle	25	50	75	100	125	150	175	200
Velocity • fps	915	893	873	854	836	818	802	786	770
Energy • ft-lbs	372	355	339	324	310	298	286	274	264
Taylor KO Index	10.7	10.5	10.2	10.0	9.8	9.6	9.4	9.2	9.0
Mid-Range Trajectory Height • Inches	0.0	0.3	1.3	3.1	5.7	9.1	13.4	18.6	24.9
Drop • Inches	0.0	-1.3	-5.4	-12.2	-22.1	-35.1	-51.3	-70.9	-94.0

(Left to right) .40 Smith & Wesson, 10mm Auto, .45 Auto.

10mm Auto

First chambered in the Bren Ten pistol in 1983, the 10mm Auto was a powerful pistol. The first ammo was manufactured by Norma. The Bren Ten went nowhere, and it wasn't until 1989 when the FBI announced the selection of the 10mm Auto as its officially favored sidearm that the 10mm Auto took off. The best performance overall as a pistol round came at velocities near 1,000 fps. The cartridge case has lots more volume than is needed for that kind of performance, leaving room for a considerable jump in performance as a submachine cartridge. The .40 Smith & Wesson Auto is a spin-off design that seems to be taking over the law-enforcement market.

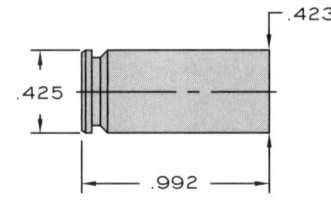

Relative Recoil Factor = 0.96

Specifications:

Controlling Agency for Standardization of this Ammunition: SAAMI

Bullet Weight Grains	Velocity fps	Maximum Average Pressure	
		Copper Crusher	Transducer
155 HP/XP	1,410	N/S	37,500 psi
155 FMJ	1,115	N/S	37,500 psi
170 HP/XP	1,320	N/S	37,500 psi
175 STHP	1,275	N/S	37,500 psi
200 FMJ/FP	1,150	N/S	37,500 psi
200 SXT	985	N/S	37,500 psi

Standard barrel for velocity testing: 5 inches long— 1 turn in 16-inch twist

Availability:

Cor-Bon 115-grain Glaser Safety Slug (03400/03600)
G1 Ballistic Coefficient = 0.1

Distance • Yards	Muzzle	25	50	75	100	125	150	175	200
Velocity • fps	1650	1504	1373	1258	1162	1086	1027	980	940
Energy • ft-lbs	695	578	481	404	345	301	270	245	226
Taylor KO Index	10.9	9.9	9.0	8.3	7.6	7.1	6.8	6.4	6.2
Mid-Range Trajectory Height • Inches	0.0	0.1	0.5	1.2	2.3	4.0	6.2	9.2	12.9
Drop • Inches	0.0	-0.4	-1.8	-4.4	-8.3	-13.8	-21.2	-30.6	-42.3

Cor-Bon 135-grain Pow'RBall (PB10135/20)
G1 Ballistic Coefficient = 0.1

Distance • Yards	Muzzle	25	50	75	100	125	150	175	200
Velocity • fps	1400	1294	1202	1127	1066	1017	976	941	910
Energy • ft-lbs	588	502	434	381	341	310	286	266	248
Taylor KO Index	10.8	10.0	9.3	8.7	8.2	7.8	7.5	7.3	7.0
Mid-Range Trajectory Height • Inches	0.0	0.1	0.6	1.5	3.0	5.0	7.6	11.0	15.1
Drop • Inches	0.0	-0.6	-2.5	-5.8	-11.0	-18.0	-27.1	-38.4	-52.3

Cor-Bon 135-grain JHP (SD10135/20)
G1 Ballistic Coefficient = 0.1

Distance • Yards	Muzzle	25	50	75	100	125	150	175	200
Velocity • fps	1400	1294	1202	1127	1066	1017	976	941	910
Energy • ft-lbs	588	502	434	381	341	310	286	266	248
Taylor KO Index	10.8	10.0	9.3	8.7	8.2	7.8	7.5	7.3	7.0
Mid-Range Trajectory Height • Inches	0.0	0.1	0.6	1.5	3.0	5.0	7.6	11.0	15.1
Drop • Inches	0.0	-0.6	-2.5	-5.8	-11.0	-18.0	-27.1	-38.4	-52.3

Cor-Bon 150-grain JHP (SD10150/20)
G1 Ballistic Coefficient = 0.

Distance • Yards	Muzzle	25	50	75	100	125	150	175	200
Velocity • fps	1325	1232	1154	1090	1038	996	959	928	899
Energy • ft-lbs	585	506	446	396	359	330	307	287	269
Taylor KO Index	11.4	10.6	9.9	9.3	8.9	8.5	8.2	8.0	7.7
Mid-Range Trajectory Height • Inches	0.0	0.1	0.7	1.7	3.2	5.3	8.2	11.7	16.0
Drop • Inches	0.0	-0.6	-2.7	-6.4	-12.0	-19.5	-29.3	-41.4	-56.0

Hornady 155-grain JHP/XTP (9122)

G1 Ballistic Coefficient = 0.138

Distance • Yards	Muzzle	25	50	75	100	125	150	175	200
Velocity • fps	1265	1186	1119	1065	1020	983	950	921	895
Energy • ft-lbs	551	484	431	390	358	333	311	292	276
Taylor KO Index	11.2	10.5	9.9	9.4	9.0	8.7	8.4	8.2	7.9
Mid-Range Trajectory Height • Inches	0.0	0.2	0.8	1.8	3.5	5.6	8.5	12.1	16.4
Drop • Inches	0.0	-0.7	-3.0	-4.0	-12.8	-20.8	-31.0	-43.5	-58.5

Cor-Bon 155-grain DPX (DPX10155/20)

G1 Ballistic Coefficient = 0.135

Distance • Yards	Muzzle	25	50	75	100	125	150	175	200
Velocity • fps	1200	1130	1073	1026	987	953	923	896	871
Energy • ft-lbs	496	440	396	363	335	313	293	276	261
Taylor KO Index	12.3	11.6	11.0	10.6	10.2	9.8	9.5	9.2	9.0
Mid-Range Trajectory Height • Inches	0.0	0.2	0.9	2.0	3.8	6.2	9.3	13.0	17.8
Drop • Inches	0.0	-0.8	-3.3	-7.6	-14.1	-22.7	-33.8	-47.4	-63.7

Cor-Bon 165-grain JHP (SD10165/20)

G1 Ballistic Coefficient = 0.140

Distance • Yards	Muzzle	25	50	75	100	125	150	175	200
Velocity • fps	1250	1174	1110	1058	1016	976	947	919	893
Energy • ft-lbs	573	505	452	411	378	351	329	309	292
Taylor KO Index	11.8	11.1	10.5	10.0	9.6	9.2	8.9	8.7	8.4
Mid-Range Trajectory Height • Inches	0.0	0.1	0.7	1.8	3.5	5.7	8.6	12.3	16.7
Drop • Inches	0.0	-0.7	-3.0	-7.1	-13.1	-21.2	-31.6	-44.4	-59.7

PMC 170-grain Jacketed Hollow Point (10B)

G1 Ballistic Coefficient = 0.112

Distance • Yards	Muzzle	25	50	75	100	125	150	175	200
Velocity • fps	1200	1117	1052	1000	958	921	888	858	831
Energy • ft-lbs	544	471	418	378	347	321	298	278	261
Taylor KO Index	11.7	10.9	10.2	9.7	9.3	8.9	8.6	8.3	8.1
Mid-Range Trajectory Height • Inches	0.0	0.2	0.9	2.1	3.9	6.4	9.6	13.7	18.6
Drop • Inches	0.0	-0.8	-3.3	-7.8	-14.5	-23.5	-35.0	-49.2	-66.3

Winchester 175-grain Silvertip Hollow Point (X10MMSTHP)

G1 Ballistic Coefficient = 0.142

Distance • Yards	Muzzle	25	50	75	100	125	150	175	200
Velocity • fps	1290	1209	1141	1084	1037	998	965	935	908
Energy • ft-lbs	649	568	506	457	418	387	362	340	320
Taylor KO Index	12.9	12.1	11.4	10.8	10.4	10.0	9.7	9.4	9.1
Mid-Range Trajectory Height • Inches	0.0	0.2	0.7	1.8	3.3	5.4	8.2	11.7	15.9
Drop • Inches	0.0	-0.7	-2.8	-6.7	-12.4	-20.0	-29.9	-42.0	-56.5

Cor-Bon 180-grain Bonded Core SP (HT10180BC/20)

G1 Ballistic Coefficient = 0.175

Distance • Yards	Muzzle	25	50	75	100	125	150	175	200
Velocity • fps	1300	1233	1173	1122	1078	1040	1008	979	954
Energy • ft-lbs	676	607	550	503	464	433	406	384	364
Taylor KO Index	13.4	12.7	12.1	11.5	11.1	10.7	10.4	10.1	9.8
Mid-Range Trajectory Height • Inches	0.0	0.2	0.7	1.7	3.2	5.5	7.8	10.9	14.9
Drop • Inches	0.0	-0.7	-2.8	-6.4	-11.8	-19.1	-28.4	-39.8	-53.5

Hornady 180-grain JHP/XTP (9126)

G1 Ballistic Coefficient = 0.165

Distance • Yards	Muzzle	25	50	75	100	125	150	175	200
Velocity • fps	1180	1124	1077	1038	1004	974	948	923	901
Energy • ft-lbs	556	505	464	431	403	378	359	341	324
Taylor KO Index	12.1	11.6	11.1	10.7	10.3	10.0	9.8	9.5	9.3
Mid-Range Trajectory Height • Inches	0.0	0.2	0.9	2.0	3.7	6.0	9.0	12.7	17.0
Drop • Inches	0.0	-0.8	-3.3	-7.7	-14.1	-22.7	-33.5	-46.7	-62.4

Federal 180-grain Hydra-Shok JHP (P10HS1)

G1 Ballistic Coefficient = 0.169

Distance • Yards	Muzzle	25	50	75	100	125	150	175	200
Velocity • fps	1030	1000	970	950	920	898	878	859	841
Energy • ft-lbs	425	400	375	355	340	323	308	295	283
Taylor KO Index	10.6	10.3	10.0	9.8	9.5	9.2	9.0	8.8	8.7
Mid-Range Trajectory Height • Inches	0.0	0.3	1.1	2.5	4.7	7.4	10.9	15.3	20.4
Drop • Inches	0.0	-1.0	-4.3	-9.8	-17.8	-28.3	-41.5	-57.6	-76.5

UMC (Remington) 180-grain Metal Case (L10MM6)

G1 Ballistic Coefficient = 0.180

Distance • Yards	Muzzle	25	50	75	100	125	150	175	200
Velocity • fps	1150	1103	1063	1023	998	971	947	925	904
Energy • ft-lbs	529	486	452	423	398	377	359	342	327
Taylor KO Index	11.8	11.3	10.9	10.6	10.3	10.0	9.7	9.5	9.3
Mid-Range Trajectory Height • Inches	0.0	0.2	0.9	2.0	3.7	6.2	9.2	12.9	17.3
Drop • Inches	0.0	-0.8	-3.5	-8.0	-14.6	-23.4	-34.4	-47.9	-63.9

American Eagle (Federal) 180-grain FMJ (AE10A)

G1 Ballistic Coefficient = 0.169

Distance • Yards	Muzzle	25	50	75	100	125	150	175	200
Velocity • fps	1030	1000	970	950	920	898	878	859	841
Energy • ft-lbs	425	400	375	355	340	323	308	295	283
Taylor KO Index	10.6	10.3	10.0	9.8	9.5	9.2	9.0	8.8	8.7
Mid-Range Trajectory Height • Inches	0.0	0.3	1.1	2.5	4.7	7.4	10.9	15.3	20.4
Drop • Inches	0.0	-1.0	-4.3	-9.8	-17.8	-28.3	-41.5	-57.6	-76.5

Cor-Bon 200-grain RN PN (HT10200PN/20)

G1 Ballistic Coefficient = 0.20

Distance • Yards	Muzzle	25	50	75	100	125	150	175	200
Velocity • fps	1125	1086	1052	1022	996	971	949	929	910
Energy • ft-lbs	562	524	491	464	440	419	400	383	368
Taylor KO Index	11.6	11.2	10.8	10.5	10.2	10.0	9.8	9.6	9.4
Mid-Range Trajectory Height • Inches	0.0	0.2	0.9	2.1	3.9	6.3	9.4	12.9	17.5
Drop • Inches	0.0	-0.9	-3.6	-8.3	-15.0	-24.0	-35.3	-48.9	-65.1

Hornady 200-grain JHP/XTP (9129)

G1 Ballistic Coefficient = 0.20

Distance • Yards	Muzzle	25	50	75	100	125	150	175	200
Velocity • fps	1050	1020	994	970	948	928	909	891	874
Energy • ft-lbs	490	462	439	418	399	382	367	353	339
Taylor KO Index	12.0	11.7	11.4	11.1	10.8	10.6	10.4	10.2	10.0
Mid-Range Trajectory Height • Inches	0.0	0.2	1.0	2.4	4.4	7.0	10.3	14.4	19.2
Drop • Inches	0.0	-1.0	-4.1	-9.4	-17.0	-27.0	-39.5	-54.6	-72.5

Blazer (CCI) 200-grain TMJ (3597)

G1 Ballistic Coefficient = 0.1

Distance – Yards	Muzzle	25	50	75	100	125	150	175	200
Velocity – fps	1050	1015	985	957	933	910	889	869	850
Energy – Ft lbs	490	458	431	407	387	368	351	335	321
Taylor KO Index	12.0	11.6	11.3	11.0	10.7	10.4	10.2	9.9	9.7
Mid-Range Trajectory Height – Inches	0.0	0.2	1.1	2.5	4.5	7.2	14.5	14.9	20.0
Drop – Inches	0.0	-1.0	-4.1	-9.5	-17.2	-27.4	-40.3	-55.9	-74.4

PMC 200-grain Truncated Cone – Full Metal Jacket (10A)

G1 Ballistic Coefficient = 0.1

Distance – Yards	Muzzle	25	50	75	100	125	150	175	200
Velocity – fps	1050	1008	972	941	912	887	863	840	819
Energy – Ft lbs	490	451	420	393	370	349	331	314	298
Taylor KO Index	12.0	11.5	11.1	10.8	10.4	10.1	9.9	9.6	9.4
Mid-Range Trajectory Height – Inches	0.0	0.2	1.1	2.5	4.6	7.4	11.0	15.4	20.7
Drop – Inches	0.0	-1.0	-4.2	-9.6	-17.5	-28.0	-41.2	-57.4	-76.6

.41 Remington Magnum

Remington announced the .41 Remington Magnum in 1964 at the same time that Smith & Wesson announced a Model 57 pistol chambered .41 Remington Magnum. The idea in offering the .41 was a simple one. The .357 Magnum wasn't quite enough gun for some shooters, but the .44 Remington Magnum was too much. The .41 Remington Magnum was supposed to be "just right." While the .41 Remington Magnum does just what it was designed to do, a couple things happened on the way to the gun shop. The first was that the .41 Remington Magnum was a little too much for many shooters. The second was that the caliber arrived just about the time that semiautomatic pistols were experiencing a big jump in popularity. The .41 Remington Magnum is still a powerful cartridge and one that fills the hunting application very easily.

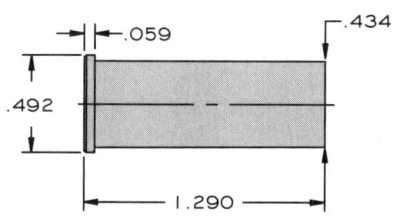

Relative Recoil Factor = 1.23

Specifications:

Controlling Agency for Standardization of this Ammunition: SAAMI

Bullet Weight Grains	Velocity fps	Maximum Average Pressure	
		Copper Crusher	Transducer
170 STHP	1,400	40,000 cup	36,000 psi
175 STHP	1,250	40,000 cup	36,000 psi
210 L	955	40,000 cup	36,000 psi
210 SP-HP	1,280	40,000 cup	36,000 psi

Standard barrel for velocity testing: 4 inches long (vented)—1 turn in 18.75-inch twist.
Alternate one-piece barrel length: 10.135 inches.

Availability:

Cor-Bon 170-grain JHP Self Defense (SD41M170/20)

G1 Ballistic Coefficient = 0.150

Distance – Yards	Muzzle	25	50	75	100	125	150	175	200
Velocity – fps	1275	1200	1136	1083	1039	1001	969	940	914
Energy – Ft lbs	614	544	487	443	407	379	355	334	316
Taylor KO Index	12.7	11.9	11.3	10.8	10.3	10.0	9.6	9.4	9.1
Mid-Range Trajectory Height – Inches	0.0	0.1	0.7	1.8	3.3	5.5	8.3	11.8	16.0
Drop – Inches	0.0	-0.7	-2.9	-6.8	-12.5	-20.3	-30.2	-42.4	-57.1

Winchester 175-grain Silvertip Hollow Point (X41MSTHP2)

G1 Ballistic Coefficient = 0.153

Distance – Yards	Muzzle	25	50	75	100	125	150	175	200
Velocity – fps	1250	1180	1120	1071	1029	994	963	936	911
Energy – Ft lbs	607	541	488	446	412	384	361	340	322
Taylor KO Index	12.8	12.1	11.5	11.0	10.5	10.2	9.9	9.6	9.3
Mid-Range Trajectory Height – Inches	0.0	0.2	0.8	1.8	3.4	5.6	8.4	11.9	16.1
Drop – Inches	0.0	-0.7	-3.0	-7.0	-12.9	-20.8	-31.0	-43.4	-58.2

Federal 180-grain Barnes Expander (P41XB1) LF

G1 Ballistic Coefficient = 0.156

Distance • Yards	Muzzle	25	50	75	100	125	150	175	200
Velocity • fps	1340	1260	1190	1130	1080	1038	1002	971	943
Energy • ft-lbs	720	635	670	515	470	431	402	377	356
Taylor KO Index	14.1	13.3	12.5	11.9	11.4	10.9	10.6	10.2	9.9
Mid-Range Trajectory Height • Inches	0.0	0.2	0.7	1.6	3.1	5.1	7.7	10.8	14.8
Drop • Inches	0.0	-0.6	-2.6	-6.2	-11.4	-18.5	-27.6	-38.9	-52.4

Cor-Bon 180-grain DPX (DPX41180/20)

G1 Ballistic Coefficient = 0.150

Distance • Yards	Muzzle	25	50	75	100	125	150	175	200
Velocity • fps	1300	1222	1155	1099	1052	1013	979	950	923
Energy • ft-lbs	676	597	533	483	443	410	383	361	340
Taylor KO Index	13.7	12.9	12.2	11.6	11.1	10.7	10.3	10.0	9.7
Mid-Range Trajectory Height • Inches	0.0	0.2	0.7	1.8	3.3	5.3	8.1	11.5	15.6
Drop • Inches	0.0	-0.7	-2.8	-6.6	-12.1	-19.6	-29.2	-41.1	-55.4

Federal 210-grain Swift A-Frame (P41SA)

G1 Ballistic Coefficient = 0.182

Distance • Yards	Muzzle	25	50	75	100	125	150	175	200
Velocity • fps	1360	1290	1220	1170	1120	1078	1042	1010	983
Energy • ft-lbs	860	775	700	635	580	542	506	476	450
Taylor KO Index	16.7	15.9	15.0	14.4	13.8	13.3	12.8	12.4	12.1
Mid-Range Trajectory Height • Inches	0.0	0.2	0.7	1.6	2.9	4.7	7.2	10.2	13.8
Drop • Inches	0.0	-0.6	-2.5	-5.9	-10.8	-17.5	-26.1	-36.6	-49.3

Remington 210-grain Jacketed Soft Point (R41MG1)

G1 Ballistic Coefficient = 0.160

Distance • Yards	Muzzle	25	50	75	100	125	150	175	200
Velocity • fps	1300	1226	1162	1108	1062	1024	991	962	935
Energy • ft-lbs	788	702	630	573	526	489	458	431	408
Taylor KO Index	16.0	15.1	14.3	13.6	13.1	12.6	12.2	11.8	11.5
Mid-Range Trajectory Height • Inches	0.0	0.2	0.7	1.7	3.2	5.2	7.9	11.2	15.2
Drop • Inches	0.0	-0.7	-2.8	-6.5	-12.0	-19.4	-28.8	-40.4	-54.4

Cor-Bon 210-grain JHP (HT41210JHP/20)

G1 Ballistic Coefficient = 0.16

Distance • Yards	Muzzle	25	50	75	100	125	150	175	200
Velocity • fps	1350	1271	1201	1141	1090	1047	1011	979	951
Energy • ft-lbs	850	753	673	607	554	511	476	447	422
Taylor KO Index	16.6	15.6	14.8	14.0	13.4	12.9	12.4	12.0	11.7
Mid-Range Trajectory Height • Inches	0.0	0.1	0.6	1.6	3.0	4.9	7.5	10.7	14.5
Drop • Inches	0.0	-0.6	-2.6	-6.1	-11.2	-18.2	-27.1	-38.2	-51.5

Federal 210-grain JHP (C41A)

G1 Ballistic Coefficient = 0.17

Distance • Yards	Muzzle	25	50	75	100	125	150	175	200
Velocity • fps	1230	1170	1120	1080	1040	1008	979	954	930
Energy • ft-lbs	705	640	585	545	505	474	447	424	404
Taylor KO Index	15.1	14.4	13.8	13.3	12.8	12.4	12.0	11.7	11.4
Mid-Range Trajectory Height • Inches	0.0	0.2	0.8	1.9	3.5	5.6	8.4	11.8	16.0
Drop • Inches	0.0	-0.7	-3.1	-7.1	-13.1	-21.0	-31.1	-43.4	-58.1

Speer 210-grain Gold Dot Hollow Point (23996)

G1 Ballistic Coefficient = 0.1

Distance • Yards	Muzzle	25	50	75	100	125	150	175	200
Velocity • fps	1280	1217	1162	1114	1073	1038	1007	980	955
Energy • ft-lbs	764	691	630	579	537	503	473	448	426
Taylor KO Index	15.7	15.0	14.3	13.7	13.2	12.8	12.4	12.1	11.7
Mid-Range Trajectory Height • Inches	0.0	0.2	0.7	1.7	3.2	5.2	7.9	11.1	15.0
Drop • Inches	0.0	-0.7	-2.9	-6.8	-12.4	-19.9	-29.4	-41.0	-54.8

Winchester 240-grain Platinum Tip (S41PTHP)

G1 Ballistic Coefficient = 0.2

Distance • Yards	Muzzle	25	50	75	100	125	150	175	200
Velocity • fps	1250	1198	1151	1111	1075	1044	1016	991	969
Energy • ft-lbs	833	765	706	658	616	581	551	524	500
Taylor KO Index	17.6	16.8	16.2	15.6	15.1	14.7	14.3	13.9	13.6
Mid-Range Trajectory Height • Inches	0.0	0.2	0.8	1.7	3.3	5.3	7.9	11.1	15.0
Drop • Inches	0.0	-0.7	-2.9	-6.8	-12.4	-19.9	-29.4	-41.0	-54.8

Cor-Bon 250-grain Hard Cast

G1 Ballistic Coefficient = 0.2

Distance • Yards	Muzzle	25	50	75	100	125	150	175	200
Velocity • fps	1325	1267	1215	1167	1125	1088	1056	1028	1002
Energy • ft-lbs	975	892	819	756	703	658	619	586	558
Taylor KO Index	19.4	18.1	17.8	17.1	16.5	15.9	15.5	15.1	14.7
Mid-Range Trajectory Height • Inches	0.0	0.1	0.6	1.5	2.9	4.7	7.1	10.1	13.6
Drop • Inches	0.0	-0.6	-2.6	-6.1	-11.2	-17.9	-26.5	-37.1	-49.7

Federal 250-grain Cast Core (P41B)

G1 Ballistic Coefficient = 0.

Distance • Yards	Muzzle	25	50	75	100	125	150	175	200
Velocity • fps	1160	1120	1090	1050	1030	1006	984	964	945
Energy • ft-lbs	745	695	655	615	585	562	538	516	496
Taylor KO Index	17.0	16.4	16.0	15.4	15.1	14.7	14.4	14.1	13.8
Mid-Range Trajectory Height • Inches	0.0	0.2	0.9	2.0	3.7	5.9	8.8	12.1	16.4
Drop • Inches	0.0	-0.8	-3.4	-7.8	-14.1	-22.5	-33.0	-45.8	-60.9

.44 Colt

This is a really oldie. The cartridge was introduced in 1870 to be used in centerfire conversions of the percussion .44 Colt. It was also used a few years later in a .44 Remington revolver. When introduced, it was a blackpowder number, which resulted in very low pressures. The cartridge dropped out of the factory catalogs in 1940. The popularity of Cowboy Action shooting revived this cartridge. A word of CAUTION here. These cartridges are now loaded with smokeless powder and should NOT to be fired in any guns designed or manufactured in the blackpowder era. They are only suitable for modern replica guns in good condition.

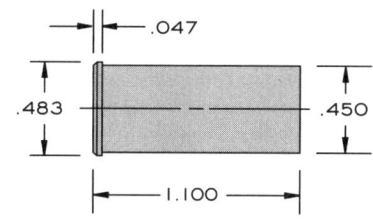

Relative Recoil Factor = 0.75

Specifications:

Controlling Agency for Standardization of this Ammunition: Factory

Availability:
Cowboy Action Loads:

Black Hills 230-grain FPL (DCB44CLTN2)

G1 Ballistic Coefficient = 0.150

Distance • Yards	Muzzle	25	50	75	100	125	150	175	200
Velocity • fps	730	715	700	685	671	657	641	630	617
Energy • ft-lbs	272	261	250	240	230	221	212	203	195
Taylor KO Index	10.4	10.1	9.9	9.7	9.5	9.3	9.1	8.9	8.8
Mid-Range Trajectory Height • Inches	0.0	0.5	2.1	4.8	8.8	14.1	20.8	29.0	38.6
Drop • Inches	0.0	-2.1	-8.4	-19.1	-34.5	-54.7	-80.0	-100.4	-146.4

.44 Smith & Wesson Russian

The only reason this old dog is still around is for Cowboy Action events. Its origin was in 1870 when it was designed by Smith & Wesson for the Russian military revolver. It obviously started life as a blackpowder number. Pressures are very mild but for the Cowboy application, who cares. Nobody wants "hot" ammunition for Cowboy Action shooting.

Relative Recoil Factor = 0.81

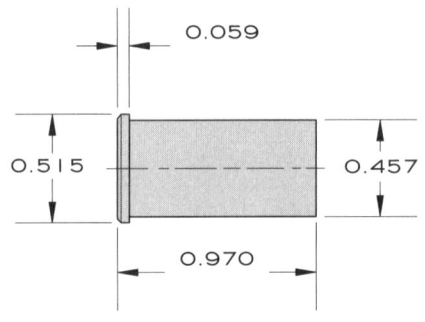

Specifications:

Controlling Agency for Standardization of this Ammunition: Factory

Availability:
Cowboy Action Loads:

Black Hills 210-grain LFN (DCB44RUSN1)

G1 Ballistic Coefficient = 0.10?

Distance • Yards	Muzzle	25	50	75	100	125	150	175	200
Velocity • fps	650	630	611	592	574	557	540	523	507
Energy • ft-lbs	197	185	174	164	154	144	138	128	120
Taylor KO Index	8.4	8.2	7.9	7.7	7.4	7.2	7.0	6.8	6.6
Mid-Range Trajectory Height • Inches	0.0	0.7	2.8	6.4	11.7	18.8	28.0	39.4	53.1
Drop • Inches	0.0	-2.6	-10.7	-24.6	-44.8	-71.5	-105.2	-146.4	-195.6

Many alligators are shot with a "shooting stick" which consists of a handgun cartridge in a very short barrel at the end of a metal pole that is slammed down with force to create a detonation. It works well and they are deadly devices. (Safari Press Archives)

.44 Smith & Wesson Special

The .44 S&W is one of the first-generation pistol cartridges designed to use smokeless powder. At the time of its introduction, the working pressures were kept very low, pretty much a duplication of blackpowder pressure levels. As a result, the performance is certainly modest when compared with .44 Remington Magnum performance levels (although very potent when compared with calibers like the .38 Special). Like the .38 Special in a .357 Magnum revolver, the .44 S&W Special can be fired in modern guns in good condition that are chambered for the .44 Remington Magnum. This provides a way for the nonreloader to get modestly powered ammo for use in the .44 Remington Magnum. Using .44 Remington Magnum ammo in a gun chambered for the .44 S&W Special is definitely NOT recommended.

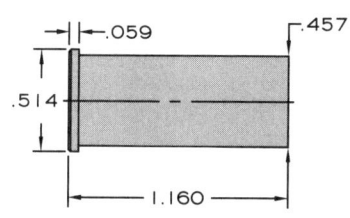

Relative Recoil Factor = 0.81

Specifications:

Controlling Agency for Standardization of this Ammunition: SAAMI

Bullet Weight Grains	Velocity fps	Maximum Average Pressure	
		Copper Crusher	Transducer
200 STHP	900	14,000 cup	N/S
200 SWCHP	1,025	14,000 cup	N/S
246 L	800	N/S	15,500 psi

Standard barrel for velocity testing: 4 inches long [vented]—1 turn in 20-inch twist. (Alternate one-piece barrel is 8.15 inches long) Several manufacturers are using 6- or 6.5-inch vented barrels to obtain their velocity data.

Availability:

Cor-Bon 135-grain Glaser Safety Slug (03800/04000)

G1 Ballistic Coefficient = 0.125

Distance • Yards	Muzzle	25	50	75	100	125	150	175	200
Velocity • fps	1350	1251	1166	1098	1043	988	961	927	898
Energy • ft-lbs	546	469	408	361	326	299	277	258	242
Taylor KO Index	11.2	10.4	9.7	9.1	8.7	8.3	8.0	7.7	7.5
Mid-Range Trajectory Height • Inches	0.0	0.1	0.7	1.7	3.2	5.3	8.1	11.6	15.9
Drop • Inches	0.0	-0.6	-2.6	-6.3	-11.7	-19.1	-28.7	-40.6	-55.1

CCI 140-grain Magnum Shotshell (3979)

Shot load using #9 shot—Muzzle Velocity = 1000 fps.

Hornady 165-grain FTX (90700)

G1 Ballistic Coefficient = 0.126

Distance • Yards	Muzzle	25	50	75	100	125	150	175	200
Velocity • fps	900	873	848	824	802	781	761	741	723
Energy • ft-lbs	297	279	263	249	235	223	212	201	191
Taylor KO Index	9.2	8.9	8.6	8.4	8.2	8.0	7.7	7.5	7.4
Mid-Range Trajectory Height • Inches	0.0	0.3	1.4	3.3	6.1	9.7	14.4	20.2	27.1
Drop • Inches	0.0	-1.4	-5.6	-12.8	-23.3	-37.1	-54.5	-75.6	-100.7

Cor-Bon 165-grain JHP (SD44S165/20)

G1 Ballistic Coefficient = 0.140

Distance • Yards	Muzzle	25	50	75	100	125	150	175	200
Velocity • fps	1050	1009	973	942	914	889	865	843	822
Energy • ft-lbs	404	373	347	325	306	289	274	260	248
Taylor KO Index	10.7	10.3	9.9	9.6	9.3	9.1	8.8	8.6	8.4
Mid-Range Trajectory Height • Inches	0.0	0.3	1.1	2.5	4.6	7.4	11.0	15.3	20.8
Drop • Inches	0.0	-1.0	-4.2	-9.6	-17.5	-28.0	-41.2	-57.3	-76.5

Hornady 180-grain JHP/XTP (9070)

G1 Ballistic Coefficient = 0.139

Distance • Yards	Muzzle	25	50	75	100	125	150	175	200
Velocity • fps	1000	965	935	907	882	959	837	816	796
Energy • ft-lbs	400	373	350	329	311	295	280	266	253
Taylor KO Index	11.1	10.7	10.3	10.0	9.8	9.5	9.3	9.1	8.8
Mid-Range Trajectory Height • Inches	0.0	0.3	1.2	2.7	5.0	7.9	11.8	16.6	22.2
Drop • Inches	0.0	-1.1	-4.6	-10.5	-19.0	-30.4	-44.6	-62.0	-82.6

PMC 180-grain JHP (44SB)
G1 Ballistic Coefficient = 0.104

Distance • Yards	Muzzle	25	50	75	100	125	150	175	200
Velocity • fps	980	938	902	869	839	811	785	761	737
Energy • ft-lbs	383	352	325	302	281	263	246	231	217
Taylor KO Index	10.9	10.4	10.0	9.7	9.3	9.0	8.7	8.5	8.2
Mid-Range Trajectory Height • Inches	0.0	0.3	1.2	2.9	5.4	8.7	13.0	18.3	24.8
Drop • Inches	0.0	-1.2	-4.8	-11.1	-20.3	-32.6	-48.1	-67.2	-90.1

Cor-Bon 200-grain DPX (DPX44S200/20)
G1 Ballistic Coefficient = 0.135

Distance • Yards	Muzzle	25	50	75	100	125	150	175	200
Velocity • fps	950	920	893	868	845	823	803	783	764
Energy • ft-lbs	401	376	354	335	317	301	286	272	254
Taylor KO Index	11.7	11.4	11.0	10.7	10.4	10.2	9.9	9.7	9.4
Mid-Range Trajectory Height • Inches	0.0	0.3	1.3	3.0	5.5	8.7	13.0	18.1	24.4
Drop • Inches	0.0	-1.2	-5.0	-11.5	-20.9	-33.4	-49.0	-68.0	-90.6

Winchester 200-grain Silvertip Hollow Point (X44STHPS2)
G1 Ballistic Coefficient = 0.16?

Distance • Yards	Muzzle	25	50	75	100	125	150	175	200
Velocity • fps	900	879	860	840	822	804	788	772	756
Energy • ft-lbs	360	343	328	313	300	287	276	265	254
Taylor KO Index	11.1	10.8	10.6	10.3	10.1	9.9	9.7	9.5	9.3
Mid-Range Trajectory Height • Inches	0.0	0.3	1.4	3.2	5.9	9.3	13.8	19.3	25.7
Drop • Inches	0.0	-1.4	-5.5	-12.7	-22.9	-36.3	-53.0	-73.3	-97.2

Blazer (CCI) 200-grain JHP (3556)
G1 Ballistic Coefficient = 0.14?

Distance • Yards	Muzzle	25	50	75	100	125	150	175	200
Velocity • fps	920	895	872	850	829	809	791	773	756
Energy • ft-lbs	376	356	337	321	305	291	278	265	254
Taylor KO Index	11.4	11.0	10.8	10.5	10.2	10.0	9.8	9.5	9.3
Mid-Range Trajectory Height • Inches	0.0	0.3	1.4	3.1	5.7	9.2	13.6	18.9	25.3
Drop • Inches	0.0	-1.3	-5.3	-12.2	-22.1	-35.1	-51.5	-71.4	-94.8

Federal 200-grain Semi-Wadcutter HP (C44SA)
G1 Ballistic Coefficient = 0.1?

Distance • Yards	Muzzle	25	50	75	100	125	150	175	200
Velocity • fps	870	850	830	810	790	773	756	739	723
Energy • ft-lbs	335	320	305	290	275	265	254	243	232
Taylor KO Index	10.7	10.5	10.2	10.0	9.8	9.5	9.3	9.1	8.9
Mid-Range Trajectory Height • Inches	0.0	0.3	1.5	3.4	6.3	10.1	14.9	20.8	27.8
Drop • Inches	0.0	-1.5	-5.9	-13.6	-24.5	-39.0	-57.0	-78.9	-104.8

Speer 200-grain GDHP (23980)
G1 Ballistic Coefficient = 0.1?

Distance • Yards	Muzzle	25	50	75	100	125	150	175	200
Velocity • fps	875	853	832	813	794	776	759	742	726
Energy • ft-lbs	340	323	308	293	280	268	256	245	234
Taylor KO Index	10.8	10.5	10.3	10.0	9.8	9.6	9.4	9.2	9.0
Mid-Range Trajectory Height • Inches	0.0	0.4	1.5	3.5	6.3	10.0	14.8	20.4	27.6
Drop • Inches	0.0	-1.4	-5.9	-13.4	-24.3	-38.6	-56.5	-78.2	-103.8

Remington 246-grain Lead Round Nose (R44SW)
G1 Ballistic Coefficient = 0.?

Distance • Yards	Muzzle	25	50	75	100	125	150	175	200
Velocity • fps	755	739	725	709	695	681	667	654	641
Energy • ft-lbs	310	299	285	275	265	253	243	233	224
Taylor KO Index	11.4	11.2	11.0	10.7	10.5	10.3	10.1	9.9	9.7
Mid-Range Trajectory Height • Inches	0.0	0.5	2.0	4.6	8.3	13.2	19.4	27.0	36.0
Drop • Inches	0.0	-1.9	-7.8	-17.9	-32.3	-51.1	-74.6	-103.1	-136.6

Winchester 246-grain Lead-Round Nose (X44SP)
G1 Ballistic Coefficient = 0.

Distance • Yards	Muzzle	25	50	75	100	125	150	175	200
Velocity • fps	755	739	725	709	695	681	667	654	641
Energy • ft-lbs	310	299	285	275	265	253	243	233	224
Taylor KO Index	11.4	11.2	11.0	10.7	10.5	10.3	10.1	9.9	9.7
Mid-Range Trajectory Height • Inches	0.0	0.5	2.0	4.6	8.3	13.2	19.4	27.0	36.0
Drop • Inches	0.0	-1.9	-7.8	-17.9	-32.3	-51.1	-74.6	-103.1	-136.6

Cowboy Action Loads:

Magtech 200-grain LFN (44E)
G1 Ballistic Coefficient = 0.152

Distance • Yards	Muzzle	25	50	75	100	125	150	175	200
Velocity • fps	722	707	693	678	665	651	638	625	612
Energy • ft-lbs	232	222	213	204	196	188	181	173	167
Taylor KO Index	8.9	8.7	8.6	8.4	8.2	8.0	7.9	7.7	7.6
Mid-Range Trajectory Height • Inches	0.0	0.5	2.1	5.0	9.0	14.4	21.2	29.6	39.4
Drop • Inches	0.0	-2.1	-8.6	-19.6	-35.3	-55.9	-81.6	-112.7	-149.4

Black Hills 210-grain FPL (DCB44SPLN1)
G1 Ballistic Coefficient = 0.100

Distance • Yards	Muzzle	25	50	75	100	125	150	175	200
Velocity • fps	700	678	657	637	618	599	580	563	545
Energy • ft-lbs	229	215	202	189	178	167	157	148	139
Taylor KO Index	9.0	8.7	8.5	8.2	8.0	7.8	7.5	7.3	7.1
Mid-Range Trajectory Height • Inches	0.0	0.6	2.4	5.5	10.1	16.1	24.1	33.9	45.7
Drop • Inches	0.0	-2.3	-9.2	-21.3	-38.6	-61.7	-90.8	-126.3	-168.7

Magtech 240-grain LFN (44B)
G1 Ballistic Coefficient = 0.100

Distance • Yards	Muzzle	25	50	75	100	125	150	175	200
Velocity • fps	761	737	714	691	670	649	629	610	592
Energy • ft-lbs	309	289	271	255	239	225	211	199	187
Taylor KO Index	11.3	10.9	10.6	10.2	9.9	9.6	9.3	9.0	8.8
Mid-Range Trajectory Height • Inches	0.0	0.5	2.0	4.6	8.5	12.6	19.0	28.8	38.9
Drop • Inches	0.0	-1.9	-7.8	-18.0	-32.8	-52.3	-77.0	-107.3	-143.3

Winchester 240-grain Cast Lead (USA44CB)
G1 Ballistic Coefficient = 0.153

Distance • Yards	Muzzle	25	50	75	100	125	150	175	200
Velocity • fps	750	734	719	704	690	676	662	649	636
Energy • ft-lbs	300	287	276	264	254	244	234	224	216
Taylor KO Index	11.0	10.8	10.6	10.4	10.2	10.0	9.8	9.6	9.4
Mid-Range Trajectory Height • Inches	0.0	0.5	2.1	4.8	8.4	13.3	19.7	27.4	36.5
Drop • Inches	0.0	-2.0	-7.9	-18.1	-32.7	-51.8	-75.7	-104.5	-138.5

.44 Remington Magnum (Pistol Data)

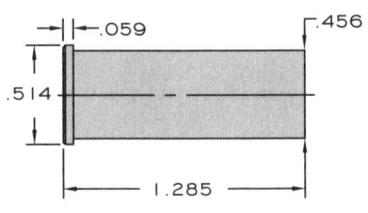

The .44 Remington Magnum cartridge was introduced in 1956. From its inception, it has been the stuff of legends. Many species of the world's dangerous game have been killed by a hunter equipped with a .44 Magnum revolver (backed up, usually, by a professional hunter equipped with a suitable large rifle). As a pistol cartridge, the .44 Magnum kills at both ends. Few shooters can get off more than 5 or 6 shots without beginning to flinch. Ammunition in .44 S&W Special caliber can be used in revolvers chambered for .44 Remington Magnum, a condition similar to firing a .38 Special in a .357 Magnum. See the data in the rifle section for further information. (Generally, .44 S&W Special ammunition can't be used in tubular magazine rifles designed for the .44 RM because the shorter cartridge promotes feeding problems that aren't present in a revolver.)

Relative Recoil Factor = 1.45

Specifications:

Controlling Agency for Standardization of this Ammunition: SAAMI

Bullet Weight Grains	Velocity fps	Maximum Average Pressure	
		Copper Crusher	Transducer
180 JHP	1,400	40,000 cup	36,000 psi
210 STHP	1,250	40,000 cup	36,000 psi
240 L-SWC	995	40,000 cup	36,000 psi
240 L	1,335	40,000 cup	36,000 psi

Standard barrel for velocity testing: 4 inches long [vented]—1 turn in 20-inch twist
Some velocities listed below are taken with "nonstandard" barrel lengths

Availability:

Cor-Bon 135-grain Glaser Safety Slug (04200/04400)

G1 Ballistic Coefficient = 0.1?

Distance • Yards	Muzzle	25	50	75	100	125	150	175	200
Velocity • fps	1600	1470	1352	1249	1162	1092	1036	991	953
Energy • ft-lbs	768	648	548	468	405	357	322	294	272
Taylor KO Index	13.3	12.2	11.3	10.4	9.7	9.1	8.6	8.3	7.9
Mid-Range Trajectory Height • Inches	0.0	0.1	0.5	1.3	2.4	4.1	6.3	9.3	12.9
Drop • Inches	0.0	-0.4	-1.9	-4.6	-8.4	-14.7	-21.7	-31.2	-42.9

CCI 140-grain Shotshell (3979)
Shotshell load using # 9 shot at 1000 fps.

Cor-Bon 165-grain JHP Self Defense (SD44M165/20)

G1 Ballistic Coefficient = 0.1

Distance • Yards	Muzzle	25	50	75	100	125	150	175	200
Velocity • fps	1300	1217	1146	1087	1040	1000	965	935	908
Energy • ft-lbs	618	542	481	433	396	366	342	320	302
Taylor KO Index	13.2	12.4	11.7	11.1	10.6	10.2	9.8	9.5	9.2
Mid-Range Trajectory Height • Inches	0.0	0.1	0.7	1.7	3.3	5.4	8.2	11.7	15.9
Drop • Inches	0.0	-0.7	-2.8	-6.6	-12.2	-19.9	-29.7	-41.8	-56.5

UMC (Remington) 180-grain Jacketed Soft Point (L44MG7)

G1 Ballistic Coefficient = 0.?

Distance • Yards	Muzzle	25	50	75	100	125	150	175	200
Velocity • fps	1610	1482	1365	1262	1175	1103	1047	1001	962
Energy • ft-lbs	1036	878	745	637	551	487	438	400	370
Taylor KO Index	17.9	16.5	15.2	14.0	13.1	12.3	11.6	11.1	10.7
Mid-Range Trajectory Height • Inches	0.0	0.1	0.5	1.2	2.3	3.9	6.1	9.0	12.5
Drop • Inches	0.0	-0.4	-1.9	-4.5	-8.4	-14.0	-21.3	-30.6	-41.9

Cor-Bon 180-grain JHP (HT44180JHP/20)

G1 Ballistic Coefficient = 0.130

Distance • Yards	Muzzle	25	50	75	100	125	150	175	200
Velocity • fps	1675	1549	1433	1327	1235	1156	1092	1040	998
Energy • ft-lbs	1122	959	821	704	609	534	477	433	398
Taylor KO Index	18.6	17.2	15.9	14.7	13.7	12.8	12.1	11.6	10.9
Mid-Range Trajectory Height • Inches	0.0	0.1	0.5	1.1	2.2	3.6	5.6	8.2	11.5
Drop • Inches	0.0	-0.4	-1.7	-4.1	-7.7	-12.7	-19.4	-27.9	-38.4

Hornady 180-grain JHP / XTP (9081)

G1 Ballistic Coefficient = 0.138

Distance • Yards	Muzzle	25	50	75	100	125	150	175	200
Velocity • fps	1550	1440	1340	1250	1173	1109	1057	1014	977
Energy • ft-lbs	960	829	717	624	550	492	446	411	382
Taylor KO Index	17.2	16.0	14.9	13.9	13.0	12.3	11.7	11.3	10.9
Mid-Range Trajectory Height • Inches	0.0	0.1	0.5	1.3	2.4	4.0	6.2	9.1	12.5
Drop • Inches	0.0	-0.5	-2.0	-4.7	-8.9	-14.6	-22.0	-31.4	-42.8

Federal 180-grain JHP (C44B)

G1 Ballistic Coefficient = 0.122

Distance • Yards	Muzzle	25	50	75	100	125	150	175	200
Velocity • fps	1460	1340	1240	1160	1090	1036	992	954	921
Energy • ft-lbs	850	720	615	535	470	429	393	364	339
Taylor KO Index	16.2	14.9	13.8	12.9	12.1	11.5	11.0	10.6	10.2
Mid-Range Trajectory Height • Inches	0.0	0.1	0.6	1.5	2.8	4.7	7.3	10.6	14.5
Drop • Inches	0.0	-0.5	-2.3	-5.4	-10.2	-16.8	-25.4	-36.2	-49.4

PMC 180-grain Jacketed Hollow Point (44B)

G1 Ballistic Coefficient = 0.107

Distance • Yards	Muzzle	25	50	75	100	125	150	175	200
Velocity • fps	1400	1270	1167	1091	1032	983	942	906	874
Energy • ft-lbs	784	653	553	479	426	387	355	328	305
Taylor KO Index	15.6	14.1	13.0	12.1	11.5	10.9	10.5	10.1	9.7
Mid-Range Trajectory Height • Inches	0.0	0.2	0.7	1.6	3.1	5.3	8.1	11.8	16.2
Drop • Inches	0.0	-0.6	-2.5	-6.0	-11.3	-18.6	-28.2	-40.3	-54.9

Hornady 200-grain JHP/XTP (9080)

G1 Ballistic Coefficient = 0.170

Distance • Yards	Muzzle	25	50	75	100	125	150	175	200
Velocity • fps	1500	1413	1333	1260	1196	1140	1092	1051	1017
Energy • ft-lbs	999	887	789	706	635	577	530	491	459
Taylor KO Index	18.5	17.4	16.5	15.6	14.8	14.1	13.5	13.0	12.6
Mid-Range Trajectory Height • Inches	0.0	0.1	0.5	1.3	2.5	4.1	6.2	8.9	12.2
Drop • Inches	0.0	-0.5	-2.1	-4.9	-9.1	-14.8	-22.2	-31.4	-42.5

Fiocchi 200-grain Hornady XTP JHP (44XTBP)

G1 Ballistic Coefficient = 0.170

Distance • Yards	Muzzle	25	50	75	100	125	150	175	200
Velocity • fps	1490	1404	1325	1253	1189	1134	1087	1047	1013
Energy • ft-lbs	986	876	780	697	628	571	525	487	456
Taylor KO Index	18.4	17.3	16.4	15.5	14.7	14.0	13.4	12.9	12.5
Mid-Range Trajectory Height • Inches	0.0	0.1	0.5	1.3	2.5	4.1	6.2	9.0	12.3
Drop • Inches	0.0	-0.5	-2.1	-5.0	-9.2	-15.0	-22.4	-31.7	-43.0

Fiocchi 200-grain SJHP (44B)

G1 Ballistic Coefficient = 0.170

Distance • Yards	Muzzle	25	50	75	100	125	150	175	200
Velocity • fps	1475	1390	1312	1241	1178	1125	1080	1041	1008
Energy • ft-lbs	968	858	764	685	617	563	518	481	451
Taylor KO Index	18.2	17.2	16.2	15.3	14.5	13.9	13.3	12.8	10.8
Mid-Range Trajectory Height • Inches	0.0	0.1	0.5	1.3	2.5	4.2	6.4	9.1	12.6
Drop • Inches	0.0	-0.5	-2.2	-5.1	-9.4	-15.3	-22.9	-32.3	-43.8

Magtech 200-grain SCHP (44D)

G1 Ballistic Coefficient = 0.220

Distance • Yards	Muzzle	25	50	75	100	125	150	175	200
Velocity • fps	1296	1242	1193	1149	1110	1076	1046	1020	996
Energy • ft-lbs	746	685	632	587	548	515	486	462	440
Taylor KO Index	16.0	15.3	14.7	14.2	13.7	13.3	12.9	12.6	12.3
Mid-Range Trajectory Height • Inches	0.0	0.1	0.7	1.6	3.0	4.9	7.4	10.4	14.1
Drop • Inches	0.0	-0.7	-2.7	-6.3	-11.6	-18.6	-27.5	-38.3	-51.3

Speer 200-grain GDHP – Short Barrel (23971)
G1 Ballistic Coefficient = 0.145

Distance • Yards	Muzzle	25	50	75	100	125	150	175	200
Velocity • fps	1075	1031	994	961	933	906	882	860	839
Energy • ft-lbs	513	472	440	410	387	365	346	328	312
Taylor KO Index	13.3	12.7	12.3	11.9	11.5	11.2	10.9	10.6	10.4
Mid-Range Trajectory Height • Inches	0.0	0.2	1.0	2.4	4.4	7.1	10.6	14.7	20.0
Drop • Inches	0.0	-1.0	-4.0	-9.2	-16.7	-26.8	-39.5	-55.0	-73.4

Speer 210-grain GDHP (23972)
G1 Ballistic Coefficient = 0.154

Distance • Yards	Muzzle	25	50	75	100	125	150	175	200
Velocity • fps	1450	1359	1276	1203	1140	1088	1044	1007	975
Energy • ft-lbs	980	861	759	675	606	552	509	473	443
Taylor KO Index	18.8	17.6	16.5	15.6	14.8	14.1	13.5	13.1	12.6
Mid-Range Trajectory Height • Inches	0.0	0.1	0.6	1.4	2.7	4.4	6.8	9.8	13.4
Drop • Inches	0.0	-0.5	-2.3	-5.3	-9.9	-16.1	-24.1	-34.2	-46.4

Winchester 210-grain Silvertip Hollow Point (X44MS)
G1 Ballistic Coefficient = 0.133

Distance • Yards	Muzzle	25	50	75	100	125	150	175	200
Velocity • fps	1250	1171	1106	1053	1010	973	941	912	886
Energy • ft-lbs	729	640	570	518	475	442	413	388	366
Taylor KO Index	16.1	15.1	14.3	13.6	13.1	12.6	12.2	11.8	11.5
Mid-Range Trajectory Height • Inches	0.0	0.2	0.8	1.9	3.7	5.7	8.7	12.4	16.7
Drop • Inches	0.0	-0.7	-3.0	-7.1	-13.2	-21.3	-31.7	-44.5	-59.8

Hornady 225-grain FTX (92782)
G1 Ballistic Coefficient = 0.15

Distance • Yards	Muzzle	25	50	75	100	125	150	175	200
Velocity • fps	1410	1320	1240	1170	1111	1063	1022	987	956
Energy • ft-lbs	993	871	768	684	617	564	522	487	457
Taylor KO Index	19.6	18.3	17.2	16.3	15.4	14.8	14.2	13.7	13.3
Mid-Range Trajectory Height • Inches	0.0	0.1	0.6	1.5	2.8	4.7	7.2	10.2	14.0
Drop • Inches	0.0	-0.6	-2.4	-5.6	-10.4	-17.0	-25.5	-36.1	-48.9

Cor-Bon 225-grain DPX (DPX44M225/20)
G1 Ballistic Coefficient = 0.15

Distance • Yards	Muzzle	25	50	75	100	125	150	175	200
Velocity • fps	1350	1266	1193	1130	1078	1035	998	967	938
Energy • ft-lbs	911	801	711	638	581	535	498	467	440
Taylor KO Index	18.7	17.6	16.6	15.7	15.0	14.4	13.9	13.4	13.0
Mid-Range Trajectory Height • Inches	0.0	0.2	0.7	1.6	3.1	5.0	7.6	10.9	14.9
Drop • Inches	0.0	-0.6	-2.6	-6.1	-11.3	-18.4	-27.5	-38.8	-52.4

Federal 225-grain Barnes Expander (P44XB1) LF
G1 Ballistic Coefficient = 0.1

Distance • Yards	Muzzle	25	50	75	100	125	150	175	200
Velocity • fps	1280	1210	1150	1100	1050	1013	981	953	927
Energy • ft-lbs	820	730	660	600	555	513	481	454	429
Taylor KO Index	17.8	16.8	16.0	15.3	14.6	14.1	13.6	13.2	12.9
Mid-Range Trajectory Height • Inches	0.0	0.2	0.7	1.8	3.3	5.4	8.1	11.5	15.7
Drop • Inches	0.0	-0.7	-2.9	-6.7	-12.3	-20.0	-29.7	-41.7	-56.0

Fiocchi 240-grain TCSP (44A)
G1 Ballistic Coefficient = 0.1

Distance • Yards	Muzzle	25	50	75	100	125	150	175	200
Velocity • fps	1310	1224	1151	1092	1042	1002	967	936	908
Energy • ft-lbs	914	799	706	635	579	535	498	467	440
Taylor KO Index	19.0	17.8	16.7	15.8	15.1	14.5	14.0	13.6	13.2
Mid-Range Trajectory Height • Inches	0.0	0.2	0.7	1.8	3.3	5.4	8.2	11.6	15.3
Drop • Inches	0.0	-0.7	-2.8	-6.5	-12.1	-19.7	-29.4	-41.4	-56.0

Sellier and Bellot 240-grain SP (V311402U)
G1 Ballistic Coefficient = 0.0

Distance • Yards	Muzzle	25	50	75	100	125	150	175	200
Velocity • fps	1181	1055	971	907	855	809	768	730	695
Energy • ft-lbs	743	593	502	439	389	349	314	284	257
Taylor KO Index	17.5	15.6	14.4	13.4	12.7	12.0	11.4	10.8	10.3
Mid-Range Trajectory Height • Inches	0.0	0.2	1.0	2.4	4.6	7.7	11.9	17.2	23.9
Drop • Inches	0.0	-0.8	-3.6	-8.7	-16.5	-27.2	-41.2	-58.9	-80.7

PMC 240-grain Truncated Cone – Soft Point (44D)

G1 Ballistic Coefficient = 0.139

Distance • Yards	Muzzle	25	50	75	100	125	150	175	200
Velocity • fps	1300	1216	1145	1086	1038	999	964	934	906
Energy • ft-lbs	900	788	699	629	575	532	496	465	438
Taylor KO Index	19.3	18.0	17.0	16.1	15.4	14.8	14.3	13.8	13.4
Mid-Range Trajectory Height • Inches	0.0	0.2	0.7	1.8	3.3	5.4	8.2	11.6	15.8
Drop • Inches	0.0	-0.7	-2.8	-6.6	-12.2	-19.9	-29.6	-41.7	-56.2

American Eagle (Federal) 240-grain Jacketed Soft Point (AE44B)

G1 Ballistic Coefficient = 0.174

Distance • Yards	Muzzle	25	50	75	100	125	150	175	200
Velocity • fps	1270	1200	1150	1100	1060	1025	994	967	942
Energy • ft-lbs	860	770	700	640	595	560	527	498	473
Taylor KO Index	18.8	17.8	17.0	16.3	15.7	15.2	14.7	14.3	14.0
Mid-Range Trajectory Height • Inches	0.0	0.2	0.8	1.8	3.3	5.4	8.1	11.4	15.4
Drop • Inches	0.0	-0.7	-2.9	-6.7	-12.4	-19.9	-29.5	-41.4	-55.5

Magtech 240-grain SJSP Flat (44A)

G1 Ballistic Coefficient = 0.173

Distance • Yards	Muzzle	25	50	75	100	125	150	175	200
Velocity • fps	1180	1127	1081	1043	1010	981	955	931	909
Energy • ft-lbs	741	677	624	580	544	513	486	462	440
Taylor KO Index	17.5	16.7	16.0	15.4	15.0	14.5	14.1	13.8	13.5
Mid-Range Trajectory Height • Inches	0.0	0.2	0.9	2.0	3.7	6.0	8.9	12.5	16.8
Drop • Inches	0.0	-0.8	-3.3	-7.7	-14.1	-22.5	-33.2	-46.3	-61.9

Remington 240-grain Soft Point (R44MG2)

G1 Ballistic Coefficient = 0.173

Distance • Yards	Muzzle	25	50	75	100	125	150	175	200
Velocity • fps	1180	1127	1081	1043	1010	981	955	931	909
Energy • ft-lbs	741	677	624	580	544	513	486	462	440
Taylor KO Index	17.5	16.7	16.0	15.4	15.0	14.5	14.1	13.8	13.5
Mid-Range Trajectory Height • Inches	0.0	0.2	0.9	2.0	3.7	6.0	8.9	12.5	16.8
Drop • Inches	0.0	-0.8	-3.3	-7.7	-14.1	-22.5	-33.2	-46.3	-61.9

USA (Winchester) 240-grain Jacketed Soft Point (Q4240)

G1 Ballistic Coefficient = 0.173

Distance • Yards	Muzzle	25	50	75	100	125	150	175	200
Velocity • fps	1180	1127	1081	1043	1010	981	955	931	909
Energy • ft-lbs	741	677	624	580	544	513	486	462	440
Taylor KO Index	17.5	16.7	16.0	15.4	15.0	14.5	14.1	13.8	13.5
Mid-Range Trajectory Height • Inches	0.0	0.2	0.9	2.0	3.7	6.0	8.9	12.5	16.8
Drop • Inches	0.0	-0.8	-3.3	-7.7	-14.1	-22.5	-33.2	-46.3	-61.9

Cor-Bon 240-grain JHP (HT44240JHP/20)

G1 Ballistic Coefficient = 0.200

Distance • Yards	Muzzle	25	50	75	100	125	150	175	200
Velocity • fps	1475	1402	1335	1272	1215	1165	1120	1082	1048
Energy • ft-lbs	1160	1048	950	863	787	723	669	624	586
Taylor KO Index	21.8	20.8	19.8	18.8	18.0	17.3	16.6	16.0	15.5
Mid-Range Trajectory Height • Inches	0.0	0.1	0.6	1.3	1.9	4.0	6.1	8.6	11.8
Drop • Inches	0.0	-0.5	-2.1	-5.0	-9.2	-14.8	-22.1	-31.0	-41.9

Speer 240-grain Gold Dot HP (23973)

G1 Ballistic Coefficient = 0.177

Distance • Yards	Muzzle	25	50	75	100	125	150	175	200
Velocity • fps	1400	1324	1255	1193	1139	1093	1054	1020	990
Energy • ft-lbs	1044	934	839	758	691	637	592	554	523
Taylor KO Index	20.7	19.6	18.6	17.7	16.9	16.2	15.6	15.1	14.7
Mid-Range Trajectory Height • Inches	0.0	0.1	0.6	1.4	2.7	4.5	6.8	9.8	13.3
Drop • Inches	0.0	-0.6	-2.3	-5.6	-10.3	-16.7	-24.9	-35.0	-47.2

Fiocchi 240-grain XTP HP (44XTP)

G1 Ballistic Coefficient = 0.205

Distance • Yards	Muzzle	25	50	75	100	125	150	175	200
Velocity • fps	1350	1288	1231	1180	1134	1095	1060	1030	1003
Energy • ft-lbs	971	884	807	742	686	639	599	566	537
Taylor KO Index	20.0	19.1	18.2	17.5	16.8	16.2	15.7	15.3	14.9
Mid-Range Trajectory Height • Inches	0.0	0.1	0.6	1.5	2.8	5.1	7.5	9.9	13.5
Drop • Inches	0.0	-0.6	-2.5	-5.9	-10.8	-17.4	-25.8	-36.2	-48.6

Hornady 240-grain JHP/XTP (9085)

G1 Ballistic Coefficient = 0.205

Distance • Yards	Muzzle	25	50	75	100	125	150	175	200
Velocity • fps	1350	1288	1231	1180	1134	1095	1060	1030	1003
Energy • ft-lbs	971	884	807	742	685	639	599	566	537
Taylor KO Index	20.0	19.1	18.2	17.5	16.8	16.2	15.7	15.3	14.9
Mid-Range Trajectory Height • Inches	0.0	0.2	0.7	1.6	2.9	4.7	7.0	10.0	13.5
Drop • Inches	0.0	-0.6	-2.5	-5.9	-10.8	-17.4	-25.9	-36.2	-48.6

Fiocchi 240-grain JHP (44D)

G1 Ballistic Coefficient = 0.172

Distance • Yards	Muzzle	25	50	75	100	125	150	175	200
Velocity • fps	1330	1258	1195	1139	1092	1052	1017	987	960
Energy • ft-lbs	943	844	761	692	636	590	552	520	492
Taylor KO Index	19.7	18.6	17.7	16.9	16.2	15.6	15.1	14.6	14.2
Mid-Range Trajectory Height • Inches	0.0	0.1	0.7	1.6	3.0	5.0	7.5	10.7	14.5
Drop • Inches	0.0	-0.6	-2.6	-6.2	-11.4	-18.4	-27.4	-38.5	-51.8

PMC 240-grain Starfire Hollow Point (44SFA)

G1 Ballistic Coefficient = 0.13

Distance • Yards	Muzzle	25	50	75	100	125	150	175	200
Velocity • fps	1300	1212	1138	1079	1030	990	956	925	897
Energy • ft-lbs	900	784	692	621	566	523	487	456	429
Taylor KO Index	19.3	18.0	16.9	16.0	15.3	14.7	14.2	13.7	13.3
Mid-Range Trajectory Height • Inches	0.0	0.2	0.7	1.8	3.3	5.5	8.3	11.8	16.0
Drop • Inches	0.0	-0.7	-2.8	-6.6	-12.3	-20.0	-29.9	-42.1	-56.8

Winchester 240-grain DUAL BOND (S44RMDB)

G1 Ballistic Coefficient = 0.19

Distance • Yards	Muzzle	25	50	75	100	125	150	175	200
Velocity • fps	1300	1238	1183	1134	1092	1055	1024	996	972
Energy • ft-lbs	901	817	745	685	635	594	559	529	502
Taylor KO Index	19.3	18.3	17.5	16.8	16.2	15.6	15.2	14.8	14.4
Mid-Range Trajectory Height • Inches	0.0	0.2	0.7	1.7	3.1	5.1	7.6	10.8	14.5
Drop • Inches	0.0	-0.7	-2.8	-6.4	-11.7	-18.8	-27.9	-39.1	-52.4

Black Hills 240-grain Jacketed Hollow Point (D44MN2)

G1 Ballistic Coefficient = 0.1

Distance • Yards	Muzzle	25	50	75	100	125	150	175	200
Velocity • fps	1260	1187	1125	1074	1031	995	964	936	910
Energy • ft-lbs	848	751	675	615	567	528	495	467	441
Taylor KO Index	18.7	17.6	16.7	15.9	15.3	14.7	14.3	13.9	13.5
Mid-Range Trajectory Height • Inches	0.0	0.2	0.8	1.8	3.4	5.5	8.4	11.9	16.1
Drop • Inches	0.0	-0.7	-3.0	-6.9	-12.8	-20.6	-30.7	-43.0	-57.8

American Eagle (Federal) 240-grain JHP (AE44A)

G1 Ballistic Coefficient = 0.1

Distance • Yards	Muzzle	25	50	75	100	125	150	175	200
Velocity • fps	1230	1170	1120	1070	1040	1008	979	954	930
Energy • ft-lbs	805	730	665	615	570	541	511	485	461
Taylor KO Index	18.2	17.3	16.6	15.8	15.4	14.9	14.5	14.1	13.8
Mid-Range Trajectory Height • Inches	0.0	0.2	0.8	1.8	3.4	5.6	8.4	11.9	16.0
Drop • Inches	0.0	-0.7	-3.1	-7.1	-13.0	-21.0	-31.1	-43.4	-58.1

Federal 240-grain JHP (C44A)

G1 Ballistic Coefficient = 0.1

Distance • Yards	Muzzle	25	50	75	100	125	150	175	200
Velocity • fps	1230	1170	1120	1070	1040	1008	979	954	930
Energy • ft-lbs	805	730	665	615	570	541	511	485	461
Taylor KO Index	18.2	17.3	16.6	15.8	15.4	14.9	14.5	14.1	13.8
Mid-Range Trajectory Height • Inches	0.0	0.2	0.8	1.8	3.4	5.6	8.4	11.9	16.0
Drop • Inches	0.0	-0.7	-3.1	-7.1	-13.0	-21.0	-31.1	-43.4	-58.1

Federal 240-grain Hydra-Shok JHP (P44HS1)

G1 Ballistic Coefficient = 0.

Distance • Yards	Muzzle	25	50	75	100	125	150	175	200
Velocity • fps	1210	1150	1100	1060	1020	989	960	935	912
Energy • ft-lbs	780	705	645	600	560	521	492	466	443
Taylor KO Index	17.9	17.0	16.3	15.7	15.1	14.6	14.2	13.8	13.5
Mid-Range Trajectory Height • Inches	0.0	0.2	0.8	2.0	3.6	5.8	8.8	12.2	16.7
Drop • Inches	0.0	-0.8	-3.2	-7.4	-13.5	-21.8	-32.2	-45.0	-60.3

Blazer (CCI) 240-grain JHP (3564)
G1 Ballistic Coefficient = 0.165

Distance • Yards	Muzzle	25	50	75	100	125	150	175	200
Velocity • fps	1200	1142	1092	1050	1015	984	956	931	908
Energy • ft-lbs	767	695	636	588	549	516	487	462	439
Taylor KO Index	17.8	16.9	16.2	15.6	15.0	14.6	14.2	13.8	13.4
Mid-Range Trajectory Height • Inches	0.0	0.2	0.5	1.9	3.3	5.8	8.8	12.4	16.7
Drop • Inches	0.0	-0.8	-3.2	-7.5	-13.7	-22.1	-32.6	-45.5	-60.9

Remington 240-grain Semi-Jacketed Hollow Point (R44MG3)
G1 Ballistic Coefficient = 0.173

Distance • Yards	Muzzle	25	50	75	100	125	150	175	200
Velocity • fps	1180	1127	1081	1043	1010	981	955	931	909
Energy • ft-lbs	741	677	624	580	544	513	486	462	440
Taylor KO Index	17.5	16.7	16.0	15.4	15.0	14.5	14.1	13.8	13.5
Mid-Range Trajectory Height • Inches	0.0	0.2	0.9	2.0	3.7	6.0	8.9	12.5	16.8
Drop • Inches	0.0	-0.8	-3.3	-7.7	-14.1	-22.5	-33.2	-46.3	-61.9

Winchester 240-grain Hollow Soft Point (X44MHSP2)
G1 Ballistic Coefficient = 0.173

Distance • Yards	Muzzle	25	50	75	100	125	150	175	200
Velocity • fps	1180	1127	1081	1043	1010	981	955	931	909
Energy • ft-lbs	741	677	624	580	544	513	486	462	440
Taylor KO Index	17.5	16.7	16.0	15.4	15.0	14.5	14.1	13.8	13.5
Mid-Range Trajectory Height • Inches	0.0	0.2	0.9	2.0	3.7	6.0	8.9	12.5	16.8
Drop • Inches	0.0	-0.8	-3.3	-7.7	-14.1	-22.5	-33.2	-46.3	-61.9

PMC 240-grain EMJ NT (44EMD) LF
G1 Ballistic Coefficient = 0.138

Distance • Yards	Muzzle	25	50	75	100	125	150	175	200
Velocity • fps	1300	1216	1144	1086	1038	998	963	933	905
Energy • ft-lbs	900	788	698	628	574	531	495	464	437
Taylor KO Index	18.9	17.6	16.6	15.8	15.1	14.5	14.0	13.5	13.1
Mid-Range Trajectory Height • Inches	0.0	0.2	0.7	1.8	3.3	5.5	8.3	11.7	16.1
Drop • Inches	0.0	-0.7	-2.8	-6.6	-12.3	-19.9	-29.8	-41.9	-56.6

Magtech 240-grain FMJ Flat (44C)
G1 Ballistic Coefficient = 0.173

Distance • Yards	Muzzle	25	50	75	100	125	150	175	200
Velocity • fps	1180	1127	1081	1043	1010	981	955	931	909
Energy • ft-lbs	741	677	624	580	544	513	486	462	440
Taylor KO Index	17.5	16.7	16.0	15.4	15.0	14.5	14.1	13.8	13.5
Mid-Range Trajectory Height • Inches	0.0	0.2	0.9	2.0	3.7	6.0	8.9	12.5	16.8
Drop • Inches	0.0	-0.8	-3.3	-7.7	-14.1	-22.5	-33.2	-46.3	-61.9

Winchester 250-grain Platinum Tip (S44PTHP)
G1 Ballistic Coefficient = 0.202

Distance • Yards	Muzzle	25	50	75	100	125	150	175	200
Velocity • fps	1250	1196	1148	1106	1070	1038	1010	985	962
Energy • ft-lbs	867	794	732	680	635	599	567	539	514
Taylor KO Index	19.3	18.5	17.7	17.1	16.5	16.0	15.6	15.2	14.8
Mid-Range Trajectory Height • Inches	0.0	0.2	0.8	1.8	3.3	5.3	8.0	11.3	15.1
Drop • Inches	0.0	-0.7	-3.0	-6.8	-12.5	-20.0	-29.6	-41.3	-55.2

Winchester 250-grain Partition Gold (S44MP)
G1 Ballistic Coefficient = 0.201

Distance • Yards	Muzzle	25	50	75	100	125	150	175	200
Velocity • fps	1230	1178	1132	1092	1057	1027	1000	975	953
Energy • ft-lbs	840	770	711	662	620	585	555	528	504
Taylor KO Index	18.9	18.1	17.4	16.8	16.2	15.8	15.4	15.0	14.7
Mid-Range Trajectory Height • Inches	0.0	0.2	0.8	1.8	3.4	5.4	8.1	11.4	15.4
Drop • Inches	0.0	-0.7	-3.0	-7.0	-12.9	-20.6	-30.4	-42.4	-56.6

Cor-Bon 260-grain Bonded Core HP (CB44260BHP/20)
G1 Ballistic Coefficient = 0.155

Distance • Yards	Muzzle	25	50	75	100	125	150	175	200
Velocity • fps	1450	1364	1286	1217	1158	1108	1066	1030	998
Energy • ft-lbs	1214	1074	955	855	774	709	656	612	576
Taylor KO Index	23.3	21.9	20.6	19.5	18.6	17.8	17.1	16.5	16.0
Mid-Range Trajectory Height • Inches	0.0	0.1	0.5	1.4	2.6	4.3	6.6	9.4	12.8
Drop • Inches	0.0	-0.5	-2.2	-5.3	-9.8	-15.9	-23.7	-33.5	-45.2

Speer 270-grain GDSP (23968)

G1 Ballistic Coefficient = 0.193

Distance • Yards	Muzzle	25	50	75	100	125	150	175	200
Velocity • fps	1250	1194	1144	1101	1064	1031	1003	977	954
Energy • ft-lbs	937	854	785	727	678	638	603	573	546
Taylor KO Index	20.4	19.5	18.7	18.0	17.4	16.8	16.4	15.9	15.6
Mid-Range Trajectory Height • Inches	0.0	0.2	0.8	1.8	3.3	5.4	8.1	11.2	15.3
Drop • Inches	0.0	-0.7	-3.0	-6.9	-12.6	-20.2	-29.8	-41.6	-55.7

Federal 280-grain Swift A-Frame (P44SA)

G1 Ballistic Coefficient = 0.142

Distance • Yards	Muzzle	25	50	75	100	125	150	175	200
Velocity • fps	1170	1110	1060	1010	980	948	920	895	871
Energy • ft-lbs	850	760	695	640	595	559	527	498	472
Taylor KO Index	20.2	19.2	18.3	17.5	16.9	16.4	15.9	15.5	15.1
Mid-Range Trajectory Height • Inches	0.0	0.2	0.9	2.1	3.9	6.3	9.4	13.3	18.1
Drop • Inches	0.0	-0.8	-3.4	-8.0	-14.6	-23.5	-34.8	-48.7	-65.3

Cor-Bon 300-grain JSP (HT44300JSP/20)

G1 Ballistic Coefficient = 0.24?

Distance • Yards	Muzzle	25	50	75	100	125	150	175	200
Velocity • fps	1250	1205	1164	1128	1095	1066	1040	1016	995
Energy • ft-lbs	1041	968	903	847	799	757	721	688	660
Taylor KO Index	22.7	21.8	21.1	20.4	19.9	19.3	18.9	18.4	18.0
Mid-Range Trajectory Height • Inches	0.0	0.2	0.8	1.8	3.2	5.2	7.7	10.7	14.5
Drop • Inches	0.0	-0.7	-2.9	-6.7	-12.2	-19.6	-28.8	-40.1	-53.4

Cor-Bon 300-grain FP Penetrator (HT44305FPPN/20)

G1 Ballistic Coefficient = 0.24

Distance • Yards	Muzzle	25	50	75	100	125	150	175	200
Velocity • fps	1250	1205	1164	1128	1095	1066	1040	1016	995
Energy • ft-lbs	1058	984	918	861	812	770	732	700	670
Taylor KO Index	23.0	22.2	21.5	20.8	20.2	19.6	19.2	18.7	18.3
Mid-Range Trajectory Height • Inches	0.0	0.2	0.8	1.8	3.2	5.2	7.7	10.7	14.5
Drop • Inches	0.0	-0.7	-2.9	-6.7	-12.2	-19.6	-28.8	-40.1	-53.4

Black Hills 300-grain Jacketed Hollow Point (D44MN3)

G1 Ballistic Coefficient = 0.2?

Distance • Yards	Muzzle	25	50	75	100	125	150	175	200
Velocity • fps	1150	1108	1071	1039	1010	985	962	940	921
Energy • ft-lbs	879	817	764	719	680	646	616	589	565
Taylor KO Index	21.3	20.5	19.8	19.2	18.7	18.2	17.8	17.4	17.1
Mid-Range Trajectory Height • Inches	0.0	0.2	0.9	2.1	3.8	6.1	9.1	12.7	17.0
Drop • Inches	0.0	-0.8	-3.5	-8.0	-14.5	-23.1	-34.0	-47.3	-63.0

Hornady 300-grain JHP/XTP (9088)

G1 Ballistic Coefficient = 0.2

Distance • Yards	Muzzle	25	50	75	100	125	150	175	200
Velocity • fps	1150	1115	1084	1056	1031	1008	987	968	950
Energy • ft-lbs	881	828	782	743	708	677	649	624	602
Taylor KO Index	21.3	20.6	20.1	19.6	19.1	18.7	18.3	17.9	17.6
Mid-Range Trajectory Height • Inches	0.0	0.2	0.9	2.0	3.7	5.9	8.8	12.2	16.3
Drop • Inches	0.0	-0.8	-3.4	-7.9	-14.2	-22.7	-33.2	-46.0	-61.1

Cor-Bon 320-grain Hardcast (HT44320HC/20)

G1 Ballistic Coefficient = 0.2?

Distance • Yards	Muzzle	25	50	75	100	125	150	175	200
Velocity • fps	1175	1134	1098	1066	1038	1012	990	969	949
Energy • ft-lbs	981	914	857	808	765	729	696	667	640
Taylor KO Index	22.7	21.9	21.2	20.6	20.1	19.6	19.1	18.7	18.4
Mid-Range Trajectory Height • Inches	0.0	0.2	0.8	2.0	3.6	5.8	8.6	11.9	16.1
Drop • Inches	0.0	-0.8	-3.3	-7.6	-13.8	-22.0	-32.4	-44.9	-59.8

.45 GAP (Glock Automatic Pistol)

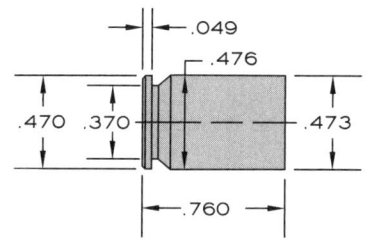

It took much longer than the shrinking of the 10mm Auto into the .40 S&W, but some bright lad finally figured out that the .45 Auto case was just a little larger than it had to be to get the desired performance. Thus, the .45 GAP. Now, in this sort of situation it is common for the proponents of the new cartridge to claim, "It'll do everything the .45 Auto will do." Not quite true. But the .45 GAP does do everything its designers wanted it to do, at the price of slightly higher working pressure. Since it is a new caliber, the guns can be given the strength necessary to withstand the higher pressure, and there are no old guns to worry about. The result is a very useful cartridge that goes into a gun with a slightly smaller grip. That's a boon to shooters with not-so-large hands. This caliber may not be an exact replacement for the .45 Auto, but it is likely to be with us for a long time.

Relative Recoil Factor = 0.92

Specifications:

Controlling Agency for Standardization of this Ammunition: SAAMI

Bullet Weight Grains	Velocity fps	Maximum Average Pressure	
		Copper Crusher	Transducer
185 JHP FMJ	1,090	N/S	23,000 psi
200 JHP FMJ	1,020	N/S	23,000 psi

Standard barrel for velocity testing: 5 inches long—1 turn in 16-inch twist

Availability:

Cor-Bon 160-grain DPX (DPX45GAP160/20)

G1 Ballistic Coefficient = 0.145

Distance • Yards	Muzzle	25	50	75	100	125	150	175	200
Velocity • fps	1075	1031	994	961	933	906	882	860	839
Energy • ft-lbs	411	378	351	328	309	292	277	263	250
Taylor KO Index	11.1	10.7	10.3	9.9	9.6	9.4	9.1	8.9	8.7
Mid-Range Trajectory Height • Inches	0.0	0.2	1.0	2.4	4.4	7.1	10.6	14.7	20.0
Drop • Inches	0.0	-1.0	-4.0	-9.2	-16.7	-26.8	-39.5	-55.0	-73.4

Cor-Bon 165-grain Pow'RBall (PB454GAP165/20)

G1 Ballistic Coefficient = 0.145

Distance • Yards	Muzzle	25	50	75	100	125	150	175	200
Velocity • fps	1075	1031	994	961	933	906	882	860	839
Energy • ft-lbs	424	390	362	339	319	301	285	271	258
Taylor KO Index	11.5	11.0	10.6	10.2	9.9	9.7	9.4	9.2	8.9
Mid-Range Trajectory Height • Inches	0.0	0.2	1.0	2.4	4.4	7.1	10.6	14.9	20.0
Drop • Inches	0.0	-1.0	-4.0	-9.2	-16.8	-26.8	-39.5	-55.0	-73.4

Magtech 185-grain JHP (GG45GA)

G1 Ballistic Coefficient = 0.192

Distance • Yards	Muzzle	25	50	75	100	125	150	175	200
Velocity • fps	1148	1104	1066	1034	1005	979	955	934	914
Energy • ft-lbs	542	501	467	439	415	394	375	358	343
Taylor KO Index	13.8	13.2	12.8	12.4	12.1	11.7	11.5	11.2	11.0
Mid-Range Trajectory Height • Inches	0.0	0.2	0.9	2.0	3.8	6.1	9.1	12.8	17.2
Drop • Inches	0.0	-0.8	-3.5	-8.0	-14.6	-23.3	-34.3	-47.7	-63.5

Federal 185-grain Hydra-Shok JHP (PD45G1H)

G1 Ballistic Coefficient = 0.157

Distance • Yards	Muzzle	25	50	75	100	125	150	175	200
Velocity • fps	1090	1050	1010	970	950	924	900	873	858
Energy • ft-lbs	490	450	415	390	370	351	333	317	302
Taylor KO Index	13.0	12.5	12.1	11.6	11.3	11.0	10.8	10.5	10.2
Mid-Range Trajectory Height • Inches	0.0	0.2	1.0	2.3	4.3	6.9	10.2	14.2	19.3
Drop • Inches	0.0	-0.9	-3.9	-8.9	-16.2	-26.0	-38.2	-53.2	-71.0

Speer 185-grain GDHP (23977)

G1 Ballistic Coefficient = 0.109

Distance • Yards	Muzzle	25	50	75	100	125	150	175	200
Velocity • fps	1090	1030	982	941	906	874	845	818	793
Energy • ft-lbs	488	436	396	364	337	314	294	275	243
Taylor KO Index	13.0	12.3	11.7	11.2	10.8	10.4	10.1	9.8	9.5
Mid-Range Trajectory Height • Inches	0.0	0.2	1.0	2.4	4.5	7.3	11.0	15.7	21.2
Drop • Inches	0.0	-1.0	-4.0	-9.2	-16.9	-27.3	-40.4	-56.7	-76.2

Winchester 185-grain Silvertip HP (X45GSHP)

G1 Ballistic Coefficient = 0.146

Distance • Yards	Muzzle	25	50	75	100	125	150	175	200
Velocity • fps	1000	967	938	911	887	864	843	823	804
Energy • ft-lbs	411	384	361	341	323	307	292	278	265
Taylor KO Index	11.9	11.6	11.2	10.9	10.6	10.3	10.1	9.8	9.6
Mid-Range Trajectory Height • Inches	0.0	0.3	1.2	2.7	4.9	7.9	11.8	16.5	22.1
Drop • Inches	0.0	-1.1	-4.5	-10.5	-19.0	-30.2	-44.4	-61.7	-82.1

Speer 200-grain GDHP (23978)

G1 Ballistic Coefficient = 0.138

Distance • Yards	Muzzle	25	50	75	100	125	150	175	200
Velocity • fps	970	939	911	885	862	839	818	798	779
Energy • ft-lbs	418	392	369	348	330	313	297	283	269
Taylor KO Index	12.6	12.2	11.8	11.5	11.2	10.9	10.6	10.4	10.1
Mid-Range Trajectory Height • Inches	0.0	0.3	1.2	2.8	5.2	8.4	12.5	17.5	23.4
Drop • Inches	0.0	-1.2	-4.8	-11.1	-20.1	-32.1	-47.1	-65.4	-87.1

Cor-Bon 200-grain JHP (SD45GAP200/20)

G1 Ballistic Coefficient = 0.17

Distance • Yards	Muzzle	25	50	75	100	125	150	175	200
Velocity • fps	950	926	904	884	864	846	828	812	796
Energy • ft-lbs	401	381	363	347	332	318	305	293	281
Taylor KO Index	12.3	12.0	11.7	11.4	11.2	10.9	10.7	10.5	10.3
Mid-Range Trajectory Height • Inches	0.0	0.3	1.3	2.9	5.3	8.5	12.5	17.5	23.3
Drop • Inches	0.0	-1.2	-5.0	-11.4	-20.6	-32.7	-47.8	-66.1	-87.7

American Eagle (Federal) 185-grain Full Metal Jacket (AE45GA)

G1 Ballistic Coefficient = 0.15

Distance • Yards	Muzzle	25	50	75	100	125	150	175	200
Velocity • fps	1090	1050	1010	970	950	924	900	873	858
Energy • ft-lbs	490	450	415	390	370	351	333	317	302
Taylor KO Index	13.0	12.5	12.1	11.6	11.3	11.0	10.8	10.5	10.2
Mid-Range Trajectory Height • Inches	0.0	0.2	1.0	2.3	4.3	6.9	10.2	14.2	19.3
Drop • Inches	0.0	-0.9	-3.9	-8.9	-16.2	-26.0	-38.2	-53.2	-71.0

Speer 185-grain TMJ FN Lawman (53979)

G1 Ballistic Coefficient = 0.0

Distance • Yards	Muzzle	25	50	75	100	125	150	175	200
Velocity • fps	1060	999	950	909	872	839	809	781	754
Energy • ft-lbs	462	410	370	339	312	290	269	251	234
Taylor KO Index	12.7	11.9	11.3	10.9	10.4	10.0	9.7	9.3	9.0
Mid-Range Trajectory Height • Inches	0.0	0.2	1.0	2.6	4.8	7.9	11.9	16.7	22.9
Drop • Inches	0.0	-1.0	-4.2	-9.8	-18.0	-29.1	-43.2	-60.7	-81.7

Speer 200-grain TMJ FN Lawman (53980)

G1 Ballistic Coefficient = 0.1

Distance • Yards	Muzzle	25	50	75	100	125	150	175	200
Velocity • fps	990	946	908	874	843	814	788	762	739
Energy • ft-lbs	435	397	365	339	316	294	276	258	242
Taylor KO Index	12.8	12.2	11.7	11.3	10.9	10.5	10.2	9.8	9.5
Mid-Range Trajectory Height • Inches	0.0	0.3	1.2	2.9	5.3	8.6	12.8	18.0	24.5
Drop • Inches	0.0	-1.1	-4.7	-10.9	-20.0	-32.1	-47.5	-66.4	-89.0

Magtech 230-grain JHP (GG45GB)

G1 Ballistic Coefficient = 0.

Distance • Yards	Muzzle	25	50	75	100	125	150	175	200
Velocity • fps	1007	980	956	934	914	894	876	859	843
Energy • ft-lbs	518	491	467	446	426	409	392	377	363
Taylor KO Index	15.0	14.6	14.3	13.9	13.6	13.3	13.1	12.8	12.6
Mid-Range Trajectory Height • Inches	0.0	0.2	1.1	2.6	4.7	7.6	11.2	15.6	20.8
Drop • Inches	0.0	-1.1	-4.4	-10.2	-18.4	-29.2	-42.7	-59.1	-78.4

Federal 230-grain Hydra-Shok JHP (P45GHS1)
G1 Ballistic Coefficient = 0.207

Distance • Yards	Muzzle	25	50	75	100	125	150	175	200
Velocity • fps	880	870	850	840	820	807	794	781	769
Energy • ft-lbs	395	380	370	355	345	333	322	312	302
Taylor KO Index	13.1	12.9	12.6	12.5	12.2	12.0	11.8	11.6	11.4
Mid-Range Trajectory Height • Inches	0.0	0.3	1.5	3.3	6.0	9.6	14.0	19.5	25.8
Drop • Inches	0.0	-1.4	-5.8	-13.1	-23.6	-37.2	-54.3	-74.7	-98.8

USA (Winchester) 230-grain Jacketed Hollow Point (USA45GJHP)
G1 Ballistic Coefficient = 0.165

Distance • Yards	Muzzle	25	50	75	100	125	150	175	200
Velocity • fps	880	860	842	824	807	790	775	759	745
Energy • ft-lbs	396	378	363	347	332	319	306	294	283
Taylor KO Index	13.1	12.8	12.5	12.2	12.0	11.7	11.5	11.3	11.1
Mid-Range Trajectory Height • Inches	0.0	0.3	1.5	3.4	6.1	9.8	14.4	20.1	26.7
Drop • Inches	0.0	-1.4	-5.8	-13.2	-23.8	-37.8	-55.2	-76.3	-101.1

American Eagle (Federal) 230-grain FMJ (AE45GB)
G1 Ballistic Coefficient = 0.207

Distance • Yards	Muzzle	25	50	75	100	125	150	175	200
Velocity • fps	880	870	850	840	820	807	794	781	769
Energy • ft-lbs	395	380	370	355	345	333	322	312	302
Taylor KO Index	13.1	12.9	12.6	12.5	12.2	12.0	11.8	11.6	11.4
Mid-Range Trajectory Height • Inches	0.0	0.3	1.5	3.3	6.0	9.6	14.0	19.5	25.8
Drop • Inches	0.0	-1.4	-5.8	-13.1	-23.6	-37.2	-54.3	-74.7	-98.8

UMC (Remington) 230-grain Metal Case (L45GAP4)
G1 Ballistic Coefficient = 0.160

Distance • Yards	Muzzle	25	50	75	100	125	150	175	200
Velocity • fps	880	860	841	822	805	788	772	756	741
Energy • ft-lbs	395	377	361	345	331	317	304	292	280
Taylor KO Index	13.1	12.8	12.5	12.2	12.0	11.7	11.5	11.2	11.0
Mid-Range Trajectory Height • Inches	0.0	0.3	1.5	3.4	6.2	9.8	14.5	20.2	26.9
Drop • Inches	0.0	-1.4	-5.8	-13.2	-23.9	-37.9	-55.4	-76.6	-101.5

Winchester 230-grain Brass Enclosed Base (WC45G)
G1 Ballistic Coefficient = 0.179

Distance • Yards	Muzzle	25	50	75	100	125	150	175	200
Velocity • fps	875	857	840	823	808	793	778	764	750
Energy • ft-lbs	391	375	360	346	333	321	309	298	287
Taylor KO Index	13.0	12.7	12.5	12.2	12.0	11.8	11.6	11.3	11.1
Mid-Range Trajectory Height • Inches	0.0	0.3	1.5	3.4	6.2	9.8	14.4	20.1	26.7
Drop • Inches	0.0	-1.4	-5.8	-13.3	-24.0	-38.0	-55.5	-76.5	-101.3

USA (Winchester) 230-grain Full Metal Jacket (USA45G)
G1 Ballistic Coefficient = 0.165

Distance • Yards	Muzzle	25	50	75	100	125	150	175	200
Velocity • fps	850	832	814	798	782	766	751	737	723
Energy • ft-lbs	369	353	338	325	312	300	288	277	267
Taylor KO Index	12.6	12.4	12.1	11.9	11.6	11.4	11.2	10.9	10.7
Mid-Range Trajectory Height • Inches	0.0	0.3	1.6	3.6	6.6	10.4	15.4	21.4	28.5
Drop • Inches	0.0	-1.5	-6.2	-14.1	-25.5	-40.4	-59.0	-81.5	-107.9

Magtech 230-grain FMJ (45GA)
G1 Ballistic Coefficient = 0.155

Distance • Yards	Muzzle	25	50	75	100	125	150	175	200
Velocity • fps	837	818	800	783	767	751	735	720	706
Energy • ft-lbs	356	342	326	313	300	288	276	265	254
Taylor KO Index	12.4	12.1	11.9	11.6	11.4	11.2	10.9	10.7	10.5
Mid-Range Trajectory Height • Inches	0.0	0.4	1.6	3.7	6.8	10.8	16.0	21.9	29.6
Drop • Inches	0.0	-1.6	-6.4	-14.6	-26.4	-41.8	-61.1	-84.4	-111.9

Sellier & Bellot 230-grain FMJ (V311145U)
G1 Ballistic Coefficient = 0.175

Distance • Yards	Muzzle	25	50	75	100	125	150	175	200
Velocity • fps	807	792	777	762	748	734	721	708	696
Energy • ft-lbs	333	320	308	297	286	275	266	256	247
Taylor KO Index	12.0	11.8	11.5	11.3	11.1	10.9	10.7	10.5	10.3
Mid-Range Trajectory Height • Inches	0.0	0.4	1.7	4.0	7.2	11.5	16.9	23.0	31.1
Drop • Inches	0.0	-1.7	-6.8	-15.6	-28.1	-44.5	-64.9	-89.5	-118.4

Cor-Bon 230-grain FMJ (PM45GAP230/50)

G1 Ballistic Coefficient = 0.175

Distance • Yards	Muzzle	25	50	75	100	125	150	175	200
Velocity • fps	750	736	723	710	697	685	673	661	649
Energy • ft-lbs	287	277	267	257	248	239	231	223	215
Taylor KO Index	11.1	10.9	10.7	10.5	10.4	10.2	10.0	9.8	9.6
Mid-Range Trajectory Height • Inches	0.0	0.5	2.0	4.6	8.3	13.2	19.4	26.9	35.9
Drop • Inches	0.0	-2.0	-7.9	-18.0	-32.5	-51.4	-74.9	-103.2	-136.5

The Alaska Peninsula is a bird shooter's paradise, but there are lots of big bears, too. Here guide Brad Adams is carrying a handgun of serious proportions whilst guiding bird hunters. Just in case! (Safari Press Archives)

.45 Auto (.45 ACP)

Born in the flurry of gun and cartridge design following the end of the Spanish-American War (1905), the .45 Auto cartridge was designed by John Browning to be fired in his automatic pistol. The first cut at the gun's design didn't do well, so the gun was modified (some might say improved) into what became the M1911 pistol. The original loading for that pistol used a 230-grain bullet at 850 fps. Here we are, over 100 years later, and the 230-grain bullet at 850 fps is still pretty close to the standard load for the .45. Like most calibers that have been adopted by the U.S. military, the .45 Auto remains a very popular caliber, and certainly a very effective defensive weapon.

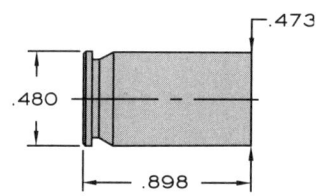

Relative Recoil Factor = 0.93

Specifications:

Controlling Agency for Standardization of this Ammunition: SAAMI

Bullet Weight Grains	Velocity fps	Maximum Average Pressure Copper Crusher	Transducer
180 JHP	995	18,000 cup	21,000 psi
180 JHP	930	18,000 cup	21,000 psi
230 FMC	830	18,000 cup	21,000 psi
+ P Loads			
185 JHP	1,130	N/S	23,000 psi

Standard barrel for velocity testing: 5 inches long—1 turn in 16-inch twist

Availability:

CCI 117-grain Shotshell (3567)
Number 9 shot at 1100 fps

Cor-Bon 145-grain Glaser Safety Slug +P (04600/04800)
G1 Ballistic Coefficient = 0.125

Distance • Yards	Muzzle	25	50	75	100	125	150	175	200
Velocity • fps	1350	1251	1166	1098	1043	998	961	927	898
Energy • ft-lbs	587	504	438	388	350	321	297	277	260
Taylor KO Index	12.6	11.7	10.9	10.3	9.8	9.3	9.0	8.7	8.4
Mid-Range Trajectory Height • Inches	0.0	0.2	0.7	1.7	3.2	5.3	8.1	11.6	15.9
Drop • Inches	0.0	-0.6	-2.7	-6.3	-11.7	-19.1	-28.7	-40.6	-55.1

MC 155-grain SinterFire/Frangible (45F) LF
G1 Ballistic Coefficient = 0.136

Distance • Yards	Muzzle	25	50	75	100	125	150	175	200
Velocity • fps	1100	1048	1006	970	939	910	884	860	838
Energy • ft-lbs	416	379	349	324	303	285	269	255	242
Taylor KO Index	11.0	10.5	10.1	9.7	9.4	9.1	8.8	8.6	8.4
Mid-Range Trajectory Height • Inches	0.0	0.2	1.0	2.3	4.3	6.9	10.4	14.6	19.7
Drop • Inches	0.0	-0.9	-3.8	-8.9	-16.2	-26.0	-38.5	-53.7	-71.8

Cor-Bon 160-grain DPX (DPX45160/20)
G1 Ballistic Coefficient = 0.145

Distance • Yards	Muzzle	25	50	75	100	125	150	175	200
Velocity • fps	1050	1010	976	945	918	893	870	848	827
Energy • ft-lbs	392	362	338	317	299	283	269	256	243
Taylor KO Index	10.8	10.4	10.1	9.6	9.5	9.2	9.0	8.8	8.5
Mid-Range Trajectory Height • Inches	0.0	0.3	1.1	2.5	4.6	7.4	11.0	15.2	20.6
Drop • Inches	0.0	-1.0	-4.2	-9.6	-17.4	-27.9	-41.0	-57.0	-76.1

Cor-Bon 165-grain JHP +P (SD45165/20)
G1 Ballistic Coefficient = 0.140

Distance • Yards	Muzzle	25	50	75	100	125	150	175	200
Velocity • fps	1250	1180	1121	1073	1032	997	967	939	914
Energy • ft-lbs	573	510	461	422	390	365	342	323	306
Taylor KO Index	13.3	12.6	11.9	11.4	11.0	10.6	10.3	10.0	9.7
Mid-Range Trajectory Height • Inches	0.0	0.1	0.7	1.8	3.4	5.6	8.4	11.9	16.1
Drop • Inches	0.0	-0.7	-3.0	-7.0	-12.9	-20.8	-30.9	-43.2	-58.0

Cor-Bon 165-grain Pow'RBall +P (PB45165/20)

G1 Ballistic Coefficient = 0.145

Distance • Yards	Muzzle	25	50	75	100	125	150	175	200
Velocity • fps	1225	1155	1097	1050	1010	975	945	918	893
Energy • ft-lbs	550	489	441	404	378	349	327	309	292
Taylor KO Index	13.1	12.3	11.7	11.2	10.8	10.4	10.1	9.8	9.5
Mid-Range Trajectory Height • Inches	0.0	0.2	0.8	1.9	3.6	5.9	8.8	12.6	17.0
Drop • Inches	0.0	-0.8	-3.1	-7.3	-13.5	-21.8	-32.3	-45.3	-60.9

Federal 165-grain EFMJ (PD45CSP2H)

G1 Ballistic Coefficient = 0.127

Distance • Yards	Muzzle	25	50	75	100	125	150	175	200
Velocity • fps	1140	1080	1030	990	950	919	890	864	840
Energy • ft-lbs	475	425	390	360	335	309	291	274	259
Taylor KO Index	12.1	11.5	11.0	10.5	10.1	9.8	9.5	9.2	8.9
Mid-Range Trajectory Height • Inches	0.0	0.2	0.9	2.2	4.1	6.7	10.1	14.1	19.2
Drop • Inches	0.0	-0.9	-3.6	-8.4	-15.4	-24.9	-36.9	-51.6	-69.3

Magtech 165-grain SCHP (FD45A)

G1 Ballistic Coefficient = 0.168

Distance • Yards	Muzzle	25	50	75	100	125	150	175	200
Velocity • fps	1100	1058	1022	991	963	937	914	893	873
Energy • ft-lbs	443	410	383	360	340	322	306	292	279
Taylor KO Index	11.8	11.3	10.9	10.6	10.3	10.0	9.8	9.6	9.3
Mid-Range Trajectory Height • Inches	0.0	0.2	0.9	2.2	4.1	6.7	9.9	13.9	18.7
Drop • Inches	0.0	-0.9	-3.8	-8.7	-15.9	-25.4	-37.4	-51.9	-69.2

Federal 165-grain Hydra-Shok JHP (PD45HS3 H)

G1 Ballistic Coefficient = 0.12

Distance • Yards	Muzzle	25	50	75	100	125	150	175	200
Velocity • fps	1060	1010	980	940	910	883	857	833	811
Energy • ft-lbs	410	375	350	325	305	285	269	255	241
Taylor KO Index	11.3	10.8	10.4	10.0	9.7	9.4	9.1	8.9	8.6
Mid-Range Trajectory Height • Inches	0.0	0.2	1.1	2.5	4.6	7.4	11.1	15.4	21.0
Drop • Inches	0.0	-1.0	-4.1	-9.5	-17.4	-27.8	-41.1	-57.3	-76.7

Fiocchi 165-grain Frangible Sinterfire (45SFNT) LF

G1 Ballistic Coefficient = 0.1

Distance • Yards	Muzzle	25	50	75	100	125	150	175	200
Velocity • fps	1125	1058	1004	961	923	891	860	833	807
Energy • ft-lbs	464	410	370	338	313	291	271	254	239
Taylor KO Index	12.0	11.3	10.7	10.3	9.9	9.5	9.2	8.9	8.6
Mid-Range Trajectory Height • Inches	0.0	0.2	0.9	2.3	4.3	7.0	10.6	15.0	20.4
Drop • Inches	0.0	-0.9	-3.7	-8.7	-16.1	-25.9	-38.6	-54.2	-72.8

Winchester 170-grain Jacketed Flat Point (SC45NT) LF

G1 Ballistic Coefficient = 0.1

Distance • Yards	Muzzle	25	50	75	100	125	150	175	200
Velocity • fps	1050	1013	982	954	928	905	883	863	843
Energy • ft-lbs	416	388	364	343	325	309	295	281	269
Taylor KO Index	11.5	11.1	10.8	10.5	10.2	9.9	9.7	9.5	9.3
Mid-Range Trajectory Height • Inches	0.0	0.2	0.9	2.4	4.5	7.2	9.7	15.0	20.0
Drop • Inches	0.0	-1.0	-4.1	-9.5	-17.3	-27.5	-40.5	-56.1	-74.7

Cor-Bon 185-grain JHP +P (SD45185/20)

G1 Ballistic Coefficient = 0.1

Distance • Yards	Muzzle	25	50	75	100	125	150	175	200
Velocity • fps	1150	1100	1058	1022	990	962	937	913	891
Energy • ft-lbs	543	497	460	429	403	381	361	343	326
Taylor KO Index	13.7	13.1	12.6	12.2	11.8	11.5	11.2	10.9	10.6
Mid-Range Trajectory Height • Inches	0.0	0.2	0.9	2.1	3.8	6.2	9.3	13.0	17.5
Drop • Inches	0.0	-0.8	-3.5	-8.1	-14.7	-23.6	-34.8	-48.4	-64.6

Hornady 185-grain FTX (90900)

G1 Ballistic Coefficient = 0.1

Distance • Yards	Muzzle	25	50	75	100	125	150	175	200
Velocity • fps	900	875	853	831	811	791	773	755	738
Energy • ft-lbs	333	315	299	284	270	257	245	234	224
Taylor KO Index	10.6	10.3	10.1	9.8	9.6	9.3	9.1	8.9	8.7
Mid-Range Trajectory Height • Inches	0.0	0.3	1.4	3.2	6.0	9.6	14.1	19.8	26.5
Drop • Inches	0.0	-1.4	-5.6	-12.8	-23.1	-36.7	-53.8	-74.6	-99.1

Magtech 185-grain GGJHP (GG45A)

G1 Ballistic Coefficient = 0.147

Distance • Yards	Muzzle	25	50	75	100	125	150	175	200
Velocity • fps	1148	1092	1046	1007	973	944	917	892	869
Energy • ft-lbs	542	490	449	417	389	366	345	327	311
Taylor KO Index	13.8	13.1	12.6	12.1	11.7	11.3	11.0	10.7	10.4
Mid-Range Trajectory Height • Inches	0.0	0.2	0.9	2.1	3.9	6.4	9.6	13.6	18.3
Drop • Inches	0.0	-0.8	-3.5	-8.2	-15.0	-24.1	-35.6	-49.7	-66.6

Remington 185-grain Golden Saber +P (GS45APC)

G1 Ballistic Coefficient = 0.150

Distance • Yards	Muzzle	25	50	75	100	125	150	175	200
Velocity • fps	1140	1086	1042	1004	971	942	916	892	870
Energy • ft-lbs	534	485	446	414	388	365	645	327	311
Taylor KO Index	13.6	13.0	12.4	12.0	11.6	11.3	10.9	10.7	10.4
Mid-Range Trajectory Height • Inches	0.0	0.2	1.0	2.2	4.0	6.4	9.6	13.5	18.2
Drop • Inches	0.0	-0.9	-3.6	-8.3	-15.1	-24.3	-35.8	-49.9	-66.8

Cor-Bon 185-grain DPX +P (DPX45185/20) + (TR45185/20)

G1 Ballistic Coefficient = 0.150

Distance • Yards	Muzzle	25	50	75	100	125	150	175	200
Velocity • fps	1075	1032	996	964	936	911	887	865	844
Energy • ft-lbs	475	438	408	382	360	341	323	307	293
Taylor KO Index	12.8	12.3	11.9	11.5	11.2	10.9	10.6	10.3	10.1
Mid-Range Trajectory Height • Inches	0.0	0.2	1.0	2.4	4.4	7.1	10.5	14.8	19.8
Drop • Inches	0.0	-1.0	-4.0	-9.2	-16.7	-26.7	-39.3	-54.7	-73.0

Speer 185-grain GDHP (23964)

G1 Ballistic Coefficient = 0.110

Distance • Yards	Muzzle	25	50	75	100	125	150	175	200
Velocity • fps	1050	998	956	919	886	856	829	803	779
Energy • ft-lbs	453	409	375	347	322	301	282	265	249
Taylor KO Index	12.5	11.9	11.4	11.0	10.6	10.2	9.9	9.6	9.3
Mid-Range Trajectory Height • Inches	0.0	0.3	1.1	2.6	4.6	7.6	11.5	16.2	22.0
Drop • Inches	0.0	-1.0	-4.2	-9.8	-18.0	-28.8	-42.6	-59.6	-79.9

Remington 185-grain Golden Saber (GS45APA)

G1 Ballistic Coefficient = 0.150

Distance • Yards	Muzzle	25	50	75	100	125	150	175	200
Velocity • fps	1015	981	951	924	899	876	855	835	815
Energy • ft-lbs	423	395	372	351	332	316	300	286	273
Taylor KO Index	12.1	11.7	11.4	11.0	10.7	10.5	10.2	10.0	9.7
Mid-Range Trajectory Height • Inches	0.0	0.3	1.1	2.6	4.5	7.6	11.4	16.0	21.4
Drop • Inches	0.0	-1.1	-4.4	-10.2	-18.4	-29.4	-43.2	-59.9	-79.7

Black Hills 185-grain Barnes TAC-XP +P (D45N7) LF

G1 Ballistic Coefficient = 0.167

Distance • Yards	Muzzle	25	50	75	100	125	150	175	200
Velocity • fps	1000	971	945	921	899	878	859	841	823
Energy • ft-lbs	411	387	367	349	332	317	303	290	278
Taylor KO Index	12.0	11.7	11.3	11.1	10.8	10.5	10.3	10.1	9.9
Mid-Range Trajectory Height • Inches	0.0	0.3	1.1	2.6	4.8	7.8	11.5	16.1	21.5
Drop • Inches	0.0	-1.1	-4.5	-10.4	-18.8	-29.9	-43.8	-60.6	-80.5

Black Hills 185-grain Jacketed Hollow Point (D45N4)

G1 Ballistic Coefficient = 0.130

Distance • Yards	Muzzle	25	50	75	100	125	150	175	200
Velocity • fps	1000	963	931	902	875	851	828	806	785
Energy • ft-lbs	411	381	356	334	315	298	282	267	254
Taylor KO Index	11.9	11.5	11.1	10.8	10.5	10.2	9.9	9.6	9.4
Mid-Range Trajectory Height • Inches	0.0	0.3	1.2	2.7	5.0	8.0	11.9	16.8	22.6
Drop • Inches	0.0	-1.1	-4.6	-10.5	-19.2	-30.6	-45.0	-62.6	-83.5

Remington 185-grain Jacketed Hollow Point (R45AP2)

G1 Ballistic Coefficient = 0.149

Distance • Yards	Muzzle	25	50	75	100	125	150	175	200
Velocity • fps	1000	968	939	913	889	866	845	826	807
Energy • ft-lbs	411	385	362	342	324	308	294	280	267
Taylor KO Index	11.9	11.6	11.2	10.9	10.6	10.3	10.1	9.9	9.6
Mid-Range Trajectory Height • Inches	0.0	0.3	1.1	2.7	4.9	7.8	11.7	16.3	21.9
Drop • Inches	0.0	-1.1	-4.5	-10.4	-18.9	-30.2	-44.3	-61.4	-81.8

Winchester 185-grain Silvertip Hollow Point (X45ASHP2)

G1 Ballistic Coefficient = 0.148

Distance • Yards	Muzzle	25	50	75	100	125	150	175	200
Velocity • fps	1000	967	938	912	888	866	845	826	806
Energy • ft-lbs	411	384	362	342	324	308	293	279	267
Taylor KO Index	11.9	11.6	11.2	10.9	10.6	10.3	10.1	9.9	9.6
Mid-Range Trajectory Height • Inches	0.0	0.3	1.2	2.7	4.9	7.8	11.7	16.3	21.9
Drop • Inches	0.0	-1.1	-4.5	-10.4	-19.0	-30.2	-44.3	-61.5	-81.8

Hornady 185-grain JHP/XTP (9090)

G1 Ballistic Coefficient = 0.136

Distance • Yards	Muzzle	25	50	75	100	125	150	175	200
Velocity • fps	970	938	910	884	860	837	816	796	777
Energy • ft-lbs	387	362	340	321	304	288	274	260	248
Taylor KO Index	11.6	11.2	10.9	10.6	10.3	10.0	9.7	9.5	9.3
Mid-Range Trajectory Height • Inches	0.0	0.3	1.2	2.9	5.3	8.4	12.5	17.3	23.5
Drop • Inches	0.0	-1.2	-4.8	-11.1	-20.2	-32.1	-47.2	-65.6	-87.3

Black Hills 185-grain JHP (D45SCN4)

G1 Ballistic Coefficient = 0.145

Distance • Yards	Muzzle	25	50	75	100	125	150	175	200
Velocity • fps	950	922	897	873	852	831	811	792	775
Energy • ft-lbs	371	349	331	313	299	284	270	258	247
Taylor KO Index	11.3	11.0	10.7	10.4	10.2	9.9	9.7	9.5	9.3
Mid-Range Trajectory Height • Inches	0.0	0.3	1.3	3.0	5.4	8.7	12.8	17.9	24.0
Drop • Inches	0.0	-1.2	-5.0	-11.5	-20.8	-33.1	-48.6	-67.4	-89.6

Federal 185-grain JHP (C45C)

G1 Ballistic Coefficient = 0.16

Distance • Yards	Muzzle	25	50	75	100	125	150	175	200
Velocity • fps	950	920	900	880	860	841	823	806	789
Energy • ft-lbs	370	350	335	320	300	291	278	267	256
Taylor KO Index	11.3	11.0	10.8	10.5	10.3	10.0	9.8	9.6	9.4
Mid-Range Trajectory Height • Inches	0.0	0.4	1.6	3.7	5.4	8.5	12.6	17.5	23.4
Drop • Inches	0.0	-1.2	-5.0	-11.4	-20.7	-32.8	-48.0	-66.4	-88.2

PMC 185-grain Jacketed Hollow Point (45B)

G1 Ballistic Coefficient = 0.10

Distance • Yards	Muzzle	25	50	75	100	125	150	175	200
Velocity • fps	900	867	836	805	776	758	735	712	691
Energy • ft-lbs	339	311	290	270	253	238	223	210	197
Taylor KO Index	10.8	10.4	10.0	9.7	9.4	9.1	8.8	8.5	8.3
Mid-Range Trajectory Height • Inches	0.0	0.3	1.4	3.3	6.2	9.9	14.8	20.9	28.2
Drop • Inches	0.0	-1.4	-5.6	-12.9	-23.9	-37.7	-55.5	-77.4	-103.5

UMC (Remington) 185-grain Metal Case (L45API)

G1 Ballistic Coefficient = 0.1

Distance • Yards	Muzzle	25	50	75	100	125	150	175	200
Velocity • fps	1015	983	955	930	907	885	864	845	827
Energy • ft-lbs	423	397	375	355	338	322	307	294	281
Taylor KO Index	12.1	11.7	11.4	11.1	10.8	10.6	10.3	10.1	9.9
Mid-Range Trajectory Height • Inches	0.0	0.3	1.1	2.6	4.8	7.6	11.3	15.8	21.1
Drop • Inches	0.0	-1.1	-4.4	-10.1	-18.3	-29.2	-42.8	-59.3	-101.7

Speer 185-grain TMJ FN (53654)

G1 Ballistic Coefficient = 0.0

Distance • Yards	Muzzle	25	50	75	100	125	150	175	200
Velocity • fps	1050	991	943	903	867	835	805	777	751
Energy • ft-lbs	453	404	365	335	308	286	266	248	232
Taylor KO Index	12.5	11.8	11.3	10.8	10.4	10.0	9.6	9.3	9.0
Mid-Range Trajectory Height • Inches	0.0	0.3	1.1	2.6	4.9	8.0	12.0	17.1	23.2
Drop • Inches	0.0	-1.0	-4.3	-9.9	-18.3	-29.5	-43.8	-61.5	-82.8

Winchester 185-grain Brass Enclosed Base (WC451)

G1 Ballistic Coefficient = 0.0

Distance • Yards	Muzzle	25	50	75	100	125	150	175	200
Velocity • fps	1000	947	902	864	829	796	764	739	712
Energy • ft-lbs	411	368	334	306	282	261	241	224	208
Taylor KO Index	11.9	11.3	10.8	10.3	9.9	9.5	9.2	8.8	8.5
Mid-Range Trajectory Height • Inches	0.0	0.3	1.2	2.9	5.3	8.7	13.1	18.7	25.4
Drop • Inches	0.0	-1.1	-4.7	-10.9	-20.0	-32.3	-48.0	-67.5	-90.9

USA (Winchester) 185-grain FMJ – FN (USA45A)

G1 Ballistic Coefficient = 0.140

Distance • Yards	Muzzle	25	50	75	100	125	150	175	200
Velocity • fps	910	885	861	839	818	799	780	762	744
Energy • ft-lbs	340	322	304	289	275	262	250	238	228
Taylor KO Index	10.9	10.6	10.3	10.0	9.8	9.5	9.3	9.1	8.9
Mid-Range Trajectory Height • Inches	0.0	0.3	1.3	3.2	5.8	9.3	13.8	19.3	25.9
Drop • Inches	0.0	-1.3	-5.5	-12.5	-22.6	-36.0	-52.7	-73.0	-97.1

Federal 185-grain FMJ – SWC Match (GM45B)

G1 Ballistic Coefficient = 0.069

Distance • Yards	Muzzle	25	50	75	100	125	150	175	200
Velocity • fps	770	740	700	670	640	612	585	560	535
Energy • ft-lbs	263	240	219	199	182	167	152	139	127
Taylor KO Index	9.9	9.6	9.0	8.7	8.3	7.9	7.6	7.2	6.9
Mid-Range Trajectory Height • Inches	0.0	0.5	2.0	4.8	8.9	14.5	21.9	31.1	42.8
Drop • Inches	0.0	-1.9	-7.8	-18.2	-33.3	-53.8	-80.1	112.7	-152.2

Magtech 200-grain LSWC (45C)

G1 Ballistic Coefficient = 0.075

Distance • Yards	Muzzle	25	50	75	100	125	150	175	200
Velocity • fps	950	899	854	815	799	746	715	685	653
Energy • ft-lbs	401	359	324	295	270	247	227	209	192
Taylor KO Index	12.3	11.7	11.1	10.6	10.1	9.7	9.3	8.9	8.5
Mid-Range Trajectory Height • Inches	0.0	0.3	1.3	3.2	5.9	9.8	14.8	21.1	28.9
Drop • Inches	0.0	-1.2	-5.2	-12.1	-22.3	-36.1	-53.7	-75.7	-102.2

Black Hills 200-grain Match Semi-Wadcutter (D45N1)

G1 Ballistic Coefficient = 0.150

Distance • Yards	Muzzle	25	50	75	100	125	150	175	200
Velocity • fps	875	854	833	814	796	778	762	745	729
Energy • ft-lbs	340	324	309	294	280	269	258	247	236
Taylor KO Index	11.3	11.0	10.8	10.5	10.3	10.0	9.8	9.6	9.4
Mid-Range Trajectory Height • Inches	0.0	0.4	1.5	3.4	6.2	9.9	14.7	21.8	27.4
Drop • Inches	0.0	-1.4	-5.9	-13.4	-24.2	-38.5	-56.3	-77.9	-103.4

Speer 200-grain GDHP +P (23969)

G1 Ballistic Coefficient = 0.139

Distance • Yards	Muzzle	25	50	75	100	125	150	175	200
Velocity • fps	1080	1033	994	960	930	903	878	855	833
Energy • ft-lbs	518	474	439	410	384	362	343	325	308
Taylor KO Index	13.9	13.3	12.8	12.4	12.0	11.7	11.3	11.0	10.8
Mid-Range Trajectory Height • Inches	0.0	0.2	1.0	2.3	4.3	7.0	10.5	14.8	19.9
Drop • Inches	0.0	-1.0	-4.0	-9.1	-16.7	-26.7	-39.4	-54.9	-73.3

Hornady 200-grain JHP/XTP +P (9113)

G1 Ballistic Coefficient = 0.151

Distance • Yards	Muzzle	25	50	75	100	125	150	175	200
Velocity • fps	1055	1015	982	952	925	900	877	856	836
Energy • ft-lbs	494	458	428	402	380	360	342	326	310
Taylor KO Index	13.6	13.1	12.7	12.3	11.9	11.6	11.3	11.1	10.8
Mid-Range Trajectory Height • Inches	0.0	0.2	1.0	2.4	4.5	7.2	10.7	15.0	20.2
Drop • Inches	0.0	-1.0	-4.1	-9.5	-17.2	-27.5	-40.4	-56.2	-74.9

Cor-Bon 200-grain JHP +P (SD45200/20)

G1 Ballistic Coefficient = 0.150

Distance • Yards	Muzzle	25	50	75	100	125	150	175	200
Velocity • fps	1050	1015	984	957	932	909	887	866	847
Energy • ft-lbs	490	458	430	407	386	367	349	333	319
Taylor KO Index	13.6	13.1	12.7	12.4	12.0	11.7	11.5	11.2	10.9
Mid-Range Trajectory Height • Inches	0.0	0.2	1.0	2.4	4.4	7.1	10.6	14.9	19.9
Drop • Inches	0.0	-1.0	-4.1	-9.5	-17.2	-27.4	-40.2	-55.8	-74.3

Fiocchi 200-grain XTP JHP (45XTPB)

G1 Ballistic Coefficient = 0.151

Distance • Yards	Muzzle	25	50	75	100	125	150	175	200
Velocity • fps	925	900	878	856	837	817	799	781	764
Energy • ft-lbs	380	360	342	326	311	296	283	271	259
Taylor KO Index	11.9	11.6	11.3	11.1	10.8	10.6	10.3	10.1	9.9
Mid-Range Trajectory Height • Inches	0.0	0.3	1.4	3.1	5.7	9.0	13.4	18.6	24.9
Drop • Inches	0.0	-1.3	-5.3	-12.1	-21.8	-34.7	-50.8	-70.3	-93.4

Hornady 200-grain JHP/XTP (9112)

G1 Ballistic Coefficient = 0.149

Distance • Yards	Muzzle	25	50	75	100	125	150	175	200
Velocity • fps	900	877	855	835	815	797	779	762	746
Energy • ft-lbs	358	342	325	310	295	282	270	258	247
Taylor KO Index	11.6	11.3	11.0	10.8	10.5	10.3	10.1	9.8	9.6
Mid-Range Trajectory Height • Inches	0.0	0.3	1.4	3.2	5.9	9.4	14.0	19.5	26.1
Drop • Inches	0.0	-1.4	-5.6	-12.7	-23.0	-36.5	-53.5	-74.0	-98.2

Fiocchi 200-grain Jacketed Hollowpoint (45B)

G1 Ballistic Coefficient = 0.120

Distance • Yards	Muzzle	25	50	75	100	125	150	175	200
Velocity • fps	890	862	835	813	790	768	748	728	709
Energy • ft-lbs	350	330	310	293	275	262	248	236	223
Taylor KO Index	11.5	11.1	10.8	10.5	10.2	9.9	9.7	9.4	9.2
Mid-Range Trajectory Height • Inches	0.0	0.3	1.5	3.4	6.2	10.0	14.8	20.6	27.9
Drop • Inches	0.0	-1.4	-5.7	-13.2	-23.9	-38.0	-55.9	-77.7	-103.5

Hornady 200-grain TAP-FPD +P (91128)

G1 Ballistic Coefficient = 0.15

Distance • Yards	Muzzle	25	50	75	100	125	150	175	200
Velocity • fps	1055	1016	982	953	926	901	879	857	837
Energy • ft-lbs	494	458	428	403	380	361	343	326	311
Taylor KO Index	13.6	13.1	12.7	12.3	12.0	11.6	11.4	11.1	10.8
Mid-Range Trajectory Height • Inches	0.0	0.2	1.1	2.5	4.5	7.3	10.8	14.9	20.3
Drop • Inches	0.0	-1.0	-4.1	-9.5	-17.2	-27.5	-40.5	-56.2	-75.0

Speer 200-grain TMJ FN Lawman (53655)

G1 Ballistic Coefficient = 0.10

Distance • Yards	Muzzle	25	50	75	100	125	150	175	200
Velocity • fps	975	933	897	864	834	807	781	756	733
Energy • ft-lbs	422	387	357	332	309	289	271	254	239
Taylor KO Index	12.6	12.0	11.6	11.2	10.8	10.4	10.1	9.8	9.5
Mid-Range Trajectory Height • Inches	0.0	0.3	1.3	2.9	5.4	8.8	13.1	18.4	25.0
Drop • Inches	0.0	-1.2	-4.8	-11.2	-20.5	-32.9	-48.6	-67.9	-91.0

Magtech 230-grain Hornady GGJHP (GG45B)

G1 Ballistic Coefficient = 0.16

Distance • Yards	Muzzle	25	50	75	100	125	150	175	200
Velocity • fps	1007	977	949	925	902	880	860	842	824
Energy • ft-lbs	518	487	460	437	415	396	378	362	346
Taylor KO Index	15.0	14.6	14.2	13.8	13.5	13.1	12.8	12.6	12.3
Mid-Range Trajectory Height • Inches	0.0	0.2	1.1	2.6	4.8	7.7	11.4	16.0	21.4
Drop • Inches	0.0	-1.1	-4.5	-10.2	-18.6	-29.6	-43.4	-60.1	-79.9

Black Hills 230-grain Jacketed Hollow Point +P (D45N6)

G1 Ballistic Coefficient = 0.1

Distance • Yards	Muzzle	25	50	75	100	125	150	175	200
Velocity • fps	950	925	903	882	862	843	825	808	792
Energy • ft-lbs	460	438	418	397	379	363	348	334	320
Taylor KO Index	14.1	13.7	13.4	13.1	12.8	12.5	12.3	12.0	11.8
Mid-Range Trajectory Height • Inches	0.0	0.3	1.2	2.9	5.3	8.5	12.5	17.5	23.3
Drop • Inches	0.0	-1.2	-5.0	-11.4	-20.6	-32.8	-48.0	-66.3	-88.0

Cor-Bon 230-grain JHP +P (SD45230/20)

G1 Ballistic Coefficient = 0.1

Distance • Yards	Muzzle	25	50	75	100	125	150	175	200
Velocity • fps	950	928	908	889	870	853	836	820	805
Energy • ft-lbs	461	440	421	403	387	372	357	344	331
Taylor KO Index	14.1	13.8	13.5	13.2	12.9	12.7	12.4	12.2	12.0
Mid-Range Trajectory Height • Inches	0.0	0.3	1.2	2.8	5.2	8.3	12.3	17.2	22.9
Drop • Inches	0.0	-1.2	-5.0	-11.3	-20.5	-32.5	-47.4	-65.5	-86.8

Hornady 230-grain TAP-FPD +P (90958)

G1 Ballistic Coefficient = 0.

Distance • Yards	Muzzle	25	50	75	100	125	150	175	200
Velocity • fps	950	929	908	890	872	855	839	824	809
Energy • ft-lbs	461	440	421	405	388	374	360	347	334
Taylor KO Index	14.1	13.8	13.5	13.2	13.0	12.7	12.5	12.2	12.0
Mid-Range Trajectory Height • Inches	0.0	0.3	1.3	2.9	5.3	8.4	12.3	17.2	22.9
Drop • Inches	0.0	-1.2	-5.0	-11.3	-20.4	-32.4	-47.4	-65.4	-86.6

Hornady 230-grain JHP/XTP +P (9096)

G1 Ballistic Coefficient = 0.190

Distance • Yards	Muzzle	25	50	75	100	125	150	175	200
Velocity • fps	950	929	908	890	872	855	839	824	809
Energy • ft-lbs	461	440	422	405	389	374	360	347	334
Taylor KO Index	14.1	13.8	13.5	13.2	13.0	12.7	12.5	12.2	12.0
Mid-Range Trajectory Height • Inches	0.0	0.3	1.3	2.9	5.3	8.4	12.3	17.2	22.9
Drop • Inches	0.0	-1.2	-5.0	-11.3	-20.5	-32.4	-47.4	-65.4	-86.6

Winchester 230-grain Bonded PDX1TM (S45PDB)

G1 Ballistic Coefficient = 0.185

Distance • Yards	Muzzle	25	50	75	100	125	150	175	200
Velocity • fps	920	900	881	864	847	831	815	800	786
Energy • ft-lbs	432	414	397	381	366	352	339	327	315
Taylor KO Index	13.7	13.4	13.1	12.8	12.6	12.3	12.1	11.9	11.7
Mid-Range Trajectory Height • Inches	0.0	0.4	1.4	3.1	5.6	8.9	13.1	18.2	24.3
Drop • Inches	0.0	-1.3	-5.3	-12.1	-21.8	-34.5	-50.3	-69.5	-92.0

Federal 230-grain Hydra-Shok JHP (P45HS1)

G1 Ballistic Coefficient = 0.183

Distance • Yards	Muzzle	25	50	75	100	125	150	175	200
Velocity • fps	900	880	870	850	830	814	799	785	771
Energy • ft-lbs	415	395	380	365	355	339	326	315	303
Taylor KO Index	13.4	13.1	12.9	12.6	12.3	12.1	11.9	11.7	11.5
Mid-Range Trajectory Height • Inches	0.0	0.3	1.4	3.2	5.9	9.3	13.7	18.7	25.3
Drop • Inches	0.0	-1.4	-5.5	-12.6	-22.7	-36.0	-52.5	-72.4	-95.9

Fiocchi 230-grain Hornady XTP JHP (45XTP)

G1 Ballistic Coefficient = 0.151

Distance • Yards	Muzzle	25	50	75	100	125	150	175	200
Velocity • fps	900	877	856	836	817	798	781	764	748
Energy • ft-lbs	414	393	374	357	341	326	311	298	286
Taylor KO Index	13.4	13.1	12.8	12.5	12.2	11.9	11.7	11.4	11.2
Mid-Range Trajectory Height • Inches	0.0	0.3	1.4	3.2	5.9	9.5	14.0	19.5	26.1
Drop • Inches	0.0	-1.4	-5.6	-12.7	-23.0	-36.5	-54.3	-73.9	-98.1

Speer 230-grain GDHP (23966)

G1 Ballistic Coefficient = 0.192

Distance • Yards	Muzzle	25	50	75	100	125	150	175	200
Velocity • fps	890	867	845	825	805	786	768	751	735
Energy • ft-lbs	405	384	365	347	331	316	302	288	276
Taylor KO Index	13.2	12.9	12.5	12.3	12.0	11.7	11.4	11.2	10.9
Mid-Range Trajectory Height • Inches	0.0	0.3	1.4	3.3	6.0	9.7	14.3	20.0	26.8
Drop • Inches	0.0	-1.4	-5.7	-13.0	-23.5	-37.4	-54.8	-75.8	-100.7

USA (Winchester) 230-grain Jacketed Hollow Point (USA45JHP)

G1 Ballistic Coefficient = 0.168

Distance • Yards	Muzzle	25	50	75	100	125	150	175	200
Velocity • fps	880	861	842	825	808	792	776	761	747
Energy • ft-lbs	396	378	363	347	334	320	308	296	285
Taylor KO Index	13.1	12.8	12.5	12.3	12.0	11.8	11.5	11.3	11.1
Mid-Range Trajectory Height • Inches	0.0	0.3	1.4	3.4	6.1	9.7	14.3	19.9	26.6
Drop • Inches	0.0	-1.4	-5.8	-13.2	-23.8	-37.7	-55.1	-76.1	-100.9

Remington 230-grain Golden Saber (GS45APB)

G1 Ballistic Coefficient = 0.148

Distance • Yards	Muzzle	25	50	75	100	125	150	175	200
Velocity • fps	875	853	833	813	795	777	760	744	728
Energy • ft-lbs	391	372	355	338	323	309	295	283	271
Taylor KO Index	13.0	12.7	12.4	12.1	11.8	11.5	11.3	11.0	10.8
Mid-Range Trajectory Height • Inches	0.0	0.3	1.5	3.4	6.1	9.9	14.7	20.6	27.5
Drop • Inches	0.0	-1.4	-5.9	-13.4	-24.3	-38.5	-56.4	-78.0	-103.5

Black Hills 230-grain Jacketed Hollow Point (D45N5)

G1 Ballistic Coefficient = 0.165

Distance • Yards	Muzzle	25	50	75	100	125	150	175	200
Velocity • fps	850	832	814	798	782	766	751	737	723
Energy • ft-lbs	368	353	339	325	312	300	288	277	267
Taylor KO Index	12.6	12.4	12.1	11.9	11.6	11.4	11.2	10.9	10.7
Mid-Range Trajectory Height • Inches	0.0	0.4	1.6	3.6	6.6	10.4	15.4	21.4	28.5
Drop • Inches	0.0	-1.5	-6.2	-14.1	-25.5	-40.4	-59.0	-81.5	-107.9

Federal 230-grain Hi-Shok JHP (C45D)

G1 Ballistic Coefficient = 0.189

Distance • Yards	Muzzle	25	50	75	100	125	150	175	200
Velocity • fps	850	830	820	800	790	776	763	750	737
Energy • ft-lbs	370	355	345	330	320	308	297	287	277
Taylor KO Index	12.6	12.3	12.2	11.9	11.7	11.5	11.3	11.1	10.9
Mid-Range Trajectory Height • Inches	0.0	0.4	1.6	3.6	6.5	10.3	15.2	20.7	27.9
Drop • Inches	0.0	-1.5	-6.2	-14.0	-25.3	-40.0	-58.4	-80.4	-106.4

Fiocchi 230-grain Jacketed Hollowpoint (45T)

G1 Ballistic Coefficient = 0.295

Distance • Yards	Muzzle	25	50	75	100	125	150	175	200
Velocity • fps	850	840	830	820	810	801	792	783	774
Energy • ft-lbs	370	360	350	343	335	328	320	313	306
Taylor KO Index	12.6	12.5	12.3	12.2	12.0	11.9	11.8	11.6	11.5
Mid-Range Trajectory Height • Inches	0.0	0.4	1.6	3.5	6.3	10.0	14.6	19.7	26.5
Drop • Inches	0.0	-1.5	-6.1	-13.9	-24.8	-39.1	-56.8	-77.9	-102.6

PMC 230-grain Starfire Hollow Point (45SFA)

G1 Ballistic Coefficient = 0.150

Distance • Yards	Muzzle	25	50	75	100	125	150	175	200
Velocity • fps	850	830	810	790	770	753	736	719	703
Energy • ft-lbs	370	350	335	320	305	289	276	264	252
Taylor KO Index	12.6	12.3	12.0	11.7	11.4	11.2	10.9	10.7	10.4
Mid-Range Trajectory Height • Inches	0.0	0.4	1.6	3.6	6.6	10.6	15.7	21.9	29.3
Drop • Inches	0.0	-1.5	-6.2	-14.2	-25.8	-40.9	-59.9	-82.9	-110.1

Remington 230-grain Jacketed Hollow Point (Subsonic) (R45AP7)

G1 Ballistic Coefficient = 0.16

Distance • Yards	Muzzle	25	50	75	100	125	150	175	200
Velocity • fps	835	817	800	783	767	751	736	722	708
Energy • ft-lbs	356	341	326	313	300	289	277	266	256
Taylor KO Index	12.4	12.1	11.9	11.6	11.4	11.2	10.9	10.7	10.5
Mid-Range Trajectory Height • Inches	0.0	0.4	1.6	3.7	6.8	10.8	15.9	22.1	29.5
Drop • Inches	0.0	-1.6	-6.4	-14.6	-26.4	-41.9	-61.2	-84.5	-112.0

UMC (Remington) 230-grain Jacketed Hollow Point (L45AP7)

G1 Ballistic Coefficient = 0.16

Distance • Yards	Muzzle	25	50	75	100	125	150	175	200
Velocity • fps	835	817	800	783	767	751	736	722	708
Energy • ft-lbs	356	341	326	313	300	289	277	266	256
Taylor KO Index	12.4	12.1	11.9	11.6	11.4	11.2	10.9	10.7	10.5
Mid-Range Trajectory Height • Inches	0.0	0.4	1.6	3.7	6.8	10.8	15.9	22.1	29.5
Drop • Inches	0.0	-1.6	-6.4	-14.6	-26.4	-41.9	-61.2	-84.5	-112.0

American Eagle (Federal) 230-grain Full Metal Jacket (AE45A)

G1 Ballistic Coefficient = 0.1?

Distance • Yards	Muzzle	25	50	75	100	125	150	175	200
Velocity • fps	890	870	860	840	820	804	789	774	760
Energy • ft-lbs	405	390	375	360	345	330	318	306	295
Taylor KO Index	13.2	12.9	12.8	12.5	12.2	11.9	11.7	11.5	11.3
Mid-Range Trajectory Height • Inches	0.0	0.3	1.4	3.3	6.0	9.5	14.0	19.2	25.9
Drop • Inches	0.0	-1.4	-5.6	-12.9	-23.2	-36.8	-53.7	-74.2	-98.2

Winchester 230-grain Brass Enclosed Base (WC452)

G1 Ballistic Coefficient = 0.1

Distance • Yards	Muzzle	25	50	75	100	125	150	175	200
Velocity • fps	875	857	840	824	808	793	778	765	751
Energy • ft-lbs	391	375	360	347	334	321	310	298	288
Taylor KO Index	13.0	12.7	12.5	12.2	12.0	11.8	11.6	11.4	11.2
Mid-Range Trajectory Height • Inches	0.0	0.4	1.5	3.4	6.2	9.8	14.4	20.0	26.6
Drop • Inches	0.0	-1.4	-5.8	-13.3	-24.0	-38.0	-55.4	-76.5	-101.2

Federal 230-grain FMJ Match (GM45A)

G1 Ballistic Coefficient = 0.?

Distance • Yards	Muzzle	25	50	75	100	125	150	175	200
Velocity • fps	860	840	830	810	800	787	774	761	748
Energy • ft-lbs	380	365	350	340	325	316	306	296	286
Taylor KO Index	12.8	12.5	12.3	12.0	11.9	11.7	11.5	11.3	11.1
Mid-Range Trajectory Height • Inches	0.0	0.4	1.5	3.5	6.3	10.1	14.8	20.1	27.2
Drop • Inches	0.0	-1.5	-6.0	-13.7	-24.7	-39.0	-56.9	-78.4	-103.7

Fiocchi 230-grain FMJ (45A)

G1 Ballistic Coefficient = 0.205

Distance • Yards	Muzzle	25	50	75	100	125	150	175	200
Velocity • fps	860	845	830	816	803	790	777	765	753
Energy • ft-lbs	377	365	352	340	329	319	308	299	289
Taylor KO Index	12.8	12.6	12.4	12.2	12.0	11.8	11.6	11.4	11.2
Mid-Range Trajectory Height • Inches	0.0	0.3	1.5	3.4	6.3	10.0	14.6	20.3	27.0
Drop • Inches	0.0	-1.5	-6.0	-13.7	-24.6	-38.9	-56.7	-78.1	-103.2

Fiocchi 230-grain FMJ-Enclosed Base (45TCEB)

G1 Ballistic Coefficient = 0.125

Distance • Yards	Muzzle	25	50	75	100	125	150	175	200
Velocity • fps	860	836	809	791	771	750	731	713	695
Energy • ft-lbs	378	357	334	319	303	287	273	259	247
Taylor KO Index	12.8	12.4	12.0	11.7	11.5	11.1	10.9	10.6	10.3
Mid-Range Trajectory Height • Inches	0.0	0.4	1.6	3.6	6.6	10.6	15.7	21.9	29.4
Drop • Inches	0.0	-1.5	-6.1	-14.0	-25.4	-40.4	-59.3	-82.3	-109.5

Sellier & Bellot 230-grain FMJ (V311252U)

G1 Ballistic Coefficient = 0.125

Distance • Yards	Muzzle	25	50	75	100	125	150	175	200
Velocity • fps	853	829	806	785	764	745	726	708	690
Energy • ft-lbs	371	351	331	315	298	283	269	256	243
Taylor KO Index	12.7	12.3	12.0	11.7	11.3	11.1	10.8	10.5	10.2
Mid-Range Trajectory Height • Inches	0.0	0.4	1.5	3.6	6.5	10.4	15.3	21.4	28.6
Drop • Inches	0.0	-1.5	-6.2	-14.2	-25.8	-41.1	-60.2	-83.5	-111.2

American Eagle (Federal) 230-grain TMJ (AE45N1)

G1 Ballistic Coefficient = 0.190

Distance • Yards	Muzzle	25	50	75	100	125	150	175	200
Velocity • fps	850	830	820	800	790	776	763	750	737
Energy • ft-lbs	370	365	345	330	320	308	297	287	278
Taylor KO Index	12.6	12.3	12.2	11.9	11.7	11.5	11.3	11.1	10.9
Mid-Range Trajectory Height • Inches	0.0	0.4	1.6	3.6	6.5	10.3	15.1	20.7	27.9
Drop • Inches	0.0	-1.5	-6.2	-14.0	-25.3	-40.0	-58.3	-80.4	-106.3

Black Hills 230-grain Full Metal Jacket (D45N3)

G1 Ballistic Coefficient = 0.150

Distance • Yards	Muzzle	25	50	75	100	125	150	175	200
Velocity • fps	850	830	811	793	775	759	742	727	711
Energy • ft-lbs	368	352	336	321	307	294	282	270	258
Taylor KO Index	12.6	12.3	12.0	11.8	11.5	11.3	11.0	10.8	10.6
Mid-Range Trajectory Height • Inches	0.0	0.4	1.6	3.6	6.6	10.5	15.5	21.6	28.9
Drop • Inches	0.0	-1.5	-6.2	-14.2	-25.6	-40.7	-59.5	-82.3	-109.1

Hornady 230-grain FMJ RN ENC (9097)

G1 Ballistic Coefficient = 0.183

Distance • Yards	Muzzle	25	50	75	100	125	150	175	200
Velocity • fps	850	834	818	803	788	774	760	747	734
Energy • ft-lbs	369	355	342	329	317	306	295	285	275
Taylor KO Index	12.6	12.4	12.1	11.9	11.7	11.5	11.3	11.1	10.9
Mid-Range Trajectory Height • Inches	0.0	0.4	1.6	3.6	6.5	10.3	15.2	21.1	28.0
Drop • Inches	0.0	-1.5	-6.2	-14.1	-25.3	-40.1	-58.5	-80.7	-106.7

MC 230-grain SFHP (45SFA)

G1 Ballistic Coefficient = 0.148

Distance • Yards	Muzzle	25	50	75	100	125	150	175	200
Velocity • fps	850	830	811	792	775	758	741	725	710
Energy • ft-lbs	369	352	336	321	307	293	281	269	258
Taylor KO Index	12.6	12.3	12.0	11.8	11.5	11.3	11.0	10.8	10.5
Mid-Range Trajectory Height • Inches	0.0	0.4	1.6	3.6	6.6	10.6	15.6	21.5	29.0
Drop • Inches	0.0	-1.5	-6.2	-14.2	-25.6	-40.7	-59.6	-82.4	-109.3

Blazer (CCI) 230-grain Clean-Fire – Totally Metal Jacket (3570)

G1 Ballistic Coefficient = 0.153

Distance • Yards	Muzzle	25	50	75	100	125	150	175	200
Velocity • fps	845	826	807	790	773	756	740	725	710
Energy • ft-lbs	365	348	333	318	305	292	280	269	258
Taylor KO Index	12.5	12.3	12.0	11.7	11.5	11.2	11.0	10.8	10.5
Mid-Range Trajectory Height • Inches	0.0	0.4	1.6	3.7	6.7	10.6	15.7	21.6	29.2
Drop • Inches	0.0	-1.6	-6.3	-14.3	-25.9	-41.1	-60.1	-83.0	-110.1

Blazer (CCI) 230-grain Clean-Fire – Totally Metal Jacket (3480)
G1 Ballistic Coefficient = 0.153

Distance • Yards	Muzzle	25	50	75	100	125	150	175	200
Velocity • fps	845	826	807	790	773	756	740	725	710
Energy • ft-lbs	365	348	333	318	305	292	280	269	258
Taylor KO Index	12.5	12.3	12.0	11.7	11.5	11.2	11.0	10.8	10.5
Mid-Range Trajectory Height • Inches	0.0	0.4	1.6	3.7	6.7	10.6	15.7	21.6	29.2
Drop • Inches	0.0	-1.6	-6.3	-14.3	-25.9	-41.1	-60.1	-83.0	-110.1

CCI 230-grain Full Metal Jacket Blazer (3570)
G1 Ballistic Coefficient = 0.153

Distance • Yards	Muzzle	25	50	75	100	125	150	175	200
Velocity • fps	845	826	807	790	773	756	740	725	710
Energy • ft-lbs	365	348	333	318	305	292	280	269	258
Taylor KO Index	12.5	12.3	12.0	11.7	11.5	11.2	11.0	10.8	10.5
Mid-Range Trajectory Height • Inches	0.0	0.4	1.6	3.7	6.7	10.6	15.7	21.6	29.2
Drop • Inches	0.0	-1.6	-6.3	-14.3	-25.9	-41.1	-60.1	-83.0	-110.1

Speer 230-grain Clean-Fire – TMJ RN Lawman (53885)
G1 Ballistic Coefficient = 0.153

Distance • Yards	Muzzle	25	50	75	100	125	150	175	200
Velocity • fps	845	826	807	790	773	756	740	725	710
Energy • ft-lbs	365	348	333	318	305	292	280	269	258
Taylor KO Index	12.5	12.3	12.0	11.7	11.5	11.2	11.0	10.8	10.5
Mid-Range Trajectory Height • Inches	0.0	0.4	1.6	3.7	6.7	10.6	15.7	21.6	29.2
Drop • Inches	0.0	-1.6	-6.3	-14.3	-25.9	-41.1	-60.1	-83.0	-110.1

Speer 230-grain TMJ RN (53967)
G1 Ballistic Coefficient = 0.153

Distance • Yards	Muzzle	25	50	75	100	125	150	175	200
Velocity • fps	830	811	794	776	760	744	728	713	699
Energy • ft-lbs	352	336	322	308	295	283	271	260	250
Taylor KO Index	12.3	12.0	11.8	11.5	11.3	11.0	10.8	10.6	10.4
Mid-Range Trajectory Height • Inches	0.0	0.4	1.7	3.8	6.9	11.0	16.3	22.6	30.2
Drop • Inches	0.0	-1.6	-6.5	-14.8	-26.8	-42.5	-62.2	-85.9	-113.9

Magtech 230-grain FMJ (45A) + (MP45A)
G1 Ballistic Coefficient = 0.15

Distance • Yards	Muzzle	25	50	75	100	125	150	175	200
Velocity • fps	837	818	800	783	767	751	735	720	706
Energy • ft-lbs	356	342	326	313	300	288	276	265	254
Taylor KO Index	12.4	12.1	11.9	11.6	11.4	11.2	10.9	10.7	10.5
Mid-Range Trajectory Height • Inches	0.0	0.4	1.6	3.7	6.8	10.8	15.9	22.1	29.5
Drop • Inches	0.0	-1.6	-6.4	-14.6	-26.4	-41.8	-61.1	-84.4	-111.9

Magtech 230-grain FEB CleanRange (CR45A)
G1 Ballistic Coefficient = 0.15

Distance • Yards	Muzzle	25	50	75	100	125	150	175	200
Velocity • fps	837	818	800	783	767	751	735	720	706
Energy • ft-lbs	356	342	326	313	300	288	276	265	254
Taylor KO Index	12.4	12.1	11.9	11.6	11.4	11.2	10.9	10.7	10.5
Mid-Range Trajectory Height • Inches	0.0	0.4	1.6	3.7	6.8	10.8	15.9	22.1	29.5
Drop • Inches	0.0	-1.6	-6.4	-14.6	-26.4	-41.8	-61.1	-84.4	-111.9

Remington 230-grain Metal Case (R45AP4)
G1 Ballistic Coefficient = 0.1

Distance • Yards	Muzzle	25	50	75	100	125	150	175	200
Velocity • fps	835	817	800	783	767	751	736	722	708
Energy • ft-lbs	356	341	326	313	300	289	277	266	256
Taylor KO Index	12.4	12.1	11.9	11.6	11.4	11.2	10.9	10.7	10.5
Mid-Range Trajectory Height • Inches	0.0	0.4	1.6	3.7	6.8	10.8	15.9	22.1	29.5
Drop • Inches	0.0	-1.6	-6.4	-14.6	-26.4	-41.9	-61.2	-84.5	-112.0

UMC (Remington) 230-grain Metal Case (L45AP4)
G1 Ballistic Coefficient = 0.1

Distance • Yards	Muzzle	25	50	75	100	125	150	175	200
Velocity • fps	835	817	800	783	767	751	736	722	708
Energy • ft-lbs	356	341	326	313	300	289	277	266	256
Taylor KO Index	12.4	12.1	11.9	11.6	11.4	11.2	10.9	10.7	10.5
Mid-Range Trajectory Height • Inches	0.0	0.4	1.6	3.7	6.8	10.8	15.9	22.1	29.5
Drop • Inches	0.0	-1.6	-6.4	-14.6	-26.4	-41.9	-61.2	-84.5	-112.0

UMC (Remington) 230-grain FNEB Leadless (LL45APB)
G1 Ballistic Coefficient = 0.160

Distance • Yards	Muzzle	25	50	75	100	125	150	175	200
Velocity • fps	835	817	800	783	767	751	736	722	708
Energy • ft-lbs	356	341	326	313	300	289	277	266	256
Taylor KO Index	12.4	12.1	11.9	11.6	11.4	11.2	10.9	10.7	10.5
Mid-Range Trajectory Height • Inches	0.0	0.4	1.6	3.7	6.8	10.8	15.9	22.1	29.5
Drop • Inches	0.0	-1.6	-6.4	-14.6	-26.4	-41.9	-61.2	-84.5	-112.0

USA (Winchester) 230-grain Full Metal Jacket (Q4170)
G1 Ballistic Coefficient = 0.160

Distance • Yards	Muzzle	25	50	75	100	125	150	175	200
Velocity • fps	835	817	800	783	767	751	736	722	708
Energy • ft-lbs	356	341	326	313	300	289	277	266	256
Taylor KO Index	12.4	12.1	11.9	11.6	11.4	11.2	10.9	10.7	10.5
Mid-Range Trajectory Height • Inches	0.0	0.4	1.6	3.7	6.8	10.8	15.9	22.1	29.5
Drop • Inches	0.0	-1.6	-6.4	-14.6	-26.4	-41.9	-61.2	-84.5	-112.0

Aguila 230-grain Full Metal Jacket (1E452110)
G1 Ballistic Coefficient = 0.130

Distance • Yards	Muzzle	25	50	75	100	125	150	175	200
Velocity • fps	830	809	789	769	749	730	712	695	679
Energy • ft-lbs	352	334	317	301	286	272	259	247	235
Taylor KO Index	12.3	12.0	11.7	11.4	11.1	10.8	10.6	10.3	10.1
Mid-Range Trajectory Height • Inches	0.0	0.4	1.6	3.8	7.0	11.2	16.6	23.2	31.0
Drop • Inches	0.0	-1.6	-6.5	-15.0	-27.1	-43.1	-63.1	-87.5	-116.3

Blazer 230-grain FMJ RN (5230)
G1 Ballistic Coefficient = 0.153

Distance • Yards	Muzzle	25	50	75	100	125	150	175	200
Velocity • fps	830	811	794	776	760	744	728	713	699
Energy • ft-lbs	352	336	322	308	295	283	271	260	250
Taylor KO Index	12.3	12.0	11.8	11.5	11.3	11.0	10.8	10.6	10.4
Mid-Range Trajectory Height • Inches	0.0	0.4	1.7	3.8	6.9	11.0	16.3	22.6	30.2
Drop • Inches	0.0	-1.6	-6.5	-14.8	-26.8	-42.5	-62.2	-85.9	-113.9

MC 230-grain EMJ NT (45EMA) LF
G1 Ballistic Coefficient = 0.131

Distance • Yards	Muzzle	25	50	75	100	125	150	175	200
Velocity • fps	830	809	789	769	749	731	713	696	680
Energy • ft-lbs	352	334	317	301	287	273	260	248	236
Taylor KO Index	12.3	12.0	11.7	11.4	11.1	10.9	10.6	10.3	10.1
Mid-Range Trajectory Height • Inches	0.0	0.4	1.7	3.9	7.0	11.2	16.6	22.9	31.1
Drop • Inches	0.0	-1.6	-6.5	-15.0	-27.1	-43.1	-63.1	-87.4	-116.2

MC 230-grain Full Metal Jacket (45A)
G1 Ballistic Coefficient = 0.130

Distance • Yards	Muzzle	25	50	75	100	125	150	175	200
Velocity • fps	830	809	789	769	749	730	712	695	679
Energy • ft-lbs	352	334	317	301	286	272	259	247	235
Taylor KO Index	12.3	12.0	11.7	11.4	11.1	10.8	10.6	10.3	10.1
Mid-Range Trajectory Height • Inches	0.0	0.4	1.6	3.8	7.0	11.2	16.6	23.2	31.0
Drop • Inches	0.0	-1.6	-6.5	-15.0	-27.1	-43.1	-63.1	-87.5	-116.3

Speer 230-grain GDHP-Short Barrel (23975)
G1 Ballistic Coefficient = 0.148

Distance • Yards	Muzzle	25	50	75	100	125	150	175	200
Velocity • fps	820	801	783	766	749	733	717	702	687
Energy • ft-lbs	343	328	313	300	287	274	263	252	241
Taylor KO Index	12.2	11.9	11.6	11.4	11.1	10.9	10.6	10.4	10.2
Mid-Range Trajectory Height • Inches	0.0	0.4	1.7	3.9	7.1	11.3	16.7	22.9	31.0
Drop • Inches	0.0	-1.6	-6.7	-15.2	-27.5	-43.6	-63.8	-88.2	-117.0

Black Hills 230-grain FMJ (D45SCN3)
G1 Ballistic Coefficient = 0.150

Distance • Yards	Muzzle	25	50	75	100	125	150	175	200
Velocity • fps	800	782	765	749	733	717	702	688	674
Energy • ft-lbs	327	313	299	286	274	263	252	242	232
Taylor KO Index	11.9	11.6	11.4	11.1	10.9	10.6	10.4	10.2	10.0
Mid-Range Trajectory Height • Inches	0.0	0.4	1.8	4.1	7.4	11.8	17.5	24.3	32.4
Drop • Inches	0.0	-1.7	-7.0	-16.0	-28.8	-45.7	-66.8	-92.4	-122.5

Magtech 230-grain FMJ SWC (45B)

G1 Ballistic Coefficient = 0.160

Distance • Yards	Muzzle	25	50	75	100	125	150	175	200
Velocity • fps	780	764	749	734	719	705	691	678	665
Energy • ft-lbs	311	298	286	275	264	254	244	235	226
Taylor KO Index	11.6	11.4	11.2	10.9	10.7	10.5	10.3	10.1	9.9
Mid-Range Trajectory Height • Inches	0.0	0.4	1.8	4.2	7.7	12.3	18.1	25.2	33.6
Drop • Inches	0.0	-1.8	-7.3	-16.8	-30.2	-47.8	-69.8	-96.4	-127.7

Cor-Bon 230-grain FMJ (PM45230/50)

G1 Ballistic Coefficient = 0.150

Distance • Yards	Muzzle	25	50	75	100	125	150	175	200
Velocity • fps	750	734	718	703	689	675	661	647	634
Energy • ft-lbs	287	275	264	253	242	232	223	214	205
Taylor KO Index	11.1	10.9	10.7	10.4	10.2	10.0	9.8	9.6	9.4
Mid-Range Trajectory Height • Inches	0.0	0.5	2.0	4.6	8.4	13.4	19.8	27.5	36.7
Drop • Inches	0.0	-2.0	-8.0	-18.2	-32.7	-51.9	-75.8	-104.8	-138.8

.45 Auto Rim

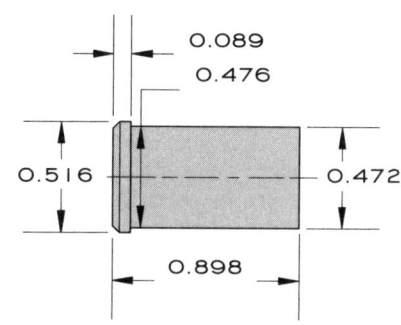

In WWI there weren't enough of the Model 1911 pistols to fill the supply needs, so the government contracted with both Colt and Smith & Wesson to produce revolvers that fired .45 Auto ammunition. The problem was that in order to fire the rimless .45 Auto round reliably in a revolver, the ammo had to be loaded into half-moon clips. That probably made sense during the war because having two types of pistol ammo in the inventory is guaranteed chaos. After the war the civilian market didn't want to mess with the half-moon clips (Frank Barnes says), so the Peters Cartridge Company introduced a new round with a thick rim to eliminate the need for the clips. The new round was named .45 Auto Rim, and it has been with us on a hit-and-miss basis ever since. Today, Black Hills is producing this ammo for the Cowboy Action application.

Relative Recoil Factor = 0.85

Specifications:

Controlling Agency for Standardization of this Ammunition: SAAMI

Bullet Weight Grains	Velocity fps	Maximum Average Pressure	
		Copper Crusher	Transducer
230 L	825	15,000 cup	N/S

Barrel used for velocity testing: 4 inches long (vented)—1 turn in 16-inch twist

Availability:
Cowboy Action Loads:

Black Hills 255-grain Semi-Wadcutter (D45AUTORIMN1)

G1 Ballistic Coefficient = 0.160

Distance • Yards	Muzzle	25	50	75	100	125	150	175	200
Velocity • fps	750	735	720	706	693	679	666	653	641
Energy • ft-lbs	318	306	294	283	272	261	251	242	233
Taylor KO Index	12.3	12.1	11.9	11.6	11.4	11.2	11.0	10.8	10.6
Mid-Range Trajectory Height • Inches	0.0	0.5	2.0	4.6	8.4	13.3	19.6	27.3	36.3
Drop • Inches	0.0	-2.0	-7.9	-18.1	-32.6	-51.8	-75.4	-104.1	-137.8

.45 Smith & Wesson Schofield

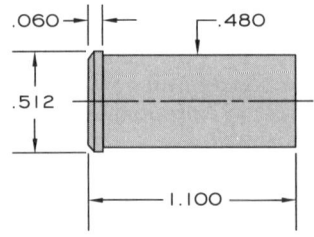

The .45 Schofield is a cartridge that had nearly disappeared into the mists of obsolescence, only to have been given a new life by the introduction of Cowboy Action shooting. It was introduced in 1875 as a competitor for the .45 Colt. The Schofield is about 0.2 inches shorter than the .45 Colt, but otherwise it is nearly identical in dimensions. For a while in the 1870s, the army had both guns and both calibers of ammunition in its inventory. Can you imagine how many units got the wrong ammunition? See .45 Auto Rim—maybe we learned something.

Relative Recoil Factor = 0.76

Specifications:

Controlling Agency for Standardization of this Ammunition: Factory
Barrel used for velocity testing: 6 inches long—1 turn in 12-inch twist

Availability:

Cor-Bon 135-grain Glaser Safety Slug (03800)

G1 Ballistic Coefficient = 0.120

Distance • Yards	Muzzle	25	50	75	100	125	150	175	200
Velocity • fps	1300	1204	1126	1063	1013	972	936	904	875
Energy • ft-lbs	507	435	390	339	308	283	263	245	230
Taylor KO Index	11.4	10.6	9.9	9.3	8.9	8.5	8.2	8.0	7.7
Mid-Range Trajectory Height • Inches	0.0	0.2	0.8	1.8	3.4	5.7	8.6	12.4	16.9
Drop • Inches	0.0	-0.7	-2.8	-6.7	-12.6	-20.5	-30.7	-43.4	-58.8

Cowboy Action Loads:

Black Hills 180-grain FNL (DCB45SCHON2)

G1 Ballistic Coefficient = 0.13

Distance • Yards	Muzzle	25	50	75	100	125	150	175	200
Velocity • fps	730	713	696	680	665	650	635	620	606
Energy • ft-lbs	213	203	195	185	177	169	161	154	147
Taylor KO Index	8.6	8.4	8.2	8.0	7.8	7.6	7.4	7.2	7.0
Mid-Range Trajectory Height • Inches	0.0	0.5	2.1	4.9	9.0	14.2	21.1	29.4	39.4
Drop • Inches	0.0	-2.1	-8.4	-19.2	-34.8	-55.2	-80.7	-111.7	-148.3

Black Hills 230-grain RNFP (DCB45SCHON1)

G1 Ballistic Coefficient = 0.15

Distance • Yards	Muzzle	25	50	75	100	125	150	175	200
Velocity • fps	730	715	700	685	671	657	644	630	617
Energy • ft-lbs	272	261	250	240	230	221	212	203	195
Taylor KO Index	10.9	10.7	10.5	10.3	10.1	9.8	9.6	9.4	9.2
Mid-Range Trajectory Height • Inches	0.0	0.5	2.1	4.9	8.9	14.1	20.8	28.9	38.6
Drop • Inches	0.0	-2.1	-8.4	-19.2	-34.5	-54.7	-80.0	-110.4	-146.4

.45 Colt (often called .45 Long Colt)

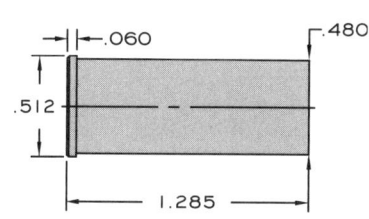

The name of this cartridge is interesting. It is often called the .45 Long Colt or the .45 Colt Long. Neither name is correct because there never was a .45 Short Colt. The name that gets used today relates to the .45 S&W Schofield. The .45 Colt was adopted by the army in 1873 for the legendary Colt Single Action Army revolver. That made it the first centerfire pistol round in the U.S. inventory. A couple years later the army adopted the .45 S&W Schofield in a Smith & Wesson revolver as an alternate. That put the army in the position of having two very similar rounds in the inventory at the same time. This guaranteed chaos! The .45 Schofield would fit the Colt revolver, but the Colt ammo wouldn't fit the Smith. Besides, the quartermasters in outfits that used the Colt revolver didn't want any of that shorter ammo, which was thought of as inferior. They were careful to specify .45 Colt (LONG) when they ordered ammo. While the standard factory ammo is pretty mild by today's standards (because there are still a lot of very old guns in circulation), the advent of Cowboy Action shooting is adding new life to this great old cartridge.

Caution: Some ammunition in this caliber has been, and still is, loaded to high pressures and these rounds are NOT intended for use in firearms of weaker receiver/frame strength such as the Colt Single Action Army revolver. Please consult your firearm manufacturer specifications.

Relative Recoil Factor = 0.90

Specifications:

Controlling Agency for Standardization of this Ammunition: SAAMI

Bullet Weight Grains	Velocity fps	Maximum Average Pressure	
		Copper Crusher	Transducer
225 STHP	915	14,000 cup	14,000 psi
225 SWC	950	14,000 cup	14,000 psi
250–255 L	900	14,000 cup	14,000 psi

Standard barrel for velocity testing: 4 inches long [vented]—1 turn in 16-inch twist

Availability:

Cor-Bon 145-grain Glaser Safety Slug (05000)

G1 Ballistic Coefficient = 0.125

Distance • Yards	Muzzle	25	50	75	100	125	150	175	200
Velocity • fps	1250	1166	1097	1043	998	960	927	898	870
Energy • ft-lbs	503	438	388	350	321	297	277	259	244
Taylor KO Index	11.8	11.0	10.4	9.9	9.4	9.1	8.8	8.5	8.2
Mid-Range Trajectory Height • Inches	0.0	0.2	0.8	1.9	3.6	5.9	9.0	12.8	17.4
Drop • Inches	0.0	-0.7	-3.1	-7.2	-13.3	-21.6	-32.3	-45.5	-61.3

CI 150-grain Magnum Shotshell (3972)

Shot load using #9 shot—Muzzle Velocity = 1000 fps.

Hornady 185-grain FTX (92790)

G1 Ballistic Coefficient = 0.140

Distance • Yards	Muzzle	25	50	75	100	125	150	175	200
Velocity • fps	920	894	870	847	826	806	787	768	751
Energy • ft-lbs	348	328	311	295	280	267	254	243	232
Taylor KO Index	11.1	10.8	10.5	10.2	10.0	9.7	9.5	9.3	9.1
Mid-Range Trajectory Height • Inches	0.0	0.3	1.4	3.2	5.8	9.2	13.6	19.1	25.5
Drop • Inches	0.0	-1.3	-5.3	-12.2	-22.2	-35.3	-51.7	-71.7	-95.3

Cor-Bon 200-grain JHP +P (SD45C200/20)

G1 Ballistic Coefficient = 0.150

Distance • Yards	Muzzle	25	50	75	100	125	150	175	200
Velocity • fps	1100	1058	1022	991	962	937	913	891	871
Energy • ft-lbs	537	497	464	436	412	390	371	353	337
Taylor KO Index	14.3	13.8	13.3	12.9	12.5	12.2	11.9	11.6	11.3
Mid-Range Trajectory Height • Inches	0.0	0.2	0.9	2.2	4.1	6.6	9.9	13.9	18.6
Drop • Inches	0.0	-0.9	-3.8	-8.7	-15.9	-25.4	-37.3	-51.8	-69.1

Blazer (CCI) 200-grain JHP (3584)

G1 Ballistic Coefficient = 0.139

Distance • Yards	Muzzle	25	50	75	100	125	150	175	200
Velocity • fps	1000	965	935	907	882	859	837	816	796
Energy • ft-lbs	444	414	388	366	345	328	311	296	282
Taylor KO Index	12.9	12.5	12.1	11.7	11.4	11.1	10.8	10.5	10.3
Mid-Range Trajectory Height • Inches	0.0	0.3	1.2	2.7	5.0	8.0	11.9	16.5	22.4
Drop • Inches	0.0	-1.1	-4.6	-10.5	-19.0	-30.4	-44.7	-62.1	-82.8

Hornady 225-grain FTX (92792)

G1 Ballistic Coefficient = 0.120

Distance • Yards	Muzzle	25	50	75	100	125	150	175	200
Velocity • fps	960	926	895	867	841	817	794	772	751
Energy • ft-lbs	460	428	400	376	354	333	315	298	282
Taylor KO Index	14.0	13.5	13.1	12.7	12.3	11.9	11.6	11.3	11.0
Mid-Range Trajectory Height • Inches	0.0	0.3	1.2	2.9	5.4	8.7	13.0	18.3	24.6
Drop • Inches	0.0	-1.2	-5.0	-11.4	-20.8	-33.2	-48.8	-68.0	-90.7

Remington 225-grain Semi-Wadcutter (R45C1)

G1 Ballistic Coefficient = 0.110

Distance • Yards	Muzzle	25	50	75	100	125	150	175	200
Velocity • fps	960	923	890	859	832	806	782	759	737
Energy • ft-lbs	460	425	395	369	346	325	305	288	271
Taylor KO Index	14.1	13.5	13.0	12.6	12.2	11.7	11.4	11.0	10.7
Mid-Range Trajectory Height • Inches	0.0	0.3	1.3	3.0	5.5	8.8	13.2	18.6	25.0
Drop • Inches	0.0	-1.2	-5.0	-11.5	-20.9	-33.5	-49.4	-66.8	-92.0

American Eagle (Federal) 225-grain Jacketed Soft Point (AE45LC)

G1 Ballistic Coefficient = 0.19

Distance • Yards	Muzzle	25	50	75	100	125	150	175	200
Velocity • fps	860	840	830	810	800	786	773	760	747
Energy • ft-lbs	370	355	345	330	320	309	298	288	279
Taylor KO Index	12.6	12.3	12.2	11.9	11.7	11.5	11.3	11.1	10.9
Mid-Range Trajectory Height • Inches	0.0	0.4	1.5	3.5	6.3	10.1	14.8	20.5	27.2
Drop • Inches	0.0	-1.5	-6.0	-13.7	-24.7	-39.1	-57.0	-78.5	-103.8

Federal 225-grain Jacketed Soft Point (AE45LC)

G1 Ballistic Coefficient = 0.19

Distance • Yards	Muzzle	25	50	75	100	125	150	175	200
Velocity • fps	860	840	830	810	800	787	774	761	748
Energy • ft-lbs	370	355	345	330	320	309	299	289	280
Taylor KO Index	12.5	12.3	12.1	11.8	11.7	11.5	11.3	11.1	10.9
Mid-Range Trajectory Height • Inches	0.0	0.3	1.5	3.5	6.3	10.0	14.7	20.4	27.1
Drop • Inches	0.0	-1.5	-6.0	-13.7	-24.7	-39.0	-56.9	-78.4	-103.7

Cor-Bon 225-grain DPX +P + (DPX45C225/20)

G1 Ballistic Coefficient = 0.16

Distance • Yards	Muzzle	25	50	75	100	125	150	175	200
Velocity • fps	1200	1140	1089	1047	1011	979	951	926	903
Energy • ft-lbs	720	349	593	548	511	479	542	429	407
Taylor KO Index	17.4	16.6	15.8	15.2	14.7	14.2	13.8	13.5	13.1
Mid-Range Trajectory Height • Inches	0.0	0.2	0.8	2.0	3.7	5.9	8.9	12.6	17.0
Drop • Inches	0.0	-0.8	-3.2	-7.5	-13.8	-22.2	-32.8	-45.9	-61.4

Winchester 225-grain Silvertip Hollow Point (X45CSHP2)

G1 Ballistic Coefficient = 0.1

Distance • Yards	Muzzle	25	50	75	100	125	150	175	200
Velocity • fps	920	898	877	857	839	821	804	788	772
Energy • ft-lbs	423	403	384	367	352	337	323	310	298
Taylor KO Index	13.5	13.2	12.9	12.6	12.3	11.9	11.7	11.4	11.2
Mid-Range Trajectory Height • Inches	0.0	0.3	1.4	3.1	5.6	9.0	13.2	18.5	24.6
Drop • Inches	0.0	-1.3	-5.3	-12.1	-21.9	-34.8	-50.8	-70.3	-93.2

Winchester 225-grain Bonded PDX1 (S45CPDB)

G1 Ballistic Coefficient = 0.1

Distance • Yards	Muzzle	25	50	75	100	125	150	175	200
Velocity • fps	850	832	815	799	784	768	754	740	726
Energy • ft-lbs	361	345	332	319	307	295	284	273	263
Taylor KO Index	12.5	12.2	11.9	11.7	11.5	11.3	11.1	10.8	10.6
Mid-Range Trajectory Height • Inches	0.0	0.4	1.6	3.6	6.6	10.4	15.3	21.3	28.4
Drop • Inches	0.0	-1.5	-6.2	-14.1	-25.4	-40.8	-58.8	-81.2	-107.6

Federal 225-grain JHP (C45LCA)

G1 Ballistic Coefficient = 0.153

Distance • Yards	Muzzle	25	50	75	100	125	150	175	200
Velocity • fps	830	810	790	780	760	744	728	713	699
Energy • ft-lbs	345	330	315	300	290	277	265	254	244
Taylor KO Index	12.1	11.8	11.5	11.3	11.0	10.8	10.6	10.4	10.2
Mid-Range Trajectory Height • Inches	0.0	0.4	1.7	3.8	6.9	11.0	16.3	22.3	30.2
Drop • Inches	0.0	-1.6	-6.5	-14.8	-26.8	-42.5	-62.2	-85.9	-113.9

Speer 250-grain GDHP (23984)

G1 Ballistic Coefficient = 0.165

Distance • Yards	Muzzle	25	50	75	100	125	150	175	200
Velocity • fps	750	735	721	707	694	681	668	358	644
Energy • ft lbs	312	300	289	278	268	257	248	239	230
Taylor KO Index	12.2	11.9	11.7	11.5	11.3	11.0	10.8	10.7	10.4
Mid-Range Trajectory Height • Inches	0.0	0.5	2.0	4.6	8.3	13.3	19.5	27.1	36.1
Drop • Inches	0.0	-2.0	-7.9	-18.1	-32.6	-51.6	-75.2	-103.8	-137.2

Winchester 255-grain Lead-Round Nose (X45CP2)

G1 Ballistic Coefficient = 0.153

Distance • Yards	Muzzle	25	50	75	100	125	150	175	200
Velocity • fps	860	838	820	798	780	762	745	728	712
Energy • ft-lbs	420	398	380	361	344	329	314	300	237
Taylor KO Index	14.3	13.9	13.6	13.3	13.0	12.5	12.3	12.0	11.7
Mid-Range Trajectory Height • Inches	0.0	0.4	1.5	3.5	6.5	10.3	15.3	21.4	28.4
Drop • Inches	0.0	-1.5	-6.1	-13.9	-25.2	-39.9	-58.5	-80.9	-107.5

Cor-Bon 265-grain BCHP +P (HT45C265BHP/20)

G1 Ballistic Coefficient = 0.170

Distance • Yards	Muzzle	25	50	75	100	125	150	175	200
Velocity • fps	1200	1143	1095	1054	1019	988	961	936	913
Energy • ft-lbs	848	769	706	654	611	575	543	516	491
Taylor KO Index	20.5	19.6	18.7	18.0	17.4	16.9	16.4	16.0	15.6
Mid-Range Trajectory Height • Inches	0.0	0.2	0.8	2.0	3.6	5.9	8.8	12.3	16.7
Drop • Inches	0.0	-0.8	-3.2	-7.5	-13.7	-22.0	-32.5	-45.4	-60.7

Cor-Bon 300-grain JSP +P (HT45C300JSP/20)

G1 Ballistic Coefficient = 0.180

Distance • Yards	Muzzle	25	50	75	100	125	150	175	200
Velocity • fps	1300	1234	1176	1126	1082	1045	1013	985	959
Energy • ft-lbs	1126	1015	922	844	781	728	684	646	613
Taylor KO Index	25.2	23.9	22.8	21.8	21.0	20.2	19.6	19.1	18.6
Mid-Range Trajectory Height • Inches	0.0	0.2	0.7	1.7	3.2	5.1	7.7	10.8	14.8
Drop • Inches	0.0	-0.7	-2.8	-6.4	-11.8	-19.0	-28.2	-39.6	-53.1

Cor-Bon 335-grain HC +P (HT45C335HC/20)

G1 Ballistic Coefficient = 0.195

Distance • Yards	Muzzle	25	50	75	100	125	150	175	200
Velocity • fps	1050	1020	993	968	946	925	906	888	871
Energy • ft-lbs	820	773	733	697	666	637	611	587	564
Taylor KO Index	22.7	22.1	21.5	20.9	20.5	20.0	19.6	19.2	18.8
Mid-Range Trajectory Height • Inches	0.0	0.2	1.1	2.4	4.4	7.0	10.4	14.3	19.4
Drop • Inches	0.0	-1.0	-4.1	-9.4	-17.0	-27.0	-39.6	-54.8	-72.8

Cowboy Action Loads:

Magtech 200-grain LFN (45F)

G1 Ballistic Coefficient = 0.165

Distance • Yards	Muzzle	25	50	75	100	125	150	175	200
Velocity • fps	705	692	679	666	653	641	629	617	606
Energy • ft-lbs	220	212	205	197	190	183	176	169	163
Taylor KO Index	9.1	8.9	8.8	8.6	8.4	8.3	8.1	8.0	7.8
Mid-Range Trajectory Height • Inches	0.0	0.5	2.3	5.2	9.5	15.0	19.4	30.3	40.8
Drop • Inches	0.0	-2.2	-9.0	-20.4	-36.8	-58.3	-85.0	-117.2	-155.2

Magtech 250-grain LFN (45D)

G1 Ballistic Coefficient = 0.177

Distance • Yards	Muzzle	25	50	75	100	125	150	175	200
Velocity • fps	761	747	735	720	708	695	683	671	659
Energy • ft-lbs	313	310	301	288	280	268	259	250	241
Taylor KO Index	12.3	12.1	11.9	11.6	11.4	11.2	11.0	10.8	10.6
Mid-Range Trajectory Height • Inches	0.0	0.5	2.0	4.5	8.1	12.8	18.9	26.1	34.8
Drop • Inches	0.0	-1.9	-7.7	-17.5	-31.5	-49.9	-72.7	-100.2	-132.5

Winchester 250-grain Lead Flat Nose (X45CBTR)

G1 Ballistic Coefficient = 0.158

Distance • Yards	Muzzle	25	50	75	100	125	150	175	200
Velocity • fps	750	735	720	706	692	678	665	652	639
Energy • ft-lbs	312	300	288	277	266	255	245	236	227
Taylor KO Index	12.1	11.9	11.6	11.4	11.2	10.9	10.7	10.5	10.3
Mid-Range Trajectory Height • Inches	0.0	0.4	2.0	4.6	8.4	13.3	19.6	27.2	36.3
Drop • Inches	0.0	-2.0	-7.9	-18.1	-32.6	-51.7	-75.5	-104.2	-138.0

Black Hills 250-grain RNFP (DCB45CLTN1)

G1 Ballistic Coefficient = 0.140

Distance • Yards	Muzzle	25	50	75	100	125	150	175	200
Velocity • fps	725	709	693	677	663	648	634	620	606
Energy • ft-lbs	292	279	267	255	244	233	223	213	204
Taylor KO Index	11.8	11.5	11.3	11.0	10.8	10.5	10.2	10.0	9.8
Mid-Range Trajectory Height • Inches	0.0	0.5	2.2	5.0	9.1	14.4	21.3	29.7	39.6
Drop • Inches	0.0	-2.1	-8.5	-19.5	-35.2	-55.8	-81.6	-112.8	-149.6

Hornady 255-grain Cowboy (9115)

G1 Ballistic Coefficient = 0.13

Distance • Yards	Muzzle	25	50	75	100	125	150	175	200
Velocity • fps	725	708	692	676	660	645	630	615	601
Energy • ft-lbs	298	284	271	259	247	235	225	215	205
Taylor KO Index	12.0	11.8	11.5	11.2	11.0	10.6	10.4	10.1	9.9
Mid-Range Trajectory Height • Inches	0.0	0.5	2.2	5.0	9.1	14.5	21.4	29.9	39.9
Drop • Inches	0.0	-2.1	-8.5	-19.5	-35.2	-56.0	-81.9	-113.3	-150.4

.45 Winchester Magnum

In 1979 Winchester introduced its .45 Winchester Magnum cartridge. The round was intended for the gas-operated Wildey pistol. The buying public never took to the Wildey gun, but the cartridge was a real performer. At the time of its introduction, it was the most powerful pistol cartridge in the inventory. While the .45 Casull and the .50 Action Express have pushed it out of first place in the power derby, the .45 Winchester Magnum is still a very potent pistol round. It makes a great hunting cartridge in single-shot, long-barrel pistols.

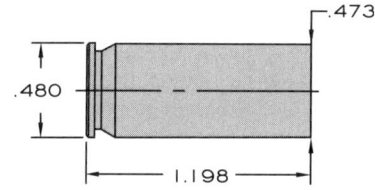

Relative Recoil Factor = 1.40

Specifications:

Controlling Agency for Standardization of this Ammunition: SAAMI

Bullet Weight Grains	Velocity fps	Maximum Average Pressure	
		Copper Crusher	Transducer
230 MC	1,380	40,000 cup	N/S

Standard barrel for velocity testing: 5 inches long—1 turn in 16-inch twist

Availability:

Winchester 260-grain Jacketed Hollow Point (X45MWA)

G1 Ballistic Coefficient = 0.180

Distance • Yards	Muzzle	25	50	75	100	125	150	175	200
Velocity • fps	1200	1146	1099	1060	1026	996	969	945	923
Energy • ft-lbs	831	758	698	649	607	573	543	516	492
Taylor KO Index	20.4	19.5	18.7	18.0	17.5	16.7	16.3	15.9	15.5
Mid-Range Trajectory Height • Inches	0.0	0.2	0.8	1.9	3.5	5.7	8.6	12.1	16.3
Drop • Inches	0.0	-0.8	-3.2	-7.4	-13.6	-21.8	-32.2	-44.8	-59.9

.454 Casull

If you look carefully at their dimensions, the .454 Casull is really a long version of the .45 Colt (which is often called Long Colt). The Casull is 0.100 inch longer, but the real performance improvement comes from using chamber pressures that approach 50,000 psi. The .454 Casull is one of just a few cartridges factory manufactured in the U.S. that have not been standardized by either SAAMI or CIP (the European agency). If a milder load is desired, .45 Colt ammunition can be fired in .454 Casull chambers. The result is pretty much the same as firing a .38 Special in a .357 Magnum. The Casull factory does not recommend this.

The .454 Casull, which captured the pistol power championship from the .44 Remington Magnum, has recently been slightly surpassed by the .460 Smith & Wesson and completely overwhelmed by the .500 Smith & Wesson Magnum. None of these guns are all that much fun to shoot.

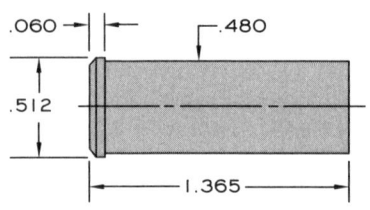

Relative Recoil Factor = 2.20

Specifications:

Controlling Agency for Standardization of this Ammunition: Factory
Standard barrel for velocity testing: 7.5 inches long [vented]—1 turn in N/S-inch twist

Availability:

Magtech 225-grain SCHP (454C)

G1 Ballistic Coefficient = 0.23

Distance • Yards	Muzzle	25	50	75	100	125	150	175	200
Velocity • fps	1640	1569	1501	1437	1375	1318	1264	1215	1171
Energy • ft-lbs	1344	1230	1126	1031	945	868	799	738	685
Taylor KO Index	23.9	22.9	21.9	21.0	20.1	19.2	18.4	17.7	17.1
Mid-Range Trajectory Height • Inches	0.0	0.1	0.4	1.0	1.9	3.1	4.7	6.8	9.3
Drop • Inches	0.0	-0.4	-1.7	-4.0	-7.3	-11.8	-17.5	-24.6	-33.1

Magtech 240-grain SJSP (454D)

G1 Ballistic Coefficient = 0.24

Distance • Yards	Muzzle	25	50	75	100	125	150	175	200
Velocity • fps	1771	1700	1632	1566	1502	1442	1384	1330	1279
Energy • ft-lbs	1672	1541	1419	1307	1203	1108	1021	942	872
Taylor KO Index	27.6	26.5	25.4	24.4	23.4	22.4	21.5	20.7	19.9
Mid-Range Trajectory Height • Inches	0.0	0.0	0.3	0.8	1.6	2.6	4.0	5.7	7.8
Drop • Inches	0.0	-0.4	-1.5	-3.4	-6.2	-10.0	-14.8	-20.7	-27.9

Hornady 240-grain XTP-MAG (9148)

G1 Ballistic Coefficient = 0.1

Distance • Yards	Muzzle	25	50	75	100	125	150	175	200
Velocity • fps	1900	1786	1678	1575	1478	1388	1305	1231	1137
Energy • ft-lbs	1923	1701	1500	1322	1163	1026	908	808	725
Taylor KO Index	29.4	27.7	26.0	24.4	22.9	21.5	20.2	19.1	18.1
Mid-Range Trajectory Height • Inches	0.0	0.1	0.3	0.8	1.6	2.6	4.0	5.8	8.1
Drop • Inches	0.0	-0.3	-1.3	-3.1	-5.7	-9.4	-14.1	-20.2	-27.6

Cor-Bon 240-grain JHP (HT454240JHP/20)

G1 Ballistic Coefficient = 0.1

Distance • Yards	Muzzle	25	50	75	100	125	150	175	200
Velocity • fps	1450	1362	1282	1210	1148	1096	1053	1016	983
Energy • ft-lbs	1120	989	876	781	703	641	591	550	516
Taylor KO Index	22.6	21.2	20.2	18.8	17.9	17.1	16.4	15.8	15.3
Mid-Range Trajectory Height • Inches	0.0	0.1	0.5	1.4	2.6	4.3	6.6	9.6	13.1
Drop • Inches	0.0	-0.5	-2.2	-5.3	-9.8	-16.0	-23.9	-33.9	-45.9

Cor-Bon 250-grain DPX (DPX454250/20)

G1 Ballistic Coefficient = 0.1

Distance • Yards	Muzzle	25	50	75	100	125	150	175	200
Velocity • fps	1650	1542	1441	1348	1265	1191	1129	1077	1034
Energy • ft-lbs	1512	1320	1153	1009	888	788	708	644	594
Taylor KO Index	26.6	24.9	23.3	21.8	20.4	19.2	18.2	17.4	16.7
Mid-Range Trajectory Height • Inches	0.0	0.1	0.5	1.1	2.1	3.5	5.4	8.0	11.0
Drop • Inches	0.0	-0.4	-1.8	-4.1	-7.7	-12.6	-19.1	-27.3	-37.3

Federal 250-grain Barnes Expander (P454XB1) LF

G1 Ballistic Coefficient = 0.142

Distance • Yards	Muzzle	25	50	75	100	125	150	175	200
Velocity • fps	1530	1420	1330	1240	1170	1107	1057	1015	979
Energy • ft-lbs	1300	1125	975	850	755	681	620	572	533
Taylor KO Index	24.7	22.9	21.5	20.0	18.9	17.9	17.1	16.4	15.8
Mid-Range Trajectory Height • Inches	0.0	0.1	0.6	1.3	2.5	4.2	6.4	9.2	12.7
Drop • Inches	0.0	-0.5	-2.0	-4.8	-9.0	-14.8	-22.4	-31.9	-43.5

Winchester 250-grain Jacketed Hollow Point (X454C3)

G1 Ballistic Coefficient = 0.146

Distance • Yards	Muzzle	25	50	75	100	125	150	175	200
Velocity • fps	1300	1220	1151	1094	1047	1008	974	944	917
Energy • ft-lbs	938	826	735	665	608	564	526	495	467
Taylor KO Index	21.0	19.7	18.6	17.7	16.9	16.3	15.7	15.2	14.8
Mid-Range Trajectory Height • Inches	0.0	0.2	0.8	0.8	3.2	5.3	8.1	11.5	15.6
Drop • Inches	0.0	-0.7	-2.8	-6.6	-12.2	-19.7	-29.4	-41.2	-55.5

Magtech 260-grain SJHP (454A)

G1 Ballistic Coefficient = 0.266

Distance • Yards	Muzzle	25	50	75	100	125	150	175	200
Velocity • fps	1800	1734	1669	1607	1547	1488	1433	1380	1330
Energy • ft-lbs	1871	1736	1609	1491	1381	1279	1186	1100	1021
Taylor KO Index	30.4	29.2	28.1	27.1	26.1	25.1	24.2	23.3	22.4
Mid-Range Trajectory Height • Inches	0.0	0.0	0.3	0.8	1.5	2.5	3.8	5.4	7.3
Drop • Inches	0.0	-0.3	-1.4	-3.3	-5.9	-9.5	-14.1	-19.8	-26.5

Magtech 260-grain FMJ Flat (454B)

G1 Ballistic Coefficient = 0.266

Distance • Yards	Muzzle	25	50	75	100	125	150	175	200
Velocity • fps	1800	1734	1669	1607	1547	1488	1433	1380	1330
Energy • ft-lbs	1871	1736	1609	1491	1381	1279	1186	1100	1021
Taylor KO Index	30.4	29.2	28.1	27.1	26.1	25.1	24.2	23.3	22.4
Mid-Range Trajectory Height • Inches	0.0	0.0	0.3	0.8	1.5	2.5	3.8	5.4	7.3
Drop • Inches	0.0	-0.3	-1.4	-3.3	-5.9	-9.5	-14.1	-19.8	-26.5

Winchester 260-grain Dual Bond (S454DB)

G1 Ballistic Coefficient = 0.176

Distance • Yards	Muzzle	25	50	75	100	125	150	175	200
Velocity • fps	1800	1700	1605	1515	1430	1352	1279	1215	1158
Energy • ft-lbs	1870	1670	1488	1326	1181	1055	945	852	774
Taylor KO Index	30.4	28.7	27.1	25.5	24.1	22.8	21.6	20.5	19.5
Mid-Range Trajectory Height • Inches	0.0	0.1	0.3	0.9	1.7	2.8	4.3	6.2	8.6
Drop • Inches	0.0	-0.4	-1.4	-3.4	-6.3	-10.2	-15.4	-21.8	-29.8

Winchester 260-grain Partition Gold (SPG454)

G1 Ballistic Coefficient = 0.175

Distance • Yards	Muzzle	25	50	75	100	125	150	175	200
Velocity • fps	1800	1700	1605	1513	1427	1349	1276	1211	1154
Energy • ft-lbs	1871	1668	1486	1322	1178	1051	941	847	769
Taylor KO Index	30.4	28.7	27.1	25.5	24.1	22.7	21.5	20.4	19.5
Mid-Range Trajectory Height • Inches	0.0	0.1	0.3	0.9	1.6	2.8	4.3	6.2	8.6
Drop • Inches	0.0	-0.4	-1.4	-3.4	-6.3	-10.2	-15.4	-21.9	-29.9

Cor-Bon 265-grain Bonded Core HP (HT454265BHP/20)

G1 Ballistic Coefficient = 0.170

Distance • Yards	Muzzle	25	50	75	100	125	150	175	200
Velocity • fps	1725	1625	1531	1442	1360	1285	1271	1158	1107
Energy • ft-lbs	1751	1555	1380	1224	1088	971	872	789	722
Taylor KO Index	29.5	27.8	26.2	24.7	23.3	22.0	20.8	19.8	18.9
Mid-Range Trajectory Height • Inches	0.0	0.1	0.4	1.0	1.9	3.1	4.8	6.9	9.5
Drop • Inches	0.0	-0.4	-1.6	-3.7	-6.9	-11.2	-16.9	-24.0	-32.8

Cor-Bon 285-grain Bonded Core SP (HT454285BC/20)

G1 Ballistic Coefficient = 0.180

Distance • Yards	Muzzle	25	50	75	100	125	150	175	200
Velocity • fps	1625	1536	1451	1373	1300	1285	1176	1126	1082
Energy • ft-lbs	1672	1493	1334	1193	1070	965	876	802	742
Taylor KO Index	29.9	28.3	26.7	25.3	23.9	22.7	21.6	20.7	19.9
Mid-Range Trajectory Height • Inches	0.0	0.1	0.5	1.1	2.1	3.4	5.2	7.5	10.4
Drop • Inches	0.0	-0.4	-1.8	-4.2	-7.7	-12.5	-18.8	-26.6	-36.1

Cor-Bon 300-grain JSP (HT454300JSP/20)

G1 Ballistic Coefficient = 0.185

Distance • Yards	Muzzle	25	50	75	100	125	150	175	200
Velocity • fps	1650	1562	1478	1399	1326	1260	1200	1147	1102
Energy • ft-lbs	1814	1625	1456	1305	1172	1057	959	877	809
Taylor KO Index	32.1	30.4	28.8	27.2	25.8	24.5	23.3	22.3	21.4
Mid-Range Trajectory Height • Inches	0.0	0.1	0.4	1.0	2.0	3.2	5.0	7.2	9.9
Drop • Inches	0.0	-0.4	-1.7	-4.0	-7.4	-12.1	-18.1	-25.6	-34.8

Federal 300-grain Swift A-Frame (P454SA)

G1 Ballistic Coefficient = 0.135

Distance • Yards	Muzzle	25	50	75	100	125	150	175	200
Velocity • fps	1520	1410	1310	1220	1150	1089	1039	998	963
Energy • ft-lbs	1540	1325	1145	995	880	790	720	664	618
Taylor KO Index	29.4	27.3	25.4	23.6	22.3	21.1	20.1	19.3	18.7
Mid-Range Trajectory Height • Inches	0.0	0.1	0.6	1.4	2.6	4.3	6.6	9.5	13.1
Drop • Inches	0.0	-0.5	-2.1	-4.9	-9.2	-15.2	-23.0	-32.8	-44.7

Hornady 300-grain XTP-MAG (9150)

G1 Ballistic Coefficient = 0.200

Distance • Yards	Muzzle	25	50	75	100	125	150	175	200
Velocity • fps	1650	1568	1490	1417	1348	1285	1227	1175	1089
Energy • ft-lbs	1813	1639	1480	1338	1210	1100	1003	919	849
Taylor KO Index	32.0	30.4	28.9	27.4	26.1	24.9	23.8	22.8	21.1
Mid-Range Trajectory Height • Inches	0.0	0.1	0.4	1.1	2.0	3.2	4.9	7.1	9.7
Drop • Inches	0.0	-0.4	-1.7	-4.0	-7.3	-11.9	-17.8	-25.1	34.0

Speer 300-grain Gold Dot Hollow Point (23990)

G1 Ballistic Coefficient = 0.23

Distance • Yards	Muzzle	25	50	75	100	125	150	175	200
Velocity • fps	1625	1556	1489	1426	1366	1310	1258	1210	1167
Energy • ft-lbs	1759	1612	1477	1355	1243	1143	1054	975	907
Taylor KO Index	31.5	30.1	28.8	27.6	26.5	25.4	24.4	23.4	22.6
Mid-Range Trajectory Height • Inches	0.0	0.1	0.4	1.1	2.0	3.2	4.8	6.9	9.4
Drop • Inches	0.0	-0.4	-1.8	-4.1	-7.4	-12.0	-17.8	-24.9	-33.6

Cor-Bon 325-grain FPPN (HT454320FPPN/20)

G1 Ballistic Coefficient = 0.19

Distance • Yards	Muzzle	25	50	75	100	125	150	175	200
Velocity • fps	1550	1469	1393	1323	1258	1200	1149	1105	1066
Energy • ft-lbs	1708	1534	1380	1244	1125	1024	938	867	808
Taylor KO Index	32.0	30.4	28.8	27.3	26.0	24.8	23.7	22.8	22.0
Mid-Range Trajectory Height • Inches	0.0	0.1	0.5	1.2	2.3	3.7	5.6	8.1	11.1
Drop • Inches	0.0	-0.5	-1.9	-4.5	-8.4	-13.6	-20.3	-28.7	-38.8

Cor-Bon 335-grain Hard Cast (HT454335HC/20)

G1 Ballistic Coefficient = 0.19

Distance • Yards	Muzzle	25	50	75	100	125	150	175	200
Velocity • fps	1550	1471	1397	1328	1265	1207	1156	1112	1074
Energy • ft-lbs	1788	1611	1452	1312	1190	1085	995	920	858
Taylor KO Index	33.5	31.8	30.2	28.7	27.4	26.1	25.0	24.1	23.2
Mid-Range Trajectory Height • Inches	0.0	0.1	0.5	1.2	2.2	3.7	5.6	8.0	11.0
Drop • Inches	0.0	-0.5	-1.9	-4.5	-8.3	-13.5	-20.2	-28.5	-38.5

Cor-Bon 360-grain FPPN (HT454360FPPN/20)

G1 Ballistic Coefficient = 0.2

Distance • Yards	Muzzle	25	50	75	100	125	150	175	200
Velocity • fps	1300	1241	1187	1140	1099	1063	1032	1004	979
Energy • ft-lbs	1351	1231	1127	1039	965	904	851	806	767
Taylor KO Index	30.2	28.8	27.6	26.5	25.5	24.7	24.0	23.3	22.8
Mid-Range Trajectory Height • Inches	0.0	0.2	0.7	1.7	3.1	5.0	7.5	10.6	14.4
Drop • Inches	0.0	-0.7	-2.7	-6.4	-11.6	-18.7	-27.7	-38.8	-52.0

.460 Smith & Wesson

Introduced in 2005, the .460 Smith & Wesson is currently second on the list of most powerful handgun cartridges. This cartridge is an example of name inflation. Despite the .460 designation, there is nothing 0.460 about this cartridge—it's really a .45 that uses the same size bullets as the .45 Colt and the .454 Casull. The power is possible because of its large volume, and the 8.375-inch test barrel doesn't hurt anything either. It is the longest revolver cartridge in the current inventory, even a little bit longer than the .500 Smith & Wesson Magnum. The current fascination with huge handgun cartridges is interesting since these guns are too large and powerful for any application except hunting, and hunting very large game at that.

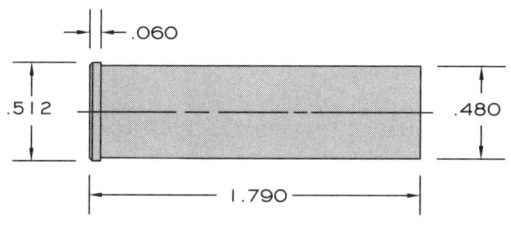

Relative Recoil Factor = 2.46

Specifications:

Controlling Agency for Standardization of this Ammunition: SAAMI
Standard barrel for velocity testing: 8.375 inches long [vented]—1 turn in 10-inch twist

Availability:

Hornady 200-grain FTX (9152)
G1 Ballistic Coefficient = 0.155

Distance • Yards	Muzzle	25	50	75	100	125	150	175	200
Velocity • fps	2200	2072	1948	1829	1715	1606	1504	1409	1322
Energy • ft-lbs	2149	1906	1685	1485	1305	1146	1005	882	777
Taylor KO Index	28.4	26.8	25.2	23.6	22.1	20.7	19.4	18.2	17.1
Mid-Range Trajectory Height • Inches	0.0	0.1	0.3	0.6	1.2	1.9	3.0	4.4	6.1
Drop • Inches	0.0	-0.2	-1.0	-2.3	-4.2	-7.0	-10.5	-15.0	-20.6

Cor-Bon 200-grain DPX (HT460SW200/20)
G1 Ballistic Coefficient = 0.155

Distance • Yards	Muzzle	25	50	75	100	125	150	175	200
Velocity • fps	2300	2168	2041	1918	1801	1688	1581	1481	1387
Energy • ft-lbs	2350	2089	1851	1635	1440	1265	1110	874	855
Taylor KO Index	29.7	28.0	26.4	24.8	23.3	21.8	20.4	19.1	17.9
Mid-Range Trajectory Height • Inches	0.0	0.1	0.2	0.6	1.1	1.8	2.7	4.0	5.5
Drop • Inches	0.0	-0.2	-0.9	-2.1	-3.9	-6.3	-9.6	-13.6	-18.7

Magtech 225-grain SCHP (460A)
G1 Ballistic Coefficient = 0.230

Distance • Yards	Muzzle	25	50	75	100	125	150	175	200
Velocity • fps	2132	2047	1963	1882	1803	1726	1652	1581	1513
Energy • ft-lbs	2272	2093	1926	1770	1625	1489	1364	1249	1143
Taylor KO Index	31.0	29.7	28.5	27.3	26.2	25.1	24.0	23.0	21.9
Mid-Range Trajectory Height • Inches	0.0	0.1	0.3	0.6	1.2	1.9	2.8	4.0	5.4
Drop • Inches	0.0	-0.2	-1.0	-2.3	-4.3	-6.9	-10.2	-14.4	-19.4

Winchester 250-grain Jacketed Hollow Point (X460SW)
G1 Ballistic Coefficient = 0.146

Distance • Yards	Muzzle	25	50	75	100	125	150	175	200
Velocity • fps	1450	1354	1267	1192	1127	1075	1031	995	962
Energy • ft-lbs	1167	1018	891	789	705	642	591	549	514
Taylor KO Index	23.4	21.9	20.5	19.2	18.2	17.4	16.6	16.1	15.5
Mid-Range Trajectory Height • Inches	0.0	0.1	0.3	0.6	1.2	1.9	2.8	4.0	5.4
Drop • Inches	0.0	-0.5	-2.3	-5.4	-10.0	-16.3	-24.5	-34.7	-47.1

Winchester 260-grain Dual Bond (S460SWDB)
G1 Ballistic Coefficient = 0.176

Distance • Yards	Muzzle	25	50	75	100	125	150	175	200
Velocity • fps	2000	1893	1790	1691	1596	1506	1422	1344	1273
Energy • ft-lbs	2309	2069	1849	1651	1470	1311	1168	1043	935
Taylor KO Index	33.6	31.8	30.1	28.4	26.8	25.3	23.9	22.6	21.4
Mid-Range Trajectory Height • Inches	0.0	0.1	0.3	0.7	1.4	2.3	3.5	5.0	7.0
Drop • Inches	0.0	-0.3	-1.2	-2.7	-5.1	-8.2	-12.4	-17.6	-24.0

Magtech 260-grain FMJ (460B)

G1 Ballistic Coefficient = 0.250

Distance • Yards	Muzzle	25	50	75	100	125	150	175	200
Velocity • fps	2001	1925	1851	1779	1710	1642	1576	1514	1453
Energy • ft-lbs	2312	2141	1979	1829	1688	1557	1435	1323	1220
Taylor KO Index	33.6	32.3	31.1	29.9	28.7	27.6	26.5	25.4	24.4
Mid-Range Trajectory Height • Inches	0.0	0.1	0.3	0.7	1.3	2.1	3.1	4.4	6.0
Drop • Inches	0.0	-0.3	-1.1	-2.6	-4.8	-7.8	-11.5	-16.1	-21.6

Cor-Bon 275-grain DPX (HT460SW275/20)

G1 Ballistic Coefficient = 0.160

Distance • Yards	Muzzle	25	50	75	100	125	150	175	200
Velocity • fps	1825	1715	1610	1510	1418	1333	1256	1188	1130
Energy • ft-lbs	2034	1796	1582	1393	1228	1085	963	862	779
Taylor KO Index	32.4	30.5	28.6	26.8	25.2	23.7	22.3	21.1	20.1
Mid-Range Trajectory Height • Inches	0.0	0.1	0.4	0.9	1.7	2.8	4.4	6.3	8.8
Drop • Inches	0.0	-0.3	-1.4	-3.3	-6.2	-10.2	-15.3	-21.9	-30.0

Federal 275-grain Barnes Expander (P460XB1) LF

G1 Ballistic Coefficient = 0.222

Distance • Yards	Muzzle	25	50	75	100	125	150	175	200
Velocity • fps	1800	1720	1640	1570	1500	1483	1370	1311	1256
Energy • ft-lbs	1980	1810	1650	1505	1370	1255	1147	1050	964
Taylor KO Index	32.0	30.5	29.1	27.9	26.6	25.3	24.3	23.3	22.3
Mid-Range Trajectory Height • Inches	0.0	0.1	0.4	0.9	1.6	2.7	4.0	5.7	7.8
Drop • Inches	0.0	-0.4	-1.4	-3.3	-6.1	-9.8	-14.6	-20.5	-27.7

Cor-Bon 300-grain JSP (HT460SW300/20)

G1 Ballistic Coefficient = 0.14?

Distance • Yards	Muzzle	25	50	75	100	125	150	175	200
Velocity • fps	1750	1632	1522	1419	1325	1242	1169	1109	1058
Energy • ft-lbs	2041	1776	1543	1342	1170	1027	911	820	748
Taylor KO Index	33.9	31.6	29.5	27.5	25.7	24.1	22.6	21.5	20.5
Mid-Range Trajectory Height • Inches	0.0	0.1	0.4	1.0	1.9	3.2	4.9	7.2	10.1
Drop • Inches	0.0	-0.4	-1.6	-3.7	-6.9	-11.3	-17.2	-24.6	-33.8

Federal 300-grain Swift A-Frame (P460SA)

G1 Ballistic Coefficient = 0.13?

Distance • Yards	Muzzle	25	50	75	100	125	150	175	200
Velocity • fps	1750	1630	1510	1400	1300	1217	1144	1085	1038
Energy • ft-lbs	2040	1760	1510	1300	1125	981	868	780	712
Taylor KO Index	33.9	31.6	29.3	27.1	25.2	23.6	22.2	21.0	20.1
Mid-Range Trajectory Height • Inches	0.0	0.1	0.4	1.0	2.0	3.3	5.1	7.5	10.4
Drop • Inches	0.0	-0.4	-1.6	-3.7	-7.0	-11.6	-17.6	-25.3	-34.8

Cor-Bon 325-grain BC (HT460SW325/20)

G1 Ballistic Coefficient = 0.1?

Distance • Yards	Muzzle	25	50	75	100	125	150	175	200
Velocity • fps	1650	1542	1441	1348	1265	1191	1129	1077	1034
Energy • ft-lbs	1965	1716	1499	1312	1154	1024	920	838	772
Taylor KO Index	34.6	32.4	30.2	28.3	26.5	25.0	23.7	22.6	21.7
Mid-Range Trajectory Height • Inches	0.0	0.1	0.5	1.1	2.1	3.5	5.4	7.9	11.0
Drop • Inches	0.0	-0.4	-1.8	-4.1	-7.7	-12.6	-19.1	-27.3	-37.3

Cor-Bon 325-grain FPPN (HT460SW325FPPN/20)

G1 Ballistic Coefficient = 0.1?

Distance • Yards	Muzzle	25	50	75	100	125	150	175	200
Velocity • fps	1500	1402	1313	1234	1165	1107	1059	1019	984
Energy • ft-lbs	1624	1419	1245	1098	979	884	809	749	699
Taylor KO Index	31.5	29.4	27.6	25.9	24.4	23.2	22.2	21.4	20.6
Mid-Range Trajectory Height • Inches	0.0	0.1	0.6	1.4	2.6	4.2	6.5	9.3	12.8
Drop • Inches	0.0	-0.5	-2.1	-5.0	-9.3	-15.2	-22.9	-32.5	-44.2

Cor-Bon 395-grain HC (HT460SW395/20)

G1 Ballistic Coefficient = 0.1?

Distance • Yards	Muzzle	25	50	75	100	125	150	175	200
Velocity • fps	1525	1434	1350	1273	1205	1146	1096	1054	1018
Energy • ft-lbs	2040	1804	1599	1422	1274	1152	1054	974	909
Taylor KO Index	38.9	36.6	34.4	32.5	30.7	29.2	28.0	26.9	26.0
Mid-Range Trajectory Height • Inches	0.0	0.1	0.5	1.3	2.4	4.0	6.1	8.7	12.0
Drop • Inches	0.0	-0.5	-2.0	-4.8	-8.8	-14.4	-21.6	-30.7	-71.7

.475 Linebaugh

This cartridge is very similar in both size and performance to the .480 Ruger. Both cartridges use .475 diameter bullets. The Linebaugh case is bit longer (0.115 inch), but both use the same dies for reloading. Judging by the factory numbers, the Linebaugh is loaded to slightly higher pressures. That suggests you might be able to shoot the .480 Ruger ammo in the .475 Linebaugh chamber, although the only reason I can see for doing that would be for the sake of availability.

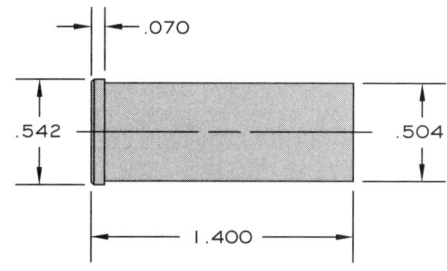

Relative Recoil Factor = 2.30

Specifications:

Controlling Agency for Standardization of this Ammunition: SAAMI
Standard barrel for velocity testing: 7.5 inches long [vented]—1 turn in 18-inch twist

Availability:

Hornady 400-grain XTP-MAG (9140)

G1 Ballistic Coefficient = 0.182

Distance • Yards	Muzzle	25	50	75	100	125	150	175	200
Velocity • fps	1300	1235	1177	1127	1084	1047	1015	987	961
Energy • ft-lbs	1501	1355	1231	1129	1043	974	915	865	821
Taylor KO Index	35.3	33.5	31.9	30.6	29.4	28.4	27.6	26.8	26.1
Mid-Range Trajectory Height • Inches	0.0	0.2	0.7	1.7	3.1	5.1	7.7	10.9	14.7
Drop • Inches	0.0	-0.7	-2.8	-6.4	-11.8	-19.0	-28.2	-39.5	-53.0

.480 Ruger

Here's another entrant in the powerful pistol category. Hornady and Ruger cooperated to produce a totally new round, and everything about this cartridge is big. It has slightly less relative recoil than either the .454 Casull or the .50 Action Express, but it is far from wimpy. This cartridge will find some immediate favor with pistol hunters, but it has no application to any of the current competition games. This cartridge should be an excellent performer in lever-action, carbine-style rifles.

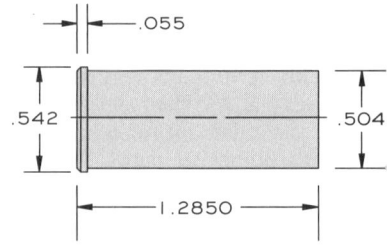

Relative Recoil Factor = 1.95

Specifications:

Controlling Agency for Standardization of this Ammunition: SAAM1

Bullet Weight Grains	Velocity fps	Maximum Average Pressure Copper Crusher	Transducer
N/S	N/S	N/S	48,000 psi

Standard barrel for velocity testing: 7.5 inches long [vented]—1 turn in 15-inch twist

Availability:

Cor-Bon 275-grain DPX (DPX480275/20)
G1 Ballistic Coefficient = 0.15

Distance • Yards	Muzzle	25	50	75	100	125	150	175	200
Velocity • fps	1550	1452	1361	1278	1206	1143	1090	1047	1009
Energy • ft-lbs	1467	1287	1131	998	888	799	726	669	622
Taylor KO Index	29.8	27.1	25.4	23.8	22.5	21.3	20.3	19.5	18.8
Mid-Range Trajectory Height • Inches	0.0	0.1	0.5	1.3	2.4	3.9	6.0	8.8	12.0
Drop • Inches	0.0	-0.5	-2.0	-4.7	-8.7	-14.2	-21.3	-30.3	-41.8

Speer 275-grain GDHP (23988)
G1 Ballistic Coefficient = 0.15

Distance • Yards	Muzzle	25	50	75	100	125	150	175	200
Velocity • fps	1450	1363	1284	1213	1152	1100	1056	1019	987
Energy • ft-lbs	1284	1135	1007	899	810	739	682	635	595
Taylor KO Index	27.1	25.1	24.0	22.6	21.5	20.5	19.7	19.0	18.4
Mid-Range Trajectory Height • Inches	0.0	0.1	0.6	1.4	2.7	4.4	6.7	9.6	13.1
Drop • Inches	0.0	-0.5	-2.2	-5.3	-9.8	-15.9	-23.8	-33.7	-45.7

Federal 275-grain Barnes Expander (P480XB1) LF
G1 Ballistic Coefficient = 0.15

Distance • Yards	Muzzle	25	50	75	100	125	150	175	200
Velocity • fps	1350	1270	1190	1130	1080	1036	1000	968	940
Energy • ft-lbs	1115	980	870	780	710	656	611	572	539
Taylor KO Index	25.2	23.7	22.2	21.1	20.2	19.3	18.7	18.1	17.5
Mid-Range Trajectory Height • Inches	0.0	0.2	0.7	1.6	3.1	5.0	7.6	10.9	14.9
Drop • Inches	0.0	-0.6	-2.6	-6.1	-11.3	-18.4	-27.4	-38.7	-52.3

Hornady 325-grain XTP-MAG (9138)
G1 Ballistic Coefficient = 0.1

Distance • Yards	Muzzle	25	50	75	100	125	150	175	200
Velocity • fps	1350	1265	1191	1129	1076	1033	997	965	936
Energy • ft-lbs	1315	1156	1023	919	835	771	717	672	633
Taylor KO Index	29.8	28.2	26.5	25.2	24.0	23.0	22.2	21.5	20.9
Mid-Range Trajectory Height • Inches	0.0	0.1	0.6	1.6	3.0	5.0	7.6	10.9	14.7
Drop • Inches	0.0	-0.6	-2.6	-6.1	-11.3	-18.4	-27.5	-38.7	-52.2

Speer 325-grain GDHP (23985)
G1 Ballistic Coefficient = 0.1

Distance • Yards	Muzzle	25	50	75	100	125	150	175	200
Velocity • fps	1350	1283	1223	1169	1122	1082	1047	1016	989
Energy • ft-lbs	1315	1189	1078	987	908	845	791	745	706
Taylor KO Index	29.8	28.3	27.0	25.8	24.7	23.9	23.1	22.4	21.8
Mid-Range Trajectory Height • Inches	0.0	0.2	0.7	1.6	2.9	4.8	7.2	10.1	13.8
Drop • Inches	0.0	-0.6	-2.6	-5.9	-10.9	-17.6	-26.2	-36.7	-49.4

Hornady 400-grain XTP-MAG (9144)
G1 Ballistic Coefficient = 0.1

Distance • Yards	Muzzle	25	50	75	100	125	150	175	200
Velocity • fps	1100	1061	1027	997	971	947	925	904	885
Energy • ft-lbs	1075	1000	937	884	838	797	760	726	696
Taylor KO Index	29.9	28.8	27.9	27.1	26.4	25.7	25.1	24.5	24.0
Mid-Range Trajectory Height • Inches	0.0	0.2	1.0	2.3	4.1	6.6	9.8	13.8	18.4
Drop • Inches	0.0	-0.9	-3.8	-8.7	-15.8	-25.2	-37.0	-51.3	-68.4

.50 Action Express

The .50 Action Express, which was introduced in 1991, is the king of the auto pistol hill. It packs about the same muzzle energy as a .30-30 with much higher Taylor KO Index values. This is because the .50 AE pushes a much heavier bullet but at a lower velocity. This is one of those calibers for the macho pistol shooter. Like the .500 Smith & Wesson Magnum and the .454 Casull, there are few shooters who can shoot the .50 AE really well for more than a few shots at a time. The .50 AE cartridge chambered into a carbine-style gun would make an interesting deer rifle for brushy hunting conditions where long shots aren't normally encountered.

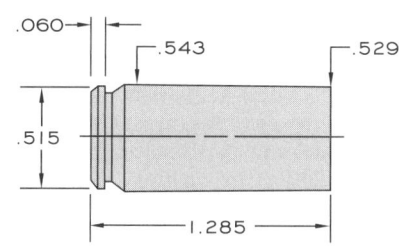

Relative Recoil Factor = 2.05

Specifications:

Controlling Agency for Standardization of this Ammunition: SAAMI

Bullet Weight Grains	Velocity fps	Maximum Average Pressure	
		Copper Crusher	Transducer
300 PHP	1,400	N/S	35,000 psi

Standard barrel for velocity testing: 6 inches long—1 turn in 20-inch twist

Availability:

Speer 300-grain GDHP (23995)

G1 Ballistic Coefficient = 0.155

Distance • Yards	Muzzle	25	50	75	100	125	150	175	200
Velocity • fps	1550	1452	1361	1278	1205	1143	1090	1047	1009
Energy • ft-lbs	1600	1404	1233	1089	968	870	792	730	679
Taylor KO Index	33.2	31.1	29.2	27.4	25.8	24.5	23.4	22.4	21.6
Mid-Range Trajectory Height • Inches	0.0	0.1	0.5	1.3	2.4	3.9	6.0	8.7	12.0
Drop • Inches	0.0	-0.5	-2.0	-4.7	-8.7	-14.2	-21.3	-30.3	-41.3

Hornady 300-grain XTP-HP (9245)

G1 Ballistic Coefficient = 0.121

Distance • Yards	Muzzle	25	50	75	100	125	150	175	200
Velocity • fps	1475	1358	1253	1167	1095	1040	995	956	923
Energy • ft-lbs	1449	1228	1046	907	799	721	659	609	567
Taylor KO Index	31.7	29.2	27.0	25.1	23.6	22.4	21.4	20.6	19.9
Mid-Range Trajectory Height • Inches	0.0	0.1	0.6	1.5	2.8	4.7	7.2	10.4	14.4
Drop • Inches	0.0	-0.5	-2.2	-5.3	-10.0	-16.5	-25.0	-35.7	-48.8

Speer 325-grain DCHP (3977)

G1 Ballistic Coefficient = 0.169

Distance • Yards	Muzzle	25	50	75	100	125	150	175	200
Velocity • fps	1450	1367	1291	1221	1162	1110	1067	1030	998
Energy • ft-lbs	1517	1348	1202	1077	974	889	821	765	718
Taylor KO Index	34.2	32.2	30.4	28.8	27.4	26.2	25.2	24.3	23.5
Mid-Range Trajectory Height • Inches	0.0	0.1	0.6	1.4	2.6	4.3	6.6	9.4	12.9
Drop • Inches	0.0	-0.5	-2.2	-5.2	-9.7	-15.8	-23.6	-33.4	-45.2

.500 Smith & Wesson Special

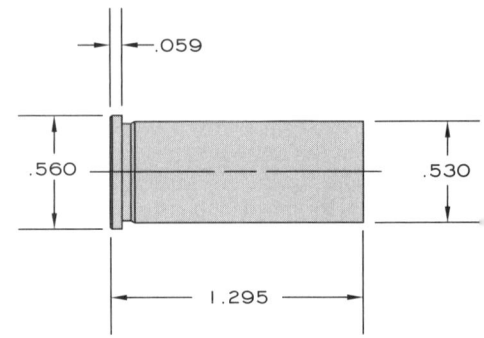

It's not a really new idea. When you have a revolver that is a bit much, using a shorter (and lower-power) cartridge is a simple way to make shooting the big gun a bit more pleasant. Cor-Bon has done this with the .500 Smith and Wesson Special, introduced in late 2006. The cartridge is 0.150 inches shorter than the .500 S&W Magnum. It is a bit nicer to shoot, but that's a relative thing. This is still a very potent cartridge as can be seen from the energy and Taylor Index values.

Relative Recoil Factor = 2.01

Specifications:

Controlling Agency for Standardization of this Ammunition: Factory

Bullet Weight Grains	Velocity fps	Maximum Average Pressure	
		Copper Crusher	Transducer

No standardization data available at this time.

Standard barrel for velocity testing: 10 inches long—1 turn in 18.75-inch twist

Availability:

Cor-Bon 275-grain DPX (DPX500S275/12)

G1 Ballistic Coefficient = 0.14

Distance • Yards	Muzzle	25	50	75	100	125	150	175	200
Velocity • fps	1250	1174	1111	1059	1016	980	948	920	894
Energy • ft-lbs	954	842	754	685	631	586	549	516	488
Taylor KO Index	24.6	23.1	21.8	20.8	20.0	19.3	18.6	18.1	17.6
Mid-Range Trajectory Height • Inches	0.0	0.2	0.8	1.9	3.5	5.8	8.7	12.4	16.8
Drop • Inches	0.0	-0.7	-3.0	-7.1	-13.1	-21.2	-31.5	-44.3	-59.7

Cor-Bon 350-grain JHP (HT500S350/12)

G1 Ballistic Coefficient = 0.15

Distance • Yards	Muzzle	25	50	75	100	125	150	175	200
Velocity • fps	1250	1179	1119	1069	1027	991	960	933	907
Energy • ft-lbs	1215	1080	973	888	820	764	717	676	640
Taylor KO Index	31.4	29.6	28.1	26.8	25.8	24.9	24.1	23.4	22.8
Mid-Range Trajectory Height • Inches	0.0	0.2	0.8	1.9	3.5	5.7	8.6	12.0	16.4
Drop • Inches	0.0	-0.7	-3.0	-7.0	-13.0	-20.9	-31.7	-43.7	-58.7

Cor-Bon 350-grain FMJ (HT500S350FMJ/12)

G1 Ballistic Coefficient = 0.1

Distance • Yards	Muzzle	25	50	75	100	125	150	175	200
Velocity • fps	1100	1053	1014	980	950	923	899	876	855
Energy • ft-lbs	941	862	799	747	702	663	628	596	568
Taylor KO Index	27.9	26.7	25.8	24.9	24.1	23.4	22.8	22.3	21.7
Mid-Range Trajectory Height • Inches	0.0	0.3	1.0	2.3	4.2	6.8	10.1	14.3	19.2
Drop • Inches	0.0	-0.9	-3.8	-8.8	-16.1	-25.7	-37.9	-52.8	-70.5

.500 Smith & Wesson Magnum

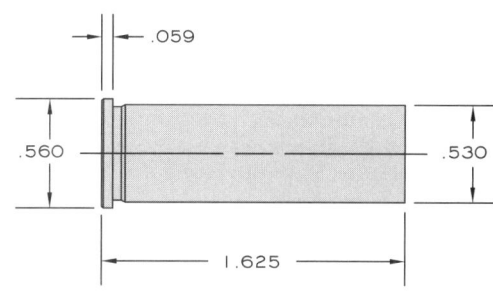

Introduced in 2004, this is the most potent handgun cartridge in the commercial inventory. By many accounts it is a beast to shoot, but those stories are a bit exaggerated. It isn't the thing to take for an afternoon's plinking, but if the idea of a hugely powerful handgun turns you on, this is the gun for you. A glance at the energy and Taylor numbers shows some loadings of this round to be more potent than the .308 Winchester. But the numbers don't matter a bit if you can't shoot the gun accurately; I wonder just how many shooters can. Like several other large, rimmed, pistol cartridges, the .500 S&W would make an excellent cartridge for lever-action carbines.

Relative Recoil Factor = 2.95

Specifications:

Controlling Agency for Standardization of this Ammunition: SAAMI

Bullet Weight Grains	Velocity fps	Maximum Average Pressure	
		Copper Crusher	Transducer
N/A	1,400	N/S	N/A

Standard barrel for velocity testing: 10 inches long—1 turn in 18.75-inch twist

Availability:

Cor-Bon 275-grain DPX (HT500SW275/20)

G1 Ballistic Coefficient = 0.140

Distance • Yards	Muzzle	25	50	75	100	125	150	175	200
Velocity • fps	1665	1548	1440	1341	1253	1176	1113	1061	1018
Energy • ft-lbs	1693	1465	1267	1099	959	845	756	687	633
Taylor KO Index	33.2	30.9	28.7	26.8	25.0	23.5	22.2	21.2	20.3
Mid-Range Trajectory Height • Inches	0.0	0.1	0.4	1.1	2.1	3.5	5.5	8.0	11.2
Drop • Inches	0.0	-0.4	-1.7	-4.1	-7.7	-12.6	-19.2	-27.4	-37.7

Magtech 275-grain SCHP (500C)

G1 Ballistic Coefficient = 0.120

Distance • Yards	Muzzle	25	50	75	100	125	150	175	200
Velocity • fps	1667	1555	1443	1354	1267	1192	1128	1075	1032
Energy • ft-lbs	1696	1477	1272	1120	964	868	778	706	650
Taylor KO Index	32.7	30.5	28.3	26.6	24.9	23.4	22.2	21.1	20.3
Mid-Range Trajectory Height • Inches	0.0	0.1	0.5	1.1	2.1	3.5	5.4	7.9	10.9
Drop • Inches	0.0	-0.4	-1.7	-4.1	-7.6	-12.5	-18.9	-27.0	-37.0

Federal 275-grain Barnes Expander (P500XB1) LF

G1 Ballistic Coefficient = 0.140

Distance • Yards	Muzzle	25	50	75	100	125	150	175	200
Velocity • fps	1660	1540	1440	1340	1250	1174	1110	1059	1016
Energy • ft-lbs	1680	1455	1255	1090	950	841	753	685	630
Taylor KO Index	32.6	30.3	28.3	26.3	24.6	23.1	21.8	20.8	19.9
Mid-Range Trajectory Height • Inches	0.0	0.1	0.5	1.1	2.1	3.6	5.5	8.1	11.2
Drop • Inches	0.0	-0.4	-1.7	-4.1	-7.7	-12.7	-19.3	-27.6	-37.9

Hornady 300-grain FTX (9249)

G1 Ballistic Coefficient = 0.200

Distance • Yards	Muzzle	25	50	75	100	125	150	175	200
Velocity • fps	1950	1857	1767	1681	1598	1518	1443	1373	1307
Energy • ft-lbs	2533	2298	2080	1882	1700	1536	1388	1256	1139
Taylor KO Index	42.5	40.4	38.5	36.6	34.8	33.0	31.4	29.9	28.5
Mid-Range Trajectory Height • Inches	0.0	0.0	0.3	0.7	1.4	2.3	3.5	5.0	6.9
Drop • Inches	0.0	-0.3	-1.2	-2.8	-5.2	-8.5	-12.7	-17.9	-24.3

Magtech 325-grain SJSP (500B)

G1 Ballistic Coefficient = 0.197

Distance • Yards	Muzzle	25	50	75	100	125	150	175	200
Velocity • fps	1801	1712	1626	1544	1467	1394	1326	1263	1206
Energy • ft-lbs	2341	2115	1909	1722	1553	1402	1268	1151	1050
Taylor KO Index	42.5	40.4	38.4	36.4	34.6	32.9	31.3	29.8	28.4
Mid-Range Trajectory Height • Inches	0.0	0.0	0.3	0.9	1.6	2.7	4.1	5.9	8.1
Drop • Inches	0.0	-0.4	-1.4	-3.3	-6.2	-10.0	-15.0	-21.1	-28.7

Federal 325-grain Swift A-Frame (P500SA)

G1 Ballistic Coefficient = 0.140

Distance • Yards	Muzzle	25	50	75	100	125	150	175	200
Velocity • fps	1800	1680	1560	1450	1350	1260	1183	1118	1065
Energy • ft-lbs	2340	2030	1755	1515	1315	1146	1010	902	819
Taylor KO Index	41.8	39.0	36.2	33.7	31.3	29.3	27.5	26.0	24.7
Mid-Range Trajectory Height • Inches	0.0	0.1	0.4	1.0	1.8	3.1	4.7	7.0	9.7
Drop • Inches	0.0	-0.4	-1.5	-3.5	-6.6	-10.8	-16.4	-23.6	-32.5

Magtech 325-grain SJSP – Flat Light Recoil (500L)

G1 Ballistic Coefficient = 0.197

Distance • Yards	Muzzle	25	50	75	100	125	150	175	200
Velocity • fps	1378	1311	1250	1195	1146	1103	1066	1034	1006
Energy • ft-lbs	1371	1241	1127	1030	948	879	821	772	731
Taylor KO Index	32.5	30.9	29.5	28.2	27.0	26.0	25.1	24.4	23.7
Mid-Range Trajectory Height • Inches	0.0	0.1	0.6	1.5	2.8	4.5	6.8	9.7	13.2
Drop • Inches	0.0	-0.6	-2.4	-5.7	-10.5	-16.9	-25.1	-35.2	-47.4

Cor-Bon 325-grain DPX (DPX500SW325/12)

G1 Ballistic Coefficient = 0.130

Distance • Yards	Muzzle	25	50	75	100	125	150	175	200
Velocity • fps	1800	1666	1541	1425	1321	1229	1151	1088	1037
Energy • ft-lbs	2338	2004	1714	1467	1259	1090	957	855	776
Taylor KO Index	41.8	38.7	35.8	33.1	30.7	28.5	26.7	25.3	24.1
Mid-Range Trajectory Height • Inches	0.0	0.1	0.4	1.0	1.9	3.2	4.9	7.4	10.2
Drop • Inches	0.0	-0.4	-1.5	-3.5	-6.7	-11.0	-16.8	-24.3	-33.6

Magtech 325-grain FMJ – Flat (500D)

G1 Ballistic Coefficient = 0.11

Distance • Yards	Muzzle	25	50	75	100	125	150	175	200
Velocity • fps	1801	1655	1520	1396	1286	1192	1115	1054	1005
Energy • ft-lbs	2341	1978	1667	1407	1194	1025	897	802	730
Taylor KO Index	41.8	38.4	35.3	32.4	29.9	27.7	25.9	24.5	23.3
Mid-Range Trajectory Height • Inches	0.0	0.1	0.4	1.0	1.9	3.3	5.1	7.6	10.7
Drop • Inches	0.0	-0.4	-1.5	-3.6	-6.8	-11.3	-17.4	-25.2	-34.9

Hornady 350-grain XTP MAG (9250)

G1 Ballistic Coefficient = 0.14

Distance • Yards	Muzzle	25	50	75	100	125	150	175	200
Velocity • fps	1700	1585	1478	1379	1289	1210	1143	1087	1041
Energy • ft-lbs	2246	1954	1697	1478	1291	1138	1015	919	842
Taylor KO Index	43.2	40.3	37.5	35.0	32.7	30.7	29.0	27.6	26.4
Mid-Range Trajectory Height • Inches	0.0	0.1	0.4	1.0	2.0	3.3	5.2	7.6	10.6
Drop • Inches	0.0	-0.4	-1.6	-3.9	-7.3	-12.0	-18.2	-26.1	-35.8

Cor-Bon 350-grain JHP (HT500SW350/12)

G1 Ballistic Coefficient = 0.1

Distance • Yards	Muzzle	25	50	75	100	125	150	175	200
Velocity • fps	1600	1492	1391	1300	1220	1151	1094	1047	1007
Energy • ft-lbs	1990	1730	1505	1314	1157	1030	930	852	788
Taylor KO Index	40.0	37.3	34.8	32.5	30.5	28.8	27.4	26.2	25.2
Mid-Range Trajectory Height • Inches	0.0	0.1	0.5	1.2	2.3	3.8	5.8	8.5	11.7
Drop • Inches	0.0	-0.4	-1.9	-4.4	-8.2	-13.5	-20.5	-29.2	-39.9

Winchester 350-grain Jacketed Hollow Point (X500SW)

G1 Ballistic Coefficient = 0.1

Distance • Yards	Muzzle	25	50	75	100	125	150	175	200
Velocity • fps	1350	1266	1192	1129	1077	1034	997	965	937
Energy • ft-lbs	1416	1245	1104	991	902	831	773	724	683
Taylor KO Index	33.6	31.7	29.8	28.2	26.9	25.9	24.9	24.1	23.4
Mid-Range Trajectory Height • Inches	0.0	0.2	0.7	1.6	3.1	5.1	7.7	11.0	14.9
Drop • Inches	0.0	-0.6	-2.6	-6.1	-11.3	-18.4	-27.5	-38.8	-52.4

Winchester 375-grain Dual Bond (S500SWDB)

G1 Ballistic Coefficient = 0.2

Distance • Yards	Muzzle	25	50	75	100	125	150	175	200
Velocity • fps	1725	1648	1573	1502	1435	1371	1311	1256	1206
Energy • ft-lbs	2477	2261	2061	1879	1713	1565	1432	1313	1211
Taylor KO Index	46.2	44.1	42.1	40.2	38.4	36.7	36.1	33.6	32.3
Mid-Range Trajectory Height • Inches	0.0	0.1	0.4	1.0	1.8	2.9	4.4	6.3	8.5
Drop • Inches	0.0	-0.4	-1.6	-3.6	-6.6	-10.7	-15.9	-22.4	-30.3

Cor-Bon 385-grain BC (HT500SW385/12)

G1 Ballistic Coefficient = 0.150

Distance • Yards	Muzzle	25	50	75	100	125	150	175	200
Velocity • fps	1700	1589	1485	1388	1300	1222	1155	1099	1052
Energy • ft-lbs	2471	2159	1885	1648	1446	1278	1141	1033	947
Taylor KO Index	46.9	43.9	41.0	38.3	35.9	33.7	31.9	30.3	29.0
Mid-Range Trajectory Height • Inches	0.0	0.1	0.4	1.1	2.0	3.3	5.1	7.5	10.4
Drop • Inches	0.0	-0.4	-1.6	-3.9	-7.3	-11.9	-18.0	-25.8	-35.3

Magtech 400-grain SJSP (500A)

G1 Ballistic Coefficient = 0.242

Distance • Yards	Muzzle	25	50	75	100	125	150	175	200
Velocity • fps	1608	1542	1478	1418	1361	1307	1257	1211	1169
Energy • ft-lbs	2297	2112	1942	1786	1645	1517	1403	1302	1214
Taylor KO Index	46.7	44.8	42.9	41.2	39.5	37.9	36.5	35.2	33.9
Mid-Range Trajectory Height • Inches	0.0	0.1	0.4	1.0	2.0	3.2	4.9	6.9	9.4
Drop • Inches	0.0	-0.4	-1.8	-4.1	-7.5	-12.1	-18.0	-25.2	-34.0

Winchester 400-grain Platinum Tip (S500PTHP)

G1 Ballistic Coefficient = 0.226

Distance • Yards	Muzzle	25	50	75	100	125	150	175	200
Velocity • fps	1675	1602	1531	1464	1400	1340	1284	1232	1185
Energy • ft-lbs	2491	2279	2092	1903	1741	1595	1464	1348	1249
Taylor KO Index	47.9	45.8	43.7	41.8	40.0	38.3	36.7	35.2	33.9
Mid-Range Trajectory Height • Inches	0.0	0.1	0.4	1.0	1.9	3.1	4.6	6.6	9.0
Drop • Inches	0.0	-0.4	-1.6	-3.8	-7.0	-11.3	-16.8	-23.6	-31.9

Cor-Bon 400-grain JSP (HT500SW400SP/12)

G1 Ballistic Coefficient = 0.154

Distance • Yards	Muzzle	25	50	75	100	125	150	175	200
Velocity • fps	1625	1521	1424	1335	1255	1185	1125	1075	1034
Energy • ft-lbs	2346	2056	1802	1584	1400	1247	1125	1027	949
Taylor KO Index	46.4	43.5	40.7	38.1	35.9	33.9	32.1	30.7	29.5
Mid-Range Trajectory Height • Inches	0.0	0.1	0.5	1.1	2.2	3.6	5.5	8.1	11.1
Drop • Inches	0.0	-0.4	-1.8	-4.2	-7.9	-12.9	-19.5	-27.8	-38.0

Cor-Bon 440-grain HC (HT500SW440HC/12)

G1 Ballistic Coefficient = 0.160

Distance • Yards	Muzzle	25	50	75	100	125	150	175	200
Velocity • fps	1625	1525	1431	1345	1267	1197	1138	1087	1045
Energy • ft-lbs	2580	2272	2002	1767	1568	1401	1265	1156	1068
Taylor KO Index	51.1	47.9	45.0	42.3	39.8	37.5	35.8	34.2	32.8
Mid-Range Trajectory Height • Inches	0.0	0.1	0.5	1.1	2.1	3.5	5.4	7.9	10.9
Drop • Inches	0.0	-0.4	-1.8	-4.2	-7.8	-12.8	-19.3	-27.5	-37.5

Hornady 500-grain FT – XTP (9252)

G1 Ballistic Coefficient = 0.185

Distance • Yards	Muzzle	25	50	75	100	125	150	175	200
Velocity • fps	1300	1236	1179	1126	1087	1050	1018	990	964
Energy • ft-lbs	1876	1698	1543	1416	1310	1224	1151	1088	1033
Taylor KO Index	47.2	44.8	42.8	41.0	39.4	38.1	36.9	35.9	35.0
Mid-Range Trajectory Height • Inches	0.0	0.1	0.7	1.7	3.1	5.1	7.6	10.8	14.8
Drop • Inches	0.0	-0.7	-2.8	-6.4	-11.8	-18.9	-28.1	-39.4	-52.8

Cor-Bon 500-grain HC (HT500SW500HC/12)

G1 Ballistic Coefficient = 0.175

Distance • Yards	Muzzle	25	50	75	100	125	150	175	200
Velocity • fps	1500	1416	1338	1266	1203	1147	1099	1059	1024
Energy • ft-lbs	2499	2226	1987	1781	1606	1461	1342	1245	1165
Taylor KO Index	53.8	50.8	48.0	45.4	43.1	41.1	39.4	38.0	36.7
Mid-Range Trajectory Height • Inches	0.0	0.1	0.5	1.3	2.4	4.0	6.1	8.8	12.1
Drop • Inches	0.0	-0.5	-2.1	-4.9	-9.0	-14.7	-22.0	-31.1	-42.2

Stopping Power

What You Need To Know

by Dave Spaulding

A bullet removed at autopsy is the best indicator of performance in the author's opinion. This 127 grain +P+ 9mm was removed from an arm robber at autopsy.

I have long been interested in the topic of handgun stopping power. For almost thirty years, I've read everything I could on the topic. I've gone to autopsies, spoken with coroners and emergency-room physicians, interviewed people who had shot others in self-defense, and pursued shooting reports from many law-enforcement agencies. I even wrote my master's thesis on the topic, which was a huge mistake. After all, part of a thesis is to draw a conclusion and defend it. How do you defend something that seems to defy a definitive pattern?

My agency allowed me to use department letterhead to solicit shooting reports for my thesis, and major agencies across the country were very good to me. I received written reports, including autopsy data, from across the nation. The only

thing these agencies requested in return were r results. And what did I discover or uncover? No darn thing! For every shooting in which a particul caliber and ammo combination worked well, I fou another in which it failed. After all of the time a effort I spent on the subject, you'd think I'd have good handle on the subject, but I don't.

Many look to the ballistics laboratory for definitive answer, but a human's resilience can't re-created in a block of ballistic gelatin. Keep in mi that gelatin is calibrated against swine tissue, a even though swine tissue is very similar to a humar it is not the same. Furthermore, no human being a 100 percent consistent, homogeneous substan We are muscle, fat, bone, and so forth. What definitive regarding the subject? We know handg

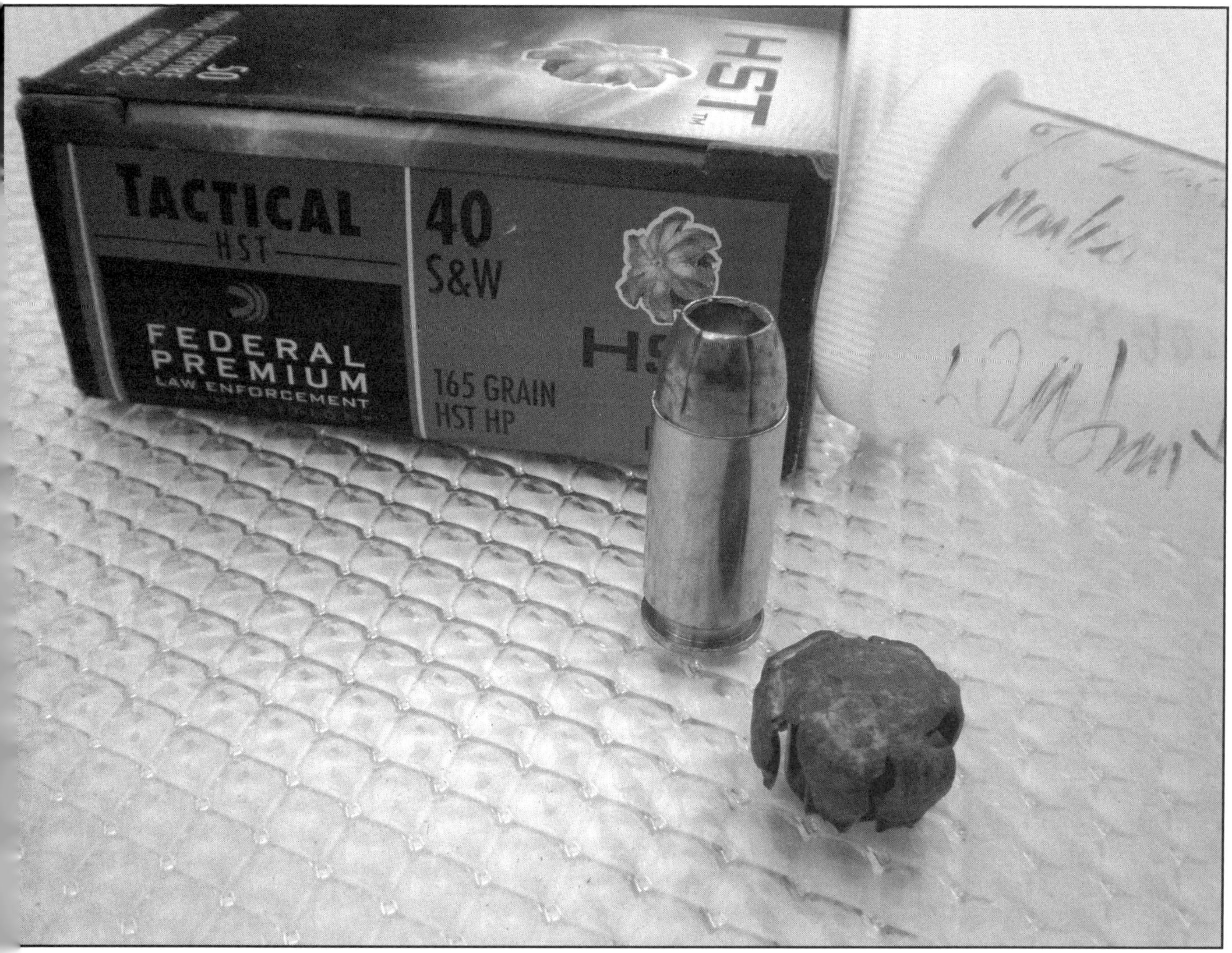

...e Federal HST series of ammo is excellent, as this 165-grain .40 reveals. This round was also removed at autopsy.

...mmo is not long gun ammo. Rifle ammo travels ...uch faster than does handgun ammo, creating a ...drostatic shock wave—an effect that can damage ...sue beyond the area the round actually comes in ...ntact with. Shotguns launch rounds that strike ...e body with multiple projectiles at once, creating ... overload shock to the system. Or they can deliver ...e large, heavy chunk of lead weighed in ounces ...stead of grains. Multiple hits at once or overly large ...ojectiles cause incapacitation, to say the least.

...Unfortunately, no handgun round will perform ...this level. Typical handgun ammunition must ...ntact tissue or a vital organ in order to disrupt or ...mage it, which means that for a handgun projectile ...cause physiological incapacitation, it must hit ...nething important. "Something important" ...ludes the brain, heart, or major vessels that will ...idly leak blood and lower blood pressure, which

While few shootings have been recorded with Corbon DPX, it has all the right "tools" for a superior combative load.

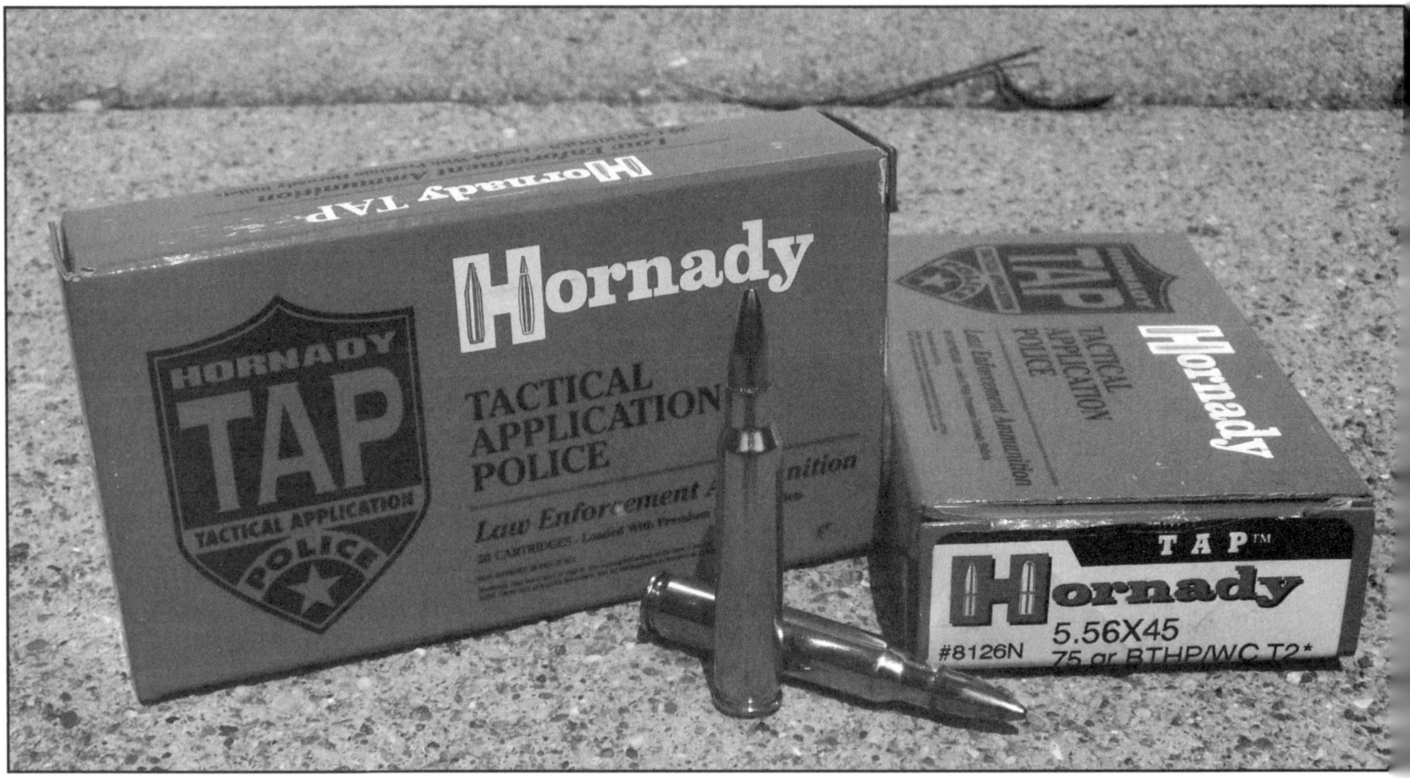

Rifle ammunition is not pistol ammo and the two cannot be viewed the same.

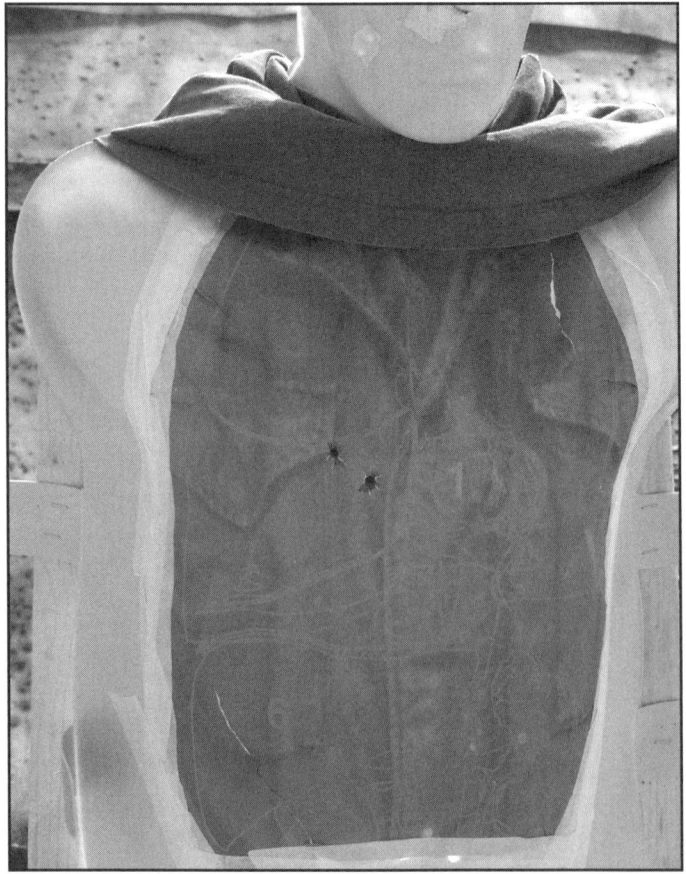

Shot placement always has . . . and always will be the single best indicator of handgun stopping-power potential.

will eventually result in loss of consciousness-"eventually" being the operative word. This mean pistol fire must be more accurate than rifle fir which is a tall order considering the handgun short barrel, limited sight radius, and fewer poin of body contact that help stabilize the gun when i fired. The truth is the handgun is portable but n really effective. Some say to "use the handgun fight your way to a long gun," but in the time fran of a typical armed confrontation, how likely is tha And don't expect the handgun to work with only o or two rounds, regardless of caliber and especia against an opponent high on drugs or pumped f of adrenaline.

What about caliber? Is one better than anothe Common sense dictates a bigger bullet is a bett bullet, but a bigger bullet is also a heavier bull which means it will deliver more felt recoil to t shooter. Yes, recoil can be controlled with prop training, but how often do you train? I know th the majority of law enforcement officers train w their sidearms only when required, but what abc the legally armed citizen? Even if he or she wa to train more often, gun clubs and public ran; are justifiably concerned with liability and saf issues, and that limits what skills a person c practice. As a general rule, rapid fire, draw from the holster, movement, and other related skills are banned. Most of the time, the gun is lifted, fired slowly, and then "benched" in the interest of safety. What good is that? This being the case, we need to take recoil into account. Multiple shots that miss the target are like not shooting at all, and additionally these misses pose the risk of traveling down the street and hitting an innocent bystander or passing through a wall and injuring a neighbor. In the eyes of the community, those unfortunate errant rounds will be the only rounds fired—regardless of how dire the situation was for the shooter.

Can added speed make a smaller bullet more effective? Sure it can. Picture a pickup truck hitting a wall at 25 mph and a VW Beetle hitting the same wall at 50 mph. A smaller, faster bullet can do considerable damage, but only if it slows inside the body to deliver its energy to the surrounding tissue. If you're restricted to using full-metal-jacket ammo (as the military is), using a fat, slow-moving bullet makes more sense because it has a greater chance

of slowing in the body, thereby "dumping" energy and reducing overpenetration hazards. A smaller projectile traveling at great speed can zip in and out of the body like someone stabbing tissue with an ice pick, damaging very little along the way. Such a

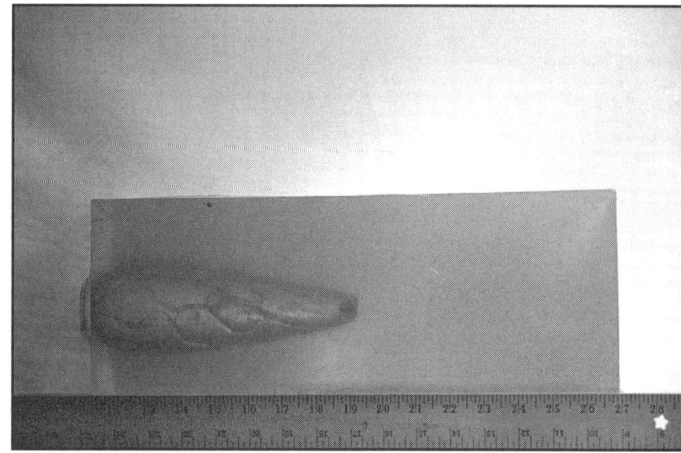

Ballistic gelatin is an excellent "apples to apples" test, but the consistent substance is not the same as the inside of a human body.

author likes to use clothing and meat-covered torsos to test ammo. While not perfect, the tests reveal some noteworthy results.

bullet must be designed to deform, stop, or at least slow dramatically before exiting the torso.

Take the 9mm, for example. It can be a good choice for some, but history has clearly shown that it is load dependent—requiring a hollowpoint bullet of reasonable weight driven at a high velocity. For many, the 9mm is a wise option due to the reduced recoil it offers over the larger bullets of calibers .40 and higher, but only with carefully selected ammunition.

What if we remove the human element—one's ability to hit while in conflict—and look at the science alone? Compare the best 9mm load with the best .45, and you will find the final wound cavity is between 15 to 20 percent larger for the .45. This would settle the argument that a bigger bullet is better than a smaller bullet when it comes to doing damage. Now for the bad news: This 15 to 20 percent is not enough to make up for poor shot placement. Remember, the larger and heavier the bullet, the more felt recoil will be generated. This will result in greater shot-to-shot time unless you are well trained and practice regularly. Only the individual can decide if this added felt recoil is important when selecting a carry gun. So you see, when it comes to a realistic discussion of handgun stopping power, the human element can't be removed.

As I mentioned earlier, I began to collect shooting reports as part of my master's thesis, and I have continued to do so over the years. It was never my intention to use them to create some type of stopping power "formula"; I merely wanted to know what worked and what didn't. So please do not read this as "Dave Spaulding's Manifesto on Bullet

"Backyard" ballistic testing is a worthwhile endeavor as long as the tester understands he/she is comparing one bullet to another. (See top picture.) It not help determine how a bullet will interact with tissue (above).

Top Stopping Power Choices

Here are the handgun ammunition formulas I've seen work well in the street over time.

.38 Special 158-grain all-lead hollowpoint. Made by all of the major manufacturers, this may be the most proven load in law-enforcement history. Another good choice is the Speer 135-grain Gold Dot hollowpoint +P as specified by the NYPD for the two-inch revolver. Available commercially, this round has proven to be very good in a number of shootings.

9mm 124- to 127-grain hollowpoint loaded to 1,250 fps approximately. This bullet is heavy enough to penetrate deeply and moves fast enough to ensure expansion (at least as much as anything like this can be assured) and deliver energy to the body.

The most proven 9mm load: the Winchester 127-grain +P+ SXT hollowpoint moving at 1,300 fps. This is a "police-only" load, but I have seen it sold online and at gun shows. Friends in the NYPD tell me they are very happy with their Speer 124-grain +P Gold Dot hollowpoint duty load while Winchester's newer 124-grain +P hollowpoint is also doing well in actual shootings and is for sale to the public. Federal offers its HST hollowpoint in a 124-grain +P version, which has also done well.

.40 S&W 155- to 165-grain jacketed hollowpoint moving at 1,100 fps or faster. All of the major manufacturers make a load like this. The U.S. Border Patrol has conducted extensive testing with this load using both the Federal and Remington versions in the lab and have had good results on the street—and in the desert, which is good to know considering the considerable threat these men and women face daily. Speer Gold Dot and Federal HST appear to lead the way.

.357 SIG 125-grain hollowpoint moving at 1,350 fps. The Speer Gold Dot and Federal JHP loads are the most proven in the field. Others are available, but I do not have any hard data on them.

.45 ACP 200- to 230-grain JHP. The Federal 230-grain Hydra Shok hollowpoint is the most proven. The Remington 230-grain Golden Sabre hollowpoint and Speer 200- and 230-grain Gold Dot hollowpoints have also worked very well.

fectiveness." It is not, and I wouldn't even try formulate such a document. What I have seen er the years, though, is that certain "formulas" of ndgun ammunition work better than others, and ese are listed in the accompanying sidebar. These nds or formulas have proven successful time d again. No, they are not absolutes as there are t too many outside factors that skew any type of apacitation prediction. However, if the shooter es his or her job and places the round in a vital a, these formulas have proven to work.

In addition to these formulas, my research and perience has led me to draw some conclusions arding caliber and ammunition choices. First, ing on anything smaller than a .38 Special is less n wise. Yes, I know that the .380 is enjoying a high l of popularity, and there are a few good ammo ices in this caliber. But while smaller calibers certainly prove lethal, reports have shown they

cause less incapacitation than those of .38 Special caliber or larger. People shot with a good quality .38 (and larger) rounds tend to react to their impact: They notice. Those shot with lesser loadings may not even know they are hit unless the bullet violates a vital organ. I have seen suspects shot with .22, .25, and .32 rounds and stay mobile as if nothing has happened. Over the years, I've seen a number of law-enforcement agencies change ammunition because of a single incident, which is not a good idea if the original-issue ammo was extensively tested and evaluated. No single event is a good indicator of overall performance because, again, there are just too many factors involved in any shooting. History is the best indicator of future performance.

What does it all mean? Ammunition that fails to stop an adversary is usually due to poor shot placement and not poor ammo performance. Expansion is a means to an end—not the end in itself. You're better off with a bullet that failed to expand but hit a vital organ than to have a fully expanded bullet that did not hit anything important. Remember, in order to be effective, handgun bullets must strike a vital location, and no amount of expansion or bullet diameter will change this. The best thing anyone can do to ensure his or her ammo incapacitates a dangerous attacker is to train hard to hit vital areas of the body while bobbing, moving,

and weaving as would happen in a real fight. This means training time and ammunition, and no super-duper, thermonuclear, +P+, thunder-flash, wonder hollowpoint will make up for this.

It also means using targets that represent real human anatomy and how the vital zones will change as the body twists and turns. It is wise to remember that an opponent will not always be facing you head on as targets do on the shooting range. As Col. Dave Grossman so wisely stated, "Amateurs talk hardware (equipment) while professionals talk software (training and mindset)." If you don't believe this is the case, take a few moments and look at some of the gun forums on the Internet. You can't help but notice that the threads related to gear are full of commentary and debate while the training threads remain largely unused. Let's face it: Gear is fun while training, and practice requires time and effort.

In the end, it's all about having the correct information—no "fables" or individual bias—and then making an informed decision. One of my favorite stories comes from Bert DuVernay, chief of police in New Braintree, Massachusetts, and the former director of the Smith & Wesson Academy. Bert had the opportunity to speak with Dr. Vincent DiMaio, one of the nation's foremost experts on wound ballistics, in regards to what constituted handgun stopping power. What Dr. DiMaio told him was the best piece of advice on the subject that I have ever heard: "The secret to handgun stopping power remains where you shoot your opponent and how many times you shoot them."

A training target needs to restrict the score zone to the vital areas and display the various positions a human body can be in during conflict.

Data For Rimfire Cartridges

This section covers rimfire ammunition in its various forms. While some of this ammunition is used primarily in pistols, the SAAMI specifications for all .22 Rimfire variations call for testing in 24-inch barrels. Some manufacturers also report performance in pistol-length barrels, but because these data are somewhat unstandardized in terms of barrel lengths used, the results can't always be used for detailed comparison. In the final analysis, if knowing the exact velocity is really critical to your shooting application, you have to chronograph a sample of the ammunition in question in your own gun.

There's an interesting situation regarding some brands of match and training ammunition. These manufacturers make their match and training ammo on the same production line, using identical components, the same everything. In fact, they don't know which grade they are making until it is tested, then the best becomes the match grade. Quite often the ammo coming off the production line is so good that there's more match-quality ammo than can be sold so the excess supply simply becomes the training grade. It could be just as good as the match-quality grade, but, of course, there's no guarantee. There is at least one manufacturer that makes three grades on the same production line.

.17 Aguila

Small calibers have always held a special fascination with shooters. It was pretty much a sure thing that when the .17 HMR took hold that a .17 caliber rimfire based on the .22 Long Rifle case was not going to be far behind. Early in 2005 there were actually two slightly different, new .17s introduced. This version was initially developed by Aguila. It is based on the standard length .22 LR case. With a 20-grain jacketed bullet at or in excess of 1,800 fps, it provides a flat shooting, light varmint capability out to at least 100 yards. This round can be adapted to conversions of Ruger's 10-22 semiauto with nothing more than a switch of barrel (a five-minute job.)

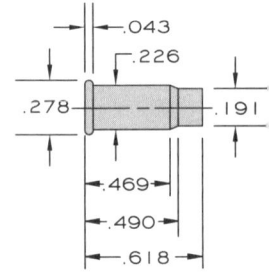

Relative Recoil Factor = 0.16

Specifications:

Controlling Agency for Standardization of this Ammunition: SAAMI (Pending)

Bullet Weight Grains	Velocity fps	Maximum Average Pressure	
		Copper Crusher	Transducer
No data available.			

Standard barrel for velocity testing: 24 inches long—1 turn in 16-inch twist

Availability:

Aguila 20-grain Jacketed Soft Point (Discontinued in 2007)

G1 Ballistic Coefficient = 0.1(

Distance • Yards	Muzzle	25	50	75	100	125	150	175	200
Velocity • fps	1850	1682	1527	1387	1267	1163	1084	1023	975
Energy • ft-lbs	152	126	104	86	71	60	52	46	42
Path • Inches	-1.5	+0.2	+1.1	+1.1	0.0	-2.5	-6.6	-12.5	-20.6
Wind Drift • Inches	0.0	0.4	1.5	3.4	6.2	10.0	14.6	20.0	26.1

Prairie dogs are a handful for most rimfires, but if shots are placed well they succumb very quickly. (Safari Press Archives)

.17 Mach 2

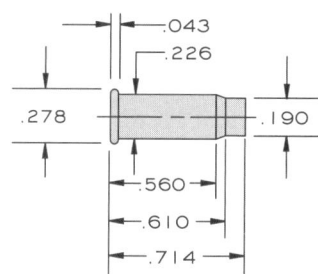

In 2005, Hornady followed its .17 HMR with the .17 Mach 2. This is another .17 rimfire based on the .22 Long Rifle case (well sort of.) This particular round seems to be based on the Stinger case, which is slightly longer than the regular .22 LR. Overall length is the same as for .22 LR ammo so gun conversion problems are somewhat minimized. Not all blowback-style .22 semiauto rifles can be converted to this cartridge without altering the bolt or mainspring along with a new barrel. That's not surprising since this round is somewhat more energetic than the .17 Aguila.

Relative Recoil Factor = 0.17

Specifications:

Controlling Agency for Standardization of this Ammunition: SAAMI (Pending)

Bullet Weight Grains	Velocity fps	Maximum Average Pressure	
		Copper Crusher	Transducer
N/S	N/S	N/S	24,000 psi

Standard barrel for velocity testing: 24 inches long—1 turn in 16-inch twist

Availability:

Hornady 15.5-grain NTX (83176) LF

G1 Ballistic Coefficient = 0.115

Distance • Yards	Muzzle	25	50	75	100	125	150	175	200
Velocity • fps	2050	1885	1729	1584	1450	1330	1225	1139	1071
Energy • ft-lbs	149	122	103	86	75	61	52	45	41
Path • Inches	-1.5	-0.1	+0.7	+0.8	0.0	-1.8	-4.9	-9.4	-15.6
Wind Drift • Inches	0.0	0.3	1.2	2.7	5.0	8.0	12.0	16.7	22.2

Eley 17-grain Hunting (None)

G1 Ballistic Coefficient = 0.125

Distance • Yards	Muzzle	25	50	75	100	125	150	175	200
Velocity • fps	2100	1946	1801	1660	1532	1411	1304	1212	1134
Energy • ft-lbs	166	143	122	104	88	75	64	55	49
Path • Inches	-1.5	-0.2	+0.7	+0.7	0.0	-1.6	-4.4	-8.3	-13.8
Wind Drift • Inches	0.0	0.2	1.0	2.4	4.4	7.1	10.3	14.7	19.7

CCI 17-grain V-MAX (0048)

G1 Ballistic Coefficient = 0.128

Distance • Yards	Muzzle	25	50	75	100	125	150	175	200
Velocity • fps	2010	1863	1724	1594	1471	1360	1262	1178	1108
Energy • ft-lbs	152	131	112	96	82	70	60	52	46
Path • Inches	-1.5	-0.1	+0.7	+0.8	0.0	-1.8	-4.8	-9.1	-15.0
Wind Drift • Inches	0.0	0.2	1.0	2.4	4.5	7.3	10.8	15.1	20.0

Hornady 17-grain V-Max (83177)

G1 Ballistic Coefficient = 0.125

Distance • Yards	Muzzle	25	50	75	100	125	150	175	200
Velocity • fps	2100	1946	1799	1660	1530	1411	1304	1212	1134
Energy • ft-lbs	166	143	122	104	88	75	64	55	49
Path • Inches	-1.5	-0.2	+0.7	+0.7	0.0	-1.6	-4.4	-8.3	-13.8
Wind Drift • Inches	0.0	0.2	1.0	2.4	4.4	7.1	10.3	14.7	19.7

CCI 17-grain V-Max (0048)

G1 Ballistic Coefficient = 0.148

Distance • Yards	Muzzle	25	50	75	100	125	150	175	200
Velocity • fps	2010	1883	1759	1645	1535	1434	1341	1257	1184
Energy • ft-lbs	153	134	117	102	89	78	68	60	53
Path • Inches	-1.5	-0.1	+0.7	+0.7	0.0	-1.6	-4.4	-8.3	-13.6
Wind Drift • Inches	0.0	0.2	0.9	2.1	3.8	6.2	9.1	12.7	17.0

.17 Hornady Magnum Rimfire (HMR)

Hornady introduced its first entry into the rimfire field in 2002 with a .17 caliber that fires a jacketed, ballistic-tip bullet. Performance is not unlike the wildcat centerfires based on the .22 Hornet case. The ballistic performance is good. The fact that there are so many loadings available today illustrates the growing popularity of this little screamer.

Relative Recoil Factor = 0.19

Specifications:

Controlling Agency for Standardization of this Ammunition: SAAMI (Pending)

Bullet Weight Grains	Velocity fps	Maximum Average Pressure	
		Copper Crusher	Transducer
N/S	N/S	N/S	24,000 psi

Standard barrel for velocity testing: 24 inches long—1 turn in 16-inch twist

Availability:

Winchester 15.5-grain NTX (S17HMR1LF) LF
G1 Ballistic Coefficient = 0.12

Distance • Yards	Muzzle	25	50	75	100	125	150	175	200
Velocity • fps	2550	2378	2212	2053	1901	1757	1620	1494	1378
Energy • ft-lbs	224	195	168	145	124	106	90	77	65
Path • Inches	-1.5	-0.5	+0.2	+0.3	0.0	-0.9	-2.6	-5.0	-8.5
Wind Drift • Inches	0.0	0.2	0.8	1.8	3.3	5.3	8.0	11.3	15.3

Hornady 15.5-grain NTX (83171) LF
G1 Ballistic Coefficient = 0.1

Distance • Yards	Muzzle	25	50	75	100	125	150	175	200
Velocity • fps	2525	2339	2161	1991	1829	1677	1535	1406	1291
Energy • ft-lbs	219	188	161	136	115	97	81	68	59
Path • Inches	-1.5	-0.5	+0.2	+0.4	0.0	-1.0	-2.8	-5.5	-9.4
Wind Drift • Inches	0.0	0.2	0.8	2.0	3.7	6.0	9.0	12.8	17.3

CCI 17-grain JHP (0053)
G1 Ballistic Coefficient = 0.1

Distance • Yards	Muzzle	25	50	75	100	125	150	175	200
Velocity • fps	2550	2375	2199	2046	1892	1745	1608	1480	1364
Energy • ft-lbs	246	213	184	158	135	115	98	83	70
Path • Inches	-1.5	-0.5	+0.2	+0.3	0.0	-0.9	-2.6	-5.1	-9.9
Wind Drift • Inches	0.0	0.2	0.8	1.8	3.4	5.4	8.1	11.5	15.6

CCI 17-grain V-MAX (0049)
G1 Ballistic Coefficient = 0.1

Distance • Yards	Muzzle	25	50	75	100	125	150	175	200
Velocity • fps	2550	2382	2220	2064	1915	1773	1639	1514	1399
Energy • ft-lbs	245	214	186	161	138	119	101	87	74
Path • Inches	-1.5	-0.5	+0.1	+0.3	0.0	-0.9	-2.5	-5.0	-8.3
Wind Drift • Inches	0.0	0.2	0.7	1.7	3.2	5.2	7.8	11.0	14.8

Hornady 17-grain V-Max Bullet
G1 Ballistic Coefficient = 0.

Distance • Yards	Muzzle	25	50	75	100	125	150	175	200
Velocity • fps	2550	2378	2212	2053	1902	1757	1621	1494	1380
Energy • ft-lbs	245	213	185	159	136	116	99	84	72
Path • Inches	-1.5	-0.5	+0.2	+0.3	0.0	-0.9	-2.6	-5.0	-8.5
Wind Drift • Inches	0.0	0.2	0.8	1.8	3.3	5.3	8.0	11.3	15.3

Remington 17-grain AccuTip-V (PR17HM1)
G1 Ballistic Coefficient = 0

Distance • Yards	Muzzle	25	50	75	100	125	150	175	200
Velocity • fps	2550	2378	2212	2053	1901	1757	1620	1494	1378
Energy • ft-lbs	245	213	185	159	136	117	99	84	72
Path • Inches	-1.5	-0.5	+0.1	+0.3	0.0	-0.9	-2.6	-5.0	-8.5
Wind Drift • Inches	0.0	0.2	0.8	1.8	3.3	5.3	8.0	11.3	15.3

Winchester 17-grain Poly Tip V-Max (S17HMR)

G1 Ballistic Coefficient = 0.128

Distance • Yards	Muzzle	25	50	75	100	125	150	175	200
Velocity • fps	2550	2382	2220	2064	1915	1773	1639	1514	1399
Energy • ft-lbs	245	214	186	161	138	119	101	86	74
Path • Inches	-1.5	-0.5	+0.1	+0.3	0.0	-0.9	-2.5	-4.9	-8.3
Wind Drift • Inches	0.0	0.2	0.7	1.6	2.9	4.7	7.0	9.9	13.4

Federal 17-grain Hornady V-Max (P771)

G1 Ballistic Coefficient = 0.125

Distance • Yards	Muzzle	25	50	75	100	125	150	175	200
Velocity • fps	2530	2359	2194	2036	1884	1741	1606	1480	1366
Energy • ft-lbs	240	210	182	156	134	114	97	83	70
Path • Inches	-1.5	-0.5	+0.2	+0.3	0.0	-1.0	-2.6	-5.2	-8.7
Wind Drift • Inches	0.0	0.2	0.7	1.6	3.0	4.9	7.3	10.3	14.0

CCI 17-grain TNT HP (0053)

G1 Ballistic Coefficient = 0.100

Distance • Yards	Muzzle	25	50	75	100	125	150	175	200
Velocity • fps	2550	2336	2132	1938	1758	1588	1435	1300	1187
Energy • ft-lbs	245	206	172	142	116	95	78	64	53
Path • Inches	-1.5	-0.4	+0.2	+0.4	0.0	-1.1	-3.1	-6.1	-10.4
Wind Drift • Inches	0.0	0.2	1.0	2.3	4.3	7.0	10.6	15.1	20.5

Federal 17-grain Speer TNT JHP (P770)

G1 Ballistic Coefficient = 0.110

Distance • Yards	Muzzle	25	50	75	100	125	150	175	200
Velocity • fps	2530	2331	2150	1973	1805	1648	1502	1371	1256
Energy • ft-lbs	240	206	175	147	123	102	85	71	60
Path • Inches	-1.5	-0.4	+0.2	+0.4	0.0	-1.0	-2.9	-5.7	-9.7
Wind Drift • Inches	0.0	0.2	0.8	1.9	3.5	5.7	8.5	12.1	16.5

CCI 17-grain TNT Green HP (0951) LF

G1 Ballistic Coefficient = 0.090

Distance • Yards	Muzzle	25	50	75	100	125	150	175	200
Velocity • fps	2500	2265	2043	1835	1642	1466	1313	1186	1090
Energy • ft-lbs	222	182	148	120	96	76	61	50	42
Path • Inches	-1.5	-0.4	+0.3	+0.5	0.0	-1.3	-3.6	-7.2	-12.3
Wind Drift • Inches	0.0	0.3	1.1	2.7	5.0	8.2	12.5	17.8	24.1

CCI 20-grain Game Point (0052)

G1 Ballistic Coefficient = 0.125

Distance • Yards	Muzzle	25	50	75	100	125	150	175	200
Velocity • fps	2375	2210	2051	1898	1754	1618	1492	1376	1274
Energy • ft-lbs	250	217	187	160	137	116	99	84	72
Path • Inches	-1.5	-0.4	+0.3	+0.4	0.0	-1.2	-3.1	-6.1	-10.2
Wind Drift • Inches	0.0	0.2	0.8	2.0	3.7	5.9	8.9	12.5	17.0

Hornady 20-grain XTP (833172)

G1 Ballistic Coefficient = 0.130

Distance • Yards	Muzzle	25	50	75	100	125	150	175	200
Velocity • fps	2375	2216	2063	1916	1776	1644	1520	1406	1304
Energy • ft-lbs	250	218	189	163	140	120	103	88	75
Path • Inches	-1.5	-0.4	+0.3	+0.4	0.0	-1.1	-3.1	-5.9	-9.9
Wind Drift • Inches	0.0	0.2	0.8	1.9	3.5	5.7	8.5	11.9	16.1

Winchester 20-grain JHP (X17HMR2)

G1 Ballistic Coefficient = 0.130

Distance • Yards	Muzzle	25	50	75	100	125	150	175	200
Velocity • fps	2375	2216	2063	1916	1776	1644	1520	1406	1304
Energy • ft-lbs	250	218	189	163	140	120	103	88	66
Path • Inches	-1.5	-0.4	+0.3	+0.4	0.0	-1.1	-3.1	-5.9	-9.9
Wind Drift • Inches	0.0	0.2	0.7	1.7	3.1	5.1	7.6	10.7	14.5

CCI 20-grain TMJ (0055)

G1 Ballistic Coefficient = 0.130

Distance • Yards	Muzzle	25	50	75	100	125	150	175	200
Velocity • fps	2375	2216	2063	1915	1776	1644	1520	1406	1304
Energy • ft-lbs	250	218	189	163	140	120	103	88	76
Path • Inches	-1.5	-0.4	+0.3	+0.4	0.0	-1.1	-3.1	-5.9	-9.9
Wind Drift • Inches	0.0	0.2	0.8	1.9	3.5	5.7	8.5	11.9	16.1

5mm Remington Rimfire Magnum

Remington introduced two new rimfire rifles in 1968 chambered for this cartridge. As a rimfire, it was a screamer, pushing a 38-grain bullet to 2,100 fps. Unfortunately, after about 50,000 rifles were produced, Remington dropped the whole idea, including the ammunition support.

In 2007, Aguila announced that it was producing a reengineered version driving a 30-grain hollowpoint bullet to 2,300 fps. Maybe the second chance for this cartridge will turn out better than the first try.

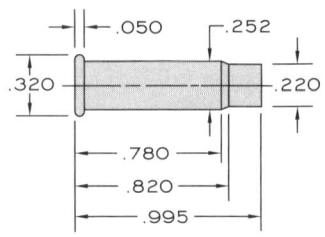

Relative Recoil Factor = 0.19

Specifications:

Controlling Agency for Standardization of this Ammunition: Factory

Standard barrel for velocity testing: N/A

Availability:

Aguila 30-grain Solid (1B222405)

G1 Ballistic Coefficient = 0.1:

Distance • Yards	Muzzle	25	50	75	100	125	150	175	200
Velocity • fps	2300	2155	2014	1880	1752	1630	1516	1410	1314
Energy • ft-lbs	352	309	270	235	204	177	153	132	115
Path • Inches	-1.5	-0.4	+0.3	+0.5	0.0	-1.2	-3.2	-6.1	-10.2
Wind Drift • Inches	0.0	0.2	0.8	1.8	3.4	5.4	8.1	11.4	15.4

Aguila 30-grain JHP (1B222406)

G1 Ballistic Coefficient = 0.1

Distance • Yards	Muzzle	25	50	75	100	125	150	175	200
Velocity • fps	2300	2155	2014	1880	1752	1630	1516	1410	1314
Energy • ft-lbs	352	309	270	235	204	177	153	132	115
Path • Inches	-1.5	-0.4	+0.3	+0.5	0.0	-1.2	-3.2	-6.1	-10.2
Wind Drift • Inches	0.0	0.2	0.8	1.8	3.4	5.4	8.1	11.4	15.4

Nothing gets your eye-hand coordin going better than a day of shootir varmints. The value of such session pay off well during hunts and sh matches. (Safari Press Archives)

.22 Short

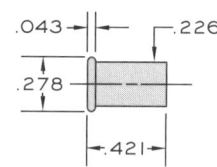

The .22 Short has the distinction of being the cartridge that has been in continuous production for longer than any other. Introduced in 1857 (loaded with blackpowder, of course), the short today has been relegated for the most part to specialized applications.

Relative Recoil Factor = 0.13

Specifications:

Controlling Agency for Standardization of this Ammunition: SAAMI

Bullet Weight Grains	Velocity fps	Maximum Average Pressure	
		Copper Crusher	Transducer
29 CB	710	N/S	21,000 psi
29 SV	1,035	N/S	21,000 psi
27 HP	1,105	N/S	21,000 psi
29 HV	1,080	N/S	21,000 psi

Standard barrel for velocity testing: 24 inches long—1 turn in 16-inch twist

Availability:

CCI 27-grain CPHP (0028)
G1 Ballistic Coefficient = 0.080

Distance • Yards	Muzzle	25	50	75	100	125	150	175	200
Velocity • fps	1105	1022	961	910	868	829	794	762	732
Energy • ft-lbs	73	63	55	50	45	41	38	35	32
Path • Inches	-1.5	+2.3	+4.0	+3.3	0.0	-6.2	-15.5	-28.3	-44.8
Wind Drift • Inches	0.0	0.4	1.7	3.7	6.3	9.6	13.5	18.0	23.2

CWS 28-grain Short R25 (None)
G1 Ballistic Coefficient = 0.047

Distance • Yards	Muzzle	25	50	75	100	125	150	175	200
Velocity • fps	560	524	490	457	427	398	371	346	322
Energy • ft-lbs	20	17	15	13	11	10	9	8	6
Path • Inches	-1.5	+12.0	+17.5	+13.9	0.0	-25.9	-65.5	-120.9	-194.5
Wind Drift • Inches	0.0	0.7	3.0	6.9	12.5	20.2	29.8	41.8	56.3

Aguila 29-grain Short High Velocity (1B222110)
G1 Ballistic Coefficient = 0.105

Distance • Yards	Muzzle	25	50	75	100	125	150	175	200
Velocity • fps	1095	1032	981	939	903	870	841	813	787
Energy • ft-lbs	77	69	62	57	52	49	46	43	40
Path • Inches	-1.5	+2.2	+3.8	+3.1	0.0	-5.7	-14.3	-25.9	-40.8
Wind Drift • Inches	0.0	0.4	1.4	3.2	5.4	8.3	11.7	15.6	20.0

Remington 29-grain Plated Lead Round Nose (1022)
G1 Ballistic Coefficient = 0.105

Distance • Yards	Muzzle	25	50	75	100	125	150	175	200
Velocity • fps	1095	1032	981	939	903	870	841	813	787
Energy • ft-lbs	77	69	62	57	52	49	46	43	40
Path • Inches	-1.5	+2.2	+3.8	+3.1	0.0	-5.7	-14.3	-25.9	-40.8
Wind Drift • Inches	0.0	0.4	1.4	3.2	5.4	8.3	11.7	15.6	20.0

Winchester 29-grain Short Standard Velocity Solid (X22S)
G1 Ballistic Coefficient = 0.105

Distance • Yards	Muzzle	25	50	75	100	125	150	175	200
Velocity • fps	1095	1032	981	939	903	870	841	813	787
Energy • ft-lbs	77	69	62	57	52	49	46	43	40
Path • Inches	-1.5	+2.2	+3.8	+3.1	0.0	-5.7	-14.3	-25.9	-40.8
Wind Drift • Inches	0.0	0.4	1.4	3.2	5.4	8.3	11.7	15.6	20.0

CCI 29-grain CPRN (0027)

G1 Ballistic Coefficient = 0.080

Distance • Yards	Muzzle	25	50	75	100	125	150	175	200
Velocity • fps	1080	1004	946	898	857	819	785	754	724
Energy • ft-lbs	75	65	58	52	47	43	40	37	34
Path • Inches	-1.5	+2.4	+4.1	+3.4	0.0	-6.4	-16.0	-29.1	-46.0
Wind Drift • Inches	0.0	0.4	1.6	3.5	6.1	9.3	13.1	17.5	22.6

CCI 29-grain Short Target Solid (0037)

G1 Ballistic Coefficient = 0.080

Distance • Yards	Muzzle	25	50	75	100	125	150	175	200
Velocity • fps	830	795	763	733	704	677	651	626	602
Energy • ft-lbs	44	41	37	35	32	30	27	25	23
Path • Inches	-1.5	+4.3	+6.7	+5.4	0.0	-9.8	-24.3	-43.9	-69.1
Wind Drift • Inches	0.0	0.3	1.3	2.8	5.1	8.0	11.6	15.9	20.9

Eley 29-grain Short Rapid Fire Match

G1 Ballistic Coefficient = 0.05

Distance • Yards	Muzzle	25	50	75	100	125	150	175	200
Velocity • fps	750	703	661	621	583	548	515	483	453
Energy • ft-lbs	36	32	28	25	22	19	17	15	13
Path • Inches	-1.5	+6.1	+9.2	+7.4	0.0	-13.8	-34.8	-64.1	-102.6
Wind Drift • Inches	0.0	0.6	2.3	5.4	9.7	15.4	22.7	31.6	42.2

CCI 29-grain Short CB (0026

G1 Ballistic Coefficient = 0.08

Distance • Yards	Muzzle	25	50	75	100	125	150	175	200
Velocity • fps	710	683	656	631	607	584	562	540	520
Energy • ft-lbs	32	30	28	26	24	22	20	19	17
Path • Inches	-1.5	+6.3	+9.4	+7.4	0.0	-13.3	-33.0	-59.5	-93.5
Wind Drift • Inches	0.0	0.3	1.4	3.1	5.5	8.8	12.8	17.6	23.3

.22 Long

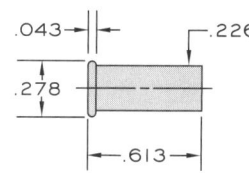

The .22 Long has just about dropped from the inventory. There was a time when not all the .22-caliber rifles were chambered for .22 Long Rifle. Since today's .22 Long Rifle ammunition will do everything the .22 Long cartridges will do, and most of it a lot better, the .22 Long is sliding quickly down the chute to the scrap pile of history.

Relative Recoil Factor = 0.14

Specifications:

Controlling Agency for Standardization of this Ammunition: SAAMI

Bullet Weight Grains	Velocity fps	Maximum Average Pressure	
		Copper Crusher	Transducer
29 HV	1,215	N/S	24,000 psi

Standard barrel for velocity testing: 24 inches long—1 turn in 16-inch twist

Availability:

CI 29-grain GLRN (0029)

G1 Ballistic Coefficient = 0.080

Distance • Yards	Muzzle	25	50	75	100	125	150	175	200
Velocity • fps	1215	1100	1019	958	908	865	827	792	760
Energy • ft-lbs	95	78	67	59	53	48	44	41	37
Path • Inches	-1.5	+1.9	+3.4	+2.9	0.0	-5.5	-14.0	-25.6	-40.6
Wind Drift • Inches	0.0	0.6	2.2	4.7	8.0	12.1	16.8	22.3	28.4

Winchester 29-grain Super-X CB-Match (X22LRCBMA)

G1 Ballistic Coefficient = 0.104

Distance • Yards	Muzzle	25	50	75	100	125	150	175	200
Velocity • fps	770	746	724	702	681	661	641	622	604
Energy • ft-lbs	38	36	34	32	30	28	26	25	24
Path • Inches	-1.5	+5.0	+7.6	+6.0	0.0	-10.7	-26.3	-47.2	-73.8
Wind Drift • Inches	0.0	0.2	1.0	2.2	4.0	6.3	9.1	12.5	16.4

CI 29-grain CB (0038)

G1 Ballistic Coefficient = 0.062

Distance • Yards	Muzzle	25	50	75	100	125	150	175	200
Velocity • fps	710	675	642	610	581	552	525	499	474
Energy • ft-lbs	32	29	27	24	22	20	18	16	14
Path • Inches	-1.5	+6.6	+9.8	+7.8	0.0	-14.3	-35.7	-65.0	-103.1
Wind Drift • Inches	0.0	0.5	2.0	4.5	8.0	12.8	18.7	25.9	34.5

.22 Standard Velocity and Match Ammunition

This is what the .22 Rimfire does best. Even the most expensive .22 Long Rifle match ammunition is inexpensive when compared to centerfire ammo. It's not reloadable, but that's not a factor since the .22 Rimfire case is so inexpensive that there just isn't anything worth saving. On an accuracy basis, it is sometimes very difficult to tell the difference in performance between "regular" standard velocity ammo and match ammo, especially since match ammo comes in at least three grades: club or practice, match grade, and the super match grade. That's a good class of problem. Unless you have a high-quality target rifle, you probably won't notice any improvement if you buy better than the standard velocity or the practice grade. In competition, you should use whatever you think works best for you. Mental attitude is an elemental part of competition.

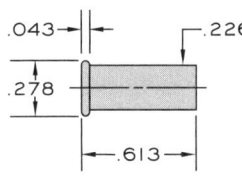

Relative Recoil Factor = 0.19

Specifications:

Controlling Agency for Standardization of this Ammunition: SAAMI

Bullet Weight Grains	Velocity fps	Maximum Average Pressure	
		Copper Crusher	Transducer
40 SV	1,135	N/S	24,000 psi

Standard barrel for velocity testing: 24 inches long—1 turn in 16-inch twist

Availability:

Winchester 29-grain RN (X22LRCBMA) LF
G1 Ballistic Coefficient = 0.1

Distance • Yards	Muzzle	25	50	75	100	125	150	175	200
Velocity • fps	770	746	724	702	681	661	641	622	604
Energy • ft-lbs	38	36	34	32	30	28	26	25	24
Path • Inches	-1.5	+5.0	+7.6	+6.0	0.0	-10.7	-26.3	-47.2	-73.8
Wind Drift • Inches	0.0	0.3	1.1	2.5	4.4	7.0	10.1	13.9	18.3

Remington 38-grain Long Rifle Subsonic Hollow Point (SUB22HP)
G1 Ballistic Coefficient = 0.

Distance • Yards	Muzzle	25	50	75	100	125	150	175	200
Velocity • fps	1050	1004	965	932	901	874	849	825	802
Energy • ft-lbs	93	85	79	73	69	64	61	57	54
Path • Inches	-1.5	+2.3	+3.9	+3.2	0.0	-5.9	-14.5	-26.2	-41.0
Wind Drift • Inches	0.0	0.3	1.1	2.5	4.3	6.6	9.4	12.6	16.2

Aguila 38-grain LR Subsonic Hollow Point (1B222286)
G1 Ballistic Coefficient = 0.

Distance • Yards	Muzzle	25	50	75	100	125	150	175	200
Velocity • fps	1025	978	939	904	873	845	818	794	770
Energy • ft-lbs	88	81	74	69	64	60	57	53	50
Path • Inches	-1.5	+2.5	+4.2	+3.4	0.0	-6.2	-15.5	-28.0	-43.9
Wind Drift • Inches	0.0	0.3	1.2	2.7	4.6	7.1	10.1	13.6	17.6

Fiocchi 38-grain HP (22HPSUB)
G1 Ballistic Coefficient = 0.

Distance • Yards	Muzzle	25	50	75	100	125	150	175	200
Velocity • fps	950	919	891	866	842	819	798	778	758
Energy • ft-lbs	76	71	67	63	60	57	54	51	49
Path • Inches	-1.5	+2.9	+4.7	+3.8	0.0	-6.9	-17.0	-30.5	-47.6
Wind Drift • Inches	0.0	0.2	0.9	2.1	3.6	5.6	8.1	10.9	14.2

RWS 40-grain R100 (213 4195)
G1 Ballistic Coefficient = 0

Distance • Yards	Muzzle	25	50	75	100	125	150	175	200
Velocity • fps	1175	1107	1065	1007	970	937	907	880	855
Energy • ft-lbs	123	109	98	90	84	78	73	69	65
Path • Inches	-1.5	+1.7	+3.2	+2.6	0.0	-4.9	-12.3	-22.3	-35.0
Wind Drift • Inches	0.0	0.4	1.4	2.0	5.1	7.7	10.8	14.3	18.3

Remington 40-grain Long Rifle Target Solid (6122) (6100)

G1 Ballistic Coefficient = 0.149

Distance • Yards	Muzzle	25	50	75	100	125	150	175	200
Velocity • fps	1150	1094	1048	1009	976	946	919	895	872
Energy • ft-lbs	117	106	98	90	85	80	75	71	68
Path • Inches	-1.5	+1.8	+3.2	+2.7	0.0	-4.9	-12.3	-22.1	-34.7
Wind Drift • Inches	0.0	0.3	1.2	2.5	4.4	6.6	9.3	12.4	15.8

Winchester 40-grain Long Rifle Standard Velocity Solid (XT22LR)

G1 Ballistic Coefficient = 0.149

Distance • Yards	Muzzle	25	50	75	100	125	150	175	200
Velocity • fps	1150	1094	1048	1009	976	946	919	895	872
Energy • ft-lbs	117	106	98	90	85	80	75	71	68
Path • Inches	-1.5	+1.8	+3.2	+2.7	0.0	-4.9	-12.3	-22.1	-34.7
Wind Drift • Inches	0.0	0.3	1.2	2.5	4.4	6.6	9.3	12.4	15.8

Fiocchi 40-grain Training (22MAXAC)

G1 Ballistic Coefficient = 0.150

Distance • Yards	Muzzle	25	50	75	100	125	150	175	200
Velocity • fps	1148	1089	1042	1001	967	937	909	884	861
Energy • ft-lbs	117	105	96	89	83	78	73	69	66
Path • Inches	-1.5	+1.8	+3.2	+2.7	0.0	-5.0	-12.5	-22.7	-35.6
Wind Drift • Inches	0.0	0.3	1.2	2.6	4.6	6.9	9.8	13.0	16.6

Fiocchi 40-grain Exacta Biathlon Match (22SM340)

G1 Ballistic Coefficient = 0.140 (Est.)

Distance • Yards	Muzzle	25	50	75	100	125	150	175	200
Velocity • fps	1120	1066	1022	985	953	924	897	873	850
Energy • ft-lbs	111	101	93	86	81	76	72	68	64
Path • Inches	-1.5	+1.9	+3.4	+2.8	0.0	-5.2	-12.9	-23.3	-36.5
Wind Drift • Inches	0.0	0.3	1.2	2.5	4.4	6.7	9.4	12.5	16.1

Norma 40-grain Biathlon (15616)

G1 Ballistic Coefficient = 0.150

Distance • Yards	Muzzle	25	50	75	100	125	150	175	200
Velocity • fps	1115	1066	1024	989	958	931	906	882	861
Energy • ft-lbs	110	101	93	87	82	77	73	69	66
Path • Inches	-1.5	+1.9	+3.4	+2.8	0.0	-5.2	-12.8	-23.2	-36.3
Wind Drift • Inches	0.0	0.3	1.1	2.4	4.1	6.2	8.8	11.7	15.0

Lapua 40-grain Polar Biathlon (420166)

G1 Ballistic Coefficient = 0.109

Distance • Yards	Muzzle	25	50	75	100	125	150	175	200
Velocity • fps	1106	1042	992	950	914	881	852	824	799
Energy • ft-lbs	109	97	87	80	74	69	64	60	57
Path • Inches	-1.5	+2.1	+3.6	+3.0	0.0	-5.6	-14.1	-25.5	-40.1
Wind Drift • Inches	0.0	0.3	1.3	2.8	4.8	7.3	10.3	13.7	17.6

Eley 40-grain Tenex (None)

G1 Ballistic Coefficient = 0.150

Distance • Yards	Muzzle	25	50	75	100	125	150	175	200
Velocity • fps	1085	1041	1003	971	941	915	891	869	848
Energy • ft-lbs	105	96	89	84	79	74	71	67	64
Path • Inches	-1.5	+2.0	+3.6	+2.9	0.0	-5.4	-13.3	-24.0	-37.5
Wind Drift • Inches	0.0	0.3	1.0	2.2	3.9	5.9	8.4	11.2	14.4

Eley 40-grain Match (None)

G1 Ballistic Coefficient = 0.150

Distance • Yards	Muzzle	25	50	75	100	125	150	175	200
Velocity • fps	1085	1041	1003	971	941	915	891	869	848
Energy • ft-lbs	105	96	89	84	79	74	71	67	64
Path • Inches	-1.5	+2.0	+3.6	+2.9	0.0	-5.4	-13.3	-24.0	-37.5
Wind Drift • Inches	0.0	0.3	1.0	2.2	3.9	5.9	8.4	11.2	14.4

Eley 40-grain (None)

G1 Ballistic Coefficient = 0.150

Distance • Yards	Muzzle	25	50	75	100	125	150	175	200
Velocity • fps	1085	1041	1003	971	941	915	891	869	848
Energy • ft-lbs	105	96	89	84	79	74	71	67	64
Path • Inches	-1.5	+2.0	+3.6	+2.9	0.0	-5.4	-13.3	-24.0	-37.5
Wind Drift • Inches	0.0	0.3	1.0	2.2	3.9	5.9	8.4	11.2	14.4

Eley 40-grain Target (None)

G1 Ballistic Coefficient = 0.150

Distance • Yards	Muzzle	25	50	75	100	125	150	175	200
Velocity • fps	1085	1041	1003	971	941	915	891	869	848
Energy • ft-lbs	105	96	89	84	79	74	71	67	64
Path • Inches	-1.5	+2.0	+3.6	+2.9	0.0	-5.4	-13.3	-24.0	-37.5
Wind Drift • Inches	0.0	0.3	1.0	2.2	3.9	5.9	8.4	11.2	14.4

Eley 40-grain Team

G1 Ballistic Coefficient = 0.150

Distance • Yards	Muzzle	25	50	75	100	125	150	175	200
Velocity • fps	1085	1041	1003	971	941	915	891	869	848
Energy • ft-lbs	105	96	89	84	79	74	71	67	64
Path • Inches	-1.5	+2.0	+3.6	+2.9	0.0	-5.4	-13.3	-24.0	-37.5
Wind Drift • Inches	0.0	0.3	1.0	2.2	3.9	5.9	8.4	11.2	14.4

Remington/Eley 40-grain Match EPS (RE22EPS)

G1 Ballistic Coefficient = 0.14?

Distance • Yards	Muzzle	25	50	75	100	125	150	175	200
Velocity • fps	1085	1040	1006	970	941	915	891	868	847
Energy • ft-lbs	105	96	90	84	79	74	70	67	64
Path • Inches	-1.5	+2.0	+3.6	+2.9	0.0	-5.4	-13.3	-24.0	-37.5
Wind Drift • Inches	0.0	0.3	1.0	2.2	3.9	6.0	8.4	11.3	14.5

Remington/Eley 40-grain Club Extra (RE22CX)

G1 Ballistic Coefficient = 0.14?

Distance • Yards	Muzzle	25	50	75	100	125	150	175	200
Velocity • fps	1085	1040	1006	970	941	915	891	868	847
Energy • ft-lbs	105	96	90	84	79	74	70	67	64
Path • Inches	-1.5	+2.0	+3.6	+2.9	0.0	-5.4	-13.3	-24.0	-37.5
Wind Drift • Inches	0.0	0.3	1.0	2.2	3.9	6.0	8.4	11.3	14.5

Remington/Eley 40-grain Target Rifle (RE22T)

G1 Ballistic Coefficient = 0.14?

Distance • Yards	Muzzle	25	50	75	100	125	150	175	200
Velocity • fps	1085	1040	1006	970	941	915	891	868	847
Energy • ft-lbs	105	96	90	84	79	74	70	67	64
Path • Inches	-1.5	+2.0	+3.6	+2.9	0.0	-5.4	-13.3	-24.0	-37.5
Wind Drift • Inches	0.0	0.3	1.0	2.2	3.9	6.0	8.4	11.3	14.5

Aguila 40-grain Match Rifle (1B222503)

G1 Ballistic Coefficient = 0.1

Distance • Yards	Muzzle	25	50	75	100	125	150	175	200
Velocity • fps	1080	1026	981	943	910	880	853	827	803
Energy • ft-lbs	104	93	86	79	74	69	65	61	57
Path • Inches	-1.5	+2.2	+3.8	+3.1	0.0	-5.7	-14.2	-25.8	-40.4
Wind Drift • Inches	0.0	0.3	1.3	2.8	4.8	7.3	10.4	13.9	17.8

Aguila 40-grain Target (1B222500)

G1 Ballistic Coefficient = 0.?

Distance • Yards	Muzzle	25	50	75	100	125	150	175	200
Velocity • fps	1080	1026	981	943	910	880	853	827	803
Energy • ft-lbs	104	93	86	79	74	69	65	61	57
Path • Inches	-1.5	+2.2	+3.8	+3.1	0.0	-5.7	-14.2	-25.8	-40.4
Wind Drift • Inches	0.0	0.3	1.3	2.8	4.8	7.3	10.4	13.9	17.8

Aguila 40-grain Long Rifle Standard Velocity (1B222332)

G1 Ballistic Coefficient = 0.?

Distance • Yards	Muzzle	25	50	75	100	125	150	175	200
Velocity • fps	1080	1016	965	922	885	852	821	793	766
Energy • ft-lbs	104	92	83	76	70	64	60	56	52
Path • Inches	-1.5	+2.3	+3.9	+3.2	0.0	-6.0	-15.0	-27.2	-42.9
Wind Drift • Inches	0.0	0.4	1.5	3.3	5.7	8.7	12.2	16.4	21.1

Federal 40-grain Premium Gold Medal (711B)

G1 Ballistic Coefficient = 0.

Distance • Yards	Muzzle	25	50	75	100	125	150	175	200
Velocity • fps	1080	1031	990	955	924	896	870	846	823
Energy • ft-lbs	105	94	87	81	76	71	67	64	60
Path • Inches	-1.5	+2.1	+3.7	+3.0	0.0	-5.6	-13.9	-25.0	-39.2
Wind Drift • Inches	0.0	0.3	1.0	2.3	3.9	6.0	8.4	11.3	14.5

Federal 40-grain Solid (UM22) (922A)

G1 Ballistic Coefficient = 0.093

Distance • Yards	Muzzle	25	50	75	100	125	150	175	200
Velocity • fps	1080	1013	960	917	880	845	814	785	757
Energy • ft-lbs	105	91	80	75	70	63	59	55	51
Path • Inches	-1.5	+2.3	+4.0	+3.3	0.0	-6.1	-15.2	-27.6	-43.5
Wind Drift • Inches	0.0	0.4	1.6	3.4	5.9	9.0	12.7	17.0	21.9

RWS 40-grain L.R. Target Rifle (213 2478)

G1 Ballistic Coefficient = 0.109

Distance • Yards	Muzzle	25	50	75	100	125	150	175	200
Velocity • fps	1080	1022	990	935	900	869	841	814	789
Energy • ft-lbs	100	93	85	78	70	67	63	59	55
Path • Inches	-1.5	+2.2	+3.8	+3.1	0.0	-5.8	-14.4	-26.1	-41.0
Wind Drift • Inches	0.0	0.4	1.4	3.0	5.1	7.8	11.1	14.8	19.0

Lapua 40-grain X-Act (420161)

G1 Ballistic Coefficient = 0.161

Distance • Yards	Muzzle	25	50	75	100	125	150	175	200
Velocity • fps	1073	1025	985	951	920	892	867	843	821
Energy • ft-lbs	102	93	86	80	75	71	67	63	60
Path • Inches	-1.5	+2.1	+3.7	+3.1	0.0	-5.6	-14.0	-25.2	-39.5
Wind Drift • Inches	0.0	0.2	1.0	2.2	3.9	5.9	8.4	11.2	14.4

Lapua 40-grain Midas + (4201621)

G1 Ballistic Coefficient = 0.161

Distance • Yards	Muzzle	25	50	75	100	125	150	175	200
Velocity • fps	1073	1025	985	951	920	892	867	843	821
Energy • ft-lbs	102	93	86	80	75	71	67	63	60
Path • Inches	-1.5	+2.1	+3.7	+3.1	0.0	-5.6	-14.0	-25.2	-39.5
Wind Drift • Inches	0.0	0.2	1.0	2.2	3.9	5.9	8.4	11.2	14.4

Lapua 40-grain Center-X (420163)

G1 Ballistic Coefficient = 0.161

Distance • Yards	Muzzle	25	50	75	100	125	150	175	200
Velocity • fps	1073	1025	985	951	920	892	867	843	821
Energy • ft-lbs	102	93	86	80	75	71	67	63	60
Path • Inches	-1.5	+2.1	+3.7	+3.1	0.0	-5.6	-14.0	-25.2	-39.5
Wind Drift • Inches	0.0	0.2	1.0	2.2	3.9	5.9	8.4	11.2	14.4

CI 40-grain Long Rifle Standard Velocity Solid (0032)

G1 Ballistic Coefficient = 0.120

Distance • Yards	Muzzle	25	50	75	100	125	150	175	200
Velocity • fps	1070	1019	977	940	908	879	852	827	804
Energy • ft-lbs	102	92	85	78	73	69	64	61	57
Path • Inches	-1.5	+2.2	+3.8	+3.1	0.0	-5.8	-14.3	-25.9	-40.6
Wind Drift • Inches	0.0	0.3	1.1	2.4	4.2	6.4	9.0	12.0	15.5

CI 40-grain Long Rifle Green Tag Comp. Solid (0033)

G1 Ballistic Coefficient = 0.120

Distance • Yards	Muzzle	25	50	75	100	125	150	175	200
Velocity • fps	1070	1019	977	940	908	879	852	827	804
Energy • ft-lbs	102	92	85	78	73	69	64	61	57
Path • Inches	-1.5	+2.2	+3.8	+3.1	0.0	-5.8	-14.3	-25.9	-40.6
Wind Drift • Inches	0.0	0.3	1.1	2.4	4.2	6.4	9.0	12.0	15.5

CI 40-grain Long Rifle Pistol Match Solid (0051)

G1 Ballistic Coefficient = 0.120

Distance • Yards	Muzzle	25	50	75	100	125	150	175	200
Velocity • fps	1070	1019	977	940	908	879	852	827	804
Energy • ft-lbs	102	92	85	78	73	69	64	61	57
Path • Inches	-1.5	+2.2	+3.8	+3.1	0.0	-5.8	-14.3	-25.9	-40.6
Wind Drift • Inches	0.0	0.3	1.1	2.4	4.2	6.4	9.0	12.0	15.5

WS 40-grain LR R50 (21304187)

G1 Ballistic Coefficient = 0.105

Distance • Yards	Muzzle	25	50	75	100	125	150	175	200
Velocity • fps	1070	1012	970	926	890	859	830	803	778
Energy • ft-lbs	100	91	80	76	70	66	61	57	54
Path • Inches	-1.5	+2.3	+3.9	+3.2	0.0	-5.9	-14.8	-26.7	-42.0
Wind Drift • Inches	0.0	0.4	1.4	3.0	5.2	8.0	11.3	15.1	19.5

Norma 40-grain Match 1 LRN (15613)

G1 Ballistic Coefficient = 0.150

Distance • Yards	Muzzle	25	50	75	100	125	150	175	200
Velocity • fps	1066	1025	989	959	930	906	882	861	840
Energy • ft-lbs	101	93	87	82	77	73	69	66	63
Path • Inches	-1.5	+2.1	+3.7	+3.0	0.0	-5.5	-13.7	-24.7	-38.6
Wind Drift • Inches	0.0	0.2	1.0	2.2	3.8	5.8	8.1	10.9	14.0

Norma 40-grain Training 2 LRN (15614)

G1 Ballistic Coefficient = 0.150

Distance • Yards	Muzzle	25	50	75	100	125	150	175	200
Velocity • fps	1066	1025	989	959	930	906	882	861	840
Energy • ft-lbs	101	93	87	82	77	73	69	66	63
Path • Inches	-1.5	+2.1	+3.7	+3.0	0.0	-5.5	-13.7	-24.7	-38.6
Wind Drift • Inches	0.0	0.2	1.0	2.2	3.8	5.8	8.1	10.9	14.0

Winchester 40-grain LHP (X222LRSUBA)

G1 Ballistic Coefficient = 0.13

Distance • Yards	Muzzle	25	50	75	100	125	150	175	200
Velocity • fps	1065	1021	984	951	922	896	871	849	827
Energy • ft-lbs	101	93	86	80	76	71	68	64	61
Path • Inches	-1.5	+2.2	+3.7	+3.1	0.0	-5.6	-14.0	-25.1	-39.3
Wind Drift • Inches	0.0	0.3	1.1	2.3	4.0	6.2	8.7	11.6	15.0

CCI 40-grain Sub-sonic HP (0056)

G1 Ballistic Coefficient = 0.1

Distance • Yards	Muzzle	25	50	75	100	125	150	175	200
Velocity • fps	1050	1002	963	928	897	869	843	818	796
Energy • ft-lbs	98	89	82	76	72	67	63	60	56
Path • Inches	-1.5	+2.3	+3.9	+3.2	0.0	-5.9	-14.7	-26.6	-41.7
Wind Drift • Inches	0.0	0.3	1.1	2.3	4.0	6.2	8.8	11.7	15.2

Fiocchi 40-grain Training (22MAXAC)

G1 Ballistic Coefficient = 0.1

Distance • Yards	Muzzle	25	50	75	100	125	150	175	200
Velocity • fps	1050	1011	978	948	921	897	874	853	833
Energy • ft-lbs	98	91	85	80	75	72	68	65	62
Path • Inches	-1.5	+2.2	+3.8	+3.1	0.0	-5.7	-14.0	-25.2	-39.4
Wind Drift • Inches	0.0	0.2	1.0	2.1	3.6	5.6	7.9	10.6	13.8

Fiocchi 40-grain Long Rifle Match Training Solid (22M320)

G1 Ballistic Coefficient = 0.1

Distance • Yards	Muzzle	25	50	75	100	125	150	175	200
Velocity • fps	1050	1006	968	935	906	879	854	831	809
Energy • ft-lbs	98	90	83	78	73	69	65	61	58
Path • Inches	-1.5	+2.3	+3.9	+3.2	0.0	-5.8	-14.4	-26.0	-40.6
Wind Drift • Inches	0.0	0.3	1.1	2.4	4.2	6.4	9.1	12.2	15.7

RWS 40-grain LR Rifle Match (213 4225)

G1 Ballistic Coefficient = 0.

Distance • Yards	Muzzle	25	50	75	100	125	150	175	200
Velocity • fps	1035	980	945	895	860	829	800	772	747
Energy • ft-lbs	95	85	80	71	65	61	57	53	50
Path • Inches	-1.5	+2.5	+4.2	+3.4	0.0	-6.4	-15.9	-28.7	-45.2
Wind Drift • Inches	0.0	0.4	1.4	3.1	5.4	8.3	11.7	15.8	20.4

Eley 40-grain Tenex Pistol

G1 Ballistic Coefficient = 0.

Distance • Yards	Muzzle	25	50	75	100	125	150	175	200
Velocity • fps	1030	993	960	932	906	881	859	838	818
Energy • ft-lbs	94	88	82	77	73	69	66	62	59
Path • Inches	-1.5	+2.3	+4.0	+3.2	0.0	-5.9	-14.5	-26.1	-40.7
Wind Drift • Inches	0.0	0.2	1.0	2.1	3.6	5.6	8.0	10.7	13.8

Eley 40-grain Match OSP

G1 Ballistic Coefficient = 0

Distance • Yards	Muzzle	25	50	75	100	125	150	175	200
Velocity • fps	1030	993	960	932	906	881	859	838	818
Energy • ft-lbs	94	88	82	77	73	69	66	62	59
Path • Inches	-1.5	+2.3	+4.0	+3.2	0.0	-5.9	-14.5	-26.1	-40.7
Wind Drift • Inches	0.0	0.2	1.0	2.1	3.6	5.6	8.0	10.7	13.8

Aguila 40-grain Subsonic Solid Point (1B222269)

G1 Ballistic Coefficient = 0.110

Distance • Yards	Muzzle	25	50	75	100	125	150	175	200
Velocity • fps	1025	978	938	904	873	844	817	792	769
Energy • ft-lbs	93	85	78	73	68	63	59	56	52
Path • Inches	-1.5	+2.5	+4.2	+3.4	0.0	-6.3	-15.6	-28.2	-44.2
Wind Drift • Inches	0.0	0.3	1.2	2.7	4.7	7.2	10.2	13.7	17.8

Eley 40-grain Match Pistol (None)

G1 Ballistic Coefficient = 0.140

Distance • Yards	Muzzle	25	50	75	100	125	150	175	200
Velocity • fps	1000	966	935	908	883	859	838	817	797
Energy • ft-lbs	89	83	78	73	69	66	62	59	56
Path • Inches	-1.5	+2.5	+4.2	+3.4	0.0	-6.2	-15.3	-27.5	-43.0
Wind Drift • Inches	0.0	0.2	0.9	2.1	3.6	5.6	7.9	10.7	13.8

Fiocchi 40-grain Super Match (22SM300)

G1 Ballistic Coefficient = 0.130

Distance • Yards	Muzzle	25	50	75	100	125	150	175	200
Velocity • fps	980	946	915	888	862	839	817	795	775
Energy • ft-lbs	85	80	75	70	66	63	59	56	53
Path • Inches	-1.5	+2.7	+4.4	+3.6	0.0	-6.5	-16.1	-29.0	-45.3
Wind Drift • Inches	0.0	0.3	1.0	2.1	3.8	5.8	8.3	11.2	14.6

Lapua 40-grain Pistol King (420164)

G1 Ballistic Coefficient = 0.131

Distance • Yards	Muzzle	25	50	75	100	125	150	175	200
Velocity • fps	958	927	898	872	848	826	804	784	764
Energy • ft-lbs	82	76	72	68	64	60	57	55	52
Path • Inches	-1.5	+2.8	+4.6	+3.7	0.0	-6.8	-16.7	-30.0	-46.9
Wind Drift • Inches	0.0	0.3	0.8	1.9	3.3	5.1	7.2	9.8	12.8

Fiocchi 40-grain Rapid Fire Pistol (22SM280)

G1 Ballistic Coefficient = 0.130

Distance • Yards	Muzzle	25	50	75	100	125	150	175	200
Velocity • fps	950	919	891	866	842	819	798	778	758
Energy • ft-lbs	80	75	71	67	63	60	57	54	51
Path • Inches	-1.5	+2.9	+4.7	+3.8	0.0	-6.9	-17.0	-30.5	-47.6
Wind Drift • Inches	0.0	0.2	0.9	2.1	3.6	5.6	8.1	10.9	14.2

Aguila 40-grain Match Pistol (1B222502-1)

G1 Ballistic Coefficient = 0.124

Distance • Yards	Muzzle	25	50	75	100	125	150	175	200
Velocity • fps	925	895	868	843	819	797	776	756	736
Energy • ft-lbs	76	71	67	63	59	56	54	51	48
Path • Inches	-1.5	+3.1	+5.0	+4.0	0.0	-7.3	-18.0	-32.3	-50.4
Wind Drift • Inches	0.0	0.2	0.9	2.1	3.7	5.8	8.3	11.3	14.7

Fiocchi 40-grain Rapid Fire Pistol (22SM280)

G1 Ballistic Coefficient = 0.130

Distance • Yards	Muzzle	25	50	75	100	125	150	175	200
Velocity • fps	920	892	866	842	820	799	778	759	740
Energy • ft-lbs	75	71	67	63	60	57	54	51	49
Path • Inches	-1.5	+3.1	+5.0	+4.0	0.0	-7.3	-18.0	-32.2	-50.3
Wind Drift • Inches	0.0	0.2	0.8	1.8	3.2	5.0	7.1	9.7	12.6

Lapua 40-grain Pistol OSP (420165)

G1 Ballistic Coefficient = 0.131

Distance • Yards	Muzzle	25	50	75	100	125	150	175	200
Velocity • fps	909	882	857	834	812	792	772	753	734
Energy • ft-lbs	73	69	65	62	59	56	53	50	48
Path • Inches	-1.5	+3.2	+5.2	+4.1	0.0	-7.4	-18.3	-32.9	-51.3
Wind Drift • Inches	0.0	0.2	0.8	1.8	3.1	4.9	7.0	9.5	12.4

Winchester 40-grain Long Rifle Dyna-Point (W22LRB)

G1 Ballistic Coefficient = 0.150

Distance • Yards	Muzzle	25	50	75	100	125	150	175	200
Velocity • fps	1150	1094	1048	1010	976	947	920	896	873
Energy • ft-lbs	117	106	98	91	85	80	75	71	68
Path • Inches	-1.5	+1.8	+3.2	+2.7	0.0	-4.9	-12.2	-22.1	-34.6
Wind Drift • Inches	0.0	0.3	1.2	2.5	4.3	6.6	9.2	12.3	15.7

Eley 40-grain Subsonic Hollow Point

G1 Ballistic Coefficient = 0.115

Distance • Yards	Muzzle	25	50	75	100	125	150	175	200
Velocity • fps	1065	1013	970	932	900	870	843	817	794
Energy • ft-lbs	100	91	83	77	72	67	63	59	56
Path • Inches	-1.5	+2.2	+3.9	+3.2	0.0	-5.8	-14.5	-26.2	-41.2
Wind Drift • Inches	0.0	0.3	1.3	2.8	4.8	7.3	10.3	13.8	17.8

CCI 40-grain Segmented HP Sub-Sonic (0074)

G1 Ballistic Coefficient = 0.120

Distance • Yards	Muzzle	25	50	75	100	125	150	175	200
Velocity • fps	1050	1002	963	928	897	869	843	818	796
Energy • ft-lbs	98	89	82	76	72	67	63	60	56
Path • Inches	-1.5	+2.3	+3.9	+3.2	0.0	-5.9	-14.7	-26.6	-41.7
Wind Drift • Inches	0.0	0.3	1.1	2.3	4.0	6.2	8.8	11.7	15.2

RWS 40-grain LR Subsonic HP (213 2494)

G1 Ballistic Coefficient = 0.09?

Distance • Yards	Muzzle	25	50	75	100	125	150	175	200
Velocity • fps	1000	949	915	869	835	805	776	749	723
Energy • ft-lbs	90	80	75	67	60	58	53	50	46
Path • Inches	-1.5	+2.7	+4.5	+3.7	0.0	-6.8	-16.9	-30.6	-48.1
Wind Drift • Inches	0.0	0.4	1.4	3.1	5.4	8.2	11.8	15.9	20.6

Aguila 60-grain SSS Sniper Subsonic (1B222112)

G1 Ballistic Coefficient = 0.09?

Distance • Yards	Muzzle	25	50	75	100	125	150	175	200
Velocity • fps	950	906	868	834	802	773	746	719	695
Energy • ft-lbs	120	109	100	93	86	80	74	69	64
Path • Inches	-1.5	+3.1	+5.0	+4.1	0.0	-7.4	-18.5	-33.4	-52.5
Wind Drift • Inches	0.0	0.3	1.3	3.0	5.2	8.1	11.6	15.7	20.5

.22 Long Rifle High Velocity and Hyper Velocity

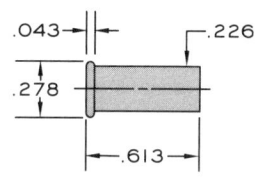

This ammunition is the generic form of the .22 Long Rifle caliber. There is more of this ammo made than any other basic performance category. The bullets are available in a variety of forms and weights. In general, the high-speed ammunition is not quite as accurate as the standard velocity form, but from time to time you will run across a gun and ammo combination that shoots extremely well with high-velocity ammunition.

Relative Recoil Factor = 0.22

Specifications:

Controlling Agency for Standardization of this Ammunition: SAAMI

Bullet Weight Grains	Velocity fps	Maximum Average Pressure Copper Crusher	Transducer
36 HVHP	1,260	N/S	24,000 psi
37 HVHP	1,260	N/S	24,000 psi
40 HV	1,235	N/S	24,000 psi
33 Hyper HP	1,465	N/S	24,000 psi
36 Hyper	1,385	N/S	24,000 psi

Standard barrel for velocity testing: 24 inches long—1 turn in 16-inch twist

Availability:

CCI 21-grain GREEN HP (0952) LF
G1 Ballistic Coefficient = 0.070

Distance • Yards	Muzzle	25	50	75	100	125	150	175	200
Velocity • fps	1650	1427	1243	1106	1013	945	891	845	804
Energy • ft-lbs	127	95	72	57	48	42	37	33	30
Path • Inches	-1.5	+0.8	+2.1	+1.9	0.0	-4.0	-10.5	-19.7	-32.0
Wind Drift • Inches	0.0	0.6	2.5	5.8	10.3	15.8	22.2	29.5	37.5

Winchester 26-grain THP TIN (X22LRHLF) LF
G1 Ballistic Coefficient = 0.073

Distance • Yards	Muzzle	25	50	75	100	125	150	175	200
Velocity • fps	1650	1433	1252	1115	1023	954	900	854	813
Energy • ft-lbs	157	119	91	72	60	53	47	42	39
Path • Inches	-1.5	+0.8	+2.0	+1.9	0.0	-3.9	-10.2	-19.2	-31.1
Wind Drift • Inches	0.0	0.6	2.4	5.6	9.9	15.3	21.5	28.5	36.2

Aguila 30-grain Super Maximum Solid Point (1B222298)
G1 Ballistic Coefficient = 0.110

Distance • Yards	Muzzle	25	50	75	100	125	150	175	200
Velocity • fps	1700	1550	1414	1293	1191	1109	1045	994	952
Energy • ft-lbs	193	160	133	111	95	82	73	66	60
Path • Inches	-1.5	+0.4	+1.4	+1.4	0.0	-2.9	-7.6	-14.2	-23.1
Wind Drift • Inches	0.0	0.4	1.5	3.5	6.4	10.2	14.7	19.9	25.7

Aguila 30-grain Super Maximum Hollow Point (1B222297)
G1 Ballistic Coefficient = 0.110

Distance • Yards	Muzzle	25	50	75	100	125	150	175	200
Velocity • fps	1700	1550	1414	1293	1191	1109	1045	994	952
Energy • ft-lbs	193	160	133	111	95	82	73	66	60
Path • Inches	-1.5	+0.4	+1.4	+1.4	0.0	-2.9	-7.6	-14.2	-23.1
Wind Drift • Inches	0.0	0.4	1.5	3.5	6.4	10.2	14.7	19.9	25.7

Federal 31-grain Long Rifle Hyper Velocity Hollow Point Copper Plated (724)
G1 Ballistic Coefficient = 0.110

Distance • Yards	Muzzle	25	50	75	100	125	150	175	200
Velocity • fps	1430	1307	1202	1118	1052	1000	957	920	888
Energy • ft-lbs	140	118	100	86	76	69	63	58	54
Path • Inches	-1.5	+1.0	+2.3	+2.0	0.0	-4.0	-10.1	-18.6	-29.6
Wind Drift • Inches	0.0	0.4	1.6	3.5	6.2	9.5	13.3	17.7	22.5

CCI 32-grain Stinger CPHP (0050)
G1 Ballistic Coefficient = 0.085

Distance • Yards	Muzzle	25	50	75	100	125	150	175	200
Velocity • fps	1640	1453	1292	1162	1066	997	943	896	858
Energy • ft-lbs	191	150	119	96	81	71	63	57	52
Path • Inches	-1.5	+0.7	+1.9	+1.7	0.0	-3.6	-9.5	-17.8	-28.7
Wind Drift • Inches	0.0	0.5	1.9	4.4	7.8	12.1	17.1	22.8	29.1

CCI 32-grain Segmented HP (0064)
G1 Ballistic Coefficient = 0.085

Distance • Yards	Muzzle	25	50	75	100	125	150	175	200
Velocity • fps	1640	1453	1292	1162	1066	997	943	896	858
Energy • ft-lbs	191	150	119	96	81	71	63	57	52
Path • Inches	-1.5	+0.7	+1.9	+1.7	0.0	-3.6	-9.5	-17.8	-28.7
Wind Drift • Inches	0.0	0.5	1.9	4.4	7.8	12.1	17.1	22.8	29.1

Winchester 32-grain LHP Lubaloy Plated (S22LRUHV)
G1 Ballistic Coefficient = 0.08?

Distance • Yards	Muzzle	25	50	75	100	125	150	175	200
Velocity • fps	1640	1459	1302	1174	1078	1008	953	908	868
Energy • ft-lbs	191	151	120	98	83	72	65	59	54
Path • Inches	-1.5	+0.7	+1.8	+1.7	0.0	-3.6	-9.3	-17.4	-28.1
Wind Drift • Inches	0.0	0.5	2.0	4.7	8.4	13.0	18.4	24.6	31.4

Remington 33-grain Long Rifle Yellow Jacket Hollow Point (1722)
G1 Ballistic Coefficient = 0.10

Distance • Yards	Muzzle	25	50	75	100	125	150	175	200
Velocity • fps	1500	1365	1247	1151	1075	1018	971	931	896
Energy • ft-lbs	165	137	114	97	85	76	69	64	59
Path • Inches	-1.5	+0.9	+2.0	+1.8	0.0	-3.7	-9.4	-17.4	-27.8
Wind Drift • Inches	0.0	0.4	1.8	4.0	7.0	10.9	15.4	20.4	26.1

Remington 36-grain Long Rifle Viper Solid (1922)
G1 Ballistic Coefficient = 0.11

Distance • Yards	Muzzle	25	50	75	100	125	150	175	200
Velocity • fps	1410	1296	1198	1119	1056	1006	965	926	897
Energy • ft-lbs	159	134	115	100	89	81	74	69	64
Path • Inches	-1.5	+1.0	+2.3	+2.0	0.0	-3.9	-10.0	-18.2	-29.0
Wind Drift • Inches	0.0	0.4	1.6	3.7	6.5	9.9	14.0	18.6	23.7

Remington 36-grain Long Rifle Cyclone Hollow Point (CY22HP) + (GL22HP)
G1 Ballistic Coefficient = 0.1

Distance • Yards	Muzzle	25	50	75	100	125	150	175	200
Velocity • fps	1280	1190	1117	1057	1010	970	935	905	877
Energy • ft-lbs	131	113	100	89	82	75	70	65	61
Path • Inches	-1.5	+1.4	+2.7	+2.3	0.0	-4.4	-11.1	-20.3	-32.0
Wind Drift • Inches	0.0	0.4	1.5	3.4	5.9	8.9	12.4	16.5	21.0

Remington 36-grain Long Rifle Hollow Point (1622/1600+)
G1 Ballistic Coefficient = 0.1

Distance • Yards	Muzzle	25	50	75	100	125	150	175	200
Velocity • fps	1280	1190	1117	1057	1010	970	935	905	877
Energy • ft-lbs	131	113	100	89	82	75	70	65	61
Path • Inches	-1.5	+1.4	+2.7	+2.3	0.0	-4.4	-11.1	-20.3	-32.0
Wind Drift • Inches	0.0	0.4	1.5	3.4	5.9	8.9	12.4	16.5	21.0

Winchester 36-grain Long Rifle Hollow Point (XPERT22)
G1 Ballistic Coefficient = 0.

Distance • Yards	Muzzle	25	50	75	100	125	150	175	200
Velocity • fps	1280	1173	1089	1026	975	933	897	864	834
Energy • ft-lbs	131	110	95	84	76	70	64	60	56
Path • Inches	-1.5	+1.5	+2.9	+2.5	0.0	-4.8	-12.0	-22.0	-34.8
Wind Drift • Inches	0.0	0.5	1.9	4.0	6.9	10.5	14.6	19.3	24.5

CCI 36-grain Mini-Mag HP (0031)
G1 Ballistic Coefficient = 0.

Distance • Yards	Muzzle	25	50	75	100	125	150	175	200
Velocity • fps	1260	1174	1104	1048	1003	964	931	901	873
Energy • ft-lbs	127	110	97	88	80	74	69	65	61
Path • Inches	-1.5	+1.4	+2.8	+2.4	0.0	-4.6	-11.4	-20.8	-32.9
Wind Drift • Inches	0.0	0.4	1.4	3.0	5.2	7.8	10.9	14.5	18.4

Federal 36-grain Copper Plated HP (745)
G1 Ballistic Coefficient = 0.120

Distance • Yards	Muzzle	25	50	75	100	125	150	175	200
Velocity • fps	1260	1171	1099	1042	996	957	923	892	865
Energy • ft-lbs	125	110	97	87	79	73	68	64	60
Path • Inches	-1.5	+1.5	+2.8	+2.4	0.0	-4.6	-11.6	-21.1	-33.4
Wind Drift • Inches	0.0	0.4	1.4	3.1	5.3	8.1	11.3	15.0	19.1

Winchester 37-grain FHP (S22LRFSP)
G1 Ballistic Coefficient = 0.118

Distance • Yards	Muzzle	25	50	75	100	125	150	175	200
Velocity • fps	1435	1319	1219	1136	1070	1018	975	939	906
Energy • ft-lbs	169	143	122	106	94	85	78	72	68
Path • Inches	-1.5	+1.0	+2.2	+1.9	0.0	-3.8	-9.7	-17.9	-28.6
Wind Drift • Inches	0.0	0.4	1.6	3.6	6.4	9.9	13.9	18.5	23.7

Winchester 37-grain Long Rifle Hollow Point (X22LRH)
G1 Ballistic Coefficient = 0.128

Distance • Yards	Muzzle	25	50	75	100	125	150	175	200
Velocity • fps	1280	1193	1120	1062	1015	976	941	911	883
Energy • ft-lbs	135	117	103	93	85	78	73	68	64
Path • Inches	-1.5	+1.4	+2.7	+2.3	0.0	-4.4	-11.0	-20.1	-31.6
Wind Drift • Inches	0.0	0.4	1.5	3.3	5.7	8.7	12.1	16.1	20.5

Aguila 38-grain L.R. High Velocity Hollow Point (1B222335)
G1 Ballistic Coefficient = 0.124

Distance • Yards	Muzzle	25	50	75	100	125	150	175	200
Velocity • fps	1280	1190	1116	1057	1010	970	935	905	877
Energy • ft-lbs	138	120	105	94	86	79	74	69	65
Path • Inches	-1.5	+1.4	+2.7	+2.3	0.0	-4.4	-11.1	-20.3	-21.0
Wind Drift • Inches	0.0	0.4	1.5	3.4	5.9	8.9	12.4	16.5	21.0

Fiocchi 38-grain CPHP (22HVHP)
G1 Ballistic Coefficient = 0.130

Distance • Yards	Muzzle	25	50	75	100	125	150	175	200
Velocity • fps	1280	1194	1123	1065	1065	979	945	915	887
Energy • ft-lbs	138	120	106	96	88	81	75	71	66
Path • Inches	-1.5	+1.4	+2.7	+2.3	0.0	-4.4	11.1	20.2	31.8
Wind Drift • Inches	0.0	0.3	1.3	2.9	5.0	7.7	10.8	14.2	18.2

Federal 38-grain Copper Plated HP (712)
G1 Ballistic Coefficient = 0.130

Distance • Yards	Muzzle	25	50	75	100	125	150	175	200
Velocity • fps	1260	1177	1110	1054	1009	971	943	908	882
Energy • ft-lbs	135	117	105	94	86	80	75	70	66
Path • Inches	-1.5	+1.4	+2.8	+2.4	0.0	-4.5	-11.5	-20.6	-32.5
Wind Drift • Inches	0.0	0.3	1.3	2.9	5.0	7.6	10.6	14.0	17.9

American Eagle (Federal) 38-grain Copper Plated HP (AE22)
G1 Ballistic Coefficient = 0.130

Distance • Yards	Muzzle	25	50	75	100	125	150	175	200
Velocity • fps	1260	1177	1110	1054	1009	971	943	908	882
Energy • ft-lbs	135	117	105	94	86	80	75	70	66
Path • Inches	-1.5	+1.4	+2.8	+2.4	0.0	-4.5	-11.5	-20.6	-32.5
Wind Drift • Inches	0.0	0.3	1.3	2.9	5.0	7.6	10.6	14.0	17.9

Aguila 40-grain Interceptor Solid (1B222320)
G1 Ballistic Coefficient = 0.140

Distance • Yards	Muzzle	25	50	75	100	125	150	175	200
Velocity • fps	1470	1368	1277	1197	1130	1074	1029	991	958
Energy • ft-lbs	192	166	145	127	113	103	94	87	82
Path • Inches	-1.5	+0.8	+1.9	+1.7	0.0	-3.4	-8.7	-16.1	-25.6
Wind Drift • Inches	0.0	0.3	1.2	2.7	4.9	7.6	10.8	14.5	18.6

Aguila 40-grain Long Rifle High Velocity (1B222328)
G1 Ballistic Coefficient = 0.139

Distance • Yards	Muzzle	25	50	75	100	125	150	175	200
Velocity • fps	1255	1178	1113	1061	1017	981	949	920	894
Energy • ft-lbs	139	123	110	100	92	86	80	75	71
Path • Inches	-1.5	+1.4	+2.8	+2.3	0.0	-4.4	-11.1	-20.2	-31.9
Wind Drift • Inches	0.0	0.4	1.4	3.0	5.2	7.9	11.1	14.7	18.7

Fiocchi 40-grain Copper Plated Solid Point (22HVCRN)

G1 Ballistic Coefficient = 0.135

Distance • Yards	Muzzle	25	50	75	100	125	150	175	200
Velocity • fps	1255	1176	1110	1057	1013	976	943	919	888
Energy • ft-lbs	140	123	109	99	91	85	79	74	70
Path • Inches	-1.5	+1.4	+2.8	+2.4	0.0	-4.5	-11.2	-20.4	-32.2
Wind Drift • Inches	0.0	0.3	1.3	2.8	4.8	7.3	10.2	13.5	17.3

PMC 40-grain LRN (22SC)

G1 Ballistic Coefficient = 0.139

Distance • Yards	Muzzle	25	50	75	100	125	150	175	200
Velocity • fps	1255	1178	1113	1061	1017	981	949	920	894
Energy • ft-lbs	140	123	110	100	92	86	80	75	71
Path • Inches	-1.5	+1.4	+2.8	+2.3	0.0	-4.4	-11.1	-20.2	-31.9
Wind Drift • Inches	0.0	0.4	1.4	3.0	5.2	7.9	11.1	14.7	18.7

Remington 40-grain Long Rifle Solid (1522)

G1 Ballistic Coefficient = 0.13

Distance • Yards	Muzzle	25	50	75	100	125	150	175	200
Velocity • fps	1255	1167	1113	1061	1017	979	947	918	892
Energy • ft-lbs	140	123	110	100	92	85	80	75	71
Path • Inches	-1.5	+1.4	+2.8	+2.3	0.0	-4.4	-11.1	-20.1	-31.7
Wind Drift • Inches	0.0	0.4	1.4	3.0	5.2	8.0	11.2	14.8	18.9

Remington 40-grain Long Rifle Thunderbolt Solid (TB22A)

G1 Ballistic Coefficient = 0.13

Distance • Yards	Muzzle	25	50	75	100	125	150	175	200
Velocity • fps	1255	1177	1110	1060	1017	979	949	920	892
Energy • ft-lbs	140	123	109	100	92	85	80	75	71
Path • Inches	-1.5	+1.4	+2.8	+2.3	0.0	-4.4	-11.1	-20.1	-31.7
Wind Drift • Inches	0.0	0.4	1.4	3.0	5.2	8.0	11.2	14.8	18.9

Winchester 40-grain Long Rifle Wildcat Solid (WW22LR)

G1 Ballistic Coefficient = 0.1

Distance • Yards	Muzzle	25	50	75	100	125	150	175	200
Velocity • fps	1255	1177	1110	1060	1017	979	949	920	892
Energy • ft-lbs	140	123	109	100	92	85	80	75	71
Path • Inches	-1.5	+1.4	+2.8	+2.3	0.0	-4.4	-11.1	-20.1	-31.7
Wind Drift • Inches	0.0	0.4	1.4	3.0	5.2	8.0	11.2	14.8	18.9

Winchester 40-grain Long Rifle Solid (X22LR)

G1 Ballistic Coefficient = 0.1

Distance • Yards	Muzzle	25	50	75	100	125	150	175	200
Velocity • fps	1255	1177	1110	1060	1017	979	949	920	892
Energy • ft-lbs	140	123	109	100	92	85	80	75	71
Path • Inches	-1.5	+1.4	+2.8	+2.3	0.0	-4.4	-11.1	-20.1	-31.7
Wind Drift • Inches	0.0	0.4	1.4	3.0	5.2	8.0	11.2	14.8	18.9

Fiocchi 40-grain Lead Round Nose (22CRN)

G1 Ballistic Coefficient = 0.1

Distance • Yards	Muzzle	25	50	75	100	125	150	175	200
Velocity • fps	1250	1172	1107	1054	1011	974	942	913	886
Energy • ft-lbs	139	122	109	99	91	84	79	74	70
Path • Inches	-1.5	+1.4	+2.8	+2.4	0.0	-4.5	-11.3	-20.5	-32.4
Wind Drift • Inches	0.0	0.3	1.3	2.8	4.8	7.2	10.2	13.5	17.2

Federal 40-grain Long Rifle Solid (510)

G1 Ballistic Coefficient = 0.

Distance • Yards	Muzzle	25	50	75	100	125	150	175	200
Velocity • fps	1240	1163	1100	1047	1005	968	937	908	882
Energy • ft-lbs	137	120	107	98	90	83	78	73	69
Path • Inches	-1.5	+1.5	+2.8	+2.4	0.0	-4.6	-11.4	-20.8	-32.8
Wind Drift • Inches	0.0	0.3	1.3	2.8	4.8	7.2	10.1	13.4	17.1

American Eagle (Federal) 40-grain Long Rifle Solid (AE5022)

G1 Ballistic Coefficient = 0.

Distance • Yards	Muzzle	25	50	75	100	125	150	175	200
Velocity • fps	1240	1163	1100	1047	1005	968	937	908	882
Energy • ft-lbs	137	120	107	98	90	83	78	73	69
Path • Inches	-1.5	+1.5	+2.8	+2.4	0.0	-4.6	-11.4	-20.8	-32.8
Wind Drift • Inches	0.0	0.3	1.3	2.8	4.8	7.2	10.1	13.4	17.1

Federal 40-grain Solid (710) + (810)
G1 Ballistic Coefficient = 0.139

Distance • Yards	Muzzle	25	50	75	100	125	150	175	200
Velocity • fps	1240	1165	1100	1052	1010	975	943	915	889
Energy • ft-lbs	137	121	108	98	91	84	79	74	70
Path • Inches	-1.5	+1.5	+2.8	+2.4	0.0	-4.5	-11.3	-20.6	-32.4
Wind Drift • Inches	0.0	0.3	1.4	3.0	5.1	7.8	10.9	14.5	18.5

Blazer 40-grain LRN (0021)
G1 Ballistic Coefficient = 0.150

Distance • Yards	Muzzle	25	50	75	100	125	150	175	200
Velocity • fps	1235	1166	1104	1060	1026	985	955	927	902
Energy • ft-lbs	135	121	108	100	93	86	81	76	72
Path • Inches	-1.5	+1.4	+2.8	+2.4	0.0	-4.4	-11.1	-20.2	-31.8
Wind Drift • Inches	0.0	0.3	1.2	2.8	4.8	7.3	10.2	13.5	17.3

CCI 40-grain Mini-Mag CPRN (0030)
G1 Ballistic Coefficient = 0.130

Distance • Yards	Muzzle	25	50	75	100	125	150	175	200
Velocity • fps	1235	1156	1092	1040	998	961	929	901	874
Energy • ft-lbs	135	119	106	97	88	82	77	72	68
Path • Inches	-1.5	+1.5	+2.9	+2.4	0.0	-4.6	-11.6	-21.1	-33.3
Wind Drift • Inches	0.0	0.3	1.3	2.8	4.9	7.4	10.3	13.7	17.5

CCI 40-grain Small Game Bullet (0058)
G1 Ballistic Coefficient = 0.125

Distance • Yards	Muzzle	25	50	75	100	125	150	175	200
Velocity • fps	1235	1153	1088	1035	992	955	922	893	866
Energy • ft-lbs	135	118	105	95	87	81	76	71	67
Path • Inches	-1.5	+1.5	+2.9	+2.5	0.0	-4.7	-11.7	-21.4	-33.7
Wind Drift • Inches	0.0	0.3	1.3	2.9	5.0	7.6	10.7	14.2	18.0

CCI 40-grain AR Tactical GREEN CPRN (0953) LF
G1 Ballistic Coefficient = 0.116

Distance • Yards	Muzzle	25	50	75	100	125	150	175	200
Velocity • fps	1200	1120	1056	1006	964	929	897	868	841
Energy • ft-lbs	128	111	99	90	82	77	71	67	63
Path • Inches	-1.5	+1.7	+3.1	+2.6	0.0	-5.0	-12.5	-22.6	-35.7
Wind Drift • Inches	0.0	0.4	1.6	3.4	5.8	8.7	12.2	16.2	20.6

CCI 40-grain SELECT LRN (0045)
G1 Ballistic Coefficient = 0.116

Distance • Yards	Muzzle	25	50	75	100	125	150	175	200
Velocity • fps	1200	1120	1056	1006	964	929	897	868	841
Energy • ft-lbs	128	111	99	90	82	77	71	67	63
Path • Inches	-1.5	+1.7	+3.1	+2.6	0.0	-5.0	-12.5	-22.6	-35.7
Wind Drift • Inches	0.0	0.4	1.6	3.4	5.8	8.7	12.2	16.2	20.6

Aguila 40-grain Interceptor HP (1B222321)
G1 Ballistic Coefficient = 0.140

Distance • Yards	Muzzle	25	50	75	100	125	150	175	200
Velocity • fps	1470	1368	1277	1197	1130	1074	1029	991	958
Energy • ft-lbs	192	166	145	127	113	103	94	87	82
Path • Inches	-1.5	+0.8	+1.9	+1.7	0.0	-3.4	-8.7	-16.1	-25.6
Wind Drift • Inches	0.0	0.3	1.2	2.7	4.9	7.6	10.8	14.5	18.6

CCI 40-grain Velocitor CPHP (0047)
G1 Ballistic Coefficient = 0.125

Distance • Yards	Muzzle	25	50	75	100	125	150	175	200
Velocity • fps	1435	1325	1230	1149	1184	1032	989	952	920
Energy • ft-lbs	183	156	134	117	104	95	87	81	75
Path • Inches	-1.5	+0.9	+2.1	+1.9	0.0	-3.7	-9.5	-17.5	-27.9
Wind Drift • Inches	0.0	0.3	1.4	3.1	5.5	8.4	11.9	15.9	20.3

Winchester 40-grain Super-X LHP (XHV22LR)
G1 Ballistic Coefficient = 0.140

Distance • Yards	Muzzle	25	50	75	100	125	150	175	200
Velocity • fps	1435	1336	1249	1173	1110	1058	1016	979	948
Energy • ft-lbs	183	159	139	122	109	100	92	85	80
Path • Inches	-1.5	+0.9	+2.0	+1.8	0.0	-3.6	-9.1	-16.7	-26.6
Wind Drift • Inches	0.0	0.3	1.2	2.8	4.9	7.6	10.8	14.4	18.5

RWS 40-grain High Velocity Hollow Point (213 2494)

G1 Ballistic Coefficient = 0.104

Distance • Yards	Muzzle	25	50	75	100	125	150	175	200
Velocity • fps	1310	1199	1120	1043	990	947	909	875	845
Energy • ft-lbs	150	128	110	97	85	80	73	68	63
Path • Inches	-1.5	+1.4	+2.8	+2.4	0.0	-4.5	-11.4	-20.9	-33.1
Wind Drift • Inches	0.0	0.5	1.8	4.0	7.0	10.5	14.7	19.4	24.7

Winchester 40-grain Super-X Power-Point (X22LRPP)

G1 Ballistic Coefficient = 0.118

Distance • Yards	Muzzle	25	50	75	100	125	150	175	200
Velocity • fps	1280	1186	1110	1049	1001	961	926	894	866
Energy • ft-lbs	146	125	109	98	89	82	76	71	67
Path • Inches	-1.5	+1.4	+2.8	+2.4	0.0	-4.5	-11.3	-20.6	-32.5
Wind Drift • Inches	0.0	0.4	1.6	3.6	6.1	9.3	13.0	17.2	21.9

More and more rimfire ammo is now available in lead free or "green" ammo. This is the CCI 21-grain hollowpoint load. (Safari Press Archives)

506

.22 Winchester Magnum Rimfire (WMR)

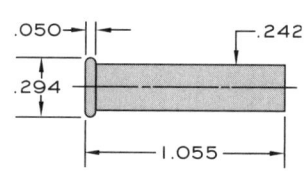

The .22 WMR cartridge can be thought of as a .22 Long Rifle on steroids. The bad news is that the .22 WMR is a significantly larger cartridge and will NOT fit into standard .22 guns chambered for .22 Long Rifle—nor for that matter can the .22 Long Rifle cartridge be fired in a .22 WMR chamber. The good news is that in the fastest loadings the .22 WMR begin to approach the performance of the .22 Hornet. That's a lot of performance from a rimfire cartridge.

Relative Recoil Factor = 0.34

Specifications:

Controlling Agency for Standardization of this Ammunition: SAAMI

Bullet Weight Grains	Velocity fps	Maximum Average Pressure	
		Copper Crusher	Transducer
40	1,875	N/S	24,000 psi

Standard barrel for velocity testing: 24 inches long—1 turn in 16-inch twist

Availability:

Winchester 28-grain JHP Tin (X22MLF) LF
G1 Ballistic Coefficient = 0.086

Distance • Yards	Muzzle	25	50	75	100	125	150	175	200
Velocity • fps	2200	1972	1759	1564	1394	1243	1128	1044	981
Energy • ft-lbs	301	242	192	152	121	96	79	68	60
Path • Inches	-1.5	-0.1	+0.7	+0.8	0.0	-1.9	-5.2	-10.3	-17.3
Wind Drift • Inches	0.0	0.3	1.4	3.4	6.4	10.4	15.6	21.8	28.8

Winchester 30-grain V-MAX (S22M2PT)
G1 Ballistic Coefficient = 0.089

Distance • Yards	Muzzle	25	50	75	100	125	150	175	200
Velocity • fps	2250	2026	1817	1623	1450	1296	1171	1078	1009
Energy • ft-lbs	337	274	220	175	140	112	91	77	68
Path • Inches	-1.5	-0.2	+0.6	+0.7	0.0	-1.8	-4.8	-9.4	-15.9
Wind Drift • Inches	0.0	0.3	1.3	3.2	5.9	9.7	14.6	20.4	27.3

Winchester 30-grain JHP (S22M2)
G1 Ballistic Coefficient = 0.090

Distance • Yards	Muzzle	25	50	75	100	125	150	175	200
Velocity • fps	2250	2029	1821	1629	1455	1303	1179	1084	1015
Energy • ft-lbs	337	274	221	177	141	113	93	78	69
Path • Inches	-1.5	+0.2	+0.6	+0.7	0.0	-1.7	-4.8	-9.3	-15.8
Wind Drift • Inches	0.0	0.3	1.2	2.8	5.2	8.6	12.9	18.2	24.2

CI 30-grain 22MAG V-MAX (0073)
G1 Ballistic Coefficient = 0.116

Distance • Yards	Muzzle	25	50	75	100	125	150	175	200
Velocity • fps	2200	2030	1866	1714	1571	1439	1321	1219	1135
Energy • ft-lbs	322	274	232	196	164	138	116	99	86
Path • Inches	-1.5	-0.2	+0.5	+0.6	0.0	-1.5	-4.1	-7.9	-18.2
Wind Drift • Inches	0.0	0.2	1.0	2.4	4.5	7.2	10.8	15.2	20.5

CI 30-grain WMR Maxi-Mag. + V Hollow Point (0059)
G1 Ballistic Coefficient = 0.084

Distance • Yards	Muzzle	25	50	75	100	125	150	175	200
Velocity • fps	2200	1967	1749	1551	1375	1228	1114	1032	971
Energy • ft-lbs	322	258	204	160	126	100	83	71	63
Path • Inches	-1.5	+0.1	+0.7	+0.8	0.0	-2.0	-5.4	-10.5	-17.7
Wind Drift • Inches	0.0	0.4	1.5	3.5	6.5	10.7	16.0	22.4	29.6

CI 30-grain WMR TNT Hollow Point (0063)
G1 Ballistic Coefficient = 0.088

Distance • Yards	Muzzle	25	50	75	100	125	150	175	200
Velocity • fps	2200	1977	1768	1577	1405	1258	1142	1055	991
Energy • ft-lbs	322	260	208	166	131	106	87	74	66
Path • Inches	-1.5	+0.1	+0.7	+0.8	0.0	-1.9	-5.1	-10.0	-16.9
Wind Drift • Inches	0.0	0.3	1.4	3.3	6.2	10.1	15.2	21.2	28.1

Federal 30-grain WMR Speer TNT HP (P765)

G1 Ballistic Coefficient = 0.090

Distance • Yards	Muzzle	25	50	75	100	125	150	175	200
Velocity • fps	2200	1982	1780	1589	1420	1273	1155	1067	1002
Energy • ft-lbs	320	262	210	168	134	108	89	76	67
Path • Inches	-1.5	-0.2	+0.6	+0.8	0.0	-1.8	-5.0	-9.8	-16.5
Wind Drift • Inches	0.0	0.3	1.4	3.2	6.0	9.8	14.7	20.6	27.4

Hornady 30-grain WMR V-Max (83202)

G1 Ballistic Coefficient = 0.090

Distance • Yards	Muzzle	25	50	75	100	125	150	175	200
Velocity • fps	2200	1982	1780	1589	1420	1273	1155	1067	1002
Energy • ft-lbs	320	262	210	168	134	108	89	76	67
Path • Inches	-1.5	-0.2	+0.6	+0.8	0.0	-1.8	-5.0	-9.8	-16.5
Wind Drift • Inches	0.0	0.3	1.4	3.2	6.0	9.8	14.7	20.6	27.4

CCI 30-grain WMR Speer TNT Green HP (0060) LF

G1 Ballistic Coefficient = 0.090

Distance • Yards	Muzzle	25	50	75	100	125	150	175	200
Velocity • fps	2050	1841	1648	1471	1317	1190	1092	1021	965
Energy • ft-lbs	280	226	181	144	116	94	80	69	62
Path • Inches	-1.5	0.0	+0.8	+0.9	0.0	-2.2	-5.9	-11.5	-19.1
Wind Drift • Inches	0.0	0.4	1.5	3.5	6.6	10.7	15.9	22.0	28.8

Remington 33-grain WMR Premier AccuTip-V (PR22M1)

G1 Ballistic Coefficient = 0.13

Distance • Yards	Muzzle	25	50	75	100	125	150	175	200
Velocity • fps	2000	1836	1730	1609	1495	1388	1292	1208	1138
Energy • ft-lbs	293	254	219	190	164	141	122	107	95
Path • Inches	-1.5	-0.1	+0.7	+0.8	0.0	-1.8	-4.6	-8.8	-14.5
Wind Drift • Inches	0.0	0.2	1.0	2.3	4.2	6.8	10.0	14.0	18.7

Winchester 34-grain WMR Jacketed Hollow Point (S22WM)

G1 Ballistic Coefficient = 0.10

Distance • Yards	Muzzle	25	50	75	100	125	150	175	200
Velocity • fps	2120	1931	1753	1537	1435	1304	1192	1104	1037
Energy • ft-lbs	338	282	232	190	155	132	113	100	90
Path • Inches	-1.5	-0.1	+0.7	+0.8	0.0	-1.8	-5.0	-9.6	-16.1
Wind Drift • Inches	0.0	0.3	1.2	2.9	5.5	8.9	13.3	18.6	24.7

Fiocchi 40-grain WMR JSP (22WMA)

G1 Ballistic Coefficient = 0.1

Distance • Yards	Muzzle	25	50	75	100	125	150	175	200
Velocity • fps	1910	1770	1638	1515	1402	1300	1211	1136	1076
Energy • ft-lbs	324	278	238	204	175	150	130	115	103
Path • Inches	-1.5	0.0	+0.9	+0.9	0.0	-2.0	-5.3	-10.1	-16.6
Wind Drift • Inches	0.0	0.2	1.0	2.3	4.2	6.8	10.1	14.0	18.6

Remington 40-grain WMR Pointed Soft Point Solid (R22M2)

G1 Ballistic Coefficient = 0.1

Distance • Yards	Muzzle	25	50	75	100	125	150	175	200
Velocity • fps	1910	1751	1600	1466	1340	1235	1147	1076	1021
Energy • ft-lbs	324	272	227	191	159	136	117	103	93
Path • Inches	-1.5	+0.1	+1.0	+1.0	0.0	-2.2	-5.8	-11.1	-18.2
Wind Drift • Inches	0.0	0.3	1.3	3.0	5.5	8.8	13.0	18.0	23.7

Winchester 40-grain WMR Full Metal Jacket Solid (X22M)

G1 Ballistic Coefficient = 0.

Distance • Yards	Muzzle	25	50	75	100	125	150	175	200
Velocity • fps	1910	1746	1592	1452	1326	1218	1130	1061	1007
Energy • ft-lbs	324	271	225	187	156	132	113	100	90
Path • Inches	-1.5	+0.1	+1.0	+1.0	0.0	-2.2	-6.0	-11.4	-18.7
Wind Drift • Inches	0.0	0.3	1.3	3.1	5.7	9.2	13.6	18.7	24.6

Federal 40-grain WMR Full Metal Jacket Solid (737)

G1 Ballistic Coefficient = 0.

Distance • Yards	Muzzle	25	50	75	100	125	150	175	200
Velocity • fps	1880	1719	1570	1432	1310	1205	1121	1055	1003
Energy • ft-lbs	315	263	219	182	152	129	112	99	89
Path • Inches	-1.5	+0.1	+1.0	+1.0	0.0	-2.3	-6.1	-11.7	-19.2
Wind Drift • Inches	0.0	0.3	1.3	3.1	5.8	9.2	13.6	18.7	24.6

CCI 40-grain WMR Gamepoint JSP (0022)
G1 Ballistic Coefficient = 0.133

Distance • Yards	Muzzle	25	50	75	100	125	150	175	200
Velocity • fps	1875	1740	1614	1494	1385	1287	1202	1131	1073
Energy • ft-lbs	312	269	231	198	170	147	128	114	102
Path • Inches	-1.5	+0.1	+0.9	+1.0	0.0	-2.1	-5.5	-10.4	-17.0
Wind Drift • Inches	0.0	0.2	1.0	2.3	4.2	6.8	10.0	13.9	18.4

CCI 40-grain WMR Maxi-Mag Solid (0023)
G1 Ballistic Coefficient = 0.127

Distance • Yards	Muzzle	25	50	75	100	125	150	175	200
Velocity • fps	1875	1734	1603	1478	1366	1266	1180	1110	1054
Energy • ft-lbs	312	267	228	194	166	142	124	109	99
Path • Inches	-1.5	+0.1	+1.0	+1.0	0.0	-2.1	-5.6	-10.7	-17.5
Wind Drift • Inches	0.0	0.2	1.0	2.4	4.5	7.2	10.6	14.6	19.3

RWS 40-grain WMR Hollow Point
G1 Ballistic Coefficient = 0.116

Distance • Yards	Muzzle	25	50	75	100	125	150	175	200
Velocity • fps	2020	1858	1710	1563	1430	1314	1213	1130	1065
Energy • ft-lbs	360	307	260	217	180	153	131	114	101
Path • Inches	-1.5	0.0	+0.8	+0.8	0.0	-1.9	-5.0	-9.7	-16.0
Wind Drift • Inches	0.0	0.3	1.2	2.7	5.0	8.1	12.0	16.8	22.3

Fiocchi 40-grain WMR JHP (22WMB)
G1 Ballistic Coefficient = 0.120

Distance • Yards	Muzzle	25	50	75	100	125	150	175	200
Velocity • fps	1910	1759	1617	1486	1366	1261	1172	1100	1043
Energy • ft-lbs	324	275	232	196	166	141	122	107	97
Path • Inches	-1.5	0.1	+0.9	+1.0	0.0	-2.1	-5.6	-10.7	-17.6
Wind Drift • Inches	0.0	0.3	1.1	2.5	4.6	7.5	11.0	15.3	20.2

Remington 40-grain WMR Jacketed Hollow Point
G1 Ballistic Coefficient = 0.116

Distance • Yards	Muzzle	25	50	75	100	125	150	175	200
Velocity • fps	1910	1754	1610	1472	1350	1244	1155	1084	1028
Energy • ft-lbs	324	273	230	193	162	137	118	104	94
Path • Inches	-1.5	+0.1	+1.0	+1.0	0.0	-2.2	-5.8	-11.0	-18.0
Wind Drift • Inches	0.0	0.3	1.2	2.9	5.4	8.7	12.8	17.7	23.3

Winchester 40-grain WMR Jacketed Hollow Point (X22MH)
G1 Ballistic Coefficient = 0.110

Distance • Yards	Muzzle	25	50	75	100	125	150	175	200
Velocity • fps	1910	1746	1592	1452	1326	1218	1130	1061	1007
Energy • ft-lbs	324	271	225	187	156	132	113	100	90
Path • Inches	-1.5	+0.1	+1.0	+1.0	0.0	-2.2	-6.0	-11.4	-18.7
Wind Drift • Inches	0.0	0.3	1.3	3.1	5.7	9.2	13.6	18.7	24.6

CCI 40-grain WMR Maxi-Mag Hollow Point (0024)
G1 Ballistic Coefficient = 0.115

Distance • Yards	Muzzle	25	50	75	100	125	150	175	200
Velocity • fps	1875	1720	1574	1442	1320	1219	1134	1068	1015
Energy • ft-lbs	312	263	220	185	155	132	114	101	92
Path • Inches	-1.5	+0.1	+1.0	+1.0	0.0	-2.3	-6.0	-11.4	-18.8
Wind Drift • Inches	0.0	0.3	1.2	2.7	5.0	8.0	11.8	16.3	21.4

CCI 40-grain JHP (0069)
G1 Ballistic Coefficient = 0.120

Distance • Yards	Muzzle	25	50	75	100	125	150	175	200
Velocity • fps	1300	1204	1125	1063	1013	972	936	904	875
Energy • ft-lbs	169	145	126	113	103	94	88	82	77
Path • Inches	-1.5	+1.3	+2.7	+2.3	0.0	-4.4	-11.1	-20.3	-32.2
Wind Drift • Inches	0.0	0.4	1.6	3.5	6.1	9.3	13.0	17.2	21.8

Speer 40-grain GDHP-SB (954)
G1 Ballistic Coefficient = 0.087

Distance • Yards	Muzzle	25	50	75	100	125	150	175	200
Velocity • fps	1150	1061	995	943	898	860	826	794	764
Energy • ft-lbs	117	100	88	79	72	66	61	56	52
Path • Inches	-1.5	+2.0	+3.6	+3.0	0.0	-5.7	-14.4	-26.2	-41.5
Wind Drift • Inches	0.0	0.5	1.9	4.0	6.9	10.5	14.6	19.5	25.0

Winchester 40-grain FMJ (S22M)

G1 Ballistic Coefficient = 0.086

Distance • Yards	Muzzle	25	50	75	100	125	150	175	200
Velocity • fps	2200	1972	1759	1564	1394	1243	1128	1044	981
Energy • ft-lbs	430	345	275	217	172	137	113	97	86
Path • Inches	-1.5	-0.1	+0.7	+0.8	0.0	-1.9	-5.2	-10.3	-17.3
Wind Drift • Inches	0.0	0.3	1.4	3.4	6.4	10.4	15.6	21.8	28.8

Winchester 45-grain Dynapoint (USA22M)

G1 Ballistic Coefficient = 0.126

Distance • Yards	Muzzle	25	50	75	100	125	150	175	200
Velocity • fps	1550	1430	1322	1227	1147	1083	1032	989	953
Energy • ft-lbs	240	204	175	150	131	117	106	98	91
Path • Inches	-1.5	+0.7	+1.8	+1.6	0.0	-3.2	-8.3	-15.5	-24.8
Wind Drift • Inches	0.0	0.4	1.4	3.3	5.9	9.2	13.2	17.8	22.9

Federal 50-grain WMR Jacketed Hollow Point (757)

G1 Ballistic Coefficient = 0.160

Distance • Yards	Muzzle	25	50	75	100	125	150	175	200
Velocity • fps	1530	1436	1350	1270	1200	1141	1090	1047	1011
Energy • ft-lbs	260	229	202	179	160	144	132	122	114
Path • Inches	-1.5	+0.6	+1.6	+1.5	0.0	-3.0	-7.7	-14.2	-22.6
Wind Drift • Inches	0.0	0.3	1.1	2.6	4.7	7.3	10.5	14.3	18.5

.22 Rimfire Shotshells

In the 1940s my father took me to some outdoor shows. One of the booths that attracted my interest was a trap-shooting game that used .22 rimfire shotshells and miniature clay targets. The backstop was nothing more than a sheet of heavy canvas hung from the ceiling of the hall. That was enough to stop the very fine shot. It's hard to imagine anyone getting away with that today, but I never heard of any accident resulting from those shows. The .22 rimfire shotshells also make a fine load for a snake gun.

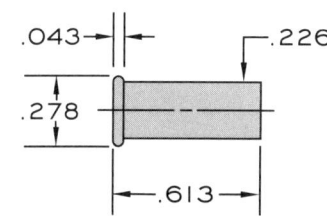

Relative Recoil Factor = 0.14

Specifications:

Controlling Agency for Standardization of this Ammunition: SAAMI

Bullet Weight Grains	Velocity fps	Maximum Average Pressure	
		Copper Crusher	Transducer
25 #12 Shot	1,000	N/S	24,000 psi

Standard barrel for velocity testing: 24 inches long—1 turn in 16-inch twist

Availability:

Federal 25-grain Long Rifle Bird Shot (716)

This load uses #12 shot. No performance data are given.

CCI 31-grain Long Rifle Shotshell (0039)

This load uses #12 shot.
Muzzle Velocity is listed at 1000 fps.

Winchester Long Rifle Shot (X22LRS)

This load uses #12 shot. No performance data are given.

CCI 52-grain WMR Shotshell (0025)

This load uses #12 shot.
Muzzle velocity is listed at 1000 fps.

Choosing a Big-Game Bullet

Your Guide to Modern Hunting Bullets and How to Know Which Ones Are Best for the Job at Hand

by John Barsness

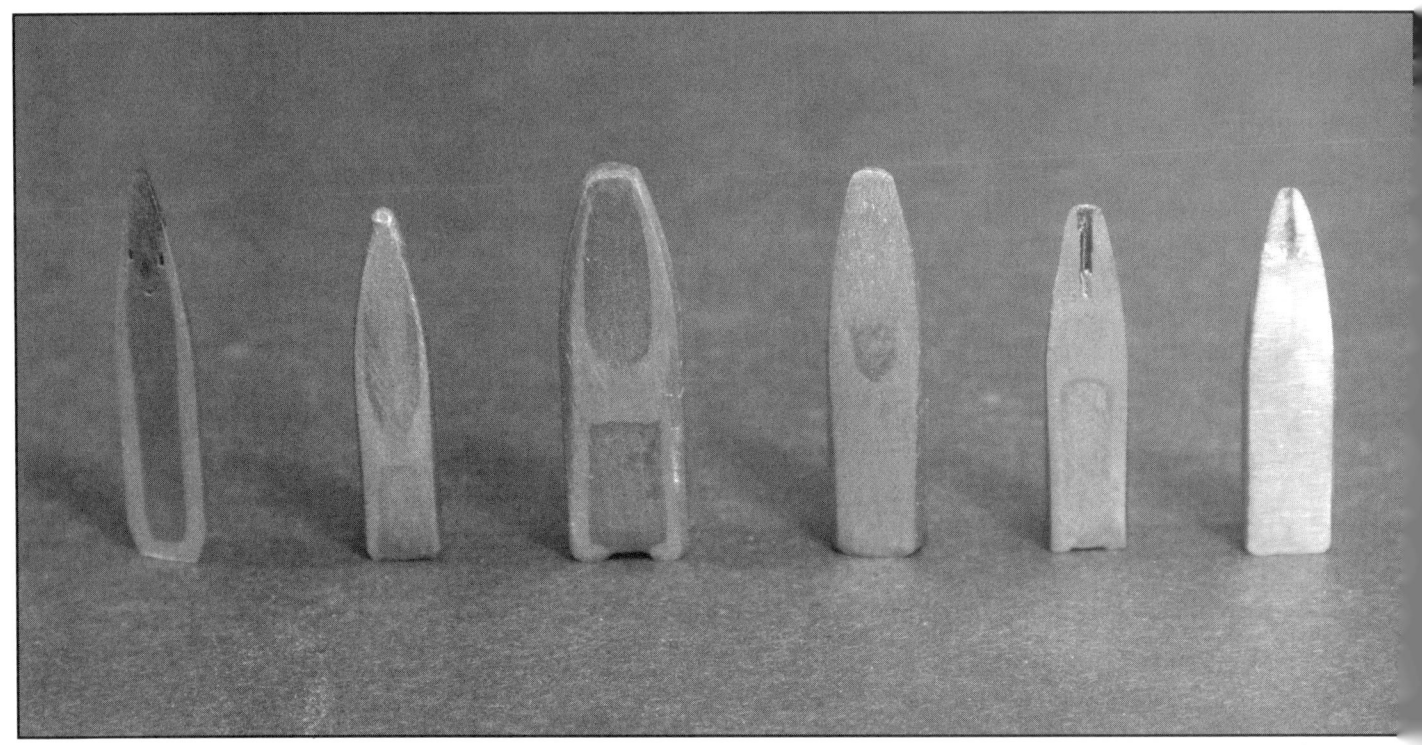

Many different designs have been used for premium bullets over the years. This lineup includes six designs as shown: the Swift Scirocco II, Nosler Partitio Swift A-Frame, Combined Technology Fail Safe, and Barnes X-Bullet.

In recent years many hunters have started judging all big-game bullets by how much weight they retain after expansion. This probably started in 1978 because of a book by Bob Hagel with a title almost as long as a Nile crocodile: *Game Loads and Practical Ballistics for the American Hunter.*

In 1978 there were only two truly "premium" bullets available in America: the Nosler Partition and the Bitterroot Bonded Core (BBC). The Nosler was available almost everywhere, and it featured a two-part lead core separated by a "partition" of its copper-alloy jacket that prevented the bullet from expanding beyond a certain point. The BBC was a rather obscure bullet made by a guy from Idaho named Bill Steigers, with a lead core essentially soldered to a heavy copper jacket. Hagel noted that the BBC often retained 90 percent or more of its weight after expansion, while the Nosler Partition usually retained around 65 to 70 percent. A lot of hunters said *"Aha!"* and jumped on retained weight as the secret to bullet performance on big game.

What these hunters apparently missed (sinc it was plainly stated in Hagel's book) was the fa that the BBC bullets tended to open up widely, s they didn't penetrate as deeply as Nosler Partition The frontal area of an expanded bullet has a f greater influence than retained weight on hc deeply bullets penetrate, something that mar hunters still don't get despite the vast selection premium bullets available today. Both "bonde core" and "high weight retention" became catc phrases for high quality. Some years ago a g phoned me to ask questions about various bulle but in reality he wanted to brag about his hunti accomplishments and great taste in fine equipme Eventually he said, "I won't even use a bullet tha not bonded." I was tempted to ask about licens and insured as well, but kept my mouth shut.

Today's hunting bullets offer almost as ma options as cable TV, and almost as much copy of what sells: bonded-core, controlled-expansi monometal, nontoxic, partitioned, plastic-tipp etc. This long list of jargon implies that bullets are complex, but the goal is simple: a hole through the vitals of the animal we're hunting. This is the bottom line according to the foremost forensic ballisticians, the people who study bullet wounds. They even have a technical term for the hole, calling it the "permanent wound channel."

All hunting bullets perform somewhere between two extremes. At one extreme are bullets that fragment completely upon impact, creating very wide but relatively shallow wound channels. At the other end of the spectrum are nonexpanding bullets, generally known as "solids." These penetrate very deeply, creating long, narrow wound channels.

Penetration isn't the only criterion for a big-game bullet. If it were, we'd all be using solids, even on doe whitetails. The big advantage of a bullet like the Bitterroot Bonded Core is not extreme penetration, but the bigger permanent wound channel created by its wider mushroom.

With some bullets the rear end also tends to expand. This often happens with the Swift A-Frame and the Winchester XP3. Rear-end expansion functions much like front-end expansion, reducing penetration but creating a wider hole. This is the reason that Swift A-Frames usually don't penetrate any deeper than Nosler Partitions of the same size, even though the A-Frames usually retain more weight—and make a bigger hole.

Today's tendency to equate weight retention with "killing power" is so pronounced that some hunters are offended when a super-bullet like a Barnes X loses even one petal. In my recovered bullet collection are both an X-Bullet and a Fail Safe with all their petals missing. Yet the animals (an axis deer and a gemsbok) died very quickly. Several years ago I shot a Cape buffalo with a 300-grain Fail Safe through both lungs. We found one of the petals from the bullet in the skin at the edge of the exit hole. The bullet may have lost more petals, but since it disappeared into the Okavango Delta, we'll never know. The bull still died within 75 yards.

In fact, a bullet that at least partially fragments, sending chunks of jacket or core spinning off along the bullet's path, tends to make a bigger hole. This

Northwest Territories caribou was taken with a 150-grain Nosler AccuBond from a .308 Winchester, dropping the bull on impact. The AccuBond is **a bullet** *that both expands violently yet penetrates pretty well, a good combination on animals this size.*

The Right Bullet for the Job

Most expanding big-game bullets fall into certain categories. Here's a rough guide.

Deer Bullets: Berger VLD, Federal Fusion, Hornady SST and Interlock, Nosler Ballistic Tip, Remington Core-Lokt, Sierra GameKing and ProHunter, Speer Hot-Cor, Winchester Power Point.

In heavier weights and at lower velocities, all of these can be used on bigger game, especially when placed broadside through the ribs, but they work perfectly on lighter game. The Berger VLD is the only bullet I've encountered with "delayed expansion." All other expanding bullets start to open as soon as they encounter skin, but the VLD penetrates about two inches before expanding violently.

Another exception found in this general list is the .30-caliber, 180-grain Nosler Ballistic Tip, recently redesigned with a very heavy jacket (about 60 percent of the bullet's weight) because so many hunters insisted on using it on elk. Now it really is an elk bullet. The Hornady Interlock is also a pretty tough bullet, called by some "the poor hunter's premium." Speer boattails are not Hot-Cors but are swaged with a softer lead core, so they tend to fragment somewhat more than Hot-Cors.

Wide Expanders: Hornady Interbond, Norma Oryx, Remington Ultra Core-Lokt, Speer Grand Slam, Swift Scirocco II, Woodleigh Weldcore.

These are all lead-cored, bonded bullets that don't have any feature to stop expansion, except perhaps a thickening of the jacket. Consequently, they tend to open up even wider at faster impact velocities. They all retain most of their weight but open widely enough that penetration normally isn't super-deep unless the bullets are very heavy and started at moderate velocities. Woodleighs, for instance, penetrate quite well when their super-heavy models are started at under 2,500 fps.

Deep Penetrators: Barnes X, Hornady GMX and DGX, North Fork, Nosler AccuBond, Partition, and E-Tip, Swift A-Frame, Winchester XP3, Trophy Bonded Bear Claw.

These only open up so far, typically with less frontal area than the wide expanders, so they penetrate deeper. The oddball in this group is the Nosler AccuBond, a "tipped" bonded bullet designed to expand similarly to the Nosler Partition, losing about one-third of its weight. Since most of the front end blows off, it can't be classed with the wide expanders. Like Partitions, the heavier AccuBonds are designed around their expected use, retaining more weight. I have seen both the 250-grain 9.3mm, and 260-grain penetrate close to lengthwise through heavy game like kudu and elk. The smaller AccuBonds act more like bonded-core deer bullets.

is exactly the function of the front end of a Nosler Partition. With Partitions, the wound channel starts out widely, with fragments of core creating what could be called "side channels," then narrows down considerably.

Today the ideal for good bullet performance is apparently complete penetration. The other day I read a thread on an Internet chat room, started by a guy asking about a newer brand of bonded bullet. Most of the replies noted "complete pass-throughs" or "good exit hole," all meant as compliments. Very few mentioned how quickly the animal died.

This is also common. Instead of judging bullets by how quickly they dispatch game animals, hunters talk about exit holes, or how much weight the bullet retained, or how pretty the bullet looked after expansion. Unfortunately, animals don't weigh bullets or admire the perfect shape of their expanded form. In the end, even exit holes have

nothing to do with how well a bullet kills becau: skin isn't a vital organ.

The Laws of Physics

Despite what many hunters believe, bullets a: also subject to the laws of physics. There's no w the same bullet can both create a wider hole a: penetrate deeper. Bullets that penetrate deeply ha smaller frontal areas and, hence, create smal wound channels. It's that simple. This is good wh hunting really big game. On animals weighing ei; hundred pounds or more we want a long wou: channel, in order to put a hole all the way throu the animal's big chest. This tears apart more of lungs and other pulmonary essentials than a wic hole through only one lung—the possible result w a bullet that expands wider, or loses half its weig

However, deer-size animals usually die quicl when hit with a more violently expanding bullet deer's chest isn't very thick, and a widely expanding bullet that loses some weight tears up far more of its vital organs than a bullet designed to penetrate deeply. I know this not because of any theory but because for some years I've paced off how far big-game animals traveled before dropping after solid chest hits. (And most do go a little way, unless hit in the spine or both shoulders.)

The deepest penetrating expanding bullets are what might be called "petal" bullets that usually retain almost all of their weight, including the Barnes X, Nosler E-Tip, the late lamented Combined Technology Fail Safe, and the new Hornady GMX. These open up not into a mushroom, but a "flower" with four petals. (Isn't it odd how something used to collect meat inspires images of mushrooms and flower petals?) These bullets penetrate deeply partly because of retained weight but primarily because they present less frontal area than most other bullets. They don't open as widely as a bonded-core bullet that expands halfway down its shank, and there are spaces between the petals.

Hunters have been using such bullets for over twenty years now, and I have been pacing off "traveling" distances for almost that long. Now, any animal can drop at the shot sometimes, no matter what the bullet. (This is known in statistics as an "example of one," which never proves anything.) The average traveling distance of the animals I've seen killed with petal bullets is a little over fifty yards—and some have gone over 50 yards before falling.

"Petal" bullets have less overall frontal area when expanded than most other bullets because of the spaces between the petals, which is one reason they penetrate so deeply.

On average, animals hit with bullets that expand widely or violently don't go nearly as far. Bullets with a quick-killing reputation share the tendency to lose quite a bit of their initial weight, up to half or even more. This basically describes any "conventional" bullet made with a copper-alloy jacket that isn't bonded to the lead core. Put one of these in the lungs of any "deer-size" game animal, and the animal will usually go down relatively quickly.

These 7mm bullets originally weighed between 154 and 160 grains, and were shot into dry newspaper at an impact velocity of around 2,600 fps, the approximate impact velocity from a 7mm magnum at 200 yards. All retained at least 70 percent of their weight and penetrated between 8 and 13 inches. They're arranged according to the depth of penetration, demonstrating that frontal area affects penetration more than retained weight. From left: Norma Oryx, Hornady Interbond, Kodiak, Swift Scirocco II, North Fork, Remington Core-Lokt Ultra Bonded, Swift A-Frame, Nosler Partition, Barnes Triple Shock X-Bullet.

The exception is when the impact (not muzzle) velocity of this type of bullet exceeds about 2,800 fps, causing the bullet to break up on shoulder bones. However, even a 130-grain bullet from a .270 Winchester, started at 3,100 fps or so, slows down to around 2,800 fps at 100 yards. While some hunters insist on the "insurance" of using a premium bullet on deer-size game, my hunting notes (from several hundred big-game animals) show that all the deer that traveled more than 100 yards after solid lung hits were shot with so-called "premium" bullets.

Yes, bullets that retain almost all their weight help us take larger game, but they aren't ideal for every kind of hunting. Europeans have known this for a long time, partly because they invented rifles. The German ammunition firm of RWS, for example, offers five big-game bullets of varying performance. The front ends of their "deer" bullets are designed to fragment precisely into several pieces because this has been proven by lots of field-testing to kill smaller animals more quickly. This is important in countries like Germany where the game belongs to the landowner. A lung-shot deer that runs off your 100-hectare hunting lease before dropping isn't your deer anymore.

While many American hunters continue to argue that one particular bullet (their favorite, naturally) is the absolutely perfect bullet for all game anywhere, more American bullet companies are recognizing that different bullets work better for some jobs. That's why Nosler offers an array from the Ballistic Tip (about 50 to 60 percent weight retention) through the E-Tip (a petal-type bullet that usually retains 100 percent of its weight). Hornady offers everything from the violently expanding SST to the new monometal, petal-type GMX, plus the steel-jacketed DGS and DGX dangerous-game bullets. All the major ammunition companies do the same thing, whether with their own designs or bullets purchased from companies like Barnes, Hornady, and Nosler.

Some companies also design the same basic bullet to perform somewhat differently, depending on the intended use. The lighter Nosler Partitions, for instance, usually lose about one-third of their

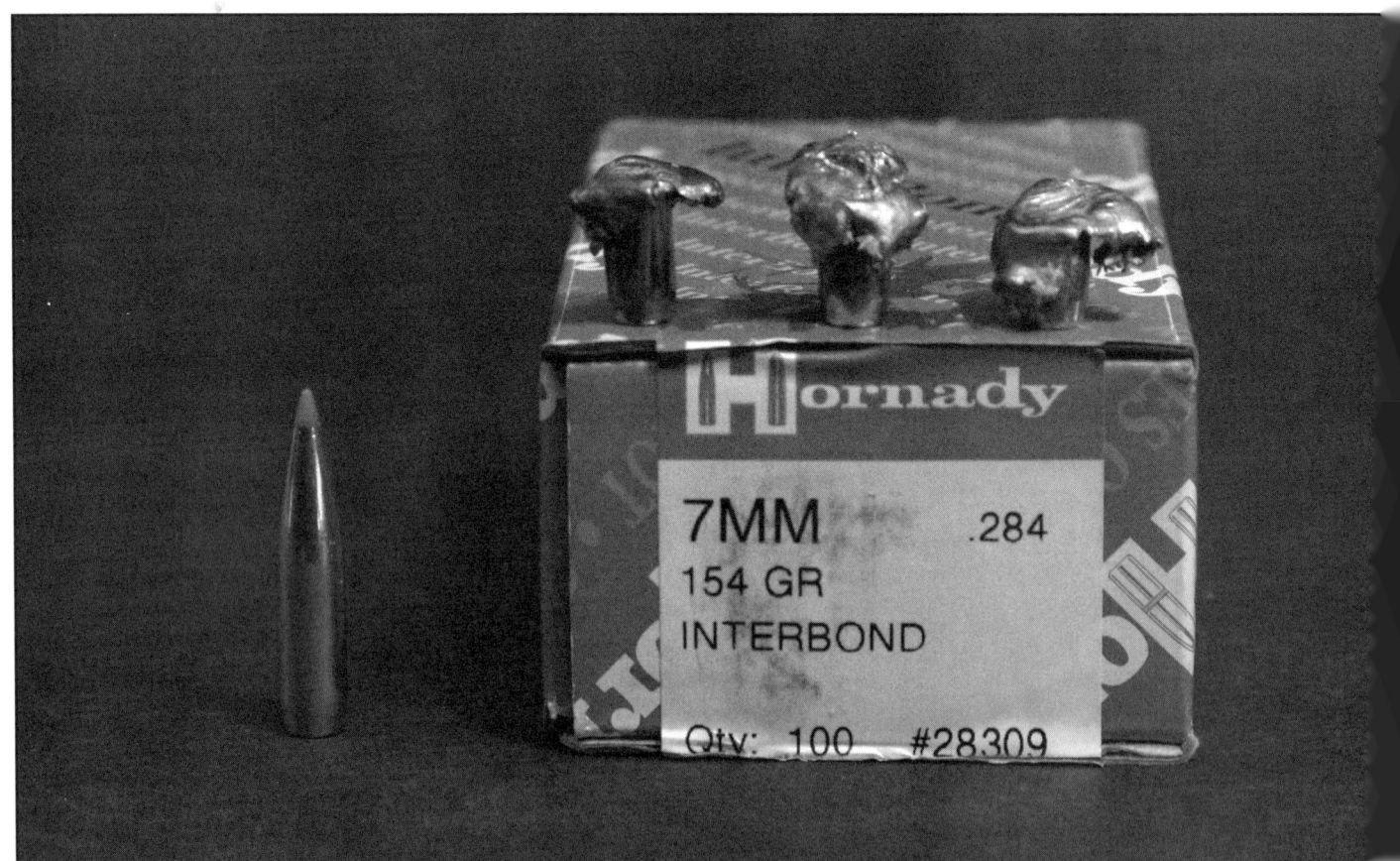

Impact velocity makes a difference in expansion in most bonded-core bullets. These three Hornady Interbonds were shot into newspaper at 2,2 2,500, and 2,650 fps.

"Conventional" bullets work very well, especially when heavier. This big Colorado buck was taken with a 175-grain Hornady Interlock (often called the "poor hunter's premium") from a 7mm Weatherby Magnum. A day or two later another 175-grain Hornady also worked just as well on a 5-point bull elk.

eight in order to work reasonably well on both deer nd larger game. But the biggest Partitions have e partition shifted forward and typically retain at ast 85 percent and sometimes 90 to 95 percent their weight since they're most likely to be used ostly on big animals. All the expanded Partitions e recovered from 9.3mm to .416 have retained 85 95 percent of their weight.

Plastic Tips

One feature seen on many modern bullets is a astic tip, widely advertised to increase ballistic efficient. This helps when shooting at ranges m 500 yards out, but really doesn't make any nificant difference at shorter ranges. The tip o tends to protect the point of the bullet from ttering in the magazine during recoil.

However, the plastic tip's most important role may lie in bullet expansion. On impact, the light plastic tip tends to wedge the bullet open more quickly. This not only helps bullet expansion at very long range, where velocity has dropped off, but it also makes hollowpoint bullets expand every time.

This was first noticed on small varmints such as prairie dogs. Anybody who's shot many prairie dogs from long range has experienced a few hollowpoints that didn't open up. At first the shot appears to be a miss, but a few moments later the prairie dog falls over, the delayed victim of a very tiny hole. When varmint-size Nosler Ballistic Tips first appeared, this ceased—and a Ballistic Tip is just a hollowpoint with a plastic tip.

Hollowpoint big-game bullets also occasionally fail to open, especially when the hollowpoint is smaller. I have seen this a few times myself with both Combined Technology Fail Safes and Barnes X-Bullets, and other experienced hunters have as well. The phenomenon appears to be almost entirely limited to bullets in calibers from .30 down. This makes sense as these tend to have the smallest hollowpoints, which can also be easily battered, and maybe closed, in a rifle's magazine during recoil. This is one reason some hunters (including me) tend to prefer "tipped hollowpoints" in smaller calibers.

Of course, even if today's hunter can choose the "perfect" big-game bullet for any game, sometimes we're hunting different animals at the same time. In my native Montana I'm often hunting both deer and elk, and in Canada or Africa an even wider variety of game can be encountered. The obvious strategy is to use a bullet sufficient for the largest game we might find, but some bullets are very good compromises. The classic is the Nosler Partition, of course, but any bullet designed to open up relatively widely will also work well. I've had very good luck on a variety of North American and African game with the Norma Oryx, a bonded-core

bullet designed to open up easily and widely yet retain most of its weight.

Another solution is to work up a load with two bullets of the same weight, a "deer" bullet and tougher "big-game" bullet. In Africa, I've spent one day hunting the plains for springbok with Sierra GameKings, and the next day up in the hills hunting kudu with a Nosler Partition or Barnes Triple Shock. (It's also much cheaper to check your rifle's zero with the "deer" bullet.)

I've known elk hunters who've heard about how such-and-such bullet retained 90 percent of its weight, yet were disappointed in its penetration—and other hunters who chose a deep-penetrating bullet for a Dall sheep hunt and were even more disappointed when their ram ran over a cliff with a small hole through its lungs. If you're unsure, shoot some bullets into an 18-inch stack of dry (not wet) newspaper, placed against a backstop. This will stop 95 percent of the expanding bullets made and it provides a very good indication of relative expansion and penetration. Such knowledge will help a lot when choosing the right bullet for your next hunt.

Big, deep-penetrating bullets make a real difference on heavy game. This Cape buffalo fell within 75 yards to a single 400-grain Swift A-Frame from .416 Remington Magnum.